ENCYCLOPEDIA OF STATISTICAL SCIENCES

VOLUME 8

Regressograms
to St. Petersburg Paradox, The

ENCYCLOPEDIA OF STATISTICAL SCIENCES

VOLUME 8

REGRESSOGRAMS to ST. PETERSBURG PARADOX, THE

A WILEY-INTERSCIENCE PUBLICATION

John Wiley & Sons

NEW YORK · CHICHESTER · BRISBANE · TORONTO · SINGAPORE

Library of Congress Cataloging-in-Publication Data:

Encyclopedia of statistical sciences.

"A Wiley-Interscience publication."
Includes bibliographies.
Contents: v. 1. A to Circular probable error—
v. 3. Faà di Bruno's formula to Hypothesis testing—
[etc.]—v. 8. Regressograms to St. Petersburg paradox, the.
1. Mathematical statistics—Dictionaries.
2. Statistics—Dictionaries. I. Kotz, Samuel.
II. Johnson, Norman Lloyd. III. Read, Campbell B.

QA276.14.E5 1982 519.5′03′21 81-10353
ISBN 0-471-05556-5 (v.8)

Printed in the United States of America

10 9 8 7 6 5 4 3 2 1

CONTRIBUTORS

J. Abrahams, *Carnegie Mellon University, Pittsburgh, Pennsylvania.* Slepian Process

I. A. Amara, *Quintiles, Inc., Chapel Hill, North Carolina.* Repeated Measurements: Design and Analysis

S. Amari, *Tokyo University, Tokyo, Japan.* Statistical Curvature.

O. D. Anderson, *Pennsylvania State University, University Park, Pennsylvania.* Serial Correlation; Serial Dependence

D. Baird, *University of South Carolina, Columbia, South Carolina.* Significance Tests, History and Logic

G. Bamberg, *Universität Augsburg, Augsburg, Federal Republic of Germany.* Statistische Hefte

T. A. Bancroft,* *Iowa State University, Ames, Iowa.* Snedecor, George Waddel

O. E. Barndorff-Nielsen, *Aarhus Universitet, Aarhus, Denmark.* Reproductive Models.

V. Barnett, *The University of Sheffield, Sheffield, England.* Relationship

D. J. Bartholomew, *London School of Economics, London, England.* Social Statistics.

A. P. Basu, *University of Missouri, Columbia, Missouri.* Reliability, Probabilistic

L. A. Baxter, *State University of New York, Stony Brook, New York.* Relevation

J. M. Begun, *University of North Carolina, Chapel Hill, North Carolina.* Ryan's Multiple Comparisons Procedure

D. R. Bellhouse, *University of Western Ontario, London, Ontario, Canada.* Spatial Sampling

S. Berg, *University of Lund, Lund, Sweden.* Snowball Sampling; Stirling Distributions; Stirling Numbers

J. O. Berger, *Purdue University, West Lafayette, Indiana.* Stein Effect, The

J. E. Besag, *University of Durham, Durham, England.* Resistant Techniques

U. N. Bhat, *Southern Methodist University, Dallas, Texas.* Semi-Markov Processes

P. J. Bickel, *University of California, Berkeley, California.* Robust Estimation

P. Blaesild, *Aarhus Universitet, Aarhus, Denmark.* Reproductive Models

L. Bondesson, *The Swedish University of Agricultural Sciences, Umeå, Sweden.* Shot-Noise Processes and Distributions

E. F. Borgatta, *University of Washington, Seattle, Washington.* Social Network Analysis; Sociometry

B. M. Brown, *University of Tasmania, Hobart, Tasmania, Australia.* Spatial Median

M. Browne, *University of South Africa, Pretoria, South Africa.* Rotation Techniques

D. S. Burdick, *Duke University, Durham, North Carolina.* Statistics in Parapsychology

S. Cambanis, *University of North Carolina, Chapel Hill, North Carolina.* Separable Space

P. Cazes, *Centre D'Enseignement et Recherche de Statistique Appliquée, Paris, France.* Revue de Statistique Appliquée

*Deceased

I. M. Chakravarti, *University of North Carolina, Chapel Hill, North Carolina.* Room's Squares

C.-S. Cheng, *University of California, Berkeley, California.* Regular Graph Designs

S. L. Cheuk, *Louisiana State University Medical Center, New Orleans, Louisiana.* Statistics in Dentistry

L. Cobb, *Medical University of South Carolina, Charleston, South Carolina.* Statistical Catastrophe Theory

J. E. Cohen, *Rockefeller University, New York, New York.* Stochastic Demography

R. D. Cook, *University of Minnesota, St. Paul, Minnesota.* Residuals

P. Coughlin, *University of Maryland, College Park, Maryland.* Single-Peakedness and Median Voters

C. D. Cowan, *Bureau of the Census, Washington, D.C.* Selection Bias

L. H. Cox, *Bureau of the Census, Washington, D.C.* Statistical Confidentiality

P. R. Cox, *The Level House, Mayfield, Sussex, England.* Reproduction Rates

E. M. Cramer, *University of North Carolina, Chapel Hill, North Carolina.* Statistics in Psychology

E. B. Dagum, *Statistics Canada, Ottawa, Canada.* Seasonality

J. Davidson, *London School of Economics, London, England.* Sampling Theory with Dependent Observations, Asymptotic

M. H. A. Davis, *Imperial College, London, England.* Stochastics

M. L. Deaton, *Kansas State University, Manhattan, Kansas.* Simulation Models, Validation of

M. H. DeGroot, *Carnegie Mellon University, Pittsburgh, Pennsylvania.* Regret; Statistical Sciences

T. de Wet, *I. M. T., Simon's Town, Cape Province, South Africa.* South African Statistical Journal

W. J. Dixon, *UCLA Medical Center, Los Angeles, California.* Staircase Method

N. R. Draper, *University of Wisconsin, Madison, Wisconsin.* Response Surface Designs; Runs

M. S. Dueker, *University of Connecticut, Stamford, Connecticut.* Saturated Designs

J. D. Elashoff, *B.M.D.P. Statistics Software, Los Angeles, California.* Repeated Measurements: Design and Analysis

E. L. Etnier, *Oak Ridge National Laboratory, Oak Ridge, Tennessee.* Risk Analysis, Society for

W. G. Faris, *University of Arizona, Tucson, Arizona.* Stochastic Mechanics

T. S. Ferguson, *University of California, Los Angeles, California.* Stochastic Games

L. T. Fernholz, *Princeton Forecasting Associates, Inc., Princeton, New Jersey.* Statistical Functionals

S. E. Fienberg, *Carnegie Mellon University, Pittsburgh, Pennsylvania.* Rotation Group Bias; Statistics on Crime and Criminal Justice

P. D. Finch, *Monash University, Clayton, Victoria, Australia.* Standardization

P. C. Fishburn, *AT & T Bell Laboratories, Murray Hill, New Jersey.* Risk Measurement, Foundations of

N. I. Fisher, *CSIRO, Lindfield, Australia.* Spherical Medians

M. A. Fligner, *Ohio State University, Columbus, Ohio.* Scale Tests

J. D. Flora, Jr., *Midwest Research Institute, Kansas City, Missouri.* Ridit Analysis

B. L. Foote, *University of Oklahoma, Norman, Oklahoma.* Robust Inspection Sampling Plans

M. R. Frankel, *National Opinion Research Center, New York, New York.* Response Bias

D. A. S. Fraser, *University of Toronto, Ontario, Canada.* Rotation Group; Statistical Inference

L. S. Freedman, *Medical Research Centre, Cambridge, England.* Statistics in Medicine

G. H. Freeman, *National Vegetable Research Station, Warwick, England.* Reversal Designs; Row and Column Designs

J. S. Gardenier, *Washington, D.C.* Risk Management, Statistical Aspects of

T. K. Gardenier, *Washington, D.C.* Risk Management, Statistical Aspects of

D. W. Gaylor, *Center for Toxicological Research, Jefferson, Arizona.* Satterthwaite's Formula

E. J. Gbur, Jr., *Bureau of the Census,*

Washington, D.C. Rotation Sampling

W. Gersch, *University of Hawaii, Honolulu, Hawaii.* Smoothing Priors

B. K. Ghosh, *Virginia Polytechnic Institute, Blacksburg, Virginia.* Sequential Analysis

M. Ghosh, *University of Florida, Gainesville, Florida.* Sequential Rank Estimators

S. P. Ghosh, *IBM Research Laboratory, San José, California.* Statistics Metadata

J. D. Gibbons, *University of Alabama, University, Alabama.* Selection Procedures; Sign Tests; Steel Statistics

I. J. Good, *Virginia Polytechnic Institute and State University, Blacksburg, Virginia.* Scientific Method and Statistics; Stable Estimation; Statistical Evidence

A. D. Gordon, *University of St. Andrews, St. Andrews, Scotland.* Sequence Comparison Statistics

P. J. Green, *University of Durham, Durham, England.* Sharpening Data

R. S. Greenberg, *Atlanta Cancer Surveillance Center, Decatur, Georgia.* Retrospective Studies (Including Case-Control)

W. S. Griffith, *University of Kentucky, Lexington, Kentucky.* Shock Models

R. R. Harris, *Sheffield City Polytechnic, Sheffield, England.* Statistician, The

A. Hart, *Lancashire Polytechnic, Preston, England.* Standard Deviation

A. R. Hayes, *University of the District of Columbia, Washington, D.C.* Statistical Software

A. M. Herzberg, *Imperial College, London, England.* Robustness in Experimental Design; Rotatable Designs

I. D. Hill, *Clinical Research Centre, Middlesex, England.* Royal Statistical Society

A. E. Hoerl, *University of Delaware, Newark, Delaware.* Ridge Regression

H.-K. Hsieh, *University of Massachusetts, Amherst, Massachusetts.* Savage Test; Schafer-Sheffield Test

H. Huyhn, *University of South Carolina, Columbia, South Carolina.* Sphericity, Tests of

J. T. Hwang, *Cornell University, Ithaca, New York.* Stochastic and Universal Domination

D. R. Jensen, *Virginia Polytechnic Institute and State University, Blacksburg, Virginia.* Semi-Independence

M. A. Johnson, *The Upjohn Company, Kalamazoo, Michigan.* Screening Designs

R. A. Johnson, *University of Wisconsin, Madison, Wisconsin.* Statistics and Probability Letters

G. K. Jones, *National Center for Health Statistics, Hyattsville, Maryland.* Sampling Errors (Computation of)

K. Kafadar, *Hewlett Packard Corporation, Palo Alto, California.* Robust-Resistant Line; Slash Distribution; Stem-and-Leaf Display

A. Kahnert, *U.N. Economic Commission for Europe, Geneva, Switzerland.* Statistical Journal of the U.N. Economic Commission for Europe

M. I. Kamien, *Northwestern University, Evanston, Illinois.* Stochastic Differential Equations: Applications

G. K. Kanji, *Sheffield City Polytechnic, Sheffield, England.* Statistician, The

R. Karandikar, *University of North Carolina, Chapel Hill, North Carolina.* Stochastic Integrals

A. F. Karr, *Johns Hopkins University, Baltimore, Maryland.* Stochastic Processes, Point

B. Kedem, *University of Maryland, College Park, Maryland.* Sinusoidal Limit Theorem

D. G. Kendall, *University of Cambridge, Cambridge, England.* Seriation; Shape Statistics

R. W. Kennard, *Groveland, Florida.* Ridge Regression

K. Kercher, *University of Washington, Seattle, Washington.* Social Network Analysis; Sociometry

R. A. Khan, *Cleveland State University, Cleveland, Ohio.* Rényi-Anscombe Theorem

G. Kimeldorf, *University of Texas at Dallas, Richardson, Texas.* Setwise Dependence

G. G. Koch, *University of North Carolina, Chapel Hill, North Carolina.* Repeated Measurements: Design and Analysis

W. S. Krasker, *Harvard University, Boston, Massachusetts.* Robust Regression; Schweppe-Type Estimators

W. H. Kruskal, *The University of Chicago, Chicago, Illinois.* Representative Sampling

R. G. Krutchkoff, *Virginia Polytechnic Institute and State University.* Statistical Computation and Simulation, Journal of

R. V. Kulkarni, *University of North Carolina, Chapel Hill, North Carolina.* Sequential Procedures, Adaptive

G. S. Ladde, *University of Texas, Arlington, Texas.* Stochastic Analysis and Applications

N. Laird, *Harvard University, Boston, Massachusetts.* Self-Consistency

V. Lakshmikantham, *University of Texas, Arlington, Texas.* Stochastic Analysis and Applications

H. O. Lancaster, *University of Sydney, Sydney, New South Wales, Australia.* Statistics, History of

J. Lawless, *University of Waterloo, Ontario, Canada.* Reliability, Nonparametric Methods in

E. L. Lehmann, *University of California, Berkeley, California.* Statistics: An Overview

H. H. Lemmer, *Rand Afrikaans University, Johannesburg, South Africa.* Shrinkage Estimators

D. V. Lindley, *Minehead, England.* Savage, Leonard J.

T. J. Lorenzen, *General Motors Research Laboratories, Warren, Michigan.* Snowflakes

C. Löschke, *Deutsche Statistische Gesellschaft, Federal Republic of Germany.* Statistische Gesellschaft, Deutsche

L. Y. Low, *Wright State University, Dayton, Ohio.* Resampling Procedures

L. Lyberg, *Statistiska Centralbyran, Stockholm, Sweden.* Statistisk Tidskrift

V. C. Mavron, *The University College, Aberystwyth, Wales.* Residuation

I. McKeague, *Florida State University, Tallahassee, Florida.* Sieves, Method of

I. M. McKinney, *Central Statistical Office, London, England.* Statistical News

A. I. McLeod, *University of Western Ontario, London, Ontario, Canada.* Simple Random Sampling

R. E. Miles, *The Australian National University, Canberra, Australia.* Sports, Scoring Systems in; Stereology

D. Monrad, *University of Illinois, Urbana, Illinois.* Stable Distributions

R. Morton, *CSIRO, Canberra, Australia.* Spearman Estimators

J. E. Mosimann, *National Institutes of Health, Bethesda, Maryland.* Size and Shape Analysis

F. Mosteller, *Harvard University, Cambridge, Massachusetts.* Representative Sampling

A. Mukherjea, *University of South Florida, Tampa, Florida.* Semi-Dissipative and Non-Dissipative Markov Chains; Sojourn Time

R. J. Myers, *Silver Spring, Maryland.* Social Security Statistics

H. N. Nagaraja, *Ohio State University, Columbus, Ohio.* Selection Differentials

N. K. Namboodiri, *University of North Carolina, Chapel Hill, North Carolina.* Sociology, Statistics in

B. Natvig, *University of Oslo, Oslo, Norway.* Reliability Importance

J. I. Naus, *East Brunswick, New Jersey.* Scan Statistics

J. Neter, *University of Georgia, Athens, Georgia.* Statistics in Auditing

J. B. Neuhardt, *Ohio State University, Columbus, Ohio.* Resolution

G. Neuhaus, *University of Hamburg, Hamburg, Federal Republic of Germany.* Repeated Chi-Square Testing

M. F. Neuts, *University of Arizona, Tucson, Arizona.* Stochastic Models

H. Niederhausen, *Florida Atlantic University, Boca Raton, Florida.* Rényi-Type Distributions; Sheffer Polynomials

D. Oakes, *University of Rochester, Rochester, New York.* Semi-Parametric Models

J. K. Ord, *Pennsylvania State University, University Park, Pennsylvania.* Spatial Process; Spectral Analysis

F. O'Sullivan, *University of California, Berkeley, California.* Robust Smoothing

D. B. Owen, *Southern Methodist University, Dallas, Texas.* Screening by Correlated Variates

S. C. Pearce, *University of Kent, Canterbury, England.* Split Plots; Strip Plots

S. S. Perng, *Springfield, Virginia.* Simple

Expansions

J. D. Petruccelli, *Worcester Polytechnic Institute, Worcester, Massachusetts.* Secretary Problem

W. V. Philipp, *University of Illinois, Urbana, Illinois.* Skorokhod Embeddings

K. C. S. Pillai,* *Purdue University, West Lafayette, Indiana.* Roy's Characteristic Root Statistic

W. R. Pirie, *Virginia Polytechnic Institute and State University, Blacksburg, Virginia.* Spearman Rank Correlation Coefficient

S. Portnoy, *University of Illinois, Urbana, Illinois.* Statistics in Religious Studies

R. F. Potthoff, *Burlington Industries, Greensboro, North Carolina.* Run Lengths, Tests of

N. U. Prabhu, *Cornell University, Ithaca, New York.* Stochastic Processes and Their Applications

J. E. Prather, *Georgia State University, Atlanta, Georgia.* Spurious Correlation

D. A. Preece, *East Malling Research Station, East Malling, Kent, England.* Semi-Latin Squares

I. Prucha, *University of Maryland, College Park, Maryland.* Seemingly Unrelated Regression

J. Putter, *The Volcani Center, Bet-Dagan, Israel.* Selective Inference

B. R. Rao, *University of Pittsburgh, Pittsburgh, Pennsylvania.* SSR_2; STP_2

C. B. Read, *Southern Methodist University, Dallas, Texas.* Scheffé's Simultaneous Comparison Procedure; Siegel-Tukey Test; Spacings; Sports, Statistics in

H. T. Reynolds, *University of Delaware, Newark, Delaware.* Statistics in Political Science

M. G. Ribe, *Statistika Centralbyran, Stockholm, Sweden.* Statistics Sweden; Statistisk Tidskrift

H. R. Richardson, *Daniel H. Wagner Associates, Paoli, Pennsylvania.* Search Theory

B. D. Ripley, *University of Strathclyde, Glasgow, Scotland.* Spatial Data Analysis

W. L. Rodgers, *University of Michigan, Ann Arbor, Michigan.* Statistical Matching

B. Rosén, *Royal Institute of Technology, Stockholm, Sweden.* Scandinavian Journal of Statistics

W. D. Rowe, *The American University, Washington, D.C.* Risk Assessment

L. Rowebottom, *Statistics Canada, Ottawa, Canada.* Statistics Canada

J. P. Royston, *The Clinical Research Centre, Middlesex, England.* Shapiro-Wilk W Statistics

D. Ruppert, *University of North Carolina, Chapel Hill, North Carolina.* Robustification and Robust Substitutes

A. R. Sampson, *University of Pittsburgh, Pittsburgh, Pennsylvania.* Setwise Dependence; Stochastic Approximation

I. Scardovi, *Istituto di Statistica, Bologna, Italy.* Statistica

R. L. Scheaffer, *University of Florida, Gainesville, Florida.* Statistical Education

W. Schlee, *Technische Universität München, München, Federal Republic of Germany.* Regressograms

W. R. Schucany, *Southern Methodist University, Dallas, Texas.* Sample Reuse

Z. Schuss, *Northwestern University, Evanston, Illinois.* Stochastic Differential Equations

R. C. Scott, *University of Durham, Durham, England.* Smear-and-Sweep

H. L. Seal,* *La Mottaz, Switzerland.* Risk Theory

A. Seheult, *University of Durham, Durham, England.* Resistant Techniques

A. R. Sen, *University of Calgary, Alberta, Canada.* Sequential Estimation of the Mean in Finite Populations; Statistics in Animal Science

P. K. Sen, *University of North Carolina, Chapel Hill, North Carolina.* Robust Tests for Change-Point Models; Sequential Analysis; Signed Rank Statistics; Statistics and Decisions; Stopping Numbers and Stopping Times

E. Seneta, *University of Sydney, Sydney, New South Wales, Australia.* Slutsky, Evgenii Evgenievich

G. Shafer, *University of Kansas, Lawrence,*

*Deceased

*Deceased

Kansas. Sharp Null Hypotheses; St. Petersburg Paradox, The

J. P. Shaffer, *University of California, Berkeley, California.* Simultaneous Testing

G. Shorack, *University of Washington, Seattle, Washington.* Shorack Estimators

D. Singh, *New Delhi, India.* Sequential Sampling

N. Singpurwalla, *George Washington University, Washington, D.C.* Software Reliability

C. J. Skinner, *University of Southampton, Southampton, England.* Rejective Sampling

A. F. M. Smith, *University of Nottingham, Nottingham, England.* Schwarz Criterion

H. Smith, *Durham, North Carolina.* Stepwise Regression

W. L. Smith, *University of North Carolina, Chapel Hill, North Carolina.* Renewal Theory

F. Steen, *Allegheny College, Meadville, Pennsylvania.* Scan Diagrams

F. W. Steutel, *Technisch Hogeschool Eindhoven, Eindhoven, The Netherlands.* Statistica Neerlandica

W. F. Stout, *University of Illinois, Urbana, Illinois.* Stable Distributions

G. P. H. Styan, *University of Alberta, Edmonton, Alberta, Canada.* Samuelson-Nair Inequality

R. Syski, *University of Maryland, College Park, Maryland.* Stochastic Processes

M. Taqqu, *Cornell University, Ithaca, New York.* Self-Similar Processes (Non-Gaussian)

R. E. Tarone, *National Cancer Institute, Bethesda, Maryland.* Score Statistics

K. K. Tatsuoka, *University of Illinois, Champaign, Illinois.* Rule Space

M. M. Tatsuoka, *University of Illinois, Champaign, Illinois.* Rule Space

G. Taylor, *Colorado State University, Fort Collins, Colorado.* Remez Algorithm

L. A. Thibodeau, *Applied Management Sciences, Silver Spring, Maryland.* Sensitivity and Specificity

P. Thyregod, *Technical University of Denmark, Lyngby, Denmark.* Sampling Plans

H. T. Tigelaar, *Tilburg University, Tilburg, The Netherlands.* Sample Size, Informative and Predictive

R. D. Tortora, *U.S. Department of Agriculture, Washington, D.C.* Respondent Burden

C. C. Travis, *Oak Ridge National Laboratory, Oak Ridge, Tennessee.* Risk Analysis, Journal of

R. L. Tweedie, *Siromath Pty. Ltd., Sydney, Australia.* Return State

I. Vincze, *Hungarian Academy of Sciences, Budapest, Hungary.* Rényi, Alfréd

J. W. E. Vos, *International Statistical Institute, Voorburg, The Netherlands.* Statistical Theory and Methods Abstracts

K. W. Wachter, *University of California, Berkeley, California.* Statistics in Historical Studies

W. G. Warren, *Environmental Protection Agency, Corvallis, Oregon.* Statistics in Forestry

E. J. Wegman, *George Mason University, Fairfax, Virginia.* Reproducing Kernel Hilbert Spaces; Sobolev Spaces; Statistical Software

L. J. Wei, *University of Michigan, Ann Arbor, Michigan.* Selection Bias

R. Weinberg, *Louisiana State University, New Orleans, Louisiana.* Statistics in Dentistry

G. H. Weiss, *National Institutes of Health, Bethesda, Maryland.* Saddle Point Approximations; Statistics in Crystallography

L. Weiss, *Cornell University, Ithaca, New York.* Runs; Sequential Estimation

F. S. Whaley, *University of Vermont, Burlington, Vermont.* Runs Test, Multidimensional

W. Whitt, *AT & T Bell Laboratories, Murray Hill, New Jersey.* Stochastic Ordering

H. Wold, *University of Uppsala, Uppsala, Sweden.* Specification, Predictor

H. Wolkowicz, *Emory University, Atlanta, Georgia.* Samuelson-Nair Inequality

M. Woodroofe, *Rutgers University, New Brunswick, New Jersey.* Repeated Significance Tests.

R *continued*

REGRESSOGRAMS

The regressogram (RGM) is a nonparametric estimator* for the regression function, that is the conditional expectation of a random variable Y with respect to X. It is constructed analogously to the histogram*.

A verbal description of the RGM is as follows. The x domain is partitioned into intervals. The value of the RGM is constant in each of these intervals. This value is zero for those intervals in which no sample values of X occur. In the other intervals the value of the RGM is equal to the arithmetic mean of those y values whose corresponding x values lie in the considered interval.

Besides the visual significance of the RGM in the case of X and Y being one-dimensional there is also an importance for hypothesis testing if X and Y are multidimensional.

MATHEMATICAL DEFINITION

Let (X_l, Y_l), $l = 1,2,\ldots, n$, be random variables with $E[Y_l|X_l] = g(X_l)$, X_l p-dimensional and Y_l $\bar{p}$-dimensional. The x domain is assumed to be an interval $[A, B] \subset \mathbb{R}^p$. $[A, B]$ is a priori partitioned into intervals $I(i) = I_1(i_1) \times \cdots \times I_p(i_p)$. $J_i(\cdot)$ is the indicator function of the interval $I(i)$, that is $J_i(z) = 1$ if $z \in I(i)$ and 0 otherwise.

The function $\hat{g}_n : \mathbb{R}^p \to \mathbb{R}^{\bar{p}}$, defined by

$$\hat{g}_n(z) = \sum_i Y_i J_i(z),$$

is called a *regressogram*.

An obvious generalization of this definition is the RGM with random partition. Here $[A, B]$ is chosen as the smallest interval containing all X_l, $l = 1,2,\ldots, n$. $[A, B]$ is then partitioned by means of order statistics of the x sample.

HISTORICAL REMARK

The RGM (with a priori chosen interval partition) was introduced and denominated by Tukey [12]. Its statistical properties were first investigated by Bosq [2]. Pearson and Lee [8] also used a RGM, but only as an aid for estimation of linear regression; they neither denominated it nor pointed out its significance for nonparametric estimation. The data they used are only given up to a very restricted number of digits. Consequently the possible x values repeated themselves within the sample. Herein an intuitive background of the RGM becomes apparent: It is not necessary to distinguish between the x values within a small interval; therefore, the corresponding y values can be reduced to their arithmetic mean*. Independently of Tukey, Bhattacharya and Parthasarathy [1] investigated the RGM with random partition. They presented it as a method of fractile graphical analysis*, which was introduced by Mahalanobis [6].

STATISTICAL PROPERTIES

According the intention to use the RGM as a nonparametric estimator, only consistency and asymptotic probability laws have been investigated. This makes it necessary to connect the interval diameter with the sample size n and let this diameter converge to 0 if n tends to infinity. The various results are then obtained by assuming the appropriate speed of convergence besides weak conditions on the underlying probability distribution. The published results are further restricted to $p = \bar{p} = 1$ and (X_l, Y_l), $l = 1, 2, \ldots, n$, i.i.d. random variables.

Bosq [2] gives conditions for almost sure convergence to zero of the maximal deviation

$$M_n = \sup_z |\hat{g}_n(z) - g(z)|$$

if $[A, B] = [0, 1]$. Sabry [9] extends this result to the case of $[0, B]$, where B depends on the sample size n and tends to infinity with n. Geffroy [4] gives further related results. Major [7] evaluates the asymptotic distribution of a maximal deviation quantity modified by the values of the marginal density of X and the conditional variance of Y on X. For the random partition case with $[A, B] = [0, 1]$, Bhattacharya and Parthasarathy [1] give the convergence (in probability and almost sure) of some modified M_n and an asymptotic distribution result. The results in refs. 1 and 7 also yield asymptotic tests for the true regression function. Lecoutre [5] extends the convergence of M_n to the case $[A, B] = \mathbb{R}$.

RECENT GENERALIZATIONS

Collomb [3] applies the method of RGMs to Markov processes* $(X_l)_l$ of order k^* with values in $[0, 1]^p$ to estimate

$$E[h(X_{l+s}) | (X_{l-k+1}, \ldots, X_l)]$$

for some measurable function h. L^2 consistency is proved and a rate of convergence is given. He calls his nonparametric predictor a *predictogram*.

Schlee [11] points out the possibility of RGM-like estimators for more complex characteristics of the joint probability distribution of X and Y than the conditional expectation of Y. He considers estimators of the kind

$$\tilde{g}_n(z) = \sum_i h\left(\bar{V}_{(i)1}, \ldots, \bar{V}_{(i)d}\right) J_i(z)$$

with some function $h: \mathbb{R}^d \Rightarrow \mathbb{R}^{\bar{d}}$ that corresponds to the quantity to be estimated. $\bar{V}_{(i)j}$, $j = 1, 2, \ldots, d$, are quantities that use only the y values whose corresponding x lie in the interval $I(i)$. They are computed according to certain prescribed instructions. $\tilde{g}_n(\cdot)$ is called a *functiogram*. It especially includes the RGM, the *quantilogram* of Schlee [10], and an estimator of the conditional covariance matrix of Y on X; see ref. 11. In ref. 11 weak and strong consistency* results are given in the case of (X_l, Y_l), $l = 1, 2, \ldots, n$, i.i.d. random variables and p, $\bar{p} \geqslant 1$ for a smoothed version of the functiogram.

RELATED NONPARAMETRIC ESTIMATORS

In some sense nearest-neighbor estimators*, kernel, and spline estimators* may be regarded as related (compare also ref. 11).

References

[1] Bhattacharya, P. K. and Parthasarathy, K. R. (1961). *Sankhyā A*, **23**, 91–102.

[2] Bosq, D. (1970). *Publ. Inst. Statist. Univ. Paris*. **19**, 97–177 (in French).

[3] Collomb, G. (1982). *C. R. Acad. Sci. Paris*, Sér, I, **294**, 59–62 (in French).

[4] Geffroy, J. (1980). *Publ. Inst. Statist. Univ. Paris*, **25**, I–II, 41–56 (in French).

[5] Lecoutre J. P. (1980). *C. R. Acad. Sci. Paris*, Sér. A, **291**, 355–358 (in French).

[6] Mahalanobis, P. C. (1961). *Sankhyā A*, **23**, 41–64.

[7] Major, P. (1973). *Studia Sci. Math. Hung.*, **8**, 347–361.

[8] Pearson, K. and Lee, A. (1903). *Biometrika*, **2**, 357–462.

[9] Sabry, H. (1978). *C. R. Acad. Sci. Paris*, Sér. A, **286**, 941–944 (in French).

[10] Schlee, W. (1982). *Statistique et Analyse des Données*, **7**, 32–47 (in French).

[11] Schlee, W. (1985). In *Limit Theorems in Probability and Statistics*, Veszprém, Hungary, 1982. *Colloq. Math. Soc. János Bolyai*, **36** (P. Révész, ed.) North-Holland, Amsterdam, The Netherlands.

[12] Tukey, J. W. (1961). *Proc. 4th Berkeley Symp. Math. Statist.*, 681–694.

(HISTOGRAM
KERNEL ESTIMATORS
MARKOV PROCESSES
NEAREST NEIGHBOR METHODS
QUANTILES)

WALTER SCHLEE

REGRET

In Webster's *Third New International Dictionary, regret* is defined as "sorrow aroused by circumstances beyond one's control or power to repair: grief or pain tinged with disappointment, dissatisfaction, longing, remorse, or comparable emotion." In the theory of decision making, regret is usually caused by comparing the actual result of a decision with a more favorable result that could have been obtained if a different decision had been made. In particular, consider a problem in which a decision maker must choose a decision a from some given set A; the consequences of the decision depend on a parameter whose unknown value θ lies in the parameter space Ω; and $L(a, \theta)$ denotes the loss or cost to the decision maker if the decision a is chosen when the true value of the parameter is θ (see DECISION THEORY and MINIMAX DECISION RULES). The following function L^* is often called the *regret function* in this problem:

$$L^*(a, \theta) = L(a, \theta) - \min_{a' \in A} L(a', \theta). \quad (1)$$

The function L^* was explicitly introduced by Savage [9] who suggested that it was more appropriate to select a decision $a \in A$ that was *minimax** with respect to L^* rather than to select one that was minimax with respect to the original loss function* L. Savage felt that the use of L^* was implicit in the original development of decision theory by Wald [12]. However, Savage [10, p. 163] did not approve of calling L^* "regret" because, as he stated, "that term seems... charged with emotion and liable to lead to such misinterpretation as that the loss necessarily becomes known to the person."

The principle of choosing a decision in accordance with the criterion of minimax regret has several undesirable features. The following criticisms are paraphrased from Chernoff [5]: (a) It has never been demonstrated that the difference between the realized loss and the minimum possible attainable loss does in fact measure what one may call regret. (b) There are examples where an arbitrarily small advantage of a_1 over a_2 for one value of the parameter outweighs a considerable advantage of a_2 over a_1 for another value. This same shortcoming is true of the minimax loss criterion. (c) There are examples in which the decision a_1 will be chosen when only the decisions a_1, a_2, and a_3 are available, but the decision a_2 will be chosen when an additional decision a_4 is made available. An example of this type is given in MINIMAX DECISION RULES.

Because of these undesirable features, many statisticians believe that it is more effective to assign a prior density function or weight function ξ to the values of θ and to consider the *risk*

$$R(a, \xi) = \sum_{\theta \in \Omega} L(a, \theta)\xi(\theta). \quad (2)$$

A *Bayes decision*, that is, a decision a for which the risk (2) is minimized, is then chosen. Bayes decisions do not suffer from any of the undesirable features just described. Furthermore, the Bayes decision in a given problem will be the same regardless of whether the loss function L or the regret function L^* is used in (2).

In a famous experiment, Allais [1] found that the preferences of many persons among gambles involving monetary payoffs seemed to violate the principle of the maximization of *subjective expected utility**. The systematic nature of these violations has been termed the *Allais paradox* (Allais and Hagen [2]) and explanations of the paradox are often based on notions of potential regret by the decision makers (Bell [3], DeGroot [6, pp. 93–94]). In light of these findings, there has been intensive reconsideration of traditional utility theory (Stigum and Wenstøp [11]). Bell [3, 4] has incorporated regret into a multiattribute utility function, and Loomes and Sugden [7, 8] discuss an alternate development of regret theory.

References

[1] Allais, M. (1953). *Econometrica*, **21**, 503–546.

[2] Allais, M. and Hagen, O. eds. (1979). *Expected Utility and the Allais Paradox*. Reidel, Dordrecht, Netherlands.

[3] Bell, D. E. (1982). *Operat. Res.*, **30**, 961–981.

[4] Bell, D. E. (1983). *Manag. Sci.*, **29**, 1156–1166.

[5] Chernoff, H. (1954). *Econometrica*, **22**, 422–443.

[6] DeGroot, M. H. (1970). *Optimal Statistical Decisions*. McGraw-Hill, New York.

[7] Loomes, G. and Sugden, R. (1983). *Econ. Lett.*, **12**, 19–21.

[8] Loomes, G. and Sugden, R. (1983). *Amer. Econ. Rev.*, **73**, 428–432.

[9] Savage, L. J. (1951). *J. Amer. Statist. Ass.*, **46**, 55–67.

[10] Savage, L. J. (1954). *The Foundations of Statistics*. Wiley, New York.

[11] Stigum, B. P. and Wenstøp, F., eds. (1983). *Foundations of Utility and Risk Theory with Applications*. Reidel, Dordrecht, Netherlands.

[12] Wald, A. (1950). *Statistical Decision Functions*. Wiley, New York.

(DECISION THEORY
MINIMAX)

MORRIS H. DEGROOT

REGULAR EXPONENTIAL FAMILY *See* EXPONENTIAL FAMILIES

REGULAR GRAPH DESIGNS

Regular graph designs were introduced by John and Mitchell [10] as a class of efficient incomplete block designs. A regular graph design with v varieties and b blocks of size k is an incomplete block design ($k < v$) such that

(a) Each variety is replicated $r = bk/v$ times.

(b) Each variety appears in each block at most once.

(c) $|\lambda_{ij} - \lambda_{i'j'}| \leqslant 1$ for all $i \neq j$, $i' \neq j'$, where λ_{ij} is the number of blocks in which both varieties i and j appear.

Condition **(c)** implies that $\lambda_{ij} = \lambda$ or $\lambda + 1$ for all $i \neq j$, where λ is the largest integer not exceeding $r(k-1)/(v-1)$. Let $\alpha = r(k-1) - \lambda(v-1)$ and $n = v - 1 - \alpha$. Then for any fixed i_0, $1 \leqslant i_0 \leqslant v$, there are n λ_{i_0j}'s equal to λ, and α λ_{i_0j}'s equal to $\lambda + 1$. If we consider the v varieties as the vertices of a graph in which there is a line between vertex i and vertex j if and only if $\lambda_{ij} = \lambda$, then the resulting graph is a regular graph with degree n, i.e., each vertex is adjacent to n other vertices. This is why such a design is called a regular graph design. For convenience, we shall denote the above graph corresponding to a regular graph design d by $G(d)$.

As an example, the following is a regular graph design with $v = b = 5$ and $k = 3$: (124), (235), (341), (452), (513). Notice that this design has $r = 3$, $\lambda = 1$, $\alpha = 2$, and $G(d)$ is a circuit on five vertices.

According to the above definition, a balanced incomplete block design is a regular graph design with $n = v - 1$ and the corresponding graph being a complete graph. A balanced incomplete block* design is known to have strong optimum properties (see refs. 14 and 15). The rationale for studying regular graph designs is that these designs are combinatorially close to balanced incomplete block designs and therefore are expected to be very efficient; *see* NEARLY BALANCED DESIGNS. Furthermore, the stringent condition in the definition of a balanced incomplete block design that all the λ_{ij}'s be equal is relaxed a bit to keep the high efficiencies, yet the resulting flexibility is enough to make the regular graph designs far more available than the balanced incomplete block designs; a regular graph design can be found for most of the parameter sets. This makes regular graph designs very attractive from a practical point of view.

John and Mitchell [10] conjectured that for given values of v, b, and k, if there exists a regular graph design, then there must be a regular graph design which is D-, A- and E-optimal (*see* OPTIMUM DESIGN OF EXPERIMENTS); also see ref. 11. If their conjecture is true, then the search for optimal designs can be reduced to the set of regular graph designs. Using the relation to graphs, John and Mitchell carried out a computer search for A-, D-, and E-optimal regular graph designs (or their duals) for 209 parameter sets in the practical range ($v \leqslant 12$ and $r \leqslant 10$). A complete list of these optimal regular graph designs can be found in the technical report of Mitchell and John [18]. Other uses of this conjecture in design construction have also appeared in refs. 9, 12, 19, and 22.

For convenience, let $S_{v,b,k}$ be the collection of all the designs with v varieties in b blocks of size k. Jones and Eccleston [13] obtained, for the three parameter sets $(v, b, k) = (10, 10, 2), (11, 11, 2), (12, 12, 2)$, designs with unequal replications which were A and E better than the best regular graph designs in $S_{v,b,k}$ found by John and Mitchell. In spite of these findings, regular graph designs are indeed very efficient, if not optimal. For instance, Mitchell [16] used the algorithm DETMAX (see ref. 17) to construct D-optimal incomplete block designs for $v + b \leqslant 31$ and $bk \leqslant 44$. The algorithm does not guarantee D-optimality, but it matched or bettered the best design previously available in 55 of the 66 distinct cases considered. All the designs obtained by Mitchell were regular graph designs or duals of regular graph designs. In recent years, many classes of regular graph designs have been shown to be optimal over $S_{v,b,k}$ with respect to various criteria. See, e.g., refs. 2, 4–6, 8, and 21.

The performance of a regular graph design d under various optimality criteria depends on the structure of the graph $G(d)$. We shall illustrate this on the E criterion. The C matrix (in terms of which a typical optimality criterion is defined) of an equally replicated design in $S_{v,b,k}$ is of the form

$$\mathbf{C} = r\mathbf{I}_v - k^{-1}\mathbf{N}\mathbf{N}',$$

where $\mathbf{I}_v$ is the identity matrix of order v and $\mathbf{N}$ is the variety-block incidence matrix. For a regular graph design, all the diagonal elements of $\mathbf{NN}'$ are equal to r and the off-diagonal elements are λ or $\lambda + 1$. Thus the C matrix of a regular graph design can be written as

$$\mathbf{C} = k^{-1}\left[\{(k-1)r + \lambda + 1\}\mathbf{I}_v - (\lambda + 1)\mathbf{J}_v + \mathbf{T}\right],$$

where $\mathbf{J}_v$ is the $v \times v$ matrix of 1's and $\mathbf{T}$ is a $(0, 1)$ matrix such that all the diagonal elements are zero and the (i, j)th entry is equal to 1 if and only if $\lambda_{ij} = \lambda$. Therefore $\mathbf{T}$ is the adjacency matrix of $G(d)$ whose smallest eigenvalue δ is related to the next-to-least eigenvalue μ of $\mathbf{C}$ by

$$\mu = k^{-1}\{(k-1)r + \lambda + 1 + \delta\}.$$

Thus the search for an E-optimal regular graph design is equivalent to maximizing the smallest eigenvalue of the adjacency matrix of $G(d)$. It is well known in graph theory* that the smallest eigenvalue of the adjacency matrix of a nonempty graph is always less than or equal to -1 and this upper bound is achieved if and only if the graph is a complete graph or is a disconnected graph in which all the connected components are cliques. The corresponding regular graph designs are balanced incomplete block designs or group-divisible designs* with $\lambda_2 = \lambda_1 + 1$, whose E-optimality can be proven along this line.

The regular graphs with the smallest eigenvalue less than -1 but greater than or equal to -2 have also been characterized (see, e.g., refs. 1 and 7). For instance, it is known that if all the connected components of $G(d)$ are line graphs, then $\delta \geqslant -2$, and all the other regular graphs have $\delta \leqslant -2$. This shows that a line graph is a desirable structure. Regular graph designs such that all the connected components of $G(d)$ are line graphs include, e.g., balanced incomplete block designs, group-divisible designs with $\lambda_2 = \lambda_1 + 1$, triangular-type partially balanced* incomplete block (PBIB) designs with $\lambda_2 = \lambda_1 + 1$, L_2-type PBIB designs with $\lambda_2 = \lambda_1 + 1$, and many others. It also follows from the result of Doob and Cvetković [7] that $-2 < \delta < -1$ is possible only if the degree (n) of $G(d)$ is 2. Thus if $n \neq 2$ and a balanced incomplete block design or a group-divisible design with $\lambda_2 = \lambda_1 + 1$ does not exist, then any regular graph design such that the smallest eigenvalue of $\mathbf{T}$ is -2 [or all the connected components of $G(d)$ are line graphs] is E-optimal over the regular graph designs. It is interesting to note that among the 209 parameter sets considered by John and Mitchell, all of the 176 designs known to be E-optimal over equally replicated designs in $S_{v,b,k}$ (see ref. 4) have $\delta \geqslant -2$ or are duals of designs with $\delta \geqslant -2$. Even among the D- and A-optimal designs found by John and Mitchell, 173 have $\delta \geqslant -2$ or are duals of such designs. As a matter of fact, all the 209 designs satisfy $\delta \geqslant -3$ or are duals of designs with $\delta \geqslant -3$. For details of the above discussion, see ref. 5.

A general rule for determining D- and A-optimal regular graph designs is still lacking, but as long as the number of blocks is not too small, one would expect little difference among the regular graph designs; any of them is highly efficient if not optimal. The efficiencies of regular graph designs are discussed in ref. 3.

We close with a remark about the relationship between regular graph designs and PBIB designs with two associate classes. Although a regular graph design has at most two distinct λ_{ij}'s, it is not necessarily a PBIB design with two associate classes. A necessary and sufficient condition for a regular graph design d to be a PBIB design with two associate classes is that the regular graph $G(d)$ be strongly regular. (See ref. 20 for a definition of strongly regular graphs.) This is also equivalent to the C matrix of d having two distinct nonzero eigenvalues.

References

[1] Cameron, P. J., Goethals, J. M., Seidel, J. J., and Shult, E. E. (1976). *J. Algebra*, **43**, 305–327. (A beautiful work characterizing the graphs with the smallest eigenvalue $\geqslant -2$.)

[2] Cheng, C. S. (1978). *Ann. Statist.*, **6**, 1239–1261.

[3] Cheng, C. S. (1978). *Commun. Statist., A*, **7**, 1327–1338.

[4] Cheng, C. S. (1980). *J. R. Statist. Soc. B*, **42**, 199–204.

[5] Cheng, C. S. and Constantine, G. M. (1986). *J. Statist. Plann. Inf.* **15**, 1–10.

[6] Conniffe, D. and Stone, J. (1975). *Biometrika*, **62**, 685–686.

[7] Doob, M. and Cvetković, D. (1979). *Linear Algebra Appl.*, **27**, 17–26.

[8] Jacroux, M. A. (1980). *J. R. Statist. Soc. B*, **42**, 205–209.

[9] John, J. A. (1966). *J. R. Statist. Soc. B*, **28**, 345–360.

[10] John, J. A. and Mitchell, T. J. (1977). *J. R. Statist. Soc. B*, **39**, 39–43. (Introduces regular graph designs.)

[11] John, J. A. and Williams, E. R. (1982). *J. R. Statist. Soc. B.*, **44**, 221–225. (Discusses several outstanding conjectures on optimal design.)

[12] John, J. A., Wolock, F. W., and David, H. A. (1972). *Cyclic Designs*. Appl. Math. Ser. **62**, National Bureau of Standards, Washington, DC.

[13] Jones, B. and Eccleston, J. A. (1980). *J. R. Statist. Soc. B*, **42**, 238–243.

[14] Kiefer, J. (1958). *Ann. Math. Statist.*, **29**, 675–699.

[15] Kiefer, J. (1975). In *A Survey of Statistical Design and Linear Models*, J. N. Srivastava, ed. North-Holland, Amsterdam, pp. 333–353.

[16] Mitchell, T. J. (1973). *Proc. 39th Session ISI*, 199–205.

[17] Mitchell, T. J. (1974). *Technometrics*, **16**, 203–210.

[18] Mitchell, T. J. and John, J. A. (1976). *Report No. ORNL/CSD-8*, Oak Ridge National Laboratory. (Contains an extensive table of optimal regular graph designs.)

[19] Patterson, H. D. and Williams, E. R. (1976). *Biometrika*, **63**, 83–92.

[20] Raghavarao, D. (1971). *Constructions and Combinatorial Problems in Design of Experiments*. Wiley, New York.

[21] Takeuchi, K. (1961). *Rep. Statist. Appl. Res. Union Japan Sci. Eng.*, **8**, 140–145.

[22] Williams, E. R., Patterson, H. D., and John, J. A. (1977). *Biometrics*, **33**, 713–717.

(BLOCKS, BALANCED INCOMPLETE
DESIGN OF EXPERIMENTS
GRAPH THEORY
GROUP DIVISIBLE DESIGNS
NEARLY BALANCED DESIGNS
OPTIMUM DESIGN OF EXPERIMENTS
PARTIALLY BALANCED DESIGNS
SUPERSATURATED DESIGNS
WITT DESIGNS)

CHING-SHUI CHENG

REINFORCED INCOMPLETE BLOCK DESIGNS

These designs were introduced by Das [1]. They are augmented *incomplete block* designs* obtained by including in each block of an ordinary balanced incomplete block design a certain number of additional treatments and then taking some more blocks in each of which all the treatments are present. In some cases this results in balanced block designs with a smaller number of replications than a balanced incomplete design having the same design parameters. See Das [1] and Giri [2] for more details.

References

[1] Das, M. N. (1956). *J. Indian Soc. Agric. Statist.*, **10**, 73–77.

[2] Giri, N. C. (1958). *J. Indian Soc. Agric. Statist.*, **12**, 41–51.

(PARTIALLY BALANCED DESIGNS)

REJECTABLE PROCESS LEVEL (RPL)

Rejectable process level (RPL) [also known as limiting quality level (LQL), unacceptable quality level (UQL), and lot tolerance percent defective (LTPD)] is the percentage or proportion of "variant units" (namely nonconforming or defective units) in a lot for which—for purposes of acceptance sampling—the consumer wishes the probability of acceptance to be restricted to a specified *low* value. [Compare with acceptance process level* (APL).] *Sampling plans** indexed by RPL are designed for use when emphasis is placed on the quality of individual lots.

(ACCEPTANCE SAMPLING
QUALITY CONTROL, STATISTICAL)

REJECTION

The term "rejection," in regard to a statistical hypothesis*, has a highly sophisticated meaning. It means that observed values indicate—according to a predetermined rule, called a test—that a statistical hypothesis is not valid. It *need not* mean that the hypothesis is rejected, in the sense that it is no longer worthy of consideration.

Attempts have been made to produce "action rules" of a more concise nature, usually based on cost consideration (*see* DECISION THEORY).

(HYPOTHESIS TESTING)

REJECTIVE SAMPLING (INCLUDING SAMPFORD–DURBIN SAMPLING)

Rejective sampling is a type of scheme for sampling a finite population. In its most general sense [9] it involves first selecting a sample by a given randomised procedure; if the sample does not obey a given criterion it is rejected and a second sample is then selected independently by the same procedure. This sample is also rejected if it does not obey the criterion and this process is continued until a sample is obtained that does obey the criterion. This forms the final sample to be used. For example, Royall and Cumberland [19] suggest a rejective scheme for *balanced sampling* where the randomised procedure is *simple random sampling** and the criterion is that the sample mean of an auxiliary variable x differs from the population mean of x by less than a specified amount.

The idea of rejective sampling goes back at least as far as Yates and Grundy in 1953 [22], although Hájek [11] was the first to coin the term in 1964. In both papers the context is without replacement *probability proportional to size** (PPS) sampling. Here the randomised procedure consists of a series of n with replacement random draws from the population and the criterion is that the n units in the sample are distinct. It is in this restricted sense that the term rejective sampling is usually understood and this article will only be concerned with this use. The case $n = 2$ is especially important because it corresponds to the common choice of sampling by PPS with two primary sampling units per stratum. Rao [17] suggested a simple rejective scheme for $n = 2$ that is equivalent to the method of Durbin [10]. This was generalised by Sampford [20], for $n > 2$, to a scheme known as Sampford–Durbin sampling. Before describing these methods we outline the associated estimation theory. *See also* PROBABILITY PROPORTIONAL TO SIZE SAMPLING and HORVITZ–THOMPSON ESTIMATOR.

ESTIMATION THEORY FOR WITHOUT REPLACEMENT PPS SAMPLING

We consider a finite population of N units with value y_i associated with the ith unit, $i = 1, \ldots, N$. A classical problem of survey sampling (see, e.g., ref. 8) is to choose a sampling design and an estimator for the population total

$$Y = \sum_{i=1}^{N} y_i.$$

A without replacement sampling design is defined as a probability measure $p(s)$ on the set of samples s consisting of subsets of the population, usually of fixed size n. An estimator is a function of the sampled y values.

Sometimes for each unit i in the population an auxiliary variable x_i is known that is related approximately proportionately to y_i. This variable is usually a *size* measure such that $x_i > 0$, $i = 1, \ldots, N$. For example, if the units are primary sampling units, then x_i may be the number of second stage units in the primary unit. In this case a common strategy is to use a *PPS sampling design*; that is a design $p(s)$ for which the inclusion probabilities Π_i are proportional to x_i,

$$\Pi_i = \sum_{s \ni i} p(s) = np_i, \qquad (1)$$

where $p_i = x_i/X$, $X = \sum_{i=1}^{N} x_i$ (and $x_i \leqslant X/n$ is assumed; $i = 1, \ldots, N$), and to use the *Horvitz–Thompson estimator*,

$$\hat{Y}_{\text{HT}} = \sum_{i \in s} y_i/\Pi_i.$$

This estimator is unbiased and (providing the sample size n is fixed) has variance

$$\operatorname{var}(\hat{Y}_{\text{HT}}) = \sum_{i<j}^{N} (\Pi_i \Pi_j - \Pi_{ij}) \times (\Pi_i^{-1} y_i - \Pi_j^{-1} y_j)^2,$$

where Π_{ij} is the joint inclusion probability of units i and j,

$$\Pi_{ij} = \sum_{s \ni i, j} p(s).$$

The standard estimator of $\operatorname{var}(\hat{Y}_{\text{HT}})$, proposed by Sen [21] and Yates and Grundy [22], is

$$v(\hat{Y}_{\text{HT}}) = \sum_{i<j \in s} (\Pi_i \Pi_j - \Pi_{ij}) \times (\Pi_i^{-1} y_i - \Pi_j^{-1} y_j)^2 \Pi_{ij}^{-1}. \qquad (2)$$

The problem then is to find a PPS design that has desirable properties such as:

(a) it is easily implemented;

(b) the Π_{ij} are positive and easy to compute for use in (2);

(c) $\Pi_i \Pi_j - \Pi_{ij} \geqslant 0$ for all i, j so that v is always nonnegative;

(d) the Π_{ij} are such that $\hat{Y}_{\text{HT}}$ is an efficient estimator of Y;

(e) the Π_{ij} are such that v is an efficient estimator of $\text{var}(\hat{Y}_{\text{HT}})$.

Other considerations can be added, such as ease of use in rotating samples (see ref. 6).

Given the multiplicity of the preceding objectives, it is perhaps not surprising that many solutions have been offered to this problem. Hansen and Hurwitz [13] were the first to propose a PPS scheme *with* replacement, whilst Madow [15] proposed the first without replacement PPS scheme using *systematic sampling**. Many papers followed and are reviewed in ref. [6]. Simple schemes for $n = 2$ were proposed by Brewer [4] and Durbin [10] for which properties **(b)** and **(c)** hold [for systematic sampling **(b)** may not hold since the Π_{ij} may be zero]. These schemes do extend to $n > 2$ (e.g., 5) although the Π_{ij} become more difficult to compute.

An algorithmic approach for $n > 2$ that has properties **(a)**, **(b)**, and **(c)** is given by Chao [7]. Rejective schemes are described in the next section.

REJECTIVE SAMPLING WITH FIXED DRAW PROBABILITIES

The simplest type of rejective sampling is now described. A set of fixed probabilities $\alpha_1, \ldots, \alpha_N$ is chosen:

$$\sum_{i=1}^{N} \alpha_i = 1, \qquad \alpha_i > 0, \quad i = 1, \ldots, N.$$

On each repeated (with replacement) draw from the population unit i is selected into the sample with probability α_i. If the selected unit already forms part of the sample, the whole sample is rejected and the procedure is begun again until a sample s of distinct units is obtained.

The aim is to choose the α_i such that $\Pi_i = np_i$ as in (1). The most obvious choice is to set $\alpha_i = p_i$. This is referred to as the *Yates–Grundy rejective procedure* in ref. 6 although Yates and Grundy [22] only mention it briefly, discarding it as inferior to a simple draw by draw approach. As an example of the Π_i produced, suppose $N = 4$, $n = 2$, and $p_i = i/10$, $i = 1, \ldots, 4$ (see ref. 22). For any rejective scheme with $n = 2$ and fixed draw probabilities α_i the inclusion probabilities are easily obtained as

$$\Pi_i = 2\alpha_i(1 - \alpha_i)\left(1 - \sum_{1}^{N} \alpha_j^2\right)^{-1}$$

In our example setting $\alpha_i = p_i$ the implied values of $\frac{1}{2}\Pi_i$ are 0.129, 0.229, 0.300, and 0.343 as compared with the desired values of 0.1, 0.2, 0.3, and 0.4.

Instead of taking $\alpha_i = p_i$ we would like to seek a function $\alpha_i = g(p_i)$, which implies $\Pi_i = np_i$. Unfortunately no closed-form expression for g is available although iterative solutions seem possible [16]. Some approximations are, however, given by Hájek [11, 12]. As a best approximation he suggests setting α_i proportional to

$$p_i + (\gamma + \delta - 1)p_i(1 - np_i)^{-1},$$

where

$$\gamma = \sum_{j=1}^{N} np_j(1 - np_j),$$

$$\delta = \sum_{j=1}^{N} n^2 p_j^2 (1 - np_j)/\gamma.$$

As a crude approximation he suggests taking α_i proportional to

$$p_i(1 - np_i)^{-1}$$

For the preceding example with $N = 4$, the best approximation gives α_i values of 0.070, 0.153, 0.268, and 0.510 with implied values of $\frac{1}{2}\Pi_i$ as 0.102, 0.202, 0.306, and 0.390, which are very close to the desired values. For the crude approximation the α_i values are 0.039, 0.104, 0.234, and 0.623 with the values of $\frac{1}{2}\Pi_i$ being 0.069, 0.171, 0.329, and 0.431, which are no better than for the Yates–Grundy procedure. Hájek's approximations, however, assume n large. For the case $N = 10$, $n = 4$, and $p_i = i/55$, $i = 1, \ldots, 10$, the values of $\frac{1}{4}\Pi_i/p_i$ vary between 0.82 and 1.47 for the Yates–Grundy procedure, between 0.85 and 1.06 for Hájek's crude procedure, and between 0.998 and 1.001 for Hájek's best procedure [11].

For $n = 2$ the joint inclusion probabilities are easily obtained as

$$\Pi_{ij} = 2\alpha_i\alpha_j\left(1 - \sum_1^N \alpha_k^2\right)^{-1}, \qquad i \neq j. \quad (3)$$

In this case it is easy to verify that property **(c)** holds. For $n > 2$ expressions for the Π_{ij} become more complicated and Hájek [12] gives the approximation

$$\Pi_{ij} = \Pi_i\Pi_j\left[1 - (1 - \Pi_i)(1 - \Pi_j)\gamma^{-1} + o(\gamma^{-1})\right]. \quad (4)$$

Hence property **(c)** again holds, at least approximately. For the $N = 4$ example using Hájek's best approximation to the α_i, the ratios of the $\Pi_i\Pi_j - \Pi_{ij}$ as predicted by (4) compared with their true values vary between 0.60 and 0.98. For the $N = 10$ example the range is 0.87 to 1.07 [11].

One practical problem with rejective sampling is that one may have to reject samples many times before obtaining a sample of distinct units. The probability of the event E that n consecutive draws will not be distinct clearly depends on n and the sampling fraction $f = nN^{-1}$. It also depends on the α_i and $\Pr(E)$ will be large if any of the α_i are close to unity. This can be seen for $n = 2$, where $\Pr(E) = \Sigma\alpha_i^2$. Hájek [12] gives an approximation for general n,

$$\Pr(E) = 1 - \exp\left[-\tfrac{1}{2}f(n - 1)(1 + V^2)\right],$$

where V^2 is a measure of dispersion of the α_i.

Rejective sampling is only one scheme that implements a given sampling design $p(s)$. The sampling design induced by a rejective sampling scheme with fixed draw probabilities α_i may alternatively be implemented by a conditional (rejective) Poisson sampling scheme (see ref. 12). In this approach the N population units are independently included in the sample with probabilities β_i where

$$\alpha_i = \beta_i(1 - \beta_i)^{-1} \Big/ \sum_{j=1}^N \beta_j(1 - \beta_j)^{-1}.$$

If the sample contains n distinct units it is retained; otherwise, it is rejected and the whole process repeated. This scheme proves to be less practical than the original scheme because it will generally lead to many more rejections. However, this characterisation of the sampling design is theoretically useful for obtaining asymptotic distributional results for $\hat{Y}_{\mathrm{HT}}$ (see ref. 11).

REJECTIVE SAMPLING WITH NONFIXED DRAW PROBABILITIES

We now allow the draw probabilities to vary over different draws. Otherwise the procedure remains as before, with the whole sample being rejected when repeat units are obtained and the identical procedure being begun again. In the *Sampford–Durbin* procedure the ith unit is selected with probability $\alpha_{1i} = p_i$ at the first draw and with probability α_{2i} proportional to $p_i(1 - np_i)^{-1}$ at the $n - 1$ subsequent draws. It may be shown [12, 20] that this scheme yields inclusion probabilities exactly equal to np_i. The scheme was first suggested by Rao [17] for $n = 2$ and was extended for $n > 2$ by Sampford [20]. For $n = 2$ the sampling design implied by this scheme may alternatively be implemented by the following nonrejective scheme of Durbin [10]. Select the first unit i with probability p_i and the second from the remaining $n - 1$ units with probability

$$p_{j,i} = \frac{p_j\left[(1 - 2p_i)^{-1} + (1 - 2p_j)^{-1}\right]}{\left[1 + \Sigma_{k=1}^N p_k(1 - 2p_k)^{-1}\right]}, \qquad j \neq i.$$

Sampford [20] also gives corresponding nonrejective schemes for $n > 2$ although these quickly become very complicated. For $n = 2$ second-order inclusion probabilities are given simply as

$$\Pi_{ij} = 2p_i p_{j,i}.$$

For $n > 2$ the Π_{ij} become much more complicated to compute, although formulae are given in ref. 20. Some approximate formulae that seem to work well for some numerical examples are given in refs. 2 and 12. The formula in ref. 12 is in fact identical to (5).

For $n = 2$ the rejective scheme using general unequal draw probabilities α_{1i} and α_{2i} is discussed in ref. 1, where it is shown that property **(c)** always holds. Even more generally we may allow the procedure to change following each rejection. In Hanurav's scheme A [14] two units are selected with probability p_i with replacement. If the two units are the same, the sample is rejected and two units are selected with probability proportional to p_i^2. The process is continued using probabilities proportional to $p_i^{2^{K-1}}$ on the Kth procedure. Hanurav shows that for this scheme $\Pi_i = np_i$ and properties **(b)** and **(c)** hold provided that the two largest p_i values are equal. If they are unequal a modified procedure is suggested.

CONCLUSION

Some empirical and model-based comparisons of different PPS strategies are given in refs. 18 ($n = 2$) and [3] ($n > 2$). Amongst rejective sampling schemes there seems very little difference in terms of criteria **(e)** and **(f)** between a fixed draw-probability scheme and the Sampford–Durbin scheme, and indeed little difference between them and the scheme of ref. 14. In comparison with nonrejective schemes, the rejective methods do seem to stand up well although a nonrejective method due to Murthy seems preferable when both criteria **(e)** and **(f)** are important. Whilst rejective methods are fairly straightforward to implement, simpler methods do exist, for example that of ref. 7 for general n. Certainly the major users of unequal probability sampling, governmental agencies conducting large scale surveys, seem to prefer systematic sampling for its simplicity even if it creates problems in variance estimation.

References

[1] Agarwal, S. K., Kumar, P., and Dey, A. (1982). *J. R. Statist. Soc. B*, **44**, 43–46.

[2] Asok, C. and Sukhatme, B. V. (1976). *J. Amer. Statist. Ass.*, **71**, 912–913.

[3] Bayless, D. L. and Rao, J. N. K. (1970). *J. Amer. Statist. Ass.*, **65**, 1645–1667.

[4] Brewer, K. R. W. (1963) *Aust. J. Statist.*, **5**, 5–13.

[5] Brewer, K. R. W. (1975) *Aust. J. Statist.*, **17**, 166–172.

[6] Brewer, K. R. W. and Hanif, M. (1983). *Sampling with Unequal Probabilities*. Springer-Verlag, New York. (A comprehensive reference book on designs and estimators. The best place to start after this article.)

[7] Chao, M. T. (1982). *Biometrika*, **69**, 653–656.

[8] Cochran, W. G. (1977). *Sampling Techniques*, 3rd ed. Wiley, New York. (A classic text on sampling.)

[9] Deshpande, M. N. and Prabhu-Ajgaonkar, S. G. (1977). *Biometrika*, **64**, 422–424.

[10] Durbin, J. (1967). *Appl. Statist.*, **16**, 152–164.

[11] Hájek, J. (1964). *Ann. Math. Statist.*, **35**, 1491–1523. (The paper that coined the term "rejective sampling." Complicated algebraically —try ref. 12 first.)

[12] Hájek, J. (1981). *Sampling From a Finite Population*. Dekker, New York. (A major reference on rejective sampling. Chapter 7 gives various approximate results for fixed-probability rejective sampling and Chapter 8 gives corresponding results for Sampford–Durbin sampling.)

[13] Hansen, K. H. and Hurwitz, W. N. (1943). *Ann. Math. Statist.*, **14**, 333–362.

[14] Hanurav, T. V. (1967). *J. R. Statist. Soc. B*, **29**, 374–391.

[15] Madow, W. G. (1949). *Ann. Math. Statist.*, **20**, 333–354.

[16] Rao, J. N. K. (1963). *J. Amer. Statist. Ass.*, **58**, 202–215.

[17] Rao, J. N. K. (1965). *J. Indian Statist. Ass.*, **3**, 173–180.

[18] Rao, J. N. K. and Bayless, D. L. (1969). *J. Amer. Statist. Ass.*, **64**, 540–559.

[19] Royall, R. M. and Cumberland, W. G. (1981). *J. Amer. Statist. Ass.*, **76**, 66–88.

[20] Sampford, M. R. (1967). *Biometrika*, **54**, 499–513. (The first three sections provide a very clear introduction to Sampford–Durbin sampling.)

[21] Sen, A. R. (1953). *J. Indian Soc. Agric. Statist.*, **5**, 119–127.

[22] Yates, F. and Grundy, P. M. (1953). *J. R. Statist. Soc. B*, **15**, 253–261.

(FINITE POPULATIONS, SAMPLING FROM
HORVITZ–THOMPSON ESTIMATOR
PROBABILITY PROPORTIONAL TO SIZE SAMPLING
SIMPLE RANDOM SAMPLING
SURVEY SAMPLING
SYSTEMATIC SAMPLING)

C. J. Skinner

RELATIONSHIP

The notion of *relationship* is one of the most fundamental factors in statistical methodology, occurring at all levels from simple descriptive study of pattern among univariate random variables to aspects of complex structure in extensive multivariate data sets. Many other, more specific, concepts are subsumed under the general heading of relationship: These include association, causation*, classification*, correlation*, dependence*, functional and structural (law-like) models, goodness-of-fit*, linear (and generalised linear*) models, regression*, and components of time series* analysis. The principles and specific statistical techniques for many of these topics are discussed elsewhere in these volumes. We shall consider here only some of the conceptual and philosophical principles underlying the statistical manifestation of the notion of relationship (eschewing the broader nonstatistical philosophical, sociological, and semantic aspects ably covered in more general encyclopedias). The study of relationships exhibited through statistical data is one of the earliest aspects of statistical enquiry, as is evidenced, for example, in the work of Lagrange, Gauss, Laplace, and others around the turn of the eighteenth century.

We can conveniently trace different aspects of "relationship" by examining a hypothetical practical situation of ever-increasing complexity.

Suppose we are interested in the yield of a particular variety of wheat and take measurements of yield for many sites of similar size. We can express the results as a (grouped) frequency distribution. But we might be able to usefully fit some distributional model to the data; perhaps a normal distribution. Already we encounter a primitive notion of relationship: The observed frequencies are *not unrelated*, they follow (with acceptable levels of random variation) a pattern anticipated under the normal model we have fitted. Thus relationship is embodied in simple model fitting* processes (for univariate or multivariate data) and associated parametric or nonparametric procedures for examining the goodness-of-fit of the models.

Consider two extensions of the practical problem: For each yield measurement we might note which one of four levels (or types) of fertilizer treatment was applied and we might also record the altitude of the site.

It would not be surprising to find that mean yield varies with (is not unrelated to) the level of fertilizer and in this respect analysis of variance* techniques (and nonparametric equivalents) are used to investigate relationship. Examination of contrasts* in designed experiment data enables us to proceed from the *rejection of lack of relationship* between mean yield and fertilizer level to an examination of *the nature of such a relationship* (especially where the fertilizer levels are quantitative, when orthogonal polynomials* analysis may be relevant).

Alternatively we may observe that the wheat yield tends to vary with the altitude measurements: a tendency perhaps to find lower yields at higher altitudes. We do not expect to find a deterministic relationship, more a propensity for one variable to change as does the other, but with a degree of scatter about the relationship. The extent to (and direction in) which any linearity of relationship is present in the data is represented by the idea of correlation with the various sample correlation coefficients (product-moment, rank, etc.) providing corresponding statistical estimates.

Extensive data sets (for example, with very many wheat yields and associated altitudes) may be available only in frequency form. Here we will have a contingency table* where the two margins represent yield and altitude grouped in convenient ways. A test of independence (or in other contexts, of homogeneity of distribution) provides a basis for examining the prospect of no relationship between the two variables. The extent and direction of any relationship is also estimable by means of the coefficient of contingency* or (for ordered marginal groupings) modified rank correlation coefficients* or a variety of other measures of association such as those due to Goodman and Kruskal [1].

Rather than studying the symmetric correlation relationship between yield and altitude we may wish to examine how knowledge of altitude (as a predictor or regressor variable) provides information about the value of yield (as a response variable) that may be encountered. We are now concerned with the important field of regression*; where relationship is modelled in terms of the conditional expectation $\mu_{Y|x}$ of the response variable Y given the value of a predictor (or regressor) variable X. The variables now have differing roles, the regressor variable serving as an input variable and the response as an output variable whose characteristics we wish to infer. If $\mu_{Y|x} = \alpha + \beta x$ we have a linear regression model* as a special case, but in practice we will often need more complicated models (e.g., polynomial regression*).

In our agricultural example there are various regression prospects. We can examine if some declared regression model makes sense in describing how yield varies with altitude. Inversely, we might ask what altitude ranges might be expected to lead to particular yield values (this inverse regression problem relates to calibration*). If a linear regression model seems appropriate with constant β values for different levels of fertilizer we can then use analysis of covariance* methods to allow for the *relationship between yield and altitude* in examining the *relationship between yield and fertilizer level.*

Suppose even more information is available, with latitude and rainfall figures as additional regressor variables and straw yields as a further response variable. Many other forms of relationship can now be investigated. In the field of correlation we might examine the multiple correlation* between the three environmental variables or partial correlations* (e.g., between rainfall and altitude having allowed for latitude). Canonical correlation* is also informative in determining which linear combinations of the two yield variables and the three environmental variables are most highly correlated. In the regression context we might study the multiple regression* of yield on all three environmental variables.

The classical linear model wherein a vector of n observations $\mathbf{y}$ is related to a vector of p parameters $\boldsymbol{\beta}$ in the form

$$\mathbf{y} = \mathbf{A}\boldsymbol{\beta} + \boldsymbol{\epsilon}$$

(where $\mathbf{A}$ is a known $n \times p$ design matrix and $\boldsymbol{\epsilon}$ a vector of n observations of zero-mean residual random variables), has long played a central role in modelling statistical relationships. It encompasses multiple regression problems and, with $\mathbf{A}$ having components 0 or 1 (and normal residuals), it underlies the wide field of analysis of variance for designed experiments. More recently (see, e.g., McCullagh and Nelder [2]), the generalized linear model* has come into prominence, where $\mathbf{y}$ has expected value $\boldsymbol{\mu}$ with $\boldsymbol{\mu} = g(\mathbf{A}\boldsymbol{\beta})$ and the link function $g(\ )$ is a monotonic differentiable function and where the distribution of residuals may be any member of the exponential family. This goes beyond the classical linear model that is included as a special case (as are probit* and logit* models for proportional responses and log-linear models* for counts or frequencies).

Many aspects of relationship have been illustrated by means of this simple agricultural example, but it is necessary to draw some basic distinctions and to indicate some refinements and extensions.

The difference between correlation and regression exemplifies a distinction drawn by Kendall and Stuart [3] between *interdependence* (a symmetric relationship between two or more variables) and *dependence* (of one or more variables on other variables). It is important to recall that statistical relationship does not imply direct causation*; it can and usually does reflect concomitant factors. We must beware of G. B. Shaw's tongue-in-cheek conclusion "that the wearing of tall hats . . . enlarges the chest" (preface to *The Doctor's Dilemma* (1906); see also Kendall and Stuart [3]). Conversely, causal relationships need not show up as statistical relationships; it will depend on how we model the relationship. Thus a correlation coefficient is typically a measure of *linear* relationship and need not reasonably reflect more complex relationships.

We need to consider the status of the variables. They may be *random variables** or *mathematical (nonrandom) variables* and this constitutes a further basic distinction. The categorization is not always clear, however. In relating wheat yield to fertilizer level the former variable is random whilst the latter variable is typically nonrandom (although if we contemplate choosing the four levels at random from some larger set of possibilities even this apparently clear distinction disappears). In comparing wheat yield with altitude, the altitude variable may or may not be random. In correlation studies we assume the latter—we relate two random variables. In regression, we assume either that the regressor variable is predetermined (nonrandom) or draw inferences *conditional on* observed values of the regressor variable.

An intermediate class of problems where both (all) variables have the *same* status but we are interested in explicitly modelling interrelationship is of some importance. Suppose we want to estimate the density d of moon rock from specimens where weight W and volume V have been determined. We know that the variables are related, $W = dV$. But we can only observe W and V with superimposed errors of measurement ϵ and η. So our data constitute a sample of values $(y_1, x_1) \cdots (y_n, x_n)$ of random variables (Y, X) where $Y = W + \epsilon$ and $X = V + \eta$. We are now concerned with so-called lawlike relationships*; depending on assumptions we use linear functional models or linear structural models.

We have only illustrated aspects of statistical relationship and mainly in the context of experimental data. Relationships exist, of course, and merit study in survey data or even with full enumeration. The field of time series* also yields many relationship problems particularly of an interesting intercorrelational style (see, for example, Priestley [4]).

References

[1] Goodman, L. A. and Kruskal, W. H. (1963). *J. Amer. Statist. Ass.*, **58**, 310–364.

[2] McCullagh, P. M. and Nelder, J. A. (1983). *The Generalised Linear Model*. Chapman and Hall, London, England.

[3] Kendall, M. G. and Stuart, A. (1979). *The Advanced Theory of Statistics*, 4th ed., Vol. 2. Griffin, London, England.

[4] Priestley, M. B. (1981). *Spectral Analysis and Time Series*. Academic, London, England.

(CAUSATION
CORRELATION
DEPENDENCE, CONCEPTS OF
LAWLIKE RELATIONSHIPS
STRUCTURAL INFERENCE)

VIC BARNETT

RELATIVE ASYMPTOTIC VARIANCE

See EFFICIENCY, ASYMPTOTIC RELATIVE (ARE)

RELATIVE BETTING ODDS

See CLINICAL TRIALS

RELEVATION

The Stieltjes convolution* of two distribution functions with nonnegative support*, F

and G say, is

$$F * G(t) = \int_0^t F(t-u)\, dG(u).$$

It is the distribution function of the time to the failure of the second of two components when the second component (with life distribution G) is placed in service on the failure of the first (with life distribution F); the replacement component is assumed to be new on installation and to be independent of the first. Suppose, however, that the failed component is replaced by one of equal age. Then the survivor function of the time until the failure of both components is said to be the *relevation* of $\bar{F} = 1 - F$ and $\bar{G} = 1 - G$, and is denoted

$$\bar{F}\#\bar{G}(t) = \bar{F}(t) + \bar{G}(t)\int_0^t dF(u)/\bar{G}(u).$$

This definition is due to Krakowski [8].

One can interpret $\bar{F}\#\bar{G}(t)$ in the following manner: at time 0, a component with life distribution F is set into operation. Simultaneously, a standby component with life distribution G, independent of the first, is also installed. Should the first component fail before the standby component, the latter instantaneously assumes the function of the first component. Then the survivor function of the time to the second failure is $\bar{F}\#\bar{G}(t)$.

The asymmetry of the definition suggests, correctly, that relevation is not a commutative operation; indeed, $\bar{F}\#\bar{G}(t) \equiv \bar{G}\#\bar{F}(t)$ if and only if there exists a real constant α such that $\bar{F}(t) = [\bar{G}(t)]^{\alpha}$ for all t. It can also be shown that relevation is not associative. Further, relevation is left-distributive, but not right-distributive.

Suppose, now, that the life lengths of the two components are not independent. The survivor function of the time to failure of the second of the two components is now

$$\bar{F}_T(t) = \bar{F}(t) + \int_0^t \frac{S(t|u)}{S(u|u)}\, dF(u),$$

where $S(x|y)$ is the probability that the lifetime of the second component exceeds x given that the life length of the first component equals y and F is the marginal distribution of the life length of the first component.

We now give some examples of special cases where X and Y are exchangeably distributed with common marginal distribution function F; this would be the case, for example, when both components come from the same batch. If the joint distribution of X and Y is of the Fairlie–Gumbel–Morgenstern* form with parameter $\alpha \in (-1, 1)$, then

$$\bar{F}_T(t) = \bar{F}(t)\left\{1 + \int_{\bar{F}(t)}^{t} \frac{1 + \alpha F(t)(1-2u)}{u[1 + \alpha(1-u)(1-2u)]}\, du\right\}.$$

If X and Y follow a bivariate Burr* distribution with parameters $k > 0$ and $c > 0$, then

$$\bar{F}_T(t) = (1 + t^c)^{-k}\left\{1 + k\int_0^{t^c}\left[\frac{2u+1}{(u+1)(u+1+t^c)}\right]^{k+1} du\right\}.$$

If X and Y follow a bivariate Pareto* distribution with parameters $a > 0$ and $\theta > 0$, then

$$\bar{F}_T(t) = (\theta/t)^a\left\{1 + a\int_1^{t/\theta}\left[\frac{2u-1}{u(u+t/\theta-1)}\right]^{a+1} du\right\}.$$

See Johnson and Kotz [6] for further details.

It is clear that, if G is an exponential distribution*, $1 - \bar{F}\#\bar{G}(t) = F * G(t)$ for all t. Grosswald et al. [4] show that the converse holds provided that $G(t)$ admits of a power-series expansion. An alternative proof [10] assumes only that G is continuous.

In the characterization of many optimal maintenance and replacement policies, it is not necessary to know the functional form of the life distribution; it is sufficient to know the class in which the distribution lies (*see* HAZARD RATE AND OTHER CLASSIFICATIONS OF DISTRIBUTIONS). Important examples of such classes include the increasing failure rate* (IFR), increasing failure rate average* (IFRA), and new better than used* (NBU) distributions. *See also* RELIABILITY THEORY, PROBABILISTIC. Since relevation is not symmetric in F and G, closure theorems analogous to those concerning the preservation of class of life distribution under Stieltjes convolution cannot be found: different conditions on F and G are required. The following results are proved by Shanthikumar and Baxter [9]. Let $\Lambda_F(t) = -\log \bar{F}(t)$ denote the cumulative hazard function corresponding to the distribution F.

(1) If Λ_F/Λ_G is nondecreasing and G is NBU (IFRA), then $1 - \overline{F\#G}$ is NBU (IFRA).

(2) If $\bar{F}/\bar{G}$ is nonincreasing (or, alternatively, if Λ'_F and Λ'_G both exist and Λ'_F/Λ'_G is nondecreasing) and G is IFR, then $1 - \overline{F\#G}$ is IFR.

Result (2) replaces Theorem 2 of ref. 2, which is incorrect.

Let $\bar{F}_{(n)}$ denote the n-fold recursive relevation of $\bar{F}$ with itself ($n \geqslant 1$) where $\bar{F}_{(0)}(t) = 0(1)$ if $t \geqslant (<)0$. We can interpret $\bar{F}_{(n)}$ as the survivor function of either

(a) the time to the nth failure given that a failed component is replaced by an identical component of equal age

or

(b) the time to the nth failure if, after each breakdown, instantaneous repair restores the component to its condition immediately prior to failure, cf. Ascher's "bad-as-old" concept [1].

A straightforward induction argument [2, 8] shows that

$$\bar{F}_{(n)}(t) = \bar{F}(t) \sum_{k=0}^{n-1} [\Lambda_F(t)]^k/k!$$

for all n, i.e., that the sequence of component failures comprises a nonhomogeneous Poisson process* [2, 3]. Thus, relevation generates a nonhomogeneous Poisson process in a manner analogous to the way in which Stieltjes convolution generates a renewal process*. Note that $\bar{F}_{(n)}(t)$ is a generalization of the distribution function of the gamma distribution [8].

Relevation has been applied to warranty analysis: an item with life distribution F is required by a consumer for a length of time L; on failure before L, the item is replaced (or repaired) free if it fails during the warranty period of length W; otherwise it is replaced (or repaired) at a fixed cost K. Assuming that the sequence of failures follows a nonhomogeneous Poisson process, it can be shown that the total expected outlay by the consumer is $K[1 + \sum_{n=1}^{\infty} A^{(n)}(L)]$, where $A(L) = [F(L + W) - F(W)]/\bar{F}(W)$ and where the superscript denotes n-fold recursive Stieltjes convolution. See ref. 2 for details.

A further application is in hierarchical replacement [5, 7], where failed components are replaced from stock that has been subject to deterioration while in storage.

References

[1] Ascher, H. E. (1968). *IEEE Trans. Rel.*, **R-17**, 103–110.

[2] Baxter, L. A. (1982). *Naval Res. Logist. Quart.*, **29**, 323–330.

[3] Girault, M. (1976). *Rev. Française Aut. Inf. Rech. Opérat.* Sér. V-2, **10**, 71–73.

[4] Grosswald, E., Kotz, S., and Johnson, N. L. (1980). *J. Appl. Prob.*, **17**, 874–877.

[5] Johnson, N. L. and Kotz, S. (1979). *IEEE Trans. Rel.*, **R-28**, 292–299.

[6] Johnson, N. L. and Kotz, S. (1981). *Amer. J. Math. Manag. Sci.*, **1**, 155–165.

[7] Kotz, S. and Johnson, N. L. (1984). *Rel. Eng.*, **7**, 151–157.

[8] Krakowski, M. (1973). *Rev. Française Aut. Inf. Rech. Opérat.* Sér. V-2, **7**, 107–120.

[9] Shanthikumar, J. G. and Baxter, L. A. (1985). *Naval Res. Logist. Quart.*, **32**, 185–189.

[10] Westcott, M. (1981). *J. Appl. Prob.*, **18**, 568.

(POISSON PROCESS
RELIABILITY, IMPORTANCE OF COMPONENTS
SURVIVAL ANALYSIS)

L. A. BAXTER

RELIABILITY COEFFICIENTS (TUCKER–LEWIS)

Tucker and Lewis [2] proposed a *reliability coefficient* designed to measure goodness-of-fit for a factor analysis* model fitted by maximum likelihood*.

The reliability coefficient for a fitted factor model with k factors is

$$\hat{\rho}_k = (M_k - M_{k+1})/(M_k - 1),$$

where M_k is defined (with $k = p$) in FACTOR ANALYSIS (Vol. 3, page 4). When k is sufficiently large, $\hat{\rho}_k$ should approximate to 1. Empirical investigations (e.g., ref. 1) indicate that values of $\hat{\rho}_k$ of 0.90 or greater correspond to a satisfactory fit.

References

[1] Bohrnstedt, G. W. (1983). Measurement. In *Handbook for Survey Research*, P. H. Rossi, J. D. Wright, and A. B. Anderson, eds. Academic, New York.

[2] Tucker, L. R. and Lewis, C. (1973). *Psychometrika*, **38**, 1–10.

(FACTOR ANALYSIS
LISREL)

RELIABILITY: IMPORTANCE OF COMPONENTS

In reliability theory* (*see* COHERENT STRUCTURE THEORY and MULTISTATE COHERENT SYSTEMS) a key problem is to find out how the reliability of a complex system can be determined from knowledge of the reliabilities of its components. However, trying to apply this theory on a large technological system, seems often almost impossible. This is due to a poor and often irrelevant data base, to little knowledge on human components, and vague information on the dependencies coming into play. This was clearly demonstrated in the Reactor Safety Study [9] on the safety of nuclear reactors in the USA (*see also* NUCLEAR MATERIAL SAFEGUARDS). Hence the use of risk analysis and reliability theory to back political decisions on some controversial safety issues, may at least be doubtful.

If, however, a political decision is already made, these disciplines can contribute essentially to *improve* the safety of a system. This seems to be the present philosophy for instance in both the existing nuclear industry and in offshore engineering. When aiming at such improvements, measures of relative importance of each component to system reliability are basic tools. First, it permits the analyst to determine which components merit the most additional research and development to improve overall system reliability* at minimum cost or effort. Second, it may suggest the most efficient way to diagnose system failure by generating a repair checklist for an operator to follow.

SOME MEASURES OF IMPORTANCE OF SYSTEM COMPONENTS

Consider a system consisting of n components. As is true for most of the theory in this field, we shall here restrict to the case where the components and hence the system cannot be repaired. We shall also assume that we have a binary description of system and component states as in classical COHERENT STRUCTURE THEORY. Let $(i = 1, \ldots, n)$

$$X_i(t) = \begin{cases} 1, & \text{if } i\text{th component functions at time } t, \\ 0, & \text{if } i\text{th component is failed at time } t. \end{cases}$$

For mathematical convenience the stochastic processes $\{X_i(t),\ t \geqslant 0\}$, $i = 1, \ldots, n$, are assumed to be mutually independent. Introduce

$$\underline{X}(t) = (X_1(t), \ldots, X_n(t))$$

and let

$$\phi(\underline{X}(t)) = \begin{cases} 1, & \text{if system functions at time } t, \\ 0, & \text{if system is failed at time } t, \end{cases}$$

where the system's structure function ϕ is assumed to be coherent.

Let now the ith component have an absolutely continuous life distribution $F_i(t)$ with density $f_i(t)$. Then the reliability of this component at time t is given by

$$P(X_i(t) = 1) = 1 - F_i(t) \overset{\text{def}}{=} \bar{F}_i(t)$$

Introduce

$$\underline{\bar{F}}(t) = (\bar{F}_1(t), \ldots, \bar{F}_n(t)).$$

Then the reliability of the system at time t is given by

$$P(\phi(\underline{X}(t)) = 1) = h(\underline{\bar{F}}(t)),$$

where h is the system's reliability function.

The following notation will be used

$$(\cdot_i, \underline{x}) = (x_1, \ldots, x_{i-1}, \cdot, x_{i+1}, \ldots, x_n).$$

Birnbaum [3] defines the importance of the ith component at time t by

$$I_B^{(i)}(t) = P[\phi(1_i, \underline{X}(t)) - \phi(0_i, \underline{X}(t)) = 1],$$

which in fact is the probability that the system is in a state at time t in which the functioning of the ith component is critical for system functioning. As in ref. 1 it is not hard to see that

$$I_B^{(i)}(t) = \partial h(\underline{\bar{F}}(t))/\partial \bar{F}_i(t),$$

which is the rate at which system reliability improves as the reliability of the ith component improves. Vesely [10] and Fussell [4] suggest the following definition of the importance of the ith component at time t:

$$I_{\mathrm{VF}}^{(i)}(t) = P(X_i(t) = 0 | \phi(\underline{X}(t)) = 0).$$

Hence this definition takes into account the fact that a failure of a component can be contributing to system failure without being critical. However, also a failure of the ith component after system failure but before time t is contributing to this measure. Another objection is that according to this measure all components in a parallel system are equally important at any time irrespective of their life distributions.

One objection against both measures above, when applied during the system development phase, is that they both give the importance at fixed points of time, leaving it for the analyst to determine which points are important. This is not the case for the definition by Barlow and Proschan [2], who give the (time-independent) importance of the ith component by

$I_{\mathrm{BP}}^{(i)} = P$(the failure of the ith component coincides with the failure of the system).

Now

$$I_{\mathrm{BP}}^{(i)} = \int_0^\infty I_{\mathrm{B}}^{(i)}(t) f_i(t)\, dt = \int_0^\infty [h(1_i, \underline{\bar{F}}(t)) - h(0_i, \underline{\bar{F}}(t))] f_i(t)\, dt,$$

implying that the Barlow–Proschan measure is a weighted average of the Birnbaum measure, the weight at time t being $f_i(t)$.

Intuitively it seems that components that by failing strongly reduce the remaining system lifetime are the most important. This seems at least true during the system development phase. However, even when setting up a repair checklist for an operator to follow, one should just not try to get the system functioning. Rather one should try to increase the time until the system breaks down next. Introduce the random variable

$Z_i =$ reduction in remaining system lifetime due to the failure of the ith component.

Natvig [5] suggests the following measure of the importance of the ith component:

$$I_{N_1}^{(i)} = EZ_i \Big/ \sum_{j=1}^{n} EZ_j.$$

In ref. 6, Z_i is given the following representation:

$$Z_i = Y_i^1 - Y_i^0, \qquad (1)$$

where

$Y_i^1 =$ remaining system lifetime just *after* the failure of the ith component, which, however, immediately undergoes a mini mal repair; i.e., it is repaired to have the same distribution of remaining lifetime as it had just before failing;

$Y_i^0 =$ remaining system lifetime just *after* the failure of the ith component.

Also the distribution of Z_i is arrived at.

Let now T be the lifetime of a new system and T_i the lifetime of a new system where the life distribution of the ith component is replaced by the corresponding distribution where exactly one minimal repair of the component is allowed. As in ref. 7 it follows from (1) that

$$Z_i = T_i - T,$$

which leads to

$$EZ_i = \int_0^\infty \bar{F}_i(t)(-\ln \bar{F}_i(t)) I_B^{(i)}(t)\, dt.$$

If instead a total repair of the ith component is allowed, i.e., the component is repaired to have the same distribution of remaining lifetime as originally, the expected increase in system lifetime is given by

$$EU_i = \int_0^\infty \int_0^t f_i(t-u)\bar{F}_i(u)\, du\, I_B^{(i)}(t)\, dt.$$

Finally, the expected increase in system lifetime by replacing the ith component by a perfect one, i.e., $\bar{F}_i(t)$ is replaced by 1, is given by

$$EV_i = \int_0^\infty F_i(t) I_B^{(i)}(t)\, dt.$$

Now let the components have proportional hazards, i.e.,

$$\bar{F}_i(t) = \exp(-\lambda_i R(t)),$$
$$\lambda_i > 0;\ t \geqslant 0,\ i = 1, \ldots, n,$$

where λ_i, $i = 1, \ldots, n$, are the proportional hazard rates*. In ref. 6 the following measure is suggested

$$I_{N_2}^{(i)} = \frac{\partial ET}{\partial \lambda_i^{-1}} \Bigg/ \sum_{j=1}^n \frac{\partial ET}{\partial \lambda_j^{-1}}.$$

At least for the special case where components are exponentially distributed* this measure is easily motivated since λ_i^{-1} is the expected lifetime of the ith component. As in ref. 7 it is not hard to see that

$$\frac{\partial ET}{\partial \lambda_i^{-1}} = \lambda_i EZ_i.$$

We now define the measures

$$I_{N_3}^{(i)} = EU_i \Bigg/ \sum_{j=1}^n EU_j,$$

$$I_{N_4}^{(i)} = EV_i \Bigg/ \sum_{j=1}^n EV_j.$$

Hence we see that all measures $I_{N_k}^{(i)}$, $k = 1, 2, 3, 4$ and $I_{BP}^{(i)}$ are weighted averages of the Birnbaum measure. In ref. 7 one is comparing the different weight functions. A preliminary conclusion seems to be that the $I_{N_1}^{(i)}$ measure is advantageous.

As a very simple example from ref. 5 consider a series system of two components where

$$\bar{F}_i(t) = \exp(-\lambda_i t^{\alpha_i}),$$
$$\lambda_i > 0,\ i = 1, 2;\ \alpha_1 = 2,\ \alpha_2 = 1;\ t \geqslant 0.$$

For instance for $\lambda_2 \big/ \sqrt{2\lambda_1} = 0.6$ we have

$$I_{BP}^{(2)} = 0.494 < 0.506 = I_{BP}^{(1)},$$

whereas

$$I_{N_1}^{(2)} = 0.539 > 0.461 = I_{N_1}^{(1)}.$$

Hence the ordering of importance is different using the two measures, illustrating the need for a theory behind the choice of measures.

Finally, the measures suggested in refs. 2, 3, and 5 are generalized to the multistate case in ref. 8. As a concluding remark it should be admitted that the costs of improving the components are not entering into the measures reviewed here. Hence continued research in this important field is needed.

References

[1] Barlow, R. E. and Proschan, F. (1975). *Statistical Theory of Reliability and Life Testing: Probability Models*. Holt, Rinehart and Winston, New York.

[2] Barlow, R. E. and Proschan, F. (1975). *Stoch. Processes Appl.*, **3**, 153–173.

[3] Birnbaum, Z. (1969). In *Multivariate Analysis II* (P. R. Krishnaiah, ed.), 581–592.

[4] Fussell, J. B. (1975). *IEEE Trans. Rel.*, **24**, 169–174.

[5] Natvig, B. (1979). *Stoch. Processes Appl.*, **9**, 319–330.

[6] Natvig, B. (1982). *J. Appl. Prob.*, **19**, 642–652 (correction, **20**, 713).

[7] Natvig, B. (1985). *Scand. J. Statist.* **12**, 43–54.

[8] Natvig, B. (1985). In *Probabilistic Methods in the Mechanics of Solids and Structures*. Springer-Verlag, Berlin.

[9] Reactor Safety Study (1975). An Assessment of Accident Risks in U.S. Commercial Nuclear Power Plants. *Report No. Wash-1400*, Nuclear Regulatory Commission, Washington, D.C.

[10] Vesely, W. E. (1970). *Nuclear Eng. Design*, **13**, 317–360.

(COHERENT STRUCTURE THEORY
MULTISTATE COHERENT SYSTEMS
NUCLEAR MATERIAL SAFEGUARDS
RELEVATION
SURVIVAL ANALYSIS)

B. NATVIG

RELIABILITY, NONPARAMETRIC METHODS IN

Reliability* deals with the study of the proper functioning of equipment and systems. In discussing statistical problems in reliability, it is convenient to think of two main, but related, types of situations: (a) those where the emphasis is on the lifetime or failure-free operating time of a system or piece of equipment, and (b) those where the emphasis is on broader aspects of a system's performance over time, the possibility of repeated failure and repair or of varying levels of performance being allowed for. The first case stresses modelling and estimation of lifetime distributions. Statistical methods for independent observations from univariate or multivariate distributions are relevant here. In the second case, stochastic processes such as renewal and Markov processes* are generally used to model system performance, and related statistical methods are important. Both areas employ nonparametric methods.

NONPARAMETRIC METHODS FOR LIFETIME DATA

Consider a nonnegative continuous random variable T representing a lifetime of some kind. Let $S(t) = \Pr(T \geqslant t)$ denote the survival distribution function (SDF) of T and denote the probability density function (PDF) and hazard rate* function (HRF) by $f(t) = -dS(t)/dt$ and $\lambda(t) = f(t)/S(t)$, respectively. The cumulative hazard function (CHF) is

$$\Lambda(t) = \int_0^t \lambda(s)\,ds = -\log S(t).$$

A main application of nonparametric methods is in the estimation of $S(t)$ or other characteristics of T's distribution.

Estimation of a Lifetime Distribution*

If a random sample $t_1, \dots, t_n$ from the distribution of T is observed, the empirical SDF given by $\hat{S}(t) = (\#\text{ of lifetimes} \geqslant t)/n$ is a nonparametric estimator of $S(t)$. For any fixed t, $n\hat{S}(t)$ has a binomial distribution* with parameters n and $S(t)$, and confidence limits for $S(t)$ are easily obtained (e.g., refs. 22 and 24). Lifetime data are, however, often right censored*. That is, for some observations the exact lifetime is not known, but only a lower bound on it, called a censoring time. In this case $\hat{S}(t)$ as just defined cannot be used, since we may not know exactly how many lifetimes exceed a given t. Instead, suppose that $t_1 < t_2 < \cdots < t_k$ are the distinct observed lifetimes in a sample consisting of k lifetimes and $n - k$ censoring times. Let $n(t)$ be the number of individuals still at risk, i.e., known to be alive and uncensored just prior to time t. The *product-limit estimator* of $S(t)$ is

$$\hat{S}(t) = \prod_{i:\, t_i < t} \left(\frac{n(t_i) - 1}{n(t_i)} \right), \tag{1}$$

with the proviso that, if there is a censoring time t^* larger than t_k, $\hat{S}(t)$ is undefined past t^*.

This estimator, also known as the *Kaplan–Meier estimator** after the authors who first studied it extensively [18], is motivated by the observation that for any constants $0 < a_1 < a_2 < \cdots < a_k$, we can decompose $\Pr(S \geqslant a_k)$ as

$$\Pr(S \geqslant a_1)\Pr(S \geqslant a_2|S \geqslant a_1) \cdots \times \Pr(S \geqslant a_k|S \geqslant a_{k-1}).$$

In (1), the term $(n(t_i) - 1)/(n(t_i))$ can be thought of as estimating the probability of surviving past t_i, given survival up to t_i. In general $n(t_{i+1}) \leqslant n(t_i) - 1$, with equality only when there are no censoring times between t_i and t_{i+1}. When no observations in the sample are censored, $\hat{S}(t)$ reduces to the previous empirical SDF. The estimator has been studied extensively under various censorship models (e.g., ref. 5). A variance estimator for $\hat{S}(t)$ is available and can be used to give large sample confidence limits or confidence bands for $S(t)$.

An alternative estimator of $S(t)$ is the empirical CHF estimator

$$\tilde{H}(t) = -\log \tilde{S}(t) = \sum_{i:\, t_i < t} \frac{1}{n(t_i)}. \quad (2)$$

This has similar asymptotic properties to $\hat{S}(t)$. Besides providing nonparametric estimates of $S(t)$ and $H(t)$, these estimators are useful in both graphical and significance test procedures for assessing parametric lifetime distributions (e.g., ref. 23) and in obtaining nonparametric estimates of other characteristics of the lifetime distribution, such as quantiles.

When data are uncensored or singly censored (i.e., only the r largest lifetimes are censored), expressions such as (1) and (2) can be written in terms of the order statistics of the sample; the censored-data procedures are then usually just truncated versions of the corresponding complete-data procedures.

Many other nonparametric procedures are also available: for example, bootstrap* or jackknife* methods can be used in conjunction with quantile estimators to obtain confidence limits. Nonparametric estimation of the PDF and HRF can be approached through smoothed versions of $\hat{S}(t)$ or $\tilde{H}(t)$ or through kernel* estimates [29]. Another related set of procedures uses nonparametric estimators in testing whether a distribution has characteristics such as a monotone increasing (or decreasing) HRF; for example, ref. 20 discusses numerous nonparametric properties of lifetime distributions and ref. 16 provides examples of related tests. Such methods are also useful in obtaining confidence bands for $S(t)$.

Tests of Equality of Distributions

There are many nonparametric tests for the equality of two or more distributions. Tests such as the Wilcoxon, Kruskal–Wallis, and exponential ordered scores tests were the first to be extended for use with censored data, making them especially useful with lifetime data. Since the mid 1970s there have been extensive further developments, particularly in the area of rank tests*; refs. 2, 17, and 22 contain numerous results. For examples and discussion of other types of tests, see ref. 15.

Regression Methods

Regression* analysis of failure or lifetime data is important, for example in handling data from life tests carried out under varying conditions or in analyzing observational lifetime data in relation to concomitant variables. Many nonparametric approaches to regression are actually semiparametric: they involve nonparametric assumptions about the underlying distribution of T, given the (vector of) regressor variables $\mathbf{x}$, but the dependence of T's distribution on $\mathbf{x}$ is specified parametrically. Two approaches are in wide use and involve what are referred to as proportional hazards and accelerated failure time models, respectively.

The proportional hazards model takes the effect of regressor variables to be multiplicative on the HRF of T. That is, the HRF is assumed to be of the form

$$\lambda(t|\mathbf{x}) = \lambda_0(t)g(\mathbf{x};\boldsymbol{\beta}), \quad (3)$$

where $\boldsymbol{\beta}$ is a vector of regression coefficients, $g(\mathbf{x};\boldsymbol{\beta})$ is $\geqslant 0$, and $\lambda_0(t)$ is a baseline HRF, for an individual with $g(\mathbf{x};\boldsymbol{\beta}) = 1$. The most widely used model (3) is that for which $g(\mathbf{x};\boldsymbol{\beta}) = \exp(\mathbf{x}'\boldsymbol{\beta})$, where $\mathbf{x}$ and $\boldsymbol{\beta}$ are both

$p \times 1$ vectors; then $\lambda_0(t)$ is the HRF for an individual with $\mathbf{x} = \mathbf{0}$. An approach to estimating and testing $\boldsymbol{\beta}$ without having to specify $\lambda_0(t)$ parametrically was given by Cox [9]. This is based on an idea similar to that used to motivate (1), and yields a pseudo likelihood* for $\boldsymbol{\beta}$ on which inferences can be based. It is also possible to estimate $\lambda_0(t)$ nonparametrically, along lines similar to those used to obtain (1) or (2). This approach is very powerful and has undergone extensive development; ref. 17 is a key source.

Accelerated failure time models take the effect of covariates to be multiplicative on the time scale or, in other words, $\mathbf{x}$ acts additively on the scale of $Y = \log T$. The distribution of Y given $\mathbf{x}$ is then of general location–scale regression form,

$$Y = \eta(\mathbf{x}; \boldsymbol{\beta}) + \sigma e, \qquad (4)$$

where e has error PDF $f_0(e)$, $-\infty < e < \infty$, and $\sigma > 0$ is a scale parameter. Nonparametric approaches to analysis under (4) avoid assumptions about the form of $f_0(e)$. Rank regression methods (e.g., ref. 17, Chap. 6) or least squares, sometimes combined with nonparametric SDF estimation (e.g., refs. 7 and 28), are both used.

Discrete Models

Binomial or multinomial models are frequently used with lifetime data, in particular when the time axis has been partitioned into intervals $[a_{j-1}, a_j)$, where $0 = a_0 < a_1 < \cdots < a_k < a_{k+1} = \infty$. Interest then focusses on estimating survival probabilities $S(a_j)$ and related quantities. Analogs of the methods discussed are available here. For example, life-table estimation (e.g., refs. 22 and 12) corresponds to product-limit estimation (1), and binomial and multinomial response models [22] provide methods of regression analysis. In a sense these methods can be termed parametric, since they involve specific distributional assumptions (i.e., multinomial), but they are nonparametric in spirit and provide robust methodology that does not assume a specific family of continuous distributions for T.

Multivariate Methods

Multivariate distributions frequently arise with lifetime problems in reliability. One important instance concerns multiple failure modes, where equipment or a system can fail in different ways. If each observation has associated with it a pair of variables (T, C), where T represents failure time and $C \in \{1, \ldots, k\}$ the mode of failure, then models can be specified in terms of the sub-SDF's $S_j(t) = \Pr(T \geqslant t,\ C = j)$, $j = 1, \ldots, k$. When there is no censoring, $\hat{S}_j(t) = (\#$ of observations with $T \geqslant t$ and $C = j)/n$ is a nonparametric estimator. With censoring, product-limit-type estimators of the $S_j(t)$'s and related quantities are available; see, e.g., ref. 22, Chap. 10.

Multivariate lifetime distributions are also important: for example, a system may have k components, with lifetimes $T_1, \ldots, T_k$. A complicating factor in estimating the joint SDF, $S(t_1, \ldots, t_k) = \Pr(T_1 \geqslant t_1, \ldots, T_k \geqslant t_k)$, is that many types of censoring schemes are possible in multivariate situations, some of which are difficult to handle. Product-limit-type estimators are available in some instances; e.g., see ref. 8. With some types of data only certain aspects of the joint distribution of $T_1, \ldots, T_k$ are estimable. The best known such situation is the classical competing risks* problem, where only $T = \min(T_1, \ldots, T_k)$ and $C = j : T_j = t$ are observed.

In system reliability work the lifetime of a system may be expressed in terms of the lifetimes $T_1, \ldots, T_k$ of its components. If the T_i's are independent, nonparametric methods for use with their separate distributions can be applied to yield inferences about system lifetime. In this and other connections, nonparametric tests of independence (e.g., of $T_1, \ldots, T_k$) are useful. Only a few procedures (e.g., ref. 6) are available for censored data*.

NONPARAMETRIC METHODS FOR SYSTEMS OPERATION DATA

Reliability problems involving the operation of a system over time often employ stochastic processes* to represent random aspects of the system's performance, such as failure and repair, or differential levels of performance. In this context point process* models are especially useful [3, 10]. A realization of such a process consists of events (failures, say) occurring at times $T_1 < T_2 < \cdots$. The random variable $W_i = T_i - T_{i-1}$ (with $T_0 = 0$) is called the ith interarrival time. Statistical questions concern things like the existence of a trend (that is, a tendency for failures to occur more closely or less closely together as time passes) and the effect of concomitant variables on the process. The most widely used models are Poisson processes*, renewal processes, and generalizations of them.

The main types of nonparametric procedures used with point process models in reliability are (a) tests for trend in a process, (b) nonparametric estimation of the intensity function of a nonhomogeneous Poisson process* (NHPP), (c) nonparametric methods used with the interarrival time distribution in a renewal process (RP), and (d) regression analysis. Category (a) includes procedures such as tests for independence of interarrival times, tests for stationarity of a process, and so on; see for example refs. 3 and 11. Under category (b), the failure intensity function for a NHPP, defined as

$$\lambda(t) = \lim_{h \downarrow 0} \frac{1}{h} \Pr\{\text{an event occurs in } (t, t+h)\},$$

and the associated cumulative intensity function, $\Lambda(t) = \int_0^t \lambda(s)\, ds$, can be estimated nonparametrically, and asymptotic properties can be obtained along lines similar to those leading to (2); e.g., see ref. 1. Other nonparametric approaches to the estimation of $\lambda(t)$ are also available; e.g., see ref. 4. Category (c) involves the use of nonparametric methodology for estimating a univariate distribution discussed earlier, since in a RP interarrival times are independent and identically distributed. A complicating twist occurs when observation of a process may start or end between two events [30]. Category (d) includes regression methods for the NHPP and more complicated processes ([25, 26]) similar to those involved with the proportional hazards lifetime model. Estimation of $\boldsymbol{\beta}$ is via a pseudo likelihood function analogous to that used for (3). For RPs, methods for lifetime data that deal with models (3) or (4) can be used in connection with analysis of the interarrival times.

Multistate models are potentially of much use, for example in dealing with systems that may be in any one of a set of distinct operational states at any point in time. Nonparametric methods are available for estimating transition intensities and for regression analysis of certain types of processes; methods for Markov and semi-Markov processes (e.g., refs. 13, 21, and 22) are, for example, well developed. At the time of this writing, experience in analyzing reliability data under such models is, however, rather limited.

SOME FURTHER READING

There are several books dealing with the analysis of lifetime data, and most discuss nonparametric methods. References 12, 17, 22, and 24 provide broad coverage; it will be noted that these techniques are also widely used in other areas such as biostatistics. References 14 and 19 provide general background on nonparametric methods, and ref. 27 provides special results useful in lifetime and reliability problems. For system operation and reliability applications, refs. 3 and 11 are useful.

References

[1] Aalen, O. O. (1978). *Ann. Statist.*, **6**, 701–726.

[2] Andersen, P. K., Borgan, O., Gill, R., and Keiding, N. (1982). *Int. Statist. Rev.*, **50**, 219–258.

[3] Ascher, H. and Feingold, H. (1984). *Repairable Systems Reliability: Modelling, Inference, Misconceptions and Their Causes*. Dekker, New York.

[4] Bartoszynski, R., Brown, B. W., McBride, C. M., and Thompson, J. R. (1981). *Ann. Statist.*, **9**, 1050–1060.

[5] Breslow, N. E. and Crowley, J. (1974). *Ann. Statist.*, **2**, 435–453.

[6] Brown, B. W., Hollander, M., and Korwar, R. M. (1974). In *Reliability and Biometry: Statistical Analysis of Life Length*, SIAM, Philadelphia, pp. 327–354.

[7] Buckley, J. and James, I. (1975). *Biometrika*, **66**, 429–436.

[8] Campbell, G. (1981). *Biometrika*, **68**, 417–422.

[9] Cox, D. R. (1972). *J. R. Statist. Soc. B*, **34**, 187–220.

[10] Cox, D. R. and Isham, V. (1980). *Point Processes*. Chapman and Hall, London, England.

[11] Cox, D. R. and Lewis, P. A. W. (1966). *The Statistical Analysis of Series of Events*. Methuen, London, England.

[12] Elandt-Johnson, R. and Johnson, N. L. (1980). *Survival Models and Data Analysis*. Wiley, New York.

[13] Fleming, T. R. (1978). *Ann. Statist.*, **6**, 1057–1070.

[14] Gnedenko, B. V., Puri, M. L., and Vincze, I., eds. (1982). *Nonparametric Statistical Inference*, Vols. 1 and 2. North-Holland, New York.

[15] Harrington, D. and Fleming, T. R. (1982). *Biometrika*, **69**, 533–546.

[16] Hollander, M. and Proschan, F. (1975). *Biometrika*, **62**, 585–593.

[17] Kalbfleisch, J. D. and Prentice, R. L. (1980). *The Statistical Analysis of Failure Time Data*. Wiley, New York.

[18] Kaplan, E. L. and Meier, P. (1958). *J. Amer. Statist. Ass.*, **53**, 457–481.

[19] Krishnaiah, P. R. and Sen, P. K., eds. (1985). *Handbook of Statistics*, Vol. 4, Elsevier, New York.

[20] Langberg, N. A., Leon, R. V., and Proschan, F. (1980). *Ann. Prob.*, **8**, 1163–1170.

[21] Lagakos, S. W., Sommer, C. J., and Zelen, M. (1978). *Biometrika*, **65**, 311–317.

[22] Lawless, J. F. (1982). *Statistical Models and Methods for Lifetime Data*. Wiley, New York.

[23] Nair, V. N. (1981). *Biometrika*, **68**, 99–104.

[24] Nelson, W. (1982). *Applied Life Data Analysis*. Wiley, New York.

[25] Oakes, D. (1981). *Int. Statist. Rev.*, **49**, 235–264.

[26] Prentice, R. L., Williams, B. J., and Peterson, A. V. (1981). *Biometrika*, **68**, 373–379.

[27] Sen, P. K. (1981). *Sequential Nonparametrics*. Wiley, New York.

[28] Shaked, M. and Singpurwalla, N. D. (1982). *IEEE Trans. Rel.*, **R-31**, 69–74.

[29] Singpurwalla, N. D. and Wong, M. (1983). *Commun. Statist. A*, **12**, 559–588.

[30] Vardi, Y. (1982). *Ann. Statist.*, **10**, 772–785.

(CENSORED DATA
COX'S REGRESSION MODEL
RANK TESTS
RELIABILITY, PROBABILISTIC
STOCHASTIC PROCESSES, POINT
SURVIVAL ANALYSIS)

J. F. Lawless

RELIABILITY, PROBABILISTIC

Probabilistic reliability theory deals with solution of problems useful in predicting, estimating, or optimizing a suitable objective function. The objective function could be mean life, survival probability, or more generally, life distribution of components or systems.

There is a considerable literature available. For a bibliography see Barlow and Proschan [4, 5], Gnedenko et al. [26], Mann et al. [34], Bain [2], Nelson [38], Ascher and Feingold [1], and Martz and Waller [36]. For applications of the theory in *survival analysis** see Johnson and Johnson [31], Lawless [33], and Gross and Clark [27]. For applications in *risk analysis* see McCormick [37] and in the area of power system see Dhillon [22]. To get a flavor of current problems in reliability see conference proceedings edited by Proschan and Serfling [39], Barlow et al. [3], Tsokos and Shimi [41], Crowley and Johnson [21], and Basu [10].

We describe a few important concepts. Let the nonnegative random variable X denote the lifetime of a system, let the cumulative distribution function (cdf) of X be $F(x)$, and let its mean be μ. The *reliability function* or survival function of X is given by

$$\bar{F}(x) = P(X > x) = 1 - F(x). \quad (1)$$

For simplicity throughout assume X to be absolutely continuous with density function $f(x)$ and $\bar{F}(x) > 0$. The failure rate function

$$r(x) = f(x)/\bar{F}(x) \qquad (2)$$

is variously called the intensity function, hazard rate* function, and the force of mortality*. Since

$$\bar{F}(x) = \exp\left[-\int_0^x r(u)\,du\right], \qquad (3)$$

there is a one to one relationship between $r(x)$ and $\bar{F}(x)$.

PARAMETRIC MODELS

A number of lifetime distributions have been considered as parametric models. We list the most important among these.

(a) Exponential distribution*. Here the cdf is given by

$$F(x) = 1 - \exp(-x/\theta), \qquad x \geqslant 0,\ \theta > 0.$$

The mean life is θ, the scale parameter, and

$$\bar{F}(x) = e^{-x/\theta}, \qquad r(x) = 1/\theta.$$

(b) The Weibull distribution*. The cdf of the two-parameter Weibull distribution is given by

$$F(x) = 1 - \exp(-x^{\alpha}/\theta),$$

$$x \geqslant 0,\ \alpha, \theta > 0.$$

Here $r(x) = \alpha x^{\alpha-1}/\theta$, $x > 0$. Note that if the shape parameter $\alpha > 1$, $r(x)$ is an increasing function of x; if $\alpha < 1$, $r(x)$ is decreasing; and if $\alpha = 1$, $r(x)$ is a constant and the Weibull distribution reduces to the exponential distribution.

(c) The gamma distribution*. Here

$$F(x) = \int_0^x f(u;\ \theta, \alpha)\,du;$$

the two-parameter gamma density is given by

$$f(u;\ \theta, \alpha) = [\theta^{\alpha}/\Gamma(\alpha)]\,e^{-u/\theta}\,u^{\alpha-1},$$

$$u \geqslant 0,\ \theta, \alpha > 0.$$

Like the Weibull distribution, the gamma distribution reduces to the exponential distribution when the shape parameter $\alpha = 1$. The failure rate function $r(x)$ is an increasing function of x for $\alpha > 1$ and a decreasing function of x for $\alpha < 1$.

The preceding models have been found suitable in many cases, especially when a product is subject to "wear out" with time. Other distributions have been considered as models. Among these are the preceding distributions with additional parameters, and the lognormal distribution*. Mixtures of some of those distributions have also been considered as models.

Since the cdf can be obtained given the failure rate function, a number of attempts have been made to model the failure rate function. Since products are subject to "break in" at first and subject to "wear out" with time, a distribution for which $r(x)$ is "bathtub" shaped is considered desirable in many cases; see Hjorth [29].

For modelling cdfs for multicomponent systems several multivariate distributions have been proposed. Among these are the multivariate exponential distributions* of Marshall and Olkin [35] and Block and Basu [16], based on physical considerations.

DISTRIBUTIONS WITH AGING CONCEPTS

Since many of the inference procedures using specific parametric distributions as models are not robust, broader classes of distributions using various concepts of aging or wear out have been proposed to study lifetimes of systems and components. The 12 most commonly studied classes of life distributions in the univariate case are the following:

1. increasing failure rate (IFR);
2. decreasing failure rate (DFR);
3. increasing failure rate in average (IFRA);

4. decreasing failure rate in average (DFRA);
5. new better than used (NBU);
6. new worse than used (NWU);
7. decreasing mean residual life (DMRL);
8. increasing mean residual life (IMRL);
9. new better than used in expectation (NBUE);
10. new worse than used in expectation (NWUE);
11. harmonic new better than used in expectation (HNBUE);
12. harmonic new worse than used in expectation (HNWUE).

The dual of IFR is DFR, the dual of IFRA is DFRA and so forth. Assume the mean $\mu = \int_0^\infty \bar{F}(x)\,dx$ to be finite and let $S = \{x : \bar{F}(x) > 0\}$.

(i) F is said to be IFR (DFR) if the conditional survival function (at age t), i.e., $\bar{F}(x+t)/\bar{F}(x)$, is a decreasing (increasing) function of x for $t \geqslant 0$.

(ii) F is said to be IFRA (DFRA) if $-\log \bar{f}(x)/x$ is increasing (decreasing) on S.

(iii) F is said to be NBU (NWU) if

$$\bar{F}(x)\bar{F}(y) \geqslant (\leqslant) \bar{F}(x+y)$$

for x, y in S.

(iv) F is said to be NBUE (NWUE) if

$$\mu \bar{F}(x) \geqslant (\leqslant) \int_0^\infty \bar{F}(x+y)\,dy$$

for x in S.

(v) F is said to be DMRL (IMRL) if

$$\int_x^\infty \bar{F}(u)\,du / \bar{F}(x)$$

is decreasing (increasing) on S.

(vi) F is said to be HNBUE (HNWUE) if

$$\int_x^\infty \bar{F}(u)\,du \leqslant (\geqslant) \mu \exp(-x/\mu)$$

for x in S.

Note that all 12 classes include the exponential distribution as a member. The following chain of implication exists among the six classes of distributions (and among their duals):

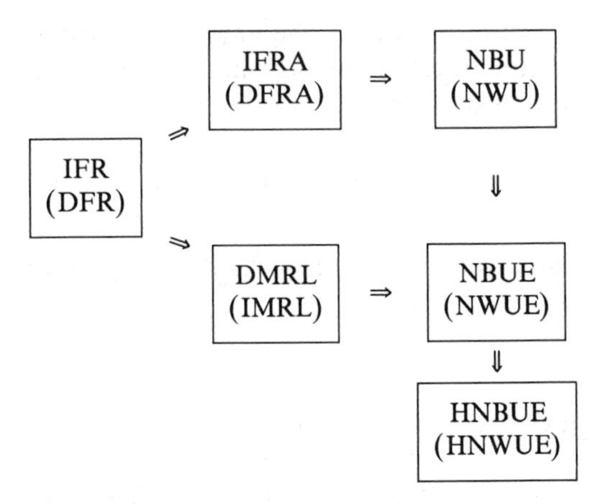

Properties of these classes have been studied extensively. For a bibliography see Barlow and Proschan [5], Basu and Ebrahimi [11, 12], Basu and Kirmani [15], Bryson and Siddiqui [19], Haines [28], Klefsjö [32], and Hollander and Proschan [30].

As in the case of parametric models, considerable work has been done toward extending the concept of univariate aging to the multivariate case. See, for example, Basu [6], Basu et al. [13], Block and Savits [17], Brindley and Thompson [18], Buchanan and Singpurwalla [20], Ghosh and Ebrahimi [25], and El-Neweihi [22]. In particular, Basu et al. [13] studied the multivariate extension of HNBUE and HNWUE.

RELIABILITY OF COMPLEX SYSTEMS

A physical system is called *simple* if it consists of a single component. Otherwise it is a *complex* system. Assume that a system can be in two states only, a functioning state and a failed state. The reliability of the system can be defined in two ways. Let X and Y be two random variables with cdfs $f(x)$ and $G(y)$, respectively. Suppose Y is the strength of a component subject to a random stress X. Then the component fails if at any moment the applied stress (or load) is greater than its strength or resistance. The reliability of the component in this case is given by

$$R_1 = P(X > Y).$$

This is the *stress-strength model*. As an example, consider a solid propellant rocket engine successfully fired provided the chamber pressure (X) generated by ignition stays below the burst pressure (Y) of the rocket chamber.

A second definition of reliability R_2 is the probability that the system will be in operating condition and function satisfactorily at mission time t.

For a survey of some recent results see Basu [7, 8, 9] and Basu and El Mawaziny [14].

An important problem is to determine the reliability of a complex system given the reliabilities of its components. Explicit expressions for reliability can be obtained for a number of cases. As an example, consider a p-component *series* system, which functions if and only if all components function. Let R_i be the reliability of the ith component. The reliability of the system in this case, assuming that the components are operating independently, is

$$R = R_1 R_2 \cdots R_p = \prod_{i=1}^{p} R_i$$

Assume next that the components are in *parallel*; that is, the system fails only when all components fail. The reliability of the system in this case, again assuming independence, is

$$1 - (1 - R_1)(1 - R_2) \cdots (1 - R_p).$$

A system with p components is called a *k-out-of-p system* if it functions only when k or more components operate successfully. As an example of a two-out-of-three system, consider an airplane that can function satisfactorily if and only if at least two out of its three engines are functioning. When $k = p$ ($k = 1$) we obtain the series (parallel) system.

Consider a k-out-of-p system S. Let X be the system lifetime, and X_i and R_i be the lifetime and reliability function of the ith component ($i = 1, 2,, \ldots, p$). Then the system reliability at time t is

$$\begin{aligned} R(t) &= P(X > t) \\ &= P(k \text{ or more components fail} \\ &\qquad\qquad \text{at time} > t) \\ &= \sum_{j=k}^{m} \sum_{\alpha_i} \sum_{i=1}^{j} \left(R_{\alpha_i}(t)\right) \\ &\quad \times \prod_{i=j+1}^{m} \left(1 - R_{\alpha_i}(t)\right). \end{aligned}$$

Here Σ is the sum over all $\binom{m}{j}$ distinct α_i combinations of the integers $1, 2, \ldots, m$ taken j at a time such that exactly j of the X_i's are greater than t.

Using the combinatorial theory, expressions for the reliability of complex systems can be derived in many cases. In other cases approximate bounds might be available.

A number of problems in reliability theory deal with *repairable systems*. That is, a system is repaired on failure. Barlow and Proschan [4, 5] discuss repairable systems and associated maintenance policies. Ascher and Feingold [1] point out misconceptions and problems relative to repairable systems.

References

[1] Ascher, H. and Feingold, H. (1984). *Repairable Systems Reliability*. Dekker, New York. (First book to seriously discuss issues related to repairable systems.)

[2] Bain, L. J. (1978). *Statistical Analysis of Reliability and Life Testing Models*. Dekker, New York.

[3] Barlow, R. E., Fussell, J. B., and Singpurwalla, N. D. (1975). *Reliability and Fault-tree Analysis*. Ser. Appl. Math., SIAM, Philadelphia.

[4] Barlow, R. E. and Proschan, F. (1965). *Mathematical Theory of Reliability*. Wiley, New York.

[5] Barlow, R. E. and Proschan, F. (1981). *Statistical Theory of Reliability and Life Testing*. To Begin With, Silver Spring, MD. (An excellent book in the area.)

[6] Basu, A. P. (1971). *J. Amer. Statist. Ass.*, **60**, 103–104. (Defines bivariate failure rate.)

[7] Basu, A. P. (1977). *Proc. 22nd Conference on Design of Experiments in Army Research and Testing. U.S. Army Research Office, Research Triangle Park, NC,* pp. 97–110.

[8] Basu, A. P. (1981). *Naval Res. Logist. Quart.*, **28**, 383–392.

[9] Basu, A. P. (1985). In *The Frontiers of Modern Statistical Inference Procedures*, E. J. Dudewicz, ed. American Sciences Press, Columbus, OH, pp. 271–282.

[10] Basu, A. P., ed. (1986). *Reliability and Quality Control*. North-Holland, Amsterdam, Netherlands.

[11] Basu, A. P. and Ebrahimi, N. (1985). *Ann. Inst. Statist. Math. A*, **37**, 347–359.

[12] Basu, A. P. and Ebrahimi, N. (1986). In *Reliability and Quality Control*, A. P. Basu, ed. North-Holland, Amsterdam, Netherlands. (Surveys properties of HNBUE distributions.)

[13] Basu, A. P., Ebrahimi, N., and Klefsjö, B. (1983). *Scand. J. Statist.*, **10**, 19–25. (Introduces multivariate HNBUE distributions.)

[14] Basu, A. P. and El Mawaziny, A. H. (1978). *J. Amer. Statist. Ass.*, **73**, 850–854. (Studies reliability of *k*-out-of-*m* systems.)

[15] Basu, A. P. and Kirmani, S. N. U. A. (1986). *J. Appl. Prob.* **23**, 1038–1044.

[16] Block, H. W. and Basu, A. P. (1974). *J. Amer. Statist. Ass.*, **69**, 1031–1037.

[17] Block, H. W. and Savits, T. H. (1980). *Ann. Prob.*, **8**, 793–801. (Studies multivariate IFRA distributions.)

[18] Brindley, E. C., Jr. and Thompson, W. A., Jr. (1972). *J. Amer. Statist. Ass.*, **67**, 822–830. (Considers multivariate version of IFR.)

[19] Bryson, M. C. and Siddiqui, M. M. (1969). *J. Amer. Statist. Ass.*, **64**, 1472–1483. (Introduces a number of concepts for aging.)

[20] Buchanan, W. B. and Singpurwalla, N. D. (1977). In *The Theory and Applications of Reliability*, Vol. 1, C. P. Tsokos and I. N. Shimi, eds. Academic, New York.

[21] Crowley, J. and Johnson, R. A. (1982). *Survival Analysis. IMS Lecture Notes—Monograph Series,* Vol. 2.

[22] Dhillon, B. S. (1983). *Power System Reliability, Safety and Management*. Ann Arbor Science, Ann Arbor, MI. (Considers applications in power systems.)

[23] El-Neweihi, E. (1981). *Commun. Statist. A,* **10**, 1655–1672. (Discusses multivariate NBU distributions.)

[24] Epstein, B. and Sobel, M. (1953). *J. Amer. Statist. Ass.*, **48**, 486–502. (Basic paper in life testing.)

[25] Ghosh, M. and Ebrahimi, N. (1980). *Egypt Statist. J.*, **25**, 36–55.

[26] Gnedenko, B. V., Belyayev, Yu. K., and Solovyev, A. D. (1969). *Mathematical Methods of Reliability Theory*. Academic, New York.

[27] Gross, A. J. and Clark, V. A. (1975). *Survival Distributions: Reliability Applications in the Biomedical Sciences*. Wiley, New York.

[28] Haines, A. (1973). Some Contributions to the Theory of Restricted Classes of Distributions with Applications to Reliability. Ph.D. dissertation, The George Washington University, Washington, DC.

[29] Hjorth, V. (1980). *Technometrics*, **22**, 99–107. (Considers "bathtub" distributions.)

[30] Hollander, M. and Proschan, F. (1984). In *Handbook of Statistics*, Vol. 4, P. R. Krishnaiah and P. K. Sen, eds. North-Holland, Amsterdam, Netherlands, Chap. 27.

[31] Johnson, R. C. E.- and Johnson, N. L. (1979). *Survival Models and Data Analysis*. Wiley, New York.

[32] Klefsjö, B. (1982). *Naval Res. Logist. Quart.*, **29**, 331–344. (Studies HNBUE and HNWUE distributions.)

[33] Lawless, J. F. (1982). *Statistical Models and Methods for Lifetime Data*. Wiley, New York.

[34] Mann, N. R., Schafer, R. E., and Singpurwalla, N. D. (1974). *Methods for Statistical Analysis of Reliability and Life Data*. Wiley, New York.

[35] Marshall, A. W. and Olkin, I. (1967). *J. Amer. Statist. Ass.*, **62**, 30–44.

[36] Martz, H. F. and Waller, R. A. (1982). *Bayesian Reliability Analysis*. Wiley, New York.

[37] McCormick, N. J. (1981). *Reliability and Risk Analysis*. Academic, New York. (Considers nuclear power applications.)

[38] Nelson, W. (1982). *Applied Life Data Analysis*. Wiley, New York.

[39] Proschan, F. and Serfling, R. J. (1974). *Reliability and Biometry*. SIAM, Philadelphia, PA.

[40] Shooman, M. L. (1968). *Probabilistic Reliability: An Engineering Approach*. McGraw-Hill, New York.

[41] Tsokos, C. P. and Shimi, I. N. (1977). *The Theory and Applications of Reliability*, Vols. I, II. Academic, New York.

[42] Weibull, W. (1951). *J. Appl. Mech.*, **18**, 293–297. (Proposes Weibull distributions as a model.)

Bibliography

Bergman, B. (1985). *Scand. J. Statist.*, **12**, 1–41. (A survey of some selected results and their usefulness.)

(EXPONENTIAL DISTRIBUTION
GAMMA DISTRIBUTION

HAZARD RATE AND OTHER
CLASSIFICATIONS OF DISTRIBUTIONS
RELEVATION
RELIABILITY: IMPORTANCE OF
COMPONENTS
SHOCK MODELS
SURVIVAL ANALYSIS
WEIBULL DISTRIBUTION)

ASIT D. BASU

RELIABILITY, SOFTWARE *See* SOFTWARE RELIABILITY

RELVARIANCE

The square of the coefficient of variation*. The term is used mainly in finite population sampling* theory; see, for example, refs. 1 and 2.

References

[1] Hansen, M. H., Hurwitz, W. N., and Madow, W. G. (1953). *Sample Survey Methods and Theory*. Wiley, New York, p. 124.

[2] Kish, L. (1965). *Survey Sampling*. Wiley, New York, p. 47.

(FINITE POPULATIONS, SAMPLING FROM)

REMEZ ALGORITHM

This is an iterative algorithm that calculates a minimax (L_∞, uniform, or Chebyshev) approximation to a function of a single variable or a one-dimensional data set. Let $f \in C(X)$ be the set of continuous real-valued functions defined on X a compact subset of the real line. Fix V as an n-dimensional subspace of $C[a, b]$, where $X \subset [a, b]$. (For a data set $\{(x_i, y_i)\}_{i=1}^N$, with each x_i and y_i real and the x_i all distinct, one sets $X = \{x_i\}_{i=1}^N$ and defines $f(x_i) = y_i$, $i = 1, \dots, N$ in order to use the functional description.) Under the assumptions that card$(X) \geqslant n + 1$ and V is a Haar subspace of $C[a, b]$ (i.e., if $v \in V$ vanishes at more than $n - 1$ points of $[a, b]$ then $v \equiv 0$), the Remez algorithm calculates $v^* \in V$ satisfying

$$\|f - v^*\|_X = \inf\{\|f - v\|_X : v \in V\},$$

where

$$\|h\|_X = \max\{|h(x)| : x \in X\}$$

for each $h \in C(X)$. Under these assumptions a unique best uniform approximation, v^*, to f from V on X exists and is characterized by $n + 1$ points $t_1 < \cdots < t_{n+1}$ in X on which

$$f(t_i) - v^*(t_i) = (-1)^{i-1}\sigma\|f - v^*\|_X, \qquad i = 1, \dots, n + 1,$$

with a $\sigma = \text{sgn}(f(t_1) - v^*(t_1))$ [2].

Using this characterization, Remez [9] developed the basic ideas for this algorithm for algebraic polynomials. (See ref. 8 for a complete discussion of this algorithm.) Basically, at the kth iteration a best uniform approximation, v^k, to f from V on $X_k = \{t_1^k < \cdots < t_{n+1}^k\} \subset X$ is calculated by solving an $(n + 1) \times (n + 1)$ linear system. If

$$e_k \triangleq \max\{|f(t_i^k) - v^k(t_i^k)| : i = 1, \dots, n + 1\} = \|f - v^k\|_X,$$

then v^k is the unique best approximation and the algorithm terminates. Otherwise, an exchange procedure finds a new reference set

$$X_{k+1} = \{t_1^{k+1} < \cdots < t_{n+1}^{k+1}\} \subset X$$

on which

$$e_k \leqslant (-1)^{i-1}\text{sgn}\left(f\left(t_1^{k+1}\right) - v^k\left(t_1^{k+1}\right)\right)\left(f\left(t_i^{k+1}\right) - v^k\left(t_i^{k+1}\right)\right), \qquad i = 1, \dots, n + 1,$$

and

$$\left|f\left(t_m^{k+1}\right) - v^k\left(t_m^{k+1}\right)\right| = \|f - v^k\|_X$$

for some m, $1 \leqslant m \leqslant n + 1$ hold. Normally, either a single point exchange (essentially equivalent to the simplex method* [10]) or a multiple point exchange (essentially equivalent to Newton's method* [12]) is used. In both cases, a globally convergent (independent of initial reference set used) iteration results. When the set X is not finite, this algorithm is applied to a sufficiently dense finite subset of X to find a nearly best approximation [4].

This algorithm has also been extended to algebraic rational approximation [3]. Here it may fail to converge due to various difficulties including nonexistence of best approximations on reference sets and the existence of poles between reference set points [6]. A more robust algorithm for rational approximation is the differential correction algorithm [5, 11]. (Reference [11] is a survey of this algorithm.) The differential correction algorithm also applies to multidimensional problems and generalized rational fits subject to linear constraints. Its cost over the Remez algorithm (when the latter runs) is greater storage and time requirements [6, 7]. Finally, the L_∞ norm treats all data values as known with essentially equal accuracy.

For statistical applications, see ref. 1.

References

[1] Chambers, J. M. (1977). *Computational Methods for Data Analysis*. Wiley, New York.

[2] Cheney, E. W. (1966). *Introduction to Approximation Theory*. McGraw-Hill, New York.

[3] Fraser, W. and Hart, J. F. (1962). *Commun. ACM*, **5**, 401–403.

[4] Golub, G. H. and Smith, L. B. (1971). *Commun. ACM*, **14**, 737–746.

[5] Kaufman, E. H. and Taylor, G. D. (1981). *Int. J. Numer. Meth. Eng.*, **17**, 1273–1280.

[6] Kaufman, E. H., Leeming, D. G., and Taylor, G. D. (1980). *Comp. Math Appl.*, **6**, 155–160.

[7] Lee, C. M. and Roberts, F. D. K. (1973). *Math. Comp.*, **27**, 111–121.

[8] Meinardus, G. (1967). *Approximation of Functions: Theory and Numerical Methods*. Springer, Berlin, Germany.

[9] Remez, E. Ja. (1934). *C. R. Acad. Sci. Paris*, **199**, 337–340.

[10] Stiefel, E. (1960). *Numer. Math.*, **2**, 1–17.

[11] Taylor, G. D. (1985). *Int. Ser. Numer. Math.*, **74**, 288–303.

[12] Veidinger, L. (1960). *Numer. Math.*, **2**, 99–105.

(MATHEMATICAL FUNCTIONS, APPROXIMATIONS TO)

GERALD D. TAYLOR

RENEWAL THEORY

INTRODUCTION

Renewal theory originated in the discussion of self-renewing aggregates and in the non-stochastic treatment of population growth*. The publications of Lotka [7] and Fréchet [6] contain full bibliographies of this early work, some of which (to a reader of today) is surprisingly controversial. The rigorous and general treatment of the subject owes much to papers by Feller [3–5]; the first of these, amongst other important results, cleared up some of the early controversies.

In this article the term *renewal process* will mean a sequence of nonnegative random variables $\{X_0, X_1, X_2, \ldots\}$ that are mutually independent, with all except X_0 identically distributed in accordance with a CDF $F(x) = P\{X_n \leqslant x\}$, $n \leqslant 1$. Where necessary we shall write $G(x) = P\{X_0 \leqslant x\}$. The physical model behind the mathematics is most simple: We are imagining an item subject to wear, failure, and replacement from an infinite pool of "similar" items (similar, that is, except for their statistically distributed lifetimes). X_0 is the residual lifetime of the item in use at the initial instant ($t = 0$). At time $S_0 = X_0$ this initial item fails and is immediately replaced with a "new" item (from the hypothetical infinite pool of such items). This constitutes the "zeroth" renewal. At time $S_1 = X_0 + X_1$ a further failure occurs and instantly the "first" renewal. It is plain to see how this simple process continues. Thus the nth renewal occurs at time $S_n = X_0 + X_1 + \cdots + X_n$.

We shall write $N(t)$ for the number of renewals to occur in the half-open time interval $(0, t]$. Thus, when $X_0 > 0$, $N(S_k) = k + 1$; if $X_0 = 0$ [which is the case, in particular, when the choice of our initial time $(t = 0)$ coincides with a renewal], then $N(S_k) = k$. It may reasonably be argued that *renewal theory* is the study of the random variable $N(t)$, which we shall call the *renewal count*.

In much published theory it is supposed that $X_0 = 0$; this makes matters somewhat simpler. We shall, except in one or two places that will be noted, follow this simpler practice. It is usually a straightforward matter to obtain results for the more general case from results for this simpler procedure.

In some applications of the theory it may happen that every X_n $(n \geqslant 1)$ is (almost surely) an integral multiple of some constant equal to 1; the lifetimes $\{X_n\}$ $(n \geqslant 1)$ are then (almost surely) all nonnegative integers, and we are now studying what is more usually called a *recurrent event process*. When the renewal process is not thus periodic we say it is *aperiodic*. For this article we shall, in the interest of brevity, assume that we are always dealing with an aperiodic process.

It is conceivable to have a general sequence of lifetimes $\{X_0, X_1, X_2, \dots\}$ such that $\{X_1, X_2, \dots\}$ is periodic but X_0 is, let us say, absolutely continuous* in its distribution. There are, conceivably, applications of such a process but the literature seems to contain no commentary on this Janus-like object.

Since the $\{X_n\}$ are nonnegative, all the moments $\mu'_k = \mathbb{E}(X_n^k)$ $(n \geqslant 1,\ k = 1, 2, \dots)$ are unambiguously defined, though possibly infinite. A standard convention, which we follow, is that when $(1/\mu'_1)$ occurs in any formula or theorem it is to be taken as zero when $\mu'_1 = \infty$.

BASIC RESULTS

If $F(0+) = 1$ all the lifetimes are (almost surely) zero and the renewal process is a worthless object of study. Henceforth, therefore, we shall reasonably assume $F(0+) < 1$. It can then be shown that, no matter what CDF $F(x)$ is chosen, there is necessarily a real $\sigma > 0$ [depending, of course, on $F(x)$] such that $\mathbb{E}\exp[\sigma N(t)] < \infty$ for every finite $t > 0$. Thus every ordinary moment of $N(t)$ is finite. Of preeminent interest is $H(t) \equiv \mathbb{E}N(t)$, called the *renewal function*. Knowledge of the behavior of $H(t)$ is important in many theoretical investigations related to renewal processes as well as in many real-world applications of the theory.

Blackwell's renewal theorem tells us, for any fixed real α, that as $t \to \infty$,

$$H(t+\alpha) - H(t) \to \alpha/\mu'_1. \quad (1)$$

The *key renewal theorem** is technically equivalent to (1) but is in a form more convenient for applications. Essentially it states that for a useful class $\mathbb{K}$ of "filter functions"* $K(x)$ in $L_1(0, \infty)$, as $t \to \infty$,

$$\int_0^t K(t-u)H(du) \to (1/\mu'_1)\int_0^\infty K(u)\,du. \quad (2)$$

Much work has gone into obtaining more and more refined estimates of $H(t)$ for large values of t. For example, when μ_2 is finite, a suitable choice of the filter $K(x)$ in (2) yields the *second renewal theorem*: As $t \to \infty$,

$$H(t) - \frac{t}{\mu'_1} \to \frac{\mu'_2}{2\mu_1'^2} - 1. \quad (3)$$

When $F(x)$ is absolutely continuous there will be a *renewal density* $h(t)$ such that $H(t) = \int_0^t h(u)\,du$. Blackwell's renewal theorem suggests that, as $t \to \infty$, we should have

$$h(t) \to 1/\mu'_1. \quad (4)$$

This is the *renewal density theorem*, but it is only valid if $f(x)$, the PDF of $F(x)$, satisfies suitable conditions. For instance, (4) holds if (a) $f(x) \to 0$ as $x \to \infty$ and (b) for some constant $p > 1$, $\int_0^\infty |f(x)|^p\,dx > \infty$. For the most general results on the *renewal density theorem* see Smith [10], where necessary and sufficient conditions are discussed.

THE INTEGRAL EQUATION* OF RENEWAL THEORY

This is the easily derived equation

$$H(t) = F(t) + \int_0^t H(t-u)F(du). \quad (5)$$

When densities exist there is the parallel *renewal density equation*

$$h(t) = f(t) + \int_0^t h(t-u)f(u)\,du. \quad (6)$$

A more general form of (5) corresponding to the case when X_0 is not identically zero is

$$H_G(t) = G(t) + \int_0^t H_G(t-u)F(du). \quad (7)$$

However, there is a simple expression for $H_G(t)$ in terms of $H(t)$:

$$H_G(t) = G(t) + \int_0^t H(t-u)G(du). \quad (8)$$

Thus it is not too difficult to learn about the solution of (7) once we know the solution of (5).

Both (5) and (6) are easy to solve formally in terms of Laplace transforms (*see* INTEGRAL TRANSFORMS). For instance, if for real $s > 0$ we write

$$f^o(s) = \int_0^\infty e^{-sx}f(x)\,dx,$$

then it follows that

$$h^o(s) = f^o(s)/\{1 - f^o(s)\}. \quad (9)$$

In general, of course, there is no convenient inversion of the Laplace transform (9). In some cases, however, particularly when $f^o(s)$ is a rational function, inversion is possible.

It is interesting to note that $H(t)$, although ostensibly merely the first moment of $N(t)$, contains information sufficient to determine all higher moments thereof. As an easy example, if we write $H_2(t) = \mathbb{E}(N(t))^2$, then one can show that

$$H_2(t) = H(t) + 2\int_0^t H(t-u)H(du).$$

Similar equations hold for higher moments of $N(t)$. These facts explain in some part the overriding interest in determining $H(t)$ as precisely as possible in any application of renewal theory.

In various research areas not obviously related to renewal theory, such as the theory of branching processes*, integral equations arise that are similar to some of those quoted immediately above. Knowledge of the behavior of the renewal function can thus translate into helpful information in these other research areas.

CUMULANTS OF $N(T)$

When $\mu_2' < \infty$ it has been shown that, as $t \to \infty$,

$$\operatorname{Var} N(t) \sim \left[\left(\mu_2' - \mu_1'^2\right)/\mu_1'^3\right]t. \quad (10)$$

To obtain asymptotic results for higher cumulants of $N(t)$ it has been necessary to make a technical assumption about $F(x)$, an assumption that $F(x)$ belongs to a class $\mathbb{S}$. Let us write $F_n(x)$ for the CDF of S_n, where $F_n(x)$ is the n-fold Stieltjes convolution of $F(x)$ with itself. Then $F(x)$ belongs to $\mathbb{S}$ if, for some $n \geqslant 1$, $F_n(x)$ possesses a nonnull absolutely continuous component. It is possible for $F(x)$ to have no absolutely continuous component and yet for $F_3(x)$, say, to have one. Of course, in practical applications we are usually dealing with a totally absolutely continuous $F(x)$ and this technical assumption presents no obstacle.

Once it is assumed that $F(x)$ belongs to $\mathbb{S}$ somewhat stronger conclusions are possible than (10). Let us write $K_n(t)$ for the nth *cumulant** of $N(t)$, $n \geqslant 1$. Suppose that for some integer $p \geqslant 0$ we have $\mu'_{n+p+1} < \infty$. Then there exist constants a_n and b_n such that

$$K_n(t) = a_n t + b_n + \lambda(t)/(1+t)^p, \quad (11)$$

where $\lambda(t)$ is of bounded total variation in $(0, \infty)$ and $\lambda(t) \to 0$ as $t \to \infty$. The constant a_n is a rational function of $\mu_1', \mu_2', \dots, \mu_n'$; the constant b_n is a rational function of $\mu_1', \mu_2', \dots, \mu'_{n+1}$.

As a particular example of (11), if $\mu_3' < \infty$, then (10) can be strengthened to

$$\operatorname{Var} N(t) = \frac{\mu_2' - \mu_1'^2}{\mu_1'^3} + \left(\frac{5\mu_2'^2}{4\mu_1'^4} - \frac{2\mu_3'}{3\mu_1'^3} - \frac{\mu_2'}{2\mu_1'^2}\right) + o(1). \quad (12)$$

This particular formula is useful in studying the so-called *variance-time curve*, sometimes a useful inferential tool in dealing with problems of superposed renewal processes.

SOME GENERALIZATIONS

There is an alternative way of regarding the problems of renewal theory, and this leads naturally to certain generalizations. Imagine a random process $X(t)$, say, for $t \geqslant 0$, defined on each time interval $[n, n+1)$, $n = 0, 1, 2, \ldots$, by $X(t) = S_n$. Thus the process $X(t)$ is a step function with a saltus of amount X_n at every integer "time" point $t = n$, $n = 0, 1, 2, \ldots$; $X(t)$ is defined so as to be continuous to the right. From this new viewpoint $N(x)$ can be equated to the total time that $X(t)$ spends in $(0, x]$; we can thus regard $N(x)$ as a *sojourn time*. It then becomes interesting to ask what happens if the $\{X_n\}$ are no longer restrained to be nonnegative, provided only that $\mathbb{E}X_n > 0$ (so that, as the law of large numbers* easily shows, the $\{S_n\}$ will display an overall drift towards $+\infty$). From this generalized standpoint it is no longer sensible to regard the $\{X_n\}$ as "lifetimes," since some of them may now have negative values. However, it still makes sense to study a function $\tilde{H}(I)$, say, the expected sojourn time of $X(t)$ in the interval I. Many of the results already described in this article (but *not* those referring to cumulants) have been proved by various authors in suitably generalized versions. In particular the key renewal theorem holds in the more general setting.

The notion of regarding $N(x)$ as a *sojourn time* rather than a *count* can be pushed further. Suppose $X(t)$ to be a time-homogeneous additive process in continuous time such that $X(0) = 0$ and $\mathbb{E}[X(t+1) - X(t)] = \mu_1' > 0$. In a familiar way let $\chi_I(\ )$ be the indicator function of the interval I. Then we can define a sojourn time $\tilde{N}(I)$ by the equation

$$\tilde{N}(I) = \int_0^\infty \chi_I(X(t))\, dt. \qquad (13)$$

If we write $F(I, t) = P\{X(t) \in I\}$ then (13) yields an expression for the renewal function $\tilde{H}(I) = \mathbb{E}\tilde{N}(I)$:

$$\tilde{H}(I) = \int_0^\infty F(I, t)\, dt. \qquad (14)$$

This formula corresponds to a well-known one in conventional renewal theory:

$$H(a) - H(b) = \sum_{n=0}^{\infty} \{F_n(a) - F_n(b)\}. \qquad (15)$$

In the present context we may call $X(t)$ an *infinitesimal renewal process*. It transpires that to many, if not to most, of the theorems that hold for the conventional renewal process there exist corresponding parallel theorems for the infinitesimal renewal process. For example, if

$$\mathbb{E}[X(t+1) - X(t)]^2 = \mu_2' < \infty,$$

and if we write $\tilde{H}(t)$ for $\tilde{H}((-\infty, t])$, then a result holds for $\tilde{H}(t)$ precisely like the second renewal theorem (3). It should be mentioned, however, that even here we must assume the infinitesimal process to be *aperiodic*. This means we assume that there is no $\tilde{\omega} > 0$ such that $F(I, t) = 1$ for $t > 0$ when I is the lattice set $\{\ldots, -2\tilde{\omega}, -\tilde{\omega}, 0, \tilde{\omega}, 2\tilde{\omega}, \ldots\}$. For more details of the infinitesimal process, see Smith [9].

QUASI-POISSON PROCESSES

The *forward delay* at time t ($\geqslant 0$) of a renewal process is the time that will elapse from t to the next renewal instant; thus, if we write $Y_+(t)$ for this forward delay, then

$$Y_+(t) = S_{N(t)+1} - t \qquad (16)$$

(we are assuming $X_0 = 0$). The *backward delay*, $Y_-(t)$ is the time that has elapsed by time t since the immediately preceding renewal instant. Evidently this backward delay is the *age* of the item in use at time t; the forward delay is the residual future life of that item. Conditional on having observed the process throughout $[0, t]$, the backward delay would be a known quantity (except in certain somewhat artificial cases). When μ_1

$< \infty$ it has long been known that, as $t \to \infty$,

$$P\{Y_+(t) \leqslant x\} \to (1/\mu_1')\int_0^x \{1 - F(u)\}\, du. \quad (17)$$

The integrand $\mu_1'^{-1}\{1 - F(u)\}$ in this result is a PDF, the *first derived* PDF, and it arises in various other contexts (such as when sampling lengths of fibres with the probability of a fibre's *inclusion* in the sample proportional to the length of that fibre).

The *Poisson process** can be defined in a variety of ways. One way is to regard it as a renewal process such that, for some fixed $\lambda > 0$,

$$F(x) = 1 - e^{-\lambda x}. \quad (18)$$

For the Poisson process many pleasant properties hold *exactly*, whereas they hold only approximately or in a limiting sense for an arbitrary renewal process. For instance, for the Poisson process we have *exactly*, for all $t \geqslant 0$,

$$H(t) = \lambda t, \qquad \operatorname{Var} N(t) = \lambda t. \quad (19)$$

These are exact versions of the second renewal theorem (3) and the variance result (12). Much more than this is true, of course. In particular, let us notice that the cumulant generating function* of $N(t)$ for a Poisson process is

$$\log \mathbb{E} e^{TN(t)} = \lambda t(e^T - 1), \quad (20)$$

so that *all* the cumulants are simple multiples of t.

There is an interesting class of renewal processes called *quasi-Poisson*; members of this class exhibit in a partial way some of the pleasing characteristics of the Poisson process. A renewal process is deemed to be quasi-Poisson with index τ ($\geqslant 0$) if $H(t)$ is exactly linear in t for all $t > \tau$. It can be shown that (as is the case for the Poisson process) all the moments $\{\mu_n'\}$ are necessarily finite for a quasi-Poisson process. Thus, for all $t > \tau$ the renewal function $H(t)$ of the quasi-Poisson process satisfies *exactly* the equation

$$H(t) = \frac{t}{\mu_1'} + \frac{\mu_2'}{2\mu_1'^2} - 1. \quad (21)$$

A necessary and sufficient condition for a renewal process to be quasi-Poisson with index $\tau > 0$ is that, for all $t > \tau$,

$$P\{Y_+(t) \leqslant x\} = (1/\mu_1')\int_0^x \{1 - F(u)\}\, dx \quad (22)$$

exactly (for all $x \leqslant 0$).

It can also be shown that, for a quasi-Poisson process, the cumulant $K_n(t)$ is *exactly* linear for all $t > n\tau$. Thus the variance of $N(t)$ is exactly a linear function for $t > 2\tau$.

The class of quasi-Poisson processes is not empty! Such processes arise in a natural way, for example, when one considers certain procedures for *censoring* an underlying Poisson process, such as happens in certain *electronic counter* processes. See Smith [8]. There is an interesting parallel to the quasi-Poisson process in the theory of infinitesimal renewal processes.

ASYMPTOTIC DISTRIBUTIONS AND APPROXIMATIONS

It was Feller [5] who drew attention to the fruitful identity

$$P\{S_k \leqslant t\} = P\{N(t) \geqslant k\}. \quad (23)$$

Using this he was able to deduce, when $\mu_2' < \infty$, the asymptotic normality of $N(t)$ from the more familiar central limit theorem* for iid variables. If we set $\sigma^2 = \mu_2' - \mu_1'^2$, then, as $t \to \infty$, he found that

$$P\left\{\frac{N(t) - (t/\mu_1')}{\sigma t^{1/2}/\mu_1'^{3/2}} \leqslant x\right\} \to \frac{1}{\sqrt{(2\pi)}}\int_{-\infty}^x e^{-u^2/2}\, du. \quad (24)$$

If one has a precise analytic form for $F(x)$ and can determine the Laplace-Stieltjes transform

$$F^*(s) = \int_{0-}^\infty e^{-sx} F(dx),$$

then, sometimes, the method of steepest descents* will yield approximations to the distribution of $N(t)$ more useful than those provided by (24). Let us write $K(s) = \log F^*(s)$ and, for large real t and large integer n, suppose there to be a unique real root α of the equation

$$K'(s) = -(t/n).$$

Then $P\{N(t) = n\}$ is given approximately by

$$\frac{e^{\alpha t}[1 - F^*(\alpha)][F^*(\alpha)]^n}{\alpha[2\pi n K''(\alpha)]^{1/2}}. \qquad (25)$$

This approximation will be good if $nK''(\alpha)$ is very large, and this requirement dictates, in each special case, suitable ranges for t and n.

For an example, if, for some fixed $\nu > 0$, $f(t) = e^{-t}t^{\nu-1}/\Gamma(\nu)$, then (25) becomes

$$\frac{e^{-t}t^{n\nu}}{e^{-n\nu}(n\nu)^{n\nu}\sqrt{(2\pi n\nu)}}\left[\frac{1 - (t/n\nu)^{\nu}}{1 - (t/n\nu)}\right]. \qquad (26)$$

This approximation to $P\{N(t) = n\}$ will be good if $t^2/(n\nu)$ is large.

More recently the simple normality result (24) has been greatly improved by the methods of *weak convergence* theory. Billingsley [1] showed, under the condition $\mu_2 < \infty$, that the following is true. Let us set

$$Z(t) = \frac{N(t) - (t/\mu_1')}{\sigma t^{1/2}/\mu_1'^{3/2}},$$

and, for each $t > 0$, define a process $W_t(\tau)$ on the interval $0 \leqslant \tau \leqslant 1$ by the relation $W_t(\tau) = Z(\tau t)$. Then $W_t(\tau)$ converges weakly to a Brownian process as $t \to \infty$. This is a much more powerful result than (23) and it has generated several applications of weak convergence to renewal theory or related processes.

Dall'Aglio [2] obtained a different kind of central limit theorem* for a renewal process. Suppose that at each renewal instant one will be paid \$1 and suppose the force of interest to be $\alpha > 0$. Then the present value of this "random perpetuity" is \$$Z(\alpha)$, say, where

$$Z(\alpha) = \sum_{n=1}^{\infty} e^{-\alpha S_n}.$$

Dall'Aglio showed that as α decreases to zero, the distribution of $Z(\alpha)$ more and more closely approaches normal. (At the time Dall'Aglio did this research, low interest rates were the norm!)

Sometimes one wants bounds to $H(t) - (t/\mu_1')$. A number of results have been published in this connection, but we shall merely mention one, an elegant device owing to Feller [4]. If $B(t)$ is any function $\geqslant F(t)$ and if $A(t)$ is the solution to

$$A(t) = B(t) + \int_0^t A(t-z)F(dz), \qquad (27)$$

then $A(t) \geqslant H(t)$. Similarly if $B(t) \leqslant F(t)$, then $A(t) \leqslant H(t)$. Feller sets $A(t) = \mu_1'^{-1}t + C$ and determines $B(t)$ from (27), adjusting C so that $B(t) \geqslant F(t)$. Let C_1 be the value so obtained for C. Similarly we adjust C so that $B(t) \leqslant F(t)$ and get C_2. Then $C_2 \leqslant H(t) - \mu_1'^{-1}t \leqslant C_1$.

FURTHER GENERALIZATIONS

We have, of necessity, been selective and allusive in this article. In this final short section, nonetheless, we refer to certain other generalizations.

What happens if the $\{X_n\}$ are independent but not identically distributed? The theory becomes enormously more complicated. One line of inquiry has been to discover under what general circumstances the key renewal theorem (2) generalizes. Although a few papers have tackled this matter and, in some cases, displayed the impressive mathematical prowess of their authors, much has still to be done before results are available in a sufficiently useful form to interest those who would apply them to other areas of investigation, theoretical or applied. A second line has been to find suitable generalizations of the elementary renewal theorem. This is an easier line of inquiry, perhaps; it *has* led to some results in a more easily comprehended form.

We quote the following result due to Smith [11]. Let $\{X_n\}$ be an infinite sequence of independent nonnegative random variables such that, for some regularly varying nondecreasing function $\lambda(n)$, with exponent $1/\beta$, $0 < \beta < \infty$, as $n \to \infty$,

$$\mathbb{P}\{(X_1 + X_2 + \cdots + X_n)/\lambda(n) \leqslant x\} \to K(x)$$

at all continuity points of some DF $K(x)$. Let $\Lambda(x)$ be the function inverse to $\lambda(n)$, and let $R(x)$ be any other regularly varying function of exponent $\alpha > 0$. Then if $N(x)$ is the maximum k for which $X_1 + X_2 + \cdots + X_n \leqslant x$, it follows that

$$\mathbb{E}R(N(x)) \sim I(\alpha\beta)R(\Lambda(x)), \quad \text{as } x \to \infty,$$

where

$$I(\alpha\beta) = \int_0^\infty u^{-\alpha\beta}K(du),$$

and the latter integral may diverge.

References

[1] Billingsley, P. (1968). *Convergence of Probability Measures*. Wiley, New York.

[2] Dall'Aglio, G. (1964). *Ann. Math. Statist.*, **35**, 1326–1331.

[3] Feller, W. (1941). *Ann. Math. Statist.*, **12**, 243–267.

[4] Feller, W. (1948). *Courant Anniversary Volume*. pp. 105–115.

[5] Feller, W. (1949). *Trans. Amer. Math. Soc.*, **67**, 98–119.

[6] Fréchet, M. (1949). *Statistical Self-Renewing Aggregates*. Fouad I University Press, Cairo, Egypt.

[7] Lotka, A. (1939). *Ann. Math. Statist.*, **10**, 1–25.

[8] Smith, W. L. (1957). *Proc. Camb. Phil. Soc.*, **53**, 175–193.

[9] Smith, W. L. (1960). In *Contributions to Probability and Statistics*, Stanford University Press, Stanford, CA, pp. 396–413.

[10] Smith, W. L. (1962). *Trans. Amer. Math. Soc.*, **104**, 79–100.

[11] Smith, W. L. (1968). *Ann. Math. Statist.*, **39**, 139–154.

Bibliography

Baxter, L. A., Scheuer, E. M., Blischke, W. R., and McConalogue (1981). *Renewal Tables: Tables of Functions Arising in Renewal Theory*, Tech. Rep., Dept. Management and Policy Sciences, University of Southern California, Los Angeles, CA.

(ABSOLUTE CONTINUITY
POISSON PROCESSES
QUEUEING THEORY
RETURN STATE
SEMI-MARKOV PROCESSES
STOCHASTIC PROCESSES)

W. L. SMITH

RÉNYI, ALFRÉD

Born: March 20, 1921, in Budapest, Hungary

Died: February 1, 1970, in Budapest, Hungary

Contributed to: Probability, statistical theory, information theory, number theory, combinatorial theory.

CAREER

Alfréd Rényi came from an intellectual family. After finishing his secondary school, he was not admitted into the university immediately because of racial laws, but one year later, after winning a mathematical competition, he gained admittance and studied mathematics and physics at the University of Budapest until 1944. Professor L. Fejér was a great influence on his formative career years, but Rényi also profited very much from the young, but already productive, mathematicians P. Turán and T. Gallai.

In 1944, Rényi was interned in a labour camp but escaped and, in the guise of a uniformed soldier, was able to help other persecuted persons. After the liberation of Budapest, he went to Szeged and earned a doctoral degree with F. Riesz. Then he was employed for one year by a social security institution. After marriage to the able mathematician Catherine Schulhof (a very lively person), he traveled to Leningrad with her on scholarship. Under the guidance of

Yu. V. Linnik, he obtained the candidate degree of mathematical sciences in one year instead of the prescribed three.

Beginning in 1947, again in Budapest, he acted as assistant lecturer at the University of Budapest; at the same time (between 1948 and 1950) he regularly visited the University of Debrecen as a professor. In 1950, the Mathematical Institute of the Hungarian Academy of Sciences was established and he soon became the director of the Institute and head of the Department of Probability. From this time on he was also chairman of the department of Probability Theory at the University of Budapest. He performed both these duties until the end of his life.

Around 1950, the Hungarian School of Probability was formed; it included P. Medgyessy, Ilona Palásti, A. Prékopa, G. Székely, I. Vincze, Margaret Ziermann, and L. Takács, who already had been active in this field as a student of the distinguished Hungarian statistician Ch. Jordan.

During the next 20 years Rényi played an important role in the Mathematical and Physical Department of the Hungarian Academy of Sciences (HAS) on the Committee for Scientific Qualification, and in the János Bolyai Mathematical Society. In the frame of this activity, he organized the first and second Hungarian Mathematical Congresses in 1950 and 1960, respectively. In almost every year conferences and colloquia in different fields of probability theory and its applications were held. He had a considerable role in improving teaching in the field of mathematics in secondary schools and in the university among others, organizing several kinds of mathematical competitions. He initiated and founded the *Publications of the Mathematical Institute of HAS* (now *Studia Scientiarum Mathematicarum Hungaricae*).

In recognition of his scientific activity he became corresponding member of the HAS in 1949 and ordinary member in 1956. He received the silver grade of the Kossuth Prizein 1949 and the golden grade in 1956. As a member of the International Statistical Institute* he was elected vice president in 1965. He was a member of the editorial boards of ten (Hungarian and foreign) periodicals. As a visiting professor he lectured at Michigan State University (1961), the University of Michigan (1964), Stanford University (1966), Cambridge University (1968), and the University of North Carolina (1968).

Following the unexpected death of Catherine Rényi in August 1969, physicians diagnosed an inoperable lung carcinoma in Alfréd; he died on February 1, 1970.

SCIENTIFIC ACTIVITY

Rényi wrote his doctoral dissertation on Cauchy–Fourier series and his candidate thesis on the quasi-Goldbach problem, giving a remarkable development of the large-sieve method. Then he published a sequence of articles on the theory and applications of probability theory and mathematical statistics. He collaborated with L. Jánossy and J. Aczél on compound Poisson distributions, and with L. Pukánszky on measurable functions. He has also written on mixing sequences and on algebras of distributions. In 1953, his basic paper on order statistics* appeared, establishing a method for the determination of limiting distribution laws; he determined the limiting law of the relative deviation between empirical and theoretical distributions, giving a new version of the Kolmogorov test. In 1954, his axiomatic foundation of the theory of probability based on conditional probability*, was published. Since 1945, he published on analytic functions of a complex variable, on geometry, on algebra, and on Newton's method of approximation; he continued his work on varying topics for the remainder of his life.

Many joint papers with his friends P. Turán and P. Erdős appeared: papers on combinatorics*, and particularly on random graphs*; and also on the theory of functions and in the theory of numbers, in which topic Catherine Rényi was also sometimes his coauthor. Beginning in 1950, with the stimulus of problems posed by industrial and other experts, he solved practical problems and published on storage problems, breaking processes, energy needs of plants, rational dimensioning of compressors, chemical reactions, and replacement policy in stocks. In 1956, his paper on information theory*, where he dealt mainly with the concept of entropy* in the several interpretations and applications of this concept. In a sequence of papers he considered entropy and statistical physics, entropy and mathematical statistics, and finally established a theory of search*. He collaborated with J. Hájek on a generalization of the Kolmogorov inequality*, with R. Sulanke on geometrical probabilities, and with J. Neveu on inequalities in connection with probabilities of events. We cannot list all the topics he considered, as his papers number about 300, all achieved in a period of 25 years.

The most complete collection of Rényi's articles appears in *Selected Papers of Alfréd Rényi*, edited by P. Turán, in three volumes (Akadémiai Kiadó [Publishing House of the HAS], Budapest, 1979), which include the English translation of several works written in Hungarian.

BOOKS

Rényi's first book, *Theory of Probability* (Tankőnyvkiadó [Publisher of Textbooks], Budapest, 1954), is a monumental synthesis of theory, applications, and practice, still not translated into any foreign language. Its theoretical part, in a revised and extended form, with an addendum on information theory, appeared in German as *Wahrscheinlichkeitstheorie mit einem Anhang über Informationstheorie*, which was translated, each time in revised form, into Hungarian, English, French, and Czech. He wrote a book on the foundation of the theory of probability based on his conditional probability concept, which posthumously appeared in English. The character of his other books takes the form of popularization or essays. It can be a notable event for the mathematician or an interested layman to read them. They include *Dialogues* (1967) (Socratic dialogues on mathematics and applications), *Die Sprache des Buches der Natur* (1968), *Ars Mathematica* (1970) (includes discussion of information theory), and *Letters on Probability* (1972) (a fictional correspondence between Fermat and Pascal).

ISTVÁN VINCZE

RÉNYI–ANSCOMBE THEOREM

The nature of the Rényi–Anscombe theorem justifies a few words on the central limit theorem* (CLT). Let $X_1, X_2, \ldots$ be independent and identically distributed (iid) random variables with mean μ and finite variance σ^2. Write $Z_n = \sqrt{n}(\bar{X}_n - \mu)/\sigma$, where $\bar{X}_n = (X_1 + \cdots + X_n)/n$, $n \geqslant 1$. Then Z_n converges in distribution to the standard normal law as $n \to \infty$ (*see* LIMIT THEOREM, CENTRAL). The result is of great importance in statistical problems such as interval estimation and hypothesis testing*, etc.

Motivated by an asymptotic theory of sequential estimation*, Anscombe [1] extended the CLT when n is replaced by a suitable random number. The original Anscombe theorem, however, is a much more general limit theorem for a randomly indexed sequence of random variables under appropriate conditions (see ref. 1). An elegant proof of Anscombe's theorem in the case of independent random variables was given by Rényi [6]. This CLT for random sums can be formally described as follows.

Theorem (Rényi–Anscombe). Let $X_1, X_2, \ldots$ be iid random variables with mean μ and finite variance σ^2, and set $\bar{X}_n = (X_1 + \cdots + X_n)/n$. Let $N(c)$ be a positive integer-valued random variable such that $N(c)/c \to 1$ in probability as $c \to \infty$. Then $Z_{N(c)} = \sqrt{N(c)}\,(\bar{X}_{N(c)} - \mu)/\sigma$ converges in distribution to the standard normal law as $c \to \infty$.

The following example shows an interesting application.

Example. Let $X_1, X_2, \dots$ be iid random variables with unknown mean μ and variance σ^2. We wish to estimate μ by a confidence interval of length $2d$ and coverage probability γ. Chow and Robbins [2] developed a sequential procedure with good asymptotic properties. Let s_n^2 be the sample variance and let a be the γ fractile of a standard normal distribution. Define the stopping variable

$$N(d) = \inf\{n \geqslant m : n \geqslant a^2 s_n^2/d^2\},\quad m \geqslant 2,$$

and let $I_{N(d)} = [\overline{X}_{N(d)} \pm d]$. Then $I_{N(d)}$ is a confidence interval of length $2d$, and, among other things, it has been shown by Chow and Robbins [2] via the Rényi–Anscombe theorem that

$$P\left(|\overline{X}_{N(d)} - \mu| \leqslant d\right) \to \gamma \quad \text{as } d \to 0.$$

See also FIXED-WIDTH AND BOUNDED-LENGTH CONFIDENCE INTERVALS.

References

[1] Anscombe, F. J. (1952). *Proc. Camb. Phil. Soc.*, **48**, 600–607.

[2] Chow, Y. S. and Robbins, H. (1965). *Ann. Math. Statist.*, **36**, 457–462.

[3] Chow, Y. S. and Teicher, H. (1978). *Probability Theory*. Springer-Verlag, New York. (Excellent treatment of Rényi–Anscombe theorem at graduate level.)

[4] Doeblin, W. (1938). *Bull. Soc. Math. France*, **66**, 210–220.

[5] Mogyorodi, J. (1962). *Magyar. Tud. Akad. Mat. Kutato Int. Közl.*, **7**, 409–424.

[6] Rényi, A. (1960). *Acta Math. Acad. Sci. Hung.*, **11**, 97–102.

(LIMIT THEOREM, CENTRAL
SEQUENTIAL ESTIMATION)

RASUL A. KHAN

RENYI-TYPE DISTRIBUTIONS

In his famous paper "On the Theory of Order Statistics" [40], Rényi derived the asymptotic distribution of

$$\sqrt{n}\,\sup_{a \leqslant F(x) \leqslant b}[(F_n(x) - F(x))/F(x)]$$

and

$$\sqrt{n}\,\sup_{a \leqslant F(x) \leqslant b}|(F_n(x) - F(x))/F(x)|.$$

Here $F_n(x)$ means the empirical distribution function of an independent and identically distributed (iid) sample $X_1, \dots, X_n$ with continuous CDF $F(x)$.

This statistic is used in situations similar to the Kolmogorov–Smirnov-type tests of fit*. For critical comparisons see refs. 32 and 33.

There are two levels of generalizations of Rényi's statistic. First, let $T_n(x; c, d, \delta) = (F_n(x) - cF(x))/(\delta - dF(x))$. A *Rényi-type distribution* is of the form

$$\Pr[-r \leqslant T_n(x; c, d, \delta) \leqslant s \text{ for all } x \in I],$$

where $I = \{x : a \leqslant F(x) \leqslant b\}$ or $I = \{x : A/n \leqslant F_n(x) \leqslant B/n\}$.

Equivalently

$$\Pr\left[\frac{F_n(x) - s\delta}{c - sd} \leqslant F(x) \leqslant \frac{F_n(x) + r\delta}{c + rd} \text{ for all } x \in I\right], \quad (1)$$

if $d < \delta$, $c + rd > 0$, and $c - sd > 0$. Thus, in a second step the general Rényi-type distribution is defined by

$$\Pr[f(F_n(x)) \leqslant F(x) \leqslant g(F_n(x)) \text{ for all } x \in I]. \quad (2)$$

It is clear from this derivation that the functions f and g are determined by the values of the underlying Rényi-type statistic. This statistic is distribution-free, because (2) is equal to

$$\Pr[f(U_n(u)) \leqslant u \leqslant g(U_n(u)) \text{ for all } u \in I'], \quad (3)$$

where $U_n(u)$ is the empirical distribution function (*see* EDF STATISTICS) of n independent uniform $(0, 1)$ random variables $U_1, \dots, U_n$. Finally, (3) can be written as

$$\Pr[\nu_i \leqslant U_{i:n} \leqslant \mu_{i-1} \text{ for } i = 1, \dots, n] \quad (4)$$

$$= \Pr[1 - \mu_{n-i} < U_{i:n} < 1 - \nu_{n+1-i} \text{ for } i = 1, \dots, n], \quad (4')$$

where $\nu_1, \ldots, \nu_n$ and $\mu_1, \ldots, \mu_{n-1}$ depend on f, g, and I (see ref. 36 for details).

In general, (4) can only be recursively calculated as in Wald and Wolfowitz [50], Epanechnikov [25], Noé and Vandewiele [38, 39], and Steck's determinant [46, 42]. Asymptotically, the Brownian bridge and the Poisson process* determine (4). See ref. 37 and Durbin's monograph [21, eq. (3.5)]. Siegmund [43] surveyed extensively the theory and applications of boundary crossing probabilities.

The distribution (2) [or (4)] is one-sided if $g(x) = 1$ for all $x \in I$ (or $\mu_i \geqslant 1$ for all $i = 0, \ldots, n-1$). This means no preference between f and g (or ν_i and μ_i) because of (4′). The algorithms are simpler in this case [50; 15; 46, eq. (2.3); 36, eq. (2.9)].

Examples. (References providing tables for the given distribution are marked by a prime, e.g., [5′].)

(a) Weighted Kolmogorov–Smirnov Statistics. $\text{Sup}_{x \in I}|T_n(x; c, d, \delta)|$ is a special case of a weighted statistic $\sup_{x \in I}\{w(F(x), F_n(x))|F_n(x) - F(x)|\}$. Borovkov and Sycheva [5] showed the sense in which

$$V_n = \sup_{x \in I}\left\{\frac{|F_n(x) - F(x)|}{[F(x)\{1 - F(x)\}]^{1/2}}\right\}$$

is asymptotically uniformly optimal among those weighted statistics. For more results about V_n see: *Asymptotics*, refs. 1, 5′, 37, 9, 24 and 27; *Exact*, refs. 38′, 36′ and 28′; *Monte Carlo*, ref. 6′.

(b) Composed Statistics. Several Rényi-type statistics together constitute a multivariate Rényi-type statistic, which has a distribution of form (2). The same is true for the maximum of such statistics. Examples are refs. 33 and 19′; for truncated samples, refs. 14 and 10.

(c) $T_n(x; c, d, \delta)$. To this large class also belong those distributions where the $F(x)$ in the denominator of T_n is replaced by $F_n(x)$, because

$$\Pr\left[-r \leqslant \frac{F_n(x) - cF(x)}{\delta - dF_n(x)} \leqslant s \text{ for all } x \in I\right]$$
$$= \Pr\left[-\frac{r}{1 - rd} \leqslant T_n(x; c, cd, \delta) \leqslant \frac{s}{1 + sd} \text{ for all } x \in I\right] \quad (5)$$

if $0 \leqslant d < \delta$, $cd \leqslant \delta$, and $rd - 1 \leqslant 0 \leqslant 1 + sd$. Similar identities hold for other restrictions on the parameters.

The general recursions can be simplified for the computation of (5) (Durbin's matrix method [21, eq. (2.44); 36, p. 933]).

For one-sided distributions, closed forms are available for some special cases of (4). How complicated these formulas are (owing to a multiplicity of summations) depends on how many lines are needed to cover the points (i, ν_i) (*see* SHEFFER POLYNOMIALS). If $I = \{a \leqslant F(x) \leqslant b\}$ one obtains, for $d < \delta$ and $c > sd$,

$$\Pr[T_n(x; c, d, \delta) \leqslant s \text{ for all } a \leqslant F(x) \leqslant b]$$
$$= \Pr[f(F_n(x)) \leqslant F(x) \text{ for all } a \leqslant F(x) \leqslant b]$$
$$= \sum_{j=0}^{b_f}\binom{n}{j}(1-b)^{n-j}\left(b - f\left(\frac{j}{n}\right)\right) \times \sum_{i=0}^{a_f}\binom{j}{i} f\left(\frac{i}{n}\right)^i\left(b - f\left(\frac{i}{n}\right)\right)^{j-i-1} \quad (6)$$

if $b_f = \lfloor nb(c - sd) + ns\delta \rfloor < n$ and $f(x) = (x - s\delta)/(c - sd)$. Replace b by a to define a_f. If $b_f \geqslant n$, (6) reduces to

$$(1 - f(1))\sum_{i=0}^{a_f}\binom{n}{i} f(i/n)^i \times (1 - f(i/n))^{n-i-1}. \quad (7)$$

If $I = \{A/n \leqslant F_n(x) \leqslant B/n\}$ replace a_f by $A - 1$, b_f by B, and b by $f(B/n)$ in (6) and (7). Table 1 gives an introduction to the rich literature about these distributions.

Table 1 Summary and Literature Guide for Rényi-type Distributions

T_n	I	
$T_n(x; 0, -1, 0)$	$0 \leqslant F(x) \leqslant 1$	*Ex(act)*, [15]; *As(ymptotic)*, [52]
$= F_n(x)/F(x)$	$0 \leqslant F(x) \leqslant b$	*Ex*, [8, Th. 4] and [49, Th. 2] can be simplified—see [36, eq. (4.1)]; *As*, [8′]
	$\frac{1}{n} \leqslant F_n(x) \leqslant 1$	*As*, [41]
	$\frac{1}{n} \leqslant F_n(x) \leqslant \frac{B}{n}$	*Ex*, [8, Th. 7; 49, Th. 4]
$T_n(x, 1, 0, 1)$	$0 \leqslant F(x) \leqslant b$	*Ex*, [3, 4′]; *As*, [31]
$= F_n(x) - F(x)$	$a \leqslant F(x) \leqslant 1$	*Ex*, [44], [3]; *As*, [31]
	$a \leqslant F(x) \leqslant b$	*Ex*, [3]; *Power*, [2]
	$\frac{A}{n} \leqslant F_n(x) \leqslant 1$	*Ex*, [3, 4′]
	$0 \leqslant F_n(x) \leqslant \frac{B}{n}$	*Ex*, [3]
$T_n(x, c, 0, 1)$	$0 \leqslant F(x) \leqslant 1$	*Ex*, [22, 16, 20]; *As*, [22, 20]
$= F_n(x) - cF(x)$	$a \leqslant F(x) \leqslant b$	*Ex*, [48, 23] very detailed
$T_n(x; 1, -1, 0)$	$0 \leqslant F(x) \leqslant 1$	*Ex*, [15]
$= (F_n(x) - F(x))/F(x)$	$a \leqslant F(x) \leqslant 1$	*Ex*, [26, Th. 1] corrected in [12, p. 1116]; *As*, [40]
$= T_n(x; 0, -1, 0) - 1$	$a \leqslant F(x) \leqslant b$	*Ex*, [3]; *As*, [40, Th. 7] corrected in [14, p. 553], [7′] computer program, [13] k-sample analogs; *Power*: [53].
	$\frac{A}{n} \leqslant F_n(x) \leqslant 1$	*Ex*, [12, 3, 4′]
	$\frac{1}{n} \leqslant F_n(x) \leqslant \frac{B}{n}$	*Ex*, [3]
$T_n(x; c, -1, 0)$	$0 \leqslant F(x) \leqslant 1$	*Ex*, [16]
$= T_n(x; 0, -1, 0) - c$	$a \leqslant F(x) \leqslant 1$	*Ex*, [48]
$T_n(x; 1, 1, 1)$	$0 \leqslant F(x) \leqslant b$	*Ex*, [26, Th. 2] corrected in [3, eq. (2.33)].
$= (F_n - F)/(1 - F)$	$a \leqslant F(x) \leqslant 1$	*Ex*, [12, 3]
	$a \leqslant F(x) \leqslant b$	*Ex*, [3]; *As*, [11]; *Power*, [53]
	$\frac{A}{n} \leqslant F_n(x) \leqslant 1$	*Ex*, [12, 3]
	$0 \leqslant F_n(x) \leqslant \frac{B}{n}$	*Ex*, [12, 3]
$T_n(x; c, d, \delta)$	$a \leqslant F(x) \leqslant b$	*Ex*, [36]; *As*, [29] bounds, [30] law of iterated logarithm

TWO-SAMPLE TESTS

Let $Y_1, \ldots, Y_m$ be a second iid sample from the same continuous distribution as $X_1, \ldots, X_n$. Denote the empirical distribution function of the second sample by G_m, and let $H_{n+m} = (nF_n + mG_m)/(n+m)$ be the empirical distribution function of the combined sample. Analogous to the one sample case, define

$$T_{n,m}(x; c, d, \delta) = \frac{F_n(x) - cH_{n+m}(x)}{\delta - dH_{n+m}(x)}.$$

The Rényi-type distributions are

$$\Pr[-r \leqslant t_{n,m}(x; c, d, \delta) \leqslant s \text{ for all } x \in I],$$

where

$$I = \left[\frac{a}{n+m} \leqslant H_{n+m}(x) \leqslant \frac{b}{n+m}\right]$$

$$\text{or} \quad I = \left[\frac{A}{n} \leqslant F_n(x) \leqslant \frac{B}{n}\right].$$

Again, this leads to the general Rényi-type distributions

$$\begin{aligned} &\Pr[f(F_n(x)) \leqslant H_{n+m}(x) \\ &\quad \leqslant g(F_n(x)) \text{ for all } x \in I] \qquad (8) \\ &= \Pr[\nu_i < R_i - i \\ &\quad \leqslant \mu_{i-1} \text{ for all } i = 1, \ldots, n], \qquad (9) \end{aligned}$$

where R_i is the relative rank of $X_{i:n}$ in the combined sample. How to derive ν_i and μ_i from f, g, and I is shown in ref. 36.

In general, (8) can be evaluated only by use of recursion formulas. Steck's determinant [45′] was derived from the rank distribution (9). He also considered the special case of Lehmann alternatives [47′].

Equation (8) can be computed by counting restricted lattice paths. For more details see Mohanty's monograph [34]. The asymptotic results are very similar to those in the one-sample case.

Example (a). A general concept of weighted Kolmogorov–Smirnov* two-sample statistics is discussed in ref. 35. The standardized statistic

$$\sup\left\{\frac{|F_n(x) - G_n(x)|}{\sqrt{H_{n+m}(x)(1 - H_{n+m}(x))}} : A/n \leqslant F_n(x) \leqslant B/n\right\}$$

was thoroughly studied by Doksum and Sievers [17′]. See also refs. 36′ (exact) and 6′ (simulation). This statistic is used as a test for symmetry in ref. 18′.

Example (b). $T_{n,m}(x; c, d, \delta)$. The Kolmogorov–Smirnov* distribution ($I = \{0 \leqslant x \leqslant 1\}$) is the only one in this class where closed formulas exist for the two-sided case. For one-sided distributions, closed forms can be derived using Sheffer polynomials, Steck's determinant [45], or lattice path combinatorics [34]. The multiplicity of the resulting summation depends (as in the one-sample case) on the minimal number of lines covering the points (i, ν_i), but now ν_i has to be integer-valued. The following example is the counterpart to the one-sample case.

Define a_f and b_f as before. If $d < \delta$, $c - sd > 0$ and if $\rho_i = [(m+n)(i/n - s\delta)/(c - sd)]$ is affine (i.e., $\rho_i = ui - v$, for some integers u and v) for all $i = 0, \ldots, b_f$, then

$$\begin{aligned} &\Pr\left[-s \leqslant T_{m,n}(x; c, d) \right. \\ &\quad \left. \text{for all } \frac{a}{m+n} \leqslant H_{m+n}(x) \leqslant \frac{b}{m+n}\right] \\ &= \sum_{j=0}^{b_f} \binom{n - j + m - b + b_f}{n - j} \sum_{i=0}^{a_f} \binom{\rho_i - 1}{i} \\ &\quad \times \frac{b - b_f - \rho_j + j}{b - b_f - \rho_i + i} \binom{j - 1 + b - b_f - \rho_i}{j - i}. \end{aligned}$$

Shou-Jen Wang [51] derived asymptotic results for $\sup_x\{(F_m(x) - G_n(x))/G_n(x)\}$. Further asymptotic distributions for this statistic and for $\sup\ \{(F_m(x) - G_n(x))/H_{m+n}(x)\}$ are given in ref. 11.

References

[1] Anderson, T. W. and Darling, D. A. (1952). *Ann. Math. Statist.*, **23**, 193–212.

[2] Birnbaum, Z. W. (1953). *Ann. Math. Statist.*, **24**, 484–489.

[3] Birnbaum, Z. W. and Lientz, B. P. (1969). *Appl. Math.*, **10**, 179–192.

[4] Birnbaum, Z. W. and Lientz, B. P. (1969). *J. Amer. Statist. Ass.*, **64**, 870–877.

[5] Borovkov, A. A. and Sycheva, N. M. (1968). *Theory Prob. Appl.*, **13**, 359–393.

[6] Canner, P. L. (1975). *J. Amer. Statist. Ass.*, **70**, 209–211.

[7] Chamayou, J. M. F. (1976). *Comp. Phys. Commun.*, **12**, 173–178.

[8] Chang, Li-Chien (1955). *Acta Math. Sinica*, **5**, 347–368 [*IMS & AMS Sel. Transl. Math. Statist. Prob.*, **4**, 17–38 (1963)].

[9] Csàki, E. (1977). *Zeit. Wahrscheinl.*, **38**, 147–167.

[10] Csörgő, M. (1965). *Ann. Math. Statist.*, **36**, 322–326.

[11] Csörgő, M. (1965). *Ann. Math. Statist.*, **36**, 1113–1119.

[12] Csörgő, M. (1965). *Proc. Amer. Math. Soc.*, **16**, 1158–1167.

[13] Csörgő, M. (1965). *Bull. Amer. Math. Soc.*, **71**, 616–618.

[14] Csörgő, M. (1967). *Canad. J. Math.*, **19**, 550–558.

[15] Daniels, H. E. (1945). *Proc. R. Soc., Lond. Ser. A*, **183**, 405–435.

[16] Dempster, A. P. (1959). *Ann. Math. Statist.*, **30**, 593–597.

[17] Doksum, K. A. and Sievers, G. L. (1976). *Biometrika*, **63**, 421–434. (Confidence bands, applications.)

[18] Doksum, K. A., Fenstad, G., and Aaberge, R. (1977). *Biometrika*, **64**, 473–487. (Confidence bands, applications.)

[19] Dufour, R. and Maag, U. R. (1978). *Technometrics*, **20**, 29–32. (Truncated or censored samples.)

[20] Durbin, J. (1968). *Ann. Math. Statist.*, **39**, 398–411.

[21] Durbin, J. (1973). Distribution theory for tests based on the sample distribution function. *Regional Conference Series in Mathematics*, **9**, SIAM, Philadelphia. (Good review.)

[22] Dwass, M. (1959). *Ann. Math. Statist.*, **30**, 1024–1028.

[23] Eicker, F. (1970). *Ann. Math. Statist.*, **41**, 2075–2092.

[24] Eicker, F. (1979). *Ann. Statist.*, **7**, 116–138.

[25] Epanechnikov, V. A. (1968). *Theory Prob. Appl.*, **13**, 686–690.

[26] Ishii, G. (1959). *Ann. Inst. Statist. Math. Tokyo*, **11**, 17–24.

[27] Jaeschke, D. (1979). *Ann Statist.*, **7**, 108–115.

[28] Kotel'nikova, V. F. and Chmaladze, E. V. (1983). *Theory Prob. Appl.*, **27**, 640–648.

[29] Krumbholz, W. (1976). *J. Multivariate Anal.*, **6**, 644–652.

[30] Krumbholz, W. (1976). *J. Multivariate Anal.*, **6**, 653–658.

[31] Maniya, G. M. (1949). *Dokl. Akad. Nauk SSSR*, **69**, 495–497.

[32] Mantel, N. (1968) *Biometrics*, **24**, 1018–1023.

[33] Mason, D. M. and Schuenemeyer, J. H. (1983). *Ann. Statist.*, **11**, 933–946.

[34] Mohanty, S. G. (1979). *Lattice Path Counting and Applications*. Academic, New York.

[35] Nair, V. N. (1981). *Biometrika*, **68**, 99–103.

[36] Niederhausen, H. (1981). *Ann. Statist.*, **9**, 923–944.

[37] Nikitin, J. J. (1972). *Sov. Math. Dokl.*, **13**, 1081–1084.

[38] Noé, M. and Vandewiele, G. (1968). *Ann. Math. Statist.*, **39**, 233–241.

[39] Noé, M. (1972). *Ann. Math. Statist.*, **43**, 58–64.

[40] Rényi, A. (1953). *Acta Math. Acad. Sci. Hung.*, **4**, 191–227.

[41] Rényi, A. (1968). *Magy. Tud. Akad. III Oszt. Közl.*, **18**, 23–30 [*IMS & AMS Sel. Transl. Math. Statist. Prob.*, **13**, 289–298 (1973)].

[42] Ruben, H. (1976). *Commun. Statist. Theor. Meth. A*, **5**, 535–543.

[43] Siegmund, D. (1986). *Ann. Statist.*, **14**, 361–404. (Boundary crossing probabilities.)

[44] Smirnov, N. V. (1961). The probability of large values on non-parametric one-sided criteria of fit. *Trudy Mat. Inst. Steklov*, **64**, 185–210 (in Russian).

[45] Steck, G. P. (1969). *Ann. Math. Statist.*, **40**, 1449–1466.

[46] Steck, G. P. (1971). *Ann. Math. Statist.*, **42**, 1–11.

[47] Steck, G. P. (1974). *Ann. Prob.*, **2**, 155–160.

[48] Takàcs, L. (1964). *J. Appl. Prob.*, **1**, 389–392.

[49] Tang, S. C. (1962). *Pacific J. Math.*, **12**, 1107–1114.

[50] Wald, A. and Wolfowitz, J. (1939). *Ann. Math. Statist.*, **10**, 105–118.

[51] Wang, S-J. (1955). *Acta Math. Sinica*, **5**, 253–267.

[52] Wellner, J. A. (1978). *Zeit. Wahrscheinl.*, **45**, 73–88.

[53] Yu, G. C. S. (1975). *J. Amer. Statist. Ass.*, **70**, 233–237.

(CHI-SQUARED TESTS
EDF STATISTICS
GOODNESS-OF-FIT
KOLMOGOROV–SMIRNOV-TYPE TESTS OF FIT)

H. NIEDERHAUSEN

REPAIRABILITY

A term used in reliability* theory for the probability that a failed system can be restored to operating condition within a specified period of repair time.

REPEATABILITY *See* MEASUREMENT ERROR

REPEATED CHI-SQUARE TESTING

Repeated significance testing is a technique for sequentializing fixed sample tests in cases where there is a need for rejecting a false null hypothesis as soon as possible. For example, in medical trials, ethical reasons may prevent carrying out a trial as soon as it becomes clear that the treatment under experiment is worse than a standard one. *See also* REPEATED SIGNIFICANCE TESTING. In this entry this sequentializing technique is described for certain chi-square* (χ^2) goodness-of-fit* tests.

Consider Pearson's [4] goodness-of-fit χ^2 statistic $Q(n) = (1/n)\Sigma_{j=1}^{r}(n_j - np_j^0)^2/p_j^0$, where $n_1, \ldots, n_r$ are the observed counts of a multinomial* $(n, \mathbf{p})$ distribution with $n = n_1 + \cdots + n_r$ and $\mathbf{p} = (p_1, \ldots, p_r)$. In order to test the simple hypothesis $H_0 : \mathbf{p} = \mathbf{p}^0$, $\mathbf{p}^0$ being a theoretical probability vector with strictly positive components, Pearson [4] suggested rejection of H_0 at significance level α if $Q(n)$ exceeds the critical value $\chi_{1-\alpha}^2(r-1)$.

The easiest way of sequentializing the fixed sample χ^2 test based on $Q(n)$ is to plot, after an initial sample of size n_0 ($\geqslant 1$), the successive values of $Q(n)$ against $n_0, n_0 + 1, \ldots, N$, where N ($\geqslant n_0$) is the target sample size, n_0 and N being given in advance. The procedure stops with rejection of H_0 at the first member n, $n_0 \leqslant n \leqslant N$, where $Q(n)$ exceeds some prescribed critical value c (> 0); if no such n exists, the procedure stops at N with acceptance of H_0. The resulting so-called repeated χ^2 test (for goodness-of-fit) rejects H_0 if and only if $Q(n_0, N) \equiv \max\{Q(n) : n_0 \leqslant n \leqslant N\}$ exceeds c. The random number N^* (say), where the above procedure stops, may be substantially smaller than the target sample size. In order to compute asymptotically critical values c, one may use the following result of Neuhaus and Kremer [3]: Assume that the target sample size N is proportional to N_0, i.e., $N = [n_0 T]$ for some fixed $T \geqslant 1$, $[x]$ denoting the integer part of x. Then $\lim_{N\to\infty}\Pr\{Q(n_0, N) \leqslant c | H_0\}$ equals

$$\Pr\left\{\sup_{1\leqslant t\leqslant T}(1/t)\|\mathbf{B}(t)\|^2 < c\right\}, \tag{1}$$

where $\|\mathbf{B}(t)\| \equiv (B_1^2(t) + \cdots + B_k^2(t))^{1/2}$, $k = r - 1$, is the so-called *Bessel process* built up by k independent standard Brownian motions* $B_1(t), \ldots, B_k(t)$. Though (1) is not available in an analytic form, De Long [1] expressed (1) as an infinite series $\Sigma_{i=1}^{\infty}\alpha_i(c, k)T^{-\beta_i(c,k)}$, where the β_i are roots of certain confluent hypergeometric functions* and the α_i can be computed in terms of β_i, c, and k. In ref. 1, tables for (1) and for critical values corresponding to (1) are given for selected significance levels α and selected values of T for $k = 1, 2, 3, 4$, while in ref. 3 critical values (obtained by Monte Carlo methods*) are given for the finite sample size distribution of $Q(n_0, N)$ for selected values of n_0, N, and α and various numbers of categories r up to $r = 20$.

If $\mathbf{p}$ ($\neq \mathbf{p}^0$) is some alternative probability vector, the error of the first kind of the repeated χ^2 test with critical value c equals approximately (1) with $B_1(t)$ replaced by $B_1(t) + \delta T^{1/2}$, δ^2 being the usual noncentrality parameter

$$\delta^2 = N\Sigma_{i=1}^{r}\left(p_i - p_i^0\right)^2/p_i^0.$$

Clearly, H_0 corresponds to $\delta = 0$. Since for $\delta^2 > 0$ a similar infinite series can be obtained as in (1) ($\delta^2 = 0$) (see De Long [1]), one may in principle compute the asymptotic power of the repeated χ^2 test.

For illustration we give an example from a simulation* study in ref. 3: For $r = 4$ cells, $n_0 = 10$ initial observations, and target sample size $N = 60$, the level $\alpha = 0.05$ critical value is $c(0.05) = 11.7$. For alternatives with corresponding $\delta = 1.0$, the repeated χ^2 test has the relatively low power $\beta(\delta) = 0.37$. In that case the expected stopping time, EN^*, is about 50, while a fixed sample χ^2 test with only $\overline{N} = 45$ observations would give (roughly) the same power 0.37. The situation changes for larger δ; e.g., for $\delta = 1.5$, the power of the repeated χ^2 test is $\beta(\delta) = 0.79$ with expected stopping time $EN^* = 37$. In the latter case a fixed sample χ^2 test would need $\overline{N} = 50$ observations to achieve the same power 0.79. The two preceding cases show a general feature of the repeated χ^2 test: For alternatives with δ leading to values of the power of $\beta(\delta)$ around 0.6 or smaller, EN^* becomes larger than the corresponding sample number $\overline{N}$, for which the fixed sample χ^2 test yields the same power under the given alternatives. On the other hand, for values of δ with $\beta(\delta)$ somewhat larger than 0.6 one gets $EN^* < \overline{N}$, thus making the repeated χ^2 test favourable to the fixed sample version.

In practice one has to judge what one wants to achieve: If one has to reject heavy deviations from the null hypothesis as soon as possible while small deviations are not so dangerous, the repeated χ^2 test should be used. On the other hand, the fixed sample version is preferable in cases where there is a need for detecting small deviations, regardless of the fact that large deviations can be rejected only at the end of the whole experiment.

An extension of the repeated χ^2 test to composite hypotheses $H_0: p \in \{p(\vartheta): \vartheta \in \theta\}$, where $p(\vartheta)$ is a prespecified set of r functions of t nonredundant parameters ϑ from the Euclidean t space, is possible by using

$$\hat{Q}(n) = (1/n)\Sigma_{j=1}^{r}\left(n_j - np_j\left(\hat{\vartheta}_n\right)\right)^2/p_j\left(\hat{\vartheta}_n\right)$$

instead of $Q(n)$, where $\hat{\vartheta}_n$ is some asymptotic efficient estimator of ϑ, e.g., a minimum χ^2 estimator or a maximum likelihood* estimator. Then it may be shown (see Neuhaus [2]) that $\Pr\{\hat{Q}(n_0, N) \leqslant c|H_0\}$ with $\hat{Q}(n_0, N) \equiv \max\{\hat{Q}(n): n_0 \leqslant n \leqslant N\}$ tends to (1) with $k = r - 1$, and this result extends to alternatives as previously described.

It should be mentioned that the same limiting probabilities as in (1) (as well as its variant for $\delta^2 > 0$) arise when considering certain nonparametric repeated significance tests; see Sen [5, Section 9.3].

Siegmund [6] uses the name *sequential χ^2 test* in a one-way analysis of variance* setting with independent, identically distributed normal random r vectors $\mathbf{X}_n = (X_{1n}, \dots, X_{rn})'$, observed sequentially for $n = 1, 2, \dots$, each having normal distribution with mean $\boldsymbol{\mu} = (\mu_1, \dots, \mu_r)$ and known variance–covariance matrix $\sigma^2\mathbf{I}$, $\mathbf{I}$ being the $(r \times r)$ unit matrix. For testing the hypothesis $H_0: \mu_1 = \cdots = \mu_r$ (equality of r treatments) he considers, instead of $Q(n)$, the statistic $\Lambda(n) = n\Sigma_{i=1}^{r}(\overline{X}_{i\cdot} - \overline{X}_{\cdot\cdot})^2/(2\sigma^2)$ and proceeds as described at the beginning of this entry, with the extension that H_0 may be rejected at the target sample size N even if $\Lambda(N)$ exceeds a prescribed value $d > 0$ smaller than c, where c has the same meaning as before. Asymptotic approximations to the significance level, the power of the test, and the expected sample size (*see* AVERAGE SAMPLE NUMBER) are given. Siegmund's asymptotics is different from the one herein described, since he assumes that $N \to \infty$ and $c \to \infty$ with $0 < c/N$ staying constant.

We cite one numerical example from ref. 6: For $n_0 = 1$, $N = 20$, and $c = d$ the critical value at level $\alpha = 0.05$ is $c = 4.96$. The power β of the corresponding test under $\boldsymbol{\mu}$ depends only on $\vartheta \equiv \sigma^{-1}(\Sigma_{i=1}^{r}(\mu_i - \bar{\mu}_{\cdot})^2)^{1/2}$, i.e., $\beta = \beta(\vartheta)$. From ref. 6 one has, e.g., $\beta(0.7) = 0.685$ and $\beta(1.0) = 0.959$, while the corresponding sample sizes $E_\vartheta N^*$ are $E_{0.7}N^* = 13.0$ and $E_{1.0}N^* = 7.8$. One notices again that the test is designed to have a small expected sample size for large ϑ, when the effects of the various treatments differ substantially from one another. If σ is unknown, an estimator of σ has to be incorporated. The resulting procedure is the *sequential F test* and has been studied in the same paper by Siegmund [6].

References

[1] De Long, D. M. (1981). *Commun. Statist.-Theor. Meth.*, **10**, 2197–2213.

[2] Neuhaus, G. (1983). *Commun. Statist.-Seq. Anal.*, **2**, 99–121.

[3] Neuhaus, G. and Kremer, E. (1981). *Commun. Statist. Simul. Comp. B*, **10**, 143–161.

[4] Pearson, K. (1900). *Philos. Mag. Ser. 5*, **50**, 157–172.

[5] Sen, P. K. (1981). *Sequential Nonparametrics*. Wiley, New York.

[6] Siegmund, D. (1980). *Biometrika*, **67**, 389–402.

(REPEATED SIGNIFICANCE TESTS
SEQUENTIAL ANALYSIS)

GEORG NEUHAUS

REPEATED MEASUREMENTS—DESIGN AND ANALYSIS

A broad range of statistical investigations can be classified as repeated measurements studies. Their essential feature is that each subject is observed under two or more conditions. Four important classes of repeated measures studies are as follows:

1. **Split-Plot Experiments in Agriculture*.** An example is the evaluation of the effects of fertilizer and crop variety on crop yield, where fertilizer types are randomly assigned to fields and crop varieties are randomly assigned to plots within fields. The fields represent the "subjects" that are randomly assigned to levels of one factor (fertilizer type), and the plots correspond to the observational conditions that are randomly assigned to levels of a second factor (crop variety).

2. **Longitudinal Studies*.** For example, cattle are randomly assigned to one of three diets and their weight is measured every week for three months. The cattle are the subjects, and the successive time intervals correspond to the observational conditions.

3. **Change-Over Design* Studies.** For example, each subject is randomly assigned either to the sequence group with treatment A first and B second or to the sequence group with treatment B first and then A. Responses to each successive treatment are measured for their corresponding periods of administration. The observational conditions are not only the successive time periods, but also the treatments, and possibly the immediately preceding treatments.

4. **Sources of Variability Studies.** An example is the study of the number of gastrin cells per unit length of rat stomach tissue via measurements for adjacent microscopic fields, in adjacent biopsies, from selected gastric sites, by two different observers. The rats constitute the subjects, and the sites, biopsies, fields, and observers can be either a fixed set or a random sample from a large population.

For studies like those described in **1**–**4**, the subjects are primary sampling units randomly selected to represent various strata or randomly assigned to levels of a grouping factor; they are often called *experimental units*. The responses measured under the respective conditions constitute the observational units; because within each subject these constitute a profile of inherently multivariate data, their covariance* structure plays an important role in the formulation of statistical methods for their analysis.

A useful strategy for a broad range of repeated measurements studies is to view the subjects as the units of analysis in the following two-stage procedure: certain measures of interest (e.g., sums for total response, differences between conditions, orthogonal contrasts* between conditions) are first constructed from the observational unit data within each subject; then this information is analyzed across experimental units by parallel, but separate, univariate methods and/or by simultaneous multivariate methods. This general strategy can appropriately account for the measurement scale of the response (categorical, ordinal, interval) and the nature of the randomization* in the study design.

DESIGN

In this section, principles pertaining to the design of repeated measurements studies are discussed through representative examples. Separate attention is given to split-plot experiments, longitudinal studies, change-over design studies, and sources of variability studies. The features that are shared by these different classes of designs are noted as well as those that characterize them; also, relative advantages and limitations are identified.

Split-Plot Experiments

The distinguishing aspect of split-plot experiments relative to other types of studies is the use of two or more stages of randomization. In the first stage, subjects are randomly allocated to treatments or randomly selected from strata; in the subsequent stages, conditions are randomly allocated within subjects. In the usual agricultural experiment, the subjects are whole plots or fields within which split plots are the observational units. Some other representative examples are:

S1. The whole plots are six batches of synthetic fiber; these correspond to three batches from each of two different combinations of ingredients. The split plots are four subsamples from each batch; they are tested for pulling strength under four different temperature conditions.

S2. The whole plots are litters (or cages) of rats that are assigned to diets containing different amounts of fat, and the split plots are the different doses of carcinogen under which the tumor levels of individual animals are observed; or, the split plots could be the different times at which the rats were sacrificed for tumor evaluation.

S3. The whole plots are school systems that are assigned to one of two different sets of materials for a science unit, and the split plots are individual schools in which different teaching strategies are used. The response measure is student examination performance.

S4. Rats are paired on the basis of weight; within each pair, one rat receives an experimental diet and the other receives the same amount of a control diet (i.e., it is a pair-fed control). Seven days later, the rats are sacrificed and the liver of each rat is divided into three parts. For each section of a rat liver, the amount of iron absorbed is determined from a solution with one of three randomly assigned pH (acidity) levels and with randomly assigned temperature for the pair to which the animal belonged. Pairs of rats are whole plots, rats within pairs are split plots, and liver thirds within a rat are split split plots. Further discussion of this *split-split-plot experiment* is given in Koch [45].

S5. Three generations of animals with large litters are used to assess the effect of hormone and diet treatments on hormonal blood levels in the offspring. In the first generation, four females from the same litter are randomly assigned to four hormone doses. When these females have offspring, three females are selected from each litter and randomly assigned to be fed one of three different diets; also, the four animals receiving the same diet are maintained in the same cage. Thus, at the second generation, there is a block of 12 females originating from the same first generation litter and residing in three cages (for diet) with four animals per cage (for hormone dose). The third generation litters from these 12 females represent whole plots, and five females within each of them correspond to split plots to which five weekly intervals for time of sacrifice are assigned in order to measure hormonal blood levels. These 60 animals are raised in 12 cages with the five animals from the same litter sharing the same cage. In all, there are 10 blocks with the (4 × 3)(mother × cage) structure described here. This type of study is called a *split-block experiment*. It can be naturally applied in agriculture by assigning one type of treatment to rows of a field (e.g., planting method) and another type to the columns (e.g., fertilizer); row × column cells are whole plots within which a third factor (e.g., varieties of a crop) can be assigned to split plots; see Federer [22] for further discussion.

A broad range of potential research designs is thus available for split-plot experiments, and so relevant aspects of a particular situation can be taken into account. The treatment groups can be based on a single factor or on the cross-classification of two or more factors; they can be assigned to whole plots according to a completely randomized design, a randomized complete blocks design, or some type of incomplete blocks design. A similar statement applies to the nature of conditions and their assignment to split plots. When incomplete blocks structures are used for cross-classified treatment or condition factors, issues concerning the *confounding** of effects require careful attention for both design specification and analysis. Finally, the preceding considerations also apply to situations where whole plots are randomly selected from strata that correspond to the groups.

Split-plot experiments have two important advantages over other research designs without their nested structure. They can be less costly or more straightforward to implement when the treatment for each whole plot can be applied to its entirety (i.e., to all of its split plots simultaneously rather than separately). The second advantage is that split plots within the same whole plot are usually more homogeneous than those from different whole plots. As a result, comparisons between conditions and (treatment × condition) interaction* effects (i.e., between treatment differences for comparisons of conditions) are estimated more precisely. The basic consideration here is that the variability of such estimates comes from within-whole plot variability for split plots rather than overall (i.e., across whole plots) variability. Thus, it corresponds to the way in which the design of split-plot experiments enables the researcher to control the extent of variability influencing comparisons between conditions. Since whole plots are the units through which such control is applied, they are often said to serve as their own controls.

Three limitations of split-plot experiments should be noted. One is that differences between treatments applied to whole plots are estimated less precisely than differences between conditions applied to split plots and also less precisely than if they had been applied to the same number of independent split plots; this occurs because split plots within whole plots usually have a positive *intraclass correlation**. Secondly, greater cost or effort might be required for the administration of split-plot experiments in order to ensure that each condition only affects the split plots to which it was assigned. Contamination of condition effects to neighboring split plots needs to be negligible so that estimated comparisons among conditions and treatments are not biased to a potentially misleading extent. The third limitation of split-plot experiments is that complexities in their structure can make the analysis of their data relatively difficult.

The statistical literature for split-plot experiments is extensive; bibliographies have been published by Federer and Balaam [24], Hedayat and Afsarinejad [32], and Federer [23]. A useful basic reference is Snedecor and Cochran [82]. Some textbooks that discuss alternative designs are Allen and Cady [2], Bennett and Franklin [5], Cochran and Cox [12], Cox [17], Federer [21], Gill [27], Kempthorne [40], Myers [64], and Winer [92].

Longitudinal Studies

The primary way in which longitudinal studies differ from split-plot experiments is that the observational units for their subjects are systematically linked to the conditions rather than being randomly assigned. The usual dimension for such linkage is time, but it can also be location in space or different components of a concept, item, or process. The design of longitudinal studies for subjects specifies parallel groups, which can be based on either random allocation to treatments, random selection from strata, or both. Some representative examples are:

L1. Two treatments for chronic pain are randomly assigned to subjects, and the extent of pain relief is evaluated at weekly visits for six weeks. Alternatively, for studies of the rapidly occurring effects of treatments, heart rate, blood pressure, or gastric pH might be measured at more frequent intervals such as every hour or every 10 minutes.

L2. Boys and girls from a cohort of one year olds are observed every six months for five years to assess their ability to perform a manual dexterity task (or measurements of height, weight, or physical fitness might be made).

L3. Two treatments for a dental problem are randomly assigned to children. The status of teeth on the upper and lower jaws is evaluated every three months for one year. Here, the eight conditions are determined by the site × time cross-classification.

L4. In a political survey, each subject is asked about the degree of trust in three political institutions (the Presidency, the Senate, and the Supreme Court). Subjects are drawn from different demographic groups. The three institutions constitute the conditions.

An important aspect of the design of longitudinal studies is the specification of the number of conditions and their nature (e.g., a set of time intervals). As the number of conditions is increased, the amount of information for the response is increased; but the cost or effort for its acquisition and management is also usually increased (a potential exception being studies with automated data collection* and processing devices). Since conditions represent situations for comprehensively observing within-subject response across some dimension(s) of interest (e.g., time course) rather than an experimental factor, which can be assigned either within or among subjects, their cost is a necessary consequence of their scope. Also, there are no alternative designs that would provide the same information. Thus, conditions are specified to encompass the observational situations of greatest interest in a manner compatible with available resources.

In some cases, the number of conditions may be too extensive for all of them to be observed on each subject (e.g., more than 100 time points over a period exceeding five years), owing to cost or subject tolerance, say. One way of dealing with this problem is to define incomplete subsets that suitably cover the range of conditions and to assign them randomly to the subjects in the respective treatment groups. Such subsets can be formed in ways that appropriately account for what is feasible in a particular application. For example, when the conditions are based on time, overlapping subsets of time points (i.e., 0–12, 6–18, 12–24, etc.) can be used. Studies with this structure are sometimes said to have a *panel* design* with *rotating groups* of new subjects entering every six months and leaving after 12 months of participation. Similarly, when age is the principal dimension for conditions, subjects with consecutive initial ages (e.g., children in the range 6–17 years) can be subsequently observed for some response (e.g., height) at the same time intervals over some specified period (e.g., every six months over three years); in this way, longitudinal information is obtained for the entire age range under study (e.g., children 6–20 years old). Studies of this nature are said to have a linked cross-sectional or mixed longitudinal design;

see Rao and Rao [72] and Woolson et al. [93].

The range of potential designs for longitudinal studies is very broad. Since the data for each subject form a multivariate profile, the relevant design principles are the same as for multivariate studies. Discussion and related references are given in Roy et al. [77]. For more specific consideration of the number of conditions to be used and the spacings between them, see Morrison [62] and Schlesselman [79]. When the number of conditions is small relative to the number of subjects and all conditions are observed on all subjects, data from longitudinal studies can often be satisfactorily analyzed by multivariate analysis of variance* procedures like those described in the Methods section of this entry; also, see textbooks dealing with multivariate analysis* such as Anderson [4], Bock [7], Gill [27], Morrison [63], Rao [71], and Timm [86]. For situations in which the number of conditions is large or which require methods specifically designed for longitudinal studies *see* GROWTH CURVES, LONGITUDINAL DATA ANALYSIS, and TIME SERIES, or refer to Cook and Ware [16], Dielman [19], Geisser [26], Grizzle and Allen [30], Laird and Ware [52], Nesselroade and Baltes [66], Rao [70], Snee et al. [83], and Ware [90]. The longitudinal studies described here do not include follow-up studies for the time until some event such as death or recurrence of disease; *see* CENSORED DATA; CLINICAL TRIALS; and SURVIVAL ANALYSIS for discussion of such data.

Change-Over Designs

Change-over designs have similarities to both split-plot and longitudinal studies. Subjects are assigned at random to groups or selected from strata. As in longitudinal studies, the response of each subject is observed at several points in time (or locations in space); and as in split-plot experiments, a treatment condition is assigned to each time period. Subjects are randomly assigned to groups that receive alternative sequences of treatments. More than one sequence is needed to allow the separate estimation of time or location effects and treatment effects (since they are completely confounded in any single sequence). For this reason, sequences are often based on the rows of a *Latin square**.

Another important aspect of change-over studies is that the preceding or neighboring treatment can influence the response to the next treatment. Such factors are called *carryover effects* or *residual effects*. When their extent cannot be presumed negligible, adjustments for their presence become necessary. In such cases, the sequences in the design of a change-over study need to be constructed so as to allow the estimation of location effects, treatments effects and carryover effects. Alternatively, it may suggest that some other research design in which subjects receive only one treatment be used. CHANGE-OVER DESIGNS provides additional discussion concerning construction and properties; also see Brown [10], Cochran and Cox [12], Constantine and Hedayat [15], Gill [27], Kershner and Federer [42], Hedayat and Afsarinejad [32, 33], Koch et al. [51], Laycock and Seiden [53], and Wallenstein and Fisher [88].

Some examples that illustrate the repeated measurements nature of change-over studies are:

C1. Information on recent smoking was obtained for each subject by two different methods; one was the subject's self-report to a direct question and the other was a biochemical determination based on carbon monoxide levels in the blood. Subjects were randomly assigned to one of two sequence groups; for one group, the self-report preceded the biochemical determination; and for the second group, the self-report followed the biochemical determination.

C2. Patients with an unfavorable skin condition on both sides of their body are randomly assigned to one of two groups; one group receives an active treatment for daily application on the right side of the body and placebo for the left side, and the other receives the active treatment for the left side and placebo for the right side. The extent of healing is evaluated separately for each side at weekly visits during a four-week period.

C3. The relative potency of two drugs that influence cardiovascular function is assessed through a change-over design. Volunteers are randomly assigned to one of two sequence groups. One group receives drug A during the first six-week study period and drug B during the second, and the other group receives the opposite regimen. A two-week washout period separates the two treatment periods. During each treatment period, three doses of the drug are tested with the drug dose being successively increased every two weeks. At the beginning of treatment and at the end of each two-week dose interval, heart rate is measured before and after a treadmill exercise test. Additional discussion of this type of example is given in Elashoff [20].

C4. Subjects are randomly assigned to one of six groups based on the cross-classification of a treatment factor (alcohol vs. no alcohol) and a sequence factor for the assignment of three drugs; the sequence factor is based on the three rows of a 3×3 Latin square. At each of the three one-day study periods, performance on a task is measured in response to one of the three drugs. The study periods are separated by one-week washout intervals in which no drugs or alcohol are to be used.

For change-over studies like those in **C1**–**C4**, note that subjects are analogous to whole plots in split-plot experiments, and treatment sequences are analogous to their treatment groups. A straightforward extension of this structure is its application within different subpopulations of subjects. These subpopulations could correspond to different strata from which subjects were randomly selected or to one or more additional experimental factors that were randomly assigned to subjects (e.g., alcohol status in **C4**). The structure of conditions in change-over studies can be extended by observing the response to each within-subject treatment at a longitudinal set of locations (e.g., several time points as in **C2** and **C3**). All of the considerations outlined here imply that change-over studies can be designed in many different ways.

Since change-over designs have a similar structure to split-plot experiments, they have similar advantages and disadvantages. The advantages are potentially reduced cost through the need for fewer subjects (because more than one treatment is tested for each) and more precise estimates for within-subject treatment comparisons. The principal disadvantage is that research design, management, and analysis strategies to minimize or deal with carryover effects may be complex and expensive. The major issue is that carryover effects may induce substantial bias in treatment comparisons if they are ignored; adjustments to eliminate such bias may lead to use of less efficient estimation methods such as analyzing only the data from each subject's first treatment.

Sources of Variability Studies

Sources of variability studies are concerned with identifying the amount of variability between responses that is attributable to each component of the sampling or measurement process. Some representative examples are:

V1. The variability in product performance is to be studied. On each of 10 randomly identified production days, two samples are obtained at randomly specified times. Each sample is divided into five subsamples, and each subsample is assigned to one of five evaluators who make two performance determinations. The same five evaluators are used on each of the 10 days.

V2. The variability between and within observers for ratings of severity of a periodontal condition is to be assessed. For each of a sample of patients, two photographs of gum status are obtained. All photographs are shown in random order to five observers for rating.

V3. The variability associated with households within clusters and with interviewers is to be assessed in a survey of a socioeconomic variable. A random sample of 288 clusters of eight households is randomly divided into three sets (A, B, C) of 96 clusters. Twenty-four interviewers are randomly assigned to each set of clusters. Each assignment in the following protocol is made at random. In set A of household clusters, each interviewer is assigned four clusters and obtains responses from all eight households in each cluster. In set B, 12 sets of two interviewers are formed and eight clusters are assigned to each pair; one member of the pair of interviewers is assigned to four households in each cluster, and the second to the other four households. In set C, six blocks of four interviewers are formed and each block is assigned 16 clusters; each interviewer in the block is assigned to two households in each cluster.

Studies of sources of variability can thus deal with fixed components of a measurement process such as the evaluators in **V1**, and random components such as days and samples in **V1**. Their structure can be straightforward or very complex. For designs where interest is focused on a fixed component of variability with few levels in a straightforward design, as in the case of the observers in **V2**, a satisfactory analysis can usually be undertaken with the procedures described in the Methods section. For other discussion *see* INTRACLASS CORRELATION COEFFICIENT; KAPPA STATISTICS; MEASURES OF AGREEMENT; and VARIANCE COMPONENTS. See also, for example, Anderson and Bancroft [3], Kempthorne [41], Scheffé [78], and Searle [81].

Combination Designs

The previous discussion has dealt with split-plot experiments, longitudinal studies, change-over studies, and variability studies as separate types of repeated measurements studies; however, many research designs combine features across these types. Longitudinal data can be obtained for each split plot or for each successive treatment in a change-over study (see **C2** and **C3**). Responses can be assessed by more than one observer for each split plot or subject. Repeated measurements studies can be designed to provide information for factors of direct interest (treatment effects or time patterns) and on background factors (place or observer effects); the latter are often called *lurking variables** so as to reflect their potential influence on the analysis of a response variable; see Joiner [39]. The examples presented here illustrate the wide variety of research designs involving repeated measurements; others are given in Federer [22], Gill [27], Koch et al. [46], and Monlezun et al. [61].

STATISTICAL METHODS

In this section, several strategies for the analysis of repeated measurements studies are described. They include:

1. Univariate analysis of within-subject functions.
2. Multivariate analysis for within-subject functions.
3. Repeated measures analysis of variance*.
4. Nonparametric rank methods.
5. Categorical data* methods.

The specification of these methods takes five important considerations into account: the measurement scale, the number of subjects, the role of randomization, the nature of the covariance structure of the observational units within subjects, and the potential influence of carryover effects. Primary attention is given to split-plot experiments since there are other entries for LONGITUDINAL DATA ANALYSIS and GROWTH CURVES; CHANGE-OVER DESIGNS; and VARIANCE COMPONENTS. Basic methods for relatively straightforward situations are emphasized. Examples 1–4 provide insight on their application and are potentially helpful to read in parallel with the discussion of methods or prior to it.

Data Structure

A general framework for repeated measurements studies involves a set of subjects for whom data are obtained for each of d conditions; these conditions can be based on a single design factor or the cross-classification of two or more factors. For each subject, the data for each condition can include a set of N determinations for each of R response variables and M concomitant variables for which potential associations with the variation between conditions are to be taken into account. Also, the information for the respective subjects can include classifications for one or more study design factors by which the subjects are partitioned into G groups, as well as K background variables for which potential associations with the variation between groups are to be taken into account. The subsequent discussion is directed at situations with $N = 1$ determination for $R = 1$ response variable and M concomitant variables for the respective conditions within each subject; when $N > 1$, the methods for $N = 1$ are usually applicable to averages across the multiple determinations. When $R > 1$, the multivariate analogs of methods for a univariate response variable are usually applicable; see Reinsel [73] and Thomas [85].

Let $i = 1, 2, \ldots, G$ index the G groups and $j = 1, 2, \ldots, d$ index the d conditions. Let $l = 1, 2, \ldots, n_i$ index the n_i subjects within the ith group. Let Y_{ijl} denote the response of the lth subject in the ith group for the jth condition and $\mathbf{z}_{ijl} = (z_{ijl1}, z_{ijl2}, \ldots, z_{ijlM})'$ be the corresponding set of observed values for M concomitant variables*. Finally, let $\mathbf{x}_{i*l} = (x_{i1l}, x_{i2l}, \ldots, x_{iKl})'$ denote the set of K background variables that represent fixed characteristics for the lth subject in the ith group. Interpretation of this notation can be clarified by considering a change-over design example like C4, where $i = 1, 2, 3, 4, 5, 6$ indexes the six groups of subjects with respect to alcohol status and drug sequence and $j = 1, 2, 3$ indexes the three periods during which they receive the three drugs; the background variables $\mathbf{x}_{i*l}$ for the lth subject represent characteristics such as age, sex, and previous medical history, for which the values are defined prior to the study (or apply to all conditions); and the concomitant variables $\mathbf{z}_{ijl}$ represent aspects of health status such as baseline performance at the time of administration of the jth drug, period effects, drug effects, or carryover effects; the response y_{ijl} represents performance after the administration of the jth drug. More specific illustrations are given in Examples 1 and 2.

Univariate Analysis of Within-Subject Functions

For many repeated measurements studies, the questions of interest can be expressed in terms of some summary functions of within-subject responses to the conditions. These summary functions could be contrasts; the general form of a contrast for the lth subject in the ith group is $f_{il} = \sum_{j=1}^{d} a_j y_{ijl}$, where $\sum_{j=1}^{d} a_j = 0$. Examples are pairwise comparisons of conditions $\{(y_{ijl} - y_{ij'l})\}$; or when the conditions represent equally spaced doses or times, the contrasts could be orthogonal polynomials (*see* ORTHOGONAL EXPANSIONS), e.g., for four doses, the linear contrast is $-3y_{i1l} - y_{i2l} + y_{i3l} + 3y_{i4l}$. Summary functions could also be the average response over conditions ($f_{il} = \bar{y}_{i \cdot l} = \sum_{j=1}^{d} y_{ijl}/d$) or nonlinear functions such as ratios.

Suppose that the variation of the $\{y_{ijl}\}$ across groups and conditions can be described by the additive model

$$E\{y_{ijl}\} = \mu + \xi_i + \theta_j + (\xi\theta)_{ij}, \quad (1)$$

where μ is a reference parameter for group 1 and condition 1, the $\{\xi_i\}$ are incremental effects for the respective groups, the $\{\theta_j\}$ are incremental effects for the respective conditions, and the $\{(\xi\theta)_{ij}\}$ are interaction effects between group and condition; also, $\xi_1 = 0$, $\theta_1 = 0$, the $\{(\xi\theta)_{1j} = 0\}$ and the $\{(\xi\theta)_{i1} = 0\}$ to avoid redundancy*. To simplify the exposition, we shall concentrate the subsequent discussion in this section on a contrast function. Two cases are considered:

1. One group or multiple groups with no (group $\times$ condition) interaction.
2. Multiple groups that possibly interact with conditions.

Case 1: One Group or No Group-By-Condition Interaction. Here the contrast functions $\{f_{il}\}$ for the respective subjects have the same expected value

$$E\{f_{il}\} = \sum_{j=1}^{d} a_j\theta_j = \theta_f, \quad (2)$$

since no (group $\times$ condition) interaction implies $\{(\xi\theta)_{ij} = 0\}$ for (1), and if there is only one group, the $\{(\xi\theta)_{ij}\}$ do not exist. The variances for the $\{f_{il}\}$ have the form

$$\text{Var}\{f_{il}\} = \sum_{j=1}^{d}\sum_{j'=1}^{d} a_j a_{j'} v_{i,jj'}, \quad (3)$$

where $v_{i,jj'} = \text{Cov}\{y_{ijl}, y_{ij'l}\}$ denotes the (jj')th element of the covariance matrix $\mathbf{V}_i$ for the d responses of subjects in the ith group. The $\{f_{il}\}$ will have homogeneous variances

$$\text{Var}\{f_{il}\} = v_f \quad (4)$$

if the elements of the $\{\mathbf{V}_i\}$ have the structure

$$v_{i,jj'} = v_{i,*} + v_{*,jj'}; \quad (5)$$

here $v_{i,*}$ represents the between-subject variance component for the ith group and $v_{*,jj'}$ the within-subject covariance component for the jth and j'th conditions. However, it is often realistic to make the stronger assumption that the d responses for all subjects have the same covariance matrix $\mathbf{V}$ (i.e., all $\mathbf{V}_i = \mathbf{V}$), and this is done henceforth.

Confidence intervals* and statistical tests concerning θ_f in (2) can be obtained by using the t-distribution* if it is also assumed that the $\{f_{il}\}$ for the $n = \sum_{i=1}^{G} n_i$ subjects have independent normal distributions. The overall mean

$$\bar{f} = \left\{ \sum_{i=1}^{G}\sum_{l=1}^{n_i} f_{il}/n \right\} \quad (6)$$

is an unbiased estimator of θ_f. An unbiased estimator of its variance is given by $(\hat{v}_f/n)$, where

$$\hat{v}_f = \sum_{i=1}^{G}\sum_{l=1}^{n_i} (f_{il} - \bar{f})^2/(n-1). \quad (7)$$

Thus, the $100(1-\alpha)\%$ confidence interval for θ_f is given by

$$\bar{f} - t\left[\hat{v}_f/n\right]^{1/2} \leqslant \theta_f \leqslant \bar{f} + t\left[\hat{v}_f/n\right]^{1/2}, \quad (8)$$

where $t = t_{1-(\alpha/2)}\{(n-1)\}$ is the $100\{1-(\alpha/2)\}$ percentile of the t-distribution with $(n-1)$ degrees of freedom. The hypothesis $H_0: \theta_f = 0$ can be tested by rejecting it if 0 is not included in the interval (8), i.e., if

$$\left|\bar{f}/\left[\hat{v}_f/n\right]^{1/2}\right| \geqslant t. \quad (9)$$

Analogous one-sided confidence intervals to (8) and hypothesis tests to (9) can be constructed by using $t = t_{1-\alpha}(n-1)$ for the specified direction only.

When the overall sample size n is sufficiently large (e.g., $n > 30$), $\bar{f}$ can have an approximately normal distribution on the basis of central limit theory. For such situations, (8) and (9) are applicable with $t = t_{1-(\alpha/2)}(\infty)$ under the more general assumption that the $\{f_{il}\}$ are independent random variables with a common distribution. Also, the common distribution assumption (and hence the homogeneous variance assumption) can be relaxed to hold within each of the respective groups if $\hat{v}_f$ in (7) is replaced by

$$\tilde{v}_f = \sum_{i=1}^{G} \left(\frac{n_i^2}{n^2} \right) \sum_{l=1}^{n_i} \frac{(f_{il} - \bar{f}_i)^2}{(n_i - 1)n_i} = \sum_{i=1}^{G} \frac{n_i}{n^2} \hat{v}_{f,i}, \qquad (10)$$

where $\bar{f}_i = \{\sum_{l=1}^{n_i} n_i f_{il}/n_i\}$ is the sample mean of the $\{f_{il}\}$ for the ith group and $\hat{v}_{f,i}$ is the sample variance. The large sample counterparts of (8) and (9) with (10) replacing (7) are of particular interest when the $\{f_{il}\}$ represent contrasts among scored values for ordinally scaled categorical data, e.g., $f_{il} = -1, 0, 1$ for pairwise differences between dichotomous observations $\{y_{ijl} = 0, 1\}$ and $f_{il} = -3, -2, -1, 0, 1, 2, 3$ for pairwise differences between integer scores $\{y_{ijl} = 1, 2, 3, 4\}$ for observations with respect to four ordinal categories. For additional discussion, see Guthrie [31], Koch et al. [50], Stanish et al. [84], and CHI-SQUARE TESTS.

Nonparametric methods provide another way to proceed when the normal distribution assumption for the $\{f_{il}\}$ is not realistic. Symmetry and independence of the distributions of the $\{f_{il}\}$ are sufficient to support usage of the *sign test*; if the $\{f_{il}\}$ can also be ranked, the *Wilcoxon signed ranks test** is applicable. Related confidence intervals can be constructed under the further assumption that the $\{f_{il}\}$ have a common, continuous distribution; see Conover [14]. Otherwise, when ratio or percent change comparisons are of interest, the symmetry condition can often be satisfied by replacing the $\{f_{il}\}$ by analogous contrasts for logarithms.

Case 2: Multiple Groups That May Interact With Conditions. The methods discussed for Case 1 are appropriate for designs where there is only one group or it can be assumed that the effect of conditions is the same in every group. In most situations, however, it will be of interest to test this assumption; frequently, in fact, the question of whether there is a (group × condition) interaction is of interest in its own right and can be assessed by comparing the $\{f_{il}\}$ among groups.

When the contrast functions $\{f_{il}\}$ are normally distributed, the hypothesis of no group differences in their expected values,

$$E\{f_{il}\} = \sum_{j=1}^{d} a_j \theta_j + \sum_{j=1}^{d} a_j (\xi\theta)_{ij} = \theta_f + (\xi\theta)_{f,i} = \Delta_{f,i}, \qquad (11)$$

can be tested by one-way analysis of variance*; relevant background variables, concomitant variables, or functions of them can be taken into account by analysis of covariance* methods. For situations where the normal distribution assumption for the $\{f_{il}\}$ is not realistic, other strategies are available. If the sample sizes for each group are sufficiently large for the within-group means $\{\bar{f}_i\}$ to have approximately normal distributions via central limit theory, then the Wald statistic methods described in CHI-SQUARE TESTS can be used; also see Koch et al. [50]. Alternatively, randomization of the subjects to groups or a common family for the distributions of the $\{f_{il}\}$ with location shifts being the only source of across-group variation is sufficient to support usage of nonparametric methods like the Kruskal–Wallis statistic; see Koch [44], Koch et al. [46], Lehmann [54], Puri and Sen [68], and CHI-SQUARE TESTS.

When there is (group × condition) interaction, the contrasts $\{f_{il}\}$ may have different expected values $\{\Delta_{f,i}\}$ for the G groups. As a result, the expected value of their overall mean $\bar{f}$ in (6) is a weighted average of the $\{\Delta_{f,i}\}$, the weights being the proportions $\{(n_i/n)\}$ of the subjects in the respective groups, i.e.,

$$E\{\bar{f}\} = \sum_{i=1}^{G} (n_i/n)\Delta_{f,i}. \qquad (12)$$

Sometimes these weights are reasonable, sometimes other weights $\{w_i\}$ are preferable [e.g., $w_i = (1/G)$ for all i], and sometimes usage of any weights at all is considered undesirable because alternative choices lead to seemingly different conclusions. When the $\{f_{il}\}$ have independent normal distributions with the expected value structure in (11) and the homogeneous variance structure in (4), a confidence interval analogous to (8) can be formed for the weighted average $\Delta_{f,w} = \Sigma_{i=1}^{G} w_i \Delta_{f,i}$. It is

$$f_w - t\{\hat{v}_{f,w}\}^{1/2} \leqslant \Delta_{f,w} \leqslant f_w + t\{\hat{v}_{f,w}\}^{1/2}, \quad (13)$$

where

$$f_w = \sum_{i=1}^{G} w_i \bar{f}_i,$$

$$\hat{v}_{f,w} = \bar{v}_f \sum_{i=1}^{G} (w_i^2/n_i)$$

with $\bar{v}_f = \Sigma_{i=1}^{G} \Sigma_{l=1}^{n_i} (f_{il} - \bar{f}_i)^2/(n - G)$, and $t = t_{1-(\alpha/2)}\{(n - G)\}$. Here, the pooled within-groups variance $\bar{v}_f$ rather than $\hat{v}_f$ in (7) is used to estimate v_f since it accounts for the potential variation between groups for the expected values $\{\Delta_{f,i}\}$. Thus, the application of (13) to $\bar{f}$ in the context of (12) can be viewed as providing a more robust interval for θ_f when (2) is viewed as approximately true rather than strictly true; a limitation of (13) is that it can give a confidence interval undesirably wider than (8) for small samples (e.g., $n - G < 10$). Otherwise, when the assumption of normal distributions is not realistic for the $\{f_{il}\}$, large sample or nonparametric methods can be formulated in ways that yield results similar in spirit to (13).

Another summary function of interest, which is not a contrast, is the within-subject mean

$$\bar{y}_{i \cdot l} = \left(\sum_{j=1}^{d} y_{ijl}/d \right) = \mathbf{1}'_d \mathbf{y}_{il}/d, \quad (14)$$

where $\mathbf{y}_{il} = (y_{i1l}, \ldots, y_{idl})'$ and $\mathbf{1}_d$ is the $(d \times 1)$ vector of 1's. It allows between group comparisons of the average responses over conditions, i.e., the

$$E\{\bar{y}_{i \cdot l}\} = \left\{ \sum_{j=1}^{d} \mu_{ij}/d \right\} = \bar{\mu}_{i \cdot}, \quad (15)$$

where $\mu_{ij} = E\{y_{ijl}\}$. When all $(\xi\theta)_{ij} = 0$ in the model (1), comparisons between the $\{\bar{\mu}_{i \cdot}\}$ are the same as comparisons between the group effects $\{\xi_i\}$ because

$$\bar{\mu}_{i \cdot} = (\mu + \bar{\theta}) + \xi_i, \quad (16)$$

with $\bar{\theta} = (\Sigma_{j=1}^{d} \theta_j/d)$, in this case. For more general situations with (group $\times$ condition) interaction*, comparisons among the $\{\bar{\mu}_{i \cdot}\}$ may need to be interpreted cautiously since tendencies for one group to have higher responses than another for some conditions and lower responses for other conditions will be absorbed in the average over conditions. The methods of analysis for the $\{\bar{y}_{i \cdot l}\}$ are similar to those described for contrasts. They are one-way analysis of variance when the $\{\bar{y}_{i \cdot l}\}$ have independent normal distributions with homogeneous variance $v_m = \mathbf{1}'_d \mathbf{V} \mathbf{1}_d/d^2$, and its large sample or nonparametric counterparts when these assumptions are not realistic. Also, extensions for covariance analysis allow relevant explanatory variables to be taken into account; see Snedecor and Cochran [82] or Neter et al. [67] for the normal distribution framework and Koch et al. [47] and Quade [69] for the nonparametric framework.

In this section, we have discussed estimation and testing of summary functions as if each were dictated by the nature of the conditions and were to be evaluated separately. However, if several summary functions are to be tested and simultaneous inference is an important issue, *multiple comparisons* methods* may be used to control the overall significance level. A useful, simple approach is the Bonferroni* method (see Miller [60]) in which for ν comparisons, the overall significance level α is obtained by testing each comparison at the nominal significance level (α/ν).

The methods described in this section have been focused on linear functions for complete data situations, but their applicability is broader. The same strategies, particularly the large sample or nonparametric approaches, can be directed at more complicated functions such as ratios or medians. When some subjects have incomplete data (i.e., data for some conditions are missing), functions can be defined to take the observed missing data* pattern into account. One approach is to restrict attention to the subset of subjects with the necessary data for determining a particular function; e.g., for a pairwise difference, restriction to those subjects with responses to both conditions. However, the populations represented by this subset of subjects and those for whom the function could not be determined must be equivalent in order for selection bias to be avoided; the justification for such equivalence can involve stratification or a statistical model (*see* INCOMPLETE DATA). Alternatively, attention can be directed at functions that have meaningful definitions for all subjects; e.g., the last observation or the median* in a longitudinal study. Since such functions may have complicated distributions, large sample or nonparametric methods may be more appropriate for their analysis; see Koch et al. [46, 47] and CHI-SQUARE TESTS.

Multivariate Methods for Within-Subject Functions

Some analysis questions involve the simultaneous consideration of $u \geqslant 2$ within-subject linear functions $\mathbf{f}_{\mathbf{A},il} = \mathbf{A}'\mathbf{y}_{il}$, where $\mathbf{A}$ is a $(d \times u)$ specification matrix. These can be addressed by the multivariate extensions of the previously discussed univariate methods. For example, when there is only one group or no (group $\times$ condition) interaction in the model (1), the overall hypothesis of no differences between condition effects is equivalent to

$$E\{(y_{ijl} - y_{idl})\} = 0 \qquad \text{for } j = 1, 2, \ldots, (d-1). \tag{17}$$

The corresponding matrix formulation is directed at $u = (d-1)$ linear functions $\mathbf{f}_{\mathbf{A},il}$ for which $\mathbf{A}' = [\mathbf{I}_u, -\mathbf{1}_u]$; here $\mathbf{I}_u$ is the uth order identity matrix and $\mathbf{1}_u$ is the $(u \times 1)$ vector of 1's. When the $\{\mathbf{f}_{\mathbf{A},il}\}$ have independent multivariate normal distributions with homogeneous covariance structure $\mathbf{A}'\mathbf{V}_i\mathbf{A} = \mathbf{V}_{\mathbf{A},i} = \mathbf{V}_{\mathbf{A}}$, the expression of the hypothesis in (17) as $\mathbf{E}\{\mathbf{f}_{\mathbf{A},il}\} = \mathbf{0}_u$, where $\mathbf{0}_u$ is a $(u \times 1)$ vector of 0's, allows it to be tested by Hotelling's T^2 statistic. More specifically, for the pooled groups, let

$$\bar{\mathbf{f}}_{\mathbf{A}} = \frac{1}{n} \sum_{i=1}^{G} \sum_{l=1}^{n_i} \mathbf{f}_{\mathbf{A},il} = \mathbf{A}'\left\{\frac{1}{n} \sum_{i=1}^{G} \sum_{l=1}^{n_i} \mathbf{y}_{il}\right\} = \mathbf{A}'\bar{\mathbf{y}} \tag{18}$$

denote the mean vector of the $\{f_{\mathbf{A},il}\}$ and let

$$\begin{aligned}\bar{\mathbf{V}}_{\mathbf{A}} &= \frac{1}{(n-G)} \sum_{i=1}^{G} \sum_{l=1}^{n_i} (\mathbf{f}_{\mathbf{A},il} - \bar{\mathbf{f}}_{\mathbf{A},i})(\mathbf{f}_{\mathbf{A},il} - \bar{\mathbf{f}}_{\mathbf{A},i})' \\ &= \mathbf{A}'\left\{\frac{1}{(n-G)} \sum_{i=1}^{G} \sum_{l=1}^{n_i} (\mathbf{y}_{il} - \bar{\mathbf{y}}_i)(\mathbf{y}_{il} - \bar{\mathbf{y}}_i)'\right\}\mathbf{A} \\ &= \mathbf{A}'\bar{\mathbf{V}}\mathbf{A},\end{aligned} \tag{19}$$

where $\bar{\mathbf{y}}_i = \sum_{l=1}^{n_i}(\mathbf{y}_{il}/n_i)$ and $\bar{\mathbf{f}}_{\mathbf{A},i} = \mathbf{A}'\bar{\mathbf{y}}_i$, denote the pooled within groups unbiased estimator of $\mathbf{V}_{\mathbf{A}}$. The rejection region with significance level α for the hypothesis (17) concerning $u = (d-1)$ linear functions is

$$\frac{(n - G - u + 1)}{(n-G)u} T^2 \geqslant \mathscr{F}, \tag{20}$$

where $\mathscr{F} = \mathscr{F}_{1-\alpha}\{u, (n - G - u + 1)\}$ is the $100(1-\alpha)$ percentile of the $\mathscr{F}$-distribution with $\{u, (n - G - u + 1)\}$ degrees of freedom, and

$$\begin{aligned}T^2 &= n\bar{\mathbf{f}}_{\mathbf{A}}'\bar{\mathbf{V}}_{\mathbf{A}}^{-1}\bar{\mathbf{f}}_{\mathbf{A}} \\ &= n\bar{\mathbf{y}}'\mathbf{A}\{\mathbf{A}'\bar{\mathbf{V}}_{\mathbf{A}}\mathbf{A}\}^{-1}\mathbf{A}'\bar{\mathbf{y}}\end{aligned} \tag{21}$$

is the Hotelling T^2 statistic*. Here, it is assumed that $\mathbf{V}_{\mathbf{A}}$ is nonsingular and $n \geqslant (G + u)$ so that $\bar{\mathbf{V}}_{\mathbf{A}}$ is almost certainly nonsingular.

An important property of T^2 is that its value stays the same across all nonsingular transformations of the $\{\mathbf{f}_{\mathbf{A},il}\}$. Thus, it is invariant for any specification of the hypothesis (17) or corresponding choice of $\mathbf{A}$ that is a basis of contrast space [i.e., any $(d \times (d-1))$ matrix $\mathbf{A}$ such that rank $(\mathbf{A}) = (d-1)$ and $\mathbf{A}'\mathbf{1}_d = \mathbf{O}_{(d-1)}$], and so it provides a well defined overall test procedure for the hypothesis of no differences among condition effects.

The application of the Roy and Bose [76] method for multiple comparisons to the distribution of T^2 in (21) allows simultaneous confidence intervals* to be constructed for all linear contrasts $\theta_f = \Sigma_{j=1}^{d} a_j \theta_j$ between condition effects at an overall $100(1-\alpha)\%$ level. Such intervals, which are analogous to those from the *Scheffé method* (*see* SCHEFFÉ'S SIMULTANEOUS COMPARISON PROCEDURE), have the form

$$\mathbf{a}'\bar{\mathbf{y}} - \mathscr{T}\{\mathbf{a}'\bar{\mathbf{V}}_{\mathbf{A}}\mathbf{a}/n\}^{1/2} \leqslant \theta_f \leqslant \mathbf{a}'\bar{\mathbf{y}} + \mathscr{T}\{\mathbf{a}'\bar{\mathbf{V}}_{\mathbf{A}}\mathbf{a}/n\}^{1/2}, \quad (22)$$

where $\mathbf{a} = (a_1, a_2, \ldots, a_d)'$ and

$$\mathscr{T} = \left\{\frac{(n-G)u}{(n-G-u+1)} \times [\mathscr{F}_{1-\alpha}\{u, (n-G-u+1)\}]\right\}^{1/2} \quad (23)$$

with $u = (d-1)$; such intervals can also be formed with respect to lower-dimensional subspaces with $u < (d-1)$.

A potential limitation of the multivariate procedures (20) and (22) is that their effectiveness for overall inference purposes requires enough subjects to provide a stable estimate of $\mathbf{V}_{\mathbf{A}}$ (e.g., $n - G - u \geqslant 15$); a somewhat smaller number is sufficient if the multivariate counterpart of $\hat{v}_f$ in (7) is used to estimate $\mathbf{V}_{\mathbf{A}}$ instead of $\bar{\mathbf{V}}_{\mathbf{A}}$ (e.g., $n - u \geqslant 15$). Also, complete response vectors are required (that is, each subject must be observed under all conditions). An advantage of these methods is their applicability under minimal covariance structure assumptions.

Large sample and nonparametric counterparts to the multivariate procedures (20) and (22) are similar in spirit to those described for the univariate procedures (8) and (9). When the overall sample size n is sufficiently large (e.g., $n \geqslant d + G + 30$), $\bar{\mathbf{f}}_{\mathbf{A}}$ has approximately a multivariate normal distribution (*see* MULTIDIMENSIONAL CENTRAL LIMIT THEOREMS). Thus, T^2 has the χ^2 distribution* with u degrees of freedom. Also, potential heterogeneity among the covariance structures $\{\mathbf{V}_{\mathbf{A},i}\}$ can be taken into account by refinements analogous to (10). As for univariate methods, such analysis is of particular interest for categorical data. Nonparametric methods for multivariate comparisons among conditions can be based on extensions of the sign test and Wilcoxon signed ranks test; see Koch et al. [46] for discussion and references.

For studies with multiple groups, the hypothesis of no (group × condition) interaction can be tested by multivariate analysis of variance methods when the $\{\mathbf{f}_{\mathbf{A},il}\}$ have independent multivariate normal distributions with homogeneous covariance structure; see Cole and Grizzle [13], Gill [27], Morrison [63], and Timm [86, 87]. Strategies for evaluating condition effects in the potential presence of (group × condition) interaction are analogous to (13). When the assumption of multivariate normality is not realistic, large sample or nonparametric procedures are available; see Koch [44], Koch et al. [46, 50], Puri and Sen [68], or CHI-SQUARE TESTS.

Repeated Measures Analysis of Variance

The number of subjects n in split-plot experiments and change-over design studies is often small (e.g., $n \leqslant 10$) because of cost constraints for the total number of observational units partitioned among them. For such situations, univariate or multivariate analysis of within-subject functions may be unsatisfactory for assessing condition effects because critical values like t in (8) or (13) or $\mathscr{F}$ in (20) may be so large that they excessively weaken the effectiveness of the corresponding confidence intervals and test procedures; also multivariate methods are not applicable if $n \leqslant u$. One way to resolve this problem is to use methods based on more stringent assumptions that enable the covariance structure of contrasts $\mathbf{f}_{\mathbf{A},il}$ to be estimated with larger degrees of freedom and thereby provide smaller critical values of t or F.

An assumption that can be realistic for many split-plot experiments and change-over studies is that the elements of the covariance matrices of the response vectors satisfy (5) with $v_{*,jj} = v_{*,0}$ for all j and $v_{*,jj'} = 0$ for all $j \neq j'$; i.e., all diagonal elements of $\mathbf{V}_i$ are equal to $(v_{i,*} + v_{*,0})$, and all other elements are equal to $v_{i,*}$. Covariance matrices with this structure are said to have the property of *compound symmetry*. An underlying model for which it holds is

$$y_{ijl} = \mu_{ij} + s_{i*l} + e_{ijl}, \qquad (24)$$

where the $\{s_{i*l}\}$ are independent subject effects with $E\{s_{i*l}\} = 0$ and $\text{Var}\{s_{i*l}\} = v_{i,*}$, the $\{e_{ijl}\}$ are independent response errors with $E\{e_{ijl}\} = 0$ and $\text{Var}\{e_{ijl}\} = v_{*,0}$, and the $\{s_{i*l}\}$ and $\{e_{ijl}\}$ are mutually independent.

For split-plot experiments, randomization of conditions within subjects and the interchangeable nature of observational units (e.g., subsamples, littermates, etc.) often provide justification for the model (24); for change-over design studies, the response process needs to be sufficiently stable that its conditional distributions within subjects are independent and have homogeneous variance for the respective periods.

When the model (24) applies, a set of contrasts $\mathbf{f}_{\mathbf{A},il} = \mathbf{A}'\mathbf{y}_{il}$ will have variance

$$\text{Var}(\mathbf{f}_{\mathbf{A},il}) = \mathbf{V}_{\mathbf{A}} = (\mathbf{A}'\mathbf{A})v_{*,0}. \qquad (25)$$

Moreover, if $\mathbf{A}$ is orthonormal (i.e., its columns are mutually orthogonal and have unit length), then $\mathbf{V}_{\mathbf{A}} = \mathbf{I}_u v_{*,0}$. This property of $\mathbf{V}_{\mathbf{A}}$ is called *sphericity** or *circularity* (see Huynh and Feldt [35] and Rouanet and Lepine [75]); it is implied by compound symmetry, which in turn is implied by the model (24), but it can arise under somewhat more general conditions for the covariance structure of the $\{\mathbf{y}_{il}\}$.

The usual analysis of variance calculations for the model (24) are displayed in Table 1. When the covariance matrix satisfies the sphericity condition with $u = (d-1)$ and the $\{s_{i*l}\}$ and $\{e_{ijl}\}$ are normally distributed, the mean square error MSE is an unbiased estimate of $v_{*,0}$, and the hypothesis of no (group $\times$ condition) interaction can be tested with the rejection region

$$\text{MS}(GC)/\text{MSE} \geqslant \mathscr{F}_{1-\alpha}[(d-1)(G-1), (d-1)(n-G)]. \qquad (26)$$

When the conditions correspond to the cross-classification of two or more factors or have some other relevant structure, there usually is interest in assessing (group $\times$ condition) interaction for one or more subsets of contrasts. For a set of u contrasts $\{\mathbf{f}_{\mathbf{A},il}\}$, the numerator of the counterpart test to (26) would be

$$\text{MS}(GC, \mathbf{A}) = \sum_{i=1}^{G} n_i(\bar{\mathbf{f}}_{\mathbf{A},i} - \bar{\mathbf{f}}_{\mathbf{A}})'(\bar{\mathbf{f}}_{\mathbf{A},i} - \bar{\mathbf{f}}_{\mathbf{A}})/[u(G-1)], \qquad (27)$$

where $\bar{\mathbf{f}}_{\mathbf{A},i} = (\sum_{l=1}^{n_i} \mathbf{f}_{\mathbf{A},il}/n_i)$ and $\bar{\mathbf{f}}_{\mathbf{A}} = \sum_{i=1}^{G}(n_i \bar{\mathbf{f}}_{\mathbf{A},i}/n)$. The denominator would be MSE if sphericity applied to the entire $(d-1)$-dimensional space for contrasts or it would be

$$\text{MS}(E, \mathbf{A}) = \sum_{i=1}^{G}\sum_{l=1}^{n_i} (\mathbf{f}_{\mathbf{A},il} - \bar{\mathbf{f}}_{\mathbf{A},i})'(\mathbf{f}_{\mathbf{A},il} - \bar{\mathbf{f}}_{\mathbf{A},i})/[u(n-G)] \qquad (28)$$

if it only applied to the u-dimensional subspace $\mathbf{A}$; in the former case, the critical value of $\mathscr{F}$ would be that appropriate to $u(G-1)$ and $(d-1)(n-G)$ degrees of freedom (d.f.), and in the latter, it would be for d.f. $= [u(G-1), u(n-G)]$. It should be noted that the test based on (28) should be used for situations that involve more than two components of variance in models analogous to (24); e.g., split-split-plot experiments or studies that involve several sources of random variation.

When there is no (group $\times$ condition) interaction, the overall null hypothesis (17) of no differences between conditions can be tested with the rejection region

$$\text{MS}(C)/\text{MSE} \geqslant \mathscr{F}_{1-\alpha}[(d-1), (d-1)(n-G)]. \qquad (29)$$

Table 1 Repeated Measures Analysis of Variance

Source of Variation	Degrees of Freedom (d.f.)	Sums of Squares[a]	Mean Square
Groups	$(G-1)$	$SS(G)=\sum_{i=1}^{G} dn_i(\bar{y}_{i\cdot\cdot}-\bar{y}_{\cdot\cdot\cdot})^2$	$MS(G)=SS(G)/(G-1)$
Subjects within groups	$(n-G)$	$SS(S)=\sum_{i=1}^{G}\sum_{l=1}^{n_i} d(\bar{y}_{i\cdot l}-\bar{y}_{i\cdot\cdot})^2$	$MS(S)=SS(S)/(n-G)$
Total among subjects	$(n-1)$	$SS(AS)=\sum_{i=1}^{G}\sum_{l=1}^{n_i} d(\bar{y}_{i\cdot l}-\bar{y}_{\cdot\cdot\cdot})^2$	
Conditions	$(d-1)$	$SS(C)=\sum_{j=1}^{d} n(\bar{y}_{\cdot j\cdot}-\bar{y}_{\cdot\cdot\cdot})^2$	$MS(C)=SS(C)/(d-1)$
Groups × conditions	$(G-1)(d-1)$	$SS(GC)=\sum_{i=1}^{G}\sum_{j=1}^{d} n_i(\bar{y}_{ij\cdot}-\bar{y}_{i\cdot\cdot}-\bar{y}_{\cdot j\cdot}+\bar{y}_{\cdot\cdot\cdot})^2$	$MS(GC)=SS(GC)/\{(G-1)(d-1$
Conditions × subjects within groups (error)	$(n-G)(d-1)$	$SSE=\sum_{i=1}^{G}\sum_{l=1}^{n_i}\sum_{j=1}^{d}(y_{ijl}-\bar{y}_{ij\cdot}-\bar{y}_{i\cdot l}+\bar{y}_{i\cdot\cdot})^2$	$MSE=SSE/\{(n-G)(d-1)\}$
Total within subjects	$n(d-1)$	$SS(WS)=\sum_{i=1}^{G}\sum_{l=1}^{n_i}\sum_{j=1}^{d}(y_{ijl}-\bar{y}_{i\cdot l})^2$	
Total among observational units	$(nd-1)$	$SST=\sum_{i=1}^{G}\sum_{l=1}^{n_i}\sum_{j=1}^{d}(y_{ijl}-\bar{y}_{\cdot\cdot\cdot})^2$	

[a] Here, $n=\Sigma_{i=1}^{G}n_i$, $\bar{y}_{ij\cdot}=(1/n_i)\Sigma_{l=1}^{n_i}y_{ijl}$, $\bar{y}_{i\cdot l}=(1/d)\Sigma_{j=1}^{d}y_{ijl}$, $\bar{y}_{i\cdot\cdot}=(1/d)\Sigma_{j=1}^{d}\bar{y}_{ij\cdot}$, $\bar{y}_{\cdot j\cdot}=(1/n)\Sigma_{i=1}^{G}n_i\bar{y}_{ij\cdot}$, an $\bar{y}_{\cdot\cdot\cdot}=(1/n)\Sigma_{i=1}^{G}n_i\bar{y}_{i\cdot\cdot}$

For the subset of u contrasts $\{\mathbf{f}_{\mathbf{A},il}\}$ the numerator is replaced by

$$MS(C,\mathbf{A})=n\bar{\mathbf{f}}_{\mathbf{A}}'\bar{\mathbf{f}}_{\mathbf{A}}/u,$$

and either MSE or (28) can be the denominator in accordance with the previous discussion for (27). If MSE is used, d.f. = $[u,(d-1)(n-G)]$, while if (28) is used, d.f. = $[u,u(n-G)]$.

When it can be assumed that there is no (group × condition) interaction and that the within-subject means $\{\bar{y}_{i\cdot l}\}$ have homogeneous variances across groups, the hypothesis of no differences between group effects can be tested by the rejection region

$$MS(G)/MS(S) \geqslant \mathscr{F}_{1-\alpha}[(G-1),(n-G)]. \quad (30)$$

Note that the denominator in (30) for the comparisons between groups is different from that used in (26) and (29) for the comparisons between conditions. This is because the common variance of the $\{\bar{y}_{i\cdot l}\}$ usually involves both between-subject and within-subject components of variation. Specifically, relative to the model (24) with all $v_{i,*}=v_s$,

$$\mathrm{Var}\{\bar{y}_{i\cdot\cdot}\}=\frac{v_s}{n_i}+\frac{v_{*,0}}{n_i d}, \quad (31)$$

whereas (25) indicates that the variances of contrasts only involve $v_{*,0}$. Typically, more powerful results are provided for comparisons between conditions than comparisons between groups; the underlying considerations are the extent to which pairwise differences between the $\{\bar{y}_{i\cdot\cdot}\}$ have larger variances than pairwise differences between the $\{\bar{y}_{\cdot j\cdot}\}$ and the extent to which the denominator degrees of freedom for (30) is smaller than those for (26) and (29); see Jensen [38] for discussion of related issues.

The previous discussion of repeated measures analysis of variance was based on several assumptions. These included complete response vectors, normal distributions with homogeneous covariance matrices, and the sphericity structure in (25). Also, the question of whether there is (group $\times$ condition) interaction is important. When all of these assumptions hold, then usage of these methods and their counterparts for subsets of within-subject contrasts is reasonable. Otherwise, alternative procedures need to be applied. Three general strategies that are of interest for this purpose are summarized as follows:

1. Regression methods for situations with normal distributions and sphericity.
2. Nonparametric methods for situations where normal distributions do not apply.
3. Methods for situations where sphericity does not apply.

Strategy 1: Regression Methods for Situations With Normal Distributions and Sphericity. When response vectors are incomplete or more complex models are required to account for period effects, carryover effects or concomitant variables, within-subject contrasts can be analyzed by least-squares* methods. Suppose the lth subject in the ith group is observed for d_{il} conditions; consideration is given to contrasts $\{\mathbf{f}_{il} = \mathbf{A}'_{il}\mathbf{y}_{il}\}$, where the $\{\mathbf{A}'_{il}\}$ are orthonormal $\{(d_{il} - 1) \times d_{il}\}$ matrices. Sphericity is assumed and so the respective covariance matrices of the $\{\mathbf{f}_{il}\}$ have the form $\{\mathbf{I}_{(d_{il}-1)}v_{*,0}\}$. Let the variation within and between the $E\{\mathbf{f}_{A,il}\} = \boldsymbol{\psi}_{il}$ be described by the linear model $\boldsymbol{\psi}_{il} = \mathbf{Z}_{il}\boldsymbol{\beta}$, where the $\{\mathbf{Z}_{il}\}$ are known $\{[(d_{il} - 1) \times t]\}$ submatrices of the full rank specification matrix $\mathbf{Z} = [\mathbf{Z}'_{11}, \ldots, \mathbf{Z}'_{1n_1}, \ldots, \mathbf{Z}'_{G,n_G}]'$ for the respective subjects and $\boldsymbol{\beta}$ is the $(t \times 1)$ vector of unknown parameters. Inferences concerning $\boldsymbol{\beta}$ are then based on multiple regression* methods; such analysis is illustrated for Example 2.

An equivalent approach involves the application of least-squares methods to the $\{y_{ijl}\}$ with the subject effects $\{s_{i*l}\}$, but no group effects or background variables for subjects, forced into the model. Linkage of the parameters of such a model to the $\boldsymbol{\psi}_{il}$ and $\boldsymbol{\beta}$ needs to be identified; for related discussion, see Schwertman [80]. This analysis strategy is not appropriate for inferences concerning between-subject sources of variation (e.g., groups) unless (24) holds with all $v_{i,*} = 0$; such questions need to be addressed either by analyses of within-subject means like (14) or in the setting of maximum likelihood* or related methods for a general multivariate linear model for the $\{y_{ijl}\}$.

Strategy 2: Nonparametric Methods. For situations where the normal distribution assumption for the $\{y_{ijl}\}$ is not realistic and the response vectors could be incomplete, randomization-based nonparametric methods are appropriate for comparisons between conditions if they have been randomly assigned within subjects or the distributions of the $\{(\mathbf{y}_{il} - \boldsymbol{\mu}_i)\}$ are invariant under within-subject permutations. Examples of such methods are the Cochran [11] statistic for dichotomous data, the Friedman [25] two-way rank analysis of variance statistic for ordinal data, and various extensions; see Darroch [18], Koch et al. [46], Myers et al. [65], White et al. [91], and CHI-SQUARE TESTS. Exact probability levels can be determined for small n, and chi-square approximations can be used when n is at least moderate.

Strategy 3: Methods Where Sphericity Does Not Apply. The sphericity structure in (25) may not be realistic for several reasons. For complex split-plot experiments with several levels of randomization with respect to a hierarchy of observational units, the model (24) may need to be extended to include additional components of variance. It is then possible for sphericity to hold for subsets of orthonormal contrasts; these can then be separately analyzed by methods analogous to (27) and (28).

On the other hand, for longitudinal studies and change-over studies, sphericity is often contradicted by the inherent tendency for closely adjacent observational units to be more highly correlated than remote ones, i.e., for there to be a simplex serial correlation pattern among responses to the conditions. A general strategy here is to use extensions of multivariate models that describe the variation of the $\{\mu_{ij}\}$ across both groups and conditions and specify patterned covariance structures $\{\mathbf{V}_i\}$ with respect to underlying variance components; see Laird and Ware [52], Cook and Ware [16], Ware [90], and GROWTH CURVES. This approach can be particularly advantageous for longitudinal studies with an at least moderately large number of subjects whose response vectors are possibly incomplete.

When n is small, d is moderate, and response vectors are complete, corrections can be applied to the test procedures (26) and (29) to adjust for lack of sphericity. The F critical values are applied, but the degrees of freedom of both numerator and denominator are reduced by multiplying by the factor

$$\epsilon = \{\text{trace}(\mathbf{V}_{\mathbf{A}})\}^2/[(d-1)\{\text{trace}(\mathbf{V}_{\mathbf{A}}^2)\}] \tag{32}$$

for orthonormal contrasts $\mathbf{A}$ with rank $(d-1)$; see Box [9] and Greenhouse and Geisser [29]. Since $[1/(d-1)] \leqslant \epsilon \leqslant 1$, use of $1/(d-1)$ for ϵ will provide a conservative test; similarly, if attention were restricted to $u = \text{Rank}(\mathbf{A})$ orthonormal contrasts, then $(d-1)$ would be replaced by u in (32) and $(1/u)$ would be the conservative lower bound. For longitudinal studies with correlation ρ^{δ} for observational units δ units apart, Wallenstein and Fleiss [89] discuss the use of less conservative lower bounds for ϵ. Greenhouse and Geisser [29], Huynh and Feldt [36], and Rogan et al. [74] all suggest estimators of ϵ based on sample estimates of $\mathbf{V}_{\mathbf{A}}$; also, see Huynh [34] and Maxwell and Arvey [57] for further consideration of approximate tests.

Since lack of sphericity can cause the results of repeated measures analysis of variance to be potentially misleading (see Boik [8] and Maxwell [56]), statistical tests concerning it are of some interest. However, methods for this purpose (see Anderson [4], Mauchley [55], and Mendoza [59]) are sensitive to nonnormal distributions for the data (particularly outliers*), and their power is relatively weak for small or moderate n (see Boik [8] and Keselman et al. [43]). Thus, usage of repeated measures analysis of variance is only recommended when sphericity can be presumed on the basis of subject matter knowledge and research design structure (e.g., split-plot experiments); otherwise, univariate analysis of within-subject functions is preferable.

Example 1: A Split-Plot Experiment With Litters of Baby Rats. This experiment was undertaken to compare plasma fluoride concentrations (PFC) for $G = 6$ groups of litters of baby rats. These groups corresponded to two age strata (six day old and 11 day old) within which three doses for intraperitoneal injection of fluoride (0.10, 0.25, 0.50 micrograms per gram of body weight) were investigated. For each age stratum, $n_0 = 3$ litters of baby rats were fortuitously assigned

Table 2 Average Logarithms of Plasma Fluoride Concentrations from Pairs of Baby Rats in Split-Plot Study

Age (days)	Dose (microgram)	Litter	Minutes Post-Injection 15	30	60
6	0.50	1	4.1	3.9	3.3
6	0.50	2	5.1	4.0	3.2
6	0.50	3	5.8	5.8	4.4
6	0.25	4	4.8	3.4	2.3
6	0.25	5	3.9	3.5	2.6
6	0.25	6	5.2	4.8	3.7
6	0.10	7	3.3	2.2	1.6
6	0.10	8	3.4	2.9	1.8
6	0.10	9	3.7	3.8	2.2
11	0.50	1	5.1	3.5	1.9
11	0.50	2	5.6	4.6	3.4
11	0.50	3	5.9	5.0	3.2
11	0.25	4	3.9	2.3	1.6
11	0.25	5	6.5	4.0	2.6
11	0.25	6	5.2	4.6	2.7
11	0.10	7	2.8	2.0	1.8
11	0.10	8	4.3	3.3	1.9
11	0.10	9	3.8	3.6	2.6

Table 3 Means and Estimated Standard Errors for Logarithms of Plasma Fluoride Concentrations and Orthonormal functions for Groups of Baby Rats in Split-Plot Study

Age (days)	Dose (micrograms)	Minutes Post-Injection			Orthonormal Function		
		15	30	60	Average	Trend	Curvature
6	0.50	5.02	4.53	3.65	7.62	0.97	0.16
6	0.25	4.63	3.89	2.83	6.56	1.27	0.13
6	0.10	3.44	2.97	1.89	4.79	1.09	0.25
11	0.50	5.85	4.36	2.85	7.54	2.12	0.01
11	0.25	4.89	3.65	2.29	6.25	1.84	0.05
11	0.10	3.65	2.99	2.12	5.06	1.08	0.08
Estimated s.e. for all groups		0.41	0.53	0.36	0.70	0.20	0.20

to each of the three doses. The $n = 18$ litters are the subjects for this study. Six rats from each litter were assessed; two fortuitously selected rats were sacrificed at each of the three post-injection time conditions (15, 30, 60 minutes) for PFC measurement. The response values for the respective conditions are the averages of the natural logarithms of the PFCs for the corresponding pairs of baby rats; thus, pairs of baby rats are the observational units. The data are displayed in Table 2.

The means for each of the $d = 3$ conditions are shown in Table 3 for each group. They indicate that PFC tends to decrease over time for each (age × dose) group and tends to decrease with lower doses for each (age × time). The 11-day old rats show a greater decrease over time in response to the higher doses than the 6-day old rats. These aspects of the data provide the motivation for considering three orthonormal summary functions of the data from each litter. Function 1 reflects the average response for the three time periods; function 2 is the linear trend across log time; and function 3 is a measure of any lack of linearity (or curvature) in the trend. The functions are specified by

$$\mathbf{U}' = \begin{bmatrix} (1,1,\ \ 1)/\sqrt{3} \\ (1,0,-1)/\sqrt{2} \\ (-1,2,-1)/\sqrt{6} \end{bmatrix} = \begin{bmatrix} \mathbf{1}'_3/\sqrt{3} \\ \mathbf{A}' \end{bmatrix}; \tag{33}$$

and their means are shown on the right side of Table 4.

Table 4 Analysis of Variance Mean Squares and Significance Indicators for Orthonormal Functions

Source of Variation	d.f.	Separate Functions		
		Average	Trend	Curvature
Overall mean	1	715.349[a]	35.036[a]	0.231
Age	1	0.008	1.460[a]	0.081
Dose	2	10.618[a]	0.431[c]	0.012
Age × dose	2	0.124	0.500[b]	0.003
Within-groups error	12	1.487	0.115	0.117

[a] Significant results with $p < 0.01$.
[b] Results with $0.01 < p < 0.05$.
[c] Suggestive results with $0.05 < p < 0.10$.

Separate analyses of variance can be undertaken for the three functions as long as it is reasonable to assume normality and across-group homogeneity of variances for each of them. Results are shown in the corresponding columns of Table 4. There is a significant effect of dose on the "average response," a significant overall linear trend, and a nearly significant effect of dose on the size of the linear trend. In addition, there is a significant (age × dose) interaction for the linear trend; this corresponds to the tendency for the difference between ages to increase with dose. There is no evidence of curvature in the time trend or of differences in curvature due to age or dose.

The application of repeated measures analysis of variance to these data presumes that the response vectors $\{\mathbf{y}_{il}\}$ for the respective litters have multivariate normal distributions with the same covariance matrix $\mathbf{V}$ and that sphericity holds. Normality and equality of covariance matrices seem reasonable for this example. The pooled within-groups estimate for $\mathbf{V}$ is

$$\bar{\mathbf{V}} = \frac{1}{12}\sum_{i=1}^{6}\sum_{l=1}^{3}(\mathbf{y}_{il} - \bar{\mathbf{y}}_i)(\mathbf{y}_{il} - \bar{\mathbf{y}}_i)'$$

$$= \begin{bmatrix} 0.5040 & 0.5318 & 0.3279 \\ 0.5318 & 0.8337 & 0.5116 \\ 0.3279 & 0.5116 & 0.3816 \end{bmatrix}.$$

Its counterpart for the orthonormal functions defined by $\mathbf{U}'$ in (33) is

$$\bar{\mathbf{V}}_{\mathrm{U}} = \mathbf{U}'\bar{\mathbf{V}}\mathbf{U}$$

$$= \begin{bmatrix} 1.4873 & 0.0582 & 0.2756 \\ 0.0582 & 0.1148 & -0.0237 \\ 0.2756 & -0.0237 & 0.1171 \end{bmatrix}.$$

The lower right hand (2 × 2) block of $\bar{\mathbf{V}}_{\mathrm{U}}$ is the estimated covariance matrix $\bar{\mathbf{V}}_{\mathbf{A}} = \mathbf{A}'\bar{\mathbf{V}}\mathbf{A}$ for the within-subject contrasts $\{\mathbf{A}'\mathbf{y}_{il}\}$. Its compatibility with the assumption of sphericity can be confirmed in the following way: diagonal structure is supported by the nonsignificance ($p > 0.100$) of the correlation of -0.20 between the "linear trend" vs. "curvature" (t-test with 11 d.f.); then equality of diagonal elements is supported by the nonsignificance ($p > 0.100$) of the F-test with (12, 12) d.f. for their ratio. An overall test of sphericity is provided by BMDP2V [37] (*see* SPHERICITY, TESTS OF); it was nonsignificant for $\bar{\mathbf{V}}_{\mathbf{A}}$ with $p = 0.79$. As an additional consideration, the significance ($p = 0.014$) of the correlation of 0.66 between "average response" vs. "curvature" (t-test with 11 d.f.) can be interpreted as contradicting compound symmetry for $\mathbf{V}$.

Table 5 Analysis of Variance Mean Squares and Significance Indicators for Repeated Measures Analysis of Variance

Source	d.f.	MS
Age	1	0.008
Dose	2	10.618[a]
Age × dose	2	0.124
Litter within group	12	1.487
Time	2	17.633[a]
Time × age	2	0.770[a]
Time × dose	4	0.222
Time × age × dose	4	0.252[b]
Within litters	24	0.116

[a] Significant results with $p < 0.01$.
[b] Suggestive results with $p = 0.104$.

Because the assumption of sphericity is reasonable, the repeated measures analysis of variance results are shown in Table 5. Note that the between-litters part of the table reproduces the information for the "average response" in Table 5; also, the within-litters part of the table could have been obtained by adding the sums of squares from the orthonormal linear and curvature contrasts. The time and (age × time) effects are significant because of the significance of the linear trend function.

Other methods of analysis could be applied to this example. If the assumption of sphericity were unrealistic, time effects and their interactions with other sources of variation could be assessed by multivariate analysis of variance methods. If the PFC data for the litters did not appear to have a multivariate normal distribution, nonparametric rank methods could be used; see Koch et al. [46] and CHI-SQUARE TESTS.

Table 6 Means and Covariance Matrices for Peak Heart Rate of Subjects in Two Sequence Groups of a Change-Over Study

Evaluation period	Sequence A : B ($n_1 = 9$) Mean	Covariance Matrix				Sequence B : A ($n_2 = 11$) Mean	Covariance Matrix			
Pretreatment	104	142	133	73	137	117	403	160	206	148
1st treatment	108		392	61	134	95		173	191	152
Drug-free	105	Symmetric		131	113	116	Symmetric		362	168
2nd treatment	92				174	115				168

Example 2: A Multiperiod Change-Over Study. Peak heart rate responses were obtained during four evaluation periods for $G = 2$ sequence groups of subjects; the data are shown in Table 2 of CHI-SQUARE TESTS, NUMERICAL EXAMPLES. One group with $n_1 = 9$ subjects received drug A during the first treatment period and drug B during the second treatment period, while the other group with $n_2 = 11$ subjects received the opposite regimen; a pretreatment period preceded the first treatment period and a drug-free period occurred between it and the second treatment period. The response values for the drug-free period were based on one visit for each subject while those for the other three periods were based on averages for two visits. The mean vectors $\{\bar{\mathbf{y}}_i\}$ and estimated covariance matrices $\{\hat{\mathbf{V}}_i\}$ for each sequence group are shown in Table 6. This information is the underlying framework for the subsequent discussion.

A preliminary statistical model of interest here has the structure in Table 7 where μ denotes a common reference parameter for the pretreatment status of both treatment groups; π_1, π_2, π_3 are period effects for the first treatment period, the drug-free period, and the second treatment period, respectively; τ is the direct drug A vs. drug B differential effect, γ_1 is the drug A vs. drug B differential carryover effect for the drug-free period; and γ_2 is the drug A vs. drug B differential carryover effect for the second treatment period. Also π_1 includes the effect of drug B as the reference treatment, π_2 includes its carryover effect to the drug-free period, and π_3 includes its carryover effect to the second treatment period. Four orthogonal within-subject functions that are useful for analyses pertaining to this model have the following specifications with respect to the seven visits of the study:

F1. The average over all seven visits.

F2. The difference between drug-free and the pretreatment average.

F3. The difference between the second treatment period average and the first treatment period average.

F4. The difference between the average over the first treatment period and the second treatment period vs. the average over pretreatment and the drug-free period.

These four functions are obtained from the four period summary framework for the data

Table 7 Preliminary Model for Change-Over Study

Sequence Group	Pretreatment	1st Treatment	Drug-Free	2nd Treatment
A : B	μ	$\mu + \pi_1 + \tau$	$\mu + \pi_2 + \gamma_1$	$\mu + \pi_3 + \gamma_2$
B : A	μ	$\mu + \pi_1$	$\mu + \pi_2$	$\mu + \pi_3 + \tau$

Table 8 Expected Value Structure of Orthogonal Functions for Change-Over Study

Function	Sequence A : B	Sequence B : A	A : B vs. B : A Difference
Overall mean	$\mu + (2\pi_1 + 2\tau + \pi_2 + \gamma_1 + 2\pi_3 + 2\gamma_2)/7$	$\mu + (2\pi_1 + \pi_2 + 2\pi_3 + 2\tau)/7$	$(\gamma_1 + 2\gamma_2)/7$
Drug-free vs. pretreatment	$\pi_2 + \gamma_1$	π_2	γ_1
1st vs. 2nd treatment	$(\pi_3 - \pi_1) - \tau + \gamma_2$	$(\pi_3 - \pi_1) + \tau$	$\gamma_2 - 2\tau$
1st and 2nd treatment vs. drug-free and pretreatment	$(3\pi_1 + 3\tau + 3\pi_3 + 3\gamma_2 - 2\pi_2 - 2\gamma_1)/6$	$(3\pi_1 + 3\tau + 3\pi_3 - 2\pi_2)/6$	$(3\gamma_2 - 2\gamma_1)/6$

by the linear transformation matrix

$$\mathbf{U}' = \begin{bmatrix} \frac{2}{7} & \frac{2}{7} & \frac{1}{7} & \frac{2}{7} \\ -1 & 0 & 1 & 0 \\ 0 & -1 & 0 & 1 \\ -\frac{2}{3} & \frac{1}{2} & -\frac{1}{3} & \frac{1}{2} \end{bmatrix}. \quad (34)$$

The application of the transformation $\mathbf{U}'$ to the model in Table 7 yields the expected value structure shown in Table 8 for the functions **F1**–**F4** for each sequence group and the difference between them.

UNIVARIATE AND MULTIVARIATE ANALYSES OF WITHIN-SUBJECT FUNCTIONS. If differential carryover effects are present, the estimate of the direct differential effect τ for the two treatments can have substantially larger variance than if they are negligible. For this reason, the usage of change-over designs typically presumes that carryover effects are negligible. We assess this assumption for this example through tests of hypotheses about the parameters γ_1, γ_2. If $\gamma_1 = \gamma_2 = 0$, then the difference between the sequence groups should have expected value zero for functions **F1**, **F2**, and **F4**. The two-sample t-tests for these functions each have 18 d.f.; their p-values* are 0.18, 0.72, and 0.14, respectively. Since all of these results are nonsignificant, the assumption that $\gamma_1 = \gamma_2 = 0$ is supported.

The hypothesis of no difference between treatments in carryover effects can also be tested in a more comprehensive way with the two-sample Hotelling's T^2. The comparison between the two sequence groups of all three functions **F1**, **F2**, and **F4** is nonsignificant ($p = 0.34$) relative to the F-distribution with (3, 16) d.f. However, **F1** incorporates subject effects, and hence can have substantially greater variability than **F2** and **F4**, which are within-subject contrasts. For this reason, the comparison of the two groups for functions **F2** and **F4** potentially provides a more effective, multivariate assessment of carryover effects; the corresponding two-sample Hotelling T^2 yields $p = 0.35$ relative to the F-distribution with (2, 17) d.f. Thus, the overall tests also support the conclusion that the carryover effects are equivalent for the two treatments.

Given that $\gamma_1 = \gamma_2 = 0$, we can test the hypothesis $\tau = 0$ of no difference in effect between drug A and drug B by comparing function **F3** for the two sequence groups with a two-sample t-test with 18 d.f. Since the resulting $p < 0.01$, it can be concluded that drug B lowers peak heart rate significantly more than drug A.

Regression analysis under sphericity assumption. Aspects of repeated measures analysis of variance for this example can be illustrated through the orthonormal counterparts of the functions **F1**–**F4** with respect to the seven visits of the study. For these functions, the linear transformation matrix is

$$\begin{bmatrix} \sqrt{7} & 0 & 0 & 0 \\ 0 & \sqrt{2/3} & 0 & 0 \\ 0 & 0 & 1 & 0 \\ 0 & 0 & 0 & \sqrt{12/7} \end{bmatrix} \mathbf{U}' = \begin{bmatrix} \sqrt{7} & \mathbf{u}_1' \\ & \mathbf{A}' \end{bmatrix}, \quad (35)$$

where $\mathbf{U}'$ is defined in (34), $\mathbf{u}_1'$ is its first row, and $\mathbf{A}'$ is the specification matrix for the three orthonormal contrasts that correspond to the last three rows of $\mathbf{U}'$. Under the model (24) for the seven visits of the study, the covariance matrix of the linear functions $\{\mathbf{A}'\mathbf{y}_{il}\}$ has the sphericity structure $\mathbf{V_A} = \mathbf{I}_3 v_{*,0}$. Relative to this background, the pooled within-groups estimate for $\mathbf{V_A}$ is

$$\overline{\mathbf{V}}_{\mathbf{A}} = \mathbf{A}'\overline{\mathbf{V}}\mathbf{A} = \begin{bmatrix} 169 & 12 & -52 \\ 12 & 153 & 65 \\ -52 & 65 & 201 \end{bmatrix}, \quad (36)$$

where $\overline{\mathbf{V}} = (8\hat{\mathbf{V}}_1 + 10\hat{\mathbf{V}}_2)/18$. Its compatibility with the assumption of sphericity can be confirmed in several ways: diagonal structure for $\mathbf{V_A}$ is supported by the nonsignificance ($p \geqslant 0.10$) of t-tests with 17 d.f. for each of the pairwise correlation estimates from $\overline{\mathbf{V}}_{\mathbf{A}}$; equality of the diagonal elements of $\mathbf{V_A}$ is supported by the nonsignificance ($p \geqslant 0.10$) of F-tests with (18, 18) d.f. for the pairwise ratios of their estimates from $\overline{\mathbf{V}}_{\mathbf{A}}$. The overall sphericity test from BMDP2V [37] was nonsignificant with $p = 0.51$.

Since sphericity is considered realistic for $\mathbf{V_A}$, the effect parameters $\boldsymbol{\beta} = (\pi_1, \pi_2, \pi_3, \tau, \gamma_1, \gamma_2)'$ of the model in Table 8 can be estimated on a within-subject basis by the application of multiple regression methods to the $\{\mathbf{A}'\mathbf{y}_{il}\}$. The corresponding specification matrix $\mathbf{Z}$ has respective components,

$$\mathbf{Z}_{1l} = \begin{bmatrix} (0, 2, 0, 0, 2, 0)/\sqrt{6} \\ (-1, 0, 1, -1, 0, 1) \\ (-3, 2, -3, -3, 2, -3)/\sqrt{21} \end{bmatrix}, \quad (37)$$

for the subjects from the A : B sequence group and respective components,

$$\mathbf{Z}_{2l} = \begin{bmatrix} (0, 2, 0, 0, 0, 0)/\sqrt{6} \\ (-1, 0, 1, 1, 0, 0) \\ (-3, 2, -3, -3, 0, 0)/\sqrt{21} \end{bmatrix}, \quad (38)$$

for the subjects from the B : A sequence group. The least-squares* estimates for $\boldsymbol{\beta}$ from this framework, their corresponding estimated standard errors, and p-values for t-tests with 54 d.f. for 0 values are

Parameter	π_1	π_2	π_3
Estimate	−22.3	−0.8	−28.7
s.e.	4.0	4.9	8.2
p	< 0.01	0.87	< 0.01
Parameter	τ	γ_1	γ_2
Estimate	26.5	2.6	16.7
s.e.	5.9	7.3	10.3
p	< 0.01	0.72	0.11

(39)

The error mean square estimate for the within-subject variance component $v_{*,0}$ is $\hat{v}_{*,0} = 174$; it is the average (28) of the diagonal elements of $\overline{\mathbf{V}}_{\mathbf{A}}$.

The results in (39) indicate that the carryover effect parameters γ_1 and γ_2 can be removed from the model in Table 7 because of their nonsignificance ($p \geqslant 0.10$), that the difference τ between direct treatment effects is significant ($p < 0.01$), and suggest that period effects are compatible with the constraints $\pi_1 - \pi_3 = 0$, $\pi_2 = 0$. Thus, peak heart rates during treatment periods with drug A were essentially the same as at pretreatment while those during treatment periods with drug B were significantly lower by about 27 beats per minute; any potential carryover effects of the two treatments were equivalent. These conclusions were also provided by the previously discussed univariate analyses of functions **F1**–**F4**. The advantage of the univariate function approach is that its use does not require the assumption of sphericity. However, it lacks comprehensiveness since each function is analyzed separately. For this example, the assumption of sphericity seems reasonable, and so repeated measures model fitting methods are applicable. They have the advantage of providing an effective estimation framework that encompasses the variation of response both across conditions within subjects and across groups of subjects.

NONPARAMETRIC ANALYSIS. The application of nonparametric rank methods is illustrated for this example in CHI-SQUARE

TESTS, NUMERICAL EXAMPLES. These methods are of interest because they are based on randomization in the research design rather than on assumptions concerning distributions and covariance structure for the data. It is possible that the covariance matrices for the two sequence groups are not homogeneous; the F-test for the comparison of the estimated variances for functions **F3** is significant ($p = 0.004$), although such tests for the other functions and for the responses during each period are not. The conclusions from nonparametric analyses agree with those reported here; for related discussion *see* INFERENCE, DESIGN BASED VS. MODEL BASED.

Example 3: A Study to Compare Two Diagnostic Procedures. One thousand subjects were classified according to both a standard version and a modified version of a diagnostic procedure with four ordinally scaled categories: strongly negative, moderately negative, moderately positive, and strongly positive. The resulting 4×4 contingency table is shown in Table 1 of HIERARCHICAL KAPPA STATISTICS.

Let $\mathbf{n}$ denote the vector of frequencies for the $n = 1000$ subjects and $\mathbf{p} = (\mathbf{n}/n)$ the vector of sample proportions for the 16 possible outcomes in the (4×4) table. If the 1000 subjects in this study can be considered a simple random sample, $\mathbf{p}$ has approximately a multivariate normal distribution. A consistent estimate of the covariance matrix is $\mathbf{V}_\mathbf{p} = [\mathbf{D}_\mathbf{p} - \mathbf{pp}']/n$, where $\mathbf{D}_\mathbf{p}$ is a diagonal matrix with elements $\mathbf{p}$ on the diagonal.

In the repeated measures context, this study involves one group with two conditions, and so it could be analyzed with a t-test for the pairwise difference between conditions if the two responses were normally distributed rather than just ordinal. Thus, attention needs to be given to the analogous comparison of the response category distributions for the two diagnostic procedures. Let $\mathbf{f}$ be the vector of the six marginal proportions for strongly negative, moderately negative, and moderately positive for each diagnostic procedure; proportions for the strongly positive category are not needed because they are linear functions of the others. The vector $\mathbf{f}$ can be obtained from $\mathbf{p}$ by constructing a (6×16) matrix $\mathbf{A}$ such that $\mathbf{f} = \mathbf{Ap}$. The estimates $\mathbf{f}$ and their estimated covariance matrix $\mathbf{V}_\mathbf{f} = \mathbf{AV}_\mathbf{p}\mathbf{A}'$ are shown in Table 9.

Under the hypothesis that the marginal distributions for the two diagnostic procedures are the same (i.e., marginal homogeneity), the Wald statistic

$$Q = \mathbf{f}'\mathbf{W}'[\mathbf{WV}_\mathbf{f}\mathbf{W}']^{-1}\mathbf{Wf} \qquad (40)$$

with $\mathbf{W} = [I_3, -I_3]$ approximately has the chi-square distribution with 3 d.f.; this criterion is analogous to the Hotelling T^2 statistic in (21). Since $Q = 6.68$ approaches significance with $p = 0.083$, the two diagnostic procedures potentially have somewhat different marginal distributions; additional evaluation of their agreement is given in HIERARCHICAL KAPPA STATISTICS. Wald statistics are discussed generally in CHI-SQUARE TESTS and for repeated measurements of categorical data in Guthrie [31] and Koch et al.

Table 9 Estimated Marginal Proportions and Covariance Matrix for Two Diagnostic Procedures

Diagnosis	Response Category	Estimate	Estimated Covariance Matrix × 10^6					
Modified	Strongly negative	0.362	231	−122	−34	144	−39	−31
Modified	Moderately negative	0.339		224	−32	−36	99	5
Modified	Moderately positive	0.093			84	−30	1	28
Standard	Strongly negative	0.394		Symmetric		239	−119	−37
Standard	Moderately negative	0.302					211	−28
Standard	Moderately positive	0.094						85

Table 10 Means and Estimated Covariance Matrices for Responses to Two Treatments for a Medical Condition

Treatment Group	Observer	Time	Mean	Estimated Covariance Matrix × 10^6					
A	1	1 month	2.2	5020	2902	566	203	856	1727
A	1	Final	2.4		8183	1727	2191	3497	4048
A	2	1 month	2.8			4498	638	−1219	73
A	2	Final	2.7		Symmetric		5049	3772	1872
A	3	1 month	2.2					9779	5296
A	3	Final	2.5						7284
B	1	1 month	1.9	6081	−39	849	1007	−197	2448
B	1	Final	2.2		7068	1441	928	2468	3001
B	2	1 month	2.4			6633	1599	4284	3415
B	2	Final	2.7		Symmetric		5923	2448	2369
B	3	1 month	1.9					1015	4778
B	3	Final	2.3						9713

[50]. Other methods for assessing marginal homogeneity and the related hypotheses of symmetry and quasisymmetry* are given in Bishop et al. [6] and Gokhale and Kullback [28]; methods for ordinal data* are given in Agresti [1] and McCullagh and Nelder [58].

Example 4: A Study to Compare Two Psychiatric Drugs. A randomized clinical trial* was undertaken to compare two drugs for a psychiatric condition. One group of $n_1 = 37$ patients received drug A for two months and another group of $n_2 = 37$ patients received drug B for two months. Each patient's mental condition was evaluated by three observers at the end of one month and again at the end of the trial. The responses were scored using the ordinal values of (1) unsatisfactory, (2) satisfactory, or (3) good. The mean vectors $\{\bar{\mathbf{y}}_i\}$ relative to the values 1, 2, 3 and their estimated covariance matrices $\hat{\mathbf{V}}_i$ for each treatment group are shown in Table 10.

The sampling framework is considered sufficient for the composite mean vector $\bar{\mathbf{y}} = [\bar{\mathbf{y}}_1', \bar{\mathbf{y}}_2']'$ to have an approximately multivariate normal distribution, and so Wald statistics can be used to test hypotheses concerning groups, observers, time, and their interactions. These statistics are computed via

$$Q = \bar{\mathbf{y}}'\mathbf{W}'\left[\mathbf{W}\hat{\mathbf{V}}_{\bar{\mathbf{y}}}\mathbf{W}'\right]^{-1}\mathbf{W}\bar{\mathbf{y}}, \qquad (41)$$

where $\hat{\mathbf{V}}_{\bar{\mathbf{y}}}$ is a block diagonal matrix with $\hat{\mathbf{V}}_1$ and $\hat{\mathbf{V}}_2$ as the respective blocks and $\mathbf{W}$ is the specification matrix. Results shown in Table 11 indicate significant variation ($p < 0.01$) between treatment groups, between observers, and between evaluation times; also the (group × time) interaction is significant ($p = 0.045$). The other sources of variation are interpreted as essentially random since no interaction of observers with group or time was expected on a priori grounds [although the (observer × time) interaction is recognized to be suggestive with $p = 0.060$].

A model that reflects the stated conclusions has the form $E\{\bar{\mathbf{y}}\} = \mathbf{X}\boldsymbol{\beta}$, where

$$\mathbf{X} = \begin{bmatrix} 1 & 1 & 1 & 1 & 1 & 1 & 1 & 1 & 1 & 1 & 1 & 1 \\ 0 & 0 & 0 & 0 & 0 & 0 & 1 & 1 & 1 & 1 & 1 & 1 \\ 0 & 0 & 1 & 1 & 0 & 0 & 0 & 0 & 1 & 1 & 0 & 0 \\ 0 & 0 & 0 & 0 & 1 & 1 & 0 & 0 & 0 & 0 & 1 & 1 \\ 0 & 1 & 0 & 1 & 0 & 1 & 0 & 1 & 0 & 1 & 0 & 1 \\ 0 & 0 & 0 & 0 & 0 & 0 & 0 & 1 & 0 & 1 & 0 & 1 \end{bmatrix} \qquad (42)$$

and $\boldsymbol{\beta} = (\beta_1, \beta_2, \beta_3, \beta_4, \beta_5, \beta_6)'$. For this specification, β_1 is a reference value for the expected response score for the classification of observer 1 at one month for treatment A, β_2 is the increment for treatment B, β_3 and

Table 11 Results of Wald Statistics for Preliminary Assessment of Group, Observer, and Time Sources of Variation

Source of Variation	W Matrix												Q	d.f
Group (G)	1	1	1	1	1	1	−1	−1	−1	−1	−1	−1	9.34[a]	1
Observer (O)	1	1	0	0	−1	−1	1	1	0	0	−1	−1	106.90[a]	2
	0	0	1	1	−1	−1	0	0	1	1	−1	−1		
Time (T)	1	−1	1	−1	1	−1	1	−1	1	−1	1	−1	40.22[a]	1
G × O	1	1	0	0	−1	−1	−1	−1	0	0	1	1	0.83	2
	0	0	1	1	−1	−1	0	0	−1	−1	1	1		
G × T	1	−1	1	−1	1	−1	−1	1	−1	1	−1	1	4.02[b]	1
O × T	1	−1	0	0	−1	1	1	−1	0	0	−1	1	5.62[c]	2
	0	0	1	−1	−1	1	0	0	1	−1	−1	1		
G × O × T	1	−1	0	0	−1	1	−1	1	0	0	1	−1	2.11	2
	0	0	1	−1	−1	1	0	0	−1	1	1	−1		

[a] Significant results with $p < 0.01$.
[b] Significant results with $0.01 < p < 0.05$.
[c] Suggestive results with $0.05 < p < 0.10$.

β_4 the increments for observers 2 and 3, β_5 the increment for the final evaluation time, and β_6 the increment for the interaction between treatment B and the final evaluation time. As noted in CHI-SQUARE TESTS, weighted least-squares* methods can be used to determine the asymptotically unbiased and asymptotically efficient estimator

$$\mathbf{b} = \left(\mathbf{X}'\hat{\mathbf{V}}_{\bar{\mathbf{y}}}^{-1}\mathbf{X}\right)^{-1}\mathbf{X}'\hat{\mathbf{V}}_{\bar{\mathbf{y}}}^{-1}\bar{\mathbf{y}} \qquad (45)$$

for $\boldsymbol{\beta}$; a consistent estimator for its covariance matrix is $\hat{\mathbf{V}}_{\mathbf{b}} = (\mathbf{X}'\hat{\mathbf{V}}_{\bar{\mathbf{y}}}^{-1}\mathbf{X})^{-1}$. The goodness of fit of the model specified by (42) is supported by the nonsignificance ($p = 0.16$) of the Wald goodness-of-fit statistic*,

$$Q = (\bar{\mathbf{y}} - \mathbf{Xb})'\hat{\mathbf{V}}_{\bar{\mathbf{y}}}^{-1}(\bar{\mathbf{y}} - \mathbf{Xb}) = 9.30, \quad (44)$$

relative to the chi-square distribution with 6 d.f.

Predicted values for the mean response of each group for the $d = 6$ (observer × time) conditions are obtained via $\hat{\mathbf{y}} = \mathbf{Xb}$; these quantities and their estimated standard errors (via square roots of the diagonal elements of $\mathbf{X}\hat{\mathbf{V}}_{\mathbf{b}}\mathbf{X}'$) are shown in Table 12. They indicate that the response is more favorable for treatment A, and that this tendency is larger at one month. The classifications of observer 2 were higher than those for the other two observers.

Finally, if the covariance matrices for the two groups were not substantially different from each other, the standard multivariate analysis of variance methods discussed in the Methods section could be applied here in

Table 12 Model Predicted Values (and Standard Errors) for Mean Response to Treatments A and B for (Observer × Time) Conditions

	Observer 1		Observer 2		Observer 3	
Group	1 Month	Final	1 Month	Final	1 Month	Final
Treatment A	2.17	2.32	2.68	2.83	2.28	2.44
	(0.06)	(0.06)	(0.05)	(0.05)	(0.06)	(0.07)
Treatment B	1.88	2.18	2.39	2.69	1.99	2.29
	(0.05)	(0.06)	(0.06)	(0.06)	(0.07)	(0.07)

a large sample context. Additional discussion of the methods illustrated with this example is given in Koch et al. [49, 50] and Stanish et al. [84].

References

[1] Agresti, A. (1984). *Analysis of Ordinal Categorical Data*. Wiley, New York.

[2] Allen, D. M. and Cady, F. B. (1982). *Analyzing Experimental Data by Regression*. Lifetime Learning Publications, Belmont, CA.

[3] Anderson, R. L. and Bancroft, T. A. (1952). *Statistical Theory in Research*. McGraw-Hill, New York.

[4] Anderson, T. W. (1984). *An Introduction to Multivariate Statistical Analysis*, 2nd ed. Wiley, New York.

[5] Bennett, C. A. and Franklin, N. L. (1954). *Statistical Analysis in Chemistry and the Chemical Industry*. Wiley, New York.

[6] Bishop, Y. M. M., Fienberg, S. E., and Holland, P. W. (1975). *Discrete Multivariate Analysis*. MIT Press, Cambridge, MA.

[7] Bock, R. D. (1975). *Multivariate Statistical Methods in Behavioral Research*. McGraw-Hill, New York.

[8] Boik, R. J. (1981). *Psychometrika*, **46**, 241–255. (A priori tests in repeated measures designs: effects of nonsphericity.)

[9] Box, G. E. P. (1954). *Ann. Math. Statist.*, **25**, 484–498.

[10] Brown, B. W. (1980). *Biometrics*, **36**, 69–79.

[11] Cochran, W. (1950). *Biometrika*, **37**, 256–266.

[12] Cochran, W. G. and Cox, G. M. (1957). *Experimental Designs*. Wiley, New York.

[13] Cole, J. W. L. and Grizzle, J. E. (1966). *Biometrics*, **22**, 810–828. (Applications of multivariate analysis of variance to repeated measurements experiments.)

[14] Conover, W. J. (1971). *Practical Nonparametric Statistics*. Wiley, New York.

[15] Constantine, G. and Hedayat, A. (1982). *J. Statist. Plann. Infer.*, **6**, 153–164.

[16] Cook, N. and Ware, J. H. (1983). *Ann. Rev. Public Health*, **4**, 1–24.

[17] Cox, D. R. (1958). *Planning of Experiments*. Wiley, New York.

[18] Darroch, J. N. (1981). *Int. Statist. Rev.*, **49**, 285–307. (The Mantel–Haenszel test and tests of marginal symmetry, fixed-effects, and mixed models for a categorical response.)

[19] Dielman, T. E. (1983). *Amer. Statist.*, **37**, 111–122. (Pooled cross-sectional and time series data: A survey of current statistical methodology.)

[20] Elashoff, J. D. (1985). Analysis of Repeated Measures Designs. *BMDP technical report no. 83*, BMDP Software, Los Angeles, CA.

[21] Federer, W. T. (1955). *Experimental Design*. Macmillan, New York.

[22] Federer, W. T. (1975). In *Applied Statistics*, R. P. Gupta, ed. North-Holland, Amsterdam, pp. 9–39.

[23] Federer, W. T. (1980, 1981). *Int. Statist. Rev.*, **48**, 357–368; **49**, 95–109, 185–197. (Some recent results in experiment design with a bibliography.)

[24] Federer, W. T. and Balaam, L. N. (1972). *Bibliography on Experiment and Treatment Design Pre-1968*. Oliver and Boyd, Edinburgh, Scotland.

[25] Friedman, M. (1937). *J. Amer. Statist. Ass.*, **32**, 675–699.

[26] Geisser, S. (1980). *Handbook of Statistics*, Vol. 1, P. R. Krishnaiah, ed. North-Holland, Amsterdam, pp. 89–115.

[27] Gill, J. L. (1978). *Design and Analysis of Experiments in the Animal and Medical Sciences*. Iowa State University Press, Ames, IA.

[28] Gokhale, D. V. and Kullback, S. (1978). *The Information in Contingency Tables*. Dekker, New York.

[29] Greenhouse, S. W. and Geisser, S. (1959). *Psychometrika*, **24**, 94–112. (On methods in the analysis of profile data.)

[30] Grizzle, J. E. and Allen, D. (1969). *Biometrics*, **25**, 357–381.

[31] Guthrie, D. (1981). *Psychol. Bull.*, **90**, 189–195.

[32] Hedayat, A. and Afsarinejad, K. (1975). In *A Survey of Statistical Design and Linear Models*, J. N. Srivastava, ed. North-Holland, Amsterdam, pp. 229–242.

[33] Hedayat, A. and Afsarinejad, K. (1978). *Ann. Statist.*, **6**, 619–628.

[34] Huynh, H. (1978). *Biometrika*, **43**, 161–175.

[35] Huynh, H. and Feldt, L. S. (1970). *J. Amer. Statist. Ass.*, **65**, 1582–1589. (Conditions under which mean square ratios in repeated-measurement designs have exact *F*-distributions.)

[36] Huynh, H. and Feldt, L. S. (1976). *J. Educ. Statist.*, **1**, 69–82.

[37] Jennrich, R., Sampson, P., and Frane, J. (1981). In *BMDP Statistical Software*, W. J. Dixon et al., eds. University of California Press, Los Angeles, CA, Chap. 15.2.

[38] Jensen, D. R. (1982). *Biometrics*, **38**, 813–825. (Efficiency and robustness in the use of repeated measurements.)

[39] Joiner, B. L. (1981). *Amer. Statist.*, **35**, 227–233.

[40] Kempthorne, O. (1952). *Design and Analysis of Experiments*. Wiley, New York.

[41] Kempthorne, O. (1969). *An Introduction to Genetic Statistics*. Wiley, New York.

[42] Kershner, R. P. and Federer, W. T. (1981). *J. Amer. Statist. Ass.*, **76**, 612–619.

[43] Keselman, H. J., Rogan, J. C., Mendoza, J. L., and Breen, L. J. (1980). *Psychol. Bull.*, **87**, 479–481.

[44] Koch, G. G. (1969). *J. Amer. Statist. Ass.*, **64**, 485–505.

[45] Koch, G. G. (1970). *Biometrics*, **26**, 105–128.

[46] Koch, G. G., Amara, I. A., Stokes, M. E., and Gillings, D. B. (1980). *Int. Statist. Rev.*, **48**, 249–265. (Some views on parametric and nonparametric analysis for repeated measurements and selected bibliography.)

[47] Koch, G. G., Amara, I. A., Davis, G. W., and Gillings, D. B. (1982). *Biometrics*, **38**, 563–595.

[48] Koch, G. G., Imrey, P. B., and Reinfurt, D. W. (1972). *Biometrics*, **28**, 663–692.

[49] Koch, G. G., Imrey, P. B., Singer, J. M., Atkinson, S. S., and Stokes, M. E. (1985). *Analysis of Categorical Data*. University of Montreal Press, Montreal, Canada.

[50] Koch, G. G., Landis, J. R., Freeman, J. L., Freeman, D. H., and Lehnen, R. G. (1977). *Biometrics*, **33**, 133–158. (A general methodology for the analysis of experiments with repeated measurement of categorical data.)

[51] Koch, G. G., Gitomer, S. L., Skalland, L., and Stokes, M. E. (1983). *Statist. Med.*, **2**, 397–412.

[52] Laird, N. M. and Ware, J. H. (1982). *Biometrics*, **38**, 963–974. (Random effects models for longitudinal data.)

[53] Laycock, P. J. and Seiden, E. (1980). *Ann. Statist.*, **8**, 1284–1292.

[54] Lehmann, E. L. (1975). *Nonparametrics: Statistical Methods Based on Ranks*. Holden-Day, San Francisco, CA.

[55] Mauchley, J. W. (1940). *Ann. Math. Statist.*, **11**, 204–209.

[56] Maxwell, S. E. (1980). *J. Educ. Statist.*, **5**, 269–287.

[57] Maxwell, S. E. and Arvey, R. D. (1982). *Psychol. Bull.*, **92**, 778–785.

[58] McCullagh, P. and Nelder, J. A. (1983). *Generalized Linear Models*. Chapman and Hall, London, England.

[59] Mendoza, J. L. (1980). *Psychometrika*, **45**, 495–498.

[60] Miller, R. (1981). *Simultaneous Statistical Inference*, 2nd ed. McGraw–Hill, New York.

[61] Monlezun, C. J., Blouin, D. C., and Malone, L. C. (1984). *Amer. Statist.*, **38**, 21–27.

[62] Morrison, D. (1970). *Biometrics*, **26**, 281–290.

[63] Morrison, D. F. (1976). *Multivariate Statistical Methods*, 2nd ed. McGraw-Hill, New York.

[64] Myers, J. L. (1979). *Fundamentals of Experimental Design*, 3rd ed. Allyn and Bacon, Boston, MA.

[65] Myers, J. L., DiCecco, J. V., White, J. B., and Borden, V. M. (1982). *Psychol. Bull.*, **92**, 517–525.

[66] Nesselroade, J. R. and Baltes, P. B., eds. (1979). *Longitudinal Research in the Study of Behavior and Development*. Academic, New York.

[67] Neter, J., Wasserman, W., and Kutner, M. H. (1985). *Applied Linear Statistical Models*, 2nd ed. Irwin, Homewood, IL.

[68] Puri, M. L. and Sen, P. K. (1971). *Nonparametric Methods in Multivariate Analysis*. Wiley, New York.

[69] Quade, D. (1982). *Biometrics*, **38**, 597–611.

[70] Rao, C. R. (1965). *Biometrika*, **52**, 447–458. (The theory of least squares when the parameters are stochastic and its application to the analysis of growth curves.)

[71] Rao, C. R. (1965). *Linear Statistical Inference and Its Application*. Wiley, New York.

[72] Rao, M. N. and Rao, C. R. (1966). *Sankhyā Ser. B*, **28**, 237–258.

[73] Reinsel, G. (1982). *J. Amer. Statist. Ass.*, **77**, 190–195.

[74] Rogan, J. C., Keselman, H. J., and Mendoza, J. L. (1979). *Brit. J. Math. Statist. Psychol.*, **32**, 269–286.

[75] Rouanet, H. and Lepine, D. (1970). *Brit. J. Math. Statist. Psychol.*, **23**, 147–163.

[76] Roy, S. N. and Bose, R. C. (1953). *Ann. Math. Statist.*, **24**, 513–536. (Simultaneous confidence interval estimation.)

[77] Roy, S. N., Gnanadesikan, R., and Srivastava, J. N. (1971). *Analysis and Design of Certain Quantitative Multiresponse Experiments*. Pergamon, Oxford, England.

[78] Scheffé, H. (1959). *The Analysis of Variance*. Wiley, New York.

[79] Schlesselman, J. (1973). *J. Chronic Disease*, **26**, 561–570.

[80] Schwertman, N. C. (1978). *J. Amer. Statist. Ass.*, **73**, 393–396.

[81] Searle, S. R. (1971). *Linear Models*. Wiley, New York.

[82] Snedecor, G. W. and Cochran, W. G. (1967). *Statistical Methods*, 7th ed. Iowa State University Press, Ames, IA.

[83] Snee, R. D., Acuff, S. K., and Gibson, J. R. (1979). *Biometrics*, **35**, 835–848.

[84] Stanish, W. M., Gillings, D. B., and Koch, G. G. (1978). *Biometrics*, **34**, 305–317.

[85] Thomas, D. R. (1983). *Psychometrika*, **48**, 451–464.

[86] Timm, N. H. (1975). *Multivariate Analysis with Applications in Education and Psychology*. Brooks/Cole, Monterey, CA.

[87] Timm, N. H. (1980). *Handbook of Statistics*, Vol. 1, P. R. Krishnaiah, ed. North-Holland, Amsterdam, pp. 41–87.

[88] Wallenstein, S. and Fisher, A. C. (1977). *Biometrics*, **33**, 261–269.

[89] Wallenstein, S. and Fleiss, J. L. (1979). *Psychometrika*, **44**, 229–233.

[90] Ware, J. A. (1985). *Amer. Statist.*, **39**, 95–101. (Linear models for the analysis of longitudinal studies.)

[91] White, A. A., Landis, J. R., and Cooper, M. M. (1982). *Int. Statist. Rev.*, **50**, 27–34.

[92] Winer, B. J. (1971). *Statistical Principles in Experimental Design*, 2nd ed. McGraw-Hill, New York.

[93] Woolson, R. F., Leeper, J. D., and Clarke, W. R. (1978). *J. R. Statist. Soc. Ser. A*, **141**, 242–252.

Acknowledgments

The authors would like to thank James Bawden for providing the data in Example 1 and William Shapiro for providing the data in Example 2. They would also like to express appreciation to Keith Muller for helpful comments with respect to the revision of an earlier version of this entry and to Ann Thomas for editorial assistance. This research was partially supported by the U.S. Bureau of the Census through Joint Statistical Agreements JSA 83-1, 84-1, and 84-5 and by grant AM 17328 from NIADDK.

(ANALYSIS OF COVARIANCE
ANALYSIS OF VARIANCE
CHANGE-OVER DESIGNS
CHI-SQUARE TESTS
CONFOUNDING
DESIGN OF EXPERIMENTS
GROWTH CURVES
HIERARCHICAL KAPPA STATISTICS
INTERACTION
INTRACLASS CORRELATION
LONGITUDINAL DATA ANALYSIS
MULTIVARIATE ANALYSIS OF VARIANCE (MANOVA)
REGRESSION ANALYSIS (VARIOUS ENTRIES)
SPHERICITY, TESTS OF
VARIANCE COMPONENTS)

GARY G. KOCH
JANET D. ELASHOFF
INGRID A. AMARA

REPEATED SIGNIFICANCE TESTS

ORIGINS

Suppose that random variables (or vectors) $X_1, X_2, \ldots$ are observed sequentially; let H_0 denote a hypothesis about the joint distribution of $X_1, X_2, \ldots$; and let $\Lambda_n = \Lambda(X_1, \ldots, X_n)$ denote a test statistic for each $n \geqslant n_0 \geqslant 1$. Suppose that Λ_n has been so constructed that large values constitute evidence against H_0 and so normalized that it has a limiting distribution G, when H_0 is true. Let a be a critical level and let $\alpha = 1 - G(a)$. If a sample of a large predetermined size n is taken, then the test that rejects H_0 iff (if and only if) $\Lambda_n > a$ has type I error,

$$P_{H_0}[\Lambda_n > a] \approx 1 - G(a) = \alpha. \quad (1)$$

On the other hand, since the data accumulate sequentially, one might perform these tests repeatedly over time; that is, letting $m > n_0$ and $M > m$ denote the minimum and maximum sample sizes that might be taken, one might determine whether $\Lambda_n > a$ for each n, $m \leqslant n \leqslant M$, stopping with the least such n, if any, and rejecting H_0 iff $\Lambda_n > z$ for some n, $m \leqslant n \leqslant M$. Such a procedure is called a *repeated significance test*. The actual sample size is the minimum of M and

$$t = t_a = \inf\{n \geqslant m : \Lambda_n > a\}; \quad (2)$$

and the actual probability of a type I error is

$$\alpha^* = P_{H_0}[\Lambda_n > a, \text{for some } m \leqslant n \leqslant M], \quad (3)$$

which may exceed α substantially, even if m and M are large.

For example, suppose that $X_1, X_2, \ldots$ are independent normally distributed random variables with unknown mean θ and known variance σ_0^2, that $H_0 : \theta = 0$, and that Λ_n is the log likelihood ratio statistic

$$\Lambda_n = S_n^2/(2n\sigma_0^2), \tag{4}$$

where $S_n = X_1 + \cdots + X_n$ for $n \geqslant 1$. Then Λ_n has a chi-squared distribution* on one degree of freedom for each $n \geqslant 1$, so that α may be computed for any $a > 0$. However, formally setting $M = \infty$ in (3) yields

$$\alpha^* = P_0\{|S_n| \geqslant \sigma_0\sqrt{2an}, \text{ for some } n \geqslant m\}$$
$$= 1 \tag{5}$$

for any $m \geqslant 1$ and any $a > 0$ by the law of the iterated logarithm*.

That repeated testing could increase the type I error* was noted by Feller [7] in connection with some experiments on extrasensory perception; and the importance of computing α^* was recognized by Robbins [16]. The observation reflects an importance difference between the classical (frequentist) approach to testing and approaches based on the likelihood principle* and/or Bayesian* analysis. For the latter, the likelihood* function and posterior distributions* are unaffected by optional stopping; for the former, the effect on the type I error may be huge, as in (5). See refs. 1 and 5 for statements of the two points of view.

Subsequently, Armitage [1, 2], Miller [15], and others have suggested the use of repeated significance tests (with a finite M) in clinical trials* (medical experiments on human subjects) to reduce the ethical problems inherent in such experiments. To understand why, suppose that a test of a new treatment calls for testing $N = 100$ pairs of subjects with one of each pair to receive the new treatment and one to receive an old treatment. Suppose also that halfway through the test the experimenters observe a large value of Λ_n and strongly suspect that the new treatment is better. Then they confront the following ethical dilemma: continuing the experiment requires giving the old treatment, which is thought to be inferior, to half of the remaining subjects; stopping violates the experimental design, so that the results might not be believed by colleagues. Using a repeated significance test or other sequential test provides a compromise by building the option of early termination into the experimental design. See ref. 2, Sections 1.3 and 2.1–2.4 for a detailed discussion of this point.

At a more technical level, Schwarz [17, 18] has shown that optimal sequential tests of two (appropriately) separated hypotheses, H_0 and H_1 say, may be approximated by performing repeated significance tests. In Schwarz's formulation, $X_1, X_2, \ldots$ are independent and identically distributed (i.i.d.) with a common density that depends on unknown parameters; there is a positive loss for a wrong decision whenever either of the two hypotheses is true; there is a cost $c > 0$ for each observation; and there is a prior distribution* with full support and positive probability for both of the hypotheses. The result takes the form of a limit theorem as $c \to 0$. Let Λ_n^0 and Λ_n^1, $n \geqslant 1$, denote the log likelihood ratio statistics for testing H_0 and H_1. It is shown that sampling until $\max(\Lambda_n^0, \Lambda_n^1) > a = (\log l)/c$ approximates the optimal Bayesian sequential test. Thus, the approximation consists of performing repeated significance tests on both hypotheses.

PROPERTIES

There are several approaches to computing α^* and such related quantities as power and expected sample size.

In ref. 3, α^* was computed numerically for likelihood ratio tests of $\omega = \frac{1}{2}$, 1, or 0 when $X_1, X_2, \ldots$ are i.i.d. Bernoulli*, exponential*, or normal* with unknown mean θ (and unit variance in the normal case). In these cases, α^* is the last of the sequence of M recursively defined sums or integrals that may be computed exactly in the Bernoulli case and by numerical integration in the exponential and normal cases. For the exponential and normal cases with $m = 1$, α^*/α varied between 6 and 11 for selected values of α, $0.01 \leqslant \alpha \leqslant 0.05$, and M, $50 \leqslant M \leqslant 100$. For the Bernoulli case, $3 \leqslant \alpha^*/\alpha \leqslant 5$ in the same range. Later ref. 14 studied the power of repeated significance tests for the same distributions. While limited to special cases, these two papers have provided important checks on the theoretical approximations in the descriptions to follow.

In many cases α^* may be approximated by using an invariance principle*. Suppose that $m, M \to \infty$ in such a manner that $m/M \to \epsilon$, $0 < \epsilon < 1$ and that a remains fixed. Let λ_M denote the stochastic process*

$$\lambda_M(s) = \Lambda_{[Ms]}, \qquad m/M \leqslant s \leqslant 1,$$

where $[x]$ denotes the integer part of x. In many problems, λ_M converges in distribution to a continuous stochastic process $\lambda(s)$, $\varepsilon \leqslant s \leqslant 1$, as $M \to \infty$ when H_0 is true, and

$$\alpha^* \to P\left[\max_{\epsilon \leqslant s \leqslant 1} \lambda(s) > a\right]. \quad (6)$$

For example, if $X_1, X_2, \ldots$ are i.i.d. $N(\theta, \sigma_0^2)$, $H_0: \theta = 0$, and $\Lambda_n = S_n^2/(2n\sigma_0^2)$, $n \geqslant 1$, then λ_M converges in distribution to $\lambda(s) = (1/2s)b^2(s)$, $\epsilon \leqslant s \leqslant 1$, where b is a standard Brownian motion*. Refinements give approximations to the power against local alternatives. The details of this approach are described in refs. 4 and 26 for parametric models (when Λ_n is a log likelihood ratio statistic), in ref. 20 for many nonparametric models, and in refs. 19 and 21 for semiparametric models involving censored data. Tables of the asymptotic distribution are given in ref. 6, and an approximation in ref. 22 (see also ref. 9). REPEATED CHI-SQUARE TESTING and ref. 20 contain further references.

Observe that the type I error does not approach zero in (6). In most examples, the type II error does approach zero at fixed alternatives, but little is known about the exact rate.

If Λ_n, $n \geqslant n_0$, denote log likelihood ratio* statistics, then Schwarz's derivation suggests that a should approach ∞ with m and M, say $m \sim a/\delta_1$ and $M \sim a/\delta_0$ for suitable values of $0 < \delta_0 < \delta_1 < \infty$. If the common distributions of $X_1, X_2, \ldots$ form an exponential family*, then a detailed description of the asymptotic properties of the repeated significance test is possible. For example, If $X_1, X_2, \ldots$ are independent normally distributed random variables with unknown mean θ and unit variance $\sigma_0^2 = 1$, then

$$\alpha^* \sim K\sqrt{a}e^{-a}, \quad (7)$$

$$E_\theta(t) = \frac{2}{\theta^2}\left[a + \rho_\theta - \frac{1}{2}\right] + o(1),$$

$$0 < |\theta| < \sqrt{2\delta}, \quad (8)$$

as $a \to \infty$, where $K = K(\delta_0, \delta_1)$ and ρ_θ are constants. Refinements yield approximations to the power function at fixed alternatives.

The derivations of (7) and (8) make essential use of the nonlinear theorem of Lai and Siegmund [10, 11], described in NONLINEAR RENEWAL THEORY. For details in the case of one-parameter exponential families, including tables of K and ρ_θ, see refs. 25, Chap. 4, or 29, Chap. 7. For (7) in the case of several parameters, see refs. 12 and 25, Chap. 5. The values reported in [28] are too small by about 10%. References [22] and [31] study the normal case and report good agreement between the asymptotic formulas and the numerical calculations of ref. 14. A nonparametric case is considered in ref. 30 and a nonregular one in ref. 27. There is recent work on models with sequential allocation and censoring in refs. 8 and 13.

Repeated significance testing complicates the sampling distributions of estimators that might be used after the test and may introduce a substantial bias. Siegmund [23–25] shows how to use the relationship between tests of hypotheses and confidence sets to form confidence intervals following sequential testing.

Approximations by Brownian motion implicitly neglect the overshoot $\Lambda_t - a$, while those using nonlinear renewal theory include it. The latter are more complicated and model dependent. One may expect the latter to be more accurate when the model is correct; and this expectation is realized in the (few) special cases for which the answer is known.

Reference 25 is a Masters' level text that describes the use of repeated significance tests and the derivation of their properties. It also attempts a synthesis of the two methods of approximation, called *corrected diffusion approximations.* I recommend it highly.

References

[1] Armitage, P. (1967). In *Proc. 5th Berkeley Symp.*, Vol. 4. University of California Press, pp. 791–804.

[2] Armitage, P. (1975). *Sequential Medical Trials.* Halsted Press, New York.

[3] Armitage, P., McPherson, C. K., and Rowe, B. C. (1969). *J. R. Statist. Soc. A*, **132**, 235–244.

[4] Barbour, A. (1979). *Proc. Camb. Philos. Soc.*, **86**, 85–90.

[5] Cornfield, J. and Greenhouse, S. (1967). In *Proc. 5th Berkeley Symp.*, Vol. 4. University of California Press, pp. 813–829.

[6] De Long, D. M. (1981). *Commun. Statist. Theor. Meth. A*, **10**, 2197–2213.

[7] Feller, W. (1940). *J. Parapsych.*, **4**, 271–298.

[8] Heckman, N. (1982). *Two Treatment Comparisons with Random Allocations.* Ph.D. Thesis, The University of Michigan.

[9] Jennen, C. and Lerche, H. R. (1981). *Zeit. Wahrsch. Ver. Geb.*, **55**, 133–148.

[10] Lai, T. L. and Siegmund, D. (1977). *Ann. Statist.*, **5**, 946–954.

[11] Lai, T. L. and Siegmund D. (1979). *Ann. Statist.*, **7**, 60–76.

[12] Lalley, S. (1983). *Zeit. Wahrsch. Ver. Geb.*, **63**, 293–322.

[13] Lalley, S. (1984). *Ann Prob.*, **12**, 1113–1148.

[14] McPherson, C. K. and Armitage, P. (1971). *J. R. Statist. Soc. A*, **134**, 15–26.

[15] Miller, R. G. (1970). *J. Amer. Statist. Ass.*, **65**, 1554–1561.

[16] Robbins, H. (1952). *Bull. Amer. Math. Soc.*, **58**, 527–535.

[17] Schwarz, G. (1962). *Ann. Math. Statist.*, **33**, 224–236.

[18] Schwarz, G. (1968). *Ann. Math. Statist.*, **39**, 2038–2043.

[19] Sellke, T. and Siegmund D. (1983). *Biometrika*, **70**, 315–326.

[20] Sen, P. K. (1981). *Sequential Nonparametrics: Invariance Principle and Statistical Inference.* Wiley, New York.

[21] Sen, P. K. (1981). *Ann. Statist.*, **9**, 109–121.

[22] Siegmund, D. (1977). *Biometrika*, **64**, 177–189.

[23] Siegmund, D. (1978). *Biometrika*, **65**, 341–349.

[24] Siegmund, D. (1979). *Biometrika*, **67**, 389–402.

[25] Siegmund, D. (1985). *Sequential Analysis*, Springer, New York.

[26] So, C. and Sen, P. K. (1982). *Commun. Statist. Seq. Anal.*, **1**, 101–120.

[27] Swanepoel, J. (1980). *S. Afr. Statist. J.*, **14**, 31–41.

[28] Woodroofe, M. (1979). *Biometrika*, **66**, 453–463.

[29] Woodroofe, M. (1982). *Non-linear Renewal Theory in Sequential Analysis.* S.I.A.M., Philadelphia, PA.

[30] Woodroofe, M. (1984). *Sankhyā* , *A* **46**, 233–252.

[31] Woodroofe, M. and Takahashi, H. (1982). *Ann. Statist.*, **10**, 895–908.

(CLINICAL TRIALS
HYPOTHESIS TESTING
PARAPSYCHOLOGY, STATISTICS IN
REPEATED CHI-SQUARE TESTING
SEQUENTIAL ANALYSIS)

MICHAEL B. WOODROOFE

REPORTS OF STATISTICAL APPLICATIONS RESEARCH, UNION OF JAPANESE SCIENTISTS AND ENGINEERS

This journal is sponsored by the Statistical Applications Research Group of the Union of Japanese Scientists and Engineers (JUSE). JUSE was founded in 1946.

The journal is devoted mainly to applicable statistics, but contains a substantial number of papers describing new methods of analysis and of constructing designs for experiments and surveys. Articles are classified as follows:

A Section: Original papers dealing with the development of statistical theories and methods.

B Section: Original papers dealing with the application of statistical theories and methods.

C Section: Reports on activities in quality control and other statistical applications.

Short Note: Brief reports on creative studies, which include, short as they may be, important discoveries or conclusions, or persuasive data.

The contents of the March, 1986 issue (Vol. 33, No. 1) are:

(A) "Comparison of two highly multivariate small samples" by Shunichi Tani, Mitsugu Kanboa, and Masashi Goto.

(C) "Marketing and TQC activities at Komatsu Zenoah Co." by Toshiro Suzuki.

The Editor (March 1986) is Kaoru Ishikawa. All papers are in English. The journal publishes reports by nonmembers of the Statistical Application Research Group, JUSE, if they are recommended by a member of the group. Communications should be addressed to

Union of Japanese Scientists and Engineers (JUSE),
5-10-11, Sendagaya, Shibuya-ku,
Tokyo, Japan.

REPRESENTATIVE SAMPLING

The attractive concept of a representative sample advanced the development of sample surveys (*see* SURVEY SAMPLING), although no single satisfactory definition of the concept has been forthcoming. In studies of usage, Kruskal and Mosteller [18–21] found eight distinct senses of "representative sample" in lay, scientific, and statistical literature. We first explain these meanings and then present a short history of representative sampling.

(1) Rhetorical Usage; General Acclaim; Puffery. A journalist or other author may speak of interviewing a representative sample of people when no process designed to produce representativeness stands back of the sampling. C. D. Wright says

> We have aimed to make our investigations of such a degree of comprehensiveness that our deductions would bear the impress of true representative character, and seem founded upon a tangible basis. [25, p. 201]

Others criticize reports of sampling by complaining about nonrepresentativeness, often with no specific biases in mind and sometimes even when the sample has been a carefully executed probability sample. These rhetorical usages should be avoided: if a sample has been well drawn, its procedure should be reported; if its drawing deserves criticism, specifics should accompany the criticism.

(2) Absence or Presence of Selective Forces. This meaning ordinarily arises in criticisms of sampling. Our sample should make possible an inference to a specific population rather than one badly distorted. For example, Campbell and Stanley [2, p. 19] say

> Consider ... an experiment on teaching in which the researcher has been turned down by nine school systems and is finally accepted by a tenth. This tenth almost certainly differs from the other nine, and from the universe of schools to which we would like to generalize. ... It is, thus, non-representative. Almost certainly its staff has higher morale, less fear of being inspected, more zeal for improvement than does that of the average school.

(3) Miniature of the Population. The inviting idea of the sample as a miniature or mirror of the population usually fails because of selection, because of sampling fluctuations, or because a sample cannot represent all the traits at once. Human populations often have too many traits for adequate miniaturization. In some physical situations, especially for miscible liquids, a drop adequately mirrors a whole beaker. Stephan and McCarthy [24, p. 31] say

> The first aim of most sampling procedures is to obtain a sample ... that will *represent* the population from which it is selected. In other words ... results ... from the sample [should] ... agree "closely" with results

> that would have been obtained ... [by studying] the entire population .... This ... idea has frequently been stated by saying that a sample should be a miniature population or universe.

Gilbert et al. [8, pp. 218–220] explain the problem of many variables.

> When we sample from a population, we would like ideally a sample that is a microcosm or replica or mirror of the target population—the population we want to represent. For example, for a study of types of families, we might note that there are adults who are single, married, widowed, and divorced. We want to stratify our population to take proper account of these four groups and include members from each in the sample.... [To take account of more variables we might] have four marital statuses, two sexes, say five categories for size of immediate family (by pooling four or over), say four regions of the country, and six sizes and types of city, say five occupation groups, four levels of education, and three levels of income. This gives us in all $4 \times 2 \times 5 \times 4 \times 6 \times 5 \times 4 \times 3 =$ 57,600 different possible types, if we are to mirror the population or have a small microcosm; and one observation per cell may be far from adequate. We thus may need hundreds of thousands of cases!

(4) Typical or Ideal Cases. Attempts to select individuals or even sets of individuals as typical of the population have failed in agriculture (Cochran and Watson [4] and Yates [26, 27]) and in government (Gini [9] and Gini and Galvani [10]), even though the idea of such "purposive" sampling carries enormous appeal. Stratified sampling can ensure that some strata in a population are represented, but the number of strata grows exponentially with the number of variables.

Hodges and Lehmann illustrate [11, pp. 35–36]:

> Again, in trying to forecast an election, we might select counties that have voted like the country as a whole in recent elections, and conduct interviews with voters in these counties. While this method of "purposive" sampling is superficially attractive, it has not worked well in practice. Biases of the selector creep in too easily when the items for the sample are selected purposively by the exercise of judgment. Also, items that are typical of the population in some respects need not be typical in others. For instance, counties that vote like the country on the last election may well not do so on the next, when the issues will be different.

Statisticians may throw up their hands at Emerson's [7] ideal case: the representative item for him is the superlative or quintessential in its class; Plato, the ideal or representative philosopher; Newton, the ideal physicist; the finest cabbage in the patch.

(5) Coverage of the Population. The idea of coverage makes sense when one item can represent its entire class. Except for occasional blemishes, one mass-produced necktie does represent well all those with the same pattern, and collecting one example of each pattern covers the population. Coverage reveals what categories are available when frequency may be of little interest. For example, a study may try to learn what techniques are available to handle cases of alleged fraud within research organizations without trying to learn how often a technique is adopted.

The early work in sampling emphasized coverage as a feature of representative samples intended to reflect properly the heterogeneity of the population. For example, Kiaer* says [14, p. 180; 21, p. 176 (this is a free translation)]:

> By representative investigation I understand a partial exploration with observations on a large number of scattered localities, distributed over the whole territory so that they form a miniature of that whole. The localities are not chosen arbitrarily, but according to a rational grouping based on census results; and the results should be controlled by comparison with those censuses.

Kiaer's definition represents a mix of ideas of coverage and the miniature.

(6) Representative as a Vague Term to Be Made Precise. In the early stages of exposition of the results of a study, one may want to speak of a specially designed sample as a "representative sample" without going into detail. Then "representative" becomes a placeholder for the actual description of the sample. This usage seems satisfactory, especially if a description of the sampling procedure is included in the same document.

(7) Representative as a Specific Sampling Method. At some times and for some people, specific sampling plans have been assigned the name "representative sample" —for example, the simple random sample and more often stratified probability samples with sampling from each stratum proportional to the size of the stratum in the population. This usage no longer seems wise because it has not been widely adopted.

(8) Representative as Good Enough for a Particular Purpose. A sampling that can prove that something thought to be rare is frequent, or vice versa, may be adequately representative even though the sample itself is defective. For example, a random sample with 50% nonresponse might still prove that at least 15% of burn patients had a special kind of burn.

HISTORY OF REPRESENTATIVE SAMPLING

We turn now to the history of representative sampling. At the 1895 Berne meeting of the International Statistical Institute* (ISI) [12], Anders Nicolai Kiaer of Norway offered the first analytical use of the term. He pressed hard for what he called the "representative method" and what others called "partial investigation" and "indirect methods." One of his definitions of the representative method appears under usage (5) above. His proposals for sampling met with stiff opposition from other government statisticians, who believed in complete enumeration. In 1897, Kiaer [13] suggested drawing a sample by lot, though he did not actually do that. Instead, he picked people according to certain ages and initial letters of surnames. He also suggested differential sampling rates with weighting in the analysis. He noted that in many studies, it is impossible to take a census—of agricultural yields from every holding, of all letters mailed between countries, or of all geological structures in a region. He pushed his position at various conferences: St. Petersburg [14] and Stockholm in 1897 [5, p. 30; 15], Budapest in 1901 [16], and Berlin in 1903 [17].

At Budapest, Bortkiewicz* [16, discussion] discussed Kiaer's work by applying a test of goodness of fit* to the data. Bortkiewicz rejected Kiaer's proposition that the differences between sample and population could have been due to accidental causes, but Bortkiewicz may well have used variances too small to be appropriate for Kiaer's sampling methods. Chunks of the population were ordinarily used, and Bortkiewicz may have used the theory of independent random samples.

At Berlin in 1903 [17], Lucien March discussed representative sampling in a different vein. As so often happens in science, what was new and outrageous one year turns out a little later to be old and respectable after all. March reminded everyone that the French had used sampling to estimate the population of France by computing the ratio of population to numbers of births for some scattered districts, and then applying the ratio to the whole country, assuming that the numbers of births were known. Laplace* (see Cochran [3]) himself had estimated the uncertainty of the results. For the first time, March revealed the role of randomness in the sampling and appreciated that the various probabilistic calculations had been made on an "as if" basis. Thus March might be said to have developed the idea of probability sampling.

Actually carrying out a sample survey with tabled numbers to select the units was done by Bowley and Burnett–Hurst [1] in 1906 (Bowley usually used systematic samples, every kth item). An early well-documented use of physical randomization was performed by Edin [6, p. 3] in Göteborg for a 1910 study of apartments. Edin says (in Dalenius's translation [5, p. 40])

> As far as I can discover, the method used is the only one which can be called representative, in the strict meaning of the word —something which is not valid when that form of sample is used where a few suburbs are surveyed more or less completely.

This last remark is especially relevant to the so-called representative method as practiced by most of its early users, who tended to take districts or large chunks of populations and enumerate them completely. Wright [25] used such methods in the Department of Labor in the United States. His bombastic justification for representativeness is included in our usage (1). Repeatedly he merely asserted the representativeness of his samples. Nevertheless, Kiaer was pleased to report on the success the American claimed to have with the representative method.

At the Italian census bureau the major failure in 1926 of an attempt at purposive sampling exposed the weaknesses of that method. These threatened to spill over and destroy the credibility of sampling methods generally. Gini* and Galvani [10] needed to discard part of the 1921 census data to make room for those from another. They wanted to be able to recover population information later from a sample of the discarded census. By trial and error, they chose 29 administrative units out of 214 in such a way that the average values for seven important characteristics over the 29 units were very close to the values for the nation as a whole. They discovered that other characteristics over the 29 units often differed substantially from the population values, for example, variability and measures of association did not agree well. Jerzy Neyman* retold this story to great effect, arguing that we must distrust purposive sampling, not sensible probability sampling.

Neyman's 1934 paper [22] treats the problem of stratified samples of districts. Thus he clarified the problem that Gini and Galvani were concerned with, and also showed that resulting estimates from the stratified method using chunks could be very uncertain, especially if the districts were large. W. Edwards Deming brought Neyman to give a set of lectures at the U. S. Department of Agriculture, and thus ideas of sampling became much more widely spread. By 1939 Snedecor [23, p. 852] published a paper arguing for replacing censuses by sample surveys. He wrote that "... the stratified random sample is recognized by mathematical statisticians as the only practical device now available for securing a representative sample from human populations ...," in agreement with Edin [5]. Thus what started as unacceptable, ultimately became standard practice.

References

[1] Bowley, A. L. and Burnett-Hurst, A. R. (1915). *Livelihood and Poverty*. G. Bell, London, England.

[2] Campbell, D. T. and Stanley, J. C. (1963). *Experimental and Quasi-experimental Designs for Research*. Rand McNally, Chicago, IL. (A catalogue of research designs with strengths and weaknesses explained.)

[3] Cochran, W. G. (1978). In *Contributions to Survey Sampling and Applied Statistics. Papers in Honor of H. O. Hartley*, H. A. David, ed. Academic, New York, pp. 3–10.

[4] Cochran, W. G. and Watson, D. J. (1936). *Empire J. Exper. Agric.*, **4**, 69–76. (Proof that agricultural workers could not choose unbiased crop samples by eye.)

[5] Dalenius, T. (1957). *Sampling in Sweden: Contributions to the Methods and Theories of Sample Survey Practice*. Almqvist and Wiksell, Stockholm, Sweden.

[6] Edin, K. A. (1912). *De Mindre Bemedlades Bostadföhåll Anden (Första Bosladsbeskrifningin) i Göteborg. Statistik Undersökning Ulförd År 1911 på Uppdrag af Kommittén för Stadens Kommunalstatistik.* Göteborg, Sweden.

[7] Emerson, R. W. (1850). *Representative Men*. Phillips, Samson, Boston, MA. (Many later republications.)

[8] Gilbert, J. P., Light, R. J., and Mosteller, F. (1977). In *Statistics and Public Policy*, W. B. Fairley and F. Mosteller, eds. Addison-Wesley, Reading, MA, pp. 185–241.

[9] Gini, C. (1928). *Bull. Int. Statist. Inst.*, **23**, Liv. 2, 198–215.

[10] Gini, C. and Galvani, L. (1929). *Annali di Statist., Ser. 6*, **4**, 1–107.

[11] Hodges, J. L., Jr. and Lehmann, E. L. (1964). *Basic Concepts of Probability and Statistics.* Holden-Day, San Francisco, CA.

[12] Kiaer, A. N. (1895–1896). *Bull. Int. Statist. Inst.*, **9**, Liv. 2, 176–183. (The meeting was in Berne in 1895.)

[13] Kiaer, A. N. (1897). Den repraesentative Undersøgelsesmethode. *Christiania Videnskabsselskabets Skrifter.* II. Historiskfilosofiske Klasse, No. 4, 24 pages. (Translated by Svein Brenna into English, as "The representative method of statistical surveys". *Papers from the Norwegian Academy of Science and Letters.* II. The Historical, Philosophical Section. The original and the translation, together with a bibliography of Kiaer and a preface, were published in 1976 by the Norwegian Central Bureau of Statistics, No. 27 in its series *Samfunnsøkonomiske Studier.*)

[14] Kiaer, A. N. (1899). *Bull. Int. Statist. Inst.*, **11**, Liv. 1, 180–185. (The meeting was in St. Petersburg in 1897.)

[15] Kiaer, A. N. (1899). *Allg. Statist. Arch.*, **5**, 1–22. (The author's name is spelled Kiär.)

[16] Kiaer, A. N. (1903). *Bull. Int. Statist. Inst.*, **13**, Liv. 1, 66–70; discussion, 70–78. (The meeting was in Budapest in 1901.)

[17] Kiaer, A. N. (1905). *Bull. Int. Statist. Inst.*, **14**, Liv. 1, 119–126; discussion, 126–134. (The meeting was in Berlin in 1903.)

[18] Kruskal, W. H. and Mosteller, F. (1979). *Int. Statist. Rev.*, **47**, 13–24. (Usage of representative sample in nonscientific literature.)

[19] Kruskal, W. H. and Mosteller, F. (1979). *Int. Statist. Rev.*, **47**, 111–127. (Usage of representative sample in the scientific literature.)

[20] Kruskal, W. H. and Mosteller, F. (1979). *Int. Statist. Rev.*, **47**, 245–265. (Usage of representative sample in the statistical literature.)

[21] Kruskal, W. H. and Mosteller, F. (1980). *Int. Statist. Rev.*, **48**, 169–195. (History of representative sampling 1895–1939.)

[22] Neyman, J. (1934). *J. R. Statist. Soc. A*, **97**, 558–606; discussion, 607–625. (The Gini–Galvani sampling problem described in English, a solution to the purposive sampling problem and, irrelevant to this topic, the invention of confidence limits.)

[23] Snedecor, G. W. (1939). *J. Farm Economics*, **21**, 846–855.

[24] Stephan, F. F. and McCarthy, P. J. (1958). *Sampling Opinions: An Analysis of Survey Procedure.* Wiley, New York.

[25] Wright, C. D. (1875). *Sixth Annual Report of the Bureau of Statistics of Labor, March 1875.* Wright and Potter, Boston, MA.

[26] Yates, F. (1934–1935). *Ann. Eugenics*, **6**, 202–213.

[27] Yates, F. (1936–1937). *Trans. Manchester Statist. Soc.* (Articles separately paginated; this has 26 pages.)

Bibliography

David, H. T. (1978). In *International Encyclopedia of Statistics*, W. H. Kruskal and J. Tanur, eds. Free Press, New York. pp. 399–409. (For discussion of goodness-of-fit.)

Goodman, L. A. and Kruskal, W. H. (1959). *J. Amer. Statist. Ass.*, **54**, 123–163; **58**, 525–540 (1963); **67**, 415–421 (1972). (Discusses various measures of representativeness.)

Mosteller, F. (1948). *J. Amer. Statist. Ass.*, **43**, 231–242. (Using representativeness to support estimation based on samples from different populations.)

Sobel, M. and Huyett, M. J. (1958). *Bell Syst. Tech. J.* **37**, 135–161. (A special kind of representativeness assessment for samples from a multinomial distribution.)

(PROBABILITY PROPORTIONAL TO SIZE SAMPLING
PUBLIC OPINION POLLS
SURVEY SAMPLING
TARGET POPULATION)

W. H. KRUSKAL
F. MOSTELLER

REPRODUCIBILITY *See* MEASUREMENT ERROR

REPRODUCING KERNEL HILBERT SPACES

INTRODUCTION

Let $\mathscr{F}$ be a Hilbert space of real functions f defined over a set E. We suppose that a norm is defined by $\|f\|^2 = (f, f)$ for $f \in \mathscr{F}$ where $(\cdot, \cdot)$ is the inner product. Every class $\mathscr{F}$ of real functions forming a real Hilbert space determines a complex Hilbert space by considering all $f_1 + if_2$, for f_1 and f_2 in $\mathscr{F}$. The complex norm is defined by $\|f_1 + if_2\|^2 = \|f_1\|^2 + \|f_2\|^2$. The space of such complex functions, say $\mathscr{F}_c$, is a *complex Hilbert space.*

The space formed in this way has the key properties that if $f \in \mathscr{F}_c$, then its complex conjugate $\bar{f} \in \mathscr{F}_c$ and has the same norm. If $\mathscr{F}$ is a class of functions defined on E forming a Hilbert space (complex or real), a function $K(x, y)$ of x and y in E is a *reproducing kernel* of $\mathscr{F}$ if:

(a) For every y, $K(x, y)$ as a function of x belongs to $\mathscr{F}$.

(b) For every $y \in E$ and $f \in \mathscr{F}$,

$$f(y) = (f(x), K(x, y))_x.$$

The subscript x on the inner product indicates that this inner product is taken on functions of x. If $\mathscr{F}$ has a reproducing kernel K, then the corresponding complex space $\mathscr{F}_c$ has the same reproducing kernel. Hence, in general one may formulate the theory in terms of complex Hilbert spaces, an assumption we make here.

BASIC PROPERTIES

A Hilbert space of functions together with a reproducing kernel provides a structure that admits some interesting properties, summarized next.

(a) If a reproducing kernel (rk) exists it is unique.

(b) A necessary and sufficient condition for the existence of a rk K is that for every $y \in E$, $f(y)$ is a continuous functional for f running through $\mathscr{F}$.

(c) $K(\cdot, \cdot)$ is a nonnegative definite function.

(d) Conversely, for every nonnegative definite function $K(\cdot, \cdot)$, there is a unique class of functions with a uniquely determined quadratic form forming a Hilbert space and admitting $K(\cdot, \cdot)$ as a reproducing kernel.

(e) If $\mathscr{F}$ possesses a rk $K(\cdot, \cdot)$, every sequence of functions $\{f_n\}$ that converges strongly to a function f in $\mathscr{F}$ also converges pointwise in the sense $\lim_n f_n(x) = f(x)$, $x \in E$. This convergence is uniform on every subset of E in which $K(x, x)$ is uniformly bounded.

(f) F possesses a rk $K(\cdot, \cdot)$ in a subspace of a larger Hilbert space $\mathscr{H}$, then

$$f(y) = (h, K(x, y))_x$$

is the projection of $h \in \mathscr{H}$ into $\mathscr{F}$.

(g) If $\mathscr{F}$ possesses a rk $K(\cdot, \cdot)$ and if $\{g_n\}$ is an orthonormal system in $\mathscr{F}$, then for every sequence $\{a_n\}$ such that

$$\sum_1^\infty |a_n|^2 < \infty,$$

we have

$$\sum_1^\infty |a_n|^2 |g_n(x)| \leqslant K(x, x)^{1/2} \left(\sum_1^\infty |a_n|^2 \right)^{1/2}.$$

(h) If $\mathscr{F}$ possesses a rk $K(\cdot, \cdot)$, then the same is true of all closed linear subspaces by virtue of (a) and (b). If $\mathscr{F}_1$ and $\mathscr{F}_2$ are subspaces with rk K_1 and K_2, respectively, such that $\mathscr{F}_1 \cup \mathscr{F}_2 = \mathscr{F}$, then $K_1 + K_2 = K$.

Results **(c)** and **(d)** taken together immediately suggest a statistical application, in the next section. In statistical applications it is customary to define inner products in terms of Stieltjes integrals so that $(f, g) = \int f(x) g(x)\, d\mu(x)$ for some measure μ. This places the user in an L_2 setting so that the norm

$$\|f - g\|^2 = \int \{f(x) - g(x)\}^2\, d\mu(x)$$

corresponds to square error and one is poised for carrying out least-squares* inference.

HISTORY

The fundamental theory and systematic development of reproducing kernels and associated Hilbert spaces was laid out by Aronszajn [1, 2]. The discussion in the preceding sections is adapted from the 1950 paper, which is quite long. It is the fundamental reference for the theory of reproducing kernel Hilbert spaces (rkhs). Properties of rkhs are intimately bound up with properties of nonnegative definite functions; Stewart [22] provides a historical survey of such functions and an extensive bibliography.

The covariance function of a second-order stochastic process is nonnegative definite, and is thus the reproducing kernel of a function space. Loève [15] was the first to establish this link between reproducing kernels and stochastic processes*. He stated the basic isomorphism between the rkhs determined by the covariance of a second-order process and the Hilbert space of random variables spanned by a stochastic process.

Once it is realized that the Hilbert space spanned by a stochastic process $H(X_t)$ is a rkhs also, the framework for estimation, prediction, etc., becomes clear. In the prediction situation, for example, consider the sub-Hilbert space $M = H(X_s, s \leqslant t_0)$ spanned by the process up to time t_0. One need only use the result **(f)** to determine the projection of a process Z in $H(X_t)$ into M so as to determine the least-squares predictor of the process Z given the data up to time t_0. Result **(e)** gives an excellent handle on convergence results. In an approximation setting, if M is an approximating subspace, the projection into M will yield an approximator possessing properties of M. For example, if M is taken as a *Sobolov space* of smooth functions, then the approximator will possess precisely those smoothness properties. Generalized splines are the approximators determined by the rkhs technique using Sobolov spaces.

Parzen [16] was the first to apply the rkhs method to time series* problems. He demonstrated that the rkhs method provided a unified framework for least-squares estimation of random variables, for minimum variance unbiased estimation* of regression coefficients*, and for detection of known signals in Gaussian noise*. Although not cast in the framework of Aronszajn's rkhs, the Hájek [8] paper independently suggests the same framework. Parzen's 1959 technical report [16] was published as a series of papers [17–20], which are conveniently collected in ref. 21.

The line of thinking begun by Loève, Parzen, and Hájek has blossomed extensively. Notable among contributions to the applications of rkhs methods to stochastic processes are those of Kailath and his colleagues (see for example Kailath [9], Kailath and Weinert [11], Kailath and Duttweiler [10], Duttweiler and Kailath [6, 7]) and those of Kallianpur [12] and Kallianpur and Oodaira [13].

The notion that M can be tailored to cause estimators or approximations to inherit smoothness properties extends at least as far back as de Boor and Lynch [4]. Indeed, this paper discusses some fundamental theory of splines in the context of Hilbert space projections. Wahba and her colleagues have been early and widely published advocates of spline methods based on the rkhs formulation. Kimeldorf and Wahba [14] represents early work, and Wahba [24] and Wahba and Wendelberger [25] are other recent examples. Nonparametric density estimation* is a popular object of rkhs methods using splines. Key papers include Wahba [23], Boneva et al. [3], and de Montricher et al. [5].

The paper by Weinert [28] is an excellent survey of rkhs-based methods of curve estimation while Wegman and Wright [27] is a summary of spline methods. Splines are solutions to minimum norm problems in Sobolov subspaces, but similar approximation problems may be formulated with other types of rkhs. An example of this is Yao [31], who deals with spaces of band-limited functions. Tailoring of the subspace to cause the approximator/estimator to inherit desired properties was also implicitly done in Wright and Wegman [30]. An explicit formulation was given in Wegman [26]. Finally, we mention the fine volume edited by Weinert [29], which includes many of the papers mentioned here. Weinert's highly intelligent commentary provides an excellent guide to the literature as well as much of the basis for the present discussion.

References

[1] Aronszajn, N. (1943). *Proc. Camb. Philos. Soc.*, **39**, 133–153 (in French).

[2] Aronszajn, N. (1950). *Amer. Math. Soc. Trans.*, **68**, 337–404.

[3] Boneva, L., Kendall, D., and Stefanov, I. (1971). *J. R. Statist. Soc. B*, **33**, 1–70.

[4] de Boor, C. and Lynch, R. (1966). *J. Math. Mech.*, **15**, 953–969.

[5] de Montricher, G., Tapia, R. and Thompson, J. (1975). *Ann. Math. Statist.*, **3**, 1329–1348.

[6] Duttweiler, D. and Kailath, T. (1973). *IEEE Trans. Inf. Theory*, **IT-19**, 19–28.

[7] Duttweiler, D. and Kailath, T. (1973). *IEEE Trans. Inf. Theory*, **IT-19**, 29–37.

[8] Hájek, J. (1962). *J. Czech. Math.*, **12**, 404–444.

[9] Kailath, T. (1971). *IEEE Trans. Inf. Theory*, **IT-17**, 530–549.

[10] Kailath, T. and Duttweiler, D. (1972). *IEEE Trans. Inf. Theory*, **IT-18**, 730–745.

[11] Kailath, T. and Weinert, H. (1975). *IEEE Trans. Inf. Theory*, **IT-21**, 15–23.

[12] Kallianpur, G. (1970). In *Advances in Probability and Related Spaces*, P. Ney, ed. Marcel Dekker, New York, 51–83.

[13] Kallianpur, G. and Oodaira, H. (1973). *Ann. Prob.*, **1**, 104–122.

[14] Kimeldorf, G. and Wahba, G. (1971). *J. Math. Anal. Appl.*, **33**, 82–95.

[15] Loève, M. (1948). *Stochastic Processes and Brownian Motion*, P. Levy, ed. Gauthier-Villars, Paris, Appendix (in French).

[16] Parzen, E. (1959). Statistical inference on time series by Hilbert space methods, I. *Tech. Rep.*, No. *23*, Statistics Department, Stanford University, Stanford, CA.

[17] Parzen, E. (1961). *Ann. Math. Statist.*, **32**, 951–989.

[18] Parzen, R. (1961). In *Proc. 4th Berkeley Symp. Math. Statist. Prob.*, Vol. 1, J. Neyman, ed. University of California Press, Berkeley, 469–489.

[19] Parzen, E. (1962). *SIAM J. Control*, **1**, 35–62.

[20] Parzen, E. (1963). In *Mathematical Optimization Techniques*, R. Bellman, ed. University of California Press, Berkeley, 75–108.

[21] Parzen, E. (1967). *Time Series Analysis Papers*. Holden-Day, San Francisco.

[22] Stewart, J. (1976). *Rocky Mountain J. Math.*, **6**, 409–434.

[23] Wahba, G. (1975). *Ann. Statist.*, **3**, 30–48.

[24] Wahba, G. (1978). *J. R. Statist. Soc. B*, **40**, 364–372.

[25] Wahba, G. and Wendelberger, J. (1980). *Monthly Weather Rev.*, **108**, 1122–1143.

[26] Wegman, E. (1984). *J. Statist. Plann. Infer.*, **9**, 375–388.

[27] Wegman, E. and Wright, I. (1983). *J. Amer. Statist. Ass.*, **78**, 351–365.

[28] Weinert, H. (1978). *Commun. Statist. B*, **7**, 417–435.

[29] Weinert, H. (1982). *Reproducing Kernel Hilbert Spaces: Applications in Statistical Signal Processing*. Hutchinson-Ross Publishing, Stroudsburg, PA.

[30] Wright, I. and Wegman, E. (1980). *Ann. Statist.*, **8**, 1023–1035.

[31] Yao, K. (1967). *Inf. Control*, **11**, 429–444.

Acknowledgments

This paper was supported by the Naval Air Systems Command under Contract N00014-85-K-0202 with the George Washington University. We are grateful to Professor N. Singpurwalla for his assistance.

(KERNEL ESTIMATORS
SPLINE FUNCTIONS
STOCHASTIC PROCESSES)

EDWARD J. WEGMAN

REPRODUCTION RATES

A population is reckoned to reproduce itself if its numbers are maintained from generation to generation. If the number of births between time $t - x$ and time $t - x + \delta t$ is written as ${}^{t-x}B\,\delta t$, then the number of births to which they give rise in their turn x years later, between age x and age $x + \delta x$, may be written as

$${}^{t-x}B\,\delta t\left[{}^{t}_{x}p_0 . {}^{t}\phi_x\,\delta x\right],$$

where ${}^{t}_{x}p_0$ expresses the chance of survival from birth to age x and ${}^{t}\phi_x$ is the fertility rate at age x. The total number of births between times t and $t + \delta t$ is thus $\int_{x=0}^{\infty} {}^{t-x}B\,\delta t . {}^{t}_{x}\,p_0 . {}^{t}\phi_x\,dx$, but this also equals

${}^{t}B\,\delta t$. If now ${}_{x}p_0$ and ϕ_x are constant in time, and the annual geometrical rate of growth in the number of births is written as r, then

$$ {}^{t}B\,\delta t = \int_0^\infty {}^{t-x}B\,\delta t \cdot {}_{x}p_0 \cdot \phi_x\,dx $$

$$ = {}^{t}B\,\delta t \int_0^\infty e^{-rx} \cdot {}_{x}p_0 \cdot \phi_x\,dx, $$

and so

$$ 1 = \int_0^\infty e^{-rx} \cdot {}_{x}p_0 \cdot \phi_x\,dx. $$

This equation applies to one sex or the other. It has exactly one real solution, besides a number of complex roots in conjugate pairs. It can be shown that, as t increases, the behavior of ${}^{t}B$ is dominated by the real root, and the proportion of the sex in each age group tends towards a constant amount. Thus, for large t, the age distribution is constant, and it follows that so also is the rate of growth r. Such a population has been called a *stable* population.

The equation shown immediately above expresses the rate of population growth r in terms of mortality and fertility* rates. It can be solved approximately by expanding the exponential

$$ e^{-rx} = 1 - rx + r^2x^2/2! - \cdots $$

and thus arriving at

$$ 1 = \int_0^\infty {}_{x}p_0 \cdot \phi_x\,dx $$

$$ - \int_0^\infty r_x \cdot {}_{x}p_0 \cdot \phi_x\,dx + \cdots . $$

Because r is small, its second and higher powers may be ignored for an approximate answer. Writing the first two terms on the righthand side of the equation for convenience as m_1 and m_2 we have

$$ m_1 - rm_2 \doteqdot 1 $$

or

$$ m_1\{1 - (m_2/m_1)r\} \doteqdot 1. $$

Taking logarithms, this leads to $\log_e m_1 + \log_e\{1 - (m_2/m_1)r\} \doteqdot 0$. Expanding the logarithmic series and once again disregarding powers of r above the first, $\log_e m_1 - (m_2/m_1)r \doteqdot 0$. Hence

$$ m_1 \doteqdot e^{rm_2/m_1}. $$

So m_1 is a near measure of the population increase in time m_2/m_1. Now m_2/m_1 represents the average age at bearing a child, that is, the average length of a generation. Thus m_1 can be regarded as a measure of the rate of reproduction, i.e., $\int_0^\infty {}_{x}p_0 \cdot \phi_x\,dx$. Its value exceeds, equals, or falls short of unity as population is growing, steady, or falling, respectively. It represents the average number of offspring born, in a generation, of the same sex as the parent.

Both ${}_{x}p_0$ and ϕ_x differ in value between the sexes, and so reproduction rates are not the same for men and women. Attempts to find a mathematical solution common to both have not proved wholly satisfactory. Men marry, on the whole, later in life than women do, and so they are older when the children are born. Thus the male length of a generation is longer than the female. There may also be sex differences in the proportion marrying. Male reproduction rates exceed female rates, but it is not often possible to calculate them, as few national statistical systems classify births by age of father. Reproduction rates for women are, however, often available, and they are normally quoted in one of two forms: "net" rates are those assessed according to the formula $\int {}_{x}p_0 \cdot \phi_x\,dx$; "gross" reproduction rates disregard the element of mortality and thus represent $\int \phi_x\,dx$. They enable comparisons in time or between countries to be made, free from the complications of differences in survivorship. The following are some specimen figures:

Country and Period	Female Reproduction Rate	
	Net	Gross
Ghana, 1960	2.3	3.3
Ireland, 1968	1.83	1.91
France, 1931–35	0.90	1.06

The following table shows how gross reproduction rates for women have varied in some recent years.

	1950	1955	1960	1965	1970
Brazil		2.76		2.61	2.42
Finland	1.55	1.42	1.31	1.18	0.89
Luxemburg	0.90	1.01	1.13	1.19	0.97
Netherlands	1.50	1.48	1.51	1.47	1.26
Singapore	3.06	3.17	2.77	2.25	1.51
Tunisia			3.34	3.34	3.12

There is a considerable contrast in these data, not only in the size of the figures, but also in their trends. In some countries there has been a sharp downward movement; in others there has been only a slight fall, while in one the rates have risen and then fallen. The high rates are those experienced in the developing areas and are akin to those for European countries in past centuries. As high fertility is often a concomitant of high mortality, the net reproduction rates corresponding to the above would show a narrower range of variation.

Differences in reproduction rates can, by a simple algebraic process, be analysed into their component parts where the data are available. However, demographers do not now make very much use of such rates. They represent little more than a convenient index of current fertility and are not sufficiently stable or illustrative of underlying trends to have predictive value. Considerable efforts were made at one time to improve them by processes of standardization, for instance in respect to marriage and legitimacy, or by the use of fertility rates analysed by marriage duration. The variety of answers obtained demonstrated that nothing steady or fundamental was being produced to give a reliable guide to the future.

From a historical point of view, however, it is of value to measure, wherever data are available for women born in each period of time, the number of female births to which they gave rise in due course at each adult age; in other words to make a study of replacement in cohort or generation form. The following are some estimated generation *replacement rates* for England and Wales:

Year of Birth	Replacement Rate
1848–53	1.36
1868–73	1.09
1888–93	0.81
1908–13	0.70
1928–33	0.96
1948–53	1.04

These rates allow for the effects of mortality, falls in which, with the passage of time, assisted in the rise in the figures from their low point of 0.70.

Bibliography

Henry, L. (1976). *Population, Analysis and Models.* Edward Arnold, London, England. (Translated from the French, it illustrates the calculation of reproduction rates.)

Keyfitz, N. and Smith, D. (1977). *Mathematical Demography*. Springer-Verlag, Berlin, Germany. (Contains reprints of original papers on stable population theory and on the relationships between male and female reproduction rates.)

Pollard, J. H. (1973). *Mathematical Models for the Growth of Human Populations*. Cambridge University Press, London, England. (Includes a rigorous analysis of the deterministic population models of Lotka.)

Royal Commission on Population (1950). *Reports and Selected Papers of the Statistics Committee*. Her Majesty's Stationery Office, London, England. (Discusses the measurement of reproductivity and attempts to improve reproduction rates.)

(BIRTH-AND-DEATH PROCESSES
DEMOGRAPHY
FERTILITY MEASUREMENT
LIFE TABLES
MARRIAGE
POPULATION GROWTH MODELS
SURVIVAL ANALYSIS
VITAL STATISTICS)

P. R. Cox

REPRODUCTIVE MODELS

Let $\mathscr{D} = \{D(\omega) : \omega \in \Omega\}$, where $\Omega \subset R^d$, be a parametric family of probability distributions on a sample space $\mathscr{X}$ and let $\mathbf{s}$ be a statistic on $\mathscr{X}$. We write $\mathbf{s} \sim D(\omega)$ if $\mathbf{s}$ is distributed according to $D(\omega)$, and we let $\mathbf{s}_1, \mathbf{s}_2, \ldots, \mathbf{s}_n$ denote independent and identically distributed copies of $\mathbf{s}$.

The pair $(\mathscr{D}, \mathbf{s})$ is said to be *reproductive* in ω if for all $\omega \in \Omega$ and all $n = 1, 2, \cdots$ we have

(i) $n\Omega \subset \Omega$

and

(ii) If $\mathbf{s} \sim D(\omega)$ then $\bar{\mathbf{s}} = n^{-1}(\mathbf{s}_1 + \cdots + \mathbf{s}_n) \sim D(n\omega)$.

When it appears from the context which $\mathscr{D}$ or $\mathbf{s}$ is at issue we simply speak of $\mathbf{s}$ or $\mathscr{D}$ as being reproductive.

Note first, that if the vector $\mathbf{s}$ is partitioned into components $\mathbf{u}$ and $\mathbf{v}$, i.e., $\mathbf{s} = (\mathbf{u}, \mathbf{v})$, and if $\mathbf{s}$ is reproductive in ω, then $\mathbf{u}$ and $\mathbf{v}$ are both reproductive in ω; and second, that if $\kappa(\zeta; \omega)$ denotes the cumulant generating function* for $\mathbf{s}$, corresponding to the distribution $D(\omega)$ and evaluated at ζ, then condition (ii) is equivalent to

$$n\kappa(\zeta; \omega) = \kappa(n\zeta; n\omega), \qquad (1)$$

i.e., κ considered as a function of both ζ and ω has a homogeneity property of order 1.

Example. The Wishart distribution* with arbitrary scalar "degrees of freedom parameter" λ has cumulant generating function

$$\kappa(\zeta; \lambda, \Sigma) = -(\lambda/2)\ln|\mathbf{I} - 2i\zeta\Sigma^{-1}|,$$

where ζ is a symmetric $r \times r$ matrix. Thus, for fixed r, the class of these distributions is reproductive in (λ, Σ).

Example. Let $\mathscr{D}$ be the scale parameter family generated by any of the one-dimensional stable laws* with cumulant generating function

$$\kappa(\zeta) = -|\zeta|^{\alpha}\{1 + i\beta\zeta/|\zeta|\tan(\pi\alpha/2)\},$$

where $\alpha \in (0, 2)$, $\alpha \neq 1$, and $|\beta| \leqslant 1$, α and β being considered as known. Writing the scale parameter as $\omega^{\alpha-1}$ we have

$$\kappa(\zeta; \omega) = -\omega^{1-\alpha}|\zeta|^{\alpha}\{1 + i\beta\zeta/|\zeta|\tan(\pi\alpha/2)\}$$

and hence $\mathscr{D}$ is reproductive in ω (but note that $\mathscr{D}$ is not reproductive in the scale parameter itself).

The Cauchy distribution*, which corresponds to $\alpha = 1$ and $\beta = 0$, is peculiar in constituting a reproductive model on its own.

The exponential family* generated by $(D(\omega), s)$ has a cumulant generating function of the form

$$\kappa(\zeta; \omega, \boldsymbol{\theta}) = \kappa(\zeta + \theta; \omega) - \kappa(\boldsymbol{\theta}; \omega), \quad (2)$$

where $\boldsymbol{\theta}$ denotes the canonical parameter of the exponential family. Denoting the distribution given by (2) as $D(\omega, \boldsymbol{\theta})$ we see from (1) that the class of all the distributions $D(\omega, \boldsymbol{\theta})$ will be reproductive in $(\omega, \boldsymbol{\theta})$, provided the domain for $(\omega, \boldsymbol{\theta})$ contains $n(\omega, \theta) = (n\omega, n\boldsymbol{\theta})$ for any $n = 1, 2, \ldots$ and any $(\omega, \boldsymbol{\theta})$ of the domain.

Of particular interest are those cases where $\mathbf{s} = \mathbf{x}$, the identity mapping on the sample space $\mathscr{X}$, and where $\mathscr{D}$ is a (steep) exponential model of the form

$$\exp\{\chi \cdot \mathbf{H}(\mathbf{x}) + \psi \cdot \mathbf{x} - \kappa(\chi, \psi) - d(\mathbf{x})\}, \qquad (3)$$

the reproductivity parameter being the canonical parameter $\boldsymbol{\theta} = (\chi, \psi)$. The discussion in the sequel is based mainly on Barndorff-Nielsen and Blæsild [3, 4]; cf. also Bar-Lev and Reiser [2].

The two-parameter normal, gamma, and inverse Gaussian models are of this type. For instance, the inverse Gaussian distribution* has probability density function

$$\varphi^{-}(x; \chi, \psi) = \frac{\sqrt{\chi}}{\sqrt{2\pi}} e^{\sqrt{\chi\psi}} x^{-3/2} e^{-\{\chi x^{-1} + \psi x\}/2}$$
$$(x > 0,\ \chi > 0,\ \psi \geqslant 0) \qquad (4)$$

and x is reproductive in (χ, ψ). Here $H(x)$ of (3) equals x^{-1}, while in the case of the normal model and the gamma model we have that $H(x)$ equals x^2 and $\ln x$, respectively. Further examples may be constructed by suitable combination of these, and in discussing the more complicated combinations, graph theoretic concepts are helpful; cf. Barndorff-Nielsen and Blæsild [3–5]. Here we just present a single instance of such a combination.

Example. Let u_1, u_2, u_3, and u_4 be random variables with joint distribution

$$\begin{aligned}&\varphi^-(u_1;\chi_1,\psi_1)\varphi^-(u_2;u_1^2\chi_2,\psi_2)\\&\quad\times\varphi^-\left(u_3;(u_1+u_2)^2\chi_3,\psi_3\right)\\&\quad\times\varphi\left(u_4;(u_1+u_2)\xi,(u_1+u_2)\sigma^2\right),\end{aligned}$$

where φ^- is given by (4) while φ denotes the normal probability density function. By construction, the conditional distribution of u_2 given u_1 is inverse Gaussian, etc. The class $\mathscr{D}$ of distributions of (u_1,u_2,u_3,u_4) obtained by letting the parameters χ_i, ψ_i $(i=1,2,3)$, ξ, and σ vary freely is then reproductive of type (3) with $x=(u_l,\ldots,u_4)$ and $H(x)=(u_1^{-1},u_1^2u_2^{-1},(u_1+u_2)^2u_3^{-1},(u_1+u_2)^{-1}u_4)$.

In particular, the marginal distribution of (u_1,u_2) is reproductive in $(\chi_1,\psi_1,\chi_2,\psi_2)$ and if $(u_{11},u_{12}),\ldots,(u_{n1},u_{n2})$ is a random sample from this distribution then $\bar{u}=(\bar{u}_1,\bar{u}_2)=n^{-1}(u_{11}+\cdots+u_{n1},u_{12}+\cdots+u_{n2})$ follows the same type of distribution but with parameter $(n\chi_1,n\psi_1,n\chi_2,n\psi_2)$. Furthermore, it follows from general results mentioned below that if we define r and s by

$$\begin{aligned}r&=n^{-1}\sum u_{i1}^{-1}-\bar{u}_1^{-1},\\s&=n^{-1}\sum u_{i1}^2u_{i2}^{-1}-\bar{u}_1^2\bar{u}_2^{-1},\end{aligned}$$

then $\bar{u}$, r, and s are independent and

$$\begin{aligned}r&\sim\Gamma((n-1)/2,n\chi_1/2),\\s&\sim\Gamma((n-1)/2,n\chi_2/2),\end{aligned}$$

where Γ indicates the gamma distribution*. Thus the marginal model for (u_1,u_2) is a two-dimensional analog of the inverse Gaussian model.

On the assumption that the vector-valued function $\mathbf{H}$ in (3) is continuous, $(\mathscr{D},\mathbf{x})$ is reproductive in $\boldsymbol{\theta}=(\chi,\psi)$ if and only if $c\,\text{int}\,\Theta\subset\text{int}\,\Theta$, where Θ is the domain of $\boldsymbol{\theta}$ and ψ is of the form

$$\psi=-\chi\mathbf{h}(\xi),\qquad(5)$$

ξ denoting the mean value of $\mathbf{x}$. In fact, $\mathbf{h}(\xi)=\partial\mathbf{H}(\xi)^*/\partial\xi$, where * indicates matrix transposition. Furthermore, in this case, the mean value η of $\mathbf{H}(\mathbf{x})$ is of the form

$$\eta=\mathbf{m}(\chi)+\mathbf{H}(\xi)\qquad(6)$$

for some one-to-one function $\mathbf{m}$.

Of more direct statistical relevance are the following consequences of the reproductivity of (3). Let

$$\mathbf{p}(\mathbf{x})=\mathbf{H}(\mathbf{x})-\mathbf{x}\mathbf{h}(\xi)^*+\check{\mathbf{H}}(\xi),$$

where $\check{\mathbf{H}}(\xi)=\xi\mathbf{h}(\xi)^*-\mathbf{H}(\xi)$ is the Legendre transform of $\mathbf{H}(\xi)$; set $\overline{\mathbf{H}}=n^{-1}(\mathbf{H}(\mathbf{x}_1)+\cdots+\mathbf{H}(\mathbf{x}_n))$, $\bar{\mathbf{p}}=n^{-1}(\mathbf{p}(\mathbf{x}_1)+\cdots+\mathbf{p}(\mathbf{x}_n))$, and

$$\begin{aligned}\acute{\mathbf{q}}&=\mathbf{p}(\bar{\mathbf{x}})=\mathbf{H}(\bar{\mathbf{x}})-\bar{\mathbf{x}}\mathbf{h}(\xi)^*+\check{\mathbf{H}}(\xi),\\\acute{\mathbf{w}}&=\overline{\mathbf{H}}-\mathbf{H}(\bar{\mathbf{x}})=\bar{\mathbf{p}}-\mathbf{p}(\bar{\mathbf{x}}).\end{aligned}$$

The model $\mathscr{D}$ of (3) may be parametrised by the mixed parameter (χ,ξ). Assuming that the maximum likelihood* estimate $(\hat{\chi},\hat{\xi})$ of (χ,ξ) exists with probability 1, we have

(i) $\hat{\chi}$ and $\hat{\xi}$ are independent.

(ii) In the identity

$$\bar{\mathbf{p}}=\acute{\mathbf{w}}+\acute{\mathbf{q}},\qquad(7),$$

the quantities $\acute{\mathbf{w}}$ and $\acute{\mathbf{q}}$ are independent.

(iii) The statistic $\acute{\mathbf{w}}$ follows a linear exponential model with canonical parameter $n\chi$. Thus, in particular, the distribution of $\acute{\mathbf{w}}$ depends on $\boldsymbol{\theta}$ through χ only.

(iv) For fixed ξ, the quantities $\bar{\mathbf{p}}$ and $\acute{\mathbf{q}}$ each follow a linear exponential model with canonical parameter $n\chi$.

(v) In terms of cumulant generating functions, (7) takes the form

$$\begin{aligned}&\exp\left(n\{\mathbf{M}(\chi+n^{-1}\zeta)-\mathbf{M}(\chi)\}\right)\\&\quad=\exp\left(n\{\mathbf{M}(\chi+n^{-1}\zeta)-\mathbf{M}(\chi)\}\right.\\&\qquad\left.-\{\mathbf{M}(n\chi+\zeta)-\mathbf{M}(n\chi)\}\right)\\&\quad\times\exp(\mathbf{M}(n\chi+\zeta)-\mathbf{M}(n\chi)),\end{aligned}$$

where ζ is the argument of the transform and $\mathbf{M}$ is an indefinite integral of $\mathbf{m}$ in (6).

The decomposition result embodied in (7) constitutes a generalisation of the standard decomposition theorem for χ^2-distributed quadratic forms* in normal variates, and the independence and distributional statements in **(i)**–**(v)** include well known and important properties of the normal, gamma, and inverse Gaussian distributions.

Various properties, somewhat weaker than reproductivity, each suffices for some of the preceding conclusions. One such weaker property is that

$$\bar{\mathbf{x}} \text{ follows an exponential model with canonical parameter } n\boldsymbol{\theta} = (n\chi, n\psi) \text{ and with corresponding canonical statistic } (\mathbf{H}(\bar{\mathbf{x}}), \bar{\mathbf{x}}), \tag{8}$$

and it implies, in particular, that $\overline{\mathbf{H}} - \mathbf{H}(\bar{\mathbf{x}})$ and $\bar{\mathbf{x}}$ are independent (in fact, it is equivalent to this; see Barndorff-Nielsen and Blæsild [3, 4] and Bar-Lev [1]). It also implies, and is conceivably equivalent to, η being of the form (6).

The latter relation is of interest from the viewpoint of the geometry of exponential models and the connections between geometrical and statistical properties. Thus (6) is equivalent to ψ being of the form (compare with (5))

$$\psi = -\chi\mathbf{h}(\xi) + \mathbf{k}(\xi), \tag{9}$$

which means that the mean-affine submodels of (3) corresponding to fixed values of ξ are also affine in the canonical parameter space. Furthermore, (9) is equivalent to nonrandomness of the observed profile information function for χ. If $\hat{\chi}$ and $\hat{\xi}$ are independent then (9) holds and the converse is true under the condition (8).

The dual versions of (6) and (9), i.e., respectively,

$$\psi = \mathbf{H}(\chi) + \mathbf{m}(\xi) \tag{10}$$

and

$$\eta = -\xi\mathbf{h}(\chi) + \mathbf{k}(\chi), \tag{11}$$

are each equivalent to $\mathbf{x}$ being a cut, whence $\mathbf{x}$ is S-sufficient for ξ and S-ancillary with respect to χ. (For a detailed discussion of models satisfying (6) and (9) or (10) and (11) see Barndorff-Nielsen and Blæsild [3].)

Finally, it should be noted that for exponential models there is a difference of terminology between the present article and Barndorff-Nielsen and Blæsild [4]. In the latter paper the term strong reproductivity was used for the property discussed above, while reproductivity referred to models satisfying (8) for all $n = 1, 2, \ldots$.

References

[1] Bar-Lev, S. K. (1983). *Ann Statist.*, **11**, 746–752.

[2] Bar-Lev, S. K. and Reiser, B. (1982). *Ann Statist.*, **10**, 979–989.

[3] Barndorff-Nielsen, O. E. and Blæsild, P. (1983). *Ann Statist.*, **11**, 753–769.

[4] Barndorff-Nielsen, O. E. and Blæsild, P. (1983). *Ann Statist.*, **11**, 770–782.

[5] Barndorff-Nielsen, O. E. and Blæsild, P. (1987). *Ann. Statist.*, **15**.

(CONVOLUTION
EXPONENTIAL FAMILY
NATURAL EXPONENTIAL FAMILY)

O. E. BARNDORFF-NIELSEN
P. BLÆSILD

REPRODUCTIVE PROPERTIES OF DISTRIBUTIONS

A linear function of *independent* normal* random variables again has a normal distribution. Thus the normal distribution "reproduces" itself through a linear function. This is an example of a reproductive property. (Under certain conditions this property characterizes the normal distribution law. *See* NORMAL DISTRIBUTION.) Several other important distributions reproduce themselves through summation. In particular:

(a) If X_i $(i = 1, \ldots, k)$ are independent binomial* random variables with parameter n_i and p (the same p for all X_i, $i = 1, \ldots, k$), then ΣX_i also has a binomial distribution, with parameters Σn_i and p.

(b) If X_i $(i = 1, \ldots, k)$ are independent negative binomial* random variables each with parameters r_i and p then ΣX_i is a negative binomial random variable with parameters Σr_i and p.

(c) If X_i $(i = 1, \ldots, k)$ are independent Poisson* random variables with scale parameters λ_i, then ΣX_i is also a Poisson variable, with parameters $\Sigma \lambda_i$.

(d) If X_i $(i = 1, \ldots, k)$ are independent gamma* random variables with scale parameters α_i and the same shape parameter β, then ΣX_i is also a gamma random variable with parameters $\Sigma \alpha_i$ and β. [In particular a sum of independent χ^2 (chi-squared) random variables is again a χ^2 (chi-squared) random variable (with degrees of freedom of the sum equal to the sum of degrees of freedom of the summand). This property of the chi-squared distribution is of special importance in statistical inference.]

(BINOMIAL DISTRIBUTION
CHI-SQUARE DISTRIBUTION
F DISTRIBUTION
GAMMA DISTRIBUTION
NEGATIVE BINOMIAL DISTRIBUTION
NORMAL DISTRIBUTION
POISSON DISTRIBUTION)

RERANDOMIZATION TESTS *See* RANDOMIZATION TESTS

RESAMPLING PROCEDURES (INCLUDING BOOTSTRAPPING)

Many statistical techniques are derived by assuming that the sample has a specified distributional form. When such assumptions hold, the appropriate technique should be used. However, data from exploratory situations, sampling, or highly nonlinear processes may not be known to fit any standard asumptions. Statistics may be biased. There will be no reference values for testing hypotheses, no tabled values for obtaining confidence intervals, and no sufficient statistics to give concise summaries of the data. The data must be allowed to "speak for itself" in order to tell more about the underlying population and the statistic of interest. This is done by recomputing the statistic many times with reweighted sample values.

Resampling techniques have been used in testing and in estimation. Randomization tests* were introduced more than a half-century ago. However, computational complexity limited their use to the smallest samples. The jackknife* was introduced more than a quarter-century ago as a bias reduction technique. Later, the jackknife and other subsampling techniques, such as half-sampling*, were used to obtain variance estimates and confidence intervals. The explosive growth of computing power has allowed statistics with no closed distributional forms or variance expressions to be analyzed using resampling methods such as the bootstrap* and its recently introduced variations.

TESTS OF HYPOTHESES

Randomization or permutation tests* date back to the 1930s and have been described by Kempthorne ([19] and elsewhere) as truly nonparametric tests. In his book *Randomization Tests* [9], Edgington points out that *however* the data were obtained, the inference is valid for the given sample. He gives an excellent survey and further comments about inference and historical origins in RANDOMIZATION THEORY.

A very brief summary of the procedure is this: To test whether a term belongs in a statistical model, the observations are permuted with respect to that term, and the test statistic is computed for all possible configurations of the data. This gives an empirical distribution of the test statistic values. The actual value is compared with these, and the null hypothesis is rejected if the actual value has small probability of being exceeded. Computations are massive if the sample is large. Thus, large data sets are usually handled by asymptotic considerations or sampling.

Gabriel [14] and coauthors have written a series of papers on weather modification* using a closely related technique he calls *rerandomization* theory. Here the treatments must be randomly assigned to the experimental units, thus broadening the scope of the inference.

Puri and Sen [23] give insight into underlying theoretical foundations.

JACKKNIFING, BOOTSTRAPPING*, AND SUBSAMPLING METHODS

Jackknifing was introduced in the mid-1950s by Quenouille as a bias reduction technique and further investigated by Tukey. The most used version of the jackknife consists of recomputing the sample statistic n times with a different sample value deleted each time.

Much interest in resampling theory in the 1960s and 1970s was in the field of sampling. The emphasis was on estimation of variance. In JACKKNIFE METHODS, Hinkley describes a fraction-sampling approach to variance estimation and suggests that those interested in further reading consult Kish and Frankel [20] and Krewski and Rao [21]. Efron, in *The Jackknife, the Bootstrap and Other Resampling Plans* [11] mentions work by McCarthy [22], Hartigan ([16] and later), and others.

The term "bootstrapping" was introduced by Efron [10] in the late 1970s. He has said that he wanted the word to be "euphonious with jackknifing." While the previous methods (except for permutation tests*) reduce the size of the sample in each of the recomputations of the statistic, the bootstrap sample may be of any size. This is accomplished by resampling from the data values, residuals, or other functions of the data *with replacement* and then recomputing the original statistic. The recomputed values are used to estimate empirical distributions, confidence intervals, variances, etc. *See* JACKKNIFE METHODS for an example. Efron also presents some simulation studies comparing subsampling techniques with the jackknife and the bootstrap. The results are mixed, but bias and relatively large variances are evident for half-sampling and random subsampling methods. The quantity of research in this area in recent years following Efron [10] has been growing, as indicated, for example, in the *Current Index to Statistics* [15].

FURTHER RESULTS ON BOOTSTRAPPING

Bickel and Freedman [3, 4] have written a series of theoretical papers on the bootstrap. For certain statistics they have shown that the variance of the bootstrap realizations is a consistent estimator of the variance of the bootstrapped statistic. They have obtained laws of large numbers and central limit theorems* for means, U-statistics*, stratified simple random sample, "t" statistics, etc. (Here a "t" statistic is the quotient of a bootstrap value minus the realized sample value and the standard deviation of the bootstrapped statistic.)

Babu and Singh [2] show that the bootstrapped t statistic converges a.s. to the correct distribution assuming finiteness of the sixth moment. Hinkley and Wei [17] give an example where a bootstrapped t statistic of a ratio estimator is preferable to jackknifed values.

A novel application resulted in reduced computation costs for Moore [7] and coauthors in a series of papers. In their double Monte Carlo technique they simulated a large sample, and then drew bootstrap samples from it for subsequent Monte Carlo trials.

Robinson [26] gives an illuminating example.

Some "popular"-level treatises of bootstrapping are Diaconis and Efron [8] and Efron and Gong [12]. Rey [25] lends support to Efron's conjecture that the empirical distribution of the bootstrapped statistic is a discrete approximation to the true distribution.

Wu [28] has conducted an investigation of quadratic regression with homoscedastic and heteroscedastic errors. He gives a general representation for resampling in a regression model. He also compares coverage probabilities and lengths of empirical confidence intervals for a nonlinear parameter $\Theta = -\beta_1/(2\beta_2)$ that maximizes the quadratic function $f(x) = \beta_0 + \beta_1 x + \beta_2 x^2$ over x. The model is $y_i = f(x_i) = e_i$, $i = 1(1)12$, $x_i = 1(0.5)4(1)10$, where $e_i \sim N(0, 1)$ or $N(0, 1)\sqrt{x_i}/2$. He used 3000 simulation samples with $\beta_0 = 0$, $\beta_1 = 4$, and $\beta_2 = -0.25$, -0.35, -0.5, and -1.0. He investigated nine interval estimates, which included Fieller's (*see* FIELLER'S THEOREM), five jackknife and two bootstrap variations, and the least-squares estimator with a linearization procedure for variance estimation. For all methods, bias was negligible except for the unequal variance, $\beta_2 = -0.25$ case.

For the bootstrap, Wu fits the least-squares* estimator of β, computes the residuals, and draws e_i^* randomly with replacement from the residuals divided by the proper scaling factor, $\sqrt{1 - k/n}$ (here $k = 3$ and $n = 12$). Then $y_i^* = f(x_i) = e_i^*$, and the least-squares estimator is recomputed to obtain the bootstrapped estimator. Table 1 gives the coverage probabilities and (median lengths) of five of the estimators: Fieller's, a percentile method, and $\Theta_i \pm t_\alpha\sqrt{V\Theta_i}$, where $V\Theta_i$ is the appropriate variance estimator as described by Wu. In the heteroscedastic case, Fieller's method gave intervals of infinite length in 199 and 7 cases for $\beta_2 = -0.25$ and -0.35, respectively, and these are separated from the finite length case.

RESAMPLING METHODS WITH U-STATISTICS

Reduced computational effort for *U*-statistics* first defined by Hoeffding [28] in the 1940s was the motivation of resampling-related methods introduced in the mid-1970s. Brown and Kildea's reduced *U*-statistics [6] are analogous to McCarthy's balanced half-samples [22]. Blom's incomplete *U*-statistics

Table 1 Average Coverage Probabilities and (Median Lengths) for Four Interval Estimation Methods. Nominal level = 0.95, $\beta_0 = 0$, $\beta_1 = 4$. (3,000 Simulation Samples[a].)

	Unequal Variances β_2				Equal Variances	
Method	-0.25	-0.35	-0.5	-1	-0.25	-1
Fieller	0.858 (∞, 3.81)	0.866 (∞, 1.10)	0.968 (0.98)	0.952 (0.92)	0.947 (2.48)	0.950 (0.64)
Jackknife	0.887 (29.08, 3.87)	0.848 (8.92, 1.04)	0.961 (0.91)	0.950 (0.89)	0.904 (2.03)	0.935 (0.62)
Linearization	0.865 (14.75, 2.91)	0.891 (5.82, 1.02)	0.969 (0.93)	0.952 (0.90)	0.949 (2.18)	0.948 (0.64)
Bootstrap	0.866 (313.17, 3.73)	0.902 (86.63, 1.07)	0.973 (0.97)	0.955 (0.91)	0.956 (2.42)	0.946 (0.64)
Bootstrap percentile	0.829 (55.05, 3.07)	0.814 (17.78, 0.93)	0.940 (0.84)	0.921 (0.79)	0.912 (2.05)	0.916 (0.56)

[a] For unequal variance, the $n = 3000$ trials are presented for $\beta_2 = -0.25$ as ($n = 199$, $n = 2801$) and for $\beta_2 = -0.35$ as ($n = 7$, $n = 2993$) trials. There were no occurrences of the interval $(-\infty, \infty)$.

Source. Wu [28].

[5] are also closely related to some of the previous methods. The Blom [5] and Brown and Kildea [6] efforts were directed to using only some of the subsamples in point estimation, but the techniques could be used in variance estimation.

Enqvist [13] obtained asymptotic results for incomplete *U*-statistics. Bickel and Freedman [3] obtained some asymptotic results for bootstrapped *U*-statistics. Athreya, et al. [1] obtained further results, including a martingale* representation for bootstrapping, and applied it to bootstrapped *U*-statistics.

References

[1] Athreya, K. B., Ghosh, M., Low, L. Y., and Sen, P. K. (1984). *J. Statist. Plann. Inf.*, **9**, 185–194. [Obtains a strong law for the bootstrap assuming a $(1 + \delta)$ moment.]

[2] Babu, C. J. and Singh, K. (1983). *Ann. Statist.*, **11**, 999–1003.

[3] Bickel, P. J. and Freedman, D. A. (1981). *Ann. Statist.*, **9**, 1196–1217.

[4] Bickel, P. J. and Freedman, D. A. (1984). *Ann. Statist.*, **12**, 470–482.

[5] Blom, G. (1976). *Biometrika*, **63**, 573–580.

[6] Brown, B. M. and Kildea, D. G. (1978). *Ann. Statist.*, **6**, 828–835.

[7] Depuy, K. M., Hobbs, J. R., Moore, A. H., and Johnston, J. W., Jr. (1982). *IEEE Trans. Rel.*, **R-31**, 474–477.

[8] Diaconis, P. and Efron, B. (1983). *Scientific American*, May, 116–130.

[9] Edgington, E. S. (1980). *Randomization Tests*. Marcel Dekker, New York.

[10] Efron, B. (1979). *Ann. Statist.*, **7**, 1–26.

[11] Efron, B. (1982). *The Jackknife, the Bootstrap, and Other Resampling Plans*. CBMS Monograph No. 38, SIAM, Philadelphia. (This discusses and compares several resampling techniques.)

[12] Efron, B., and Gong, G. (1983). *Amer. Statist.*, **38**, 36–48.

[13] Enqvist, E. (1978). On sampling from sets of random variables with applications to incomplete *U*-statistics. *Dissertation*, Lund Univ., Sweden.

[14] Gabriel, K. R. and Hsu, C. F. (1983). *J. Amer. Statist. Ass.*,**78**, 766–775.

[15] Gentle, J. E. (1982). *Current Index Statist*, **8**.

[16] Hartigan (1969). *J. Amer. Statist. Ass.*, **64**, 1303–1317.

[17] Hinkley, D. V. and Wei, B. C. (1984). *Biometrika*, **71**, 331–339.

[18] Hoeffding, W. (1948). *An. Math. Statist.*, **19**, 293–325.

[19] Kempthorne, O. (1955). *J. Amer. Statist. Ass.*, **50**, 946–967.

[20] Kish, L. and Frankel, M. R. (1974). *J. R. Statist. Soc. B*, **36**, 1–37. (An empirical study of resampling methods. They conclude that half-sampling provides more reliable confidence intervals of the methods studied.)

[21] Krewski, D. and Rao, J. N. K. (1981). *Ann. Statist.*, **9**, 1010–1019.

[22] McCarthy, P. J. (1969). *Rev. Int. Statist. Inst.*, **37**, 239–264. (This describes half-sampling methods.)

[23] Puri, M. L. and Sen, P. K. (1971). *Nonparametric Methods in Multivariate Analysis*, Wiley, New York.

[24] Rao, J. N. K. and Wu, C. F. J. (1984). Bootstrap inference with stratified samples. In *Proc. Survey Res. Methods Sect., American Statistical Association* (to appear).

[25] Rey, W. J. J. (1983). *Introduction to Robust and Quasi-Robust Statistical Methods*, Springer-Verlag, New York.

[26] Robinson, J. A. (1983). *Technometrics*, **25**, 179–187. (The bootstrap approximation is used to estimate the distribution of the pivot when censoring is progressive in a location-scale model.)

[27] Rubin, D. B. (1981). *Ann. Statist.*, **9**, 130–134. (Suggests caution in the use of the bootstrap and the Bayesian bootstrap, which he applies only when all distinct values of the population appear in the sample.)

[28] Wu, C. F. J. (1984). Jackknife and Bootstrap Inference in Regression and a Class of Representations for the L.S.E., *MRC Tech. Summary Rep. No. 2675*, Univ. of Wisconsin, Madison, WI.

(BOOTSTRAPPING
FIELLER'S THEOREM
HALF-SAMPLES
JACKKNIFE METHODS
MARTINGALES
MONTE CARLO METHODS
RANDOMIZATION TESTS
REGRESSION (VARIOUS ENTRIES)
RERANDOMIZATION TESTS
SAMPLE REUSE
U-STATISTICS
WEATHER MODIFICATION)

LEONE Y. LOW

RESEARCH HYPOTHESIS

This is an alternative term for "alternative hypothesis*." It is often used in social sciences, where it represents a hypothesis in which the researcher is interested.

RESIDUAL CHANGE

This is another name for deviation from a value predicted by a regression formula. If (x_1, x_2) are the observed values of random variables (X_1, X_2) corresponding to values of a character X on two successive occasions, the *residual change* in X between the first and second occasions is

$$x_2 - E[X_2|X_1 = x_1].$$

Use of the residual change as a measure of change is intended to eliminate the effect of regression to the mean*.

The regression function is usually taken to be linear. It should be noted that if the linear regression function is fitted by least squares* to data, the sum of residual changes for all individuals contributing to the data must be zero. In such cases, therefore, one is really comparing each individual with all other individuals contributing to that particular data set.

The method has been used in psychometry [2] and more recently in epidemiology [1].

References

[1] Glynn, R. J., Rosner, B., and Silbert, J. E. (1982). *Circulation*, **66**, 724–731.

[2] Lord, F. M. (1966). In *Problems in Measuring Change*, 2nd ed., C. W. Harris, ed. University of Wisconsin, Madison, WI.

(REGRESSION TO THE MEAN)

RESIDUAL WAITING TIME *See* WAITING TIMES

RESIDUALS

Residuals contain important information on the often idealized assumptions that underlie statistical analyses, and thereby form a common foundation for many diagnostic methods that are designed to detect disagreement between the observed data and the postulated model. There is generally one residual for each observation on the random response variable.

Residuals have been studied most extensively in the context of the standard linear model,

$$\mathbf{Y} = \mathbf{X}\beta + \epsilon, \tag{1}$$

in combination with the tentative assumption that the $n \times 1$ vector of unobservable random errors ϵ follows a multivariate normal distribution with mean vector $\mathbf{0}$ and variance–covariance matrix $\sigma^2\mathbf{I}$. Some important implications of this assumption are (a) the model displays the correct form of the mean of the observable random vector of responses $\mathbf{Y}$ (expectation $(\mathbf{Y}) = E(\mathbf{Y}) = \mathbf{X}\beta$), (b) the individual observations y_i have constant variance σ^2, and (c) the individual errors ϵ_i are independent identically distributed normal random variables. Analyses based on such assumptions can be misleading and inefficient if the assumptions are not in fact appropriate. Residuals are used to identify corresponding model inadequacies such as curvature ($E(\mathbf{Y}) \neq \mathbf{X}\beta$), outliers*, heteroscedasticity*, nonnormality, and serial correlation*, and can also be used to guide further analysis.

Both graphical and nongraphical residual-based diagnostic methods are available to aid in analyses founded on model (1). Nongraphical methods usually involve a formal statistical comparison of model (1) with a larger alternative model in which a selected assumption is relaxed. Examples included the well-known Durbin–Watson [12] test* for serial correlation* and the more recent tests for heteroscedasticity proposed by Cook and Weisberg [9]. Such tests are often sensitive to several alternatives and should be used with care. Graphical counterparts of formal methods provide a degree of robustness by allowing the investigator to distinguish between several alternatives.

Anscombe [2] gives an excellent introduction to residual-based graphical methods.

Several types of residuals have been investigated in connection with various diagnostic methods. The *ordinary residuals* e_i, $i = 1, 2, \ldots, n$, are the easiest to construct and are defined informally as $e_i = i$th observed value y_i minus ith fitted value $\hat{y}_i$. When ordinary least squares is used to estimate the $p \times 1$ vector of unknown parameters $\boldsymbol{\beta}$, the $n \times 1$ vector of ordinary residuals $\mathbf{e}$ is defined as

$$\mathbf{e} = \mathbf{Y} - \hat{\mathbf{Y}}, \tag{2}$$

where $\hat{\mathbf{Y}} = \mathbf{HY}$ is the $n \times 1$ vector of fitted values and $\mathbf{H} = \mathbf{X}(\mathbf{X}^T\mathbf{X})^{-}\mathbf{X}^T$ is the orthogonal projection operator for the column space of $\mathbf{X}$. In the remainder of this discussion, $\mathbf{e}$ will be defined as in (2).

If, as displayed in (1), $E(\mathbf{Y}) = \mathbf{X}\boldsymbol{\beta}$, then $E(\mathbf{e}) = \mathbf{0}$; otherwise $E(\mathbf{e}) \neq \mathbf{0}$. For example, if the postulated model is missing a single independent variable Z then $E(\mathbf{Y}) = \mathbf{X}\boldsymbol{\beta} + \mathbf{Z}\boldsymbol{\gamma}$ and $E(\mathbf{e}) = (\mathbf{I} - \mathbf{H})\mathbf{Z}\boldsymbol{\gamma}$. A useful implication of this result, which illustrates the motivation behind the use of residuals in diagnostic methods, is that certain systematic patterns in plots of $\mathbf{e}$ vs. $\mathbf{Z}$ or functions of $\mathbf{Z}$ indicate a model inadequacy, namely $E(\mathbf{Y}) \neq \mathbf{X}\boldsymbol{\beta}$. In routine analyses useful diagnostic information can be obtained by setting $\mathbf{Z}$ equal to $\hat{\mathbf{Y}}$ or a selected independent variable (a column of the known $n \times p$ matrix $\mathbf{X}$); different model inadequacies are indicated by different systematic patterns in the corresponding plots of $\mathbf{e}$ vs. $\mathbf{Z}$. Such patterns are nicely illustrated by Weisberg [21].

An added variable plot of $\mathbf{e}$ vs. $(\mathbf{I} - \mathbf{H})\mathbf{Z}$, the ordinary residuals from the regression of $\mathbf{Z}$ on $\mathbf{X}$, can be used to assess the appropriateness of adding the new independent variable $\mathbf{Z}$ to the current model (1). Added variable plots can also be used as diagnostics for assessing the need to transform $\mathbf{Y}$ or a selected independent variable. The various uses of these and related plots are discussed by Atkinson [4, 5], Cook and Weisberg [8], and Belsley et al. [6], who call them partial-regression leverage* plots.

Normal probability plots of residuals are used to detect substantial deviations from normality. The Shapiro–Wilk [19] test* of normality can be viewed as a formal counterpart of this graphical method.

The ith *internally Studentized residual* r_i (often called a standardized or a Studentized residual) is defined as

$$r_i = e_i/s(1 - h_i)^{1/2} \tag{3}$$

where $s^2 = \mathbf{e}^T\mathbf{e}/(n-p)$ and h_i is the ith diagonal element of H. In contrast to the ordinary residuals, these residuals are scaled to have constant variance under model (1), a property that may facilitate interpretation of diagnostic plots (*see also* STUDENTIZATION).

The ith *externally Studentized residual* t_i (also referred to as a Studentized residual, a deletion residual, and a jackknife residual) is defined as

$$t_i = r_i s/s_{(i)}, \tag{4}$$

where $s_{(i)}^2 = \mathbf{e}_{(i)}^T\mathbf{e}_{(i)}/(n-p-1)$ and $\mathbf{e}_{(i)}$ is the $(n-1) \times 1$ vector of ordinary residuals based on the data with the ith observation removed. The residual t_i is often used to assess the evidence that the ith observation is an outlier. Under model (1), t_i follows a Student's t-distribution* with $n - p - 1$ degrees of freedom. The residuals shown in (3) and (4) are related, $t_i^2 = (n-p-1)r_i^2/(n-p-r_i^2)$. Further information on externally and internally Studentized residuals, and related concerns is available in Atkinson [4], Cook and Weisberg [8], and Hoaglin and Welsch [15].

Recursive residuals are useful for studying model inadequacies that occur along a selected ordering (e.g., by time) of the data. For ordered observations the first p *ordinary recursive residuals* play no important role and are often taken to be zero. The remaining ordinary recursive residuals w_i are constructed sequentially by using the data on only the first i observations. Let r_i^* denote the ith unscaled Studentized residual, $r_i^* = e_i/(1-h_i)^{1/2}$. Then $w_i = r_{ii}^*$, where r_{ii}^* is the ith (last) unscaled Studentized residual from the analysis based on the first i observations, $i = p + 1, \ldots, n$. Under model (1), the w_i's are independent normal random variables with mean 0 and variance σ^2. *Studentized recursive residuals* u_i are constructed similarly: $u_i = t_{ii}$, where t_{ii} is the ith externally Studentized residual from the analysis based on the first i observation, $i = p + 2, \ldots, n$. Under model (1) the u_i's are independent, a property that is not shared by the externally Studentized residuals t_i, and u_i follows a Students t-distribution with $i - p - 1$ degrees of freedom.

Brown et al. [7] give a comprehensive account of the use of recursive residuals in methods for investigating the stability of models over time. Such concerns occur frequently in economics and quality control, for example. Recursive residuals have also been studied in connection with tests for serial correlation [17] and heteroscedasticity [14]. Although interest in recursive residuals is relatively recent, equivalent versions appear as early as 1891 [13].

Various other types of residuals have been studied in the framework of model (1). The *predicted residuals*, for example, are defined as $p_i = y_i - \mathbf{x}_i^T\hat{\boldsymbol{\beta}}_{(i)}$, where $\mathbf{x}_i^T$ is the ith row of $\mathbf{X}$ and $\hat{\boldsymbol{\beta}}_{(i)}$ is the ordinary least-squares* estimator of $\boldsymbol{\beta}$ based on the data with the ith observation removed. The predicted residuals are related to the ordinary residuals, $p_i = e_i/(1 - h_i)$, and the predicted residual sum of squares PRESS $= \Sigma p_i^2$ is used as a criterion for model selection [16]. *BLUS* (best linear unbiased scalar) residuals are uncorrelated and have certain optimality properties as described by Thiel [20].

Beyond linear models, relatively little is known about appropriate ways to construct and use residuals. Cox and Snell [10] investigate residuals in a fairly general class of models that covers many frequently encountered problems such as nonlinear and exponential regression. Residuals in logistic regression* are discussed by Pregibon [18].

Anscombe [1] and Anscombe and Tukey [3] are classic works on residuals. Beginning to intermediate material on residuals can be found in most comprehensive texts on regression, such as Draper and Smith [11] and Weisberg [21]. More advanced treatments are available in Belsley et al. [6] and Cook and Weisberg [8]; these books contain many references and should serve as useful guides to the substantial literature on residuals.

References

[1] Anscombe, F. J. (1961). *Proc. Fourth Berkeley Symp.*, Vol. 1, 1–36. University of California Press, Berkeley, CA. (Advanced, establishes many fundamental results.)

[2] Anscombe, F. J. (1973). *Amer. Statist.*, **27**, 17–21. (Elementary, provides clear motivation for the use of residual graphs.)

[3] Anscombe, F. J. and Tukey, J. (1963). *Technometrics*, **5**, 141–160. (Intermediate, with emphasis on two-way arrays; contains a good discussion of fundamental ideas.)

[4] Atkinson, A. C. (1981). *Biometrika*, **68**, 13–20. (Intermediate, emphasizes plots for outliers and influential observations.)

[5] Atkinson, A. C. (1982). *J. R. Statist. Soc. B*, **44**, 1–36.

[6] Belsley, D. A., Kuh, E., and Welsch, R. (1980). *Regression Diagnostics*. Wiley, New York. (Intermediate to advanced, emphasizes influence and collinearity.)

[7] Brown, R. L., Durbin, J., and Evans, J. M. (1975). *J. R. Statist. Soc. B.*, **37**, 149–163. (Intermediate to advanced, a good first source for recursive residuals.)

[8] Cook, R. D. and Weisberg, S. (1982). *Residuals and Influence in Regression*. Chapman and Hall, London. (Intermediate to advanced, contains an overview of many topics with references, emphasizes influence.)

[9] Cook, R. D. and Weisberg, S. (1983). *Biometrika*, **70**, 1–10.

[10] Cox, D. R. and Snell, E. (1968). *J. R. Statist. Soc. B*, **30**, 248–275. (Advanced, promotes a general definition of residuals.)

[11] Draper, N. R. and Smith, A. (1981). *Applied Regression Analysis*, 2nd ed. Wiley, New York.

[12] Durbin, J. and Watson, G. (1971). *Biometrika*, **58**, 1–19.

[13] Farebrother, R. W. (1978). *J. R. Statist. Soc. B.*, **40**, 373–375.

[14] Hedayat, A. and Robson, D. S. (1970). *J. Amer. Statist. Ass.*, **65**, 1573–1581.

[15] Hoaglin, D. C. and Welsch, R. (1978). *Amer. Statistician*, **32**, 17–22.

[16] Mosteller, F. and Tukey, J. W. (1977). *Data Analysis and Linear Regression*. Addison–Wesley, Reading, MA.

[17] Phillips, G. D. A. and Harvey, A. C. (1974). *J. Amer. Statist. Ass.*, **69**, 935–939.

[18] Pregibon, D. (1981). *Ann. Statist.*, **9**, 705–724. (Covers diagnostics for logistic regression.)

[19] Shapiro, S. S. and Wilk, M. (1965). *Biometrika*, **52**, 591–611.

[20] Theil, H. (1968). *J. Amer. Statist. Ass.*, **63**, 242–251.

[21] Weisberg, S. (1980). *Applied Regression Analysis*. Wiley, New York. (A good source for beginning to intermediate material on residuals.)

(EXPLORATORY DATA ANALYSIS
LEVERAGE
REGRESSION DIAGNOSTICS
RESISTANT TECHNIQUES)

R. DENNIS COOK

RESIDUATION

In design theory, residuation essentially means the process of obtaining a (residual) design D_0 by deleting a block and its points from a design D. In practice, D is a symmetric $2-(v, k, \lambda)$ and then D_0 is a two-design with parameters $b_0 = v - 1$, $v_0 = v - k$, $r_0 = k$, $k_0 = k - \lambda$, and $\lambda_0 = \lambda$, so that $r_0 = k_0 + \lambda_0$.

A two-design satisfying $r = k + \lambda$ is said to be quasiresidual (QR). Not every QR design is a residual of a symmetric two-design (e.g., see refs. 2, 6, 7).

QR designs with $\lambda = 1$ are the finite affine planes and they are all residuals of projective planes. In general, affine two-designs are QR and it is conjectured that all are residual. Certainly affine three-designs (the extensions of symmetric Hadamard three-designs) are all residual. See [13] for details.

For given λ, a QR design with sufficiently large k is residual [1, 11]. This is true for all k if $\lambda < 3$ (see ref. 3 or 4).

A theorem of Mann asserts that if a block in a $2-(v, k, \lambda)$ D is repeated ρ times, then $\rho \leqslant b/v$. If $\rho = b/v$, one gets from D a generalized quasiresidual design D_0 by deleting the block, its repetitions, and their points [15] (see also ref. 8). When D is symmetric, $\rho = 1$ and D_0 is quasiresidual. This idea is further generalized in ref. 10 to group divisible designs*. Further information on properties of residuation can be found in refs. 5, 9, 13, and 15.

References

[1] Bose, R. C., Shrikhande, S. S., and Singhi, N. M. (1976). Edge regular multigraphs and partial geometric designs with an application to the embedding of quasiresidual designs. In *Colloq. Int. Sul. Teo. Combin.*, Tom I, 49–81. Acc. Naz. Lincei, Roma, Italy.

[2] Brown, R. B. (1975). *J. Comb. Theory A*, **19**, 115–116.

[3] Hall, M., Jr. (1967). *Combinatorial Theory*. Blaisdell, Waltham, MA.

[4] Hall, M., Jr. and Connor, W. S. (1953). *Canad. J. Math.*, **6**, 35–41.

[5] Kelly, G. (1982). *Discrete Math.*, **39**, 153–160.

[6] van Lint, J. H. (1978). *Proc. Kon. Ned. Akad. Wet.*, **81**, 269–275.

[7] van Lint, J. H. and Tonchev, V. D. (1984). *J. Combin. Theory A*, **37**, 359–362.

[8] Lloyd, C. E. (1986). A method for constructing generalised residual designs. *J. Combin. Theory A* (to appear).

[9] Mann, H. B. (1969). *Ann. Math. Statist.*, **40**, 679–680.

[10] Mavron, V. C. (1980). *Residuals of group divisible designs. Combinatorics and its Applications*. Indian Statistical Institute, Calcutta, India.

[11] Neumaier, A. (1982). *Geometriae Dedicata*, **12**, 351–366.

[12] Shrikhande, S. S. (1976). *Aequationes Mathematicae*, **14**, 251–269.

[13] Shrikhande, S. S. and Singhi, N. M. (1973). *Utilitas Math.*, **4**, 35–43.

[14] Shrikhande, S. S. and Singhi, N. M. (1972). *Utilitas Math.*, **1**, 191–201.

[15] Tonchev, V. D. (1981). *Coll. Math. Soc. Janos Bolyai*, **37**, 685–695.

(BLOCKS
DESIGN OF EXPERIMENTS
FACTORIAL DESIGNS)

VASSILI C. MAVRON

RESISTANCE

A term used in exploratory data analysis* to denote a property of measures of location or spread, or of some smoothing techniques that makes them "relatively unaffected" (or impervious) by the presence of outliers*. The median and midspread (interquartile range) are simple examples of resistant measures of location and scale, respectively.

RESISTANT TECHNIQUES

In recent years, much attention has been paid to the development of resistant techniques for data analysis, primarily under the influence of J. W. Tukey and his co-workers. Useful accounts, at somewhat differing levels, are given by McNeil [11], Mosteller and Tukey [13], Tukey [17], Velleman and Hoaglin [18], and, particularly, Hoaglin et al. [8], henceforth referred to as HMT.

Formally, a summary statistic is called *resistant* if it is insensitive to any change in any small part of the data and to any small change in a large part of the data. In practice, however, resistance is usually interpreted as resistance to outliers*. Thus, consider the mean* and median* as measures of location in a single batch of data. The former is proportionately influenced by moving a single observation away from the main body of the data, whereas the latter remains essentially unchanged: the median but not the mean is resistant.

Resistance and robustness are often thought of as synonymous. However, it is useful to differentiate between the terms, particularly at a philosophical level. Consider, for example, a typical regression problem. A robust analysis usually aims to produce a single set of parameter estimates and associated confidence intervals, as with classical statistical analysis but employing a procedure that is highly efficient across a plausible range of error distributions; *see* ROBUST ESTIMATION and the references therein. Resistant techniques are often used in a more informal manner with emphasis on exploration and no consideration of an underlying probability distribution or optimality. Estimation at more than one level of resistance may be involved and results compared with those of a standard analysis. If only one resistant analysis were contemplated, it would usually be preferable to opt for high resistance. The examination of residuals takes on a special importance, a primary objective being the detection of discrepant observations.

Some possible consequences of resistant analysis include:

- (i) Positive suggestions for the collection of further data.
- (ii) Respecification of an underlying model, perhaps through transformation or the addition of further explanatory variables.
- (iii) Careful questioning of the experimenter or observer concerning particular elements of the data.
- (iv) Outright rejection of some observations as mavericks, perhaps followed by a standard analysis of the remainder.
- (v) Sometimes, acceptance of the resistant analysis itself.
- (vi) Further research into interesting anomalies, which a standard analysis may not have revealed.

Thus, in the last case, an atypical region may become the focus of geographical attention or a special month or year may engage the economist or forecaster. However, especially in routine situations, a resistant analysis may be carried out as no more than a check on the standard procedure, to be subsequently discarded if discrepancies are small.

We now consider some specific areas of application.

LOCATION AND DISPERSION

Regarding the location of a single batch of data, the most common resistant estimates are the median, mentioned above, and the $\alpha\%$ trimmed mean (*see* TRIMMING AND WINSORIZATION), for which the arithmetic mean* of the observations is taken after removal of (approximately) $\alpha\%$ of the data in each tail; α might vary between 5 (mildly resistant) and 25 (highly resistant). Note that the median is the location estimate that minimizes the sum of absolute deviations. Corresponding estimates of scale or dispersion, alternative to the nonresistant sample standard deviation, include the interquartile range* and the median absolute deviation (MAD) of the observations about the location estimate. Chapters 9–12 of ref. [8] review more sophisticated measures of particular interest for robustness. Gnanadesikan [6, Chaps. 5 and 6] considers some multivariate generalizations; see also Brown [3].

Resistant summaries of location and dispersion are useful in graphical methods, particularly in the construction of box plots; *see* EXPLORATORY DATA ANALYSIS and Tukey [17, Chaps. 2 and 4]. Displays of parallel box plots form a powerful tool in comparing several batches of data: dominant patterns, the need for transformation, and the existence of outliers are often forcibly highlighted.

y VERSUS x DATA

Special techniques for fitting straight lines to simple y vs. x data are reviewed by HMT (Chap. 5); *see also* ROBUST-RESISTANT LINE. These usually involve splitting the data into three groups and fitting a line to measures of their centres. Straight-line fits also provide a convenient focus when assessing the performance of general purpose resistant regression procedures. For example, one may consider the influence of highly discrepant x values as well as y values. The former are termed points of high leverage*; see, for example, Cook and Weisberg [5]. Note that minimizing the sum of absolute residuals* is *not* resistant to such points.

For nonparametric curve fitting and smoothing of y vs. x data, see Cleveland [4] and Tukey [17].

TWO-WAY TABLES

Simple resistant techniques are particularly appealing in the analysis of two-way tables, where the response variable y is classified by both row and column factors (*see* CONTINGENCY TABLES). It is usually assumed that the data follow an additive model of the form

$$y_{ij} = \alpha + \rho_i + \gamma_j + z_{ij};$$

the classical least-squares* approach uses row and column means to estimate the corresponding effects. An analogous but resistant alternative [8, Chap. 6], known as median polish* (Tukey [17]), involves the successive sweeping out of row and column medians from the body of the table until further changes become negligible. The fitting procedure, which for small tables may be quickly carried out by hand, will ignore occasional anomalies in the data, which are then highlighted by large residuals. A few missing observations can also be accommodated. If required, polishing can be based on a different resistant measure of location, such as a trimmed mean. The results of the analysis can be used to diagnose flaws in the model and to suggest a transformation of the response variable that may achieve better additivity ([8], Chap. 6].

In practice, median polish usually comes close to minimizing the sum of absolute residuals, a procedure that *is* resistant in this context but that is rarely worth the extra computational effort. For direct resistant fitting of multiplicative two-way tables, see McNeil and Tukey [12].

MULTIPLE REGRESSION*

Consider the estimation of $\boldsymbol{\theta}$ in the usual linear model $y = \mathbf{X}\boldsymbol{\theta} + z$. Let $\tilde{\boldsymbol{\theta}}$ denote a current estimate of $\boldsymbol{\theta}$ and write $\tilde{z}_i \equiv z_i(\tilde{\boldsymbol{\theta}})$ for the ith component of the corresponding residual vector $\tilde{z}$. Then it is possible to re-estimate θ resistantly by the method of weighted least squares* using weights $w_i \equiv w(\tilde{z}_i)$ determined by the current residuals.

Suitable choices of weight function w include Tukey's bisquare

$$w(z) = \begin{cases} \left[1 - (z/(cs))^2\right]^2, & |z| < cs, \\ 0, & |z| \geqslant cs, m \end{cases}$$

where s is a resistant measure of dispersion among the $\tilde{z}_i$'s, usually the MAD (see above), and c is a tuning constant that determines the level of resistance. The procedure may be iterated, using the updated residuals at each stage, and generally converges to a value $\boldsymbol{\theta}^*$, say, that may depend on the initial choice of $\tilde{\boldsymbol{\theta}}$. A detailed description and interpretation in terms of the normal equations* is in Mosteller and Tukey [13, Chap. 14]. For the robustness viewpoint and other techniques see Huber [9] and ROBUST ESTIMATION. Note that, in general, $\boldsymbol{\theta}^*$ does *not* seek to minimize $\Sigma w(z_i) z_i^2$, unless $w(z) \propto |z|^{-p}$ and $p < 2$.

An important extension of the preceding procedure is to the large class of maximum likelihood* problems that can be solved by iteratively reweighted least squares; see, for example, Green [7] and Nelder and Wedderburn [14]. As indicated by Besag [1], resistant variants can be readily obtained by including a second weight function, such as the bisquare, acting on the current standardized residuals; see also Green [7] and Pregibon [15].

DESIGNED EXPERIMENTS

In designed experiments, there is an additional problem concerning resistance. As the simplest example, consider r replicates of a two-factor experiment. In a classical framework, estimates of main effects* and interactions are obtained from the associated two-way table of means over replicates. Suppose the data are well behaved and that main effects alone provide an adequate fit, except at one particular combination of levels; this is reflected in the relevant r observations and in the corresponding cell of the two-way table. It is scientifically desirable that the analysis isolates this single peculiarity. However, in a least-squares analysis, its influence is smeared among all estimates of main effects and interaction* terms. The fit is mathematically correct but the proper interpretation is likely to be missed. The same comment applies to the results of all-purpose resistant regression procedures when applied in such situations. Compare this with the use of median polish on the two-way table, for which the effect will generally be isolated, provided each factor appears at three or more levels; of course, in practice, the two-way table for a resistant analysis would not be constructed from means, but this is a separate issue. The experiments considered by Brown [2] exemplify the general problem.

Considerations, such as those above, suggest that special estimation procedures should be developed for the analysis of designed experiments; some implications for the analysis of variance are discussed briefly in Mallows and Tukey [10]. An example of extended median polish for Latin squares* is given by Besag [1]. Note that median polish will not be helpful for resistant analysis of 2^n experiments, for which see Seheult and Tukey [16].

References

[1] Besag, J. E. (1981). *Biometrika*, **68**, 463–469.

[2] Brown, M. B. (1975). *Appl. Statist.*, **24**, 288–298.

[3] Brown, B. M. (1983). *J. R. Statist. Soc. B*, **45**, 25–30.

[4] Cleveland, W. S. (1979). *J. Amer. Statist. Ass.*, **74**, 829–836.

[5] Cook, R. D. and Weisberg, S. (1982). *Residuals and Influence in Regression*. Chapman and Hall, London.

[6] Gnanadesikan, R. (1977). *Methods for Statistical Data Analysis of Multivariate Observations*. Wiley, London.

[7] Green, P. J. (1984). *J. R. Statist. Soc. B*, **46**, 149–192.

[8] Hoaglin, D. C., Mosteller, F., and Tukey, J. W. (1983). *Understanding Robust and Exploratory Data Analysis*. Wiley, New York.

[9] Huber, P. (1981). *Robust Statistics*. Wiley, New York.

[10] Mallows, C. L. and Tukey, J. W. (1982). An overview of techniques of data analysis, emphasising its exploratory aspects. In *Some Recent Advances in Statistics*. Academic Press, London.

[11] McNeil, D. R. (1977). *Interactive Data Analysis*. Wiley, New York.

[12] McNeil, D. R. and Tukey, J. W. (1975). *Biometrics*, **31**, 487–510.

[13] Mosteller, F. and Tukey, J. W. (1977). *Data Analysis and Regression*. Addison-Wesley, London, England and Reading, MA.

[14] Nelder, J. A. and Wedderburn, R. W. M. (1972). *J. R. Statist. Soc. A.*, **135**, 370–384.

[15] Pregibon, D. (1982). *Biometrics*, **38**, 485–498.

[16] Seheult, A. H. and Tukey, J. W. (1982). *Utilitas Math.*, **21B**, 57–98.

[17] Tukey, J. W. (1977). *Exploratory Data Analysis*. Addison-Wesley, London, England and Reading, MA.

[18] Velleman, P. F. and Hoaglin, D. C. (1981). *Applications, Basics, and Computing of Exploratory Data Analysis*. Duxbury Press, Boston, MA.

(EXPLORATORY DATA ANALYSIS
GRAPHICAL REPRESENTATION OF DATA
LEVERAGE
MEAN, MEDIAN, MODE, AND SKEWNESS
MEDIAN POLISH
OUTLIERS
RESIDUALS
ROBUST ESTIMATION
ROBUST-RESISTANT LINE
TRIMMING AND WINSORIZATION)

JULIAN BESAG
ALLAN SEHEULT

RESOLUTION

The development of fractional factorial* designs was followed by the creation of a scheme to classify these designs according to a characteristic called "resolution." Resolution is defined in the following way: A fractional factorial design is of resolution R if all factorial effects (main and interaction*) up to order k are estimable (k being the largest integer less than $R/2$) under the assumption that all effects of order greater than or equal to $R - k$ are zero (Box and Hunter [5]; the concept was introduced earlier with different terminology by Box and Wilson [7], and later expanded upon by Webb [28]). Clearly, a design of resolution R (greater than 1) is also of resolution $Q < R$. Construction methods surveyed by Addelman [2, 3], Raktoe et al. [20], and Raghavarao [23], and tables for "R" classes (e.g., refs. 4 and 8) aid the experimenter in design selection provided certain "smoothness" assumptions are made about the factorial model that is to be used in analyzing the data (see Addelman [2], Box and Hunter [5], and Plackett and Burman [19] for brief discussions on this point). Issues related to the existence of resolution R designs and methods of construction are, perhaps, as extensive as the issues related to the existence and methods of constructing fractional designs [24] [such methods are listed in FRACTIONAL FACTORIAL DESIGNS (FFD)]. However, most methods are based on (1) group theory owing to the relationship between resolution and the minimum number of nonzero elements in the "generators" of the design [5]; (2) properties of orthogonal arrays* [21, Chap. 13] owing to the relationship of the strength of such an array and the resolution of an associated design (e.g., refs. 24 and 9); and more recently, (3) properties of block designs [14].

Early writings on resolution were partitioned into orthogonal or nonorthogonal (depending on whether the covariance matrix of the effects is diagonal or not, the latter sometimes referred to as "irregular") and symmetric or asymmetric (depending on whether the number of levels of each factor are equal or not). These works were followed by a further classification called "balancedness" (special cases of nonorthogonal designs; see, e.g., refs. 13 and 26 and FFD*) as well as "search" designs [11, 21, and 25] and "compromise" plans, which allow subsets of high-order effects to be nonnegligible (i.e., refs. 1 and 23, p. 279). These represented elaborations of the basic resolution framework the experimenter can adopt in determining suitable experimental design alternatives.

The greater the resolution, generally the greater the total number of required observations, N. If R is odd, the minimum possible value of N is the number of parameters to be estimated. See Margolin [16] for R = IV minimum "N" requirements and Seiden [24] for a discussion on such requirements deduced from the properties of arrays. It is also of interest to consider how many unique fractions exist in the set of R = III designs. This number has been calculated for up to seven factors, with the cardinality of the set equal to 858,240,222,176 for exactly seven factors and $N = 8$ (*see* FFD). If one restricts the feasible (R = III) set of fractions to those that are orthogonal and generated using group theory methods, the number of unique fractions of the seven factor problem is 30 for $N = 8$ and 12,840 for $N = 16$ [18].

Example. An experiment was planned to study the effects of nine factors on the error rate, per unit time, of experienced data entry operators encoding written information at a computer terminal. The nine factors were related to workplace enhancements and included (a) a tiltable terminal, (b) a filter screen on the terminal, (c) an adjustable chair, (d) overhead indirect lighting, etc. Each factor could be present or absent, so there was a total of 2^9 possible combinations of factors that could be studied in the complete factorial. Since time did not permit such an exhaustive study, it was decided to select a fractional design that would allow estimation of the mean and first-order (main) effects with the additional assumption that all third-order and higher factor effects were negligible. A decision was made to restrict the study to orthogonal fractions with $N = 2^r$. Three options now remained, depending on further assumptions regarding the second-order effects:

1. If all second-order effects are assumed negligible, then a resolution III design could be selected with $N = 16$ [8].
2. If any nonnegligible second-order effect is composed of terms containing only those from a set of four prespecified factors, a "compromise" R = V may be selected with $N = 16$ [1].
3. If no second-order effect may be assumed negligible, then a resolution IV design is needed with $N = 32$ [8].

An extended list of possibilities for this design is shown in Table 1, which lists N and a corresponding reference, depending upon resolution and whether the design is orthogonal.

Raktoe et al. [21, p. 88] have added the convention that "when R is odd, the mean effect (order zero) is estimable, while if R is even, the mean effect is of no interest for estimation." Traditionally, most fractional designs have been constructed with the implicit requirement that the mean be estimable. Because a design of even resolution does permit "nuisance" effects (i.e., those that need not be estimated separately from each

Table 1 The Number of Observations and Bibliographic Reference for Some Fractional Designs of a 2^9 Factorial

	Resolution		
	III	IV	V
Orthogonal	16 [8]	32 [8]	128 [8]
	12 [2, 18]	24 [16, 7]	16 [1][a]
Nonorthogonal	11 [10, 17]	18 [15, 16]	80 [12]
	10 [19, 21, 28]	20 [26]	46 [9, 20]

[a] "Compromise" resolution V.

other but cannot be assumed negligible), allowing the mean to be such an effect (1) increases the number of feasible designs, (2) may introduce more realistic restrictions, and (3) makes the mathematical problem formulation "less cumbersome" (particularly for nonorthogonal designs; *see* FFD).

References

[1] Addelman, S. (1962). *Technometrics*, **4**, 50.

[2] Addelman, S. (1963). *J. Amer. Statist. Ass.*, **58**, 45–71.

[3] Addelman, S. (1972). *J. Amer. Statist. Ass.*, **67**, 103–111. (This and ref. 2 provide comprehensive surveys as well as an extensive bibliography.)

[4] Addelman, S. and Kempthorne, O. (1961). *Aeronautical Research Tech. Rep. No. 79*, NTIS Order No. AD 272-250. (Tables for R = III, asymmetric designs.)

[5] Box, G. E. P. and Hunter, J. S. (1961). *Technometrics*, **3**, 311–351.

[6] Box, G. E. P. and Hunter, J. S. (1961). *Technometrics*, **3**, 449–458. (This and ref. 5 are basic introductory references.)

[7] Box, G. E. P. and Wilson, K. B. (1951). *J. R. Statist. Soc. B*, **13**, 1–45. [Classic article on "response surface" methodology (all factors quantitative); idea of resolution introduced.]

[8] Box, G. E. P., Hunter W. G., and Hunter, J. S. (1978). *Statistics for Experimenters*. Wiley, New York. (A good basic text in design for initial reading.)

[9] Chopra, D. V. (1977). *J. Indian Statist. Ass.* **15**, 179–186. (Specific results related to the 2^9, resolution V, problem.)

[10] Galil, Z. and Kiefer, J. (1980). *Ann. Statist.*, **8**, 1293–1306.

[11] Ghosh, S. (1981). *J. Statist. Plann. Inf.*, **5**, 381–389. (Examination of search designs with small number of observations; two-level factors.)

[12] John, P. W. M. (1971). *Statistical Design and Analysis of Experiments*. Macmillan, New York. (Several chapters on fractional designs; good bibliography.)

[13] Kuwada, M. (1979). *J. Statist. Plann. Inf.*, **3**, 347–360. [Relationships of balanced arrays and symmetric resolution V (three-level) designs; references for the two-level problem.]

[14] Lewis, S. M. (1979). *J. R. Statist. Soc. B*, **41**, 352–357. (Cyclic design method, with the constraint of R = III imposed.)

[15] Margolin, B. H. (1969). *J. R. Statist. Soc. B*, **31**, 514–523.

[16] Margolin, B. H. (1969). *Technometrics*, **11**, 431–444. [Minimum total observation requirements, R = IV (see also ref. 15).]

[17] Mitchell, T. J. (1974). *Technometrics*, **16**, 211–220. (Good introduction to R = III construction, along with a computer algorithm and results.)

[18] Mount-Campbell, C. A. and Neuhardt, J. B. (1981). *Commun. Statist. A*, **10**, 2101–2111. (Enumeration of the number of R = III designs using group theory, up to 20 factors.)

[19] Plackett, R. L. and Burman, J. P. (1946). *Biometrika*, **33**, 305–325. [Classic paper on R = III orthogonal designs (although the term resolution was introduced much later).]

[20] Raktoe, B. L. and Federer, W. T. (1973). *Ann. Statist.*, **1**, 924–932. (Investigation of equivalence classes of two-level, $n = 4m - 1$ (m integer) observations; saturated R = III designs; suggestions for other n.]

[21] Raktoe, B. L., Hadayat, A., and Federer, W. T. (1981). *Factorial Designs*. Wiley, New York. [A must for the serious student wishing to bridge the gaps in (analysis of) linear models and (construction of) factorial designs; extensive bibliography.]

[22] Rechstaffer, R. L. (1967). *Technometrics*, **9**, 569–575. (A treatment of the saturated design problem; two- and three-level factors.)

[23] Raghavarao, D. (1971). *Constructions and Combinatorial Problems in Design of Experiments*. Wiley, New York. (Emphasis on Latin square; block design methods of constructing fractional designs; Chaps. 13–16 cover basic aspects of design; excellent references.)

[24] Seiden, E. (1973). In *A Survey of Combinatorial Theory*, J. N. Srivastava, F. Harary, C. R. Rao, G. C. Rota, and S. S. Shrikande, eds. North-Holland, Amsterdam, pp. 397–401. (Concise description of results dealing with relationship of strength of orthogonal arrays and resolution.)

[25] Srivastava, J. N. (1965). In *A Survey of Statistical Designs and Linear Models*, J. N. Srivastava, ed. North-Holland, Amsterdam, pp. 507–519.

[26] Srivastava, J. N. (1970). In *S. N. Roy Memorial Volume*. University of North Carolina and Indian Statistical Institute, pp. 227–241.

[27] Srivastava, J. J. and Anderson, D. A. (1970). *J. Amer. Statist. Ass.*, **65**, 828–843. [A thorough treatment of the two-level (covariance) balanced R = IV problem, with extensive tables.]

[28] Webb, S. R. (1965). *Tech. Document No. 65-116*, Aerospace Research Laboratories p. 38.

[29] Webb, S. R. (1968). *Technometrics*, **10**, 535–550.

[30] Whitwell, J. C. and Morbey, G. K. (1961). *Technometrics*, **3**, 459–478. [An early discussion dealing with the selection of a subset of higher-order interactions for estimation (two-level); group theory design method.]

(DESIGN OF EXPERIMENTS
ESTIMABILITY
FRACTIONAL FACTORIAL DESIGNS
INTERACTION
MAIN EFFECTS
ORTHOGONAL ARRAYS
PARTIALLY BALANCED DESIGNS)

JOHN B. NEUHARDT

RESOLVABLE DESIGNS *See* RESOLUTION

RESPONDENT BURDEN, REDUCTION OF

Providing data for administrative or statistical purposes imposes a burden on the participant. The United States Government [7] defines *burden* as "...the time, effort, or financial resources expended by persons to provide information... ." It has been hypothesized that respondent burden can increase the nonresponse* rate or decrease the accuracy of the data supplied by individual respondents, in both cases increasing nonsampling errors. Thus there is interest in reducing the respondent burden in censuses* and surveys*. The magnitude of this respondent burden can depend on several factors, some quantitative in nature, like the length of time required to provide the information and the frequency of contacts, to factors subjective in nature, like the respondent's opinion of the value of the data and the effort needed to provide the data. Bradburn [1] generalizes the concept of respondent burden to more precisely address its subjective aspects. He suggests that perceived burden is composed of four factors: frequency of contact, length of contact, required respondent effort, and the stress of disturbing questions.

One way to reduce respondent burden is to make fewer contacts. Fewer contacts may come about by conducting fewer surveys or reducing sample sizes, that is, reducing overall burden, or by simply not contacting the same sampling unit too often, that is, reducing the individual burden. Individual burden can also be reduced by asking fewer questions of the respondent. An example of this approach is the use of a long and short form for a survey or census. One part of the sample would receive a long questionnaire and the other part receive the short form of the questionnaire. The U.S. Bureau of the Census* (Cole and Altman [2] and Makovak [6]) has tested long and short forms for the Annual Survey of Manufacturers and the decennial census, respectively. This work has compared return rates and item nonresponse rates for mail questionnaires. While no statistically significant differences were found for the return rates, there was some evidence of a higher item nonresponse rate for the long forms, conditional on the fact that questionnaire design does not impact on item nonresponse.

Ford et al. [3] suggest sample designs and estimation procedures to reduce the average number of questions per respondent for a particular survey. They show that by systematically deleting groups of questions and administering those deleted groups to random subsamples of the sample that the average respondent burden can be decreased without significantly increasing the sampling errors of the survey. The correlation* structure between questions on the long and short form is used in the estimation process. However, it may not always be possible to use these alternatives for reducing burden. Data needs often dictate the frequency of surveys and sample sizes usually provide the maximum acceptable sampling errors for a given survey cost. Even though the data obtained for all questions for a given instrument are not published directly, they often appear on the questionnaire in order to provide internal consistency checks. The survey design can also be altered to reduce individual burden.

The most common method proposed is the use of rotation sampling*, as opposed to a fixed panel* of respondents, for surveys that are repeated over time. Here the idea is to spread the burden out to a larger group of respondents over the survey period. While

Table 1 Decrease in Expected Number of Contacts for Selected Large Farms in South Dakota

Type of Large Farm or Ranch	Percent Decrease in Burden
Cattle on feed	−8.7
Cattle, including cattle on feed	−8.3
Hogs	−13.4
Chickens	−9.8

there are several other reasons to use rotation sampling, reduction of burden may not be achieved if the total sample size is nearly the population size for the period considered. This may be the case for relatively small populations, for example, a specific industry type, and for relatively frequent surveys, say monthly during a calendar year. Further research on burden has been concerned with developing innovative sampling procedures to reduce the quantitative factors of respondent burden and attempting to better understand the nature of respondent burden in terms of the relationship between the quantitative and subjective factors.

Sunter [9] and Tortora [10] independently generalize the concept of reducing burden for all surveys conducted by an organization. The collection of all surveys targeted at a particular population represents a system of surveys where the respondent burden can be reduced by controlling and coordinating sample selection in all surveys, including rotational surveys. Both studies propose the use of unequal probability sampling where the probability of selection is based on some function of respondent burden, for example, the total number of contacts over the last calendar year. Tortora and Crank [11, 12] demonstrate, in simulation studies, that respondent burden for the large sampling units can be greatly decreased using this approach, without loss of efficiency, when compared to equal probability sampling. Table 1 shows the percent decrease in expected number of contacts over a survey year using probabilities of selection based on past number of contacts for large or extreme farm operators in South Dakota. This decrease in burden for the large farm operators corresponds with an increase in burden of 1.2% for the nonextreme farmer or rancher. In addition, Tortora and Crank estimated the relative efficiency of this stratified unequal probability of selection scheme to the operational stratified equal probability of selection scheme for the cattle and hog surveys. These design effects were 1.02 and 1.03, respectively.

On the other hand, Jones et al. [5] and Frankel [4] and Frankel and Sharp [8] have studied burden, as received by survey participants, for farm operators in North and South Dakota and for a white middle-class suburb of Philadelphia, respectively. They found for these subpopulations that burden is largely reflected by the respondents interpretation of the usefulness and value of the survey, rather than the survey contact(s) itself. While generalizations to all surveys and all populations cannot be made from these results, they demonstrate the importance of the interviewer being able to convey the uses, needs, and benefits of the survey data.

References

[1] Bradburn, N. (1979). Respondent burden. In *Health Survey Research Methods: Second Biennial Conference, Williamsburg, Virginia*, L. Reeder, ed. U.S. Government Printing Office, Washington, DC.

[2] Cole, S. and Altman, M. (1977). *An Experiment with Short Forms in the Annual Survey of Manufacturers*. Unpublished U.S. Census Bureau Report, Washington, DC.

[3] Ford, B., Hocking, R., and Coleman, A. (1978). Reducing Respondent Burden on an Agriculture Survey. *Proc. Survey Res. Methods Sect.*, American Statistical Association, Washington, DC, pp. 341–345.

[4] Frankel, J. (1980). Measurement of Respondent Burden: Study Designs and Early Findings.

Bureau of Social Science Research Rep. No. 0529-8, BSSR, Washington, DC.

[5] Jones, C., Sheatsley, P., and Stinchcombe, A. (1979). Dakota Farmers and Ranchers Evaluate Crop and Livestock Surveys. *Report No. 128*, National Opinion Research Center, Chicago, Il.

[6] Makovak, W. (1982). Analysis of the Effect of Questionnaire Length on Item Nonresponse. *Preliminary Evaluation Results Memorandum No. 25*, U.S. Bureau of the Census, Washington, DC.

[7] Office of Management and Budget (1980). The Paperwork Reduction Act of 1980. *Public Law 96-511*, U.S. Government Printing Office, Washington, DC.

[8] Sharp, L. and Frankel, J. (1983). *Public Opinion Quarterly*, **47**, 36–53.

[9] Sunter, A. (1977). *Int. Statist. Rev.*, **45**, 209–222.

[10] Tortora, R. (1977). *Agric. Econ. Res.*, **30**, 100–107.

[11] Tortora, R. and Crank, K. (1978). The Use of Unequal Probability Sampling to Reduce Respondent Burden. *Tech. Rep.*, Statistical Reporting Service, Washington, DC.

[12] Tortora, R. and Crank, K. (1980). *Estadística*, **34**, 71–90.

(CENSUS
DATA COLLECTION
NONRESPONSE IN SAMPLE SURVEYS
PROBABILITY PROPORTIONAL TO SIZE SAMPLING
QUESTION WORDING EFFECTS IN SURVEYS
SURVEY SAMPLING)

ROBERT D. TORTORA

RESPONSE BIAS

The term 'response bias' is used to describe the systematic (nonvariable) component of the difference between information provided by a survey respondent and the "appropriate" information [1]. In the psychological literature, an individual's response to a certain stimulus is often viewed as a single observation from a random process that has a mean and standard deviation. Under the assumption of repeated applications of the stimulus, the difference between the expected value of these repeated measurements* and the correct or true value is defined as the response bias. Each element in a population may have its own degree of response bias.

Given the difficulties associated with repeated measurements for the same individual, the survey* research literature has tended to recognize individual level response bias, but measurement of this bias and its reporting is carried out in terms of average (over the entire sample or population) response bias. For example, if one-half of the respondents to a survey provide answers that are one unit too high $(+1)$ and the other half provide responses that are one unit too low (-1), the response bias over the entire sample would be zero. The impact of individual level bias would be subsumed under the heading of response *variation*.

The survey research literature contains examples of response bias that may either inflate or deflate estimated parameters. Some of the factors that may affect magnitude and direction of response bias include questionnaire wording (*see* QUESTION WORDING EFFECT IN SURVEYS), question context, and type of interview (face-to-face, telephone, and self-administered–mail).

Suggested causes or sources of response bias include (a) forgetting, (b) telescoping, (c) social desirability, prestige or threat, (d) misunderstanding, (e) fatigue, and (f) gravitating or heaping.

Forgetting with the passage of time is probably the most commonly cited source of response bias. Attempts to correct for this type of bias have lead to various forms of "aided recall" devices. For example, in a survey designed to measure television viewing by one-week recall, a respondent may first be shown a list of television shows that aired during the past week. In surveys designed to estimate trial of consumer products, the respondent may be shown pictures of the particular products in question.

The phenomenon of telescoping is used to describe the incorrect placement of events in a specific time period. Telescoping may be either forward or backward in time, but the most commonly reported form of telescoping is forward. Forward telescoping involves the incorrect placement of a past event in a more recent time period. The resulting overreporting or estimate inflation that occurs has been documented in consumer expenditure and crime victimization surveys. Elimination of forward telescoping involves bounding by the use of repeated interviews. In the first interview a respondent may be asked to recall all of the events of a certain type that occurred during the past six months. A second interview, conducted six months after the first, will ask a similar six-month recall question, but any reported events will be cross-checked to make certain that they were not reported during the first interview. In surveys of this type, data from the first interview are typically not used for estimation, but rather to establish a time boundary.

Response bias that results from social desirability, prestige or threat, describes situations where respondents report participation in socially desirable activities and do not report participation in socially undesirable activities. Comparison of stated participation in elections with actual official counts attest to the overreporting of socially desirable behavior such as voting participation. It is hypothesized that income surveys suffer from significant underreporting of income from illegal and nontaxed sources.

Attempts to reduce overreporting have led to the use of "filter" questions, and attempts to reduce underreporting have lead to the use of randomized response* techniques. Filter questions are often used in surveys designed to measure average daily newspaper readership. The respondent is first asked whether or not a particular newspaper was read during the past week. This is followed by a question to establish the last time the newspaper was read. A respondent is counted as a reader only if reading is reported on the day prior to the interview. The question concerning readership during the past week allows nonreaders the opportunity to claim readership without influencing the survey outcome.

Misunderstanding of a question can be a serious source of response bias. Much of this bias can be eliminated by questionnaire pretesting.

Fatigue bias is used to describe bias resulting from respondent fatigue in overly long interviews. This type of bias may result in ritualistic repetition of either positive or negative answers to a series of questions.

(NONRESPONSE IN SAMPLE SURVEYS
PUBLIC OPINION POLLS
QUESTION WORDING EFFECT IN SURVEYS
SURVEY SAMPLING)

MARTIN R. FRANKEL

RESPONSE ERROR *See* RESPONSE BIAS

RESPONSE, QUANTAL *See* QUANTAL RESPONSE

RESPONSE SURFACE DESIGNS

Suppose we have a set of observations $y_u, \xi_{1u}, \xi_{2u}, \ldots, \xi_{ku}$, $u = 1, 2, \ldots, n$, taken on a response variable y and on k predictor variables $\xi_1, \xi_2, \ldots, \xi_k$. A response surface model is a mathematical model fitted to y as a function of the ξ's in order to provide a summary representation of the behaviour of y. Two basic types of models can be fitted to data arising from a response-predictor relationship:

(a) **Empirical Models**. These are typically models linear in the parameters, often polynomials, either in the basic predictor variables or in transformed entities constructed from these basic predictors. The purpose of fitting empirical models is to provide a mathematical French curve that will summarize the data. (The mechanism that produced these data is, in this context, either unknown or poorly understood.) This article will be concerned only with design of experiments for such empirical models.

(b) Mechanistic Models. When knowledge of the underlying mechanism that produced the data is available, it is often possible to construct a model that, reasonably well, represents that mechanism. Such a model is preferable to an empirical one, because it usually contains fewer parameters, fits the data better, and extrapolates more sensibly. (Polynomial models often extrapolate poorly.) However, mechanistic models are often nonlinear in the parameters, and more difficult to fit and evaluate. Also the choice of an experimental design presents intricate problems. For basic information on designs for mechanistic models see NONLINEAR MODELS and NONLINEAR REGRESSION and the references therein. A cornerstone article on nonlinear experimental design is by Box and Lucas [15].

We now continue with (a). Typically, when little is known of the nature of the true underlying relationship, the model fitted will be a polynomial in the ξ's. (The philosophy is that we are approximating the true but unknown surface by low-order terms in its Taylor's series* expansion. The words "order" and "degree" are interchangeable in response surface work, and the choice of one word over the other is a matter of personal preference.) Most used in practice are polynomials of first and second order. The first-order model is

$$y_u = \beta_0' + \beta_1'\xi_{1u} + \beta_2'\xi_{2u} + \cdots + \beta_k'\xi_{ku} + \epsilon_u, \quad (1)$$

where it is usually tentatively assumed that the errors $\epsilon_u \sim N(0, \sigma^2)$ and are independent. The second-order model contains additional terms

$$\beta_{11}'\xi_{1u}^2 + \beta_{22}'\xi_{2u}^2 + \cdots + \beta_{kk}'\xi_{ku}^2 + \beta_{12}'\xi_{1u}\xi_{2u} + \cdots + \beta_{k-1,k}'\xi_{k-1,u}\xi_{ku}. \quad (2)$$

Polynomial models of order higher than 2 are rarely fitted, in practice. This is partially because of the difficulty of interpreting the form of the fitted surface, which, in any case, produces predictions whose standard errors are greater than those from the lower-order fit, and partly because the region of interest is usually chosen small enough for a first- or second-order model to be a reasonable choice. Exceptions occur in meteorology*, where quite high-order polynomials have been fitted, but there are only two or three ξ's commonly used. When a second-order polynomial is not adequate, and often even when it is, the possibility of making a simplifying transformation in y or in one or more of the ξ's would usually be explored before reluctantly proceeding to higher order, because more parsimonious representations involving fewer terms are generally more desirable. In actual applications, it is common practice to code the ξ's via $x_{iu} = (\xi_{iu} - \xi_{i0})/S_i$, $i = 1, 2, \ldots, k$, where ξ_{i0} is some selected central value of the ξ_i range to be explored, and S_i is a selected scale factor. For example, if a temperature (T) range of 140–160°C is to be covered using three levels 140, 150, 160°C, the coding $x = (T - 150)/10$ will code these levels to $x = -1, 0, 1$, respectively. The second-order model would then be recast as

$$\begin{aligned} y_{iu} = \beta_0 &+ \beta_1 x_{1u} + \cdots + \beta_k x_{ku} \\ &+ \beta_1 x_{1u}^2 + \cdots + \beta_k x_{ku}^2 \\ &+ \beta_{12} x_{1u} x_{2u} + \cdots \\ &+ \beta_{k-1,k} x_{k-1,u} x_{ku} + \epsilon_u \end{aligned} \quad (3)$$

and would usually be fitted by least squares* in that form. Substitution of the coding formulas into (3) enables the β''s to be expressed in terms of the β's.

What sorts of surfaces are representable by a model of form (3)? Figure 1 shows, for $k = 2$, one basic type that occurs frequently in practice. The three-dimensional upper portion of Fig. 1 shows a rising ridge, while the lower portion shows the contours of that ridge in the (x_1, x_2) plane. The details of the specific example used are in the figure caption. A change of origin and a rotation of axes brings the fitted equation into the so-called canonical form (in X_1 and X_2) in which the nature of the surface may be im-

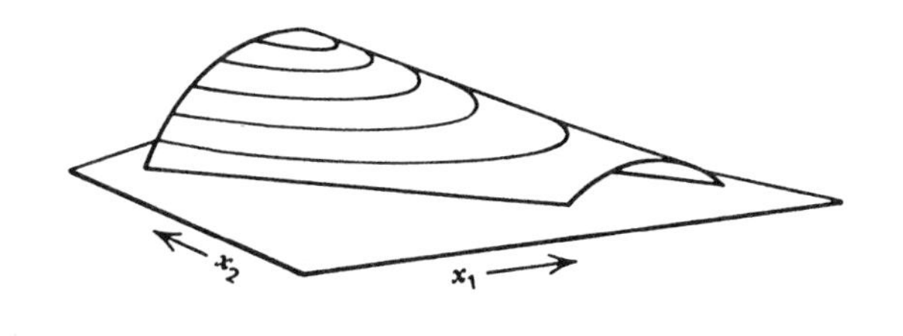

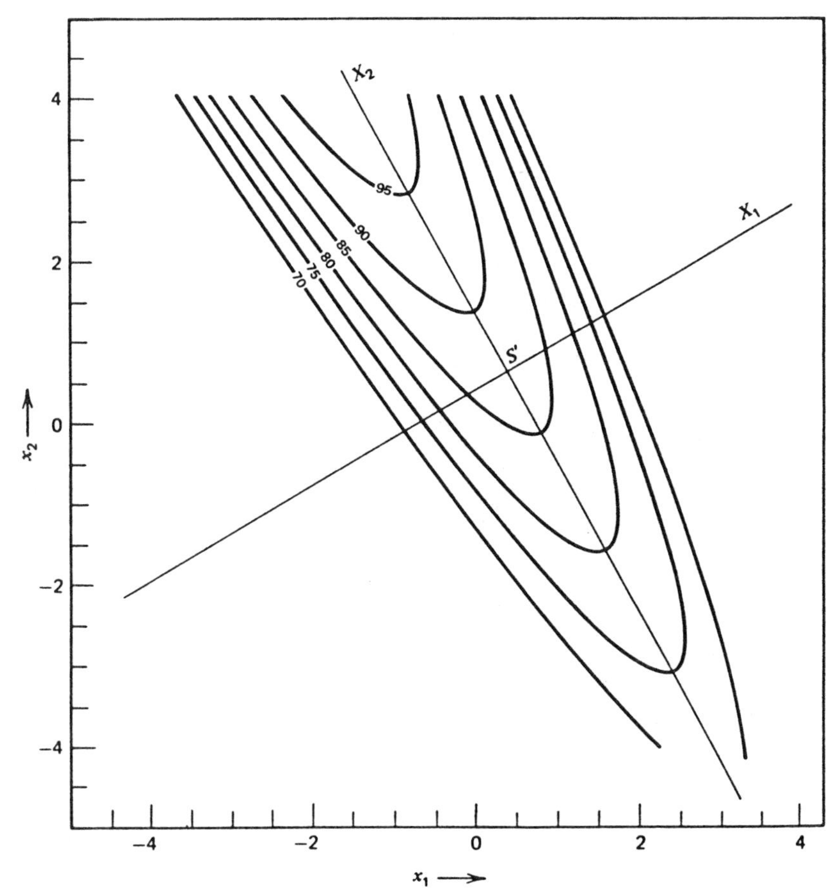

Figure 1 Example of a second-degree equation representing a rising ridge. $\hat{y} = 82.71 + 8.80x_1 + 8.19x_2 - 6.95x_1^2 - 2.07x_2^2 - 7.59x_1x_2$, $\hat{y} - 87.69 = -9.02X_1^2 + 2.97X_2$.

mediately appreciated. (See Davies [19, Chap. 11] or Box and Draper [12].) Figure 2 shows three other surface types representable by (3), the simple maximum, the saddle, and the stationary ridge. For additional details here, see Box et al. [14, pp. 526–534], the source from which Figs. 1 and 2 were adapted.

The n sets of values $(x_{1u}, x_{2u}, \ldots, x_{ku})$ are the coded experimental design points, and may be regarded as a pattern of n points in a k-dimensional space. A response surface design is simply an experimental arrangement of points in **x**-space that permits the fitting of a response surface to the corresponding observations y_u. We thus speak of first-order designs (if a first-order surface can be fitted), second-order designs,

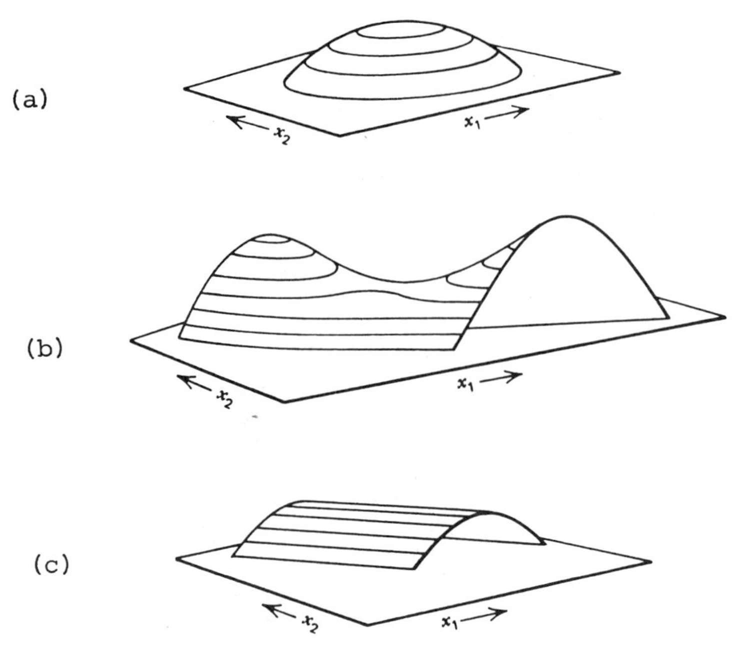

Figure 2 Examples of surfaces representable by a second-degree equation: (a) simple maximum, (b) saddle or col, and (c) stationary ridge.

and so on. Obviously, a design of a particular order is also necessarily a design of lower order.

The choice of a response surface design is thus one of selecting a set of suitable points in k-dimensional x space according to some preselected criterion or criteria of goodness. The technical literature of experimental design contains many discussions of so-called "optimal designs." However, wary skepticism is called for in reading many of these papers, because their authors usually concentrate on one criterion only (and sometimes one that by practical experimental standards is inappropriate) and then derive the best designs under that single criterion. While this often provides interesting mathematical and/or computational exercises and throws light on the behaviour of the examined criterion, it does not necessarily lead to sound practical advice. There are many possible desirable characteristics for a "good" response surface design. Box and Draper [10] gave 14 such characteristics, and, all, or some of which might in different circumstances be of importance. The design should:

1. generate a satisfactory distribution of information about the behaviour of the response variable throughout a region of interest, R;
2. ensure that the fitted value at $\mathbf{x}$, $\hat{y}(\mathbf{x})$, be as close as possible to the true value at $\mathbf{x}$, $\eta(\mathbf{x})$;
3. give good detectability of lack of fit;
4. allow transformations to be estimated;
5. allow experiments to be performed in blocks;
6. allow designs of increasing order to be built up sequentially;
7. provide an internal estimate of error;
8. be insensitive to wild observations and to violation of the usual normal theory assumptions;
9. require a minimum number of experimental points;

10. provide simple data patterns that allow ready visual appreciation;

11. ensure simplicity of calculation;

12. behave well when errors occur in the settings of the predictor variables, the x's;

13. not require an impractically large number of predictor variable levels;

14. provide a check on the "constancy of variance" assumption.

Part of the art of the good practising statistician is his ability to assess the special needs of a given situation and to choose a design that comes close to meeting them. To aid this choice, it would be helpful, where possible, to have appropriate numerical measures of a design's desirability in relation to the various criteria. It would also be helpful to know which criteria are in conflict and which in harmony. Much work remains to be done along these lines.

No design satisfies all the criteria simultaneously. However, there are types of designs that do satisfy many of them. Before discussing any particular designs we briefly elaborate on some of the features mentioned, using the same numbering.

1. In order that $\hat{y}(\mathbf{x})$ should be estimable for all $\mathbf{x}$ belonging to R, the main requirements are

   (a) There must be enough design points to estimate all the coefficients, and preferably additional runs* to cover points **3** and **7**.

   (b) The number of levels of each x_i must exceed the order of the model; otherwise the $\mathbf{X}'\mathbf{X}$ matrix used in the least-squares procedure will be singular.

6. It is an advantage if the observations used to fit, say, a first-order model can be combined with some additional observations and reused to fit a second-order model, especially if a blocking scheme (see point **5**) can be arranged so that differences in levels between the various blocks of the complete design do not affect the final estimates. Such an arrangement allows very economical use of experimental facilities. A design that can be built up in this way is called a *sequentially blocked* response surface design. Randomization of run order would be made only *within* blocks of the design.

7. The provision of an internal estimate of variance error can be achieved by using repeat (replicated) design points. These would often be repeats at the *center* of the design but, where the allowable number of runs permits it, noncentral points could also be replicated. This latter course might be advisable if (i) it were known that the magnitudes of the errors were fairly large in relation to the average size of the observations to be used, and/or (ii) it was desired to measure the error variance at a number of $\mathbf{x}$ locations (see **14**), and/or (iii) some noncentral region were of special interest.

A GENERAL PHILOSOPHY OF SEQUENTIAL EXPERIMENTATION

The center of the experimental design is usually the point representing current "best" (whatever that is defined to mean) conditions, and the objective in empirically fitting a response surface may be:

1. To examine the local nature of the relationship of the response and the predictors and so "explain" the response's behavior. It may, for example, be desired to keep the response within specifications requested by a customer, and/or to see if predictor variable settings are critical and sensitive.

2. To proceed from the current "best" conditions to better conditions (lower cost, higher yield, improved tear resistance, and so on).

Table 1 The Rows are the Coordinates of the $(k+1)$ Points of a Simplex Design in k Dimensions

x_1	x_2	x_3	$\cdots$	x_i	$\cdots$	x_k
$-a_1$	$-a_2$	$-a_3$	$\cdots$	$-a_i$	$\cdots$	$-a_k$
a_1	$-a_2$	$-a_3$	$\cdots$	$-a_i$	$\cdots$	$-a_k$
0	$2a_2$	$-a_3$	$\cdots$	$-a_i$	$\cdots$	$-a_k$
0	0	$3a_3$	$\cdots$	$-a_i$	$\cdots$	$-a_k$
.	.	0		.		.
.	.	.		.		.
.	.	.		.		.
				ia_i		
				0		
				.		
				.		
				.		
0	0	0		0		ka_k

3. To use the fitted surface as a stepping stone to mechanistic understanding of the underlying process.

A more detailed list of possible objectives is given by Herzberg [27].

We would usually first consider the possibility that a first-order model might be satisfactory and perform a first-order design. A simple but good choice (see Box [3]) would be a simplex design with one or more center points. The general simplex in k dimensions has $n = k + 1$ points (runs) and can be oriented to have its coordinates given as in Table 1, where $a_i = \{cn/[i(i+1)]\}^{1/2}$, and c is a scaling constant to be selected. Alternatively, a two-level factorial or fractional factorial*, or a Plackett and Burman design*, with added center point(s) would be excellent. In all cases, the center point(s) average response can be compared to the average response at the noncentral points to give a measure of nonplanarity. For additional details, see Box et al. [14, p. 516] or Box and Draper [12].

If the first-order surface fitted well, one would either interpret its nature if the local relationship were being sought, or else move out along a path of steepest ascent* (or descent) if improved conditions were sought; see Box and Draper [12]. If the first-order surface were an inadequate representation of the local data, either initially or after one or more steepest ascent(s) (or descent(s)), it would be sensible to consider transformations of the response and/or predictor variables that *would* allow a first-order representation. When the possibilities of using first-order surfaces had been exhausted, one would then consider a second-order surface.

It would usually not be necessary at this stage to start from scratch, particularly if a two-level factorial or fractional factorial had just been used. This previous design could be incorporated as an *orthogonal block* in a larger second-order composite design. We first explain how such a design is formed and then how orthogonal blocking may be achieved.

THE CENTRAL COMPOSITE DESIGN

A particular type of second-order design that has many of the desirable features listed is the *central composite* design (normally called just the *composite design**). It is constructed from three sets of points. In the coded **x** space, these three sets can be characterized as follows:

(a) the 2^k vertices $(\pm 1, \pm 1, \ldots, \pm 1)$ of a k-dimensional "cube" $(k \leqslant 4)$, or a fraction of it $(k \geqslant 5)$;

(b) the $2k$ vertices $(\pm\alpha, 0, \ldots, 0)$, $(0, \pm\alpha, \ldots, 0), \ldots, (0, 0, \ldots, 0, \pm\alpha)$ of a k-dimensional cross-polytope or "star";

(c) a number, n_0, of "center points," $(0, 0, \ldots, 0)$.

Set **(a)** is simply a full 2^k factorial design or a 2^{k-p} fractional factorial if $k \geqslant 5$. The notation $(\pm 1, \pm 1, \ldots, \pm 1)$ means that 2^k points obtained by taking all possible combinations of signs are used for full factorial cases. (In response surface applications, these points are often referred to as a "cube," whatever the number of factors.)

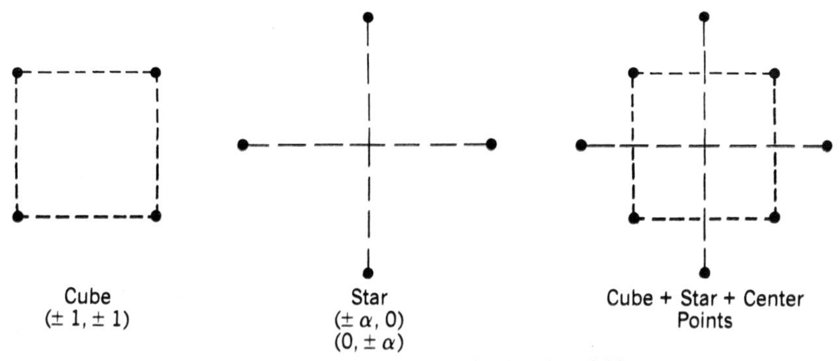

Figure 3 Composite design for $k = 2$ variables.

Set **(b)** consists of pairs of points on the coordinate axes all at a distance α from the origin. (The quantity α has yet to be specified; according to its value the points may lie inside or outside the cube.) In three dimensions the points are the vertices of an octahedron and this word is sometimes used for other values of $k \neq 3$. However, a more convenient name for such a set of points in k dimensions is "star" or, more formally, cross-polytope.

These sets and the complete design (the n_0 center points represented by a single center point) are shown diagrammatically in Figs. 3 and 4 for the cases $k = 2$ and 3.

Fractionation of the cube is possible whenever the resulting design will permit individual estimation of all the coefficients in Eq. (3). This is guaranteed for fractions of resolution $\geqslant 5$. (*See* RESOLUTION and PLACKETT AND BURMAN DESIGNS and references therein for further information.) The smallest useable fraction is then a 2^{k-1} design (a half-fraction) for $k = 5, 6, 7$, a 2^{k-2} design (a quarter-fraction) for $k = 8, 9$, a 2^{k-3} for $k = 10$, and so on. (See Box et al. [14, p. 408].) Table 2, adapted from Box and Hunter [13, p. 227], shows the number of parameters in Eq. (3) and the number of noncentral design points in the corresponding composite design for $k = 2, \ldots, 9$. The values to be substituted for p are $p = 0$ for $k = 2$, 3, and 4; $p = 1$ for $k = 5$, 6, and 7; and $p = 2$ for $k = 8$ and 9; they correspond

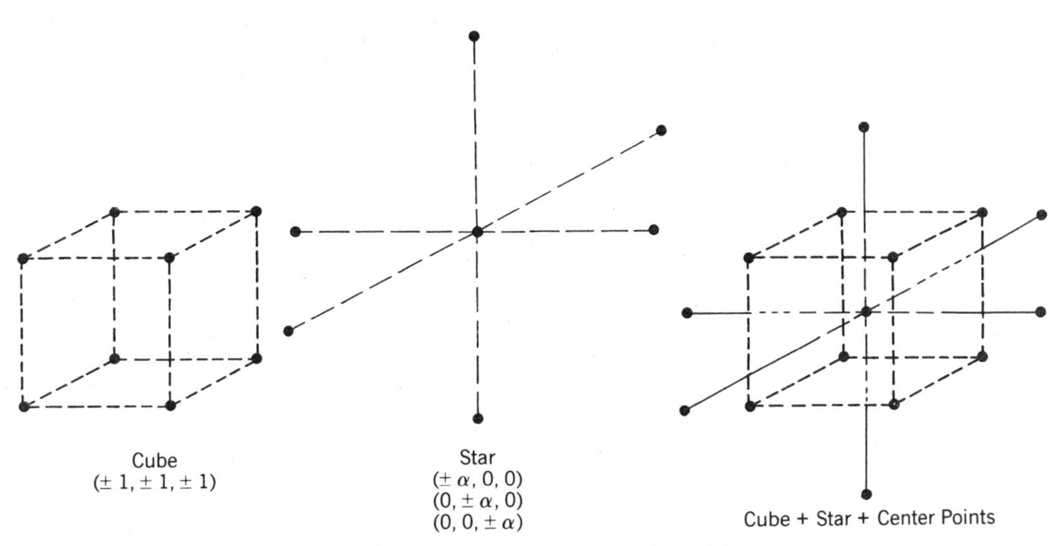

Figure 4 Composite design for $k = 3$ variables.

Table 2 Features of Certain Composite Designs

No. of variables	k	2	3	4	5	6	7	8	9
No. of parameters	$(k+1)(k+2)/2$	6	10	15	21	28	36	45	55
Cube + star	$2^k + 2k$	8	14	24	—	—	—	—	—
$\frac{1}{2}$(cube) + star	$2^{k-1} + 2k$	—	—	—	26	44	78	—	—
$\frac{1}{4}$(cube) + star	$2^{k-2} + 2k$	—	—	—	—	—	—	80	130
α (rotatable)	$2^{(k-p)/4}$	1.414	1.682	2	2	2.378	2.828	2.828	3.364
Suggested n_0		2–4	2–4	2–4	0–4	0–4	2–4	2–4	2–4

to the fraction, $1/2^p$, of the cube used for the design.

We can check immediately that the composite designs have at least some of the 14 desirable features. For example, there are enough points and enough levels (three if $\alpha = 1$, five if $\alpha \neq 1$) to satisfy points 1 and 13. The designs can be performed sequentially; the cube or factorial portion plus center points can be used as a first-order design and the additional star points, plus center points, complete the second-order design. Thus points **5** and **6** are achieved. (If a block effect changes the response level between the running of the two sections, it will usually be detected through the center point readings. The block effect could be estimated as the difference between the average responses at the center levels in each of the two blocks and the observations in one or the other block could be appropriately adjusted, if desired. Alternatively, the design can be *orthogonally blocked*, that is, blocked in such a way that block effects are orthogonal to model estimates and so do not affect them; see below.)

The n_0 repeated center points allow the internal (pure error) estimation of error as in point **7**. The number of design points is reasonable in relation to the number of coefficients if not minimal (point 9). The pattern of the design (point 10) is clearly excellent, and the least-squares* calculations are simple (point **11**). The designs are also robust to small errors in the settings of the x's since a slight displacement of the design points will not materially affect the fitted surface (point **12**). However, a wild observation may cause an erroneous displacement of the fitted surface (point **8**). This can often be detected from the patterns exhibited by the standard residuals plots if the effect is serious. The size of a possible displacement may be reduced if all or some of the noncentral design points are replicated, because, in a set of repeats, a single wild observation will be "muted" by its "correct" replicates. Points 2 and 3 can be satisfied by choice of α, n_0, and by shrinking or expanding all the design points relative to the region R (see Box and Draper [8, 9]; see also, Welch [37] and Houck and Myers [30]). Point 4 is also satisfied (see Box and Draper [11]). Overall then, the composite design is an excellent choice.

What values should be chosen for α and n_0? The value of α determines if the star points fall inside the cube ($\alpha < 1$), outside the cube ($\alpha > 1$), or on the faces of the cube ($\alpha = 1$). Note that when $\alpha = 1$ only three experimental levels $(-1, 0, 1)$ are required, which may be an advantage or necessity in some experimental situations. For additional comments and specific designs see De Baun [20] and Box and Behnken [7].

If three levels are not essential, what value of α should be selected? One criterion that can be applied to decide this is that of *rotatability*. A design (of any order) is rotatable* when the contours of the variance function $V\{\hat{y}(\mathbf{x})\}$ are spheres about the origin in the k-dimensional factor space defined by variables $x_1, x_2, \ldots, x_k$. Box and Hunter [13] showed that the required values (given in Table 2) are $\alpha = 2^{(k-p)/4}$, where $p = 0$, 1, or 2 according to the fraction of the cube used in the design.

Note that the rotatability property is specifically related to the codings chosen for the x's. It is usually assumed that these codings have been chosen in a manner that anticipates (roughly speaking) that one unit of change in any x will have about the same effect on the response variable. In such a case, obtaining equal information at the same radial distance in any direction (which is what rotatability implies) is clearly sensible. Codings are rarely perfect; the codings are adjusted in future designs as a result of information gained in current and past experiments. Exact rotatability is not a primary consideration. However, knowledge of the tabulated values provides a target to aim at, while attempting to satisfy other desirable design features.

How large a value should be selected for n_0? There are many possible criteria to apply; these are summarized by Draper [22, 23]. The suggested values in the table appear to be sensible with respect to many criteria, the overall message being that only a few center points are usually needed. (Whenever α is chosen so that all the design points lie on a sphere, at least one center point is essential; otherwise not all of the coefficients can be individually estimated.) A few additional center points will do no harm. Nevertheless, additional runs are probably better used to duplicate selected noncentral design points, unless special considerations apply, as below. Repeated points spread over the design provide a check of the usual "homogeneous variance" assumption; see Box [5] and Dykstra [25].

For a numerical example of a second-order response surface fitting for $k = 3$, see Draper and Smith [24, pp. 390–403]. For a wide variety of examples, see Box and Draper [12].

ORTHOGONAL BLOCKING

Another criterion (previously mentioned) that may be applied to the choice of α and n_0 in the composite design is that of *orthogonal blocking*. This requires division of the runs into two or more blocks in such a manner that this division does not affect the estimates of the second-order model obtained via the standard least-squares regression analysis*. The basic approach was given by Box and Hunter [13]; see also DeBaun [20] and Box [5]. Two conditions must be satisfied:

1. Each block must itself be a first-order orthogonal design. Thus $\Sigma_u x_{iu} x_{ju} = 0$, $i \neq j$, for each block.
2. The fraction of the total sum of squares of each variable x_i contributed by every block must be equal to the fraction of the total observations allotted to the block. Thus, for each block,

$$\frac{\Sigma_u x_{iu}^2}{\Sigma_{u=1}^n x_{iu}^2} = \frac{n_b}{n}, \tag{4}$$

where n_b denotes the number of runs in the block under consideration, Σ_u denotes summation *only in that block*, and the denominators of (4) refer to the entire design.

The simplest orthogonal block division of the composite design is into the orthogonal design pieces:

Block 1. Cube portion (2^{k-p} points) plus c_0 center points.

Block 2. Star portion ($2k$ points) plus s_0 center points.

Application of (4) then implies that

$$\alpha = \left\{2^{k-p-1}(2k + s_0)/\left(2^{k-p} + c_0\right)\right\}^{1/2}. \tag{5}$$

For example, if $k = 3$ and $p = 0$, so that the first block is a 2^3 factorial plus c_0 center points and the second block is a six-point octahedron plus s_0 center points, then

$$\alpha = \{4(6 + s_0)/(8 + c_0)\}^{1/2}. \tag{6}$$

If $c_0 = 4$ center points are added to the cube and no center points are added to the star ($s_0 = 0$), then $\alpha = 2^{1/2} = 1.414$. This design is orthogonally blocked but is *not* rotatable. However, values of α closer to the rotatable value 1.682 are possible. For example, if $c_0 = 0$, $s_0 = 0$, and $\alpha = (24/8)^{1/2} = 1.732$, or if $c_0 = 4$, $s_0 = 2$, and $\alpha = (32/12)^{1/2} = 1.633$. The choices are, of course, limited by the fact that c_0 and s_0 must be integers. Generally, orthogonal blocking [α from (5)] takes precedence over rotatability, for which $\alpha = 2^{(k-p)/4}$ is needed. In certain cases, both can be achieved simultaneously. This requires

$$2^{k-p} + c_0 = 2^{0.5(k-p)-1}(2k + s_0) \quad (7)$$

to be satisfied for integer (k, p, c_0, s_0). Some possibilities are $(2, 0, s_0, s_0)$, $s_0 \geqslant 1$; $(4, 0, 2s_0, s_0)$, $(5, 1, (4 + 2s_0), s_0)$, $(7, 1, 4(s_0 - 2), s_0)$, $s_0 \geqslant 2$; and $(8, 2, 4s_0, s_0)$, where $s_0 = 0, 1, 2, \ldots$, unless otherwise specified. (Note that some of these arrangements call for more center points than recommended in the table, an example of how applications of different criteria can produce conflicting conclusions.)

Further division of the star will not lead to an orthogonally blocked design. However, it is possible to divide the cube portion into smaller blocks and still maintain orthogonal blocking if $k > 2$. As long as the pieces that result are fractional factorials of resolution III or more (see Box et al. [14, p. 385]), each piece will be an orthogonal design. All fractional factorial pieces *must* contain the same number of center points or else (4) cannot be satisfied. Thus c_0 must be divisible by the number of blocks.

Table 3 A 24-Run Second-Order Rotatable Response Surface Design for Three Factors, Orthogonally Blocked into Four Blocks of Equal Size

Block	x_1	x_2	x_3	Design
I	-1	-1	1	2^{3-1} design,
	1	-1	-1	$x_1x_2x_3 = 1$
	-1	1	-1	plus two
	1	1	1	center points
	0	0	0	
	0	0	0	
II	-1	-1	-1	2^{3-1} design,
	1	-1	1	$x_1x_2x_3 = -1$
	-1	1	1	plus two
	1	1	-1	center points
	0	0	0	
	0	0	0	
III	$-2^{1/2}$	0	0	star,
	$2^{1/2}$	0	0	$\alpha = 2^{1/2}$
	0	$-2^{1/2}$	0	
	0	$2^{1/2}$	0	
	0	0	$-2^{1/2}$	
	0	0	$2^{1/2}$	
IV	$-2^{1/2}$	0	0	star,
	$2^{1/2}$	0	0	$\alpha = 2^{1/2}$
	0	$-2^{1/2}$	0	
	0	$2^{1/2}$	0	
	0	0	$-2^{1/2}$	
	0	0	$2^{1/2}$	

AN ATTRACTIVE THREE-FACTOR DESIGN

In a composite design, replication of either the cube portion or the star portion, or both can be chosen if desired. As an example of such possibilities, we now provide, in Table 3, a 24-run second-order design for three factors that is rotatable *and* orthogonally blocked into four blocks of equal size. It consists of a cube (fractionated via $x_1x_2x_3 = \pm 1$) plus replicated (doubled) star plus four center points, two in each 2^{3-1} block. This design provides an illustration of the fact that center points in *different* blocks of the design are no longer comparable due to possible block effects. Thus, the sum of squares for pure error must be obtained by pooling the separate sums of squares for pure error from each block.

GENERAL COMMENT

A second-order response surface design will be very effective if the underlying surface being examined is roughly quadratic. If it is an attenuated or distorted quadratic, transformations on the x variables will often be needed. In practice, one usually discovers the need for such transformations by observing the nonquadratic curvature in the data *after* a second-order design has been used and finding that the fitted quadratic surface cannot properly handle that curvature.

QUALITATIVE VARIABLES

Our discussion so far has effectively assumed that all the ξ's are quantitative variables able to assume any value in some specified range limited only by the practicalities of the experimental situation. In some experimentation, some of the predictor variables are qualitative, that is, able to take only distinct values. For example, three different catalysts might constitute three qualitative levels of one factor. So might three fertilizers, unless they were constructed, for example, by altering the level of an ingredient; in such a case, the three fertilizers would usually be regarded as constituting three levels of a continuous variable. Variables such as shifts, reactors, operators, machines, and railcars would, typically, be qualitative variables. When qualitative variables occur in a response surface study, surfaces in the quantitive variables are fitted separately for each combination of qualitative variables. For illustrative commentary see Box et al. [14, pp. 296–299].

ANALYSIS FOR ORTHOGONALLY BLOCKED DESIGNS

When a second-order design is orthogonally blocked:

1. Estimate the β coefficients of the second-order model in the usual way, ignoring blocking.
2. Calculate pure error from repeated points *within* the same block only, and then combine these contributions in the usual way. Runs in different blocks cannot be considered as repeats.
3. Place an extra term

$$\text{SS(blocks)} = \sum_{w=1}^{m} \frac{B_w^2}{n_w} - \frac{G^2}{n},$$

with $(m - 1)$ degrees of freedom in the analysis of variance* table, where B_w is the total of the n_w observations in the wth block and G is the grand total of all the observations in all the m blocks.

FURTHER READING

The literature of response surface methodology is very extensive. For readers who would like to know more, we encourage the following course of action.

1. Obtain an overview of the field from the excellent review papers of Mead and Pike [33] and Morton [34]. (Although the former paper is "from a biometric viewpoint," it will also serve the nonbiometric viewpoint reader extremely well.) Then look at the earlier review papers of Hill and Hunter [29] and Herzberg and Cox [28] for additional broadening.
2. Read the succession of key papers by Box and Wilson [16], Box [3, 4], Box and Youle [17], and Box and Hunter [13]. They contain the basic ideas and philosophy essential to a full understanding of the field. (Although the fundamental ideas of response surface methodology (RSM) existed before his series of contributions began, George E. P. Box is justifiably regarded as the founding father of modern RSM. His succession of publications, written alone and with co-authors, extends over several decades.)
3. For a definitive textbook account of response surface methodology, see Box and Draper [12]. Less extensive accounts may also be found in other texts, for example, Box et al. [14], Davies [19], Guttman et al. [26], and Myers [35]. For response surface applications in experiments with mixtures* of ingredients, see Cornell [18].
4. Readers interested in the mathematical and computational problems of "optimal design"* theory with its various alphabetic optimality criteria should read

the ingenious contributions of J. Kiefer, his co-authors, and others. For references see, for example, St. John and Draper [36], Herzberg [27], and Kiefer [31]. For a discussion of the problems, see Box [6], summarized in Box and Draper [12, Chap. 14]. See also Lucas [32] and Atkinson [2]. For the application of optimality criteria for compromise purposes, for example, to obtain designs that can be reasonably efficient both for model testing and parameter estimation, see Atkinson [1].

References

[1] Atkinson, A. C. (1975). Planning experiments for model testing and discrimination. *Math. Oper. und Statist.*, **6**, 253–267.

[2] Atkinson, A. C. (1982). Developments in the design of experiments. *Int. Statist. Rev.*, **50**, 161–177.

[3] Box, G. E. P. (1952). Multifactor designs of first order. *Biometrika*, **39**, 49–57.

[4] Box, G. E. P. (1954). The exploration and exploitation of response surfaces: some general considerations and examples. *Biometrics*, **10**, 16–60.

[5] Box, G. E. P. (1959). Answer to query: Replication of non-center points in the rotatable and near-rotatable central composite design. *Biometrics*, **15**, 133–135.

[6] Box, G. E. P. (1982). Choice of response surface design and alphabetic optimality. *Utilitas Math.*, **21B**, 11–55.

[7] Box, G. E. P. and Behnken, D. W. (1960). Some new three level designs for the study of quantitative variables. *Technometrics*, **2**, 455–475.

[8] Box, G. E. P. and Draper, N. R. (1959). A basis for the selection of a response surface design. *J. Amer. Statist. Ass.*, **54**, 622–654.

[9] Box, G. E. P. and Draper, N. R. (1963). The choice of a second order rotatable design. *Biometrika*, **50**, 335–352.

[10] Box, G. E. P. and Draper, N. R. (1975). Robust designs. *Biometrika*, **62**, 347–352.

[11] Box, G. E. P. and Draper, N. R. (1982). Measures of lack of fit for response surface designs and predictor variable transformations. *Technometrics*, **24**, 1–8.

[12] Box, G. E. P. and Draper, N. R. (1987). *Empirical Model-Building with Response Surfaces*. Wiley, New York.

[13] Box, G. E. P. and Hunter, J. S. (1957). Multifactor experimental designs for exploring response surfaces. *Ann. Math. Statist.*, **28**, 195–241.

[14] Box, G. E. P., Hunter, W. G., and Hunter, J. S. (1978). *Statistics for Experimenters. An Introduction to Design, Data Analysis and Model Building*. Wiley, New York.

[15] Box, G. E. P. and Lucas, H. L. (1959). Design of experiments in non-linear situations. *Biometrika*, **46**, 77–90.

[16] Box, G. E. P. and Wilson, K. B. (1951). On the experimental attainment of optimum conditions. *J. R. Statist. Soc. B*, **13**, 1–38 (discussion 38–45).

[17] Box, G. E. P. and Youle, P. V. (1955). The exploration and exploitation of response surfaces: an example of the link between the fitted surface and the basic mechanism of the system. *Biometrics*, **11**, 287–322.

[18] Cornell, J. A. (1981). *Experiments with Mixtures: Designs, Models, and the Analysis of Mixture Data*. Wiley, New York.

[19] Davies, O. L., ed. (1978). *Design and Analysis of Industrial Experiments*. Longman Group, New York and London.

[20] De Baun, R. M. (1956). Block effects in the determination of optimum conditions. *Biometrics*, **12**, 20–22.

[21] De Baun, R. M. (1959). Response surface designs for three factors at three levels. *Technometrics*, **1**, 1–8.

[22] Draper, N. R. (1982). Center points in response surface designs. *Technometrics*, **24**, 127–133.

[23] Draper, N. R. (1984). Schläflian rotatability. *J. R. Statist. Soc. B*, **46**, 406–411.

[24] Draper, N. R. and Smith, H. (1981). *Applied Regression Analysis*, 2nd ed. Wiley, New York.

[25] Dykstra, O. (1959). Partial duplication of response surface designs. *Technometrics*, **2**, 185–195.

[26] Guttman, I., Wilks, S. S., and Hunter, J. S. (1971). *Introductory Engineering Statistics*. Wiley, New York (see Chap. 17).

[27] Herzberg, A. M. (1982). The robust design of experiments: a review. *Serdica Bulgaricae Math. Publ.*, **8**, 223–228.

[28] Herzberg, A. M. and Cox, D. R. (1969). Recent work on the design of experiments: a bibliography and a review. *J. R. Statist. Soc. A*, **132**, 29–67.

[29] Hill, W. J. and Hunter, W. G. (1966). A review of response surface methodology: a literature survey. *Technometrics*, **8**, 571–590.

[30] Houck, E. C. and Myers, R. H. (1978). A sequential design procedure for the estimation of graduating polynomials of order one. *Proc. Tenth Annual Mtg. Amer. Inst. Decision Sciences*, **1**, 275–277.

[31] Kiefer, J. C. (1984). The publications and writings of Jack Kiefer. *Ann. Statist.*, **12**, 424–430. ("The complete bibliography... was prepared through the efforts of Roger Farrell and Ingram Olkin." The quotation is from the cited journal, p. 403.) See Kiefer, J. C. (1985). *Collected Papers III, Design of Experiments* (L. D. Brown, I. Olkin, J. Sacks, and H. P. Wynn, eds.) Springer-Verlag, New York.

[32] Lucas, J. M. (1976). Which response surface design is best? *Technometrics*, **18**, 411–417.

[33] Mead, R. and Pike, D. J. (1975). A review of response surface methodology from a biometric viewpoint. *Biometrics*, **31**, 803–851.

[34] Morton, R. H. (1983). Response surface methodology. *Math. Sci.*, **8**, 31–52.

[35] Myers, R. H. (1971). *Response Surface Methodology*. Edwards Bros, Ann Arbor, MI.

[36] St. John, R. C. and Draper, N. R. (1975). Optimality for regression designs: a review. *Technometrics*, **17**, 15–23.

[37] Welch, W. J. (1983). A mean squared error criterion for the design of experiments. *Biometrika*, **70**, 205–213.

Acknowledgments

Helpful comments and suggestions from A. C. Atkinson, J. A. Cornell, D. R. Cox, A. M. Herzberg, W. J. Hill, J. S. Hunter, W. G. Hunter, R. Mead, R. H. Myers, D. J. Pike, R. Snee, and the editors are gratefully acknowledged.

(ANALYSIS OF VARIANCE
COMPOSITE DESIGNS
DESIGN OF EXPERIMENTS
FRACTIONAL FACTORIAL DESIGNS
OPTIMAL DESIGN OF EXPERIMENTS
ORTHOGONAL DESIGNS
RIDGE REGRESSION
ROTATABLE DESIGNS)

NORMAN R. DRAPER

RESTRICTED RANDOMIZATION *See* RANDOMIZATION, CONSTRAINED

RESTRICTION ERROR

This term, introduced by Anderson [1] (see also Anderson and McLean [2]), refers to the effect of restrictions on the randomization* used in designing an experiment. It seems to refer to the choice of model on which to analyze the experimental data, though in its discussion there appears to be a lack of clearcut distinction between randomization models and fixed, mixed, and random effects models*.

As an example to clarify ideas, consider an experiment in which there are bt experimental units to each of which will be assigned one of the $b \times t$ possible combinations of levels of two factors, one (blocks) having b levels, and the other (treatments). If each of the $(bt)!$ possible arrangements is equally likely to be chosen by the randomization procedure, the design is said to be *completely randomized*. If, on the other hand the t units in each of the b blocks are prespecified, and randomization consists only in random assignment of the t treatments among the t units in each block, the randomization is then restricted. Such a situation arises, for example, if the units are plots in blocks on a field. The appropriate randomization distribution is different in the two cases (see, e.g., Johnson [3]). Common models for the two cases, based on observed values $\{x_{ij}\}$ for combination of ith level of blocks and jth level of treatments would be:

(a) For complete randomization

$$X_{ij} = \xi + \beta_i + \tau_j + V_{ij},$$

$$i = 1, \ldots, b; \; j = 1, \ldots, t,$$

where $V_{11}, \ldots, V_{bt}$ is a rearrangement of the quantities $\{x_{ij} - \bar{x}..\}$, where $x.. = (bt)^{-1}\Sigma_{i=1}^{b}\Sigma_{j=1}^{t}x_{ij}$, each of the $(bt)!$ possible rearrangements being equally likely.

(b) For restricted randomization (as described above)

$$X_{ij} = \xi + \bar{x}_{i\cdot} + \tau_j + U_{ij},$$

$$i = 1, \ldots, b; \; j = 1, \ldots, t,$$

where $U_{i1}, \ldots, U_{it}$ is a rearrangement of the quantities $x_{i1} - \bar{x}_{i\cdot}, \ldots, x_{it} - \bar{x}_{i\cdot}$ $(i = 1, \ldots, b)$, each one of the $(t!)^b$ possible rearrangements being equally likely (and $\bar{x}_{i\cdot} = t^{-1}\Sigma_{j=1}^{t}x_{ij}$).

Anderson [1], on the other hand, proposes use of a model of the form

$$X_{ij} = \xi + \beta_i + \delta_{(i)} + \tau_j + \epsilon_{(ij)},$$

$$i = 1, \ldots, b;\ j = 1, \ldots, t,$$

where $\delta_{(i)}$ represents the *restriction error*. The quantities $\delta_{(i)}$ and $\delta_{(ij)}$ are supposed to be independent random variables, each normal with expected value zero and variances $\sigma_\delta^2, \sigma_\epsilon^2$ (not depending on i or j), respectively. It seems that it is not expected that randomization theory will be used in the analysis. Anderson gives a formal analysis of variance*, based on the model in which the "restriction error" sum of squares is zero and has zero degrees of freedom.

References

[1] Anderson, V. L. (1970). *Biometrics*, **26**, 255–268.

[2] Anderson, V. L. and McLean, R. A. (1974). *Design of Experiments*. Dekker, New York.

[3] Johnson, N. L. (1959). *Biometrika*, **45**, 265–266.

(ANALYSIS OF VARIANCE
BLOCKS, RANDOMIZED COMPLETE
CROSS-CLASSIFICATION
RANDOMIZATION
RANDOMIZATION,
CONSTRAINED)

RESUBMITTED LOT

A lot that has previously been designated as "nonacceptable" but has been again submitted for acceptance inspection* after (presumably) having been further checked and corrected.

Reference

[1] Freund, R. A., Tsiakals, J. J., and Murphy, T. D. (1983). *Glossary and Tables for Statistical Quality Control*, 2nd edn. American Society for Quality Control. Milwaukee, WI.

(QUALITY CONTROL, STATISTICAL)

RETROSPECTIVE STUDIES (INCLUDING CASE-CONTROL)

In nonexperimental research, historical information can be used to evaluate the relationship between two or more study factors. When all of the phenomena under investigation occur prior to the onset of a study, the research may be described as retrospective in design. That is to say, the study findings are based upon a look backward in time. The retrospective approach has proven useful for research in the social, biological, and physical sciences. Regardless of the field of inquiry, there are certain contexts in which a retrospective study is appropriate. The most obvious reason to conduct a retrospective investigation is to evaluate the effects of an event that occurred in the past and is not expected to recur in the future. Similarly, it may be necessary to resort to historical information for the study of events that occur infrequently, or are difficult to predict in advance. For example, much of the information on the health effects of ionizing radiation was obtained through retrospective studies of persons exposed to the bombings of Hiroshima and Nagasaki [3]. Retrospective studies are often performed to evaluate deleterious exposures, such as ionizing radiation, when an experiment with human subjects cannot be morally justified (*see* CLINICAL TRIALS and HISTORICAL CONTROLS).

A retrospective approach also offers logistical advantages for the study of factors that are separated by a prolonged period of time. With the use of historical information, the investigator does not have to wait for the occurrence of the outcome variable. As an illustration, consider research on a suspected cause of cancer in humans. For many carcinogens, the median time span between first exposure and the subsequent clinical detection of cancer is more than a decade [1]. A study that begins at the time of exposure to a suspected carcinogen would require years to evaluate the subsequent risk of cancer. In contrast, a retrospective investigation could be undertaken after the occurrence of the clinical outcomes, thereby saving time and resources.

TYPES OF RETROSPECTIVE STUDIES

Retrospective research may be subclassified according to the sequence of observations on the factors of interest. In a retrospective cohort study, (also known as historical cohort study), the subjects are entered by level of past exposure to the antecedent variable. In the usual situation, two groups of subjects are defined: exposed and unexposed subjects (Fig. 1). Then, the subsequent frequencies of the outcome variable in these groups are compared. Although the subjects are traced forward in time, the study is retrospective because all of the phenomena under consideration occurred prior to the onset of investigation.

Example of a Retrospective Cohort Study. The evaluation of the long-term health effects of occupational exposures often is performed with the retrospective cohort method. As a typical example, consider a study conducted by McMichael and colleagues [8] on mortality in workers in the rubber industry. These authors used employment records to identify a group of active and retired employees of a tire factory who were alive on January 1, 1964. Through various record sources, the subsequent mortality experience of these 6678 workers was determined for a nine-year period of time. The age and sex-specific mortality rates of the U.S. population were used for comparison. The authors found that the rubber workers had excessive numbers of deaths from cancers of the stomach, prostate, and hematopoietic tissues [8].

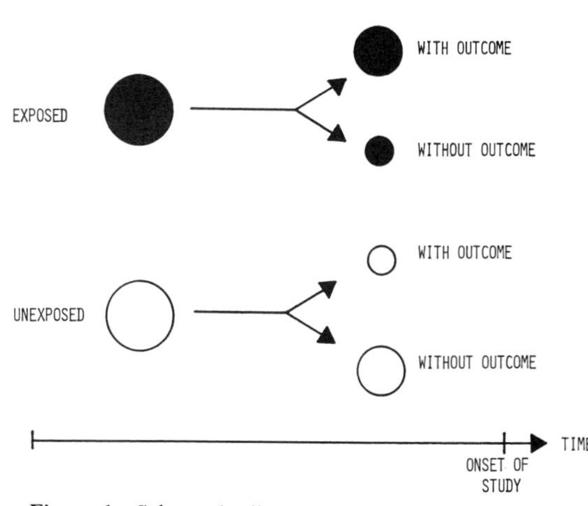

Figure 1 Schematic diagram of a retrospective cohort study. Shaded areas represent subjects exposed to the antecedent factor; unshaded areas correspond to unexposed subjects.

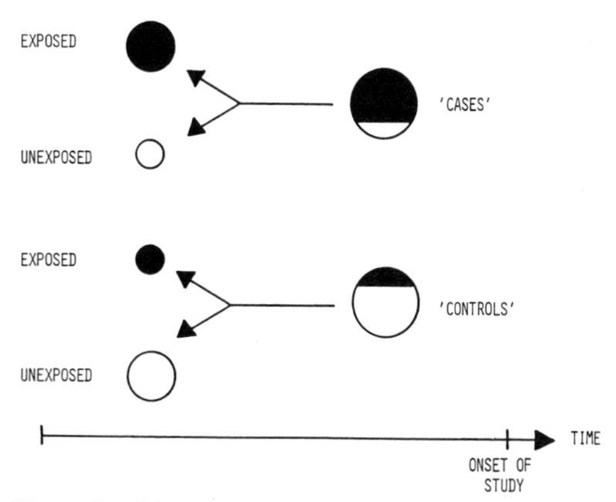

Figure 2 Schematic diagram of a case-control study. Shaded areas represent subjects exposed to the antecedent factor; unshaded areas correspond to unexposed subjects.

A second type of retrospective study is referred to as the case-control method, (also known as the case-compeer, or case-referent study). In a case-control investigation, the subjects are entered according to the presence or absence of the outcome variable (Fig. 2). By convention, persons who possess the outcome are labelled as "cases" and persons who do not possess the outcome are termed "controls." The case and control groups then are compared with respect to previous exposure to the antecedent factor. Thus, in a case-control study, the natural sequence of events is reversed and subjects are traced from outcome to a preceding exposure.

Example of a Case-Control Study. Often, the evaluation of a suspected cause of disease is performed with the case-control method. As a typical example, consider a study conducted by Clarke and colleagues [4] to evaluate the purported association between cigarette smoking and cancer of the cervix. A total of 181 women with cervical cancer (i.e., cases) and 905 women without cervical cancer (i.e., controls) were interviewed to determine previous smoking history. The authors found that a significantly greater proportion of cases had smoked cigarettes, when compared with controls [4]. Of course, this finding does not prove that cigarette smoking causes cervical cancer. One cannot exclude the possibility that this association occurred because cigarette smoking is related to a correlate of cervical cancer, such as sexual practices.

THE FREQUENCY OF STUDY FACTORS AND RETROSPECTIVE RESEARCH

In nonexperimental research, the choice of a study design often is dictated by the characteristics of the factors under investigation. As indicated previously, a retrospective approach is useful when the study factors are separated by a prolonged period of time. Also, consideration must be given to the relative frequencies of the study factors in the source population. When exposure to the putative causal factor is rare in the general population, a study sample that is selected randomly will have few exposed subjects. For this situation, disproportionate sampling rates of exposed and unexposed subjects are desirable. In a retrospective cohort study, the investigator can fix the prevalence of exposure, because the subjects are sampled contingent upon the presence or absence of the antecedent factor.

Example of a Retrospective Cohort Study of a Rare Exposure. The polybrominated biphenyls (PBBs) are chemicals that are used commercially as fire retardants. Although heavy exposure to these compounds is uncommon, concern has been raised about the health effects of such exposures. Bahn and co-workers [2] used historical information to identify a group of 86 men that were employed at a firm that manufactured PBBs. A separate group of 89 unexposed persons was chosen from two other industries and the community. Evidence of primary thyroid dysfunction was found in four exposed subjects, as compared with none of the unexposed persons [2].

When the effect of interest is rare in the general population, a randomly selected sample will have few subjects with the outcome under study. For this situation, disproportionate sampling rates for the outcomes are desirable. In a case-control study, the investigator can fix the prevalence of the effect of interest, because the subjects are sampled contingent upon the presence or absence of the outcome.

Example of a Case-Control Study of a Rare Outcome. The Toxic Shock Syndrome (TSS) is a life-threatening illness that is related to certain types of staphylococcal infections. Among menstruating women, the population at greatest risk of TSS, this disease is still uncommon, with an estimated annual incidence of 144 cases per million women [6]. To evaluate the relationship between use of tampons and subsequent risk of TSS, Kehrberg and colleagues [6] conducted a case-control study. These authors found that the prior use of tampons was significantly more frequent among the 29 women with TSS, as compared to the 91 women without TSS [6].

When both the supposed cause and effect of interest are rare in the general population, the standard retrospective methods often lack sufficient statistical power to evaluate the association of these factors. In this situation, the investigator may choose a hybrid design, which combines the cohort and case-control sampling procedures. Although a variety of hybrid studies might be envisioned, consider one particular approach. First, historical information is used to identify a group with a moderate baseline rate of exposure to the suspected causal factor. Then, subjects within this cohort are sampled contingent upon the presence or absence of the outcome of interest. This hybrid approach has been described as a case-control study nested within a cohort [7].

Example of a Case-Control Study within a Cohort. To evaluate the relationship between heavy exposure to certain heavy metal oxides and subsequent risk of prostate cancer, Goldsmith and colleagues [5] first identified an occupational group with a moderate baseline rate of exposure to these compounds. Then, within this cohort, 88 men with prostate cancer (i.e., cases) and 258 men without prostate cancer (i.e., controls) were sampled. The authors subsequently found that the level of exposure to heavy metal oxides and organic accelerators was significantly more frequent among cases, as compared with controls [5].

OTHER CONSIDERATIONS IN RETROSPECTIVE RESEARCH

Aside from the temporal pattern and frequencies of the study factors, the choice of a retrospective study design may be influenced by other considerations. For instance, case-control studies are relatively inexpensive and can be conducted in a short period of time. Also, the case-control approach allows the simultaneous evaluation of multiple suspected causal factors. However, these advantages must be weighed against the following limitations of case-control studies:

1. Only one outcome can be evaluated.
2. The rates of the outcomes within exposure groups cannot be estimated.
3. The method is especially susceptible to certain types of systematic errors.
4. The manner in which cases and controls are selected can introduce bias into the study results.
5. A causal relationship cannot be established by a single study.

With the retrospective cohort method, more than one outcome can be evaluated. Also, the rates of the various outcomes within exposure groups can be estimated with this design. However, the retrospective cohort approach has the following limitations:

1. Only one suspected causal factor can be evaluated.
2. The task of determining outcomes within cohorts can be tedious and time-consuming.
3. The reliance upon recall or historical records may limit the type of information available and even lead to erroneous conclusions.
4. A causal relationship cannot be established by a single study.

Ultimately, the choice of a research strategy is affected by the goals of the study and the nature of the factors under investigation. For many scientific questions, a retrospective design provides an expedient and appropriate method of evaluation.

References

[1] Armenian, H. K. and Lilienfeld, A. M. (1974). *Amer. J. Epidemiol.*, **99**, 92–100.

[2] Bahn, A. K., Mills, J. L., Snyder, P. J., et al. (1980). *New Engl. J. Med.*, **302**, 31–33.

[3] Beebe, G. W., Kato, H., and Land, C. E. (1978). *Radiat. Res.*, **75**, 138–201.

[4] Clarke, E. A., Morgan, R. W., and Newman, A. M. (1982). *Amer. J. Epidemiol.*, **115**, 59–66.

[5] Goldsmith, D. F., Smith, A. H., and McMichael, A. J. (1980). *J. Occupat. Med.*, **22**, 533–541.

[6] Kehrberg, M. W., Latham, R. H., Haslam, B. T., et al. (1981). *Amer. J. Epidemiol.*, **114**, 873–879.

[7] Kupper, L. L., McMichael, A. J., and Spirtas, R. (1975). *J. Amer. Statist. Ass.*, **70**, 524–528.

[8] McMichael, A. J., Spirtas, R., and Kupper, L. L. (1974). *J. Occupat. Med.*, **16**, 458–464.

Bibliography

Breslow, N. E. and Day, N. E. (1980). *Statistical Methods in Cancer Research. Volume 1. The Analysis of Case-Control Studies.* International Agency for Research on Cancer, Lyon, France. (A technical treatment of analytical methods for case-control studies.)

Ibrahim, M. A. (1979). *The Case-Control Study. Consensus and Controversy*. Pergamon Press, Oxford, England. (A compendium of invited papers on various aspects of case-control research.)

Kleinbaum, D. G., Kupper, L. L., and Morgenstern, H. (1982). *Epidemiologic Research. Principles and Quantitative Methods*. Lifetime Learning Publications, Belmont, CA. (An excellent textbook on epidemiological methods, with a thorough coverage of study design and analysis.)

Schlesselman, J. J. (1982). *Case-Control Studies. Design, Conduct, Analysis*. Oxford University Press, New York. (A highly readable reference on the principles and methods of case-control research.)

Acknowledgments

The preparation of this entry was supported in part by National Cancer Institute Contract N01-CN-61027. The manuscript was typed by Ms. Vickie Thomas.

(BIOSTATISTICS
CLINICAL TRIALS
COX'S REGRESSION MODEL
EPIDEMIOLOGICAL STATISTICS
HISTORICAL CONTROLS
PROSPECTIVE STUDIES
SURVIVAL ANALYSIS)

RAYMOND S. GREENBERG

RETURN PERIOD

The most common use of this term is to denote the expected number of observations of a variable needed to obtain one observation in excess of a specified quantity θ, say. If the observed values $X_1, X_2, \ldots$ are mutually independent, with common cumulative distribution function (CDF) $F(x)$, then the return period is

$$\{1 - F(\theta)\}^{-1}.$$

In particular, the term is used in connection with flood flows, with X representing the maximum flow in a specified period, such as a year (*see* HYDROLOGY, STOCHASTIC).

The term is also used to denote the expected time elapsing until a value of X exceeding a specified value is exceeded. If $X_1, X_2, \ldots$ are values observed in a discrete process, with constant expected value (τ, say) for time between observations of successive X's ($X_2 - X_1$, $X_3 - X_2$, and so on) then the return period would be

$$\tau\{1 - F(\theta)\}^{-1}.$$

(RECORDS)

RETURN STATE

The notion of a return state is of major importance in the use of renewal theory* in evaluating the limiting properties of a general stochastic process*. Also known as a regeneration state or renewal state, a return state is a value such that at the times when the value is attained, the future of the process has a probability law governing its evolution that is independent of the past history of the process. There is a large amount of literature seeking to make this notion precise [2, 3], and these references also give many consequences flowing from the existence of a return state.

For Markovian processes*, all states have this renewal property, whilst in many other non-Markovian contexts, selected states serve as renewal states. For example, in queueing theory*, with exponential interarrival times for customers, the times at which the queue becomes empty have the renewal property.

In order to use renewal theory effectively, the random variables T_x denoting the time taken by the processes to return state x must be proper, i.e., $\Pr[T_x < \infty] = 1$; and, in practical contexts, $E[T_x]$ needs to be finite also. Various conditions for this are known (*see*, e.g., RECURRENCE CRITERION).

When $E[T_x] < \infty$, the blocks of time spent by the process between entries to x form a sequence of finite mean i.i.d. variables and a variety of asymptotic results follow from this observation.

The most detailed study of conditions for a return state x to have $\Pr[T_x < \infty] = 1$ have been carried out for random walks [5]. Essentially, for a random walk* on an integer lattice, the origin is a return state for which $\Pr[T_0 < \infty] = 1$, provided the increment distribution of the random walk has zero mean and the space is one or two dimensional; in three and more dimensions, there are no states for which return is certain.

The existence of a particular state to which a discrete-time process returns is usually only of relevance to discrete-space-valued processes. One of the more interesting recent developments in stochastic processes is a technique for the construction, for Markovian processes in discrete time and general state space, of "artificial" return states [4] (see also [1, Chaps. 5 and 6]). This idea, which involves augmenting the state space and probabilistic structure of the process, enables renewal arguments to be applied in very much more complex situations.

In continuous time, processes such as Brownian motion* (in one or two dimensions) with continuous paths may also return to fixed states with probability 1. In general, such processes are of considerable mathematical interest but the application of the results can be difficult.

References

[1] Breiman, L. (1968). *Probability*. Addison-Wesley, Reading, MA.

[2] Feller, W. (1968). *An Introduction to Probability Theory and Its Applications*, 3rd ed., Vol. 1, Wiley, New York.

[3] Kingman, J. F. C. (1972). *Regenerative Phenomena*. Wiley, London.

[4] Nummelin, E. (1978). *Zeit. Wahrsch. verw. Geb.*, **43**, 309–318.

[5] Spitzer, F. (1964). *Principles of Random Walk*. Van Nostrand, Princeton, NJ.

(BROWNIAN MOTION
MARKOV PROCESSES
RANDOM WALK
REGENERATION PROCESSES
RENEWAL THEORY
STOCHASTIC PROCESSES)

R. L. TWEEDIE

REVERSAL DESIGNS

In some trials an experimental subject may receive different treatments in successive time periods, such trials using what are known as *changeover designs**. The simplest possible changeover design has two groups of subjects, one receiving treatment A in period 1 and treatment B in period 2, and the other receiving B in period 1 and A in period 2. If there are two groups of subjects and p time periods, with one group receiving the sequence A, B, A, ... and the other the sequence B, A, B, ..., then the design is called a *reversal* design; an alternative name is a *switchback* design. The first reversal designs had only two treatments, but Taylor and Armstrong [9] described a generalization of the design to more treatments; with three treatments, six groups are needed so that all possible pairs can be compared and, in general, comparing n treatments requires $n(n-1)$ groups.

Reversal designs with three periods were first described by Brandt [2], their original use being in trials with lactating cows. During a cow's period of lactation her milk yield rises initially then declines, but the rate of decline varies markedly from cow to cow. Thus, in the simple changeover design described above, the difference between the daily yields with treatments A and B might depend critically on the rate of decline that happened to be exhibited by the cows with those treatments. The reversal design overcomes this because the estimated difference between treatments A and B is orthogonal to the linear component of the period effects, and a standard analysis of variance* of the results is possible. However, treatment comparisons are confounded with second-order period effects and these cannot always safely be neglected [4].

A fully worked example with three periods and three treatments on 12 cows in one block of six and two blocks of three was given by Lucas [7], though this in fact used uniformity trial* data. A more realistic example of a design with three periods and three treatments on six groups of lactating cows was described by Smith [8]. The treatments here were three methods of preparing alfalfa for feeding to the cows, and the records were vitamin A potency per pound of butter fat and milligrams of carotene intake, vitamin A being adjusted by carotene in an analysis of covariance*.

Reversal designs may have more than three periods, and Brandt also considered four periods. An experiment to compare the responses of two groups of rabbits to injection with different insulin mixtures used a four-period reversal design [3]. Blood samples were taken at various times after injection and blood sugar levels determined for each sample.

Reversal designs may be of use in any situation where living subjects receive a sequence of treatments. Changeover designs are commonly used in clinical trials*, most often with only two treatments and two periods; however, more than two treatments or periods, or both, can be used. Although reversal designs are available in this context, recent reviews [1, 6] do not explicitly mention them. As has been shown above, the theory of the designs has been known for very many years, but practical application appears to have lagged behind theory. Taylor and Armstrong [9] compared the results from reversal designs and other changeover designs for dairy husbandry trials conducted at various stages during lactation. They concluded that reversal designs with p periods were more efficient than other changeover designs when the yields of milk or fat from successive periods closely conformed to a curve of degree $(p-2)$, but that otherwise some other form of changeover design was more efficient. Kershner and Federer [5] compared three- and four-period reversal designs for two treatments with extra-period changeover designs for the same number of periods and treatments. Under a model allowing for direct and residual treatment effects they showed that reversal designs gave higher estimates to the variances of contrasts and so did not recommend them.

References

[1] Bishop, S. H. and Jones, B. (1984). *J. Appl. Statist.*, **11**, 29–50.

[2] Brandt, A. E. (1938). *Res. Bull. 234*, Iowa State Agricultural Experiment Station. (The paper that introduced these designs, and the leading reference.)

[3] Ciminera, J. L. and Wolfe, E. K. (1953). *Biometrics*, **9**, 431–446.

[4] Cox, C. P. (1958). *Biometrics*, **14**, 499–512.

[5] Kershner, R. P. and Federer, W. T. (1981). *J. Amer. Statist. Ass.*, **76**, 612–619.

[6] Koch, G. G., Amara, I. A., Stokes, M. E., and Gillings, D. B. (1980). *Int. Statist. Rev.*, **48**, 249–265.

[7] Lucas, H. L. (1956). *J. Dairy Sci.*, **39**, 146–154.

[8] Smith, H. F. (1957). *Biometrics*, **13**, 282–308.

[9] Taylor, W. B. and Armstrong, P. J. (1953). *J. Agric. Sci., Camb.*, **43**, 407–412. (Best comparison of reversal designs with other changeover designs.)

(CHANGEOVER DESIGNS)

G. H. Freeman

REVERSAL TESTS (FOR INDEX NUMBERS) *See* INDEX NUMBERS

REVERSE MARTINGALE *See* MARTINGALES

REVERSION, COEFFICIENT OF

This term was coined by Galton* [1] to describe the "exceedingly simple law connecting parent and offspring seeds." The current name for it is *regression coefficient**.

Reference

[1] Galton, F. (1877). *Proc. R. Inst.*, **8**, 282–301.

Bibliography

MacKenzie, D. A. (1981). *Statistics in Britain 1865–1930*. Edinburgh University Press, Edinburgh, Scotland.

(GALTON, FRANCIS
LINEAR REGRESSION
REGRESSION (VARIOUS ENTRIES))

RÉVÉSZ ESTIMATOR OF REGRESSION FUNCTION

Let (X_i, Y_i) $(i = 1, 2, \dots, n)$ be n mutually independent pairs of random variables with ranges of variation $0 \leqslant X_i \leqslant 1$ and $-\infty \leqslant Y_i \leqslant +\infty$, respectively. The regression function* of Y on X is

$$r(x) = E[Y|X = x].$$

Révész [2] suggested the following nonparametric estimator of this function:

Take an arbitrary function $r_0(x)$, transforming the interval $[0, 1]$ into R (i.e., $-\infty$ to ∞), and define $\{r_n(x)\}$ by the recursive relation

$$r_{n+1}(x) = r_n(x) - \frac{1}{(n+1)a_{n+1}} + K\left(\frac{x - X_{n+1}}{a_{n+1}}\right)(Y_{n+1} - r_n(x)),$$

where $a_n = n^{-\alpha}$ $(\frac{1}{2} \leqslant \alpha < 1)$ and

$$K(x) = \begin{cases} 1 & \text{if } -\frac{1}{2} \leqslant x < \frac{1}{2}, \\ 0 & \text{otherwise.} \end{cases}$$

Koronacki [1] extended Révész's original studies of these estimators of $r_0(x)$ to the case of weakly dependent (so-called ϕ-mixed) random variables $\{X_i\}$. (*See* MARTINGALES and SUBMARTINGALES.)

References

[1] Koronacki, J. (1984). *Math. Operationsforsch. Statist. Ser. Statist.*, **15**, 195–203 (new title *Statistics*).

[2] Révész, P. (1977). *Math. Operationsforsch. Statist. Ser. Statist.*, **8**, 119–126.

(MARTINGALES
STOCHASTIC APPROXIMATION)

REVUE DE STATISTIQUE APPLIQUÉE (R.S.A.)

The *R.S.A.* is edited by CERESTA (Centre d'Enseignement et de Recherche de Statistique Appliquée) and publishes, in the French language, quarterly issues constituting each year a volume of about 350 pages that present various articles dealing with theory, methods, and actual applications of statistics. It was founded in 1953 by the Centre de Formation des Ingénieurs et Cadres aux Applications Industrielles de la Statistique, renamed CERESTA in 1972, which Professor G. Darmois established within the Institut de Statistique de l'Université de Paris.

From the very beginning, the *R.S.A.* was intended to serve as a link between this Centre de Formation and former trainees (upper and middle management, engineers, and technicians) who had attended its statistical training courses.

The first editor of the *R.S.A.* was E. Morice, who served from 1953–1968; he was followed by G. Morlat (1969–1978) and P. Cazes (1979–). The editorial policy of this journal aims at covering the widest field of statistical interests and techniques: quality control, reliability, experimental designs, multivariate analysis, data analysis, econometrics, inventory control research, demography, etc., and also practical applications of statistical methods in industry, agriculture, medicine, meteorology, psychology, etc. As an example, the contents of a recent issue [Vol. 31, No. 3 (1983)] include:

"Probabilité mathématique et inférence statistique. Quelques souvenirs personnels sur une important étape du progrès scientifique" by H. Cramér (translated by A. Vessereau).

"Introduction de la loi de probabilité du coefficient de variation dans les applications de la méthode 'résistancecontrainte' " by A. Desroches and M. Neff.

"Calcul de blocs diagrammes 'complexes' de fiabilité par la méthode dite des 'matrices de condition' " by D. Turinetti.

"Etude de la persistance d'un champ météorologique" by M. Deque.

"Représentations optimales des matrices imports-exports" by Y. Le Foll and B. Burtschy.

Each issue also includes an extensive review of the latest books dealing with statistics, probability, and applied mathematics; it also gives the contents of the main journals, both French and foreign, devoted to quality control, statistics,... .

Authors desiring to submit papers must send a copy (from about 10–20 pages) written in French, to the Editor, who will submit it to several referees before possible publication.

Interest in the journal has continually expanded and it has subscribers in more than 30 countries.

PIERRE CAZES

RICE DISTRIBUTION

Another name for the noncentral chi-square distribution*, used in mathematical physics, and especially in communication theory*. The names *generalized Rayleigh* and *Rayleigh–Rice distribution* are also used.

(RAYLEIGH DISTRIBUTION)

RICHARDS FAMILY OF GROWTH CURVES

This flexible growth function for empirical use proposed by Richards [1] can be written in terms of four parameters (α, β, ρ, and λ):

$$f(t) = \alpha(1 + s\beta\rho^t)^\lambda, \qquad t > 0. \qquad (1)$$

The parameter λ determines the point of inflection of the curve and the constant s ($s = 1, -1$) determines the sign of the term $\beta\rho^t$ (see, e.g., Du Toit and Gonin [2]). If the function increases monotonically in t, then $s = -1$ for $\lambda \geqslant 0$ and $s = +1$ for $\lambda < 0$. The constraints imposed on the parameters are:

$$\begin{aligned} &\alpha \geqslant 0, \\ &0 \leqslant \beta, &&\text{if } s = 1, \\ &0 \leqslant \beta \leqslant 1, &&\text{if } s = -1, \\ &0 \leqslant \rho \leqslant 1. \end{aligned}$$

Under these constraints $f(t)$ is monotonic and possesses a uniquely determined inverse.

Special cases of the Richards family are:

$$f(t) = \alpha(1 - \beta\rho^t)$$

(modified exponential)

$$f(t) = \alpha/(1 + \beta\rho^t) \quad \text{(logistic curve*)}$$

$$f(t) = \alpha \exp(-\beta^x\rho^t)$$

(Gompertz* curve; obtained from (1) for $|\lambda| \to \infty$ and $\beta^x = |\beta\lambda|$).

This family is closed under linear transformations. Properties of this family were investigated by Du Toit and Gonin [2], who introduce the *extended* Richards curve

$$f(t) = f(g(x)) = \alpha(1 + s\beta\rho^{g(x)})^\lambda,$$

where $t = g(x)$ is a general monotonic transformation and $g(x)$ is a continuous function in x. They also developed a method for determination of the parameters and in particular the transformation $t = g(x)$ based on observed data.

References

[1] Richards, F. J. (1954). *J. Exper. Botany*, **10**, 290–300.

[2] Toit, S. H. C. du and Gonin, R. (1984). *S. Afr. Statist. J.*, **18**, 161–176.

(GOMPERTZ DISTRIBUTION
LOGISTIC DISTRIBUTION)

RIDGE REGRESSION

INTRODUCTION

The theory of regression* analysis and least-squares* estimation was fairly well developed by 1950. However, the tremendous amount of manual computation associated with the solution of the normal equations* precluded its broad use except for small problems. The digital computer changed all this and the use of regression analysis grew rapidly in all fields—engineering, science, business, etc.—where data collection* and analysis are regular operational procedures. This increased use almost immediately pointed out deficiencies in least-squares estimation. In particular, it was observed that the estimated regression coefficients could be inflated in magnitude, could have the wrong sign, and could be unstable in that radical changes in their values could result from small changes or additions to the data. These deficiencies are worse, the poorer the conditioning of the $\mathbf{X}'\mathbf{X}$ matrix of the normal equations. It was found (Hoerl and Kennard [2, 3]) that these estimation difficulties could be overcome by adding small positive quantities to the diagonal of the normal equations, that is, with estimation based on $[\mathbf{X}'\mathbf{X} + k\mathbf{I}]$, $k \geqslant 0$, which is called *ridge regression*. This method of estimation allows the effects of correlations among the predictor variables to be portrayed graphically and leads to point estimates of the regression coefficients that are stable and closer to their true values.

The sections that follow include: a standard form of the model for multiple linear regression; the properties of least-squares estimation that lead to its deficiencies; the definition of ridge regression; the ridge trace for displaying the effects of predictor variable correlations; a summary of the theorems that guarantee the performance of ridge regression; and algorithms to choose the parameter k; alternative ways to view ridge regression; a bibliography of selected theoretical and applied papers that have appeared in the period 1970–1983.

THE GENERAL LINEAR MODEL—STANDARD FORM

Model in Observed Units

The general linear model* in multiple regression* can be expressed in matrix form as

$$\mathbf{Y} = \gamma_0 \mathbf{1} + \mathbf{W}\gamma + \epsilon, \tag{1}$$

where $\mathbf{W}$ is a known matrix $n \times p$ and rank p; $\mathbf{1}$ is an $n \times 1$ vector of 1's; γ_0 is an unknown scalar and γ is $p \times 1$ and unknown; and $E[\epsilon] = \mathbf{0}$, $E[\epsilon\epsilon'] = \sigma^2\mathbf{I}_n$ with σ^2 unknown. In this form it is assumed that the values of the response $\mathbf{Y}$ and the values of the factors or predictor variables $\mathbf{W} = [\mathbf{W}_1\mathbf{W}_2 \cdots \mathbf{W}_p]$ are given in the units in which they are measured. After the unknown parameters are estimated, the predicting function may be reported as $\hat{\mathbf{Y}} = \hat{\gamma}_0\mathbf{1} + \mathbf{W}\hat{\gamma}$, where $\hat{\gamma}_0$ and $\hat{\gamma}$ are estimates of γ_0 and γ, respectively. However, for a number of important reasons it is imperative to center and scale the factors, $\mathbf{W}_j$, to have zero means and unit vector lengths (correlation form). Likewise, it is useful to center the response to have zero mean. The model (1) will be transformed to correlation units. Then the important reasons for their transformation will be discussed.

The Standard Form Model with Factors Scaled to Correlation Units

In this version of the model each factor is mean-corrected and divided by its root sum of squares of deviations, i.e.,

$$x_{ij} = (w_{ij} - \bar{w}_j)/S_j,$$

$$\text{where} \quad S_j^2 = \sum_j (w_{ij} - \bar{w}_j)^2.$$

Hence

$$\mathbf{y} = \mathbf{X}\beta + \epsilon, \tag{2}$$

where $\mathbf{y}$ has zero mean. The relationship

between γ_j and β_j is defined by

$$\beta_j = S_j \gamma_j, \qquad j = 1, 2, \ldots, p. \qquad (3)$$

The most important reason for using the standard model (2) is to be able to compare the magnitudes of the regression coefficients* directly. Without the centering and scaling of the factors to correlation* units, such comparisons are not possible. In the original units, the factors $\mathbf{W}_j$ can, and usually do, have quite different dimensions and ranges of values and these differences are necessarily reflected in the regression coefficients. The magnitude of the values associated with a factor is not necessarily related to the size of the effect that it has on predicting the values of the response. The $\mathbf{X}_j$ are dimensionless, span the same ranges, and therefore result in β_j estimates that are comparable.

Another important reason for use of the standard model concerns the domain of definition for the factors. Technically, this domain should be stated with the specification of the model (1), but this is rarely, if ever, done. Rather the domain is given implicitly as not being much larger than that defined by the n points in p-space tabulated in $\mathbf{W}$. In other words, the model is not meant to be global; it is not meant to apply to all possible values for the factors. The origin in the real number system can be far from the center of the observations for the factors. In such cases, the regression coefficients estimated from uncentered data will provide a predictor, but they are not readily interpretable as derivatives within the domain of the factors if some of the factors are functions, such as powers of primary predictor variables. Furthermore, even though the overall F-ratio is not affected, F- or t-tests on individual coefficients can be meaningless.

As a matter of standard practice the primary factors should be centered first; then the functions of these factors to be used in the model are generated from the centered primary factors. Then *all* factors should be centered prior to scaling. An important reason for doing this is to reduce the superficial correlations among the primary factors and the factors generated from them. It removes one unnecessary cause of poor conditioning of the normal equations.

PROPERTIES OF BEST LINEAR UNBIASED ESTIMATION

Using unbiased linear estimation with minimum variance or maximum likelihood estimation when the error ϵ is normal form (2) results in the least-squares estimate,

$$\hat{\beta} = (\mathbf{X}'\mathbf{X})^{-1}\mathbf{X}'\mathbf{y}, \qquad (4)$$

with a minimum sum of squares of residuals,

$$\phi_{\min} = \phi(\hat{\beta}) = (\mathbf{y} - \mathbf{X}\hat{\beta})'(\mathbf{y} - \mathbf{X}\hat{\beta}). \quad (5)$$

Ridge regression is primarily directed toward applications for which $\mathbf{X}'\mathbf{X}$ is not nearly a unit matrix. To demonstrate the effects of this condition on the estimate of β, consider two properties of $\hat{\beta}$—its variance–covariance matrix and the squared distance from its expected value or mean square error (MSE). The variance–covariance matrix is

$$\mathbf{COV}(\hat{\beta}) = \sigma^2(\mathbf{X}'\mathbf{X})^{-1} = \sigma^2(c_{ij}). \quad (6)$$

Then the variance of the estimate of the jth regression coefficient is

$$\text{var}(\hat{\beta}_j) = \sigma^2 c_{jj} = \sigma^2(1 - R_j^2)^{-1}, \quad (7)$$

where R_j^2 is the coefficient of determination when factor j is regressed on the remaining p-1 factors. If factor j is orthogonal to the other factors, then $c_{jj} = 1$, $R_j^2 = 0$, and factor j values cannot be predicted from the values of other factors. When there are significant interfactor correlations the c_{jj} are large and the variances of the estimates of the β_j will be large.

Let L_1 be the distance from $\hat{\beta}$ to β. Then

$$L_1^2 = (\hat{\beta} - \beta)'(\hat{\beta} - \beta) \qquad (8)$$

and

$$\text{MSE} = E[L_1^2] = \sigma^2 \text{Trace}(\mathbf{X}'\mathbf{X})^{-1}, \quad (9)$$

or equivalently,

$$E[\hat{\beta}'\hat{\beta}] = \hat{\beta}'\beta + \sigma^2 \text{Trace}(\mathbf{X}'\mathbf{X})^{-1}. \quad (10)$$

When the error is normally distributed

$$\operatorname{Var}\left[L_1^2\right] = 2\sigma^4 \operatorname{Trace}(\mathbf{X}'\mathbf{X})^{-2}. \quad (11)$$

These related properties show the uncertainty in $\hat{\boldsymbol{\beta}}$ as a function of the conditioning of $\mathbf{X}'\mathbf{X}$. If the eigenvalues of $\mathbf{X}'\mathbf{X}$ are denoted by

$$\lambda_1 \geqslant \lambda_2 \geqslant \cdots \geqslant \lambda_p, \quad (12)$$

then the MSE is given by

$$\text{MSE} = E\left[L_1^2\right] = \sigma^2 \Sigma(1/\lambda_j) \quad (13)$$

and the variance of L_1^2 when the error is normal is

$$\operatorname{Var}\left[L_1^2\right] = 2\sigma^4 \Sigma\left(1/\lambda_j^2\right). \quad (14)$$

Hence, if the shape of the factor space is such that reasonable data collection results in $\mathbf{X}'\mathbf{X}$ with one or more small eigenvalues, the distance from $\hat{\boldsymbol{\beta}}$ to $\boldsymbol{\beta}$ will tend to be large. Also the squared norm $\hat{\boldsymbol{\beta}}'\hat{\boldsymbol{\beta}}$ has an expectation $(\boldsymbol{\beta}'\boldsymbol{\beta} + \sigma^2\Sigma(1/\lambda_j^2))$ that can be substantially larger than $\boldsymbol{\beta}'\boldsymbol{\beta}$ and hence the point $\hat{\boldsymbol{\beta}}$ in p-space is farther away from the origin on the average than the point $\boldsymbol{\beta}$. Hence, shrinking $\boldsymbol{\beta}$ toward the origin is a viable alternative.

RIDGE REGRESSION BASICS

The ridge estimator for $\boldsymbol{\beta}$ in the standard model (2) is defined as

$$\hat{\boldsymbol{\beta}}(k) = \left[\hat{\mathbf{X}}\mathbf{X} + k\mathbf{I}_p\right]^{-1}\mathbf{X}'\mathbf{y}, \qquad k \geqslant 0. \quad (15)$$

Using the normal equation (4), this is readily shown to be a linear transform of the least-squares estimator, namely,

$$\hat{\boldsymbol{\beta}}(k) = \left[\mathbf{X}'\mathbf{X} + k\mathbf{I}_p\right]^{-1}\mathbf{X}'\mathbf{X}\hat{\boldsymbol{\beta}}. \quad (16)$$

To further display the character of the estimator, it is convenient to use the singular value decomposition of $\mathbf{X}$, namely, $\mathbf{X} = \mathbf{P}\Lambda^{1/2}\mathbf{Q}'$, where Λ is the matrix of eigenvalues of $\mathbf{X}'\mathbf{X}$ and $\mathbf{P}'\mathbf{P} = \mathbf{Q}'\mathbf{Q} = \mathbf{Q}\mathbf{Q}' = \mathbf{I}_p$, and rewrite the standard model in canonical form thus:

$$\mathbf{y} = \mathbf{X}\boldsymbol{\beta} + \boldsymbol{\epsilon} = \mathbf{P}\Lambda^{1/2}\mathbf{Q}'\boldsymbol{\beta} + \boldsymbol{\epsilon}.$$

Using the transformations $\mathbf{Z} = \mathbf{P}\Lambda^{1/2} = \mathbf{X}\mathbf{Q}$ and $\boldsymbol{\alpha} = \mathbf{Q}'\boldsymbol{\beta}$, the canonical form of the standard model is

$$\mathbf{y} = \mathbf{Z}\boldsymbol{\alpha} + \boldsymbol{\epsilon}. \quad (17)$$

Using the singular value decomposition $\hat{\boldsymbol{\beta}}(k)$ can be written as

$$\hat{\boldsymbol{\beta}}(k) = \mathbf{Q}\left[\Lambda + k\mathbf{I}_p\right]^{-1}\mathbf{Z}'\mathbf{y}, \quad (18)$$

and since $\boldsymbol{\alpha} = \mathbf{Q}'\boldsymbol{\beta}$, the canonical form of the ridge estimator is

$$\hat{\boldsymbol{\alpha}}(k) = \left[\Lambda + k\mathbf{I}_p\right]^{-1}\mathbf{Z}'\mathbf{y}$$
$$= \left[\Lambda + k\mathbf{I}_p\right]^{-1}\Lambda\hat{\boldsymbol{\alpha}}. \quad (19)$$

Thus, the individual components of $\hat{\boldsymbol{\alpha}}(k)$ are

$$\hat{\alpha}_j(k) = \left[\lambda_j/(\lambda_j + k)\right]\hat{\alpha}_j = \delta_j\hat{\alpha}_j, \qquad j = 1, 2, \ldots, p. \quad (20)$$

Since $\delta_j \leqslant 1$ for all values of j, the ridge estimator is a shrinkage estimator; it shrinks the least-squares estimator toward the origin. Other shrinkage estimators* that have different characteristics are common or Stein shrinkage where $\delta_j = \delta \leqslant 1$ for all j and principal components that set $\delta_j = 0$ for the $\hat{\alpha}_j$ associated with the smallest eigenvalues.

To make a comparison with least squares,

$$\mathbf{COV}\left[\hat{\boldsymbol{\beta}}(k)\right] = \sigma^2\left[\mathbf{X}'\mathbf{X} + k\mathbf{I}_p\right]^{-1} \times \mathbf{X}'\mathbf{X}\left[\mathbf{X}'\mathbf{X} + k\mathbf{I}_p\right]^{-1} \quad (21)$$

and the residual sum of squares is

$$\phi(k) = \left[\mathbf{y} - \mathbf{X}\hat{\boldsymbol{\beta}}(k)\right]'\left[\mathbf{y} - \mathbf{X}\hat{\boldsymbol{\beta}}(k)\right]$$
$$= \phi_{\min} + k^2\hat{\boldsymbol{\beta}}'(k)(\mathbf{X}'\mathbf{X})^{-1}\hat{\boldsymbol{\beta}}(k) \quad (22)$$
$$= \phi_{\min} + k^2\sum_1^p \hat{\alpha}_j^2\lambda_j/(\lambda_j + k)^2. \quad (23)$$

THE RIDGE TRACE

The ridge estimator

$$\hat{\boldsymbol{\beta}}(k) = \left[\mathbf{X}'\mathbf{X} + k\mathbf{I}_p\right]^{-1}\mathbf{X}'\mathbf{y}$$

is really a family of estimators indexed by the parameter k. This family of estimators can be used in graphical form as a data

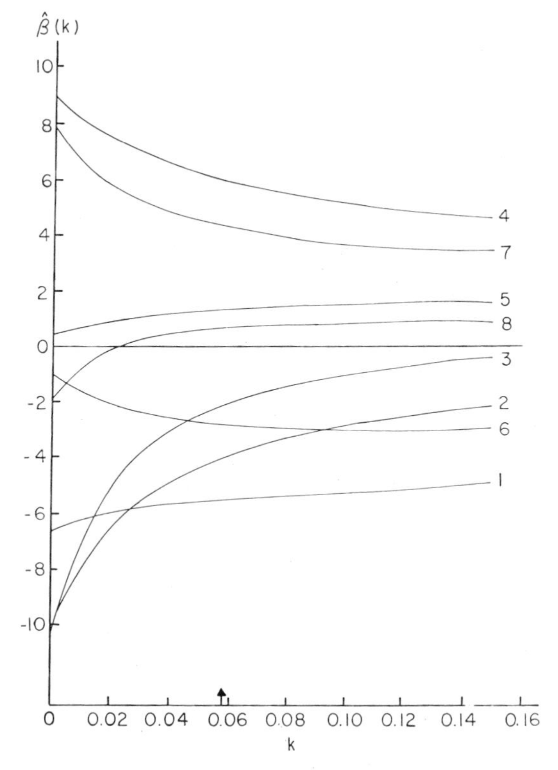

Figure 1 Ridge trace.

analysis tool to show the relative importance of the individual factors and to show the effects of the conditioning of $\mathbf{X'X}$ on the sensitivity of the estimates. The graphical tool is the *ridge trace*, which is a two-dimensional plot of the individual coefficients, $\hat{\beta}_j(k)$, as functions of k (see Fig. 1).

The ridge trace is best demonstrated with an example. In Table 1 the estimation data for a problem with $n = 17$, $p = 8$ are shown. The problem exhibits a moderate degree of ill-conditioning; trace$(\mathbf{X'X})^{-1} = 109.46$. An examination of the ridge trace in Fig. 1 shows these characteristics of the system:

(i) The least-squares solution is unstable. There are significant changes in values of the coefficients as k is increased from zero. This is especially true for factors 2, 3, 4, and 7.

(ii) There is a change in sign for factor 8.

(iii) There is an apparent underestimate of the magnitude for factor 6.

In the region from $k = 0.04$ to $k = 0.10$ there is much more stability to the system. A set of coefficients chosen at a point in this region for k should provide a better estimate for the coefficients than does least squares. Algorithms for choosing a particular value of k will be given under a separate heading.

MEAN SQUARE ERROR PROPERTIES OF RIDGE

The mean square error (MSE) for the ridge estimator is

$$\begin{aligned}\text{MSE}(k) &= E\left[L_1^2(k)\right]\\ &= E\left[(\hat{\beta}(k) - \beta)'(\hat{\beta}(k) - \beta)\right]\\ &= \sigma^2\sum_1^p \lambda_j/(\lambda_j + k)^2 + k^2\sum_1^p \alpha_j^2/(\lambda_j + k)^2 \quad (24)\\ &= \gamma_1(k) + \gamma_2(k). \quad (25)\end{aligned}$$

The meanings of the two components $\gamma_1(k)$ and $\gamma_2(k)$ are readily established. Using (21), the variance of $\hat{\beta}(k)$,

$$\gamma_1(k) = \text{Trace}\left(\sigma^2\left[\mathbf{X'X} + k\mathbf{I}_p\right]^{-1} \times \mathbf{X'X}\left[\mathbf{X'X} + k\mathbf{I}_p\right]^{-1}\right). \quad (26)$$

Hence, $\gamma_1(k)$ is the sum of the variances of the components of $\hat{\beta}(k)$.

If the bias of the estimator $\hat{\beta}(k)$ is defined as

$$\begin{aligned}\text{Bias} &= E\left[\hat{\beta}(k) - \beta\right]\\ &= \left[(\mathbf{X'X} + k\mathbf{I}_p)^{-1}\mathbf{X'X} - \mathbf{I}_p\right]\beta, \quad (27)\end{aligned}$$

then $\gamma_2(k)$, using the singular value decomposition of $\mathbf{X}$ and the canonical form of the general model (17), is the square of the bias.

Table 1 Sample Data Set

$W1$	$W2$	$W3$	$W4$	$W5$	$W6$	$W7$	$W8$	Response
29.2	275	12.7	6.6	24.7	14.6	2.1	430	35.569
28.3	313	10.0	8.4	19.6	14.0	1.7	496	38.815
28.6	244	15.2	6.2	34.1	14.4	2.3	400	40.596
28.7	323	8.9	10.0	21.0	13.0	1.2	492	39.814
29.5	329	9.2	9.0	18.5	15.2	1.4	492	35.117
28.0	254	13.3	5.8	23.4	12.6	1.8	426	40.170
28.0	253	12.5	5.8	25.7	11.2	1.5	416	38.089
28.3	245	15.4	7.0	33.0	13.0	2.6	397	42.374
28.4	240	15.9	5.6	30.5	13.4	2.5	374	39.335
28.4	236	14.6	6.6	31.4	13.0	2.3	392	41.681
28.6	288	9.5	7.0	18.2	13.6	1.4	465	38.767
29.1	289	11.2	6.4	24.8	15.0	1.5	430	33.693
28.3	240	13.4	5.8	29.5	13.4	1.9	388	39.382
28.6	336	9.1	8.2	17.6	12.8	1.5	470	38.003
28.1	264	12.2	6.2	23.2	12.8	1.7	401	40.296
28.1	280	11.6	8.2	25.6	12.0	1.9	453	44.263
28.1	268	10.7	6.2	21.2	11.0	1.7	436	41.583

The foundation for ridge regression is contained in the following theorem:

Existence Theorem. There always exists a $k > 0$ such that

$$E\left[L_1^2(k)\right] < E\left[L_1^2(0)\right] = \sigma^2\Sigma(1/\lambda_j). \tag{28}$$

Theobald [8] showed that the existence is guaranteed if $k \leqslant 2\sigma^2/(\boldsymbol{\beta}'\boldsymbol{\beta})$.

The existence theorem is the result of two other theorems.

Theorem 1. The variance function $\gamma_1(k)$ is a continuous monotonically decreasing function of k. Its first derivative $\gamma_1(k) < 0$ at $k = 0$.

Theorem 2. The squared bias function $\gamma_2(k)$ is a continuous monotonically increasing function of k, and approaches $\boldsymbol{\beta}'\boldsymbol{\beta}$ as an upper limit. Its first derivative $\gamma_2'(k)$ is zero at $k = 0$ for any bounded $\boldsymbol{\beta}'\boldsymbol{\beta}$.

Theorems 1 and 2 guarantee that $E[L_1^2(k)]$ attains a minimum value for some $k < 0$, i.e., there exists a value of k that will produce a decrease in the variance that more than offsets the resulting squared bias.

ALGORITHMS FOR DETERMINING THE BIASING PARAMETER

To obtain a point estimate $\hat{\boldsymbol{\beta}}(k)$ with a potentially smaller mean square error than least squares, it is necessary to obtain a value for k from the data for a given problem. Many different algorithms have been proposed (Hoerl and Kennard [5]). However, when a value for k is obtained from the data, closed mathematical forms cannot be derived to compare the ridge estimate with least squares. Statistical performance has to be assessed by simulation*.

The need to simulate the performance of an algorithm raises an important question in regression methodology and simulation. How is an algorithm validated for the class of all regression problems? No formal procedure exists for validation. However, the elements that must be considered to characterize the set of all regression problems can be delineated. They are: (i) the dimension of the problem, that is, the number of factors, p, and the number of observations, n; (ii) the structure of the factor space—the distribution of points in the factor space must be quantified, e.g., using one or more measures of the conditioning of **X**; and (iii) the signal-to-noise ratio, the ability to predict changes in the response in the presence of random errors, must be quantified, e.g., with $\beta'\beta/\sigma^2$.

Two algorithms that have been assessed by simulations that considered all of the elements necessary to validate a regression algorithm are k_{at} (Hoerl and Kennard [4]) and k_b (Lawless and Wang [7]). These algorithms are defined as follows:

1. k_{at}: A sequence of estimates of β and k is constructed, viz.

$$\begin{aligned}
&\hat{\beta}\\
&k_{a0} = p\hat{\sigma}^2/(\hat{\beta}'\hat{\beta})\\
&\hat{\beta}(k_{a0})\\
&k_{a1} = p\hat{\sigma}^2/([\hat{\beta}(k_{a0})]'[\hat{\beta}(k_{a0})])\\
&\hat{\beta}(k_{a1})\\
&k_{a2} = p\hat{\sigma}^2/([\hat{\beta}(k_{a1})]'[\hat{\beta}(k_{a1})])\\
&\cdots\\
&\hat{\beta}(k_{at}).
\end{aligned}$$

The sequence is terminated when

$$[(k_{a,i+1} - k_{a,i})/k_{a,i}] \leqslant \delta = 20T^{-1.30}, \tag{29}$$

where $T = \text{trace}(\mathbf{X}'\mathbf{X})^{-1}/p$.

2.

$$\begin{aligned}
k_b &= 1/F = p\hat{\sigma}^2/(\hat{\beta}'\mathbf{X}'\mathbf{X}\hat{\beta})\\
&= p\hat{\sigma}^2/(\hat{\alpha}'\Lambda\hat{\alpha}).
\end{aligned} \tag{30}$$

The extensive simulations used to assess these algorithms show that for any given n, p, factor space structure, or signal-or-noise ratio the estimates have the following properties:

(i) With probability greater than 0.5 the mean square error, $E[l_1^2(k_{at})]$ or $E[L_1^2(k_b)]$, is smaller than that for least squares.

(ii) The standard deviations of the $L_1^2(k_{at})$ and $L_1^2(k_b)$ distributions are smaller than that for least squares.

Two other algorithms are worthy of mention. They are k_a (Hoerl et al. [6]) and k_d (Dempster et al. [1]). They are defined as

$$k_a = p\hat{\sigma}^2/(\hat{\beta}'\hat{\beta}) \tag{31}$$

and k_d is the solution to the nonlinear equation

$$(1/\hat{\sigma}^2)\sum_{j=1}^{p}\hat{\alpha}_j[(1/k_d) + (1/\lambda_j)]^{-1} - p = 0. \tag{32}$$

The algorithm k_a has been subjected to the same assessment as k_{at} and k_b, and has the same performance characteristics. However, it is more conservative, and k_{at} and k_b can give greater improvement. The k_d algorithm was derived using Bayesian arguments. It has not been assessed to the detail of k_{at} and k_b, but does seem to exhibit the same general properties.

In Table 1, values of a data set are given for a problem with $n = 17$ and $p = 8$. The data were generated using a known set of coefficients. Table 2 gives the known coefficient values, the least-squares estimates, and the estimates from two of the algorithms, k_{at} and k_b.

The k values chosen by the algorithms are in the stable region of the ridge trace. As can be seen from the L_1^2 values, the coefficients are much closer to the true values than are the least-squares coefficients. Two primary deficiencies have been corrected, namely, the large magnitudes and the incorrect sign.

ALTERNATIVE VIEWS

Ridge regression can be given a Bayesian* interpretation. Let the prior information for the regression coefficients β_j be modeled by independent normal random variables with

Table 2 Regression Coefficients

Coeff. No.	True	Least Squares	$k_{at} = 0.0587$	$k_b = 0.0976$
$W1$	−5.7844	−6.5315	−5.5002	−5.1850
$W2$	−2.3933	−10.1687	−3.9273	−2.8672
$W3$	−1.6437	−10.5596	−2.0398	−1.0023
$W4$	5.0397	9.0092	5.9344	5.2043
$W5$	1.2218	0.4895	1.3429	1.5031
$W6$	−2.4431	−1.0175	−2.7188	−2.9536
$W7$	5.0642	7.8824	4.4410	3.8748
$W8$	0.3306	−1.8814	0.6956	0.8600
$L_1^2 = \Sigma(\hat{\beta}_j - \beta_j)^2$		171.67	4.00	3.06

parameters

$$E[\boldsymbol{\beta}] = \boldsymbol{\beta}_0 \quad \text{and} \quad \text{Var}[\boldsymbol{\beta}] = \gamma^2. \tag{33}$$

Then for the general linear model $\mathbf{y} = \mathbf{X}\boldsymbol{\beta} + \boldsymbol{\epsilon}$ with normally distributed error the likelihood, L, of the sample is

$$L = (2\pi\sigma^2)^{-n/2}\exp\left[-\tfrac{1}{2}(\mathbf{y} - \mathbf{X}\boldsymbol{\beta})'(\mathbf{y} - \mathbf{X}\boldsymbol{\beta})\sigma^{-2}\right] \times \prod_{j=1}^{p}(2\pi\gamma_j^2)^{-1}\exp\left[-(\beta_j - \beta_{j0})^2/(2\gamma_j^2)\right]. \tag{34}$$

The maximum likelihood* estimate, $\boldsymbol{\beta}$, or the mean of the posterior distribution is

$$\bar{\boldsymbol{\beta}} = \left[\mathbf{X}'\mathbf{X} + \sigma^2\gamma^{-2}\right]^{-1}\left[\mathbf{X}'\mathbf{y} + \sigma^2\gamma^{-2}\boldsymbol{\beta}_0\right]. \tag{35}$$

In (35) the unknown σ^2 appears inside the inverse matrix. However, if the uncertainty in the β_j's is made relative to σ^2, then σ^2 can be eliminated. Take $\gamma_j^2 = \sigma^2/k_j$. Then, if $\mathbf{K}$ is the diagonal matrix of parameters, k_j, the estimate is

$$\bar{\boldsymbol{\beta}} = [\mathbf{X}'\mathbf{X} + \mathbf{K}]^{-1}[\mathbf{X}'\mathbf{y} + \mathbf{K}\boldsymbol{\beta}_0]. \tag{36}$$

Now if it is assumed a priori that all regression coefficients are zero, i.e., $\boldsymbol{\beta}_0 = \mathbf{0}$, and the uncertainty is the same for all p coefficients, viz. $\gamma_j^2 = \sigma^2/k$ for all j, then

$$\bar{\boldsymbol{\beta}} = \left[\mathbf{X}'\mathbf{X} + k\mathbf{I}_p\right]^{-1}\mathbf{X}'\mathbf{y}. \tag{37}$$

which is the same form as the ridge estimator. Thus, the Bayesian character of the estimate $\hat{\boldsymbol{\beta}}(k)$ is displayed. It is equivalent to an a priori assumption that each β_j has expectation zero and variance σ^2/k.

References

[1] Dempster, A. P., Schatzoff, M., and Wermuth, N. (1977). A simulation study of alternatives to ordinary least squares. *J. Amer. Statist. Ass.*, **72**, 77–106.

[2] Hoerl, A. E. and Kennard, R. W. (1970). Ridge regression: Biased estimation for nonorthogonal problems. *Technometrics*, **12**, 55–67.

[3] Hoerl, A. E. and Kennard, R. W. (1970). Ridge regression: Applications to nonorthogonal problems. *Technometrics*, **12**, 69–82.

[4] Hoerl, A. E. and Kennard, R. W. (1976). Ridge regression: Iterative estimation of the biasing parameter. *Commun. Statist. A*, **5**, 77–88.

[5] Hoerl, A. E. and Kennard, R. W. (1981). Ridge regression—1980: Advances, algorithms, and applications. *Amer. J. Math. Mgmt. Sci.*, **1**, 5–83.

[6] Hoerl, A. E., Kennard, R. W., and Baldwin, K. F. (1975). *Commun. Statist. A*, **4**, 105–124.

[7] Lawless, J. F. and Wang, P. (1976). A simulation study of ridge and other regression estimators. *Commun. Statist. A*, **5**, 307–323.

[8] Theobald, C. M. (1974). Generalizations of mean square error applied to ridge regression. *J. R. Statist. Soc. B*, **36**, 103–105.

Bibliography

Casella, C. (1980). Minimax ridge regression estimation. *Ann. Statist.*, **6**, 1036–1056.

Farebrother, R. W. (1976). Further results on the mean square error of ridge regression. *J. R. Statist. Soc. B*, **38**, 248–250.

Hemmerle, W. J. (1975). An explicit solution for generalized ridge regression. *Technometrics*, **17**, 309–314.

Hocking, R. R., Speed, F. M., and Lynn, M. J. (1976). A class of biased estimators in linear regression. *Technometrics*, **18**, 425–438.

Marquardt, D. W. and Snee, R. D. (1975). Ridge regression in practice. *Amer. Statist.*, **29**, 3–19.

Obenchain, R. L. (1977). Classical *F*-tests and confidence regions for ridge regression. *Technometrics*, **19**, 429–439.

Obenchain, R. L. (1978). Good and optimal ridge estimators. *Ann. Statist.*, **6**, 1111–1121.

Vinod, H. D. and Ullah, A. (1981). *Recent Advances in Regression Methods*. Marcel Dekker, New York.

(GENERAL LINEAR MODEL
LINEAR REGRESSION
MULTIPLE REGRESSION
REGRESSION (various entries)
SHRINKAGE ESTIMATORS)

A. E. HOERL
R. W. KENNARD

RIDIT ANALYSIS

Ridit analysis refers to a technique introduced by Bross [3] for analyzing data with responses measured on an ordered categorical (semiquantal) scale. The acronym RIDIT stands for relative to an identified distribution integral transformation. An empirical distribution function is used in a probability integral transformation* to assign scores called ridits to each category of the ordered response variable. These ridit scores are then used in the analysis.

Suppose that responses can be considered to occur along an unobservable continuum. The actual data that are recorded, however, consist only of membership in a number of categories along the underlying continuum. The order of the categories is known, but no numerical scale is available. As an example consider the data in Table 1 adapted from Scott et al. [14]. Passengers in the front seats of 1974 cars involved in a sample of crashes were classified according to whether or not safety belts were used. The response variable is an injury severity scale. To apply ridit analysis, a reference distribution over the same response categories must be identified. This could be a standard population from an external source, but in practice it has usually been taken to be one of the observed groups. For illustration, take the distribution of injuries to unrestrained passengers as the reference distribution for calculation of the ridits. Let P_{ij} be the proportion in injury category $j = 1, \ldots, k$ of the group i and define the ridits by

$$R_j = \sum_{n=1}^{j-1} P_{2n} + P_{2j}/2.$$

These are also shown in the table. Kantor et al. [10] note that these can be interpreted as percentile ranks*. If X denotes the injury severity for an occupant selected at random from the unrestrained population, and Y denotes the injury severity for a restrained occupant, then the mean ridit,

$$\overline{R} = \sum_{j=1}^{k} R_j P_{1j},$$

can be interpreted as an estimate of $\Pr[X \leqslant Y]$, the probability that an unrestrained occupant would be less seriously injured than a restrained occupant. (More precisely, $\overline{R}$ estimates $\Pr[X < Y] + (\frac{1}{2})\Pr[X = Y]$.) Because the ridit scores are bounded by zero and one, and because they are linear combinations of the multinominal* response vector, mean ridits will have asymptotic normal distributions. This asymptotic normality was used by Bross [3] in suggesting analytical procedures. If the ridits are estimated from the data, then different limiting theorems must be used, for example Theorem 4.1 of Conover [5] or Theorem 29C of Hájek [8].

Mantel [13] has criticized the use of ridit analysis. Responses by Bross [4] and Selvin [15] follow his article. Much of Mantel's criticism is directed against the automatic use of ridit analysis for problems with ordinal data*. He argues against the use of the ridit scale particularly in the case where some numerical information about the underlying continuum is available for the categories. The point is that the ridit scale is based on a probability integral transformation and is not an attempt to reconstruct the numerical scale. Consequently ridit analysis must be interpreted in probabilistic terms, not in numerical terms on the original scale. If numerical scores for the categories based on the original or natural scale are available, then, as Mantel says, these may be preferred in the analysis. However, in their absence, or even in their presence, the ridit analysis provides an additional interpretation that may be useful.

THEORETICAL BASIS

Brockett and Levine [2] provide a theoretical basis for choosing the ridit score: Assume that a scoring function $h_k(i, \mathbf{P})$ is desired to assign a score to category i of k, based on a vector of proportions $\mathbf{P} = (p_1, \ldots, p_k)$. They prove the following theorem.

Theorem (Brockett and Levine). Let h be a real-valued function satisfying

(a) h is continuous;

(b) (branching property)

$$h(x, y) = \lambda h(x, y + (1 - \lambda)c) + (1 - \lambda)h(x + \lambda c, y)$$

for any x, y, c, and λ.

Then $h(x, y) = a + b(x - y)$ for some choice of a and b.

Application of this to the assignment function

$$h_k(i, \mathbf{P}) = h\left(\sum_{j<i} p_j, \sum_{j>i} p_j\right)$$

yields Bross's ridit scores. The branching property **(b)** requires that if two adjacent categories are combined, scores in the other categories are not affected and the resulting score for the combined category is the weighted average of the scores of the two categories that were combined. Under a slightly different set of assumptions, Brockett and Levine also derive by functional equations a slightly more general transformation. Dennis et al. [6] use the general transformation defined in ref. 2, but allow a reference distribution to be continuous, discrete, or a mixture and derive the mean and variance of the transformed variable.

Lynch [11, 12] considers the loss function

$$L(F; \hat{F}) = \int [F(x) - \hat{F}(x)]^2 \times h(F(x))\, dF(x),$$

for estimating an unknown distribution function. He shows that for $h(F) = [F(1 - F)]^{-1}$, ridits form a best invariant estimate of F.

Vigderhous [17] has summarized relationships of ridits to several measures of association*. These include Somers d* and Kendall's tau* for ordered $2 \times k$ contingency tables. He also notes the relationship to the Mann–Whitney–Wilcoxon* U and relates ridits to a Markov model in the case of a 2×2 contingency table.

Ridit analysis is related to rank tests* with many ties (refs. 6, 7, and 13), and to U-statistics*. The mean ridit can also be viewed as a linear function of the proportion of a sample falling in the ordered response categories. As such, ridit analysis is a special case of linear model approaches to categorical data* as illustrated in ref. 18.

The mean ridit, $\bar{R}$, was suggested by Bross [3] as an estimate of $\Pr[X \leqslant Y]$. Flora [6] noted that this probability is the consistency parameter for the Wilcoxon rank sum test (Mann–Whitney U). Flora suggested that testing in the two-sample case be based on the Mann–Whitney U adjusted for many ties and showed the relationship to the mean ridit and the normal theory tests suggested by Bross. Let $\Pi^0 = \Pr[X = Y]$, $\Pi^+ = \Pr[X > Y]$, and $\Pi^- = \Pr[X < Y]$. Then Flora showed that $\bar{R}$ actually estimates $\Pi^- + \frac{1}{2}\Pi^0$. He suggested a modification to estimate Π^0, Π^+, and Π^- separately, since in many cases Π^0 is large and may convey important information. For example, in Table 1, Π^0 represents the probability that a restrained driver

Table 1 Injury Severity for Belted and Nonbelted Occupants of 1974 Cars

Restraint Use	Injury				
	None	Minor	Moderate	Serious	Total
Yes	417	330	33	17	797
No	357	540	53	35	985
Ridit score	0.181	0.637	0.938	0.982	

would receive the same injury severity as an unrestrained driver. Hochberg [7] gave U-statistic estimators for the variance of $(\hat{\Pi}^+ - \hat{\Pi}^-)$ under the general (nonnull hypothesis) case, and extended the estimation to confidence intervals*. Selvin [15] independently derived the algebraic relation of $\overline{R}$ to Wilcoxon's rank sum statistic.

Illustration of Ridit Analysis for Two-Sample Case

Let f_i denote the frequencies of the restrained (Y) group and g_i the frequencies of the unrestrained (X) group, and $T_i = f_i + g_i$. Then form

$$W = \sum_{i=2}^{k} f_i\left(\sum_{j=1}^{i-1} g_j\right) - \sum_{i=1}^{k-1} f_i\left(\sum_{j=i+1}^{k} g_j\right)$$

and

$$\hat{\Pi}^0 = \left(\sum_{i=1}^{k} f_i g_i\right)\Big/mn.$$

W/mn estimates $(\Pi^+ - \Pi^-)$. Then the estimate of Π^+ is found from $\hat{\Pi}^+ = (1 - \hat{\Pi}^0 + W/mn)/2$, and $\hat{\Pi}^- = 1 - \hat{\Pi}^0 - \hat{\Pi}^+$. Under H_0; $P^+ = P^-$, the variance of W conditioned on the ties is that of the Mann–Whitney U:

$$\operatorname{Var}(W) = \frac{mn(N+1)}{3} \times \left(1 - \frac{\sum_{i=1}^{k}(T_i^3 - T_i)}{N^3 - N}\right).$$

Application to the data in Table 1 gives the values $\hat{\Pi}^0 = 0.420$, $\hat{\Pi}^- = 0.372$, and $\hat{\Pi}^+ = 0.208$. The statistic $Z = W/\sqrt{\operatorname{Var}(W)} = -6.645$ is related to the standard normal distribution to test H_0. The mean ridits for the two groups are $\overline{R} = 0.418$ for restrained and $\overline{R} = 0.5$ for unrestrained. (The mean ridit calculated for the reference distribution will always equal 0.5, although the variance will change with the sample size and distribution.) Note that $\overline{R} = \Pi^+ + \frac{1}{2}\Pi^0$. Application of the two-sample t test gives $t = -6.646$. The estimated standard errors of $\overline{R}_x$ and $\overline{R}_y$ are 9.19×10^{-3} and 8.19×10^{-3}, respectively.

Uses

Ridit analysis has been used in analyzing data from medical research, with a response variable representing some degree of disease. Usually comparisons on the basis of mean ridits have used t-tests or analysis of variance. Williams and Grizzle [18] give several examples of the use of ridit analysis. In their work, ridit scores are viewed as a particular case of functions defined over the response categories and are analyzed by the linear-models approach to categorical data*. Agresti [1] has applied ridit analysis in testing for marginal homogeneity in tables with ordered categories. Dennis et al. [6] discuss the use of ridit analysis in discriminant problems.

Relations to Probit Analysis

The ridit transformation was described an an alternative to the probit transformation [3], being based on an empirical rather than a theoretical distribution. The usual use of the two is different. The ridit is applied in cases where there are at least three ordered response categories, while the most frequent use of probits or logits is for a dichotomous response. Ridit analysis is primarily a test of differences in location. The application of probits would be analysis of a trend in proportions. Suppose that the columns, instead of the rows, in Table 1 were viewed as fixed. Then probit analysis could be used to analyze trends in the proportion using seat belts in each injury severity category.

Ridit analysis represents an interface between categorical data analysis and nonparametric statistical analysis. It transforms ordinal data to a probability scale. The transformed data are then useful in a wide variety of statistical problems, including those of location, association, and discrimination. The results are interpreted on a probabilistic scale rather than on the original scale of the measurements.

References

[1] Agresti, A. (1983). *Biometrics*, **39**, 505–510.

[2] Brockett, P. and Levine, A. (1977). *Ann. Statist.*, **5**, 1245–1248.

[3] Bross, I. D. J. (1958). *Biometrics.*, **14**, 18–38.

[4] Bross, I. D. J. (1979). *Amer. J. Epidemiol.*, **109**, 29–30.

[5] Conover, W. J. (1973). *Ann. Statist.*, **1**, 1105–1125.

[6] Dennis, T. B., Pore, M. D., and Terrell, G. R. (1980). *Amer. Statist. Ass. Proc. Statist. Comput.*, pp. 303–308.

[7] Flora, J. D. (1974). A Note on Ridit Analysis. *Technical Report No. 3*, Department of Biostatistics, University of Michigan, Ann Arbor, Michigan.

[8] Hájek, J. (1969). *Nonparametric Statistics*. Holden Day, San Francisco, p. 133.

[9] Hochberg, Y. (1975). *On the Variance Estimate of a Mann–Whitney Statistics for Ordered Grouped Data*. Highway Safety Research Center, University of North Carolina, Chapel Hill, North Carolina.

[10] Kantor, S., Winklestein, W. J., and Ibrahim, M. A. (1968). *Amer. J. Epidemiol.*, **87**, 609–615.

[11] Lynch, G. W. (1978). *Commun. Statist. A*, **7**, 607–614.

[12] Lynch, G. W. (1980). *Commun. Statist. A*, **9**, 1207.

[13] Mantel, N. (1979). *Amer. J. Epidemiol.*, **109**, 25–29.

[14] Scott, R. E., Flora, J. D., and Marsh, J. C. (1976). *An Evaluation of the 1974 and 1975 Restraint Systems*. Highway Safety Research Institute, University of Michigan, Ann Arbor, Michigan.

[15] Selvin, S. (1977). *Amer. J. Epidemiol.*, **105**, 16–20.

[16] Selvin, S. (1979). *Amer. J. Epidemiol.*, **109**, 30–31.

[17] Vigderhous, G. (1979). *Quality and Quantity*, **13**, 187–201.

[18] Williams, O. D. and Grizzle, J. E. (1972). *J. Amer. Statist. Ass.*, **67**, 29–30.

(BIOASSAY, STATISTICAL METHODS IN
CATEGORICAL DATA
ORDINAL DATA
QUANTAL RESPONSE ANALYSIS
RANK TESTS)

JAIRUS D. FLORA, JR.

RIEDWYL ONE SAMPLE TEST *See* V-STATISTIC

RIESZ–FISCHER THEOREM

If the sequence of functions $\{f_n(x)\}$, $(n = 1, 2, \dots)$ satisfies the condition

$$\lim_{\substack{m\to\infty \\ n\to\infty}} \int_a^b |g_m(x) - g_n(x)|^p \, dx = 0, \quad p \geqslant 1,$$

then there is a uniquely defined (apart from sets of measure zero) function $g(x)$ such that

$$\lim_{n\to\infty} \int_a^b |g_n(x) - g(x)|^p \, dx = 0.$$

(It is supposed that all the functions belong to the class L^p—that is, the Lebesgue integrals of the pth power of the functions exist.)

Bibliography

Fischer, E. (1907). *C. R. Acad. Sci. Paris*, **144**, 1022–1024.

Riesz, F. (1907). *Göttingen Nachr.*, 116–122.

RING-WIDTH INDEX *See* DENDROCHRONOLOGY

RIORDAN IDENTITY

A combinatorial identity

$$n^n = \sum_{j=1}^{n} \binom{n-1}{j-1} n^{n-j} j!.$$

It is useful in partitioning and selection problems.

(COMBINATORICS)

RISK ANALYSIS, JOURNAL OF

Risk Analysis, an international journal, was founded by the Society for Risk Analysis in 1981. The journal is published quarterly by Plenum Publishing Corporation, 233 Spring Street, New York, NY 10013. The founding editor of the journal was Robert B. Cumming, who served from 1981–1983; the current editor (1986) is Curtis C. Travis. The editorial address is:

The Editor,
Risk Analysis, An International Journal,
Health and Safety Research Division,
P.O. Box X,
Oak Ridge National Laboratory,
Oak Ridge, TN 37831.

Risk Analysis provides a focal point for new developments in risk analysis for scientists from a wide range of disciplines. The journal covers topics of potential interest to researchers, regulators, and scientific administrators. It deals with such issues in health risk analysis as dose–response functions, intraspecies extrapolation, and pharmacokinetics. Engineering topics treated include reliability analysis, probabilistic risk analysis, and nuclear and nonnuclear plant safety. The social and psychological aspects of risk addressed include such topics as risk perception, acceptability, economics, and ethics. The editor tries to maintain a balance between the biological, engineering, and social sciences. The journal publishes articles on a wide range of risk-related issues and most volumes contain discussions of statistical techniques and applications in risk analysis. Following is a brief selection of articles that have appeared in recent issues:

"On judging the frequency of lethal events: A replication."

"Facts and values in risk analysis for environmental toxicants."

"A social decision analysis of the earthquake safety problem: The case of existing Los Angeles buildings."

"How unique are the Price–Anderson limitations on nuclear accident liability?"

"A comparative potency approach for cancer risk assessment: Application to diesel particulate emissions."

"Decision making under risk: A comparison of Bayesian and fuzzy set methods."

"A methodology for seismic risk analysis of nuclear power plants."

"The use of landscape chemical cycles for indexing the health risks of toxic elements and radionuclides."

"Specifying risk goals: Inherent problems with democratic institutions."

"How close is close enough?: Public perceptions of the risks of industrial facilities."

Certain issues have been devoted entirely to special topics of concern such as health effects of diesel emissions, lessons learned from probabilistic risk analysis for nuclear plants, and the effect of toxic materials on human reproduction.

The journal currently (1983–1987) receives about 125 manuscripts per year, with an acceptance rate of about 25%. All scientific articles in *Risk Analysis* are fully peer-reviewed by three to four individuals. Each quarterly issue contains eight to ten articles, plus editorials, letters, and comments. The editor solicits expert comment and author response on selected articles in each issue.

CURTIS C. TRAVIS

RISK ANALYSIS, SOCIETY FOR

The Society for Risk Analysis was formed in 1980 in response to a perceived need to address the broad questions of risk as they span traditional disciplines. Society members represent many diverse fields, including

- engineering
- health sciences
- physical sciences
- social sciences
- economics

The purpose of the Society is to promote interdisciplinary exchanges between these fields, and to insure equal representation of all groups in Society activities and publications. Professionals interested in risk analysis, as it pertains to their area of interest or to other areas, are urged to join the Society.

The Society for Risk Analysis is an international body that holds annual meetings each summer or fall. The first meeting of the Society was the "International Workshop on the Analysis of Actual vs. Perceived Risks," held in June, 1981. Successive meetings have included a "Workshop on Low-Probability/High Consequence Risk Analysis," "Risk in the Private Sector," "Uncertainty in Risk Analysis," "New Risks: Issues in Management," and "Risk Assessment: Setting National Priorities." Attendance at the annual meetings reflects the diversity of persons associated with the Society: about one-third of the attendees have been from universities and policy institutions, one-third from government, and one-third from business and industry.

The Society sponsors an annual 3-day course on Carcinogen Risk Assessment, held at the National Academy of Sciences, Washington, DC.

The affairs of the Society are conducted by its members through an elected council composed of four officers, the past president, and nine councilors. They serve varying terms of from one to four years. A slate of officers is presented to the membership by a nominating committee, and efforts are made to insure representation of as many of the membership disciplines as possible. The presidents who have served since the inception of the Society are:

1981–82	Robert B. Cumming Oak Ridge National Laboratory
1982–83	Christopher G. Whipple Electric Power Research Institute
1983–84	Paul Slovic Decision Research
1984–85	Elizabeth L. Anderson U.S. Environmental Protection Agency
1985–86	Lester B. Lave Carnegie-Mellon University
1986–87	Paul F. Deisler, Jr. Shell Oil Company
1987–88	Vincent T. Covello National Science Foundation

There are four classes of membership available: regular, student, sustaining, and honorary. Membership in the Society entitles one to receive the quarterly journal, *Risk Analysis: An International Journal*, published by Plenum Press. The Society also distributes a quarterly newsletter including committee reports and information on national and international meetings of interest. For membership information contact: The Secretariat, SRA, 1340 Old Chain Bridge Road, Suite 300, McLean, Virginia 22101; telephone (703) 790-1745.

The Society is undertaking a series of volumes on topical issues in risk analysis in conjunction with Plenum Press. Suggested titles include: "Health and Environmental Risk Analysis," "Probabilistic Risk Analysis for Nuclear Power Plants," and "Contemporary Issues in Risk Analysis: The Social and Behavioral Sciences."

ELIZABETH L. ETNIER

RISK, CONSUMER'S *See* ACCEPTANCE SAMPLING

RISK MANAGEMENT, STATISTICS OF

Risk may be defined quite elaborately [27] or quite briefly. A concise summary is "adverse chance." *Risk management* is here defined as "any systematic process for limiting risk." *Statistics of risk management* is interpreted fairly broadly to involve any tools or methods from the discipline of statistics that have been, or logically should be, used in a systematic process for limiting risk.

GENERAL CONCEPT OF RISK MANAGEMENT AND MONITORING

As a general concept, risk management blends four crucial types of discipline. First is the substantive area discipline of economics/finance or engineering or medicine in which the primary risk control technology is embedded. The second is psychology and human factors analysis, which is necessary to help/induce people to work effectively with the risk control technology. The third is management/policy science needed to organize, coordinate, and monitor the implementation of the risk control measures. None of these would be sufficient without the fourth, statistical evaluation of the risk levels and monitoring of the effective impact of the risk management measures.

SCOPE

Risk management is not treated here as a subdiscipline of statistics having specialized models, distributions, or approaches. Rather it is covered as a broad family of applications to which many forms of statistical analysis apply. The reader should consult other sections of this encyclopedia and/or listed references for more detail on the mathematical structure of such processes as actuarial statistics*, logistic regression*, ARIMA models, or cumulative transition state scores. The applications range is illustrated, rather than exhausted, by brief discussions on finance, games, insurance, and developmental project management; on floods, chemical and oil spills, and nuclear power; on epidemiology and environmental regulation.

EVOLUTION OF RISK MANAGEMENT

The statistical aspects of risk management are twentieth century additions to historical principles of managing risks. Early humans who lived in caves not only found shelter from the elements, but also limited the possible approaches by an attacker. Fires at the cave entrance not only ventilated the smoke, but also lowered the probability of predatory animals attempting to enter the cave. Even primitive societies made organized efforts to limit risks of war, plague, crop failures, flood, and storm damage. Diplomacy, elementary public hygiene, systematic agriculture (or hunting), careful village siting, and storm resistant structural design were (and remain) necessary achievements to the survival of most cultures. Impressions of the subjective probability* of various hazards must have played a role in risk management, as did religious practices and subject matter expertise.

Further social evolution broadened risk management concerns from basic survival to include technological and economic risks. Accumulation of wealth beyond survival needs allowed venture capital undertakings. Selection of the few affordable risk ventures to be funded involved implicit probabilistic analysis. Ferdinand V and Isabella I of Spain decided that Christopher Columbus's expedition to find a shorter trade route to the East Indies by sailing west was an acceptable risk as a three-ship expedition. Fewer vessels were considered to involve excessive risk. High subjective confidence that at least one vessel would survive to report the results was the crux of the risk management decision.

INSURANCE

The meetings of shipowners and merchants at Edward Lloyd's coffeehouse in London, starting in 1688, led to formalization of insurance as a commercial risk management process. Mortality and injury life tables* were developed in the eighteenth and nineteenth centuries to support the determination of fair premiums for individual life insurance. The functions represented by these tables were not subjected to true statistical analysis until this century. Today, there is a highly developed field of actuarial statistics dealing with life and injury probabilities. Similar methods are applied to other common hazards such as fires. The original maritime focus of Lloyd's of London, however, remains nonstatistical. Ship hull insurance underwriters work by professional judgment of numerous particulars regarding a vessel and its owners, trade routes, cargoes, crew, and damage experience to manage risk. Ship classification societies inspect and measure ships and the results of their surveys are important to the setting of the premiums. Statistical methods are not employed because the ship populations at risk are not considered to be sufficiently homogeneous. Risk management using insurance principles may be explored further in several references, such as refs. 2 and 32.

GAMBLING*, CAPITAL BUDGETING, AND INVESTMENTS

Very likely, the earliest quantitative analysis explicitly designed to limit risk in monetary decisions may have been game theory* for gambling. Given fair dice or roulette wheels and honest game management, games are formal, closed probabilistic systems. Betting strategies can be formulated so that, over a large number of homogeneous plays, the statistically astute bettor is likely to be a net winner—unless equally astute management limits betting opportunities to prevent this.

There is little conceptual difference between betting decisions in games of chance* and capital budgeting decisions (selection and development of new products to market) in industry. In both cases, money is advanced on the basis of perceived possibilities of loss (risk) and potential gain (return). The major difference from gambling is that the "game" of industry is not a closed formal system. Production and marketing of a new product not only risks consumers' preferences relative to competing products, but also risks technological breakthroughs by competitors, breakdowns in distribution systems, strikes, recessions, and many other phenomena. Primitive quantitative risk management, therefore, leads to the simplistic notion of minimizing the payback period (the time when the cumulative discounted stream of returns equals the initial investment). This, in turn, leads to preference for the cheapest capital investments and/or those that return capital in the shortest number of years, based on an idea that the less money at risk for the shortest time assures the lowest risk. That principle ignores the opportunity costs in avoiding larger, longer term investments that may yield substantially greater returns.

Substantially improved capital budgeting decisions can be made by employing statistical decision theory* [22] to perform a more detailed analysis of several aspects of a number of capital budgeting alternatives. In principle, the same benefits can be achieved using decision analysis methods from operations research* [23] or econometrics* [1], which operate on single points of probability distributions—the expected mean value or a lower confidence limit of returns. The advantage of statistical decision theory is that it preserves the full probability distributions in the computations, thereby using more of the available information.

Choosing a group of capital budgeting ventures for a large corporation is similar to choosing a portfolio of investments for an individual or a financial institution. Two additional tools useful in this setting are utility theory* and portfolio optimization models. Utility theory deals with the concept that investors or risk capital managers attach varying weights ("utiles") to equivalent dollar amounts depending on whether those amounts are potential losses or gains, and gains of just satisfactory amounts or spectacular gains. Furthermore, a mixture of product lines or a portfolio of securities will not be optimized by a simple linear combination of individual items, each having acceptable risk-return patterns. Markowitz has shown that a quadratic programming model has desirable properties for efficient risk management over a diverse portfolio [16]. Risk management in capital budgeting or investments addresses market risk, assuming that technological risk is minimal.

TECHNOLOGICAL RISK MANAGEMENT

For development of major new technological systems, statistical analysis is appropriate when key technical parameters have a homogeneous development history over time. One of the most common examples is the development of successive generations of aircraft or missiles in terms of such parameters as thrust-to-weight ratio, maneuvering properties, combat radius, and payload. Various analyses by the RAND Corporation and other think tanks have shown that logistic (or logit) regression over time can be extrapolated usefully to evaluate whether a proposed new aircraft design is conservative, "on line," or highly ambitious in terms of its technological risk [12]. Using several such logit analyses* on different aspects of the system can help the project management team to focus early attention toward those aspects of the development where the technological risk is the greatest. Other forms of regression* and of statistical decision theory are also applicable to project management of cost and schedule risk.

LOW-PROBABILITY–HIGH-CONSEQUENCE RISK MANAGEMENT

There is a class of social risks such that periods between successive occurrences (return periods) span several years or several tens of years, but which are massively destructive when they occur. Examples include volcanic eruptions, floods, or hurricanes striking large cities. The statistically interesting property of this class of phenomena is that the occurrences of greatest interest occur in the far tails of skewed probability distributions, rather than around central tendencies.

The first major statistical work in this topic was in extreme value* statistics by Gumbel [10]. In studying very destructive floods, with return periods of 50–200 years, he noted that their probability distributions are transforms of more established distributions. Flood risk management is somewhat complex because increasing protection, say, from "25-year" floods to "50-year" floods will tend to induce more development of the flood plain than would otherwise occur. The potential destructiveness of a more-than-50-year flood may be heightened by the very management steps that reduce the risk due to 50-or-less-year floods. Thus, greater precision in defining the tail of the distribution of maximal floods as a function of return period can be of significant value in defining how far to go in flood protection schemes.

The risk of maritime toxic or flammable chemical spills that endanger shore populations can be evaluated by a vulnerability model [7]. Spilled material mixing with air may disperse simply according to a Gaussian puff or plume model, or buoyancy factors may also be important [11]. Probability of ignition of flammable clouds has been postulated to follow a probabilistic error function [28]. Probit analysis of dose response functions has been applied to lethal dose (LD) or lethal concentration (LC) experimental data to determine percentages of exposed populations expected to suffer various levels of morbidity or death [21].

In the case of oil spills from offshore drilling, pipelines, or tanker transport, major statistical problems were noted in 1975 by the U.S. National Academy of Sciences [17] because few data points were available, their precision was suspect, and the properties of their theoretical distributions were relatively unknown. The focus of effort was on risk estimation using stable law* (or Pareto*) distributions [20] or Bayesian* variations on parametric statistics to estimate spill frequencies and magnitudes independently [16]. From a risk management perspective, however, all that is required from the estimative process is a determination as to whether major new efforts to reduce the spill risk in a given project or operation are warranted. Effective risk reduction depends on improving the engineering, work structure and/or training, and the management processes for oil drilling or oil transport. Studies of these processes can be aided by such diverse statistical processes as reliability/maintainability/availability analysis of equipment [33], cluster analysis of tasks performed by personnel so as to improve functional work allocation and training [3], and even statistical design of experiments* [6] for work site or simulator experiments on how alternative operational procedures affect the spill risk [4].

The first major attempt to unify the risk estimation and risk management processes was the Reactor Safety Study, also known as WASH 1400 or the "Rasmussen report" [31]. The systems safety engineering technique called fault-tree analysis [5] was utilized for detailed insight into the designs, strengths, and weaknesses of two types of light water fission reactors. Lognormal distributions* of failure possibilities were estimated for thousands of independent potential accident-initiating factors. These were then combined by the Boolean algebraic logic of the fault-tree structure to try to arrive at net overall probabilities of serious accidents formulated in a manner that could be used directly for risk management. Systems engineers would then try to improve reliabilities of critical components (including human beings as system components with similarly structured failure probability functions to the mechanical components). Resolving double counting tendencies at the upper levels of the fault trees led the analytic team to formulate very useful additions to the technique—event trees. In retrospect, the attempt to combine risk estimation and risk management in a single massive model may not be the best way to achieve either goal. The Lewis Committee review of the study [15] found that the claimed confidence bounds could not be supported based on the data or theory available; indeed, later estimates revised the calculated risks markedly. Similarly, while the fault-tree analysis approach was useful, no single tool or technique is sufficient to identify or evaluate alternative risk management steps.

Factor analysis* has been used to show that many aspects of public perception of risk can be resolved into two orthogonal components—degree of dread of the hazard and degree to which the hazard and its management are perceived to be understood by science as well as by the public [29].

ENVIRONMENTAL RISK AND MONITORING SYSTEMS

Statistical aspects of health risk management need to be distinguished from health risk assessment. Implementing regulations under the Food and Drug Act of 1901 demanded monitoring, inspection, and control. Among the many laws that mandate the control of chemicals that are hazardous to humans, other living species, and the environment are the following: Toxic Substances Control Act (TSCA), Resource Conservation and Recovery Act (RCRA), the Federal Insecticide, Fungicide, and Rodenticide Act (FIFRA), The Safe Drinking Water Act, Water Pollution Control Act, the Clean Air Act, and the Endangered Species Act. Most of these were promulgated after 1970. The U.S. Environmental Protection Agency was given the responsibility by the U.S. Congress to determine, implement, and monitor regulations that prevent environmental health risks. Other agencies of the U.S. government also have the authority to control the use of chemicals and/or food additives that are hazardous for use or human consumption. Implementing their decision, however, usually involves banning the use of the substance. In contrast, environmentally oriented regulations establish "threshold limits," which defined the concentrations or levels that hazardous substances should not exceed in air, water, soil, and agents such as pesticides.

Establishing tolerable levels for toxic pollutants (hazardous substances) demands a knowledge of the statistical distribution properties of the following:

1. The concentration of the toxic pollutant in the medium in which it will be measured. For example, carbon monoxide levels in air have been assumed to follow the lognormal distribution*.
2. The relationship between the environmental concentration level and the biological concentration that will elicit the health hazard. For example, a linear regression* line has been fitted between the logarithms of ambient air lead levels and the logarithm of blood lead levels in determining the ambient air lead standard.
3. The concentration of the toxic substance over time. For example, the decision to penalize industrial sources if the second highest measured level during the year exceeded the allowable standard was based upon a model using the exponential distribution*.

The last issue has been particularly relevant to risk managment in terms of detecting trends or changes in environmental concentrations of hazardous pollutants. Box–Jenkins* time series* models [30, 13, 25] have been used to develop the best statistical fit to the time series data. Modified statistical quality control charts* [9] have been developed that account for the time dependence between successive values–autocorrelation. Typical industrial quality control* models [26] assume that successive observations are independent.

Determining standards for monitoring compliance demands attention to statistical criteria such as:

1. Whether the samples are obtained at intermittent discrete time intervals or whether they represent averages of continuous measurements.
2. Whether averages incorporate any calculations based upon the use of moving averages.

Either or both of these factors will introduce autocorrelation and require compensatory statistical processing.

EPIDEMIOLOGY AND DISEASE DETECTION

A nonregulatory oriented framework for health risk management pertains to the identification of exposure cohorts from population and disease registries. The U.S. Center for Disease Control maintains registries to identify the incidence of various untoward health effects. Potentially correlated environmental* surveillance data based on air and water quality are kept by the U.S. Environmental Protection Agency and the U.S. Geological Survey. Bivariate maps of morbidity*/mortality rates with air or water pollution levels at the national, state, county, and city level have been useful in targeting areas that need more remedial assistance, control, or intervention than the average [20]. After such maps have helped to focus on areas and environmental factors of interest, more numerical treatment can be provided through contingency table* analysis. This two-step process is a useful method for prioritizing budget and for generating hypotheses for further monitoring.

PRINCIPLES OF STATISTICAL MONITORING

A management-oriented aspect of statistical monitoring is determining the frequency of audits or intermittent samples that will establish conformance to standards. For this, general principles for selecting sample size are used, but an adjustment is often necessary for autocorrelation. When autocorrelation exists, sample sizes calculated may be too small to detect ongoing compliance, especially when pollutant source managers are aware that they will be subject to noncompliance penalties. Another management-oriented decision for establishing compliance has been to determine if sources have used best available control technology (BACT) for emission control. That procedure is more technological than statistical.

The statistical principles of monitoring use type I and type II errors. In the risk regulation and monitoring settings, type I error means deciding that the source is in violation when it actually is in compliance; type II error lies in deciding the source is in compliance when it actually is not. Such analysis assumes that the long-range emission probability distribution and its mean are known and also that the standard is set at an upper bound above the mean under a specific probability distribution. The expected exceedances (ExEx) method for establishing emission control standards for air quality has used this concept. Using principles borrowed from industrial quality control methods, including Shewhart control charts*, cumulative sums* (CUSUMS) and V masks, these techniques use the concepts of run length to define the upper control limit and to define a violation.

An innovation in regulatory monitoring was the "bubble" concept, where industrial and other pollutant sources work together with local or state authorities in monitoring area-wide compliance to a health or environmental standard. This has led to "controlled trading" of emission credits and thereby introduced new vistas for statistical exploration. While a binomial (pass/fail) framework would be sufficient in deciding whether a source was in violation or in compliance, a trinomial (pass/fail/credit) framework is needed when credits (periods of being better than the standard) are to be accumulated. A trinomial-based statistic for integrating changes in these states (pass/fail/credit) over time is the Cumulative Transitional State Score, CTSS [8]. CTSS integrates trinomial statistical principles with the transition probability concepts of Markov chain analysis to formulate a "benefit" score suitable to the accumulation and controlled trading of emission credits.

References

[1] Baumol, W. J. (1963). *Manag. Sci.*, **10**, 174–182.

[2] Chernick, P. L., Fairley, W. B., Meyer, M. B., and Scharff, L. C. (1981). *Design Costs and Acceptability of an Electric Utility Self-Insurance Pool for Assuring the Adequacy of Funds for Nuclear Power Plant Decommissioning Expense*, Nuclear Regulatory Commission, Washington, DC.

[3] Christal, R. E. and Ward, J. H., Jr. (1968). *The MAXOF Clustering Model.* Personnel Research Laboratory, Lackland Air Force Base, Texas.

[4] Cook, R. C., Marino, K. L., and Cooper, R. B. *A Simulator Study of Deepwater Port Shiphandling and Navigation Problems in Poor Visibility*, NTIS, Springfield, VA.

[5] Delong, T. W. (1971). *A Fault Tree Manual.* National Technical Information Service (NTIS), Springfield, VA.

[6] Diamond, W. J. (1981). *Practical Experiment Designs for Engineers and Scientists.* Lifetime Learning, Belmont, CA.

[7] Eisenberg, N. A., Lynch, C. J., and Breeding, R. J. *Vulnerability Model: A Simulation System for Assessing Damage Resulting from Marine Spills.* NTIS, Springfield, VA.

[8] Gardenier, T. K. (1979). Profiling for efficiency in long-range clinical trials. In *Advanced Medical Systems: An Assessment of the Contributions*, E. Hinman, ed. Yearbook Publications, Chicago.

[9] Gardenier, T. K. (1982). *Simulation*, **39**, 49–58.

[10] Gumbel, E. J. (1958). *Statistics of Extremes.* Columbia University Press, New York.

[11] Havens, J. A. (1977). *Predictability of LNG Vapor Dispersion from Catastrophic Spills Onto Water: An Assessment.* National Technical Information Service, Springfield, VA.

[12] Hutzler, W. P., Nelson, J. R., Pei, R. Y., and Francisco, C. M. (1983). *Non-nuclear Air to Surface Ordnance for the Future: An Approach to Propulsion Technology Risk Assessment.* RAND Corp., Santa Monica, CA.

[13] Ledolter, J. and Tiao, G. C. (1979). A statistical analysis of New Jersey CO data, *Proc. APCA*

ASQC Conference on Quality Assurance in Air Pollution Measurements. pp. 282–293.

[14] Lewis, H. W. et al. (1978). *Risk Assessment Review Group Report to the U. S. Nuclear Regulatory Commission*, Washington, DC.

[15] Markowitz, H. (1952). *J. Finance*, **8**, 77–91.

[16] Massachusetts Institute of Technology (1974). An analysis of oil spill statistics. *OCS Oil and Gas—An Environmental Assessment*, Council on Environmental Quality, Washington, DC.

[17] National Academy of Sciences (1975). *Petroleum in the Marine Environment*.

[18] Office of Federal Statistical Policy and Standards (1979). *Domestic Information for Decision-Making (DIDS): A New Alternative*, Washington, DC.

[19] Page, E. S. (1961). *Technometrics*, **3**, 1–9.

[20] Paulson, A. S., Schumaker, A. D., and Wallace, W. A. (1975). Risk-analytic approach to control of large volume oil spills. *Proc. Conf. on Prevention and Control of Oil Spills*, American Petroleum Association, Washington, DC.

[21] Perry, W. W. and Articola, W. P. (1980). *Study to Modify the Vulnerability Model of the Risk Management System*. NTIS, Springfield, VA.

[22] Raiffa, H. and Schlaifer, R. (1961). *Applied Statistical Decision Theory*. Harvard University Press, Cambridge, MA.

[23] Raiffa, H. (1968). *Decision Analysis: Introductory Lectures on Choices under Uncertainty*. Addison-Wesley, Reading, MA.

[24] Rasmussen, J. and Rouse, W. B., eds. (1981). *Human Detection and Diagnosis of System Failures*. Plenum, New York.

[25] Reinsel, G., Tiao, G. C., and Lewis, R. (1981). Statistical analysis of stratospheric ozone data for trend detection. *Proc. Environmetrics '81 Conference*, pp. 215–236.

[26] Roberts, S. W. (1966). *Technometrics*, **8**, 411–430.

[27] Rowe, W. D. (1977). *An Anatomy of Risk*. Wiley, New York.

[28] Simmons, J. A., Erdmann, R. C., and Naft, B. N. (1973). *The Risk of Catastrophic Spills of Toxic Chemicals*. NTIS, Springfield, VA.

[29] Slovic, P., Fischhoff, B., and Lichtenstein, S. (1983). Characterizing perceived risk. *Technological Hazard Management*, R. W. Kater and C. Hohenemser eds. Oelgeschlager, Gunn and Hain, Cambridge, MA.

[30] Tiao, G. C., Box, G. E. P., and Hamming, W. J. (1975). *J. Air Pollution Control Ass.*, **25**, 260–268.

[31] U. S. Nuclear Regulatory Commission (1975). *Reactor Safety Study No. WASH 1400*, Washington, DC.

[32] Williams, G. A. and Heins, R. M. (1976). *Risk Management and Insurance*, 3d ed. McGraw-Hill, New York.

[33] Willoughby, W. G. (1983). *Design and Manufacturing Fundamentals for Reliability*. Navy Materiel Command., Washington, DC.

(ACTUARIAL STATISTICS
CONTROL CHARTS
DECISION THEORY
EXTREME VALUE DISTRIBUTIONS
GAMBLING, STATISTICS IN
GAME THEORY
MANAGEMENT SCIENCE, STATISTICS IN
PHARMACEUTICAL INDUSTRY, STATISTICS IN
RISK MEASUREMENT, FOUNDATIONS OF
QUALITY CONTROL, STATISTICS IN
UTILITY THEORY)

J. S. Gardenier
T. K. Gardenier

RISK MEASUREMENT, FOUNDATIONS OF

Many fields—including medicine, safety, insurance, strategic planning, diplomacy, farming, banking, and investment management—view risk as a chance of something bad happening. In specific practical contexts, risk is often dealt with through two related activities: first, risk assessment estimates likelihoods of undesirable outcomes and evaluates the severity or cost of such outcomes; second, risk management* structures and implements courses of action that achieve a reasonable balance between risks and potential benefits.

The foundations of risk measurement focus on efforts to get behind specific contextual referents to consider aspects of risk that apply to many situations. These efforts adopt the perspective of measurement theory* [12, 17] by proposing axioms for a binary relation $\succsim$ (is at least as risky as) on a set of real or hypothetical probability distributions defined over a spectrum of possible decision outcomes. Typical axioms include ordering conditions and monotonicity postulates, such as $p \succsim q$ if distribution p assigns more probability to bad outcomes than does q.

The axioms usually permit the construction of a real-valued risk function on the distributions that orders them in the same way as the qualitative risk relation $\succsim$. The resultant risk measure often has special characteristics, as will be illustrated shortly.

This approach does not stand in isolation from empirical and practical concerns. Axioms are shaped by conventional wisdom about the meaning of risk and by empirical studies of perceived risk as well as by previous theoretical research. Moreover, applicability to risk assessment motivates the formation of risk measures that are as simple and elegant as possible, consistent with present understanding of risk. In the final analysis, any set of axioms that purports to characterize risk is subject to empirical verification or refutation.

Theories of risk involve several primitive notions on which the axioms are built. These generally include a set X of decision outcomes, a preference relation on X, a set P of probability measures defined on X or on a suitable algebra of subsets of X, and a binary risk relation $\succsim$ on P. It is assumed that X is completely ordered by the preference relation, and that P is completely ordered by $\succsim$. These are the basic ordering axioms for risk measurement.

For convenience, we shall presume that elements in X are numbers and that x is preferred to y if $x > y$. This is quite natural in monetary contexts and in several other settings. In less-structured situations we can think of x in X as a numerical value or utility of an underlying outcome [6].

Most theoretical and empirical work within the preceding formulation has taken outcomes as amounts of money to be won ($x > 0$) or lost ($x < 0$) so that P can be viewed as a set of monetary gambles. One of the first axiomatizations of $\succsim$ in this setting is ref. 16. This study shows that four axioms imply the following risk measure R on P that preserves $\succsim$ and is additive under convolutions:

$$p \succsim q \quad \text{if and only if} \quad R(p) \geqslant R(q),$$
$$R(p * q) = R(p) + R(q),$$

for all p and q in P. Here $*$ denotes convolution*:

$$(p * q)(x) = \sum_y p(y)q(x - y).$$

One of the four axioms—an independence condition, which says that, for all p, q, and r in P, $p \succsim q$ if and only if $p * r \succsim q * r$—seems empirically doubtful. For example, many people feel that p is riskier than q when p has probabilities 0.99 of a \$10,000 loss and 0.01 of a \$1000 gain, and q is an even-chance gamble between a \$12,000 loss and a \$2000 gain. Let r be a sure-thing \$11,000 gain. Since $p * r$ gives at least a \$1000 gain while $q * r$ yields a loss of \$1000 with probability $\frac{1}{2}$, it seems likely that most people would consider $q * r$ riskier than $p * r$.

Three more axioms imply that R has a linear decomposition in mean E and variance V; thus,

$$R(p) = \theta V(p) - (1 - \theta)E(p), \qquad 0 < \theta \leqslant 1.$$

The common but inaccurate notion of risk as variance obtains when $\theta = 1$. Although some of the seven axioms in ref. 16 are very appealing, empirical research [3] shows that factors besides mean and variance affect perceived risk.

A subsequent axiomatization [10] of risk uses the axioms of expected utility [6] to derive an order-preserving risk measure R that is linear under convex combinations. With $\lambda p + (1 - \lambda)q$ the convex ($0 \leqslant \lambda \leqslant 1$) combination of measures p and q, linearity means that

$$R(\lambda p + (1 - \lambda)q) = \lambda R(p) + (1 - \lambda)R(q).$$

Although there are settings in which this measure appears to be more viable [1] than an alternative theory [2] of preference over P that takes risk as an implicit determinant of preference, *its* independence condition—given p, q, and r in P, and $0 < \lambda < 1$,

$$p \succsim q \quad \text{if and only if}$$
$$\lambda p + (1 - \lambda)r \succsim \lambda q + (1 - \lambda)r$$

—also seems questionable.

For example, if a \$10,000 loss would be uncomfortable but a \$50,000 loss nearly ruinous, a person may perceive p as riskier than q when p yields a \$50,000 loss with probability 0.1 or a \$20,000 gain with probability 0.9, and q is certain to result in a \$10,000 loss. However, when r yields a \$50,000 loss with probability 0.2 or a \$100,000 gain with probability 0.8, and $\lambda = 0.1$, the same individual may perceive that $\lambda q + (1-\lambda)r$ is riskier than $\lambda p + (1-\lambda)r$:

$\lambda q + (1-\lambda)r$ yields:

a \$50,000 loss with probability 0.18,

a \$10,000 loss with probability 0.10,

a \$100,000 gain with probability 0.72;

$\lambda p + (1-\lambda)r$ yields:

a \$50,000 loss with probability 0.19,

a \$20,000 gain with probability 0.09,

a \$100,000 gain with probability 0.72.

Empirical observations of preference assessments [9, 11] and risk perception [4, 15, 18], coupled with common ideas about risk, have motivated theories of risk that differentiate favorable ("nonrisky") from unfavorable ("risky") outcomes. In ref. 13, aggregate axioms are used to derive the following multiplicative and additive measures in the monetary context:

$$R(p) = a_1\int_{-\infty}^{0} |x|^{\theta}\, dp(x) + a_2\int_{0}^{\infty} x^{\theta}\, dp(x), \qquad \theta > 0,$$

$$R(p) = b_1\int_{-\infty}^{0} dp(x) + b_2\int_{0}^{\infty} dp(x) + b\int_{x\neq 0} \log|x|\, dp(x), \qquad b > 0.$$

In the first form, risk is expressed as a weighted expectation of a power function whose parameter θ governs the curvature of risk away from 0. Normally, $a_1 > 0 > a_2$ and $a_1 > |a_2|$, with risk negative and variable when only gains are involved. In the latter form, $\int_{-\infty}^{0} dp(x)$ is the loss probability, and $\int_{0}^{\infty} dp(x)$ is the gain probability. It is noted that more data are needed to assess the empirical tractability of these forms.

Differential effects on risk of gains and losses lead to a general formulation that can apply to all contexts once a preference order over the outcomes is established. The first task in the general case is to identify a target outcome, which might be thought of as a point of no loss and no gain, a safe level, an aspiration level, or an outcome of minimal acceptable risk. For convenience, let the target be 0. Nonzero outcomes—or the numerically ordered utilities of outcomes—are then partitioned into unfavorable $(-)$ and favorable $(+)$ sets:

$$X^- = \{x : x < 0\}, \qquad X^+ = \{x : x > 0\}.$$

Sets of probability measures P^- and P^+ are defined on X^- and X^+, respectively. Then the general set P of measures can be taken as

$$P = \{(\alpha, p; \beta, q) : \alpha \geqslant 0, \beta \geqslant 0, \alpha + \beta \leqslant 1, p \text{ in } P^-, q \text{ in } P^+\}.$$

In $(\alpha, p; \beta, q)$, α is the loss probability, p is the loss distribution given a loss, β is the gain probability, q is the gain distribution given a gain, and $1 - \alpha - \beta$ is the probability for the target outcome 0.

For example, if g is a gamble that yields either -30, -20, 0, $+10$, or $+40$ with respective probabilities 0.06, 0.09, 0.35, 0.30, and 0.20, then the $(\alpha, p; \beta, q)$ way of writing g has

$$\begin{aligned} \alpha &= 0.15 \text{ (loss probability)}, \\ p &= \text{a gamble with probability } \tfrac{1}{3} \\ &\quad \text{for } -30 \text{ and } \tfrac{2}{3} \text{ for } -20, \\ 1 - \alpha - \beta &= 0.35 \text{ (target probability)}, \\ \beta &= 0.50 \text{ (gain probability)}, \\ q &= \text{a gamble with probability } \tfrac{3}{5} \\ &\quad \text{for } +10 \text{ and } \tfrac{2}{5} \text{ for } +40. \end{aligned}$$

The risk measure in this general approach satisfies

$$(\alpha, p; \beta, q) \succsim (\gamma, r; \delta, s) \quad \text{if and only if} \quad R(\alpha, p; \beta, q) \geqslant R(\gamma, r; \delta, s).$$

The preceding formulation is adopted in a two-part study [7, 8] that distinguishes "pure risk" from "speculative risk." The "pure risk" part [7] fixes β at 0, so there is no chance of gains, and considers the abbreviated risk measure $\rho(\alpha, p) = R(\alpha, p; 0, \cdot)$. In this setting, α is the loss probability, $1 - \alpha$ is the target-outcome probability, and p is the distribution of losses given a loss. Monotonicity, separability, and other axioms are used to obtain special forms for ρ, including

$$\rho(\alpha, p) = \rho_1(\alpha)\rho_2(p),$$

$$\rho(\alpha, p) = \int_{X^-} \rho(\alpha, x)\, dp(x),$$

and

$$\rho(\alpha, p) = \rho_1(\alpha) \int_{X^-} \rho_2(x)\, dp(x),$$

with ρ nonnegative and $\rho(\alpha, p) = 0$ only when $\alpha = 0$. The third form is similar to the expectation measure [10] when $\rho_1(\alpha) = \alpha$. Specializations of this form when $\rho_1(\alpha) = \alpha$ are discussed elsewhere: $\rho_2(x) = |x|^\theta$ is from ref. 13; $\theta = 1$ gives a weighted-losses measure of risk [5]; and $\theta = 2$ is below-target semivariance [14].

The first and third forms for ρ are separable in loss probability α and loss distribution p. This requires, among other things, that if α and α' are positive and $(\alpha, p) \succ (\alpha, p')$, then $(\alpha', p) \succ (\alpha', p')$. It is by no means obvious that this will hold for most people. Suppose p gives a sure loss of 200, and p' yields a loss of 250 with probability 0.8, and a loss of 1 otherwise. Then some people will see $(1, p)$ as riskier than $(1, p')$ because of the large sure loss for $(1, p)$. However, the risk direction between p and p' could reverse for small α. At $\alpha = 0.2$,

$(0.2, p')$ yields:

a loss of 250 with probability 0.16,
a loss of 1 with probability 0.04,
0 with probability 0.80;

$(0.2, p)$ yields:

a loss of 200 with probability 0.20,
0 with probability 0.80.

Because of the change in loss probability, it could happen that $(0.2, p') \succ (0.2, p)$ and $(1, p) \succ (1, p')$, and in such a case ρ cannot be separable.

The full scope of the $(\alpha, p; \beta, q)$ formulation is investigated in ref. 8 under the commitment that R is nonnegative, with $R = 0$ if and only if $\alpha = 0$. That is, it is presumed that there is no risk unless a loss is possible. This rules out additive forms like $R(\alpha, p; \beta, q) = R_1(\alpha, p) + R_2(\beta, q)$, but allows forms that are multiplicative in losses and gains. Examples include

$$R(\alpha, p; \beta, q) = \rho(\alpha, p)\tau(\beta, q)$$

and

$$R(\alpha, p; \beta, q) = \left[\rho_1(\alpha) \int_{X^-} \rho_2(x)\, dp(x)\right] \times \left[1 - \tau_1(\beta) \int_{X^+} \tau_2(y)\, dq(y)\right].$$

If τ is constant in the first form, gains have no effect on risk [7], but we would usually expect τ to decrease as β increases or as q shifts weight to larger outcomes. In the second form, where $1 > \tau_1(\beta) \int \tau_2(y)\, dq(y) > 0$, we expect ρ_2 to decrease in its argument and ρ_1, τ_1, and τ_2 to increase in their arguments [15, 18]. There is of course no surety that losses and gains will be separable in the general case, but separability may be a reasonable approximation in many situations.

Empirical studies of the proposals in the preceding paragraphs could be valuable in guiding further research.

References

[1] Aschenbrenner, K. M. (1978). *Acta Psych.*, **42**, 343–356.

[2] Coombs, C. H. (1975). In *Human Judgment and Decision Processes*, M. K. Kaplan and S. Schwartz, eds. Academic, New York. (Exposition of

Coombs's portfolio theory of choice between gambles.)

[3] Coombs, C. H. and Bowen, J. N. (1971). *Acta Psych.*, **35**, 15–28.

[4] Coombs, C. H. and Lehner, P. E. (1981). *Evaluation of Two Alternative Models for a Theory of Risk: II. The Bilinear Model and Its Conjoint Analysis.* Dept. of Psychology, University of Michigan, Ann Arbor, MI. (Important empirical study of risk.)

[5] Domar, E. V. and Musgrave, R. A. (1944). *Quart. J. Econ.*, **58**, 389–422. (Dated but valuable example of risk in economic analysis.)

[6] Fishburn, P. C. (1970). *Utility Theory for Decision Making.* Wiley, New York. (Standard reference for axiomatic utility theories.)

[7] Fishburn, P. C. (1984). *Manag. Sci.*, **30**, 396–406. (Risk as probable loss.)

[8] Fishburn, P. C. (1982). *J. Math. Psychol.*, **25**, 226–242. (Effects of gain on risk.)

[9] Fishburn, P. C. and Kochenberger, G. A. (1979). *Decision Sci.*, **10**, 503–518. (Argues that assessed utility functions behave differently below and above a target point.)

[10] Huang, L. C. (1971). The Expected Risk Function. *Report No. MMPP 71-6*, Dept. of Psychology, University of Michigan, Ann Arbor, MI.

[11] Kahneman, D. and Tversky, A. (1979). *Econometrica*, **47**, 263–291. (Presents a significant alternative to traditional theory of preference between gambles.)

[12] Krantz, D. H., Luce, R. D., Suppes, P., and Tversky, A. (1971). *Foundations of Measurement. Volume I: Additive and Polynomial Representations.* Academic, New York. (The standard reference on measurement theory.)

[13] Luce, R. D. (1980). *Theor. Decision*, **12**, 217–228; correction (1981), **13**, 381. (Establishes new theoretical directions.)

[14] Markowitz, H. (1959). *Portfolio Selection.* Wiley, New York. (Initiator of significant work in financial analysis.)

[15] Payne, J. W. (1975). *J. Exp. Psych.: Human Percep. Perform.*, **104**, 86–94. (Empirical study of factors that affect perceived risk among gambles.)

[16] Pollatsek, A. and Tversky, A. (1970). *J. Math. Psychol.*, **7**, 540–553. (Good background reading on risk measurement.)

[17] Roberts, F. S. (1979). *Measurement Theory.* Addison-Wesley, Reading, MA. (Lighter and more comprehensive than ref. 12.)

[18] Slovic, P. (1967). *Psychonomic Sci.*, **9**, 223–224. (Only two pages, but a powerful and valuable empirical report on factors that affect preference and risk perception.)

(DECISION THEORY
GAMBLING, STATISTICS IN
MEASUREMENT THEORY
NUCLEAR MATERIALS SAFEGUARDS
RISK MANAGEMENT, STATISTICAL ASPECTS OF)

PETER C. FISHBURN

RISK, PRODUCER'S *See* ACCEPTANCE SAMPLING

RISK THEORY

Some 50 years ago Cramér [4] wrote: "The object of the Theory of Risk is to give a mathematical analysis of the random fluctuations in an insurance business and to discuss the various means of protection against their inconvenient effects." One of these effects is the complete depletion of the company's "free reserves" and, because its mathematical formulation has much in common with certain early problems of probability theory, this is known as the ruin of the company. A more modern definition of risk theory is implied by the following introductory sentence from Beard et al. [1], which provides a convenient explanation of some technical insurance terms: "... the following analysis is ... restricted to the study of claims and to that part of the premiums which remains when loadings for expenses of management have been deducted, i.e. risk (net) premiums increased by a safety loading." Even more recently what the foregoing authors have called risk theory has been retitled ruin theory and described as only part of a risk theory that includes such subjects as premium calculation, experience rating, and reinsurance—matters to which we can give no attention (Gerber [8]).

The earliest attempt to give a mathematical form to the risk run by an insurance company, specifically a fund to pay pensions to widows, was that of Tetens [20], and for nearly a century all technical articles on the theory were written in German (Wagner [23]).

Broadly the idea underlying these early researches was to calculate the variance of a life insurance contract and, assuming that each contract in a portfolio was independent of any other, to sum these variance and apply the central limit theorem* to determine the probabilities of different aggregate losses (i.e., claims minus net premiums) over a shorter or longer period. A classic summary of the theory as it had developed by the turn of the century is that of Bohlmann [2].

A completely novel approach to risk theory and at the same time the birth of discontinuous stochastic processes*, in particular the Poisson process* (Cramér [5]), is due to the brilliant but idiosyncratic Swedish actuary Filip Lundberg [10, 11]. While the "individual" contract theory was essentially applicable to life insurance and the new "collective" theory was introduced with life and annuity policies in view, the enormous reserves held by such companies to meet the "savings" element in these policies militated against the urgent need for a mathematical theory of risk. On the contrary, in fire, accident, and other branches of nonlife insurance where no savings element is present in the premiums, risk theory plays (or should play) an important role. Lundberg's risk theory is based on assumptions that appear to be well suited to such nonlife companies.

Briefly, a portfolio of, e.g., fire insurance policies can be supposed to be subject to two independent sets of random variations: (i) The stochastic (point) process of time epochs at which claims occur, and (ii) the amount of the claim (supposed paid immediately) that the fire has caused. Point processes and their mathematical properties form an independent subject of study (Cox and Isham, [3]), but risk theory has so far limited itself to consideration of stationary point processes*, which can be defined as those in which $p_n(t)$, the probability of n events (claims) in the following interval of time of length t, is unchanged when the epoch of commencement is moved along with time axis (McFadden [13]). The simplest of all stationary point processes is the Poisson process in which the intervals between successive events are independent variates, the probability distribution of which is (negative) exponential. There have been several mathematically sophisticated papers extending the Poisson process to a "renewal process" in which the independent interval lengths are distributed arbitrarily (Thorin [21]), but as yet no statistical example of such a process of the occurrence of claims has been published. What has been shown in a number of practical papers is that (a) the Poisson process only provides a broad agreement with the actual claim process, (b) successive claim interval lengths are *not* independent, and (c) the observed distribution of claim intervals can always be fitted successfully by a member of the family of so-called mixed Poisson processes.

The mixed Poisson process* was introduced by Dubourdieu [6] and has the following properties, each of which can be shown to imply the others (Dubourdieu, [7, note]; McFadden [14]):

(A) If n claims have occurred in an interval of length t, they are distributed uniformly and independently over the interval.

(B) If $p_n(t)$ denotes the probability of n claims in an interval of length t,

$$p_n(t) \equiv (-1)^n \frac{t^n}{n!} p_0^{(n)}(t),$$

the parenthetical index denoting differentiation of the order indicated.

(C)

$$p_0(t) = \int_0^{\infty} e^{-\lambda t}\, dM(\lambda) = e^{\theta}(t),$$

$$\theta(0) = 0, \qquad (-1)^n \theta^{(n)}(t) \geqslant 0, \qquad n = 1, 2, \ldots,$$

where $M(\cdot)$ is a distribution function over the positive axis.

Two members of the mixed Poisson family that have been successfully fitted to nonlife claim statistics are the negative binomial*

$$p_n(t) = \binom{n+h-1}{n} \left(\frac{h}{h+t}\right)^h \left(\frac{t}{h+t}\right)^n,$$

$$h > 0,\ n = 0, 1, 2, \ldots,$$

where the $\theta(t) = h \ln(h/h + t)$, and the Hofmann distribution

$$p_0(t) = \exp\left\{\frac{1}{c(1-a)}\left[1 - (1 + ct)^{1-a}\right]\right\},$$

$$a, c > 0,\ a \neq 1.$$

In both cases $E(N) = t$.

With regard to the distribution of claim sizes, assumed to be continuous for convenience, any distribution over the positive axis would seem possible. Nevertheless, statistical observations have indicated a clear preference for the Pareto*, namely,

$$B(y) = 1 - (1 + y/b)^{-\nu},$$

$$\nu, b > 0;\ 0 \leqslant y < \infty,$$

or the lognormal*,

$$B(y) = \frac{1}{\sigma\sqrt{2\pi}} \int_0^y x^{-1} \exp\left[-\frac{(\ln x - \zeta)^2}{2\sigma^2}\right] dx,$$

$$\zeta, \sigma > 0;\ 0 < y < \infty,$$

where $B(\cdot)$ is the distribution function of claim sizes. By writing $b = \nu - 1$ and $\zeta = -\sigma^2/2$ the mean claim size is unity.

Having specified the claim occurrence process and the claim size distribution we turn to the two problems that have occupied actuaries working with Lundberg's theory of risk. The first of these is the dual of a problem in queueing theory*: Given the point process of claims (arrivals at the queue) and the distribution of their sizes (the times they occupy the single server), find the distribution function of aggregate claims (work load offered to the server) during an interval of length t. Writing $F(x, t)$ for this distribution function and $B^{n*}(\cdot)$ for the distribution function of the aggregate claim stemming from n independent claims we have

$$F(x, t) = \sum_{n=0}^{\infty} p_n(t) B^{n*}(x), \qquad x \geqslant 0, \tag{1}$$

where the distribution function of the sum of n independent claims is available iteratively from the standard relation

$$B^{n*}(x) = \int_0^x B^{(n-1)*}(x - y)\, dB(y),$$

$$n = 1, 2, 3, \ldots,$$

$$B^{0*}(x) = \begin{cases} 0, & x < 0, \\ 1, & x \geqslant 0. \end{cases}$$

While this formula for $F(x, t)$ is easily written down it is much harder to get numerical values when $B(\cdot)$ is of Pareto or lognormal form. In the former case the relatively simple formula for the Laplace transform or characteristic function* of $F(x, t)$ can be used (Seal [17]), but the lognormal has no explicit characteristic function and its numerical values must be obtained by quadrature.

The second important problem of risk theory is to calculate numerically the probability of ruin, or its complement the probability of survival, through an epoch t. This, too, can be written down quite easily, but only for the practical case of mixed Poisson claim occurrences (Seal [18]). Suppose that n independent claims have occurred in $(0, t)$; then standard probability theory states that the distribution function of the length of any subinterval between uniformly distributed claims (including the time between the origin and the first claim) is

$$A(z) = 1 - (1 - z/t)^n, \qquad 0 \leqslant z \leqslant t.$$

Write $g_n(x)$ for the density of a loss of x to the company on any one claim, namely a claim of $x + z$ following a subinterval of z since the prior claim during which a uniform premium has been paid at a unit rate. In terms of the density of claim sizes and intervals between claims,

$$g_n(x) = \int_{\max(0,\,-x)}^{\infty} b(x + z)\, dA(z),$$

$$-t \leqslant x < \infty.$$

Now let $W_{jn}(w)$ be the probability of company survival past the jth of the n claims given that the company's initial risk reserve (or free surplus) was w. [It is customary to ignore interest on a usually growing risk reserve, but it can be included (Gerber [9]).]

The latter probability is equivalent to the probability that an initial loss of x, with density $g_n(x)$, reduced w to $w - x$ and left $j - 1$ claims (out of n) to be survived. Summing this recursive relation over all x, we get

$$W_{jn}(w) = \int_{-t}^{w} W_{j-1,n}(w - x) g_n(x)\, dx; \qquad j = 1, 2, 3, \ldots, n,$$

with

$$W_{0n}(w) = 1, \qquad w \geqslant 0.$$

If $U(w, t)$ denotes the probability of company survival throughout an interval $(0, t)$, which it started with a risk reserve of w, we may write

$$U(w, t) = \sum_{n=0}^{\infty} p_n(t) W_{nn}(w, t), \qquad (2)$$

a relation analogous to (1).

The two principal functions of risk theory can thus be evaluated numerically for any chosen values of t, w, and x and for selected members of the mixed Poisson family together with suitable Pareto or lognormal size distributions.

In Seal [15] there is a numerical example based on an actual two-component compound Poisson distribution of claim occurrences, namely,

$$p_n(t) = ae^{-\alpha_1 t}\frac{(\alpha_1 t)^n}{n!} + (1 - a)e^{-\alpha_2 t}\frac{(\alpha_2 t)^n}{n!},$$

$$a = 0.9403, \qquad \alpha_1 = 0.7553, \qquad \alpha_2 = 4.8547,$$

and a "fitted" workmen's compensation claims size density

$$B'(y) = b(y) = b\beta_1 e^{-\beta_1 y} + (1 - b)\beta_2 e^{-\beta_2 y},$$

$$b = 0.00663, \qquad \beta_1 = 0.09026, \qquad \beta_2 = 1.0722.$$

Both these distributions have unit means.

The following results were obtained by inversion of Laplace transforms in a manner described in the article cited. Notice the relatively small decrease in the first function to obtain the last for given small t.

t	$F(10 + 1.1t, t)$	$U(0, t)$ $1 + \eta = 1.1$	$U(10, t)$ $1 + \eta = 1.1$
1	0.9954	0.5990	0.995
2	0.9821	0.4969	0.980
3	0.9637	0.4494	0.958
4	0.9484	0.4212	0.940
5	0.9385	0.4023	0.928
6	0.9328	0.3886	0.922
7	0.9295	0.3781	0.918
8	0.9275	0.3698	0.916
9	0.9260	0.3631	0.914
10	0.9247	0.3575	0.912
20	0.9174	0.3291	0.896
30	0.9150	0.3177	0.884
40	0.9150	0.3113	0.876

(Recent research indicates small third-place corrections in the last column.)

If larger companies are to be considered, asymptotic methods are available. For the Poisson point process of claims we have Cramér's [4] integral equation*

$$\lim_{t\to\infty} U(w, t) \equiv U(w) = \frac{\eta}{1 + \eta} + \int_0^w U(w - x)\frac{1 - B(x)}{1 + \eta}\, dx,$$

where $\eta \neq 0$ is the risk loading on the unit premium per expected claim. When w is large we may use Lundberg's [12, p. 26] asymptotic value derived from this, namely,

$$1 - U(w) \sim \frac{\eta}{\int_0^\infty xe^{\kappa x}\, dB(x) - (1 + \eta)} e^{-\kappa w},$$

where κ is given by

$$\int_0^\infty e^{\kappa y} b(y)\, dy = 1 + (1 + \eta)\kappa.$$

Writing **(C)** in the form

$$p_n(t) = \int_0^\infty e^{-\lambda t}\frac{(\lambda t)^n}{n!}\, dM(\lambda),$$

(2) becomes

$$U(w, t) = \int_0^\infty dM(\lambda) \sum_{n=0}^\infty e^{-\lambda t} \frac{(\lambda t)^n}{n!} W_{nn}(w, t),$$

and

$$\begin{aligned} U(w) &= \lim_{t \to \infty} U(w, t) \\ &= \lim_{t, k \to \infty} \int_0^\infty dM(\lambda) \sum_{n=0}^\infty e^{-k} \frac{k^n}{n!} W_{nn}(w, t) \\ &= \lim_{t, k \to \infty} \sum_{n=0}^\infty e^{-k} \frac{k^n}{n!} W_{nn}(w, t) \end{aligned}$$

so that mixed Poisson ruin may be evaluated by Cramér's equation.

References

Books and articles dated before 1965 can be regarded as having mainly historical importance.

[1] Beard, R. E., Pentikäinen, T., and Pesonen, E. (1969, 1977). *Risk Theory*. Chapman and Hall, London. ("... a basic text for universities which offer courses in actuarial science.... ")

[2] Bohlmann, G. (1909). Die Theorie des mittleren Risikos in der Lebensversicherung. *Berichte VI Int. Kong. Versich. Wiss.*, **1**, 593–683.

[3] Cox, D. R. and Isham, V. (1980). *Point Processes*. Chapman and Hall, London. ("... emphasis on results and methods directly useful in applications.")

[4] Cramér, H. (1930). On the mathematical theory of risk. *Festskrift Skand.* 1855–1930, Stockholm, Sweden.

[5] Cramér, H. (1969). Historical review of Filip Lundberg's works on risk theory. *Skand. Aktuarie-tidskr. Suppl.*, **52**, 6–12.

[6] Dubourdieu, J. (1938). Remarques relatives à la théorie mathématique de l'assurance-accidents. *Bull. Inst. Actu. Française*, **44**, 79–126.

[7] Dubourdieu, J. (1952). *Théorie Mathématique du Risque dans les Assurances de Répartition*. Gauthier-Villars, Paris.

[8] Gerber, H. (1971). Der Einfluss von Zins auf die Ruinwahrscheinlichkeit. *Mitt. Ver. Schweiz. Versich.-Math.*, **71**, 63–70.

[9] Gerber, H. (1979). *An Introduction to Mathematical Risk Theory*. University of Pennsylvania, Philadelphia, PA. ("... several semesters of college mathematics are required.")

[10] Lundberg, F. (1903). *I. Approximerad Framställning af Sannolikhetsfunktionen. II. Återförsäkring af Kollektivrisker*. Almqvist and Wiksell, Uppsala, Sweden.

[11] Lundberg, F. (1909). Über die Theorie der Rückversicherung. *Berichte VI Int. Kong. Versich. Wiss.*, **1**, 877–948.

[12] Lundberg, F. (1926). *Försäkringsteknisk Riskutjämning. I Theori*. Englund, Stockholm, Sweden. (Lundberg's pathbreaking work has been regarded as conceptually and mathematically impenetrable but Cramér's historical review above provides a partial interpretation.)

[13] McFadden, J. A. (1962). On the lengths of intervals in a stationary point process. *J. R. Statist. Soc. B*, **24**, 364–382, 500.

[14] McFadden, J. A. (1965). The mixed Poisson process. *Sankhyā A*, **27**, 83–92. (The latter relatively inaccessible paper is easily understood and very important for practical risk theory.)

[15] Seal, H. L. (1974). The numerical calculation of $U(w, t)$, the probability of non-ruin in an interval $(0, t)$. *Scand. Actu. J.*, **57**, 121–139.

[16] Seal, H. L. (1978). *Survival Probabilities: The Goal of Risk Theory*. Wiley, Chichester, England.

[17] Seal, H. L. (1980). Survival probabilities based on Pareto claim distributions. *Astin Bull.*, **11**, 61–71.

[18] Seal, H. L. (1980). Ruin probabilities for mixed Poisson claim numbers without Laplace transforms. *Mitt. Ver. Schweiz. Versich.-Math.*, **80**, 297–306.

[19] Seal, H. L. (1983). Numerical probabilities of ruin when expected claim numbers are large. *Mitt. Ver. Schweiz. Versich.-Math*, **83**, 89–104. (Works principally concerned with the numerical evaluation of ruin probabilities.)

[20] Tetens, J. N. (1786). *Einleitung zur Berechnung der Leibrenten und Anwartschaften*. Zweyter Teil. Weidmanns, Erben and Reich, Leipzig, Germany.

[21] Thorin, O. (1977). Ruin probabilities prepared for numerical calculation. *Scand. Actu. J.*, **60**, 7–17. (Mathematically difficult.)

[22] Thyrion, P. (1969). Extension of the collective risk theory. *Skand. Aktuarietidskr. Suppl.*, **52**, 84–98. (Calculation of $F(x, t)$ when $p_n(t)$ is mixed Poisson.)

[23] Wagner, K. (1898). *Das Problem vom Risiko in der Lebensversicherung*. Fischer, Jena, Germany.

(ACTUARIAL STATISTICS, LIFE
DECISION THEORY)

Hilary L. Seal

R-MATRIX

A matrix of which the (i, j)th element is the correlation between X_i and X_j. For a set of m variables $X_1, \ldots, X_m$ it is an $m \times m$ matrix. Each diagonal element is 1.

(CORRELATION
MULTIPLE CORRELATION
PARTIAL CORRELATION
VARIANCE–COVARIANCE MATRIX)

ROBBINS–MONRO PROCEDURE *See* STOCHASTIC APPROXIMATION

ROBUST ESTIMATION

Robustness is a desirable but vague property of statistical procedures. A robust procedure, like a robust individual, performs well not only under ideal conditions, the model assumptions that have been postulated, but also under departures from the ideal. The notion is vague insofar as the type of departure and the meaning of "good performance" need to be specified. Robust estimation has been studied primarily when gross errors are the departure from assumptions we are concerned with. We begin by discussing this departure in the simplest and best understood situation.

THE NORMAL MEASUREMENT (ONE-SAMPLE) MODEL

We want to estimate an unknown constant θ using measurements $X_1, \ldots, X_n$, where we represent $X_i = \theta + e_i$, e_i being measurement error*. If, as usual, we assume the measurement errors are independent with a common normal distribution with mean 0 and variance σ^2, the sample mean $\bar{X}$ is the estimate of choice on a variety of grounds. It is the maximum likelihood* estimate, best unbiased, minimax, and asymptotically efficient. But it is not robust against even small departures from the assumption of normality. In particular suppose that the measuring instrument, which normally produces normal errors, malfunctions on each observation with probability ϵ (independent of what the measurement error might have been without malfunction) and produces e_i distributed according to an irrelevant distribution H. This leads to the gross error model of Tukey [35] and Huber [17]. The e_i have common distribution $G(x)$, where

$$G(x) = (a - \epsilon)\Phi(x/\sigma) + \epsilon H(x) \quad (1)$$

and $F(x) = G(x - \theta)$ is the distribution of X_i. Experience suggests that G has heavier tails than the normal component, i.e., the "bad" e_i tend to be "gross errors," larger than the "good" ones in absolute value, and the corresponding X_i tend to be outliers*. We would expect that:

- **(a)** When present and large enough, the outliers could determine the value of $\bar{X}$ leading to entirely inaccurate estimates of θ.
- **(b)** Unless G is symmetric about 0, $\bar{X}$ would be biased.
- **(c)** Even if G is symmetric about 0, the variance of $\bar{X}$ might be much higher than in the absence of gross errors and $\bar{X}$ might be highly inefficient.

Experience with data has suggested to Tukey, Huber, and other writers that ϵ of the order of 1–10% is common.

This lack of robustness of the sample mean has been recognized in practice for a long time and has been dealt with by more or less ad hoc rejection-of-outliers procedures or by the use of estimates less affected by outliers, such as the median* or, more generally, trimmed means. The α trimmed mean $\bar{X}_\alpha$ is defined as the average of the inner $n - 2[n\alpha]$ order statistics*, $0 \leqslant \alpha < \frac{1}{2}$, $X'_{[n\alpha]+1}, \ldots, X'_{n-[n\alpha]}$. For a history of the early use of such procedures and other aspects of the history of robust estimation see Stigler [33].

Following an important paper by Tukey [35], in which he discussed quantitatively the serious effect of gross errors on estimation of location and scale in the normal one-sample model, theories of robust estimation (against gross errors) were developed by various workers. We sketch some of these approaches, emphasizing the central contributions of Huber and Hampel. A full and very valuable account may be found in Huber's [19] monograph, on which we draw heavily. We hereafter refer to this work as [H].

HAMPEL'S CRITERIA

In his fundamental thesis, Hampel [14] proposed some useful criteria that an estimate should satisfy to be called robust against gross errors (in general, one-sample models). As is true of most criteria in the theory of robust estimation they are based on large-sample approximations since analysis (as opposed to simulation) for small samples is difficult. We shall indicate fixed sample size analogs where they exist. So an estimate $\{T_n\}$ is to be thought of as a sequence of related procedures, T_n being appropriate to sample size n. In all cases of interest, if the common distribution of the X_i is F (and regularity conditions hold), $T_n = T_n(X_1, \ldots, X_n)$ stabilizes in probability as $n \to \infty$ to a value $T(F)$, which defines a functional on as broad a class of distributions as the regularity conditions allow. For example, $T_n = \bar{X}$ leads to $T(F) = \int x\, dF(x)$, the population mean; $T_n = \text{median}(X_1, \ldots, X_n)$ leads to $T(F) = F^{-1}(\frac{1}{2})$, the population median (if it is uniquely defined). In most cases of interest, including the above, $T_n = T(F_n)$, where F_n is the empirical distribution* of the sample.

Qualitative Robustness

$T(\cdot)$ should be continuous for convergence in law at distributions F that are of interest. That is, if $F_m \to F$ in law then $T(F_m) \to T(F)$. The notion that small changes in F should lead to small changes in T clearly catches some of what we would like robustness to mean. An appealing argument that continuity for convergence in law is the right notion is given in [H, Sec. 1.3]. The notion of resistance* in small samples, owing to Mosteller and Tukey [29], is closely related. The sample mean is ruled out by this criterion.

Breakdown

As we have noted, a single very large outlier can essentially determine the value of $\bar{X}$ no matter how large n is. This is not the case with the median*. A single outlier can only move the median one order statistic over. The empirical breakdown point is essentially the smallest fraction of outliers that the estimate can tolerate before being determined by the outliers. For $\bar{X}$ the proportion is 0, for $\bar{X}_\alpha$ essentially α, for the median $\frac{1}{2}$. In general, the empirical breakdown point depends on the configuration of the data and so it is preferable to give a population (large-sample) version of this notion. So Hampel defines the breakdown point of $\{T_n\}$ at F by

$$\epsilon(f) = \sup\left\{\epsilon : \sup_Q |T((1-\epsilon)F + \epsilon Q)| < \sup_Q |T(Q)|\right\}.$$

A positive, preferably high, breakdown point is a reasonable requirement for a robust estimate.

The Influence Function* and Gross Error Sensitivity

For estimates of interest (under regularity conditions), not only does T_n tend to $T(F)$ in probability but also,

$$T_n = T(F) + n^{-1} \sum_{i=1}^{n} \text{IC}(X_i, F; \{T_n\}) + o_p(n^{-1/2}), \tag{2}$$

where IC, Hampel's *influence curve* or the first von Mises derivative is uniquely defined by $\{T_n\}$ and F and has the properties

$$\int \mathrm{IC}(x, F; \{T_n\})\, dF(x) = 0,$$

$$\int \mathrm{IC}^2(x, F; \{T_n\})\, dF(x) < \infty.$$

Thus,

$$\mathrm{IC}(x, F: \{\bar{X}\}) = x - \int x\, dF(x),$$

$$\mathrm{IC}(x, F; \{\mathrm{median}(X_1, \dots, X_n)\}) = \frac{\mathrm{sgn}(x - F^{-1}(\frac{1}{2}))}{2f(F^{-1}(\frac{1}{2}))},$$

where $\mathrm{sgn}(x) = \pm 1$ as $x \gtrless 0$, as was originally shown by Bahadur [2]. Calculation of IC is facilitated by Hampel's remark

$$\mathrm{IC}(x, F; \{T_n\}) = \lim_{\epsilon \to 0} \frac{T((1-\epsilon)F + \epsilon\delta_x) - T(F)}{\epsilon}, \tag{3}$$

where δ_x is point mass at x. The notion of expansion of statistics in this sense is due to von Mises [27]. Hampel noted that if $\{T_n\}$ is an estimate, $\mathrm{IC}(X_i, F; \{T_n\})$ can be interpreted as the contribution that X_i makes to the value of the estimate. To be more precise, consider the following natural estimate of IC:

$$\mathrm{SC}(x; X_1, \dots, X_{n-1}) = \frac{T((1 - 1/n)F_{n-1} + (1/n)\delta_x) - T(F_{n-1})}{1/n}.$$

This is the sensitivity curve proposed by Mosteller and Tukey, which if $T_n = T(F_n)$ can be written as

$$\mathrm{SC}(x; X_1, \dots, X_{n-1}) = n(T_n(X_1, \dots, X_{n-1}, x) - T_{n-1}(X_1, \dots, X_{n-1})).$$

In this form, SC is clearly proportional to the change in the estimate when one observation with value x is added to a sample $X_1, \dots, X_{n-1}$. There is also a simple relation to the jackknife*—see [H] and Efron [13].

In line with his interpretation, Hampel defines the gross error sensitivity of $\{T_n\}$ at F,

$$\gamma^*(F, \{T_n\}) = \sup_x |\mathrm{IC}(x, F, \{T_n\})|.$$

For robustness it seems reasonable to require that γ^* be finite.

The influence curve has taken on a major role in robust estimation, proving to be a very important heuristic tool for construction of estimates. It can sensibly be generalized to the linear model and other multiparameter problems.

Hampel's criteria give some guidelines as to what we should expect of robust estimates. To pick members out of the large classes of procedures that fit his requirements requires optimality considerations of some sort. The most fruitful considerations so far have been those of Huber's asymptotic minmax* theory.

HUBER'S ASYMPTOTIC MINMAX* APPROACH FOR LOCATION

Huber [17] proposed as a goal minimization of the maximum asymptotic variance over the gross error model (with σ^2 provisionally assumed known). This goal can be easily justified if G is in addition assumed symmetric about 0, in which case asymptotic biases are negligible compared to the variance. His solution is one of an important class of procedures he proposes, the (M) estimates*, defined as solutions T_n of equations of the form

$$\sum_{i=1}^{n} \psi\left(\frac{X_i - T_n}{\sigma}\right) = 0.$$

These are generalizations of maximum likelihood* estimates, for which $\psi(x/\sigma) = -g'(x)/g(x)$, where g is the common density of the e_i. Huber's solution has

$$\psi_k(x) = \begin{cases} x, & \text{if } |x| \leqslant k, \\ k \,\mathrm{sgn}\, x, & \text{if } |x| > k, \end{cases}$$

where k is related to the hypothesized probability of a gross error ϵ by

$$2\frac{\Phi(k)}{k} - 2\phi(-k) = \frac{\epsilon}{1-\epsilon}.$$

If σ is not known, it too is to be estimated.

Popular choices are $k = 1.0$–1.5, corresponding to ϵ ranging from 5–10% and

$$\hat{\sigma} = \text{MAD}/0.6745,$$

where

$$\text{MAD} = \text{median}\{|X_i - \text{median}(X_1, \dots, X_n)|\},$$

although this choice of $\hat{\sigma}$ may not preserve the minmax property of the estimate. To contrast the behaviour of this estimate with $\bar{X}$, see [H, p. 145]. Huber's estimates satisfy Hampel's criteria.

It can be shown (Jaeckel [20]) that if $\alpha = \Phi(-k)$, the trimmed mean $\bar{X}_\alpha$ is also asymptotically minmax in Huber's sense and satisfies Hampel's criteria. Since $\bar{X}_\alpha$ is scale equivariant (a common change of scale in the data changes $\bar{X}_\alpha$ accordingly), there is no problem of internal estimation of scale. The trimmed mean is one of a class of scale equivariant* estimates, the linear combinations of order statistics* ((L) estimates), which have been extensively studied. Yet another class, which also provides an explicitly calculable solution to Huber's problem is that of the (R) estimates obtained by inverting rank tests*. These procedures were introduced by Hodges and Lehmann [16]. The most commonly used member of this class, obtained by inverting the Wilcoxon test is the Hodges–Lehmann estimate*, the median of all averages of pairs $(X_i + X_j)/2$, $i \leqslant j$. See Jaeckel [21] for a discussion of these estimates and their properties.

Other robust choices of ψ and $\hat{\sigma}$ are current (particularly of ψ), which vanish outside an interval (*see* R-DESCENDING M-ESTIMATORS), although they are not solutions to Huber's problem. The rationale for such ψ is that they entirely delete the influence of outliers rather than merely bounding them. Some theoretical justification for their use may be found in Collins [12].

An important comparative Monte Carlo study of estimates of location including much valuable numerical and theoretical information is Andrews et al. [1]. Interesting applications to real data where the answer is known are Stigler ([34]) and Relles and Rogers [30].

EXTENSION OF HUBER'S APPROACH TO THE LINEAR MODEL

These considerations and methods have been extended to the general linear model. In regression form we suppose we observe

$$\mathbf{y}_{n\times 1} = \mathbf{X}_{n\times p}\boldsymbol{\theta}_{p\times 1} + \mathbf{e}_{n\times 1}.$$

The gross error mechanism is assumed to operate on the components $e_1, \dots, e_n$ of $\mathbf{e}_{n\times 1}$ so that, rather than being mean 0 identically distributed normal variates, the e_i are independent identically distributed with distribution G as in (1). Maximum likelihood for such a model suggests the class of (M) estimates of $\boldsymbol{\theta}$, so that $\hat{\boldsymbol{\theta}}_{p\times 1}$ solves

$$\mathbf{X}^T\psi\left\{\frac{\mathbf{y} - \mathbf{X}\hat{\boldsymbol{\theta}}}{\sigma}\right\}_{n\times 1} = \mathbf{0}_{p\times 1}, \qquad (4)$$

where $\psi(\mathbf{z}) = (\psi(z_1), \dots, \psi(z_n))^T$ and $\mathbf{z} = (z_1, \dots, z_n)$.

If we embed the stage n model into a sequence of models keeping p fixed, under suitable regularity conditions (Huber [18]), $\hat{\theta}$ has approximately a normal distribution with mean θ and covariance matrix proportional to $[X^TX]^{-1}$ with the proportionality constant

$$\frac{\sigma^2 \int \psi^2\left\{\dfrac{x - T(G)}{\sigma}\right\} dG(x)}{\left[\int \psi'\left\{\dfrac{x - T(G)}{\sigma}\right\} dG(x)\right]^2}, \qquad (5)$$

where $\int\psi\{[x - T(G)/\sigma\} dG(x) = 0$. The expression in (5) does not depend on the design and in particular agrees with the standardized asymptotic variance in the one-sample model. So the minmax criterion applied to the asymptotic covariance matrix now leads to the same ψ_k as before.

Again σ has to be estimated and if G is symmetric the one-sample model results apply. As Carroll [10] notes, if G is not symmetric, the asymptotic behaviour of the estimate where $\hat{\sigma}$ is itself an estimate may be complex. The applicability of the p fixed large-sample theory can be questioned if, as is common, the ratio p/n is not negligible. Moreover, design effects on the approximation do not appear in the first-order fixed p theory.

Huber has developed a large p, large n theory that points up the importance of the hat matrix* in robustness and suggests a modification of (4); see [H, Sec. 7.9].

The (L) and (R) methods may also be generalized to the linear model: see Bickel [4] and Koenker and Bassett [24] for (L); Jaeckel [22] for (R). However, they lose their advantage—closed form rather than implicit definition and ease of interpretation over the (M) method. Computational algorithms for Huber's estimates are given in [H]. From a large-sample point of view it is in fact enough to take just one Newton-Raphson* step towards the solution of (2) (Bickel [5]). The same holds true for the (L) and (R) generalizations.

The estimates $\hat{\boldsymbol{\theta}}$ defined by (4) are ordinary least-squares* estimates for the pseudo-observations

$$y^*_i = t_i + \psi(r_i/\hat{\sigma})\hat{a},$$

where $(t_1, \ldots, t_n) = \mathbf{X}\hat{\boldsymbol{\theta}}$, $r_i = y_i - t_i$, for any choice of $\hat{a}$. If $\hat{a}$ is chosen suitably, for instance,

$$\hat{a} = \hat{\sigma}\left[n^{-1}\sum_{i=1}^{n}\psi'(r_i/\hat{\sigma})\right]^{-1},$$

it can be argued that if $\mathbf{G}$ is symmetric or p is large, ordinary analysis of variance* (ANOVA) procedures for subhypotheses applied to these pseudo-observations should behave approximately as ordinary ANOVA for the robustified model (with bounded errors),

$$\mathbf{y} = \mathbf{X}\boldsymbol{\theta} + \alpha\psi(e/\sigma),$$

where

$$\alpha = \sigma\left[\int\psi'(x/\sigma)\,dG(x)\right]^{-1}.$$

See Bickel [6], Hettmansperger and Schrader [15], and [H, Sec. 7.10]. So these estimates can be accompanied by F tests, C_p criteria, etc., even as for least squares.

For some successful applications of these methods see Launer and Wilkinson [26]. There are important extensions to multivariate analysis* in [H, Chap. 8] and in Kleiner et al. [23], which also contains significant applications.

ALTERNATIVES TO HUBER'S MINMAX APPROACH

Huber's minmax approach poses problems. The asymptotic variance is only a compelling criterion if the asymptotic bias is negligible. We would expect this in the location problem if F is symmetric about 0 and in the linear model if the nature and frequency of gross errors are independent of the value of the independent variables (the rows of X), but not otherwise. Neither of these assumptions is generally plausible.

An approach with an alternative rationale has been developed by Hampel [14]. He finds that estimate that minimizes the asymptotic variance *at the model* within the class of all estimates that are Fisher consistent at the model with a specified bound on the sensitivity. Hampel's rationale is that once quantitative robustness control has been achieved by bounding the sensitivity, it is enough to consider behaviour at the model. Since Fisher consistency yields estimates with asymptotically negligible bias it is enough to consider asymptotic variance. It can be shown (Rieder [31] and Bickel [7]) that Hampel's approach is essentially equivalent to minimizing maximum asymptotic MSE over infinitesimal $(1/\sqrt{n})$ contamination neighbourhoods.

Reassuringly, for estimation of location and of scale, the form of Hampel's solutions coincides with Huber's. For the linear model this continues to be valid, provided that the distribution of the gross errors that can occur in the dependent variables may depend on the value of the independent variable; see

Huber [19]. However, if the gross errors can affect the independent variables as well, we are led to different solutions studied by Hampel [14] and Krasker and Welsch [25].

There is an important conceptual difference between Hampel's original point of view and Huber's, which highlights a difficult issue. If gross errors have occurred, or more generally when a model is (as always!) only approximately valid, what should we be estimating? A parameter can be well defined at the model in many ways (e.g., the mean of the normal distribution is also its median and its mode*). What definition should be adopted when the distribution is perturbed? The Huber formulation implicitly postulates the existence of an unobservable ideal population whose parameters we wish to estimate. Hampel's point of view is, at least in part, that any estimator that is Fisher consistent, i.e., estimates correctly under the ideal model and is stable, is reasonable. For a variant of this position, see Bickel and Lehmann [8].

ROBUST ESTIMATION AND ADAPTATION

In Huber's minmax formulation for the location and linear models it is possible to construct so called adaptive estimates whose asymptotic variance is no larger than that of Huber's estimate and is strictly smaller for all but one F (Huber's least favorable distribution). However, it is not clear whether they satisfy Hampel's robustness criteria and if their convergence is sufficiently uniform. *See* adaptive methods* and [H] for further references.

ROBUST ESTIMATION AND REJECTION OF OUTLIERS*

An alternative to robust estimation (against gross errors) is the application of outlier rejection procedures to "identify" the gross errors followed by classical estimation procedures. For a discussion of the relative merits of these techniques, see [H, p. 4] and Barnett and Lewis [3].

Robustness in a Bayesian Context

A Bayesian parametric approach to robust estimation has been taken by Box and Tiao in a series of papers. An excellent reference is Box [9], which contains a very interesting general discussion of the process of model building and the role of robustness in that process.

Other Kind of Robustness

Other departures from assumptions can be important and have been studied, such as heteroscedasticity (primarily for tests) and dependence (Portnoy [29]). A rather different type of robustness centering on unbiasedness* has been investigated in the sampling context (see Cassel et al. [11] for a description).

References

[1] Andrews, D. F., Bickel, P. J., Hampel, F. R., Huber, P. J., Rogers, W. H., and Tukey, J. W. (1972). *Robust Estimates of Location: Survey and Advances*. Princeton University Press, Princeton, NJ.

[2] Bahadur, R. R. (1966). *Ann. Math. Statist.*, **37**, 577–580.

[3] Barnett, V. M. and Lewis, T. (1977). *Rejection of Outliers*. Wiley, New York.

[4] Bickel, P. J. (1973). *Ann. Statist.*, **1**, 597–616.

[5] Bickel, P. J. (1975). *J. Amer. Statist. Ass.*, **70**, 428–434.

[6] Bickel, P. J. (1976). *Scand. J. Statist.*, **3**, 145–168.

[7] Bickel, P. J. (1981). *Quelques Aspects de la Statistique Robuste*. Springer Lecture Notes in Mathematics No. 876. Springer, Berlin.

[8] Bickel, P. J. and Lehmann, E. L. (1975–1976). *Ann Statist*. **3**, 1038–1069; **4**, 1139–1158.

[9] Box, G. E. P. (1980). *J. R. Statist. Soc. A*, **143**, 383–430.

[10] Carroll, R. J. (1978). *Ann Statist*, **2**, 314–318.

[11] Cassel, C. M., Sarndal, C. E., and Wretman J. (1977). *Foundations of Inference in Survey Sampling*, Wiley, New York.

[12] Collins, J. R. (1976). *Ann. Statist.*, **4**, 68–85.

[13] Efron, B. (1980). *The Jackknife, the Bootstrap and Other Resampling Plans*. In *C.B.M.S. Lecture Notes*, SIAM, Philadelphia.

[14] Hampel F. R. (1968). *Contributions to the Theory of Robust Estimation*. Ph.D. Thesis, University of California, Berkeley.

[15] Hettmansperger, T. and Schrader, R. M. (1980). *Biometrika*, **67**, 93–101.

[16] Hodges, J. L. and Lehmann, E. L. (1963). *Ann. Math. Statist.*, **34**, 598–611.

[17] Huber, P. J. (1964). *Ann. Math. Statist.*, **35**, 73–101.

[18] Huber, P. J. (1973). *Ann. Statist.*, **1**, 799–821.

[19] Huber, P. J. (1981). *Robust Statistics*. Wiley, New York.

[20] Jaeckel, L. A. (1971). *Ann. Math. Statist.*, **42**, 1020–1034.

[21] Jaeckel, L. A. (1971). *Ann. Math. Statist.*, **42**, 1540–1552.

[22] Jaeckel, L. A. (1972). *Ann. Math. Statist.*, **43**, 1449–1458.

[23] Kleiner, B., Martin, R. D., and Thomson, D. J. (1979). *J. R. Statist. Soc. B*, **41**, 313–351.

[24] Koenker, R. G. and Bassett, G. B. (1978). *Econometrica*, **46**, 33–50.

[25] Krasker, W. J. and Welsch, R. E. (1982). *J. Amer. Statist. Ass.*, **77**, 595–604.

[26] Launer, R., and Wilkinson, G., (eds.) (1979). *Robustness in Statistics*. Academic Press, New York.

[27] von Mises, R. (1947). *Ann. Math. Statist.*, **18**, 309–348.

[28] Mosteller, F. and Tukey, J. W. (1977). *Data Analysis and Regression*. Addison-Wesley, Reading MA.

[29] Portnoy, S. (1979). *Ann. Statist.*, **5**, 22–43, 224–231.

[30] Relles, D. A. and Rogers, W. H. (1977). *J. Amer. Statist. Ass.*, **72**, 107–111.

[31] Rieder, H. (1981). *Ann. Statist.*, **9**, 266–277.

[32] Spjötvoll, E. and Aastveit, A. H. (1980). *Scand. J. Statist.*, **7**, 1–13.

[33] Stigler, S. M. (1973). *J. Amer. Statist. Ass.*, **68**, 872–879.

[34] Stigler, S. M. (1977). *Ann. Statist.*, **5**, 1055–1098.

[35] Tukey, J. W. (1960). A survey of sampling from contaminated distributions. In *Contributions to Probability and Statistics*, I. Olkin, ed. Stanford University Press, Stanford, CA, pp. 445–485.

(INFLUENCE FUNCTIONS
INFLUENTIAL DATA
M-ESTIMATORS
R-DESCENDING M-ESTIMATORS
RESISTANCE)

PETER BICKEL

ROBUST INSPECTION SAMPLING PLANS

The term "robust" in statistics is not a term defined with mathematical rigor. It refers (in statistics and quality control*) to procedures that can be used even when their basic assumptions are violated with negligible error. The degree of violation is usually not specified and the size of error that is negligible is not stated quantitatively. There is a good example of a demonstration of robustness in confidence interval* determination in Barrett and Goldsmith [2], where simulation is used to test the impact of violating the underlying assumption of normality. Wright [12] approaches the demonstration in an analytic way for the problem of developing estimators with a minimum variance.

INSPECTION SAMPLING PLANS

A procedure used frequently in industry is the single sampling plan:

(1) draw a random sample of size n;
(2) find the number of defectives X;
(3) accept the whole lot if $X \leqslant c$;
(4) reject the lot if $X > c$.

Inspection of Lots to Control Lot Product Quality

In quality control administration inspection* is deemed necessary to control the quality of a process when quality cannot be assured by the processes and workmanship producing the item. Standards are set to define a quality product and items not meeting these standards are labeled "defective." The role of an inspector is to measure the item against the standards and determine if the item is defective. Usually the labor expense is prohibitive to inspect all items produced, so a sample of items is selected (hopefully at random). Because the sample is random, it may contain more or less defectives per hundred items than the production lot as a whole. Thus there is a chance for a statistical decision error to be made when the lot is of good quality or of bad quality.

No Cost Structure Procedures

Good quality lots are usually defined by a subjective assessment of a value p_1, which is the proportion defective in a production lot that is definitely acceptable, and a value p_2, which is a lot proportion defective that is definitely unacceptable. Management hopes that lots with proportion defective p_1 or better will be accepted and lots with proportion defective p_2 or greater will be rejected.

The following diagram illustrates the types of decisions that can be made and the probabilities of their occurrence due to random variation in sample proportion defective.

	Condition of Lot	
	Acceptable	Unacceptable
Decision	(p_1)	(p_2)
Accept	$1 - \alpha$ (1)	β (2)
Reject	α (3)	$1 - \beta$ (4)

Decision–condition combinations (2) and (3) are called type II and type I errors*, respectively. They are falsely accepting and falsely rejecting errors, respectively.

Standard textbooks such as Duncan [6] describe the basic procedure for determining n and c to come nearest to the desired levels of α and β while keeping n as small as possible.

Inspection Error Problems

The preceding formulas hold when the inspector is perfect. But if the inspector makes errors then the plan will not be executed and α and β will not even be approximately achieved. Let the inspector error be defined in tabular form as

	State of Item	
	Defective (D)	Nondefective (ND)
Inspector D	$1 - \phi$	Θ
Decision ND	ϕ	$1 - \Theta$

Where $\phi = \Pr[\text{called nondefective}|\text{defective}]$ and $\Theta = \Pr[\text{called defective}|\text{nondefective}]$.

The inspector's average errors are Θ and ϕ, corresponding to type I and type II decision errors. If the average proportion defective submitted to the inspector is p, then $\pi = p(1 - \phi) + (1 - p)\Theta$ will be reported as defective.

HISTORY OF INSPECTOR ERROR RESEARCH

The preceding formula for π was first observed by Lavin [9] in 1946. No further work in a quality control context was done until Ayoub et al. [1] in 1970. Collins et al. [4] gave the first rigorous developments of the impact of inspection error on quality parameters for single sampling plans. Hoag et al. [7] extended this type analysis to sequential* sampling. Dorris and Foote [5] published a comprehensive survey of the inspection error literature impacting quality control and related work. An assessment of the impact of ergonomic information on attempts to compensate for inspector error was done by Rahali and Foote [10].

COMPENSATION FOR INSPECTION ERROR

In theory, inspector errors can be compensated for (i.e., the design α and β achieved) by using what the inspector will report in each case (acceptable and unacceptable quality) as a basis for choosing a plan. Thus, to compensate for known inspector errors, compute $\pi_1 = p_1(1-\phi) + (1-p_1)\Theta$ and $\pi_2 = p_2(1-\phi) + (1-p_2)\Theta$ and use these values to determine the single sampling plan using normal procedure. This is a robust procedure that will also compensate for inspector error when choosing a sequential sampling plan depending on p_1 and p_2.

However, when the new plans are designed using π_1 and π_2 in place of p_1 and p_2, an increase in sample size occurs. The increased sample size can be gigantic. A value of ϕ observed in inspection of shell casings was 0.6. For $\Theta = 0$, $p_2 = 0.0406$, and $p_1 = 0.01$, n is 493 compared to 197 in the no error case. As Θ increases $\pi_2 \to \pi_1$ and n further increases; eventually the sample size computed will exceed the lot size N.

Other quality control parameters such as the average outgoing quality limit* (AOQL) will be affected by inspector error. What happens depends on the hypotheses of the individual cases and the use of π in designing the plan will not compensate. In some cases AOQL may not exist (Case et al. [3]).

Further problems occur when ergonomic factors are accounted for and problems in determining Θ and ϕ are considered. According to Dorris and Foote [5], Θ, ϕ will vary over the day due to fatigue, biorhythm, work pace, and incoming quality variation. Further, no one has been able to show how to determine Θ and ϕ without inserting known defectives and known good items in the line. However, the inspector may detect this by observing the insertions or the identifications on the items so they can be retrieved and respond with increased alertness, biasing the estimate of (Θ, ϕ). Thus when a compensating plan is implemented it will probably be invalidated within an hour.

Rahali and Foote [10] proposed a method of coping with this variation. They posed the problem as a game, with the QC analyst choosing plans $P_1, P_2, \ldots, P_n$ according to a random strategy and nature choosing a pair (Θ, ϕ) according to a random strategy.

If $(\Theta, \phi)_i$ occurs, plan i compensates exactly. If $(\Theta, \phi)_j$ occurs, plan i will have an associated α_r, β_r different from the design (α, β). Let $\Delta_{ij} = |\alpha - \alpha_r| + |\beta - \beta_r|$ measure this deviation.

For example if $n = 138$ and $c = 21$, Δ is 0.0299 for $(\phi, \Theta) = (0.1, 0.1)$, but Δ is 1.04334 for $(\phi, \Theta) = (0.25, 0.16)$. Δ_{ij} is the payoff if nature chooses strategy i and the QC analyst chooses plan j. Rahali and Foote solve games of this type and show that indeed a random use of plans according to a given strategy will reduce the Δ obtained, but this reduction is small.

To see what reduction is possible, (ϕ, Θ) combinations were formed from $\Theta = 0, 0.05, 0.1$ and $\phi = 0, 0.1, 0.2, 0.3$ and the compensating plans computed. Then the game matrix

	P_1	P_2	$\cdots$	P_{12}
$(\phi, \Theta)_1$	Δ_{11}	Δ_{12}	$\cdots$	$\Delta_{1,12}$
$(\phi, \Theta)_2$	Δ_{21}	Δ_{22}	$\cdots$	$\Delta_{2,12}$
$\vdots$	$\vdots$	$\vdots$		$\vdots$
$(\phi, \Theta)_{12}$	$\Delta_{12,1}$	$\Delta_{12,2}$	$\cdots$	$\Delta_{12,12}$

was formed and solved using a linear programming* approach demonstrated by Taha [11]. The solution uses five of the twelve plans, with the following random strategy:

Probability of Use	Plan n	c	Number
0.55	27	1	1
0.05	39	1	2
0.15	99	16	3
0.05	154	23	4
0.20	196	28	5

The value of the game was 0.5 compared to 0.9737 if only $n = 39$, $c = 1$ is used. Thus each hour a random two digit number y should be drawn. If $0 \leqslant y \leqslant 54$ use plan 1, $55 \leqslant y \leqslant 59$ use plan 2, $60 \leqslant y \leqslant 74$ use plan 3, etc.

Notice that the best Δ is 0.5, a total deviation from design so great as to render the idea of α and β protection to be meaningless. In these cases where the inspector error values are uncertain, compensation is not possible. It is clear that resources should be focused on reducing inspector error and *most* important of all, implement plans to motivate employees and improve processes to eliminate production of defectives in the first place, as no alternatives to the Rahali and Foote approach have been developed.

UNSOLVED PROBLEMS IN ANALYSIS

The major problem in applying inspector error models is in determining the three parameters p, ϕ, and Θ by examining the output of the inspector without a perfect inspector. An exact form of the distribution of the number of defectives an inspector will report in terms of p, ϕ, and Θ has been developed by Johnson and Kotz [8], which may lead to estimators of p, ϕ, and Θ that could be rapidly redetermined if conditions change. Several procedures such as skip-lot sampling plans* remain unanalyzed.

References

[1] Ayoub, M. M., Lambert, B., and Walvekar, A. G. (1970). Human Factors Society Conference, October, 1970, San Francisco, CA.

[2] Barrett, J. P. and Goldsmith, L. (1976). *Amer. Statist.*, **30**, 67–70. (A simulation approach to sensitivity of t-statistic confidence intervals to the normality assumption. Three shapes including a bimodal are investigated.)

[3] Case, K. E., Bennett, G. K., and Schmidt, J. W. (1975). *J. Qual. Tech.*, **7**, 28–33. (Nine different rectification policies are analyzed for single sampling plans and it is shown that AOQL may not exist in some cases.)

[4] Collins, R. D., Case, K. E., and Bennett, G. K. (1973). *Int. J. Prod. Res.*, **11**, 289–298. (Analyzes probability of acceptance, average total inspection, and average outgoing quality under inspector error for single sampling plans.)

[5] Dorris, A. L. and Foote, B. L. (1978). *AIIE Trans.*, **10**, 183–192. (Comprehensive survey of inspector error research, which includes related ergonomic and human factors work.)

[6] Duncan, A. J. (1986). *Quality Control and Industrial Statistics*, 5th ed. Richard D. Irwin, Inc., Homewood, IL. (A basic reference for quality control concepts and procedures.)

[7] Hoag, L. L., Foote, B. L., and Mount-Campbell, C. (1975). *J. Qual. Tech.*, **7**, 157–164. (Details of computing true α and β for sequential sampling under inspector error.)

[8] Kotz, S. and Johnson, N. L. (1982). *Commun. Statist. Theor. Meth.*, **11**, 1997–2016. (Summarizes their work on obtaining the distribution of the number of "called" defectives by an inspector who makes type I and type II errors.)

[9] Lavin, M. (1946). *J. Amer. Statist. Ass.*, **41**, 432–438. (First basic observations in the field of apparent defective rate.)

[10] Rahali, B. and Foote, B. L. (1982). *J. Qual. Tech.*, **14**, 190–195. (Assesses the impact of ergonomic information on the attempts to compensate for inspector error.)

[11] Taha, H. A. (1971). *Operations Research*. McMillan, NY. (Basic text in the field.)

[12] Wright, R. L. (1980). *Amer. Statist. Ass., Proc. Bus. Econ. Statist.*, 580–583. (Extends previous work of others on obtaining robust confidence intervals in large scale population surveys.)

(ACCEPTANCE SAMPLING
GAME THEORY
INSPECTION SAMPLING
QUALITY CONTROL, STATISTICS IN
SAMPLING PLANS)

B. L. FOOTE

ROBUST REGRESSION

Robust regression refers to the application of robust estimation to regression models (*see* ROBUST ESTIMATION). The model is generally taken to be the classical linear regression* model

$$y_i = x_i\beta + u_i, \qquad i = 1, \ldots, n, \qquad (1)$$

where y_i is the ith observation on the dependent variable, x_i is the ith row of the $n \times p$ matrix X of observations on the explanatory variables, β is the p vector of unknown parameters, and u_i is the ith disturbance. Under the classical assumptions the u's are taken to be independently and identically distributed as $N(0, \sigma^2)$, given X. In principle there are many kinds of violations of the assumptions against which one would like an estimator for β to be robust. However, the term robust regression has come to refer to estimators that maintain a high efficiency in the presence of heavy-tailed disturbance distributions. Estimators that are resistant* to more general violations of the assumptions, such as gross errors in either the x or y data, are called *bounded-influence estimators* (*see* REGRESSION, BOUNDED INFLUENCE).

Robust regression was motivated by the fact that ordinary least squares* (OLS), though efficient under the assumption that the disturbances are Gaussian, can be very inefficient if the disturbance distribution has heavy tails. The most common robust alternatives come from the family of M-estimators, which are direct generalizations of the estimators of the same name for the "location" problem (*see* M-ESTIMATORS). The estimate $\hat{\beta}$ is defined as the value of β that minimizes

$$\sum_{i=1}^{n} \rho(y_i - x_i\beta), \qquad (2)$$

where ρ is a (convex) function that is symmetric around zero. M-estimators were so named because they correspond to maximum likelihood* estimators, under the assumption that the disturbances have density proportional to $\exp(-\rho(u))$.

Two important special cases of M-estimators are OLS and least absolute residuals* (LAR), with $\rho(u) = u^2/2$ and $\rho(u) = |u|$, respectively. OLS is of course not robust in the sense used here. LAR was one of the earliest robust alternatives to OLS. LAR can be computed as a linear program (*see* METHOD OF LEAST ABSOLUTE VALUES for discussion and references), and its asymptotic properties have been studied by many authors (see ref. 3).

As an example, we consider the regression of the logarithm of the price of avocados (adjusted for changes in the Consumer Price Index) on the logarithm of California avocado production, for the years 1960–1969. The yearly price data (in cents per pound, adjusted for inflation) are 6.3, 15.8, 11.9, 14.8, 14.1, 28.0, 13.9, 10.4, 23.0, and 14.4, while the corresponding production data (in millions of pounds) are 140, 71, 100, 80, 93.6, 48, 116, 149, 74.8, and 122.2. In the logarithms of these variables, the OLS regression is $\hat{y} = 7.11 - 0.99x$, with standard errors on the intercept and slope coefficients of 2.84 and 0.22, respectively, and a standard error of the regression of 0.23. By contrast, the LAR estimates are $\hat{y} = 5.24 - 0.57x$, with the difference due in large part to the (outlying) first observation.

Generally one expresses the $\hat{\beta}$ that minimizes (2) as the solution to the first-order condition that the derivative of (2) with respect to β be zero; this gives

$$0 = \sum_{i=1}^{n} \psi(y_i - x_i\hat{\beta})x_i^t, \qquad (3)$$

where $\psi = \rho'$, and where a superscript t indicates transpose. Note that an M-estimator can be expressed as a weighted least-squares* estimator

$$0 = \sum_{i=1}^{n} w_i(y_i - x_i\hat{\beta})x_i^t, \qquad (4)$$

where $w_i = \psi(y_i - x_i\hat{\beta})/(y_i - x_i\hat{\beta})$.

The most common choice of ψ is the Huber ψ-function indexed by $c > 0$: $\psi_c(t) = \min\{1, c/|t|\}t$ (*see* M-ESTIMATORS). The constant c is chosen by the statistician to achieve the right balance between robustness against heavy tails and efficiency if the disturbances are truly Gaussian. As $c \to \infty$, the Huber M-estimator converges to OLS, whereas $c \to \infty$ it converges to LAR. If $\sigma^2 = 1$, then $c = 1.5$ yields an estimator with 95% of the OLS efficiency if the disturbances are Gaussian and (as in the "location" problem) maintains a relatively high efficiency even if the disturbance distribution has heavy tails (see ref. 2). Huber estimates inherit from the "location" problem a certain minimax robustness property.

Under mild regularity conditions an M-estimator will be consistent and asymptotically normal, with asymptotic covariance matrix proportional to that for least squares (see ref. 7). The asymptotic covariance matrix can be consistently estimated by

$n^2 k(X^t X)^{-1}$, where

$$k = \sum_i \psi(e_i)^2 \Big/ \Big[\sum_i \psi'(e_i)\Big]^2 \text{ and } e_i = y_i - x_i\hat{\beta}.$$

Usually the scale σ is unknown and must be estimated along with β. In this case the M-estimator (3) can be generalized to $0 = \sum_{i=1}^n \psi((y_i - x_i\hat{\beta})/\hat{\sigma})x_i^t$, where $\hat{\sigma}$ is determined by a side condition. Alternatively, one can choose β and σ simultaneously to minimize $d\sigma + \sum_{i=1}^n \rho((y_i - x_i\beta)/\sigma)\sigma$, where d is a constant chosen to make $\hat{\sigma}$ consistent for σ when the disturbance distribution is Gaussian (see ref. 7).

Though less commonly used than M-estimators, various other types of robust regression estimators have been proposed. Zeckhauser and Thompson [10] proposed estimating β by the maximum likelihood estimator corresponding to the assumption that the disturbances have a Subbotin distribution*, with density function

$$f(u; \sigma, \theta) \propto \exp\{-|u/\sigma|^{\theta}\}. \quad (5)$$

This density reduces to the Gaussian if $\theta = 2$. However, leaving θ as an extra parameter to be estimated allows the maximum-likelihood estimator to adapt to the tail length of the disturbance distribution.

Using the ordered residuals from a regression obtained from some consistent preliminary estimator, one can generalize to regression the idea of linear combinations of order statistics, called L-estimators* (see Bickel [4]). So called R-estimates of regression are obtained by letting E_i be the rank of the residual e_i in $(e_1, \ldots, e_n)$, and choosing β to minimize $\sum_{i=1}^n a(E_i)e_i$, where $a(\cdot)$ is some monotone scores function satisfying $\sum_i a(i) = 0$ (see Jaeckel [8]).

Koenker and Bassett [9] have proposed an alternative generalization of location L-estimators*, based on a notion called the "θth regression quantile"* $(0 < \theta < 1)$, defined as the vector $b^*(\theta)$ minimizing

$$\sum_{\{i: y_i \geqslant x_i b\}} \theta|y_i - x_i b| + \sum_{\{i: y_i < x_i b\}} (1-\theta)|y_i - x_i b|.$$

One then forms an estimator for β by setting $\hat{\beta} = \sum_{\theta} \pi(\theta) b^*(\theta)$, where π is some discrete symmetric probability distribution on $(0, 1)$.

Other robust estimators have been proposed primarily as starting values for the iterations required to compute most robust regression estimators (such as M-estimators). Harvey [5] has compared two such "preliminary estimators" to LAR. The first, suggested by Hinich and Talwar [6], divides the sample into m nonoverlapping subsamples, computes OLS estimates for each subsample, and sets $\hat{\beta}$ equal to the median of the m OLS estimates. The second, developed by Andrews [1] and called "regression by medians," exploits the property of OLS that it is possible to estimate the coefficients by a sequence of regressions, each having only one explanatory variable. In this sequence, Andrews replaces the OLS estimates by resistant estimates of slope.

References

[1] Andrews, D. F. (1974). *Technometrics*, **16**, 523–531.

[2] Andrews, D. F., Bickel, P. J., Hampel, F. R., Huber, P. J., Rogers, W. H., and Tukey, J. W. (1972). *Robust Estimates of Location*. Princeton University Press, Princeton, NJ.

[3] Bassett, G. and Koenker, R. (1978). *J. Amer. Statist. Ass.*, **73**, 618–622.

[4] Bickel, P. J. (1973). *Ann. Statist.*, **1**, 597–616.

[5] Harvey, A. C. (1977). *J. Amer. Statist. Ass.*, **72**, 910–913.

[6] Hinich, M. J. and Talwar, P. P. (1975). *J. Amer. Statist. Ass.*, **70**, 113–119.

[7] Huber, P. J. (1981). *Robust Statistics*. Wiley, New York.

[8] Jaeckel, L. A. (1972). *Ann. Math. Statist.*, **43**, 1149–1158.

[9] Koenker, R. and Bassett, G. (1978). *Econometrica*, **46**, 33–50.

[10] Zeckhauser, R. and Thompson, M. (1970). *Rev. Econ. Statist.*, **52**, 280–286.

(HEAVY-TAILED DISTRIBUTIONS
INFLUENCE FUNCTIONS
LEAST SQUARES
METHOD OF LEAST ABSOLUTE VALUES
REGRESSION, BOUNDED INFLUENCE

RESISTANT TECHNIQUES
ROBUST-RESISTANT LINE)

WILLIAM S. KRASKER

ROBUST-RESISTANT LINE

A robust-resistant line is a special case of robust regression* in which a line is fitted to data in such a way that the parameters for slope and intercept are not unduly influenced by high leverage* points. That is, the equation for the line is determined to fit data from a broad class of error distributions ("robust") and not be influenced by any small subset of the data, either in the dependent or independent variables ("resistant").

Several methods for fitting a line robustly have been proposed. A very simple procedure is explained in ref. 7 (Chap. 5). Denoting the independent variable by X_i and the dependent variable Y_i, $i = 1, \dots, n$:

(1) Divide the data into thirds (approximately), according to the X values;

(2) Find the (median X, median Y) in each third [say, (X_L, Y_L), (X_M, Y_M), (X_R, Y_R)]. The slope is calculated from the two points in the outer thirds [i.e., $\hat{b} = (Y_R - Y_L)/(X_R - X_L)$];

(3) The intercept is calculated as the mean of the intercepts determined from the three pairs of points from each third [i.e., $\hat{a} = \{(Y_L - \hat{b}X_L) + (Y_M - \hat{b}X_M) + (Y_R - \hat{b}X_R)\}/3$].

In ref. 7, this fitting procedure is illustrated in Fig. 1 for the data in Table 1. For well-behaved data sets (i.e., no obvious outliers), the resulting line is very similar to the familiar least-squares line. For this reason, it is wise to evaluate both straight line fits. The example in Fig. 1 shows a case where the robust-resistant line fits the bulk of the data, whereas the least-squares line is heavily influenced by two points having extreme Y values. Notice in Fig. 2 [7] that the residuals from the robust-resistant line are more heavily concentrated about 0.

Modifications to the above procedure to simplify the calculation include:

(3′) $\hat{a} = \text{med}\{Y_i - \hat{b}X_i\}$ or

(3″) move the line with slope $\hat{b}$ until half the points are above, half below.

In addition, one may determine a closer fit by iterating as follows:

(4) Compute the residuals* $\{R_i\} = \{Y_i - \hat{a} - \hat{b}X_i\}$, and repeat steps (1)–(3) using the points (X_i, R_i), updating the slope and intercept accordingly until there is no further change.

Ramifications of this iteration are discussed in ref. 2a. Several authors have proposed robust-resistant lines for purposes of quick analyses or in connection with the errors-in-variables problem (e.g., refs. 1, 2, 4, 5, and 8). The above procedure is based on that proposed in ref. 6, which itself is based on ref. 4. Its theoretical properties and generalizations are discussed in ref. 3.

References

[1] Bartlett, M. S. (1949). *Biometrics*, **5**, 207–212.

[2] Brown, G. W. and Mood, A. M. (1951). *Proc. 2nd Berkeley Symp.*, University of California Press, Berkeley, CA, pp. 159–166.

[2a] Emerson, D. and Hoaglin, D. C. (1983). "Resistant lines for y versus x." In *Understanding Robust and Exploratory Data Analysis*, D. C. Hoaglin, F. Mosteller, and J. W. Tukey, eds. Wiley, New York.

[3] Johnstone, I. and Velleman, P. (1981). *Proc. Statist. Computing Sec., Amer. Statist. Ass.*, Meeting, Detroit, MI, pp. 218–223.

[4] Nair, K. R. and Shrivastava, M. P. (1942). *Sankhyā*, **6**, 121–132.

[5] Quenouille, M. H. (1959). *Rapid Statistical Calculations*. Hafner, New York, p. 33.

[6] Tukey, J. W. (1977). *Exploratory Data Analysis*. Addison-Wesley, Reading, MA.

[7] Velleman, P. F. and Hoaglin, D. C. (1981). *Applications, Basics, and Computing of Exploratory Data Analysis*. Duxbury Press, Boston, MA.

[8] Wald, A. (1940). *Ann. Math. Statist.*, **11**, 284–300.

Acknowledgment

The author wishes to thank P. Velleman for pointing out several references related to this article.

(EXPLORATORY DATA ANALYSIS
GRAPHICAL REPRESENTATION OF DATA
RESISTANT TECHNIQUES
ROBUST REGRESSION)

K. KAFADAR

ROBUST SMOOTHING

Think of data, (y_i, x_i), being generated according to the model

$$y_i = \theta(x_i) + \epsilon_i, \qquad i = 1, \ldots, n, \quad (1)$$

where θ is a smooth function of x and the measurement errors* ϵ_i are independently and identically distributed. The goal is to estimate the function θ. Unfortunately, standard kernel and spline smoothing methods (*see* KERNEL ESTIMATION AND SPLINE FUNCTIONS) may behave poorly if the data contain gross outliers. For example, in Fig. 1 one can see that for data generated according to

$$y_i = \sin(\pi e^{-5i/101}) + 0.05\epsilon_i,$$

$$i = 1, \ldots, 100, \quad (2)$$

with $e_i \sim n(0,1)$, the addition of one gross outlier $y_i = 4$ at $i = 60$ dramatically degrades the performance of the standard kernel and spline smooths. In robust estimation (*see* ROBUSTNESS), resistance to outliers is achieved by replacing a least-squares* fitting criterion by one that is less sensitive to extreme deviations. Applying this idea in the smoothing context gives rise to robust versions of the kernel and spline methods. Figure 1*c* shows that the robust smooths work well on our outlier contaminated data. A brief description of these robust smooths is given in the next two sections.

ROBUST KERNEL SMOOTHING

In the usual kernel smoother, observations are *averaged* with respect to a kernel or weighting function. The kernel estimate of the regression function at t is of the form

$$\hat{\theta}(t) = \frac{(1/n)\sum_{j=1}^{n} K(\lambda^{-1}(t - x_j))\, y_j}{(1/n)\sum_{j=1}^{n} K(\lambda^{-1}(t - x_j))}, \quad (3)$$

where K is a kernel or weighting function. Equivalently, $\hat{\theta}(t)$ is the minimizer with respect to μ of the weighted residual sum of squares,

$$\sum_{j=1}^{n} [y_j - \mu]^2 K(\lambda^{-1}(t - x_j)). \quad (4)$$

Smoothness of the estimate is controlled by adjusting the band-width parameter λ. Typically, large values of λ make the estimate more smooth and in the limit the estimate is a constant.

Robustification* of the kernel smoother is accomplished by replacing the least-squares criterion in eq. (4) by one of the form

$$\sum_{j=1}^{n} \rho[(y_j - \mu)/\sigma] K(\lambda^{-1}(t - x_j)), \quad (5)$$

where ρ is a function that is less responsive to extremely large residuals and σ is the scale of the errors. Differentiable and nondifferentiable ρ functions are admissible. With ρ taken to be the absolute value function, one gets an important class of the running median-type smoothers. A general discussion including computational aspects of robust kernel methods can be found in Cleveland [1] and Tukey [12]. Asymptotic convergence properties are considered by Hardle and Gasser [5]. Further theoretical discussion may be found in Mallows [9, 10] and Velleman [14].

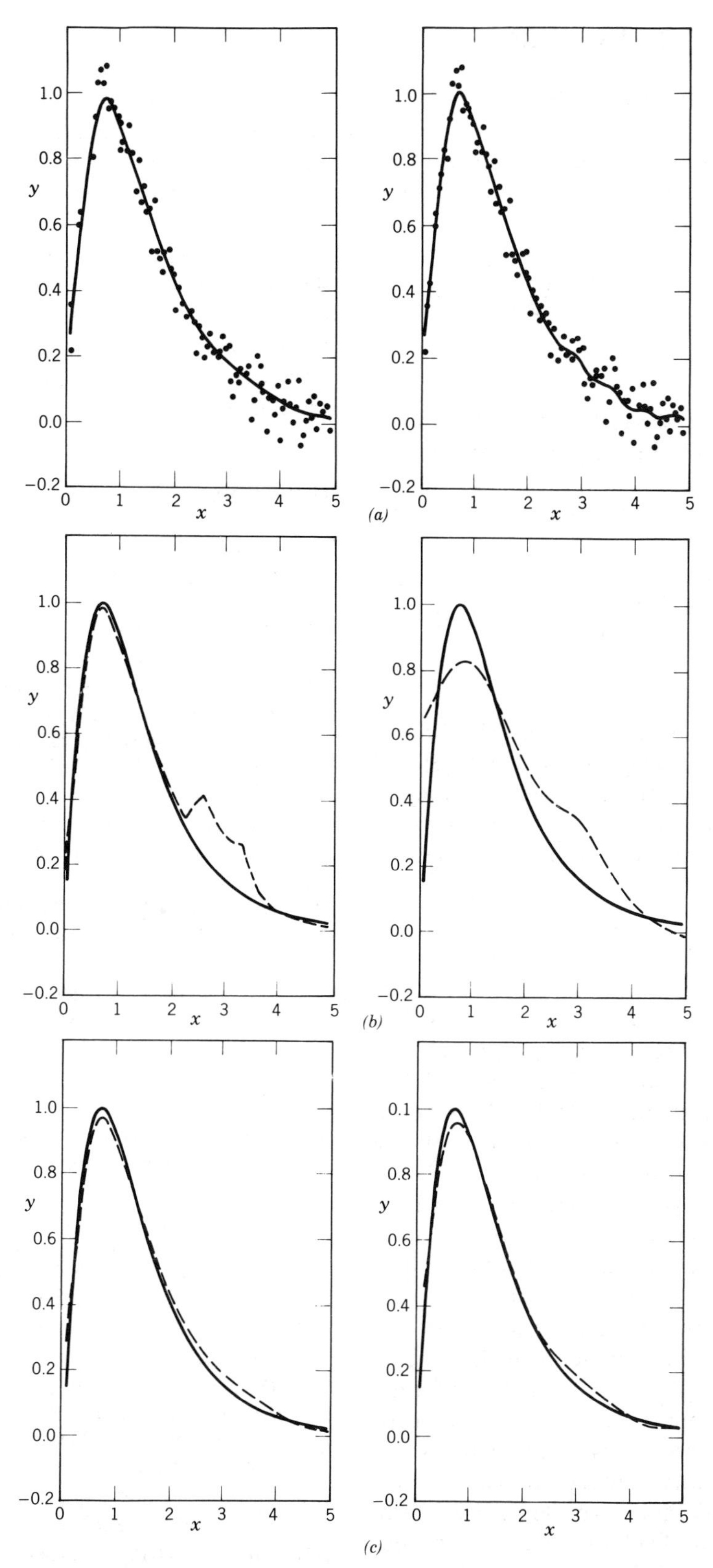

Figure 1. (*a*) Standard smooths on original data. (*b*) Standard smooths on data with added outlier. (*c*) Robust smooths on data with added outlier.

ROBUST SPLINE SMOOTHING

The least-squares smoothing spline estimate of the function θ is defined to be the minimizer of the quantity

$$(1/n) \sum_{i=1}^{n} [y_i - \theta(x_i)]^2 + \lambda \int [\theta^{(2)}(s)]^2 \, ds, \quad \lambda > 0, \tag{6}$$

over a general class of twice differentiable functions. As in the kernel case, large values of λ imply smoother estimates. Robust smoothing splines are obtained as the minimizers of

$$\frac{1}{n} \sum_{i=1}^{n} \sigma \rho \left[\frac{y_i - \theta(x_i)}{\sigma} \right] + \lambda \int [\theta^{(2)}(s)]^2 \, ds, \tag{7}$$

where σ and ρ are as in eq. (5). Robust splines are originally due to Lenth [8] and Huber [6]. Asymptotic convergence properties are given by Cox [2], while the computational aspects are covered by Utreras [13].

DISCUSSION

Choices for the function ρ are discussed in many parts of the robustness literature. The ρ function used in Fig. 1*c* corresponded to Huber's ψ-function; see Huber [7]. When ρ is differentiable the computation of the smoothers is carried out by iteratively reweighted least-squares (IRLS) kernel or spline smoothing algorithms. In least-squares or weighted least-squares smoothing there are a variety of methods available for choosing an appropriate value for the smoothing parameter. Most of these methods combine residuals, weights, and hat-matrix information (the hat-matrix is the matrix that maps data $\underline{y}$ into predictions $\hat{\underline{y}}$) to obtain assessments for the smoothing parameter. When an IRLS smoothing algorithm is used then, as outlined in O'Sullivan et al. [11], residuals, weights, and the hat-matrix* from the final iteration can be used to obtain an appropriate assessment for the smoothing parameter. Cross-validation* methods (see Friedman and Stuetzle [4] for the kernel smoother and Craven and Wahba [3] for the spline smoother) were used to select the degree of smoothing in Fig. 1*a* and *b*. The degree of smoothness in the robust smooths of Fig. 1*c* used the same cross-validation assessments except it was based on residuals, weights, and hat-matrix information from the final iteration of the IRLS algorithm.

References

[1] Cleveland, W. S. (1979). *J. Amer. Statist. Ass.*, **74**, 829–836. (Robust locally weighted regression and smoothing scatterplots.)

[2] Cox, D. D. (1983). *Ann. Statist.*, **11**, 530–551. (Asymptotics for *M*-type smoothing splines.)

[3] Craven, P. and Wahba, G. (1979). *Numer. Math*, **31**, 377–403. (Smoothing noisy data with spline functions: estimating the correct degree of smoothing by the method of generalized cross-validation.)

[4] Friedman, J. H. and Stuetzle, W. (1982). *Tech. Rep. Orion 3*, Statistics Dept., Stanford University, Stanford, CA. (Smoothing of scatterplots.)

[5] Hardle, W. and Gasser, Th. (1984). *J. R. Statist. Soc. B*, **14**, 42–51. (Robust nonparametric function fitting.)

[6] Huber, P. J. (1979). In *Robustness in Statistics*, R. L. Launer and G. H. Wilkinson, eds. Academic Press, New York, pp. 33–47. (Robust smoothing.)

[7] Huber, P. J. (1981). *Robust Statistics*. Wiley, New York.

[8] Lenth, R. V. (1977). *Commun. Statist. A*, **6**, 847–854. (Robust splines.)

[9] Mallows, C. L. (1979). In *Smoothing Techniques for Curve Estimation. Lecture Notes in Mathematics No. 757*, Th. Gasser and M. Rosenblatt, eds. Springer-Verlag, New York. (Some theoretical results on Tukey's 3*R* smoother.)

[10] Mallows, C. L. (1980). *Ann. Statist.*, **8**, 695–715. (Some theory of nonlinear smoothers.)

[11] O'Sullivan, F., Yandell, B., and Raynor, W. J. (1986). *J. Amer. Statist. Ass.*, **81**, 96–103. (Automatic smoothing of regression functions in generalized linear models.)

[12] Tukey, J. W. (1977). *Exploratory Data Analysis*. Addison-Wesley, Reading, MA.

[13] Utreras, F. I. (1981). *SIAM J. Sci. Stat. Comput.*, **2**, 153–163. (On computing robust splines and applications.)

[14] Velleman, P. (1975). *Tech. Rep. No. 89*, Series 2, Dept. of Statistics, Princeton University, Princeton, NJ. (Robust nonlinear data smoothing.)

(DATA ANALYSIS
GRADUATION
INTERPOLATION
ROBUST REGRESSION
SMOOTHING)

FINBARR O'SULLIVAN

ROBUST TESTS FOR CHANGE-POINT MODELS

Consider n independent random variables $X_1, \ldots, X_n$, taken at (ordered) time-points $t_1, \ldots, t_n$, respectively, where X_i has the distribution F_i for $i = 1, \ldots, n$. In a continuous *sampling inspection plan** (*see also* SAMPLING PLANS) and in some related problems in statistical quality control*, one may want to test for the identity of $F_1, \ldots, F_n$ against a (composite) alternative that at some (unknown) time-point τ (the *change-point**) there is a change of the distribution [or its parameter(s)]. Thus, the null hypothesis is

$$H_0: F_1 = \cdots = F_n = F \quad \text{(generally, unknown)}, \qquad (1)$$

and the alternative hypothesis H is the union of $H_1, \ldots, H_{n-1}$, where

$$H_m: F_1 = \cdots = F_m \neq F_{m+1} = \cdots = F_n, \quad 1 \leqslant m \leqslant n-1. \qquad (2)$$

[Note that $\tau \in (t_m, t_{m+1})$ and m is unknown.] See Page [9] and Bartholomew [1] for some nice statistical treatments.

A *sequential version* arises in a *sequential detection problem* (viz. Shiryayev [23, 24]), where for a sequence $\{X_i;\ i \geqslant 1\}$ of independent random variables, one assumes that for some positive integer m (possibly $+\infty$), $X_1, \ldots, X_m$ have a common distribution (F_1) and the $X_{m+j} (j \geqslant 1)$ have a common distribution (F_{m+1}) with $F_{m+1} \neq F_1$: The problem is to raise an alarm if $m < \infty$, while not stopping unnecessarily if $m = +\infty$. A *stopping time* N (a positive integer-valued random variable) needs to be so chosen that $P\{N < \infty | m = +\infty\}$ (the probability of a *false alarm*) and $E\{\max(n - m, 0) | m < \infty\}$ are both small (*quickest detection*); in a change-point model, n is prefixed. A test for (1) against (2) may or may not be sequential in nature.

In a typical parametric setup, assuming the F_i's normal with means μ_i and common variance σ^2, Chernoff and Zacks [5] considered a likelihood-ratio-based Bayes test for (1) (i.e., the equality of the μ_i) against a *shift alternative* (i.e., $\mu_1 = \cdots = \mu_m = \mu_{m+1} - \Delta$ and $\mu_{m+1} = \cdots = \mu_n$, $\Delta \neq 0$, $1 \leqslant m \leqslant n-1$). Their test statistic is

$$T_n^{(1)} = \sum_{i=1}^{n} (i-1)(X_i - \overline{X}_n)/\sigma, \qquad (3)$$

where $\overline{X}_n = n^{-1}\Sigma_{i=1}^n X_i$. Under a similar Bayesian* setup, for some shift alternative with the F_i not necessarily normal, Bhattacharyya and Johnson [3] proposed some *robust rank tests*. With a set $q_1, \ldots, q_n$ of nonnegative weights (which may be interpreted as prior probabilities for τ to be equal to $t_1, \ldots, t_n$, respectively), their test statistic is of the form

$$T_n^{(2)} = \sum_{i=1}^{n} (Q_i - \overline{Q}_n) a_n(R_{ni}), \qquad (4)$$

where $Q_i = \Sigma_{j \leqslant i} q_j$ $(1 \leqslant i \leqslant n)$, $\overline{Q}_n = n^{-1}\Sigma_{i=1}^n Q_i$, R_{ni} = *rank* of X_i among $X_1, \ldots, X_n$ $(1 \leqslant i \leqslant n)$, and $a_n(1), \ldots, a_n(n)$ are suitable *scores*. Typically, $q_1 = 0$, $q_2 = \cdots = q_n = (n-1)^{-1}$, and $T_n^{(2)}$ reduces to

$$(n-1)^{-1} \sum_{i=1}^{n} \{(i-1)\,[a_n(R_{ni}) - \bar{a}_n]\},$$

$$\bar{a}_n = n^{-1} \sum_{i=1}^{n} a_n(i);$$

in this form (3) and (4) are very comparable. Also typically, one may use *Wilcoxon scores* $[a_n(k) = k/(n+1),\ 1 \leqslant k \leqslant n]$, normal scores* $[a_n(k)$ = expected value of the kth order statistic of a sample of size n from the standard normal distribution, $1 \leqslant k \leqslant n$], or log-rank scores* [$a_n(k) = -1 + \sum_{j=1}^{k}(n-j+1)^{-1}$, $1 \leqslant k \leqslant n$]. For a given F, they have also characterized the choice of *optimal scores* for which the corresponding $T_n^{(2)}$ is *locally most powerful rank-invariant*. Under H_0, $T_n^{(2)}$ is distribution-free. For a specified process median μ_0, they considered an alternative statistic

$$T_n^{(3)} = \sum_{i=1}^{n} Q_i \operatorname{sign}(X_i - \mu_0) a_n^*(R_{ni}^+), \quad (5)$$

where $a_n^*(1), \dots, a_n^*(n)$ are suitable *scores* and $R_{ni}^+ =$ *rank* of $|X_i - \mu_0|$ among $|X_1 - \mu_0|, \dots, |X_n - \mu_0|$ for $i = 1, \dots, n$. Under H_0 (when F is symmetric about μ_0), $T_n^{(3)}$ is distribution-free with local optimality properties. Earlier, for the same model, Page [9] proposed the statistics

$$T_n^{(4)} = \max_{0 \leqslant k \leqslant n} \left\{ S_k - \min_{0 \leqslant i \leqslant k} S_i \right\} \quad \text{(one sided)}, \quad (6)$$

$$T_n^{(5)} = \max_{0 \leqslant k \leqslant n} \left\{ \max_{0 \leqslant i \leqslant k} S_i - \min_{0 \leqslant i \leqslant k} S_i \right\} \quad \text{(two-sided)}, \quad (7)$$

where $S_0 = 0$ and $S_k = \sum_{i \leqslant k} \operatorname{sign}(X_i - \mu_0)$, $k \geqslant 1$. These are also distribution-free under H_0.

While for (3), (4), and (5), the test for the null hypothesis H_0 can only be made when all the X_i have been observed, recursive (or quasisequential) tests may be based on (6) or (7). At each time-point t_k, $S_k - \min_{0 \leqslant i \leqslant k} S_i$ (or $\max_{0 \leqslant i \leqslant k} S_i - \min_{0 \leqslant i \leqslant k} S_i$) is computed. If this exceeds the critical value of $T_n^{(4)}$ (or $T_n^{(5)}$), sampling may be curtailed at t_k along with the rejection of H_0. If not, one makes a similar test at time-point t_{k+1}. In this setup, an *early stopping* is plausible for the recursive tests, without increasing the overall risk for rejecting H_0 when it is really true.

For μ_0 unspecified, other robust statistics have been proposed. Let $U_{k,n-k}$ be the Mann–Whitney–Wilcoxon* statistic based on $(X_1, \dots, X_k)$ and $(X_{k+1}, \dots, X_n)$ (two samples of sizes k and $n-k$, respectively) for $k = 1, \dots, n-1$. Sen and Srivastava [12] considered the statistic

$$T_n^{(6)} = \max_{1 \leqslant k \leqslant n-1} \left\{ \frac{(U_{k,n-k} - E_0 U_{k,n-k})/r}{\sqrt{V_0(U_{k,n-k})}} \right\} \quad (8)$$

(along with a two-sided version), while Pettitt [11] proposed

$$T_n^{(7)} = \max_{1 \leqslant k \leqslant n-1} \{(U_{k,n-k} - E_0 U_{k,n-k})\} \quad (9)$$

(and a two-sided version too); E_0 and V_0 denote the mean and variance under H_0. Though these statistics are distribution-free under H_0, the associated tests are not recursive in nature. For large n, $T_n^{(6)}$ may have a very complicated distribution: for some Monte Carlo studies*, see Sen and Srivastava [12] and Wolfe and Schechtman [25]. See also Sen [14].

For the case where μ_0 is unspecified, Sen [13] considered some aligned rank statistics of the form

$$T_n^{(8)} = \sum_{i=1}^{n} (\operatorname{sign} \hat{X}_i) a_n^*(\hat{R}_{ni}^+), \quad (10)$$

where $\hat{X}_i = X_i - \hat{\mu}_n$ $(i = 1, \dots, n)$, $\hat{R}_{ni}^+ =$ rank of $|\hat{X}_i|$ among $|\hat{X}_1|, \dots, |\hat{X}_n|$ $(i = 1, \dots, n)$, and the scores $a_n^*(k)$ are defined as in (5); $\hat{\mu}_n$ is the estimator of μ_0 (under H_0) based on the same rank statistic [but on observations $X_i - b$ $(i = 1, \dots, n)$, where b is so chosen that the statistic is closest possible to 0], i.e., $\hat{\mu}_n$ is the classical R-estimator of μ_0. The test is also nonrecursive and only asymptotically distribution-free.

Recursive rank tests for H_0 in (1), owing to Bhattacharya and Frierson [2], are based on statistics of the form

$$T_n^{(9)} = \sum_{i=1}^{n} (c_i - \bar{c}_i) a_i(R_{ii}), \quad (11)$$

where $R_{ii} =$ rank of X_i among $X_1, \dots, X_i$ $(i \geqslant 1)$, the c_i are suitable constants $\bar{c}_i = i^{-1} \sum_{j=1}^{i} c_j$ $(i \geqslant 1)$, and the scores $a_i(j)$ $(1 \leqslant j$

$\leqslant i,\ i \geqslant 1)$ are defined as in (4). A recursive test may be based on the partial sequence $\{T_k^{(9)};\ k \geqslant 1\}$, distribution-free under H_0. See also Lombard [7, 8].

A more general form of the change-point model relates to the *constancy of the regression relationships over time*. Typically, for the random variables X_i $(i \geqslant 1)$ observed at time-points t_i $(i \geqslant 1)$, one considers a linear regression model

$$X_i = \boldsymbol{\beta}_i'\mathbf{c}_i + e_i, \qquad i = 1, \dots, n, \quad (12)$$

where the e_i are assumed to have a common distribution F, $\mathbf{c}_i = (c_{i1}, \dots, c_{ip})'$ are known regression constants (vectors), and $\boldsymbol{\beta}_i = (\beta_{i1}, \dots, \beta_{ip})'$ are unknown parameters (vectors) for some $p \geqslant 1$. The shift model is a special case of (12), where $\mathbf{c}_i = 1$ and the $\boldsymbol{\beta}_i$ are location parameters. The null hypothesis is

$$H_0 : \boldsymbol{\beta}_1 = \cdots = \boldsymbol{\beta}_n = \boldsymbol{\beta} \ \text{(unknown)} \quad (13)$$

(constancy of regression relationships over time), and one is interested in a change-point alternative

$$\boldsymbol{\beta}_1 = \cdots = \boldsymbol{\beta}_m \neq \boldsymbol{\beta}_{m+1} = \cdots = \boldsymbol{\beta}_n,$$
$$\text{for some } m: 1 \leqslant m \leqslant n-1. \quad (14)$$

See Brown et al. [4] for the case where F is assumed to be normal. Let

$$\mathbf{Q}_k = \sum_{i=1}^{k} \mathbf{c}_i\mathbf{c}_i', \qquad k \geqslant 1,$$

$$\hat{\boldsymbol{\beta}}_k = \mathbf{Q}_j^{-1} \sum_{i=1}^{k} \mathbf{c}_i X_i, \qquad k \geqslant p,$$

$$w_k = \frac{X_k - \hat{\boldsymbol{\beta}}_{k-1}'\mathbf{c}_k}{\left(1 + \mathbf{c}_k'\mathbf{Q}_{k-1}^{-1}\mathbf{c}_k\right)^{1/2}}, \qquad k \geqslant p+1; \quad (15)$$

conventionally, $w_k = 0$ for $k \leqslant p$. Based on these *recursive residuals*, consider the cumulative sums $W_k = \Sigma_{i \leqslant k} w_i$, $k \geqslant 1$. The test for (13) against (14) is then based on

$$T_n^{(10)} = \max\left\{(n-p)^{-1/2} W_k : 0 \leqslant k \leqslant n\right\}. \quad (16)$$

In a two-sided version, the W_k are replaced by $|W_k|$, $k \geqslant 0$. These tests are generally recursive in nature and early stopping is plausible; see also Sen [17].

For the model (12), Sen [21] considered recursive residual rank tests for H_0 in (13) against the alternative in (14). Under H_0 in (13), the X_i need not be identically distributed, and hence (4) or (5) may not be applicable. For a general class of recursive estimators $\hat{\boldsymbol{\beta}}_k$ $(k \geqslant 1)$, consider recursive residuals $\hat{X}_i = X_i - \hat{\boldsymbol{\beta}}_{i-1}'\mathbf{c}_i$ $(i \geqslant 1)$, and similar to (11), recursive signed rank statistics of the form

$$T_n^{(11)} = \sum_{i=1}^{n} \left(\text{sign } \hat{X}_i\right) a_i^*\left(\hat{R}_{ii}^+\right), \quad (17)$$

where the $a_i^*(j)$ $(1 \leqslant j \leqslant i,\ i \geqslant 1)$ are defined as in (5) and $\hat{R}_{ii}^+$ is the rank of $|\hat{X}_i|$ among $|\hat{X}_1|, \dots, |\hat{X}_i|$ for $i \geqslant 1$. Recursive tests are based on the partial sequence $\{T_k^{(11)};\ k \leqslant n\}$ and are asymptotically distribution-free. Sen [21, 22] similarly constructed recursive M-tests [for (13) vs. (14)] using M-statistics of the type

$$\sum_{i=1}^{k} \psi\left(X_i - \hat{\boldsymbol{\beta}}_{i-1}'\mathbf{c}_i\right), \qquad k \geqslant 1; \quad (18)$$

here $\psi(\cdot)$ is a nondecreasing and skew-symmetric score function. These robust tests are also only asymptotically distribution-free.

In a more general setup, one may consider a *regular functional* or *estimable parameter* $\theta = \theta(F) = E_F g(X_1, \dots, X_k)$, where $g(\cdot)$ is a *kernel* of *degree* k $(\geqslant 1)$. For $g(x) = x$, $\theta(F) = E_F X$ (the population mean) and, for $g(x, y) = \frac{1}{2}(x-y)^2$, $\theta(F) = \text{Var}(X)$ are notable examples of such parameters. In testing for (1) against (2), one may be interested in some change with respect to the $\theta(F_i)$ [instead of the shift model as in (3)]. Optimal estimators of $\theta(F)$ under H_0 in (1) are given by

$$U_r = \{r \cdots (r-k+1)\}^{-1} \times \sum_r^{*} g\left(X_{i_1}, \dots, X_{i_k}\right), \qquad r \geqslant k, \quad (19)$$

where Σ_r^* extends over all $1 \leqslant i_1 \neq \cdots \neq i_k$

$\leqslant r$; see Hoeffding [6] for detailed study of these U-statistics*. A *recursive* U-statistic U_r^* may be defined by

$$U_r^* = \{(r-1)\cdots(r-k+1)\}^{-1} \times \sum_r^0 g(X_r, X_{i_2}, \ldots, X_{i_i}), \qquad r \geqslant k, \tag{20}$$

where Σ_r^0 extends over all $1 \leqslant i_2 \neq \cdots \neq i_k \leqslant r-1$. With these, define the recursive residuals as

$$w_r = U_r^* - U_{r-1}, \qquad r \geqslant k+1;$$
$$w_r = 0, \qquad \forall r \leqslant k. \tag{21}$$

Then, the change-point tests are based on the cumulative sums $W_r = \Sigma_{i \leqslant r} w_i$, $r \geqslant 0$, in a manner similar to that in Brown et al. [4]; see Sen [19].

Excepting the rank statistics, in the other cases we need suitable estimates of variability; for U-statistics this is no problem. For M-statistics too, these estimators are studied in detail in Sen ([16], Chap. 8). Finally, one may also conceive of a general change-point model involving possible multiple change-points. The tests considered here remain applicable, though their performance picture may become more complicated; see Sen [22, Chap. 3]. For a nice review of some parametric procedures, see Zacks [26].

References

[1] Bartholomew, D. J. (1959). *Biometrika*, **46**, 36–48, 328–335.

[2] Bhattacharya, P. K. and Frierson, D. (1981). *Ann. Statist.*, **9**, 544–554.

[3] Bhattacharyya, G. K. and Johnson, R. A. (1968). *Ann. Math. Statist.*, **39**, 1731–1743.

[4] Brown, R. L., Durbin, J., and Evans, J. M. (1975). *J. R. Statist. Soc. B*, **37**, 149–192.

[5] Chernoff, H. and Zacks, S. (1964). *Ann. Math. Statist.*, **35**, 999–1018.

[6] Hoeffding, W. (1948). *Ann. Math. Statist.*, **19**, 293–325.

[7] Lombard, F. (1981). *S. Afr. Statist. J.*, **15**, 129–152.

[8] Lombard, F. (1983). *S. Afr. Statist. J.*, **17**, 83–105.

[9] Page, E. S. (1954). *Biometrika*, **41**, 100–115.

[10] Page, E. S. (1957). *Biometrika*, **44**, 248–252.

[11] Pettitt, A. N. (1979). *Appl. Statist.*, **28**, 126–135.

[12] Sen, A. and Srivastava, M. S. (1975). *Ann. Statist.*, **3**, 98–108.

[13] Sen, P. K. (1977). *Ann. Statist.*, **5**, 1107–1123.

[14] Sen, P. K. (1978). *Sankhyā*, *A.*, **40**, 215–236.

[15] Sen, P. K. (1980). *Zeit. Wahrsch. Verw. Geb.*, **52**, [illegible]-218.

[16] Sen, P. K. (1981). *Sequential Nonparametrics*. Wiley, New York.

[17] Sen, P. K. (1982). *Ann. Statist.*, **10**, 307–312.

[18] Sen, P. K. (1982). *Math. Operat. Forschung, Statist. Ser. Statist.*, **13**, 21–32.

[19] Sen, P. K. (1982). *Commun. Statist.-Sequential Anal.*, **1**, 263–284.

[20] Sen, P. K. (1983). *Chernoff Festschrift*, Rizvi et al. eds., Academic Press, New York, pp. 371–391.

[21] Sen, P. K. (1983). *Proc. Int. Statist. Inst.*, 44th Sess. **1**, 206–209.

[22] Sen, P. K. (1985). *Sequential Nonparametrics: Theory and Applications*. SIAM, Philadelphia.

[23] Shiryayev, A. N. (1963). *Teor. Verojatnost. Primen.*, **8**, 22–46.

[24] Shiryayev, A. N. (1978). *Optimal Stopping Rules*. Springer-Verlag, New York.

[25] Wolfe, D. A. and Schechtman, E. (1984). *J. Statist. Plann. Inf.*, **9**, 389–396.

[26] Zacks, S. (1983). *Chernoff Festschrift*, Rizvi et al. eds. Academic Press, New York, pp. 245–269.

(CHANGE-POINT MODEL
DISTRIBUTION-FREE METHODS
LOCALLY OPTIMAL STATISTICAL TESTS
LOG RANK SCORES, STATISTICS AND TESTS
NORMAL SCORES TESTS
ROBUSTNESS OF TESTS
SAMPLING PLANS
STOPPING RULES AND TIMES
U-STATISTICS)

P. K. SEN

ROBUSTIFICATION AND ROBUST SUBSTITUTES

A statistical method is said to be *robust* if its behavior is relatively insensitive to slight departures from the assumptions that justify that method. There are two reasons for these assumptions: to calculate certain probabilities such as confidence coefficients* and type I error rates, and to prove that the method has high efficiency. If the calculated probabilities are approximately correct under slight violations of the assumptions then the method is called *validity robust*. The method is said to be *efficiency robust* if its efficiency, for example the power* of a test or the mean square error* of an estimator, can be only moderately degraded by such violations.

Many classical methodologies, especially those appropriate for normally distributed data, are not robust. In fact, the term "robustness" was coined by Box [5], who found that tests for equality of several variances have under nonnormality substantially higher type I error rates than under normality, and so they are not validity robust. Although many inferential procedures about population means, for example t-tests and F-tests* for fixed effects linear models, are validity robust (Scheffé [31, Chap. 10]), since a seminal paper by Tukey [35] they have become notorious for their lack of efficiency robustness. The nonrobustness of classical statistical procedures has caused statisticians to make them robust by modification, a process called *robustification*, or to find alternative robust procedures, that is, robust substitutes.

Nonrobustness is usually caused by high sensitivity to outliers*. "Outlier" is a term used in a variety of ways in the statistics literature. Beckman and Cook [2] suggest calling any observation that appears unusual a *discordant* value and calling any observation that is not from the target population a *contaminant*. They define an outlier as any observation that is either discordant or contaminant. There are two reasons why a robust procedure must not be too sensitive to outliers: there may be contaminants in the sample and the target population may have a higher frequency of discordant values than the assumed distribution.

It is convenient to divide methods of robustification into four categories: (1) trimming*, Winsorization, and other methods based on order statistics, (2) M-estimation* and minimum distance estimation*, (3) rank statistics*, and (4) outlier tests and diagnostics.

TRIMMING, WINSORIZATION, AND OTHER METHODS BASED ON ORDER STATISTICS

Trimming and Winsorization are, respectively, the processes of removing extreme values from the sample and of changing the extreme values by setting each equal to the values of less extreme observations. For example, suppose that one has a univariate sample $X_1, \ldots, X_N$, let k be a positive integer less than $N/2$, and define $\alpha = k/N$. Then the α-symmetrically trimmed sample is the original sample after the k smallest-and k largest-order statistics have been removed. The α-symmetrically Winsorized sample is obtained by replacing the k smallest- and k largest-order statistics by, respectively, $X_{k+1:N}$ and $X_{N-k:N}$.

The idea behind trimming and Winsorization is quite simple. Any grossly discordant outliers will be among the extreme values. If the fraction of outliers is less than α, then any statistic calculated from the modified sample will be robust. Tukey and McLaughlin [36] advocated the trimmed mean as an estimate of location. It has a number of desirable properties, including being easily understood, simple to compute, and robust. As α changes from 0 to $\frac{1}{2}$, the trimmed mean changes along a continuum from the arithmetic mean* to the median*. For samples from a symmetric population, the symmetrically trimmed mean is an unbiased estimate of the population mean. Otherwise, the sample trimmed mean estimates the corresponding population trimmed mean. The latter can be understood as a compromise between the mean and the median.

By trial-and-error, Tukey and McLaughlin [36] found that the Winsorized sample variance is a suitable estimate of the variance of the trimmed mean. They suggested a "t-statistic," the ratio of the trimmed mean to the

Winsorized variance, for purposes of hypothesis testing and interval estimation. Their ideas have since been supported by asymptotic theory (Stigler [34], Huber [25], and deWet and Venter [15]) and Monte Carlo simulation* (Gross [18]).

Trimming has been extended to linear models by Koenker and Bassett [26]. They introduce the concept of regression quantiles*, an extension of ordinary quantiles for univariate samples. Their suggestion is to trim all observations "below" the αth regression quantile or "above" the $(1-\alpha)$th regression quantile and to compute the least-squares estimate with the remaining data. Ruppert and Carroll [31] establish the asymptotic distribution of the trimmed least-squares estimator, and they also investigate trimming based on residuals* from a preliminary estimate.

There have been several proposals for trimming multivariate samples. One method, called *peeling**, is the removal of the extreme points of the convex hull of the data; this process can, of course, be repeated. Another method, *depth trimming*, owing to Tukey is described in Donoho and Huber [15]. Gnanadesikan and Kettenring [17] use trimming to define robust estimates of bivariate and multivariate correlation.

The trimmed and Winsorized means of a univariate sample are both examples of linear combinations of order statistics, called L-statistics. A general L-statistic is of the form

$$\sum_{i=1}^{N} \omega_i X_{i:N},$$

where $\omega_1, \ldots, \omega_N$ are fixed weights not depending upon the data. L-estimators of location and scale are discussed by Huber [25]. Both this reference and Serfling [33] gives a good treatment of the asymptotic properties of L-statistics.

M-ESTIMATORS AND MINIMUM DISTANCE ESTIMATORS

M-estimators*, which were introduced by Huber [24], are generalizations of the maximum likelihood* estimator (MLE). Let $X_1, \ldots, X_N$ be independent, identically distributed from a density of $f(x, \theta)$ depending upon the parameter θ. The MLE is found by maximizing

$$\sum_{i=1}^{N} \log f(x_i, \theta)$$

or solving

$$\sum_{i=1}^{N} \dot{l}(x_i, \theta) = 0,$$

where $\dot{l}(x, \theta)$ is the gradient of $\log f(x, \theta)$ with respect to θ. M-estimators are obtained by replacing $\log f(x, \theta)$ by another function, say $\rho(x, \theta)$, or replacing $\dot{l}(x, \theta)$ by say $\psi(x, \theta)$. For simplicity only the second possibility will be considered. Using Hampel's [19] influence curve* it can be shown that the MLE is not robust when $\dot{l}(x, \theta)$ is not bounded. Thus $\psi(x, \theta)$ should be bounded.

Hampel [19] shows that for univariate θ the good choice of ψ that maintains the highest possible efficiency subject to a specified degree of robustness is to let ψ be $\dot{l}$ suitably truncated. More precisely, let

$$\psi(x, \theta) = \min\{1, k/|\dot{l}(x, \theta) - a|\} \times (\dot{l}(x, \theta) - a),$$

where k is the truncation point and a is chosen so that

$$\int \psi(x, \theta) f(x, \theta)\, dx \equiv 0. \qquad (1)$$

Equation (1) is necessary for $\hat{\theta}$ to be consistent. As k decreases, the truncation becomes more severe and the estimator becomes more robust but less efficient when $f(x, \theta)$ is the true density of the data.

Theoretical properties of M-estimators are treated by Huber [25] and Serfling [33]. Hoaglin et al. [22] contains more elementary material and is more suitable for those with primary interest in applications.

Minimum distance estimation was introduced by Wolfowitz [37] as a particularly powerful method of constructing consistent estimators. Parr and Schucany [29] advocate minimum distance estimation for its ability to produce robust estimators. Let $\delta(F_1, F_2)$ be a distance measure between the distribution functions F_1 and F_2. For example,

$$\delta(F_1, F_2) = \sup_{x \in R} |F_1(x) - F_2(x)| \psi(F_2(x))$$

is a weighted Kolmogorov distance and

$$\delta(F_1, F_2) = \int_{-\infty}^{\infty} [F_1(x) - F_2(x)]^2 \times \psi(F_2(x))\, dF_2(x)$$

is a weighted Cramér-von Mises distance. Further let $\{F_\theta : \theta \in \theta\}$ be a parametric family of distribution functions. Suppose the sample comes from a distribution G and that G_N is an estimate of G, for example, the empirical distribution function. Then the minimum distance estimator is the value of θ that minimizes $\delta(G_N, F_\theta)$. This method of estimator is intuitively appealing, and the minimum value $\delta(G_N, F_{\hat{\theta}})$ can be used to test goodness of fit* (Boos [4]). Parr and Schucany [29] study the large sample properties of minimum distance estimators and they compare them with M-estimators in a Monte Carlo* simulation.

Rank Statistics

An effective way to robustify is to use statistics based only on the ranks of observations. Rank tests are well known as competitors to normal-theory tests. For example, the two-sample Wilcoxon test* is an excellent substitute for the classical t-test. For normal populations the Wilcoxon has efficiency 0.955 relative to the t-test and for heavy-tailed distributions the Wilcoxon can be considerably more powerful than the t-test (Lehmann [28, Appendix 6]). Moreover, the type I error probability of the Wilcoxon can be computed exactly under the null hypothesis that the two populations have a common continuous distribution, regardless of what that distribution may be. However, the type I error rate of the t-test is reasonably stable as the populations deviate from the normal distribution, so the real advantage of the Wilcoxon is its efficiency robustness.

Lehmann [28] gives an excellent introduction to rank statistics. Conover [11] and Hollander and Wolfe [23] are also good introductory texts. Randles and Wolfe [30] provide an intermediate level theoretical treatment. Huber [25] and Serfling [33] discuss the theoretical properties of estimates based on rank tests, which are usually called R-estimates.

OUTLIER TESTS AND OTHER METHODS

Significance tests and informal diagnostics can be used to detect outliers for possible removal. Rather than removing a fixed number of extreme values, as with trimming, only sufficiently outlying observations are removed and the decision to remove an observation may be based on subjective factors as well as objective criteria.

Significance tests for outliers test the null hypothesis that all observations come from the parent population of known parametric form against the alternative that one or more of the cases are contaminants. Diagnostics are methods, often graphical, for highlighting discordant data with no formal means of testing for contaminants.

Beckman and Cook [2] give a nice survey of the literature on outliers*. Barnett and Lewis [1] and Hawkins [21] present a variety of tests for outliers. Hampel [20] studies the robustness of the sample mean when used with outlier rejective rules.

Since discordant cases in a univariate sample are easily located, diagnostics are mostly used in linear models and multivariate analysis*. The best known regression diagnostics* are plots of the least-squares* residuals against the fitted values, explanatory variables both in and excluded from the model, and normal quantiles. See Draper and Smith [16, Chap. 3]. However, the effectiveness of traditional residual analysis is somewhat mitigated by the nonrobustness of the least-squares estimator, and this has led to new diagnostics that measure the effects of deleting the observations one at a time. Another possibility is to use residuals from a robust fit, perhaps the Krasker–Welsch [27] estimator. See Cook and Weisberg [13] or Belsley et al. [3] for a survey of recent developments.

Diagnostics can locate both difficulties with the data and weaknesses in the model. In the first situation, one can delete discordant data or downweight them using an M-estimator. In the second case, one should consider modifying the model, and model revision can be considered a robustification. In regression analysis, possible modifications include power transformations of the response and/or explanatory variables, additional explanatory variables, or heteroscedastic models. However, even with an improved model a robust analysis may be needed. Carroll and Ruppert [8, 9] investigate M-estimators for heteroscedastic linear models. Carroll [7] and Carroll and Ruppert [10] study M-estimators for the Box–Cox [6] transformation model, and Cook and Wang [11] propose diagnostics for this model.

References

[1] Barnett, V. D. and Lewis, T. (1978). *Outliers in Statistical Data*. Wiley, New York. (Comprehensive discussion of outliers including significance testing, trimming, and M-estimation.)

[2] Beckman, R. J. and Cook, R. D. (1983). *Technometrics*, **25**, 119–149. (A very readable survey article.)

[3] Belsley, D. A., Kuh, E., and Welsch, R. E. (1980). *Regression Diagnostics*. Wiley, New York.

[4] Boos, D. D. (1981). *J. Amer. Statist. Ass.*. **76**, 663–670.

[5] Box, G. E. P. (1953). *Biometrika*, **40**, 318–335.

[6] Box, G. E. P. and Cox, D. R. (1964). *J. R. Statist. Soc., B*, **26**, 211–252.

[7] Carroll, R. J. (1980). *J. R. Statist. Soc. B*, **42**, 71–78.

[8] Carroll, R. J. and Ruppert, D. (1982). *Ann. Statist.*, **10**, 429–441.

[9] Carroll, R. J. and Ruppert, D. (1983). In *Contributions to Statistics: Essays in Honor of Norman L. Johnson*, P. K. Sen, ed. North-Holland, Amsterdam, pp. 81–96.

[10] Carroll, R. J. and Ruppert, D. (1985). *Technometrics*, **27**, 1–12.

[11] Conover, W. J. (1971). *Practical Nonparametric Statistics*. Wiley, New York. (Good text for the beginning user of rank statistics.)

[12] Cook, R. D. and Wang, P. C. (1983). *Technometrics*, **25**, 337–343.

[13] Cook, R. D. and Weisberg, S. (1980). *Technometrics*, **22**, 495–508.

[14] deWet, T. and Venter, J. H. (1974). *S. Afr. Statist. J.*, **8**, 127–134.

[15] Donoho, D. L. and Huber, P. J. (1983). In *A Festschrift for Erich L. Lehmann*, P. J. Bickel, K. A. Doksum, and J. L. Hodges, Eds. Wadsworth, Belmont, CA, pp. 157–184.

[16] Draper, N. and Smith, H. (1981). *Applied Regression Analysis*, 2nd ed. Wiley, New York.

[17] Gnanadesikan, R. and Kettenring, J. R. (1972). *Biometrics*, **29**, 81–124.

[18] Gross, A. M. (1976). *J. Amer. Statist. Ass.*, **71**, 409–416.

[19] Hampel, F. R. (1974). *J. Amer. Statist. Ass.*, **69**, 1179–1186. Introduces important concepts.

[20] Hampel, F. R. (1985). *Technometrics*, **27**, 95–107.

[21] Hawkins, D. M. (1980). *Identification of Outliers*. Chapman and Hall, London.

[22] Hoaglin, D. C., Mosteller, F., and Tukey, J. W. (1983). *Exploring Data Tables, Trends and Shapes*, Wiley, New York.

[23] Hollander, M. and Wolfe, D. A. (1973). *Nonparametric Statistical Inference*. Wiley, New York.

[24] Huber, P. J. (1964). *Ann. Math. Statist.*, **35**, 73–101. (Technical; seminal paper.)

[25] Huber, P. J. (1981). *Robust Statistics*. Wiley, New York. (Terse style requires careful reading. Introductory chapters require considerable mathematical knowledge but can be skipped or only perused.)

[26] Koenker, R. and Bassett, G., Jr. (1978). *Econometrica*, **46**, 33–50.

[27] Krasker, W. S. and Welsch, R. E. (1982). *J. Amer. Statist. Ass.*, **77**, 595–604.

[28] Lehmann, E. L. (1975). *Nonparametrics: Statistical Methods Based on Ranks*. Holden-Day, San Francisco. (Excellent text, but not as suitable for browsing as Conover's.)

[29] Parr, W. C. and Schucany, W. R. (1980). *J. Amer. Statist. Ass.*, **75**, 616–624.

[30] Randles, R. H. and Wolfe, D. A. (1979). *Introduction to the Theory of Nonparametric Statistics*. Wiley, New York.

[31] Ruppert, D. and Carroll, R. J. (1980). *J. Amer. Statist. Ass.*, **75**, 828–838.

[32] Scheffé, H. (1959). *The Analysis of Variance*. Wiley, New York. (A classic text but somewhat technical. One of the first books to give a detailed discussion of the violations of the assumptions.)

[33] Serfling, R. J. (1980). *Approximation Theorems of Mathematical Statistics*. Wiley, New York. (Very readable among books at this mathematical level. Covers parametric and nonparametric statistics.)

[34] Stigler, S. M. (1973). *Ann. Statist.*, **1**, 472–477.
[35] Tukey, J. W. (1960). In *Contributions to Probability and Statistics*, I. Olkin, ed. Stanford University Press, Stanford, CA, pp. 448–485.
[36] Tukey, J. W. and McLaughlin, D. H. (1963). *Sankhyā A*, **25**, 331–352.
[37] Wolfowitz, J. (1957). *Ann. Math. Statist.*, **28**, 75–88.

(CENSORED DATA
MASKING AND SWAMPING
M-ESTIMATORS
MINIMUM DISTANCE ESTIMATION
ORDER STATISTICS
OUTLIERS
RANKING PROCEDURES
REGRESSION DIAGNOSTICS
ROBUST ESTIMATION
TRIMMING AND WINSORIZATION)

DAVID RUPPERT

ROBUSTNESS IN EXPERIMENTAL DESIGN

The term robust* was introduced into statistics by Box [4], although the idea involved is much older; see, for example, the studies of E. S. Pearson*, in particular ref. 19. There are broadly two senses in which robust is used in the statistical design of experiments*. In the first, a design that is optimal in relation to a particular criterion is examined to see how well it performs in relation to other criteria; see, for example, Kiefer [16], Galil and Kiefer [9, 10], and Kiefer and Studden [17]. Often there remains some freedom in the choice of design points of an optimal design and further subsidiary optimality criteria may be incorporated; see Herzberg and Cox [14]. In the second, designs are constructed that guard against particular shortcomings; see, for example, Box and Draper [7] and Herzberg and Andrews [12]. Such robust designs will be slightly less than optimal under ideal conditions, but will be more efficient under more realistic conditions.

Box and Draper [7] introduced a criterion for the construction of designs to minimize the effect of outliers* or spurious values on the least-squares* estimate of the predicted response function. Thus they wanted to obtain designs that made the analysis insensitive to outliers. For the comparison of designs, they used the value of the population variance* of the variances of the estimated response at the design points, the smaller the value implying the more robust the design.

Let

$$\mathbf{Y} = \mathbf{X}\boldsymbol{\theta} + \boldsymbol{\epsilon}, \qquad (1)$$

where $\mathbf{Y}$ is the $n \times 1$ vector of observations, $\mathbf{X}$ is the $n \times p$ matrix of "independent" variables, $\boldsymbol{\theta}$ is the $n \times p$ vector of unknown parameters to be determined by least squares, and $\boldsymbol{\epsilon}$ is the $n \times 1$ vector of errors, which have zero mean, variance σ^2, and are independent. Let

$$\mathbf{V} = \mathbf{X}(\mathbf{X}'\mathbf{X})^{-1}\mathbf{X}' = \{v_{ij}\}.$$

Then

$$\sigma^2 v_{ii} = \operatorname{var}\{\hat{y}(\mathbf{x}_i)\},$$

the variance of the estimated response at the ith observation. Box and Draper showed that in order to minimize the population variance of the v_{ii}'s a design should be chosen for which $\Sigma_{i=1}^{n} v_{ii}^2$ is minimized. For example, first-order designs that have this property are two-level factorial and fractional factorial designs*.

Herzberg and Andrews [12] and Andrews and Herzberg [1] developed robust measures for assessing designs in situations where outliers, missing values, or non-Gaussian distributions are contemplated. Cox and Hinkley [8] showed that in asymptotic theory a non-Gaussian distribution of errors in a location family* had no effect on optimality if a maximum likelihood estimator* were employed. Let $\alpha(\mathbf{x})$ be the probability of losing an observation at the point $\mathbf{x}$, the losses at different points being independent. Missing observations may be modelled in the following way. The formulation does not explain missing values, but is a way to investigate their influence. Outliers and non-Gaussian distributions may be formulated similarly.

Let $\mathbf{D} = \{d_{ii}(i = 1, \ldots, n)\}$ be a diagonal matrix, where

$$d_{ii} = \begin{cases} 0 & \text{with probability } \alpha(\mathbf{x}_i), \\ 1 & \text{with probability } 1 - \alpha(\mathbf{x}_i), \end{cases}$$

the value 0 being associated with a missing observation.

If a sufficient number of observations are missing, not all the elements of $\boldsymbol{\theta}$ given in (1) may be estimated. This occurs when $|\mathbf{X}'\mathbf{D}\mathbf{X}| = 0$. The probability that this occurs is

$$\Pr[|\mathbf{X}'\mathbf{D}\mathbf{X}| = 0],$$

the probability of breakdown of the design, which may be used to compare designs. Robust designs have a small value for the probability of breakdown.

Another measure of robustness is

$$E\left[|\mathbf{X}'\mathbf{D}\mathbf{X}|^{1/p}\right],$$

which is related to the D-optimality design criterion. A further measure is

$$E\left[\max_{\mathbf{x}}\left[\operatorname{var}\{\hat{y}(\mathbf{x})\}\,\big||\mathbf{X}'\mathbf{D}\mathbf{X}| \neq 0\right]\right],$$

where $\hat{y}(\mathbf{x})$ is the least-squares estimate of $y(\mathbf{x})$, which is related to the G-optimality design criterion.

Examples are given in Herzberg and Andrews [12] and Andrews and Herzberg [1]. In particular, Herzberg and Andrews [13] compared the robustness of chain block and coat-of-mail designs, two special types of incomplete block designs*.

Bickel and Herzberg [2] and Bickel et al. [3] investigated the problem of the robustness of design against autocorrelation in time. The assumption that observations over time are independent is usually artificial.

Suppose that an observation at time t can be written as

$$Y(t) = \beta_1 f_1(t) + \cdots + \beta_p f_p(t) + \epsilon(t), \tag{2}$$

where the $f_j(t)$ are known functions, the β_j ($j = 1, \ldots, p$) are unknown parameters, and $\epsilon(t)$ is a random error with "center" 0. For example, $f_i(t)$ may be a function of temperature at time t, this being assumed to be outside the investigator's control, but the choice of the time at which the observations are taken is under the investigator's control. Suppose that n observations are taken in the interval $[-T, T]$, $-T < t_1 < \cdots < t_n < T$, and

$$\operatorname{var}\{Y(t_i)\} = \sigma^2, \qquad i = 1, \ldots, n,$$

$$\operatorname{cov}\{Y(t_i), Y(t_j)\} = \gamma\rho(t_i - t_j), \qquad i \neq j,$$

where $0 < \gamma < 1$ and $\rho(\cdot)$ is the correlation function of a nondegenerate stationary process, i.e., $\rho(\cdot)$ is a positive symmetric function and $\rho(0) = 1$.

When ρ is known, the variance–covariance matrix* of the least-squares estimates is proportional to $n\Sigma$, where

$$\Sigma = (\mathbf{F}'\mathbf{U}^{-1}\mathbf{F})^{-1};$$

when ρ is unknown, the variance–covariance matrix of the least squares is proportional to $n\Sigma^*$, where

$$\Sigma^* = (\mathbf{F}'\mathbf{F})^{-1}\mathbf{F}'\mathbf{U}\mathbf{F}(\mathbf{F}'\mathbf{F})^{-1}.$$

Further

$$\mathbf{U} = \{\gamma\rho(t_i - t_j) + (1 - \gamma)\delta_{ij}\ (i, j = 1, \ldots, n)\}$$

is an $n \times n$ matrix and

$$\mathbf{F}' = \{f_i(t_j)\ (i = 1, \ldots, p;\ j = 1, \ldots, n)\}$$

is a $p \times p$ matrix. Bickel and Herzberg [2] and Bickel et al. [3] restricted attention to (2) with $p = 1$ and $f_1(t) = 1$ or $f_1(t) = t$, i.e., estimation of location and regression through the origin and simple linear regression.

Designs were obtained for the case $\rho(t) = e^{-\lambda|t|}$. In this case $n\Sigma$ and $n\Sigma^*$ have asymptotically the same limit. The uniform design was shown to be asymptotically optimal in a strong sense for estimating location and in a weaker sense for estimating the slope of a straight line regression*. For the regression case, the asymptotically optimal design for the estimation of the slope parameter has positive mass in the intervals $[-T, -a]$ and

$[a, T]$, where a is a function of γ and λ. When $\lambda \to \infty$, $a \to T$; this is the D-optimal and G-optimal situation.

Hedayat and John [11] and John [15] have discussed robustness for balanced incomplete block (BIB) designs*; in particular they have looked at those that remain BIB designs when one treatment is missing.

Rotatable designs* can be considered variance robust and have model robustness in some situations; see Box and Draper [5, 6]. Kiefer and Studden [17] discussed model robustness in one dimension for design measures with n points of support in interpolation and extrapolation problems. Kiefer and Wynn [18] looked at robustness of balanced incomplete block and Latin square* designs for simple nearest neighbor* correlation models.

References

[1] Andrews, D. F. and Herzberg, A. M. (1979). *J. Statist. Plann. Inf.*, **3**, 249–257.

[2] Bickel, P. J. and Herzberg, A. M. (1979). *Ann. Statist.*, **7**, 77–95.

[3] Bickel, P. J., Herzberg, A. M., and Schilling, M. F. (1981). *J. Amer. Statist. Ass.*, **76**, 870–877.

[4] Box, G. E. P. (1953). *Biometrika*, **40**, 318–335.

[5] Box, G. E. P. and Draper, N. R. (1959). *J. Amer. Statist. Ass.*, **54**, 622–654.

[6] Box, G. E. P. and Draper, N. R. (1963). *Biometrika*, **50**, 335–352.

[7] Box, G. E. P. and Draper, N. R. (1975). *Biometrika*, **62**, 347–352.

[8] Cox, D. R. and Hinkley, D. V. (1968). *J. R. Statist. Soc. B*, **30**, 284–289.

[9] Galil, Z. and Kiefer, J. (1977). *J. Statist. Plann. Inf.*, **1**, 121–132.

[10] Galil, Z. and Kiefer, J. (1977). *J. Statist. Plann. Inf.*, **1**, 27–40.

[11] Hedayat, A. and John, P. W. M. (1974). *Ann. Statist.*, **2**, 148–158.

[12] Herzberg, A. M. and Andrews, D. F. (1976). *J. R. Statist. Soc. B*, **38**, 284–289.

[13] Herzberg, A. M. and Andrews, D. F. (1978). *Commun. Statist. A*, **7**, 479–485.

[14] Herzberg, A. M. and Cox, D. R. (1972). *Biometrika*, **59**, 551–561.

[15] John, P. W. M. (1975). *Ann. Statist.*, **4**, 960–962.

[16] Kiefer, J. (1975). *Biometrika*, **62**, 277–288.

[17] Kiefer, J. and Studden, W. J. (1976). *Ann. Statist.*, **4**, 1113–1123.

[18] Kiefer, J. and Wynn, H. P. (1981). *Ann. Statist.*, **9**, 737–757.

[19] Pearson, E. S. (1929). *Biometrika*, **21**, 259–286.

(BLOCKS, BALANCED INCOMPLETE
NEAREST NEIGHBOR METHODS
OPTIMAL DESIGN OF EXPERIMENTS
OUTLIERS
ROBUST ESTIMATION
ROBUST REGRESSION
ROTATABLE DESIGNS)

A. M. HERZBERG

RODRIGUES' FORMULA

The Legendre polynomial of degree n denoted by $P_n(x)$ is defined by

$$P_n(x) = \sum_{m=1}^{N} (-1)^m \times \frac{(2n-2m)!}{2^n m!(n-m)!(n-2m)!} x^{n-2m},$$

where $N = n/2$ or $(n-1)/2$, whichever is an integer. These polynomials ($P_0(x) = 1$, $P_1(x) = x$, $P_2(x) = \frac{1}{2}(3x^2 - 1), \dots$) form an orthogonal* sequence on $[-1, 1]$ and are used extensively in the theory and applications of statistical distributions. They satisfy Rodrigues' formula [2]

$$P_n(x) = \frac{1}{2^n n!} \frac{d^n}{dx^n}\left[(x^2-1)^n\right],$$

which is also valid in the complex domain. (*In general*, Rodrigues' formula expresses the nth function of a set of special functions in terms of the nth derivative of some polynomial.)

There are tables of values of Legendre functions in Abramowitz and Stegun [1].

References

[1] Abramowitz, M. and Stegun, I. A., eds. (1964). *Handbook of Mathematical Functions*. Nat. Bur. Standards, U.S. Government Printing Office, Washington, DC.

[2] Rodrigues, I. O. (1816). *Corresp. Ecole Poly.*, **3**, 361–385.

(JACOBI POLYNOMIALS
LAGUERRE POLYNOMIALS
ORTHOGONAL EXPANSIONS)

ROMANOVSKIĬ, V. I.

Born: December 4, 1879, in Alma-Ata, (now) Kazakhstan, USSR.

Died: October 6, 1954, in Tashkent, Uzbekhistan, USSR.

Contributed to: Markov chain theory, urn models, applications of mathematical statistics in quality control.

V. I. Romanovskiĭ graduated in 1906 from St. Petersburg University, where he later also served as a lecturer. He worked as a professor at Warsaw University (1911–1915), Donsk University (1915–1918), and the Central Asia University (1918 onwards). He became a Doctor of Physics–Mathematical Sciences in 1935 and an Academician of the Academy of Sciences of the Uzbek SSR in 1943.

His textbook *Mathematical Statistics* has had a substantial influence on the development of Soviet mathematical statistics since the 1960s and his work on quality control* has found applications in Soviet industry throughout the same period.

Apart from contributions to statistical theory and practice, Romanovskiĭ also produced research results in mathematics, especially in the theory of systems of partial differential equations.

ROMANOVSKIĬ R-TEST

A modification of Fisher's *F*-test* of the equality of two variances under normal assumptions was suggested by Romanovskiĭ in 1928 [2]. (Fisher's [1] test was proposed in 1924.) Instead of using the test statistic

$$F = S_1^2/S_1^2,$$

where $S_1^2 = \sum_{i=1}^{n_1}(x_i - \bar{x})^2/(n_1 - 1)$ and $S_2^2 = \sum_{i=1}^{n_2}(y_i - \bar{y})^2/(n_2 - 1)$, based on two independent random samples $(x_1, \dots, x_{n_1})$ and $(y_1, \dots, y_{n_2})$ from normal populations, Romanovskiĭ defines the quantity

$$\theta = [(\nu_2 - 2)/\nu_2]F,$$

where $\nu_2 = n_2 - 1$. Since $E(\theta) = 1$ and $\sigma_\theta = \sqrt{2(\nu_1 + \nu_2 - 2)/\{\nu_1(\nu_2 - 4)\}}$, where $\nu_1 = n_1 - 1$, the Romanovskiĭ test statistic is $R = |\theta - 1|/\sigma_\theta$ and the null hypothesis $\sigma_1^2 = \sigma_2^2$ is rejected if $R \geqslant 3$. Note that to apply this test, *one* of the degrees of freedom (to be chosen as ν_2) should exceed four, i.e., the corresponding sample size should be at least five.

Romanovskiĭ claims that this test statistic is more efficient than Fisher's, since

$$\sigma_\theta^2 = \left(\frac{n_1 - 3}{n_1 - 1}\right)^2 \sigma_F^2 < \sigma_F^2.$$

This, of course, reflects a misunderstanding of the nature of the tests.

References

[1] Fisher, R. A. (1924). *Proc. Int. Math. Congr., Toronto, Canada*, pp. 805–813. (Reprinted in Fisher, R. A. (1950). *Contributions to Mathematical Statistics*. Wiley, NY, pp. 12, 804a–813.)

[2] Romanovskiĭ, V. (1928). *Metron*, **7**, 1–54.

(*F*-TESTS)

ROOM'S SQUARES

A Room's square of order $2n$ [9, 1] is an arrangement of $\binom{2n}{2} = n(2n-1)$ unordered pairs of symbols formed from $2n$ distinct symbols, in a square of $2n - 1$ rows and columns such that (a) a cell of the square either contains an unordered pair of symbols or is empty, (b) each of the $2n$ symbols appears precisely once in each row and in each column, and (c) every unordered pair of distinct symbols occurs in exactly one cell of the square.

It has been established that a Room's square of order $2n$ exists for every integer n except for $n = 2, 3$. A brief survey of various techniques, some quite involved and ingenious, which were used to demonstrate the result stated above, is given in ref. 8.

If, in a Room's square, every unordered pair of symbols is replaced by one of the ordered pairs and from every row of this *ordered Room's square*, two blocks are formed—one consisting of the first elements of the ordered pairs and the other consisting of the second elements—one gets an *incomplete block design** in $2(2n - 1)$ blocks that is self-complementary. If the self-complementary incomplete block design is a *balanced* incomplete block* design, then the corresponding ordered Room's square is called a *balanced* Room's square [11]. A *doubly balanced* or a *three-design* is an arrangement of elements in blocks of k distinct elements such that every unordered triple of distinct elements occurs in precisely μ blocks. It has been shown [11] that every self-complementary balanced incomplete block design is a doubly balanced or a three-design. Thus a balanced Room's square furnishes a three-design. A balanced Room's square of order $2n$ can only exist for n an even integer [2, 11].

In a duplicate bridge tournament for $2n$ teams, a *complete Howell rotation* is an arrangement such that (i) each board is played by at most one pair of teams on any round, (ii) every team plays one board every round and every team plays each board precisely once, and (iii) every team opposes every team precisely once.

A Room's square of order $2n$ is equivalent to a complete Howell rotation of $2n$ teams. Suppose now that every board has a NS direction and an EW direction. Team 1 is said to compete against Team 2 on a particular board if they play the board in the same direction. A complete Howell rotation of $2n$ teams is called a *complete balanced Howell rotation* if the arrangement also satisfies the condition (iv) each team competes equally often, say λ times, with every other team.

It has been shown [11] that a balanced Room's square is equivalent to a complete balanced Howell rotation.

Complete balanced Howell rotations for $4t = p^r + 1$ (with $p^r > 3$ a prime power) and hence balanced Room's squares of order $4t$, have been shown in ref. 2 to exist. Other constructions are given in refs. 11 and 6.

A survey of various generalizations of Room's squares is given in ref. 10. One such combinatorial arrangement, called a *generalized Room's square* of degree k and order n, is a square array satisfying (i) every cell of the array either contains a k subset of an n-set N or is empty, (ii) every element of N appears precisely once in each row and each column, and (iii) every k subset of N is contained in exactly one cell of the square. It follows that k must divide n and the square is of side $\binom{n-1}{k-1}$.

Room's squares furnish various types of experimental designs*; see for instance, refs. 1 and 5.

If the columns of a Room's square of order $2n$ are regarded as blocks, the rows as a set of treatments $a_1, \ldots, a_{2n-1}$, and the symbols in the square as a second set of treatments $b_0, \ldots, b_{2n-1}$, then the design is an incomplete block design with respect to the first set of treatments (in special cases a balanced incomplete block design) and a complete randomized block* design with respect to the second set of treatments. Further, the design is balanced with respect to combinations of levels of treatment a with levels of treatment b. The additive parametric model for this design would be $y_{tij} = A + B_t + \alpha_i + \beta_j + z_{tij}$, where y_{tij} denotes the observation on the treatment combination $a_i b_j$ in the tth block, $\Sigma B_t = \Sigma\alpha_i = \Sigma\beta_j = 0$, and z_{tij}'s are independent random variables with common variance and mean zero. The analysis of variance appropriate to this model is given in ref. 1.

Alternatively, the design may also be regarded as an incomplete block design for treatments $\{a_1, \ldots, a_{2n-1}\}$ with main plots split for treatments $\{b_0, \ldots, b_{2n-1}\}$. The model for this design [1] then becomes

$$y_{tij} = A + B_t + \alpha_i + \beta_j + u_{ti} + z_{tij},$$

where the u_{ti}'s are independent random variables, with zero mean and common variance, which are also independent of the z_{tij}'s. The correct analysis of variance* is given in ref. 12.

Table 1 Room's Square of Order $2n = 8$

	1	2	3	4	5	6	7
1			24		56	37	01
2		35		67	41	02	
3	46		71	52	03		
4		12	63	04			57
5	23	74	05			61	
6	15	06			72		34
7	07			13		45	26

Source: Ref. 1.

The parameters of a balanced incomplete block design corresponding to a *balanced* Room's square of order $2n$ are $v = 2n$, $b = 2(2n - 1)$, $r = 2n - 1$, $k = n$, and $\lambda = n - 1$. Since this is a doubly balanced or a three design, it can be used to estimate residual effects of treatments where the same experimental units are used in a serial experiment; see, for instance, refs. 3 and 13. Again a doubly balanced design gives rise to a partially balanced array [4], which is used in factorial experiments*. Construction of incomplete paired-comparisons* designs for Room's squares and statistical analysis of observations from such designs are given in ref. 5. Optimality of generalized Room's squares based on a model given in ref. 1, has been considered in ref. 7.

A Room's square of order $2n = 8$ taken from [1] is given in Table 1. This happens to be a balanced Room's square. It also gives rise to a symmetrical balanced incomplete block design $v = b = 7$, $r = k = 4$, and $\lambda = 2$, when the rows are considered as treatments and the columns as blocks.

References

[1] Archbold, J. W. and Johnson, N. L. (1958). *Ann. Math. Statist.*, **29**, 219–225.

[2] Berlekamp, E. R. and Hwang, F. K. (1972). *J. Comb. Theory A*, **12**, 159–166.

[3] Calvin, L. D. (1954). *Biometrics*, **10**, 61–88.

[4] Chakravarti, I. M. (1961). *Ann. Math. Statist.*, **32**, 1181–1185.

[5] Chakravarti, I. M. (1976). Statistical Designs from Room's Squares with Applications. In *Exp. Suppl. 22, Contrib. Appl. Statist. Dedicated to A. Linder*, W. J. Ziegler, ed. Birkhäuser Verlag, Basel, Switzerland, pp. 223–231.

[6] Du, Ding-Zhu and Hwang, F. K. (1982). *Trans. Amer. Math Soc.*, **271**, 409–413.

[7] Ghosh, S. and Shah, K. R. (1983). *Commun. Statist. A*, **12**, 2119–2125.

[8] Mullin, R. C. and Wallis, W. D. (1975). *Aequationes Math.*, **13**, 1–7.

[9] Room, T. G. (1955). *Math. Gazette*, **39**, 307.

[10] Rosa, A. (1980). *Ann. Discrete Math.*, **8**. *Combinatorics* 79, part 1, M. Deza and I. G. Rosenberg, eds., pp. 43–57.

[11] Schellenberg, P. J. (1973). *Aequationes Math.*, **9**, 75–90.

[12] Shah, K. R. (1970), *Ann. Math. Statist.*, **41**, 743–745.

[13] Williams, E. J. (1949). *Aust. J. Sci. Res. A*, **2**, 149–168.

(ANALYSIS OF VARIANCE
BLOCKS, BALANCED INCOMPLETE
DESIGN OF EXPERIMENTS
LATIN SQUARES, LATIN CUBES,
LATIN RECTANGLES, ETC.
YOUDEN SQUARES)

I. M. CHAKRAVARTI

ROOTAGON

This is a frequency polygon* with cell counts reexpressed on a square root scale to stabilize the variance of the counts.

ROOTISTIC DISTRIBUTION

A particular case of (Tukey's) lambda distribution* corresponding to the parameter value $\lambda = \frac{1}{2}$.

ROOTOGRAM *See* GRAPHICAL REPRESENTATION OF DATA

ROOT MEAN SQUARE

The root mean square of values $x_1, x_2, \ldots, x_n$ is defined as

$$\sqrt{\left(\sum_{j=1}^{n} x_j^2/n\right)}.$$

For a grouped frequency distribution with values $y_1, y_2, \dots, y_k$ with frequencies $f_1, f_2, \dots, f_k$, respectively, the root mean square is

$$\sqrt{\left(\sum_{i=1}^{k} f_i y_i^2/n\right)},$$

where $n = f_1 + f_2 + \cdots + f_k$.

ROOT MEAN SQUARE DEVIATION (RMSD)

The RMSD of a set of values $x_1, x_2, \dots, x_n$ from a fixed value ξ is

$$\sqrt{\left\{\sum_{j=1}^{n} (x_j - \xi)^2/n\right\}},$$

that is, it is the root mean square* of the deviations $(x_1 - \xi), (x_2 - \xi), \dots, (x_n - \xi)$.

For a grouped frequency distribution of values $y_1, y_2, \dots, y_k$ with frequencies $f_1, f_2, \dots, f_k$, respectively, the RMSD about ξ is

$$\sqrt{\left\{\sum_{i=1}^{k} f_i (y_i - \xi)^2/n\right\}},$$

where $n = f_1 + f_2 + \cdots + f_k$.

For a sample of size n, with observed values $X_1, \dots, X_n$ the sample standard deviation is

$$\left(\frac{n}{n-1}\right)^{1/2} (\text{RMSD about the arithmetic mean}).$$

ROSE DIAGRAM *See* DIRECTIONAL DATA

ROSENTHAL'S MOMENT INEQUALITY

Let $\{X_n\}$, $n = 1, 2, \dots,$ be a sequence of independent random variables on probability space (Ω, A, P), $p \geqslant 1$, and

$$S_n = \sum_{i=1}^{n} X_i - E(X_i).$$

Assume, without loss of generality, that $E[X_i] = 0$. Rosenthal [2] has shown that there exists a constant C_p such that for any integer n,

$$E\left(|S_n|^{2p}\right) \leqslant C_p \max\left\{\sum_{i=1}^{n} E\left(|X_i|^{2p}\right), \sum_{i=1}^{n} \left(E\left(X_i^2\right)\right)^p\right\}.$$

This result was utilized by Wittman [3] to prove the strong law of large numbers* under somewhat less restrictive assumptions than those in classical texts such as Loève [1].

References

[1] Loève, M. (1977). *Probability Theory*, 4th ed., Vol. 1. Springer-Verlag, New York.

[2] Rosenthal, H. P. (1970). *Israel J. Math.*, **8**, 273–303.

[3] Wittmann, R. (1985). *Statist. Prob. Lett.*, **3**, 131–33.

(LAWS OF LARGE NUMBERS)

ROTATABLE DESIGNS

A design for the investigation of response surfaces* involving k factors can be considered as a set of N, not necessarily distinct, points in k-dimensional space. Each point in k-dimensional space represents a particular combination of the levels of the k factors.

Let $x_1, \dots, x_k$ denote the standardized levels of the k factors. It is assumed that the response surface may be represented in a given region by a polynomial of degree d. A design that allows all the coefficients of a polynomial of degree d in k variables to be estimated by the method of least squares is called a design of order d or a dth-order design.

In cases where little, if anything, is known about the nature of the response surface, it is reasonable to use an experimental design that will provide approximately the same amount of information at all points in the factor space that are equidistant from the center of the design.

Rotatable designs were introduced by Box and Hunter (1957); these designs are such that the variance of the estimated response is constant for all points equidistant from the center of the design, i.e., the variance of the estimated response at a point is a function of the distance of the point from the center, origin, of the design. Distance is in terms of Euclidean distance and presupposes meaningful scales for the x_i's. Thus equal information may be obtained in all directions at any specified distance from the center.

The necessary and sufficient condition that a dth-order design is rotatable or equivalently that the variance is constant at points equidistant from the center of the design is that

$$\mathbf{X}'\mathbf{X} = \mathbf{R}'\mathbf{X}'\mathbf{X}\mathbf{R}, \qquad (1)$$

where $\mathbf{X}$ is the matrix of "independent" variables of the model, $(\mathbf{X}'\mathbf{X})^{-1}$ being proportional to the variance–covariance matrix* of the estimated coefficients of the dth degree polynomial, and

$$\mathbf{R} = \begin{bmatrix} 1 & 0 & \cdots & 0 \\ 0 & & & \\ \vdots & & \mathbf{W} & \\ 0 & & & \end{bmatrix},$$

$\mathbf{W}$ being an arbitrary orthogonal matrix. The logic of (1) is that the moment structure of the design remains unaltered when the design is rotated about the origin. From (1), the moments of a dth-order rotatable design in k dimensions are

$$\frac{1}{N}\sum_{u=1}^{N} x_{1u}^{\alpha_1} \cdots x_{ku}^{\alpha_k} = \begin{cases} \dfrac{\lambda_\alpha \Pi_{i=1}^{k} \alpha_i!}{2^{\alpha/2}\Pi_{i=1}^{k}\left(\frac{1}{2}\alpha_i\right)!}, & \text{all } \alpha_i \text{ even}, \\ 0, & \text{one or more of the } \alpha_i \text{ odd}, \end{cases}$$

where $\alpha = 2d$ and λ_α is constant for any design and any α. These moments are identical to the moments of order up to and including $2d$ of a spherical distribution.

There are many rotatable designs. A first-order rotatable design can always be formed from an orthogonal set of $k+1$ points in k dimensions; see Box [2]. The k-dimensional point sets $(\pm a, \ldots, \pm a)$, $(\pm 2^{k/4}a, 0, \ldots, 0)$, $(0, \pm 2^{k/4}a, 0, \ldots, 0), \ldots, (0, \ldots, 0, \pm 2^{k/4}a)$ plus points at the center always form a second-order rotatable design, for all a, $a > 0$. Various authors present methods of construction of second- and third-order rotatable designs. Herzberg and Cox [14] give a large set of references; see, in particular, Bose and Draper [1] and Draper [8–10]. Sometimes the number of design points required is prohibitive.

Box and Draper [3, 4] considered designs that minimize the averaged integrated mean-squared error of the response function over a spherical region of the factor space, the error arising from bias due to the use of a polynomial model of too small degree and from random errors. They showed that when bias alone was considered, the bias was minimized when rotatable designs were used.

Cylindrically rotatable designs, or group-divisible* rotatable designs, were introduced by Herzberg [11, 12] and Das and Dey [7] in order to find designs that were in some dimensions rotatable but required a smaller number of points than rotatable designs.

When a rotatable design is used, the variance of the difference of the estimated responses at any two points in the factor space is a function of the distance of the two points from the center of the design and the angle subtending the points at the center of the design; see Herzberg [13]. Box and Draper [5] have extended and unified the results.

While the notion of rotatability for a response surface design is not essential, it is generally better to use a rotatable design rather than a nonrotatable design, other things being equal.

References

[1] Bose, R. C. and Draper, N. R. (1959). *Ann. Math. Statist.*, **30**, 1097–1112.

[2] Box, G. E. P. (1952). *Biometrika*, **39**, 49–57.

[3] Box, G. E. P. and Draper, N. R. (1959). *J. Amer. Statist. Ass.*, **54**, 622–654.

[4] Box, G. E. P. and Draper, N. R. (1963). *Biometrika*, **50**, 335–352.

[5] Box, G. E. P. and Draper, N. R. (1980). *J. R. Statist. Soc. B*, **42**, 79–82.

[6] Box, G. E. P. and Hunter, J. S. (1957). *Ann. Math. Statist.*, **28**, 195–241.

[7] Das, M. N. and Dey, A. (1967). *Ann. Inst. Math. Statist, Tokyo*, **19**, 331–347. (Correction, **20**, 337.)

[8] Draper, N. R. (1960). *Ann. Math. Statist.*, **31**, 875–877.

[9] Draper, N. R. (1960). *Ann. Math. Statist.*, **31**, 23–33.

[10] Draper, N. R. (1960). *Ann. Math. Statist.*, **31**, 865–874.

[11] Herzberg, A. M. (1966). *Ann. Math. Statist.*, **37**, 242–247.

[12] Herzberg, A. M. (1967). *Ann. Math. Statist.*, **38**, 167–176.

[13] Herzberg, A. M. (1967). *J. R. Statist. Soc. B*, **29**, 174–179.

[14] Herzberg, A. M. and Cox, D. R. (1969). *J. R. Statist. Soc. A*, **132**, 29–67.

(ANALYSIS OF VARIANCE
COMPOSITE DESIGNS
DESIGN OF EXPERIMENTS
ORTHOGONAL DESIGNS
RESPONSE SURFACE DESIGNS)

A. M. HERZBERG

ROTATION GROUP

A rotation through the angle θ on the plane produces new coordinates

$$y_1 = \cos\theta \cdot x_1 + \sin\theta \cdot x_2,$$

$$y_2 = -\sin\theta \cdot x_1 + \cos\theta \cdot x_2$$

from initial coordinates x_1, x_2. These new coordinates can be viewed as coordinates with respect to the orthonormal unit vectors $(\cos\theta, \sin\theta)$ and $(-\sin\theta, \cos\theta)$ that define new axes. Designate such a transformation of the plane by $T(\theta)$.

If the transformation $T(\theta_1)$ is followed by $T(\theta_2)$, then the combined transformation $T(\theta_2)T(\theta_1)$ is given by $T(\theta_2 + \theta_1)$. If the transformation $T(\theta)$ is reversed, we obtain $T^{-1}(\theta) = T(-\theta)$ and $T(\theta)T(-\theta) = T(0)$, which is the identity transformation. This closure of the class $\{T(\theta) : \theta \in R\}$ of transformation under product and inverse means the class is a *group* of transformations. Here it suffices to work with θ in $[\theta, 2\pi)$ and replace other θ values by the equivalents in that range.

The density function

$$(2\pi\sigma^2)^{-1}\exp\{-(x_1^2 + x_2^2)/2\sigma^2\}$$

is unchanged by a rotation $T(\theta)$: the density is *rotationally symmetric*. The same applies for the density $c \cdot h(r^2)$, where $r^2 = x_1^2 + x_2^2$. The rotation group is the formal mathematical way of expressing rotational symmetry: under the rotations the density does not change.

More generally on R^n, consider the transformation $\mathbf{y} = \mathbf{A}\mathbf{x}$, where the $n \times n$ matrix $\mathbf{A}$ satisfies $\mathbf{A}\mathbf{A}' = \mathrm{I}$ and $|\mathbf{A}| = 1$, and is called a rotation matrix. The class of such transformations forms the rotation group on R^n. If the condition $|\mathbf{A}| = 1$ is relaxed to $|\mathbf{A}| = \pm 1$, then the orthogonal group is obtained (it has twice as many elements and includes transformations such as the reversal of sign of a single coordinate). The density function $f = (2\pi\sigma^2)^{-n/2}\exp\{-\Sigma x_i^2/(2\sigma^2)\}$, or more generally $c \cdot h(r^2)$, is invariant under the rotations on R^n $(r^2 = \Sigma x_i^2)$. Thus if $x_1, \ldots, x_n$ are independent and identically distributed (IID) Normal(0, σ^2), then $y_1, \ldots, y_n$ are IID Normal(0, σ^2), where $\mathbf{y} = \mathbf{A}\mathbf{x}$ and $\mathbf{A}$ is a *rotation matrix* (or even is an orthogonal matrix).

A statistical problem can be invariant under the rotation group. As a simple example, let $x_1, \ldots, x_n$ be a sample from Normal(μ_0, σ^2), where μ_0 is known. This can be recast in terms of an origin at μ_0 on each axis: $x_1 - \mu_0, \ldots, x_n - \mu_0$ is a sample from Normal(0, σ^2). In this centered form the problem is invariant under the rotation group on R^n: the rotation matrix $\mathbf{A}$ changes $(x_1 - \mu_0, \ldots, x_n - \mu_0)'$ into $\mathbf{A}(x_1 - \mu_0, \ldots, x_n - \mu_0)'$. Considering the set of all rotations $\mathbf{A}$ generates all points on the sphere of radius $r = (\Sigma(x_i - \mu_0)^2)^{1/2}$: in fact r is the only unalterable characteristic; it is the *maximal invariant* function with respect to this rotation group. The invariance principle* then says that statistical inference should be based on r; for example $r^2/n = \Sigma(x_i - \mu_0)^2/n$ is the invariant unbiased estimate of σ^2.

More generally the sample space* R^n may factor as a Cartesian product $X_1 \times X_2$ and a statistical problem can be invariant under rotations on say X_1; this can typically be recast as invariance on R^n with respect to a subgroup of the full rotation group. As a simple example, consider $x_1, \dots, x_n$ IID $N(\mu, \sigma^2)$. The sample space can be factored as $L(1) \times L^{\perp}(1)$ and the problem is invariant under the rotation group on $L^{\perp}(1)$. This reduces the problem to the maximal invariant $(\bar{x}, \Sigma(x_i - \bar{x})^2)$.

(INVARIANCE CONCEPTS IN STATISTICS)

D. A. S. FRASER

ROTATION GROUP BIAS

ROTATING PANEL SURVEYS

In sample surveys involving periodically repeated interviews with the same individuals, many investigators have reported that the number of times respondents have been interviewed seems to affect their answers to certain questions (e.g., see Kemsley [10], Lehnen and Reiss [11], Mooney [12], Neter and Waksberg [13], Pearl [14], and Turner [17]). This phenomenon has evoked considerable concern amongst survey statisticians because several major government household surveys in Canada, Great Britain, and the United States involve repeated interviewing.

Because of the sheer size of the sample of many large-scale government household surveys, and the frequency of data collection and reporting, some segments of the population of the United States and Canada would be exhausted were new nonoverlapping samples drawn each month for each survey. For example, the U.S. Current Population Survey (CPS), which collects and reports information on employment and unemployment, is based on monthly interviews with respondents in 60,000 households, while its Canadian counterpart, the Labour Force Participation Survey (LFPS), involves monthly interviews with respondents in over 50,000 households (*see* BUREAU OF THE CENSUS, U.S. and BUREAU OF LABOR STATISTICS). For this and other reasons, such as the administrative costs of drawing new samples and locating respondents, the cheapness of telephone reinterviews, and the desire for longitudinal information, statisticians have developed elaborate multiple interview and rotation schedules. For example, the U.S. National Crime Survey (NCS), which collects information about *victimization*, has seven rotation groups of about 9000 housing units each, with each rotation group remaining in the sample for seven interviews, once every six months (*see* JUSTICE STATISTICS, BUREAU OF and STATISTICS OF CRIMINAL JUSTICE). Every six months a new rotation group enters the sample and the oldest group is rotated out. The LFPS with its monthly interviews has a six-month rotation schedule, and the CPS currently uses a 4-8-4 rotation scheme in which each of the eight rotation groups is interviewed for four months, is dropped from the sample for the next eight months, and is reincluded for the final four months. In this CPS scheme the month-to-month overlap is 75%, while there is a 50% overlap in the sample location for the same month in successive years.

For characteristics such as household expenditures and victimization, some of the differential reporting of information in panel surveys (*see* PANEL DATA) has been attributed to memory errors. For example, events, purchases, or victimizations are often reported as happening more recently than they actually did happen. This moving of events to more recent times is referred to as "telescoping." There is some evidence, however, that respondents may sometimes move events into the more distant past, and thus telescoping can go either forward or backward. To control telescoping a technique called "bounded recall" has been useful in panel studies where respondents are interviewed repeatedly (Neter and Waksberg [13]). At the start of the second and subsequent interviews, respondents are reminded of what they have previously reported and asked what has occurred since those events. The NCS uses this kind of bounded recall, going so far as to disregard the victimization data from the first interview at a household location for the computation of victimization rates and using such data only for bounding purposes. Despite this form of control, at many of the subsequent interviews there is no boundary information available (Fienberg [4]).

Even after attempts to control for various types of response errors that manifest themselves in rotating panel surveys such as the CPS, LFPS, and the NCS, estimates from different panels relating to the same periods of time have differing values (e.g., see Bailar [1] and Woltman et al. [19]). Because these differences do not appear to depend appreciably on the sample size the resulting phenomenon has come to be known as "rotation group bias." The exact causes of rotation bias in most surveys are unknown; it is a reflection of various nonsampling and measurement errors* and is typically complicated by the fact that the actual number of interviews for individuals and households in a rotation group is not constant because of the replacement rules used (Fienberg [4] and Reiss [15]).

EMPIRICAL EVIDENCE

Each rotation group in a rotating panel survey, such as the CPS, is designed as a national sample. Thus, one might expect that the eight estimates for the eight different rotation groups present each month in the CPS would have differences in measures of employment and unemployment that are explainable primarily in terms of sampling error. Nevertheless, Bureau of the Census studies have consistently documented variation among the estimates from the different rotation groups in the CPS in excess of that explainable by sampling error.

The problem with rotation group bias is that we do not know which rotation group should be used as a standard from which the bias of the others can be determined. Moreover, the effect of rotation group bias differs depending on the quantity being measured, and the statistical estimator being used (Bailar [1]). For example, in the CPS for the five-year period from 1968 to 1972 the average unemployment rate expressed as a percentage of the overall rate for eight rotations groups was

Rotation Group	Bias Index
1	110.4
2	100.6
3	97.9
4	98.7
5	104.1
6	97.4
7	94.6
8	96.4

(These figures are based on values reported in Brooks and Bailar [3]. More extensive data and details are available in Bailar [1, 2].)

The bias index data for the CPS might be thought of as suggesting that the second rotation group would be a reasonable standard, but that would be true only if the average of all eight groups is unbiased, and there is no way to check such an assumption. Indeed, it may well be more reasonable to assume that estimates of all the rotation groups are biased, and thus, unless the biases cancel out, the resulting overall estimate based on all rotation groups will be biased as well (Bailar [1]).

One justification for the use of panel rotation structures is that the overlap from one interview to another can be used to decrease the sampling error of period-to-period changes if there is a positive correlation between responses over time. Part of the folklore often cited in connection with panel surveys such as the CPS and NCS both orally and in print, is that, although estimates of level may suffer from rotation group bias, estimates of change are unbiased. This conventional wisdom about change estimates is true only if

(1) rotation group biases are constant from interview period to interview period;

(2) rotation group biases are additive.

In the context of the accumulation of many years of CPS data, there is apparently some evidence that (1) may not be true. In addition, Bailar [2] provides considerable empirical evidence to support the contention that certain estimates of change are affected by rotation group biases to a greater extent than are estimates of level.

EXPLANATIONS

The causes of rotation group bias are not really known, but it must come from the nonsampling and measurement errors referred to above. For example, it is often suggested that at least part of the bias in the CPS unemployment rates is due to added stimulation of respondents toward looking for jobs, and if procedures change (as they continually do), nonsampling errors change. Thus, even if we had been able to sort out and adjust for the contributions of the various sources of nonsampling error to rotation group bias in data from the years prior to the existing studies of the problem, changes that have occurred in the intervening decade, such as the more extensive use of telephone interviews (e.g., see the discussion in Fienberg and Tanur [5]), are all likely to have made matters again uncertain.

A special problem of rotating panel household surveys (*see* SURVEYS, HOUSEHOLD) that needs to be considered in the context of a discussion of rotation group bias is the selective attrition of individual respondents and families—something that cuts across the various sources of nonresponse error. The rotation design in most of these surveys holds the housing unit constant, but individuals and families come and go. In fact, individuals can "drop out" even though their families remain; a particular problem arises because departures from families and, hence, panel attrition at the individual level may well be related to the phenomenon of interest such as unemployment or victimization.

For the NCS, Reiss [15] has noted that households moving out of sample locations had higher rates of victimization prior to moving than did those houses that remained in the sample. Moreover, replacement households appear to have different characteristics and victimization experiences than the movers, although the matching of victimization levels for movers and stayers is compounded by a host of other response-related biases. These differences are in part due to compositional changes in NCS households after they enter the survey. Thus we see that the attrition problem is related to the phenomenon under study, i.e., victimization (see also Griffin [6]).

Estimation problems associated with panel attrition, or self-selection as the phenomenon is also known, have been the subject of considerable study. For example, Williams [18] estimates the biases in panel designs associated with attrition and self-selection. But to estimate the impact of selected attrition and to adjust for it, we require longitudinal records for individuals and households. Only through such longitudinal analyses (e.g., see Griffin [6], Griliches et al. [7], Heckman [8], Heckman and Singer [9], and Stasny [16]) and through supplementary studies of movers from sample locations can we hope to gain a deeper knowledge of the sources of rotation group bias.

References

[1] Bailar, B. A. (1975). *J. Amer. Statist. Ass.*, **70**, 23–30.

[2] Bailar, B. A. (1979). In *Proc. Conf. Recent Developments in Statistical Methods and Application.* Academia Sinica, Taipei, Taiwan, pp. 13–29.

[3] Brooks, C. A. and Bailar, B. A. (1978). An Error Profile: Employment as Measured by the Current Population Survey. *Statistical Policy Working Paper 3*, U.S. Department of Commerce, Washington, DC.

[4] Fienberg, S. E. (1980). *Statistician, London*, **29**, 313–350.

[5] Fienberg, S. E. and Tanur, J. M. (1983). *Behav. Sci.*, **28**, 135–153.

[6] Griffin, D. L. (1984). *Estimation of Victimization Prevalence Using Data From the National Crime Survey. Lecture Notes in Statistics*. Springer-Verlag, New York.

[7] Griliches, Z., Hall, B. H., and Hausman, J. A. (1977). In *INSEE Conf. Economics of Panel Data, Paris, France*.

[8] Heckman, J. D. (1976). *Ann. Econ. Social Meas.*, **5**, 475–492.

[9] Heckman, J. D. and Singer, B. (1982). In *Advances in Econometrics*, W. Hildenbrand, ed. Cambridge Univ. Press, Cambridge, England, pp. 39–77.

[10] Kemsley, W. F. F. (1961). *Appl. Statist.*, **10**, 117–135.

[11] Lehnen, R. G. and Reiss, A. J., Jr. (1978). *Victimology*, **3**, 110–124.

[12] Mooney, H. (1962). Methodology in Two California Health Surveys. *Public Health Monograph No. 70*, U.S. Public Health Service, U.S. Government Printing Office, Washington, DC.

[13] Neter, J. and Waksberg, J. (1964). *J. Amer. Statist. Ass.*, **59**, 18–55.

[14] Pearl, R. B. (1963). *Employment and Earnings*, **9**, iv–x.

[15] Reiss, A. J., Jr. (1977). Unpublished Report, Institute for Social and Policy Studies, Yale University, New Haven, CT.

[16] Stasny, E. (1983). Estimating Gross Flows in Labor Force Participation Using Data From the Canadian Labour Force Survey. Ph.D. Dissertation, Department of Statistics, Carnegie-Mellon University, Pittsburgh, PA.

[17] Turner, R. (1961). *Appl. Statist.*, **10**, 136–146.

[18] Williams, W. H. (1978). In *Contributions to Survey Analysis and Applied Statistics*, H. A. David, ed. Academic Press, New York, pp. 89–112.

[19] Woltman, H., Bushery, J., and Carstensen, L. (1975). Internal Bureau of the Census Memorandum.

Acknowledgments

The preparation of this entry was supported in part by the National Science Foundation under Grant No. SES-8406952 to Carnegie-Mellon University.

(BUREAU OF LABOR STATISTICS
BUREAU OF THE CENSUS, U.S.
JUSTICE STATISTICS, BUREAU OF
PANEL DATA
ROTATION SAMPLING
SURVEY SAMPLING
TELEPHONE SURVEYS,
COMPUTER-ASSISTED
TELESCOPING)

STEPHEN E. FIENBERG

ROTATION SAMPLING

Many important surveys* are repeated at regular time intervals. Data from repeated sampling of a population over time can be used to estimate parameters associated with the characteristics of interest for the current time period, changes in the parameters from the previous time period, and parameters such as averages or totals defined over several time periods. Data from the current period can also be combined with past data to revise estimates for previous periods. In addition to information on the current status of the population, repeated surveys can provide insight into the forces acting on the population over time.

In designing a repeated survey there are two considerations that do not arise for one time surveys, viz., the frequency of sampling and the composition of the sample at each time point relative to previous time periods. In the selection of a sampling frequency, the need for periodic estimates and the rate at which the parameters of interest are changing over time must be balanced against the cost and time required to produce estimates having the desired precision. The sample at each time point may be composed of

1. the same set of units that were sampled in the preceding time period;
2. an entirely different set of units than were sampled in the preceding time period;
3. a set of units that were sampled in the preceding period (matched units) and a set of units that were not sampled in the preceding period (unmatched units).

The first type of sample composition is referred to as *panel data** and repeated sampling with the third type of sample composition at each time point is referred to as rotation sampling. The definition of rotation sampling does not preclude units remaining in the sample for several consecutive time periods or units being resampled at a future time point. For example, the *Current Population Survey* conducted by the U.S. Bureau of the Census* to estimate monthly labor force characteristics uses a rotation design in which a selected unit is sampled for four consecutive months, deleted from the sample for eight months, and then resampled for four consecutive months. Each month approximately 75% of the sample consists of matched units and 25% consists of unmatched units.

The selection of a good rotation pattern is dependent on the correlation between the responses from a unit sampled in two consecutive time periods as well as on other factors that are not as easily quantifiable. Among these other factors is the burden imposed on respondents who are sampled several times. Cases have been reported (e.g., ref. 2) where data from groups of individuals sampled in the same time period but having differing numbers of previous interviews have varied greatly. The only apparent difference in the groups was the number of previous interviews. (This effect is called *rotation group bias**.) In extreme cases, repeated sampling can affect the normal response pattern, making the sample no longer representative of the population. On the other hand, data collected during the initial interview may be less reliable than that collected at subsequent interviews. This is especially true when sensitive information is requested and respondents may be reluctant to provide the necessary information to an interviewer who is a complete stranger. From an administrative viewpoint there are additional costs and burdens associated with the sampling of a unit for the first time that do not arise for subsequent sampling of the unit. These include listing, identifying, and locating first time units. Because factors such as those listed are not easily quantifiable, much of the research on the selection of a rotation pattern has not directly included them.

Three general approaches to estimation have been considered in rotation sampling. The approach found in most sampling textbooks (e.g., refs. 3 and 5) utilizes composite estimators. One form of the composite estimator for a parameter at time t is a weighted linear combination of an estimator formed from the unmatched units at time t and an estimator based on the matched units and the composite estimator at time $t-1$. For example, suppose that the parameter of interest is the population mean. Let $\bar{y}_{tm}$ and $\bar{y}_{tu}$ be the sample means of the matched and unmatched units, respectively, at time t. Then the composite estimator is of the form

$$\hat{\mu}_t = w_u \bar{y}_{tu} + w_m \bar{y}^*,$$

where the weights w_u and w_m are usually functions of the reciprocal of the variances of the estimators $\bar{y}_{tu}$ and $\bar{y}^*$. Examples of the estimator $\bar{y}^*$ are

1. a regression-type estimator (e.g., ref. 3)

$$\bar{y}^* = \bar{y}_{tm} + b(\hat{\mu}_{t-1} - \bar{y}_{(t-1)m});$$

2. a ratio-type estimator (e.g., ref. 1)

$$\bar{y}^* = (\bar{y}_{tm}/\bar{y}_{(t-1)m})\hat{\mu}_{t-1};$$

3. a difference-type estimator (e.g., ref. 9)

$$\bar{y}^* = (\bar{y}_{tm} - \bar{y}_{(t-1)m}) + \hat{\mu}_{t-1}.$$

When a unit is sampled in more than two consecutive periods, the correlation between responses is assumed to decrease as the number of intermediate time periods increases. For this reason, data collected on a unit prior to time $t - 1$ are usually only utilized indirectly through the composite estimator for time $t - 1$. Most descriptions of composite estimation assume a product correlation structure, i.e., if ρ is the correlation between the responses from a unit in two consecutive time periods, then ρ^k is the correlation between responses that are k time periods apart. References 7 and 8 discuss more general correlation structures.

An optimal rotation pattern may be defined as one that minimizes the variance of the composite estimator for a specified sample size. Formulas for the proportion of matched units at each time point for an optimal design as a function of the correlation between responses from a matched unit can be found in ref. 3 for simple random sampling. For a given sample size, the optimal number of matched units to be included in the sample at time t is inversely related to the correlation.

The development of composite estimators in refs. 3 and 5 assumes that the sampled population is infinite. Results for composite estimators that explicitly account for sample selection from a finite population can be found in ref. 10.

The composite estimation approach assumes a relationship between observations on a unit over time. However, it does not assume a corresponding relationship for the values of the parameter of interest over time. An alternative approach to estimation is based on modeling the population parameter using time-series* models and methods. The data y_t at time t are represented as the sum of the parameter of interest θ_t and a sampling error e_t; i.e., $y_t = \theta_t + e_t$ for all t, where θ_t and e_t are usually assumed to be uncorrelated stationary processes. References 6 and 11 describe methods of estimating the parameters θ_t.

Linear models provide a third approach to estimation in rotation sampling. An example of this approach can be found in ref. 12. For related models *see* PANEL DATA.

In the preceding discussion it has been tacitly assumed that at each time point the sampled units report information only for that time period. Higher level rotation schemes in which units sampled at each point in time report information not only for that time period but for one or more of the immediately preceding time periods is discussed in refs. 4 and 12.

References

[1] Avadhani, M. S. and Sukhatme, B. V. (1970). *Appl. Statist.*, **19**, 251–259. (Discusses ratio-type composite estimators.)

[2] Bailar, B. A. (1975). *J. Amer. Statist. Ass.*, **70**, 23–30. (Basic reference on rotation group bias.)

[3] Cochran, W. G. (1977). *Sampling Techniques*, 3rd ed. Wiley, New York, pp. 344–355.

[4] Eckler, A. R. (1955). *Ann. Math. Statist.*, **26**, 644–685. (Early paper on rotation sampling.)

[5] Hansen, M. H., Hurwitz, W. N., and Madow, W. G. (1953). *Sample Survey Methods and Theory*, Vol. 2, Wiley, New York, pp. 268–279.

[6] Jones, R. G. (1980). *J. R. Statist. Soc. B*, **42**, 221–226. (Unifies the time-series and composite estimation approaches.)

[7] Kulldorf, G. (1979). *Proc. Int. Statist. Inst.*, **48**, 417–445. (A study of optimal allocation.)

[8] Prabhu-Ajgaonkar, S. G. (1968). *Aust. J. Statist.*, **10**, 56–63. (Contains estimation results under general correlation structures.)

[9] Raj, D. (1965). *J. Amer. Statist. Ass.*, **60**, 270–277. (Discusses difference-type composite estimators.)

[10] Rao, J. N. K. and Graham, J. E. (1964). *J. Amer. Statist. Ass.*, **59**, 492–509. (Contains results for composite estimation assuming a finite population.)

[11] Scott, A. J., Smith, T. M. F., and Jones, R. G. (1977). *Int. Statist. Rev.*, **45**, 12–28. (Describes a time-series approach to estimation.)

[12] Wolter, K. M. (1979). *J. Amer. Statist. Ass.*, **74**, 604–613. (Discusses a linear model approach to estimation.)

(PANEL DATA
ROTATION GROUP BIAS
SURVEY SAMPLING)

EDWARD E. GBUR, JR.

ROTATION TECHNIQUES

The factor analysis* model (*see* FACTOR ANALYSIS; MINIMUM RANK FACTOR ANALYSIS) implies that the observed variable correlation matrix $\mathbf{P}$, of order $p \times p$ say, can be expressed in the form

$$\mathbf{P} = \Lambda\Phi\Lambda' + \Psi, \tag{1}$$

where the elements of the $p \times k$ factor pattern matrix Λ are the factor loadings or standardized multiple regression weights of the observed variables on the factors, Φ is the $k \times k$ factor correlation matrix, and therefore must be positive definite with unit diagonal elements, and Ψ is a nonnegative definite diagonal matrix. While the factor analysis model can be expressed as a covariance structure, the standardized form in (1) is convenient for the interpretation of the relative magnitudes of elements of Λ because both the dependent (observed) variables and independent (factor) variables are standardized to comparable units of one standard deviation. We observe that the component matrices in (1) may represent either population parameters to be estimated or estimates obtained under the assumption of the model. Since certain indeterminacies that apply both to the population parameters and the estimates will be discussed here it will not be essential to distinguish between the two.

In exploratory factor analysis, the factor pattern matrix Λ and factor correlation matrix Φ in (1) are not unique when $k > 1$. The aim of rotation techniques is to select an interpretable factor pattern matrix and a corresponding factor correlation matrix from the infinite set satisfying (1).

Given an arbitrary initial factor matrix $\mathbf{A}$ of order $p \times k$ satisfying

$$\mathbf{AA}' = \mathbf{P} - \Psi, \tag{2}$$

a $k \times k$ nonsingular transformation matrix $\mathbf{T}$ is required such that the new "factor pattern" matrix $\Lambda = \mathbf{AT}$ has characteristics that are desirable for interpretability and the factor correlation matrix $\Phi = (\mathbf{T}'\mathbf{T})^{-1}$ has unit diagonal elements. It follows from (2) that (1) will be satisfied.

In principal component analysis* the matrix of standardized regression weights of the original variables on the first k principal components is sometimes rotated (see ref. 21 for an application in meteorology). While this destroys the maximum variance property (cf. COMPONENT ANALYSIS) of the individual components, the total amount of variance explained (cf. ref. 20, Secs. 3–6) by the set of k transformed components remains constant.

There are two main classes of rotation technique—orthogonal rotation, where the factor correlation matrix Φ is required to be an identity matrix so that $\mathbf{T}$ must be an orthogonal matrix, and oblique rotation, where $\mathbf{T}$ is required only to result in unit variances for the factors so that

$$\text{Diag}\left[(\mathbf{T}'\mathbf{T})^{-1}\right] = \mathbf{I}. \tag{3}$$

The elements of Λ, being standardized regression weights of the observed variables on the factors, are generally interpreted as being representative of the magnitude of the effect of the factors on the observed variables. Interpretation therefore is simplified if any observed variable has substantial loadings on few factors so that the corresponding row of Λ has several near zero elements. A factor is named according to the subset of observed variables it affects so that each column of Λ should also have several negligible elements.

Graphical rotation procedures [cf. refs. 24 (Chaps. IX–XI), 11 (pp. 248–277), 17 (pp. 68–72), and 18 (Chap. 9)] were first developed and are sometimes still advocated as a final step after a computerized analytic rotation. The disposition of negligible loadings sought in graphical rotation is that specified by Thurstone [24, p. 335] in his five criteria for a "simple structure."

The "structure" referred to by Thurstone is his "reference structure" Γ, which is related to $\mathbf{A}$ by $\Gamma = \mathbf{AM}$, where $\mathbf{M}$ is a $k \times k$ transformation matrix satisfying

$$\text{Diag}[\mathbf{M}'\mathbf{M}] = \mathbf{I}. \tag{4}$$

Since the required factor pattern Λ is obtained by rescaling the columns of the reference structure Γ, i.e.,

$$\Lambda = \Gamma \operatorname{Diag}^{1/2}\left[(\mathbf{M}'\mathbf{M})^{-1}\right],$$

the disposition of (identically) zero loadings of Λ will be the same as that of Γ (cf. refs. 11, 18, and 24). In orthogonal rotation Γ and Λ are identical.

Rotation to a reference structure was introduced by Thurstone in graphical oblique rotation and was retained for many years as intermediate step in computerized analytic rotation techniques. It is not really clear why the reference structure was ever employed, but this may have been because methods for imposing the constraints (4) were more immediately obvious than were methods for imposing (3).

Graphical methods are subjective in nature. Attention was therefore given to the development of "analytic" rotation procedures involving the maximization or minimization of a function $f(\Lambda)$ of the factor loadings. Initially, far greater progress was made in orthogonal rotation than in oblique rotation. The first function employed was the *Quartimax criterion*, $f_Q(\Lambda) = \Sigma_{i=1}^{p}\Sigma_{j=1}^{k}\lambda_{ij}^4$. Equivalent versions of this criterion were derived independently by several authors using different rationales. [See refs. 11 (Chap. 14) or 18 (Chap. 10) for historical background.] The Quartimax criterion proved only partially successful and tended to yield a "general factor" or column of Λ with no negligible loadings.

Kaiser [15] then developed the *raw Varimax* criterion

$$f_V(\Lambda) = p^{-2}\sum_{j=1}^{k}\left\{p\sum_{i=1}^{p}\lambda_{ij}^4 - \left(\sum_{i=1}^{p}\lambda_{ij}^2\right)^2\right\}. \tag{6}$$

This is the sum of within-column variances of squared loadings. Since a variance is large when observations are concentrated at extremes, maximizing the Varimax criterion will tend to yield a rotated factor matrix with both large and small loadings in each column. Because the raw Varimax criterion of (6) will be more greatly influenced by rows of Λ with large loadings than by rows with small loadings, Kaiser [15] suggested the use of the weighted normal Varimax criterion in practice. This merely involves the replacement of the λ_{ij} in (6) by $w_i\lambda_{ij}$, where the weight w_i is given by $w_i = (\Sigma_{j=1}^{k}\lambda_{ij}^2\}^{-1/2}$. This normal Varimax criterion frequently yields an acceptable solution, but can prove unsatisfactory in certain situations in which a number of observed variables are factorially complex (i.e., substantial positive or negative loadings of the observed variable on several factors are unavoidable). A different weighting system for Varimax rotation, which has given good results in a situation of this type, has been proposed by Cureton and Mulaik [6].

It is possible for rotation criteria to have more than one maximum, but this does not seem to occur frequently in practical applications (see ref. 7).

Other criteria for orthogonal rotations have been proposed. One family of criteria to be minimized, suggested by Crawford and Ferguson [5], is

$$f(\Lambda|\kappa) = (1-\kappa)\sum_{i=1}^{p}\sum_{j=1}^{k}\sum_{s\neq j}\lambda_{ij}^2\lambda_{is}^2 + \kappa\sum_{j=1}^{k}\sum_{i=1}^{p}\sum_{r\neq i}\lambda_{ij}^2\lambda_{rj}^2 \tag{7}$$

with $0 \leqslant \kappa \leqslant 1$. The first term in (7) when minimized tends to yield near zero loadings in rows of Λ, and the second, near zero loadings in columns of Λ. Minimizing $f(\Lambda|\kappa = 0)$ is equivalent to maximizing the Quartimax criterion $f_Q(\Lambda)$ and minimizing $f(\Lambda|\kappa = 1/p)$ is equivalent to maximizing the Varimax criterion $f_V(\Lambda)$ in orthogonal rotation. Taking values of κ greater than $1/p$ will sometimes, but not always, give slightly preferable solutions to Varimax. The Varimax criterion is still most widely used in practical applications.

The algorithms for orthogonal rotation that were first employed, and that are still widely used, involve a sequence of plane rotations involving two columns of the factor matrix at a time (cf. ref. 16). An alternative iterative procedure, which simultaneously provides new values of all elements of **T** on each iteration, was suggested by Horst [12, Secs. 18.4 and 18.7.2]; also see refs. 17 (Sec. 6.3) and 18 (Sec. 10.5). Another possibility [3] is to make use of a reparameterisation to ensure that the transformation matrix **T** is orthogonal. If **S** is a skew-symmetric matrix*, then $\mathbf{T} = 2(\mathbf{I} + \mathbf{S})^{-1} - \mathbf{I}$ is orthogonal and a gradient method may be employed to optimize the rotation criterion with respect to the $\frac{1}{2}k(k-1)$ elements of **S** above the main diagonal.

The methods for analytic oblique rotation that were proposed initially [cf. refs. 11 (Chap. 15) or 18 (Chap. 11)] proved unsatisfactory because they operated on the reference structure Γ instead of on the factor pattern Λ. A transformation matrix **M** to a reference structure can be singular and still satisfy (4). This happened fairly frequently in practice, resulting in linearly dependent rotated factors and, consequently, an unsatisfactory rotated factor matrix. The first really effective method for oblique analytic rotation was developed by Jennrich and Sampson [14], who found a method for rotating directly to a factor pattern. The imposition of (3) on **T** prevented the occurrence of singular transformation matrices. Successive transformations operating on pairs of columns of the factor matrix were employed to minimize the *Quartimin criterion*, defined by (7) with $\kappa = 0$. This algorithm is easily adapted to minimize other members of the family of criteria in (7). The Quartimin criterion, however, does not appear to be inferior to any other presently available criterion in oblique rotation (cf. refs. 4 and 10). An alternative algorithm [3] operates on all columns of the factor matrix simultaneously. The reparameterisation $\mathbf{T} = \mathbf{X}\,\mathrm{Diag}^{1/2}[\mathbf{X}^{-1}\mathbf{X}^{-1\prime}]$, where **X** has unit diagonals, is employed to impose (3), and a gradient method is employed to optimize the rotation criterion with respect to the $k(k-1)$ free elements of **X**.

Methods are available for rotating a factor matrix obliquely or orthogonally to a least-squares fit to a target matrix whose elements need not all be specified. A method for orthogonal rotation to a fully specified target (e.g., refs. 8 and 19) may be employed for rotating several factor matrices to maximum similarity [23]. The mean factor matrix may then be rotated orthogonally to simple structure (e.g., Varimax) and the same transformation matrix applied to each individual factor matrix without affecting the similarity criterion. In certain situations prior information is available as to the disposition of negligible loadings. Algorithms for orthogonal [1] or oblique [2; 9; 18, pp. 308–314] rotation to a partially specified target may then be employed, specifying only the positions of zero elements of the target. These methods are also useful for "cleaning up" an initial rotation, such as Varimax or direct Quartimin, by reducing the magnitudes of all loadings regarded as negligible. While some oblique target rotation procedures regard the target as a reference structure and impose (4), methods of this type can sometimes give poor results (cf. ref. 9). The use of a method that regards the target as a factor pattern and imposes (3) is to be recommended.

Methods for obtaining standard errors of factor loadings after rotation are available (see ref. 13 and references therein).

An illustration of how an oblique rotation may be employed to simplify the interpretation of a factor analysis is given in Table 1. Templer [22] conducted a study to investigate the relationship between anxiety and extroversion–introversion. Four tests purporting to measure extroversion–introversion (EI1, EI2, EI3, EI4) and three tests for anxiety (A1, A2, A3) were applied to a sample of 105 university students. A factor matrix was obtained from the correlation matrix using an unweighted least-squares procedure (*see* FACTOR ANALYSIS). This factor matrix **A** is shown in columns (a). The two sets of test have loadings of opposite sign on the first factor but no loadings are

Table 1 Rotation Example

Variable	(a) Original factor Matrix **A**		(b) Direct Quartimin Factor Pattern Λ	
	1	2	I	II
EI1	0.86	0.35	0.93	0.01
EI2	0.69	0.35	0.80	0.08
EI3	0.93	0.29	0.95	−0.07
EI4	0.86	0.28	0.88	−0.06
A1	−0.45	0.61	0.05	0.78
A2	−0.60	0.63	−0.05	0.85
A3	−0.60	0.66	−0.03	0.88
	Factor Correlation Matrix		Factor Correlation Matrix **Φ**	
	1.00	0.00	1.00	−0.32
	0.00	1.00	−0.32	1.00

negligible and interpretation of the results is difficult. Columns (b) show the result of a direct Quartimin rotation. The first factor now clearly represents introversion–extroversion and the second factor, anxiety. Examination of the factor correlation matrix **Φ** shows that introversion–extroversion and anxiety are negatively correlated.

References

[1] Browne, M. W. (1972). *Brit. J. Math. Statist. Psychol.*, **25**, 115–120.

[2] Browne, M. W. (1972). *Brit. J. Math. Statist. Psychol.*, **25**, 207–212.

[3] Browne, M. W. (1974). *Brit. J. Math. Statist. Psychol.*, **27**, 115–121.

[4] Crawford, C. B. (1975). *Brit. J. Math. Statist. Psychol.*, **28**, 201–213.

[5] Crawford, C. B. and Ferguson, G. A. (1970). *Psychometrika*, **35**, 321–332.

[6] Cureton, E. E. and Mulaik, S. A. (1975). *Psychometrika*, **40**, 183–195.

[7] Gebhardt, F. (1968). *Psychometrika*, **33**, 35–36.

[8] Green, B. F. (1952). *Psychometrika*, **17**, 429–440.

[9] Gruvaeus, G. T. (1970). *Psychometrika*, **34**, 493–505.

[10] Hakstian, A. R. and Abell, R. A. (1974). *Psychometrika*, **39**, 429–444.

[11] Harman, H. H. (1976). *Modern Factor Analysis*, 3rd ed. University of Chicago Press, Chicago. (Comprehensive review of rotation techniques in Part III.)

[12] Horst, P. (1965). *Factor Analysis of Data Matrices*. Holt, Rinehart and Winston, New York.

[13] Jennrich, R. I. and Clarkson, D. B. (1980). *Psychometrika*, **45**, 237–247.

[14] Jennrich, R. I. and Sampson, P. F. (1966). *Psychometrika*, **31**, 313–323. (Key article on oblique rotation.)

[15] Kaiser, H. F. (1958). *Psychometrika*, **23**, 187–200. (Original presentation of the most widely used method of orthogonal rotation.)

[16] Kaiser, H. F. (1959). *Educ. Psychol. Meas.*, **19**, 413–420. [Details of successive plane algorithm for orthogonal rotation (Varimax).]

[17] Lawley, D. N. and Maxwell, A. E. (1971). *Factor Analysis as a Statistical Method*. Butterworth, London.

[18] Mulaik, S. A. (1972). *The Foundations of Factor Analysis*. McGraw-Hill, New York. (Comprehensive review of rotation techniques in Chaps. 9–12.)

[19] Schönemann, P. H. (1966). *Psychometrika*, **31**, 1–10.

[20] Rao, C. R. (1964). *Sankhyā A*, **26**, 329–358.

[21] Richman, M. B. (1981). *J. Appl. Meteorology*, **20**, 1145–1159.

[22] Templer, A. J. (1971). *Psychologia Africana*, **14**, 20–31.

[23] Ten Berge, J. M. F. (1977). *Psychometrika*, **42**, 267–276.

[24] Thurstone, L. L. (1947). *Multiple Factor Analysis*. University of Chicago Press, Chicago.

(FACTOR ANALYSIS
PSYCHOLOGY, STATISTICS IN)

M. W. BROWNE

ROUGH

A term used in exploratory data analysis* to denote the deviations from the smooth*. Also, and more commonly, referred to as residuals*. The fundamental model of exploratory analysis can be given in the form: smooth + rough, or, more conventionally, (simple) mathematical formula + residual.

ROUGHNESS PENALTY *See* PENALIZED MAXIMUM LIKELIHOOD

ROUNDED DISTRIBUTIONS

If values of a continuous random variable X are rounded off (e.g., to the nearest 0.01), the resulting variable X', say, has a discrete distribution. This is sometimes called a *rounded distribution*. An alternative name is a *grouped distribution*. If the groups have boundaries

$$\ldots,(a-1)w, aw, (a+1)w, (a+2)w, \ldots$$

(the *rounding lattice*),

the rounded distribution is defined by

$$\Pr\left[X' = \left(a + h + \tfrac{1}{2}\right)w\right] = \Pr\left[(a+h)w < X < (a+h+1)w\right].$$

The ratio of w to the standard deviation of X is called the *degree of rounding*.

An early paper on this topic by Eisenhart [1] gives a clear account of the basic ideas. Detailed results for specific distributions are given by Lowell [2] (for normal) and Tricker [3, 4] (for normal, Laplace, gamma, and exponential distributions). From these investigations it appears that the degree of skewness* of a distribution is crucial in determining how its moments can be distorted by rounding.

References

[1] Eisenhart, C. (1947). Effects of rounding on grouped data. In *Selected Techniques of Statistical Analysis*, C. Eisenhart, M. W. Hastay, and W. A. Wallis, eds. McGraw-Hill, New York, pp. 185–223.

[2] Lowell, N. S. (1980). *Anal. Chem.*, **52**, 1141–1147.

[3] Tricker, A. R. (1984). *J. Appl. Statist.*, **11**, 54–87.

[4] Tricker, A. R. (1984). *Statistician, Lond.*, **33**, 381–390.

(GROUPED DATA
ROUND-OFF ERROR)

ROUNDING RULES *See* ROUND-OFF ERROR

ROUND-OFF ERROR

In itself, this is simply the error in a recorded value arising from giving the nearest value on some grid rather than an actual value. A common example is in calculations where the results are given to a certain number—m, say—of decimal places. The round-off error in such circumstances can be between $-\frac{1}{2} \times 10^{-m}$ and $+\frac{1}{2} \times 10^{-m}$. Somewhat more subtle examples arise from the necessary limitations on reading instruments.

While a single round-off error is easily allowed for, more complicated considerations arise when round-off occurs in many stages of lengthy computations, with the possibility of accumulation of such errors. This possibility is very real, especially in the use of *computer programs*. A useful, elementary discussion of some of the prinicples involved is contained in ref. 1.

Reference

[1] Wilkinson, J. H. (1963). *Rounding Errors in Algebraic Processes*. Prentice-Hall, Englewood Cliffs, NJ.

Bibliography

Berry, H. (1933). *Math. Gazette*, **17**, 112–113.

Clarke, M. R. (1984). *J. Appl. Statist.*, **11**, 12–20.

Inman, S. (1932). *Math. Gazette*, **16**, 306–315.

Larson, J. L., Pasternale, M. E., and Wisniewski, J. A. (1983). Algorithm 594, Software for relative error analysis, *ACM Trans. Math. Software*, **9**, 125–130.

(MEASUREMENT ERROR
NUMERACY)

ROUND ROBIN TOURNAMENTS *See* KNOCK-OUT TOURNAMENTS; SPORTS, STATISTICS IN

ROUTE SAMPLING

A method of selecting samples from a population covering a specified area. A "route" (or routes) (usually straight lines, though not necessarily so) is chosen at random, and all units within a specified distance of a route are included in the sample.

(AREA SAMPLING)

ROW AND COLUMN DESIGNS

Many practical experiments are conducted on material arranged in the form of a grid, often a physical two-dimensional structure such as a glasshouse bench or a field. These two dimensions may then be described as *rows* and *columns*. In the present article the discussion of row and column designs will be confined to those in which the intersection of any row and any column does not result in more than one experimental unit or plot. The same principles of experimental design* can also be used without the physical structure, for example, if there is an existing set of treatments and blocks to which a further set of treatments is to be added [11, 26]. Also, there are situations where a set of experimental treatments is applied to a number of units, not just once but on several occasions, so that the units are "rows" and the occasions "columns." Row and column designs with time as one classifying factor are used in, amongst other areas, animal experimentation, chemical engineering, and market research. An important practical consideration in all these row and column designs is that the effects of rows, columns, and treatments do not interact. If there are such interactions*, it is still possible to conduct an experiment using rows and columns and also to carry out a formal analysis of the results. However, the interpretation of the analysis is more difficult and invalid conclusions may be drawn. Despite these drawbacks, row and column designs have a place in practical experimentation: many examples of their use, in the contexts described and others, are given in standard textbooks [4, 5].

SOME SIMPLE DESIGNS

If there are the same number of treatments, rows, and columns, and each treatment occurs just once in each row and each column, then the simplest type of row and column design, a *Latin square**, is possible. These have been known for very many years, and discussed for over 200, while the first experiment using these principles appears to be that of Cretté de Palluel [6]. However, the two requirements for Latin squares, that the numbers of rows and columns shall both be equal to n, say, and further that the numbers of treatments and replicates shall also be n, are so restrictive that other row and column designs are often needed. The principle of lack of interaction between the effects of treatments, rows, and columns may still be preserved by having several Latin squares side by side. Also, two or more of the treatments in a Latin square may be the same, though possibly without equal replication of all treatments. Any block design with n treatments, replicates, blocks, and units within a block may have its units arranged in the form of an $n \times n$ Latin square. Indeed, Smith and Hartley [32] proved the more general result that a block design with t treatments and blocks and n replicates and units to a block can be arranged as an $n \times t$ row and column design. This result is of great value in considering the relation of row and column designs to block designs.

When the numbers of treatments, rows, and columns are not all equal, but it is still desired to maintain equal replication of all treatments, there are a number of possible designs. The earliest of these to be used in practice seems to be the incomplete Latin square of Yates [37], consisting of the omission of one row from a Latin square. This principle was taken further by Youden [40], who showed how to rearrange the plots of balanced incomplete block* designs with equal numbers of treatments and blocks in rows and columns. Such designs, though rectangular rather than square, have come to be known as *Youden squares**. Whereas the omission of one row of a Latin square always gives a statistically valid design, it is not in general possible to omit more than one row and obtain a design with the balance required for a Youden square. Shrikhande [31] obtained general conditions for designs with treatments having one of the three possibilities, orthogonality, total balance, and partial balance with respect to rows and another with respect to columns.

MORE GENERAL DESIGNS

A more general description of some types of row and column design was given by Hoblyn et al. [17], who introduced the following notation. If two classifications are related to each other orthogonally (O), with total balance (T) or with partial balance (P) then, if X, Y, Z are any of O, T, P, a design of type $X: YZ$ is one in which the relation of columns to rows is X, that of treatments to rows is Y, and that of treatments to columns is Z. Most row and column designs have rows and columns orthogonal, so are of type $O: YZ$. In particular, Latin squares are of type $O: OO$ and Youden squares of type $O: OT$. Pearce and Taylor [26] gave the only $O: TT$ design of reasonable size, with nine replicates of four treatments on six rows and columns. Kiefer [19] defined a class of generalised Youden designs, which are effectively the same as $O: TT$ designs, and a comprehensive listing of such designs has been given by Ash [1].

Designs of type $O: OP$ may be obtained by rearranging the units in some partially balanced* incomplete block designs in the manner indicated by Smith and Hartley, while some designs of type $O: TP$ are found by a suitable arrangement of the units of balanced incomplete block designs with unequal numbers of treatments and blocks. Many families of $O: PP$ designs were described by Freeman [10]. Designs with slightly less balance were given by Potthoff [27], who described the use of one for a trial on traffic counts. There are also some examples of designs with rows and columns not orthogonal, Hoblyn et al. giving a $T: TT$ design.

If there is a block design with n replicates of t treatments on h blocks with k units each and another block design with n replicates of t treatments on k blocks with h units each, then they can usually, but not always, be combined into a row and column design [9]. When the block designs can be so combined, they may be referred to as the component designs of the row and column design [25]. The row and column design may then be such that any loss of efficiency in one component is compensated for by greater efficiency in the other component. One particular $O: PP$ design, described by Kshirsagar [20] (and also in ref. 10), illustrates this concept very well. The design has four replicates of nine treatments on six rows and columns: each component design is of type P, but those treatments concurring three times in the same row concur twice in the same column and vice versa. The row and column design thus has total balance, so may be said to be of type $O: (T)$ [24]. Potthoff's designs, though not even partially balanced, are such that the combined row and column design has greater efficiency than either component block design.

Row and column designs with unequal replication of treatments may also be useful. In addition to O, T, and P, Hoblyn et al. [17] had another category, S, for supplemented balance, in which one treatment, possibly a control, has a different replication from all the others. The most obvious examples of designs of type $O:SS$ are Latin squares with (a) a row and column missing [37], or (b) a row and column added, the supplementing treatment occurring on the unit common to both the added row and column, or (c) a row added and a column missing [22]. Pearce [23] gave examples of designs of type $O:SS$ and $O:OS$; he also found that designs of type $O:TS$ and $O:PS$ exist, though without any practical use. Row and column designs with two groups of treatments having different replications were considered in more detail by Freeman [12], who gave examples of designs in which both groups contain more than one treatment. Again, some of these row and column designs have better properties of balance and efficiency than the component block designs. Another generalisation of supplemented balance is to multipartite designs, where there are many groups; these were called type M by Pearce [24], who gave examples of type $O:OM$ and $O:MM$.

ANALYSIS OF ROW AND COLUMN DESIGNS

As has been seen, there are many combinatorially possible row and column designs, with varying degrees of balance. Not all of these possibilities, however, have been much used for the practical design of experiments. When a row and column design has been used, the analysis of the results may, but need not, be more complex than that for a block design. For Latin and Youden squares the analyses follow the same lines as those for randomised blocks and balanced incomplete blocks, respectively. The first general method of analysis for row and column designs was given by Tocher [33], who assumed that rows and columns were orthogonal but made no assumption about the disposition of treatments with respect to rows and columns. Consider a row and column design with h rows, k columns, and N plots in all, where the incidence matrices of treatments with respect to rows and columns are $\mathbf{n}$ and $\mathbf{m}$, respectively, and the vector of replications is $\mathbf{r}$. If $\mathbf{r}^{\delta}$ is the diagonal matrix with elements those of $\mathbf{r}$ and $\mathbf{X}'$ is the transpose of $\mathbf{X}$ for any matrix $\mathbf{X}$, then the variance–covariance matrix* of the design is $\boldsymbol{\Omega}\sigma^2$, where σ^2 is the residual mean square and

$$\boldsymbol{\Omega}^{-1} = \mathbf{r}^{\delta} - (1/k)\mathbf{n}\mathbf{n}' - (1/h)\mathbf{m}'\mathbf{m} + (2/N)\mathbf{r}\mathbf{r}'.$$

If the vectors of treatment, row, and column totals are, respectively, $\mathbf{T}$, $\mathbf{R}$, and $\mathbf{C}$, and the grand total is G, while $\mathbf{1}$ is a unit vector, then let

$$\mathbf{Q} = \mathbf{T} - (1/k)\mathbf{n}\mathbf{R} - (1/h)\mathbf{m}\mathbf{C} + (G/N)\mathbf{1}.$$

The vector of estimated treatment parameters is $\mathbf{t}^* = \boldsymbol{\Omega}\mathbf{Q}$, the treatment sum of squares is $\mathbf{Q}'\mathbf{t}^* - G^2/N$, and the residual sum of squares is

$$\mathbf{y}'\mathbf{y} - \mathbf{Q}'\mathbf{t}^* - (1/k)\mathbf{R}'\mathbf{R} - (1/h)\mathbf{C}'\mathbf{C} + 2G^2/N,$$

where $\mathbf{y}$ is the vector of observations.

Tocher's method was generalised by Freeman and Jeffers [15], who showed that if rows and columns are not orthogonal, up to three matrix inversions may be necessary. Thus, instead of there being k plots in each row and h in each column, suppose that the numbers of plots in rows and columns may be represented, respectively, by the vectors $\mathbf{k}$ and $\mathbf{h}$, while $\mathbf{w}$ is the row–column incidence matrix. Other symbols are the same as before. Then, introducing new matrices $\mathbf{K}$, $\mathbf{H}$, $\mathbf{U}$, $\mathbf{V}$, where

$$\mathbf{K}^{-1} = \mathbf{k}^{\delta} + (1/N)\mathbf{k}\mathbf{k}' - \mathbf{w}\mathbf{h}^{-\delta}\mathbf{w}',$$
$$\mathbf{H}^{-1} = \mathbf{h}^{\delta} + (1/N)\mathbf{h}\mathbf{h}' - \mathbf{w}'\mathbf{k}^{-\delta}\mathbf{w},$$
$$\mathbf{U} = (\mathbf{m}\mathbf{h}^{-\delta}\mathbf{w}' - \mathbf{n})\mathbf{K},$$
$$\mathbf{V} = (\mathbf{n}\mathbf{k}^{-\delta}\mathbf{w} - \mathbf{m})\mathbf{H},$$

allows us to write

$$\boldsymbol{\Omega}^{-1} = \mathbf{r}^{\delta} + \mathbf{U}\mathbf{n}' + \mathbf{V}\mathbf{m}'.$$

$\mathbf{Q} = \mathbf{T} + \mathbf{U}\mathbf{R} + \mathbf{V}\mathbf{C}$, and $\mathbf{t}^*$ and the treatment sum of squares are as before. The residual sum of squares is now

$$\mathbf{y}'\mathbf{y} - \mathbf{Q}'\mathbf{t}^* - \mathbf{R}'\mathbf{K}(\mathbf{R} - \mathbf{w}\mathbf{h}^{-\delta}\mathbf{C}) - \mathbf{C}'\mathbf{H}(\mathbf{C} - \mathbf{w}'\mathbf{k}^{-\delta}\mathbf{R}).$$

Freeman and Jeffers gave an example of a Youden square with a missing treatment: In an experiment where the amount of copper was determined at different leaf positions on apple shoots on a number of occasions, it was not possible to carry out the planned last determination, thus destroying any orthogonality in the intended design. Very general methods of analysis for designs with many nonorthogonal classification were given by Bradu [3] and Rees [29], these methods being applicable to row and column designs. A pattern in the design simplifies the analysis and methods for analysis for particular classes of design have been described [8, 12, 22, 27, 20]. Pearce [25] showed the relation between the matrices that have to be inverted in row and column designs and in their component block designs. With modern computing methods there is no difficulty in obtaining the formal analysis by any general technique, though the pattern present in many of the more balanced designs makes the analysis easier.

OTHER ROW AND COLUMN DESIGNS

So far, the treatments in a row and column design have been regarded as having no structure, but there are situations where factorial* designs may be used with rows and columns. The first example is a combination of Latin squares and factorial designs called *quasi-Latin squares*, in which some of the interactions between the factors are confounded with rows and some with columns [38]. There may be special problems of randomisation with these designs, because comparisons between parts of a square can be associated with important effects, so that some restriction of randomisation is necessary [16]. Even without equal numbers of rows and columns, factorial experiments may be conducted using generalised cyclic designs* [18].

A type of design outside the standard pattern of row and column designs is the *lattice square* design, which is an arrangement of a balanced lattice design* to take out variation in two dimensions within a set of squares [39]. These designs compare p^2 treatments, and require $(p + 1)$ squares if p is even, but only $(p + 1)/2$ if p is odd. They have been most used in practice for field trials to compare new crop varieties.

Another nonstandard type of row and column design is the *semi-Latin square*, which is a rectangular arrangement of t treatments, where $t = nk$, in n rows and t columns. The columns are grouped into sets each containing k consecutive columns, so that each treatment occurs exactly once in each row and each set of columns. This design has long been known to be statistically defective when considered as a design with just one residual term in its analysis [36], but has still been frequently rediscovered. A closely related design is the *Trojan square** [7], which requires the existence of k mutually orthogonal $n \times n$ Latin squares, $k > 2$, and there are then $t = kn$ treatments. There are n^2 plots, arranged in n rows and n columns, and each plot is split into k subplots, which have the treatments specified by the appropriate entries in the k Latin squares. Semi-Latin and Trojan squares are both valid experimental designs if there are two residual terms in the analysis. Many semi-Latin and Trojan squares are closely related to some semiregular group divisible* incomplete block designs* [28].

Finally, row and column designs are useful for field experiments in which it is required to assess the effects of nearest neighbours. A method of analysis adjusting for nearest neighbours proposed by Papadakis [21] was not much used for many years until fresh impetus was given to this topic by Bartlett [2] and the discussants of his paper. Most of the row and column designs suitable for nearest-neighbour experiments are, as usual, in Latin squares, the particular type being *complete* Latin squares in which every ordered pair of treatments occurs equally often in both rows and columns [35]. Other types of design not necessarily using Latin squares may also be useful [13, 14]. At the time of writing there remains much interest in this whole area, as evinced by the response to Wilkinson et al. [34], who gave both a method of analysis that they claim to be superior to that of Papadakis and suggestions for row and column designs with a high degree of balance.

References

[1] Ash, A. (1981). *J. Statist. Plann. Inf.*, **5**, 1–25.

[2] Bartlett, M. S. (1978). *J. R. Statist. Soc. B*, **40**, 147–174. (The paper, with discussion, that provided the stimulus for the use of row and column designs for nearest-neighbour analysis.)

[3] Bradu, D. (1965). *Ann. Math. Statist.*, **36**, 88–97.

[4] Cochran, W. G. and Cox, G. M. (1957). *Experimental Designs*, 2nd ed. Wiley, New York. (*The* standard textbook on experimental designs.)

[5] Cox, D. R. (1958). *Planning of Experiments.* Wiley, New York. (Still the best general book on experimental design with a minimum of mathematics and a full description of underlying principles.)

[6] Cretté de Palluel (1788). *Ann. Agric.*, **14**, 133–139. [English version by Young, A. (1790).]

[7] Darby, L. A. and Gilbert, N. (1958). *Euphytica*, **7**, 183–188.

[8] Freeman, G. H. (1957). *J. R. Statist. Soc. B*, **19**, 154–162.

[9] Freeman, G. H. (1957). *J. R. Statist. Soc. B*, **19**, 163–167.

[10] Freeman, G. H. (1958). *Ann. Math. Statist.*, **29**, 1063–1078.

[11] Freeman, G. H. (1959). *Appl. Statist.*, **8**, 13–20.

[12] Freeman, G. H. (1975). *J. R. Statist. Soc. B*, **37**, 114–128.

[13] Freeman, G. H. (1979). *J. R. Statist. Soc. B*, **41**, 88–95.

[14] Freeman, G. H. (1979). *J. R. Statist. Soc. B*, **41**, 253–262.

[15] Freeman, G. H. and Jeffers, J. N. R. (1962). *J. R. Statist. Soc. B*, **24**, 435–446.

[16] Grundy, P. M. and Healy, M. J. R. (1950). *J. R. Statist. Soc. B*, **12**, 286–291.

[17] Hoblyn, T. N., Pearce, S. C., and Freeeman, G. H. (1954). *Biometrics*, **10**, 503–515.

[18] John, J. A. and Lewis, S. M. (1983). *J. R. Statist. Soc. B*, **45**, 245–251.

[19] Kiefer, J. (1975). *Ann. Statist.*, **3**, 109–118.

[20] Kshirsagar, A. M. (1957). *Calcutta Statist. Ass. Bull.*, **7**, 469–476.

[21] Papadakis, J. S. (1937). *Bull. Inst. Amel. Plantes à Salonique*, **23**. (The paper that introduced the idea of nearest-neighbour analysis.)

[22] Pearce, S. C. (1952). *J. R. Statist. Soc. B*, **14**, 101–106.

[23] Pearce, S. C. (1960). *Biometrika*, **47**, 263–271.

[24] Pearce, S. C. (1963). *J. R. Statist. Soc. A*, **126**, 353–377. (Full description of many types of nonorthogonal row and column designs.)

[25] Pearce, S. C. (1975). *Appl. Statist.*, **24**, 60–74.

[26] Pearce, S. C. and Taylor, J. (1948). *J. Agric. Sci. Camb.*, **38**, 402–410.

[27] Potthoff, R. F. (1962). *Technometrics*, **4**, 187–208.

[28] Preece, D. A. and Freeman, G. H. (1983). *J. R. Statist. Soc. B*, **45**, 267–277.

[29] Rees, D. H. (1966). *J. R. Statist. Soc. B*, **28**, 110–117.

[30] Shah, K. R. (1977). *J. Statist. Plann. Inf.*, **1**, 207–216.

[31] Shrikhande, S. S. (1951). *Ann. Math. Statist.*, **22**, 235–247.

[32] Smith, C. A. B. and Hartley, H. O. (1948). *J. R. Statist. Soc. B*, **10**, 262–263. (The paper that showed how many incomplete block designs could be made to form row and column designs with the same degree of balance.)

[33] Tocher, K. D. (1952). *J. R. Statist. Soc. B*, **14**, 45–100. (The first attempt to produce general methods of analysis for row and column designs with minimal restrictions.)

[34] Wilkinson, G. N., Eckert, S. R., Hancock, T. W., and Mayo, O. (1983). *J. R. Statist. Soc. B*, **45**, 151–211. (A major attempt to improve nearest-neighbour analysis, with a wide ranging discussion.)

[35] Williams, E. J. (1949). *Aust. J. Sci. Res. A*, **2**, 149–168. (An early paper on designs for balancing neighbour effects.)

[36] Yates, F. (1935). *J. R. Statist. Soc. Suppl.*, **2**, 181–247.

[37] Yates, F. (1936). *J. Agric. Sci. Camb.*, **26**, 301–315. (The first paper to use a nonorthogonal row and column design.)

[38] Yates, F. (1937). *Ann. Eugen.* (Lond.), **7**, 319–331. (An early paper combining the ideas of factorial and row and column designs.)

[39] Yates, F. (1940). *J. Agric. Sci. Camb.*, **30**, 672–687.

[40] Youden, W. J. (1937). *Cont. Boyce Thompson Inst.*, **9**, 41–48. (The introduction of "Youden squares.")

Bibliography

Eccleston, J. A. and McGilchrist, C. A. (1985). *J. Statist. Plann. Inf.*, **12** (305–310). (An efficiency bound for row and column designs is found.)

(DESIGN OF EXPERIMENTS

INCOMPLETE BLOCK DESIGNS
LATIN SQUARES, LATIN CUBES, LATIN RECTANGLES, ETC.
YOUDEN SQUARES)

G. H. FREEMAN

ROYAL STATISTICAL SOCIETY

The Royal Statistical Society was founded, as the Statistical Society of London, in 1834, "for the purposes of procuring, arranging and publishing Facts calculated to illustrate the Condition and Prospects of Society ... and, as far as it may be found possible, ... facts which can be stated numerically and arranged in tables." The subject, at that time, was regarded almost exclusively as dealing with affairs of the State. Its remit nowadays includes these original functions, but is extended to cover also all theoretical and practical matters connected with variability, uncertainty, probability, and sampling, in whatever subject matter. In 1887, the Society was granted a Royal Charter by Queen Victoria and was renamed Royal Statistical Society. Her Majesty Queen Elizabeth II is the present Patron of the Society.

The Society is governed by an elected council, consisting of the President, three Honorary Secretaries, the Honorary Treasurer, and 27 other members, four of whom are nominated each year by the President to be Vice Presidents. The three most recent past Presidents also serve on the Council without further election.

In addition to the main Society, there are six sections, two study groups, and 18 local groups, each with its own semiautonomous committee to run its affairs. The sections are called Business and Industrial, General Applications, Medical, Research, Statistical Computing, and Social Statistics. The study groups are subsidiary to the General Applications Section and are called Government Statistical Service and Multivariate. The local groups take their names from their geographical locations.

The Society itself has only one grade of membership, namely Fellow of the Society, but membership of sections and local groups also exists. Fellowship is open to all, aged 21 years or over, who are interested in statistical method and concerned for its useful and ethical application. A candidate must normally be proposed by two or more existing Fellows. The number of Fellows, at the end of 1984, was 4815. Honorary Fellows, limited in number to not more than 70 at any one time, are proposed by the Council, elected by the Fellows, and normally consist only of statisticians resident abroad. Corporate bodies and institutions may also join the Society as Associated Members.

The Society's main activities are found in its meetings and conferences, its journals, and its library. It also prepares comments on matters of public interest, where statistics is relevant, sometimes in response to a request for evidence from a Royal Commission or other committee of enquiry, sometimes on its own initiative when comments seem desirable.

Its main meetings are normally held in London, but one such meeting each year is held elsewhere in association with one of the local groups. About half the main meetings are organized by the Council, through its Programme Committee, the other half by the Research Section.

Additionally there are many meetings of a less formal nature, arranged and conducted by the sections, study groups, and local groups. These are not normally published in the journals, but summaries of them frequently appear in *News and Notes*. During the 1983/84 session, there were nine main meetings (of which four were organised by the Research Section), 46 meetings of the other sections and study groups, and 80 meetings of local groups (though this includes multiple counting of repetitions of a talk in various parts of the country).

Conferences have no set pattern but are organized from time to time as seems appropriate. The 1983/84 session included the 150th Anniversary Conference held at Imperial College, London University.

Publications consist of (1) *Journal of the Royal Statistical Society* Series A*, first published in 1838, four issues per year; (2) *Journal of the Royal Statistical Society Series B*, first published in 1934, three issues per year; (3) *Applied Statistics* (also known as *JRSS Series C*), first published in 1952, three issues per year; (4) *News and Notes*, first published in 1974, ten issues per year, a newsletter for items of ephemeral interest.

The library goes back to the earliest days of the Society and contains a valuable and unique collection of books and journals. Agreement has now been reached to transfer it to be housed with the Library of University College London.

For the first 141 years of its existence the Society's offices were housed in rented accommodation in various locations, but in 1975 it purchased its present offices (25 Enford Street, London, W1H 2BH).

The staff consist of the Executive Secretary, Assistant Secretary, Executive Editor, Librarian, Fellowship Assistant, Subscriptions Assistant, and Clerical Assistant.

The Society is affiliated with the International Statistical Institute*.

Bibliography

Bonar, J. and Macrosty, H. W. (1934). *Annals of the Royal Statistical Society* Royal Statistical Society, London.

Tippett, L. H. C. (1972). Annals of the Royal Statistical Society 1934–1971. *J. R. Statist. Soc. A*, **135**, 545–568.

Hill, I. D. (1984). Statistical Society of London–Royal Statistical Society. The first 100 years: 1834–1934. *J. R. Statist. Soc. A*, **147**, 130–139.

Plackett, R. L. (1984). Royal Statistical Society. The last 50 years: 1934–1984. *J. R. Statist. Soc. A*, **147**, 140–150.

Rosenbaum, S. (1984). The growth of the Royal Statistical Society. *J. R. Statist. Soc. A*, **147**, 375–388.

I. D. Hill

ROY'S CHARACTERISTIC ROOT STATISTIC

INTRODUCTION

Roy's largest or maximum characteristic root statistic has been proposed [see Roy [21] and (1957)] for the test of the same three hypotheses as stated in the Introduction of HOTELLING'S TRACE and in PILLAI'S TRACE, namely: (a) equality of covariance matrices of two p-variate normal populations; (b) equality of p-dimensional mean vectors of l p-variate normal populations having a common unknown covariance matrix, known as MANOVA (alternately general linear hypothesis*); and (c) independence between a p-set and a q-set ($p \leqslant q$) in a ($p + q$)-variate normal population.

As discussed in HOTELLING'S TRACE and PILLAI'S TRACE, tests proposed for these three hypotheses are generally invariant tests that, under the null hypotheses, depend only on the characteristic (ch.) roots of matrices based on samples. Thus Roy's largest root test is such a test and the statistic is denoted by b_s, where the ch. roots $b_1, \ldots, b_s$ of each of the three matrices defined below have the same form of the joint density given by

$$f(b_1, \ldots, b_s) =$$

$$C(s, m, n) \begin{vmatrix} b_s^{m+s-1}(1 - b_s)^n & \cdots & b_s^m(1 - b_s)^n \\ \vdots & \vdots & \vdots \\ b_1^{m+s-1}(1 - b_1)^n & \cdots & b_1^m(1 - b_1)^n \end{vmatrix},$$

$$0 < b_1 < \cdots < b_s < 1.$$

[See Pillai (1960), and [17] for $C(s, m, n)$.] The density function $f(b_1, \ldots, b_s)$ is the same as the one given in PILLAI'S TRACE except that it is now expressed as a determinant. Here s, m, and n are to be understood differently for the three hypotheses. (See the following text, as well as HOTELLING'S TRACE and PILLAI'S TRACE.)

For (a), the b_i's are the ch. roots, of $\mathbf{S}_1(\mathbf{S}_1 + \mathbf{S}_2)^{-1}$, where $\mathbf{S}_1$ and $\mathbf{S}_2$ are sums of products (SP) matrices with n_1 and n_2 degrees of freedom (d.f.); for (b), of $\mathbf{S}^*(\mathbf{S}^* + \mathbf{S})^{-1}$, where $\mathbf{S}^*$ is the between-SP matrix and $\mathbf{S}$, the within-SP matrix with $l - 1$ and $N - l$ d.f., respectively, and N is the total of l sample sizes; and for (c), of $\mathbf{S}_{11}^{-1}\mathbf{S}_{12}\mathbf{S}_{22}^{-1}\mathbf{S}_{12}'$, where $\mathbf{S}_{ij}$ ($i, j = 1, 2$) is the SP matrix of the ith set with the jth set (1 denoting p-set and 2 denoting q-set). $\mathbf{S}_{12}\mathbf{S}_{22}^{-1}\mathbf{S}_{12}'$ and $\mathbf{S}_{11} - \mathbf{S}_{12}\mathbf{S}_{22}^{-1}\mathbf{S}_{12}'$ have q and $n' - 1 - q$ d.f., respectively, where n' is the sample size. If ν_1 and ν_2 denote the d.f. in each case in the order given, then $m = \frac{1}{2}(|\nu_1 - p| - 1)$ and $n = \frac{1}{2}(\nu_2 - p - 1)$. In (a), if $p \leqslant n_1, n_2$, then $s = p$; in (b), $s = \min(p, l - 1)$; and in (c) if $p + q < n'$, then $s = p$ (*see also* HOTELLING'S TRACE and PILLAI'S TRACE). Further, if $f_i = b_i/(1 - b_i)$, $i = 1, \dots, s$, where $0 < f_1 < \cdots < f_s < \infty$ are the ch. roots defined in HOTELLING'S TRACE, f_s is the largest root statistic in this context. However, because $0 < f_s < \infty$, but $0 < b_s < 1$, tables of percentiles have been computed by various authors for b_s (see below under Null Distribution for details).

In regard to the test of hypothesis (a), under two-sided alternatives, Roy has proposed the largest–smallest roots test [see Roy (1957) and Optimum Properties below]. Further, for the test of hypothesis (d), specifying the covariance matrix of a p-variate normal population, Roy has proposed the largest root of the sample covariance matrix as the test statistic, while for the test of the same hypothesis against two-sided alternatives, he has proposed the largest–smallest roots test statistic [see Roy (1957) and Optimum Properties below].

If $g_i = nb_i$, $i = 1, \dots, s$, as $n \to \infty$ the joint density of $g_1, \dots, g_s$ is obtained [see Nanda (1948b)] from $f(b_1, \dots, b_s)$ by replacing in the determinant the element $b_i^{m+j-1}(1 - b_i)^n$ by $g_i^{m+j-1}e^{-g_i}$, $i, j = 1, \dots, s$, and changing $C(s, m, n)$ to a constant $K(s, m)$ (see Pillai [17]). $f(g_1, \dots, g_s)$ thus obtained, where $0 < g_1 < \cdots < g_s < \infty$, is the same as the joint density of $\frac{1}{2}l_1, \dots, \frac{1}{2}l_s$, where $l_1, \dots, l_s$ are the ch. roots of a sample SP matrix having a central Wishart distribution* with $s \times s$ identity covariance matrix and ν_1 d.f., $s \leqslant \nu_1$.

Four standard tests statistics are available in the literature for tests of hypotheses (a)–(c), namely, Hotelling's trace* Pillai's trace*, Wilks' lambda criterion*, and Roy's largest root statistic, all of which have several optimal properties. Such properties of the largest root include: (i) the union–intersection* character for tests of (b) and (c) [for (a), under two-sided alternatives, largest–smallest roots tests has that character], (ii) good power for tests of (a)–(c) when there is only one nonzero deviation parameter, i.e., in the linear case, and (iii) smallest confidence sets in MANOVA under certain conditions (see below under Confidence Bounds and Optimum Properties).

DISTRIBUTION

The null and nonnull distributions may be considered separately; *see also* LATENT ROOT DISTRIBUTIONS.

Null Distribution

The distribution of the largest root has been studied by various authors through four different methods:

1. Roy–Pillai reduction formulae.
2. Mehta–Krishnaiah pfaffian method.
3. Davis differential equation method.
4. Pillai–Fukotomi–Sugiyama zonal polynomial series.

ROY–PILLAI REDUCTION FORMULAE. In the determinant given in the Introduction, if $b_i^{m+j-1}(1 - b_i)^n$ is integrated with respect to b_i in the interval 0 to b_{i+1}, $i = 1, \dots, s - 1$, $j = 1, \dots, s$, and $b_s^{m+j-1}(1 - b_s)^n$ in the interval 0 to $x < 1$, $j = 1, \dots, s$, and if the resulting determinant is denoted by $V(x; m + s - 1, \dots, m; n)$, then the CDF of the largest root b_s is given by $C(x, m, n)V(s; m + s - 1, \dots, m; n)$. A reduction formula in this connection, given by Pillai (1954, 1956b), is

$$\begin{aligned}(m + n + s)&V(x; m + s - 1, \dots, m; n)\\ &= -I_0(x; m + s - 1; n + 1)\\ &\quad \times V(x; mt + s - 2, \dots, m; n)\\ &\quad + 2\sum_{j=1}^{s-1} (-1)^{s-j-1}\\ &\quad \times I(x; 2m + s + j - 2; 2n + 1)\\ &\quad \times V(x; m + s - 2, \dots, m + j,\\ &\qquad m + j - 2, \dots, m; n),\end{aligned}$$

where $I_0(x'; a, b) = x^a(1-x)^b|_0^{x'}$ and $I(x'; a; b) = \int_0^{x'} x^a(1-x)^b\,dx$. Using the above reduction formula, Pillai (1954, 1956a) has obtained explicit expressions for the CDF of b_s ($s = 2, \ldots, 8$) in terms of incomplete beta functions*. However, computation becomes extremely prohibitive for larger values of s and hence Phillai has suggested an approximation (to be stated later) of the CDF of b_s at the upper end [see Pillai (1954, 1956a) and [15]].

The CDF of the largest root f_s was first studied by Roy [[21] and (1957)] who gave explicit expressions of its CDF for $s = 2$, 3, and 4. Nanda (1948a) also presented explicit expressions of its CDF for individual values of $s = 2(1)5$ without giving any general expression.

Pillai (1954, 1956b) has also given a parallel reduction formula for the largest root f_s starting with the joint density of $f_1, \ldots, f_s$ given in HOTELLING'S TRACE.

Now the approximation at the upper end of the CDF of the largest root obtained by Phillai [15] is given below. These expressions are the only means for computing the CDF of the b_s for larger values of s.

s even: The approximation to $\Pr(b_s \leqslant x)$ for s even is given by

$$1 + \sum_{i=1}^{s-1} (-1)^i k_{m+s-i} \times I_0(x; m+s-i; n+1),$$
$$k_{m+s} = 0,$$

where

$$k_{m+s-i} = [C(s, m, n)V(1; m+s-1, \ldots$$
$$m+s-i+1, m+s-i-1, \ldots, m; n)$$
$$-(m+s-i+1)k_{m+s-i+1}]/(m+n+s-i+1),$$

and

$$C(s-1, m, n)V(1; m+s-1, \ldots$$
$$m+s-i+1, m+s-i-1, \ldots, m; n)$$
$$= \left[\binom{s-1}{i-1} \prod_{j=1}^{s-1} \frac{(2m+s-j+1)}{(2m+2n+2s-j+1)}\right].$$

s odd: The approximation to $\Pr(b_s \leqslant x)$ for s odd is given by

$$\frac{I(x; m, n)}{\beta(m+1, n+1)} + \sum_{i=1}^{s-1} (-1)^i k_{m+s-i} \times I_0(x; m+s-i, n+1).$$

The approximation neglects terms involving $(1-x)^{rn+1}$, $r > 1$, in the development of the CDF obtained by repeated application of the Roy–Pillai reduction formulae (see Pillai [15]).

Using the approximation formulae, Pillai (1954, 1956a, 1957) gave the upper 5 and 1% points of b_s for $s = 2(1)5$ for $m = 0(1)4$ and n varying from 5 to 1000. Further, Pillai (1960) published in *Statistical Tables for Tests of Multivariate Hypotheses* similar tables for $s = 2(1)6$ for $m = 0(1)4, 5, 7, 10, 15$, and values of n as before, and Pillai (1964a) gave such tables for $s = 7$. Approximations were obtained for each s separately for the computations for s up to 7. Pillai [15] obtained the general expressions for the approximation given above and computed upper 5 and 1% points for $s = 8$, 9, and 10 for arguments as above. Further, similar tables for $s = 11$ and 12 were given in Pillai (1970) and for $s = 14(2)20$ in Pillai (1967b). In addition, Pillai and Flury [19] have computed extensive tables of the largest root, giving upper 10, 5, 2.5, 1, 0.5% points for $s = 2(1)10(2)20$, $m = 20(5)40(10)70$, 85, 100, 120, 140, 170, 200, 300, and $n = 5$, 7, 9, 12, 15(5)40(10)70, 85, 100, 120, 140, 170, 200, 300. These extensive tables facilitate the test of hypothesis (a) for which both m and n could be large. Further percentage points for $s = 13(1)20$ are available in a mimeograph report in the Department of Statistics at Purdue University [Pillai (1966a), No. 72].

Nanda (1951) has tabulated upper 5 and 1% points of the largest root for $s = 2$ for $m = 0(\frac{1}{2})2$ and $n = \frac{1}{2}(\frac{1}{2})10$. Foster and Rees (1957) have given upper 20, 15, 10, 5, and 1% points of the largest root for $s = 2$, $m = -0.5, 0(1)9$, and $n = 1(1)19(5)49, 59, 79$. Foster (1957, 1958) has extended the tabulations for $s = 3$ and 4. They have used d.f. as arguments for the tabulations. Heck (1960) has obtained some charts of upper 5, 2.5, and 1% points for $s = 2(1)5$, $m = -0.5, 0(1)10$, and $n \geqslant 5$.

In the one-sample case the reduction formulae for individual roots could be derived from those for the b_i's by transforming, as before, $g_i = nb_i$ $(i = 1, \ldots, s)$, $n \to \infty$. Nanda (1948b) has obtained explicit expressions for the CDF of g_s for $s = 2(1)5$ and so also for the smallest root. However, in view of the prohibitive computation for larger values of s, Pillai and Chang (1968, 1980), and independently Hanumara and Thompson (1968), derived an approximation for the CDF of g_s at the upper end from Pillai's approximation for the CDF of b_s given above and transforming $g_s = nb_s$, $n \to \infty$. (See also Pillai [17]). Pillai and Chang (1968, 1970) have given tables of the upper 10, 5, 2.5, 1, and 0.5% of g_s for $s = 2(1)20$ and d.f. ranging from 2 to 20. Hanumara and Thompson (1968) have also given upper percentage points as above for $s = 2(1)10$ except for the upper 10%, and have given lower percentage points for the smallest root. Earlier Thompson (1962) had given upper percentage points for $s = 2$. Pearson and Hartley (1972) have published tables of upper percentage points of b_s prepared from Pillai sources and papers of Foster and Rees, and those of g_s based on Pillai and Chang (1968, 1970) and Hanumara and Thompson (1968).

MEHTA–KRISHNAIAH PFAFFIAN METHOD. Mehta (1960, 1967) has developed a method of reduction of the V-determinant for g_s involving gamma functions [obtained in the same manner from $f(g_1, \ldots, g_s)$ as the V-determinant for the CDF of b_s from $f(b_1, \ldots, b_s)$] in terms of double integrals through pfaffians (if $\mathbf{T}$ is a skew-symmetric matrix* of even order, $|\mathbf{T}|^{1/2}$ is called a *pfaffian*). The method was extensively used with extensions to the V-determinant for b_s involving beta functions by Krishnaiah and associates [Krishnaiah and Chang (1970, 1971b); Krishnaiah and Waikar (1970, 1971b); and Krishnaiah [9]]. Chang [2] has tabulated the exact distribution of b_s for $s \leqslant 5$.

DAVIS DIFFERENTIAL EQUATION METHOD. Davis (1972a, 1972b) has shown that the marginal density functions of b_i's (g_i's) satisfy a system of ordinary differential equations of Fuchsian type. The marginal density of the largest root has been obtained as a power series that is the same as that derived by Pillai (1967a) and, independently, by Fukutomi and Sugiyama (1967) (see Zonal Polynomial Series below). Pillai's approximations to the CDF of the largest root given above for even and odd s are interpreted as exact solutions, neglecting the contributions of higher-order solutions.

PILLAI–FUKUTOMI–SUGIYAMA ZONAL POLYNOMIAL SERIES. A method, though not very useful in the null case for computational purposes in view of the better methods given above, that is quite useful in the nonnull case, is the series method with terms involving zonal polynomials* [James (1961, 1964); Hua (1959), and Muirhead [11]], which are homogeneous symmetric polynomials in the ch. roots of a real symmetric matrix (Hermitian matrix in the complex case). Pillai (1967b), and, independently, Fukutomi and Sugiyama (1967) have obtained the density of b_s as an infinite series of zonal polynomials. Sugiyama (1967b), further, has given the CDF of $b_s(b_1)$ as a series in powers of $x((1-x))$. The distributions of $g_s(g_1)$ follow by the transformation $g_i = nb_i$, $n \to \infty$, given above. For the one-sample case, the density of $l_s = 2g_s$ and that of the vector corresponding to l_s were obtained by Sugiyama (1966, 1967a) using some results of Tumura (1965), which are of interest in principal component analysis*. Sugiyama has also given an approximation (1972b) to the distribution of l_s from the zonal polynomial approach. Using series of zonal polynomials, he has also computed [24] a table of 5 and 1% points of the largest root of a sample covariance matrix for d.f. $\nu = 2(2)50$, $s = 2, 3$; for $s = 4$ and $\nu = 2(2)10(10)30$. The convergence of the zonal polynomial series is so slow that more than 100 terms were needed [see also Johnson and Kotz (1972)]. Further, Krishnaiah and Chang (1971a) have derived the distribution of l_1 in a finite series of zonal polynomials when m is a nonnegative integer.

Nonnull Distribution

The nonnull distribution of the largest root has been studied by various authors, mainly using zonal polynomial series. Pillai [[16], and (1970)] has derived the exact CDF of the largest root for (b) and (c) for $s = 2$, and $s = 3$ in the linear case, and computed powers of the tests. Pillai and Jayachandran (1967) have obtained an approximation to the CDF of the largest root in the linear case for $s = 2$, 3, and 4, and further they have (1968) obtained the CDF for $s = 2$ and computed the powers. Pillai and Dotson (1969) have derived the CDF of the largest root for (b) and (c) for $s = 2$ and 3 and Pillai and Al-Ani (1970) for (a) with power tabulations in all cases. In all the above derivations, zonal polymonials up to sixth degree have been used. For hypothesis (b), Hayakawa (1967) and, independently, Khatri and Pillai (1968a) have obtained the density of b_s in series of beta functions. The density involves coefficients of the two types: $g^{\delta}_{\kappa,\tau}$ and $b^{\delta}_{\kappa,\tau}$ given by

$$C_{\kappa}(\mathbf{B})C_{\tau}(\mathbf{B}) = \sum_{\delta} g^{\delta}_{\kappa,\tau} C_{\delta}(\mathbf{B}),$$

where $\kappa = (k_1, \ldots, k_p)$, $\tau = (\tau_1, \ldots, \tau_p)$, and $\delta = (d_1, \ldots, d_p)$ such that $d = k + t$, where κ, τ, and δ are partitions of k, t, and d, respectively. Further, if $\mathbf{G}_1 = \text{diag}(\mathbf{G}, 1)$, where $\mathbf{G} = \text{diag}(g_1, \ldots, g_{p-1})$, the g_i's being the ch. roots of $\mathbf{G}$, then

$$C_{\kappa}(\mathbf{G}_1) = \sum_{t=0}^{k} \sum_{\tau(t)} b_{\kappa,\tau} C_{\tau}(\mathbf{M}).$$

Khatri and Pillai (1968a) have given tables of the $g^{\delta}_{\kappa,\tau}$ coefficients for $d \leqslant 7$ and $b_{\kappa,\tau}$ coefficients for $k \leqslant 6$. Hayakawa (1967) has tabulated, for $d \leqslant 4$, the $h^{\delta}_{\kappa,\tau}$ coefficients, which are $g_{\kappa,\tau}$ times a factor, and also $a_{\kappa,\tau}$ coefficients corresponding to the $b_{\kappa,\tau}$ coefficients of Khatri and Pillai. Further, Khatri (1967) has derived the density of b_s for (a) as a ${}_3F_2$ hypergeometric* function.

Pillai and Sugiyama (1969) have obtained the density of the CDF of b_s in simpler power series form than obtained before for (b), and also derived those of b_s for (a) and (c). DeWaal (1969a) has also studied the distribution of b_s for (c) and Troskie and Money (1972) have obtained the distribution of the smallest root for (c). Further, Waikar (1973) has derived the joint distribution of f_1 and f_s as well as g_1 and g_s. Khatri (1972) has studied the distribution of b_s as well as b_1 for all three tests in exact zonal polynomial finite series form when $n = \frac{1}{2}(\nu_2 - p - 1)$ is a nonnegative integer. Using Khatri's finite series, Venebles (1973) has presented an algorithm for the numerical evaluation of the null distribution of b_is. Krishnaiah and Chattopadhyay (1975) have considered the extensions of the pfaffian method to the noncentral case.

Hayakawa (1969) has derived the density of the largest root of the noncentral Wishart* matrix with known covariance matrix as a series in Hermite polynomials*. For the central Wishart matrix with unknown covariance matrix, Khatri (1972) has given the distribution of the largest root as well as the smallest root in finite zonal polynomial series for nonnegative integral values of m.

The nonnull distributions of Khatri for the three tests or one-sample case are generally extremely complex, although they are in finite series form and hence are of limited use. Hence Pillai and Saweris (1974b) have derived simpler forms, in finite or infinite series, based on Khatri's results, which are generally more rapidly convergent than those derived by Pillai and Sugiyama (1969). They have also obtained some approximations to these distributions.

Anderson has studied (1951, 1963) the limiting distribution of the ch. roots and vectors in the one-sample case in view of principal component analysis. If the ch. roots, $\lambda_1, \ldots, \lambda_p$, of the covariance matrix $\Sigma(p \times p)$ are different, the deviations of the ch. roots of the sample covariance matrix from the corresponding λ_i's are asymptotically independently normally distributed with variance $2\lambda_i^2/\nu$, $i = 1, \dots, p$, where $\nu \geqslant p$ is the d.f., and are also asymptotically independently distributed with the corresponding ch. vectors. Further, the asymptotic distribution of sample ch. roots and vectors also has been studied when some of the ch. roots of $\mathbf{\Sigma}$ are equal.

Muirhead [(1970b), and [11]] derived an asymptotic expansion up to order ν^{-2} for the largest root in the one-sample case based on his expansion for ${}_1F_1$ function. But it is of limited interest because it is valid only over the range of values less than the smallest ch. root of $\mathbf{\Sigma}$. Further, using Muirhead's system of partial differential equations [(1970a), and [10]] Muirhead and Chiguse (1975) have obtained an asymptotic expansion* up to order ν^{-1} of the distribution of $x_p = (\nu/2)^{1/2}(l_p/\nu\lambda_p - 1)$ and similarly that of x_1.

Waternaux (1976) [25] has studied the asymptotic distribution of the individual ch. root of a sample covariance matrix from a nonnormal population with finite fourth cumulant. Muirhead and Waternaux [12] have further obtained a similar asymptotic distribution of the individual ch. root in the canonical correlation case, which for the special case of the elliptical distribution has a surprisingly simple form.

Further, for a survey of similar distributional aspects in the complex multivariate normal* see the review paper by Krishnaiah [8].

OPTIMUM PROPERTIES

For (a), let $\gamma_1, \dots, \gamma_p$ be the ch. roots of $\mathbf{\Sigma}_1\mathbf{\Sigma}_2^{-1}$, where $\mathbf{\Sigma}_h(p \times p)$ is the covariance matrix of the hth population ($h = 1, 2$). Similarly for (b), let $\omega_1, \dots, \omega_p$ be the ch. roots of the noncentrality matrix $\mathbf{\Omega}$, and for (c), let $\rho_1^2, \dots, \rho_p^2$ be the ch. roots of $\mathbf{\Sigma}_{11}^{-1}\mathbf{\Sigma}_{12}\mathbf{\Sigma}_{22}^{-1}\mathbf{\Sigma}_{12}'$, where $\mathbf{\Sigma}_{ij}$ is the covariance matrix of the ith set with the jth ($i, j = 1, 2$). Roy and Mikhail (1961), Mikhail (1962), Anderson and Das Gupta (1964a, 1964b), Das Gupta et al. (1964), and J. N. Srivastava (1964) have studied the monotonicity of power of the largest root test with respect to each population ch. root for (b) and (c), and for (a) for one-sided alternatives: $H_1: \gamma_i \geqslant 1$, $i = 1, \dots, p$, $\sum_{i=1}^p \gamma_i > p$. Anderson and Das Gupta (1964b) have also discussed similar results for the smallest root and individual roots for (a) and for the one-sample case for the test of $\mathbf{\Sigma} = \mathbf{I}$. In addition, Roy (1957) has given the conditions for local unbiasedness of the largest-smallest roots tests for two-sided alternatives for (a), and in the one-sample case for two-sided alternatives for hypothesis (d) the covariance matrix is a specified one. Mudholkar (1965) has shown the monotonicity of power of the largest root for (b) and (c) through the study of symmetric gauge functions and Eaton and Perlman (1974) have shown such a property of the largest root test for (b) using Schur convexity. Perlman and Olkin [14] have demonstrated the unbiasedness of all invariant tests with acceptance regions monotone in the maximal invariant statistics (ch. roots) that include the largest root. (For a discussion of powers of standard multivariate tests, see Anderson [1], Morrison (1976), Muirhead [11], Srivastava and Khatri [23], Seber [22], and others.) Pillai and Jayachandran (1967, 1968) have made exact numerical power comparisons in the two-roots case for each of the three hypotheses (a), (b), and (c) for tests based on the four statistics, Hotelling's trace ($U^{(s)}$), Pillai's trace ($V^{(s)}$), Wilks' Lambda ($W^{(s)}$), and the largest root. While the first three tests compare favorably and behave somewhat in the same manner in regard to the three hypotheses, the largest root has generally lower power than the other three when the number of nonzero deviation parameters is greater than one, as is observed from Table 1. Further, Schatzoff (1966) has made a Monte Carlo* study for (b) (also for a larger number of roots), and his findings are similar. While the power comparisons for (a) above concerns one-sided alternatives, Chu and Pillai [3] have observed

Table 1 Deviations of Powers of Tests of Hypotheses (a)–(c) by $U^{(2)}$, $V^{(2)}$, and $W^{(2)}$ from that by b_2 (Ratios of Deviations to the Power of b_2 in Parentheses), $n = 30$

			$m = 0$			$m = 2$		
			$U^{(2)}$	$V^{(2)}$	$W^{(2)}$	$U^{(2)}$	$V^{(2)}$	$W^{(2)}$
(a)	γ_1	γ_2						
	1	1.5	0.0012	0.0005	0.0009	0.002	−0.001	0.001
	1.25	1.25	0.0045	0.0047	0.0047	0.014	0.015	0.014
			(0.04)	(0.04)	(0.04)	(0.10)	(0.11)	(0.10)
(b)	ω_1	ω_2						
	0	8	−0.003	−0.015	−0.013	−0.003	−0.017	−0.10
	4	4	0.072	0.082	0.075	0.055	0.062	0.061
			(0.17)	(0.19)	(0.18)	(0.20)	(0.23)	(0.22)
(c)	ρ_1^2	ρ_2^2						
	0	0.1	−0.005	−0.012	−0.010	−0.003	−0.016	−0.010
	0.05	0.05	0.060	0.068	0.062	0.048	0.053	0.050
			(0.16)	(0.18)	(0.17)	(0.19)	(0.21)	(0.20)

that for two-sided alternatives, i.e., $\Sigma_1 \neq \Sigma_2$, in the two-roots case, all four tests are biased but the largest root generally has the least bias. They have also made comparisons of the four tests under locally unbiased conditions, including in the study the likelihood ratio test* and Roy's largest-smallest roots test and its three variations (see also Pillai [17]), and have concluded that if a single test is desired for the two-sided case over the whole parameter space, Roy's largest root is the proper candidate. Since the largest root test is the least biased, even equal tail areas could be adequate.

Roy (1953, 1957) has shown the *union–intersection* character of the largest root for tests of (b) and (c) and that of largest–smallest roots for tests of (a) and (d) against two-sided alternatives. The derivation of the largest root test based on the union–intersection principle for (b) is discussed in MULTIVARIATE ANALYSIS OF VARIANCE (MANOVA) and for (c) and (d) in UNION–INTERSECTION PRINCIPLE.

For (a), using the union–intersection principle, the hypothesis $\Sigma_1 = \Sigma_2$ vs. $\Sigma_1 \neq \Sigma_2$, the acceptance region for the hypothesis is given by $b_{1\alpha_1} \leqslant b_1 < b_p \leqslant b_{p,1-\alpha_2}$, where $b_{1\alpha_1}$ and $b_{p,1-\alpha_2}$ are percentiles from the joint distribution of the smallest and the largest roots such that $\alpha_1 + \alpha_2 = \alpha$. As regards *robustness**, the largest root comes behind the other three standard test statistics, namely, Pillai's trace, Wilks' lambda, and Hotelling's trace. For the test of (b), i.e., MANOVA, the robustness aspects are of two types, namely, (i) against nonnormality and (ii) against heteroscedasticity*. A detailed discussion on robustness of the largest root is available in the MANOVA entry. In brief, the Monte Carlo study of Olson [13] recommends Pillai's trace as the most robust and that the largest root, which produces excessive rejection of H_0 under both kurtosis and heteroscedasticity, may be dropped from consideration. Earlier, Korin [7] made a Monte Carlo study of Hotelling's trace, Wilks' lambda, and the largest root for MANOVA under heteroscedasticity, with the conclusion that the largest root does exhibit somewhat greater departures from the significance level than the other two. Further, Pillai and Sudjana (1975) have carried out an exact robustness study in the two-roots case based on Pillai's distribution of ch. roots of $\mathbf{S}_1\mathbf{S}_2^{-1}$ under violations and computing the ratio $e = (p_1 - p_0)/(p_0 - \alpha)$, where p_1 is the power under violation, p_0 is the power without violation, and $\alpha = 0.05$; the order of robustness was observed to be Pillai's trace $\succ$ Wilks' lambda $\succ$ Hotelling's trace $\succ$ Roy's largest root. Davis [4, 5] has studied the effects of nonnormality on Wilks' lambda and the largest root in multivariate Edgeworth populations. For both tests, increasing kurtosis was found to lower the type I error while increasing skewness was found to raise it (see the MANOVA entry for more details). Further, Pillai and Sudjama (1975) have made exact robustness studies as above for the test of (a) and Pillai and Hsu [20] for the test of (c), both against nonnormality, and from the numerical values of e defined earlier, observed the same order of robustness as given above.

CONFIDENCE BOUNDS

Simultaneous confidence interval estimation using the extreme ch. roots has been made by Roy and Bose (1953) and Roy (1957) in connection with multivariate location and dispersion and regression problems. In the one-sample case, *confidence bounds connected with* Σ are given [see Roy (1957)] as follows:

$$l_{1\alpha_1}^{-1} l_p \geqslant \text{all ch.}(\Sigma) \geqslant l_{p,1-\alpha_2}^{-1} l_1,$$

where the statement has confidence coefficient $\geqslant 1-\alpha$, $\alpha_1+\alpha_2=\alpha$, ch.(Σ) = ch. root of Σ, and $l_{1\alpha_1}$ and $l_{p,1-\alpha_2}$ are the appropriate percentiles obtained from the joint distribution of l_1 and l_p, the smallest and largest ch. roots of the SP matrix with ν d.f.

Further, *confidence bounds on the ch. roots of* $\Sigma_1\Sigma_2^{-1}$ are given [see Roy (1957)] by $f_{1\alpha_1}^{-1} f_p \geqslant \text{all ch.}(\Sigma_1\Sigma_2^{-1}) \geqslant f_{p,1-\alpha_2}^{-1} f_1$, where the statement has confidence coefficient $\geqslant 1-\alpha$, $\alpha_1+\alpha_2=\alpha$, and $f_i = b_i/(1-b_i)$ $(i=1,p)$, where b_i's are the ch. roots for (a).

Again, a *confidence bound on the regression matrix** $\boldsymbol{\beta} = \Sigma_{12}\Sigma_{22}^{-1}$ based on $\mathbf{B} = \mathbf{S}_{12}\mathbf{S}_{22}^{-1}$ is given [see Roy (1957)] by

$$\text{all ch.}(\mathbf{B}-\boldsymbol{\beta})(\mathbf{B}-\boldsymbol{\beta})' \leqslant b_{p,1-\alpha}(1-b_1)\text{ch.}_p(\mathbf{S}_{11})\text{ch.}_1(\mathbf{S}_{22}),$$

where ch.$_i(\mathbf{A})$ denotes the ith ordered ch. root of $\mathbf{A}$ and b_1, b_p are the smallest and largest ch. roots for (c). The confidence coefficient is not less than $1-\alpha$. [See Roy (1957) for further discussion on confidence bounds.] Wijsman [26, 27] has shown that the confidence sets based on the largest root are smallest in MANOVA under certain conditions.

APPLICATION

Two examples are available in the MANOVA entry which illustrate the test procedure based on the largest root as well as the other three standard test statistics. In Example 1 there [see Ventura (1957), and Pillai (1960)] on four variables in six groups, all the four tests accepted the hypothesis of equality of mean vectors using $\alpha = 0.05$. But in Example 2 [Rao (1952), p. 263] on equality of three variables in six groups, the two traces and Wilks' lambda rejected H_0 at the 5% level, while the largest root rejected only at the 15% level. The reason for this behavior of the largest root test may be partially explained by the power and robustness aspects described above.

Further, two more examples are given below; one on (a) and the other on (c).

Example 1. Flury and Riedwyl [6] made measurements in millimeters of the following six variables on 85 forged and 100 real Swiss bank notes: length, left and right-side widths, lower- and upper-margin widths, and length of print diagonal. The data were found to be normally distributed based on tests. (See Pillai and Flury [19] for more details and the two-sample covariance matrices.)

For the test of (a), $H_0: \Sigma_1 = \Sigma_2$ vs. $H_1: \gamma_i \geqslant 1$; $i=1,\ldots,6$, $\Sigma_{i=1}^6 \gamma_i > 6$, reject H_0 if $b_6 > b_{6,0.95}$; otherwise do not reject. Here $m = \frac{1}{2}(n_1-p-1) = \frac{1}{2}(84-6-1) = 38.5$ and $n = \frac{1}{2}(n_2-p-1) = \frac{1}{2}(99-6-1) = 46$. Now $b_6 = 0.679 > 0.657 = b_{6,0.95}$ (see Pillai and Flury [19] for extended tables of the largest root, which enables one to do the test for the first time for larger values of both degrees of freedom). Hence H_0 may be rejected with a caution for further experimentation.

Example 2. Consider the largest root test of (c) concerning the independence between a set of two variates (1) systolic pressure, (2) diastolic pressure and a set of three variates (3) weight (pounds), (4) chest (inches), and (5) waist (inches) based on measurements on 60 male reserve officers of the Armed Forces of the Philippines in civilian status, belonging to the age group 29 to 31. (The data were found to be normally distributed based on earlier tests.) Here the hypothesis is $H_0 : \mathbf{\Sigma}_{12} = \mathbf{0}$ vs. $H_1 : \mathbf{\Sigma}_{12} \neq \mathbf{0}$, where

$$\mathbf{\Sigma} = \begin{matrix} 2 \\ 3 \end{matrix}\begin{pmatrix} \mathbf{\Sigma}_{11} & \mathbf{\Sigma}_{12} \\ \mathbf{\Sigma}_{12}' & \mathbf{\Sigma}_{22} \end{pmatrix} \quad \text{in } N_{2+3}(\boldsymbol{\mu}, \mathbf{\Sigma}).$$

The SP matrix of the two-set $\mathbf{S}_{11}$, of the three-set $\mathbf{S}_{22}$, and that of the two-set on the three-set $\mathbf{S}_{12}$, are given below [see Pillai (1960) and Sen (1957)]:

$$\mathbf{S}_{11} = \begin{matrix} \\ 1 \\ 2 \end{matrix}\begin{matrix} 1 & 2 \\ \begin{pmatrix} 5632 & 2824 \\ & 3017 \end{pmatrix} \end{matrix},$$

$$\mathbf{S}_{22} = \begin{matrix} \\ 3 \\ 4 \\ 5 \end{matrix}\begin{matrix} 3 & 4 & 5 \\ \begin{pmatrix} 10171 & 904 & 1291 \\ & 169 & 140 \\ & & 300 \end{pmatrix} \end{matrix},$$

and

$$\mathbf{S}_{12} = \begin{matrix} \\ 1 \\ 2 \end{matrix}\begin{matrix} 3 & 4 & 5 \\ \begin{pmatrix} 2151 & 189 & 330 \\ 1402 & 156 & 198 \end{pmatrix} \end{matrix}.$$

Now

$$\mathbf{S}_{11}^{-1}\mathbf{S}_{12}\mathbf{S}_{22}^{-1}\mathbf{S}_{12}' = \begin{pmatrix} 0.0666624 & 0.0363025 \\ 0.0372663 & 0.0340762 \end{pmatrix}$$

with latent roots, $b_1 = 0.010141$, and $b_2 = 0.090598$. Further, $b_2 < b_{2,0.95}$ as $m = \frac{1}{2}(q - p - 1) = 0$ and $n = \frac{1}{2}(n' - 1 - q - p - 1) = 26.5$. Hence do not reject H_0. Now, Pillai's trace, $V^{(2)} = \operatorname{tr} \mathbf{S}_{11}^{-1}\mathbf{S}_{12}\mathbf{S}_{22}^{-1}\mathbf{S}_{12}' = b_1 + b_2 = 0.100739 < V_{0.95}^{(2)}$. Also, Hotelling's trace, $U^{(2)} = \Sigma_{i=1}^{2} b_i/(1 - b_i) = 0.109868 < U_{0.95}^{(2)}$. Further, Wilks' lambda, $W^{(2)} = \Pi_{i=1}^{2}(1 - b_i) = 0.900180 > W_{0.05}^{(2)}$. Hence the test results agree in all cases.

INDIVIDUAL ROOTS

Roy [21] has provided an expression for the CDF of the ith root ($i = 1, \ldots, s$). However, the expression is incorrect and Pillai and Dotson (1969) have given the correct reduction formulae for the smallest, second largest, and second smallest roots. They have also noted in the null case the relation

$$\Pr(b_i \leqslant x; m, n) = 1 - \Pr(b_{s-i+1} \leqslant 1 - x; n, m)$$

and computed upper 5 and 1% points for the smallest root for $s = 2$ and 3 and the middle root for $s = 3$. Further, they have observed for tests of (b) and (c) and Pillai and Al-Ani (1970) for the tests of (a), that the power of the smallest or intermediate root can be larger in some instances than that of the largest root. In the noncentral case, Al-Ani (1970) has obtained the CDF of the second largest roots for (a), (b), and (c) in terms of zonal polynomials* and Pillai and Al-Ani (1969) the noncentral distributions of the smallest and second smallest roots for all three cases and the one-sample case.

Krishnaiah and associates have used the pfaffian method to obtain the CDF of the ith root in the null case; Krishnaiah and Chang (1980, 1981b) obtained the joint density of b_1 and b_s and Schurmann et al. (1972) obtained tables of values of x such that $\Pr(1 - x \leqslant b_1 < b_s \leqslant x) = 1 - \alpha$; Clemm et al. (1982) obtained values of u such that $\Pr(u^{-1} \leqslant l_1 < l_s \leqslant u) = 1 - \alpha$, both tables prepared in view of Roy's largest–smallest root tests, the latter in the wake of the table of Thompson (1962) for $s = 2$ [see Pearson and Hartley (1972) also in this connection.] Further, Schurmann and Waikar (1972) have also given upper percentage points for the smallest root and Clemm et al. (1972), those for individual roots l_i ($i = 1, \ldots, s - 1$) for $s = 2(1)10$. Also, Krishnaiah and Waikar (1970, 1971b) have derived the joint density of any subset of consecutive ordered roots. Davis (1972b) has obtained explicit expressions of marginal distributions of the b_i's for $s \leqslant 5$: A differential equation is applied recursively to develop the distributions for s from those obtained from $s - 1$. Results also have been obtained for the l_i's. The system of differential equations above is related to those for Hotelling's trace [Davis (1968, 1970a, 1970c)] and Pillai's trace [Davis (1970b)].

Muirhead and Chiguse (1975) have derived the marginal density of $x_i = (\nu/2)^{1/2}(l_i/\nu\lambda_i - 1)$, $i = 1,\dots, p$, from the subasymptotic expansion of Anderson (1965) in terms of normal densities, which agrees with the result of Sugiura (1973).

References

For references not listed here see Pillai [18]. Both references 17 and 18 are annotated in HOTELLING'S TRACE.

[1] Anderson, T. W. (1984). *Introduction to Multivariate Statistical Analysis*. Wiley, New York.

[2] Chang, T. C. (1974). *Ann. Inst. Statist. Math. Suppl.*, **8**, 59–66.

[3] Chu, S. S. and Pillai, K. C. S. (1979). *Ann. Inst. Statist. Math.*, **31**, 185–205.

[4] Davis, A. W. (1980). *Biometrika*, **67**, 419–427.

[5] Davis, A. W. (1982). *J. Amer. Statist. Ass.*, **77**, 896–990.

[6] Flury, B. and Riedwyl, H. (1983). *Angewandte Multivariate Statistik*. Gustav Fischer, New York.

[7] Korin, B. P. (1972). *Biometrika*, **59**, 215–216.

[8] Krishnaiah, P. R. (1976). *J. Multivariate Analy.*, **6**, 1–30.

[9] Krishnaiah, P. R. (1978). *Developments in Statistics*, Vol. 1, P. R. Krishnaiah, ed. Academic, New York, pp. 135–169.

[10] Muirhead, R. J. (1978). *Ann. Statist.*, **6**, 5–33.

[11] Muirhead, R. J. (1982). *Aspects of Multivariate Statistical Theory*. Wiley, New York.

[12] Muirhead, R. J. and Waternaux, C. M. (1980). *Biometrika*, **67**, 31–43.

[13] Olson, C. L. (1974). *J. Amer. Statist. Ass.*, **69**, 894–908.

[14] Perlman, M. D. and Olkin, I. (1980). *Ann. Statist.*, **8**, 1326–1341.

[15] Pillai, K. C. S. (1965). *Biometrika*, **52**, 405–414. [The paper gives an approximation to the CDF of the largest root at the upper end under the null hypotheses (a)–(c) using the Roy–Pillai reduction formulae. Since the exact distribution is a determinant of order s, this approximation is the only method available for computing the upper percentiles for larger values of s. Also note that $\Pr(b_1 \leqslant x; m, n) = 1 - \Pr(b_s \leqslant 1 - x; n, m)$.]

[16] Pillai, K. C. S. (1966). *Multivariate Analysis*, P. R. Krishnaiah, ed. Academic, New York, pp. 237–251.

[17] Pillai, K. C. S. (1976). *Can. J. Statist.*, **4**, 157–183.

[18] Pillai, K. C. S. (1977). *Can. J. Statist.*, **5**, 1–62.

[19] Pillai, K. C. S. and Flury, B. N. (1984). *Commun. Statist. A*, **13**, 2199–2237.

[20] Pillai, K. C. S. and Hsu, Y. S. (1979). *Ann. Inst. Statist. Math.*, **31**, 85–101.

[21] Roy, S. N. (1945). *Sankhyā*, **7**, 133–158. [In this paper Roy studies for the first time the distributions of the largest, smallest, and intermediate roots under the null hypothesis of tests (a) and (b) in terms of $f_1, \dots, f_s$ and obtains reduction formulae for the same. However, the expression for the CDF of the ith root ($i = 1, \dots, s$) is not correct, as shown by Pillai and Dotson (1969).]

[22] Seber, G. A. F. (1984). *Multivariate Observations*. Wiley, New York.

[23] Srivastava, M. S. and Khatri, C. G. (1979). *An Introduction to Multivariate Statistics*. North-Holland, New York.

[24] Sugiyama, T. (1968). *Mimeo Series No. 590*, Institute of Statistics, University of North Carolina, Chapel Hill, NC.

[25] Waternaux, C. M. (1976). *Biometrika*, **63**, 639–645.

[26] Wijsman, R. A. (1979). *Ann. Statist.*, **7**, 1003–1018.

[27] Wijsman, R. A. (1980). *Multivariate Analysis*, Vol. 5, P. R. Krishnaiah, ed. Academic, New York, pp. 483–498.

(HOTELLING'S TRACE
LAMBDA CRITERION, WILKS'
LATENT ROOT DISTRIBUTIONS
MULTIVARIATE ANALYSIS OF VARIANCE (MANOVA)
PILLAI'S TRACE
UNION–INTERSECTION PRINCIPLE
ZONAL POLYNOMIALS)

K. C. S. PILLAI

RUIN, GAMBLER'S *See* GAMBLING, STATISTICS IN

RULE OF SUM

The following relation between binomial coefficients,

$$\binom{n}{k} = \binom{n-1}{k-1} + \binom{n-1}{k},$$
$$k = 0, 1, \ldots, n-1,$$

is sometimes called the "rule of sum." It is the basis for Pascal's triangle*.

RULE SPACE

The rule space model was developed (Tatsuoka [4]) to solve a specific classification problem that arose in the context of computerized educational testing. The entities to be classified are called "rules of operation" or simply "rules" for short. A rule is a description of a set of procedures or operations that one can use in solving a problem in some well-defined procedural domain such as arithmetic, algebra, and the like. For a sufficiently delimited subdomain (e.g., the addition and subtraction of positive and negative integers, the addition and subtraction of common fractions) there will be one, or a few, correct rules and any number of incorrect rules that a student may use in solving a set of problems on a test. A correct rule, by definition, will yield the right answer to every one of the problems. But an incorrect rule may fortuitously lead to the right answer for some subset of the problems. Each particular incorrect rule, if consistently applied to a given set of problems, will yield the right answers to some specific subset of the problems.

To illustrate this in the simple context of the subtraction of positive and negative integers (or "signed numbers"), consider a test comprising the 10 items listed in the first column of Table 1—without the answers, of course. The table also displays four blocks of results that accrue from using four incorrect rules in solving the 10 problems. These rules, labeled i, ii, iii, and iv are described at the bottom of the table. The first column of each block shows the answer that results from using the indicated rule for each problem. The second column, headed by x_j, indicates whether the answer in the first column is right ($x_j = 1$) or wrong ($x_j = 0$).

Note here that Rules i and ii both yield the correct answers in two of the 10 problems, but not the same two. Similarly, consistent use of either Rule iii or iv gives correct answers to eight of the 10 problems (again, not the same eight). It is clear, then, by comparing the natures of the pairs of rules that yield identical total scores (Rules i and ii on the one hand and Rules iii and iv on the other), that the total score (number right) is a poor measure of the ability or achievement level of a pupil in the subtraction of signed numbers. Pupils may earn the same test score for different reasons. In order to get a better understanding of what sorts of misconception lead pupils to perform as they do on a test in earning a given score, we need to know not just how many items they got correct, but which ones.

On the other hand, if we were literally to seek the information just suggested, we would be inundated with much more than we could handle. For instance, there are $10!/(5!)^2 = 252$ ways in which an examinee could get five out of 10 times right. In general, on an n-item test there are $\binom{n}{m}$ different response patterns that would give an examinee a total of m points. Thus, we need some index that will give us more information than just the total score, but which is not quite so overwhelming as trying to make interpretations based on which of $\binom{n}{m}$ possible response patterns an examinee has. This is precisely what the index ζ, which is one of the bases of rule space, is designed to do.

To define ζ and explain how it works, we need to go into some background material. Denote the response pattern of the ith examinee on an n-item test by the binary vector

$$\mathbf{x}_i = [x_{i1}, x_{i2}, \ldots, x_{in}]', \quad i = 1, 2, \ldots, N,$$

where $x_{ij} = 1$ if examinee i gets the jth item right and 0 otherwise. Let Θ be a random variable representing the hypothetical ability, or the "latent trait," that determines

Table 1 The Binary Response Vectors for a Set of Ten Items Responded to by Four Incorrect Rules

	Rule i		Rule ii		Rule iii		Rule iv	
Items	Response	x_j^*	Response	x_j	Response	x_j	Response	x_j
$-3-(-7)=+4$	-4	0	$+4$	1	$+4$	1	$+4$	1
$-2-8=-10$	$+6$	0	$+6$	0	$+6$	0	-10	1
$5-(-12)=+17$	-7	0	$+7$	0	$+17$	1	-7	0
$-11-+8=-19$	-3	0	$+3$	0	-19	1	-19	1
$9-4=+5$	$+5$	1	$+5$	1	$+5$	1	$+5$	1
$-15-(-9)=-6$	-6	1	$+6$	0	-6	1	-6	1
$-13-5=-18$	-8	0	$+8$	0	-8	0	-18	1
$8-(-6)=+14$	$+2$	0	$+2$	0	$+14$	1	$+2$	0
$-5-+11=-16$	$+6$	0	$+6$	0	-16	1	-16	1
$1-10=-9$	$+9$	0	$+9$	0	-9	1	-9	1

Rule i: The student subtracts the smaller absolute values from the larger absolute value and takes the sign of the number with the larger absolute value in his/her answer.
Rule ii: The two numbers are always subtracted as seen in Rule i but the + sign is always taken in the answers.
Rule iii: The student converts $-2-8$ and $-13-5$ into $-2+8$ and $-13+5$, respectively, but the other eight items are converted to addition correctly. Then the right addition rule is used to answer them.
Rule iv: The student has a strange idea about the parentheses. Converts operation sign $-$ to $+$ first. Then he/she follows the rule: if the signs of the two numbers are minus, then change the sign of the second number to a +; if the signs of the two numbers are not alike, then the sign of the second number becomes a minus.

*x_j is the score for the jth item in the binary response vector **x**.

(stochastically) the quality of an examinee's performance on the test. The distribution of Θ is not crucial for our purpose here, but it is often assumed to be $N(\mu, \sigma^2)$. Realizations of Θ associated with particular individuals i (although unobservable) are denoted θ_i.

The unobservable θ_i and observable x_{ij} are linked by one form or another of latent trait theory or item response theory (IRT), which postulates how the probability of a person's getting an item correct is related to his/her θ and one or more parameters characterizing each item. Among the most commonly used forms of IRT is the two-parameter logistic model:

$$\Pr\left[x_{ij}=1|\Theta=\theta_i\right] = P_j(\theta_i) = \frac{1}{1+\exp\left[-1.7a_j(\theta_i-b_j)\right]}, \quad (1)$$

where a_j and b_j are item parameters representing discriminating power and difficulty, respectively. For discussions of IRT see Lord [2], Lord and Novick [3], and PSYCHOLOGICAL TESTING THEORY. Here we only present some results that are needed for our purpose. On observing a suitably large sample of response vectors $\mathbf{x}_i$ and invoking the assumption of local independence (which asserts that, in a subpopulation of examinees with fixed θ, the events of getting any two items right are probabilistically independent), we can obtain maximum likelihood estimates of θ, a_j, and b_j. The MLE of θ_i is determined by setting its sufficient statistic*,

$$\theta_i^* = \sum_{j=1}^{n} a_j x_{ij},$$

equal to an expression in which x_{ij} is replaced by $P_j(\theta_i)$, which—as noted below—is its expected value. That is, θ_i is obtained by solving

$$\sum_{j=1}^{n} a_j x_{ij} = \sum_{j=1}^{n} a_j P_j(\theta_i). \quad (2)$$

The equations for the MLEs of a_j and b_j are complicated, and are not stated here.

The important thing to note is that, because the elements x_{ij} of any response vector $\mathbf{x}_i$ take only the values 1 and 0 with probabilities $P_j(\theta_i)$ and $1 - P(\theta_i)$, respectively, the expectation of x_{ij} for any fixed θ_i is $P_j(\theta_i)$. Because we are not concerned with any particular examinee but with the entire subpopulation of examinees with some fixed θ, we may drop the subscript i and write $\mathrm{E}(x_j|\theta) = P_j(\theta)$ or, for the whole response vector $\mathbf{x}$,

$$E(\mathbf{x}|\theta) = [P_1(\theta), P_2(\theta), \ldots, P_n(\theta)]' = \mathbf{P}(\theta). \quad (3)$$

Next, denote the mean of the $P_j(\theta)$ over the n items by $T(\theta)$,

$$T(\theta) = (1/n) \sum_{j=1}^{n} P_j(\theta),$$

and form an n-dimensional vector with $T(\theta)$ as its repeated elements,

$$\mathbf{T}(\theta) = [T(\theta), T(\theta), \ldots, T(\theta)]'. \quad (4)$$

[In practice we would replace θ by $\hat{\theta}$ from eq. (2), but we write θ to simplify the notation and also so that some subsequent statements will be strictly true that are only approximately true for $\hat{\theta}$.]

Then the inner product

$$f(\mathbf{x}; \theta) = [\mathbf{P}(\theta) - \mathbf{x}]'[\mathbf{P}(\theta) - \mathbf{T}(\theta)] \quad (5)$$

of the two residuals is, for any fixed θ, a measure of the *atypicality* of a given $\mathbf{x}$ among the sample from which the a_j and b_j were estimated. Briefly, this can be seen by rewriting (5) as

$$f(\mathbf{x}; \theta) = \sum_{j=1}^{n} P_j(\theta)[P_j(\theta) - T(\theta)] - \sum_{j=1}^{n} \{x_j[P_j(\theta) - T(\theta)]\} \quad (6)$$

and assuming without loss of generality that the items have been numbered in ascending order of difficulty so that for each θ,

$$P_1(\theta) \geqslant P_2(\theta) \geqslant \cdots \geqslant P_n(\theta).$$

(A separate ordering may be needed for different θ's.) Then, since $T(\theta)$ is the mean of the $P_j(\theta)$, there must be some k such that

$$P_j(\theta) - T(\theta) > 0 \quad \text{for all } j \leqslant k,$$

$$P_j(\theta) - T(\theta) < 0 \quad \text{for all } j > k.$$

This, plus the fact that the first sum in (6) is constant for any fixed θ, tells us that for any response pattern that has a *small* number of $x_j = 1$ for $j \leqslant k$ (i.e., the easier items) and a *large* number of $x_j = 1$ for $j > k$ (the harder items) the quantity $f(\mathbf{x}|\theta)$ will have a larger numerical value than for those response patterns in which the opposite is true. But such a response pattern (with a small number of corrects among the easier items and a large number among the harder items) is clearly "atypical"; it signals an anomaly.

Thus, the quantity $f(\mathbf{x}; \theta)$ defined in (5) serves as an index that flags unusual response patterns among those of examinees with a fixed θ—unusual, that is, for the group in which the item parameters were calibrated. (The unusualness may be the result of some peculiar misunderstanding, or the response pattern may be that of a pupil who had been taught the subject matter by a different method and who was transferred into the group just before the test was given, etc.) But what about comparisons across pupils with different θ's? For this we need only standardize $f(\mathbf{x}; \theta)$ by dividing it by its (estimated) standard deviation over the entire population of pupils with no restriction on θ. This, finally, is the index ζ, whose description was one of the objectives of this article,

$$\zeta = f(\mathbf{x}; \theta)[\mathrm{var} f(\mathbf{x}; \theta)]^{-1/2} \quad (7)$$

with $f(\mathbf{x}; \theta)$ as defined by (5), and whose variance we need not go into here; it is given, along with other properties of ζ, in Tatsuoka [5] and Tatsuoka and Linn [7].

Going back to (5), we see that $f(\mathbf{x}; \theta)$ is a linear mapping function of each response vector $\mathbf{x}$ into a real number (6). This, together with the θ that was already associated with each $\mathbf{x}$ through (2), generates a two-dimensional Cartesian product space with bases θ and ζ, and each response pattern $\mathbf{x}$ is now mapped into a point (θ, ζ). To make it explicit that each response pattern was generated by some rule R, we may use R as a subscript for the coordinates of the point and write (θ_R, ζ_R). The space comprising all such points is the rule space.

So far, we have not said anything about classification, which was the stated purpose of the rule space model. The reason is that, up to this point, we have assumed that when a pupil uses some rule R (correct or incorrect), he/she uses it consistently throughout the test, and hence the pupil (or his/her performance) is represented by a "pure" rule point (θ_R, ζ_R) in rule space. Clearly, however, this assumption of consistency is unrealistic. There are bound to be some lapses or "slips" on one or more problems, reflecting random errors (see Brown and VanLehn [1]). This introduces a "fuzz factor" or a perturbation. A pupil will almost never be represented by a pure rule point (θ_R, ζ_R), but by a point somewhat removed from one—close to it if only one or a few slips were made; far from it if many slips were committed.

It was shown (Tatsuoka [6] and Tatsuoka and Tatsuoka [8]) that these perturbed points cluster around the pure rule points from which they are perturbations. Moreover, under plausible assumptions, each cluster was shown approximately to follow a bivariate normal distribution $N[(\theta_R, \zeta_R), \Sigma]$. Here Σ is a diagonal matrix since $\sigma(\theta, \zeta) \to 0$, as shown in Tatsuoka [6]; $\sigma_\theta^2 = 1/I(\theta)$, the reciprocal of the Fisher information* function, as shown by Lord [2]; and $\sigma_\zeta^2 = 1$ by definition (7).

Thus the rule space model becomes a tool for solving a special kind of classification problem: Given a response pattern (or vector) $\mathbf{x}$, we want to determine the rule point to which it is the closest. Otherwise stated, we can, with the help of this model translate the question, "What misconception, leading to what incorrect rule of operation did this pupil most likely have?", into a statistical classification problem, "From which of the bivariate normal populations $N[(\theta_{R_k}, \zeta_{R_k}), \Sigma]$ $(k = 1, 2, \ldots, K)$ was this pupil's observed point (θ, ζ) most likely drawn?" The original question is an important one in educational diagnostics; the derived question is a well-known one in statistical classification* and pattern recognition theory. Implications are discussed in Tatsuoka and Tatsuoka [8].

References

[1] Brown, J. S. and VanLehn, K. (1980). *Cognitive Science*, **2**, 155–192.

[2] Lord, F. M. (1980). *Application of Item Response Theory to Practical Testing Problems*. Lawrence Erlbaum, Hillsdale, NJ.

[3] Lord, F. M. and Novick, M. R. (1968). *Statistical Theories of Mental Test Scores*. Addison-Wesley, Reading, MA.

[4] Tatsuoka, K. K. (1983). *J. Educ. Meas.*, **20**, 345–354.

[5] Tatsuoka, K. K. (1984). *Psychometrika*, **49**, 95–110.

[6] Tatsuoka, K. K. (1985). *J. Educ. Stat.*, **10**, 55–73.

[7] Tatsuoka, K. K. and Linn, R. L. (1983). *Applied Psychol. Meas.*, **7**, 81–96.

[8] Tatsuoka, K. K. and Tatsuoka, M. M. (1985). Bug Distribution and Pattern Classification. *Tech. Rep. No. 85-3*, ONR Contract No. N00014-82-K-0604, Computer-based Education Research Laboratory, University of Illinois at Urbana-Champaign, Urbana, IL.

(CLASSIFICATION
EDUCATIONAL STATISTICS
PSYCHOLOGICAL TESTING THEORY)

MAURICE M. TATSUOKA
KIKUMI K. TATSUOKA

RUN

A *run* (or experimental run) consists of one performance of an experimental setup with all the predictor (input) variables held (as far as is practically possible) at some prechosen levels. One or more response variables may be measured. (Note that another meaning is described in RUNS.)

Repeat runs occur when two or more separate runs are made at exactly the same selected set of predictor variable conditions. Note that genuine repeat runs must be subjected to all the possible sorts of random error that occur naturally, so that they reflect the appropriate experimental error. Thus, for example, prolonging a run to take additional response readings generally will *not* provide a genuine repeat run and may well give a false appearance of precision in the conduct of the experiment. Usually only *measurement* error* will be observed in such circumstances. Genuine repeat runs are used to estimate the full natural variation (only part of which is measurement error) in the responses observed. Because repeat runs should differ only in this natural variation, they are said to determine *pure error*.

Many writers do not define "run" specifically, but leave the reader to pick up the meaning in context. For example, Daniel [2, p. 2] writes:

> The research worker is often able to see the results of one run or trial before making another. He may guess that he can improve his yield, say, by a slight increase in temperature, by a considerable increase in pressure, by using a little more emulsifier, *and* adding a little more catalyst. He will act on all four guesses at once in his next run.

A formal definition given by Box et al. [1, pp. 24–25] reads as follows:

> We shall say that an experimental *run* has been performed when an apparatus has been set up and allowed to function under a specific set of experimental conditions. For example, in a chemical experiment a run might be made by bringing together in a reactor specific amounts of the chemical reactants, adjusting temperature and pressure to the desired levels, and allowing the reaction to proceed for a particular time. In a psychological experiment a run might consist of subjecting a human subject to controlled stress.

John [3, p. 2] makes a useful but not universally agreed distinction when he writes that

> ... the common terminology of experimental design comes from agriculture, where they sow varieties in or apply treatments to plots and often have the plots arranged in blocks. We shall use this terminology except when we shall talk about points or runs. The terms *point* and *run* will be used interchangeably; we make a run on the plant at a predetermined set of conditions, and the result will be an observation or a data point (pedants would say datum point).

(The pedants are correct!)

A selected set of runs is called an *experimental design** or an *experiment*. John [3, p. 3] points out that, in agriculture, "the emphasis has to be on designing relatively complete experiments; if anything is omitted it will have to wait until next year," whereas in industrial work "small experiments carried out in sequence, with continual feedback of information" are typically more sensible.

Broadly speaking, run is an industrial term. The statistical literature contains many words used as synonyms for run, however. Some of these are: blend, factor combination, formulation, mixture, point, recipe, test, treatment, treatment combination, and trial.

References

[1] Box, G. E. P., Hunter, W. G., and Hunter, J. S. (1978). *Statistics for Experimenters: An Introduction to Design, Data Analysis, and Model Building*. Wiley, New York.

[2] Daniel, C. (1976). *Applications of Statistics to Industrial Experimentation*. Wiley, New York.

[3] John, P. W. M. (1971). *Statistical Design and Analysis of Experiments*. Macmillan, New York.

Acknowledgments

Helpful comments from J. A. Cornell, C. Daniel, J. S. Hunter, W. G. Hunter, P. W. M. John, R. H. Myers, R. D. Snee, and the Editors, are gratefully acknowledged.

(DESIGN OF EXPERIMENTS
QUALITY CONTROL, STATISTICAL)

NORMAN R. DRAPER

RUN LENGTHS, TESTS OF *See* SUPPLEMENT

RUNNING MEDIAN

This statistic is computed in a similar manner to repeated summation formulas. Initially (first-order) medians of each successive m_1 observations are calculated; then "second-order" medians, which are medians of m_2 successive first-order medians, and so on.

As an example consider the sequence 5, 7, 4, 3, 5, 5, 6. In the table below, at the left, the calculation of the smoothed values $\tilde{u}_t = \frac{1}{9}[3]^2 u_t$ is set out; at the right, the corresponding running medians u_t^* are calculated. (Here $m_1 = m_2 = 3$.)

u_t	$[3]u_t$	$[3]^2u_t$	$\tilde{u}_t = \frac{1}{9}[3]^2u_t$	Medians: First order	Medians: Second order (u_t^*)
5					
7	16			5	
4	14	42	$4\frac{2}{9}$	4	4
3	12	39	$4\frac{1}{9}$	4	4
5	13	41	$4\frac{5}{9}$	5	5
5	16			5	
6					

Running medians give less weight to outlying observations than do summation formulas. For example if the last value of u_t were 10 instead of 6, the second-order medians would be unchanged, but the last $\tilde{u}_t$ would be 5 instead of $4\frac{5}{9}$.

Bibliography

Mosteller, F. and Tukey, J. W. (1977). *Data Analysis and Regression*. Addison-Wesley, Reading, MA.

(EXPLORATORY DATA ANALYSIS
GRADUATION
SUMMATION $[n]$)

RUNS

One of the nontechnical meanings of the noun "run" is "an uninterrupted sequence," and that is essentially its meaning in probability and statistics: a *run* of a certain type of element is an uninterrupted sequence of such elements bordered at each end by other types of elements or by the beginning or end of the complete sequence of all types of elements. As an example, the sequence 00011010111 starts with a run of three 0's, then there is a run of two 1's, a run of one 0, etc., a total of six runs. (Note that another meaning is described in RUN.)

Runs have been of interest in probability theory from the very beginning. Todhunter [12] describes the computation by de Moivre* (1667–1754) of the probability of a run of r heads in n tosses of a coin (not necessarily a fair coin). As a historical curiosity, K. Marbe [6, 7] used observations on runs to support his theory that if a coin comes up heads very often, the probability of getting a tail on the next toss increases. In this article we confine attention to the use of runs in statistical analyses.

THE WALD–WOLFOWITZ TWO-SAMPLE TEST

An early application of runs in statistics is the two-sample test of Wald and Wolfowitz [13], which is so easy to describe and so intuitively appealing that it appears in some elementary textbooks. Suppose $X_1, \ldots, X_m, Y_1, \ldots, Y_n$ are mutually independent scalar random variables to be observed. $X_1, \ldots, X_m$ have common cumulative distribution function (CDF) $F(x)$, and $Y_1, \ldots, Y_n$ have joint CDF $G(y)$. $F(x), G(y)$ are known to be continuous, but are otherwise unknown. The null hypothesis (H_0, say) to be tested is that $F(x) = G(x)$ for all x. Order all $m + n$ observations into a sequence $\{S_1, S_2, \ldots, S_{m+n}\}$ with $S_1 < S_2 < \cdots < S_{m+n}$. (Since the random variables are continuous, we ignore the possibility of ties.) Replace each S_j that was originally an X by 0, and each S_j that was originally a Y by 1. Let $\{Z_1, Z_2, \ldots, Z_{m+n}\}$ denote the resulting sequence of 0's and 1's, and let R denote the total number of runs in this sequence. If H_0 is true, we would expect the X's and Y's to be mixed up with each other, and therefore R to be relatively high. The Wald–Wolfowitz two-sample test rejects H_0 if $R < c(m, n, \alpha)$, where $c(m, n, \alpha)$ is chosen to give the desired level of significance α. (Since R is a discrete random variable, we cannot always achieve a desired α exactly, unless we are willing to use randomization*.)

To compute $c(m, n, \alpha)$, the probability distribution of R under H_0 must be found. This is done very easily. The total number of different possible sequences $\{Z_1, Z_2, \dots, Z_{m+n}\}$ is $\binom{m+n}{m}$, the number of ways to choose m positions in which to place 0's out of the $m + n$ positions available. Under H_0, each of these sequences is equally likely. The number of sequences for which R equals any given integer r can be found by elementary counting methods, as in Feller [4]. Using these formulas, Swed and Eisenhart [11] have tabulated the CDF for R under H_0 for all m, n with $m \leqslant 20$ and $n \leqslant 20$, thus enabling us to find $c(m, n, \alpha)$ for all such m, n.

Let $\mu(m, n)$ and $\sigma^2(m, n)$ denote, respectively, the mean of R and the variance of R when H_0 is true. $\mu(m, n)$ can be found as follows. $R = 1 + \sum_{j=2}^{m+n}(Z_j - Z_{j-1})^2$; $(Z_j - Z_{j-1})^2$ can be either 0 or 1, so under H_0,

$$\begin{aligned} E\{(Z_j - Z_{j-1})^2\} &= P[(Z_j - Z_{j-1})^2 = 1] \\ &= P[Z_j = 0 \cap Z_{j-1} = 1] \\ &\quad + P[Z_j = 1 \cap Z_{j-1} = 0] \\ &= 2\left(\frac{mn}{(m+n)(m+n-1)}\right), \end{aligned}$$

and thus

$$\begin{aligned} E\{R\} &= 1 + (m+n-1) \\ &\quad \times \frac{2mn}{(m+n)(m+n-1)} \\ &= 1 + \frac{2mn}{m+n}. \end{aligned}$$

Also

$$\sigma^2(m, n) = \frac{2mn(2mn - m - n)}{(m+n)^2(m+n-1)}.$$

Under H_0, $[R - \mu(m, n)]/[\sigma(m, n)]$ converges in distribution to a standard normal random variable as m and n both increase, and thus if m and n are large, $c(m, n, \alpha)$ is approximately $\mu(m, n) + \sigma(m, n)\Phi^{-1}(\alpha)$, where $\Phi^{-1}(\alpha)$ is the αth quantile* of the standard normal distribution [13]. The approximation is quite good if m, n are both at least 20. Wald and Wolfowitz also showed that the two-sample run test is consistent. That is, for fixed level of significance α $(0 < \alpha < 1)$ and fixed $F(x)$, $G(x)$ with $F(x) \neq G(x)$ for at least one value of x, the probability that the test rejects H_0 approaches 1 as m and n both increase.

A more demanding way to examine the large-sample properties of a two-sample test is to let $F(x)$ and $G(x)$ depend on m and n, and so be written as $F_{m,n}(x)$, $G_{m,n}(x)$, and examine the limiting probability of rejection of H_0 under a sequence of alternatives to H_0 such that

$$\lim_{m,n\to\infty} \max_x |F_{m,n}(x) - G_{m,n}(x)| = 0.$$

Under such alternatives, Weiss [15] constructed a test with a higher limiting probability of rejecting H_0 than the test based on R. This test is given as follows. Let $Y_1' < Y_2' < \cdots < Y_n'$ denote the ordered values of $Y_1, Y_2, \dots, Y_n$. Let $T(1)$ denote the number of X's that are less than Y_1', $T(j)$ the number of X's that fall between Y_{j-1}' and Y_j', for $j = 2, \dots, n$, and $T(n+1)$ the number of X's that are greater than Y_n'. (Most two-sample tests that have been proposed can be expressed in terms of $T(1), T(2), \dots, T(n+1)$. For example, the number of runs R can be expressed as follows: let T^* denote the number of nonzero values among $T(1), T(2), \dots, T(n+1)$, and then $R = 2T^* + 1$ if $T(1) + T(n+1) = 0$, $R = 2T^* - 1$ if $T(1)T(n+1) > 0$, and $R = 2T^*$ in all other cases.) Fix a value δ slightly greater than $\frac{3}{4}$, and let $\bar{n}_n$ denote the largest integer not greater than n^δ. Let $k(n)$ denote the largest integer such that $k(n)\bar{n}_n < n$. Define $\bar{T}(j)$ as $\sum_{i=1}^{\bar{n}_n} T((j-1)\bar{n}_n + i)$, for $j = 1, \ldots, k(n)$, and $\bar{T}(k(n)+1)$ as $m - \sum_{j=1}^{k(n)} \bar{T}(j)$. The test that rejects H_0 when

$$\sum_{j=1}^{k(n)+1} \left[\bar{T}(j) + \bar{n}_n - 1\right]^{-1} > \bar{c}_{m,n}(\alpha),$$

where $\bar{c}_{m,n}(\alpha)$ is a nonrandom value chosen to give level of significance α, has higher large-sample power than the test based on R.

A MULTIVARIATE GENERALIZATION

Suppose $X_1, \ldots, X_m, Y_1, \ldots, Y_n$ are mutually independent p-dimensional random variables to be observed. $X_1, \ldots, X_m$ have common unknown p-variate CDF F, and $Y_1, \ldots, Y_n$ have common unknown p-variate CDF G. The problem is to test the null hypothesis H_0 that F and G are identical. Friedman and Rafsky [5] have constructed a test that is a generalization of the Wald–Wolfowitz runs test, and reduces to it when $p = 1$. Construct the edge-weighted graph in p dimensions consisting of the $m + n$ points $X_1, \ldots, X_m, Y_1, \ldots, Y_n$ as nodes, with edges linking all $\binom{m+n}{2}$ pairs of nodes, and with weight assigned to an edge equal to the Euclidean distance between the nodes defining that edge (*see* GRAPH THEORY). The minimum spanning tree of this graph (*see* DENDRITES) is the subgraph of minimum total edge weight that provides a path between every pair of nodes. Remove all edges of the minimum spanning tree for which the defining nodes originate from different samples (that is, one node was an X and the other a Y). The test statistic R is the number of disjoint subtrees that result from this removal of edges. The hypothesis H_0 is rejected if R is too small. When $p = 1$, R is the number of runs in the Wald–Wolfowitz two-sample test. For general p, the properties of the test based on R are closely analogous to the properties of the Wald–Wolfowitz test.

TESTS OF RANDOMNESS

Suppose $(X_1, \ldots, X_n)$ is a vector of random variables to be observed, with a joint CDF $F(x_1, \ldots, x_n)$, which is known to be continuous but is otherwise unknown. The null hypothesis H_0 to be tested is that $X_1, \ldots, X_n$ are independent and identically distributed: that is, that $F(x_1, \ldots, x_n) = \prod_{i=1}^{n} G(x_i)$ for some unspecified continuous univariate CDF $G(x)$.

Since $X_1, \ldots, X_n$ are continuous, the probability of ties among the X's is zero. Let S denote the sequence of signs of the differences $X_2 - X_1, X_3 - X_2, \ldots, X_n - X_{n-1}$. A run of plus signs is called a *run up*, a run of minus signs is called a *run down*. Various tests of H_0 can be based on the numbers and lengths of runs up and down in S. The critical region* depends on the alternatives of interest. If the alternatives imply a trend in the sequence $(X_1, \ldots, X_n)$, we reject H_0 if there are too few runs up and down. If the alternatives imply many short cycles in the sequence $(X_1, \ldots, X_n)$, we reject H_0 if there are too many runs up and down (*see* RANDOMNESS, TESTS FOR).

QUALITY CONTROL*

Suppose a production line is turning out a product with some crucial measurable characteristic X. In any physical process, X will be a random variable. Often there are specification limits; a unit of output with a value of X that does not fall between the lower and upper specification limits is less valuable, or perhaps completely useless. The production process is "in control" if it is turning out a satisfactorily large proportion of units with values of X between the specification limits. Processes are often monitored as follows. Let $X_1, X_2, \ldots, X_n$ denote the measurements of n successive units of output. Deming [3] suggests that a run up of length above 6, or a run down of length above 6, suggests that a process is drifting out of control and should be checked.

Another type of monitoring is based on runs above and below the median, which are constructed as follows. Let M_n denote the median of the set of values $(X_1, \ldots, X_n)$. Replace each X_i that is above M_n by the symbol A, and each X_i that is below the median by the symbol B. A run of A's is called a run above the median, and a run of B's is called a run below the median. Deming suggests that a run of either type of length above 6 indicates that a process may be drifting out of control, and should be checked.

TESTS OF FIT

Suppose $X_1, \ldots, X_n$ are independent and identically distributed continuous random variables with unknown CDF $F(x)$, and the problem is to test the null hypothesis H_0 that $F(x)$ is $F_0(x)$, where $F_0(x)$ is a completely specified continuous CDF. Weiss [16] defines $k(n)$ functions $W_1, \ldots, W_{k(n)}$ of $X_1, \ldots, X_n$, such that $\lim_{n \to \infty} k(n) = \infty$, and as n increases these $k(n)$ quantities can be considered mutually independent, normally distributed, with variances 1 and means 0 if and only if H_0 is true. Weiss suggests replacing each W_i by its sign, and rejecting H_0 if there are too few runs of plus and minus signs.

An auxiliary test of fit has also been based on runs. Suppose $X_1, \ldots, X_k$ have a joint multinomial distribution*, with $X_1 + \cdots + X_k = n$, and the cell probabilities $p_1, \ldots, p_k$ unknown. The problem is to test the null hypothesis H_0 that $p_i = p_i^0$, for $i = 1, \ldots, k$, where the $\{p_i^0\}$ are specified values. The familiar chi-square test is based on the quantity $S = \sum_{i=1}^{k} (X_i - np_i^0)^2/(np_i^0)$. David [2] suggested a modification. In the sequence $(X_1 - np_1^0, X_2 - np_2^0, \ldots, X_k - np_k^0)$, replace each negative value by a minus sign and each positive value by a plus sign, ignore zero values, and denote by R the total number of runs of plus and minus signs in the resulting sequence. David suggests using both S and R to construct a critical region. This is made easier because David showed that as n increases, S and R become independent random variables.

Runs have also been used to check the adequacy of fitted models. Stigler [10] reports a remarkably early example of this use by DeForest in 1876. DeForest suggested analyzing the adequacy of a curve fitted to data by studying the runs in the sequence of signs of the residuals (a residual is the observed value minus the fitted value). Nowadays this type of checking is sometimes called *diagnostics*: for a recent illustration see Mullet [9].

RUNS OF MORE THAN TWO KINDS

In all the cases discussed, there were only two types of elements: $(0, 1)$ or $(+, -)$ or (A, B). For the theory and applications of runs of more than two kinds of elements see Barton and David [1], Mood [8], and Wallis and Roberts [14]. One important application discussed in Ref. 1 is to the several-sample problem: testing whether several samples are from the same population (a generalization of the two-sample problem). An application discussed in [14] is to testing whether several types of elements may be considered to be randomly arranged.

References

[1] Barton, D. E. and David, F. N. (1957). *Biometrika*, **44**, 168–178.

[2] David, F. N. (1947). *Biometrika*, **34**, 299–310.

[3] Deming, W. E. (1972). Making Things Right. *Statistics: A Guide to the Unknown*, J. Tanur et al., eds. Holden-Day, San Francisco, CA.

[4] Feller, W. (1968). *Probability Theory and Its Applications*, 3rd ed., Vol. 1. Wiley, New York.

[5] Friedman, J. H. and Rafsky, L. C. (1979). *Ann. Statist.*, **7**, 697–717.

[6] Marbe, K. (1916). *Die Gleichförmigkeit in der Welt*. C. H. Beck, Munich, W. Germany.

[7] Marbe, K. (1934). *Grundfragen der Angewandten Wahrscheinlichkeitsrechnung*. C. H. Beck, Munich, W. Germany.

[8] Mood, A. M. (1940). *Ann. Math. Statist.*, **11**, 367–392.

[9] Mullet, G. M. (1977). *J. Quality Tech.*, **9**, 1–5.

[10] Stigler, S. M. (1978). *Ann. Statist.*, **6**, 239–265.

[11] Swed, F. S. and Eisenhart, C. (1943). *Ann. Math. Statist.*, **14**, 66–87.

[12] Todhunter, I. (1865). *History of the Mathematical Theory of Probability*. Macmillan, London.

[13] Wald, A. and Wolfowitz, J. (1940). *Ann. Math. Statist.*, **11**, 147–162.

[14] Wallis, W. A. and Roberts, H. V. (1956). *Statistics: A New Approach*. The Free Press, Glencoe, IL.

[15] Weiss, L. (1976). *Commun. Statist. A*, **5**, 1275–1285.

[16] Weiss, L. (1977). Asymptotic Properties of Bayes Tests of Nonparametric Hypotheses. *Statistical Decision Theory and Related Topics*, Vol. II, S. S. Gupta and D. S. Moore, eds. Academic Press, New York.

Bibliography

David, F. N. and Barton, D. E. (1962). *Combinatorial Chance*. Hafner, New York. (Contains much material on the probability distributions associated with runs.)

Guenther, W. C. (1978). *Amer. Statist.*, **32**, 71–73. (Shows how to use tables of the hypergeometric distribution to construct critical values for the Wald–Wolfowitz test.)

Levene, H. (1952). *Ann. Math. Statist.*, **23**, 34–56. (A basic paper on tests of randomness.)

Shaughnessy, P. W. (1981). *J. Amer. Statist. Ass.*, **76**, 732–736. (Contains tables of critical values for tests based on runs of more than two kinds of elements.)

(QUALITY CONTROL, STATISTICAL
RANDOMNESS, TESTS OF
RUN
RUNS TEST, MULTIDIMENSIONAL
TIME SERIES)

L. WEISS

RUNS-UP-AND-DOWN TEST *See* RANDOMNESS, TESTS FOR

RUNS TEST, MULTIDIMENSIONAL

The runs* test proposed by Wald and Wolfowitz [9] is easy to apply as long as there is only one variable per observation, but if there is more than one, an ordering scheme must be determined. Friedman and Rafsky [3] were the first to explicitly generalize the runs test to test the null hypothesis of homogeneity of samples from two multidimensional populations, suggesting that the data be partially ordered by using orthogonal minimum spanning trees (MSTs) to link observations (*see* DENDRITES). They noted that, although any method or arranging links can be used to analyze data, this method will link observations that are close together or "similar" if the weight assigned to each possible link is related to Euclidean distance.

Friedman and Rafsky then defined a test statistic S to be the number of links connecting observations sampled from different populations. H_0 is rejected if S is "too small." The expected value and variance of S under H_0 can be expressed as

$$E(S) = 2Ln_1n_2/n^{(2)}$$

and

$$V(S) = 2(L + C)n_1n_2/n^{(2)} + 4\{L(L-1) - 2C\}n_1^{(2)}n_2^{(2)}/n^{(4)} - \{E(S)\}^2,$$

where

n_j = number of observations sampled from population j,

$n_j^{(k)} = n_j!/(n_j - k)!$,

$n = n_1 + n_2$,

$n^{(k)} = n!/(n-k)!$,

$L = \Sigma L_i$ = number of links, $i = 1, \dots, n$,

$C = \frac{1}{2}\Sigma L_i(L_i - 1)$ = number of pairs of links with one observation in common,

L_i = number of observations linked to observation i.

Regardless of the arrangement of links, the expected value of $S + 1$ is equal to the expected value of the Wald–Wolfowitz runs statistic, while the variance of S is equal to the variance of this statistic if $L = n - 1$ and $C = n - 2$. Friedman and Rafsky indicated that this occurs when unidimensional observations are linked using one MST,

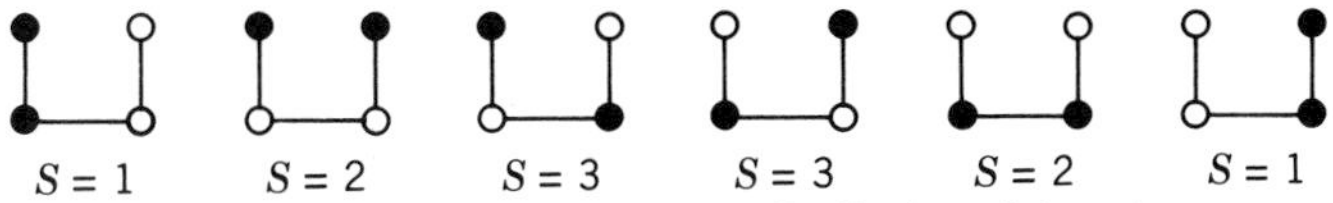

S = 1 S = 2 S = 3 S = 3 S = 2 S = 1

Figure 1 Computing a distribution of S.

which is equivalent to a rank ordering of the data.

Other investigators have independently developed statistics that are equivalent to S. These include the BW statistic of Cliff and Ord [1, 2] and a modification of the multiresponse permutation procedure* (MRPP) statistic of Mielke et al. [5, 6, 7]. Whaley [10] discussed the relationship between these statistics in more detail.

Cliff and Ord [1, p. 41] and Whaley and Quade [12] used computer simulation to investigate the fit of the normal distribution to the distribution of the multidimensional runs statistic under H_0. Cliff and Ord found that a slight adjustment to the normal distribution provided a reasonably good fit. Whaley and Quade found that the adequacy of the fit was questionable.

DISTRIBUTION PROPERTIES

The distribution of the multidimensional runs statistic under H_0 depends on the number and pattern of links, so to find the exact distribution of S, the value of the test statistic must be computed for all $\binom{n}{n_1}$ permutations of the data. Figure 1 shows that S equals 1, 2, or 3 with probability $\frac{1}{3}$ for a U-shaped arrangement of links when $n_1 = n_2 = 2$. In general, finding the exact distribution of S is rather unwieldy, even for small samples, and quite impractical for large samples.

Wald and Wolfowitz showed that the runs statistic is asymptotically normal, but Friedman and Rafsky, Cliff and Ord, and Mielke et al. reached different conclusions about the asymptotic normality* of the multidimensional runs statistic. In particular, see O'Reilly and Mielke [8]. Friedman and Rafsky [3, 4] claimed that S is asymptotically normal except for arrangements where L_i varies greatly from observation to observation. Cliff and Ord [1, p. 36] claimed that BW is asymptotically normal for lattice-shaped and some irregular link arrangements. Using the arguments of Mielke [5, 6], the modification of the MRPP statistic that is equivalent to S or BW is asymptotically nonnormal. Whaley [11] also looked into this problem, but his results were inconclusive.

POWER OF THE TEST

Friedman and Rafsky [3] and Whaley and Quade also used computer simulation to investigate the power* of the multidimensional runs test to detect differences between samples from two multidimensional normal populations. Friedman and Rafsky found that for detecting location differences, the power increased as the number of dimensions and MSTs increased, and could exceed the power of a test using normal theory. However, for detecting scale differences, no clear pattern emerged. Instead of using MSTs to link data, Whaley and Quade linked all pairs of observations within a specified Euclidean distance or "tolerance." They found that the power to detect location differences depended on the direction of the shift, and that the tolerance should usually be set so that about 70–90% of all possible pairs of observations from the same sample are linked. They also found that the power compared favorably with the power of Hotelling's T^2* under certain conditions.

It is also possible to link observations using other intuitively reasonable methods. A "city block" method of measuring distance could be used to set a tolerance, i.e., defining the distance between two observations as the sum of the distance in each dimension. Another approach is to set a tolerance for each variable and consider two observations linked if they are within each of the tolerances. Neither of these methods have been explored.

References

[1] Cliff, A. D. and Ord, J. K. (1973). *Spatial Autocorrelation*. Pion, London (Chaps. 1 and 2 primarily).

[2] Cliff, A. D. and Ord, J. K. (1981). *Spatial Processes: Models and Applications*. Pion, London. (An update and expansion of ref. 1. Some relevant formulas are presented in a less straightforward manner, however.)

[3] Friedman, J. H. and Rafsky, L. C. (1979). *Ann. Statist.*, **7**, 697–717. (The best place to start reading on this subject. Defines minimum spanning trees.)

[4] Friedman, J. H. and Rafsky, L. C. (1983). *Ann. Statist.*, **11**, 377–391, Secs. 7 and 9.

[5] Mielke, P. W. (1978). *Biometrics*, **34**, 277–282.

[6] Mielke, P. W. (1979). *Commun. Statist. A*, **8**, 1541–1550. Errata: **10**, 1795; **11**, 847.

[7] Mielke, P. W., Berry, K. J., and Johnson, E. S. (1976). *Commun. Statist. A*, **5**, 1409–1424. (The notation in refs. 5, 6, and 7 is rather intimidating.)

[8] O'Reilly, F. J. and Mielke, P. W. (1980). *Commun. Statist. A*, **9**, 629–637.

[9] Wald, A. and Wolfowitz, J. (1940). *Ann. Math. Statist.*, **15**, 147–162.

[10] Whaley, F. S. (1983). *Biometrics*, **39**, 741–745.

[11] Whaley, F. S. (1983). Some Properties of the Two-Sample Multidimensional Runs Statistic. Ph.D. dissertation, University of North Carolina, Chapel Hill, NC. [Everything you always wanted to know (and more) about the multidimensional runs statistic.]

[12] Whaley, F. S. and Quade, D. (1985). *Commun. Statist.-Comput. Simul.*, **14**, 1–11.

(CLIFF–ORD TEST
HOTELLING'S T^2
MULTIRESPONSE PERMUTATION PROCEDURES
POWER
RUNS)

FREDRICK S. WHALEY

RV-COEFFICIENT

Let $\mathbf{X}' = (\mathbf{x}_1, \dots, \mathbf{x}_n)$ and $\mathbf{Y}' = (\mathbf{y}_1, \dots, \mathbf{y}_n)$ be two d-dimensional configurations of n points (the $\mathbf{x}$'s and $\mathbf{y}$'s are $d \times 1$ vectors). The *RV-coefficient* is defined by Escoufier and Robert [1, 2] as

$$\mathrm{RV}[\tilde{\mathbf{X}}, \tilde{\mathbf{Y}}] = \frac{\mathrm{tr}[\tilde{\mathbf{Y}}'\tilde{\mathbf{X}}\tilde{\mathbf{X}}'\tilde{\mathbf{Y}}]}{\left\{\mathrm{tr}\left[(\tilde{\mathbf{X}}\tilde{\mathbf{X}}')^2\right] \mathrm{tr}\left[(\tilde{\mathbf{Y}}\tilde{\mathbf{Y}}')^2\right]\right\}^{1/2}},$$

where $\tilde{\mathbf{X}} = (\mathbf{x}_1 - \bar{\mathbf{x}}, \dots, \mathbf{x}_n - \bar{\mathbf{x}})$; $\tilde{\mathbf{Y}} = (\mathbf{y}_1 - \bar{\mathbf{y}}, \dots, \mathbf{y}_n - \bar{\mathbf{y}})$ with $\bar{\mathbf{x}} = n^{-1}\sum_{i=1}^n \mathbf{x}_i$; $\bar{\mathbf{y}} = n^{-1}\sum_{i=1}^n \mathbf{y}_i$.

Among other applications, this coefficient is used in an algorithm for selecting the "best" set of predicting variables in multiple linear regression.

References

[1] Escoufier, Y. and Robert, P. (1980). In *Optimization Methods in Statistics*, J. S. Rustagi, ed. Academic, New York, pp. 205–219.

[2] Robert, P. and Escoufier, Y. (1976). *Appl. Statist.*, **25**, 257–265.

(ASSOCIATION, MEASURES OF
CORRELATION
MEASURES OF AGREEMENT)

RYAN'S MULTIPLE COMPARISONS PROCEDURE

In 1960, Ryan [4] introduced a stepwise multiple comparisons* procedure for comparing several different populations. His method may be used for comparing means of several populations, or comparing medians, proportions, variances, correlation coefficients, or any other parameter. Each comparison of a pair of populations involves a significance test. In Ryan's procedure, the tests are performed in a stepwise fashion using "adjusted significance levels" [4]. Consequently, the probability of one or more false rejections among the entire collection of significance tests, known as the experimentwise type I error rate, is controlled at a predetermined level. See Ryan [4] for illustrations of multiple comparisons of proportions and variances.

To focus attention on Ryan's method of multiple comparisons of means, suppose that $\overline{Y}_1, \dots, \overline{Y}_k$ denotes k independent sample means, where $\overline{Y}_i$, $i = 1, \dots, k$, is based on n independent observations from a normal population $N(\mu_i, \sigma^2)$. Fix the experimentwise type I error rate at level α; let s^2 denote the pooled estimate of σ^2 with ν degrees of freedom (d.f.); let $t_{\nu,\alpha}$ denote the $\alpha 100$th percentile of Student's t-distribution* with ν d.f.; and set $\alpha_p = 2\alpha/\{k(p-1)\}$ and $W_p = t_{\nu,\alpha_p}\sqrt{2s^2/n}$ for $p = 2, \dots, k$. After arranging the sample means in descending order, begin testing with the entire set of k means and proceed stepwise, through successively smaller sets of adjacent means, to sets of adjacent pairs of sample means. At each step, a two-sample t-test is performed: the difference between the largest and the smallest of the p adjacent sample means is compared with W_p. The "adjusted significance level" α_p of the test indicates that the two means being compared are the extremes of a group of p means. If the difference is nonsignificant, i.e., smaller than W_p, the corresponding p population means are declared equal and there is no further testing among these means. Otherwise the difference is declared statistically significant and testing proceeds to the next step. When all remaining differences are declared nonsignificant, testing stops.

Einot and Gabriel [1] formulate Ryan's method of multiple comparisons of means based on range tests. The difference between the largest and smallest of the p adjacent sample means is relabeled as the sample range; and α_p and W_p are redefined as $\alpha_p = \alpha p/k$ and $W_p = q_{p,\nu,1-\alpha_p}\sqrt{s^2/n}$, where $q_{p,\nu,1-\alpha_p}$ denotes the $(1-\alpha_p)100$th percentile of the distribution of the Studentized range* of p means with ν d.f.

To clarify Ryan's method using range tests, consider the data used in MULTIPLE RANGE AND ASSOCIATED TEST PROCEDURES to illustrate several other multiple comparison procedures. The $k = 6$ sample means (each based on $n = 10$ observations) in descending order are

i	$\bar{y}_i$
1	11.11
2	10.76
3	10.68
4	9.76
5	7.94
6	6.25

The pooled estimate of variance is $s^2 = 7.6793$ with $\nu = 6 \times 9 = 54$ d.f. Using tables of the Studentized range* distribution found in Harter [2] and $\alpha = 0.05$, we have

p	α_p	$q_{p,54,1-\alpha_p}$	W_p
2	0.0167	3.55	3.11
3	0.0250	3.81	3.34
4	0.0333	4.01	3.51
5	0.0417	4.12	3.61
6	0.0500	4.18	3.67

The means are grouped into (1, 2, 3, 4), (2, 3, 4, 5), and (5, 6), with five significant pairs (15; 16; 26; 36; 46).

When the sample sizes from each population are unequal, Ryan's procedure may be implemented using F tests* instead of range tests. Suppose the sample mean $\overline{Y}_i$ ($i = 1, \dots, k$) is based on n_i independent observations. Replace the range of p ($p = 2, \dots, k$) adjacent sample means by the sum of squares of those sample means, namely $\sum_{i \in P} n_i \overline{Y}_i^2 - ((\sum_{i \in P} n_i \overline{Y}_i)^2 / \sum_{i \in P} n_i)$, where P denotes the set of indices of the p adjacent sample means. Redefine W_p by

$$W_p = (p-1)F_{p-1,\nu,1-\alpha_p}s^2,$$

where $F_{k,\nu,1-\alpha_p}$ denotes the $(1-\alpha_p)100$th percentile of the F distribution with k and ν d.f. (See ref. 1.)

For either range or sum of squares tests, Einot and Gabriel [1] modify Ryan's method by using the slightly less stringent adjusted significance levels $\alpha_p = 1 - (1 - \alpha)^{p/k}$ for $p = 2, \ldots, k$. They show analytically that, with this modification, Ryan's method is more powerful than either the Tukey–Scheffé simultaneous test procedures or Duncan's procedure using "protection levels"

$$\alpha_p^D = 1 - (1 - \alpha)^{(p-1)/(k-1)} \text{ for } p = 2, \ldots, k.$$

On the basis of Monte Carlo studies, Einot and Gabriel conclude that the power advantages are small and recommend use of Tukey's honestly significant difference* method for its "elegant simplicity." Welsch [5] modifies Ryan's adjusted significance levels by using $\alpha_{k-1} = \alpha$. The modifications of both Einot and Gabriel [1] and Welsch [5] may be combined to slightly increase the power of Ryan's procedure, while still controlling the experimentwise type I error rate at level α.

Ramsey [3] introduced a "model-testing" method for applying F tests to composite hypotheses concerning means of several different populations. Ramsey's Monte Carlo* studies for $k = 4$ and 6 means show that the model-testing method has greater power than Ryan's method. However, Ramsey's model-testing method is extremely cumbersome for even a moderately large number of means (e.g., $k = 8$ or 10).

References

[1] Einot, I. and Gabriel, K. R. (1975). *J. Amer. Statist. Ass.*, **70**, 574–583.

[2] Harter, H. L. (1960). *Ann. Math. Statist.*, **31**, 1122–1147.

[3] Ramsey, P. H. (1981). *Psychol. Bull.*, **90**, 352–366.

[4] Ryan, T. A. (1960). *Psychol. Bull.*, **57**, 318–328.

[5] Welsch, R. E. (1977). *J. Amer. Statist. Ass.*, **72**, 566–575.

(MULTIPLE COMPARISONS
MULTIPLE RANGE AND ASSOCIATED TEST PROCEDURES
NEWMAN–KEULS PROCEDURE
SCHEFFÉ'S SIMULTANEOUS TEST PROCEDURE
STUDENTIZED RANGE TESTS)

JANET BEGUN

S

SADDLE POINT APPROXIMATIONS

The saddle point method, or steepest descent method, is a technique for asymptotic evaluation of integrals of the form

$$I_n = \int f(t) e^{-ng(t)}\, dt \qquad (1)$$

in the limit of large n [10]. Integrals of this form frequently appear in probability theory, the most common source of such integrals being the inversion formula for the distribution of a sum of identically distributed random variables, in which case $f(t) = (2\pi)^{-1/2}\exp(-itx)$ and $g(t) = -\ln C(t)$, where $C(t)$ is the characteristic function*. For this case the saddle point approximation (SPA) differs from the central limit result in that, when the result is normalized so that the integrated density is equal to unity, the lowest-order correction term is $O(n^{-3/2})$, in contrast to the central limit theorem's $O(n^{-1/2})$ [6]. Another advantage of the SPA suggested by several examples is that it often leads to a roughly uniform error, in contrast to the normal approximation in which errors generally increase in the tail of the distribution. A disadvantage of central limit results is that in lowest order, it can only mimic a unimodal distribution. The SP method can, in lowest order, furnish a nonunimodal approximation when appropriate [21].

The SPA was first used in the context of probability by Cramér [4], but its general power and scope were first suggested to the statistical community in an elegant paper by Daniels [5], which contains in outline most of the major advantages of the method. Since Daniels' original paper many applications have appeared in the literature. On the theoretical side, Good discussed multidimensional generalizations [14, 15] and Barndorff-Nielsen and Cox presented in some detail the theory of SP expansions [2]. These are expansions in a series of polynomials orthogonal with respect to the lowest-order SPA and can be regarded as an analog of Edgeworth expansions. Lugannani and Rice developed a systematic approach to the calculation of correction terms in the univariate case [17].

The mathematical idea behind the asymptotic evaluation of I_n in (1) is to take the integral in the complex plane and to deform the contour so that it passes through the saddle point, i.e., that value t_0 for which $g'(t) = 0$. When such a root exists and $g(t_0) \neq 0$, it can be shown that I_n has the asymptotic expansion*

$$I_n \sim \left\{ \frac{2\pi}{n|g''(t_0)|} \right\}^{1/2} \times f(t_0) e^{-ng(t_0)} \left[1 + \sum_{j=1}^{\infty} \frac{a_j}{n^j} \right], \qquad (2)$$

where the a_j's are constants calculable in terms of $f(t)$, $g(t)$, and their derivatives. More complicated cases can occur, for example, with multiple saddle points, but these have so far not arisen in statistical applications.

In his original paper, Daniels calculated an approximation for the probability density function for the mean of n independent identically distributed random variables [5]. Let $C(t) = E\{\exp(tx)\} = \exp(K(t))$ be the characteristic function and $K(t)$ be the cumulant generating function* for an individual random variable. Then, provided that a probability density $h(x)$ exists for the underlying random variables, the density for the sample mean x can be approximated in terms of the quantities

$$\lambda_j = K^{(j)}(t_0)/|K''(t_0)|^{j/2} \tag{3}$$

by

$$\rho_n(x) = \left\{\frac{n}{2\pi|K''(t_0)|}\right\}^{1/2} e^{-n[K(t_0)-t_0 x]} \times\left\{1 + n^{-1}\left[\tfrac{1}{8}\lambda_4 - \tfrac{5}{24}\lambda_3^2\right] + O(n^{-2})\right\}. \tag{4}$$

The normalized approximation

$$\tilde{\rho}_n(x) = \rho_n(x)\Big/\int_{-\infty}^{\infty}\rho_n(u)\,du \tag{5}$$

is more accurate. If only the leading term in (4) is used then (5) has a relative error that is $O(n^{-3/2})$ [6]. As an example suppose that

$$h(x) = \frac{b^\alpha}{\Gamma(\alpha)}x^{\alpha-1}e^{-bx}, \qquad x > 0. \tag{6}$$

The associated cumulant generating function is

$$K(t) = -\alpha\ln(1 - t/b), \tag{7}$$

for which $t_0 = b - (\alpha/x)$, and $K''(t_0) = x^2/\alpha$. Thus the lowest-order SPA is

$$\rho_n(x) = \left(\frac{n\alpha}{2\pi}\right)^{1/2}\left(\frac{b}{\alpha}\right)^{n\alpha} \times x^{n\alpha-1}\exp(n\alpha - nbx), \tag{8}$$

which differs from the exact result by a replacement of factorials by their corresponding Stirling approximation. However, if one uses the lowest-order properly normalized approximation indicated in (4), the exact result is obtained. Daniels has shown that the normalized SPA leads to an exact result for the normal, gamma, and inverse normal distributions and for no others [7].

The systematic development of an asymptotic series by Lugannani and Rice is for the tail probabilities of a univariate function. If $y_n = v_1 + v_2 + \cdots + v_n$, their series is for $Q_n(y) = \Pr\{y_n > Y\}$. As before, t_0 will denote the root of $K'(t_0) = y/n$. Let $\lambda(y)$ be the quantity

$$\lambda(y) = t_0 y - nK(t_0). \tag{9}$$

The Lugannani–Rice approximation for $Q_n(y)$ can then be expressed as

$$Q_n(y) \sim 1 - \phi\left(\sqrt{2\lambda(y)}\right) + \sum_{n=0}^{\infty}(A_n - B_n), \tag{10}$$

where $1 - \phi(x)$ is the complementary error function

$$1 - \phi(v) = \left(1/\sqrt{2\pi}\right)\int_v^{\infty}\exp(-x^2/2)\,dx, \tag{11}$$

the A_n can be determined recursively, and the B_n are known explicitly. If we define the quantities

$$\mu = \left(t|K''(t_0)|^{1/2}\right)^{-1}, \qquad \theta_n = \lambda_n/n!, \tag{12}$$

where λ_n is defined in (3), the first two of the A_n required for (10) are

$$A_0 = \mu(2\pi n)^{-1/2}\exp[-\lambda(y)],$$
$$A_1 = -3A_0 n^{-1} \times\left(\mu^2/3 + \mu\theta_3 + \left(\tfrac{1}{2}\right)\left(5\theta_3^2 - 2\theta_4\right)\right) \tag{13}$$

and the B_n are

$$B_n = \frac{1}{(4\pi\lambda(y))^{1/2}} \times\frac{(-1)^n}{\lambda^n(y)}\cdot\frac{(2n)!}{4^n n!}\exp\{-\lambda(y)\}. \tag{14}$$

The authors also give an error analysis as well as a comparison of results of the SPA with several competing approximation methods [19], showing that the latter are not nearly as accurate. Daniels [9] has also compared results of the Lugannani–Rice series to approximations developed by Field and Hampel [13] for *M*-estimators*, and by Robinson [20] for permutation tests, finding that in general the Lugannani–Rice series gives the most accurate results.

Although the results just given are phrased in terms of continuous random variables, they are valid as well for discrete variables, as pointed out by Daniels and later, in more detail, by Good [14]. Good considered both the univariate and multivariate cases. Specifically, if $h(z)$ is the generating function

$$h(z) = \sum_{n=-\infty}^{\infty} h_n z^n \qquad (15)$$

and the coefficient of z^n in $h^r(z)$ is $h_n(r)$, then the saddle point z_0 is the root of

$$z\frac{d \ln h(z)}{dz} = \frac{n}{r}. \qquad (16)$$

The first two terms in the SPA to $h_n(r)$ are

$$h_n(r) = \frac{[h(z_0)]^r}{\alpha z_0^n \sqrt{2\pi r}} \times \left\{1 + \frac{1}{24r}(3\lambda_4 - 5\lambda_3^2) + O\left(\frac{1}{r^2}\right)\right\}, \qquad (17)$$

where the parameters σ and the λ's are definable in terms of cumulants* $\kappa_s(z_0)$ that are

$$\kappa_s(z_0) = \frac{\partial^s}{\partial u^s} \ln h(e^u)\bigg|_{u=\ln z_0}. \qquad (18)$$

The constants in (17) are

$$\sigma = \kappa_2^{1/2}(z_0), \qquad \lambda_s = \kappa_s(z_0)/\kappa_2^{s/2}(z_0). \qquad (19)$$

The generalization of (17) to higher dimensions is quite straightforward. In two dimensions, for example, if

$$h(z_1, z_2) = \sum_{n_1=-\infty}^{\infty} \sum_{n_2=-\infty}^{\infty} h_{n_1 n_2} z_1^{n_1} z_2^{n_2}, \qquad (20)$$

the analogous approximation to $h_{n_1 n_2}(r)$, valid for large r, can be expressed in terms of the saddle point (z_{10}, z_{20}). These are the roots of the set of equations

$$z_1 \frac{\partial}{\partial z_1} \ln h(z_1, z_2) = \frac{n_1}{r}, \qquad z_2 \frac{\partial}{\partial z_2} \ln h(z_1, z_2) = \frac{n_2}{r}. \qquad (21)$$

The lowest-order SPA for this two-dimensional case is

$$h_{n_1 n_2}(r) \sim \frac{[h(z_{10}, z_{20})]^r}{2\pi r z_{10}^{n_1} z_{20}^{n_2} \Delta^{1/2}}(1 + O(r^{-1})), \qquad (22)$$

in which Δ is the determinant

$$\Delta = \kappa_{20}\kappa_{02} - \kappa_{11}^2, \qquad (23)$$

$$\kappa_{ij} = \left(\frac{\partial}{\partial u}\right)^i \left(\frac{\partial}{\partial v}\right)^j \ln h(e^u, e^v)\bigg|_{\substack{u=\ln z_{10} \\ v=\ln z_{20}}}. \qquad (24)$$

Approximations in higher dimensions follow the same pattern; see Good [14, 15] and Barndorff-Nielsen and Cox [2].

APPLICATIONS

The saddle point approximation is especially useful when an accurate representation is desired for the tail of a distribution. A systematic expansion of the tail probability* was given by Blackwell and Hodges [3], although Daniels was the first to have considered these probabilities in detail [5]. Bahadur and Ranga Rao [1] and others have also discussed SPAs for these probabilities. The methodology was used by Robinson to develop approximations for significance levels for one- and two-sample permutation tests* [20]. Barndorff-Nielsen and Cox have given examples of the use of the method applied to the time-dependent Poisson process*, the von Mises distribution*, and to a bioassay* test. Several interesting examples are also discussed by Pedersen [18]. Hampel [16] and Field and Hampel [13] used the SP method to approximate the probability density of an estimator defined by an estimating equation. The idea is as follows [9]: Let the estimate of a parameter θ be the root of

$$\sum_j \psi(x_j, \theta) = 0, \tag{25}$$

where $\psi(x, \theta)$ is monotone decreasing in θ for all x. Let

$$\overline{W}(a) = \frac{1}{n} \sum_{j=1}^{n} \psi(x_j, a). \tag{26}$$

Then, by the assumption of monotonicity $\Pr\{\theta > a\} = \Pr\{\overline{W}(a) > 0\}$, so that the problem of estimating $\Pr\{\theta > a\}$ is reduced to the study of an integral that can be approximately evaluated using the method of this article. Dvorak applied this approximation to certain random-walk* problems [12] (cf. also [21]) and Daniels applied the method to the theory of the general birth process* [8].

References

[1] Bahadur, R. R. and Ranga Rao, R. (1960). *Ann. Math. Statist.*, **31**, 1015–1027.

[2] Barndorff-Nielsen, O. and Cox, D. R. (1979). *J. R. Statist. Soc. B*, **41**, 279–312. (An exposition of the Edgeworth and saddle point approximation for univariate and multivariate distributions, using the notion of the conjugate distribution. The discussion cites many other applications. The theory presented is for distributions of the exponential family. For a more general treatment without this restriction, cf. the cited paper by Durbin [11].)

[3] Blackwell, D. and Hodges, J. L. (1959). *Ann. Math. Statist.*, **30**, 1113–1120.

[4] Cramér, H. (1938). *Actualités Scientifiques et Industrielles*, Vol. 736. Hermann, Paris.

[5] Daniels, H. E. (1954). *Ann. Math. Statist.*, **25**, 631–650. (A classic paper that discusses a large number of fundamental questions and applications. In particular, Daniels analyzes existence and uniqueness of the saddle point, generalizations of the Edgeworth series, accuracy of the approximation at end points of the distribution, and distributions of ratios of random variables.)

[6] Daniels, H. E. (1956). *Biometrika*, **43**, 169–185.

[7] Daniels, H. E. (1980). *Biometrika*, **70**, 89–96.

[8] Daniels, H. E. (1982). *J. Appl. Prob.*, **19**, 20–28.

[9] Daniels, H. E. (1983). *Biometrika*, **70**, 89–96.

[10] De Bruijn, N. G. (1982). *Asymptotic Methods in Analysis*. Dover, New York, reprint. (An excellent and easy to read introduction to a number of asymptotic techniques.)

[11] Durbin, J. (1980). *Biometrika*, **67**, 311–333.

[12] Dvorak, A. (1972). *J. Phys. A*, **5**, 85–94.

[13] Field, C. A. and Hampel, F. R. (1982). *Biometrika*, **69**, 29–46.

[14] Good, I. J. (1957). *Ann. Math. Statist.*, **28**, 861–880. (A clearly written paper applying the saddle point method to discrete multivariate distributions.)

[15] Good, I. J. (1961). *Ann. Math. Statist.*, **32**, 535–548. (A detailed account of the multivariate saddle point approximation.)

[16] Hampel, F. R. (1974). In *Proc. Second Prague Symp. Asymptotic Statist.*, Vol. 2, J. Hajek, ed. Charles University, Prague, pp. 109–126.

[17] Lugannani, R. and Rice, S. O. (1980). *Adv. Appl. Prob.*, **12**, 475–490. (Contains considerable material on the accuracy of the saddle point approximation as well as a comparison of several different approximations on the same test problem.)

[18] Pedersen, B. V. (1979). *Biometrika*, **66**, 597–604.

[19] Petrov, V. V. (1965). *Theory Prob. Appl*, **10**, 287–298.

[20] Robinson, J. (1982). *J. R. Statist. Soc. B*, **44**, 91–101.

[21] Weiss, G. H. and Kiefer, J. E. (1983). *J. Phys. A*, **16**, 489–495.

Bibliography

Daniels, H. E. (1987). *Int. Statist. Rev.*, **55**, 37–48.

Jensen, J. L. (1976). *Uniform Saddlepoint Approximation Theory*, Tech. Rep. No. 6, University of Aarhus, Aarhus, Denmark. (Extension of [5].)

(ASYMPTOTIC EXPANSIONS
CORNISH–FISHER AND EDGEWORTH EXPANSIONS
LIMIT THEOREM, CENTRAL
M-ESTIMATORS
PERMUTATION TESTS
RANDOMIZATION TESTS
RANDOM WALKS)

GEORGE H. WEISS

SAINT PETERSBURG PARADOX *See* ST. PETERSBURG PARADOX, THE

SAMPFORD–DURBIN SAMPLING *See* REJECTIVE SAMPLING

SAMPLE AUTOCORRELATION *See* SERIAL CORRELATION

SAMPLE AUTOCOVARIANCE *See* SERIAL CORRELATION

SAMPLE, RANDOM *See* SIMPLE RANDOM SAMPLING

SAMPLE REUSE

The meaning of the term *sample reuse* rests on the idea that a sample is used to evaluate a statistic and then may be used again to assess or improve the performance of the basic statistical procedure. Some applications of the concept fit quite obviously into this sample reuse framework, while some others that belong in this category are not quite so apparent. The varieties range from the overt resampling that is characteristic of the bootstrap* to more subtle manifestations of the principle such as a cross-validatory choice of bandwidth for a kernel density estimator.

STANDARD ERROR AND RERANDOMIZATION

Several of the cornerstones of statistical inference that were laid by Karl Pearson*, "Student" (*see* GOSSET, WILLIAM SEALY), R. A. Fisher*, and E. J. G. Pitman* may be viewed as early applications of the notion that the data set could profitably be used a second time. The fundamental concept of the sampling distribution of a sample mean led almost immediately to the process of reusing the sample to estimate the standard error. Strictly speaking the first use of a random sample $X_1, X_2, \ldots, X_n$ would be to compute $\overline{X}$. The reuse would be to estimate σ^2/n with the unbiased estimator s^2/n. In a very real sense this represents a return to the sample to assess an average squared deviation about the statistic that was calculated initially.

A second landmark development that fits into this same framework is the method of randomization* due to Fisher [7] and Pitman [18]. The essence of this technique, which yields exact tests of hypotheses in some settings, is a permutation argument first conceived by Fisher*. It is founded upon the idea that, given a particular random sample, it is often the case that the algebraic signs, labels, or positions in a table are features of the data set that would tend to differ under competing hypotheses even if the magnitudes of all the numerical outcomes were not different. The logical basis for the test is that, given the validity of the null hypothesis, each of these different configurations of signs or positions would be equally likely. Consequently, this classical testing procedure utilizes a null distribution generated by the set of all reevaluations of a statistic T for a specific reference set of permutations. This set of values obtained by reusing the sample provides a frame of reference that allows one to judge whether the originally observed value of T is extreme.

To illustrate this approach to the calculation of such a conditional p-value* consider testing the hypothesis H of independence of absolutely continuous random variables X and Y based upon a random sample of size n from their joint distribution. Given the observed $n \times 2$ table, the correlation coefficient r is calculated. Under H each of the $n!$ permutations of the column of y values is equally likely and corresponding to each there would be a distinct value of r. If all $n!$ outcomes were tabulated, this would represent the null distribution. Conceptually one would have enumerated all of the potential realizations that are more extreme in some predetermined sense than the actual outcome. In practice the calculations of this

p-value may be formidable if the size of the data set is large. Monte Carlo* sampling techniques to approximate the procedure were first discussed by Dwass [4]. For recent computational developments, extensions to interval estimation, and a good review see Gabriel et al. [8]. The relatively recent and appropriate nomenclature that identifies these as "rerandomization" procedures is due to Brillinger et al. [2].

JACKKNIFE AND BOOTSTRAP

In situations that are more complex than using $\overline{X}$ to estimate the population mean, the estimators of standard error are typically more complicated than the sum of squared deviations of individual observations. In many such cases it is still feasible to use a direct estimate of the standard error through expressions derived for the pertinent parametric family; occasionally, these must be only large-sample approximations. The primary function of the jackknife* and bootstrap* is to provide nonparametric estimates of standard errors. It follows that approximate tests and confidence intervals can be and are constructed from these in nonparametric settings and for some parametric problems when an analytical solution is difficult. Even though these procedures often also yield an improved estimator, for example, one with a reduced bias*, the main emphasis is on the reuse of the sample to obtain estimates of the variability that is present in the sampling distribution of a statistic.

The related topics of empirical influence function* and delta method are examined in systematic treatments by Efron [5, 6]. Still another resampling plan that was introduced by McCarthy [15] is entitled balanced half-sample pseudo-replication. This procedure is employed mainly for highly stratified survey samples. *See* JACKKNIFE METHODS and RESAMPLING PROCEDURES for detailed descriptions of jackknife, bootstrap, and pseudo-replication.

CROSS-VALIDATION AND PREDICTIVE SAMPLE REUSE

Cross-validation as a means of assessing the quality of statistical predictions is a simple idea. Stone [19] cites examples from the 1930s. The essence of the process is a partitioning of the data into two subsamples. One portion is used to construct the predictor, the other to measure the performance of the rule when it is applied to data that were not used in selecting the specific prediction rule. At this stage of development sample reuse is not in evidence. For the simple situation involving a construction sample and a separate validation sample, the approach does not actually warrant the name cross-validation either; this scheme has an advantage over the naive alternative involving direct resubstitution, which is usually optimistically biased. Clearly, any prediction rule should appear to do better with predictions of cases that were used to construct that rule than it could be expected to do generally.

The next stage of development occurred during the 1960s. Mosteller and Wallace [17] suggested the approach and the first distinct appearances of cross-validation were published by Lachenbruch and Mickey [13] and Mosteller and Tukey [16]. The heart of the procedure is that one individual case is set aside as the validation portion and the estimation or construction is carried out with the subset of $(n - 1)$ cases. After the performance of this rule is tested upon the "held-out" case, the process is repeated for each possible case. In other words, in turn the sample has been partitioned and reused in all n possible ways of applying a leave-out-one rule. The applications of these early developments were to discriminant analysis*, but the basic ingredient is that of a prediction for which there is an observable measure of error or error rate.

In the mid-1970s another extension surfaced. Both Stone [19] and Geisser [9] raised the issue of the choice of the predictor in addition to assessing predictor performance (*see* SPECIFICATION, PREDICTOR and PARTIAL LEAST SQUARES). They then proposed a merger of the cross-validation assessment component and the choice component. Subsequent reference to cross-validation tends to pertain to this major refinement, which has shifted the methodology from assessment to construction. Wahba and Wold [20] formulated and used cross-validation to select the degree of smoothing in spline* fitting. Some asymptotic optimality results have been established for spline smoothing with generalized cross-validation of Craven and Wahba [3]. The application of cross-validation to kernel density estimation* has yielded a mixture of encouraging results with occasional failures. Hall [10, 11] summarizes these findings. Properties of cross-validated nearest-neighbor* nonparametric regression are given by Li [14]. Early independent application of this same principle, called PRESS in the regression setting, can be found in Allen [1]. This alternative to least squares basically selects predictor variables that minimize the prediction error sum of squares. For more discussion of predictive aspects in general *see* PREDICTIVE ANALYSIS. These sample reuse techniques originated from attempts to validate assumptions or particular statistical models. As such, the calculation and display of residuals* qualify as among the earliest and most widely used applications. Yet, residuals as regression diagnostics* do not exhibit the two primary ingredients of cross-validation, namely a summary assessment and a choice of the specific prediction model. However, deleted residuals do require calculations that parallel those in the cross-validation methodology.

ADAPTIVE PROCEDURES

As another distinct application of sample reuse the topic of adaptive estimation actually appears to warrant the term preuse instead (*see* ADAPTIVE METHODS). Again, the general idea can be discussed in the context of estimating the mean of a distribution. If one were reasonably certain that the data were drawn from a normal distribution then $\overline{X}$ would be an efficient* estimator. On the other hand, if the probability model were known to be the Laplace distribution*, the median would be optimal in this sense. Clearly, if one could reliably decide about the population actually sampled, assuming for the sake of simplicity that it is one or the other of these two parametric families, then one could select the better measure of location. So in its most elementary form, adaptive* estimation uses the sample first to make a choice and then uses it a second time to evaluate the appropriate estimator. Hogg [12] summarizes numerous developments along these lines.

Example applications range from a dichotomous choice, e.g., a preliminary-test estimator, to continuously adaptive estimators. An example of the latter is a trimmed mean in which the fraction to be trimmed is estimated from the same data. Both of these procedures have a characteristic two-stage nature. This is in contrast with the simultaneous character of the cross-validatory choice paradigm.

References

[1] Allen, D. M. (1974). *Technometrics*, **16**, 125–127.

[2] Brillinger, D. R., Jones, L. V., and Tukey J. W. (1978). *Report to the Secretary of Commerce, Statistical Task Force to the Weather Modification Advisory Board*. U.S. Government Printing Office, Washington, DC.

[3] Craven, P. and Wahba, G. (1979). *Numer. Math.*, **31**, 377–404. (Introduces the method of generalized cross-validation to estimate the degree of smoothing with splines.)

[4] Dwass, M. (1957). *Ann. Math. Statist.*, **28**, 181–187.

[5] Efron, B. (1981). *The Jackknife, the Bootstrap and Other Resampling Plans*, CBMS Monograph No. 38, SIAM, Philadelphia.

[6] Efron, B. and Gong, G. (1983). *Amer. Statist.*, **37**, 36–48. (An expository article explaining relationships among several resampling schemes.)

[7] Fisher, R. A. (1935). *The Design of Experiments*. Oliver and Boyd, London.

[8] Gabriel, K. R. and Hall, W. J. (1983). *J. Amer. Statist. Ass.*, **78**, 827–836. (This and two other articles coauthored by Gabriel in the same issue of the journal all contain the term "rerandomization" in the title.)

[9] Geisser, S. (1975). *J. Amer. Statist. Ass.*, **70**, 320–328.
[10] Hall, P. (1982). *Biometrika*, **69**, 383–390.
[11] Hall, P. (1983). *Ann. Statist.*, **11**, 1156–1174.
[12] Hogg, R. V. (1982). *Commun. Statist. A*, **11**, 2531–2542.
[13] Lachenbruch, P. A. and Mickey, M. R. (1968). *Technometrics*, **10**, 1–11.
[14] Li, K. C. (1984). *Ann. Statist.*, **12**, 230–240.
[15] McCarthy, P. J. (1969). *Rev. Inst. Statist. Int.*, **37**, 239–264.
[16] Mosteller, F. and Tukey, J. W. (1968). In *Handbook of Social Psychology*, G. Lindzey and E. Aronson, eds. Addison-Wesley, Reading, MA.
[17] Mosteller, F. and Wallace, D. L. (1963). *J. Amer. Statist. Ass.*, **58**, 275–309.
[18] Pitman, E. J. G. (1937). *J. R. Statist. Soc. B*, **4**, 119–130.
[19] Stone, M. (1974). *J. R. Statist. Soc. B*, **36**, 111–147. Corrigendum (1976). **38**, 102.
[20] Wahba, G. and Wold, S. (1975). *Commun. Statist. A*, **4**, 1–17.

(ADAPTIVE METHODS
BOOTSTRAP
CROSS-VALIDATION
JACKKNIFE METHODS
PREDICTIVE ANALYSIS
RANDOMIZATION TESTS
RESAMPLING PROCEDURES)

WILLIAM R. SCHUCANY

SAMPLE SIZE DETERMINATION

This phase is used to describe a part of the planning stages of an enquiry—that is, the choice of size of sample, subject to various constraints. These are described more fully in DECISION THEORY; they may be summarized under

1. What is expected from the enquiry—accuracy of estimate(s), power of test, for example
2. Costs of identifying and measurement of individuals in a sample
3. Available resources (such as time and money).

There are, of course, other, nonquantifiable constraints, such as personal prejudices and idiosyncracies, but these are often neglected in preliminary analyses of possibilities.

"Optimum" sample sizes usually correspond to maximizing (or minimizing) some more or less arbitrary, though plausible, "objective function(s)".

Sample size determination is to be distinguished from sample size *estimation* in which one attempts to estimate the size of a complete sample from the evidence provided by a (possibly) incomplete sample.

SAMPLE SIZE ESTIMATION *See* POPULATION OR SAMPLE SIZE ESTIMATION

SAMPLE SIZE, INFORMATIVE AND PREDICTIVE

INFORMATIVE SAMPLE SIZE

The problem of the determination of finite informative sample sizes originated in econometrics*, where usually a relatively small number of mutually dependent observations are available for the estimation of complicated multivariate models.

Suppose a statistician is sampling from a (possibly multivariate) stochastic process* $\{X_t\}$ in discrete time, which has a probability law depending on some unknown parameter θ running over the parameter space Θ. In general, the statistician is not primarily interested in θ itself, but merely in the value taken by some function φ on Θ, e.g., the mean $\mu_t(\theta)$ of X_t or coefficients in a model.

When the statistician observes the process during $t = 1, 2, \dots, n$, and the distribution of the sample is denoted by $P_\theta^{(n)}$, then inferences about $\varphi(\theta)$ only make sense when φ is identifiable* with respect to the class of probability measures $\{P_\theta^{(n)},\ \theta \in \Theta\}$; *see* IDENTIFIABILITY. Any value of n with this property is called an *informative sample size* (i.s.s.) for φ. Formally the sample size n (or the sample itself) is said to be *informative for* φ if for all $\theta_1, \theta_2 \in \Theta$ the implication

$$\varphi(\theta_1) \neq \varphi(\theta_2) \Rightarrow P_{\theta_1}^{(n)} \neq P_{\theta_2}^{(n)} \qquad (1)$$

holds. Clearly, if n is an i.s.s. for φ, all $N \geqslant n$ are informative for φ and there exists a minimum i.s.s. The existence of an i.s.s. depends on the space Θ. Furthermore, it is important to realize that if $n = \infty$ is an i.s.s., there does not necessarily exist a finite i.s.s. (for an example see ref. 4, p. 26).

In practice, the inequality of distributions in (1) can often be proved by considering the set of first and second moments of $P_\theta^{(n)}$. When it suffices to consider the first moment, the sample size is said to be *first-order informative*; if also the second moments are needed, it is called *second-order informative*. For example, in the standard regression model the sample size n is first-order informative for the regression coefficients if the $n \times k$ matrix of regressor values has full column rank. The concept of informative sample size is of some special interest when stationary time series* are considered. As example we shall give some results on m-variate mixed autoregressive moving average* processes [ARMA (p, q)] given in ref. 4. They generalize the classical result of Hannan on the identification of ARMA (p, q) models (see ref. 2). Thus we consider the model

$$\sum_{k=0}^{p} A_k X_{t-k} = \sum_{j=0}^{q} B_j \epsilon_{t-j}, \qquad t = 0, \pm 1, \ldots,$$

where $\{\epsilon_t\}$ is m-variate white noise* with nonsingular covariance matrix Ω and $A_0 = B_0 = I_m$ (the $m \times m$ unit matrix). The integers p and q are supposed to be known. If the generating functions* $A(z) = \sum_{k=0}^{p} A_k z^k$ and $B(z) = \sum_{j=0}^{q} B_j z^j$ satisfy the conditions

1. $\det A(z) \neq 0 \quad \text{for } |z| \leqslant 1,$

   $\det B(z) \neq 0 \quad \text{for } |z| < 1;$

2. $\text{rank}[A(z), B(z)] = m \quad \text{for all } z;$

3. if $m > 1$ then $\text{rank}\left[A_p, B_q\right] = m,$

then we have for $\varphi(\theta) = (A_1, \ldots, A_p; B_1, \ldots, B_q; \Omega)$ the following informative sample sizes:

Case		Informative Sample Size
Univariate ($m = 1$)		$p + q + 1$
Multivariate ($m > 1$)	$p = 0$	$q + 1$
	$q = 0$	$p + 1$
	$pq \neq 0$	$q + (m + 1)p$

A proof that the given i.s. sizes are the minimum is known only for the moving average case ($p = 0$) (see ref. 4, pp. 43–47 and 76–81). Because the proofs of these results rely on an analysis of the second-order properties of the observable process, the sample sizes are second-order informative. They are the basis of finding i.s. sizes for a variety of models involving lagged* variables, e.g., transfer function* models or dynamic simultaneous equation systems under certain stationarity assumptions.

PREDICTIVE SAMPLE SIZE

As in the preceding section, let $P_\theta^{(n)}$, $\theta \in \Theta$, denote the probability law of the sample $X_1, \ldots, X_n$ from the process $\{X_t\}$. In order that prediction of future values X_N, $N \geqslant n + 1$, based on $X_1, \ldots, X_n$ makes sense, a necessary condition is that the sample size n is informative for the distribution $P_\theta^{(n)}$ for $N \geqslant n + 1$. Therefore the sample size n is said to be *predictive* if for all $N \geqslant n + 1$ and for all $\theta_1, \theta_2 \in \Theta$ the implication

$$P_{\theta_1}^{(N)} \neq P_{\theta_2}^{(N)} \Rightarrow P_{\theta_1}^{(n)} \neq P_{\theta_2}^{(n)} \qquad (2)$$

holds. Notice that predictivity of the sample size n does not necessarily imply that the best predictor depends on $X_1, \ldots, X_n$, but merely that "good estimates" of $P_\theta^{(N)}$ depend on $X_1, \ldots, X_n$. For example, if $\{X_t\}$ is a sequence of i.i.d. variables with zero mean and unknown variance θ, then the best predictor for X_{n+1} is zero; however, statements like a 95% interval for X_{n+1}, in general, depend on the sample since they require an estimate of the variance.

It is seen from (1) and (2) that if the sample size $N \leqslant \infty$ is informative for some function φ, and the sample size $n < N$ is predictive, then the sample size n is also informative for φ.

Clearly, if n is a predictive sample size, all sample sizes $M \geqslant n + 1$ are predictive. Therefore there may exist a minimum predictive sample size, say n_0. A simple but curious theorem states that, for any function φ of θ for which a minimum i.s.s. N_φ exists, we have $n_0 \geqslant N_\varphi$, where equality holds if φ is $1 - 1$ [4, p. 29].

Conditions under which predictive sample sizes can be obtained are in general less restrictive than those for identification of the model. For example, the sample sizes given in the preceding section for the m-variate ARMA (p, q) process turn out to be predictive under the single condition det $A(z) \neq 0$ for $|z| \leqslant 1$ [4, p. 91].

Again, we can restrict ourselves to the second-order properties of the process and obtain second-order predictive sample sizes. In the case of a weakly stationary process with covariance function $\Gamma_s = EX_tX_{t-s}$ $(s = 0, \pm 1, \dots)$, a second-order predictive sample size n is then an integer such that the infinite sequence $\{\Gamma_s\}$ is uniquely determined by $\Gamma_0, \Gamma_1, \dots, \Gamma_{n-1}$. Since there is a one-to-one correspondence between covariance functions and spectral measures, the sample size n is then second-order informative for the spectral measure (which in general depends on θ).

References

[1] Deistler, M. (1975). *Metrika*, **12**, 13–25. (Treats the identification problem in econometrics for infinite samples.)

[2] Hannan, E. J. (1969). *Biometrika*, **57**, 223–225. (Unique factorization of spectral density matrix.)

[3] Hannan, E. J. (1971). *Econometrica*, **39**, 751–765. (Also identification w.r.t. infinite samples.)

[4] Tigelaar, H. H. (1982). *Identification and Informative Sample Size*. Math. Centre Tracts 147, Amsterdam. (Rigorous treatment covering the whole topic.)

(IDENTIFICATION PROBLEMS)

H. H. TIGELAAR

SAMPLE SPACE

A basic concept of probability theory. It has been defined as "the collection of all possible outcomes of a random phenomenon." It is analogous to the concept of a set, which is a primitive mathematical notion, usually undefined, but understood to mean something like a "collection of things or items." Indeed, the sample space is commonly represented in terms of sets in a mathematical model.

(PROBABILITY MEASURE
PROBABILTY THEORY (OUTLINE))

SAMPLE SURVEYS *See* SURVEY SAMPLING

SAMPLE VARIANCE *See* CHI-SQUARE DISTRIBUTION; STANDARD DEVIATION

SAMPLING DISTRIBUTION

This term is usually applied to the distribution of a statistic* $T(X_1, \dots, X_n)$, which is a function of directly observed random variables $X_1, \dots, X_n$. It could equally well be applied to the distribution of the X's or of any one or any subset of them, but this is not commonly done. The distinction may have some pedagogical value in introducing the idea of distribution for quantities like sample mean, sample variance, etc.

The concept "sampling distribution" should be distinguished from "*sample* distribution." The latter term often denotes the values observed in a sample, rather than a theoretical distribution.

(PROBABILITY THEORY)

SAMPLING ERRORS (COMPUTATION OF)

ESTIMATION METHODS

There are a number of methods for estimating sampling errors from complex sample surveys. Six such methods are:

1. Bootstrap [3].
2. BRR (balanced repeated replication) [10].
3. Taylor series expansion [1, pp. 319–321].
4. Keyfitz [8].
5. Jackknife [3].
6. Independent replications [2].

The bootstrap method is relatively new and there is no known generalized computer software for computing variances using this method. Also, the jackknife* procedure is very costly to compute, and only one known computer software system exists for handling it (see $REPERR below).

The Taylor series approximation method, variously referred to as expansion or linearization, is the most popular in that there are more computer programs using it than any of the other methods. The Keyfitz method is a special case of the Taylor series method where there are two PSUs (primary sampling units*) per stratum. The BRR method, which also requires two PSUs per stratum, is a replication technique based on an orthogonal subset of the replications. One drawback of some of the software for the Taylor series approximation method is that poststratification is not taken into account, and thus the variances of poststratification cells are not all zero. Another disadvantage is that it is not applicable to nonalgebraic estimates (such as medians and percentiles). One drawback to the BRR method is that the variance of an estimate based on a single case is a positive number, not equal to zero.

The format for output that is most useful is that of cross-tabulation. It is also helpful if provision is made for recoding and exclusions. Francis and Sherman [6] indicate some attributes that one would like to consider in choosing a tabulation package: those that they tried to measure include the type of package, the minimum core required, the portability, the tabulating power, and the simplicity of language; those that they did not try to measure include documentation, ease of learning, useable error messages, machine costs, and ability for handling different input formats, various data structures, and dirty data.

The next section will briefly describe a number of computer software packages that exist for computing sampling errors.

PROGRAM DESCRIPTION

HESBRR (also known as HES VAR X-TAB) was written by Gretchen Jones of the National Center for Health Statistics*. The program calculates variances using the BRR method. It was designed for use with weighted data from complex sample surveys. The output is a cross-tabulation. The tables produced contain weighted and unweighted estimates of numerator and denominator, ratios (rates, means, or percents), standard errors, relative standard errors, and relative variances of ratios, numerators, and denominators, standardized values, and also an unbiased estimate of the standard deviation. The program allows for recoding, exclusions, and labeling, and it generates totals automatically. It provides for a filter (the page dimension), stub (the row dimension, which may be nested), and spread (the column dimension, which also may be nested).

SESUDAAN was developed by B. V. Shah of the Research Triangle Institute. The program is a PROC in SAS (statistical analysis system) and as such takes advantage of all the recoding, excluding, and labeling features of SAS. SESUDAAN is an improvement on an earlier version, called STDERR, and it uses the Taylor series linearization approach for calculating sampling errors. The tables produced include the mean or proportion, the total, the standard error, the sample size, the population estimate, the design effect, and the number of PSUs and strata used in the calculations. Tables may be nested or concatenated [11].

NASSVAR is another SAS procedure, designed to compute standard errors of estimates based on sample survey data. The procedure was developed by David Morganstein and Greg Binzer of WESTAT. The BRR method is used. NASSVAR is designed to compute sampling errors and associated statistics for totals, ratios, and many other arithmetic functions of the estimates specified by the user. NASSVAR calculates the full-sample estimate, the number missing for the full sample, the variance, the relvariance (the variance divided by the estimate squared), the coefficient of variance in percent, the standard error, the lower 95% confidence limit, and the upper 95% confidence limit. NASSVAR does not produce results in a cross-tabulation.

CLUSTERS is a generalized program for computing sampling errors using the Taylor linearized approach. It was developed by Vijay Verma and Mick Pierce of the World Fertility Survey. The program provides for comprehensive recoding. Output is a printed cross-tabulation or optionally on a tape or disk. Sampling errors of proportions, means, percentages, ratios, and differences of ratios are calculated. CLUSTERS also computes the design effect (the design-based standard error divided by the standard error based on simple random sampling) and the rate of homogeneity (the design effect minus one divided by the mean cluster size minus one). Data may be subclassified by geographical region. Totals are generated automatically.

SUPER CARP is a software package developed by Michael A. Hidiroglou, Wayne A. Fuller, and Roy D. Hickman of the Department of Statistics at Iowa State University. The Taylor linearization method is used to compute sampling errors for complex sample surveys. The package estimates the sampling covariance matrices of totals, ratios, regression coefficients, and subpopulation means, totals, and proportions. Options exist for computing goodness of fit* statistics of the chi-square-type and tests of independence in two-way tables. Output is a cross-tabulation; provision for excluding records exists, but there is no provision for recoding.

The Australian Bureau of Statistics has developed several programs for computing sampling errors for complex sample surveys. These programs are VARIANCE, SPLITHALF, RATVARACROSS, RATVARWITHIN, E37 (or QUARS), and E38. They replace programs formerly called SPLITHALF, MWDVAR, and GSS. VARIANCE uses the Taylor series expansion method and SPLITHALF uses the BRR technique. Both are generalized programs that use as input files of sums, etc., produced by TPL (table producing language) and produce a file of variances to be used by other programs. RATVARACROSS and RATVARWITHIN use the Taylor series approximation method to calculate within- (RATVARWITHIN) and across-stratum (RATVARACROSS) ratio estimates from sample surveys. They are SAS procedures and thus can use SAS's data-handling facilities including recoding and exclusions. E37 and E38 are SAS macros, written in the SAS macro language. E37 calculates variances of estimated quantiles or percentiles. E 38 uses a superpopulation model* with modelled heteroscedasticity*. The method is similar to that discussed in Cochran [1, pp. 256–258, 158–160].

$PSALMS is part of the OSIRIS IV software package. It uses the Taylor series expansion method and was developed by the Survey Research Center at the University of Michigan. $PSALMS calculates estimates, sampling errors, and design effects for ratios, differences of ratios, means, and totals. An extended computational option allows the printing of the ratio variance, bias, coefficient of variation for the denominator, and covariance between the numerator and denominator; the difference of ratios variance, bias, intraclass correlation, and weight effect; and each numerator and denominator weighted and unweighted estimate, variance, sum of weights, and case count for each ratio and difference of ratios. As a command in OSIRIS IV, $PSALMS is capable of using the system features that include a filter capability to subset cases, an elaborate missing data testing feature, a powerful recode facility, and a hierarchical file structure capability [12].

$REPERR is another OSIRIS IV command, using a repeated replication procedure to compute estimates, sampling errors, design effects, and test statistics for variable means and regression statistics. Using the balanced half-sample*, jackknife*, or user specified model to automatically form replications from sets of sampling error computing units, $REPERR performs a regression analysis for each replication and the total sample. $REPERR also computes variances for variable means, correlation, regression coefficients, standardized regression coefficients, partial correlations*, and multiple R-squared (multiple correlation*) statistics by examining the departure of the replication estimates from the total sample estimate. A special feature is available to create dummy variables from categorical data* for dummy variable regressions. $REPERR can input and output replication totals and regression estimates in machine readable form for the user to interface with his/her own routines to provide a generalized repeated replication variance estimation procedure extending beyond regression statistics. $REPERR has the same system facilities for handling missing data, recording, and hierarchical files as $PSALMS does [12].

CESVP (Consumer Expenditure Survey Variance Program) was written by Ronald Lambrecht and Cathy Dippo to compute variances of estimates derived from the Bureau of Labor Statistics' Consumer Expenditure Survey. The program uses the BRR technique and calculates estimates of totals, means, ratios; variances of means, totals, and ratios; coefficients of variation, and correlations. It also performs certain significance tests [4, p. 168].

The Enumerative Survey Summary System (also known as MULTI-FRAME) was developed by the Statistical Reporting Service of the Department of Agriculture. It is a generalized series of programs for summarizing multiframe probability surveys. Variance estimates are design-based. MULTI-FRAME computes estimates and standard errors at each summary level in addition to combined estimates; it does not produce estimates in a cross-tabulation; it does provide for recoding but not exclusions.

The CLFS VAR-COV (Canadian Labor Force Survey Variance–Covariance System) is a software package developed by Barry Bogart for calculating sampling errors using the Keyfitz method. Totals, linear combinations of totals, ratios, and differences between ratios can be estimated. Estimates can be subweighted, weighted, and unweighted. The ratio estimation technique is used; variance estimates of ratios, differences of ratios, and noninterview ratios are calculated; design effects are also estimated. Statistics can be computed for specified strata or PSUs.

The KEYFITZ Variance Estimation System is a software package written by Robert Jewett of the Census Bureau. It uses a combination of methods: Taylor series expansion, BRR, and Keyfitz. It consists of three programs: the estimate, the gamma factor (a factor occurring from the Taylor series expansion of a nonlinear estimator), and variance. The estimate program produces four files as input to the variance program. The gamma factor program produces a file of factors to be used in one of the second-stage ratio estimation variance formulas in the variance program. The variance program produces variances for the various components of variance and for the different levels of estimation. Also, covariances, relative covariances, and variance totals, for self-representing (an area chosen for the sample with probability 1) and non-self-representing (an area chosen for the sample with probability less than 1) areas and by geographic regions, are produced. Additional statistics produced are quantiles*, variances of averages over time, covariances and correlations, and variances of means [4, p. 169].

GENVAR is a subroutine for computing variances using the Taylor linearization approach. It was written by Beverly Causey and Ralph Woodruff at the Bureau of the Census*. The functions must be supplied by the user and must be twice-differentiable. For each function, the output includes the first-order estimate of the function, the estimated variance and standard deviation of this estimate, the coefficient of variation, and estimates of the partial derivatives of the function [4, p. 179].

There are a number of general statistical software* systems that allow calculation of sampling errors through use of the language of the software system as long as the necessary identifiers and weights are present on the input record (Stratum and PSU codes and/or half-sample weights). Among these are RGSP (Rothamsted General Survey Program), OMNIBUS 80, DATAPLOT, FOCUS, SPEAKEASY, CS (Consistent System), IMSL Library, LSML76, TPL (Table Producing Language), and NISAN. Two software packages (FOCUS and DATAPLOT) have the facility to calculate variances using independent replications. LSML76 calculates variances using Henderson's method (see Francis [4, pp. 374–377]).

COMPARISON OF SOFTWARE

There have been a number of papers comparing software for computing sampling errors. Francis and Sedransk [5] compared CLUSTERS, STDERR, SUPERCARP, and PSALMS, all using the Taylor series approximation technique. They found some errors in calculation of some statistics, especially when there was only one case in a stratum. New versions of these programs have come out that may have corrected these errors. Maurer et al. [9] compared HESBRR and STDERR as to computational efficiency according to the number of replicates, the size of the input file, and the number of cells in the table. They found that the number of replicates and the number of input records adversely influenced the computer time for HESBRR. However, the number of cells per table increased the time for STDERR, but not for HESBRR. Jones [7] compared HESBRR with SESUDAAN according to their control language syntax, printed output format, special features, and execution efficiency. She concluded that for surveys with few replicates and a small number of input records, HESBRR is preferable. For surveys with many replicates and a large number of input records, SESUDAAN is more efficient.

Probably the most complete comparison of programs is in the book *Statistical Software: A Comparative Review* by Francis [4]. He compares 11 computer packages: HES VAR X-TAB, SPLITHALF, MULTIFRAME, MWDVAR, GSS, CLFS VARCOV, CESVP, KEYFITZ, CLUSTERS, SUPER CARP, and GENVAR. Of these, SUPER CARP is the only program that is being actively maintained and exported. Figure 1 summarizes the ratings of the developers and users for the items listed. The developer's ratings are joined by a continuous line. If the developer's rating is lower than the users' rating, a thin vertical line is drawn between the two. If the developer's rating is higher than the users' rating, a thick vertical line is drawn between the two. Thus a thick line indicates that the developer has overrated his program relative to the users [4, p. 15].

In conclusion, many computer programs for calculating sampling errors have been described and compared. Many factors must be considered in choosing a software package: availability of package, appropriate method, machines on which the package will run, cost of the package, computational efficiency, ease of use, and documentation available. Most of these factors are discussed by Francis [4].

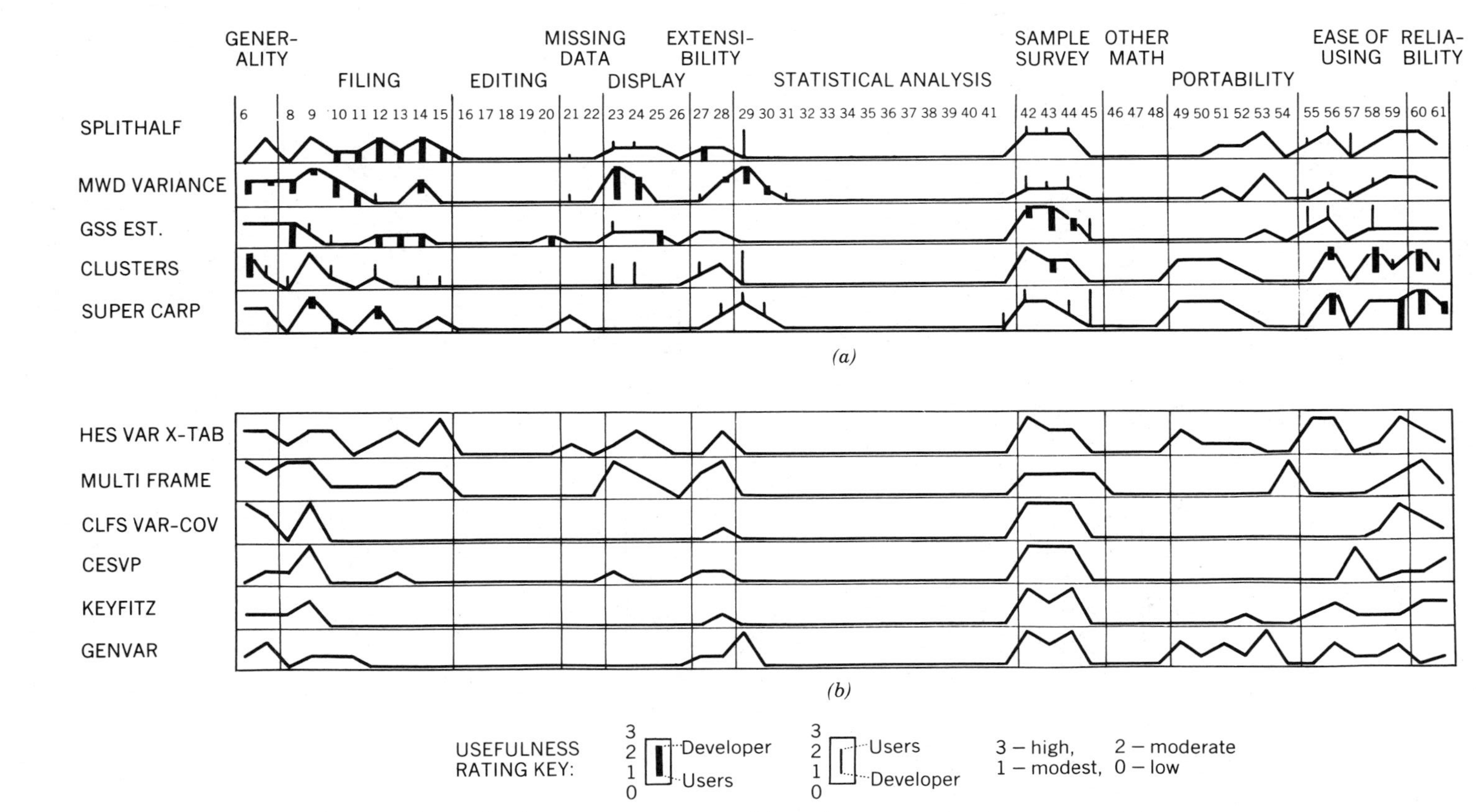

Figure 1 Ratings by developers and users on all items for survey estimation programs: (*a*) programs with users; (*b*) programs without users. Reprinted by permission of the publisher from Survey Variance-Estimation Programs by Ivor Francies, *Statistical Software, A Comparative Review*, p. 156. Copyright 1981 by Elsevier Science Publishing Co., Inc.

References

[1] Cochran, W. G. (1977). *Sampling Techniques*, 3rd ed. Wiley, New York. (This is an excellent text on sampling.)

[2] Deming, W. E. (1960). *Sample Design in Business Research*. Wiley, New York, pp. 87–97.

[3] Efron, B. (1983). *The Jackknife, the Bootstrap, and Other Resampling Plans*. SIAM, Philadelphia.

[4] Francis, I. (1981). *Statistical Software: A Comparative Review*. Elsevier North-Holland, New York. (The most comprehensive review of statistical software.)

[5] Francis, I. and Sedransk, J. (1979). *Bull. Int. Statist. Inst.*, **48**, (3), 239–272. (A comparison of software for processing and analyzing surveys.)

[6] Francis, I. and Sherman, S. P. (1976). *Proc. Comput. Sci. Statist.: 9th Annu. Symp. on the Interface*, Cambridge. (Languages and programs for tabulating data from surveys.)

[7] Jones, G. (1982). *Proc. Statist. Comput. Sect. Amer. Statist. Ass. Cincinnati*. (A comparative study of methods for calculating sampling errors.)

[8] Keyfitz, N. (1957). *J. Amer. Statist. Ass.*, **52**, 503–510. (Estimates of sampling variance where two units are selected from each stratum.)

[9] Maurer, K., Jones, G., and Bryant, E. (1978). *Proc. Survey Res. Meth. Sect. Amer. Statist. Ass., San Diego*. (Relative computational efficiency of the linearized and balanced repeated replication procedures for computing sampling variances.)

[10] McCarthy, P. J. (1966). Replication: An approach to the analysis of data from complex surveys. *Vital and Health Statistics*, Series 2, Number **14**, National Center for Health Statistics, Rockville, MD.

[11] Shah, B. V. (1981). *SESUDAAN: Standard Errors Program for Computing of Standardized Rates from Sample Survey Data*. RTI, Research Triangle Park, NC.

[12] Survey Research Center (1980). *Sampling Error Analysis in OSIRIS IV*. Institute for Social Research, University of Michigan, Ann Arbor, MI.

Bibliography

Francis, I. (1981). *Comput. Sci. Statist. Proc. 13th Symp. on the Interface, New York*. (Measuring the performance of computer software—a dual role for statisticians.)

Kaplan, B. A., Francis, I., and Sedransk, J. (1979). *Proc. 12th Annu. Symp. Interface Comput. Sci. Statist.* (Criteria for comparing programs for computing variances of estimators from complex sample surveys.)

Kaplan, B. A., Francis, I., and Sedransk, J. (1979). *Proc. Survey Res. Meth. Sect. Amer. Statist. Ass.*, Washington, DC (A comparison of methods and programs for computing variances of estimators from complex sample surveys.)

(COMPUTERS AND STATISTICS
SAMPLE REUSE
STATISTICAL SOFTWARE)

GRETCHEN K. JONES

SAMPLING FRAME

This is a list itemizing individuals in a population. The term is most commonly used in connection with planning and implementation of sample surveys*.

SAMPLING INSPECTION *See* SAMPLING PLANS; QUALITY CONTROL, STATISTICAL

SAMPLING PLANS

A sampling plan specifies the *sample size* and the associated *acceptance criterion* to be used when the acceptability of a lot or process is decided by means of sampling inspection; *see* ACCEPTANCE SAMPLING. The object of the sampling plan is to ascertain the characteristics or quality of a lot or process without inspecting each individual item, and to give a rule for judging whether the lot should be considered acceptable or not. An example is the plan: Take a random sample of 80 items and accept the lot from which the sample was taken if two or fewer items in the sample are found nonconforming with the specifications for an item; otherwise reject.

The advantage of using a sampling plan instead of inspecting each item is the reduced inspection effort. The disadvantage is that the decision regarding the acceptability of the lot or process is influenced by random variations in the sample result. The choice of a sampling plan should therefore—with due consideration of the conditions under which the plan is going to operate—express a reasonable balance between the sampling and testing effort, and the ability of the sampling plan to discriminate between satisfactory and nonsatisfactory production.

The statistical implications of a specific sampling plan depend on the distribution of sample results that, in turn, depend on the composition of item quality characteristics in the lots submitted for inspection. When assessing sampling plan performance it is customary to assume that the values of the individual item quality characteristics in the sample may be modelled as independent identically distributed random variables with a common distribution. The quality of the lots submitted for inspection is expressed through the value of a *process quality parameter* Θ, specifying the distribution of sample results.

Thus, a sampling plan may be seen as a means of testing a hypothesis concerning the process quality parameter Θ, and therefore the statistical treatment relates to the theory of hypothesis testing*. The fundamental tool for assessing sampling plan performance is the OC curve, which shows the probability of acceptance as a function of the value of Θ. The OC curve is the complement of the power* curve for the corresponding statistical test. Following the terminology by Dodge, an OC curve relating the acceptance probability to the process quality is sometimes called a *type B OC curve*. Only when the sample distribution depends on the composition of the individual lot solely through a lot quality parameter is it possible to give the acceptance probability as a function of individual lot quality (*type A OC curve*).

As an aid in the choice of a plan for a specific acceptance sampling situation, it is customary to specify the acceptable quality level* (AQL) or acceptable process level (APL) of the process quality parameter Θ. For this the value of Θ the plan should lead to acceptance with a high probability. Similarly, an unsatisfactory value of Θ, the limit-

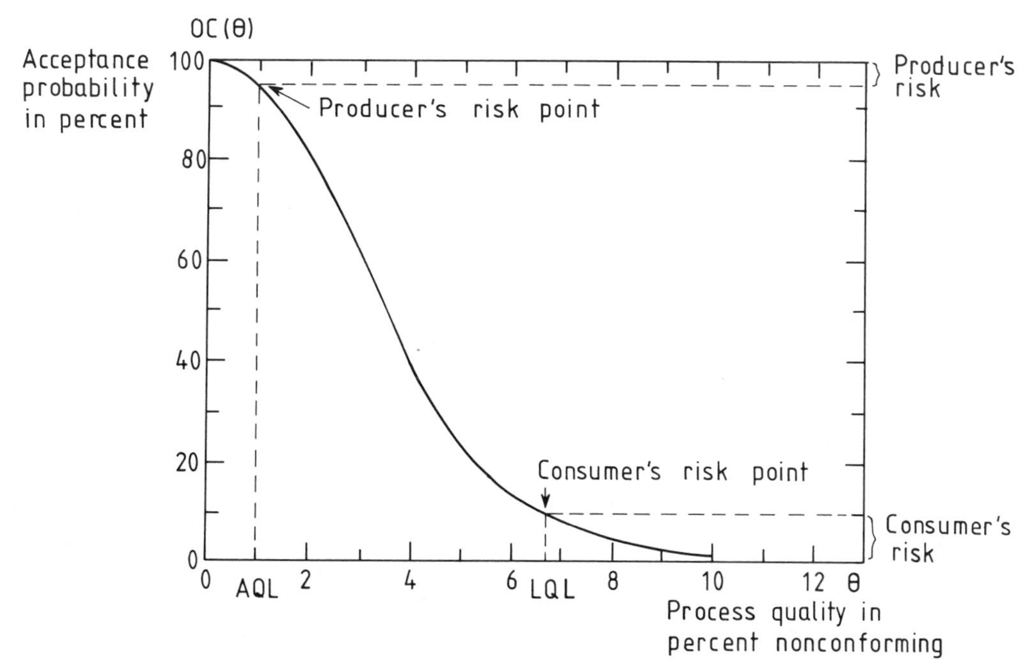

Figure 1 The OC curve for the sampling plan: take a sample of 80 items and accept the lot if two items at most are found not to conform; otherwise do not accept. The process quality parameter Θ is the fraction of nonconforming items.

ing quality level (LQL), may be specified to indicate that the plan should give a high probability of rejection for this value of the process quality parameter. When the protection against *individual lots* of inferior quality is considered important, as in safety inspection and compliance testing, an unsatisfactory lot quality, the limiting quality (LQ) or lot tolerance percent defective (LTPD), may be specified to indicate that the sampling plan should give a high probability of rejection for individual lots of this lot quality.

Often the performance of a sampling plan is summarized by two points on the OC curve: the *producer's risk point*, i.e., the producer's risk quality with the associated producer's risk, and the *consumer's risk point*, i.e., the consumer's risk quality with the associated consumer's risk. The construction of a sampling plan when these points are specified is equivalent to the construction of a statistical test with specified risks of type II and type I. Figure 1 illustrates the concepts related to the OC curve.

SINGLE SAMPLING PLANS WITH SPECIFIED RISKS

A plan for single sampling specifies the size n of a single sample, the quantity T to be determined from the sample, and the acceptance criterion: If $T \in A$ accept; otherwise reject the lot. The form of the sampling distribution, e.g., binomial*, normal*, exponential*, etc., and the parametrization in terms of suitable process quality parameter determine the sample statistic T and the form of the acceptance criterion (acceptance region). The specification of the consumer's and producer's risk points will then determine a unique single sampling plan with an OC curve satisfying the specified criteria. The following examples describe some common sampling distributions and the associated quality parameters.

When the plan is based on a simple classification of the items as conforming ($X = 0$) or not conforming ($X = 1$) to a specified criterion (an *attributes plan*), it is usual to assume that lots are formed from a process modelled by a Bernoulli sequence. The process quality parameter is the probability Θ of a nonconforming item, i.e., the *expected proportion of nonconforming items*. The attributes plan is equivalent to a test of a one-sided hypothesis of a parameter in a Bernoulli sequence. The acceptance criterion is therefore based on a comparison of the number $T = \Sigma X_i$ of nonconforming items in the sample with the acceptance number c for the plan. The distribution of T is a binomial distribution. The OC curve $\text{OC}(\Theta) = \Pr(T \leqslant c; \Theta)$ may be determined from the cumulative distribution function (CDF) of the binomial distribution.

For attributes plans one may also determine the probability of acceptance as a function of *lot quality*, i.e., the number of nonconforming items in the individual lot. Under simple random sampling* from the lot the (hypergeometric) distribution* of T only depends on the actual number of nonconforming items in the lot. Therefore, the protection against individual lots of inferior quality may be assessed by means of the (type A) OC curve giving the acceptance probability as a function of lot quality.

Example 1. An attributes plan may be as follows: Take a sample of 80 items at random from the lot. If at most two items in the sample do not conform with the item specifications, the lot is deemed acceptable; otherwise the lot is not to be accepted. For this plan $n = 80$ and $c = 2$.

When T follows the binomial distribution with sample size 80 and probability parameter $\Theta = 0.01$, one has $\Pr[T \leqslant 2] = 0.96$. Thus, the OC curve passes through the point $(0.01, 0.96)$ and, hence if lots are produced by a process in binomial control with a process average of 1% nonconforming items, then more than 95% of the lots submitted for inspection will be accepted by the sampling plan. The entire OC curve for the plan is shown in Fig. 1.

When T follows a hypergeometric distribution corresponding to a lot size of 1000 items and 100 nonconforming items in the lot, one finds for a sample size of 80 that $\Pr[T \leqslant 2] = 0.009$. Thus, the type A OC curve corresponding to lots consisting of 1000 items passes through the point (0.10, 0.009). The plan will reject more than 99% of lots containing 10% nonconforming items if such lots were submitted for inspection.

For sampling plans based on an enumeration of the number of defects or nonconformities on the individual items or units in the lot (*number of defects/nonconformities plans*) it is usually assumed that the number X of nonconformities on a unit varies over the units in accordance with a Poisson distribution*. The parameter Θ is the *expected number of nonconformities per unit*. The sampling plan is equivalent to a test of a one-sided hypothesis of the parameter of a Poisson distribution. The acceptance criterion for the number of defects plan is therefore based on a comparison of the total number $T = \Sigma X_i$ of nonconformities in the sample with the acceptance number for the plan. The distribution of T is a Poisson distribution and the OC curve may be determined from its CDF.

Plans based on quantitative measurements (variables plans) have been investigated under various assumptions. For many applications it is common to assume that the quantitative values of the quality characteristic for the individual items vary randomly over the items in accordance with a normal distribution characterized by the process mean μ and process standard deviation σ. When the process quality parameter is the *process mean*, the sampling plan is equivalent to a test of the mean in a normal distribution. The acceptance criterion is based on the standardized deviation $T = (\overline{X} - \mu_1)/\sigma$ of the sample average from a specified satisfactory process level μ_1 when σ can be assumed known. If there is no sufficient evidence for the value of the process standard deviation, it is replaced by the sample standard deviation* S, such that the acceptance criterion is based on $T' = (\overline{X} - \mu_1)/S$. The sampling distribution of T and T' is described by the normal distribution and the noncentral t-distribution*, respectively, and the OC curve may be determined from these.

When the process quality parameter is the *proportion Θ of items with quality characteristics beyond a specification limit L*, the acceptance criterion for the variables plan is based on the sample estimate of that proportion determined from the individual quantitative measurements on the sample. For normally distributed item characteristics this estimate may be determined from $T = (\overline{X} - L)/\sigma$ or $T' = (\overline{X} - L)/S$, respectively. The distribution of T and T' depend on μ and σ only through the process quality parameter Θ. The OC curve giving the acceptance probability as a function of Θ may therefore be determined from the normal distribution of T or the noncentral t-distribution related to T'.

The utilisation of the quantitative measurements in the sample results in a more discriminative sampling plan, i.e., one with a steeper OC curve than an attributes plan with the same sample size.

Example 2. Let the quality parameter of interest for an individual item be a one-dimensional real quantity, say the weight of that item. To assess the proportion of items in a lot weighing less than a specified minimum weight L, the sampling plan could be as follows: Take a sample of 35 items and determine the average item weight $\bar{x}$ and the standard deviation s of the sample weights. The lot is deemed acceptable if $\bar{x} \geqslant L + 1.89s$; otherwise the lot is not to be accepted.

For normally distributed item weights with mean μ and standard deviation σ satisfying $(\mu - L)/\sigma = 2.33$, the expected proportion of underweight items is $\Theta = \Phi(-2.33) = 0.01$. The acceptance probability corresponding to this value of the process parameter is

$$\Pr[T' \geqslant 1.89] = \Pr[\sqrt{35}\, T' \geqslant 1.89\sqrt{35}\,] = 0.94$$

since $\sqrt{35}\, T'$ follows a noncentral t-distribution with 34 degrees of freedom and noncentrality parameter $2.33\sqrt{35}$. The OC curve does not deviate much from the OC curve given in Fig. 1 for the attributes plan.

For applications in reliability* and life testing* it is often appropriate to assume that individual times to failure or lifetimes may be described by an exponential distribution, a Weibull distribution*, or an extreme-value distribution*. Relevant process quality parameters are the mean life of items, the hazard rate* after a certain elapsed time, or the reliable life, i.e., the time beyond which a specified proportion of items from the process will survive. The sample statistics pertaining, e.g., to an exponential distribution of individual lifetimes, are the total time on test until a specified number of failures has occurred (*failure terminated testing*) or the number of failures occurring during a specified time (*time terminated testing*). *See* MILITARY STANDARDS FOR FIXED LENGTH LIFE TESTS.

MULTIPLE SAMPLING PLANS

More elaborate plans have been devised in order to reduce the sample size needed to obtain a specified discriminative power (as expressed through the producer's and consumer's risk points).

A *double sampling* plan* allows for the possibility of taking a second sample from the lot before making the final decision regarding the disposition of the lot. For specified producer's and consumer's risk points there exists a multitude of double sampling plans satisfying the specification. Their OC curves for the plans do not deviate very much from the OC curve for the single sampling plan satisfying the same specification. The ASN curves (*see* AVERAGE SAMPLE NUMBER) for the various double sampling plans exhibit greater variation. When selecting such a plan with specified consumer's and producer's risk points it is customary to aim at a plan minimizing the ASN corresponding to the producer's and consumer's risk qualities. Usually the number of parameters is limited by a condition linking the size of the first sample with the size of the second sample, e.g., that they should be equal. Even then there is generally a considerable reduction of the ASN for the double sampling plan as compared with the corresponding single sampling plan.

Example 3. Consider the double sampling plan: Take a sample of 50 items: accept the lot if no nonconforming items are found in the sample; reject if three or more nonconforming items are found. If there are one or two nonconforming items in the sample, take another sample of 50 items and accept the lot if the total number of nonconforming items in the combined sample does not exceed 3, otherwise reject the lot.

If T_1 and T_2 are independent binomially distributed random variables corresponding to a sample size 50 and probability parameter $\Theta = 0.01$, one finds

$$\Pr[T_1 = 0] + \Pr[1 \leqslant T_1 \leqslant 2 \cap T_1 + T_2 \leqslant 3]$$
$$= 0.98.$$

Thus, the OC curve passes through the point (0.01, 0.98). It does not deviate much from that given in Fig. 1 for the single sampling plan.

A further reduction in the ASN may be achieved by *multiple sampling* plans*. The ultimate number of stages in such a plan is reached in a *sequential sampling* plan*, where items are inspected one at a time and a decision to accept, reject, or continue sampling is made after inspection of each item. A sequential sampling plan with specified consumer's and producer's risk points may be determined by the sequential probability ratio test*. A sampling plan determined in this way minimizes the ASN at the producer's and the consumer's risk qualities.

Double and multiple sampling plans have the psychological and legal advantage that lots that appear dubious after inspection of the first sample are rejected only after more evidence (the second sample) has been provided. This advantage is further utilized in *mixed sampling plans*, i.e., double sampling plans for checking the expected proportion of nonconforming items. In a mixed sampling plan the first sample is inspected and evaluated using a criterion based on quantitative measurements (a variables plan). The second (or combined) sample, however, is evaluated according to a criterion based on qualitative measurements (an attributes plan). In this manner the mixed sampling plan combines the efficiency of a variables plan with the robustness of an attributes plan.

OTHER PERFORMANCE MEASURES

The important balance in the trade-off between the sample size and the protection offered by the plan is generally achieved by investigating the OC curve by means of concepts from the theory of hypothesis testing*. The practical implications in the use of sampling plans have, however, led to the development of concepts for performance evaluation and design criteria that are specific to acceptance sampling. In *rectifying inspection**, rejected lots are totally inspected and nonconforming items corrected or replaced by conforming items. Thus, it is possible to take the improvement in the quality of unsatisfactory lots explicitly into consideration when evaluating the performance of a sampling plan. For attribute sampling the process quality parameter for incoming lots is Θ, the expected proportion of nonconforming items in the process. The *average outgoing quality** (AOQ) after sampling and rectification of rejected lots depends on the quality submitted and the corresponding acceptance probability (OC curve) for the plan. For a specified attributes sampling plan the AOQ curve reflects the improvement in quality as a function of incoming quality Θ, and the maximum value of the AOQ, the *average outgoing quality limit** (AOQL), is therefore used to describe the guarantee in terms of AOQ offered by the plan.

When sampling and rectification are performed by the same authority it is appropriate to consider the average amount of inspection, including the total inspection of rejected lots. The *average total inspection* (ATI) function for a specified plan depends on the process quality, the ASN, and the acceptance probability corresponding to that process quality.

PLANS FOR A CONTINUING SUPPLY

Generally, a sampling plan serves the dual purpose of verifying that the production, i.e., the level of the process quality, is satisfactory, and in the same operation the sampling plan weeds out individual lots of unsatisfactory lot quality. When lots are submitted in a continuing series of supply from a stable production process it may be appropriate to reduce the sampling effort at the expense of the protection against individual unsatisfactory lots. (This approach is a necessity when lots are not clearly defined and sampling is directly from the process; *see* CONTINUOUS SAMPLING PLANS.)

A *chain sampling plan* links the sentencing of a lot with the results of the inspection of the immediately preceding lots. In this way the individual inspection results for successive lots are combined to yield a larger effective sample size and, hence a steeper OC curve.

A *skip lot sampling plan** is similarly designed and assessed under the assumption that the individual lots may be considered as random samples from a stable process in statistical control. In a skip lot sampling plan inspection is performed only on a fraction of the lots submitted. The fraction of lots that are skipped depends on the inspection results for the preceeding lots. Generally such a plan will only be introduced when inspection results from a series of lots have provided sufficient evidence to consider the production process as operating steadily at a satisfactory level.

A more differentiated dynamical allocation of sampling effort, and consequently of the discriminatory power of the sampling system, is achieved by linking sets of sampling plans together with rules for switching between the individual plans. The switching rules give conditions for introducing a more or less severe sampling plan depending on the sample results of previously inspected lots. As an ultimate consequence, the application of such rules may lead to the verdict that the *process* is not satisfactory.

BAYESIAN PLANS

The performance evaluation of a sampling plan by means of the OC curve is based on an assessment of the properties of the plan for individual *fixed* values of the process quality parameter. In the Bayesian approach, however, the process quality parameter is modelled as a random variable with a specified prior distribution* of process quality (*see* BAYESIAN INFERENCE). The rule for sentencing the lot is based on the posterior distribution* of process quality or the predictive distribution of lot quality, calculated by Bayes' theorem* from the prior distribution and the inspection result. Thus, the structural approach is similar to that of Bayesian inference with a subjective prior distribution. In the literature on Bayesian sampling plans, however, it is most often assumed that the prior distribution of the process quality is determined from past inspection records and thus allows a direct frequency interpretation with the density function for the prior distribution of process quality given by the process curve.

The statistical specification of the prior distribution may be combined with the economic specification of the costs of sampling inspection, acceptance, and rejection to give an overall measure of the performance of a plan under stationary conditions.

A *cost optimal Bayesian plan* minimizes the expected overall cost of operating the sampling plan under specified assumptions about these costs. Thus, the determination of a cost optimal Bayesian plan uses statistical decision theory* concepts to obtain the balance between (i) the expected cost due to rejection of satisfactory lots, (ii) the expected cost due to acceptance of unsatisfactory lots, and (iii) the expected cost of sampling inspection.

SAMPLING PLANS FOR SPECIFIC PURPOSES

The performance measures of a sampling plan are useful universal tools for assessing its statistical implications when the process quality parameter has a given value. When selecting a sampling plan for a specific application the choice is made by assessing the performance of the plan under the conditions under which it is intended to operate. If operating conditions have been specified in terms of a mathematical model it is sometimes possible to reduce the problem of determining a suitable sampling plan to that of optimizing a quantity chosen to summarize the performance of the plan.

Such *optimal plans* could be those minimizing the expected amount of inspection at the anticipated quality level under the restriction that the acceptance probability for a specific unsatisfactory quality does not exceed a certain value. Other criteria of optimality could be related to models describing the various costs associated with the operation of the plan. Thus, it has been suggested to minimize the expected opportunity loss (a minimax regret* criterion) or to minimize the average cost for a specified distribution of incoming lots (a Bayes criterion).

An *effective sampling plan* is specifically tailored to its particular application and modified by knowledge of how and where nonconformities are most likely to occur. Such customized plans that concentrate the sampling effort at the weakest point of the production may often provide the desired protection at a lower cost than a sampling plan chosen according to more universal performance measures.

LITERATURE

The founder of the theory of sampling inspection was Harold F. Dodge, who originated many acceptance sampling procedures. A historical account of the development of the concepts may be found in refs. 8 and 1, which also contain discussions of basic principles. Surveys of statistical principles and results are given in refs. 2 and 12. Hald [11] has given a comprehensive exposition of the statistical theory relating to attribute sampling (including Bayesian) plans. The book is a valuable reference source, not completely restricted to attribute sampling. Schilling [30] describes, at a less demanding mathematical level, the most widely used acceptance sampling techniques. The book is a useful guide to existing tables and nomograms* of sampling plans. A survey of properties of variables sampling plans for testing the proportion of nonconforming items has been given in refs. 25 and 4. The specific problems of life testing plans are treated in ref. 21; *see also* MILITARY STANDARDS FOR SEQUENTIAL LIFE TESTING and MILITARY STANDARDS FOR FIXED LENGTH LIFE TESTS. Wetherill and Chiu [34] give a review of sampling procedures with emphasis on the economic aspect. The Bayesian approach to reliability and life testing has been reviewed in ref. 33. Methods for assessing the effect of inspection errors are given in ref. 20.

Attribute sampling plans minimizing the ATI under an LTPD or AOQL restriction have been tabulated by Dodge and Romig [7]. MIL-STD-105D [22] provides tables of individual attribute sampling plans and rules for switching between the plans. The sampling scheme in this standard has been adopted as an international standard in ISO/DIS 2859 [15, Part 1]. ISO 3951 [16] gives complementary tables for checking the fraction of nonconforming items by variables sampling. ISO/DIS 8422 [17] and ISO/DIS 8423 [18] give directions and tables for constructing sequential sampling plans for attributes and variables sampling, respectively.

A comparison of the two schemes of sampling plans in ISO/DIS 2859 (Part 1) and in ISO 3951 has been given in ref. 5. The paper also discusses the construction of double sampling plans for variables sampling. Reference 29 provides tables of pseudospecifications that may be used in connection with the plans in ISO/DIS 2859 (Part 1). Methods for evaluating mixed sampling plans have been given in ref. 9.

Sampling plans for multidimensional quality parameters were introduced in ref. 13 and further developed in ref. 19. Reference 3 describes the assessment of sampling plans for simultaneous verification of a set of one-dimensional quality parameters and ref. 23 discusses Bayesian plans for a multiple attributes and multiple criteria.

Attribute sampling plans for sorting lots in various grades are developed in ref. 26. In ref. 6 optimal attribute plans are derived using a minimax regret criterion. Reference 36 introduces the concept of effective sampling plans.

The increasing availability of computer power has simplified the task of determining sampling plans with specific properties and that of assessing sampling plans. References 10, 28, and 32 provide computer programs for evaluating single sampling attributes plans; refs. 14, 24, and 27 give programs for double and multiple plans and ref. 31 contains a program for evaluating continuous sampling plans.

Textbooks and papers on sampling plans appear annually at an ever increasing rate. A recent list of bibliographic and review references is given in ref. 35. Among the more important publication sources are: *Annual Transactions of the American Society for Quality Control*, *Communications in Statistics**, *IIE Transactions*, *Journal of Quality Technology**, *Naval Research Logistics Quarterly**, *Operations Research**, *Quality Control and Applied Statistics*, *Reports of Statistical Application Research* (Union of Japanese Scientists and Engineers, and *Technometrics*.

References

[1] American Statistical Association (1950). *Acceptance Sampling. A Symposium*. Washington, DC.

[2] Barnard, G. A. (1954). *J. R. Statist. Soc. B*, **37**, 151–165.

[3] Berger, R. L. (1982). *Technometrics*, **24**, 295–300.

[4] Bowker, A. H. and Goode, H. P. (1952). *Sampling Inspection by Variables*. McGraw-Hill, New York.

[5] Bravo, P. C. and Wetherill, G. B. (1980). *J. R. Statist. Soc. A*, **143**, 49–67.

[6] Collani, E. V. (1984). *Frontiers in Statistical Quality Control*, Vol. 2, Lenz et al., eds. Physica-Verlag, Würzburg, Germany, pp. 11–24.

[7] Dodge, H. F. and Romig, H. G. (1959). *Sampling Inspection Tables*, 2nd ed. Wiley, New York.

[8] Dodge, H. F. (1969–1970). *J. Quality Tech.*, **1**, 77–88, 155–162, 225–232; **2**, 1–8.

[9] Elder, R. S. and Muse, H. D. (1982). *Technometrics*, **24**, 207–211.

[10] Hailey, W. A. (1980). *J. Quality Tech.*, **12**, 230–235.

[11] Hald, A. (1981). *Statistical Theory of Sampling Inspection by Attributes*. Academic, London.

[12] Hamaker, H. C. (1958). *Appl. Statist.*, **7**, 149–159.

[13] Hotelling, H. (1947). *Techniques of Statistical Analysis*, C. Eisenhart, M. W. Hastay, and W. A. Wallis, eds. McGraw-Hill, New York, pp. 111–184.

[14] Hughes, H., Dickinson, P. C., and Chow, B. (1973). *J. Quality Tech.*, **5**, 39–42.

[15] International Organization for Standardization (1985). Sampling Procedures and Tables for Inspection by Attributes. *ISO/DIS 2859* (draft revision). ISO, Geneva.

[16] International Organization for Standardization (1985). Sampling Procedures and Charts for Inspection by Variables for Percent Defective. *ISO 3951*, 2nd ed. ISO, Geneva.

[17] International Organization for Standardization (1985). Sequential Sampling Plans for Inspection by Attributes (Proportion of Non-Conforming Items and Mean Number of Non-Conformities per Unit) Using the Wald Method. *ISO/DIS 8422* (draft). ISO, Geneva.

[18] International Organization for Standardization (1985). Sequential Sampling Plans for Inspection by Variables for Percent Non-Conforming (Known Standard Deviation). *ISO/DIS 8423* (draft). ISO, Geneva.

[19] Jackson, J. E. and Bradley, R. A. (1961). *Technometrics*, **3**, 519–534.

[20] Kotz, S. and Johnson, N. L. (1984). *Operat. Res.*, **32**, 575–583.

[21] Mann, N. R., Schafer, R. E., and Singpurwalla, N. D. (1974). *Methods for Statistical Analysis of Reliability and Life Data*. Wiley, New York.

[22] MIL-STD-105D (1963). *Military Standard, Sampling Procedures and Tables for Inspection by Attributes*. United States Department of Defense, U.S. Government Printing Office, Washington, DC.

[23] Moskowitz, H. (1985). *Naval Res. Logist. Quart.*, **32**, 81–94.

[24] Olorunniwo, F. O. and Salas, J. R. (1982). *J. Quality Tech.*, **14**, 166–171.

[25] Owen, D. B. (1969). *Technometrics*, **11**, 631–637.

[26] Pandey, R. J. and Chawdhary, A. K. (1972). *Sankhyā B*, **34**, 265–278.

[27] Rutemiller, H. C. and Schafer, R. E. (1985). *J. Quality Tech.*, **17**, 108–113.

[28] Schilling, E. G., Sheesley, J. H., and Nelson, P. R. (1978). *J. Quality Tech.*, **10**, 125–130; corrigenda, p. 244.

[29] Schilling, E. G. and Sommers, D. J. (1981). *J. Quality Tech.*, **13**, 83–92.

[30] Schilling, E. G. (1982). *Acceptance Sampling in Quality Control*. Marcel Dekker, New York.

[31] Sheesley, J. H. (1975). *J. Quality Tech.*, **7**, 43–45.

[32] Snyder, D. C. and Storer, R. F. (1972). *J. Quality Tech.*, **4**, 168–171.

[33] Tillman, F. A., Kuo, W., Hwang, C. L., and Grosh, D. L. (1982). *IEEE Trans. Rel.*, **R-31**, 362–372.

[34] Wetherill, G. B. and Chiu, W. K. (1975). *Int. Statist. Rev.*, **43**, 191–210.

[35] Wetherill, G. B. (1984). *Frontiers in Statistical Quality Control*, Vol. 2, Lenz et al., eds. Physica-Verlag, Würzburg, Germany, pp. 274–291.

[36] Zwickl, R. D. (1974). *Annual Technical Conference Transactions*. American Society for Quality Control, pp. 369–375.

(ACCEPTABLE QUALITY LEVEL
ACCEPTANCE SAMPLING
AVERAGE OUTGOING QUALITY
AVERAGE OUTGOING QUALITY LEVEL
CONTINUOUS SAMPLING PLANS
DOUBLE SAMPLING
MULTIPLE SAMPLING
QUALITY CONTROL, STATISTICAL
SAMPLING SCHEME)

P. THYREGOD

SAMPLING SCHEME

A set of rules indicating which sampling plans* are to be used under various conditions. For example, a sampling scheme may include a sampling plan to be used under normal conditions, a tighter (heavier) plan to use if production quality appears to be deteriorating, and a looser (lighter) plan for use if production quality has established a good record.

SAMPLING SURVEY *See* SURVEY SAMPLING

SAMPLING THEORY WITH DEPENDENT OBSERVATIONS, ASYMPTOTIC

A large proportion of the data available to econometricians are time series*. Textbooks of econometrics* often present the regression analysis of such data as basically similar to the treatment of cross-section data from the statistical point of view, apart from certain well-known "pathological" conditions such as residual autocorrelation. But the problems of statistical inference are fundamentally different when the assumption of random sampling cannot be applied to the data generation processes under study. Time series contain trends, cyclical and seasonal components, and irregular components persisting over several periods. Statistical inference in econometrics, outside the (unrealistic) classical regression model with fixed regressors and normal disturbances, is based on the application of the law of large numbers* (LLN) and the central limit theorem* (CLT) to the averaging procedures underlying regression analysis. The assumption of independent sampling on which the standard proofs of these asymptotic results depend is too strong in the time series context.

Since econometric models must explain dependence across time as well as between variables, they often take the general form of systems of stochastic linear difference equations. The sampling properties of such systems were originally investigated by Mann and Wald [7]. The simplest possible example of this type of model is the *first-order autoregressive* equation

$$x_t = \phi x_{t-1} + \epsilon_t, \qquad \epsilon_t \sim IN(0, \sigma^2). \quad (1)$$

When $|\phi| < 1$, x_t is *stationary*, meaning that it has a fixed unconditional distribution—in this case with $E(x_t) = 0$, $\text{Var}(x_t) = \sigma_x^2 = \sigma^2/(1-\phi^2)$, and $E(x_t x_{t+j}) = \phi^j \sigma_x^2$, for $j > 0$. Such a series is also *ergodic*; that is, the time average of a single realization of the series converges to the corresponding unconditional mean, so that averaging *within* a single infinite realization of the process is equivalent to averaging *across* realizations. *See* ERGODIC THEOREMS and STATIONARITY.

The maximum likelihood* estimator of ϕ from a sample $x_1, \dots, x_n$ is

$$\hat{\phi} = \left[\sum_{t=2}^{n} x_{t-1}^2\right]^{-1} \sum_{t=2}^{n} x_t x_{t-1}.$$

The proof of asymptotic normality of $n^{1/2}(\hat{\phi} - \phi)$ depends upon showing that (a) $n^{-1/2}\sum_{t=2}^{n} \epsilon_t x_{t-1}$ obeys the CLT and (b) $n^{-1}\sum_{t=2}^{n} x_{t-1}^2$ converges in probability to σ_x^2. Mann and Wald's theorem shows that these results follow when $|\phi| < 1$. While the individual terms $Z_{nt} = n^{-1/2}\epsilon_t x_{t-1}$ are not mutually independent, the dependence between Z_{nt} and $Z_{n,t+j}$ declines exponentially with $|j|$. The strategy of proof for (a) is to break up the sum $S_n = Z_{n1} + Z_{n2} + \cdots + Z_{nn}$ into pairs of partial sums of length b_n and l_n, respectively, where as $n \to \infty$, $b_n \to \infty$, and $l_n \to \infty$, but $(b_n/n) \to 0$ and also $(l_n/b_n) \to 0$. For example, set $b_n + l_n = n^\alpha$, $l_n = n^{\alpha\beta}$ for $0 < \alpha, \beta < 1$. Thus,

$$S_n = \sum_{i=1}^{r_n} (U_{ni} + V_{ni}) + \left(Z_{n, r_n(b_n + l_n)+1} + \cdots + Z_{nn}\right),$$

where r_n is the largest integer i for which $i(b_n + l_n) < n$, and

$$U_{ni} = \sum_{t=(i-1)(b_n+l_n)+1}^{(i-1)(b_n+l_n)+b_n} Z_{nt},$$

$$V_{ni} = \sum_{t=(i-1)(b_n+l_n)+b_n+1}^{i(b_n+l_n)} Z_{nt}.$$

Now, the U_{ni} approach mutual independence as $l_n \to \infty$, and satisfy the conditions of the Liapunov CLT asymptotically. But the V_{ni} are of small order relative to the U_{ni}, and it can be shown that their omission does not effect the asymptotic distribution of the sum. Hence, $S_n \xrightarrow{\mathscr{D}} N(0, \sigma_x^2\sigma^2)$ (noting that ϵ_t and x_{t-1} are independent). Then from (b), and using a theorem of Cramér [2, p. 254] by which the asymptotic distribution of $n^{1/2}(\hat{\phi} - \phi)$ is unchanged when factors that converge in probability to constants are replaced by those constants, we obtain

$$n^{1/2}(\hat{\phi} - \phi) \to N(0, 1 - \phi^2).$$

More recent results do not require linearity and stationarity. General measures of dependence in a stochastic process* are the *mixing coefficients* $\alpha(k)$ and $\phi(k)$. Let $Z_1, Z_2, \dots$ be a stochastic sequence and let $\mathscr{F}_t = \sigma(Z_1, \dots, Z_t)$ and $\mathscr{F}^{t+k} = \sigma(Z_{t+k+1}, Z_{t+k+2}, \dots)$ be the Borel subfields generated by the sequence up to time t and beyond time $t + k$, respectively. Then

$$\alpha(k) = \sup_t \left\{ \sup_{A \in \mathscr{F}_t,\, B \in \mathscr{F}^{t+k}} |P(A \cap B) - P(A)P(B)| \right\}$$

and

$$\phi(k) = \sup_t \left\{ \sup_{A \in \mathscr{F}_t,\, B \in \mathscr{F}^{t+k},\, P(A) > 0} |P(B|A) - P(B)| \right\}.$$

If $\operatorname{Lim}_{k \to \infty} \alpha(k) = 0$, the sequence is said to be *α-mixing* and if $\operatorname{Lim}_{k \to \infty} \phi(k) = 0$, the sequence is said to be *φ-mixing*, a condition which is seen to imply α-mixing. Mixing sequences have a restricted "memory," with dependence declining to zero as the time separation of the terms increases, although the rate of decline need not be exponential. If a mixing sequence is stationary, it is also ergodic. For further details see White [9] and ERGODIC THEOREMS.

CLTs generalising the Mann–Wald approach can be proved for mixing processes. For example (White and Domowitz [10]; Serfling [8]): let Z_t be an α-mixing (φ-mixing) sequence where (i) $E(Z_t) = \mu_t$, $\operatorname{Var}(Z_t) = \sigma_t^2 > 0$ and for some $r > 1$, $E|Z_t|^{2r}$ is bounded uniformly in t; (ii) $\alpha(k) = O(k^{-\lambda})$ $[\phi(k) = O(k^{-\lambda})]$ for $\lambda > r/(r-1)$; (iii) $\operatorname{Lim}_{n \to \infty} \bar{\sigma}^2_{m,n} = \bar{\sigma}^2 < \infty$, uniformly in m, where $\bar{\sigma}^2_{m,n} = \operatorname{Var}(n^{-1/2}\sum_{t=m+1}^{m+n} Z_t)$; then $n^{-1/2}\sum_{t=1}^{n}(Z_t - \mu_t)/\bar{\sigma}_{0,n} \xrightarrow{\mathscr{D}} N(0,1)$. Observe that there is a trade-off between the rate of convergence to zero of the mixing coefficients and the order of the absolute moments required to exist. When dependence dies out exponentially fast, as in the case of (1), then the existence of the $(2 + \delta)$th absolute moment is sufficient, for any $\delta > 0$, as in the regular Liapunov CLT; otherwise stronger moment conditions are needed.

A somewhat different approach to the problem is provided by martingale* theory. Here, the *type* of dependence is restricted. An analogy is provided by the Chebyshev LLN, which says that the sample mean of an *uncorrelated* sequence with finite variance (e.g., the sequence $\epsilon_t x_{t-1}$) converges in probability, regardless of more general dependence. Uncorrelatedness is not a strong enough property to yield a CLT, however.

A martingale difference sequence $Z_1, Z_2, \dots$ has the property

$$E(Z_t|\mathscr{F}_{t-1}) = 0,$$

where $\mathscr{F}_{t-1}$ is the Borel subfield generated by the previous history of the sequence. An example is the sequence of innovations relative to $\mathscr{F}_{t-1}$, $Z_t = Y_t - E(Y_t|\mathscr{F}_{t-1})$, in any stochastic process Y_t satisfying $E|Y_t| < \infty$. It carries the implication that

$$E(Z_t f(Z_1, \dots, Z_{t-1})) = 0$$

for any measurable function $f(\cdot)$ of the past values of the sequence. Thus it is a stronger property than simple uncorrelatedness, but also significantly weaker than independence.

The martingale difference property is important because it is characteristic of the sequence on which the distributions of linear and nonlinear estimators depend; the derivatives of sum-of-squares and log-likelihood functions evaluated at the true parameters, for example. This is because econometric time series modelling proceeds by formulating data densities and expectations conditioned sequentially on the past (see, e.g., Engle et al. [3]). Referring back to eq. (1), note that $E(\epsilon_t x_{t-1}|\mathscr{F}_{t-1}) = E(\epsilon_t)x_{t-1} = 0$, hence $\epsilon_t x_{t-1}$ is a martingale difference. Martingale CLTs (e.g., Macleish [6]; Serfling [8]) impose generally weaker conditions than the CLTs for mixing processes. In the theorem quoted above, if Z_t is a martingale difference (implying $\mu_t = 0$) then (ii) and (iii) can be replaced by (ii′) plim $n^{-1}\Sigma_{t=1}^{n}(Z_t^2 - \sigma_t^2) = 0$. The moment restrictions are therefore minimal. See White [9] and Hall and Heyde [5] for more on the role of martingales in estimation theory.

In practice, economic time series often have an evolutionary character; they contain trends and appear capable of "going anywhere" in the long run. One way to capture this characteristic in modelling is to assume the data generation processes are *integrated* (contain unit roots), an approach exploited in the Box–Jenkins* [1] forecasting* methodology, for example. Such processes are neither stationary nor mixing, and their memory is unrestricted; initial conditions influence the whole subsequent path. A simple example is the *random walk**, eq. (1) with $\phi = 1$. If the starting point x_0 is fixed, $\text{Var}(x_t) = E(x_t x_{t+j}) = t\sigma^2$, for $t > 0$ and $j > 0$.

Here it can be shown that $\hat{\phi}$ is "superconsistent," in that the error of estimate is of $O(n^{-1})$, not $O(n^{-1/2})$; but the limiting distribution of $n(\hat{\phi} - 1)$ is *not* normal. (See, e.g., Evans and Savin [4].) Similar properties are likely to extend to least-squares regressions involving variables generated by a multivariate integrated stochastic process. Regressions with nonstationary data should not be assumed to yield asymptotically valid normal confidence intervals, the large volume of published applied results derived from such data notwithstanding.

References

[1] Box, G. E. P. and Jenkins, G. M. (1970). *Times Series Analysis, Forecasting and Control*. Holden-Day, San Francisco.

[2] Cramér, H. (1946). *Mathematical Methods of Statistics*. Princeton University Press, Princeton, NJ.

[3] Engle, R. F., Hendry, D. F., and Richard, J-F. (1983). Exogeneity. *Econometrica*, **51**, 277–304.

[4] Evans, G. B. A. and Savin, N. E. (1981). Testing for unit roots: 1. *Econometrica*, **49**, 753–781.

[5] Hall, P. and Heyde, C. C. (1980). *Martingale Limit Theory and its Application*. Academic, New York.

[6] Macleish, D. L. (1974). Dependent central limit theorems and invariance principles. *Ann. Probab.*, **2**, 620–628.

[7] Mann, H. B. and Wald, A. (1943). On the statistical treatment of linear stochastic difference equations. *Econometrica*, **11**, 173–200.

[8] Serfling, R. J. (1968). Contributions to central limit theory for dependent variables. *Ann. Math. Statist.* **39**, 1158–1175.

[9] White, H. (1984). *Asymptotic Theory for Econometricians*. Academic, New York.

[10] White, H. and Domowitz, I. (1984). Nonlinear regression with dependent observations. *Econometrica*, **52**, 143–163.

(ECONOMETRICS
ERGODIC THEOREMS
LARGE-SAMPLE THEORY
LIMIT THEOREM, CENTRAL
MARTINGALES
TIME SERIES)

JAMES DAVIDSON

SAMPLING WITHOUT RECALL *See* DECISION THEORY

SAMPLING WITHOUT REPLACEMENT

A system of sampling wherein no individual in a population can be chosen more than once. In *simple random* sampling* without replacement from a population of N individuals, a sample of size n is equally likely to consist of any one of the $\binom{N}{n}$ possible sets of n among the N individuals.

(FINITE POPULATIONS, SAMPLING FROM)

SAMPLING ZEROS

In contrast to structural zeros*, these are cells in contingency tables* that are empty by reason of paucity of data or (if one prefers) chance. Goodman [1] discusses sampling zeros in the context of log-linear models.

Reference

[1] Goodman, L. A. (1970). *J. Amer. Statist. Ass.*, **65**, 226–256.

(STRUCTURAL ZEROS)

SAMUELSON–NAIR INEQUALITY

Given n real numbers, $x_1 \geqslant x_2 \geqslant \cdots \geqslant x_n$, the inequalities

$$x_1 \leqslant m + s(n-1)^{1/2} \tag{1}$$

and

$$x_n \geqslant m - s(n-1)^{1/2} \tag{2}$$

were obtained by Samuelson [7] and earlier by Nair [5, 6]. In (1) and (2) the mean $m = \Sigma_1^n x_i/n$ and the variance

$$s^2 = \sum_1^n (x_i - m)^2/n.$$

Thus the largest of n real numbers cannot exceed the mean by more than $(n-1)^{1/2}$ standard deviations, while the smallest cannot be less than the mean by more than that amount. Equality in (1) holds if and only if

$$x_1 = m + s(n-1)^{1/2}$$

and

$$x_2 = x_3 = \cdots x_n = m - s/(n-1)^{1/2}, \tag{3}$$

while equality in (2) holds if and only if

$$x_1 = x_2 = \cdots = x_{n-1} = m + s/(n-1)^{1/2}$$

and

$$x_n = m - s(n-1)^{1/2}. \tag{4}$$

Brunk [2] showed that

$$x_1 \geqslant m + s/(n-1)^{1/2} \tag{5}$$

and

$$x_n \leqslant m - s/(n-1)^{1/2}, \tag{6}$$

providing in addition to (1) and (2) a lower bound for the maximum and an upper bound for the minimum of n real numbers. Equality holds in (5) if and only if (4) holds, while equality holds in (6) if and only if (3) holds. It follows then that the upper bound (1) for the largest of a set of real numbers is attained if and only if the smallest attains its upper bound (6); similarly, the smallest of a set of real numbers attains its lower bound (2) if and only if the largest attains its lower bound (5).

For $k = 1, 2, \ldots, n$, the inequalities

$$m - s\left(\frac{k-1}{n-k+1}\right)^{1/2} \leqslant x_k \leqslant m + s\left(\frac{n-k}{k}\right)^{1/2}, \tag{7}$$

which were obtained by Boyd [1] and Hawkins [3], provide lower and upper bounds for the kth largest of n real numbers. When $k = 1$, the right-hand side of (7) becomes (1), and when $k = n$, the left-hand side of (7) becomes (2). Equality holds on the left of (7) if and only if

$$x_1 = x_2 = \cdots = x_{k-1} = m + s\left(\frac{n-k+1}{k-1}\right)^{1/2}$$

and

$$x_k = x_{k+1} = \cdots = x_n = m - s\left(\frac{k-1}{n-k+1}\right)^{1/2}, \tag{8}$$

and on the right if and only if

$$x_1 = x_2 = \cdots = x_k = m + s\left(\frac{n-k}{k}\right)^{1/2}$$

and

$$x_{k+1} = x_{k+2} = \cdots = x_n = m - s\left(\frac{k}{n-k}\right)^{1/2}. \tag{9}$$

The inequalities (7), as well as the inequalities (1) and (2), also follow directly from results established by Mallows and Richter [4]. Additional references are given by Wolkowicz and Styan [8].

The inequalities (1), (2), and (7) may be used to obtain bounds for the eigenvalues of an $n \times n$ matrix $\mathbf{A}$; cf. ref. 9. When the eigenvalues of $\mathbf{A}$ are all real (which is so when $\mathbf{A}$ is real and symmetric or complex and Hermitian), then (1), (2), and (7) hold with x_k denoting the kth largest eigenvalue of $\mathbf{A}$, and with

$$m = \operatorname{tr}\mathbf{A}/n \quad \text{and} \quad s^2 = (\operatorname{tr}\mathbf{A}^2/n) - m^2, \tag{10}$$

where tr denotes the trace (sum of the diagonal elements).

Another application is to the roots of an nth degree polynomial equation. If

$$x^n + c_1x^{n-1} + c_2x^{n-2} + \cdots + c_{n-1}x + c_n = 0 \tag{11}$$

has real roots, $x_1 \geqslant x_2 \geqslant \cdots \geqslant x_n$, say, then (1), (2), and (7) hold with $m = -c_1/n$ and $s^2 = [(n-1)^2c_1^2 - 2nc_2]/n$; cf. ref. 9.

References

[1] Boyd, A. V. (1971). *Publ. Faculté Électrotech. Univ. Belgrade, Sér. Math. Phys.*, **365**, 31–32.

[2] Brunk, H. D. (1959). *J. Indian Soc. Agric. Statist.*, **11**, 186–189.

[3] Hawkins, D. M. (1971). *J. Amer. Statist. Ass.*, **66**, 644–645.

[4] Mallows, C. L. and Richter, D. (1969). *Ann. Math. Statist.*, **40**, 1922–1932.

[5] Nair, K. R. (1948). *J. Indian Soc. Agric. Statist.*, **1**, 162–172. (Earliest paper on the Samuelson–Nair Inequality.)

[6] Nair, K. R. (1959). *J. Indian Soc. Agric. Statist.*, **12**, 189–190.

[7] Samuelson, P. A. (1968). *J. Amer. Statist. Ass.*, **63**, 1522–1525.

[8] Wolkowicz, H. and Styan, G. P. H. (1979, 1980). *Amer. Statist.*, **33**, 143–144, and **34**, 250–251. (Contains additional references.)

[9] Wolkowicz, H. and Styan, G. P. H. (1980). *Linear Algebra Appl.*, **29**, 471–506. (Includes applications to eigenvalues.)

(CANTELLI'S INEQUALITY
CAUCHY-SCHWARZ INEQUALITY
CHEBYSHEV'S INEQUALITY
EIGENVALUES)

HENRY WOLKOWICZ
GEORGE P. H. STYAN

SAMUELSON'S INEQUALITY *See* SAMUELSON–NAIR INEQUALITY

SANGHVI'S INDEX (SANGHVI'S DISTANCE)

Sanghvi's [4] index G^2 for the distance between two populations using *attribute* (discrete) *data* was originally given by

$$G^2 = 100 \sum_{j=1}^{r} \sum_{k=1}^{s_j+1} \left[\frac{(p_{1jk} - p_{jk})^2}{p_{jk}} + \frac{(p_{2jk} - p_{jk})^2}{p_{jk}} \right] \left(\sum_{j=1}^{r} s_j \right)^{-1},$$

where p_{1jk} and p_{2jk} are proportions in the kth class for the jth character in populations P_1 and P_2, respectively, $p_{jk} = \frac{1}{2}(p_{1jk} + p_{2jk})$ and $s_j + 1$ is the number of classes for the jth character. (Compare with the χ^2 statistic.) Alternatively

$$G_s^2 = \sum_{j=1}^{r} \sum_{k=1}^{s_j+1} \frac{(p_{1jk} - p_{2jk})^2}{p_{jk}}$$

may be used (by removing the constants). Alternative indices for measuring distances between populations using attribute data have been proposed by Edwards and Cavalli-Sforza [2], Steinberg et al. [5], Balakrishnan and Sanghvi [1], and Kurczynski [3]. The last paper presents comparisons between various definitions.

References

[1] Balakrishnan, V. and Sanghvi, L. D. (1968). *Biometrics*, **24**, 859–865.

[2] Edwards, A. W. F. and Cavalli-Sforza, L. L. (1964). *The Systematic Ass. Publ. No. 6*, The Systematic Association, London, 67–76.

[3] Kurczynski, T. W. (1970). *Biometrics*, **26**, 525–534.

[4] Sanghvi, L. D. (1953). *Amer. J. Phys. Anthrop.*, **11**, 385–404.
[5] Steinberg, A. G., Bleibtreau, M. K., Kurczynski, T. W., Martin, A. O., and Kurczynski, E. M. (1967). In *Proc. Third Int. Congress Hum. Genetics*, J. F. Crow and J. V. Neel, eds. Johns Hopkins Press, Baltimore, pp. 267–289.

(MAHALANOBIS (GENERALIZED) DISTANCE)

SANOV INEQUALITY

An inequality for the probability of large deviations in a multinomial distribution*. Let

$$\mathscr{S} = \left\{ X = (X_1, \ldots, X_k) : X_i \leqslant 0, \quad i = 1, \cdots k, \sum_{i=1}^{k} X_i \equiv 1 \right\}$$

denote a subset of a k-dimensional vector space and for each n, let $\mathbf{p}_n = (n_1/n, \ldots, n_k/n)$ be the vector of relative frequencies from a multinomial distribution with parameters $(n; \mathbf{p} \equiv (p_1, \ldots, p_k))$ belonging to the set $\mathscr{S}$. Then for any compact subset $\mathscr{A}$ of $\mathscr{S}$ and point $p \notin \mathscr{A}$, the Sanov [2] inequality asserts that

$$\Pr[p_n \in \mathscr{A} | p] \leqslant (n+1)^K e^{-nK(\mathscr{A}, p)},$$

where $K(\mathscr{A}, p) = \inf_q K(q, p)$ and $K(q, p)$ is the Kullback–Liebler information* of q with respect to p. For extensions and applications see ref 1.

References

[1] Fu, J. C. (1983). *Statist. Prob. Lett.*, **1**, 107–113.
[2] Sanov, I. N. (1957). *Math. Sbornik*, **42**, 11–44 (English transl.: *Sel. Trans. Math. Statist. Prob.*, **1**, 213–244).

(BERNSTEIN INEQUALITY
CHERNOFF INEQUALITY
LARGE DEVIATIONS)

SARGAN DISTRIBUTIONS

The Sargan distribution of order ρ has probability density function

$$f(u) = \tfrac{1}{2} K \alpha e^{-\alpha|u|} \left(1 + \sum_{j=1}^{\rho} \gamma_j \alpha^j |u|^j \right)$$

with $\alpha \geqslant 0$, $\gamma_j \geqslant 0$ $(j = 1, \ldots, \rho)$, and $K = \sum_{j=1}^{\rho} (1 + j!\gamma_j)^{-1}$. This class includes the Laplace distributions* as the distributions of order zero. The distributions are mixtures of reflected gamma distributions*; they are symmetric (about zero) and unimodal. For $\rho = 2$, the density is unimodal if and only if $\gamma_2 \leqslant \frac{1}{2}$. The rth central (and crude) moment is (see [4])

$$\mu_r = \begin{cases} 0 & \text{if } r \text{ is odd,} \\ K\alpha^{-r} \left(r! + \sum_{j=1}^{\rho} \gamma_j (j+r)! \right) & \text{if } r \text{ is even.} \end{cases}$$

Goldfield and Quandt [1] suggest the use of Sargan distributions as a "viable alternative to the normal distribution in econometric* models" because of its clear computational advantages. They attribute this suggestion to J. D. Sargan. Tse [4] compares several approximations to a unit normal distribution, and finds one with $\rho = 2$, $\gamma_1 = 1$, $\gamma_2 = 0.6888$ and $\alpha = 2.6950$ to give good results, even though the density is bimodal.

Missiakoulis [3] points out that the arithmetic mean* of $(\rho + 1)$ independent variables with a common Laplace distribution has a ρ-order Sargan distribution, and discusses the approximation to normality further. Kafaei and Schmidt [2] examined the asymptotic bias (inconsistency) of the parameter estimates using the Sargan approximation to the normal distribution in linear, censored (tobit*), and truncated regression models.

References

[1] Goldfield, S. M. and Quandt, R. E. (1981). *J. Econometrics*, **17**, 141–155.

[2] Kafaei, M.-A. and Schmidt P. (1985). *Commun. Statist. A*, **14**, 509–526.

[3] Missiakoulis, S. (1983). *J. Econometrics*, **23**, 223–233.

[4] Tse, Y. K. (1987). *J. Econometrics*, **34**, 349–354.

(APPROXIMATIONS TO DISTRIBUTIONS
LAPLACE DISTRIBUTION
NORMAL DISTRIBUTION)

SATTERTHWAITE'S FORMULA

In estimating the variance of a mean, in estimating a variance component, or in constructing an approximate *F*-test*, it is frequently necessary, particularly with unbalanced data, to form a linear function of mean squares, $s^2 = \Sigma c_i s_i^2$, where the c_i are known constants. Assuming normality, Satterthwaite [5] suggested that s^2 is distributed approximately as $\chi_f^2 \sigma^2 / f$, where $\sigma^2 = E(s^2)$. Satterthwaite's formula for the chi-square degrees of freedom is

$$f = \left(\sum c_i s_i^2\right)^2 \Big/ \left[\sum \left(c_i s_i^2\right)^2 / f_i\right],$$

where f_i is the degrees of freedom associated with s_i^2. An approximate $(1 - \alpha)100\%$ confidence interval for a mean is given by $\bar{x} \pm t_{f, \alpha/2} s$, where $\bar{x}$ is the estimate of the mean and t is the tabled value of Student's-t corresponding to the one-tailed $(\alpha/2)$ level with approximately f degrees of freedom of the estimated variance of the mean s^2 given by Satterthwaite's formula.

In comparing two means from normal populations with unequal variances, known as the Behrens–Fisher problem*, the approximate t-test is given by

$$t_f = \left(\bar{x}_1 - \bar{x}_2\right) \Big/ \sqrt{\frac{s_1^2}{n_1} + \frac{s_2^2}{n_2}},$$

where $\bar{x}_i$ is the mean of n_i observations and s_i^2 is the calculated variance, $i = 1, 2$. The approximate degrees of freedom are given by Smith [6], which is a special case of Satterthwaite's formula,

$$f = s^4 \Big/ \left[\frac{\left(s_1^2/n_1\right)^2}{n_1 - 1} + \frac{\left(s_2^2/n_2\right)^2}{n_2 - 1}\right],$$

where the variance of the difference of the two means is estimated by $s^2 = (s_1^2/n_1) + (s_2^2/n_2)$. For the difference of two means, the approximate degrees of freedom for the sum of two variances is between the minimum of $(n_1 - 1)$ and $(n_2 - 1)$ and the pooled degrees of freedom $(n_1 + n_2 - 2)$. Welch [7] showed that the approximation given by Satterthwaite was adequate for this case where mean squares are added. An approximate $(1 - \alpha)100\%$ confidence interval is given by

$$\left(\bar{x}_1 - \bar{x}_2\right) \pm t_{f, \alpha/2} \sqrt{\frac{s_1^2}{n_1} + \frac{s_2^2}{n_2}}.$$

If a variance component σ^2 is estimated in an analysis of variance* by $s^2 = \Sigma c_i s_i^2$, approximate $(1 - \alpha)100\%$ confidence limits for σ^2 are given by

$$\frac{fs^2}{\chi_2^2} \leqslant \sigma^2 \leqslant \frac{fs^2}{\chi_1^2},$$

where $\Pr[\chi^2 \geqslant \chi_2^2] = \Pr[\chi^2 \leqslant \chi_1^2] = \alpha/2$ and f is given by Satterthwaite's formula.

Estimation of variance components* generally requires subtracting some mean squares, i.e., some c_i are negative. Satterthwaite [5] noted that caution should be exercised in applying his formula for f when some of the c_i are negative. In this case, the problem can be simplified by putting the mean squares in a linear function into two groups: those with positive coefficients s_1^2 and those with negative coefficients s_2^2, so that $s^2 = s_1^2 - s_2^2$, where s_1^2 and s_2^2 may be sums of mean squares with approximately f_1 and f_2 degrees of freedom as given by Satterthwaite's formula. The approximate variance for the difference of mean squares given by Satterthwaite's formula has degrees of freedom

$$f = \left(s_1^2 + s_2^2\right)^2 \Bigg/ \left(\frac{s_1^4}{f_1} + \frac{s_2^4}{f_2}\right).$$

Gaylor and Hopper [3] show that this is an adequate approximation for the degrees of freedom when $\sigma_1^2/\sigma_2^2 \geqslant F_{f_2, f_1(0.025)}$ for $f_1 \leqslant 100$ if $f_2 \geqslant f_1/2$, where $F_{f_2, f_1(0.025)}$ is the upper 0.025 tail of Snedecor's F (note the reversal of the degrees of freedom). That is, the approximation is adequate for differences of mean squares when the mean square being subtracted is relatively small.

In the analysis of variance, exact F-tests frequently cannot be performed from the ratio of two mean squares, particularly with unequal sampling. In such cases, linear combinations of mean squares can be formed for the numerator and/or denominator to provide approximate F-tests (Anderson [1] and Eisen [2]). Suppose it is desired to test the null hypothesis that a particular component of variance is zero: $\sigma_k^2 = 0$. A denominator mean square for an approximate F-test can be constructed, $s^2 = \sum_{i=1}^{k-1} c_i s_i^2$, such that $E(s^2) = E(s_k^2)$ under the null hypothesis. An approximate test is given by $F = s_k^2/s^2$ with f_k and f degrees of freedom, where f is calculated by Satterthwaite's formula. Alternatively, the numerator mean square for the approximate F-test can be constructed or both the numerator and denominator mean squares can be constructed such that their expected values are equal under the null hypothesis. In the latter case, it is always possible to find additive combinations of mean squares for both the numerator and denominator, thereby avoiding subtracting mean squares, which may lead to poor approximations.

Mee and Owen [4] employ Satterthwaite's formula for constructing one-sided tolerance limits for observations from a balanced one-way random model.

References

[1] Anderson, R. L. (1960). *Bull. Int. Statist. Inst.*, **37**, 1–22.

[2] Eisen, E. J. (1966). *Biometrics*, **22**, 937–942.

[3] Gaylor, D. W. and Hopper, F. N. (1969). *Technometrics*, **11**, 691–706.

[4] Mee, R. W. and Owen, D. B. (1983). *J. Amer. Statist. Ass.*, **78**, 901–905.

[5] Satterthwaite, F. E. (1946). *Biometrics Bull.*, **2**, 110–114.

[6] Smith, H. F. (1936). *J. Sci. Ind. Res. India*, **9**, 211–212.

[7] Welch, B. L. (1949). *Biometrika*, **36**, 293–296.

(BEHRENS–FISHER PROBLEM
F-TESTS
WELCH TESTS)

D. W. GAYLOR

SATURATED DESIGNS

A saturated experimental design requires that the number n of trials to be run is equal to the number of parameters to be estimated. These parameters include the mean and the main effects* of each of the factors whose influence on the response variable is being studied. Saturated designs are generally used when there are many factors under consideration but only a few that will eventually be studied in detail. The preliminary experiment, a saturated (or a supersaturated*) design, will indicate which factors should be studied, since it will give estimates of the main effects of each of the factors. It will not give estimates of the possible interactions* of the factors or of the experimental standard error. Since the purpose is to search out important factors from a much larger set rather than to find the particular combination of values of factors that give an optimum result, only the high (+) and low (−) values for each factor need to be used. This allows saturated designs to be used for both quantitative and categorical variables.

The simplest possible design than can be run to allow for estimates of the main effects of each factor is a series of trials where only one of the factors is changed in each run, all other factors being held at the same level. This is known as a (one-factor-at-a-time design and, for k two-level factors, requires

Table 1

Trial Number	Factors														
	A	B	C	D	E	F	G	H	I	J	K	L	M	N	O
1	+	−	−	−	−	−	−	−	−	−	−	−	−	−	−
2	−	+	−	−	−	−	−	−	−	−	−	−	−	−	−
3	−	−	+	−	−	−	−	−	−	−	−	−	−	−	−
4	−	−	−	+	−	−	−	−	−	−	−	−	−	−	−
5	−	−	−	−	+	−	−	−	−	−	−	−	−	−	−
6	−	−	−	−	−	+	−	−	−	−	−	−	−	−	−
7	−	−	−	−	−	−	+	−	−	−	−	−	−	−	−
8	−	−	−	−	−	−	−	+	−	−	−	−	−	−	−
9	−	−	−	−	−	−	−	−	+	−	−	−	−	−	−
10	−	−	−	−	−	−	−	−	−	+	−	−	−	−	−
11	−	−	−	−	−	−	−	−	−	−	+	−	−	−	−
12	−	−	−	−	−	−	−	−	−	−	−	+	−	−	−
13	−	−	−	−	−	−	−	−	−	−	−	−	+	−	−
14	−	−	−	−	−	−	−	−	−	−	−	−	−	+	−
15	−	−	−	−	−	−	−	−	−	−	−	−	−	−	+
16	−	−	−	−	−	−	−	−	−	−	−	−	−	−	−

Table 2

Trial Number	Factors														
	A	B	C	D	E	F	G	H	I	J	K	L	M	N	O
1	+	+	+	+	+	+	+	+	+	+	+	+	+	+	+
2	+	+	+	−	−	+	−	−	−	+	+	−	+	−	−
3	+	+	−	+	−	−	+	−	−	+	−	+	−	+	−
4	+	+	−	−	+	−	−	+	+	+	−	−	−	−	+
5	+	−	+	+	−	−	−	+	−	−	+	+	−	−	+
6	+	−	+	−	+	−	+	−	+	−	+	−	−	+	−
7	+	−	−	+	+	+	−	−	+	−	−	+	+	−	−
8	+	−	−	−	−	+	+	+	−	−	−	−	+	+	+
9	−	+	+	+	−	−	−	−	+	−	−	−	+	+	+
10	−	+	+	−	+	−	+	+	−	−	−	+	+	−	−
11	−	+	−	+	+	+	−	+	−	−	+	−	−	+	−
12	−	+	−	−	−	+	+	−	+	−	+	+	−	−	+
13	−	−	+	+	+	+	+	−	−	+	−	−	−	−	+
14	−	−	+	−	−	+	−	+	+	+	−	+	−	+	−
15	−	−	−	+	−	−	+	+	+	+	+	−	+	−	−
16	−	−	−	−	+	−	−	−	−	+	+	+	+	+	+

$k + 1$ trials. The estimate of a main effect of each factor is estimated from two trials: the trial where all factors are run at the low level and the trial where the particular factor under consideration is run at the high level and all other factors are run at the low level. For example, the design for $k = 15$ factors, $n = 16$ trials is given in Table 1. Although 16 trials are being run, only two trials are used in each estimate of an effect.

A better estimate of the main effect of a factor would be obtained if all the trials could be used to get an estimate of the effect of each factor. If half the trials are run at the high level for each factor and the other half at the low level, the difference of the averages of each half would give an estimate of the effect of each factor. A factorial* design with 2^k trials is such a design. If all the trials are run, there is enough information to estimate not only the mean and the k main effects but also all two-factor interactions, three-factor interactions, etc. A saturated design can be obtained from a factorial design, where the main effects can be estimated independently, but many main effects will be completely confounded with the interaction effects. If the number n of runs is equal to $2^p(2, 4, 8, \dots)$ or if the number k of factors is $2^p - I(1, 3, 7, \dots)$, a fractional factorial design* will give estimates of the main effects. For example, if 15 factors are to be studied, a full factorial would require $2^{15} = 32{,}768$ trials. A saturated design requires $2^4 = 16$ trials, which is a 2^{-11} fraction of the 2^{15} design; see Table 2. Of course, the trials must be run in a randomly chosen order. This saturated design was obtained from the 2^4 full factorial design by substituting factor A for factor A in the original design, B for B, C for C, D for D, E for ABCD, F for ABC, G for ABD, H for ACD, I for BCD, J for AB, K for AC, L for AD, M for BC, N for BD, and O for CD. The defining relationship, therefore is Identity = ABJ = ACK = ADL = BCM = BDN = CDO = ABCF = ABDG = ACDH = BCDI = ABCDE. All the confounded interactions can be identified from these equations.

If the number k of factors is not of the form $2^p - 1$ then the next highest value can be used, with the extra runs dedicated to obtaining estimates of two-factor interactions or of the experimental standard deviation. But, as the number of factors gets large, the number of extra trials may be too great for a preliminary experiment. Plackett and Burman [1] derived saturated two-level designs for $k = 3, 7, 11, \ldots, 4m - 1, \ldots, 99$ factors in $n = 4, 8, 12, \ldots, 4m, \ldots, 100$ trials (except $n = 92$). In devising these designs, Plackett and Burman required the selection of n trials from the complete factorial that will allow the main effects to be estimated with the same accuracy as if only a single factor had been varied throughout the n trials. *See* PLACKETT AND BURMAN DESIGNS for further details and references.

A completely different approach to saturated experimental designs was suggested by Satterthwaite [2]. He proposed the use of random balance designs for the purpose of screening the important factors in an experiment using a small number of trials. Satterthwaite defined a *pure random balance design* as one for which the values of each input variable is selected by a random process. One method of satisfying this requirement is to choose a random sample of trials from a full factorial. Another method of obtaining a random balance design is to choose the level of each factor in each trial by tossing a fair coin.

In practice, additional requirements are usually imposed. It is useful to require that each level of the factor appear an equal number of times. Selection of values for one factor that cause complete confounding* with another factor should be avoided. To satisfy these additional restrictions, the experimenter may choose an appropriate fractional factorial design. Groups of variables are assigned to the design in sequence. For each group, the order of trials is chosen at random so that no factor in one group will be completely confounded with any factor in another group. The complete design is then made up of the fractional factorial, repeated as many times as necessary to provide for all the factors but with the rows in different sequences for each group of factors.

Another method for choosing a few important factors out of a large number of possibly effective factors in a small number of trials was proposed by Watson [3]. He suggested that group screening methods be used. This frequently results in saturated or supersaturated designs.

All factors to be considered are placed in to g groups, k factors to a group. Each group is then treated as a single factor with all the components in the group run at the same level. The g group factors are then tested in an appropriate fractional factorial or Plackett and Burman design. If a group factor shows no effect, it follows that all the original factors and all the interactions among those factors within that group factor can be considered to have no effect on the result. It is assumed that the factors in a group all have the same positive effect so that there is no possibility of effects cancelling each other.

If a group factor is found to be significant, then one or more original factors in the group are significant and must be studied in an additional set of trials. To keep the number of trials in the second stage to a minimum, the design for this additional set of trials should be chosen in such a way that the results of the first set of trials can be used in the analysis. Watson calculated optimal group sizes and evaluated the performance of these designs using the following assumptions:

- **(i)** All factors have, independently, the same prior probability of being effective.
- **(ii)** Effective factors have the same positive effect.
- **(iii)** There are no interactions present.
- **(iv)** The required designs exist.
- **(v)** The directions of possible effects are known.
- **(vi)** The errors of all observations are independently normal with a constant known variance.
- **(vii)** $f = gk$, where g is the number of groups, k is the number of factors in each group, and f is the number of factors being studied.

These assumptions are not very restrictive. Group screening methods have proved to be very useful when large numbers of factors are being studied.

References

[1] Plackett, R. L. and Burman J. P. (1946). The design of optimum multifactorial experiments. *Biometrika*, **33**, 305–325.

[2] Satterthwaite, F. E. (1959). Random balance experimentation. *Technometrics*, **1**, 111–137, 184–192.

[3] Watson, G. S. (1961). A study of the group screening method. *Technometrics*, **3**, 371–388.

Bibliography

Box, G. E. P. (1959). Discussion of the papers of Messrs. Satterthwaite and Budne. *Technometrics*, **1**, 174–180.

Box, G. E. P. and Hunter J. S. (1961). The 2^{k-p} fractional factorial designs, part I. *Technometrics*, **3**, 311–351.

Box, G. E. P. and Hunter, J. S. (1961). The 2^{k-p} fractional factorial designs, part II. *Technometrics*, **3**, 449–458.

Daniel, C. (1962), Sequences of fractional replicates in the 2^{p-q} series. *J. Amer. Statist. Ass.*, **57**, 403–429.

Patel, M. S. (1962). Group-screening with more than two stages. *Technometrics*, **4**, 209–217.

Rechtschaffner, R. L. (1967). Saturated fractions of 2^n and 3^n factorial designs. *Technometrics*, **9**, 569–575.

(CONFOUNDING
FACTORIAL EXPERIMENTS
FRACTIONAL FACTORIAL DESIGNS
INTERACTION
MAIN EFFECTS
PLACKETT AND BURMAN DESIGNS
SUPERSATURATED DESIGNS)

MARILYNN S. DUEKER

SAVAGE SCORES

These are used in the construction of linear rank statistics* estimating location. The scores are the coefficients $a_{n1}, a_{n2}, \ldots, a_{nn}$ of the order statistics* $X_1' \leqslant X_2' \leqslant \cdots \leqslant X_n'$ in the estimator

$$\sum_{i=1}^{n} a_{ni} X_i'.$$

For Savage scores

$$a_{ni} = \sum_{j=n-i+1}^{n} j^{-1}.$$

Reference

[1] Randles, R. H. and Wolfe, D. A. (1979). *Introduction to the Theory of Nonparametric Statistics*. Wiley, New York.

(NORMAL SCORES
WILCOXON SCORES)

SAVAGE, LEONARD J.

Born: November 20, 1917, in Detroit, Michigan.

Died: November 1, 1971, in New Haven, Connecticut.

Contributed to: probability theory, foundations of statistics, Bayesian statistics.

Leonard Jimmie Savage was the most original thinker in statistics active in America during the period from the end of World War II until his premature death. Statistics, previously a collection of useful but somewhat disconnected ideas, was given firm foundations by him: foundations that led to new ideas to replace the old. Although almost all his published work lies in the fields of statistics and probability, he was interested in and informed about so many things that he made a superb statistical consultant for those who were not satisfied with a glib response but would benefit from a thorough understanding of the situation. He was a good writer and lecturer.

He was raised in Detroit and eventually, after some difficulties largely caused by his poor eyesight, graduated in mathematics from the University of Michigan. His statistical work began when he joined the Statistical Research Group at Columbia University in 1943. Later he joined the University of Chicago and it was there that he wrote his great work, *The Foundations of Statistics* [8]. In 1960 he moved to Michigan and in 1964 to Yale where he remained until his death. A collection of his principal works has appeared [11]; this book also contains four personal tributes and an account of his technical achievements.

In the late 1940s and early 1950s, statistics was dominated by the work of R. A. Fisher* [3] and by decision-theory* ideas due to Neyman* [4], Pearson* [5, 6] and Wald* [12]. These proved to be practically useful but there was no general underlying theory, although Wald had gone some way to providing one. Savage had briefly worked with von Neumann and had seen the power of the axiomatic method in constructing a general system. Using basic probability ideas, von Neumann had shown that the only sensible decision procedure was maximization of expected utility.* Savage conceived and later executed the idea of widening the axiomatic structure and deducing, not only the utility, but the probability aspect. This he did in the first seven chapters of his first book [8].

Essentially he showed that the only satisfactory description of uncertainty is probability, of worth is utility, and that the best decision is the one that maximizes expected utility. It is the first of these that was both important and new. Now the foundations were secure; he felt he could establish the Fisherian and other results as consequences of the basic results. In the remainder of the book he attempted to do this, and gloriously failed. (He explains this in the second edition, which differs from the first only in minor details except for a new, illuminating preface [9].) Instead, gradually, over the years he developed new ideas that replaced the old ones, constructing what is now usually called Bayesian statistics. That he correctly continued to enjoy and respect the enduring values in the older work is beautifully brought out in his marvellous work on Fisher [10]. Nevertheless, he had created a new paradigm.

Savage had the attribute, unfortunately not always present in contemporary scientists, of being a true scholar, always respecting the repository of knowledge above personal aggrandizement. Having constructed new foundations, he was able to understand the work of Ramsey [7] and de Finetti [1], who earlier had developed arguments leading to essentially the same results. He spent some time in Italy with de Finetti and was largely responsible for making de Finetti's brilliant ideas available to English-speaking statisticians and for extending them. Many of the papers he wrote between 1954 and his death are concerned with his increasing understanding of the power of Bayesian methods, and to read these in sequence is to appreciate a great mind making new discoveries. His concern is usually with matters of principle. Only in his consulting work, the bulk of which was never published, does he attempt to develop techniques of Bayesian statistics to match the brilliant ones put forward by Fisher.

Savage also did important work in probability, much of it in collaboration with Lester Dubins [2]. The basic problem they considered was this: you have fortune f, $0 < f < 1$, and may stake any amount $s \leqslant f$ to win s with probability $\frac{1}{2}$ or lose it with probability $\frac{1}{2}$. Once this has been settled, you may repeat the process with $f' = f \pm s$ so long as $f' > 0$: How should the stakes be selected to maximize the probability of reaching a fortune of 1 or more? They showed that bold play is optimal: that is, stake f, or at least enough to get $f' = 1$. Generalizations of these ideas provided a rich store of results in stochastic processes*. The treatment is remarkable in using finite- rather than sigma-additivity.

Proper judgment of the importance of Savage's work can only come when the Bayesian paradigm is established as the statistical method or when it is shown to be defective. Even if the latter happens, Savage can be credited with creating a method that led to important, new ideas of lasting value. He was a true originator.

References

[1] de Finetti, B. (1974/75). *Theory of Probability*. Wiley, New York, two volumes.

[2] Dubins, L. E. and Savage, L. J. (1965). *How to Gamble If You Must: Inequalities for Stochastic Processes*. McGraw-Hill, New York.

[3] Fisher, R. A. (1950). *Contributions to Mathematical Statistics*. Wiley, New York.

[4] Neyman, J. (1967) *Early Statistical Papers*. University of California Press, Berkeley, CA.

[5] Neyman, J. and Pearson, E. S. (1967). *Joint Statistical Papers*. Univ. of California Press, Berkeley, CA.

[6] Pearson, E. S. (1966). *Selected Papers*. University of California Press, Berkeley, CA.

[7] Ramsey, F. P. (1950). *The Foundations of Mathematics and Other Logical Essays*. Humanities Press, New York.

[8] Savage, L. J. (1954). *The Foundations of Statistics*. Wiley, New York.

[9] Savage, L. J. (1972). *The Foundations of Statistics*, 2nd rev. ed. Dover, New York.

[10] Savage, L. J. (1976). *Ann. Statist.*, **4**, 441–500.

[11] Savage, L. J. (1981). *The Writings of Leonard Jimmie Savage—A Memorial Selection*. Amer. Statist. Assoc. and Inst. Math. Statist., Washington, DC.

[12] Wald, A. (1950). *Statistical Decision Functions*. Wiley, New York.

D. V. LINDLEY

SAVAGE TEST

The Savage test [21] is a nonparametric procedure for the comparison of two samples, which is especially useful for comparing lifetime distributions. The test was originally designed for complete data, and was later modified by Leonard Savage* and others to deal with censored* or grouped data*. The complete data case will be discussed first, followed by the censored case and other related topics.

Let $X_1, X_2, \ldots, X_m$ and $Y_1, Y_2, \ldots, Y_n$ be independent random samples from two populations with continuous cumulative distribution functions $F_1(x)$ and $F_2(x)$, respectively. We assume for simplicity that there are no ties. Let $R_1 < R_2 < \cdots < R_n$ denote the ordered ranks associated with the Y's in the combined sample. With $N = m + n$, the Savage statistic is defined by

$$S = \sum_{i}^{n} a_N(R_i),$$

where

$$a_N(k) = \sum_{j=1}^{k} (N - j + 1)^{-1},$$

the expected value of the kth order statistic in a random sample size N from the standard exponential distribution* with probability density function e^{-x}, $x > 0$. The statistic S is distribution-free* when $F_1 = F_2$.

Savage [21] introduced S for testing the equality of two distributions against the alternative that one is stochastically larger than the other; the null hypothesis is $H_0: F_1(x) = F_2(x)$ for all x, and the alternative $H_1: F_1(x) \geqslant F_2(x)$ for all x and $F_1(x) > F_2(x)$ for some x. The alternative says that F_2 is stochastically larger than F_1; or equivalently, that the random variable, say Y, associated with F_2 is stochastically larger than the random variable, say X, associated with F_1. Savage showed that the test that rejects H_0 for larger values of S is the locally most powerful (LMP) rank test of H_0 against the Lehmann alternative*, say, $F_2(x) = [F_1(x)]^{\Delta}$, $\Delta > 1$ (*see* LOCALLY OPTIMAL TESTS). In particular, it is the LMP rank test for testing $\theta_1 = \theta_2$ against $\theta_2 > \theta_1$ when F_1 and F_2 are exponential distributions, say, $F_i(x) = 1 - \exp(-x/\theta_i)$ for $x > 0$, $i = 1, 2$. If X and Y have exponential distributions, then the distributions of $-X$ and $-Y$ form a Lehmann alternative.

Tables for score $a_N(k)$ have been provided in Basu [1] for $N \leqslant 20$ and in Harter [7] for $N \leqslant 120$. Critical values for the Savage test for $6 \leqslant m \leqslant n \leqslant 10$ are given in Hájek [5, Tables X and XI]. If m and n are large and H_0 is true, the statistic S is approximately normal, with mean $E(S) = n$ and variance

$$\operatorname{var}(S) = mn(N-1)^{-1} \times \left[1 - N^{-1} \sum_{j=1}^{N} (1/j)\right].$$

(See, e.g., Hájek [5, Theorem 16A].) Alternatively, Cox [3] suggested an approximation using an F-distribution*,

$$\frac{mS}{n(N-S)} \sim F(2\lambda n, 2\lambda n),$$

where λ is some constant depending on n, whose value ranges from 1.242 for $n = 6$ to 1.057 for $n = 50$.

Example 1. Bickel and Doksum [2, p. 347] give failure times for the air monitors at a nuclear power plant, operated under two different maintenance policies, say, A and B:

A: 7, 26, 10, 8, 29
B: 3, 150, 40, 34, 32.

It is desired to know whether policy B is significantly better than policy A.

Here $N = 10$, $n = 5$, and the ordered ranks associated with policy B are 1, 7, 8, 9, and 10 in the combined ranking. The value of the Savage statistic is

$$\begin{aligned} S &= \tfrac{1}{10} + \left(\tfrac{1}{10} + \tfrac{1}{9} + \cdots + \tfrac{1}{4}\right) \\ &\quad + \left(\tfrac{1}{10} + \tfrac{1}{9} + \cdots + \tfrac{1}{3}\right) \\ &\quad + \left(\tfrac{1}{10} + \tfrac{1}{9} + \cdots + \tfrac{1}{2}\right) \\ &\quad + \left(\tfrac{1}{10} + \tfrac{1}{9} + \cdots + \tfrac{1}{1}\right) \\ &= 7.48. \end{aligned}$$

Since there are $\binom{10}{5} = 252$ possible ways of getting an ordered rank set $(R_1, R_2, \ldots, R_5)$, and exactly eight of them have S values exceeding 7.48, the P-value* associated with 7.48 is $\Pr[S \geqslant 7.48] = 8/252 = 0.317$. On the other hand, if one uses the Wilcoxon rank-sum statistic (*see* MANN–WHITNEY–WILCOXON STATISTIC),

$$W = R_1 + R_2 + \cdots + R_5,$$

one has $W = 1 + 7 + 8 + 9 + 10 = 35$, which has a P-value 0.0754 [2, p. 348]. The Savage test is more sensitive than the Wilcoxon test in this example.

Since ranks are preserved with increasing monotone transformations, and since there are monotone relations among the exponential, Weibull*, and the smallest extreme value distributions*, the Savage test is also the locally most powerful rank test for each of the following problems:

Table 1 Powers of the S and the T^* Test for $m = n = 5$

	δ			
Test	1	1.5	2.0	2.5
S	0.099	0.229	0.360	0.473
T^*	0.10	0.24	0.38	0.50

I. The Weibull model, where

$$F_i(x) = 1 - \exp\{-(x/b_i)^c\}$$

for $x > 0$,

with $c > 0$, $b_i > 0$, $i = 1, 2$, and

$$H_0 : b_1 = b_2,\ H_1 : b_2 > b_1.$$

II. The smallest extreme value model, where

$$F_i(x) = 1 - \exp\{-\exp[(x - \eta_i)/\xi]\},$$

$-\infty < x < \infty$,

with $\xi > 0$, $-\infty < \eta_i < \infty$, $i = 1, 2$,

and $H_0 : \eta_1 = \eta_2$, $H_1 : \eta_2 > \eta_1$.

Hence, the Savage test seems to be a reasonable distribution-free test for comparing two lifetime distributions. We note that if T_i has a Weibull distribution as in model I, then $U_i = T_i^c$ has an exponential distribution with scale $\theta_i = b_i^c$; $W_i = \log T_i$ has the smallest extreme value distribution of model II with $\eta_i = \log b_i$ and $\xi = 1/c$ (see, e.g., ref. 13, p. 141).

The power of the Savage test with respect to a specified distribution can be computed using the probability distribution of ordered ranks. With respect to the Lehmann alternative the probability of an ordered rank set has an explicit form (see, e.g., Lehmann [14, pp. 255–256]). Table 1 gives some computed values with respect to the Weibull distribution of model I, where $\delta = (b_2/b_1)^c$. Also given in Table 1 are some simulated powers corresponding to the parametric T^*-test of the same problem (model I) provided in Schafer and Sheffield [22]. For this problem the nonparametric Savage test seems to be as powerful as its parametric competitor.

James [8] computed Pitman efficiencies for the Savage test, the Wilcoxon test, and Taha's [23] test of gamma distribution scales. In the Weibull distribution model I, it can be shown that the Pitman efficiency* for the Savage test, with respect to the Wilcoxon test, is $\frac{4}{3}$; with respect to Schafer–Sheffield test, it is 1. The results remain the same for model II.

The Savage test, as discussed above, is concerned with complete data in which all X's and Y's are observed. When data are censored or grouped, some modifications are needed.

For type II censored data, modifications of the Savage statistic have been suggested by Gastwirth [4], Rao et al. [19], Basu [1], and Johnson and Mehrotra [10]. Here, we introduce Basu's generalization. Suppose two samples of items, of size m and n, respectively, are drawn from two populations with cumulative distributions $F_1(x)$ and $F_2(x)$, respectively. These N $(N = m + n)$ sample items are tested simultaneously. The observations are failure times that are naturally ordered. Suppose it is decided to terminate the experiment as soon as r $(r \leqslant N)$ failures (in the combined sample) are observed, and statistical analysis will be based on the r ordered failure times. Suppose among the r failing items, there are m_r and n_r from the first and the second samples, respectively, where m_r and n_r are random numbers $(n_r + m_r = r)$ and r is prefixed. To test the null hypothesis H_0 against H_1 based on the first r ordered observations, Basu [1] suggested the statistic $S_r^{(N)}$:

$$S_r^{(N)} = \sum_{i=1}^{r} (1 - Z_i) a_N(i) + (n - n_r)(N - r)^{-1} \times \left(\sum_{i=r+1}^{N} a_N(i) \right) - \tfrac{1}{2}(m + n),$$

where $a_N(k)$ is the ordered exponential score defined above and

$$Z_i = \begin{cases} 1, & \text{if the } i\text{th ordered failure time is from } F_1, \\ 0, & \text{otherwise.} \end{cases}$$

The null hypothesis H_0 is rejected when the value of $S_r^{(N)}$ is too large.

Tail probabilities* of $S_r^{(N)}$ under H_0 for different values of m, n, and r $[m = n = 4(1)8,\ r = r(1)8]$ are given in Basu [1]. The mean and variance of $S_r^{(N)}$ under H_0 are

$$E_0\left(S_r^{(N)}\right) = \tfrac{1}{2}(n - m),$$

$$\sigma_0^2\left(S_r^{(N)}\right) = mn\left(N(N-1)\right)^{-1} \times \left\{ \sum_{i=1}^{r} a_N^2(i) + d^2(N - r)^{-1} - N \right\},$$

where $d = \sum_{i=r+1}^{N} a_N(i)$. Basu [1, Table III] indicates that, for $m = n = r = 8$, the normal approximation to the standardized $S_r^{(N)}$ is satisfactory. Basu also showed that $S_r^{(N)}$ is asymptotically equivalent to the Gastwirth modification.

Basu's statistic $S_r^{(N)}$ becomes Savage's statistic S minus a constant when $r = N$, and it is the asymptotically most powerful rank test for censored data under Lehmann alternatives. It is also known to maximize the minimum power over IFRA (or IFR) distributions asymptotically [1]. Johnson and Mehrotra [10] showed, as a consequence of a more general result, that Basu's $S_r^{(N)}$ statistic is the LMP rank test with respect to a single parameter exponential distribution for type II censored data.

In a further generalization of the Savage statistic, Saleh and Dionne [20] considered a statistic based on the frequencies within the $k + 1$ intervals specified by k arbitrary quantiles, which yields a locally most powerful rank test against Lehmann alternatives.

The construction of rank tests* with arbitrarily censored data has been discussed by Prentice [18], Peto and Peto [17], and others. Sequential procedures for testing equality of survival distributions based on modified Savage statistics are discussed by Koziol and

Petkau [12], Jones and Whitehead [11], and Joe et al. [9].

Further references and other nonparametric testing procedures can be found in Hájek and Šidák [6], Lawless [13], Lehmann [15], and Miller [16].

References

[1] Basu, A. P. (1968). *Ann. Math. Statist.*, **39**, 1591–1604. (Modified Savage test for type II censored data with some tables.)

[2] Bickel, P. J. and Doksum, K. A. (1977). *Mathematical Statistics*. Holden-Day, San Francisco. (Excellent text for graduate students and practising statisticians.)

[3] Cox, D. R. (1964). *J. R. Statist. Soc. B*, **26**, 103–110. (*F*-statistic version of exponential ordered scores.)

[4] Gastwirth, J. L. (1965). *Ann. Math. Statist.*, **36**, 1243–1247. (Development of asymptotically most powerful rank tests for the two-sample problem with censored data.)

[5] Hájek, J. (1969). *A Course in Nonparametric Statistics*. Holden-Day, San Francisco.

[6] Hájek, J. and Šidák, Z. (1967). *Theory of Rank Tests*. Academic, New York. (Excellent treatment of the theory of rank tests.)

[7] Harter, H. L. (1969). *Order Statistics and Their Use in Testing and Estimation*, Vol. 2, U.S. Government Printing Office, Washington, DC.

[8] James, B. R. (1967). *Proc. Fifth Berkeley Symp. Math. Statist. Prob.*, University of California, Press, Berkeley, CA, pp. 389–393.

[9] Joe, H., Koziol, J. A., and Petkau, A. J. (1981). *Biometrics*, **37**, 327–340. (Sequential procedures based on the Mantel–Haenszel, the Wilcoxon, and the Savage statistics for testing the equality of two survival distributions when observations are singly censored.)

[10] Johnson, R. A. and Mehrotra, K. G. (1972). *Ann. Math. Statist.*, **43**, 823–831. (Development of locally most powerful rank tests and the unbiasedness for type II censored data.)

[11] Jones, D. and Whitehead, J. (1979). *Biometrika*, **66**, 105–113. (Sequential forms of the log rank and modified Wilcoxon tests for censored data.)

[12] Koziol, J. A. and Petkau, A. J. (1978). *Biometrika*, **65**, 615–623. (Sequential test using a modified Savage statistic.)

[13] Lawless, J. F. (1982). *Statistical Models and Methods for Lifetime Data*. Wiley, New York. (Unified treatment of censored data; a lot of references. A valuable reference for researchers and persons dealing with reliability or lifetime data.)

[14] Lehmann, E. L. (1959). *Testing Statistical Hypotheses*. Wiley, New York. (Well-known and authoritative text in this field.) (2nd. ed., 1986.)

[15] Lehmann, E. L. (1975). *Nonparametrics: Statistical Methods Based on Ranks*. Holden-Day, San Francisco. (Emphasis on proper understanding of the experimental situation and the appropriate statistical analysis. A book worth reading.)

[16] Miller, R. G., Jr. (1981). *Survival Analysis*. Wiley, New York. (A practical reference describes the latest statistical techniques pertinent to the analysis of survival data with censoring with recently developed nonparametric techniques.)

[17] Peto, R. and Peto, J. (1972). *J. R. Statist. Soc. A*, **135**, 185–206.

[18] Prentice, R. L. (1978). *Biometrika*, **65**, 167–179. (Linear rank statistics are developed for tests on regression coefficients with censored data.)

[19] Rao, U. V. R., Savage, I. R., and Sobel, M. (1960). *Ann. Math. Statist.*, **31**, 415–426. (Theoretical development of rank order statistics for the two-sample censored case.)

[20] Saleh, A. K. M. E. and Dionne, J. P. (1977). *Commun. Statist. A*, **6**, 1213–1221. (A modified Savage test when data are grouped.)

[21] Savage, I. R. (1956). *Ann. Math. Statist.*, **27**, 590–616. (Rank tests for several types of alternatives are treated. (A famous paper in dealing with exponential scores test and Lehmann alternatives).

[22] Schafer, R. E. and Sheffield, T. S. (1976). *Technometrics*, **18**, 231–235. (Test for equality of two Weibull distributions with the same shape parameter.)

[23] Taha, M. A. H. (1964). *Publ. Inst. Statist. Univ. Paris*, **13**, 169–179. (A rank test for scale parameters that is powerful for one-sided asymmetrical distributions.)

(CENSORED DATA
DISTRIBUTION-FREE METHODS
LEHMANN ALTERNATIVES
MANN–WHITNEY–WILCOXON STATISTIC
RANK TESTS
SAVAGE TEST, SEQUENTIAL)

H. K. HSIEH

SAVAGE TEST, SEQUENTIAL

The sequential Savage test is an extension of the Savage test*, proposed by Koziol and Petkau [1] and further investigated by Lesser [2]. It is a test of identity of two survival distributions, which is applied sequentially as deaths occur in two samples of sizes n_1 and n_2 from the two distributions. If the set of $(n_1 + n_2)$ times of death is arranged in ascending order then the Savage score* for the ith order statistic is

$$a_{n,i} = \sum_{j=1}^{i} (n - j + 1)^{-1} - 1,$$

where $n = n_1 + n_2$, and the Savage statistic (S_k) for the first population is the sum of $a_{n,i}$'s for all n_1 values from that population. If only the values of the first k order statistics (corresponding to the first k deaths) are available, then *modified Savage scores* are used, replacing $a_{n,i}$ by

$$a^*_{n,i} = (n - k)^{-1} \sum_{j=k+1}^{n} a_{n,j}$$

for $i \geqslant k + 1$.

The corresponding *modified Savage statistics* S^*_k are the sums of the modified Savage scores for the first population. The sequence $S^*_1, S^*_2, \ldots$ is used as test criterion.

If the two survival distributions are identical then the expected value of S^*_k is zero and the variance is

$$\operatorname{var}(S^*_k) = \frac{n_1 n_2}{n - 1} \frac{k - a_{n,n} + 1}{n}.$$

The null hypothesis of identity of the two distributions is rejected at (asymptotic) level α as soon as one observes

$$\left(\frac{n_1 + n_2}{k}\right)^{1/2} \frac{S^*_k}{\{\operatorname{var}(S_N)\}^{1/2}} > K'_\alpha$$

for a one-sided test,

$$\left(\frac{n_1 + n_2}{k}\right)^{1/2} \frac{|S^*_k|}{\{\operatorname{var}(S_N)\}^{1/2}} > K''_\alpha$$

for a two-sided test,

where

$$2\{1 - \Phi(K'_\alpha)\} = \alpha,$$

$$4 \sum_{j=0}^{\infty} (-1)^j [1 - \Phi\{(2j + 1)K''_\alpha\}] = \alpha.$$

Approximations to the power of the test under exponential alternatives $[F(x) = 1 - e^{-\lambda x}$ for various $\lambda]$ were investigated by Lesser [2] using asymptotic results developed in ref. 1.

References

[1] Koziol, J. A. and Petkau, A. J. (1978). *Biometrika*, **65**, 615–623.

[2] Lesser, M. L. (1982). *Statist. Med.*, **1**, 277–280.

(CENSORED DATA
CLINICAL TRIALS
SAVAGE SCORES
SAVAGE TEST
SEQUENTIAL ANALYSIS
SURVIVAL ANALYSIS)

S_B, S_L, S_N, S_U, S_{IJ} DISTRIBUTIONS *See* JOHNSON'S SYSTEM OF DISTRIBUTIONS

SCALE PARAMETER *See* LOCATION-SCALE PARAMETER

SCALE TESTS

INTRODUCTION

Location and scale parameters are probably the two most commonly used types of parameters for describing and comparing populations. A location parameter (*see* LOCATION TESTS) is usually chosen as a measure of "central tendency" and describes the size of a typical observation drawn at random from the population, while a scale parameter is chosen to give some indication as to the spread or variability one can expect to find in observations drawn from the population. Hypothesis tests concerning scale parameters are the subject matter of this entry.

The following situations represent instances in which comparison of scale parameters from several populations or the determination of the values of these scale parameters may be of interest. First, many tests for equality of the location parameters of several populations assume equality of some scale parameters, such as the population variances, to maintain their nominal significance level. In addition, under the assumption of normality, power calculations and sample size determinations related to tests on means require some knowledge of the values of the underlying variances. There are also situations in which one is interested in scale parameters in their own right. In order to compare the accuracy of several measuring instruments or techniques, one might compare scale parameters associated with the population of measurements for several methods [17, p. 85]. Finally, if one wishes to pool data from various sources to yield an improved estimate of some scale parameter, such as the pooled estimate of variance in the usual one-way analysis of variance, it would first be appropriate to test for equality of the scale parameters from the various sources.

Scale tests with varying assumptions on the underlying population are discussed. The next section contains a definition of scale parameter with some examples and the following section discusses procedures for testing hypotheses about the variance of a single population. The remainder of the entry deals with tests for equality of the scale parameters of several populations. Under the fourth heading, tests that assume that the underlying populations are normal are described. Unlike tests for the equality of the means, these tests are very sensitive to a failure in the assumption of normality (*see* ROBUSTNESS OF TESTS). Distribution-free* tests for equality of scale parameters are presented in the fifth section. Although distribution-free tests are valid under very general assumptions concerning the shapes of the underlying populations, they require either quite restrictive assumptions on the location parameters or random grouping of the data. The sixth section deals with robust tests for scale that can be used without such restrictive assumptions on either the location parameters or the shapes of the underlying distributions. However, the rejection regions for these procedures are based on large sample approximations to their null distributions, and some consideration must be given as to appropriate sample sizes for which these procedures should be used. Various simulation studies [5, 7, 12, 13, 20] give insight into this problem. Finally, the last section contains a discussion.

SCALE PARAMETERS

The choice of the appropriate scale test to use in a given situation depends on the types of assumptions made regarding the underlying distributions. In addition to assumptions made about the shapes of the populations, assumptions concerning location parameters become important, particularly for the distribution-free methods. Hence it will be convenient to distinguish between those scale parameter families in which a location parameter is present and those in which there is none. In both cases the effect of increasing the scale parameter is to spread out the probability distribution, although the way in which the information in the samples is used to make inferences concerning the scale parameter may be quite different in the two cases.

Case 1. Location and scale family. Let ϵ be a random variable with cumulative distribution function (CDF) $F(t) = P(\epsilon \leqslant t)$. Then the random variable $X = \eta + \sigma\epsilon$, $-\infty < \eta < \infty, 0 < \sigma < \infty$, has CDF given by

$$\begin{aligned} P(X \leqslant x) &= P(\epsilon \leqslant (x-\eta)/\sigma) \\ &= F((x-\eta)/\sigma), \end{aligned}$$

where η is a location parameter and σ is a scale parameter. The effect of increasing (decreasing) σ is to cause the probability distribution of X to become more (less) spread out. Since $\operatorname{var} X = \sigma^2 \operatorname{var} \epsilon$, increasing σ will increase the variance of X. When $\operatorname{var} \epsilon = 1$, the scale parameter σ represents the standard deviation of X. Another common scale parameter is the *interquartile range*. If $F^{-1}(\frac{3}{4}) - F^{-1}(\frac{1}{4}) = 1$, then σ represents the interquartile range.

Case 2. Scale family without a location parameter. Let ϵ be a random variable having CDF $F(t) = P(\epsilon \leqslant t)$. Then the random variable $X = \sigma\epsilon$ has CDF $P(X \leqslant x) = F(x/\sigma)$. The most common families of this type have $F(0) = 0$, so that X is a positive random variable. In this case increasing σ has the effect of increasing the size of X as well as spreading out the distribution of X.

A familiar example of Case 2 is the exponential distribution* where $F(x/\sigma) = 1 - e^{-x/\sigma}$ for $x \geqslant 0$ and $F(x/\sigma) = 0$ for $x < 0$. Inferences about the scale parameter σ are based on the sample mean, a complete sufficient statistic* for this family. However, use of the sample mean to make inferences about the standard deviation σ of a normal population, a Case 1 family, would be inappropriate. The point being made is that inference procedures appropriate for distributions in Case 2 may be inappropriate for Case 1 and vice versa.

ONE-SAMPLE TESTS OF VARIANCE

The normal distribution is that most commonly used as a model for data collected from a single population, and it is important to be able to make robust inferences about its parameters. Let $X_1, \ldots, X_n$ denote a random sample from a normal population with mean η and variance σ^2; equivalently, X_i has CDF $F((x - \eta)/\sigma)$, where F is the standard normal CDF. Inferences about η are fairly insensitive to a failure in the assumption of normality, while standard inferences about σ are quite sensitive to a failure in this assumption. To see this, let $S^2 = \Sigma(X_i - \overline{X})^2/(n - 1)$, where $\overline{X} = \Sigma X_i/n$. When sampling from a normal population, $(n - 1)S^2/\sigma^2$ has a χ^2 distribution with $n - 1$ degrees of freedom. Thus a test of $H_0 : \sigma^2 \leqslant \sigma_0^2$ vs. $H_1 : \sigma^2 \geqslant \sigma_0^2$ at level α should reject H_0 whenever $(n - 1)S^2/\sigma_0^2 > \chi_\alpha^2(n - 1)$, where $\chi_\alpha^2(n - 1)$ is the $100(1 - \alpha)$ percentile of the χ^2 distribution with $n - 1$ degrees of freedom. (A two-sided alternative would require a two-sided critical region.) When the normality assumption is not met, the probability of a type I error using this procedure is not controlled at α even for large samples, because the large sample distribution of S^2 is dependent upon another parameter, $\gamma = \{E(X - u)^4/\sigma^4\} - 3$. Specifically, $((n - 1)/2)^{1/2}(S^2/\sigma^2 - 1)$ has a limiting normal distribution with mean 0 and variance $1 + \gamma/2$, where $\gamma = 0$ for a normal distribution. Some indication of the size of the effect of different γ values on the type I error as well as further discussion on this problem can be found in Scheffé [27, pp. 336–337].

For large samples it is possible to estimate γ by

$$\hat{\gamma} = n\sum(X_i - \overline{X})^4 / \left(\sum(X_i - \overline{X})^2\right)^2 - 3$$

and create a large sample test using the fact that $((n - 1)/2)^{1/2}(S^2/\sigma_0^2 - 1)/(1 + \hat{\gamma}/2)^{1/2}$ has approximately a standard normal distribution when $\sigma^2 = \sigma_0^2$. For example, to test $H_0 : \sigma^2 \leqslant \sigma_0^2$ vs. $H_1 : \sigma^2 > \sigma_0^2$, an approximate level α test would reject H_0 whenever

$$((n - 1)/2)^{1/2}(S^2/\sigma_0^2 - 1)/(1 + \hat{\gamma}/2) > z_{1-\alpha},$$

where $z_{1-\alpha}$ is the $100(1 - \alpha)$ percentile of the standard normal distribution. If there is a question about the validity of the normality assumption and n is large, this procedure should be used as a test for variance since the type I error will be better controlled than the procedure based on the χ^2 distribution. As an alternative, Miller [22] has proposed a jackknife* procedure for testing hypotheses about σ^2, but this is also a large sample test.

In the next section, several of the classical tests for homogeneity* of variances are described. These procedures assume normality of the underlying distributions and have the same drawback as the classical one-sample test for variance based on the χ^2 distribution; the sampling distributions of the test statistics so derived are seriously disturbed if the normality assumption is not met.

TESTING HOMOGENEITY OF VARIANCES FOR NORMAL POPULATIONS

In this section, statistical procedures for testing the hypothesis that several normal populations have equal variances are given. These procedures use as their test statistics different functions of the sample variances from the populations.

Let $X_{i1}, \ldots, X_{in_i}$ denote a random sample from a population with CDF $F((x - \eta_i)/\sigma_i)$, $i = 1, \ldots, k$, where F is the standard normal CDF. Thus $X_{i1}, \ldots, X_{in_i}$ is a random sample from a normal population with mean η_i and variance σ_i^2. The procedures for testing $H_0 : \sigma_1^2 = \sigma_2^2 = \ldots = \sigma_k^2$ vs. $H_1 : \sigma_i^2$ not all equal, are usually referred to as *tests for homogeneity of variances*. All tests to be described will require the computation of the sample variances

$$S_i^2 = \Sigma_{j=1}^{n_i}\left(X_{ij} - \overline{X}_i\right)^2/(n_i - 1),$$

where $\overline{X}_i = \Sigma_{j=1}^{n_i} X_{ij}/n_i$ for $i = 1, \ldots, k$.

When $k = 2$, the testing procedure is based on the statistic $F = S_1^2/S_2^2$ which, when $\sigma_1^2 = \sigma_2^2$, has a central F-distribution* with $(n_1 - 1,\ n_2 - 1)$ degrees of freedom. Thus to test $H_0 : \sigma_1^2 = \sigma_2^2$ vs. $H_1 : \sigma_1^2 \neq \sigma_2^2$, H_0 is rejected whenever $F < F_{\alpha_1}(n_1 - 1,\ n_2 - 1)$ or $F > F_{1-\alpha_2}(n_1 - 1, n_2 - 1)$, where $\alpha = \alpha_1 + \alpha_2$ and $F_\delta(m, n)$ is the 100δ percentile of the F-distribution with (m, n) degrees of freedom. For the case $k = 2$, it is often of interest to test a one-sided hypothesis such as $H_0 : \sigma_1^2 \leqslant \sigma_2^2$ vs. $H_1 : \sigma_1^2 > \sigma_2^2$, H_0 being rejected whenever

$$F > F_{1-\alpha}(n_1 - 1, n_2 - 1).$$

A modification of the F-test*, called the F-test by Shorack [28], treats F as having an $F(d(n_1 - 1),\ d(n_2 - 1))$ distribution, where

$$d^{-1} = 1 + (b_2 - 3)/2,$$

$$b_2 = (n_1 + n_2)\frac{\sum(X_i - \overline{X})^4 + \sum(Y_j - \overline{Y})^4}{\left[\sum(X_i - \overline{X})^2 + \sum(Y_j - \overline{Y})^2\right]^2}.$$

This test is much less sensitive to a failure in the normality assumption than the classical F-test.

When $k > 2$, various tests have been proposed in the literature. The most commonly used test statistic is due to Bartlett [2], and is given by

$$B = \frac{\left(\sum \nu_i\right)\ln\left(\sum \nu_i S_i^2/\sum \nu_i\right) - \sum \nu_i \ln S_i^2}{1 + \left\{\sum(1/\nu_i) - 1/\sum \nu_i\right\}/\{3(k-1)\}},$$

where $\nu_i = n_i - 1$. Under H_0, B has an approximate χ^2 distribution with $k - 1$ degrees of freedom when $\nu_i \geqslant 3$ and thus H_0 is rejected whenever $B \geqslant \chi^2_{1-\alpha}(k - 1)$. Some exact critical values can be found in Glaser [14]. For the case where the sample sizes are equal Cochran [6] has proposed the statistic $C = \max(S_1^2, \ldots, S_k^2)/(S_1^2 + S_2^2 + \cdots + S_k^2)$, with H_0 being rejected for large values of C. Critical values for this statistic can be found in ref. 24 p. 203. Finally, Hartley [16] has proposed the statistic $H = \max(S_1^2, \ldots, S_k^2)/\min(S_1^2, \ldots, S_k^2)$ for the equal sample sizes case with H_0 being rejected for large values of H. Critical values for H can be found in refs. 9 and 18. Although these are the three most commonly used procedures for testing for homogeneity of the variances of several normal populations, other test statistics have been proposed and are described in refs. 7 and 10. *See* BARTLETT'S TEST OF HOMOGENEITY OF VARIANCES; COCHRAN'S (TEST) STATISTIC; HARTLEY'S F_{MAX} TEST.

In a simulation study comparing several procedures including the three described above, Gartside [12] recommended the use of Bartlett's test *provided* the assumption of normality holds. However, even for "small" departures from this assumption, the type I error rates of the three statistics described in this section can be seriously affected and cannot be recommended in these situations. Test procedures that do not rely on the assumption of normality are described in the next two sections.

DISTRIBUTION-FREE METHODS

The procedures to be described here are designed for testing the hypothesis of equal scale parameters when the location parameters, in this case medians, are assumed equal. The procedures are valid regardless of the shape of the underlying distributions, but their applicability is limited, due to the strong assumption made on the location parameters. A situation in which this assumption is often appropriate is in the comparison of k measuring instruments or techniques in which the measurements generated by the ith method, $i = 1, \ldots, k$, are median unbiased* for the quantity of interest. The comparison of scale parameters of the underlying distributions would then be an appropriate technique for comparing the accuracies of the methods (see ref. 17, p. 85, for an example).

Let $X_{i1}, \ldots, X_{in_i}$ denote a random sample of size n_i from a population with continuous CDF $F((x - \eta)/\sigma_i)$, $i = 1, \ldots, k$, where $F(0) = \frac{1}{2}$. The populations all have median η but may differ in the values of the scale parameters σ_i. Many distribution-free procedures for testing the hypothesis $H_0 : \sigma_1 = \cdots = \sigma_k$ vs. $H_1 : \sigma_i$ not all equal, are based on the ranks of the observations. Thus let R_{ij} denote the rank of X_{ij} in the combined sample of $N = \sum_{i=1}^{k} n_i$ observations, where ties are broken using midranks (*see* LOCATION TESTS or DISTRIBUTION-FREE METHODS for a discussion of ranking). The populations with the larger scale parameters will tend to get the extreme ranks and any test statistic based on the ranks must reflect this fact. For the comparison of the scale parameters of two populations ($k = 2$), Ansari and Bradley [1] suggested the statistic

$$A = \sum_{i=1}^{n_1} [(N+1)/2 - |R_{i1} - (N+1)/2|].$$

When $\sigma_1^2 > \sigma_2^2$ ($\sigma_1^2 < \sigma_2^2$), A will tend to be small (large), and thus to test $H_0 : \sigma_1^2 = \sigma_2^2$ vs. $H_1 : \sigma_1^2 \neq \sigma_2^2$, large or small values of A should lead to rejection of H_0; *see* ANSARI–BRADLEY W-STATISTICS. Tabled critical points as well as a large sample approximation to the null distribution of A can be found in ref. 17. In order to test a one-sided alternative, the appropriate one-sided critical region should be used. When $k > 2$, the Ansari–Bradley statistic can be modified [7, 25] as follows. Letting

$$\bar{A}_i = \sum_{j=1}^{n_i} [(N+1)/2 - |R_{ij} - (N+1)/2|] / n_i,$$

the k-sample statistic is given by

$$B = \frac{N^3 - 4N}{48(N-1)} \sum_{i=1}^{k} n_i [\bar{A}_i - \{(N+2)/4\}]^2.$$

Under H_0, B has approximately a χ^2 distribution with $k - 1$ degrees of freedom and H_0 is rejected at level α whenever $B \geqslant \chi_{1-\alpha}^2(k-1)$.

There are many distribution-free two-sample and k-sample tests that are functions of ranks and references to these can be found in survey papers [7, 10], as well as in refs. 15 and 17. To use any of the rank procedures when the populations do not have a common median, it has been suggested that the observations be replaced by $X'_{ij} = X_{ij} - \tilde{X}_i$, where $\tilde{X}_i$ is the median of the ith sample, and that the standard rank tests be applied directly to the shifted observations. These modified procedures are no longer distribution-free and require symmetric populations to guarantee that they are even asymptotically distribution-free under the null hypothesis [8, 26, 29]. In addition, when compared with the tests of the next section, they seem to compare unfavorably with regard to power [28]. When the medians are known, Sukhatme [29] has suggested a procedure.

Since the observations from the populations with the larger scale parameters tend to be farther from the common median η, an alternative ranking scheme has been suggested by Fligner and Killeen [11]. Instead of ranking the actual observations, rank their distances from the common median if η is known, or if η is unknown, rank their distances from the combined sample median. Specifically, let $V_{ij} = |X_{ij} - \eta|$ if η is known, and $V_{ij} = |X_{ij} - M|$, where M is the combined sample median when η is unknown, for $i = 1, \ldots, k$; $j = 1, \ldots, n_i$. In either case, let R'_{ij} denote the rank of V_{ij} in the combined sample of $N = \sum n_i$ V_{ij}'s. For $k = 2$, the statistic $W = \sum_{j=1}^{n_1} R'_{1j}$ can be used to test $H_0: \sigma_1^2 = \sigma_2^2$ vs. $H_1: \sigma_1^2 \neq \sigma_2^2$ with H_0 being rejected whenever W is large or small. Under H_0, W has the Wilcoxon [31] null distribution; tables can be found in ref. 17. When $k > 2$, the test statistic

$$K = \frac{12}{N(N+1)} \sum_{i=1}^{k} n_i \{\bar{R}'_{i.} - (N+1)/2\}^2$$

can be used to test $H_0: \sigma_1^2 = \cdots \sigma_k^2$ vs. $H_1: \sigma_i^2$ not all equal, H_0 being rejected for large values of K. Under H_0, K has the Kruskal–Wallis* [19] null distribution; tables can be found in ref. 17. A simulation study with symmetric populations indicates that these tests are more powerful than the Ansari–Bradley statistic or its k-sample modification in small samples [11].

None of the procedures of this section is appropriate for testing the equality of the scale parameters of positive random variables as in Case 2; namely, $X_{i1}, \ldots, X_{in_i}$ are a random sample from a population with continuous CDF $F(x/\sigma_i)$, with $F(0) = \frac{1}{2}$. In this case, since the samples from the populations with the larger scale parameters will tend to be larger as well as more spread out, it has been suggested that distribution-free location tests such as the Kruskal–Wallis be applied directly to the data; see ref. 10 (p. 1292) or 30.

ROBUST TESTS FOR SCALE

In this section, tests for equality of the scale parameters are described that are appropriate under the weakest assumptions to be considered in this entry-namely, when $X_{i1}, \ldots, X_{in_i}$ is a random sample from a population with continuous CDF $F((x - \eta_i)/\sigma_i)$, $i = 1, \ldots, k$. In this setting the shape of the underlying distributions is arbitrary and no assumptions are made regarding the location parameters. Except for a procedure due to Moses [23], all tests to be described require large sample approximations to their null distributions to determine appropriate critical regions. Simulation studies provide the only small sample information for determining how well these procedures actually maintain their nominal significance levels as well as for making power comparisons. However, from these studies several procedures appear to be "best," and these will be described. For further discussion of competing procedures see the survey papers [7, 10] or the simulation studies [5, 7, 12, 13, 20].

For testing $H_0: \sigma_1^2 = \cdots = \sigma_k^2$ vs. $H_1: \sigma_i^2$ not all equal, Levene [21] has suggested performing a one-way analysis of variance* on the variables $Z_{ij} = |X_{ij} - \bar{X}_{i.}|$, $i = 1, \ldots, k$, $j = 1, \ldots, n_i$; *see* LEVENE'S ROBUST TEST OF HOMOGENEITY OF VARIANCES. Brown and Forsythe [5] have modified Levene's procedure by considering the distances of the observations from their sample median rather than their sample mean, and this appears to have improved the small sample properties as well as making the procedure asymptotically distribution-free [22]. Specifically, let $V_{ij} = |X_{ij} - \tilde{X}_i|$, $i = 1, \ldots, k$, $j = 1, \ldots n_i$, where $\tilde{X}_i =$ median $\{X_{i1}, \ldots, X_{in_i}\}$. A one-way analysis of variance is then performed on the V_{ij}'s by computing the statistic

$$L = \frac{(N-k)\sum n_i (\bar{V}_{i.} - \bar{V}_{..})^2}{(k-1)\sum\sum (V_{ij} - \bar{V}_{i.})^2},$$

and rejecting H_0 whenever

$$L \geqslant F_{1-\alpha}(k-1, N-k).$$

An alternative procedure, which is a direct generalization of the two-sample jackknife* of Miller [22], was proposed by Layard [20]. Let $S_{i(j)}^2 = \sum_{m \neq j} (X_{im} - \bar{X}_{i(j)})^2/(n_i - 2)$, where $\bar{X}_{i(j)} = \sum_{m \neq j} X_{im}/(n_i - 1)$, $i = 1 \ldots, k$; $j = 1, \ldots, n_i$, denote the jackknife variance, i.e., the variance of the observations in the ith sample computed with the jth observation deleted. Now, form the "pseudo-observations," $U_{ij} = n_i \ln S_i^2 - (n_i - 1)\ln S_{i(j)}^2$, $i = 1, \ldots, k$; $j = 1, \ldots, n_i$. To test H_0, the one-way analysis of variance F-statistic is calculated on the U_{ij}'s, resulting in the test statistic

$$J = \frac{(N-k)\sum(\bar{U}_{i\cdot} - \bar{U}_{\cdot\cdot})^2}{(k-1)\sum\sum(U_{ij} - \bar{U}_{i\cdot})^2},$$

H_0 being rejected whenever $J \geqslant F_{1-\alpha}(k-1, N-k)$. The statistic L is better able to maintain its nominal significance level than the jackknife statistic J, particularly for small or unequal sample sizes, but it has somewhat less power [5].

The final two procedures to be described require random grouping of the data before the test statistic can be computed. For any integer $m \geqslant 2$, divide each sample into disjoint subsamples of size m in some random manner. The subsample variance S_{ij}^2 for the jth subsample in the ith sample is then computed for each subsample. Box [5], utilizing an idea of Bartlett and Kendall [3], suggested performing a one-way analysis of variance on the logarithms of the subsample variances. The reason for taking logarithms is that the logarithms tend to be more nearly normally distributed than the original subsample variances. This procedure is known as the log-ANOVA test, and although it maintains its nominal significance level quite well, it tends to have very poor power. Instead of performing an analysis of variance on the logarithms of the subsample variances, one could compute the Wilcoxon [31] statistic for $k = 2$, or the Kruskal–Wallis [19] statistic for $k > 2$, on the subsample variances resulting in an exactly distribution-free test. These procedures were suggested by Moses [23], and although the tests maintain their significance level exactly, their power tends to be quite low. Moses also suggests using functions other than the variances on the subsamples.

The next section contains a discussion which will unify the procedures contained in the preceding three sections.

DISCUSSION

Although there are a multitude of procedures for testing for homogeneity of variance, the assumptions that are valid for a particular problem will limit the number of appropriate procedures. First, as pointed out in the fourth section, the assumption of normality should not be made without some type of theoretical or empirical justification. However, when this assumption is valid, Bartlett's test should be used. The procedures of the fifth section also make fairly restrictive assumptions, this time on the equality of the medians of the populations. Again, this assumption should not be made without some justification, and when it is valid, the Ansari–Bradley or Fligner–Killeen procedures should work well. The important point is that the assumptions underlying the procedures of these two sections are not cosmetic in nature and should not be made for convenience, because a failure in the assumptions will seriously affect the validity of the inference.

In most, although not all, situations neither the assumptions of equal medians nor normality will be appropriate and the choice of a suitable test statistic will be restricted to those of the sixth section. For moderate or large sample sizes, the jackknife or Levene procedure can be used. The jackknife tends to have higher power, although the Levene test is better able to maintain its nominal significance level, particularly for unbalanced sample sizes. For small samples, if one is concerned about using a procedure whose actual significance level is quite close to the nominal value, then Moses procedures can be used, although it must be remembered that in this case the power will be quite poor and only fairly extreme alternatives will be detected.

References

[1] Ansari, A. R. and Bradley, R. A. (1960). *Ann. Math. Statist.*, **31**, 1174–1189.

[2] Bartlett, M. S. (1937). *Proc. R. Soc. A*, **160**, 268–282.

[3] Bartlett, M. S. and Kendall, D. G. (1946). *J. R. Statist. Soc. B*, **8**, 128–138.

[4] Box, G. E. P. (1953). *Biometrika*, **40**, 318–335.

[5] Brown, M. B. and Forsythe, A. B. (1974). *J. Amer. Statist. Ass.*, **69**, 364–367. (Contains a simulation study.)

[6] Cochran, W. G. (1941). *Ann. Eugenics, Lond.*, **11**, 47–52.

[7] Conover, W. J., Johnson, M. E., and Johnson, M. M. (1981). *Technometrics*, **23**, 351–361. (Contains an extensive simulation study for small samples.)

[8] Crouse, C. F. (1964). *Ann. Math. Statist.*, **35**, 1825–1826.

[9] David, H. A. (1952). *Biometrika*, **39**, 422–424.

[10] Duran, B. S. (1976). *Commun. Statist. A*, **5**, 1287–1312. (Very readable survey of nonparametric, as well as some parametric, tests for scale.)

[11] Fligner, M. A. and Killeen, T. J. (1976). *J. Amer. Statist. Ass.*, **71**, 210–213.

[12] Gartside, P. S. (1972). *J. Amer. Statist. Ass.*, **67**, 342–346 (Contains a simulation study.)

[13] Geng, S., Wang, W. J., and Miller, E. (1979). *Commun. Statist. B*, **8**, 379–389. (Contains a simulation study.)

[14] Glaser, R. E. (1976). *J. Amer. Statist. Ass.*, **71**, 488–490.

[15] Hájek, J. And Šidák, Z. (1967). *Theory of Rank Tests*, Academic, New York. (Linear rank formulation of scale tests with optimality properties.)

[16] Hartley, H. O. (1950). *Biometrika*, **37**, 308–312.

[17] Hollander, M. and Wolfe, D. A. (1973). *Nonparametric Statistical Methods*. Wiley, New York.

[18] Izenman, A. J. (1976). *Biometrika*, **63**, 185–190.

[19] Kruskal, W. H. and Wallis, W. A. (1952). *J. Amer. Statist. Ass.*, **47**, 583–621.

[20] Layard, M. W. J. (1973). *J. Amer. Statist. Ass.*, **68**, 195–198. (Contains a simulation study.)

[21] Levene, H. (1960). In *Contributions to Probability and Statistics*, I. Olkin, ed. Stanford University Press, Palo Alto, CA, pp. 278–292.

[22] Miller, R. G. (1968). *Ann. Math. Statist.*, **39**, 567–582.

[23] Moses, L. E. (1963). *Ann. Math. Statist.*, **34**, 973–983.

[24] Pearson, E. S. and Hartley, H. O., eds. (1966). *Biometrika Tables for Statisticians*, 3rd ed. Cambridge University Press, Cambridge, England.

[25] Puri, M. L. (1965) *Ann. Inst. Statist. Math.*, **17**, 323–330.

[26] Raghavachari, M. (1965). *Ann. Math. Statist.*, **36**, 1236–1242.

[27] Scheffé, H. (1959). *The Analysis of Variance*. Wiley, New York.

[28] Shorack, G. R. (1969). *J. Amer. Statist. Ass.*, **64**, 999–1013.

[29] Sukhatme, B. V. (1958). *Ann. Math. Statist.*, **29**, 60–78.

[30] Tsai, W. S., Duran, B. S., and Lewis, T. O. (1975). *J. Amer. Statist. Ass.*, **70**, 791–796. (Contains a simulation study.)

[31] Wilcoxon, F. (1945). *Biometrics Bull.*, **1**, 80–83.

(ANSARI–BRADLEY *W*-STATISTICS
BARTLETT'S TEST OF HOMOGENEITY OF VARIANCES
COCHRAN'S (TEST) STATISTIC
DISTRIBUTION-FREE METHODS
HARTLEY'S F_{MAX} TEST
HOMOGENEITY AND TESTS OF HOMOGENEITY
LEVENE'S ROBUST TEST OF HOMOGENEITY OF VARIANCES
LOCATION-SCALE PARAMETER
LOCATION TESTS
ROBUSTNESS OF TESTS)

MICHAEL A. FLIGNER

SCANDINAVIAN JOURNAL OF STATISTICS

The *Scandinavian Journal of Statistics, Theory and Applications* (*SJS*) was founded in 1973 under the auspices of statistical societies in the Nordic countries; the Danish Society for Theoretical Statistics, the Finnish Statistical Society, the Norwegian Statistical Society, and the Swedish Statistical Association.

Since the start *SJS* has appeared with one volume each year. It has gradually grown from three issues with 144 pages in Vol. 1 (1974) to four issues with 320 pages in Vol. 10 (1983).

The editorial board consists of about 10 persons, all from the Nordic countries. Bengt Rosén has acted as Editor-in-Chief since Vol. 1. The editorial policy of the journal is as follows. The main purpose of *SJS* is to publish research papers on statistical theory and its applications, including relevant aspects of probability. *SJS* has a section "Notices on Applied Statistics in Scandinavia" containing short notes of interesting applications of not necessarily original statistical methodology. There is also a section "Brief Information on Current Unpublished Research in Scandinavia." Book reviews appear occasionally.

The contents of a recent issue (Vol. 10, issue 3) were

"Some topics in regression" by S. Johansen.

"Mathematical statistics and automatic control" by T. Bohlin.

"The predictive approach and randomized population type models for finite population inference from two-stage samples" by R. Sundberg.

"Approximation to the maximum likelihood equations for some Gaussian random fields" by H. Künsch.

"The generalized Bessel process corresponding to an Ornstein–Uhlenbeck process" by B. Eie.

From the outset *SJS* has aimed at being an international journal as regards standards, contributors, and circulation. The policy in this respect is formulated as follows. While the intention of the journal is to foster research in Denmark, Finland, Norway, and Sweden, contributions from any other country are welcomed. The language of *SJS* is English. Accordingly manuscripts are judged solely on their scientific qualities. Every paper is refereed, usually by two referees. Currently approximately 70% of the published materials has authors from the Nordic countries. Volume 10 (1983) has 544 institutional subscribers from different countries and 106 individual subscribers, the latter mostly from the Nordic countries. *SJS*'s publisher is the Almqvist and Wiksell Periodical Co., Stockholm, Sweden.

BENGT ROSÉN

SCAN DIAGRAMS

Scan diagrams, suggested by F. H. Steen, are nontraditional devices for graphically and mechanically portraying probability distributions of any type.

Let X be a finite discrete random variable with probabilities $f(x_1), f(x_2), \ldots, f(x_n)$. The diagram consists of a horizontal strip one unit long divided into sections labeled $x_1, x_2, \ldots, x_n$, whose lengths are equal to the successive probabilities just mentioned (Fig. 1). In this initial elementary form it is simply the uniform-width rectangles of the distribution's histogram* laid on their sides in succession with no gaps. If this arrangement is supplied with a scanning device that can be stopped randomly, there results a physical model for the theoretical distribution. Supplied with a scale graduated from zero to one, the diagram reveals the mode, median, quartiles, deciles, and fractiles of the distribution.

The diagram may be wrapped around a cylinder of suitable size (Fig. 2), rolled to obtain an outcome, or it can be drawn on the dial of a spinner (Fig. 3), which has the disadvantage of a circular scale. In any case it serves as a single prototype, embracing coin tossing, dice rolling, card drawing, and the like.

Nonfinite discrete distributions, such as the geometric* and Poisson*, can be graphi-

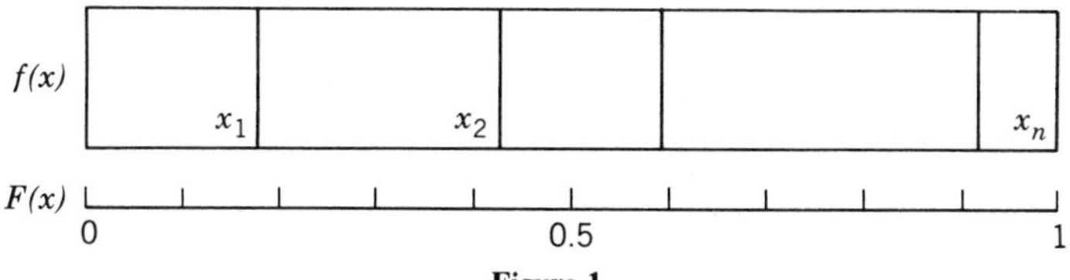

Figure 1

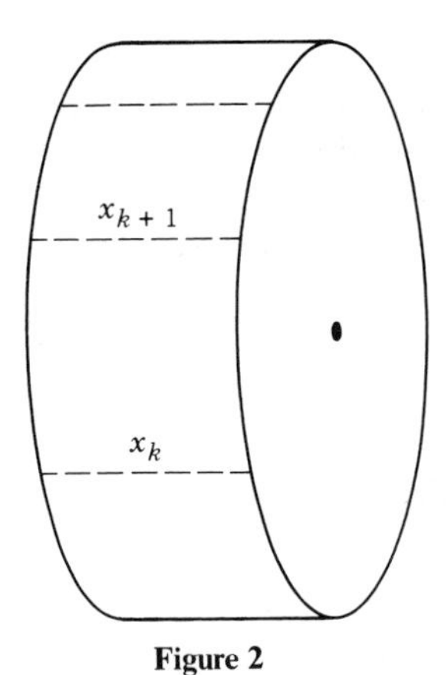

Figure 2

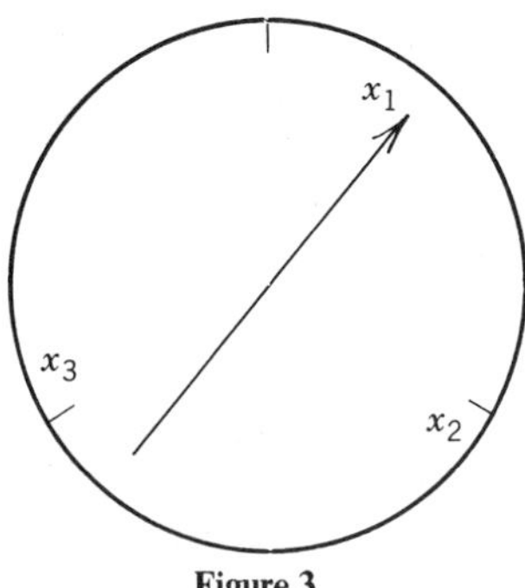

Figure 3

cally portrayed by similar means, although in this case not all of the sections can be indicated, of course (Fig. 4).

In dealing with continuous distributions such as the standard normal, the device is most conveniently constructed from tables of cumulative probabilities using selected values of the random variable and the uniform scale (Fig. 5).

The extension to discrete multivariate distributions is accomplished by forming a scan diagram for the first variable, then beneath each of its values a diagram for the joint

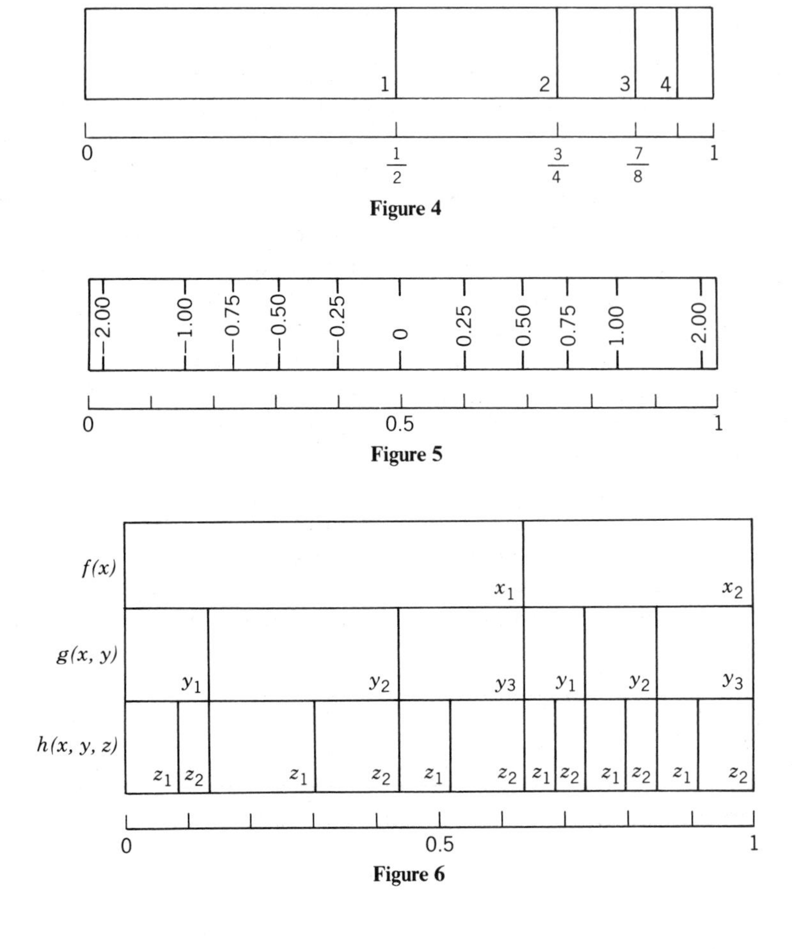

Figure 6

variation of this value and the values of the second variable, and so on (Fig. 6). The order may be varied according to the purpose at hand. The diagram itself displays graphically the various joint, marginal, and conditional probabilities. For example, $g(y_3|x_1)$ is the size of the third rectangle in the second line relative to the size of the x_1 rectangle above it in Fig. 6; $h(x_1)$ is represented by the 1st, 3rd, ..., 11th rectangles of the last line, and so on. The corresponding configuration with uniform spacing is more simply drawn and is useful for portraying various outcome possibilities as in a tree for which it is a substitute.

As in the case of a univariate distribution, the scan diagram just described can be extended to both infinite discrete and continuous multivariate distributions.

Reference

[1] Steen, Frederick H. (1982) *Elements of Probability and Mathematical Statistics*. Duxbury Press, Boston, MA, pp. 61–66, 134, 138, 152, 216.

(GRAPHICAL REPRESENTATIONS OF DATA
HISTOGRAMS
MULTIVARIATE GRAPHICS)

FREDERICK H. STEEN

SCAN STATISTICS

Professionals in a variety of disciplines scan times of occurrence of events and focus on multiple events happening close in time. Several statistics arise from scanning a time interval $(0, T)$ and observing that there exists a subinterval of length t that contains n events. Among these scan statistics are nt, the largest number of points within any subinterval of length t; pn, the length of the smallest interval containing n points, and $W(n, t)$, the waiting time until one first observes n events all occurring within a subinterval of length t. A related set of scan statistics arises from scanning the outcomes of a series of trials. We illustrate the statistics for these two cases.

Events Occurring in Time. A health official reviewing a community's records for the year is startled to find five cases of cancer occurring in the 10 day period from April 21st through April 30th. In the past the community has averaged about 36 cases per year, which is similar to other communities. Here $n10 = 5$, $p5 = 10$, and $W(5, 10) = 120$.

Events as Outcomes of Trials. A company produces motors and under normal conditions about 2% of the motors coming off the assembly line do not pass inspection. On a particular day, of 500 motors produced, 11 failed inspection. Reviewing the records, the quality control manager noted that six of the failed motors were all among 100 consecutive motors coming of the assembly line. Here $n100 = 6$, $p6 = 100$, and, if the last of the six failed motors occurred on the 280th motor off the line, $W(6, 100) = 280$.

Both the health official and the quality control manager are interested in whether the observed clusters are unlikely to have happened by chance. Scan statistics are used in such hypothesis tests.

The distributions of the scan statistics $n(t)$, $p(n)$, and $W(n, t)$ are related:

$$\begin{aligned}\Pr(n(t) \geqslant n) &= \Pr(p(n) \leqslant t)\\ &= \Pr(W(n, t) \leqslant T).\end{aligned}$$

These probabilities depend on the underlying process distribution chosen. The distributions of the scan statistics have been derived for the Poisson and Bernoulli processes and these provide simple models of chance variation for events occurring in time and as outcomes of trials, respectively. Actually each of the processes leads to several chance models appropriate for different situations.

THE POISSON PROCESS*. Given events occurring in time according to a Poisson process with mean λ, the number of events (points) k occurring in any interval $(0, T)$ has the Poisson distribution*. Conditional on there being N points from the Poisson process in $(0, T)$, the N points are independently and identically distributed according to the uniform distribution on $(0, T)$. The scan statistic probability $\Pr(n(t) \geqslant n$, given N points in $(0, T))$, is different from the unconditional probability $\Pr(n(t) \geqslant n)$. Denote and abbreviate these probabilities $P(n;\ N, t/T) = P$ and $P^*(n;\ \lambda T, t/T) = P^*$, respectively. Each of these probabilities has different testing and other applications. In a case where no comparable average is known, one would use the conditional level of significance P; in a case where the average is known and is different from the total, or in sequential monitoring over a year, which may be stopped as soon as $n(t) \geqslant n$, so that N is not known, one would use P^*.

The probabilities P and P^* are in general complex to compute, and a variety of formulas, approximations, bounds, and tables for different cases have been developed.

HISTORY AND LITERATURE

The conditional probability P was studied early in photographic science, where it was hypothesized that a silver speck on film would develop if k or more photons are absorbed within the decay time t of the nucleus. $P = P(n, N;\ t/T)$ depends on n, N, and the ratio $r = t/T$. The special cases $P(2, N;\ r)$, the CDF of the smallest gap between any two of the points, and $P(N, N;\ r)$, the CDF of the sample range for uniformly distributed variates, were known for a long time (*see* ORDER STATISTICS). The probability P could be written as an integral that is very complicated to evaluate except in simple cases; van Elteren and Gerritis [5] evaluated it for the cases $n = 3$, $N \leqslant 8$. Naus [15] uses a combinatorial approach to derive P that gives a simple formula for $n > N/2$. The approach is generalized in Naus [16] for $r = 1/L$, in Huntington and Naus [8] for all n, N, r, and in alternate form in Hwang [9]. However, for large N and small r, the general results involve the sum of a large number of determinants of large matrices. Wallenstein and Naus [28] give tables and simplified formulas (computable for large N or small r) for the case $N/3 < n \leqslant N/2$. Neff and Naus [20] give extensive tables of the probabilities and polynomials for N under 19, allowing evaluation for a full range of n and r. For $N \geqslant 19$, $2 < n \leqslant N/3$, Neff and Naus [20] and Naus [19] suggest approximations.

The probability P^* arose in certain studies of visual perception, investigations of the rate of n-fold accidental coincidences in counters, probability of overflow in queueing*, and other applications. Prior to the 1960s exact formulas for P^* were known for only a few special cases. Menon [14] develops recursion formulas but their complexity limits their use to the cases $n = 2, 3$. Newell [21], Ikeda [10], Conover et al. [1], and Cressie [4] give asymptotic results. Naus [19, Theorem I] gives exact expression for P^* for $r = \frac{1}{2}, \frac{1}{3}$, and a highly accurate approximation based on these results for all r. Janson [11] provides sharp bounds for P^*, and shows that the Naus [19] approximation falls within these bounds. Newell [21] derives asymptotic results for $\Pr(nt \geqslant n)$ for the more general process where the waiting times between points are independently and identically distributed; the special case where the interpoint distance distribution is exponential is the Poisson case. Melzak [13] gives approximations for general types of clustering and scanning regions. Samuel-Cahn [24] gives a series of approximations to the expectation of the waiting time $W(n, t)$ until the scan interval contains n points for general point processes with independent and identically distributed interarrival times of points. Ikeda [10], Mclure [12], and Glaz and Naus [6] give various asymptotic and approximate results for multiple scan statistics. Rothman [23], Wallenstein [27], and Cressie [2, 3] give various exact and approximate results for scan statistics on the circle.

For the Bernoulli trial model, Roberts [22] develops various quality control* tests and Troxell [26], various acceptance sampling* plans based on scan statistics. Saperstein [25], Huntington [7], Neff and Naus [20], and Naus [17, 18] give various results, references, and applications.

THE CONDITIONAL PROBABILITY P

The probability P is piecewise polynomial in r, with different polynomials in different ranges of r. For example, for $n > N/2$, there is one polynomial for $r \leqslant \frac{1}{2}$ and a different polynomial for $r \geqslant \frac{1}{2}$. The polynomial for $n > N/2$, $p \leqslant \frac{1}{2}$ is particularly simple and can be written in terms of binomials and sums of binomials:

$$P(n, N; r) = 2\sum_{i=n}^{N} \binom{N}{i} r^i (1-r)^{N-i} + \{(n/r) - N - 1\} \times \binom{N}{n} r^n (1-r)^{N-n}.$$

General formulas are available for all n, N, r but these involve sums of large numbers of determinants. For even moderately large N, and moderately small r, sophisticated programming is needed to use the general formula. Simplified formulas for special cases and tables of the probabilities and polynomials and approximations are available (see History and Literature).

THE UNCONDITIONAL PROBABILITY P^*

The probability P^* is important in studies of visual perception where it is assumed that photons arrive at the eye receptors according to a Poisson process. It is conjectured that several photons all arriving within a time t causes the neurons to discharge. What is of particular interest is the distribution of $W(n, t)$, the waiting time until discharge.

Except for simple cases, exact formulas for P^* involve complex computer evaluation, and various asymptotic results and bounds are available. The following approximation is highly accurate in many situations:

Let $Q^* = 1 - P^*$, $L = T/t$, $Q^*2 = Q^*(n, 2\lambda, 1/2)$, and $Q^*3 = Q^*(n, 3\lambda, 1/3)$; then $Q^*(n, L\lambda, 1/L) = Q^*2(Q^*3/Q^*2)^{L-2}$. The values of Q^*2 and Q^*3 can be read from tables or computed from simple formulas. For example, to approximate $P^*(5, 36; 10/365)$, note that $n = 5$, $\lambda L = 36$, $1/L = 10/365$, so $\lambda = 0.9863$. Then $Q^*2 = 0.98434962$, $Q^*3 = 0.9727576$, and $P^*(5, 36; 10/365)$ is approximately 0.35. The health official's observed cluster of five cancer cases within some consecutive 10-day period of the year could easily have arisen by chance.

An accurate approximation to the expectation of $W(n, t)$ is as follows: Without loss of generality define the units to make $t = 1$ and let λ denote the expected number of points per unit. Then $E(W(n, 1)) = 2 + Q^*2/\log_e(Q^*2/Q^*3)$, where Q^*2 and Q^*3 are found as above. (In the above example, $\lambda = 0.9863$ per unit of 10 days, Q^*2 and Q^*3 are as given, and the expected waiting time until getting a cluster of five cases within 10 days is 85 units of 10 days, or 850 days.)

SEQUENCE OF TRIALS: BERNOULLI MODEL

Given a sequence of T Bernoulli trials, each trial can result in a success with probability p, and in a failure with probability $1 - p = q$, and the T trials are independent. nt is the maximum number of successes within any (scanning interval consisting of) t consecutive trials. The occurrence of at least n successes within some t consecutive trials is called an *$n : t$ quota*. The special case of a $t : t$ quota is a success run of length t, and $\Pr(nt \geqslant t)$ is the probability of at least one success run of length t (*see* RUNS). An accurate approximation for $\Pr(nt \geqslant n) = 1 - Q'L$, where $L = T/t$, is $Q'L = Q'2(Q'3/Q'2)^{L-2}$, where easily computable formulas are available for $Q'2$ and $Q'3$.

Given the sequence of T Bernoulli trials results in N successes and $T - N$ failures, all $\binom{T}{N}$ arrangements of the successes and failures are equally likely. The probability of an $n:t$ quota in this conditional (on N) case is referred to as the generalized birthday probability and is known. For t and T both large relative to n, N, the conditional probability in the Bernoulli case can be approximated by P, and for the unconditional Bernoulli case by P^*. For the quality control example, the probability of $6:100$ quota given 11 defectives in 500 trials can be approximated by $P(6, 11; 100/500)$, which by simple formula ($n > N/2$, $r \leqslant \frac{1}{2}$) is 0.198. (The exact probability of the quota is 0.185.)

THE CIRCLE

Given N points distributed at random over the circle of unit circumference, let $Pc = Pc(n, N;\ t)$ denote the probability that a scanning arc of length t somewhere contains at least n points. Various exact and approximate results are known for Pc. Note that on the circle, if the arc of length t nowhere contains as many as n points, then the complementary arc of length $1 - t$ must always contain at least $N - n + 1$ points, and knowing the distribution of the maximum gives the distribution of the miminum. The minimum in a scanning arc is also a scan statistic and has been studied under coverage* problems.

References

[1] Conover, W. J., Bement, T. R., and Iman, R. L. (1979). *Technometrics*, **21**, 277–282.

[2] Cressie, N. (1977). *J. Appl. Prob.*, **14**, 272–282.

[3] Cressie, N. (1977). *Int. Statist. Bull.*, **47**, 124–127.

[4] Cressie, N. (1980). *Ann. Prob.*, **8**, 828–840.

[5] Elteren, P. H. van and Gerritis, H. J. M. (1961). *Statist. Neerlandica*, **15**, 385–401.

[6] Glaz, J. and Naus, J. I. (1983). *Commun. Statist. A*, **12**, 1961–1986.

[7] Huntington, R. J. (1976). *J. Appl. Prob.*, **13**, 604–607.

[8] Huntington, R. J. and Naus, J. I. (1975). *Ann. Prob.*, **3**, 894–896.

[9] Hwang, F. K. (1977). *Ann. Prob.*, **5**, 814–817.

[10] Ikeda, S. (1965). *Ann. Inst. Statist. Math.*, **17**, 295–310.

[11] Janson, S. (1984). *Stoch. Processes*, **18**, 313–328. (Gives several sets of bounds for P^* that are sharp in different situations; of particular interest for N large.)

[12] McLure, D. E. (1976). Extreme Nonuniform Spacings. *Pattern Analysis Report No. 44*, Brown University, Providence, RI.

[13] Melzak, Z. A. (1979). *Math. Proc. Camb. Phil. Soc.*, **86**, 313–337.

[14] Menon, M. V. (1963). Clusters in a Poisson Process. IBM Research Laboratories, San José, California.

[15] Naus, J. I. (1965). *J. Amer. Statist. Ass.*, **60**, 532–538.

[16] Naus, J. I. (1966). *J. Amer. Statist. Ass.*, **61**, 1191–1199.

[17] Naus, J. I. (1974). *J. Amer. Statist. Ass.*, **69**, 810–815.

[18] Naus, J. I. (1979). *Int. Statist. Rev.*, **47**, 47–78. (Indexed bibliography on scan statistics and related topics.)

[19] Naus, J. I. (1982). *J. Amer. Statist. Ass.*, **77**, 177–183. (Gives simple and in some cases highly accurate approximations for a variety of scan problems.)

[20] Neff, N. D. and Naus, J. I. (1980). *IMS Series of Selected Tables in Mathematical Statistics VI*. American Mathematical Society, Providence, Rhode Island. (Tables of probabilities for P and P^* and polynomials for P. Extensive introduction, review of formula and computation approaches and applications, 51 references.)

[21] Newell, G. F. (1963). In *Proc. Symp. Time Series Analysis*, M. Rosenblatt, ed. Wiley, New York, pp. 89–103.

[22] Roberts, S. W. (1958). *Bell Syst. Tech. J.*, **37**, 83–114.

[23] Rothman, E. D. (1972). *Sankhyā A*, **34**, 23–32.

[24] Samuel-Cahn, E. (1983). *Adv. Appl. Prob.*, **15**, 21–38.

[25] Saperstein, B. (1972). *J. Amer. Statist. Ass.*, **67**, 425–428.

[26] Troxell, J. R. (1972). An Investigation of Suspension Systems for Small Scale Inspections. Ph.D. Thesis, Department of Statistics, Rutgers University, New Brunswick, NJ.

[27] Wallenstein, S. R. (1971). Coincidence Probabilities Used in Nearest-Neighbor Problems on the Line and Circle. Ph.D. Dissertation, Rutgers University, New Brunswick, NJ.

[28] Wallenstein, S. R. and Naus, J. I. (1974). *J. Amer. Statist. Ass.*, **69**, 690–697.

(COVERAGE
POISSON PROCESS

RUNS
WAITING TIMES)

JOSEPH I. NAUS

SCATTERGRAM *See* TRIPLE SCATTER PLOT

SCATTER PLOT, TRIPLE *See* TRIPLE SCATTER PLOT

SCEDASTIC CURVE

The *scedastic function* of a variable Y with respect to a variable X is its conditional variance given X. The graph of

$$\text{var}(Y|X = x)$$

against x is the *scedastic curve*.

This name is also given to a graph of the ratio

$$\text{var}(Y|X = x)/\text{var}(Y).$$

The name is also applied to similar functions for standard deviation, rather than variance.

(HETEROSCEDASTICITY)

SCHAFER–SHEFFIELD TEST

The Schafer–Sheffield test [7] is a procedure for comparing two Weibull* populations. Let $x_1, x_2, \ldots, x_n$ and $y_1, y_2, \ldots, y_n$ be independent random samples, both being of size n, from two Weibull distributions

$$F_1(x) = 1 - \exp[-(x/b_1)^{c_1}], \qquad x > 0,$$
$$F_2(y) = 1 - \exp[-(y/b_2)^{c_2}], \qquad y > 0,$$

respectively. The scale parameters b_1, b_2 and the shape parameters c_1, c_2 are positive but unknown. The problem is to test the null hypothesis $H_0: b_1 = b_2,\ c_1 = c_2$ against the alternative $H_1: b_1 > b_2,\ c_1 = c_2$.

The test statistic suggested by Schafer and Sheffield is based on maximum likelihood* estimates of the parameters. Since $c_1 = c_2$ is included in both hypotheses, one may set $c_1 = c_2 = c$. With no restrictions on b_1 and b_2, the maximum likelihood estimates c^*, b_1^*, and b_2^* of c, b_1, and b_2, respectively, are given by solving the equations

$$2n/c^* - n\left[\Sigma x_i^{c^*}\ln x_i/(\Sigma x_i^{c^*}) + \Sigma y_i^{c^*}\ln y_i/(\Sigma y_i^{c^*})\right] + \Sigma \ln x_i + \Sigma \ln y_i = 0,$$

$$b_1^* = (\Sigma x_i^{c^*}/n)^{1/c^*}, \qquad b_2^* = (\Sigma y_i^{c^*}/n)^{1/c^*}.$$

Here c^* is the positive root of the first equation, and the range of the summation Σ is from $i = 1$ to n. The solutions (b_1^*, b_2^*, c^*) to the above equations uniquely maximize the joint likelihood function with $c_1 = c_2 = c$. The statistic for testing H_0 against H_1 is

$$T(c^*, b_1^*, b_2^*) = c^*[\ln b_1^* - \ln b_2^*].$$

The null hypothesis is rejected if the observed value of T is too large. The random variable $T(c^*, b_1^*, b_2^*)$ has a distribution that does not depend on b_1, b_2, and c when H_0 is true. Nevertheless, the null distribution of T is not available in an explicit form. Schafer and Sheffield obtained some quantiles of T under H_0 by the Monte Carlo method, and provided a table containing quantiles for $n = 5(1)20, 24(4)40, 50(10)100$.

In terms of power, the Schafer–Sheffield test is an improvement of a similar test considered by Thoman and Bain [8] for the same problem. But it still relies on the assumption that the populations concerned follow Weibull distributions. Any departure from this assumption may lead to a misleading significance level, should the Schafer–Sheffield test be used. Although failures of many kinds of mechanical devices can be well described by Weibull distributions [9, 2, 3], it is advisable to check whether or not data of interest are from the assumed distributions.

One method often used to determine whether or not a set of observed data came from a Weibull distribution is to perform graphical plotting on a piece of Weibull probability paper (see, e.g., Mann et al. [4, pp. 214–217]). Another method is to perform a goodness-of-fit* test, such as the test suggested by Mann et al. [5] (*see* also [4, Sec. 7.1.2] and MANN–FERTIG STATISTIC).

If data show evidence of departure from the family of Weibull distributions, one should either try some transformations to adjust them or consider nonparametric procedures, such as the Savage test* or its modifications (see Basu [1]). A discussion of recent developments in nonparametric procedures, especially for censored data*, is given in Miller [6].

References

[1] Basu, A. P. (1968). *Ann. Math. Statist.*, **39**, 1591–1604. (A modified Savage test for type II censored data.)

[2] Kao, J. H. K. (1958). *IRE Trans. Rel. Quality Control*, **13**, 15–22. (Vacuum-tube failures and the Weibull distribution.)

[3] Lieblein, J. and Zelen, M. (1956). *J. Res. Nat. Bur. Stand.* **57**, 273–316. (The fatigue life of deep-groove ball bearings, an example of the Weibull distribution.)

[4] Mann, N. R., Schafer, R. E., and Singpurwalla, N. D. (1974). *Methods for Statistical Analysis of Reliability and Life Data*. Wiley, New York. (Excellent reference book.)

[5] Mann, N. R., Scheuer, E. M., and Fertig, K. W. (1973). *Commun. Statist. A*, **2**, 383–400.

[6] Miller, R. G., Jr. (1981). *Survival Analysis*. Wiley, New York. (A book worth reading.)

[7] Schafer, R. E. and Sheffield, T. S. (1976). *Technometrics*, **18**, 231–235.

[8] Thoman, D. R. and Bain, L. J. (1969). *Technometrics*, **11**, 805–815.

[9] Weibull, W. (1951). *J. Appl. Mech.*, **18**, 293–297. (Derivation of the Weibull distribution as a probabilistic characterization for the breaking strength of materials.)

(MANN–FERTIG STATISTIC
SAVAGE TEST
WEIBULL DISTRIBUTION)

H. K. HSIEH

SCHEFFÉ'S SIMULTANEOUS COMPARISON PROCEDURE

In 1953 Henry Scheffé derived sets of confidence intervals* for the regression parameters of a general linear model* with normally distributed errors, such that the parameters jointly lie in the resulting regions (2) with predetermined levels of confidence $1 - \alpha$ [4]. The *S*-method, as it has come to be known, is now described.

Consider the general linear model*

$$\mathbf{Y} = \mathbf{X}'\boldsymbol{\beta} + \mathbf{e}, \qquad (1)$$

where $\boldsymbol{\beta}_{m\times 1}$ is unknown, $\mathbf{X}'_{n\times m}$ is known and of rank $r \leqslant m$, and $\mathbf{e}$ is a vector of n independent identically distributed normal variables with common mean 0 and unknown variance σ^2. Let L be a q-dimensional space of estimable functions ψ, i.e., of linear parametric combinations $\mathbf{a}'\boldsymbol{\beta}$ (*see* ESTIMABILITY). Then two results hold, the first of which provides a simultaneous confidence region for all ψ in L.

Theorem 1. With the preceding assumptions, the probability is $1 - \alpha$ that the least-squares* estimates $\hat{\psi}$ of ψ jointly satisfy

$$\hat{\psi} - \left[qF_{q,\nu;\alpha}\hat{\sigma}^2(\hat{\psi})\right]^{1/2} \leqslant \psi \leqslant \hat{\psi} + \left[qF_{q,\nu;\alpha}\hat{\sigma}^2(\hat{\psi})\right]^{1/2}, \qquad (2)$$

where $\hat{\sigma}^2(\hat{\psi})$ is the estimator of $\text{var}(\psi)$ with $\nu = n - r$ degrees of freedom in the analysis of variance* of (1), and $F_{q,\nu;\alpha}$ is the upper α-point of the F-distribution* with q and ν degrees of freedom.

Equation (8) of the entry MULTIPLE COMPARISONS illustrates (2) for the case in which L is the $(k-1)$-dimensional space of contrasts* of k means $\mu_1, \ldots, \mu_k$ in the one-way classification problem. The *S*-method is not restricted to contrasts in this setup, however; the form taken by (2) for joint estimation of general linear functions of $\mu_1, \ldots, \mu_k$ is given, e.g., in ref. 7. In an $s \times t$ factorial experiment*, L might be the $(s-1)(t-1)$-dimensional space of all linear combinations of the interactions* of the two factors. *See* WORKING–HOTELLING–SCHEFFÉ CONFIDENCE BANDS for the application of the *S*-method to linear and multiple regression models.

Theorem 2. With the preceding assumptions the level α test of the hypothesis

$$H: \text{for all } \psi \text{ in } L, \qquad \psi = 0,$$

will fail to reject H if and only if, for all ψ in L, the intervals (2) contain zero.

Scheffé noted the correspondence between the multiple comparisons defined in Theorem 1 and the hypothesis testing procedure in Theorem 2 [4, 5]. Olshen [1] pointed out that if the confidence region (1) is reported only when H is rejected, then the probability of simultaneous coverage of all ψ in L is not unconditional, and hence is not equal to $1 - \alpha$. In response, Scheffé [6] redefined his procedure to ensure that the multiple comparisons would be made only when they are regarded as the main objective of the experimenter, i.e., regardless of the outcome of any hypothesis testing.

The S-method is usually compared with Tukey's simultaneous comparison procedure or with extensions thereof. There are several other established methods, however; for comparisons of all of these, *see* MULTIPLE COMPARISONS and Stoline [9], whose study is restricted to pairwise comparisons of $\mu_i - \mu_j$ in the one-way analysis of variance*. The S-method is not generally recommended in the latter instance.

Roy [2] developed simultaneous confidence regions for the parameters $\mathbf{C}'\boldsymbol{\beta}$, for $\mathbf{C}_{m \times l}, \boldsymbol{\beta}_{m \times p}$ $(l \leqslant m)$ in the p-dimensional multivariate linear model given by (1), where $\mathbf{X}$ is unchanged, but $\mathbf{Y}_{n \times p}$ is a multinormal random matrix with independent columns. Roy's regions reduce to Scheffé's in the univariate setup $(p = 1)$; they are related to Roy's characteristic root statistic* in testing the hypothesis that $\mathbf{C}'\boldsymbol{\beta} = \mathbf{0}$ through the union–intersection principle*. For further details see Roy [3, Chaps. 13 and 14] and Srivastava and Khatri [8, Sec. 6.3].

References

[1] Olshen, R. A. (1973). *J. Amer. Statist. Ass.*, **68**, 692–698.

[2] Roy, S. N. (1954). *Ann. Math. Statist.*, **25**, 752–761.

[3] Roy, S. N. (1957). *Some Aspects of Multivariate Analysis*. Wiley, New York.

[4] Scheffé, H. (1953). *Biometrika*, **40**, 87–104 (Corrigenda (1969). *Biometrika*, **56**, 229).

[5] Scheffé, H. (1959). *The Analysis of Variance*. Wiley, New York.

[6] Scheffé, H. (1977). *J. Amer. Statist. Ass.*, **72**, 143–144. (This note is followed by comments by R. A. Olshen and a rejoinder by Scheffé, pp. 144–146.)

[7] Spjotvøll, E. and Stoline, M. R. (1973). *J. Amer. Statist. Ass.*, **68**, 975–978.

[8] Srivastava, M. S. and Khatri, C. G. (1979). *An Introduction to Multivariate Statistics*. North-Holland, New York.

[9] Stoline, M. R. (1981). *Amer. Statist.*, **35**, 134–141.

(MULTIPLE COMPARISONS
MULTIVARIATE MULTIPLE COMPARISONS
WORKING–HOTELLING–SCHEFFÉ CONFIDENCE BANDS)

CAMPBELL B. READ

SCHEIDEGGER–WATSON DISTRIBUTION

The density on the sphere $\mathbf{x}^t\mathbf{x} = 1$ proportional to $\exp \kappa(\mathbf{x}^t\boldsymbol{\mu})^2$, where $\mathbf{x}$ and $\boldsymbol{\mu}$ are unit vectors, is named after Scheidegger [1] and Watson [2]. If $\kappa > 0$, it is bimodal with modes at $\boldsymbol{\mu}$ and $-\boldsymbol{\mu}$. If $\kappa < 0$, it is a "girdle" distribution since it is a maximum when $\mathbf{x}$ is orthogonal to $\boldsymbol{\mu}$. The logarithm of the likelihood of a random sample from this distribution involves $\boldsymbol{\mu}$ through the term $\Sigma(\boldsymbol{x}_i^t\boldsymbol{\mu})^2 = n\boldsymbol{\mu}^t\boldsymbol{M}_n\boldsymbol{\mu}$, where $\boldsymbol{M}_n = n^{-1}\Sigma \boldsymbol{x}_i\boldsymbol{x}_i^t$. Thus if $\kappa > 0$ $(\kappa < 0)$, the maximum likelihood (ML) estimator of $\boldsymbol{\mu}$ is the eigenvector of $\boldsymbol{M}_n$ corresponding to its largest (smallest) eigenvalue. The eigenvalues of $\boldsymbol{M}_n$ sum to unity. The formula for the ML estimator of κ depends on the dimension of $\boldsymbol{x}$. Details for confidence regions* and tests in three dimensions are given in Watson [2].

Since the density is the same for $\boldsymbol{x}$ and $-\boldsymbol{x}$, it may be used to describe the distribution of undirected lines or axes. It is a special case of the Bingham distribution. Another generalization (in more than three dimensions) uses densities depending only on the length of the projection of $\boldsymbol{x}$ on a subspace V. (See Watson [3].)

References

[1] Scheidegger, A. E. (1965). On the statistics of the orientation of bedding planes, grain axes, and similar sedimentological data. *Prof. Paper, 525-C, U.S. Geol. Survey*. pp. 164–167.

[2] Watson, G. S. (1965). *Biometrika*, **52**, 193–201.

[3] Watson, G. S. (1983). *Statistics on Spheres*. Wiley, New York, p. 238.

(DIRECTIONAL DATA ANALYSIS
DIRECTIONAL DISTRIBUTIONS)

SCHEMATIC PLOT *See* NOTCHED BOX AND WHISKER PLOT

SCHOENER'S INDEX OF PROPORTIONAL SIMILARITY

Given two multinomial* populations, each with k classes and cell probability vectors $\boldsymbol{p}$ and $\boldsymbol{q}$, respectively, their *proportional similarity index* is

$$\text{PS} = \sum_{i=1}^{k} \min(p_i, q_i).$$

This was apparently first used by Schoener [2] (see also Ricklefs and Lau [1] and Socransky et al. [5]). It is used in ecological and medical studies.

An alternative formula for the index is

$$\text{PS} = 1 - \tfrac{1}{2} \sum_{i=1}^{k} |p_i - q_i|.$$

(See, for example, Smith [4].)

Given random samples of sizes m, n from the two populations, the maximum likelihood* estimator (MLE) of PS is obtained by substituting $\hat{p}_i = m_i/m$ and $\hat{q}_i = n_i/n$ (where m_i, n_i are the respective counts in the ith cell) for p_i and q_i, respectively. A fairly good approximation to the variance of the MLE of the PS, derived by Smith [3, 4], is

$$\text{var}(\widehat{PS}) = m^{-1}\left\{\sum_{i=1}^{k} p_i I_i - \left(\sum_{i=1}^{k} p_i I_i\right)^2\right\} + n^{-1}\left[\sum_{i=1}^{k} q_i(1 - I_i) - \left\{\sum_{i=1}^{k} q_i(1 - I_i)\right\}^2\right],$$

where

$$I_i = \begin{cases} 1 & \text{if } p_i < q_i, \\ \frac{1}{2} & \text{if } p_i = q_i, \\ 0 & \text{if } p_i > q_i. \end{cases}$$

Similar results can be found in ref. 1, where the index is called *coefficient of community* and compared with Morosita's index.

References

[1] Ricklefs, R. E. and Lau, M. (1980). *Ecology*, **61**, 1019–1024.

[2] Schoener, T. W. (1970). *Ecology*, **51**, 408–418. (Contains an extensive bibliography.)

[3] Smith, E. P. (1982). *The Statistical Properties of Biological Indices of Water Quality*. Ph.D. dissertation, University of Washington, Seattle, WA.

[4] Smith, E. P. (1984). *J. Statist. Comp. Simul.*, **19**, 90–94.

[5] Socransky, S. S., Tanner, A. C. R., and Goodson, J. M. (1981). *J. Dental Res. A*, **60**, 468. (special issue)

(DIVERSITY, MEASURES OF)

SCHONEMANN PRODUCT RULE *See* MATRIX DERIVATIVES: APPLICATIONS IN STATISTICS

SCHUR CONVEXITY *See* SCHUR FUNCTIONS

SCHUR DECOMPOSITION *See* LINEAR ALGEBRA, COMPUTATIONAL

SCHUR FUNCTION *See* MAJORIZATION AND SCHUR CONVEXITY

SCHUR INEQUALITY

Let $\mathbf{A} = (a_{ij})$ be a $n \times n$ matrix and let $\lambda_1, \lambda_2, \ldots, \lambda_n$ be the eigenvalues* of this matrix. The Schur inequality [1] states

$$\sum_{i=1}^{n} |\lambda_1|^2 \leqslant \sum_{i=1}^{n} |a_{ij}|^2$$

$$\text{for each } j = 1, 2, \ldots, n.$$

This inequality can be of use in multivariate analysis*.

Reference

[1] Schur, I. (1909). *Math. Ann.*, **66**, 488–510.

SCHUR PRODUCT

The Schur prouct of two (m, n) matrices of $\mathbf{A} = (a_{ij})$ and $\mathbf{B} = (b_{ij})$ is defined as

$$\mathbf{AB} = \left(a_{ij}b_{ij}\right); \quad i = 1, \ldots, m; \; j = 1, \ldots, n.$$

It is also known as the *Hadamard product*.

An elementary review of the Schur product is given in ref. 3; matrix differentiation is discussed in ref 1; applications in multivariate analysis* are described in ref 2.

References

[1] Bentler, P. M. and Lee, S. Y. (1978). *J. Math. Psychol.*, **17**, 255–262.
[2] Neudecker, H. (1981). *Psychometrika*, **46**, 343–345.
[3] Rao, C. R. and Mitra, S. K. (1971). *Generalized Inverse of Matrices and Its Applications*. Wiley, New York.

(KRONECKER PRODUCT OF MATRICES)

SCHUSTER PERIODOGRAM *See* PERIODOGRAM ANALYSIS

SCHWARZ CRITERION

The problem of selecting a model from a set of alternative statistical models has attracted particular attention in situations where the alternative models have different dimensionalities (i.e., contain different numbers of unknown estimable parameters). Examples of such problems include the choice of the order of a Markov chain, the choice of the order of an autoregressive process and the choice of degree for a polynomial regression or, more generally, the choice among multiple regression equations containing differing numbers of regression variables.

In such cases, straightforward applications of the maximum likelihood* approach or of other approaches seeking to maximize a goodness-of-fit* criterion are liable to lead automatically to the selection of the model with highest possible dimension. Such a conclusion is intuitively unsatisfactory and has motivated attempts to find alternative criteria for model choice.

Schwarz [2] has recommended the criterion: *choose the model for which the quantity*

$$\log(\text{maximized likelihood})$$

$$-\left(\tfrac{1}{2}\log n\right)(\text{number of estimable parameters})$$

is largest (where n is the sample size).

The motivation and justification for the criterion is, essentially, Bayesian (*see* BAYESIAN INFERENCE) in that, asymptotically, it corresponds to choosing the model with the largest posterior probability*, corresponding to a prior specification that assigns positive probability masses to each model, and, conditionally, appropriately dimensioned nondegenerate prior distributions* over the parameters within each model. Stone [4] has drawn attention to the systematic earlier implicit use of this criterion by Jeffreys [1] in his Bayesian approach to significance tests*. Notwithstanding the derivation, Schwarz has claimed that the criterion should have large-sample appeal outside the Bayesian framework, by virtue of being asymptotically independent of particular prior specifications.

In the general area of model choice, there is an obvious, and contrasting, competitor to the Schwarz criterion, namely, the Akaike criterion*, which takes the form: *choose the model for which the quantity*

$$\log(\text{maximized likelihood}) - (\text{number of estimable parameters})$$

is largest.

Qualitatively, both the Schwarz and Akaike criteria formalize the *principle of parsimony** in model building, in that—all other things being equal—they tend to select models with lower rather than higher dimensionality. Quantitatively, for $n \geqslant 8$ the Schwarz criterion has an increasing tendency to favour lower-dimensional models than would be selected by the Akaike criterion. In an attempt to compare and contrast these and other related criteria, Smith and Spiegelhalter [3] have considered a range of criteria based on the quantity

$$\log(\text{maximized likelihood}) - (\text{multiplier}) \times (\text{number of estimable parameters}),$$

where many possible "multiplier" factors are discussed, both in terms of the form of Bayesian prior specification from which they can be derived and also in terms of the sampling properties of the model choice procedure to which they lead. Further references are given in Smith and Spiegelhalter [3] and Stone [4].

References

[1] Jeffreys, H. (1939/1948/1961). *Theory of Probability*, 1st, 2nd, and 3rd eds. Oxford: Oxford University Press.

[2] Schwarz, G. (1978). Estimating the dimension of a model. *Ann. Statist.*, **6**, 461–464.

[3] Smith, A. F. M. and Spiegelhalter, D. J. (1980). Bayes factors and choice criteria for linear models. *J. R. Statist. Soc. B*, **42**, 213–220.

[4] Stone, M. (1979). Comments on model selection criteria of Akaike and Schwarz. *J. R. Statist. Soc. B*, **41**, 276–278.

(AKAIKE CRITERION
MODEL CONSTRUCTION: SELECTION OF DISTRIBUTIONS
PARSIMONY, PRINCIPLE OF)

A. F. M. SMITH

SCHWARZ INEQUALITY *See* CAUCHY–SCHWARZ INEQUALITY

SCHWEPPE-TYPE ESTIMATORS

The Schweppe estimator is an estimator for the linear regression* model

$$y_i = x_i\beta + u_i, \qquad i = 1, \ldots, n,$$

where y_i is the ith observation on the dependent variable, x_i is the ith row of the $n \times p$ matrix X of observations on the explanatory variables, β is the p-vector of unknown parameters to be estimated, and u_i is the ith disturbance. It was designed to be much less sensitive than ordinary least squares*, in terms of bias* and efficiency*, to both long-tailed disturbance distributions and erroneous data occurring with low probability. The Schweppe estimator is a weighted-least-squares* estimator defined implicitly by

$$0 = \sum_{i=1}^{n} \min\left\{1, \frac{c\sqrt{1-h_i}}{\left|\left(y_i - x_i\hat{\beta}\right)/\sigma\right|}\right\} \times \left(y_i - x_i\hat{\beta}\right)x_i^t,$$

where $h_i = x_i(X^tX)^{-1}x_i^t$ (the ith diagonal element of the hat matrix*), σ is the scale parameter for the disturbance distribution, c is a constant, and a superscript t denotes transpose. (Generally in applications σ is unknown and must be estimated along with β.) This estimator was developed in work by Merrill [7] and Merrill and Schweppe [8].

The term *Schweppe-type estimator* has come to refer to any estimator of the form

$$0 = \sum_{i=1}^{n} \min\left\{1, \frac{cv(x_i)}{\left|\left(y_i - x_i\hat{\beta}\right)/\sigma\right|}\right\}\left(y_i - x_i\hat{\beta}\right)x_i^t,$$

where $v(\cdot)$ is a function that downweights

extreme x-rows. Most choices for v depend on X and should properly be written $v(x_i; X)$. Moreover, they are typically "invariant" to a change in coordinates in the sense that $v(x_iB; XB) = v(x_i; X)$ for any nonsingular matrix B. The most common form for v is $v(x) = \{xA^{-1}x^t\}^{-1/2}$ where A is a symmetric positive-definite $p \times p$ matrix that depends on the marginal distribution for x; the invariance condition then translates into the requirement that if X is transformed to XB, A transforms to B^tAB. Provided $|x|v(x)$ is bounded in x, a Schweppe-type estimator is a bounded-influence estimator (*see* REGRESSION, BOUNDED INFLUENCE) and is thus qualitatively robust as defined by Hampel [1] (*see* ROBUST REGRESSION).

As $c \to \infty$, a Schweppe-type estimate always reduces to the ordinary least-squares estimate defined by

$$0 = \sum_{i=1}^{n} (y_i - x_i\hat{\beta})x_i^t.$$

More generally, whether the ith observation is downweighted depends not on the scaled residual $|(y_i - x_i\hat{\beta})/\sigma|$ or x-weight $v(x_i)$ individually, but only on their ratio. It is this feature that distinguishes the Schweppe form from other proposals for bounded-influence regression, such as that of Colin Mallows, which is of the form

$$0 = \sum_{i=1}^{n} v(x_i)\min\left\{1, \frac{c}{|(y_i - x_i\hat{\beta})/\sigma|}\right\}$$
$$\times (y_i - x_i\hat{\beta})x_i^t.$$

Maronna and Yohai [6] proved that, under general conditions, Schweppe-type estimators are consistent and asymptotically normal. However, Monte Carlo experiments have shown that the convergence to normality is slow when there are outliers* among the x variables. The sampling distribution is often more peaked than the normal distribution.

Interest in Schweppe-type estimators stems partly from their efficiency properties within the class of bounded-influence estimators. Hill [2] compared the Schweppe estimator to the Mallows estimator and others, using Monte Carlo methods, and found that the Schweppe form had a slight advantage. More general efficiency results for Schweppe-type estimators have been obtained by Krasker [4] and Krasker and Welsch [5] (*see* REGRESSION, BOUNDED INFLUENCE).

Huber [3] has analyzed the case of regression through the origin ($p = 1$) from the different standpoint of finite-sample minimax robustness. All of the estimators that emerge as minimax under his different sets of assumptions are of the Schweppe form.

References

[1] Hampel, F. R. (1973). *Ann. Statist.*, **42**, 1887–1986

[2] Hill, R. W. (1977). *Robust Regression When There Are Outliers in the Carriers.* Ph.D. dissertation, Dept. of Statistics, Harvard University, Cambridge, MA.

[3] Huber, P. J. (1983). *J. Amer. Statist. Ass.*, **78**, 66–80.

[4] Krasker, W. S. (1980). *Econometrica*, **48**, 1333–1346.

[5] Krasker, W. S. and Welsch, R. E. (1982). *J. Amer. Statist. Ass.*, **77**, 595–604.

[6] Maronna, R. A. and Yohai, V. J. (1981). *Zeit. Wahrsch. verw. Geb.*, **58**, 7–20.

[7] Merrill, H. M. (1972). *Bad Data Suppression in State Estimation, With Applications to Problems in Power.* Ph.D. thesis, Massachusetts Institute of Technology.

[8] Merrill, H. M. and Schweppe, F. C. (1971). *IEEE Trans. Power Apparatus Syst.*, **PAS-90**, 2718–2725.

(LINEAR REGRESSION
REGRESSION, BOUNDED INFLUENCE
ROBUST REGRESSION)

WILLIAM S. KRASKER

SCIENTIFIC METHOD AND STATISTICS

Statistics is a part of the "scientific method" and it makes use of it. Before elaborating on

Table 1 Some of the Main Fields of Application of Various Statistical Techniques

	Agriculture	Authorship	Artificial Intelligence	Biology	Business & Stock Market	Chemical Engineering	Clinical Trials	Credit Scoring	Commn. Th. & Cryptology	Econometrics	Engineering	Law	Manufacturing	Market Research	Medical Diagnosis	Meteorology	Oper. Res.	Pattern Recognition	Philosophy of Science	Population Genetics	Psychology	Seismology	Soc., Polit. Sc., Govt.	Statist. Mechanics
Assay				•																				
Bayes, Wt. of Evidence[a]		•	•					•	•			•			•				•					
Contingency Tables				•											•								•	
Decision Theory[a]					•														•					
Design	•										•								•					
EDA & Descriptive Statistics	•	•	•	•	•	•	•	•	•	•	•	•	•	•	•	•	•	•	•	•	•	•	•	•
Entropy & Cross-Entropy Max.									•									•	•					•
Markov chains & Renewal Theory																	•							
Multivariate (cont.)				•	•																•		•	
Nonparametrics[b]	•	•	•	•	•	•	•	•	•	•	•	•	•	•	•	•	•	•		•	•	•	•	•
Path Analysis																				•			•	
Quality Control													•											
Reg., ANOVA, Linear Models	•	•	•	•	•	•	•	•	•	•	•	•	•	•	•	•	•	•		•	•	•	•	•
Reliability													•											
Response Surfaces						•																		
Sequential Analysis							•						•											
Survey Sampling[a]														•									•	
Time Series & Stochastic Processes					•					•						•						•		

a) Bayesian Statistics, Decision Theory, and Sampling can be applied in all areas because of their fundamental nature.
b) Or "semiparametrics" or "subparametrics."

this somewhat philosophical theme, let's scan the down-to-earth Table 1, which gives a rough idea of some of the main fields of application of a variety of statistical techniques. It could be made more quantitative by entering judged degrees of applicability in the cells, but it will suffice for our purpose. It indicates that some statistical techniques are more appropriate than others in some scientific or technological fields, and, since statistics is part of the scientific method, it follows that scientific method can hardly have quite the same meaning in every field, a point to which we shall return.

The table also suggests, correctly, that the influence between statistics and the sciences flows in both directions. Most statistical techniques emerged from the sciences and are now serving them, but it is also reasonable to think of statistics as a science in its own right and capable of growth from within. This is especially true if we interpret statistics broadly to include probability, decision theory*, mathematical statistics, and their philosophical aspects. This interpretation will be adopted in this article.

A science is an organized body of knowledge (often probabilistic), concerning some field of interest, the knowledge being acquired by the scientific method, which refers more to the scientific and technological enterprise than to the methods of individual scientists and technologists, although of course the two aspects cannot be sharply separated. There is no exact definition of the scientific method, or, as the influential philosopher Popper calls it, the method of trial and error, because it has several ingredients or facets to which different weights are given by different people at different times. Thus the definite article in "The Scientific Method" is somewhat misleading but it is linguistically euphonic. It might be identified with rationality. Some of the facets are:

(i) *The interest or importance of the field of study.*

(ii) *The judgment of what is worth investigating* within that field.

(iii) *Careful and accurate observation leading to classification* of the items in the field (as in field botany). This facilitates discussion and analysis of the field.

(iv) *Repetition of one another's experiments* (sometimes in the hope of proving that the other guy is wrong).

(v) *The use of definitions and notations* that are accurate *enough* at a given stage of an investigation. Definitions can be *too* accurate, especially in exploratory stages.

(vi) *The noticing of correlations between phenomena,* leading to conjectured relationships (*induction*), including causal relationships, even when the observations are not carefully planned or not planned at all. Examples of the latter class are: the discoveries by a baby, a budding scientist, that space exists and is three-dimensional, that some things can act on other things, and later that some simple sentences represent this structure (subject, verb, object); the discovery of x-rays by Röntgen in 1895; and that of penicillin by Alexander Fleming in 1928. ("Fortune favors the prepared mind," the preparation in the case of babies being billions of years of evolution.)

(vii) *Explanation, prediction, and control, based on existing theories.* To people who do not fully understand a theory, its success in prediction and control seems to give the theory more support than is given by its role in explanation.

(viii) *The construction of new explanatory hypotheses or theories* (called *abduction* by C. S. Peirce) where a simple or elegant theory is preferred to a complex or ugly one if both explain the same facts to the same extent (*see* PARSIMONY, PRINCIPLE OF). The more speculative the theory, the more support it requires. The ideal is to attain a theory, not too complicated, that explains an entire field of experience (though the theory itself may demand an explanation). When some of a field is explained its horizons are apt to recede, thus preventing the attainment of the ideal as thereby modified, but enabling the theory to explain a larger body of empirical results, and parts of other fields. A theory is a hypothesis covering a wide field (and theories form hierarchies and networks), but here we shall usually regard theory and hypothesis as synonyms for the sake of brevity.

(ix) *Sometimes the use of mathematical or conceptual theories containing undefined terms (the axiomatic method)* but with rules of application that give the theories operational meaning. Quantum mechanics, in any of its known rigorous formulations, provides a good example at present, and most physicists are satisfied with this state of affairs. (See the later discussion of instrumentalism.)

(x) *The making of deductions from the theories or hypotheses* where the deductions might be predictive or explanatory.

(xi) *The attempt to refute or to some extent confirm these deductions by means of further experiments and observations* [which lead back to headings mentioned before, such as **(viii)**]. The evaluation of the evidence supplied by observations in favor of or against the "truth" of a hypothesis, and the provisional acceptance or rejection of hypotheses for certain purposes. *See* HYPOTHESIS TESTING.

(xii) *The attempt to avoid wishful thinking* in the evaluation of observations and in other respects, and thus to introduce a measure of objectivity into our personal and multipersonal judgments. *To use rationality rather than rationalization*. This ideal is difficult to approach closely, by individuals or by organizations, but is expedited by cooperation and free competition.

(xiii) *A striving towards a unification of the field.* This is related to simplicity, mentioned under **(viii)**.

(xiv) *The communication of results* ("publish or perish").

THE RELEVANCE OF STATISTICS TO THESE FACETS

Let us now consider the relevance of statistics and statistical philosophy to some of these facets, taking statistics first. We shall also mention some concepts that have helped or might help to unify statistics itself.

We shall append an S to the facet number when discussing statistical matters related to that facet.

(iiiS) During the last few decades, classification* [which includes cluster analysis or botryology (Good, [23, 29])] has become quite an industry among statisticians. Reduction of dimensionality, especially to two dimensions, is often useful for clustering, as in principal component analysis*, factor analysis*, and multidimensional scaling*. Also especially relevant to heading **(iii)** are descriptive statistics and exploratory data analysis*. For the philosophy of exploratory data analysis see Good [35], who argues that it has an implicit informal Bayesian component.

(viS) *See* CORRELATION, EXPLORATORY DATA ANALYSIS, REGRESSION, PATH ANALYSIS, CANONICAL ANALYSIS. For causation* see also **(viP)**.

(viiS) Any form of smoothing of data (improving the data) including regression, probability density estimation*, and time-series* analysis, is done partly in the hope of improving prediction, though smoothing can be primarily descriptive, that is, performing reduction of the data for ease of human apprehension. There may be scope here for the experimental psychologists and physiologists to find out more about what is easy for humans to apprehend and why. A somewhat exiguous sense of explanation is used when we say that a specific linear relationship explains a proportion R^2 of the total variation of a dependent variable (*see* COEFFICIENT OF DETERMINATION). See also **(viiP)** and **(viiiP)** below.

(viiiS) It is frequently said that the generation of explanatory hypotheses is a creative process for which no formal rules can be provided, though there is plenty of informal advice such as "think of analogies," "draw pictures," and "become engrossed in the details and then sleep on the problem." But there is some prospect of automatic rules. For example, there are heuristics that can be programmed for a computer and that can occasionally be successful because the computer can try many possibilities in a short time. There are strong reasons to believe that the sleeping or unconscious mind also does this (Hadamard [46], Poincaré [62] and Koestler [53]). In a more statistical context the principle in statistics of maximum entropy (see below) can also be used to generate null hypotheses (Good [24]), although the original purpose of the principle was to select maximally noncommittal priors (Jaynes [48]). Here the entropy* of mutually exclusive and exhaustive probabilities, $p_1, p_2, \ldots, p_n$, is defined as $-\Sigma_i p_i \log p_i$. Especially for continuous problems, the principle of maximum entropy is improved by using a principle of maximum cross-entropy, as used without philosophy by Kullback [54] and argued for, in a somewhat different form, simultaneously and independently by Jaynes [49] and Good [26]. Here cross-entropy is an expression of the form

$$I = -\int f(x) \log \frac{f(x)}{g(x)}\, dx,$$

where $f(x)$ is a density function to be found,

subject to assigned constraints, and where $g(x)$ is a density of special interest or is in some sense an invariant prior. When there are no constraints the principle leads to $f = g$. For a brief history of cross-entropy, which dates back in statistical mechanics to Gibbs [17, p. 136], see Good [39]. *See also* KULLBACK INFORMATION.

The principle of maximum cross-entropy, in problems of inference, selects the most conservative density if weight of evidence, $\log[f(x)/g(x)]$, in favor of the true density f (as compared with the initially assumed density g), is regarded as a "quasi-utility." (Cross-entropy is an expected weight of evidence with the sign changed, so to maximize cross-entropy is to minimize expected weight of evidence.) The principle is therefore an invariant procedure when g is in some sense invariant. For invariant distributions *see* INVARIANT PRIOR DISTRIBUTIONS, Jeffreys [51], Perks [60], Hartigan [47], and Good [26]. (In problems of experimental design, as contrasted with inference, one is concerned with *maximizing* such quasi-utilities as expected weight of evidence or Fisher information*, for a given cost or effort.)

(ixS) The axiomatic method is useful in the foundations of probability* and statistics for the usual reason that the developments from the axioms* are less controversial than their practical interpretation and because mathematically deeper arguments can be based on sharper assumptions.

(xiS) When the null hypothesis H_0 is embedded in a wider class of hypotheses having more parameters, H_0 can often be convincingly refuted, but typically only slightly confirmed. The matter is discussed below under Philosophical Issues.

(xiiS) The striving toward objectivity is one of the main reasons why scientists use statistical methods. It is also the reason why the Bayesian circumscribes his judgments by means of axioms and rules of applications. Because these are insufficient to determine all the epistemic probabilities uniquely, the Bayesian is aware of the need to avoid wishful thinking. In an attempt to overcome the problem many elementary non-Bayesian texts insist that hypotheses must be formulated before looking at a data set. This is good advice for beginners and for people with poor judgment. In exploratory data analysis*, however, one is encouraged to formulate hypotheses by looking at the data. Randomization* in the design of experiments was introduced to decrease the dangers of wishful thinking and of overlooking hidden causative agents. Total objectivity is lost if the statistician knows the outcome of the randomization (see Statistician's Stooge in the indexes of Good [33]).

PHILOSOPHICAL ISSUES

The modern Bayesian movement has increased the interest in the philosophy of science among statisticians during the last several decades, although most statisticians are still naturally more concerned with techniques such as the theory and application of linear models. Also several philosophers of science have become interested in Bayesianism.

Philosophical questions are necessarily controversial almost by definition: it's the other man's philosophy that's incorrect. But the relationship between scientific method and statistics cannot be discussed without philosophy and the writer emphasizes views that he believes are correct. Similar philosophies are espoused by Akaike, Berger, Box, C. A. B. Smith, and others.

The Bayes–Non-Bayes Compromise

To mention Hegel the Obscure may need an apology. A little Hegel goes a long way while some of what he said about science is pure drivel. If this description seems too contentious, see the hilarious examples quoted by Russell [66, p. 763] and Popper [65, p. 332n] and compare Allen [5, p. 151]. But in all the dross there is a nugget that is relevant for us, the reconciliation of opposites, also called dialectical analysis (Acton [1, p. 436]). A familiar example is the coexistence of the objective physical world on the one hand and subjective experiences on the other. Another good example that Hegel would have loved is that of wave–particle duality. The former example corresponds to the tension in statistics between physical and epistemic probability (Poisson [63]; Carnap [10]) and between non-Bayesian statistics (sampling theory) and Bayesian (Laplacian) or neo-Bayesian statistics (of which there are many varieties; see, for example, ref. 28). There is also a tension between personal and logical probability or credibility and here too a compromise is necessary in practice. Perhaps the best short definition of Bayesianism is the willingness to assume that hypotheses have probabilities. A synthesis or compromise between these seemingly opposing schools is emerging, but Hegel is seldom acknowledged, possibly because he was scientifically ignorant and is unreadable. The ancient Chinese philosophy of yin-yang (Chan [11, pp. 89–92]) partially anticipates the dialectical method. The method leads to greater unity.

Fisher*, who was generally hostile to Bayesianism, nevertheless said [15, p. 33] "... we may think of a particular throw [of a die], or of a succession of throws, as a *random* sample from the aggregate, which is in this sense subjectively homogeneous and without recognizable stratification." Thus he recognized the inevitability of some subjectivism in practice. Similarly Neyman et al. [58], having obtained a *P*-value* of 1/15, interpreted it as odds of 14 to 1 against the null hypotheses. This is surprising because it is usually assumed that Neyman* did not talk about the probabilities of hypotheses. In this case his concern was to influence an administrative decision (concerning the seeding of clouds; *see* WEATHER MODIFICATION, STATISTICS IN). Unfortunately it is a mistake to translate a *P*-value into odds against a hypothesis, and especially not odds of $(1-P)/P$, which nearly always overestimates the Bayes factor. [A Bayes factor in favor of a hypothesis H provided by evidence E is $O(H|E)/O(H)$, the ratio of the posterior to the prior odds of H, and is equal to $P(E|H)/P(E|\overline{H})$ where the bar denotes negation. The logarithm of a Bayes factor is the corresponding weight of evidence.] These two snippets of history manifest the leaders of the two main non-Bayesian schools leaning towards Bayesianism.

The prominent Bayesian, Harold Jeffreys, regards his position as closer to Fisher's than to Neyman's, presumably because he and Fisher are more concerned with inductive inference, without mention of utilities, than with "inductive behavior" (to use Neyman's expression), and also because Jeffreys was able to give a Bayesian interpretation to fiducial inference*, at least in some simple cases. A Bayes–Fisher compromise aims at a partial reconciliation of *P*-values with Bayes factors. (The arbitrary thresholds in Fisher's tables, such as 0.05, 0.01, are of no fundamental interest although they may have stimulated Neyman and Pearson to talk about errors of the first kind.) The reconciliation has several aspects. In the first place, it can come about by regarding the relevant test statistic as having a distribution, however vaguely judged, given the nonnull hypothesis (Good [18, pp. 93–94]). It is sometimes said that Fisher did not allow for the nonnull hypothesis, but how else could he have selected test statistics or distinguished between cases where single tails and double tails are appropriate? Neyman and Pearson were of course much more explicit about the nonnull hypothesis, and in *this* respect they were closer to Bayes than to Fisher.

A second reconciliation is that a Bayesian, having worked out a Bayes factor, based on a model he does not entirely trust, can treat this factor as a test criterion in the Fisherian manner (Good [19, p. 863]), though there may be difficulties in finding the distribution under the null hypothesis except by simulation.

A third reconciliation suggested, among other things, by Box [9] is the alternation between Bayesian estimation and frequentist criticism of the Bayesian model.

A fourth reconciliation consists of a method for combining statistical tests in parallel, that is, tests based on the same data. The suggested rule of thumb, which arises from an informal Bayesian argument, is to take the (possibly weighted) harmonic mean* of the individual tail probabilities (Good [20, 37] and the indexes of Good [33] under Harmonic Mean).

Yet a fifth reconciliation is the notion of standardizing a tail-area probability P, from sample size N to sample size 100, by replacing it by $\min[\frac{1}{2}, P(N/100)^{1/2}]$ (Good [31]).

Further Fisherian concepts, such as sufficiency and ancillarity, can also be discussed along Bayesian lines (Good [31]); *see* ANCILLARY STATISTICS and SUFFICIENT STATISTICS.

An example of a Bayes–Neyman compromise relates to the Neyman–Pearson–Wilks likelihood ratio or ratio of maximum likelihoods. This was first introduced as having "intuitive appeal." A Bayesian might regard it as suggested by an approximation to a Bayes factor, on the grounds that an integral is *very* roughly related to the maximum of its integrand. After making this approximation, and knowing it to be a poor one, the tolerant Bayesian might then go over to the second Bayes–Fisher reconciliation, which in effect is what Neyman, Pearson, and Wilks did (without expressing the matter in this fashion). *See* NEYMAN–PEARSON LEMMA and LIKELIHOOD RATIO TESTS.

Another example of a Bayes–Neyman compromise is the notion of the strength of a test of a hypothesis. Here strength is defined as a weighted average of power, the weights depending to some extent on a prior over the nonnull hypothesis, and perhaps on utilities (Crook and Good [12]; compare Kempthorne and Folks [52, p. 345]). Without some such concept it is difficult to obtain an intuitive grasp of the power function in multiparameter problems. Yet another aspect of the Bayes–Neyman compromise is that a Bayes factor is equal to a likelihood ratio in the special case in which the hypothesis H and its negation are both simple statistical hypotheses. A merit of a Bayes factor, or its logarithm the weight of evidence, is that it does not depend on the judgment of the initial probability of H, and therefore comes as close as a Bayesian can to measuring "what the observations are trying to say." This merit is especially valuable when the ratio of the upper to the lower initial odds of H (the ratio of the endpoints of an interval in which the initial odds are judged to lie) is large, or when the variations in the odds are large from one judge to another or from one time to another for a single judge (without change in the evidence). Thus the Bayes factor is less Bayesian than the final odds of H.

For a recent detailed justification of the expression "weight of evidence" see Good [36], who claims that the technical meaning precisely captures the most usual informal one whereas the measures $P(H|E) - P(H)$ and $P(H|E)/P(H)$ do not, although both of them are still being used by philosophers (as of 1985). *See also* STATISTICAL EVIDENCE.

For further discussion of the Bayes–non-Bayes compromise see Berger [6–8] and the indexes of Good [33].

Refutation and Corroboration

Popper [64] suggested that a demarcation between a scientific and a nonscientific or metaphysical theory is that a scientific theory can be decisively refuted if it is false, whereas degrees of corroboration do not provide such a demarcation. Moreover he claimed that the truth of basic theories is not an issue in science, and that what matters is whether, at any given time, they have "proved their mettle" by surviving honest attempts at their refutation. These views concerning scientific theories strongly resemble Fisher's views about null statistical hypotheses, so Fisher seems somewhat to have anticipated Popper although Popper [64] overlooks Fisher's partial priority. (Popper's citations to Fisher refer only to Fisher's views about likelihood, and Fisher is not indexed in Popper [65].) The concepts of refutation and degree of corroboration apply with greater difficulty in science as a whole than in statistics.

Let us consider a kind of generalization of Popperian or Fisherian refutation. Let H denote a null hypothesis that is represented by a point in a continuous space of rival hypotheses. If H is taken literally, usually $P(H)$ is extremely small or possibly zero and even when H is exactly true, if ever, it is liable to be expensive to obtain a large weight of evidence in its favor. The Fisherian cannot define any P-value against the nonnull hypothesis and it would be regarded by Popper as metaphysical because it "cannot be refuted." His position, in corresponding scientific contexts, is that the nonnull hypothesis is unscientific whereas the null hypothesis is scientific but almost certainly false! (Jeffreys [50] usually assumes that the prior probability of the null hypothesis is $\frac{1}{2}$.)

When we talk about sharp null hypotheses this is usually an abbreviated way of speech. At the back of our minds we regard a small neighborhood around H as the more practical meaning of H, and this interpretation can be formalized as, for example, in Good [18, p. 90]. This can of course be done with neighborhoods of more than one size. The smaller the neighborhood, the more expensive it is to obtain a given weight of evidence (or small tail-area probability) against $\overline{H}$. So a Bayesian and a Fisherian should both recognize that there is a continuous gradation between very scientific and very metaphysical hypotheses (compare Good [22, pp. 492–494]). The idea of considering an "onion," so to speak, of hypotheses around the null hypotheses is entirely practicable. It resembles the method of presenting a client with a whole collection of confidence intervals*, with various confidence levels, in an estimation* problem (which is not an official Neyman–Pearson procedure). Compare also the notion of a set of intervals "consonant with a definite model to a particular degree" (Kempthorne and Folks [52, pp. 364–370]; Mayo [56, pp. 324–330]).

In the following headings, a P appended to the facet number indicates a philosophical discussion of that facet.

(viP) Popper says he does not believe in scientific induction, but there must be a semantic ambiguity here, for we cannot even use language sensibly without assuming that the meanings of words are probably related to previous usage. "Induction" is unlikely to mean "cheese" next year. For a recent discussion of some aspects of induction see Good [31, 34].

Much has been written about strict causality (determinism) by philosophers, but much less about probabilistic causality though the interest in this topic is increasing. For quantitative approaches to probabilistic causality *see* PATH ANALYSIS, Wiener [69, pp. 184–190], Granger and Newbold [45], Gersch [16], Deaton [14], and Good [21, 41]; *See also* CAUSATION. Based on some desiderata, the degree to which one event F tends to cause another one E is explicated by Good as the weight of evidence against F provided by the nonoccurrence of E, given the state of the universe just before F occurred, and an application is made to the theory of regression.

(viiP), (viiiP) By using the ideas of simple probability theory, an attempt has been made to give a semiquantitative measure of the explanatory power of a hypothesis H for explaining specified observations E (Good [25, 30, 40]; Good and McMichael [44]). The measure is called *explicativity* and is denoted by $\eta(E:H)$. It is intended to measure the degree to which the reasonable degree of belief in a proposition E is increased by taking H into account together with its probability. The fact that E is known to be true is to be ignored for the calculation of this measure. Some simple and convincing desiderata lead to the formula

$$\eta(E:H) = \log P(E|H) - \log P(E) + \gamma \log P(H),$$

where γ is a number, strictly between 0 and 1, and does not depend on the probabilities. The magnitude of γ depends on how much we object to cluttering H with irrelevancies or "assuming plurality without necessity" (Ockham's razor, or the Duns–Ockham razor, which dates back at least to Aristotle, but Ockham shaved frequently; *see* PARSIMONY, PRINCIPLE OF, and, for some brief history, Good [33, p. 227]). When choosing which of the two hypotheses H and H' is more explicative for E we can write

$$\begin{aligned}\eta(E:H/H') &= \eta(E:H) - \eta(E:H') \\ &= (1-\gamma)W(H/H':E) \\ &\quad + \gamma \log[P(H|E)/P(H'|E)],\end{aligned}$$

which compromises between, on the one hand, the weight of evidence in favor of H *as compared with* H' *provided by* E and, on the other hand, the posterior log odds of H as compared with H' *given* E. If $\eta(E:H/H') > 0$, then H is more explicative than H', regarding E. This allows only for "explicativity" as a quasi-utility and, when actual utilities are allowed for, the most explicative hypothesis is not always the best one on which to base an action. But the maximization of explicativity usually leads to sensible interval estimates of parameters, and these intervals resemble confidence intervals (Good [30]). This comes about roughly because the narrower the interval that defines H, the smaller the probability of H; but the wider the interval, the smaller the conditional probability of E. So the interval should not be too wide or too narrow. The maximization of explicativity can be regarded as a sharpening of Ockham's razor.

Ockham's razor is often exemplified in statistics by a principle of parsimony* in the construction of statistical models, especially by avoiding the introduction of more parameters than necessary. A famous example outside statistics was what Einstein called his worst blunder, the Canutian introduction of the cosmological constant to prevent the universe from expanding. Sometimes in statistics new parameters are introduced one at a time by a sequence of tests of significance, whether non-Bayesian or Bayesian. The selection of regressor variables and the selection of the degree of a polynomial both depend on a principle of parsimony, and the maximization of explicativity could be applied to these and other problems, but the details of these applications have not yet been worked out. The procedure would be an alternative to the proposals of Akaike [2, 3] (*see also* AKAIKE'S CRITERION). One needs to assume a prior on the number n of parameters such as the geometrical distribution* $a^n(1-a)$ $(n = 0, 1, 2, \ldots)$. Jeffreys [50, p. 96] suggested 2^{-n-1} but was much less definite in his third edition (p. 246). All geometrical distributions are equivalent for our purpose, but with different values for γ.

In nonparametric probability density estimation, the unknown density function f is in principle determined by a countable infinity of parameters ($f^{1/2}$ is a point on the unit sphere in Hilbert space), but a principle of parsimony is nevertheless used. This is seen, for example, in the method of maximizing penalized* likelihood, where the penalty function involves one or more procedural parameters or hyperparameters (for example, Good and Gaskins [42, 43]). A "hyperrazor" is applied when one tries to minimize the number of hyperparameters, often down to only one. The hyperrazor is also applied in forms of the hierarchical Bayesian method reviewed by Good [32].

Surprisingly, some philosophers have argued that simplicity of a theory is not relevant to its probability because the world is known to be exceedingly complex. This argument misinterprets the meaning of parsimony in science. Parsimoniousness means that a specific simple theory is usually to be preferred to a *specific* complicated one (if both explain the same facts), but preferred very provisionally to the logical disjunction of all possible complicated hypotheses including ones not yet formulated. Ockham did not imply that entities should not be multiplied, only that it should not be done *unnecessarily*.

There is *some*, though unclear, relationship between simplicity and prior probability*. Perhaps, in the philosophy of inference problems, the complexity of H should be largely *replaced* by (not defined as) $-\log P(H)$, which is the amount of information obtained if H is found to be true. Although $H = A \;\&\; \bar{A}$ looks simple because it is brief, it is improbable; in fact its probability is zero and $-\log P(H) = \infty$. Note too that the simpler of two propositions is not necessarily more probable; for example, A is simpler than $A \vee B$ (usually), but not more probable. This is why the formula for explicativity makes no explicit mention of simplicity. But there are reasons, omitted here, for allowing to some extent for simplicity as well as for explicativity (Good and McMichael [44]; Good [40]).

The relationship between simplicity and probability is improved if theories are formulated without using the words "or" and "not," but "and" is acceptable because $P(A \& B) \leqslant P(A)$. (Compare the affirmative mathematics of van Dantzig [13].) Perhaps $-\log P(H)$ is an adequate measure of complexity for hypotheses restricted in this manner.

The difficulty of defining simplicity, and of estimating narrow intervals for the logarithm of the odds (initial or final, the width is the same), are reasons why scientific theories are sometimes controversial. There is a tendency for scientists to underestimate the initial credibilities of novel theories proposed by other scientists, so these theories are liable to be ignored until "the" weight of evidence in their favor is very large. In physics, especially, the weight of evidence can be enormous because of the numerical accuracy with which many measurements can be made. A theory that predicts or explains several observations to six significant figures, where there is no serious competitor, gains so large a Bayes factor that it commands respect even among those scientists who do not know they are Bayesians.

(ixP) *Instrumentalism.* There is a continuing debate among philosophers about whether all scientific theories are necessarily "instrumental." By an instrumental theory is meant, at least by some philosophers, that the theory cannot be regarded as a description of reality but only as a paper machine, so to speak, that is useful for making predictions. Some arguments against instrumentalism are given by Popper [65, pp. 111–114], Maxwell [55], and McMichael [57]. An extreme instrumentalist (a rarity) would regard all language as merely instrumental, including the words "reality" and "instrumental." This position runs into an infinite regress when we ask whether a word is really instrumental!

There is no entirely satisfactory definition of reality but sane people behave as if they believe there is a real objective world. The concept helps to explain, for example, why we often find things after we have lost them (another discovery made by babies). Scott [68] has proposed, tongue-in-cheek, that, when we lose things they leave the universe and sometimes return later. This theory seems more complicated and certainly less probable than the view that we are sometimes forgetful. Similarly nobody believes in philosophical solipsism although it cannot be refuted. In fact the impossibility of such a refutation is one way to interpret de Finetti's theorem (*see* BAYESIAN INFERENCE and Good [33], index under "solipsism"). This theorem, so fundamental to the philosophy of Bayesian statistics, is thus seen to be a contribution to philosophy as a whole although that was not de Finetti's intention.

It is remarkable that there is some indication from quantum mechanics* that a form of solipsism may be true if quantum mechanics is interpreted in a noninstrumentalist manner (Wigner [70]). It is an argument against quantum mechanics's representing "the whole of reality" (Einstein's expression).

The instrumentalist problem also arises in the foundations of probability*: the question here is whether subjective probability* and physical probability (propensity) are real. The thesis argued by Savage [67] is that certain appealing desiderata concerning the behavior of a perfectly rational woman lead her to behave *as if* she had personal (subjective) probabilities and utilities obeying the usual rules. Furthermore, de Finetti showed that the usual Bayesian arguments, which in effect assume the existence of both epistemic and physical probabilities, can be expressed in terms of epistemic probability alone (although his argument is not watertight: see Good [27]). From this it follows that it is *as if* physical probabilities exist (though de Finetti did not word it that way). Thus it seems possible to believe consistently that both epistemic and physical probabilities are only instrumental. For a more complete discussion of these matters see Good [38].

If an instrumentalist ever describes a theory as (approximately) *true*, he can only mean that the observable deductions from it are (approximately) true. In the interests of linguistic simplicity, even if one adopts the instrumentalist philosophy, there is much to be said for regarding logical and physical probabilities as real, and statements about them as being capable of being true or false. This attitude may be *mentally healthy* and stimulating of improved explanations in the future even when applied to quantum mechanics where the instrumentalist interpretation is especially rampant. But an instrumentalist interpretation of statistical hypotheses might be preferred by many statisticians.

Even if the world is thought to be deterministic, one can usefully talk about physical probabilities, as in classical statistical mechanics, and when treating pseudorandom numbers as if they were random. Such practices are in the spirit of Bentham's "fictions" (Ogden [59]). It would be more cautious to talk about "pseudoprobabilities" in these contexts.

(xP) See the previous discussion of explicativity.

(xiP) It has been recognized by several philosophers, and emphasized by Duhem (Alexander [4]), that it is not easy to refute a basic scientific theory, partly because we may have overlooked something relevant in the experimental or observational setup, partly because theories depend on other theories, and partly because a theory can usually be "patched up", and not necessarily in too ad hoc a method. The principle of conservation of energy makes a very good example; indeed it is so well entrenched as to be almost conventionally true (an opinion expressed by Poincaré [61, p. 166]). To save the principle, its interpretation was changed several times; for example, by the introduction of the concept of potential energy, by the interpretation of heat as a form of energy, by the notion of the equivalence of mass and energy, and by the suggestion of the (unobserved) neutrino. This raises the question of what is meant by an ad hoc addition J to a theory H, and whether a semiquantitative measure of the "adhockery" of J can be defined. A proposed definition is $\eta(E:H|G) - \eta(E:(H\,\&\,J)|G)$ where G denotes the background knowledge. Thus J is positively ad hoc if $H\ \&\ J$ is less explicative than H was. But the formula may usually need to be applied qualitatively because subjective probabilities do not take sharp uncontroversial values, and to a mild degree because no best value of γ is known.

(xiiiP) A theme of this article has been that a form of Bayesian theory, in which, for example, probabilities and utilities are only partially ordered, and in which there is therefore emphasis on a Bayes–non-Bayes compromise, is beginning to provide a unification of statistics.

It appears then that statistics makes use of all 14 of the listed facets of the scientific method.

References

[1] Acton, H. B. (1967). In *The Encyclopedia of Philosophy*, Vol. 3, P. Edwards, ed. Macmillan, New York, pp. 435–451. (A run-down of what Hegel seemed to be trying to say.)

[2] Akaike, H. (1973). In *2nd Int. Symp. Information Theory*, B. N. Petrov and F. Czaki, eds. Akademiai Kiado, Budapest, Hungary, pp. 267–281. (Roughly put, this penalizes likelihood by dimensionality.)

[3] Akaike, H. (1980). In *Bayesian Statistics*, J. M. Bernardo, M. H. DeGroot, D. V. Lindley, and A. F. M. Smith, eds. University of Valencia Press, Valencia, Spain, pp. 143–166. (Extension of Akaike [2] with applications to time series. Adopts Good's philosophy, which he calls the common sense approach to Bayesian statistics.)

[4] Alexander, P. (1967). In *The Encyclopedia of Philosophy*, Vol. 2, P. Edwards, ed. Macmillan, New York, pp. 423–425. (Surveys Duhem's philosophy including his ideas concerning refutation of theories.)

[5] Allen, W. (1966). *Getting Even*. Random House, New York. (Contains a passage that might have been written by Hegel.)

[6] Berger, J. O. (1984). In *Robustness of Bayesian Analyses*, J. B. Kadane, ed. North-Holland,

Amsterdam, pp. 63–144 (with discussion). (A tolerant Bayesian exposition and its relationship to robustness.)

[7] Berger, J. O. (1985). *Statistical Decision Theory and Bayesian Analysis*, Springer-Verlag, New York [2nd edition of *Statistical Decision Theory* (1980)]. (Contains a modern Bayesian course.)

[8] Berger, J. O. (1985). *J. Amer. Statist. Ass.*, **80**, 232–233. (Expresses agreement with Good's philosophy of statistics, which he calls "Doogian.")

[9] Box, G. E. P. (1983). In *Scientific Inference, Data Analysis, and Robustness*, G. E. P. Box, T. Leonard, and C.-F. Wu, eds. Academic, New York, pp. 51–84. (Suggests, among other things, the alternation between Bayesian estimation and frequentist validation of the Bayesian model.)

[10] Carnap, R. (1950). *Logical Foundations of Probability*. University of Chicago Press, Chicago, IL. (Distinguishes between logical and physical probability. Argues against the personalistic interpretation to which he was more sympathetic later.)

[11] Chan, W.-T. (1967). In *The Encyclopedia of Philosophy*, Vol. 2, P. Edwards, ed. Macmillan, New York, pp. 87–96. (Surveys the yin-yang philosophy.)

[12] Crook, J. F. and Good, I. J. (1982). *J. Amer. Statist. Ass.*, **77**, 793–802. (Develops the notion of the "strength" of a significance test for multinomials and contingency tables.)

[13] Dantzig, D. van (1947). *Ned. Akad. Wet. Proc. Ser. A*, **50**, 918–929, 1092–1103. (*Indag. Math.*, **9**, 429–440, 506–517). (Mathematics without negations.)

[14] Deaton, M. L. (1980). The Causal Analysis of Multiple Time Series: A Frequency Domain Approach Using Group Delay. Ph.D. dissertation, Dept. of Statistics, Virginia Polytechnic Institute and State University, Blacksburg, VA. (Suggests a definition of degree of causation between two time series at each frequency component.)

[15] Fisher, R. A. (1956). *Statistical Methods and Scientific Inference*. Oliver and Boyd, Edinburgh, Scotland. (Fisher's own interpretation of his statistical philosophy.)

[16] Gersch, W. (1972). *Math. Biosci.*, **14**, 177–196. (Contains a measure of causation between time series in the frequency domain.)

[17] Gibbs, J. W. (1902). *Elementary Principles in Statistical Mechanics*, Constable, London, England. [Dover reprint, (1960)]. (Not elementary. Uses cross-entropy but only for statistical mechanics.)

[18] Good, I. J. (1950). *Probability and the Weighing of Evidence*. Charles Griffin, London, England. [Proposes a theory of partially ordered or interval-valued subjective (personal) probabilities and utilities, but sympathetic to "orthodox" statistics.]

[19] Good, I. J. (1957). *Ann. Math. Statist.*, **28**, 861–881. (Mentions a compromise between *P*-values and Bayes factors.)

[20] Good, I. J. (1958). *J. Amer. Statist. Ass.*, **53**, 799–813. (Suggests the harmonic-mean rule for combining *P*-values "in parallel.")

[21] Good, I. J. (1961). *Brit. J. Philos. Sci.*, **11**, 305–318; **12**, 43–51. (Discusses the degree to which one event tends to cause another one.)

[22] Good, I. J. (1962). In *Theories of the Mind*, Jordon Scher, ed. Glencoe Free Press, New York, pp. 490–518. (For degrees of metaphysicalness. Corrections in the 2nd ed.)

[23] Good, I. J. (1962). In *The Scientist Speculates: An Anthology of Partly-Baked Ideas*, Heinemann, London, England, pp. 120–132. (Speculations about the wide variety of future applications of clustering.)

[24] Good, I. J. (1963). *Ann. Math. Statist.*, **34**, 911–934. (Shows that maximizing entropy leads to the log-linear model for contingency tables.)

[25] Good, I. J. (1968). *Brit. J. Philos. Sci.*, **19**, 123–143. (A source paper concerning explicativity.)

[26] Good, I. J. (1969). In *Multivariate Analysis II*, P. R. Krishnaiah, ed. Academic, New York, pp. 183–203. (Suggests an invariant prior that is uniquely defined in terms of a Hessian of a loss function. Reduces to the Jeffreys invariant prior when the loss function is a cross-entropy.)

[27] Good, I. J. (1970). *Synthese*, **20**, 17–24. (Expresses a view more sympathetic to the reality of physical probabilities than thought by de Finetti.)

[28] Good, I. J. (1971). *Amer. Statist.*, **25**, 62–63. (46656 kinds of Bayesians.)

[29] Good, I. J. (1977). In *Classification and Clustering*, J. van Ryzin, ed. Academic, New York, pp. 73–94. (The philosophy, purposes, classification, and 45 facets of clustering techniques.)

[30] Good, I. J. (1977). *Proc. R. Soc. Lond. A*, **354**, 303–330. (Shows how maximizing explicativity leads to significance tests and interval estimates.)

[31] Good, I. J. (1980–1985). Items numbered 67, 78, 99, 100, 140, 143, 144, 145, 165, 174, 175, 176, 191, 192, 193, 199, 200, 201, 216, 217, 232 in the column Comments, Conjectures, and Conclusions. In *J. Statist. Comput. Simul.*, **10–21**. (Various aspects of the Bayes–non-Bayes compromise are discussed, including ancillarity, sufficiency, tail-area probabilities, induction, parsimony, errors of the two kinds, Godambe's paradox, the probability of a scientific law, and "adhockery.")

[32] Good, I. J. (1981). In *Bayesian Statistics*, J. M. Bernardo, M. H. DeGroot, D. V. Lindley, and

A. F. M. Smith, eds. University of Valencia Press, Valencia, Spain, pp. 489–519 (with discussion). (Contains some history of the hierarchical Bayesian methodology especially regarding categorical data.)

[33] Good, I. J. (1983). *Good Thinking: The Foundations of Probability and its Applications*. University of Minnesota Press, Duluth, MN. (Contains 23 of the author's more philosophical articles and a well indexed bibliography.)

[34] Good, I. J. (1983). *Scientific Inference, Data Analysis, and Robustness*, G. E. P. Box, T. Leonard, and C.-F. Wu, eds. Academic, New York, pp. 191–213. (Uses mixed Dirichlet priors in a hierarchical Bayes approach to significance tests for multinomials and contingency tables. Applies the theory to the problem of induction.)

[35] Good, I. J. (1983). *Philos. Sci.*, **50**, 283–295. (The philosophy of exploratory data analysis, with a Bayesian flavor.)

[36] Good, I. J. (1984). *J. Statist. Comput. Simul.*, **19**, 294–299. (Spells out the logical reason for defining weight of evidence as the logarithm of a Bayes factor.)

[37] Good, I. J. (1984). *J. Statist. Comput. Simul.*, **20**, 173–176. (Sharpens the harmonic-mean rule of thumb.)

[38] Good, I. J. (1986). *Statist. Sci.*, **1**, 157–170. (A review of Poisson's contributions to statistics. Contains a discussion of physical probability.)

[39] Good, I. J. (1985). In *Bayesian Statistics 2*, J. M. Bernardo, M. H. DeGroot, D. V. Lindley, and A. F. M. Smith, eds. North-Holland, New York, pp. 249–269 (including discussion). (A brief survey of weight of evidence.)

[40] Good, I. J. (1985). *J. Statist. Comput. Simul.*, **22**, 184–186. (Shows that explicativity does not allow quite enough for simplicity.)

[41] Good, I. J. (1985). *Philosophy of Science Association 1984*, vol. 2. (A discussion of causal tendency with applications. A revised form of this paper will appear in the proceedings of a conference on probability and causation, held at Irvine, California in 1985. B. Skyrme and W. Harper, eds.)

[42] Good, I. J. and Gaskins, R. A. (1972). *Virginia J. Sci.*, **23**, 171–193. (An exposition of the maximum penalized likelihood method of probability density estimation. An expansion of a 1971 *Biometrika* article.)

[43] Good, I. J. and Gaskins, R. A., (1980). *J. Amer. Statist. Ass.*, **75**, 42–73, (with discussion). [A continuation of Good and Gaskins (1972).]

[44] Good, I. J. and McMichael, F. (1984). *Philos. Sci.*, **51**, 120–127. (A pragmatic modification of explicativity.)

[45] Granger, C. W. J. and Newbold, P. (1977). *Forecasting Economic Time Series*. Academic, New York. (Develops a quantitative measure of causation for pairs of time series.)

[46] Hadamard, J. (1949). *The Psychology of Invention in the Mathematical Field*. Princeton University Press, Princeton, NJ [reprinted by Dover Publications, New York (1954)]. (A famous description of how mathematicians have creative ideas.)

[47] Hartigan, J. (1964). *Ann. Math. Statist.*, **35**, 836–845. (Introduces a multiplicity of invariant priors.)

[48] Jaynes, E. T. (1957). *Phys. Rev.*, **106**, 620–630; **108**, 171–190. (Introduces the principle of maximum entropy for the purpose of choosing prior distributions, apart from new applications to statistical mechanics.)

[49] Jaynes, E. T. (1968). *IEEE Trans. Syst. Sci. Cybernetics*, **SSC-4**, 227–241. (Discusses invariant priors.)

[50] Jeffreys, H. (1939). *Theory of Probability*, 1st ed. Clarendon Press, Oxford, England. [Puts forward what was intended to be an "objectivistic Bayesian" theory, with numerous applications. The 3rd ed. (1961) moves somewhat closer to the subjectivistic Bayesian position without explicitly saying so.]

[51] Jeffreys, H. (1946). *Proc. R. Soc. A*, **186**, 453–461. (The first attempt at obtaining a general invariant prior.)

[52] Kempthorne, O. and Folks, L. (1971). *Probability, Statistics, and Data Analysis*. Iowa State University Press, Ames, IA. (Section 13.2 describes the concept of a collection of estimation intervals consistent with data to a specific degree.)

[53] Koestler, A. (1964). *The Act of Creation*. Hutchinson, London, England.

[54] Kullback, S. (1959). *Information Theory and Statistics*. Wiley, New York. (Bases significance tests and probability estimation on cross-entropy without a Bayesian flavor.)

[55] Maxwell, G. (1970). In *The Nature and Function of Scientific Theories*, R. G. Colodny, ed. University of Pittsburgh Press, Pittsburgh, PA, pp. 3–34. (Argues against instrumentalism in favor of realism.)

[56] Mayo, D. (1983). *Synthese*, **57**, 297–340. [Pages 324–330 express the idea mentioned under Kempthorne and Folks (1971) with a different slant.]

[57] McMichael, A. (1985). *Brit. J. Philos. Sci.*, **36**, 257–272. (Argues for realism against a form of instrumentalism.)

[58] Neyman, J., Scott, E. L., and Smith, J. A. (1969). *Science*, **165**, 618 (Letter concerning the Whitetop

cloud-seeding experiment. Contains a mistaken interpretation of a tail-area probability.)

[59] Ogden, C. K. (1959). *Bentham's Theory of Fictions*. Routledge and Kegan Paul, London, England. (Contains an "as if" philosophy anticipating Vaihinger by a century.)

[60] Perks, W. (1947). *J. Inst. Actuaries*, **73**, 285–334 (with discussion). [Independently of Jeffreys (1946), suggests the same invariant prior for the case of a scalar parameter.]

[61] Poincaré, H. (1905). *Science and Hypothesis*, translated from the French by "W. J. G." Walter Scott Publishing Co., London, England [reprinted by Dover Publications, New York, (1952)]. (Poincaré's philosophy of science expressed with extreme lucidity.)

[62] Poincaré, H. (1956). In *The World of Mathematics*, Vol. 4. Simon and Schuster, New York, pp. 2041–2050. [Presumably from the translation by G. B. Halstead in *The Foundations of Science*, Science Press, New York, (1913), which is cited by Hadamard (1949), p. 12. A famous description of how mathematicians have creative ideas.]

[63] Poisson, S. D. (1837). *Recherche sur la Probabilité des Jugements*. Bachelier, Paris, France. (Draws a clear distinction between two kinds of probability that are now often called epistemic and physical.)

[64] Popper. K. R. (1959). *The Logic of Scientific Discovery*. Hutchinson, London, England [based on *Logik der Forschung* (1935)]. (Emphasizes the importance of honest attempts at refutation of theories.)

[65] Popper, K. R. (1963). *Conjectures and Refutations*. Routledge and Kegan Paul, London, England.

[66] Russell, B. (1946). *History of Western Philosophy*. Allen and Unwin, London, England. (The chapter on Hegel could have been entitled "The bogosity of Hegel.")

[67] Savage, L. J. (1954). *The Foundations of Statistics*. Wiley, New York. [Argues mathematically that a rational being would behave as if it had personal (subjective) probabilities and utilities and would maximize expected utility.]

[68] Scott, C. S. O'D. (1962). *The Scientist Speculates*, I. J. Good, ed. Heinemann, London, England, pp. 149–150. (A tongue-in-cheek speculation about lost objects.)

[69] Wiener, N. (1956). In *Modern Mathematics for the Engineer*, E. F. Beckenbach, ed. McGraw-Hill, New York, pp. 165–190. (A source reference for quantitative causality measurement between time series.)

[70] Wigner, E. P. (1962). *The Scientist Speculates*, I. J. Good, ed. Heinemann, London, England, pp. 284–302. (Seems to show that quantum mechanics needs to be modified if a kind of solipsism is to be avoided.)

(BAYESIAN INFERENCE
FIDUCIAL PROBABILITY
FOUNDATIONS OF PROBABILITY
FREQUENCY INTERPRETATION IN PROBABILITY AND STATISTICAL INFERENCE
LOGIC OF STATISTICAL REASONING
PARSIMONY, PRINCIPLE OF
RANDOMIZATION
STATISTICAL EVIDENCE
SUBJECTIVE PROBABILITIES)

I. J. Good

SCORE STATISTICS

The term *efficient score* was introduced by Rao [22] (see also Fisher [9, 11]) to describe a quantity of considerable importance to statistical theory. Let the likelihood function of a vector-valued parameter $\mathbf{\Theta} = (\Theta_1, \Theta_2, \ldots, \Theta_r)$ given a data vector $\mathbf{X}$ be denoted by $L(\mathbf{\Theta}; \mathbf{X})$. In most instances, $\mathbf{X}$ represents a random sample of size n from some underlying distribution with probability density function determined by $\mathbf{\Theta}$. If $L(\mathbf{\Theta}; \mathbf{X})$ is a differentiable function of $\mathbf{\Theta}$, then the efficient score for Θ_i is defined to be

$$S_i(\mathbf{\Theta}) = \partial \log L(\mathbf{\Theta}; \mathbf{X})/\partial \Theta_i.$$

As a function of $\mathbf{\Theta}$ for fixed $\mathbf{X}$, S_i is of central importance to the theory of estimation. As a random variable for fixed $\mathbf{\Theta}$, S_i is of central importance to the theory of hypothesis testing*. *See also* LIKELIHOOD.

Let $\mathbf{S}(\mathbf{\Theta})$ denote the $r \times 1$ vector with ith entry $S_i(\mathbf{\Theta})$. For a likelihood function that is twice differentiable with respect to $\mathbf{\Theta}$, define the *observed information matrix* $\mathbf{V}(\mathbf{\Theta})$ to be the $r \times r$ matrix with (i, j)th entry $V_{ij}(\mathbf{\Theta}) = -\partial^2 \log L(\mathbf{\Theta}; \mathbf{X})/\partial \Theta_i\, \partial \Theta_j$, and assuming the V_{ij} have finite expectation, define the *information matrix* $\mathbf{I}(\mathbf{\Theta})$ to be the $r \times r$ matrix with (i, j)th entry $I_{ij}(\mathbf{\Theta}) = E[V_{ij}(\mathbf{\Theta})]$

(*see* FISHER INFORMATION). Then under general regularity conditions (e.g., see LeCam [15]), if $\boldsymbol{\Theta}_0$ is the true parameter value for the underlying distribution of $\mathbf{X}$, $E[\mathbf{S}(\boldsymbol{\Theta}_0)] = \mathbf{0}$, $\text{cov}\{\mathbf{S}(\boldsymbol{\Theta}_0)\} = \mathbf{I}(\boldsymbol{\Theta}_0)$, and, for large samples, $\mathbf{S}(\boldsymbol{\Theta}_0)$ is approximately multivariate normal with mean $\mathbf{0}$ and covariance matrix $\mathbf{I}(\boldsymbol{\Theta}_0)$.

Efficient scores were first used in forming the equations leading to the maximum likelihood estimator (e.g., see Fisher [9]). If $\sup_{\boldsymbol{\Theta}} L(\boldsymbol{\Theta}; \mathbf{X})$ is attained in the parameter space, the maximum likelihood estimator $\hat{\boldsymbol{\Theta}}$ may be obtained as a solution to the likelihood equations

$$S_i(\hat{\boldsymbol{\Theta}}) = 0 \quad \text{for all } i = 1, 2, \ldots, r.$$

Asymptotic optimality properties of the efficient scores as estimating equations were demonstrated by Fisher [9, 10] and Wilks [27], while closely related finite-sample optimality properties were presented by Godambe [13] and Barnard [2].

The likelihood equations may be solved by the *method of scoring* (*see* ITERATED MAXIMUM LIKELIHOOD ESTIMATES), an iterative method closely related to the Newton–Raphson* method. Starting with an initial value $\boldsymbol{\Theta}_1$, one finds that the $(m+1)$st iterative value by the method of scoring is

$$\boldsymbol{\Theta}_{m+1} = \boldsymbol{\Theta}_m + \{\mathbf{I}(\boldsymbol{\Theta}_m)\}^{-1}\mathbf{S}(\boldsymbol{\Theta}_m). \quad (1)$$

Although iteration may be continued until convergence, an efficient estimator can often be obtained with only two iterations [9]. The method of scoring extends easily to the case of K independent data sets, each providing a likelihood function for the same parameter $\boldsymbol{\Theta}$. Let $\mathbf{S}_k$ denote the $r \times 1$ vector of scores based on the kth likelihood function and $\mathbf{I}_k$ denote the corresponding $r \times r$ information matrix, $k = 1, 2, \ldots, K$. Then the method of scoring proceeds iteratively as described by (1), with $\Sigma \mathbf{S}_k$ substituted for $\mathbf{S}$ and $\Sigma \mathbf{I}_k$ substituted for $\mathbf{I}$. An associated test of the hypothesis of a common value of $\boldsymbol{\Theta}$ for all K data sets may be based on

$$X_H^2 = \sum_{k=1}^{K} \{\mathbf{S}_k(\hat{\boldsymbol{\Theta}})\}' \{\mathbf{I}_k(\hat{\boldsymbol{\Theta}})\}^{-1} \mathbf{S}_k(\hat{\boldsymbol{\Theta}}),$$

where $\hat{\boldsymbol{\Theta}}$ is the maximum likelihood estimator based on all K data sets. Under the null hypothesis of a common value of $\boldsymbol{\Theta}$ for all k, X_H^2 will be asymptotically distributed as a chi-squared random variable with $r(K-1)$ degrees of freedom [17, 22].

In the theory of hypothesis testing, the efficient score has a central role in the derivation of locally most powerful tests of simple null hypotheses regarding a scalar parameter Θ (*see* LOCALLY OPTIMAL STATISTICAL TESTS). Let $\beta(\Theta)$ denote the power function of a test of $H_0 : \Theta = \Theta_0$. For one-sided alternatives of the form $H_+ : \Theta > \Theta_0$ or $H_- : \Theta < \Theta_0$, the locally most powerful (LMP) test is found by maximizing $c\beta'(\Theta_0)$, where $c = 1$ for H_+ and $c = -1$ for H_-, subject to the restriction $\beta(\Theta_0) = \alpha$. By the Neyman–Pearson Lemma*, the rejection regions for the LMP tests with size α are given by $R_+ = \{\mathbf{X} \ni S(\Theta_0) > a_\alpha\}$ for H_+ and by $R_- = \{\mathbf{X} \ni S(\Theta_0) < b_\alpha\}$ for H_-, provided a_α and b_α can be determined so that $\Pr[R_+|\Theta_0] = \Pr[R_-|\Theta_0] = \alpha$ [23]. Alternatively, an approximate test may be performed by comparing the score statistic

$$Z = S(\Theta_0)/\{I(\Theta_0)\}^{1/2} \quad (2)$$

to the appropriate percentile of a standard normal distribution. The score test based on Z is the asymptotically LMP one-sided test, where asymptotic power is defined in terms of a sequence of alternatives such as $\Theta_n = \Theta_0 + \delta/\sqrt{n}$ [25]. Score tests were first used in genetic applications by Fisher [11].

Example 1. With scalar-valued parameters, uniformly most powerful (UMP) tests sometimes exists (e.g., in random sampling from distributions in the exponential family), and in such cases score test methodology is unnecessary. For a random sample $\{x_i\}_{i=1}^n$ from the logistic distribution* with density function

$$f(x; \Theta) = \exp\{-(x-\Theta)\} \times [1 + \exp\{-(x-\Theta)\}]^{-2},$$

no UMP test exists for hypotheses regarding

Θ. The LMP test of $H_0: \Theta = \Theta_0$ against $H_+ : \Theta > \Theta_0$ is based on $S(\Theta_0) = n - 2\sum_{i=1}^n y_i$, where

$$y_i = \{\exp(x_i - \Theta_0) + 1\}^{-1} \times [1 + \exp\{-(x_i - \Theta_0)].$$

Under H_0 the y_i constitute a random sample from a uniform distribution on the unit interval, and thus an exact LMP test can be performed, the critical value a_α being obtained from the distribution of the sum of n independent uniform random variables. Alternatively, since $I(\Theta_0) = n/3$, an approximate score test can be based on the statistic $Z = (n - 2\sum_{i=1}^n y_i)/(n/3)^{1/2}$ given by (2).

Example 2. Because LMP tests are equivalent to UMP tests when UMP tests exist, score tests will often have good power for alternatives other than local alternatives. That score tests need not have good power for nonlocal alternatives is demonstrated by the Cauchy distribution* with density function $f(x;\ \Theta) = [\pi\{1 + (x - \Theta)^2\}]^{-1}$. The score test for $H_0: \Theta = 0$ vs. $H_+ : \Theta > 0$ from a random sample $\{x_i\}_{i=1}^n$ is based on $S(0) = \sum_{i=1}^n\{x_i/(1 + x_i^2)\}$. Since $S(0)$ converges to 0 as all of the x_i get large, it follows that for any critical value $a_\alpha > 0$, Θ can be chosen so large that the probability of rejecting H_0 is arbitrarily close to 0. Such an anomalous result cannot occur for a convex density function [16].

In general, for the two-sided alternative $H_A: \Theta \neq \Theta_0$, no locally most powerful test exists. However, locally most powerful unbiased (LMPU) tests for H_A often exist, and may be derived by maximizing $\beta''(\Theta_0)$ subject to the restrictions $\beta(\Theta_0) = \alpha$ and $\beta'(\Theta_0) = 0$. By the generalized Neyman–Pearson lemma, the rejection region for the LMPU test is given by

$$R_A = \{\mathbf{X} \ni -V(\Theta_0) + [S(\Theta_0)]^2 \geqslant c_\alpha S(\Theta_0) + d_\alpha\},$$

provided c_α and d_α can be chosen such that $\Pr[R_A|\Theta_0] = \alpha$ and $\beta'(\Theta_0) = 0$ [20]. Alternatively, an approximate test may be performed by comparing Z^2, where Z is the score statistic defined in (2), to the appropriate percentile of a chi-squared distribution with 1 degree of freedom. The score test based on Z^2 is the asymptotically LMPU test [25]. In addition, confidence intervals based on Z^2 are asymptotically shortest in terms of expected length [27] and in terms of Neyman's criterion of shortness [26].

For vector-valued parameters, asymptotic tests of the simple hypothesis $H_0: \mathbf{\Theta} = \mathbf{\Theta}_0$ and asymptotic confidence ellipsoids for $\mathbf{\Theta}$ may be based on the fact that the score statistic

$$X^2 = \{\mathbf{S}(\mathbf{\Theta}_0)\}'\{\mathbf{I}(\mathbf{\Theta}_0)\}^{-1}\{\mathbf{S}(\mathbf{\Theta}_0)\}$$

has an asymptotic chi-squared distribution with r degrees of freedom under H_0 [22, 27, 28]. Score statistics are also useful in testing composite hypotheses* [22]. Of particular interest are tests about (or confidence regions for) one subset of $\mathbf{\Theta}$, with a second subset of $\mathbf{\Theta}$ treated as a vector of nuisance parameters*. Suppose $\mathbf{\Theta}' = (\mathbf{\Theta}_1', \mathbf{\Theta}_2')$, where $\mathbf{\Theta}_1$ is $p \times 1$ with $p < r$. Consider the corresponding partition of scores $\mathbf{S}' = (\mathbf{S}_1', \mathbf{S}_2')$, and partition $\mathbf{I}$ accordingly, with $\mathbf{I}_{11} = E[\mathbf{S}_1\mathbf{S}_1']$, $\mathbf{I}_{12} = E[\mathbf{S}_1\mathbf{S}_2']$, and $\mathbf{I}_{22} = E[\mathbf{S}_2\mathbf{S}_2']$. Let $\hat{\mathbf{\Theta}}_2^0$ be the maximum likelihood estimator of $\mathbf{\Theta}_2$ given $\mathbf{\Theta}_1 = \mathbf{\Theta}_1^0$, and define $\hat{\mathbf{\Theta}}_0' = (\mathbf{\Theta}_1^{0\prime}, \hat{\mathbf{\Theta}}_2^{0\prime})$. Then the score statistic for testing $H_0: \mathbf{\Theta}_1 = \mathbf{\Theta}_1^0$ with $\mathbf{\Theta}_2$ a vector of nuisance parameters is

$$X_{1.2}^2 = \{\mathbf{S}_1(\hat{\mathbf{\Theta}}_0)\}'\{\mathbf{I}_{11.2}(\hat{\mathbf{\Theta}}_0)\}^{-1}\{\mathbf{S}_1(\hat{\mathbf{\Theta}}_0)\}, \tag{3}$$

where $\mathbf{I}_{11.2} = \mathbf{I}_{11} - \mathbf{I}_{12}\mathbf{I}_{22}^{-1}\mathbf{I}_{12}'$. Under H_0, $X_{1.2}^2$ will be asymptotically distributed as a chi-squared random variable with p degrees of freedom. The test is asymptotically similar since the null asymptotic distribution of $X_{1.2}^2$ is independent of $\mathbf{\Theta}_2$. For $p = 1$ the test based on $X_{1.2}^2$ is the asymptotically LMPU test, and for $p > 1$ the test is the asymptotically locally most stringent test [6]. If $\hat{\mathbf{\Theta}}_2^0$ is difficult to obtain and a simpler square root

n consistent estimator of Θ_2 given $\Theta_1 = \Theta_1^0$ exists, then an asymptotically equivalent test statistic that is a modification of $X_{1.2}^2$ can be obtained using the theory of $C(\alpha)$ tests [18, 19].

Example 3. Consider J independent binomial random variables x_j, with parameters p_j and sample sizes n_j for $j = 1, 2, \ldots, J$. Suppose that for each x_j there is a corresponding value d_j of a quantitative variable, with $d_1 < d_2 < \cdots < d_J$, and that the goal is to test for an increase in the parameters p_j with increasing d_j. Modeling $p_j = H(\Theta_2 + \Theta_1 d_j)$, where H is any monotone twice differentiable function, the goal is then to test $H_0 : \Theta_1 = 0$ against $H_+ : \Theta_1 > 0$, with Θ_2 a nuisance parameter. It follows from the binomial likelihood that

$$S_1(0, \hat{\Theta}_2^0) = \frac{H'(\hat{\Theta}_2^0)}{\hat{p}\hat{q}} \left(\sum_{j=1}^{J} x_j d_j - \hat{p} \sum_{j=1}^{J} n_j d_j \right)$$

and

$$I_{11.2}(0, \hat{\Theta}_2^0) = \frac{\{H'(\hat{\Theta}_2^0)\}^2}{\hat{p}\hat{q}} \times \left\{ \sum_{j=1}^{J} n_j d_j^2 - \left(\sum_{j=1}^{J} n_j d_j \right)^2 \Big/ n. \right\},$$

where $\hat{p} = x./n.$, $\hat{q} = 1 - \hat{p}$, and $\hat{\Theta}_2^0 = H^{-1}(\hat{p})$. Thus $X_{1.2}^2$ given in (3) does not depend on the choice of H, a result which generalizes to other distributions in the exponential family* [7]. The choice of $H(y) = e^y/(1 + e^y)$ gives the natural parameterization for the binomial likelihood, and in such cases (i.e., exponential family distributions with natural parameterization) uniformly most powerful unbiased tests can be derived without resorting to score test methodology [12].

In general, for both simple and composite hypotheses, score tests are asymptotically equivalent to Wald tests based on maximum likelihood estimators and to likelihood ratio tests*. There are differences, however, among the three methods for any given application. For testing $H_0 : \Theta = \Theta_0$ based on a random sample of size n, none of the three methods is uniformly superior in terms of power, although for $H_A : \Theta \neq \Theta_0$ the likelihood ratio test is unbiased to a higher order of n than are the score or Wald tests [21]. For composite null hypotheses, tests based on all three methods are biased to order $n^{-1/2}$, and again none of the three methods uniformly dominates in terms of power [14]. The adequacy of the normal approximation to the distribution of a score statistic based on asymptotic theory is largely dependent on the shape of the corresponding likelihood function and can sometimes be improved by reparameterization [24]. Alternatively, higher-order approximations to the distribution of the score statistic have been developed [3, 4, 5, 8]. A parallel development of the asymptotic theory for score tests has evolved under the alternative terminology, Lagrange multiplier tests* [1]. In all of the applications described, asymptotic tests and confidence regions based on score statistics remain valid when the observed information matrix is substituted for the information matrix.

References

[1] Aitchison, J. and Silvey, S. D. (1958). *Ann. Math. Statist.*, **29**, 813–828.

[2] Barnard, G. A. (1973). *Sankhyā*, **35**, 133–138.

[3] Bartlett, M. S. (1953). *Biometrika*, **40**, 12–19.

[4] Bartlett, M. S. (1953). *Biometrika*, **40**, 306–317.

[5] Bartlett, M. S. (1955). *Biometrika*, **42**, 201–204.

[6] Bhat, B. R. and Nagnur, B. N. (1965). *Biometrika*, **52**, 459–468.

[7] Chen, C.-F. (1983). *J. Amer. Statist. Ass.*, **78**, 158–161.

[8] Cox, D. R. (1980). *Biometrika*, **67**, 279–286.

[9] Fisher, R. A. (1925). *Proc. Camb. Philos. Soc.*, **22**, 700–725.

[10] Fisher, R. A. (1928). *Atti Congr. Int. Mat.*, **6**, 94–100.

[11] Fisher, R. A. (1935). *Ann. Eugen.*, **6**, 187–201.

[12] Gart, J. J. and Tarone, R. E. (1983). *Biometrics*, **39**, 781–786.

[13] Godambe, V. P. (1960). *Ann. Math. Statist.*, **31**, 1208–1211.

[14] Harris, P. and Peers, H. W. (1980). *Biometrika*, **67**, 525–529.

[15] LeCam, L. (1970). *Ann. Math. Statist.*, **41**, 802–828.

[16] Lehmann, E. L. (1955). *Ann. Math. Statist.*, **26**, 399–419.

[17] Mather, K. (1935). *Ann. Eugen.*, **6**, 399–410.

[18] Moran, P.A.P. (1970). *Biometrika*, **57**, 47–55.

[19] Neyman, J. (1959). In *Probability and Statistics*, U. Grenander, ed. Wiley, New York, pp. 213–234.

[20] Neyman, J. and Pearson, E. S. (1936). *Statist. Res. Mem.*, **1**, 1–37.

[21] Peters, H. W. (1971). *Biometrika*, **58**, 577–587.

[22] Rao, C. R. (1948). *Proc. Camb. Philos. Soc.*, **44**, 50–57.

[23] Rao, C. R. and Poti, S. J. (1946). *Sankhyā*, **7**, 439.

[24] Sprott, D. A. (1975). *Sankhyā*, **37**, 259–270.

[25] Wald, A. (1941). *Ann. Math. Statist.*, **12**, 396–408.

[26] Wald, A. (1942). *Ann. Math. Statist.*, **13**, 127–137.

[27] Wilks, S. S. (1938). *Ann. Math. Statist.*, **9**, 166–175.

[28] Wilks, S. S. (1939). *Ann. Math. Statist.*, **10**, 85–86.

Bibliography

Cox, D. R. and Hinkley, D. V. (1974). *Theoretical Statistics*. Chapman and Hall, London, England. (Locally optimal tests are discussed in Sec. 4.8, and the asymptotic theory of score tests is developed in Secs. 9.1–9.3.)

(ASYMPTOTIC NORMALITY
EFFICIENT SCORE
FISHER INFORMATION
HYPOTHESIS TESTING
ITERATED MAXIMUM LIKELIHOOD ESTIMATES
LAGRANGE MULTIPLIER TEST
LIKELIHOOD
LOCALLY OPTIMUM STATISTICAL TESTS
MAXIMUM LIKELIHOOD ESTIMATION
NEYMAN–PEARSON LEMMA
WALD STATISTICS)

ROBERT E. TARONE

SCORING, METHOD OF *See* ITERATED MAXIMUM LIKELIHOOD ESTIMATES

SCORING RULES

A concept introduced by de Finetti [1] in an attempt to describe uncertainty numerically within a Bayesian framework. Suppose that in expressing *your* belief in an event A, given *your* knowledge H, a numerical value a is provided by *you*. In what sense is a a good measurement of *your* belief? De Finetti [1] had the idea of introducing a *score function*, which scores *your* measurement or, as is usually preferred, *your* assessment of *your* uncertainty of A, given H. For two functions f_0 and f_1, the score, when a is announced as the assessment, is defined to be

$f_1(a)$ if both A and H are true;
$f_0(a)$ if H is true, but A false;
zero if H is false.

De Finetti used the *quadratic* or *Brier score:* $f_0(a) = a^2$, $f_1(a) = (1-a)^2$. With the quadratic, a near 1 (0) will give a low score when A is true (false) and H is true. If H is false, the statement about A is irrelevant because it was made on the supposition of H.

Suppose now that this is done with several event pairs (A_iH_i); a is scored on each and the scores are added. De Finetti demonstrated the quadratic rule that the values of a_i must obey the rules of probability:

1. **Convexity**. $0 \leqslant p(A|H) \leqslant 1$ and $p(A|H) = 1$ if H is known by *you* logically to imply A.
2. **Addition**. $p(A_1 \cup A_2|H) = p(A_1|H) + p(A_2|H) - p(A_1 \cap A_2|H)$.
3. **Multiplication**. $p(A_1 \cap A_2|H) = p(A_1|H)p(A_2|A_1 \cap H)$.

Lindley [2] generalized the result and showed that virtually any score leads to probability: see Lindley [2] for additional details.

References

[1] de Finetti, B. (1974/1975). *Theory of Probability*, Vols. 1 and 2. Wiley, New York.

[2] Lindley, D. V. (1982). *Int. Statist. Rev.*, **50**, 1–26 (with discussion).

(BAYESIAN INFERENCE
BELIEF FUNCTIONS
COHERENCE

DEGREES OF BELIEF
FOUNDATIONS OF PROBABILITY
INFERENCE, STATISTICAL: I, II)

SCORING SYSTEMS IN SPORT *See* SPORT, SCORING SYSTEMS IN

SCOTT'S COEFFICIENT OF INTERCODER AGREEMENT

Consider data gathered to assess the agreement between two independent observers, represented in a square matrix:

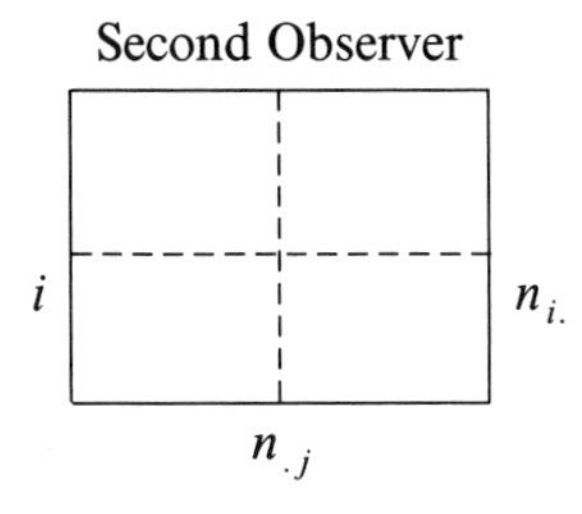

Here n_{ij} is the number of observations assigned to category (or scale value) i by the first observer and to category (or scale value) j by the second. Also, $n_{i.} = \Sigma_j n_{ij} (n_{.j} = \Sigma_i n_{ij})$ is the frequency with which the ith (jth) category (or scale value) i is used by the first (second) observer and $n = \Sigma_i n_{i.}$ is the total number of units recorded by both. The diagonal of this scale matrix, with entries n_{ii}, contains the *incidents of agreement*. Any deviation from the diagonal constitutes a disagreement. The observed proportion of agreements is

$$p_0 = \frac{1}{n} \sum_i n_{ii},$$

while the *expected* proportion of agreements is

$$p_e = \frac{1}{n^2} \sum_i \left(\frac{n_{i.} + n_{.i}}{2} \right)^2, \qquad i = 1, 2.$$

Scott's coefficient of intercoder agreement [3] is defined as

$$\frac{p_0 - p_e}{1 - p_e}.$$

The numerator is a measure of the above-chance agreement observed and the denominator is a measure of the above-chance agreement maximally possible. See, e.g., Krippendorff [2] and Andrews et al. [1] for further details.

References

[1] Andrews, F. M., Klem, L., Davidson, T. N., O'Malley, P. M., and Rodgers, W. L. (1981). *A Guide for Selecting Statistical Techniques for Analyzing Social Science Data*, 2nd ed. Institute of Social Research, University of Michigan, Ann Arbor, MI.

[2] Krippendorff, K. (1970). Bivariate Agreement Coefficients for Reliability of Data. In *Sociological Methodology*, E. F. Borgatta and G. W. Bohrnstedt, eds. Jossey-Bass, San Francisco, CA.

[3] Scott, W. A. (1955). *Public Opinion Quart.*, **19**, 321–325.

(COEFFICIENT OF CONCORDANCE)

SCREENING BY CORRELATED VARIATES

Students are chosen for entrance into college based on previous grades and standardized tests (e.g., the Scholastic Aptitude Test). The assumption is that these grades and tests are correlated with the future performance of the student in college. Similarly, the size of a crack in a manufactured item may be correlated with the lifetime of the item. Hence, items having cracks larger than a specified amount lead to rejection of those items. How do we arrive at a cutoff score on the standardized test in order to insure future college performance? How do we decide what size crack should lead to rejection of a manufactured item? A mathematical model allowing us to do this will now be discussed.

We will define

$$X = \text{a screening variate},$$

$$Y = \text{a performance variate},$$

and assume that $(X, Y)'$ have a joint bivariate normal distribution* with means $(\mu_x, \mu_y)'$ and variance–covariance matrix

$$\begin{pmatrix} \sigma_x^2 & \rho\sigma_x\sigma_y \\ \rho\sigma_x\sigma_y & \sigma_y^2 \end{pmatrix}.$$

If L is a lower specification limit on the Y variate and all items greater than L are considered satisfactory, and if the correlation* ρ is positive, then our procedure is to accept all items for which $X \geqslant \mu_x - K_\beta \sigma_x$, where K_β is a percentile of the univariate normal distribution, i.e., $\Phi(K_\beta) = \beta$. The procedure allows us to raise the proportion of acceptable items from γ to δ by selecting the upper $100\beta\%$ of the X variates. The relationship that must be solved for β is $\Pr\{Y \geqslant L | X \geqslant \mu_x - K_\beta \sigma_x\} = \delta$, where $\Pr\{Y \geqslant L\} = \gamma$.

Table 8 of Odeh and Owen [3] gives β as a function of γ, δ, and ρ. Of course, if ρ is negative, the acceptance criterion becomes: accept for $X \leqslant \mu_x + K_\beta \sigma_x$. If the specification involves an upper limit U instead of a lower limit and ρ is positive we accept if $X \leqslant \mu_x + K_\beta \sigma_x$; and if both U and a negative ρ are present we accept if $X \geqslant \mu_x - K_\beta \sigma_x$. If both upper and lower specifications apply, the procedure is that given by Li and Owen [2].

Extending screening procedures of this type to take account of several screening variables may be done in several ways. Thomas et al. [10] give tables and procedures for two screening variates that are based on the trivariate normal distribution. They also give a procedure based on a linear combination of the screening variates. The latter is easily extendable to any number of variates and uses the same table as the procedure with only one screening variate, which was given above (i.e., those given by Odeh and Owen [3]). Hence, a sketch will be given for this procedure.

Let us assume that Y, X_1, X_2 have a joint trivariate normal distribution, and (to simplify the presentation) that all three variates are standardized, i.e., have means zero and variances one. The correlations between these variables are ρ_{y1}, ρ_{y2}, ρ_{12}, where the notation indicates which variables correspond to each correlation. A linear combination $a_1x_1 + a_2x_2$ is formed and also standardized by dividing by $\sigma^* = (a_1^2 + a_2^2 + 2a_1a_2\rho_{12})^{1/2}$ to give us V. Then V and Y are used in the bivariate procedure given above, with the correlation between Y and V being $(a_1\rho_{y1} + a_2\rho_{y2})/\sigma^*$. Thomas et al. [10] show that the choice of a_1 and a_2 that maximizes the chance of obtaining items that meet the specification is given by

$$a_1 = \frac{\rho_{y1} - \rho_{y2}\rho_{12}}{\sqrt{(1 - \rho_{12}^2)(1 - \rho_{12}^2 - \Delta)}},$$

$$a_2 = \frac{\rho_{y2} - \rho_{y1}\rho_{12}}{\sqrt{(1 - \rho_{12}^2)(1 - \rho_{12}^2 - \Delta)}},$$

where

$$\Delta = 1 - \rho_{y1}^2 - \rho_{y2}^2 - \rho_{12}^2 + 2\rho_{y1}\rho_{y2}\rho_{12}.$$

If this choice is made, then $\sigma^* = 1$. Of course, a_1 and a_2 could also be chosen subjectively to reflect the relative importance that we wish to place on the two screening variates. The procedure is also immediately extendable to any number of screening variates by taking linear combinations of two variates at a time.

Up to this point, we have assumed that all of the parameters are known for the joint distribution of Y and the screening variates. Let us limit ourselves from this point on to a single screening variate. If the parameters are not known, we estimate them in the usual way, sometimes by a point estimator and sometimes by an interval estimator. Our screening procedure will be set up in the same way as tolerance limits* for a normal distribution are formed. That is, with confidence η we say that at least $100\delta\%$ of the screened population will meet specifications. Note the analogy with *β-content* tolerance regions* at confidence level γ.

We want to find a quantity k so that if we accept all items for which $X \geqslant \bar{x} - ks_x$, we can be $100\eta\%$ sure that at least $100\delta\%$ of the accepted group will have $Y \geqslant L$, where L is a lower specification limit, and $\bar{x}$ and s_x^2 are the usual unbiased estimators of μ_x and σ_x^2.

As with β-content tolerance limits we have two probabilities involved,

$$\Pr[\Pr\{Y \geqslant L | X \geqslant \bar{x} - ks_x\} \geqslant \delta] = \eta.$$

The inner probability is with respect to the conditional distribution of Y given X. The outer probability is with respect to the estimators of the unknown parameters, ρ, γ, μ_x, and σ_x. The solution to the problem of finding k from this setup is given by Chow [1].

A solution to a modified version of the problem was given by Owen and Su [9]. The modification involved moving the variation of $\bar{x}$ and s_x from the outer probability to the inner probability. A set of tables is given by Odeh and Owen [3] that enables one to implement this latter solution. In Section 3 of Owen and Boddie [4], when ρ and γ are known a problem was formulated as an approximate expected coverage problem in analogy with β-expectation tolerance limits. Hence, the solution given in Odeh and Owen [3] may be considered as an approximate expectation over $\bar{x}$ and s_x, but it involves more than an expectation since lower confidence limits on γ and ρ are used instead of their point estimates. The procedure amounts to a mixture of expected coverage and δ-content tolerance limits.

The solution given by Chow [1] more closely follows the setup for tolerance limits, but as of this writing no tables of the required constants k are available.

The implementation of either of these two procedures when all of the parameters are unknown is summarized in the following five steps. These steps are the same for both procedures, except at step **3** where **3a** applies to the Owen and Su procedure and step **3b** applies to the Chow procedure.

1. A preliminary sample of size n is taken from the joint distribution of X and Y. A lower $100\eta\%$ confidence limit on the correlation coefficient is computed and called ρ^*. If this lower limit is positive, we proceed to step **2**. If this lower limit is negative, an upper $100\eta\%$ confidence limit on ρ is computed. If this is positive, the process stops, because then X and Y could be independent. If the $100\eta\%$ upper confidence limit on ρ is also negative we call it ρ^* and proceed to step 2, but modify all of the following steps to take account of a negative correlation between X and Y.

2. A $100\eta\%$ lower confidence limit on γ is obtained and called γ^*. If $\delta < \gamma^*$ the procedure is stopped because the probability is at least η that the entire population meets the criterion $\Pr\{Y \geqslant L\} \geqslant \delta$. Table 7 of Odeh and Owen [3] or the table in Owen and Hua [6] may be used to obtain γ^*.

3a. For the Owen and Su [9] procedure we enter a table of the normal conditioned on t-distribution* to obtain a constant k for the screening procedure. Tables of the normal conditioned on t-distribution may be found in Owen and Haas [5], Owen and Ju [7], and in Odeh and Owen [3, Table 9]. These tables are functions of η, ρ, γ, and δ. Since we do not know ρ and γ, we replace them with ρ^* and γ^*. This is permissible since $\delta = \Pr[Y \geqslant L | X \geqslant \bar{x} - ks]$ is an increasing function of both ρ and γ.

3b. For the Chow [1] procedure we first obtain K_β from the normal conditioned on normal table given by Owen et al. [8] or Odeh and Owen [3, Table 8]. We enter with γ^* and ρ^* substituted for γ and ρ, respectively. Since $\delta = \Pr[Y \geqslant L | X \geqslant \mu_x - K_\beta\sigma_x]$ is an increasing function of both γ and ρ, this is permissible. The tables, which have not yet been computed, would give us k so that the $\Pr[\gamma \geqslant \gamma^*, \bar{x} - ks_x \geqslant \mu_x - K_\beta\sigma_x | \rho] = \eta'$, where ρ^* may be substituted for ρ because this probability is also an increasing function of ρ. Note that the random variables in this joint probability are γ^*, $\bar{x}$, and s_x, and that η has to be greater than η' since we found γ^* from $\Pr[\gamma \geqslant \gamma^*] = \eta$. Chow [1] derived an expression for computing η'. Because γ^* is a part of the joint probability leading to η' we no longer need to account for the $\eta = \Pr\{\gamma \geqslant \gamma^*\}$. See step **5** for the combination of these events.

4. If the ρ^* is positive, all items are accepted for which $X \geqslant \bar{x} - ks$. If ρ^* is negative, accept those items for which $X \leqslant \bar{x} + ks$.

5. We can say with confidence at least $100(\eta + \eta' - 1)\%$ that at least $100\delta\%$ of the Y's are above L in the selected group. (For the Owen and Su [9] procedure $\eta' = \eta$.) We obtain the $100(\eta + \eta' - 1)\%$ for this confidence by applying a Bonferroni inequality* to the joint event that both $\gamma \geqslant \gamma^*$ and $\rho \geqslant \rho^*$ in the Owen and Su procedure [9], and to the joint event that both $\rho \geqslant \rho^*$ and $\gamma \geqslant \gamma^*$, $\bar{x} - ks_x \geqslant \mu_x - k_\beta\sigma_x$ for the Chow [1] procedure.

The question of which procedure to use depends on several considerations: (a) A complete set of tables implementing the Owen and Su procedure [9] is given in Odeh and Owen [3], while tables for the Chow procedure have not yet been computed. (b) There is some slack in the overall probability given in step **5** due to the use of the Bonferroni inequality. The Chow procedure gives an absolute guarantee of $\eta + \eta' - 1$ confidence, while the Owen and Su procedure averages out the randomness in $\bar{x}$ and s. Simulations seem to indicate that the net effect is usually closer to the $\eta + \eta' - 1$ bound for the Owen and Su procedure than for the Chow procedure. Additional experience with these is needed before a definitive choice for all occasions can be recommended. Both will be useful in different applications.

Extension of both of the above procedures to two-sided specifications (i.e., where both L and U are specified) may be made along the lines given by Li and Owen [2].

References

[1] Chow, L. S. (1983). Application of the Bivariate Singly Noncentral t-Distribution to a Screening Problem. Dissertation, Southern Methodist University, Dallas, TX.

[2] Li, L. and Owen, D. B. (1979). *Technometrics*, **21**, 79–85. (Treats two-sided screening procedures both with known and unknown parameters.)

[3] Odeh, R. E. and Owen, D. B. (1980). *Tables for Normal Tolerance Limits, Sampling Plans, and Screening.* Dekker, New York.

[4] Owen, D. B. and Boddie, J. W. (1976). *Technometrics*, **18**, 195–199. (This is the first paper to treat screening cases with unknown parameters.)

[5] Owen, D. B. and Haas, R. (1978). In *Contributions to Survey Sampling and Applied Statistics*, H. A. David, ed. Academic, New York. (Gives tables of the normal conditioned on t-distribution.)

[6] Owen, D. B. and Hua, T. A. (1977). *Commun. Statist. B*, **6**, 285–311. (Gives tables for confidence limits on the tail area of the normal distribution when both parameters are unknown.)

[7] Owen, D. B. and Ju, F. (1977). *Commun. Statist. B*, **6**, 167–179. (Gives a table of the normal conditioned on t-distribution when the correlation is one.)

[8] Owen, D. B., McIntire, D. D., and Seymour, E. (1975). *J. Quality Tech.*, **7**, 127–138. (Gives tables for one and two screening variates in the case where all parameters are known.)

[9] Owen, D. B. and Su, Y.-L. H. (1977). *Technometrics*, **19**, 65–68. (Gives the screening procedure for special cases where certain parameters are unknown and others are known.)

[10] Thomas, J. G., Owen, D. B., and Gunst, R. F. (1977). *J. Educ. Statist.*, **2**, 55–77. (Gives the procedures for known parameters, tables for one and two screening variates, and procedures for linear combinations.)

Bibliography

Owen, D. B. (1983). In *Reliability in the Acquisitions Process*, D. J. DePriest and R. L. Launer, eds. Dekker, New York.

Owen, D. B. and Li, L. (1980). *J. of Educ. Statist.*, **5**, 157–168. (Discusses cutting scores in selection procedures.)

Owen, D. B., Li, L., and Chow, Y. M. (1981). *Technometrics*, **23**, 165–170. (Combines the idea of prediction intervals and screening procedures into a single approach.)

(CONFIDENCE INTERVALS AND REGIONS
TOLERANCE REGIONS, STATISTICAL)

DONALD B. OWEN

SCREENING DESIGNS

We screen a set of entities for those having some specified property. We might screen a population of individuals for those with a particular disease or a set of corn varieties for the one with the highest yield. Screening designs are developed to minimize the required resources. Within this broad purpose, four methodological areas stand out: disease screening, drug screening, rank screening, and group screening.

DISEASE SCREENING

Disease screening identifies those individuals who should undergo a costly diagnosis based upon a less costly predictive test result.

Let Y be a binomial variable representing the presence or absence of disease and let X denote a continuous test variable positively correlated with Y. An individual undergoes the diagnosis if $X \geqslant c$; otherwise not. The choice of c trades off the probability of a false positive ($P[Y = 0, X \geqslant c]$) against the probability of a false negative ($P[Y = 1, X < c]$). Discriminant analysis* can be used to determine a cutoff point in the multivariate case.

Recent research treats an individual as a sequence of states. In Albert et al., the individual spends time in a disease-free state, a curable state, and an incurable state when the individual seeks therapy [1]. The last two states are detected by the screen. The model includes age strata, migration rates, and false negative rates. When a policy of screening all women over 30 was compared to an "adaptive" policy, the adaptive policy performed as well or better while screening $\frac{1}{3}$ as many cases.

DRUG SCREENING

Drug screening selects compounds that are active in a test system. Screening designs are developed when more than one experimental unit* per test result is required.

A procedure in ref. 14 is illustrative. Let X_1 denote the average response at the first stage. Reject the compound if $X_1 < c_1$; otherwise go to stage 2. Let X_2 denote the mean response at stage 2. Reject the compound if $X_1 + X_2 < c_2$; otherwise declare the compound active. By properly selecting c_1 and c_2 and sample sizes at both stages, one can select an efficient design with the desired proportion of false positives and false negatives. A three-stage group sequential* procedure involving a binomial variable accomplished with 16,000 animals what would have required 40,000 animals to accomplish in a one-stage design [16].

Before a compound is studied clinically, it passes through a battery of test systems. The problem of integrating such tests has been approached from a decision theoretic viewpoint [13] and a simulation viewpoint [9].

RANK SCREENING

Rank screening selects the best or the k best members of a finite set of entities. Again, screening designs are developed when more than one experimental unit per test result is required.

Let $X_1, \ldots, X_k$ be normally distributed random variables with common known variance σ^2 and with means $\mu_1, \ldots, \mu_k$. Let $X_{1:k}, \ldots, X_{k:k}$ denote the corresponding order statistics*. Suppose we wish to determine s such that $\mu_s \geqslant \mu_j$ for $j = 1, \ldots, k$. Let $\delta = \mu_{k:k} - \mu_{k-1:k}$. We seek a subset S of $\{1, \ldots, k\}$ such that $P[s \in S] \geqslant P^*$ whenever $\delta \geqslant \delta^*$ for some specified δ^*. We call P^* the *confidence level.* (This should not be confused with the confidence coefficient of a confidence interval or region*.)

One procedure defines a random set S such that $i \in S$ whenever $X_i \geqslant X_{k:k} - d\sigma/n^{1/2}$ where d is a tabled value depending on δ^* and P^*. Another simply sets S as the index of the largest sample mean and computes P^* for specified δ^*. If δ^* is prespecified, one can determine the sample size required to assure a specified P^*.

If k is large, a multistage procedure might be used in which a sequence $S_1, \ldots, S_m$ of decreasing nested subsets is formed. Cochran [4] suggests that S_i be about $\frac{1}{5}$ the size of S_{i-1}.

GROUP SCREENING

Group screening identifies those entities with a specified property when the experimental unit is a group of entities.

Let $X_1, \ldots, X_k$ be binomial random variables. Let S denote a subset of $S_0 = \{1, \ldots, k\}$. Define $X_S = \max[X_i | i \in S]$. Group screening procedures are based on the fact that $X_S = 0$ implies that $X_i = 0$ for all $i \in S$. For example, if the pooled contents of a number of test tubes are not contaminated, then neither are the individual contents.

The Dorfman procedure [5] partitions S_0 into approximately equal subsets. If $X_S = 1$ for any of these subsets, then X_i is evaluated for all $i \in S$. Assume $P[X_i = 1] = p$. As $p \to 0$ and $k \to \infty$, the optimal size for S and the expected number of tests per entity go to $1/p^{1/2}$ and $2p^{1/2}$, respectively [15]. Allowances have been made for p varying with i [11], errors in determining X_S [12], and nonbinomial responses [2].

General binary tree procedures [17] partition S_0 into two nonempty subsets S' and S''. If $X_{S'} = 0$, then S' is not partitioned further; otherwise S' is partitioned further into two nonempty subsets each of which is tested. Proceeding iteratively in this manner, a binary tree* is generated whose terminal vertices represent either a set S, for which $X_S = 0$, or an integer i, for which $X_i = 1$.

References

[1] Albert, A., Gertman, P. M., Louis, T. A., and Liu, S. -I. (1978). *Math. Biosci.*, **40**, 61–109. (Good blend of mathematics and science.)

[2] Arnold, S. (1977). *Ann. Statist.*, **5**, 1170–1182.

[3] Bechhofer, R. E., Kiefer, J., and Sobel, M. (1968). *Sequential Identification and Ranking Procedures.* Univ. of Chicago Press, Chicago, IL. (Theoretical.)

[4] Cochran, W. G. (1951). *Proc. Second Berkeley Symp. Math. Statist. Prob.*, Vol. 2. University of California Press, Berkeley, CA, pp. 449–470.

[5] Dorfman, R. (1943). *Ann. Math. Statist.*, **14**, 436–440.

[6] Gibbons, J. D., Olkin, I., and Sobel, M. (1977). *Selecting and Ordering Populations: A New Statistical Methodology.* Wiley, New York. (For the practitioner.)

[7] Gibbons, J. D., Olkin, I., and Sobel, M. (1979), An introduction to ranking and selection. *Amer. Statist.*, **33**, 185–195. (Good introduction and bibliography.)

[8] Goldberg, J. D. and Wittes, J. T. (1981). The evaluation of medical screening procedures. *Amer. Statist.*, **35**, 4–11. (Good, short review.)

[9] Gordon, M. H. (1974). *Cancer Chemother. Rep. Part 3*, **5**, 65–78.

[10] Gupta, S. S. and Panchapakesan, S. (1979). *Multiple Decision Procedures: Theory and Methodology of Selecting and Ranking Populations.* Wiley, New York. (Theoretical.)

[11] Hwang, F. K. (1975). *J. Amer. Statist. Ass.*, **70**, 923–926.

[12] Kotz, S. and Johnson, N. (1982). *Commun. Statist. A*, **11**, 1997–2016.

[13] Marshall, A. W. and Olkin, I. (1968). *J. R. Statist. Soc. B.* 30, 407–435. Discussion 435–443. (A general formulation of drug screening.)

[14] Roseberry, T. D. and Gehan, E. A. (1964). *Biometrics*, **20**, 73–84.

[15] Samuel, S. M. (1978). *Technometrics*, **20**, 497–500. (Good introduction to the Dorfman procedure.)

[16] Schultz, J. R., Nichol, F. R., Elfring, G. L., and Weed, S. D. (1973). *Biometrics*, **29**, 293–300.

[17] Sobel, M. and Groll, P. A. (1959). Group testing to eliminate efficiently all defectives in a binomial sample. *Bell Syst. Tech. J.*, **38**, 1179–1252.

(MEDICAL DIAGNOSIS, STATISTICS IN PHARMACEUTICAL INDUSTRY, STATISTICS IN QUALITY CONTROL, STATISTICAL SEQUENTIAL ANALYSIS)

MARK JOHNSON

SEARCH THEORY

Search theory came into being during World War II with the work of B. O. Koopman and his colleagues in the Antisubmarine Warfare Operations Research Group (ASWORG). ASWORG was directed by P. M. Morse and reported to Admiral Ernest King, Chief of Naval Operations and Commander in Chief, U.S. Fleet. Inspired by Morse, many of the fundamental concepts of search theory such as sweep width and sweep rate had been established by spring of 1941. Since that time, search theory has grown to be a major discipline within the field of operations research*. Its applications range from deep-ocean search for submerged objects to deep-space surveillance for artificial satellites.

The reader interested in the origins of search theory should consult Morse [29] and Koopman [24]. Early use of it to develop naval tactics is discussed in Morse and Kimball [30]; excerpts from this work appear in ref. 31.

For a modern account, the reader should consult Stone [41]. This book provides a rigorous development of the theory and includes an excellent bibliography and notes on previous research. Washburn's monograph [52] is also recommended.

Since World War II, the principles of search theory have been applied successfully in numerous important operations. These include the 1966 search for a lost H-bomb in the Mediterranean near Palomares, Spain, the 1968 search for the lost nuclear submarine Scorpion near the Azores [39], and the 1974 underwater search for unexploded ordnance during clearance of the Suez Canal. The U.S. Coast Guard employs search theory in its open ocean search and rescue planning [37]. Search theory is also used in astronomy [47] and in radar search for satellites [34]. Numerous additional applications, including those to industry, medicine, and mineral exploration are discussed in the proceedings [15] of the 1979 NATO Advanced Research Institute on Search Theory and Applications. Applications to biology are given in refs. 17 and 18, and an application to machine maintenance and inspection is described in ref. 32.

Further references appear in the first section, followed in the second section by an illustration of how search theory can be used to solve an optimal search problem.

REVIEW OF LITERATURE

Work in search theory can be classified, at least in part, according to the assumptions made about measures of effectiveness, target motion, and the way in which search effort is characterized. This review is organized according to these criteria.

Measures of Effectiveness

Among the many such measures used in search analysis, the most common are

- probability of detection,
- expected time to detection,
- probability of correctly estimating target "whereabouts,"
- entropy of the posterior target location probability distribution.

Usually the objective of an optimal search is to maximize probability of detection with some constraint imposed on the amount of search effort available. For a stationary target (Stone [41]), when the detection function is concave or the search space and search effort are continuous, a plan that maximizes probability of detection in each of successive increments of search effort (incrementally optimal) will also be optimal for the total effort contained in the increments (totally optimal).

Moreover, for stationary targets, it is often theoretically possible to construct a *uniformly optimal search plan*, i.e., for which probability of detection is maximized at each moment during its period of application. If a uniformly optimal search plan exists, then it will (a) maximize probability of detection over any period of application (i.e., be totally and incrementally optimal) and (b) minimize the expected time to detection. An example (originally due to Koopman) is given in the last section.

In a *whereabouts search*, the objective is to correctly estimate the target's location in a collection of cells given a constraint on search cost. The searcher may succeed either by finding the target during search or by correctly guessing the target's location after search. These searches were first studied systematically by Kadane (see ref. 21). In many cases of interest, Kadane shows that the optimal whereabouts search consists of an optimal detection search among all cells *exclusive* of the cell with the highest prior target location probability. If the search fails to find the target, then one guesses that it is in the excluded highest probability cell.

More recently, Kadane and Stone [22] have considered whereabouts search in the context of moving targets. The optimal whereabouts search plan may be found by solving a finite number of optimal detection search problems, one for each cell in the grid.

Consideration of entropy* as a measure of effectiveness is useful in certain situations and can be used to draw a distinction between search and surveillance. For certain stationary target detection search problems with an exponential detection function, Barker [5] has shown that the search plan that maximizes the entropy of the posterior target location probability distribution conditioned upon search failure is the same as the search plan that maximizes probability of detection.

In a *surveillance search*, the objectives are usually more complex than in a detection search. For example, one may wish to correctly estimate target location at the end of a period of search in order to take some further action. In this case, detection before the end of the period can contribute to success but does not in itself constitute success. More general problems of this type are discussed by Tierney and Kadane [49], who obtain necessary conditions for optimality when target motion is Markovian. In surveillance problems involving moving targets and false contacts where the time of terminal action may not be known in advance, Richardson [38] suggests allocating search effort to minimize expected entropy (maximize information).

Among other measures of effectiveness used in search, those based upon minimax* criteria are of particular interest. Corwin [8] considers search as a statistical game and seeks estimates for target location. Alpern [3], Gal [14], and Isaacs [19] consider games in which minimax strategies are sought for a moving target seeking to avoid a moving searcher.

Target Motion

Assumptions about target motion have a considerable influence on the characteristics of search plans and the difficulty of computation. Until recently all but the simplest search problems involving target motion were intractable from the point of view of mathematical optimization. Results were usually obtained by considering transformations that would convert the problem into an equivalent stationary target problem (e.g., see Stone and Richardson [45], Stone [42], and Pursiheimo [35]). Representative early work on search with Markovian target motion is given in Pollock [33], Dobbie [12], and McCabe [28]. Hellman [16] investigates the effect of search upon targets whose motion is a diffusion process*.

The first computationally practical solution to the optimal search problem for stochastic target motion involving a large number of cells and time periods is due to Brown [6]. For exponential detection functions, he found necessary and sufficient conditions for discrete time and space search plans, and provided an iterative method for optimizing search for targets whose motion is described by mixtures of discrete time and space Markov chains. Washburn [51] extended Brown's necessary conditions to the case of discrete search effort. Washburn [53] also provides a useful bound on how close a plan is to the optimal plan.

Very general treatments of moving target search are provided by Stone [43] and by Stromquist and Stone [46], allowing efficient numerical solution in a wide class of practical moving target problems; these include, for example, non-Markovian motion and nonexponential detection functions.

The existence of optimal search plans for moving targets is not to be taken for granted. L. K. Arnold has shown that there are cases where no allocation function satisfies the necessary conditions given in ref. 46. In his examples, there appear to be optimal plans, but they concentrate effort on sets of measure zero and are outside the class of search allocation functions usually considered. He also shows the existence of optimal plans whenever the search density is constrained to be bounded.

Search Effort

Search effort may be either discrete (looks, scans, etc.) or continuous (time, track length, etc.).

In problems involving discrete search effort, the target is usually considered to be located in one of several cells or boxes. The search consists of specifying a sequence of looks in the cells. Each cell has a prior probability of containing the target. A detection function b is specified, where $b(j, k)$ is the conditional probability of detecting the target on or before the kth look in cell j, given that the target is located in cell j. A cost function c is also specified, where $c(j, k)$ is the cost of performing k looks in cell j. An early solution to this problem for independent glimpses and uniform cost is given by Chew [7]. In this case,

$$b(j, k) - b(j, k-1) = \alpha_j(1 - \alpha_j)^{k-1}$$

for all j and for $k > 0$;

$$c(j, k) = k \text{ for all } j \text{ and } k \geqq 0.$$

Additional important results have been obtained by Matula [27], Blackwell (see Matula [27]), Kadane [20], and Wegener ([54, 55, 56]).

Kadane's result [20] is particularly interesting since he uses a variant of the Neyman–Pearson lemma* to obtain an optimal plan for the general case where $b(j, k) - b(j, k-1)$ is a decreasing function of k for all j.

In problems involving continuous effort, the target may be located in Euclidean n-space or in cells, as in the case of discrete search. In the former case it is assumed that the search effort is "infinitely divisible" in the sense that it may be allocated as finely as necessary over the entire search space. The search problem was originally expressed in this form by Koopman (see ref. 23). The continuous effort case will be considered in greater detail in the remainder of this article.

Just as with discrete search effort, there is a detection function b, where $b(x, z)$ is the probability of detecting the target with z amount of effort applied to the point x, given the target located at x. When x is a cell index, z represents the amount of time or track length allocated to the cell. When x is a point in Euclidean n-space, then z is a density, as will be made clear in the next section.

Koopman's original solution [23] to the search problem made use of an exponential function for b of the form $b(x, z) = 1 - \exp(-\kappa z)$, where κ is a positive constant that may depend upon x. DeGuenin [9] considered a more general class of detection functions now referred to as *regular* (see ref. 41). Dobbie [11] considered sequential search with a concave detection function. Richardson and Belkin [36] have treated a special type of regular detection function obtained when the parameter κ in the exponential effectiveness function is a random variable. Such functions occur when sensor capabilities are uncertain. Tatsuno and Kisi [48] address similar problems. Stone has considered very general detection functions in ref. 41.

For differentiable detection functions, Wagner [50] obtained sufficient conditions for an optimal search plan with continuous effort using a nonlinear functional version of the Neyman–Pearson lemma.

Remarks

Search theory remains a field of active research in spite of the considerable advances made since its inception 40 years ago. A review of the current status of search theory in terms of practical applications is given in ref. 44. Many problems remain to be solved, particularly in cases involving multiple targets and false targets. Also systematic methods are needed for constructing the prior target location probability distribution from sometimes conflicting subjective opinion. More work is also needed on problems where it is essential to take exact account of search tract continuity or the switching cost of moving from one region to another. These problems remain intractable although some recent progress has been made (see, e.g., refs. 54 and 25).

Brown's innovative solution to an important class of moving target search problems has removed an impediment to progress in this area. An interesting extension of Brown's results occurs in the context of search and rescue operations carried out at sea. Here the target's state may be unknown. That is, the target may be the survivor of a mishap and may be aboard the distress vessel, afloat in a life raft, or alone in the water supported only by a life jacket. The detectability of the target varies with these alternatives as does the drift motion and the expected time of survival of the target. A solution to this important problem is given in ref. 10 under very general assumptions.

Unfortunately, Brown's approach applies only to the case where search effort is continuous. Efficient algorithms are still required in the case where search effort is discrete.

AN OPTIMAL SEARCH PROBLEM

This section shows how optimal search theory can be applied to an important class of problems where the prior (i.e., before search) probability distribution for target location is normal and the detection function b (see preceding section) is exponential. Many search problems in practice are of this form.

Definitions. Let $\mathbf{X}$ denote the plane, and A be some arbitrary region of interest. Then the prior probability that the target's location $\tilde{\mathbf{x}}$ is in A is given by

$$\Pr[\tilde{\mathbf{x}} \in A] = \int_A p(\mathbf{x})\, d\mathbf{x},$$

where

$$p(\mathbf{x}) = \frac{1}{2\pi\sigma^2} \exp\{-|\mathbf{x}|^2/(2\sigma^2)\} \quad (1)$$

is the circular normal density function, with mean at the origin and variance σ^2 in both coordinate directions. The distance of $\mathbf{x}$ from the origin is denoted by $|\mathbf{x}|$.

Let F be the class of all nonnegative functions defined on $\mathbf{X}$ with finite integral. By definition, this is the class of search allocation functions, and for $f \in F$,

$$\int_A f(\mathbf{x})\, d\mathbf{x}$$

is the amount of search effort placed in region A.

We assume that the unit of search effort is "time" and thus the "cost" C associated with the search is the total amount of time consumed. In this case $c(\mathbf{x}, z) = z$, and the cost functional C is defined by

$$C[f] = \int_{\mathbf{X}} c(\mathbf{x}, f(\mathbf{x}))\, d\mathbf{x} = \int_{\mathbf{X}} f(\mathbf{x})\, d\mathbf{x}.$$

The measure of effectiveness will be probability of detection. Hence, the "effectiveness functional" D is assumed to have the form

$$D[f] = \int_{\mathbf{X}} b(\mathbf{x}, f(\mathbf{x}))\, p(\mathbf{x})\, d\mathbf{x}, \quad (2)$$

where b is the detection function. As mentioned earlier, the detection function is assumed to be exponential, and hence, for $R > 0$,

$$b(\mathbf{x}, z) = 1 - \exp(-Rz). \quad (3)$$

The coefficient R is the *sweep rate* and measures the rate at which search is carried out (see, e.g., ref. 24).

In order to understand (2), suppose for the moment that search allocation function f is constant over A, zero outside of A, and corresponds to a finite amount of search time T. Then for $\mathbf{x} \in A$,

$$f(\mathbf{x}) = T/\text{area}(A),$$

since by definition

$$T = \int_{\mathbf{X}} f(\mathbf{x})\,d\mathbf{x}$$

$$= f(\mathbf{x}_0)\int_A d\mathbf{x} = f(\mathbf{x}_0)\ \text{area}(A)$$

for any $\mathbf{x}_0 \in A$.

The detection functional can then be written

$$D[f] = \int_{\mathbf{X}} b(\mathbf{x}, f(\mathbf{x}))p(\mathbf{x})\,d\mathbf{x}$$

$$= \int_A \{1 - \exp(-RT/\text{area}(A))\}p(\mathbf{x})\,d\mathbf{x}$$

$$= \{1 - \exp(-RT/\text{area}(A))\}\int_A p(\mathbf{x})\,d\mathbf{x}$$

$$= \{1 - \exp(-RT/\text{area}(A))\}\Pr[\tilde{\mathbf{x}} \in A].$$

This is the so-called *random-search formula*.

Sufficient Conditions for Optimal Search Plans

Using Lagrange multipliers in the fashion introduced by Everett [13], one can find sufficient conditions for optimal search plans for stationary targets that provide efficient methods for computing these plans. Define the pointwise Lagrangian l as follows:

$$l(\mathbf{x}, z, \lambda) = p(\mathbf{x})b(\mathbf{x}, z) - \lambda c(\mathbf{x}, z) \quad \text{for } \mathbf{x} \in \mathbf{X},\ z \geqq 0,\ \lambda \geqq 0. \tag{4}$$

If we have an allocation f_λ^* that maximizes the pointwise Lagrangian for some value of $\lambda \geqq 0$, i.e.,

$$l(\mathbf{x}, f_\lambda^*(\mathbf{x}), \lambda) \geqq l(\mathbf{x}, z, \lambda) \quad \text{for all } \mathbf{x} \in \mathbf{X} \text{ and } z \geqq 0, \tag{5}$$

then $D[f_\lambda^*]$ is optimal for its cost $C[f_\lambda^*]$, i.e.,

$$D[f_\lambda^*] \geqq D[f] \quad \text{for any } f \in F \text{ such that } C[f] \leqq C[f_\lambda^*]. \tag{6}$$

Equation (6) says that plan f_λ^* maximizes the detection probability over all plans using effort $C[f_\lambda^*]$ or less. To show that (6) is true, we suppose $f \in F$ and $C[f] \leqq C[f_\lambda^*]$. Since $f(\mathbf{x}) \geqq 0$, (5) yields

$$p(\mathbf{x})b(\mathbf{x}, f_\lambda^*(\mathbf{x})) - \lambda c(\mathbf{x}, f_\lambda^*(\mathbf{x})) \geqq p(\mathbf{x})b(\mathbf{x}, f(\mathbf{x})) - \lambda c(\mathbf{x}, f(\mathbf{x})) \quad \text{for } \mathbf{x} \in \mathbf{X}. \tag{7}$$

Integrating both sides of (7) over $\mathbf{X}$, we obtain

$$D[f_\lambda^*] - \lambda C[f_\lambda^*] \geqq D[f] - \lambda C[f],$$

which, along with $\lambda \geqq 0$ and $C[f] \leqq C[f_\lambda^*]$, implies

$$D[f_\lambda^*] - D[f] \geqq \lambda(C[f_\lambda^*] - C[f]) \geqq 0.$$

This proves (6) and hence that f_λ^* is an optimal search plan for its cost $C[f_\lambda^*]$.

One way to calculate optimal search plans is to choose a $\lambda > 0$ and find f_λ^* to maximize the pointwise Lagrangian for this λ. For each $\mathbf{x} \in \mathbf{X}$, finding $f_\lambda(\mathbf{x})$ is a one-dimensional optimization* problem. If the detection function is well behaved (e.g., exponential), then one can solve for f_λ^* analytically. Since $C[f_\lambda^*]$ is usually a decreasing function of λ, one can use a computer to perform a binary search to find the value of λ that yields cost $C[f_\lambda^*]$ equal to the amount of search effort available. The resulting f_λ^* is the optimal search plan.

Example. In the case of a bivariate normal target distribution (1) and exponential detection function (3), we can compute $C[f_\lambda^*]$ and the optimal plan explicitly (see ref. 23). The result is that for T amount of search time, the optimal allocation function f^* (dropping the subscript λ that depends on T) is given by

$$f^*(\mathbf{x}) = \begin{cases} \dfrac{1}{2\sigma^2}\left(r_0^2 - |\mathbf{x}|^2\right), & |\mathbf{x}| \leqq r_0, \\ 0, & \text{otherwise,} \end{cases}$$

where all search is confined to a disk of radius r_0 defined by

$$r_0^2 = 2\sigma\sqrt{RT/\pi}.$$

The probability of detection corresponding to f^* is

$$D[f^*] = 1 - \left(1 + \frac{r_0^2}{2\sigma^2}\right)\exp\left\{\frac{-r_0^2}{2\sigma^2}\right\}$$

and the expected time to detection $\bar{\tau}$ is

$$\bar{\tau} = 6\pi\sigma^2/R.$$

Thus, for a given amount of search effort T, the optimal search plan concentrates search in the disk of radius r_0, which then expands as more effort becomes available. It can be shown that the optimal search plan can be approximated by a succession of expanding and overlapping coverages. The fact that search is repeated in the high probability areas is typical of optimal search in a great many situations.

Note that probability of detection depends upon the ratio r_0/σ and increases to 1 as search effort increases without bound. The expected time to detection is finite and varies directly with σ^2 and inversely with the sweep rate R.

References

[1] Ahlswede, R. and Wegener, I. (1979). *Suchprobleme*. Teubner, Leipzig, Germany.

[2] Allais, I. M. (1957). *Manag. Sci.*, **3**, 285–347.

[3] Alpern, S. (1974). In *Differential Games and Control Theory*. Dekker, New York, pp. 181–200.

[4] Arkin, V. I. (1964). *Theory Prob. Appl.*, **9**, 674–680.

[5] Barker, W. H., II (1977). *Operat. Res.*, **25**, 304–314.

[6] Brown, S. S. (1980). *Operat. Res.*, **28**, 1275–1289.

[7] Chew, M. C. (1967). *Ann. Math. Statist.*, **38**, 494–502.

[8] Corwin, T. L. (1981). *J. Appl. Prob.*, **18**, 167–180.

[9] DeGuenin, J. (1961). *Operat. Res.*, **9**, 1–7.

[10] Discenza, J. H. and Stone, L. D. (1980). *Operat. Res.*, **29**, 309–323.

[11] Dobbie, J. M. (1963). *Naval Res. Logist. Quart.*, **4**, 323–334.

[12] Dobbie, J. M. (1974). *Operat. Res.*, **22**, 79–92.

[13] Everett, H. (1963). *Operat. Res.*, **11**, 399–417.

[14] Gal, S. (1972). *Israel J. Math.*, **12**, 32–45.

[15] Haley, K. B. and Stone, L. D. (1980). *Search Theory and Applications*. Plenum, New York.

[16] Hellman, O. B. (1970). *Ann. Math. Statist.*, **41**, 1717–1724.

[17] Hoffman, G. (1983). *Behav. Ecol. Sociobiol.*, **13**, 81–92.

[18] Hoffman, G. (1983). *Behav. Ecol. Sociobiol.*, **13**, 93–106.

[19] Isaacs, R. (1965). *Differential Games*. Wiley, New York.

[20] Kadane, J. B. (1968). *J. Math. Anal. Appl.*, **22**, 156–171.

[21] Kadane, J. B. (1971). *Operat. Res.*, **19**, 894–904.

[22] Kadane, J. B. and Stone, L. D. (1981). *Operat. Res.*, **29**, 1154–1166.

[23] Koopman, B. O. (1957). *Operat. Res.*, **5**, 613–626.

[24] Koopman, B. O. (1980). *Search and Screening, General Principles and Historical Applications*. Pergamon, New York.

[25] Lössner, U. and Wegener, I. (1982). *Operat. Res.*, **7**, 426–439.

[26] Mangel, M. (1982). *Operat. Res.*, **30**, 216–222.

[27] Matula, D. (1964). *Amer. Math. Monthly*, **71**, 15–21.

[28] McCabe, B. J. (1974). *J. Appl. Prob.*, **11**, 86–93.

[29] Morse, P. M. (1977). *In at the Beginnings: A Physicist's Life*. MIT, Cambridge, MA.

[30] Morse, P. M. and Kimball, G. E. (1946). *Operations Evaluation Group Report 54*, Center for Naval Analysis, Alexandria, VA.

[31] Newman, J. R. (1956). *The World of Mathematics*. Simon and Schuster, New York.

[32] Ozan, T. M., Chang-Tong Ng, and Saatcioglu, O. (1976). *Int. J. Prod. Res.*, **14**, 85–98.

[33] Pollock, S. M. (1970). *Operat. Res.*, **18**, 883–903.

[34] Posner, E. C. (1963). *IEEE Trans. Inf. Theory*, **IT-9**, 157–168.

[35] Pursiheimo, U. (1977). *SIAM J. Appl. Math.*, **32**, 105–114.

[36] Richardson, H. R. and Belkin, B. (1972). *Operat. Res.*, **20**, 764–784.

[37] Richardson, H. R. and Discenza, J. H. (1980). *Naval Res. Logist. Quart.*, **27**, 659–680.

[38] Richardson, H. R. and Wagner, D. H. (1973). ASW Information Processing and Optimal Surveillance in a False Target Environment. *Assoc. Rep. AD A002254*, Office of Naval Research.

[39] Richardson, H. R. and Stone, L. D. (1970). *Naval Res. Logist. Quart.*, **18**, 141–157.

[40] Stewart, T. J. (1979). *Comp. Operat. Res.*, **6**, 129–140.

[41] Stone, L. D. (1975). *Theory of Optimal Search*. Academic, New York.

[42] Stone, L. D. (1977). *SIAM J. Appl. Math.*, **33**, 456–468.

[43] Stone, L. D. (1979). *Math. Operat. Res.*, **4**, 431–440.

[44] Stone, L. D. (1982). *Operat. Res.*, **31**, 207–233.

[45] Stone, L. D. and Richardson, H. R. (1974). *SIAM J. Appl. Math.*, **27**, 239–255.

[46] Stromquist, W. R. and Stone, L. D. (1981). *Math. Operat. Res.*, **6**, 518–529.

[47] Taff, L. G. (1984). *Icarus*, **57**, 259–266.

[48] Tatsuno, K., and Kisi, T. (1982). *J. Operat. Res. Soc.*, **25**, 363–375.

[49] Tierney, L. and Kadane, J. B. (1983). *Operat. Res.*, **31**, 720–738.

[50] Wagner, D. H. (1969). *SIAM Rev.*, **11**, 52–65.

[51] Washburn, A. R. (1980). *Naval Res. Logist. Quart.*, **27**, 315–322.

[52] Washburn, A. R. (1981). *Search and Detection*. Military Applications Section, ORSA, Arlington, VA.

[53] Washburn, A. R. (1981). *Operat. Res.*, **29**, 1227–1230.

[54] Wegener, I. (1980). *Math. Operat. Res.*, **5**, 373–380.

[55] Wegener, I. (1981). *Naval Res. Logist. Quart.*, **28**, 533–543.

[56] Wegener, I. (1982). *Naval Res. Logist. Quart.*, **29**, 203–212.

(GAME THEORY
MILITARY STATISTICS
OPERATIONS RESEARCH
TARGET COVERAGE)

HENRY R. RICHARDSON

SEASONALITY

CAUSES AND CHARACTERISTICS OF SEASONALITY

A great deal of information on socioeconomic activity occurs in the form of time series* where observations are dependent and the nature of this dependence is of interest in itself. A time series is a sequence of observations ordered in time, say $X_1, X_2, \ldots, X_t, \ldots, X_T$, the interval between dates t and $t+1$ being fixed and constant throughout. The observations are generally compiled for consecutive periods, such as days, weeks, months, quarters, or years.

The analysis of time series has long distinguish different types of evolution, which may possibly be combined, namely (a) trend, (b) cycle, (c) seasonal variations, and (d) irregular fluctuations.

*Trend** is a slow variation over a long period of years, generally associated with the structural causes of the phenomenon in question. In some cases, the trend shows a steady growth; in others, it may move downward as well as upward.

Cycle is a quasiperiodic oscillation characterized by alternating periods of expansion and contraction. In most cases it is related to fluctuations in economic activity.

Seasonal variations represent the effect of climatic and institutional events that repeat more or less regularly each year.

Irregular fluctuations represent unforeseeable movements related to events of all kinds. They have a stable random appearance but, in some cases, extreme values may be present. These extreme values, or outliers*, have identifiable causes, such as strikes or floods, and, therefore, can be distinguished from the much smaller irregular variations.

The decomposition of an observed time series in several evolutionary processes is basic for time-series analysis, the study of the cycle and economic growth, and the study of seasonality. The feasibility of time-series decomposition was proved by Wold's famous theorem [29], which states that, if a time series X_t is stationary to the second order (i.e., its mean does not depend on time and its autocovariance function depends only on the time lag), then it can be uniquely represented as the sum of two mutually uncorrelated processes, one an infinite moving average n_t, and the other a deterministic process ξ_t, where the future evolution of the realization can be determined completely provided that all its previous values are known. Hence

$$X_t = \eta_t + \xi_t, \qquad (1)$$

where

$$\eta_t = \sum_{j=0}^{\infty} \beta_j U_{t-j}, \qquad \sum \beta_j^2 < \infty \qquad (2)$$

$$E(U_t U_s) = \begin{cases} \alpha^2 & \text{if } t = s, \\ 0 & \text{if } t \neq s, \end{cases}$$

$$\xi_t = \sum_{j=1}^{\infty} \alpha_j \xi_{t-j}, \qquad (3)$$

$$E(\xi_t \eta_s) = 0 \quad \text{for all } t, s.$$

There are several decomposition models that have been assumed for time-series analysis. Those more often applied are additive and multiplicative. That is,

$$X_t = C_t + S_t + U_t \quad \text{(additive model)}, \quad (4)$$

$$X_t = C_t S_t U_t \quad \text{(multiplicative model)}, \quad (5)$$

where C_t is the trend–cycle, S_t is the seasonal component, and U_t is the irregular component [in model (5), S_t and U_t are expressed as percentages]. The choice of decomposition model for a given series depends on whether the components are assumed to be independent (additive model) or dependent (multiplicative model) on one another. In a few cases, however, a mixed decomposition model, in which components are multiplicatively and additively related, may be more adequate as, for example, the one discussed in ref. 12:

$$X_t = C_t(1 + S_t) + U_t. \quad (6)$$

Among the several types of fluctuations, those due to seasonality have long been recognized. The organization of society, means of production and communication as well as social and religious events have been strongly conditioned by both climatic and conventional seasons. Seasonal variations in agriculture, the low level of winter construction, and high pre-Christmas retail sales are all well known.

It is generally accepted that there are four causes of seasonality in economic time series: the calendar, timing decisions, weather, and expectations. While the causes of seasonality are generally exogenous to the economic system, human intervention can modify their extent and nature. For example, seasonal variations in the automobile industry are affected by manufacturers' decisions regarding the extent of model changeover each year.

Another main feature of seasonality is that the phenomenon repeats with certain regularity every year, but it may evolve. Many reasons can produce changes in seasonal patterns. A decline in the importance of the primary sector in the gross national product modifies seasonal patterns in the economy as a whole, as does a change in the geographical distribution of industry in a country extending over several climatic zones. Changes in technology alter the importance of climatic factors. For most economic series, an evolving seasonality is more the rule than the exception. The assumption of stable seasonality, that is, of seasonality that repeats exactly every year, is good for a few series only. Depending on the causes of seasonal variations, their patterns can change slowly or rapidly, in a deterministic manner or in a stochastic manner.

SEASONAL MODELS

The simplest and often studied seasonal model assumes that the generating process of seasonality can be represented by strictly periodic functions of period n (e.g., 12 for monthly data, 4 for quarterly data).

For monthly series, the problem is to estimate 12 constants, one for each month, that sum to zero. That is,

$$S_t = \begin{cases} a_k & \text{for } t = k \text{ or } t - k \\ & \text{divisible by 12,} \\ 0 & \text{otherwise.} \end{cases}$$

$$\sum_{k=1}^{12} a_k = 0, \quad (7)$$

Model (7) is more often shown under its equivalent form in the frequency domain as

$$S_t = \sum_{j=1}^{6} \left(\alpha_j \cos 2\pi\lambda_j t + \beta_j \sin 2\pi\lambda_j t \right),$$

$$\lambda_j = j/12, \quad (8)$$

where the λ_j for $j = 1, 2, 3, \ldots, 6$ are the seasonal frequencies that correspond to cyclical periods of 12, 6, 4, 3, 2.4, and 2 months, respectively.

The seasonal model (8) is *deterministic* if α_j and β_j are constants, and *stochastic* if α_j

and β_j are random variables with zero mean and

$$E(\alpha_j\alpha_k) = E(\beta_j\beta_k) = \begin{cases} \sigma_j^2 & \text{if } j = k, \\ 0 & \text{if } j \neq k, \end{cases} \quad (9)$$

$$E(\alpha_j\beta_k) = 0 \quad \text{for all } j \text{ and } k.$$

Series that follow model (8) have spectra that are zero except for six spectral lines of height σ_j^2 at each λ_j. All the spectral power is concentrated at the seasonal frequencies. Although this never occurs with real data, series with very narrow seasonal peaks can be well approximated by this kind of model.

For most economic time series, however, seasonality is not stable, but changes gradually. In such cases, a more general seasonal model may be expressed as

$$S_t = \sum_{j=1}^{6} \left(\alpha_{jt}\cos 2\pi\lambda_j t + \beta_{jt}\sin 2\pi\lambda_j t\right),$$

$$\lambda_j = j/12, \quad (10)$$

where α_{jt} and β_{jt} are time-varying parameters. They can either be low degree polynomials or stochastic processes* with spectra dominated by low frequency components. For example, if the seasonal amplitudes follow a stationary autoregressive process of order 1, then

$$\alpha_{jt} = \rho\alpha_{jt-1} + U_t,$$
$$\beta_{jt} = \rho\beta_{jt-1} + V_t, \quad (11)$$

where $|\rho| < 1$ and U_t, V_t are purely random mutually uncorrelated variables with zero mean and constant variance.

If a series follows the seasonal model (10), where α_{jt} and β_{jt} are assumed to be stationary stochastic processes*, its spectrum will be characterized by peaks at the seasonal frequency bands $\lambda_s(\delta)$ of width δ, and where the value of δ depends on the rate at which α_{jt} and β_{jt} change.

The broader the bands, the less regular the seasonal variations. The set of seasonal frequency bands may be defined by

$$\lambda_s(\delta) = \{\lambda_s | \lambda_s \in (\lambda_j - \delta, \lambda_j + \delta),$$
$$j = 1, 2, 3, 4, 5, 6;$$
$$\text{where } \lambda_6 + \delta = \pi\}. \quad (12)$$

SEASONAL ADJUSTMENT METHODS

The fact that the causes of seasonality are generally exogenous to the economic system is the main reason for the removal of seasonal variations from an observed series to produce a *seasonally adjusted series*. The adjusted series thus reflects only variations attributed to the trend, the cycle, and the irregulars. The removal of seasonality from a time series, however, does not indicate how the series would have evolved had there been no seasonal variations; rather, it shows more clearly the trend–cycle abstracted from seasonality.

The estimation of seasonal variations has always posed a serious problem to statisticians because the phenomenon varies and is not directly observable.

The majority of the seasonal adjustment methods developed thus far are based on univariate time-series models where the estimation of the seasonal variations is made in a simple and automatic manner and is not based on a causal explanation of the phenomenon in question. A survey of recent developments in seasonal adjustments is given in ref. 23.

The majority of the methods for the seasonal adjustment of economic time series fall into two broad categories. The first derives from general regression and linear estimation theory; the second depends mainly on the application of moving averages* or linear smoothing filters. A very few exceptions do not fall into this broad classification, including probably the best known, the SABL (seasonal adjustment—Bell Laboratories) method [6], which uses a combination of moving means and medians.

Regression Methods

Most work on regression methods for seasonal adjustment is based on the assumption that the systematic part of a time series can be approximated closely by simple functions of time over the *entire* span of the series. In general, two types of functions of time are considered. One is a polynomial of fairly low degree that fulfills the assumption that the economic phenomenon moves slowly, smoothly, and progressively through time (the trend). The other is a linear combination of periodic functions representing oscillations that affect the total variation of the series (the cycle and seasonality). For example, a simple regression model that assumes a cubic trend and a stable deterministic seasonality is given by

$$X_t = C_t + S_t + U_t, \quad (13)$$

where

$$C_t = \sum_{i=0}^{3} \alpha_i t^i, \quad (14)$$

$$S_t = \sum_{j=1}^{12} \beta_j D_{jt}, \qquad \sum_{j=1}^{12} \beta_j = 0. \quad (15)$$

The α_i and β_j are the unknown parameters and D_{jt} are the seasonal dummy variables that take the value 1 when the tth observation concerns the jth month and zero in all other cases. If U_t is assumed to be purely random with zero mean and finite variance, the estimation of the parameters can then be made by least squares*.

Many variations are possible for the specification of the components. Thus, in the representation (15) it may be assumed that the seasonal component changes through time. The coefficient β_j is then replaced by $\beta_j + \gamma_j t$, the γ_j being restricted by the condition that their sum is zero. In other cases, a multiplicative decomposition model may be found to be more appropriate than the additive model (13), in which case the determination of the trend and the seasonal component can be carried out as before by means of a regression on the logarithms of the X_t.

Major contributions to the development of regression models for seasonal adjustment were made in the 1960s, particularly in refs. 15, 16, 18, and 20. To overcome the limitation of using a global representation of the trend–cycle, refs. 13 and 26 used local polynomials (spline* functions) for successive short segments of series. These regression moels, however, still imply a deterministic behaviour of the time-series components. More recent studies have considered the possibility of a stochastic behaviour of the components by developing mixed models [22] or regression models with time-varying parameters [14]. The effects of seasonal adjustment procedures on dynamic regression models have been discussed thoroughly in ref. 4.

Regression methods have been seldom used by statistical agencies. The main reasons for this are (a) the seasonally adjusted series are wholly revised when new observations are added; (b) all the regression models developed until very recently implied a deterministic behaviour of the components.

Moving Average Methods

The majority of the seasonal adjustment methods applied by statistical bureaus belong to the class of moving average methods, which make the assumption that, although the systematic part of a time series is a smooth function, it cannot be approximated well by simple mathematical functions over the entire range.

Since moving averages* are linear transformations, they possess the properties of scale preservation and additivity. Furthermore, they have the time-invariance property, which is not shared by the regression methods.

Methods based on moving averages techniques assume that the trend–cycle and seasonal components change through time in a stochastic manner. The majority of such methods are mainly descriptive *nonparametric* procedures in the sense that they lack explicit parametric moels for each unobserved component.

In most recent years, however, several attempts have been made to develop *model-based* procedures where univariate statistical models are explicitly assumed for each component. The explicit models mainly belong to the Gaussian ARIMA (autoregressive integrated moving average*) type developed by ref. 3 or to variations of it as developed in refs. 5, 17 and 21. Other types of models that are not ARIMA have also been studied, for example, in refs. 1, 2, and 24. All these new approaches are still in a developmental stage and the majority of the moving average procedures officially adopted by statistical bureaus belong to the *nonparametric* type (see ref. 19). Among the latter, the Method II-X-11 variant [25] and the X-11-ARIMA [7] are the most widely applied. The X-11-ARIMA was developed to produce more accurate estimates of current seasonally adjusted series when seasonality changes rapidly and in a stochastic manner, characteristics often found in main economic indicators.

These two methods follow an iterative estimation procedure where the trend–cycle is estimated first, the seasonal component next, and the irregular is derived as a residual. The properties of the combined linear filters applied to obtain a seasonal estimate have been analysed [27, 28, 30] for the X-11-variant and for the X-11-ARIMA [8].

It is inherent to all linear smoothing procedures that the end observations cannot be smoothed with the same set of symmetric filters as applied to central observations. Because of this, the estimates for current observations must be revised as more data are incorporated into the series. In the context of the X-11 and X-11-ARIMA, this means that the first and last three and a half years of a new series will be revised because their symmetric filters require seven years of data to produce a central estimate. The revisions due to differences in the moving averages applied to the same observation as it changes its positions relative to the end of the series have been studied extensively for both methods [9–11].

References

[1] Akaike, H. (1980). *J. Time Series Anal.*, **1**, 1–13.

[2] Akaike, H. and Ishiguro, M. (1980). BAYSEA, A Bayesian Seasonal Adjustment Procedure. *Monograph 13*, Inst. Statist. Math., Tokyo, Japan.

[3] Box, G. E. P. and Jenkins, G. M. (1970). *Time Series Analysis: Forecasting and Control*. Holden-Day, San Francisco, CA.

[4] Bunzel, H. ad Hylleberg, S. (1982). *J. Econometrics*, **19**, 345–366.

[5] Burman, J. P. (1980). *J. R. Statist. Soc. A*, **143**, 321–27.

[6] Cleveland, W. S., Dunn, D. M., and Terpenning, I. (1978). In *Seasonal Analysis of Economic Time Series*, A. Zellner, ed. U.S. GPO, Washington, DC, pp. 201–231.

[7] Dagum, E. B. (1980). *The X-11-ARIMA Seasonal Adjustment Method*. Statistics Canada, Ottawa, Canada.

[8] Dagum, E. B. (1983). *Canad. J. Statist.*, **11** 73–90.

[9] Dagum, E. B. (1982). *J. Forecasting*, **1**, 173–187.

[10] Dagum, E. B. (1982). *J. Amer. Statist. Ass.*, **77**, 732–738.

[11] Dagum, E. B. (1982). *Proc. Bus. Econom. Sect. Amer. Statist. Ass.*, 39–45.

[12] Durbin, J. and Kenny, P. B. (1979). In *Seasonal Analysis of Economic Time Series*, A. Zellner, ed. U.S. GPO, Washington, DC, pp. 173–187.

[13] Duvall, R. M. (1966). *J. Amer. Statist. Ass.*, **61**, 152–165.

[14] Gersovitz, M. and MacKinnon, J. G. (1978). *J. Amer. Statist. Ass.*, **73**, 264–273.

[15] Hannan, E. J. (1960). *Aust. J. Statist.*, **2**, 1–15.

[16] Henshaw, R. C. (1966). *Econometrica*, **34**, 381–395.

[17] Hillmer, S. G. and Tiao, G. C. (1982). *J. Amer. Statist. Ass.*, **77**, 63–70.

[18] Jorgenson, D. W. (1964). *J. Amer. Statist. Ass.*, **59**, 681–723.

[19] Kuiper, J. (1978). In *Seasonal Analysis of Economic Time Series*, A. Zellner, ed. U.S. GPO, Washington, DC. pp. 59–76.

[20] Lovell, M. C. (1963). *J. Amer. Statist. Ass.*, **58**, 993–1010.

[21] Nerlove, M., Grether, D. M., and Carvalho, J. L. (1979). *Analysis of Economic Time Series*. Academic, New York.

[22] Pierce, D. (1979). In *Seasonal Analysis of Economic Time Series*, A. Zellner, ed. U.S. GPO, Washington, DC. pp. 242–269.

[23] Pierce, D. (1980). *Amer. Statist.*, **34**, 125–134.

[24] Schlicht, E. (1981), *J. Amer. Statist. Ass.*, **76**, 374–378.

[25] Shiskin, J., Young, A. H., and Musgrave, J. C. (1967). *The X-11-Variant of Census Method II Seasonal Adjustment Program*. U.S. Bureau of Census, Washington, DC.

[26] Stephenson, J. A. and Farr, H. T. (1972). *J. Amer. Statist. Ass.*, **67**, 37–45.

[27] Wallis, K. F. (1974). *J. Amer. Statist. Ass.*, **69**, 18–31.

[28] Wallis, K. F. (1982). *J. R. Statist. Soc. A*, **145**, 74–85.

[29] Wold, H. (1938). *A Study in the Analysis of Stationary Time Series*. Almqvist and Wiksell, Stockholm, Sweden [2nd ed. (1954)].

[30] Young, A. H. (1968). *J. Amer. Statist. Ass.*, **63**, 445–457.

(AUTOREGRESSIVE-INTEGRATED MOVING AVERAGE (ARIMA) MODELS
FORECASTING
MOVING AVERAGES
PREDICTION AND FORECASTING
TIME SERIES
TREND
X-11)

ESTELA BEE DAGUM

SECANT METHOD *See* NEWTON–RAPHSON METHODS

SECH-SQUARED DISTRIBUTION *See* LOGISTIC DISTRIBUTIONS

SECONDARY DATA

A term used for data obtained in the course of a survey or experiment designed for some other (primary) purpose. It must be kept in mind that whatever faults there may have been in the planning and execution of the original inquiry can also affect secondary data. In addition, further errors may arise from the way in which the secondary data are extracted from the records of the inquiry.

SECRETARY PROBLEM

The Secretary Problem is one name, among many (see ref. 7), given to a collection of optimal stopping* problems. The two classical Secretary Problems are the best choice problem and the rank problem.

THE BEST CHOICE PROBLEM. It is assumed that n individuals applying for a secretarial position arrive sequentially in random order. Upon arrival each individual is interviewed and ranked (highest rank = 1) relative to all preceding arrivals. At the time of ranking an irrevocable decision must be made to hire or reject the applicant. The goal is to find a strategy to maximize the probability of selecting the best. The optimal strategy (see, e.g., refs. 2 and 7) is to reject the first $s(n)$ individuals and to choose the first, if any, among applicants $s(n) + 1, \ldots, n$ who is the best seen so far. It can be shown that $s(n)/n \to e^{-1} \doteq 0.368$ and that this is also the optimal asymptotic probability of selecting the best.

THE RANK PROBLEM. (See refs. 1 and 9.) The mechanics are the same as in the best choice problem, but here the goal is to find a strategy to minimize the expected rank of the applicant chosen. There is a nondecreasing sequence of integers $\{s(n; k), 1 \leqslant k \leqslant n\}$ such that the optimal strategy is of the form: if no selection has been made from the first $s(n; k) - 1$ applicants, select the first arrival among applicants $s(n; k), \ldots, s(n; k + 1)$ who is one of the k best seen so far. Chow et al. [1] show that the optimal asymptotic expected rank is approximately 3.87.

The origins of the Secretary Problem are unclear. According to ref. 7, in 1955

Frederick Mosteller was told the best choice problem by Andrew Gleason who had heard it from someone else. Fox and Marnie posed what is basically the best choice problem under the name "Googol" in the Mathematical Games column of the February, 1960 *Scientific American*. A solution was outlined by Moser and Pounder in the same column the following month.

Since that time the Secretary Problem has generated a substantial literature. Among the reasons for its popularity may be: it is simple to state; its solutions are intuitively appealing; it touches on the disciplines of probability, statistics, and operations research*; and it is a model for a common real-life problem that invites improvement and generalization. This last may explain why the development of the Secretary Problem up to the present has consisted primarily of generalizations designed to make the two basic models more realistic. A description of some of these follows; others are briefly described in the annotated bibliography.

FULL AND PARTIAL INFORMATION PROBLEMS

The above strategies are based only on rankings of the applicants. In some cases it may be reasonable to assume more information about the applicants—for example, one may observe scores on placement tests. In such a case the observations could be considered independent identically distributed (i.i.d.) random variables from a known distribution (full information) or from a family of distributions (partial information). Hereafter "no information" will describe problems in which only ranks are observed.

The object may be to maximize the probability of best choice, the expected quantile of the chosen observation (the analog of the rank problem), or the expectation of some other payoff function of the chosen observation.

Gilbert and Mosteller [7] solved the full information best choice and quantile problems. The asymptotic optimal probability of best choice in the former is approximately 0.58, while the asymptotic minimal expected rank in the latter equals 2.

The earliest work on the partial information problem [3, 19, 25] assumed a normal distribution with unknown mean and a payoff equal to the value of the chosen observation minus a sampling cost. Later Stewart [23] and Samuels [20] considered the best choice and quantile problems when observations are from a uniform* $U(\alpha, \beta)$ distribution with α and β unknown. The latter paper showed that for the best choice problem the minimax* rule is simply the no information rule for the classical best choice problem, while for the quantile problem the minimax rule gives an asymptotic expected rank of approximately 3.478.

For the best choice problem, Petruccelli [14] obtained sufficient conditions on location–scale parameter families of distributions for the existence of minimax rules, which asymptotically attain the asymptotic full information probability of best choice. One family that satisfies these conditions is the family of all normal distributions.

Petruccelli [10] also showed that for the family of $U[\theta - \frac{1}{2}, \theta + \frac{1}{2}]$ distributions with θ unknown, the minimax rule in the best choice problem has asymptotic probability of best choice of approximately 0.435—intermediate between the full and no information values.

Enns [4] studied problems with a different kind of partial information: the knowledge of whether or not an observation exceeds a certain value.

PROBLEMS INVOLVING BACKWARD SOLICITATION AND / OR UNCERTAINTY OF CHOICE

Yang [26] modified the classical best choice problem to allow solicitation of a past observation. He assumed that just after applicant k was interviewed, applicant $k - r$ would accept an offer with probability $q(r)$.

This model assumed $q(0) = 1$; that is, an applicant is sure to accept an immediate offer. Smith and Deely [22] assumed (in Yang's terminology) that $q(r) = 1, 0 \leqslant r \leqslant m$, and $q(r) = 0, r > m$; that is, recall memory lasts m interviews.

Smith [21] considered the case of uncertain choice without recall that corresponds to $q(0) < 1$ and $q(r) = 0, r \geqslant 1$, in Yang's terminology. Petruccelli [11] combined the models of Yang and Smith by allowing $q(0) < 1$ in Yang's formulation. In addition Petruccelli [12] extended backward solicitation and uncertainty of choice to the full information best choice problem. In this last paper, recall probabilities were functions of the quantile of the observation as well as the time since observation.

While the models described here allow for a more realistic formulation than does the classical best choice problem, there is a price to be paid: it is impossible to obtain closed-form optimal rules without further assumptions on the recall probabilities.

A RANDOM NUMBER OF OBSERVATIONS

Secretary problems with a random number of observations N are in general complicated and admit no closed form solution. For the classical best choice problem in which the number of observations is a bounded random variable, Petruccelli [13] showed that after allowing for an initial learning period, the optimal rule can take literally any form.

Among generalizations for arbitrary N, Presman and Sonin [16] obtained results for the best choice problem, while Irle [8] considered an arbitrary payoff.

For bounded N Rasmussen [17] studied general payoffs based on ranks, Rasmussen and Robbins [18] considered the best choice problem, Gianini-Pettit [6] dealt with expected rank payoff, and Petruccelli [15] added backward solicitation and uncertainty of choice to the best choice problem. Stewart [24] used a Bayesian approach to the random N problem in which arrivals were assumed to occur at i.i.d. exponential intervals.

OTHER GENERALIZATIONS

There are many other generalizations of the Secretary Problem: some are indicated in the Bibliography; a number are found in the Gilbert and Mosteller paper [7], which even after nearly two decades remains a good introduction to the topic. The best introduction to the development of the problem over the past 20 years is a review article by Freeman [5].

References

[1] Chow, Y. S., Moriguti, S., Robbins, H., and Samuels, S. M. (1964). *Israel J. Math.*, **2**, 81–90. (The first appearance of the name "Secretary Problem" in print.)

[2] Chow, Y. S., Robbins, H., and Siegmund, D. (1971). *Great Expectations: The Theory of Optimal Stopping*. Houghton-Mifflin, Boston, MA. (This is *the* reference on optimal stopping.)

[3] DeGroot, M. (1968). *J. R. Statist. Soc. B.*, **30**, 108–122.

[4] Enns, E. G. (1975). *J. Amer. Statist. Ass.*, **70**, 640–643.

[5] Freeman, P. R. (1983). *Int. Statist. Rev.*, **51**, 189–206. (Currently the best introduction to the Secretary Problem and many of its extensions.)

[6] Gianini-Pettit, J. (1979). *Adv. Appl. Prob.*, **11**, 720–736.

[7] Gilbert, J. P. and Mosteller, F. (1966). *J. Amer. Statist. Ass.*, **61**, 35–79. (Still a good introduction to the classical best choice problem and some of its variants. Contains an interesting history of the problem.)

[8] Irle, A. (1980). *Zeit. Operat. Res.*, **24**, 177–190.

[9] Lindley, D. V. (1961). *Appl. Statist.*, **10**, 39–51.

[10] Petruccelli, J. D. (1980). *Ann. Statist.*, **8**, 1171–1174.

[11] Petruccelli, J. D. (1981). *J. Appl. Prob.*, **18**, 415–425.

[12] Petruccelli, J. D. (1982). *Adv. Appl. Prob.*, **14**, 340–358.

[13] Petruccelli, J. D. (1983). *J. Appl. Prob.*, **20**, 165–171.

[14] Petruccelli, J. D. (1984). *Sankhyā, A*, **46**, 370–382.

[15] Petruccelli, J. D. (1983). *Adv. Appl. Prob.*, **16**, 111–130.

[16] Presman, E. L. and Sonin, I. M. (1972). *Theory Prob. Appl.*, **17**, 657–668.

[17] Rasmussen, W. T. (1975). *J. Optim. Theor. Appl.*, **15**, 311–325. (Lemma 3.2 is false in validating Theorem 3.1. See refs. 8 and 13 for counterexamples. See also the annotation to ref. 18.)

[18] Rasmussen, W. T. and Robbins, H. (1975). *J. Appl. Prob.*, **12**, 692–701. (The false Theorem 3.1 of ref. 17 was assumed true in this paper.)

[19] Sakaguchi, M. (1961). *J. Math. Anal. Appl.*, **2**, 446–466.

[20] Samuels, S. M. (1981). *J. Amer. Statist. Ass.*, **76**, 188–197.

[21] Smith, M. H. (1975). *J. Appl. Prob.*, **12**, 620–624.

[22] Smith, M. H. and Deely, J. J. (1975). *J. Amer. Statist. Ass.*, **70**, 357–361.

[23] Stewart, T. J. (1978). *J. Amer. Statist. Ass.*, **73**, 775–780.

[24] Stewart, T. J. (1981). *Oper. Res.*, **29**, 130–145. (The introduction contains a worthwhile discussion of the effect of various restrictive assumptions on applications of the Secretary Problem.)

[25] Yahav, J. A. (1966). *Ann. Math. Statist.*, **37**, 30–35.

[26] Yang, M. C. K. (1974). *J. Appl. Prob.*, **11**, 504–512.

Bibliography

Bartoszynski, R. and Govindarajulu, Z. (1978). *Sankhyā B*, **40**, 11–28. (A positive payoff for best or second best observation and a cost for each interview.)

Gianini, J. (1977). *Ann. Prob.*, **5**, 636–644. (Uses the infinite Secretary Problem to obtain asymptotic results for the Secretary Problem.)

Gianini, J. and Samuels, S. M. (1976). *Ann. Prob.*, **4**, 418–432. (Considers the infinite Secretary Problem.)

Gilbert, J. P. and Mosteller, F. (1966). *J. Amer. Statist. Ass.*, **61**, 35–79. (Several generalizations: (a) allowed best or second best to be a winning selection; (b) two person game in which one player orders ranked observations in an attempt to minimize the other player's probability of best choice; (c) a best choice problem with $r > 1$ choices to obtain the best.)

Guttman, I. (1960). *Canad. Math. Bull.*, **3**, 35–39. (The full information problem with payoff the value of the observation.)

Lorenzen, T. J. (1979). *Adv. Appl. Prob.*, **11**, 384–396. (Uses the infinite Secretary Problem to obtain asymptotic results for the no information problem with arbitrary payoff.)

Lorenzen, T. J. (1981). *Ann. Prob.*, **9**, 167–172. (Considers a payoff based on ranks and a cost for each interview in both the finite and infinite Secretary Problems, obtaining the infinite problem and solution as limiting cases.)

Moser, L. (1956). *Scripta Math.*, **22**, 289–292. (Assumes a known U[0, 1] distribution with payoff the value of the observation.)

Rasmussen, W. T. and Pliska, S. R. (1976). *Appl. Math. Optim.*, **2**, 279–289. (The no information case with a discounted payoff.)

Rubin, H. (1966). *Ann. Math. Statist.*, **37**, 544. (Introduces idea of an infinite Secretary Problem, in which an infinite number of rankable observations arrive at i.i.d. U[0, 1] times. This is of interest in its own right and is useful in obtaining asymptotic results.)

Tamaki, M. (1979). *J. Appl. Prob.*, **16**, 803–812. (Allows two choices in the no information case to obtain the best and second best.)

(OPTIMAL STOPPING RULES)

JOSEPH D. PETRUCCELLI

SECTOR CHART; SECTOR DIAGRAM

See PIE CHART

SECULAR TREND *See* TREND

SEEMINGLY UNRELATED REGRESSION

Consider the set of M regression models

$$y_j = X_j\beta_j + u_j, \qquad j = 1, \ldots, M, \quad (1)$$

where y_j is a $T \times 1$ vector of observations on the jth dependent variable, X_j is a $T \times K_j$ matrix of observations on K_j nonstochastic regressors assumed to be of full column rank, β_j is a $K_j \times 1$ vector of regression coefficients, and u_j is a $T \times 1$ vector of random disturbances with $E(u_j) = 0$ and $E(u_j u_i') = \sigma_{ji} I_T$ for $i, j = 1, \ldots, M$. If the observations correspond to different points in time, the specification implies that the disturbances in different equations are correlated at each point in time but are uncorrelated over time. Variances and covariances remain constant over time. We further assume that the disturbances are distributed independently over time and that the matrix of sample moments of all distinct regressors in (1) converges to a finite positive-definite matrix as T goes to infinity.

Model (1) was first considered by Zellner [7]. Even though the equations in (1) appear to be structurally unrelated, the fact that the disturbances are contemporaneously correlated across equations constitutes a link between them. For this reason, Zellner referred to (1) as a set of "seemingly unrelated regression" (SUR) equations. The SUR model differs from the structural form of a truly simultaneous equation model in that the dependent variable in one equation does not appear as a regressor in another equation. The structure of the disturbances of the SUR model is identical to that assumed for the standard linear simultaneous equation model.

Each equation in (1) by itself satisfies the assumptions of the classical linear single equation regression model. We could therefore estimate the model parameters by applying ordinary least squares* to each equation. However, this approach neglects information. By explicitly taking into account the correlation structure of the disturbances across equations and by jointly estimating the equations in (1), it is generally possible to obtain more precise estimates of the model parameters, as will be explained.

The SUR specification has been widely used; Zellner [7] used it to jointly estimate microinvestment functions for General Electric and Westinghouse. For this application Zellner also reports intermediary results of the numerical computations involved. Other applications include (i) the joint estimation of a set of consumption functions each corresponding to a different consumer unit, (ii) the joint estimation of a system of demand equations for respective factor inputs, and (iii) the joint estimation of a system of demand equations for respective consumption goods. For a review of some of the applications, see e.g., Johnston [2].

System (1) can be stacked to form the combined regression model

$$y = X\beta + u, \tag{2a}$$

where

$$y = \begin{bmatrix} y_1 \\ \vdots \\ y_M \end{bmatrix}, \qquad X = \begin{bmatrix} X_1 & & 0 \\ & \ddots & \\ 0 & & X_M \end{bmatrix},$$

$$\beta = \begin{bmatrix} \beta_1 \\ \vdots \\ \beta_M \end{bmatrix}, \qquad u = \begin{bmatrix} u_1 \\ \vdots \\ u_M \end{bmatrix}. \tag{2b}$$

We then have $E(u) = 0$ and $E(uu') = \Sigma \otimes I_T$ with $\Sigma = (\sigma_{ji})$, where $\otimes$ denotes the Kronecker product*. In a formal sense we can regard (2) as a single equation regression model. If Σ is known and positive-definite the generalized least-squares (GLS) estimator is given by

$$\begin{aligned} \hat{\beta}_{\mathrm{GLS}} &= \left[X'(\Sigma^{-1} \otimes I_T)X\right]^{-1} X'(\Sigma^{-1} \otimes I_T)y \\ &= \begin{bmatrix} \sigma^{11}X_1'X_1 & \cdots & \sigma^{1M}X_1'X_M \\ \vdots & & \vdots \\ \sigma^{M1}X_M'X_1 & \cdots & \sigma^{MM}X_M'X_M \end{bmatrix}^{-1} \\ &\quad \times \begin{bmatrix} \sum_{j=1}^{M} \sigma^{1j}X_1'y_j \\ \vdots \\ \sum_{j=1}^{M} \sigma^{Mj}X_M'y_j \end{bmatrix}, \end{aligned} \tag{3}$$

where σ^{ji} denotes the (j, i)th element of Σ^{-1}. Zellner [7] points out that the GLS estimator $\hat{\beta}_{\mathrm{GLS}}$ is the best linear unbiased estimator for β. That is, $\hat{\beta}_{\mathrm{GLS}}$ has the smallest variance within the class of estimators that are unbiased and linear in y. The variance–covariance of $\hat{\beta}_{\mathrm{GLS}}$ is

$$\left[X'(\Sigma^{-1} \otimes I_T)X\right]^{-1}.$$

If the disturbances are normally distributed, $\hat{\beta}_{\mathrm{GLS}}$ is also the maximum likelihood* estimator.

The ordinary least-squares (OLS) estimator is given by $\hat{\beta}_{\mathrm{OLS}} = (X'X)^{-1}X'y$. The OLS estimator does not take into account the correlation structure of the disturbances across equations and is generally less efficient than the GLS estimator. In the case of zero correlation over equations, i.e., $\sigma_{ji} = 0$ for $j \neq i$, or if $X_1 = X_2 = \cdots = X_M$, then $\hat{\beta}_{\mathrm{OLS}}$ and $\hat{\beta}_{\mathrm{GLS}}$ will be identical [7]. For a general set of conditions under which $\hat{\beta}_{\text{OLS}}$ and $\hat{\beta}_{\text{GLS}}$ are identical, see Dwivedi and Srivastava [1].

In typical applications Σ is unknown and has to be estimated. Zellner [7] proposed the feasible GLS estimator

$$\hat{\beta}_{\text{FGLS}} = \left[X'(\hat{\Sigma}^{-1} \otimes I_T)X\right]^{-1} \times X'(\hat{\Sigma}^{-1} \otimes I_T)y, \quad (4)$$

where $\hat{\Sigma} = (\hat{\sigma}_{ji})$ is based on OLS residuals

$$\hat{\sigma}_{ji} = \frac{1}{T}\left[y_j - X_j\hat{\beta}_{j,\text{OLS}}\right]'\left[y_i - X_i\hat{\beta}_{i,\text{OLS}}\right]. \quad (5)$$

The estimators $\hat{\sigma}_{ji}$ are biased. If instead of the divisor T the divisors $(T - K_j)^{1/2}(T - K_i)^{1/2}$ are used, the covariance estimators will be unbiased for $i = j$; if the divisors

$$T - K_j - K_i - \text{tr}\left[X_j(X_j'X_j)^{-1}X_j'X_i(X_i'X_i)^{-1}X_i'\right]$$

are used, the covariance estimators will be unbiased for all i and j. (See Zellner and Huang [9] and Stroud et al. [6].)

Asymptotically all feasible GLS estimators of the form (4) based on a consistent estimator for the disturbance variance–covariance matrix are equivalent with the true GLS estimator (3) in that the probability limit of $\sqrt{T}(\hat{\beta}_{\text{FGLS}} - \hat{\beta}_{\text{GLS}})$ is zero. In particular, $\hat{\beta}_{\text{FGLS}}$ is consistent and the limiting distribution of $\sqrt{T}(\hat{\beta}_{\text{FGLS}} - \beta)$ is normal with mean 0 and variance–covariance

$$\lim_{T\to\infty}\left[T^{-1}X'(\Sigma^{-1} \otimes I_T)X\right]^{-1}$$

(see Zellner [7]). For inference purposes we would assume that $\hat{\beta}_{\text{FGLS}}$ is approximately normal with mean β and variance–covariance matrix $[X'(\hat{\Sigma}^{-1} \otimes I_T)X]^{-1}$.

Concerning the small-sample properties of $\hat{\beta}_{\text{FGLS}}$, Bayesian interpretations, the relationship of $\hat{\beta}_{\text{FGLS}}$ with the maximum likelihood estimator, the testing of hypotheses, goodness-of-fit statistics, and the effects of specification errors, see the detailed surveys of Judge et al. [3] and Srivastava and Dwivedi [5] and the references cited therein.

The basic SUR model has been extended in various ways including: autocorrelated and heteroscedastic disturbances, random parameter and error component specifications, estimation in case of a singular variance–covariance matrix, nonlinear functional forms, and the treatment of unequal numbers of observations for different equations. See again Judge et al. [3] and Srivastava and Dwivedi [5] for helpful surveys and literature references. For recent general results on the small-sample distribution of feasible GLS etimators in SUR models see Phillips [4].

References

[1] Dwivedi, T. D. and Srivastava, V. K. (1978). *J. Econometrics*, **7**, 391–395.

[2] Johnston, J. (1984). *Econometric Methods*, 3rd ed. McGraw-Hill, New York, pp. 330–337.

[3] Judge, G. G., Griffiths, W. E., Hill, R. C., Luetkepohl, H., and Lee, T. C. (1985). *The Theory and Practice of Econometrics*, 2nd Ed. Wiley, New York, pp. 465–514.

[4] Phillips, P. C. P. (1985). *Econometrica*, **53**, 745–756.

[5] Srivastava, V. K. and Dwivedi, T. D. (1979). *J. Econometrics*, **10**, 15–32.

[6] Stroud, A., Zellner, A., and Chau, L. C. (1963). Workshop Paper No. 6803, Social Systems Research Institute, University of Wisconsin.

[7] Zellner, A. (1962). *J. Amer. Statist. Ass.*, **57**, 348–368.

[8] Zellner, A. (1963). *J. Amer. Statist. Ass.*, **58**, 977–992; (1972) **67**, 225.

[9] Zellner, A. and Huang, D. S. (1962). *Int. Econ. Rev.*, **3**, 300–313.

(LEAST SQUARES
LINEAR REGRESSION
ZELLNER ESTIMATOR)

INGMAR R. PRUCHA

SELECT LIFE TABLES

A life table* is intended to represent the survival probabilities for a specified population. Conventionally it gives the numbers of survivors l_x at exact age x to be expected from a cohort of l_0 newborn. Derived quantities, such as the probability of death within a year for a person aged x exactly

$$q_x = 1 - l_{x+1}/l_x,$$

are also given.

In a *select* table, allowance is made for differential mortality within the population, associated with some property or "event" of an individual. The commonest event has been acceptance of issuance of a life insurance policy, but it could be passing a medical examination for entering a police force, diagnosis of a specified disease, or undergoing a particular operation. The first two of these are likely to lead to lower than average mortality; the last two, to higher mortality. The former is termed *positive* selection, the latter *negative* selection.

If the event in question occurs at exact age x, one will need to replace q_x by a different value, conventionally denoted by $q_{[x]}$. For positive selection, $q_{[x]} < q_x$; for negative selection $q_{[x]} > q_x$. As time passes the effects of selection persist but usually decrease, so that if $q_{[x]+t}$ denotes the value replacing q_{x+t} for an individual for whom the event occurred at exact age x, then for positive selection

$$q_{[x+t]} < q_{[x+t-1]+1} \cdots < q_{[x]+t} < q_{x+t}.$$

(For negative selections the signs are reversed.)

Usually, the difference between $q_{[x]+t}$ and q_{x+t} becomes so small for t large enough, that it can be neglected for practical purposes and one takes $q_{[x]+t} = q_{x+t}$. If the least value of t for which this is so is d, then d is called the *select period*; it is the period over which allowance is made for differences between select and nonselect mortality.

Select life tables are usually set out with column headings:

$$x \quad l_{[x]} \quad l_{[x]+1} \quad l_{[x]+t-1} \quad l_{x+t} \quad x+t$$

and

$$x \quad q_{[x]} \quad q_{[x]+1} \quad q_{[x]+t-1} \quad q_{x+t} \quad x+t.$$

The last two columns are termed an "ultimate" life table; it is regarded and used as if it represented mortality in the population as a whole (sometimes called aggregate mortality). Strictly speaking it applies to those in the population who have not been select for a period of d years or more, but this effect is likely to be quite small.

(ACTUARIAL STATISTICS-LIFE
LIFE TABLES
SURVIVAL ANALYSIS)

SELECTION BIAS

Selection bias is the bias that occurs in a clinical trial*, experiment, or sample survey (*see* SURVEY SAMPLING) because of effects of the mechanism used to select individuals or units for inclusion in the experiment or sample survey. The bias may come about because the selection mechanism is nonrandom, is random but in some way is related to the variable(s) to be studied, or the mechanism operates differently from that intended by the investigator. Researchers tend to focus on selection bias separately in clinical trials and in sample surveys, but the phenomenon is similar in both settings.

In an experiment, suppose that eligible subjects for a study of two treatments arrive singly and must be dealt with as soon as they become available. For example, in a clinical trial, patients come to the hospital sequentially and must be treated as soon as their diseases are diagnosed; in cloud seeding experiments (*see* WEATHER MODIFICATION), it is physically impossible to collect storm clouds for simultaneous assignments of treatments. In either situation, the investigator of the study may bias the experiment through his choice of subjects. For instance, as each subject arrives, the investigator decides whether the subject is suitable for the study. If it is declared suitable, the statistician then tells the experimenter which treatment to administer, A or B. Suppose that the experimenter, in effect, attempts to bias the experiment in favor of A by his selection of a suitable subject. If he guesses or knows that the next assignment will be an A, he selects a subject with a high expected response. If he guesses B, he waits for a subject with low expected response.

A physical way to avoid the introduction of selection bias to a study is to use, for example, a double-blind experiment. In a double-blind study, neither the investigator nor the experimental subject knows what treatment is assigned. However, the double-blind experiment may not be feasible, as in a surgical procedure. Some other treatment allocation rules are usually introduced in this case to reduce the bias.

If the size of the experiment is predetermined, say $2n$, Blackwell and Hodges [1] used the expected number of correct guesses of treatment assignments by the investigator through the entire experiment as the design criterion. The minimax design in this setting is the truncated binomial; that is, the statistician assigns A or B to the subject with probability $\frac{1}{2}$ each, independently of all other assignments, until one of the treatments has been used n times. Stigler [10] considered a slightly different model and proved that the random allocation design is optimum when the investigator does not consciously bias the experiment. In the random allocation design the statistician picks n of the first $2n$ integers at random without replacement, and gives treatment A to the subjects corresponding to the integers selected. Wei [13] proposes another design criterion, which is more suitable to express the investigator's ability to bias the experiment. He also shows that the random allocation design is better than the truncated binomial design in the minimax sense under the new criterion.

In the case where the size of the experiment cannot be predetermined, Efron [4] and Wei [12, 14] introduce various restricted randomization treatment allocation rules that tend to eliminate the selection bias. Due to random entry of subjects, the problem of selection bias is not so serious in a multi-institution clinical trial with central randomization as in a single institution trial.

In observational studies* or sample surveys, selection bias can affect the estimates of certain quantities because the selection probabilities for inclusion in a survey can be related to the variables being studied. DeMets and Halperin [2] give an example from the Framingham* heart study in which a relatively large population was studied and initial serum cholesterol level was measured for the full sample. Subsequently, to study the effect of dietary cholesterol on serum cholesterol, a subsample of only those with very high or very low initial serum cholesterol was drawn. As a result of this selection procedure, the ordinary regression estimates of serum cholesterol on dietary intake were biased. The authors present an asymptotically unbiased estimator for regression coefficients to counter this sampling selection bias, and study the effects of the sampling on variance estimates for the regression coefficients. Kleinbaum et al. [7, 8] propose a model and give examples of selection bias in the measurement of ratios of proportions, risk, and prevalence in epidemiologic* studies. In their model they postulate two populations: the target population, which is the population of interest, and the actual population, which is the population observed (or sampled) in the study. If the relative proportions of disease incidence in the two populations differ, selection bias can occur in ratio estimates calculated from a survey. For example, in a study of the effects of hypertension on the incidence of skin cancer, the sample was chosen to be those persons admitted to a hospital with a condition believed to be unrelated to the variables in the study (like bone fractures). Even if the condition (for sampling purposes) is unrelated to the variables under study, but the rate of hospitilization differs for the three conditions, there can be a substantial bias in the estimate of the odds or risk ratios for the effects of hypertension of skin cancer.

Williams [15] also examines some models where selection bias was introduced in a panel survey because of differing probabilities of observing employed and unemployed individuals in the population. In a household survey, even in a self-weighting sample where each unit would have the same probability of selection, if unemployed persons are harder to find at home or have living arrangements that make it less likely to be listed as a member of a household, a bias will occur in the measurement of employment status. Williams also presents examples of the degree of selection bias in different survey estimators due to differing probabilities of response at the first and later stages of a panel* survey.

If the selection process of the participants in the study is not obvious, one usually proposes models like those mentioned above, and proposes an adjustment for selection bias. For example, Singh et al. [9] propose a model where the variable being studied is directly related to the probabilities of selection in the study, resulting in underestimates of the level of the variable (mean duration of postpartum amenorrhea). Having developed a model that describes the selection process for participants in the study, they use estimates of parameters in the model to adjust the sample estimate of the mean value in which they are interested. Dunham and Mauss [3], Trochim and Spiegelman [11], Heckman [6], and Greene [5] also propose modeling approaches to correct for selection bias.

References

[1] Blackwell, D. and Hodges, J. L. (1957). *Ann. Math Statist.*, **28**, 449–460.

[2] DeMets, D. and Halperin, M. (1977). *Biometrics*, **33**, 47–56.

[3] Dunham, R. G. and Mauss, A. L. (1979). *Eval. Quart.*, **3**, 411–426.

[4] Efron, B. (1971). *Biometrika*, **58**, 403–417.

[5] Greene, H. (1981). *Econometrica*, **49**, 795–798.

[6] Heckman, J. J. (1979). *Econometrica*, **47**, 153–161.

[7] Kleinbaum, D. G., Kupper, L., and Morgenstern, H. (1982). *Epidemiologic Research*. Lifetime Learning Publications, Belmont, CA.

[8] Kleinbaum, D. G., Morgenstern, H., and Kupper, L. L. (1981). *Amer. J. Epidemiol.*, **113**, 452–463.

[9] Singh, S. N., Bhattacharya, B. N., and Yadava, R. C. (1979). *J. Amer. Statist. Ass.*, **74**, 916–920.

[10] Stigler, S. M. (1969). *Biometrika*, **56**, 553–560.

[11] Trochim, W. M. K. and Spiegelman, C. H. (1980). *Proceedings of the Survey Research Methods Section, American Statistical Association*, 376–380.

[12] Wei, L. J. (1977). *J. Amer. Statist. Ass.*, **72**, 382–286.

[13] Wei, L. J. (1978). *Biometrika*, **65**, 79–84.

[14] Wei, L. J. (1978). *Ann. Statist.*, **6**, 92–100.

[15] Williams, W. H. (1978). *Contributions of Survey Sampling and Applied Statistics in Honor of H. O. Hartley*, H. A. David, ed. Academic, New York, pp. 89–112.

(CLINICAL TRIALS
EPIDEMIOLOGICAL STATISTICS
PANEL DATA
SURVEY SAMPLING
TARGET POPULATION)

L. J. Wei
Charles D. Cowan

SELECTION DIFFERENTIALS

Let $X_{1:n} \leqslant X_{2:n} \leqslant \cdots \leqslant X_{n:n}$ denote the order statistics* of a random sample from a continuous distribution with cumulative distribution function (CDF) $F(x)$, mean μ, and variance σ^2. Suppose we select k out of these n values. The difference between the average of the selected group and μ expressed in standard deviation (σ) units is called the *selection differential*. Usually the extreme k values are selected, resulting in directional selection. Suppose the highest k values are selected. The selection differential is then given by

$$D(k, n) = \left(\frac{1}{k} \sum_{i=n-k+1}^{n} X_{i:n} - \mu\right) \Big/ \sigma. \quad (1)$$

Without loss of generality we assume that $\mu = 0$ and $\sigma = 1$ throughout this entry. The CDF of $D(k, n)$ does not have a closed form (Nagaraja [9]). However, its mean $ED(k, n)$ and variance $\text{var}(D(k, n))$ can be computed from the means, variances, and covariances of the selected order statistics. For example, if the sample is from an exponential distribution*, $ED(k, n)$ is $\sum_{i=k+1}^{n} i^{-1}$ and $\text{var}(D(k, n))$ is $\sum_{i=k+1}^{n} i^{-2} + k^{-1}$. For a normal distribution* these moments can readily be computed from Tables 9 and 10 of Pearson and Hartley [13]. See also Schaeffer et al. [14]. Also for a normal distribution,

$$k\{1 - \text{var}(D(k, n))\}$$
$$= (n - k)\{1 - \text{var}(D(n - k, n))\}$$

(Burrows [3]). For an arbitrary CDF $F(x)$, bounds on $ED(k, n)$ are given by Nagaraja [10]. These bounds depend on the amount of restriction of $F(x)$. However, even if the sample values are dependent, $ED(k, n) \leqslant \sqrt{\{(n-k)/k\}}$ for any distribution.

If μ is unknown, it is replaced by the sample mean $\overline{X}$ and if σ is unknown, the sample standard deviation S takes its place in (1). The quantity so obtained will be called the *sample selection differential*. For example, if both μ and σ are unknown, the sample selection differential is

$$\hat{D}(k, n) = k^{-1} \sum_{i=n-k+1}^{n} (X_{i:n} - \overline{X})/S.$$

ASYMPTOTIC THEORY

The limit distribution of $D(k, n)$ as n approaches infinity depends on the behavior of k in relation to n. Recall that $D(k, n) = \sum_{i=n-k+1}^{n} X_{i:n}/k$ ($\mu = 0, \sigma = 1$). If k is held fixed while $n \to \infty$, its limit distribution is related to distributions in extreme value* theory. If there exist constants γ_n and $\delta_n > 0$ such that $(X_{n:n} - \gamma_n)/\delta_n$ has a nondegenerate limiting distribution $H(x)$, then $(D(k, n) - \gamma_n)/\delta_n$ converges in distribution to the average of the first k lower record values from $H(x)$. For possible forms of $H(x)$ *see* EXTREME-VALUE DISTRIBUTIONS. If $k = [np]$, where $0 < p < 1$ represents the proportion of individuals selected, $D(k, n)$ can be viewed as a trimmed mean (*see* TRIMMING AND WINSORIZATION), where the lower $100q\%$ of values have been trimmed. Here $[\cdot]$ is the greatest integer function and $q = 1 - p$. The limit distribution depends on whether the qth quantile* of F is unique (Stigler [15]). If ξ_q is the unique quantile, $\sqrt{k}\{D(k, n) - \mu_p\}$ is asymptotically $N(0, \sigma_p^2 + q(\mu_p - \xi_q)^2)$, where μ_p and σ_p^2 are the mean and variance of the distribution obtained by truncating F from below at ξ_q.

IMPLICATIONS IN GENETICS*

The phrase "selection differential" has been in genetics literature for a long time even though there is some arbitrariness in its definition. For example, it has been used for $D(k, n)$, $SD(k, n)$, $ED(k, n)$, or its limit value μ_p. Falconer [5, p. 192] uses the word *intensity of selection* for $D(k, n)$ whereas several others have called p, the proportion selected, the *selection intensity*. See Burrows [2, pp. 1098–1099] and references cited therein. Usually in these applications, the top $100p\%$ of a finite population is selected for breeding purposes where this population is assumed to be a random sample from a normal distribution. In any case the quantity of interest is $ED(k, n)$, representing expected improvement upon selection. This is useful in the construction of suitable breeding plans and in comparison of plans in plant as well as animal breeding problems.

If n is large and $k = [np]$, $0 < p < 1$, $ED(k, n)$ can be approximated by $\mu_p - q\{2(n+1)f(\xi_q)\}^{-1}$, where $f(x)$ is the density of $F(x)$ (Burrows [2]). This works well for a normal population.

Burrows [3] has also obtained upper bounds on $\text{var}(D(k, n))$ for the normal population. Milkman [8] has shown that the selection coefficient on a genotype is approximately equal to the product of $D(k, n)$ and standard phenotypic effect.

A situation of interest for breeders is when the sample consists of r families and the observations on members of the same family are equicorrelated. Assuming equal family size of m, let $n = mr$. Suppose that these r groups of m values form a random sample from an m-variate normal distribution where the components are identically distributed normal random variables with common correlation coefficient ρ. Under this setup, Hill

[7] has computed the exact value of, and an approximation to $ED(k, n)$ for several combinations of m and r for selected values of ρ.

A TEST FOR OUTLIERS*

Let $X_1, X_2, \ldots, X_n$ be n independent random variables, where X_i is $N(\mu_i, \sigma^2)$, $i = 1, \ldots, n$, and let $X_{1:n} \leqslant X_{2:n} \leqslant \cdots \leqslant X_{n:n}$ denote the order statistics of these X_i's. Consider the problem of testing the hypothesis $H: \mu_1 = \mu_2 = \cdots = \mu_n = \mu$ against the alternative $A: k$ of the μ_i's are equal to $\mu + \delta$ and the rest are equal to μ. If μ and σ are unknown, the test that rejects H if $\hat{D}(k, n)$ is large is the likelihood ratio test. Further it has the optimal property of being the scale- and location-invariant test of given size that maximizes the probability of identifying the k upper outliers correctly (Barnett and Lewis [1, p. 169]). If both μ and σ are known, $D(k, n)$ replaces $\hat{D}(k, n)$ above, but otherwise the appropriate version of $\hat{D}(k, n)$ is used. Barnett and Lewis [1] have compiled upper 5 and 1% points of $kD(k, n)$ and $k\hat{D}(k, n)$ for k up to 4 and selected n up to 100 (pp. 377–378, 383–385). Section 6.3.1 of [1] has a discussion of these tests and references to original sources of tables. Some of these tables are based on extensive simulation. Nagaraja [11] has compared these simulated percentage points of $D(k, n)$ with those obtained using asymptotic theory. The limiting approach taking $k = [np]$ produces results close to the simulated values. Hawkins [6] also presents some tables for $k = 1, 2$.

INDUCED SELECTION DIFFERENTIAL

Let Y be a variable highly correlated with X, having mean μ_Y and variance σ_Y^2. From a random sample of size n from the distribution of (X, Y), suppose that individuals with the top k X-values are selected. Then the induced selection differential of the selected Y-values is

$$D[k, n] = k^{-1} \sum_{i=n-k+1}^{n} \left(Y_{[i:n]} - \mu_Y\right)/\sigma_Y,$$

where $Y_{[r:n]}$ is the Y-value paired with $X_{r:n}$. Suppose direct selection based on Y-values is impossible or expensive to practice. One can do the selection based on X-values, in which case $D[k, n]$ comes naturally as a measure of improvement. In the genetics literature, this quantity is known as *response to selection* (Falconer, [5, p. 187]). Nagaraja [9, 10] has discussed some finite sample as well as asymptotic results for $D[k, n]$. For example, in the simple linear regression model* with $Y = \beta_0 + \beta_1 X + E$, $D[k, n]$ is asymptotically normal if $k = [np]$, $0 < p < 1$.

Now suppose that instead of the top k, all X-values exceeding a specified value are selected. This is known as *truncation selection*. Cochran [4] obtained the expected value of the induced selection differential for such a selection program, and also considered selection done in more than one stage.

References

[1] Barnett, V. and Lewis, T. (1984). *Outliers in Statistical Data*. 2nd ed. Wiley, New York.

[2] Burrows, P. M. (1972). *Biometrics*, **28**, 1091–1100. (Gives several references in genetics literature.)

[3] Burrows, P. M. (1975). *Biometrics*, **31**, 125–133.

[4] Cochran, W. G. (1951). *Proc. Second Berkeley Symp. Math. Statist. Prob.*, Univ. of California Press, Berkeley, CA, pp. 449–470.

[5] Falconer, D. S. (1960). *Introduction to Quantitative Genetics*. Ronald Press, New York.

[6] Hawkins, D. M. (1980). *Identification of Outliers*. Chapman and Hall, London, England, pp. 138, 153–154.

[7] Hill, W. G. (1976). *Biometrics*, **32**, 889–902.

[8] Milkman, R. (1978). *Genetics*, **88**, 391–403.

[9] Nagaraja, H. N. (1980). Contributions to the Theory of the Selection Differential and to Order Statistics. Ph.D. Thesis, Iowa State University.

[10] Nagaraja, H. N. (1981). *Ann. Inst. Statist. Math.*, **33**, 437–448.

[11] Nagaraja, H. N. (1982). *Ann. Statist.*, **10**, 1306–1310.

[12] Nagaraja, H. N. (1982). *J. Appl. Prob.*, **19**, 253–261.

[13] Pearson, E. S. and Hartley, H. O. (1972). *Biometrika Tables for Statisticians*, Vol. II. Cambridge University Press, Cambridge, England.

[14] Schaeffer, L. R., Van Vleck, L. D., and Velasco, J. A. (1970). *Biometrics*, **26**, 854–859.

[15] Stigler, S. M. (1973). *Ann. Statist.*, **1**, 472–477.

(EXTREME-VALUE DISTRIBUTIONS
ORDER STATISTICS
OUTLIERS)

H. N. NAGARAJA

SELECTION PROCEDURES

A statistical selection procedure uses sample data to select (identify) certain members of a family of k populations as "best" in such a way that we maintain control over the probability that those we select as best are indeed best. The term best is well defined and refers to the relative magnitudes of some parameter in the family of distributions. These procedures are designed to answer the question "which one (or ones) of k well-defined populations, such as drugs, makes of products, manufacturing processes, or breeds of cow, is (are) best according to a specific definition of best?" For example, a selection procedure might have the goal of selecting the one population with the largest parameter value, or selecting a subset of populations that contains the one with the largest parameter. The former goal uses the *indifference zone* approach while the latter goal uses the *subset selection* approach.

Selection procedures are one approach in multiple decision theory*, a field greatly influenced by Abraham Wald*, where we use only a simple (zero–one) loss function and our risk is an incorrect selection. A related approach in multiple decision* theory is that of ranking procedures*, where we use sample data to order or rank members of the family with respect to the relative magnitudes of some parameter in such a way that we maintain control over the probability of a correct ranking. The two kinds of problems are frequently linked together under the heading of ranking and selection procedures.

The classical and conventional statistical approach to a selection problem is to test the null hypothesis that the parameter values are all the same, a test of homogeneity*. A homogeneity test can only tell us whether the populations are equivalent or not; it cannot tell us which populations are best and hence cannot meet the goal of the experiment. Some modifications and extensions of the homogeneity test, such as multiple comparisons*, can be used to obtain additional information about the relative merits of the populations, but still cannot specifically answer the question posed.

ORIGINS OF THE PROBLEM

The theory of selection procedures originated in the 1940s when Wald [64] developed sequential analysis* and Girshick [31] modified Wald's technique and adapted it to the problem of ranking two populations. The next step in its development was study of the slippage model, where one parameter value shifts to the right or left while all other parameters remain equal, by Mosteller [48], Mosteller and Tukey [49], Bahadur [5], Bahadur and Robbins [6], Paulson [52, 53], and Truax [63]; *see also* MEAN SLIPPAGE PROBLEMS. The present formulation was introduced in Bechhofer [7], and developed more fully in Bechhofer et al. [11], Bechhofer and Sobel [9], Gupta [33], and Gupta and Sobel [39]. The first book that dealt with the theory of selection procedures was by Bechhofer et al. [13].

DESCRIPTION OF THE SELECTION PROBLEM

In the basic situation we have k populations, $\pi_1, \pi_2, \ldots, \pi_k$, each indexed by a parameter θ, where the cumulative distribution function (CDF) of π_i is $G(x; \theta_i)$ for $i = 1, 2, \ldots, k$. We assume that $G(x; \theta)$ is a stochastically increasing function of θ, i.e., $G(x; \theta') \geqslant G(x; \theta'')$ for $\theta' < \theta''$ for all x, and that the parameters can be ordered from smallest to largest. Denote the true ordered θ values by $\theta_{[1]} \leqslant \theta_{[2]} \leqslant \cdots \leqslant \theta_{[k]}$. The problem is to use parameter estimates computed from sample data to make some kind of selection from the k populations concerning the θ values, with control over the probability that the selection we make is correct. The simplest kind of selection would be to select the one best population, defined as the one with the largest θ value, $\theta_{[k]}$ (or the smallest, $\theta_{[1]}$).

Other goals concerning the θ values may be of interest. Some of the kinds of problems that have been solved using these procedures are as follows:

1. Selecting the one best population.
2. Selecting the t best populations for $t \geqslant 2$, (a) in an ordered manner or (b) in an unordered manner.
3. Selecting a random number (subset) of populations, say r, that include the t best populations (for $1 \leqslant t \leqslant r \leqslant k$).
4. Selecting a fixed number of populations, say r, that include the t best populations (for $1 \leqslant t \leqslant r \leqslant k$).
5. Selecting a random number (subset) of populations such that all populations as good as or better than a control population or known standard are included in the selected group.
6. Selecting all populations better than a control population or known standard.

The major references for the primary solution to some of the more important problems are listed under Outline of Problems and Primary References.

THE INDIFFERENCE ZONE APPROACH

Suppose we take a sample from each population and wish to use this sample data to select the population with the largest parameter $\theta_{[k]}$, i.e., to identify which of $\pi_1, \pi_2, \ldots, \pi_k$ has parameter $\theta_{[k]}$. The selection procedure here is to compute an estimate $\hat{\theta}_1, \hat{\theta}_2, \ldots, \hat{\theta}_k$ from each sample and assert that the population that produced the largest estimate $\hat{\theta}_{[k]}$ is the one with the largest parameter $\theta_{[k]}$. However, we must be concerned with the probability that this assertion is correct, called the *probability of a correct selection* (PCS).

The indifference zone approach is to guarantee a minimum probability for the PCS whenever the largest parameter value $\theta_{[k]}$ is sufficiently larger than the next largest value $\theta_{[k-1]}$, a region of the parameter space called the *preference zone* (PZ), because this is where we have a strong preference for a correct selection. The complement to this region is the *indifference zone* (IZ), because if the parameter values $\theta_{[k]}$ and $\theta_{[k-1]}$ are close, we can be indifferent about whether our selection is correct. For example, suppose we measure the proximity in the values of $\theta_{[k-1]}$ and $\theta_{[k]}$ by the difference $\delta = \theta_{[k]} - \theta_{[k-1]}$. Then the preference and indifference zones might be specified as

$$\text{PZ:} \quad \theta_{[k]} - \theta_{[k-1]} \geqslant \delta^*,$$

$$\text{IZ:} \quad \theta_{[k]} - \theta_{[k-1]} < \delta^*,$$

respectively, for some constant δ^*, as shown in Fig. 1. The probability of a correct selection in the preference zone depends on the configuration of θ-values in general, but in many cases there is a least favorable configuration (LFC) that makes the PCS a minimum for all θ in the PZ and any sample size. Then if we know that the PCS is equal to P^*, say, for the LFC in the PZ, the indifference zone approach assures us that the PCS is at least P^* for all $\theta_{[k]} - \theta_{[k-1]} \geqslant \delta^*$.

These procedures can be applied in practice as long as we have tables that relate the values of n, δ^*, and P^* for the specified distribution and parameter. The traditional approach is for the experimenter to specify δ^* and P^* and then determine the sample size n needed per population in order to

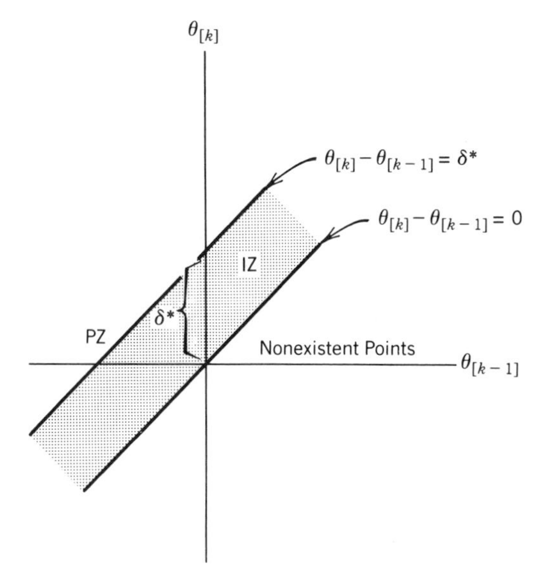

Figure 1 Graph of preference zone (PZ) and indifference zone (IZ) (shaded region) for the problem of selecting the one population with the largest θ-value when the PZ is defined as $\theta_{[k]} - \theta_{k-1} \geqslant \delta^*$ and the total parameter space is all real numbers.

satisfy the (δ^*, P^*) requirement. If n is fixed by other considerations, we could determine the preference zone threshold value δ^* for a particular P^* and the given n, or we might give the operating characteristic curve of the selection procedure as the set of all (δ^*, P^*) values that are satisfied for the given sample size. We might also wish to estimate the true probability of a correct selection.

Examples. Suppose $\pi_1, \pi_2, \ldots, \pi_k$ are k normal populations with unknown means $\mu_1, \mu_2, \ldots, \mu_k$ and common variance σ^2, and we want to select the population with the largest mean $\mu_{[k]}$. The least favorable configuration is $\mu_{[1]} = \cdots = \mu_{[k-1]} = \mu_{[k]} - \delta^*$ and the preference zone is $\mu_{[k]} - \mu_{[k-1]} \geqslant \delta^*$. The sample size needed for a specified (δ^*, P^*) requirement is computed from

$$n = \sigma^2(\tau/\delta^*)^2, \tag{1}$$

where τ is given in Table A.1 of Gibbons et al. [30], adapted from Milton [47], and reproduced here as Table 1. The sample estimates here are the sample means $\overline{X}_1, \overline{X}_2, \ldots, \overline{X}_k$, and the population selected is the one that produces the largest sample mean $\overline{X}_{[k]}$.

The classic illustration of this procedure is from Becker [14], where the problem is to select the best one out of $k = 10$ poultry stocks, best being defined as the stock with the largest mean hen-house egg production after 500 days. The data on egg production for each stock are assumed to be normally distributed with known variance $\sigma^2 = 5225$. How large a sample is needed for each stock

Table 1 Smallest Value of τ to Satisfy the P^* Requirement of k Normal Populations With Common Known Variance

	P^*					
k	0.750	0.900	0.950	0.975	0.990	0.999
2	0.9539	1.8124	2.3262	2.7718	3.2900	4.3702
3	1.4338	2.2302	2.7101	3.1284	3.6173	4.6450
4	1.6822	2.4516	2.9162	3.2220	3.7970	4.7987
5	1.8463	2.5997	3.0552	3.4532	3.9196	4.9048
6	1.9674	2.7100	3.1591	3.5517	4.0121	4.9855
7	2.0626	2.7972	3.2417	3.6303	4.0860	5.0504
8	2.1407	2.8691	3.3099	3.6953	4.1475	5.1046
9	2.2067	2.9301	3.3679	3.7507	4.1999	5.1511
10	2.2637	2.9829	3.4182	3.7989	4.2456	5.1916
15	2.4678	3.1734	3.6004	3.9738	4.4121	5.3407
20	2.6009	3.2986	3.7207	4.0899	4.5230	5.4409
25	2.6987	3.3911	3.8099	4.1761	4.6057	5.5161

Source: Adapted from Table A.1 of J. D. Gibbons, I. Olkin, and M. Sobel (1977), *Selecting and Ordering Populations: A New Statistical Methodology*, Wiley, New York, with permission.

in order that the probability of a correct selection is at least 0.90 whenever the difference between the means of the best and second-best stocks is at least 24, i.e., $\mu_{[10]} - \mu_{[9]} \geqslant 24 = \delta^*$? Table 1 gives $\tau = 2.9829$ for $k = 10$, $P^* = 0.90$, and substitution in (1) gives

$$n = 5225(2.9829/24)^2 = 80.7.$$

We round upward to 81 to obtain a conservative result.

To carry out the procedure, data on egg production for 81 chickens of each stock must be collected for 500 days and the sample means calculated. The stock with the largest sample mean is asserted to be the one with parameter $\mu_{[10]}$. We can then state with confidence 0.90 that the true mean production for the stock selected is within 24 units of the true mean production for the best stock.

In this same normal means selection problem, if the variances are assumed common but unknown, a two-stage selection procedure must be used to maintain control over the probability of a correct selection. This procedure is to take a random sample of n_0 observations from each population in the first stage and use them to calculate the pooled sample variance s^2 based on $\nu = k(n - 1)$ degrees of freedom. We then specify δ^* and P^*, with the same interpretation as before, and calculate

$$N = \max\left(n_0, \left\{2s^2h_{k,\nu}^2/\delta^{*2}\right\}^+\right),$$

where $\{X\}^+$ is the smallest integer greater than or equal to X and $h_{k,\nu}$ is a P^* quantile of the one-sided multivariate t distribution* with common correlation $\rho = 0.5$. Selected values of $h_{k,\nu}$ as a function of k and ν for $P^* = 0.95$ and $P^* = 0.99$ are given in Table A.4 of Gibbons et al. [29], adapted from Krishnaiah and Armitage [46], and reproduced here for $P^* = 0.95$ as Table 2. If $N > n_0$, a second-stage sample of $N - n_0$ additional observations must be taken from each population. The selection is then based on the sample means for all N observations. If $n_0 \geqslant N$, no additional observations are required.

THE SUBSET SELECTION APPROACH

In the subset selection approach to selecting the best population, where best is defined as the largest parameter $\theta_{[k]}$, we use the sample data to select a nonempty subset of populations (rather than a single population) and

Table 2 One-Sided Multivariate t Distribution with $\rho = 0.5$ and $P^* = 0.95$ for ν Degrees of Freedom and k Normal Populations

	k								
ν	2	3	4	5	6	7	8	9	10
5	2.01	2.44	2.68	2.85	2.98	3.08	3.16	3.24	3.30
6	1.94	2.34	2.56	2.71	2.83	2.92	3.00	3.06	3.12
7	1.89	2.27	2.48	2.62	2.73	2.81	2.89	2.95	3.00
8	1.86	2.22	2.42	2.55	2.66	2.74	2.81	2.87	2.92
9	1.83	2.18	2.37	2.50	2.60	2.68	2.75	2.81	2.86
10	1.81	2.15	2.34	2.47	2.56	2.64	2.70	2.76	2.81
20	1.72	2.03	2.19	2.30	2.39	2.46	2.51	2.56	2.60
30	1.70	1.99	2.15	2.25	2.33	2.40	2.45	2.50	2.54
60	1.67	1.95	2.10	2.21	2.28	2.35	2.39	2.44	2.48
120	1.66	1.93	2.08	2.18	2.26	2.32	2.37	2.41	2.45
∞	1.64	1.92	2.06	2.16	2.23	2.29	2.34	2.38	2.42

Source: Adapted from Table A.4 of J. D. Gibbons, I. Olkin, and M. Sobel (1977), *Selecting and Ordering Populations: A New Statistical Methodology*, Wiley, New York, with permission.

assert that this subset contains the population with parameter $\theta_{[k]}$. Here no assertion is made about which population is best within the selected subset and thus the selection is correct for any subset that includes the population with $\theta_{[k]}$. The size of the subset is random and can range between 1 and k. There is no preference and indifference zone here and we generally assume n is fixed. The experimenter specifies the minimum probability of a correct selection P^* and uses this to determine a decision rule as a function of the estimates $\hat{\theta}_1, \hat{\theta}_2, \ldots, \hat{\theta}_k$ that will indicate which populations are to be included in the selected subset. The minimum probability P^* holds for all configurations of θ-values and equality holds only if all parameters are equal.

Some variations in the subset selection approach in the literature are to select a subset containing the t best for $t \geqslant 2$, and to select a subset containing all populations as good as or better than a control population or a fixed standard value.

Subset selection procedures are useful primarily in a preliminary or screening investigation so that inferior populations can be eliminated with confidence before additional studies or comparisons are performed on the remaining populations. However, the statistical control described here applies only to the subset selection stage and not to any further tests.

Examples. Suppose $\pi_1, \pi_2, \ldots, \pi_k$ are k normal populations with unknown means $\mu_1, \mu_2, \ldots, \mu_k$, and we want to select a subset that contains the population with the largest mean $\mu_{[k]}$. A sample of n observations is taken from each population and the sample means $\bar{X}_1, \bar{X}_2, \ldots, \bar{X}_k$ are computed. The decision rule is to place population j in the selected subset if $\bar{X}_j$ satisfies

$$\bar{X}_{[k]} - c \leqslant \bar{X}_j \leqslant \bar{X}_{[k]}, \qquad (2)$$

for $j = 1, 2, \ldots, k$, where c is determined as a function of k, P^*, and either σ or the sample standard deviations. For the case where the population variances are assumed common and known, we have

Table 3 Profit Data for Five Alternative Production Plans

Plan	$\bar{X}$ ($)	s ($)	n
A	2976.40	175.83	50
B	2992.30	202.20	50
C	2675.20	250.51	50
D	3265.30	221.81	50
E	3130.90	277.04	50

$$c = \tau\sigma/\sqrt{n}, \qquad (3)$$

where τ is again found from Table 1 for the given k and P^*. Note that c here replaces δ^* in the indifference zone approach to selecting the best population. In the present problem n is specified and c is determined; in the indifference zone approach, δ^* is specified and n is determined.

If σ is assumed common but unknown, the constant c in (2) is

$$c = \sqrt{2}\, h_{k,\nu} s/\sqrt{n}, \qquad (4)$$

where $h_{k,\nu}$ is given in Table 2 and s^2 is the pooled sample variance.

A numerical example is from Naylor et al. [51], where a firm is considering five alternative operating plans for production of a product in four sequential stages. Profit is expressed as a function of six parameters that can be varied to simulate activity of the firm and thereby calculate data on total profit for a 90-day period under each of the five plans. The summary statistics shown in Table 3 are from data generated using $n = 50$ runs of each of the $k = 5$ plans and assumed to be normally distributed with common unknown variance.

To select a subset that contains the plan with the largest mean profit, we specify $P^* = 0.95$ and calculate $s^2 = 52{,}101$ with $\nu = 5(49) = 245$ degrees of freedom. Table 2 gives $h_{5,245} = 2.17$ and substitution in (4) gives $c = \sqrt{2}(2.17)\sqrt{52{,}101}/\sqrt{50} = 99.06$. The largest sample mean is $\bar{X}_{[5]} = 3265.30$, so the interval in (2) is

$$3166.24 \leqslant \bar{X}_j \leqslant 3265.30.$$

The sample mean for Plan D is the only one

that satisfies this requirement and hence our subset consists of only Plan D. We assert that this subset contains the plan with the largest profit and 0.95 is our confidence in this assertion.

OUTLINE OF PROBLEMS AND PRIMARY REFERENCES

This section gives a limited outline with references of some of the most important problems that have been solved in the area of selection procedures, including both the indifference zone approach (**I**, **II**, and **III**) and the subset selection approach (**IV**, **V**, and **VI**). All procedures given are one-stage unless noted as two-stage. This list is necessarily incomplete as the literature on the topic is vast; no sequential or Bayesian procedures are included here. See Dudewicz and Koo [22] and Gupta and Panchapakesan [38] for additional references.

I. Selecting the one best population.

A. Selecting the normal distribution with the largest (smallest) mean for:
1. variances common and known (Bechhofer [7]; Tamhane and Bechhofer [60, 61] (two-stage procedure));
2. variances common and unknown (two-stage procedure) (Bechhofer et al. [11]; Dunnett and Sobel [26]);
3. variances unequal and known (Bechhofer [7]);
4. variances unequal and unknown (two-stage procedure) (Dudewicz and Dalal [21]);
5. factorial experiments and blocking designs (Bechhofer [8]);
6. common known coefficient of variation (Tamhane [59]).

B. Selecting the normal distribution with the smallest variance (Bechhofer and Sobel [9]).

C. Selecting the binomial (or Bernoulli) distribution with the largest (smallest) probability of success (Sobel and Huyett [58]).

D. Selecting the category of the multinomial distribution* with the largest probability (Bechhofer et al. [12]) and with the smallest probability (Alam and Thompson [2]).

E. Selecting the Poisson distribution* with the largest (smallest) mean (Alam and Thompson [3]).

F. Nonparametric procedure for selecting the distribution with the largest quantile of order q (Sobel [57]).

G. Nonparametric procedure for selecting the distribution with the largest probability of producing the largest observation (Bechhofer and Sobel [10]; Dudewicz [19]).

H. Selecting the best object in a design with paired comparisons (David [17]).

I. Selecting the gamma distribution* with smallest (or largest) value of the scale parameter (Gupta [34]).

J. Selecting the multivariate normal distribution with the largest multiple correlation* (Rizvi and Solomon [55]; Alam et al. [4]).

K. Selecting the multivariate normal distribution with the largest generalized variance* (Gnanadesikan and Gupta [32]; Regier [54]).

L. Selecting the multivariate normal distribution with the largest (smallest) Mahalanobis distance (*see* MAHALANOBIS D^2) (Alam and Rizvi [1]).

M. Selecting the best multivariate normal distribution using a multivariate approach (Dudewicz and Taneja [23]).

II. Selecting the t best populations.

A. Selecting the t best populations for normal distributions with common known variance (Bechhofer [7]).

B. Selecting a subset of fixed size s that contains the t best (for $s \geqslant t$) for normal distributions with a common known variance (Desu and Sobel [18]).

III. Selecting all populations better than a control or known standard with respect to:

A. Means for normal distributions with a common unknown variance (two-stage procedure) (Tong [62]).

B. Means for normal distributions with common unknown variance (Dunnett [25a]).

C. Means for normal distributions with unknown variances (two-stage procedure) (Dudewicz et al. [24]).

D. Variances for normal distributions (Schafer [56]).

IV. Selecting a random size subset of populations that contains the one best population with respect to:

A. Means for normal distributions with

1. variances common and known (Gupta [33, 35]);
2. variances common and unknown (Gupta [33]; Gupta and Sobel [39]; Chen et al. [16]);
3. coefficient of variation common and known (Tamhane [59]).

B. Probabilities for binomial (or Bernoulli) distributions (Gupta and Sobel [41]; Gupta et al. [44]; Chen et al. [16]).

C. Variances for normal distributions (Gupta and Sobel [42, 43]).

D. Other parameters and distributions (Gupta and Panchapakesan [37]).

V. Selecting a random size subset that contains the t best populations with respect to normal population means or variances (Carroll [15]).

VI. Selecting a random size subset of populations that includes all populations as good as or better than a control or known standard with respect to:

A. Means for normal distributions with known variances (not necessarily common) (Gupta and Sobel [40]) and with common unknown variance (Gupta and Sobel [39, 40]).

B. Probabilities for binomial (or Bernoulli) distributions (Gupta and Sobel [41]; Gupta et al. [44]).

RECENT DEVELOPMENTS AND APPLICATIONS

In the last 30 years the theoretical literature on this topic has grown tremendously, and many modifications and extensions are still being developed at a rapid pace. One of the most useful recent developments is the extension of the concept from mere identification of populations with a certain characteristic to point and confidence interval estimation of certain parameters. Some procedures that combine the goals of selection and estimation are also available. Other recent useful developments are sequential adaptive procedures, procedures that combine the subset selection approach and the indifference zone approach, optimal sampling in selection problems*, and ranking* and selection in designed experiments.

A recent theoretical book on ranking and selection procedures is Gupta and Panchapakesan [38], which includes (Chap. 21) a complete description of developed estimation methods. Gupta and Huang [36] give a decision-theoretic survey of recent developments. A complete categorized bibliography is available in Dudewicz and Koo [22]. Dudewicz [20, Chap. 11] and Gibbons [29, Chap. 9] represent the only attempt to date to include these methods in a basic course on probability and statistical inference.

A limited discussion of some applications of these techniques appears in the books by

Kleijnen [45, Chaps. 5 and 6] and Naylor [50]. Gibbons et al. [29] provide a user-oriented survey of most of the methods appearing in the theoretical literature, and include all tables needed for their application. Gibbons et al. [29] is an expository survey paper including examples of data applications. Applications in marketing* are discussed in Gibbons and Gur-Arie [28], and engineering* applications are given in Gibbons [27].

References

[1] Alam, K. and Rizvi, M. H. (1966). *Ann. Inst. Statist. Math.*, **18**, 307–318.

[2] Alam, K. and Thompson, J. R. (1972). *Ann. Math. Statist.*, **43**, 1981–1990.

[3] Alam, K. and Thompson, J. R. (1973). *Technometrics*, **15**, 801–808.

[4] Alam, K., Rizvi, M. H., and Solomon, H. (1976). *Ann. Statist.*, **4**, 614–620.

[5] Bahadur, R. R. (1950). *Ann. Math. Statist.*, **21**, 362–375.

[6] Bahadur, R. R. and Robbins, H. (1950). *Ann. Math. Statist.*, **21**, 469–487.

[7] Bechhofer, R. E. (1954). *Ann. Math. Statist.*, **25**, 16–39.

[8] Bechhofer, R. E. (1977). *Proc. Winter Simulation Conference*, Vol. 1, pp. 65–70.

[9] Bechhofer, R. E. and Sobel, M. (1954). *Ann. Math. Statist.*, **25**, 273–289.

[10] Bechhofer, R. E. and Sobel, M. (1958). *Ann. Math. Statist.*, **29**, 325.

[11] Bechhofer, R. E., Dunnett, C. W., and Sobel, M. (1954). *Biometrika*, **41**, 170–176.

[12] Bechhofer, R. E., Elmaghraby, S. A., and Morse, N. (1959). *Ann. Math. Statist.*, **30**, 102–119.

[13] Bechhofer, R. E., Kiefer, J., and Sobel, M. (1968). *Sequential Identification and Ranking Procedures*, University of Chicago Press, Chicago, IL.

[14] Becker, W. A. (1961). *Poultry Science*, **40**, 1507–1514.

[15] Carroll, R. J., Gupta, S. S., and Huang, D.-Y. (1975). *Commun. Statist.*, **4**, 987–1008.

[16] Chen, H. J., Dudewicz, E. J., and Lee, Y. J. (1976). *Sankhyā B*, **38**, 249–255.

[17] David, H. A. (1963). *The Method of Paired Comparisons*. Hafner, New York.

[18] Desu, M. M. and Sobel, M. (1968). *Biometrika*, **55**, 401–410.

[19] Dudewicz, E. J. (1971). *J. Amer. Statist. Ass.*, **66**, 152–161.

[20] Dudewicz, E. J. (1976). *Introduction to Statistics and Probability*. American Sciences Press, Syracuse, New York.

[21] Dudewicz, E. J. and Dalal, S. R. (1975). *Sankhyā B*, **37**, 28–78.

[22] Dudewicz, E. J. and Koo, J. O. (1982). *The Complete Categorized Guide to Statistical Selection and Ranking Procedures*. American Sciences Press, Columbus, OH. (A comprehensive research tool and practitioners' guide to selection procedures, ranking procedures, and estimation of ordered parameters.)

[23] Dudewicz, E. J. and Taneja, V. S. (1981). *Commun. Statist. A*, **10**, 1849–1868.

[24] Dudewicz, E. J., Ramberg, J. S., and Chen, H. J. (1975). *Biometrische Zeit.*, **17**, 13–26.

[25] Dunnett, C. W. (1955). *J. Amer. Statist. Ass.*, **50**, 1096–1121.

[25a] Dunnett, C. W. (1964). *Biometrics*, **20**, 482–491.

[26] Dunnett, C. W. and Sobel, M. (1954). *Biometrika*, **41**, 153–169.

[27] Gibbons, J. D. (1982). *J. Qual. Tech.*, **14**, 80–88.

[27a] Gibbons, J. (1985). *Nonparametric Methods for Quantitative Analysis*, 2nd. ed. American Sciences Press, Syracuse, New York.

[28] Gibbons, J. D. and Gur-Arie, O. (1981). *J. Marketing Res.*, **18**, 449–455.

[29] Gibbons, J. D., Olkin, I., and Sobel, M. (1977). *Selecting and Ordering Populations: A New Statistical Methodology*. Wiley, New York. (An applications-oriented survey of selection and ranking procedures with extensive tables and examples to facilitate their use.)

[30] Gibbons, J. D., Olkin, I., and Sobel, M. (1979). *Amer. Statist.*, **33**, 185–195.

[31] Girshick, M. A. (1946). *Ann. Math. Statist.*, **17**, 123–143.

[32] Gnanadesikan, M. and Gupta, S. S. (1970). *Technometrics*, **12**, 103–117.

[33] Gupta, S. S. (1956). On a Decision Rule for a Problem in Ranking Means. Ph.D. thesis, *Mimeo Series No. 150*, Institute of Statistics, University of North Carolina, Chapel Hill, NC.

[34] Gupta, S. S. (1963). *Ann. Inst. Statist. Math.*, **14**, 199–216.

[35] Gupta, S. S. (1965). *Technometrics*, **7**, 225–245.

[36] Gupta, S. S. and Huang, D.-Y. (1981). *Multiple Statistical Decision Theory*. Springer-Verlag, New York.

[37] Gupta, S. S. and Panchapakesan, S. (1972). *Ann. Math. Statist.*, **43**, 814–822.

[38] Gupta, S. S. and Panchapakesan, S. (1979). *Multiple Decision Procedures: Theory and Methodology of Selecting and Ranking Populations*, Wiley, New York. (A comprehensive survey of the theory of selection and ranking procedures with extensive references.)

[39] Gupta, S. S. and Sobel, M. (1957). *Ann. Math. Statist.*, **28**, 957–967.

[40] Gupta, S. S. and Sobel, M. (1958). *Ann. Math. Statist.*, **29**, 235–244.

[41] Gupta, S. S. and Sobel, M. (1960). In *Contributions to Probability and Statistics, Essays in Honor of Harold Hotelling*, I. Olkin, S. G. Ghurye, W. Hoeffding, W. G. Madow, and H. B. Mann, eds. Stanford University Press, Stanford, CA., pp 224-248.

[42] Gupta, S. S. and Sobel, M. (1962). *Biometrika*, **49**, 495–507.

[43] Gupta, S. S. and Sobel, M. (1962). *Biometrika*, **49**, 509–523.

[44] Gupta, S. S., Huyett, M. J., and Sobel, M. (1957). *Transactions of the American Society for Quality Control*, 11*th Tech. Conf.*, pp. 635–644.

[45] Kleijnen, J. P. C. (1975). *Statistical Techniques in Simulation*, *Part II*. Dekker, New York.

[46] Krishnaiah, P. R. and Armitage, J. V. (1966). *Sankhyā B*, **28**, Parts 1 and 2 (Part of Aerospace Research Laboratories Report No. 65-199, Wright Patterson Air Force Base, Dayton, Ohio.)

[47] Milton, R. C. (1963). Tables of the Equally Correlated Multivariate Normal Probability Integral. *Technical Report No.* 27, University of Minnesota, Minneapolis, MN.

[48] Mosteller, F. (1948). *Ann. Math. Statist.*, **19**, 58–65.

[49] Mosteller, F. and Tukey, J. (1950). *Ann. Math. Statist.*, **21**, 120–123.

[50] Naylor, T. H. (1971). *Computer Simulation Experiments with Models of Economic Systems*. Wiley, New York.

[51] Naylor, T. H., Wertz, K., and Wonnacott, T. H. (1967). *Commun. ACM*, **10**, 703–710.

[52] Paulson, E. (1952). *Ann. Math. Statist.*, **23**, 610–616.

[53] Paulson, E. (1952). *Ann. Math. Statist.*, **23**, 239–246.

[54] Regier, M. H. (1976). *Technometrics*, **18**, 483–489.

[55] Rizvi, M. H. and Solomon, H. (1973). *J. Amer. Statist. Ass.*, **68**, 184–188; Corrigenda (1974), **69**, 288.

[56] Schafer, R. E. (1977). In *Theory and Applications of Reliability*, Vol. 1, C. P. Tsokos and I. N. Shimi, eds. Academic, New York, pp. 449–473.

[57] Sobel, M. (1967). *Ann. Math. Statist.*, **38**, 1804–1816.

[58] Sobel, M. and Huyett, M. (1957). *Bell Syst. Tech. J.*, **36**, 537–576.

[59] Tamhane, A. C. (1978). *Sankhyā B*, **39**, 334-361.

[60] Tamhane, A. C. and Bechhofer, R. E. (1977). *Commun. Statist. A*, **6**, 1003–1033.

[61] Tamhane, A. C. and Bechhofer, R. E. (1979). *Commun. Statist. A*, **8**, 337–358.

[62] Tong, Y. L. (1969). *Ann. Math. Statist.*, **40**, 1300–1324.

[63] Truax, D. R. (1953). *Ann. Math. Statist.*, **24**, 669–674.

[64] Wald, A. (1947). *Sequential Analysis*. Wiley, New York.

(MEAN SLIPPAGE PROBLEMS
MULTIPLE DECISION THEORY
OPTIMAL SAMPLING IN SELECTION PROBLEMS
RANKING PROCEDURES
SELECTIVE INFERENCE
SOMERVILLE'S MULTIPLE RANGE SUBSET SELECTION PROCEDURE)

JEAN DICKINSON GIBBONS

SELECTIVE INFERENCE

A statistical inference* procedure may be called *selective* if the identity of the object of inference (the parameter to be estimated, the hypothesis to be tested, etc.) is selected on the basis of the same sample data that are to be used in the procedure.

In a *non*selective inference procedure, the object of inference may be fixed or random, but its identity (reflecting the purpose of the inference) is assumed to be fixed and to have been determined before the data were obtained. The probabilistic properties of such a procedure (distributions, expected values, variances, risks, significance probabilities, etc.) are calculated under this assumption. Therefore, in general, these properties no longer hold when the procedure is applied to an object whose identity is selected on the basis of the data. Thus, a selective inference situation calls for devising a technique that takes into account the selective (and therefore random) identity of the object of inference. So far, such techniques have been developed only for a few types of situation.

Example 1. Estimating the Performance of a Selected Population.

(a) In one type of a selection* experiment, samples from several populations are compared, and the population that has yielded the highest sample mean is inferred to possess the highest population mean. In practice, the "populations" involved may represent alternative choices of, e.g., industrial product, agricultural technique, or medical treatment, and the selected population may then be recommended for future use. It is therefore desirable to give an estimate of how good the selected population really is.

The identity of the selected population is obviously a (random) outcome of the experiment, and in this situation the mean of the sample from the selected population is usually *not* a good estimator of the corresponding population mean. In particular, it is (on the average) an overestimator, and its positive bias may be considerable [7]. The problem of finding a better estimator has been tackled in a few papers, mainly for special cases; see refs. 2 and 3 and the references listed therein.

(b) A similar problem arises in reliability* experiments. The proposed components of a system are separately tested, and each component is accepted if its sample failure rate is low enough (and rejected otherwise). The total population failure rate of the accepted components (φ) is the sum of parameters whose identity is a (random) outcome of the experiment. Analogously to Example 1(a), the total sample failure rate of the accepted components is an underestimator of φ. Unbiased estimators of φ are available in some cases [5, 8].

Example 2. Selected Comparisons between Means. In the context of multiple comparisons* between several sample means, the inference procedures are usually formulated in terms of joint (simultaneous) inference, specifying a lower bound $(1 - \alpha)$ on the joint probability that all the comparison statements that can be made (within a given class) are simultaneously correct. In typical practice, however, only a few comparisons, selected on the basis of the data, are made [4]. If one specifies an upper bound (α) on the probability that a given actually made statement of significant difference is wrong, the random identity of the compared means should be taken into account. Also, such a significance test does not test a predetermined "null" hypothesis of no difference. Rather, it tests a selected (i.e., random) hypothesis (H), which states that a certain order relation between the populations means, suggested by the configuration of the sample means, actually holds [1]. This leads [6] to a modified definition of the significance level α, viz. Pr[H is inferred to be true but is actually false] $\leqslant \alpha$.

For the special case of comparisons between several treatments and a control, a selective testing method, analogous to the Newman–Keuls* procedure, is available [6].

In principle, the selective inference approach could be extended to cover any situation in which one (i) looks at the same data, then, perhaps, (ii) carries out some diagnostic procedures of data analysis, and only then (iii) proceeds to probabilistic inference about an object selected in stages (i) and (ii). This, however, is likely to be feasible and useful only when the set out of which the object of inference is selected is plausibly definable and not too broad. The following examples may be ripe for explicit investigation in terms of selective inference.

Example 3. Outlier* and Slippage Problems. In these problems, some sample observations or statistics are suspected of being unlike the others, and the identity of the suspects is usually selected on the basis of the data. A selective inference approach should also take into consideration the random nature of the number of suspects and of their directions of divergence.

Example 4. Regression* Problems in Which the Regressors are Selected on the Basis of the Data. In these problems, the random number and identity of the selected regressors needs to be taken into full account.

Example 5. Selected Effects in the Analysis of Variance* of a Multifactorial Experiment. In usual practice, each effect (mean effect, interaction, etc.) is tested separately. However, the conclusions from the experiment are often drawn in terms of those (selected) effects that happen to be significant. The random identity of these effects calls for modifying the testing procedure accordingly.

Example 6. Problems of Testing and Estimating Selected Correlation* Coefficients. When a correlation matrix is scanned for significant coefficients, the identity of the selected coefficients is obviously random. The significance and estimation procedures should therefore be modified accordingly.

References

[1] Bahadur, R. R. (1952). *Sankhyā*, **12**, 79–88. (Investigates the *t*-test as a test of a selected hypothesis.)

[2] Borglum, D. G. (1972). Some Estimators of Parameters from a Selected Population. *Tech. Rep. No. 195*, Dept. of Statistics, Stanford University, Stanford, CA. [Investigates Example 1(a), including a bibliography and recapitulation of previous (published and unpublished) research.]

[3] Cohen, A. and Sackrowitz, H. B. (1982). In *Statistical Decision Theory and Related Topics III, Proc. Third Purdue Symp.*, Vol. 1, S. S. Gupta and J. O. Berger, eds. Academic, New York. [Investigates Example 1(a) for the normal case.]

[4] Cox, D. R. (1965). *Technometrics*, **7**, 223–224. (Points out the selective nature of Example 2.)

[5] Kolmogorov, A. N. (1950). *Izv. Akad. Nauk SSSR, Ser. Mat.*, **14**, 303–326 (*Amer. Math. Soc. Transl.*, **98**). [Investigates a particular case of Example 1(b).]

[6] Putter, J. (1982). In *A Festschrift for E. L. Lehmann*, P. J. Bickel, K. A. Doksum and J. L. Hodges, Jr., eds. Wadsworth International Group, Belmont, CA, pp. 428–447. (Investigates Example 2.)

[7] Putter, J. and Rubinstein, D. (1968). On Estimating the Mean of a Selected Population. *Tech. Rep. No. 165*, Dept of Statistics, University of Wisconsin, Madison, WI. [Investigates particular cases of Example 1(a).]

[8] Rubinstein, D. (1965). Estimation of Failure Rates in a Dynamic Reliability Program. *Tech. Info. Ser. Rep. No. 65RG07*, General Electric Co., Ithaca, NY. [Partial abstract: Rubinstein, D. (1961). *Ann. Math. Statist.*, **32**, 924. Investigates Example 1(b).]

(MEAN SLIPPAGE PROBLEMS
MULTIPLE COMPARISONS
NEWMAN-KEULS PROCEDURE
OUTLIERS
SELECTION PROCEDURES)

JOSEPH PUTTER

SELECTIVE LEAST SQUARES *See* REGRESSION VARIABLES, SELECTION OF

SELECTIVE PROCEDURES *See* REGRESSION VARIABLES, SELECTION OF

SELECTIVE SAMPLING *See* REPRESENTATIVE SAMPLING

SELF-AFFINE PROCESSES *See* SELF-SIMILAR PROCESSES

SELF-CONSISTENCY

Nonparametric estimation of a distribution function is a common and important statistical problem. When the observed data consist of a sample of independent and identically distributed (i.i.d.) observations from the distribution to be estimated, say $F(t)$, the maximum likelihood* solution is well known to be the empirical distribution function (*see* EDF STATISTICS). When the observations are incomplete, so that the observed data no longer form an i.i.d. sample from $F(t)$, the problem is generally more complicated, although the solution is well known for certain types of incompleteness (for example, right-censoring). Other forms of incompleteness to be considered here include double censoring*, truncation*, grouping, and mixing.

Efron [3] proposed a class of estimates for the right-censored problem known as *self-consistent estimates*. The original inspiration for self-consistency was basically intuitive; it corresponds to the maximum likelihood principle in the nonparametric setting for a broad class of incomplete data* problems.

The general idea is as follows. Let $\mathbf{x} = (x_1, \dots, x_n)$ denote a random sample from $F(t)$, defined on a set χ on the real line. When $\mathbf{x}$ is observed completely, the empirical distribution function

$$\hat{F}(t) = \sum_1^n I[x_i \leqslant t]/n, \qquad t \in \chi, \quad (1)$$

maximizes the sample likelihood among the class of distribution functions on χ. Here $I[a]$ is the indicator function.

With incomplete data, we cannot observe $\mathbf{x}$ directly, but $\mathbf{y} = \mathbf{y}(\mathbf{x})$, which is a many-to-one mapping from the sample space of $\mathbf{x}$ to the sample space of $\mathbf{y}$. For example, with right-censoring, we observe $y_i = (w_i, \delta_i)$, where $w_i = x_i$ if $\delta_i = 1$; if $\delta_i = 0$, we know only that $x_i \geqslant w_i$, for some censoring time w_i. In general, we have an incomplete data problem if, for at least one i, $I[x_i \leqslant t]$ cannot be evaluated for all t in χ. In this case $\hat{F}(t)$ cannot be calculated from the available data.

To estimate $F(t)$ with right-censoring, Efron proposed a self-consistent estimate $F^*(t)$ of the distribution $F(t)$, defined as one that satisfies

$$F^*(t) = \sum_1^n P(x_i \leqslant t \,|\, \mathbf{y}, F^*)/n, \qquad t \in \chi, \quad (2)$$

where $P(x_i \leqslant t \,|\, \mathbf{y}, F^*)$ denotes the conditional probability that $x_i \leqslant t$, given $\mathbf{y}$ and F^*. The motivation for F^* is that in the absence of observing x_i directly,

$$P(x_i \leqslant t \,|\, \mathbf{y}, F) = E(I[x_i \leqslant t] \,|\, \mathbf{y}, F)$$

represents our best guess about $I[x_i \leqslant t]$, given the observed $\mathbf{y}$ and some distribution F. Notice that if $\mathbf{y}$ determines x_i exactly for some i, then $P(x_i \leqslant t \,|\, \mathbf{y}, F) = I[x_i \leqslant t]$; if $\mathbf{y}$ determines x_i exactly for all i, then (2) reduces to (1).

Equation (2) suggests an iterative procedure (the *self-consistency algorithm*) for calculating F^*: Start with an initial guess F^p; calculate $p_i(t) = P(x_i \leqslant t \,|\, \mathbf{y}, F^p)$ for each x_i and all t in χ; sum the $p_i(t)$ and divide by n to get F^{p+1}; set $F^p = F^{p+1}$; repeat the cycle until convergence. By construction, a fixed point of this algorithm is a self-consistent estimator.

Example 1. Right Censoring. For simplicity assume $n = 3$ and observe $w_1 < w_2 < w_3$, where $\delta_1 = \delta_3 = 1$ and $\delta_2 = 0$. Because w_1 and w_3 are uncensored, $\mathbf{y}$ determines x_1 and x_3 exactly; thus $p_i(t) = I[x_i \leqslant t]$ for $i = 1$ and 3. Since we know $x_2 \geqslant w_2$, $I[x_i \leqslant t]$ can be evaluated for all $t < w_2$. If $t \geqslant w_2$, we cannot know if x_2 is less than t, but given any F and y_2, we can assign it a conditional probability

$$P(x_2 \leqslant t | y_2, F) = \int_{w_2}^{t} dF(t) \Big/ \int_{w_2}^{\infty} dF(t). \quad (3)$$

Thus at each iteration $p_2(t)$ is defined by evaluating (3) at F^p if $t \geqslant w_2$; otherwise, $p_2(t) = 0$. If we start our iterations with a step function for F^p, with probability $\frac{1}{3}$ at each of w_1, w_2, and w_3, after k iterations, F^{p+k} is a step function with probability $\frac{1}{3}$ at w_1, $(\frac{1}{3})(1/2^k)$ at w_2, and the remainder at w_3. At convergence then, F^* is a step function with positive probability $\frac{1}{3}$ at w_1 and $\frac{2}{3}$ at w_3.

In any right-censoring problem, starting the iterations with the empirical CDF of all the points (censored and uncensored) will yield the nonparametric maximum likelihood (or Kaplan–Meier* [5]) estimate (NPMLE) of F. Starting with the empirical CDF of the *uncensored* points only gives faster convergence to the same solution. [If $w_{(n)}$, the largest w_i, is censored, the NPMLE is not uniquely defined beyond $w_{(n)}$. In this case the algorithm will converge to an estimate that equals the NPMLE at all points $\leqslant w_{(n)}$. For $t > w_{(n)}$, $F^*(t)$ will depend upon the starting values used for the algorithm.] Of course, an iterative procedure is not needed for computation in this case, but generally it is.

Turnbull [12, 13] extended self-consistency to include arbitrarily grouped, censored, or truncated data, showing the equivalence of Efron's principle and the nonparametric likelihood equations for these incomplete data cases. He also demonstrated convergence of the algorithm for these cases.

Example 2. Double Censoring. Here x_i can be right-censored at w_i ($\delta_i = 0$), left-censored at w_i ($\delta_i = -1$), or observed exactly ($\delta_i = 1$). The algorithm proceeds exactly as in Example 1, where now at each iteration we calculate

$$p_i(t) = I[x_i \leqslant t], \quad \text{if } \delta_i = 1,$$

$$p_i(t) = \int_{w_i}^{t} dF(t) \Big/ \int_{w_i}^{\infty} dF(t), \quad \text{if } \delta_i = 0 \text{ and } t \geqslant w_i,$$

$$p_i(t) = 0, \quad \text{if } \delta_i = 0 \text{ and } t < w_i,$$

$$p_i(t) = \int_{-\infty}^{\max(t,\, w_i)} dF(t) \Big/ \int_{-\infty}^{w_i} dF(t), \quad \text{if } \delta_i = -1.$$

Turnbull and Mitchell [14, 15] and Mitchell and Turnbull [9] have considered nonparametric estimation in serial sacrifice experiments, again showing that the NPMLE satisfies the self-consistency principle, and demonstrating convergence of the algorithm. Campbell [1] and Hanley and Parnes [4] have studied nonparametric estimation of a bivariate distribution subject to censoring on both variables using the self-consistency principle. These papers present applications to real data sets.

Laird [6] discussed self-consistent estimators in the context of nonparametric estimation of a mixing distribution. She pointed out that the principle is equivalent to the *Missing Information Principle** of Orchard and Woodbury [10], specialized to the nonparametric setting. In the case of incomplete data from parametric exponential families, the mathematical basis of both these principles reduces to *Sundberg's formulas**. She showed the self-consistency algorithm is a special case of the EM [2].

With censoring, grouping, and truncation, self-consistency uniquely defines an estimate of F that is identical to the NPMLE. In some settings, the self-consistency principle (2) may not define a unique estimate. However, the NPMLE is always self-consistent. We now give a justification for this based on the Missing Information Principle.

Suppose we desire to estimate $F(t)$ at an arbitrary fixed t_0 in χ. When we observe $\mathbf{x}$ completely, this can be recast as a simple binomial estimation problem, with sufficient statistic $s = \Sigma^n I[x_i \leqslant t_0]$, which is simply the number of $x_i \leqslant t_0$. The maximum likelihood estimate of $F(t_0)$ is well known to be

$$\hat{F}(t_0) = s/n. \tag{4}$$

Since t_0 is arbitrary, (4) is clearly true for all t_0 in χ, in agreement with (1). This extends easily to the multinomial setting when we pick a fixed grid of points in χ, the sufficient statistics* being the number of x_i below each grid point.

Now when we observe $\mathbf{y}$ rather than $\mathbf{x}$, we can apply the general theory for incomplete data from exponential family densities. This theory says that the likelihood equation for $F(t_0)$ based on $\mathbf{y}$ is found by equating expectations of the sufficient statistic, s:

$$E(s|F^*(t_0)) = E(s\,|\mathbf{y}, F^*(t_0)).$$

Here $F^*(t_0)$ is the maximum likelihood estimate based on the data $\mathbf{y}$. Substituting $\Sigma_1^n I[x_i \leqslant t_0]$ for s gives

$$nF^*(t_0) = \sum_1^n P(x_i \leqslant t_0\,|\mathbf{y}, F^*),$$

in agreement with the self-consistency principle.

The general relationship between self-consistent and nonparametric maximum likelihood (also called generalized maximum likelihood) estimates has been studied by Tsai and Crowley [11]. They show that starting the algorithm with any finite step function leads to convergence at some self-consistent estimator [13]. They give the conditions that guarantee that the solution is the NPMLE of F. Basically, one must show that the solution $F^*(t)$ maps into the empirical distribution function of the observed y_i's, which is induced by the mapping from $\mathbf{x}$ to $\mathbf{y}$. They also give conditions that guarantee that a NPMLE is self-consistent, and use the self-consistency property of the NPMLE to study its asymptotic distribution.

Example 3. Mixing. Mixing is a special, interesting type of incompleteness, since here, in contrast with all other types of incompleteness discussed in the literature, it is not possible to tell by inspection of the data where the support points of the NPMLE should be located. The NPMLE is a step function placing positive support at some finite number $K \leqslant n$ of points, but even K cannot be determined exactly by inspection of the data [7]. In this case, (2) does not uniquely define the NPMLE nor even a unique estimate; however the NPMLE does satisfy the self-consistency principle, and the self-consistency algorithm can be used to compute it.

With mixing, we observe y_i, where conditional on the unknown x_i, each y_i has known sampling density $h(y_i|x_i)$. For example y_i could be $N(x_i, 1)$, $N(0, x_i)$, binomial with probability x_i, or Poisson with mean x_i. Straightforward application of Bayes theorem* shows that self-consistency implies

$$nF^*(t) = \sum_{i=1}^{n} \left[\frac{\int_{-\infty}^{t} h(y_i|x)\, dF^*(x)}{\int_{-\infty}^{\infty} h(y_i|x)\, dF^*(x)} \right]. \quad (5)$$

If $F^*(t_0) = I[x \leqslant t_0]$ for some t_0, then (5) is trivially satisfied for all t_0. An additional set of equations, defining the location of the K support points with positive probability, is necessary to determine the NPMLE. In this case, the algorithm adjusts not only the amount of probability at each support point, but its location as well. Details of the implementation are given in ref. 6.

In a general setting, computation of the self-consistent estimate involves determining the location of and the amount of probability at a finite number K of support points. When n is large, K may be also; thus the number of parameters involved often makes conventional computational routines impractical. The EM or self-consistency algorithm is easily programmed; large numbers of parameters pose no conceptual problems. In practice, however, many iterations (500 to 1000) may be required [6, 8], especially if the support points with positive probability cannot be identified by inspecting the data.

Another type of incompleteness closely related to mixing is convolution*; it arises frequently in obtaining indirect measurements in the physical sciences. Two interesting examples that apply the EM algorithm to obtain nonparametric estimates of a convoluted distribution are given in Vardi et al. [16] and Maher and Laird [8].

References

[1] Campbell, G. (1981). *Biometrika*, **68**, 417–422. (Discusses nonparametric estimation of a bivariate distribution with censored observation.)

[2] Dempster, A. P., Laird, N. M., and Rubin, D. B. (1977). *J. R. Statist. Soc. B.*, **39**, 1–38. (With Discussion. Introduces the general form of the EM algorithm and gives many examples.)

[3] Efron, B. (1967). *Proc. Fifth Berkeley Symp. Math. Statist. Prob.*, Vol. 4, Univ. of California Press, Berkeley, CA, pp. 831–853. (Discusses the two-sample problem with censored data and introduces the self-consistency algorithm.)

[4] Hanley, J. A. and Parnes, M. N. (1983). *Biometrics*, **39**, 129–139. (Discusses nonparametric estimation of a bivariate distribution with censored observations.)

[5] Kaplan, E. L. and Meier, P. (1958). *J. Amer. Statist. Ass.*, **53**, 457–481. (Derives the nonparametric maximum likelihood estimate of a distribution based on right-censored data.)

[6] Laird, N. M. (1978). *J. Amer. Statist. Ass.*, **73**, 805–811. (Considers nonparametric estimation in mixture problems and applies the EM algorithm.)

[7] Lindsay, B. (1983). *Ann. Statist.*, **11**, 86–94. (Discusses properties of the nonparametric maximum likelihood estimate of a mixing distribution.)

[8] Maher, E. H. and Laird, N. M. (1985). *J. Aerosol Sci.*, **16**, 557–570. (Applies the EM algorithm to determine the distribution of particle sizes in a volume of air using diffusion battery data.)

[9] Mitchell, T. J. and Turnbull, B. W. (1979). *Biometrics*, **35**, 221–234. (Extends the work in ref. 14.)

[10] Orchard, T. and Woodbury, M. A. (1972). *Proc. 6th Berkeley Symp. Math. Statist. Prob.*, Vol. 1, Univ. of California Press, Berkeley, CA, pp. 697–715. (Introduces the missing information principle and considers missing data in the multivariate normal.)

[11] Tsai, W. Y. and Crowley, J. (1985). *Ann. Statist.*, **13**, 1317–1334. (Develops the general theory relating nonparametric maximum likelihood and self-consistency for incomplete data samples.)

[12] Turnbull, B. W. (1974). *J. Amer. Statist. Ass.*, **69**, 169–173. (Discusses self-consistency and nonparametric maximum likelihood estimation with doubly censored data.)

[13] Turnbull, B. W. (1976). *J. R. Statist. Soc., Ser. B*, **38**, 290–295. (Extends the results of ref. 12 to estimation with arbitrarily grouped, censored, and truncated data.)

[14] Turnbull, B. W. and Mitchell, T. J. (1978). *Biometrics*, **34**, 555–570. (Discusses nonparametric estimation for disease prevalence distributions in serial/sacrifice experiments.)

[15] Turnbull, B. W. and Mitchell, T. J. (1984). *Biometrics*, **40**, 41–50. (Discusses nonparametric estimation of time to onset for disease in serial/sacrifice experiments.)

[16] Vardi, Y., Shepp, L. A., and Kaufman, L. (1985). *J. Amer. Statist. Ass.*, **80**, 8–37. (Applies the EM algorithm to a problem in emission tomography.)

Acknowledgment

This work was supported by grant No. GM-29745 from the National Institutes of Health.

(CENSORED DATA
KAPLAN–MEIER ESTIMATOR
MAXIMUM LIKELIHOOD ESTIMATION
MISSING INFORMATION PRINCIPLE
SUNDBERG FORMULAS)

NAN M. LAIRD

SELF-CORRECTING PROCESS

This is a counting process with intensity function of the form

$$\lambda(t|H_{0t}) = \exp[\alpha + \beta\{t - \rho N(t)\}],$$
$$\beta, \rho \geqslant 0,$$

where $N(t)$ is the number of occurrences up to time t, H_{0t} is the σ-field generated by the past of the process over the interval $[0, t)$, and α, β, and ρ are real parameters. This class of processes was introduced by Isham and Westcott [1], and later developed by Ogata and Vere-Jones [2, 4]. It is related to a stress-release model used to describe the occurrence of major earthquakes in Japan (Vere-Jones [3]).

The properties of the process depend critically on the values of the parameters β and ρ. For $\beta = 0$, it is a stationary Poisson process*. (If β and ρ have opposite signs, the corresponding process is called *explosive*.)

References

[1] Isham, V. and Westcott, M. (1979). *Stoch. Processes Appl.*, **8**, 335–347.

[2] Ogata, Y. and Vere-Jones, D. (1984). *Stoch. Processes Appl.*, **17**, 337–348.

[3] Vere-Jones, D. (1978). *J. Phys. Earth*, **26**, 126–146.

[4] Vere-Jones, D. and Ogata, Y. (1984). *J. Appl. Prob.*, **21**, 335–342.

(MARKOV PROCESSES
POINT PROCESSES
STOCHASTIC PROCESSES, POINT)

SELF-ENUMERATION

A measurement or survey procedure in which each respondent is provided with a questionnaire to complete. Self-enumeration procedures eliminate interviewers and so interviewer errors are avoided. On the other hand, when a questionnaire is sent to a household or organization, there is no control over who is actually answering the questions; moreover, absence of interviewers often leads to low response rates.

(SURVEY SAMPLING)

SELF-RECIPROCAL DISTRIBUTIONS

The *reciprocal* distribution of a distribution with characteristic function* $\phi(t)$ is one with PDF proportional to $\phi(x)$. If the reciprocal distribution is the same as the original one it is called *self-reciprocal*. Among such distributions are the standard normal [with PDF $(\sqrt{(2\pi)})^{-1}\exp(-\frac{1}{2}x^2)$], the sech distribution [with PDF $(\sqrt{(2\pi)})^{-1}\operatorname{sech}(x\sqrt{(\pi/2)})$] (Bass and Levy [1]), and a special mixture of two normal distributions [with PDF

$$(\sqrt{(2\pi)})^{-1}\{\exp(-\tfrac{1}{2}x^2\sigma^{-2}) + \sigma\exp(-\tfrac{1}{2}x^2\sigma^2)\}(1+\sigma)^{-1}].$$

An extensive study of self-reciprocal PDFs and their relation to stable* and infinitely divisible* distributions has been carried out by Pastena [2], who gives the following PDF as an example of a self-reciprocal distribution that is neither infinitely divisible nor unimodal:

$$\tfrac{1}{2}\max(1-|x|,0) + \frac{1}{4\pi}\left(\frac{\sin(x/2)}{x/2}\right)^2.$$

References

[1] Bass, J. and Levy, P. (1950). *C. R. Acad. Sci. Paris*, **230**, 815–817.

[2] Pastena, D. (1983). *Publ. Inst. Statist. Univ. Paris*, **27**, 81–91.

(INFINITE DIVISIBILITY
STABLE DISTRIBUTIONS)

SELF-SIMILAR PROCESSES

A process $X(t)$, $-\infty < t < \infty$, is *self-similar* with parameter H (H-SS) if $X(at)$ and $a^H X(t)$ have identical finite-dimensional distributions for all $a > 0$. The Brownian motion* process $B(t)$, for instance, is $\frac{1}{2}$-SS.

This article illustrates how to construct non-Gaussian self-similar processes as nonlinear weighted averages of simpler processes such as Brownian motion, Lévy-stable processes, and Poisson point processes*. Increments of these non-Gaussian self-similar processes provide examples of non-Gaussian time series* exhibiting long-range dependence.

Self-similar processes are sometimes also called *semistable*, *automodel*, or *scaling processes*.

For a bibliographical guide to self-similar processes and long-range dependence, see ref. 33.

APPLICATIONS

The usefulness of self-similar processes stems in part from their connection to central limit theorems*. Recall that when random variables Y_j, $j \geqslant 1$, are independent, identically distributed, and have finite variance, the limiting distribution of the normalized $\Sigma_{j=1}^N Y_j$ is Gaussian. In that case, the process $X_N(t) = (1/N^{1/2})\Sigma_{j=1}^{[Nt]} Y_j$ belongs to the domain of attraction* of $B(t)$, i.e., the finite-dimensional distributions of $X_N(t)$ converge to those of $B(t)$. When stationarity alone is dropped, the Lindeberg condition* is necessary and sufficient to ensure convergence to a Gaussian distribution. When the finite variance assumption alone is dropped, one enters the realm of stable distributions*; the corresponding theory has been thoroughly covered in ref. 3. If we maintain the stationarity and finite variance assumptions and replace the independence assumption by various weak dependence relations such as m-dependence, ϕ-mixing, or strong mixing in the sense of Rosenblatt, then convergence to a Gaussian distribution still holds [8]. It is not likely that there exists a dependence structure or mixing condition for stationary random variables $\{Y_j,\ j \geqslant 1\}$ that precisely characterizes the convergence of the distribution of the normalized $\Sigma_{j=1}^N Y_j$ to the Gaussian. In this sense, there is little hope of ever characterizing the domain of attraction of the Gaussian distribution.

Sometimes the summands $\{Y_j,\ j \geqslant 1\}$ are too dependent for the normalized $\Sigma_{j=1}^{[Nt]} Y_j$ to be in the domain of attraction of Brownian motion. In that case, the limiting processes are still self-similar [10], but they are not necessarily Gaussian. Such sequences $\{Y_j, j \geqslant 1\}$ and the corresponding limiting self-similar processes are of interest in applications because they can model the fluctuations of random phenomena that exhibit a long-range dependence.

Long-range dependence typically manifests itself through the presence of "cycles" of all periodicities, the slowest period being roughly of the order of magnitude of the total available record or sample. This behavior is widespread in nature [18]. It is relevant in communication theory* [14], in economics [15, 31, 5], and is characteristic of many geophysical and hydrological records [21]. For the application of self-similarity to hydrology* see refs. 11, 22, and 20. Some aspects of self-similarity appear in turbulence [17, 6] and in connection to $1/f$ noises. $1/f$ noises are observed in the frequency fluctuations of quartz crystal oscillators, in the voltage fluctuations around carbon resistors, many semiconductor devices, and thin metal films [35, 27]. Self-similarity is relevant to the study of signals with ultraviolet and infrared catastrophes [29]. For a discussion of the potential relevance of self-similarity to some problems in physics see refs. 9, 24, and 1.

H-SELF-SIMILAR PROCESSES WITH STATIONARY INCREMENTS

A process $X(t)$, $-\infty < t < \infty$, is self-similar with parameter H and has stationary increments (H-SSSI) if $X(at)$ and $a^H X(t)$ have identical finite-dimensional distributions for all $a > 0$, and if the finite-dimensional distributions of $X(t_0 + t) - X(t_0)$ do not depend on t_0. (The increments $Z(t) = X(t) - X(t-1)$, $t = 0, \pm 1, \ldots,$ form a strictly stationary sequence.) This imposes some restrictions on the marginal distributions [25]. Nevertheless, H-SSSI processes may be Gaussian, non-Gaussian with finite variances, or may possess an infinite variance.

For instance, *Brownian motion* $B(t)$ is $\frac{1}{2}$-SSSI. It is Gaussian with independent increments. *Fractional Brownian motion** (FBM) $B_H(t)$ is H-SSSI with $0 < H < 1$ [21]; it is Gaussian and it reduces to Brownian motion when $H = \frac{1}{2}$. The *Lévy-stable process* $S_\alpha(t)$, parameterized by $0 < \alpha < 2$, is α^{-1}-SSSI. It has independent increments, but unlike Brownian motion its paths are not continuous. They increase by jumps, and the smaller the value of α, the higher the probability of sizable jumps. The process $S_\alpha(t)$ has infinite variance and possesses the so-called stable distribution*.

FINITE VARIANCE PROCESSES

Let $X(t)$ be an H-SSSI process with finite variance, not necessarily Gaussian. Then $0 < H \leqslant 1$ and $X(t) \equiv tX(1)$ when $H = 1$. The process has variance $EX^2(t) = \sigma^2|t|^{2H}$ and covariances

$$EX(t)X(s) = \tfrac{1}{2}\sigma^2\{|t|^{2H} + |s|^{2H} - |t-s|^{2H}\}$$

identical to those of the Gaussian fractional Brownian motion $B_H(t)$. The increments $Z(i) = X(i) - X(i-1)$, $i = \ldots, -1, 0, 1, \ldots,$ have covariances

$$r(k) = EZ(t)Z(t+k) = \tfrac{1}{2}\sigma^2\{(k+1)^{2H} - 2k^{2H} - |k-1|^{2H}\}$$

for $k \geqslant 0$. Note that $r(k) = 0$ for all $k \geqslant 1$ when $H = \frac{1}{2}$, but $r(k) \sim H(2H-1)^{2H-2}$ as $k \to \infty$ when $H \neq \frac{1}{2}$. The sequence $\{Z(i)\}$ exhibits a long-range dependence when $\frac{1}{2} < H < 1$ because $\Sigma_{k=-\infty}^{+\infty} r(k) = \infty$; when $0 < H < \frac{1}{2}$, it is negatively correlated and $\Sigma_{k=-\infty}^{+\infty} r(k) = 0$. Therefore the spectral density $f(\lambda) = \Sigma_{k=-\infty}^{+\infty} r(k)e^{ik\lambda}$ satisfies $f(0) = \infty$ when $\frac{1}{2} < H < 1$ and $f(0) = 0$ when $0 < H < \frac{1}{2}$. Warning: Many asymptotic results about time series will not apply because they typically assume that $f(0)$ is bounded away from 0 and ∞.

STATISTICAL TECHNIQUES

There is long-range dependence in a stationary finite variance time series when $r(k) \sim k^{-D}L(k)$ as $k \to \infty$; here, $0 < D < 1$ and L is a slowly varying function, that is, $\lim_{t\to\infty}[L(tx)/L(t)] = 1$ for all $x \geqslant 0$ (constants and logarithms are examples of slowly varying functions). This often implies that the spectral density $f(\lambda)$ satisfies $f(\lambda) \sim \lambda^{D-1}L_1(\lambda)$ as $\lambda \to 0$ and L_1 is another slowly varying function. Time series with long-range dependence include increments of self-similar processes (with $-D = 2H - 2$) [20] and fractional ARMA [5, 7].

How does one detect the presence of long-range dependence and estimate the exponent D? Existing methodologies include the R/S technique or variations thereof, the maximum likelihood* technique, and spectral density estimation. Theoretical properties of the R/S estimates have been investigated in refs. 16 and 20 and practical approaches are discussed in refs. 20 and 21. The maximum likelihood technique has been developed for Gaussian time series $\{Z(i)\}$. See ref. 22 for computational considerations and ref. 2 for theoretical results. The spectral density technique [23, 4] involves estimating the slope of $\log f(\lambda)$ vs. $\log \lambda$, where f denotes the spectral density.

FINITE VARIANCE PROCESSES SUBORDINATED TO BROWNIAN MOTION

Finite variance non-Gaussian H-SSSI processes subordinated to Brownian motion $B(t)$ can be constructed as sums of multiple Wiener–Itô integrals $X_m(t)$. The m-integral process $X_m(t)$, $m = 1, 2, \dots$, has a time representation

$$X_m(t) = \int \cdots \int_{\mathbb{R}^m_*} \left\{ \int_0^t g_m(s - \xi_1, \dots, s - \xi_m)\, ds \right\} dB(\xi_1) \cdots dB(\xi_m) \qquad (1)$$

and an equivalent spectral representation

$$X_m(t) = \int \cdots \int_{\mathbb{R}^m_{**}} \frac{\exp[i(\lambda_1 + \cdots + \lambda_m)t] - 1}{i(\lambda_1 + \cdots + \lambda_m)} \times A_m(\lambda_1, \dots, \lambda_m)\, d\tilde{B}(\lambda_1) \cdots d\tilde{B}(\lambda_m), \qquad (2)$$

where $B(\xi)$ is a real Brownian motion and $\tilde{B}(\lambda)$ is a complex symmetric motion [$\tilde{B}(\lambda) = B_1(\lambda) + iB_2(\lambda)$, where B_1 and B_2 are independent Brownian motions such that $\tilde{B}(-\lambda) = \tilde{B}_1(\lambda) - i\tilde{B}_2(\lambda)$ and $E|\tilde{B}(\lambda)|^2 = |\lambda|$]. The asterisk (*) [respectively, double asterisk (**)] means that the domain of integration excludes hyperplanes where any two $\xi_1, \dots, \xi_m$ are equal (respectively, any two $|\lambda_1|, \dots, |\lambda_m|$) [13, 29]. The integration over the variable s in (1) and the presence of the exponential in (2) ensure that $X_m(t)$ has stationary increments. For $X_m(t)$ to be self-similar one must also require that the functions g_m and A_m scale.

Example 1. The Hermite Processes. Let $\xi^+ = \max(\xi, 0)$ and $0 < D < 1/m$. If

$$g_m(\xi_1, \dots, \xi_m) = \prod_{j=1}^m \left(\xi_j^+\right)^{-(1+D)/2},$$

or (up to a multiplicative constant)

$$A_m(\lambda_1, \dots, \lambda_m) = \prod_{j=1}^m |\lambda_j|^{-(1-D)/2},$$

then $X_m(t)$ is H-SSSI with $H = 1 - \frac{1}{2}mD \in (\frac{1}{2}, 1)$. When $m = 1$, $X_m(t)$ reduces to the Gaussian process fractional Brownian motion.

Example 2. Again let $0 < D < 1/m$ and choose

$$A_m(\lambda_1, \dots, \lambda_m) = |\lambda_1 + \cdots + \lambda_m|^{\beta} \prod_{j=1}^m |\lambda_j|^{-(1-D)/2}.$$

The resulting process $X_m(t)$ is H-SSSI with $H = 1 - \frac{1}{2}mD - \beta$, provided β is such that $0 < H < 1$. It is not a Hermite process when $m \geqslant 2$ and $\beta \neq 0$.

Note that for given $m \geqslant 2$ and $0 < D < 1/m$, one can obtain any prescribed value for the self-similarity parameter $H \in (0,1)$ of the non-Gaussian process $X_m(t)$ by choosing $\beta = 1 - \frac{1}{2}mD - H$. For example, to get $H = \frac{1}{2}$, choose $\beta = 1 - \frac{1}{2}mD - \frac{1}{2}$. The resulting process $Z_m(t)$ is non-Gaussian but, like Brownian motion, it satisfies $EX_m(s)X_m(t) = \min(s,t)$. Its increments are uncorrelated but dependent.

INFINITE VARIANCE PROCESSES

H-SSSI processes with infinite variance are of interest because their increments can exhibit long-range dependence and high variability at the same time. Such processes can be subordinated to the Lévy-stable process $S_\alpha(t)$ or directly to a point process.

Fractional Lévy Motion (FLM)

Denoted $S_{H,\alpha}(t)$, FLM is a weighted average of increments of the Lévy-stable process $S_\alpha(t)$, and is given by

$$S_{\alpha,H}(t) = \int_{-\infty}^{0} \left((t-\xi)^{H-1/\alpha} - (-\xi)^{H-1/\alpha}\right) dS_\alpha(\xi) + \int_0^t (t-\xi)^{H-1/\alpha}\, dS_\alpha(\xi).$$

It is well defined for $0 < H < 1$, possesses a stable distribution, and hence has infinite variance [32]. $S_{\alpha,H}(t)$ becomes the FBM $B_H(t)$ when $S_\alpha(t)$ is replaced by Brownian motion and α by 2.

By analogy with FBM, view the increments of $S_{\alpha,H}(t)$ as negatively dependent when $H < 1/\alpha$ and positively dependent (long-range dependence) when $H > 1/\alpha$. Processes with either of these two types of dependence can be realized when $\alpha > 1$. The requirement $H < 1$ excludes long-range dependence when $\alpha \leqslant 1$.

Hermite-type Processes

In relation (1), replace $B(t)$ by the Lévy-stable process $S_\alpha(t)$, $1 < \alpha < 2$, and set

$$g_m(\xi_1,\ldots,\xi_m) = \prod_{i=1}^{m} (\xi_i^+)^{-D/2-1/\alpha},$$

$$0 < D < \frac{1}{m}\frac{2(\alpha-1)}{\alpha}.$$

The resulting process is well defined [28] and is H-SSSI with $H = 1 - \frac{1}{2}mD \in (1/\alpha, 1)$. It has infinite variance, is nonstable, and its increments exhibit long-range dependence.

Poincaré Processes

Poincaré processes are H-SSSI with $H > 1$. To construct them start with a Poincaré point process Π, that is, a point process on $\mathbb{R} \times \{\mathbb{R} \setminus \{0\}\}$ that is invariant in distribution under the transformation $(s,x) \to (as + b, ax)$, where $a > 0$ and $-\infty < b < \infty$. A realization of Π is a cloud of points. A point at (s,x) will indicate the presence of a jump at time s of "size" x. Let $\Pi_t = \Pi \cap \{I_t \times \mathbb{R}\}$ denote all the points of Π in the strip delineated by I_t, where $I_t = [0,t]$ if $t > 0$ and $[-t, 0]$ if $t < 0$. The Poincaré process

$$X(t) = \operatorname{sgn} t \sum_{(s,x)\in\Pi_t} |x|^H \operatorname{sgn} x$$

is an H-SSSI process with $H > 1$ whenever the sum converges absolutely. For instance, if Π is a Poisson point process, then $X(t)$ is a Lévy-stable $S_\alpha(t)$ with $\alpha = 1/H$. There are examples of Poincaré point processes Π for which the numbers of points in disjoint sets are not independent (e.g., the g-adic point process [26]). In that case the resulting $X(t)$ does not have independent increments.

SAMPLE PATH PROPERTIES

The H-SSSI process $X(t)$ is continuous in probability. Its sample paths are in fact (almost surely) continuous if it has either finite variance or if it is fractional Lévy motion with $\alpha > 1$ and $H > 1/\alpha$. Its sample paths, however, are nowhere bounded if $\alpha < 1$ [12] and if $\alpha \geqslant 1$, $H < 1/\alpha$ [27a].

The sample paths of $X(t)$ may be continuous and still very irregular. To characterize them further, choose any $-\infty < t_1 < t_2 < \infty$ and express $X(t_2) - X(t_1)$ as an arbitrary sum of its increments. If such sums always converge absolutely, then $X(t)$ has *locally bounded variation* (l.b.v.). Otherwise $X(t)$ has *nowhere bounded variation* (n.b.v.). A Poisson process for instance, while not self-similar, has l.b.v. Brownian motion has n.b.v.

The H-SSSI process $X(t)$ has n.b.v. [34] if either $H < 1$ [e.g., $S_\alpha(t)$, $1 \leqslant \alpha < 2$; $B_H(t)$, $0 < H < 1$] or if $H = 1$ and $X(t) \neq tX(1)$ [e.g., the Cauchy process $S_1(t)$]. An H-SSSI process with $H > 1$ can have either l.b.v. [e.g., $S_\alpha(t)$, $0 < \alpha < 1$ and Poincaré processes] or n.b.v. [e.g., $S_\alpha(B(t))$ with $\alpha < \frac{1}{2}$: this is an H-SSSI process with $H = (1/\alpha) \cdot \frac{1}{2} > 1$, but n.b.v. because Brownian motion $B(t)$ fluctuates infinitely often around any jump point of the Lévy-stable process S_α].

Further sample path properties are discussed in ref. 30.

References

[1] Frölich, J., ed. (1983). *Scaling and Self-Similarity in Physics: Renormalization in Statistical Mechanics and Dynamics*, *Progress in Physics*, Vol. 7. Birkhauser, Boston.

[2] Fox, R. and Taqqu, M. S. (1986). Maximum Likelihood Type Estimator for the Self-Similarity Parameter in Gaussian Sequences. *Ann. Statist.*, **14**, 517–532.

[3] Gnedenko, B. V. and Kolmogorov, A. N. (1954). *Limit Distributions for Sums of Independent Random Variables*. Addison-Wesley, Reading, MA.

[4] Graf, H. P. (1983). Long-Range Correlations and Estimation of the Self-Similarity Parameter. Thesis, ETH, Zurich, Switzerland.

[5] Granger, C. W. J. and Joyeux, R. (1980). An introduction to long-memory time series and fractional differencing. *J. Time Series Anal.*, **1**, 15–30.

[6] Helland, K. N. and Van Atta, C. W. (1978). The "Hurst phenomenon" in grid turbulence. *J. Fluid Mech.*, **85**, 573–589.

[7] Hoskings, J. R. M. (1981). Fractional differencing. *Biometrika*, **68**, 165–176.

[8] Ibragimov, I. A. and Linnik, Yu. V. (1971). *Independent and Stationary Sequences of Random Variables*. Wolters-Noordhoff, Groningen, The Netherlands.

[9] Jona-Lasinio, G. (1977). In *New Developments in Quantum Field Theory and Statistical Mechanics, Cargese, 1976* M. Levy and P. Mitter, eds. Plenum, New York, pp. 419–446.

[10] Lamperti, J. W. (1962). Semi-stable stochastic processes. *Trans. Amer. Math. Soc.*, **104**, 62–78.

[11] Lawrence, A. J. and Kottegoda, N. T. (1977). Stochastic modelling of riverflow time series. *J. R. Statist. Soc. A*, **140**, 1–47.

[12] Maejima, M. (1983). A self-similar process with nowhere bounded sample paths. *Zeit. Wahrsch. verw. Geb.*, **65**, 115–119.

[13] Major, P. (1981). *Multiple Wiener–Itô Integrals*, *Springer Lecture Notes in Mathematics*, Vol. 849. Springer Verlag, New York.

[14] Mandelbrot, B. B. (1965). Self-similar error clusters in communications systems and the concept of conditional systems and the concept of conditional stationarity. *IEEE Trans. Commun. Tech.*, **COM-13**, 71–90.

[15] Mandelbrot, B. B. (1969). Long-run linearity, locally Gaussian processes, H-spectra and infinite variances. *Int. Econom. Rev.*, **10**, 82–113.

[16] Mandelbrot, B. B. (1975). Limit theorems on the self-normalized range for weakly and strongly dependent processes. *Zeit. Wahrsch. verw. Geb.*, **31**, 271–285.

[17] Mandelbrot, B. B. (1976). In *Turbulence and Navier Stokes Equations*, *Springer Lecture Notes in Math.*, Vol. 565, R. Teman, ed. Springer Verlag, New York, pp. 121–145.

[18] Mandelbrot, B. B. (1982). *The Fractal Geometry of Nature*. Freeman, San Francisco, CA.

[19] Mandelbrot, B. B. and Van Ness, J. W. (1968). Fractional Brownian motions, fractional noises and applications. *SIAM Rev.*, **10**, 422–437.

[20] Mandelbrot, B. B. and Taqqu, M. S. (1979). Robust R/S analysis of long-run serial correlation. Proceedings of the 42nd Session of the International Statistical Institute, Manila. *Bull. Int. Statist. Inst.*, **48** (2), 69–104.

[21] Mandelbrot, B. B. and Wallis, J. R. (1969). Computer experiments with fractional Gaussian noises, Parts 1, 2, 3. *Water Resour. Res.*, **5**, 228–267.

[22] McLeod, A. I. and Hipel, K. W. (1978). Preservation of the rescaled adjusted range, Parts 1, 2, 3. *Water Resour. Res.*, **14**, 491–518.

[23] Mohr, D. (1981). Modeling Data as a Fractional Gaussian Noise. Ph.D. dissertation, Princeton University, Princeton, NJ.

[24] Newman, C. M. (1981). In *Measure Theory and Applications*, G. Goldin and R. Wheeler, eds.

[25] O'Brien, G. L. and Vervaat, W. (1983). Marginal distributions of self-similar processes with stationary increments. *Zeit. Wahrsch. verw. Geb.*, **64**, 129–138.

[26] O'Brien, G. L.and Vervaat, W. (1985). Self-similar processes with stationary increments generated by point processes. *Ann. Prob.*, **13**, 28–52.

[27] Percival, D. B. (1983). The Statistics of Long Memory Processes. Thesis, University of Washington, Seattle, WA.

[27a] Rosinski, J. (1987). On stochastic integral representation of stable processes with sample paths in Banach spaces, *J. Multiv. Anal.* to appear.

[28] Surgailis, D. (1981). In *Stochastic Differential Systems, Lecture Notes in Control and Information Sciences*, Vol. 36, Springer Verlag, New York, pp. 212–226.

[29] Taqqu, M. S. (1981). In *Random Fields: Rigorous Results in Statistical Mechanics and Quantum Field Theory, Colloquia Mathematica Societatis Janos Bolyia*, Vol. 27, Book 2. North-Holland, Amsterdam, The Netherlands, pp. 1057–1096.

[30] Taqqu, M. S. and Czado, C. (1985). A Survey of Functional Laws of the Iterated Logarithm for Self-Similar Processes. *Stochastic Models*, **1**, 77–115.

[31] Taqqu, M. S. and Levy, J. (1986). Using Renewal Processes to Generate Long-Range Dependence and High Variability. In *Dependence in Probability and Statistics*, E. Eberlein and M. S. Taqqu, eds. Birkhäuser, Boston, MA., pp. 73–89.

[32] Taqqu, M. S. and Wolpert, R. (1983). Infinite variance self-similar processes subordinate to a Poisson measure. *Zeit. Wahrsch. verw. Geb.*, **62**, 53–72.

[33] Taqqu, M. S. (1986). A bibliographical guide to self-similar processes and long-range dependence. In *Dependence in Probability and Statistics*, E. Eberlain and M. S. Taqqu, eds. Birkhäuser, Boston, MA, pp. 137–162.

[34] Vervaat, W. (1982). Sample path properties of self-similar processes with stationary increments. *Ann. Prob.*, **10**, 73–89.

[35] Wolf, D., ed. (1978). *Noise in Physical Systems, Proc. of the Fifth Int. Conf. on Noise, Bad Nauheim, March 1978*, Springer Series in Electro-Physics, Vol. 2. Springer Verlag, New York.

(BROWNIAN MOTION
DOMAIN OF ATTRACTION
FRACTIONAL BROWNIAN MOTIONS AND FRACTIONAL GAUSSIAN NOISES
STABLE (LEVY) DISTRIBUTION THEORY)

MURAD S. TAQQU

SEMI-BAYESIAN INFERENCE *See* QUASI-BAYESIAN INFERENCE

SEMI-DISSIPATIVE AND NON-DISSIPATIVE MARKOV CHAINS

Consider a homogeneous Markov chain (X_n) with a countable number of states and let P be its transition probability matrix. The chain is dissipative when the limit of the averages of the probabilities $\Pr(X_n = j)$, for all j and any initial distribution, is zero. Otherwise, the chain is semi- (or non-) dissipative. To be more specific,

$$\lim_{n\to\infty} (1/n) \sum_{k=1}^{n} (p^k)_{ij} = \pi_{ij}$$

exists, and that $\pi_{ij} \geqslant 0, \sum_{j=1}^{\infty}\pi_{ij} \leqslant 1$ (for each i). The Markov chain (X_n) is called

(i) *dissipative* if $\pi_{ij} = 0$ for all i, j;

(ii) *semi-dissipative* is the matrix π is neither a zero matrix nor a stochastic matrix;

(iii) *non-dissipative* if π is a stochastic matrix.

Kendall [2] proved that if there exists an infinite nonnegative vector (w_i) such that $\lim_{n\to\infty} w_n = \infty$ and for all i,

$$\sum_{j=1}^{\infty} p_{ij}w_j - w_i \leqslant 0, \qquad P = (p_{ij}),$$

then (X_n) is non-dissipative. Foster [1] proved that the chain (X_n) is dissipative if and only if it has no invariant distribution (that is, a probability matrix π such that $\pi P = \pi$). Mauldon [3] gave a number of sufficient conditions for a Markov chain to be non-dissipative.

It is easy to construct simple illustrative examples, e.g., any Markov chain with a finite number of states is obviously non-dissipative. A Markov chain where the transition probability matrix is given by

$$p_{ij} = 1 \quad \text{whenever } i = j - 1$$

is dissipative since $P^n \to 0$ as $n \to \infty$. If this example is modified so that $p_{ij} = 1$ whenever $i = j = 1$ or $i = j - 1 > 1$, then the limit matrix π of P^n has the first row as its only nonzero row, and consequently, the chain is semi-dissipative in this case.

As an application, consider a model for population growth. Suppose that at the beginning of the first generation, a population has X_0 members. During the first generation, each of these members has a random number of offspring, and at the end of the first generation all the original X_0 members leave the population. Let X_1 be the number of members at the end of the first generation and X_n at the end of the nth generation. Let A be the matrix given by

$$A_{ij} = \Pr(X_{n+1} = j | X_n = i), \qquad i, j = 0, 1, 2, \ldots,$$

and let A be independent of n. Let us make the following natural assumptions:

$$A_{ij} \geqslant 0, \qquad \sum_{j=0}^{\infty} A_{ij} = 1,$$

$$A_{00} = 1 \quad \text{and} \quad A_{i0} > 0 \quad \text{for each } i.$$

Then (see Foster [1]) when the matrix A is non-dissipative, the ultimate extinction of the population is almost certain; that is, for each i, $A_{i0}^{(n)} \to 1$ as $n \to \infty$.

References

[1] Foster, F. G. (1951). *Proc. Camb. Philos. Soc.*, **47**, 77–85.

[2] Kendall, D. G. (1951). *Proc. Camb. Philos. Soc.*, **47**, 633–634.

[3] Mauldon, J. G. (1957). *Proc. Camb. Philos. Soc.*, **53**, 825–835.

(MARKOV PROCESSES)

ARUNAVA MUKHERJEA

SEMI-INDEPENDENCE

Two random variables X_1 and X_2 are said to be semi-independent if each is uncorrelated with an arbitrary function of the other for which the indicated expectations are defined. From the provisional expressions

$$E[X_i f(X_j)] = E(X_i | X_j) E[f(X_j)] = E(X_i) E[f(X_j)]$$

for $i, j = 1, 2$, $i \neq j$, it follows that semi-independence is equivalent to the property that

$$E(X_i | X_j) = E(X_i), \qquad i = 1, 2,\ i \neq j.$$

Correlation* is a measure of *linear* association between pairs of random variables, and easily constructed examples give uncorrelated but perfectly dependent pairs. Such pairs are not semi-independent, however. The concept of semi-independence is stronger and thus lies somewhere between uncorrelatedness and independence.

Semi-independence assumes a natural role in prediction* theory, where the best mean-square predictor of X_0 based on $[x_1, x_2, \ldots, x_n]$ is the conditional expectation $E(X_0 | x_1, x_2, \ldots, x_n)$ given the values $[x_1, x_2, \ldots, x_n]$ of the conditioning variables. If this expectation is linear, i.e., if

$$E(X_0 | x_1, x_2, \ldots, x_n) = a_1 x_1 + a_2 x_2 + \cdots + a_n x_n,$$

then the optimal predictor is linear and the prediction error

$$E_n = X_0 - a_1 X_1 - \cdots - a_n X_n$$

is semi-independent of every random variable generated as a linear function of $[X_1, X_2, \ldots, X_n]$. Linearity of predictors is linked fundamentally to the structure of the underlying stochastic process*. In particular, every mean-square prediction problem for a second-order process has a linear solution if and only if the process is spherically invariant, i.e., its finite-dimensional distributions are essentially isotropic*. These connections between semi-independence and prediction theory were established by Vershik

[3] and were studied subsequently by Blake and Thomas [1] and Huang and Cambanis [2], for example.

In a different context, processes are described as having semi-independent increments elsewhere in the Russian literature. There are no connections apparent to this reviewer with the concept of semi-independence set forth here.

References

[1] Blake, I. F. and Thomas, J. B. (1968). *IEEE Trans. Inf. Theory*, **IT-14**, 12–16.

[2] Huang, S. T. and Cambanis, S. (1979). *J. Multivariate Anal.*, **9**, 59–83.

[3] Vershik, A. M. (1964). *Theory Prob. Appl.*, **9**, 353–356.

(ASSOCIATION, MEASURES OF
CORRELATION
DEPENDENCE, CONCEPTS OF
DEPENDENCE, MEASURES AND INDICES OF)

D. R. JENSEN

SEMI-INTERQUARTILE RANGE *See* QUARTILE DEVIATION

SEMI-INVARIANT *See* CUMULANTS; THIELE, T. N.

SEMI-INVERSE

This is another name for a reflexive generalized inverse of a matrix. *See* GENERALIZED INVERSES.

SEMI-LATIN SQUARES

The semi-Latin square is a rectangular arrangement of $t = kn$ treatments, where n and k are integers greater than 1. The arrangement has n rows and t columns, these latter being grouped into sets each containing k consecutive columns; each treatment occurs exactly once in each row and exactly once in each set of columns. The following example with $t = 8$, $n = 4$, and $k = 2$ was given in 1928 by Steven [18, p. 20], except that the rows here were his columns and vice versa:

$$\begin{array}{ccccccccccc} A & E & \vdots & F & B & \vdots & C & G & \vdots & H & D \\ G & C & \vdots & D & H & \vdots & E & A & \vdots & B & F \\ B & F & \vdots & E & A & \vdots & D & H & \vdots & G & C \\ H & D & \vdots & C & G & \vdots & F & B & \vdots & A & E \end{array}$$

One of several combinatorially distinct examples with the same parameters is

$$\begin{array}{ccccccccccc} A & B & \vdots & C & D & \vdots & E & F & \vdots & G & H \\ C & D & \vdots & F & H & \vdots & B & G & \vdots & A & E \\ E & F & \vdots & A & G & \vdots & H & D & \vdots & B & C \\ G & H & \vdots & B & E & \vdots & A & C & \vdots & D & F \end{array}$$

In 1931, such arrangements were designated as *equalized random blocks* by "Student" [19], who gave an example with $t = 10$ and $n = 5$ (rows and columns again interchanged); an example with $t = 8$ and $n = 4$ was given by Hoblyn [6, p. 14] in the same year. The name *semi-Latin square* was, however, used in 1932 by Scott [16, pp. 3, 7, 11, and 13] for his three examples with $t = 20$ and $n = 5$, and by Yates [22, 23] in 1935 and 1936; it was attributed by Ma and Harrington [9] to Professor E. J. G. Pitman*. An example with $t = 16$ and $n = 4$ was given in 1934 by Snedecor [17, p. 39].

Yates [22, Sec. 6, last paragraph and Discussion] showed that the semi-Latin square is statistically defective when used as an experimental design whose analysis of variance* has just four sources of variation, namely rows, sets of columns, treatments, and a single error. As Yates said, this design "stands condemned ... because of a biased error." Yates also commented on the matter in his 1936 paper [23, pp. 307–308]; referring first to possible experimental designs in general, he said that

> A process of randomization which will equalise the mean values of the treatment and error mean squares [if the treatments

> produce no effect] does not always exist, and some otherwise admirable experimental arrangements fail on this criterion. There does not appear, for instance, to be any correct randomization process for ... the semi-Latin square.

Validity is attained for the semi-Latin square only if the n sets of k consecutive entries in each row are regarded as constituting n whole plots each subdivided into k subplots; the design must then be randomised as a row-and-column design* having split plots* within n rows and n columns, and the analysis must have both a whole plot error and a subplot error.

In the 1930s, the use of the semi-Latin square for agricultural experimentation included its use for the Iowa Corn Yield Test in the years up to 1940 (Zuber and Robinson [24, p. 524]). The design should now be regarded as rightly obsolete, except perhaps when a specially chosen example (probably one where every row has the same n sets of k treatments) is used as a true split-plot design. Nevertheless, the semi-Latin square reappeared in the literature of agricultural experimentation in the 1950s, when it was promulgated by Behrens [2, 3], Bahn et al. [1], and Unterstenhöfer [20, 21]. Examples of its agricultural use were reported by Krüger and von Lochow [7], and it was included as an experimental design in the textbooks of Mudra [12, pp. 23–25] and von Lochow and Schuster [8, pp. 7ff and 74ff].

German and Polish workers were, however, not alone in using the design at that time; under the name *cuadro latino modificado* (modified Latin square), it was introduced into agronomic work in Mexico, as reported by Rojas and White [15]. These two authors, in a detailed statistical discussion of the design, considered the modified Latin square to be "a generalization of the semi-Latin square," as the latter term "was restricted, in the minds of many," to designs with $k = 2$; however, Yates [22, end of Sec. 6] had clearly allowed $k > 2$. The term *modified Latin square* was used by Riddle and Baker [14] in 1944.

The German authors referred to above all used the name *Lateinisches Rechteck* or *Latin rectangle* for the semi-Latin square, despite the greater generality of the two usual combinatorial definitions of Latin rectangle. Likewise, the entry *Lateinisches Rechteck* in Volume 1 of the *Biometrisches Wörterbuch* [4] describes merely the semi-Latin square. In Russian, the name Латинский квадрат с разделенными элементами (Latin square with separated elements) has been used.

The earliest known antecedent of the semi-Latin square is the row-and-column design*

$$\begin{array}{cccccccc} a & d & f & e & b & c & f & g \\ f & e & a & d & f & g & b & c \\ b & g & c & f & e & d & a & f' \\ c & f & b & g & a & f & e & d \end{array}$$

whose use for fertiliser experimentation was reported by Madsen-Mygdal [10, 11] in 1903 and 1904. This design is obtainable from a semi-Latin square by representing two of the original symbols by f.

The semi-Latin square is closely related to the *Trojan square** of Darby and Gilbert [5]. This relationship and others were discussed by Preece and Freeman [13], who also enumerated small semi-Latin squares. The enumeration procedures were similar to those that have been used for Latin squares*.

References

[1] Bahn, E., Banneick, A., Möbius, H., Müller, K. H., and Ortlepp, H. (1958). Anlageschemata für Feldversuche (2-16 Prüfglieder) erarbeitet nach den Prinzipien einer gelenkten, gerechten Verteilung der Teilstücke. *Zeit. Landwirtsch. Versuchs- und Untersuchungswesen*, **3**, 151–207.

[2] Behrens, W. U. (1956). Die Eignung verschiedener Feldversuchsanordnungen zum Ausgleich der Bodenunterschiede. *Zeit. Acker- und Pflanzenbau*, **101**, 243–278.

[3] Behrens, W. U. (1956). Feldversuchsanordnungen mit verbessertem Ausgleich der Bodenunterschiede. *Zeit. Landwirtsch. Versuchs- und Untersuchungswesen*, **2**, 176–193.

[4] (1968). *Biometrisches Wörterbuch*, Band 1. VEB Deutscher Landwirtschaftsverlag, Berlin, Germany.

[5] Darby, L. A. and Gilbert, N. (1958). The Trojan square. *Euphytica*, **7**, 183–188.

[6] Hoblyn, T. N. (1931). Field Experiments in Horticulture. *Tech. Commun. No. 2*, Imperial Bureau of Fruit Production, East Malling, Kent, England.

[7] Krüger, F. H. and von Lochow, J. (1956). Die derzeitige Anwendung der Methoden im deutschen Feldversuchswesen. *Zeit. Acker- und Pflanzenbau*, **100**, 379–387.

[8] von Lochow, J. and Schuster, W. (1961). *Anlage und Auswertung von Feldversuchen: Anleitungen und Beispiele für die Praxis der Versuchsarbeit.* DLG-Verlag, Frankfurt am Main, Federal Republic of Germany.

[9] Ma, R. H. and Harrington, J. B. (1949). A study on field experiments of semi-Latin square design. *Sci. Agric.*, **29**, 241–251.

[10] Madsen-Mygdal, A. (1903). *Beretning om Lokale Markforsøg udførte i Sommeren 1903 af de Samvirkende Landboforeninger i Fyens Stift.* Andelsbogtrikkeriet i Odense, Odense, Denmark.

[11] Madsen-Mygdal, A. (1904). *Beretning om Lokale Markforsøg undførte i Sommeren 1904 af de Samvirkende Landboforeninger i Fyens Stift.* Andelsbogtrikkeriet i Odense, Odense, Denmark.

[12] Mudra, A. (1952). *Einführung in die Methodik der Feldversuche.* Hirzel-Verlag, Leipzig, German Democratic Republic.

[13] Preece, D. A. and Freeman, G. H. (1983). Semi-Latin squares and related designs. *J. R. Statist. Soc. B*, **45**, 267–277.

[14] Riddle, O. C. and Baker, G. A. (1944). Biases encountered in large-scale yield trials. *Hilgardia*, **16**, 1–14.

[15] Rojas, B. and White, R. F. (1957). The modified Latin square. *J. R. Statist. Soc. B*, **19**, 305–317.

[16] Scott, R. A. (1932). Potato manurial trials, details of experimental work, seasons 1930–32. to *Tasmanian J. Agric.*, August 1st, 1932, supplement.

[17] Snedecor, G. W. (1934). *Calculation and Interpretation of Analysis of Variance and Covariance.* Collegiate Press, Ames, IA.

[18] Steven, H. M. (1928). Nursery Investigations. *Bull. No. 11*, Forestry Commission, HMSO, London, England.

[19] "Student" [Gosset, W. S.] (1931). Yield Trials. *Baillière's Encyclopaedia of Scientific Agriculture*, pp. 1342–1361. [Reprinted in *Student's Collected Papers*. E. S. Pearson and J. Wishart, eds. (1942). Biometrika Office, University College, London, England, pp. 150–168.]

[20] Unterstenhöfer, G. (1957). The basic principles of plant protection field trials. *Höfchen-Briefe*, **10**, 173–236.

[21] Unterstenhöfer, G. (1963). The basic principles of crop protection field trials. *Pflanzenschutz-Nachrichten "Bayer"*, **16**, 81–164.

[22] Yates, F. (1935). Complex experiments (with Discussion). *Suppl. J. R. Statist. Soc.*, **2**, 181–247. [Reprinted (without Discussion) in Yates, F. (1970). *Experimental Design: Selected Papers* Griffin, London, England, pp. 69–117.]

[23] Yates, F. (1936). Incomplete Latin squares. *J. Agric. Sci.*, **26**, 301–315.

[24] Zuber, M. S. and Robinson, J. L. (1941). The 1940 Iowa Corn Yield Test. *Bulletin P19* (New Series), Agricultural Experiment Station, Ames, IA, pp. 519–593.

(AGRICULTURE, STATISTICS IN
LATIN SQUARES
ROW AND COLUMN DESIGNS
SPLIT-PLOT EXPERIMENTS)

D. A. PREECE

SEMILOGNORMAL DISTRIBUTION

A bivariate random vector $(X, C)'$ is said to have a semilognormal distribution with parameters $\boldsymbol{\mu}$ and $\boldsymbol{\Sigma}$ if $(X, \ln C)'$ has a bivariate normal distribution with mean vector

$$\boldsymbol{\mu} = (\mu_1, \mu_2)'$$

and covariance matrix

$$\boldsymbol{\Sigma} = \begin{pmatrix} \sigma_1^2 & \sigma_{12} \\ \sigma_{12} & \sigma_2^2 \end{pmatrix}.$$

Minimum variance unbiased estimation* procedures for this distribution were studied by Suzuki et al. [1].

Reference

[1] Suzuki, M., Iwase, K., and Shimizu, K. (1984). *J. Japan Statist. Soc.*, **14**, 63–68.

(LOGNORMAL DISTRIBUTION)

SEMI-MARKOV PROCESSES

INTRODUCTION

Semi-Markov processes and Markov renewal processes are two related facets of the same phenomenon. We assume familiarity with the concepts and terminologies of Markov processes* as well as renewal processes (*see* RENEWAL THEORY).

A simple conceptualization is to consider a stochastic process* that undergoes state transitions in accordance with a Markov chain, but in which the amount of time spent in each state before a transition occurs (called the sojourn time) is random.

Let $\{X_n, n = 0, 1, 2, \dots\}$ be a stochastic process assuming values in the countable set $S = \{0, 1, 2, \dots\}$. Let $T_0, T_1, T_2, \dots$ be the transition epochs on the nonnegative half of the real line such that $0 = T_0 \leqslant T_1 \leqslant T_2 \cdots$. The two-dimensional process $(X, T) = \{X_n, T_n;\, n = 0, 1, 2, \cdots\}$ is called a *Markov renewal process* (MRP) if it has the property

$$\begin{aligned} P[X_{n+1} &= j, T_{n+1} - T_n \leqslant t \mid X_0, X_1, \dots, X_n; T_0, \dots, T_n] \\ &= P[X_{n+1} = j, T_{n+1} - T_n \leqslant t \mid X_n], \\ & j \in S,\ t \geqslant 0. \end{aligned} \tag{1}$$

We assume that it is time homogeneous. Define the probability

$$Q_{ij}(t) = P[X_{n+1} = j, T_{n+1} - T_n \leqslant t \mid X_n = i]. \tag{2}$$

Then $Q = \{Q_{ij}(t), i, j \in S, t \geqslant 0\}$ is called the *semi-Markov kernel* for the process. Let

$$P_{ij} = \lim_{t \to \infty} Q_{ij}(t)$$

and

$$F_{ij}(t) = Q_{ij}(t)/P_{ij}.$$

P_{ij} is the transition probability of the embedded Markov chain $\{X_n, n = 0, 1, 2, \dots\}$ and hence $\sum_{j \in S} P_{ij} = 1$ and $F_{ij}(t)$ [which is defined as 1 when $P_{ij} = 0$, since then $Q_{ij}(t) = 0$ for all t] is a cumulative distribution function given by

$$F_{ij}(t) = P[T_{n+1} - T_n \leqslant t \mid X_n = i, X_{n+1} = j]. \tag{3}$$

Let $N_j(t)$ be the number of visits of the process to state j during $(0, t]$ and define $N(t) = \sum_{j \in S} N_j(t)$. Let $Y(t) = X_{N(t)}$, $Y(t)$ denoting the state of the Markov renewal process (X, T) as defined in (1). The stochastic process* $\{Y(t), t \geqslant 0\}$ is called a *semi-Markov process* (SMP). The vector process $\mathbf{N(t)} = \{N_0(t), N_1(t), \dots\}$ may be identified as the Markov renewal counting process. For the MRP we have used Çinlar's [8] definition, while Pyke [19] identifies $\mathbf{N(t)}$ as the Markov renewal process. The lack of consistency in terminology exists similarly in renewal theory. To avoid confusion we have identified $\mathbf{N(t)}$ as the Markov renewal counting process.

Thus, a Markov renewal process represents the transition epoch and the state of the process at that epoch. A semi-Markov process represents the state of the Markov renewal process at an arbitrary time point and the Markov renewal counting process records the number of times the process has visited each of the states up to time t. A Markov renewal process becomes a Markov process if the sojourn times are exponentially distributed independently of the next state, it becomes a Markov chain if the sojourn times are all equal to one, and it becomes a renewal process if there is only one state.

Semi-Markov processes were introduced independently by Lévy [15] and Smith [25] in 1954. However, Takács [28] introduced essentially the same type of stochastic process in the context of problems in the theory of counters. The term "Markov renewal process" is owing to Pyke [19], who provided a formal theory including definitions, conditions for finiteness and regularity, and the classification of states of the process. An article of similar significance is by Çinlar [5], whose definition of MRP seems to eliminate some earlier confusions. In this paper he extended some renewal theory results to Markov renewal processes. For an excellent introduction to the theory, see Çinlar [8, Chap. 10]. A more elementary introduction can be found in Ross [22, Chap 5], who uses Pyke's definition of MRP. Also, see Howard [12] and Ross [24] for basic concepts of MRP and SMP. Other significant articles are by Smith [26], Pyke [20], Pyke and Schaufele [21], Feller [11], Neuts [18], and Çinlar [6]. For excellent bibliographies, see Çinlar [5, 7].

MARKOV RENEWAL PROCESS

The sojourn time $T_{n+1} - T_n$ $(n = 0, 1, 2, \dots)$ is dependent only on X_n and not any of the X_i's or the $T_{i+1} - T_i$ $(i = 0, 1, 2, \dots, n-1)$. By itself $\{X_n\}$ is Markovian. Hence the time epochs $T_n^{(j)}$ at which the process enters state j successively form a renewal process (if $X_0 \neq j$, it is a delayed renewal process). $N_j(t)$ defined earlier is the corresponding renewal counting process. Let

$$R_{ij}(t) = E\left[N_j(t) | X_0 = i\right]. \quad (4)$$

Extending renewal theory terminology we may call $R_{ij}(t)$ the *Markov renewal function*.

Let $G_{ij}(t)$ be the distribution function of the first passage time for the transition from state i to state j. Now $G_{jj}(t)$ is the recurrence time distribution for state j. Using renewal theory concepts we have

$$R_{jj}(t) = \sum_{n=0}^{\infty} G_{jj}^{(n)}(t), \quad (5)$$

where $G_{jj}^{(n)}(t)$ is the n-fold convolution of $G_{jj}(t)$ with itself. Extending (5) we also get

$$R_{ij}(t) = \int_0^t G_{ij}(ds) R_{jj}(t-s), \quad i \neq j; \quad (6)$$

other results from renewal theory follow similarly.

To capitalize on the Markovian properties of the process, define

$$Q_{ij}^{(n)}(t) = P[X_n = j, T_n \leqslant t | X_0 = i], \quad (7)$$

the extension of the n-step transition probability to the MRP. As in the case of Markov chains,

$$R_{ij}(t) = \sum_{n=0}^{\infty} Q_{ij}^{(n)}(t), \quad (8)$$

where

$$Q_{ij}^{(0)}(t) = \begin{cases} 1 & \text{if } i = j, \\ 0 & \text{if } i \neq j \end{cases}$$

and

$$Q_{ij}^{(n+1)}(t) = \sum_{k \in S} \int_0^t Q_{ik}(ds) Q_{kj}^{(n)}(t-s), \quad n \geqslant 1. \quad (9)$$

Define Laplace–Stieltjes transforms

$$\tilde{Q}_{ij}^{(n)}(\rho) = \int_0^{\infty} e^{-\rho t} Q_{ij}^{(n)}(dt), \quad Re(\rho) > 0; \quad n = 0, 1, 2, \dots;$$

and write $\tilde{Q}_{ij}^{(1)}(\rho) = \tilde{Q}_{ij}(\rho)$ and $\tilde{\mathbf{Q}}^{(\mathbf{n})}(\rho)$ as the matrix of transforms $\tilde{Q}_{ij}^{(n)}(\rho)$. Also let $\tilde{R}_{ij}(\rho)$ be the Laplace–Stieltjes transform of $R_{ij}(t)$. Then

$$\tilde{\mathbf{Q}}^{(\mathbf{n})}(\rho) = \left[\tilde{\mathbf{Q}}(\rho)\right]^{\mathbf{n}}, \quad \tilde{\mathbf{R}}(\rho) = \mathbf{I} + \tilde{\mathbf{Q}}(\rho) + \tilde{\mathbf{Q}}^2(\rho) + \cdots, \quad (10)$$

and hence $\tilde{\mathbf{R}}(\rho)$ is given by

$$\tilde{\mathbf{R}}(\rho) = \left(\mathbf{I} - \tilde{\mathbf{Q}}(\rho)\right)^{-1} \quad (11)$$

if S is finite, and by the minimal solution of

$$(\mathbf{I} - \tilde{\mathbf{Q}}(\rho))\mathbf{R} = \mathbf{I}, \quad \mathbf{R} \geqslant 0, \quad (12)$$

if S is not finite.

Thus most of the important properties of renewal processes and Markov chains carry through to Markov renewal processes with necessary modifications. A few key results are listed below.

(i) The classification of states is based on the classification of states in the embedded Markov chain or the corresponding renewal processes.

(ii) The renwal equation of renewal theory extends to MRPs as do corresponding properties.

(iii) For a given state space and the semi-Markov kernel Q, $P[N(t) < \infty$ for all $t] = 1$ for all choices of a vector of initial probability if and only if Q is regular (see Pyke [19]).

(iv) If state j is transient and $i \neq j$, then

$$\lim_{t\to\infty} R_{ij}(t) = G_{ij}{}^* R^*{}_{ij}, \qquad (13)$$

where $G_{ij}(\infty) = G^*{}_{ij}$ and $R_{ij}(\infty) = R^*{}_{ij}$. If state j is recurrent and aperiodic, then

$$\lim_{t\to\infty} \left[R_{ij}(t) - R_{ij}(t-\Delta)\right] = (\Delta/\mu_j)G^*{}_{ij}, \qquad (14)$$

where μ_j is the mean recurrence time of state j.

(v) Suppose the embedded Markov chain is irreducible and recurrent. Let $\pi = (\pi_0, \pi_1, \ldots)$ be a solution of the equations

$$\sum_{i\in S} \pi_i P_{ij} = \pi_j, \qquad j \in S,$$

and m_j be the mean sojourn time for state j; then

$$\frac{1}{\mu_j} = \frac{\pi_j}{\sum_{k\in S}\pi_k m_k}, \qquad j \in S. \qquad (15)$$

(vi) Let (X, T) be an irreducible MRP. Then for a certain class of functions h (directly Riemann integrable when the state space is finite; see Çinlar [8] for conditions in the general case)

$$\lim_{t\to\infty} \sum_{j\in S} \int_0^t R_{ij}(ds) h_j(t-s)$$

$$= \frac{1}{\sum_{k\in S}\pi_k m_k} \int_0^\infty \sum_{r\in S} \pi_r h_r(s)\, ds. \qquad (16)$$

(vii) Let (X, T) be an irreducible recurrent MRP. Then for any $i, j, k, l \in S$,

$$\lim_{t\to\infty} \left[R_{ij}(t)/R_{kl}(t)\right] = \pi_j/\pi_l. \qquad (17)$$

SEMI-MARKOV PROCESS

In the semi-Markov process $Y(t)$ defined earlier, the transition probability

$$P_{ij}(t) = P\left[Y(t) = j | X_0 = i\right] \qquad (18)$$

and related quantities are of interest. Thus

$$P_{ij}(t) = \int_0^t R_{ij}(ds) H_j(t-s), \qquad (19)$$

where

$$H_j(t) = 1 - \sum_{k\in S} Q_{jk}(t), \qquad j \in S,\ t \geqslant 0.$$

If the embedded Markov chain $\{X_n\}$ is irreducible, aperiodic, and recurrent, then as $t \to \infty$,

$$\lim_{t\to\infty} P_{ij}(t) = \pi_j m_j \Big/ \sum_{k\in S} \pi_k m_k, \qquad j \in S. \qquad (20)$$

Let $M_{ij}(t)$ be the expected occupation time of the process in state j during $(0, t]$, having initially started at i. Then

$$M_{ij}(t) = \int_0^t P_{ij}(u)\, du$$

$$= \int_0^t R_{ij}(ds) \int_0^{t-s} H_j(u)\, du. \qquad (21)$$

As $t \to \infty$ and writing $M^*{}_{ij} = M_{ij}(\infty)$,

$$M_{ij}{}^* = R^*{}_{ij} m_j, \qquad (22)$$

showing that $M_{ij}{}^* = \infty$ if j is recurrent and $G^*{}_{ij} > 0$ or if j is transient, $G^*{}_{ij} > 0$ and $m_j = \infty$. When $G_{ij}{}^* = 0$, $R^*{}_{ij} = 0$ from (13). The following two ratio results

$$\lim_{t\to\infty} \left[M_{ij}(t)/M_{kl}(t)\right] = \pi_j m_j/\pi_l m_l, \qquad (23)$$

$$\lim_{t\to\infty} \left[M_{ij}(t)/t\right] = \pi_j m_j \Big/ \sum_{k\in S} \pi_k m_k \qquad (24)$$

are direct analogs of the results for Markov processes.

SEMIREGENERATIVE PROCESS

Regenerative processes are characterized by renewal processes embedded in them. In a similar manner semiregenerative processes consist of an embedded Markov renewal process. Consequently, the histories of the process between regeneration epochs are probabilistic replicas of each other, conditional on the initial state of the embedded Markov chain.

Let $Z(t)$ be a semiregenerative process with state space $\mathscr{S}$; let (X, T) be a Markov renewal process with state space $S \in \mathscr{S}$. Define for a set $A \subset \mathscr{S}$,

$$H_{iA}(t) = P[Z(t) \in A, T_1 > t | X_0 = i],$$

$$i \in S,\ t \geqslant 0, \qquad (25)$$

$$P_{iA}(t) = P[Z(t) \in A | X_0 = i]. \qquad (26)$$

Then extending (19),

$$P_{iA}(t) = \sum_{k \in S} \int_0^t R_{ij}(ds) H_{jA}(t-s) \quad (27)$$

and

$$\lim_{t\to\infty} P_{iA}(t) = \left(\sum_{k\in S} \pi_k m_k \right)^{-1} \times \sum_{j \in S} \pi_j \int_0^\infty H_{jA}(t)\, dt$$

These generalized results prove very useful in the solution of problems in various areas of applied probability. For illustrative examples see Çinlar [5, 7, 8], Ross [22, 24], and the extensive bibliographic listings in Çinlar [5, 7] and Cheong et al. [3].

STATISTICAL INFERENCE

Inference problems on Markov renewal and semi-Markov processes are treated much the same way as for Markov processes and general point processes*. General references are Cox and Lewis [9], Snyder [27], and Basawa and Prakasa Rao [2]. Note here that these can be considered as two-dimensional Markov processes. For instance, consider the MRP (X, T) with a finite state space S. In addition to the counting processes $N_j(t)$ and $N(t)$ defined earlier, define $N_{ij}(t)$ as the number of times that $X_n = i$ and $X_{n+1} = j$ for $0 \leqslant n \leqslant N(t)$. Also let $S_k^{(i)}$ be the kth interval (between two successive transition epochs) when at the beginning of the interval the state of the process is i. Let $N_{ij}(t)$, $N_i(t)$, and $S_k^{(i)}$ be the set of observations during $(0, t]$. Moore and Pyke [17] have shown that when the MRP is irreducible recurrent and $F_{ij}(x) = F_i(x)$, $j \in S$ (which assumption incurs no loss of generality), the estimator

$$\hat{Q}_{ij}(x, t) = \hat{P}_{ij}(t) \hat{F}_i(x, t), \qquad t, x > 0,$$

where

$$\hat{P}_{ij}(t) = N_{ij}(t)/N_i(t),$$

$$\hat{F}_i(x, t) = \{N_i(t)\}^{-1} \sum_{k=1}^{N_i(t)} (x - S_k^{(i)}),$$

is uniformly strongly consistent as $t \to \infty$ in the sense that

$$\max_{i,j} \sup_x |\hat{Q}_{ij}(x, t) - Q_{ij}(x)| \to 0$$

with probability 1.

The estimator $\hat{P}_{ij}(t)$ is the same as that in Markov processes and the estimator $\hat{F}_i(x, t)$ is the ordinary empirical distribution function determined from the sample [$\hat{Q}_{ij}(x, t) = 0$ when $N_i(t) = 0$].

A computational procedure for the estimation of parameters of a semi-Markov process from censored records has been given by Thompson [28].

REMARKS

The literature on MRP and SMP includes other related aspects. Semi-Markov decision processes are studied to determine the optimality of control policies defined on them; see Ross [22–24], Howard [12], and Doshi [10]. Functionals of MRPs have been treated by Jewell [13], McLean and Neuts [16], and Pyke and Schaufele [21]. Çinlar [4], Arjas [1], and Kaplan and Sil'vestrov [14] have considered SMPs on a general state space. Major advances in techniques of analysis of MRPs and SMPs have occurred in the context of applied probability models and therefore the bibliographic listings mentioned earlier include a large number of papers from areas such as queueing theory*, reliability*, and the theory of traffic.

References

[1] Arjas, E., Nummelin, E., and Tweedie, R. L. (1980). *J. Aust. Math. Soc.*, **30**, 187–200.

[2] Basawa, I. V. and Prakasa Rao, B. L. S. (1980). *Statistical Inference for Stochastic Processes*. Academic, New York. (This is an excellent reference on this topic.)

[3] Cheong, C. K., DeSmit, J. H. A., and Teugels, J. L. (1972). Notes on Semi-Markov Processes Part II: Bibliography. *Discussion Paper No. 7207*, CORE, Université Catholique de Louvain, Belgium.

[4] Çinlar, E. (1968). *Proc. Camb. Philos. Soc.*, **66**, 381–392.

[5] Çinlar, E. (1969). *Adv. Appl. Prob.*, 1, 123–187. (Contains a good exposition on the basic theory.)

[6] Çinlar, E. (1972). *Zeit. Wahrsch. verw. Geb.*, **24**, 85–121.

[7] Çinlar, E. (1975). *Management Sci.*, **21**, 727–752. (Provides an extensive bibliography. Discussion is restricted to finite state space.)

[8] Çinlar, E. (1975). *Introduction to Stochastic Processes*. Prentice Hall, Englewood Cliffs, NJ. (Chapter 10 provides a good coverage of the basic material.)

[9] Cox, D. R. and Lewis, P. A. W. (1966). *The Statistical Analysis of Series of Events*. Methuen, London.

[10] Doshi, B. T. (1979). *J. Appl. Prob.*, **16**, 618–630.

[11] Feller, W. (1964). *Proc. Nat. Acad. Sci. USA*, **51**, 653–659.

[12] Howard, R. A. (1971). *Dynamic Probabilistic Systems Vol. II: Semi-Markov and Decision Processes*. Wiley, New York.

[13] Jewell, W. S. (1963). *Operat. Res.*, **11**, 938–948.

[14] Kaplan, E. I. and Sil'vestrov, D. S. (1979). *Theory Prob. Appl.*, **24**, 536–547.

[15] Lévy, P. (1954). *Proc. Int. Congress Math. (Amsterdam)*, **3**, 416–426.

[16] McClean, R. A. and Neuts, M. F. (1967). *SIAM J. Appl. Math.*, **15**, 726–738.

[17] Moore, E. H. and Pyke, R. (1968). *Ann. Inst. Statist. Math.*, **20**, 411–424.

[18] Neuts, M. F. (1964). *Ann. Math. Statist.*, **35**, 431–434.

[19] Pyke, R. (1961). *Ann. Math. Statist.*, **32**, 1231–1242. (Perhaps the most quoted paper on the topic.)

[20] Pyke, R. (1961). *Ann. Math. Statist.*, **32**, 1243–1259.

[21] Pyke, R. and Schaufele, R. A. (1964). *Ann. Math. Statist.*, **35**, 1746–1764.

[22] Ross, S. M. (1970). *Applied Probability Models with Optimization Applications*. Holden-Day, San Francisco. (Chapter 5 provides an excellent introduction to this topic.)

[23] Ross, S. M. (1970). *J. Appl. Prob.*, **7**, 649–656.

[24] Ross, S. M. (1982). *Stochastic Processes*. Wiley, New York.

[25] Smith, W. L. (1955). *Proc. R. Soc. Lond. A*, **232**, 6–31.

[26] Smith, W. L. (1958). *J. R. Statist. Soc. B*, **20**, 243–302.

[27] Snyder, D. L. (1975). *Random Point Processes*. Wiley, New York.

[28] Takács, L. (1954). *Magyar Tud. Akad. Mat. Kutato Int. Közl*, **3**, 115–128 [in Hungarian; English summary (1956) *Math. Rev.*, **17**, 866].

[29] Thompson, M. E. (1981). *Adv. Appl. Prob.*, **13**, 804–825.

(MARKOV PROCESSES
RENEWAL THEORY
STOCHASTIC PROCESSES)

U. NARAYAN BHAT

SEMI-MIDMEANS

The upper (lower) semi-midmean introduced by Cleveland and Kleiner [1] is defined as the midmean* of all observations above (below) the median. Under the normality assumption semi-midmeans may serve as estimators of quartiles. More precisely let $Y_1, Y_2, \ldots, Y_n$ be the ordered data. Define the sample inverse cumulative distribution function* F_n^{-1} at the points $(i - 0.5)/n$ to be Y_i, $i = 2, \ldots, n-1$, and at the points 0 and 1 to be Y_1 and Y_n, respectively. At all other points, F_n^{-1} is defined by linear interpolation*. Let

$$I(\alpha, \beta) = \frac{1}{\beta - \alpha} \int_\alpha^\beta F_n^{-1}(t)\, dt,$$

$$0 \leqslant \alpha < \beta \leqslant 1.$$

Then the lower and upper semi-midmeans are defined as $I(\frac{1}{8}, \frac{3}{8})$ and $I(\frac{5}{8}, \frac{7}{8})$, respectively. When sampling from a standard normal distribution, the asymptotic expected value of the lower (upper) semi-midmean is -0.693 $(+0.693)$; this corresponds to the quartile of order 0.244 (0.756).

Reference

[1] Cleveland, W. S. and Kleiner, B. (1975). *Technometrics*, **17**, 447–454.

(EXPLORATORY DATA ANALYSIS
MIDMEAN)

SEMI-PARAMETRIC MODELS

INTRODUCTION AND EXAMPLES

Semi-parametric statistical models include as unknowns both parameters and functions. This terminology, used by Kalbfleisch [13], Oakes [21], and others, appears to have become standard. An earlier use of the term in unpublished lectures by A. P. Dempster concerned problems with many nuisance parameters*, themselves regarded as random variables sampled from a distribution with a small number of "hyperparameters."

Interest will usually center on the parameters in the model specification, with the unspecified function(s) in the role of an infinite-dimensional nuisance parameter. A practical motivation for consideration of semi-parametric models is to avoid restrictive assumptions about secondary aspects of a problem while preserving a tight formulation for the features of primary concern. For a detailed classification of semi-parametric models, see ref. 25. The following selection illustrates the diversity of possible examples, some familiar in other guises, and the corresponding variety of inferential procedures.

Example 1. Single Sample Location. Let $f(\cdot)$ be a density function known only to be symmetric about zero. Given a sample $\{Y_i\}$ from the density $f(y-\mu)$, estimate μ.

Example 2. Proportional Hazards Regression Model [8]. Let the hazard function for the survival time T_i be

$$\begin{aligned} h_i(t) &= \lim_{\Delta\to 0}\frac{1}{\Delta}\operatorname{pr}(t < T_i < t+\Delta \mid T_i > t) \\ &= \exp(\eta)h_0(t), \end{aligned}$$

where $\eta = \beta' z$ is a linear predictor based on a vector z of explanatory variables with unknown coefficients β and $h_0(t)$ is an unknown baseline hazard function (*see* COX'S REGRESSION MODEL). Perhaps because it illustrates in striking form the separation between parametric and nonparametric components, this is the model most often described as semi-parametric.

Example 3. A Bivariate Survival Model [6]. Let (X, Y) be dependent survival times with joint survivor function

$$\begin{aligned} &\operatorname{pr}(X > x, Y > y) \\ &\quad = \left[\{G(x)\}^{-\phi} + \{H(y)\}^{-\phi} - 1\right]^{-1/\phi}, \end{aligned}$$

where G and H are continuous univariate survivor functions and ϕ is simply related to Kendall's coefficient of concordance*

$$\begin{aligned} &\operatorname{pr}\{(X_1 - X_2)(Y_1 - Y_2) > 0\} \\ &\quad = (\phi+1)/(\phi+2). \end{aligned}$$

Example 4. Linear Exponential Family* Regression [14]. Let Y_i have density

$$f_i(y, \beta) = \exp\{\beta' z_i(y) + a_i(\beta) + c(y)\},$$

where $z_i(y)$ are specified vector functions of y and β is a vector of coefficients to be estimated. If the dominating measure $c(y)$ is known, the problem has a simple parametric structure with sufficient statistic* $\Sigma z_i(y_i)$. If $c(y)$ and hence also the normalizing factors $a_i(\beta)$ are unknown, the problem is semi-parametric. This model generalizes logistic discrimination [11].

Example 5. Mixture Models. Let Y have conditional distribution $G(y|w)$ given the unobserved value of a mixing variable W

with distribution $H(w)$. A semi-parametric model can be obtained by parametric specification of one of G and H. Often qualitative restrictions on the other component will be needed to ensure identifiability*. For example, if $Y = (Y_1, Y_2, \ldots, Y_p)$ we may require the $\{Y_j\}$ to be conditionally independent given $W = w$. This gives a class of models important in empirical Bayes* analysis.

Example 6. Nonparametric Models With Constraints. Semi-parametric models can be obtained by the imposition of parametric constraints on nonparametric models. For example, in the two-sample problem (compare univariate distributions F and G given independent samples from each) we may require $1 - G(x) = \{1 - F(x)\}^\theta$, a special case of the proportional hazards regression model above, or $G(x)/\{1 - G(x)\} = \theta F(x)/\{1 - F(x)\}$, called the *proportional odds model* [3]. Such models may also be viewed as arising from the action of a group of transformations on the range space of a distribution function [17].

INFERENCE

In fully parametric models [18] subject to fairly undemanding regularity conditions, maximum likelihood* estimates are consistent, asymptotically normally distributed around the true value of the estimator, and efficient. The asymptotic variance may be estimated from the Fisher information* matrix.

For semi-parametric models no such general results obtain. Naive application of maximum likelihood techniques can give inconsistent estimates [19]. Inferential procedures for semi-parametric models have traditionally been derived ad hoc. We illustrate with the previous examples.

1. The symmetric single sample location problem has a special structure allowing *adaptive estimation*, that is, asymptotically μ may be estimated as precisely if f is unknown as if f were known [4]. The form of these adaptive estimators is quite complex.

2. Here the problem is invariant under the group of increasing differentiable transformations of the time axis, and the nuisance function can be eliminated by consideration of the marginal likelihood of the maximal invariant (*see* PSEUDO-LIKELIHOOD), namely the rank statistic [15]. In more general versions of the problem, involving right-censoring or time-dependent explanatory variables, the invariance* property may be lost, but a partial likelihood* of the same form and with the same asymptotic properties may be derived instead [8, 9, 21]. The partial or marginal likelihood has a simple explicit form.

3. In principle the marginal likelihood is available here also, but it appears to be computationally intractable. Estimators of ϕ may be obtained from unweighted or weighted estimates of the sample coefficient of concordance [6, 22, 23] or by approximating the marginal likelihood of ranks [7].

4. Here the nuisance functions may be eliminated by considering the conditional likelihood of the data given the order statistic of the $\{Y_i\}$ [14]. The resulting likelihood, over all possible permutations of the order statistic, still has $\Sigma z_j(y_j)$ as a sufficient statistic, can be written down explicitly, but requires considerable computation even for moderate sample sizes.

5. No general method for inference in mixture models is available, but maximization of the joint likelihood in the unknown parameters and functions is known to yield estimators with good properties in certain special cases [16, 24]. The EM algorithm [10] is often useful in achieving the maximization.

6. In models with constraints, the constraint itself defines a class of estimators. For example

$$\hat{\theta}(x) = \log\{1 - G(x)\}/\{1 - F(x)\}$$

and

$$\hat{\theta}(x) = G(x)\{1 - F(x)\}/[F(x)\{1 - G(x)\}]$$

are, for any x, clearly consistent for θ in the proportional hazards and proportional odds model, respectively. A large class of weighted averages of $\hat{\theta}(x)$ will also be consistent. Determination of the best such weighted average is in general difficult [1, 2].

RECENT WORK

At the time of writing, there is great interest in the development of asymptotic theory for sufficiently regular semi-parametric problems that would parallel the traditional development for fully parametric models. See for example refs. 1, 5, 12, and 25. A key idea is the local approximation of a semi-parametric model by a "least favorable" parametric model. For example, in the proportional hazards model (Example 2) the parameterization $h_0(t) = \tilde{h}_0(t)\exp\{\gamma' e(t)\}$ for the nuisance hazard [where $\tilde{h}_0(t)$ and $e(t)$ are now both assumed known] is locally least favorable if $\tilde{h}_0(t) = h_0(t)$ and $e(t) = E^*z(t)$, the expectation of the explanatory variable z on a randomly selected individual conditioned to fail at t: then the asymptotic information for β from the parametric likelihood equals that from the partial likelihood [20]. It is clear from this argument also that in this problem no valid semi-parametric procedure can improve on the partial likelihood.

References

[1] Begun, J. M., Hall, W. J., Huang, W. M., and Wellner, J. A. (1983). *Ann. Statist.*, **11**, 432–452.

[2] Begun, J. M. and Reid, N. (1983). *J. Amer. Statist. Ass.*, **78**, 337–341.

[3] Bennett, S. (1983). *Statist. Med.*, **2**, 273–277.

[4] Bickel, P. J. (1982). *Ann. Statist.*, **11**, 505–514.

[5] Bickel, P. J., Klaasen, C. A. J., Ritov, Y., and Wellner, J. A. (1987). *Efficient and Adaptive Inference in Semiparametric Models*. Johns Hopkins University Press, Baltimore, MD.

[6] Clayton, D. G. (1978). *Biometrika*, **65**, 141–151.

[7] Clayton, D. G. and Cuzick, J. (1985). *J. R. Statist. Soc. A*, **148**, 82–117.

[8] Cox, D. R. (1972). *J. R. Statist. Soc. B*, **34**, 187–220.

[9] Cox, D. R. (1975). *Biometrika*, **62**, 269–276.

[10] Dempster, A. P., Laird, N. M., and Rubin, D. B. (1977). *J. R. Statist. Soc. B*, **39**, 1–38.

[11] Efron, B. (1975). *J. Amer. Statist. Ass.*, **70**, 892–898.

[12] Gill, R. (1984). Non- and Semi-Parametric Maximum Likelihood Estimators and the Von Mises Method. Report MS-R8604. Centre for Mathematics and Computer Science, Amsterdam, Netherlands.

[13] Kalbfleisch, J. D. (1978). *J. R. Statist. Soc. B*, **40**, 214–221.

[14] Kalbfleisch, J. D. (1978). *J. Amer. Statist. Ass.*, **73**, 167–170.

[15] Kalbfleisch, J. D. and Prentice, R. L. (1973). *Biometrika*, **60**, 267–278.

[16] Lambert, D. and Tierney, D. (1984). *Ann. Statist.*, **12**, 1388–1399.

[17] Lehmann, E. L. (1953). *Ann. Math. Statist.*, **24**, 23–43.

[18] Lehmann, E. L. (1983). *Theory of Point Estimation*. Wiley, New York.

[19] Neyman, J. and Scott, E. L. (1948). *Econometrica*, **16**, 1–32.

[20] Oakes, D. (1977). *Biometrika*, **64**, 441–448.

[21] Oakes, D. (1981). *Int. Statist. Rev.*, **49**, 199–233.

[22] Oakes, D. (1982). *J. R. Statist. Soc. B*, **44**, 412–422.

[23] Oakes, D. (1986). *Biometrika*, **73**, 353–362.

[24] Tierney, L. and Lambert, D. (1984). *Ann. Statist.*, **12**, 1380–1387.

[25] Wellner, J. A. (1985). *Bull. Int. Statist. Inst.*, **51** (4), 23.1.1–23.1.20.

(COX'S REGRESSION MODEL
LINEAR EXPONENTIAL FAMILY
PARTIAL LIKELIHOOD
PSEUDO-LIKELIHOOD
SUFFICIENT ESTIMATION AND
PARAMETER-FREE INFERENCE)

DAVID OAKES

SEMI-PIE DIAGRAM

As its name implies, this is half of a complete pie chart*. It consists of a semicircle divided into sectors ("slices") proportional in size to constituents they represent.

SEMITABULAR FORM

This is a form of presentation intermediate between textual and tabular. It is also called *leader form*. It is used for material that is not sufficiently extensive to warrant formal tabular presentation, but too great in volume to be included in regular text. Usually, each item of such material is allotted a separate line. The lines are usually indented without heading, single-spaced instead of double-spaced, and there are no captions or other normal tabular form accessories.

Generally, it is inadvisable to use semitabular form if it does not completely specify the source and date of the data.

(NUMERACY)

SENSITIVE QUESTION *See* RANDOMIZED RESPONSE

SENSITIVITY ANALYSIS *See* INFLUENCE FUNCTIONS; LINEAR PROGRAMMING; ROBUSTNESS

SENSITIVITY AND SPECIFICITY

Diagnostic tests are used to discriminate between population elements that differ by some characteristic. Sensitivity and specificity are two important and interrelated concepts in the evaluation of diagnostic test accuracy [1, pp. 3–13]. These concepts are applicable to a wide range of classification* problems. The terminology used in the evaluation of diagnostic tests originated in medicine from the diagnosis of disease and therefore, serves as a useful example.

> *Sensitive Test:* If all persons with a disease have "positive" tests, we say that the test is sensitive to the presence of the disease.

> *Specific Test:* If all persons without the disease test "negative," we say that the test is specific to the absence of the disease.

If a test is both sensitive and specific, its results are clearly interpretable. Such tests are rare. Nevertheless, diagnostic tests frequently yield outcomes that are consistent with the patient's true disease condition.

While medical diagnosis is a convenient example, sensitivity and specificity are useful concepts in a wide range of applications, including decision theory*, management science*, computerized simulation modeling, and signal detection. The unifying idea across all of these applications is the classification of population elements into distinct classes based on some characteristic. Related statistical concepts include hypothesis testing*, operating characteristics, discriminant analysis*, ranking and selection*, and model fitting. For example, in hypothesis testing if we equate the null hypothesis to the absence of disease, then sensitivity is one minus the significance level and specificity is the power* of the test.

Sensitivity and specificity are defined as conditional probabilities*. Consider a population that can be partitioned into two mutually exclusive and exhaustive subsets, say D and $\bar{D}$, where $\bar{D}$ is the complement of D. A diagnostic test is a classification* method that characterizes elements of the population, without knowledge of the state D or $\bar{D}$, into one of two mutually exclusive and exhaustive sets d and $\bar{d}$, the diagnostic equivalents of D and $\bar{D}$, respectively. Henceforth, we will use $\Pr(S)$ to indicate the probability of observing an element of S, where S is a subset of a population. Yerushalmy [6] first defined $\Pr(d|D)$ as "sensitivity" and $\Pr(\bar{d}|\bar{D})$ as "specificity" and suggested their use in the evaluation of diagnostic tests.

Sensitivity and specificity are rarely known for any diagnostic test. Typically, they are both estimated by gathering test outcome information on a group of patients with the disease and another group of patients known not to have the disease. These two groups

Table 1 A Two-by-Two Table of Disease Condition by Test Outcome

		Disease Condition	
		D	$\bar{D}$
Test	d	a	b
Outcome	$\bar{d}$	c	d

represent the test population. Table 1 shows how this information is summarized in a two-by-two table* [1, pp. 15–19], from which sensitivity and specificity can be calculated directly:

$$\Pr(d|D) = \frac{a}{a+c}, \qquad \Pr(\bar{d}|\bar{D}) = \frac{b}{d+b}.$$

Sensitivity and specificity of a test typically do not depend on the underlying probability of disease. This means that test sensitivity and specificity can be applied to the population of interest even when the probability of disease in the population of interest is not equal to the probability of disease in the test population. Bayes' rule allows us to compute the probability of disease given a positive test for any probability of disease. This probability $\Pr(D|d)$ is usually referred to as the *predicted value* of a positive test. We can similarly compute the predicted value of a negative test $\Pr(\bar{D}|\bar{d})$.

Thus far, we have treated diagnostic tests as having only two outcomes, either d or $\bar{d}$. Usually, however, there are many possible test outcomes, often on a continuum. For example, suppose a test results in observing a random variable X. Further, suppose that X is distributed according to F_X if X is from D and G_X if X is from $\bar{D}$. In such instances, a critical value, say x_c, is selected such that the test is positive (d) if $x \leqslant x_c$ and negative ($\bar{d}$) if $x > x_c$. Sensitivity is then given by $F_X(x_c)$, and specificity by $1 - G_X(x_c)$. Much research goes into selecting critical values. This research is complicated by the fact that patients with a disease and those without the disease may have identical test outcomes. Critical value choice is important to the determination of sensitivity and specificity. If the critical value is chosen to maximize sensitivity, very typically it will also simultaneously minimize specificity. Therefore, there is usually a trade-off between sensitivity and specificity.

In recent years medical researchers have adopted the techniques of operating characteristics to define the trade-off between sensitivity and specificity [4]. An operating characteristic is defined by the locus of points $(F_X(x), G_X(x))$, where x ranges over all possible values. For a selected point, say $(F_X(x_c), G_X(x_c))$, sensitivity is given by $F_X(x_c)$ and specificity by $1 - G_X(x_c)$. Thus, operating characteristics* relate the dependence between sensitivity and specificity through a function over the unit square. A great deal of current research on diagnostic test evaluation is devoted to operating characteristics.

In many practical applications the state D or $\bar{D}$ is unavailable because the cost, monetary or other, is prohibitive. Hence, the determination of sensitivity and specificity must rely on a reference test that yields D^* and $\bar{D}^*$, proxies for D and $\bar{D}$.

An important issue in developing sensitivity and specificity is the choice of a reference test. Since even reference tests may not be perfectly sensitive or perfectly specific, it is usually not adequate to simply use two tests to construct the two-by-two table for calculating sensitivity and specificity. However, if the reference test and the diagnostic test are conditionally independent, as indicated by

$$\Pr(d, D^*|D) = \Pr(d|D)\Pr(D^*|D)$$

and

$$\Pr(\bar{d}, \bar{D}^*|\bar{D}) = \Pr(\bar{d}|\bar{D})\Pr(\bar{D}^*|\bar{D}),$$

maximum likelihood estimates * of sensitivity and specificity can be derived [2]. On the other hand, even small dependence between the reference test and diagnostic test will result in nonsampling errors* in the estimates of sensitivity and specificity [5].

It is rare, in medicine at least, that one diagnostic test is used without others. Therefore, sensitivity and specificity must be evaluated in the light of multiple test results [3], which are often thought of as being independent, although this is rarely true. Sensitivity and specificity can then be calculated as a product of conditional probabilities.

References

[1] Fleiss, J. L. (1973). *Statistical Methods for Rates and Proportions*. Wiley, New York. (A good basic statistical text.)

[2] Hui, S. L. and Walter, S. D. (1980). *Biometrics*, **36**, 167–171. (Estimation of error rates from independent tests. Good references.)

[3] Ingelfinger, J. A., Mosteller, F., Thibodeau, L. A., and Ware, J. H. (1983). *Biostatistics in Clinical Medicine*. Macmillan, New York. (Applications to medicine, especially Chapters 1 and 2.)

[4] Metz, C. E. (1978). *Seminars Nucl. Med.*, **4**, 283–298. (Introduction to receiver operating characteristics for diagnostic tests.)

[5] Thibodeau, L. A. (1981). *Biometrics*, **37**, 801–804.

[6] Yerushalmy, J. (1947). *Public Health Rep.*, **62**, 1432–1449.

(ACCEPTANCE SAMPLING
BIOSTATISTICS
CLASSIFICATION
DISCRIMINANT ANALYSIS
EPIDEMIOLOGICAL STATISTICS
MEDICAL DIAGNOSIS, STATISTICS IN)

L. A. THIBODEAU

SENSITIVITY CURVE

Denoting by $\hat{\theta}_n$ the estimator of a parameter θ, based on a random sample of size n, the effect of an additional random observation with value x can be measured by the *sensitivity curve*

$$\mathrm{SC}(x) = (n+1)(\hat{\theta}_{n+1}(x) - \hat{\theta}_n),$$

where $\hat{\theta}_{n+1}(x)$ represents the value of $\hat{\theta}_{n+1}$ for the original sample values combined with the new value x. More precisely

$$\mathrm{SC}(x|x_1,\ldots,x_n) = (n+1)\,(\hat{\theta}_{n+1}(x_1,\ldots,x_n,x) - \hat{\theta}_n(x_1,\ldots,x_n)).$$

This form emphasizes the fact that SC depends on $x_1,\ldots,x_n$ as well as on n.

Hampel [2] shows that if the sensitivity curve is properly normalized, in the limit (as n increases) it corresponds to the influence curve. More details are available in Andrews et al. [1].

References

[1] Andrews, D. F., Bickel, P. J., Hampel, F. R., Huber, P. J., Rogers, W. H., and Tukey, J. W. (1972). *Robust Estimates of Location: Survey and Advances*. Princeton University Press, Princeton, NJ.

[2] Hampel, F. R. (1974). *J. Amer. Statist. Ass.*, **69**, 383–393.

(INFLUENCE FUNCTIONS
ROBUST ESTIMATION)

SEPARABLE SPACE

A metric space is *separable* if it has a denumerable subset of points that approximate arbitrarily closely every point of the space, i.e., if it contains a countable dense subset.

EXAMPLES OF SEPARABLE SPACES

The space of real numbers is separable: every real can be approximated arbitrarily closely by a rational. The same is true for the space of real (or complex) vectors, as well as for the sequence spaces l_p, $1 \leqslant p < \infty$. However, the space l_∞ of bounded sequences with the supremum norm is not separable.

The space of continuous functions on the unit interval $[0,1]$ with the supremum norm is separable, while the space of continuous functions on the half line $[0,\infty)$ with the supremum norm is nonseparable. Further examples of separable spaces of functions on $[0,1]$ are the continuously differentiable functions, the absolutely continuous functions, the integrable functions, all with their natural norms, and the more general Orlicz and Sobolev spaces. Examples of nonseparable spaces of functions on [0, 1] are the Hölder-continuous functions, the functions of bounded variation, the (essentially) bounded functions, all with their usual norms, as well as the space of all measurable functions with the distance of f and g defined by

$$\int_0^1 \frac{|f(t) - g(t)|}{1 + |f(t) - g(t)|} \, dt,$$

which metrizes convergence in measure, or in probability when viewing the unit interval with Lebesgue measure as a probability space. (For most of these and related facts see Kufner et al. [10].)

An L_p space over a general measure space is separable if and only if the measure space is separable, i.e., the metric space of measurable sets of finite measure (modulo sets of measure zero), with the distance of two such sets being the measure of their symmetric difference, is separable. The same criterion for separability applies also to the more general Orlicz spaces. Even though general probability (sample) spaces are frequently required in statistics and probability, it turns out that the classical measure spaces on the real line suffice to consider all separable probability spaces. Specifically every separable probability (or σ-finite measure) space can be represented as the disjoint union of a countable set of points with positive mass, and of an interval with Lebesgue measure. For a lucid discussion of this isomorphism theorem see Halmos and Sunder [9, pp. 2–3]. For the facts stated see Halmos [8] and Rolewicz [14, p. 27].

SEPARABLE VS. NONSEPARABLE

When dealing with a nonseparable space, frequently its part within reach of usual objects of interest turns out to be separable: Borel probability measures on metric spaces usually have separable support (see Dudley [7, Sec. 5]). The image of a complete separable metric space under a Borel map into an arbitrary metric space is necessarily separable (see Cohn [6]).

What are the advantages of separability? The existence of a dense subset that is denumerable renders all kinds of desirable constructions and proofs feasible. In the absence of separability the usual approximation techniques and constructions do not work and serious technical problems arise. In technical terms one of the most significant consequences of separability is that under it the balls generate the Borel σ-field (of open sets). In nonseparable spaces one cannot generaly approximate Borel sets by balls and in certain cases measurability with respect to the (smaller) σ-field generated by the balls may be more natural than Borel measurability.

BASES

A very useful tool for analyzing problems in Hilbert and Banach spaces is provided by a basis, and whether a basis exists is closely related to the separability of the space. Every separable Hilbert space has an orthonormal basis, and if it is infinite dimensional it is isomorphic to l_2. This has been at the root of the success and ease in analyzing problems expressible in a Hilbert space setup. A Banach space with a basis is separable. A separable Banach space, however, does not necessarily have a Schauder basis, even though all those mentioned earlier do. For nonseparable Banach spaces extended notions of bases are introduced that need not form a countable set of objects (which are not elements of the space), such as Schauder decompositions. (See Singer [15, 16].)

DOMINATION

An assumption frequently made in statistics, e.g., in describing sufficient statistics*, is that a family of probability distributions is dominated (by some probability distribution). Domination is related to separability: A family of probability measures over a Euclidean space (or, more generaly, over a separable measure space) is dominated if and only if it is separable with respect to the total variation metric (see Lehmann [11]).

REGULAR CONDITIONAL DISTRIBUTIONS

Separability plays a crucial role in the construction (and thus the existence) of regular conditional probabilities and distributions for random vectors. (See, e.g., Breiman [3, Sec. 4.3] and Ash [1, Sec. 6.6].)

TIGHTNESS

The tightness of a probability measure on a complete metric space is closely related to separability. If the space is separable, every probability measure is tight. On the other hand, when the space is not required to be separable, if the probability measure has separable support*, then it is tight (and in fact this statement cannot be weakened as tightness implies separability of the support). (See Billingsley [2].)

WEAK CONVERGENCE

Separability is an important issue in the study of weak convergence of stochastic processes*. Empirical processes* have sample functions in the space $D[0, 1]$ of functions on the unit interval that are right continuous and have left limits. When $D[0, 1]$ is equipped with the (more natural and easy to handle) supremum norm it is, unfortunately, nonseparable. Still one can handle convergence in distribution to a process whose sample paths belong to a separable subset of $D[0, 1]$, such as the set of continuous functions on $[0, 1]$. This suffices to treat cases where the limiting process is Brownian motion*, Brownian bridge, or some other Gaussian process with continuous sample functions, but it excludes cases where the limiting process is Poisson, or certain other non-Gaussian processes with independent increments. In order to handle such cases the (more complex) Skorohod metric is introduced under which $D[0, 1]$ becomes separable. These issues are discussed in Pollard [12] and Billingsley [2].

STOCHASTIC PROCESSES

In a variety of problems of linear statistical inference for stochastic processes with finite second moments, a central role is played by the linear space of the process and by the reproducing kernel Hilbert space of its correlation function (which are isometrically isomorphic).

When the process is continuous in probability or in mean square, then its linear space is separable. In fact the separability of the linear space is related to the weakest smoothness assumption required of a stochastic process to be useful: that it has a measurable modification. Specifically, a measurable modification exists if and only if the linear space of the process is separable and its correlation function is measurable [5].

Under any of the smoothness assumptions just mentioned, orthonormal bases can be constructed in the observable space of the process, which is the entire linear space under continuity in probability or in mean square. They are expressed in terms of the sample functions of the process; they provide a decomposition of the process into its principal components (the Karhunen–Loève expansion); and they lead to infinite series expansions and integral approximations for (nonrecursive) linear mean square estimates based on observation of the process. (See Pugachev [13] and [4].)

Separability of the linear space of a process should not be confused with Doob's notion of a separable process, which deals with sample function regularity: Every process has a separable modification (*see* STOCHASTIC PROCESSES); separability of a process in Doob's sense, however, does not imply the separability of its linear space.

References

[1] Ash, R. B. (1972). *Real Analysis and Probability*. Academic, New York.

[2] Billingsley, P. (1968). *Convergence of Probability Measures*. Wiley, New York.

[3] Breiman, L. (1968). *Probability*. Addison-Wesley, Reading, MA.

[4] Cambanis, S. (1973). *IEEE Trans. Inf. Theory*, IT-19, 110–114.

[5] Cambanis, S. (1975). *Proc. Amer. Math. Soc.*, **47**, 467–475.

[6] Cohn, D. L. (1972). *Zeit. Wahrsch. verw. Geb.*, **22**, 161–165.

[7] Dudley, R. M. (1976). *Probabilities and Metrics: Convergence of Laws on Metric Spaces with a View to Statistical Testing.* Mathematics Institute Lecture Notes Series No. 45, Aarhus University, Aarhus, Denmark.

[8] Halmos, P. R. (1950). *Measure Theory*. Van Nostrand, Princeton, NJ.

[9] Halmos, P. R. and Sunder, V. S. (1978). *Bounded Integral Operators on L^2 Spaces*. Springer, Berlin, Federal Republic of Germany.

[10] Kufner, A., John, O., and Fučík, S. (1977). *Function Spaces*. Noordhoff, Leyden, The Netherlands.

[11] Lehmann, E. L. (1959). *Testing Statistical Hypotheses*. Wiley, New York.

[12] Pollard, D. (1984). *Convergence of Stochastic Processes*. Springer, New York.

[13] Pugachev, V. S. (1965). *Theory of Random Functions*. Pergamon, Oxford, England.

[14] Rolewicz, S. (1985). *Metric Linear Spaces*. Polish Scient. Publ., Warszawa, Poland.

[15] Singer, I. (1970). *Bases in Banach Spaces I*. Springer, New York.

[16] Singer, I. (1981). *Bases in Banach Spaces II*. Springer, Berlin, Federal Republic of Germany.

(ABSOLUTE CONTINUITY
MEASURE THEORY IN PROBABILITY AND STATISTICS
PROBABILITY SPACES, METRICS AND DISTANCES ON
PROCESSES, EMPIRICAL)

STAMATIS CAMBANIS

SEPARATING HYPERPLANE THEOREM *See* GEOMETRY IN STATISTICS; CONVEXITY

SEPARATION OF MEANS *See* k-RATIO t TESTS; MULTIPLE COMPARISONS; MULTIPLE DECISION PROCEDURES

SEQUENCE COMPARISON STATISTICS

It is often of interest to compare two or more ordered sequences of objects. Examples of such sequences are: (1) proteins, which can be described as sequences of amino acids (an object is an amino acid); (2) tree-ring sequences [an object is described by the (standardized) thickness of a single year's growth]; (3) geophysical logging data (an object is described by the values taken by several geophysical variables at a single depth down a borehole).

In these examples, it is relevant to obtain not only a measure of the overall resemblance of the sequences but also an indication of which parts of each sequence correspond with one another.

The objects in each sequence may either (i) match or not match, as in example (1) above, or (ii) merely resemble one another to a greater or lesser degree, as in examples (2) and (3). In either case, the differences between the ith and jth objects can be assessed by a measure of dissimilarity $d(i, j)$; *see* MEASURES OF SIMILARITY, DISSIMILARITY AND DISTANCE.

In example (1), the aim is to transform one sequence into the other using an optimal sequence of operations belonging to the class (deletion of an object, insertion of an object, replacement of an object by a different object). A detailed discussion of this problem is presented in [9]; extensions also considered there include replacement of an object by a string of identical objects (or vice versa), transposition of adjacent objects, and comparing different parts of a single sequence.

The data for comparison may be such that there is some relevant underlying variable, by which the objects are ordered [such as time in examples (2) and (3)]. Alternatively, such a variable may not be readily specifiable, and one might be able to assume only that the ordering of objects within each sequence is correct. In example (2), there is the further information that objects are produced at regular intervals of time, and tree-ring sequences have commonly been compared by finding the lag that maximizes the correlation coefficient* between them (e.g., ref. 3). A generalization of this approach has been used in the analysis of sequences whose objects are less precisely specified in terms of the underlying variable. The underlying variable in one sequence has been monotonely transformed (and data points interpolated) so as to ensure that the two sequences correspond as closely as possible in terms of some stated criterion (e.g., refs. 1, 7, and 8); only linear transformations have been used to date. Thus, for example, in ref. 8 measurements were available of the resistivity at regular depth intervals down two boreholes, and the underlying depth variable in the first borehole was linearly transformed (and data points interpolated at the original sampling interval) so as to ensure that the correlation coefficient between the second set of resistivity measurements and the transformed first set was maximized.

Alternatively, if one assumes only that the ordering of the objects within each sequence is correct, one can slot the sequences together so that similar objects are close together subject to the constraint that the ordering within each sequence is preserved in the joint sequence.

An illustration of the slotting together of two sequences, S_A and S_B, comprising n ($= 7$) and m ($= 6$) objects, respectively, is given in Fig. 1. For each object in each sequence, one defines a measure of the local discordance of the object in the slotting. For example [4], the local discordance of an object C could be defined as (i) the sum of the two dissimilarities between C and the pair of objects from the other sequence between which C is located in the joint slotting, e.g., the local discordance of A_2 in the slotting given in Fig. 1 is

$$d(A_2, B_1) + d(A_2, B_2).$$

In Fig. 1, the presence of a line linking two objects indicates that the dissimilarity between this pair of objects is part of the local discordance of an object: Terms included in the local discordance of objects in S_A are depicted by unbroken lines, and terms included in the local discordance of objects in S_B are depicted by dashed lines. End comparisons are treated as follows: Dummy objects A_0 and A_8, identical with A_1 and A_7, respectively, are assumed always to bracket S_B; dummy objects B_0 and B_7, bracketing S_A, are defined in a similar manner.

An alternative definition of the local discordance of an object C is (ii) the smaller of the two dissimilarities between C and the pair of objects from the other sequence between which C is located, e.g., the local discordance of A_2 is

$$\min(d(A_2, B_1), d(A_2, B_2)).$$

The choice of an appropriate measure of local discordance will depend on the nature of the expected variation in the data. Thus, the first definition would be relevant for data changing slowly with the underlying variable, whereas the second definition represents a slight relaxation of the direct matching of corresponding objects. Other possible definitions of local discordance are given in refs. 2 and 4.

The global discordance, or misfit, of two sequences is defined to be the sum, over each object in each sequence, of the local discordances. Thus, the global discordance resulting from definition (i) comprises the sum of the dissimilarities between pairs of objects

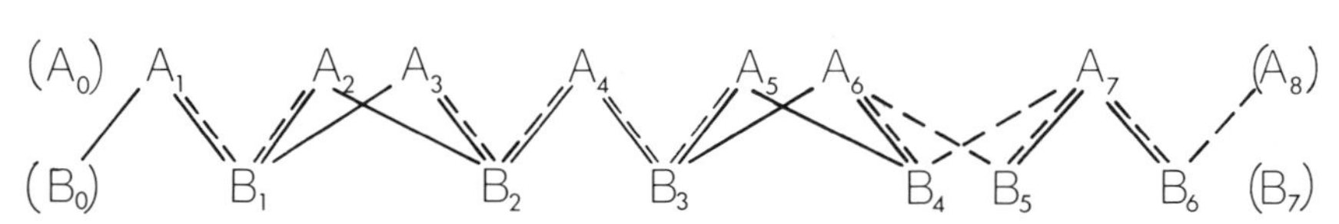

Figure 1

linked by a line in Fig. 1. One seeks the slotting(s) with minimum value of the global discordance; this can be built up recursively using the optimality principle of dynamic programming* [2].

To enable the assessment of global measures of discordance, it is helpful to standardize them. Thus, if the measure of dissimilarity d satisfies the triangle inequality, it can be shown that the first global measure σ_1 is no less than

$$\mu_1 \equiv \sum_{i=1}^{n-1} d(A_i, A_{i+1}) + \sum_{j=1}^{m-1} d(B_j, B_{j+1}),$$

suggesting the standardized statistic

$$\psi \equiv (\sigma_1 - \mu_1)/\mu_1.$$

For the second criterion, one can similarly define

$$\kappa \equiv (\sigma_2 - \mu_2)/\mu_2,$$

where

$$\mu_2 \equiv \sum_{i=1}^{n} \min_k (d(A_i, B_k)) + \sum_{j=1}^{m} \min_k (d(A_k, B_j)).$$

Since two sequences can always be slotted together, no matter how small the resemblance between them, the value taken by a discordance statistic provides a check against drawing unwarranted conclusions from a comparison. Larger values of ψ and κ cause one to be increasingly suspicious of the genuineness of any agreement that is indicated. Null models of the agreement between sequences can be investigated with the assistance of simulation studies. For example, one could envisage obtaining S_B by specifying m positions in "the period covered by" S_A and conceptually resampling from S_A at these positions, evaluating the discordance statistic between S_A and the simulated S_B [5].

It can be relevant to impose various kinds of constraint to restrict the class of allowable slottings; thus, in example (3), identifiable marker beds could be present in both sequences, corresponding to synchronous events in the past. The additive nature of the discordance statistics makes it straightforward to obtain the optimal constrained slottings by piecing together part-slottings [6].

If there are p (> 2) sequences $\{S_i\ (i = 1, \ldots, p)\}$ to be compared, one can seek the joint slotting of all the sequences that minimizes a statistic like

$$\sum_{1 \leqslant i < j \leqslant p} \psi(S_i, S_j),$$

but this soon becomes computationally infeasible. An alternative approach is to slot each sequence in turn with a "master" sequence, corresponding, for example, to a stratigraphical data set that is regarded as the standard or reference set for the region being investigated. The master sequence could be redefined before each new comparison to be the joint slotting obtained from the previous comparison. With this latter strategy, the results will in general depend on the initial choice of master sequence and the order in which the comparisons are made, but may still be informative about the general resemblances between the sequences.

References

[1] Clark, R. M. and Thompson, R. (1979). *Geophys. J. R. Astron. Soc.*, **58**, 593–607.

[2] Delcoigne, A. and Hansen, P. (1975). *Biometrika*, **62**, 661–664.

[3] Ferguson, C. W. (1968). *Science*, **159**, 839–846.

[4] Gordon, A. D. (1973). *Biometrika*, **60**, 197–200.

[5] Gordon, A. D. (1982). *Aust. J. Statist.*, **24**, 332–342.

[6] Gordon, A. D. and Reyment, R. A. (1979). *Math. Geol.*, **11**, 309–327.

[7] Kwon, B.-D., Blakeley, R. F., and Rudman, A. J. (1978). *Occasional Paper No. 26*, Indiana Geological Survey.

[8] Rudman, A. J. and Lankston, R. W. (1973). *Bull. Amer. Ass. Petroleum Geol.*, **57**, 577–588.

[9] Sankoff, D. and Kruskal, J. B., eds. (1983). *Time Warps, String Edits, and Macromolecules: The Theory and Practice of Sequence Comparison*. Addison-Wesley, Reading, MA.

(DYNAMIC PROGRAMMING
MEASURES OF SIMILARITY,
DISSIMILARITY AND DISTANCE
TIME SERIES)

A. D. GORDON

SEQUENCE DATING *See* ARCHAEOLOGY, STATISTICS IN

SEQUENTIAL ANALYSIS

BASIC IDEAS AND HISTORICAL DEVELOPMENTS

In traditional methods of statistics, which still constitute the bulk of statistical analysis, we observe a fixed set of data once and for all, and then analyze them to draw some inference about the population from which they came. The inference problem itself may be formulated in a parametric or nonparametric setting for a univariate or multivariate population in the framework of estimation* theory, hypothesis testing*, or general decision theory*. Sequential analysis, on the other hand, pertains to techniques in which the final size and composition of the data need not be predetermined but may depend, in some specified way, on the data themselves as they become available during the course of an investigation.

Two elements characterize a sequential method: a *stopping rule* (or *procedure*), which tells the statistician how to observe values from the population and when to stop sampling (*see* STOPPING NUMBERS AND STOPPING TIMES), and a *decision rule*, which tells the statistician what action should be taken about the inference problem at each stage of sampling. The *goal* of sequential analysis is to determine a stopping rule and a decision rule that satisfy certain desirable criteria prompted by the nature of the inference problem. When a stopping rule essentially instructs observation of m values from the population, no more and no less, one gets as a special case the traditional (or nonsequential) method based on a *fixed sample* of size m.

Sequential methods are motivated by questions of *solvability* of a problem, *economy* of experimentation, *ethics* of an investigation, and *intrinsic nature* of certain experiments. To illustrate these points, consider an experimental drug for controlling hypertension as measured by the diastolic blood pressure in millimeters of mercury. After the drug has been administered to a series of patients, we can record the diastolic blood pressure X_i for the ith patient; assume for present purposes that random variables $X_1, X_2, \cdots$ are independent and identically distributed (i.i.d.) according to a normal distribution with mean μ and standard deviation σ. Alternatively, we may only observe whether the patients have pressure exceeding a critical level (80 millimeters, say) or the opposite, in which case we can associate an observable random variable y_i to the ith patient such that $Y_i = 1$ if the pressure exceeds the critical level and $Y_i = 0$ if the pressure is at most the critical level.

The following examples, posed as inference problems concerning the unknown quantities μ and $p = \Pr[Y_i = 1]$, illustrate the motivations for sequential methods.

Example 1. Suppose that we wish to estimate $1/p$ unbiasedly, that is, find a function T in terms of the observed values of Y_1, Y_2 $\cdots$, such that the expected value of T is $1/p$ itself. Note that $1/p$ stands for the average number of patients needed to get the first blood pressure reading exceeding the critical value. The problem cannot be solved if one obtains a fixed sample of m patients and observes their Y values. As a second instance, suppose σ is unknown and we wish to find a confidence interval* for μ such that the interval includes μ with a specified probability and the length of the interval is at most a given number. Here, again, (see ref. 17) the problem is unsolvable if the data consist of a fixed number of patients and their X values. It becomes natural to ask in both instances whether a solution exists at all when patients are examined sequentially under suitable stopping rules. Haldane's [28] method of inverse sampling* and Stein's [53] method of two-stage sampling yield an unbiased estimate for $1/p$ and the requisite confidence interval for μ, respectively (*see* SEQUENTIAL ESTIMATION).

Example 2. Suppose σ is known and we want to test the hypothesis $H_0: \mu \leqslant \mu_0$ against the alternative hypothesis $H_1: \mu \geqslant \mu_0$ using observed values of X, where μ_0 and μ_1 are known blood pressure levels of interest to the pharmaceutical company that produces the drug. We also stipulate that the statistical test guarantee an upper limit of α and β, respectively, for the type-I error probability (i.e., of rejecting H_0 when H_0 is correct) and type-II error probability (i.e., of rejecting H_1 when H_1 is correct). For this problem, the best (uniformly most powerful) fixed-sample test observes m values on X, calculates the sample mean $\overline{X}$, and accepts H_0 or H_1 according as $\overline{X} < c$ or $\geqslant c$, where m and c are constants determined by α and β. However, if the company wants to minimize the number of patients to be examined (or the duration of the entire experiment), the best fixed-sample test need not be a good choice. It is possible to construct a sequential procedure, called the *sequential probability ratio test* (SPRT), that requires, on the average, less than m patients (see the section Hypothesis Testing). One arrives at similar conclusions for testing $H_0: p \leqslant p_0$ against $H_1: p \geqslant p_1$ using observed values of Y. This aspect of economy in cost or time renders most sequential methods more efficient than their nonsequential counterparts.

Example 3. Suppose we want to test the hypothesis $H_0: p \leqslant p_0$ at level of significance α (i.e., the maximum type-I error probability is some preassigned number α). Then the best fixed-sample test observes m patients and the value of $S_m = Y_1 + \cdots + Y_m$, which is the number of patients having blood pressure over the critical level; H_0 is rejected if $S_m \geqslant c$ for some positive integer c depending on α, and H_0 is accepted otherwise. Now, since every Y_i can be only 0 or 1, it is obvious that if the first n of m patients show that $S_n \geqslant c$, then the remaining $m - n$ patients will automatically reveal that $S_m \geqslant c$. Consequently, even if no significant cost or time is involved in carrying out the experiment, sheer medical ethics would forbid unnecessary experimentation with the additional $m - n$ patients (*see* CLINICAL TRIALS). This philosophy leads to the following sequential version of the fixed-sample procedure: Observe $Y_1, \ldots, Y_m$ one by one, and stop and reject H_0 whenever $S_n \geqslant c$ for $1 \leqslant n < m$. The sequential procedure is known as *curtailed sampling* [22]. There are many instances where a sequential procedure having otherwise nice features may itself have to be truncated, due to practical reasons, sooner than dictated by the procedure. See ref. 4 for such truncated tests and other problems in clinical medicine where sequential procedures are routinely used.

Example 4. Finally, there are investigations where the notion of sequential analysis is intrinsic. For instance, the blood pressure of a patient under intensive care may be monitored continuously and the question may be how to interpret and analyze sudden fluctuations. Similarly, the producer of a drug may want to control its quality by subjecting samples of the drug continually to chemical analysis and the question may be whether there is any changing pattern in the composition of the drug over time. In both cases, sequential aspects seem essential for understanding the factors that govern the experimental system.

Historically, the consideration of economy in cost and duration of sampling gave rise to research in sequential analysis. The rudiments originated with the double-sampling* plan of Dodge and Romig [19] to decide whether a large lot of manufactured items is of good or poor quality as measured by the proportion of defectives in the lot. In the plan itself, one inspects a fixed number of items at the first stage and tries to make a decision about the quality of the lot on the basis of the number of defectives in the sample; if the evidence is equivocal, one inspects a new sample of items and makes a final decision on the basis of the combined number of defectives in the two samples. The plan turns out to be more effective than the traditional single-sample procedure in reducing the average number of items to be inspected. Bartky [7] generalized the double-sampling plan to include more than two stages. But the idea was truly sequentialized by Wald [56] and Barnard [6] in an SPRT where items are inspected one by one and a decision (good, poor, or continue) is made about the lot at every step. Wald [57] showed that the SPRT is bound to terminate with a final decision of good or poor quality, and Wald and Wolfowitz [59] proved that the SPRT requires, on the average, the smallest number of items for inspection among all statistical procedures when the true proportion equals two critical values.

Wald's [57, p. 157] theory of the SPRT is applicable to a wide variety of situations where problems can be formulated in terms of testing hypotheses about a single parameter in the absence of nuisance parameters*. Cox [16], Bahadur [5], and Hall et al. [29] extended the theory to situations where nuisance parameters are present. Since then, numerous authors have studied various properties of the SPRT and their modified versions to meet specific goals, especially from an asymptotic viewpoint when the final sample size is large. Some of these results are described in the following section.

In the area of statistical estimation, some of the earliest attempts at sequential techniques were made by Thompson [55] to estimate proportions, Neyman [44] to estimate variances in a stratified population, and Hotelling [31] to estimate the maximum of a regression function. Their attempts were heuristic in that the properties and efficiency of the suggested methods were left unexplored. The first sophisticated sequential procedure for point estimation was developed by Haldane [28] and for interval estimation by Stein [53] (*see* FIXED-WIDTH AND BOUNDED-LENGTH CONFIDENCE INTERVALS). These were referred to in Example 1. There is at present no unified theory of sequential procedures for point or interval estimation. The reason is that, if minimization of the variance of a point estimator or attainment of the coverage probability by the interval estimator is the main criterion, which is what one often assumes in fixed-sample contexts, then for many problems optimal sequential procedures essentially reduce to fixed-sample techniques (see refs. 2, 54, and 67). As a consequence, several alternative criteria have been proposed for point and interval estimation, and numerous authors have developed and studied ad hoc sequential techniques for the corresponding problems; *see* SEQUENTIAL ESTIMATION.

It is well known that practically all problems in statistical inference can be formulated in the general framework of decision theory. Wald [58] developed a unified theory that incorporates sequential procedures. He showed that some of the sequential techniques (e.g., SPRT) used in hypothesis testing and estimation theory fall out of decision-theoretic results. More often, however, such general results answer existence or uniqueness questions for sequential procedures and do not generally provide a simple means to construct actual procedures for specific problems. A detailed account of decision-theoretic sequential procedures can be found in Berger [9], Ferguson [25], and Wald [58].

HYPOTHESIS TESTING

Let $X_1, \ldots, X_n, \ldots$ be random variables (possibly vectors) corresponding to (possibly vector) observations $x_1, \ldots, x_n, \ldots$ that can be obtained from a population. Assume that the X_i are i.i.d. according to some distribution (density or mass function) $f_\theta(x)$ at $X_i = x$, where θ is a unknown indexing parameter, possibly vector-valued, that lies in a specified set Ω. The problem is to test, under suitable criteria, a hypothesis $H_0 : \theta \in \omega_0$ against an alternative hypothesis $H_1 : \theta \in \omega_1$, where ω_0 and ω_1 are known disjoint subsets of θ (ω_1 may equal $\Omega - \omega_0$ as a special case). For specific $\theta_0 \in \omega_0$ and $\theta_1 \in \omega_1$ we denote $f_\theta(x)$ by $f_0(x)$ and $f_1(x)$, respectively.

A *sequential test* of H_0 against H_1 is defined by a random pair (N, D). The *stopping variable* (or *decisive sample number*) N is defined by a set of rules for terminating sampling with $N = n$ observations $(x_1, \ldots, x_n)$ for every $n \geqslant 1$. The *decision function* D is defined by a set of rules, after the event $N = n$ has occurred, for accepting H_0 (i.e., $D = H_0$) or rejecting H_0 (i.e., $D = H_1$) on the basis of $(x_1, \ldots, x_n)$. If $\Pr_\theta[N < \infty] = 1$ for all $\theta \in \Omega$, we say the sequential procedure (or test) *terminates surely*, and this is regarded as a desirable property (an exception occurs in the so-called tests with power one [45]). If $\Pr_\theta[N = m] = 1$ for all θ and some positive integer m, the resulting procedure becomes *fixed-sample*. We call

$$E_\theta[N] = \sum_{n=1}^{\infty} n \Pr_\theta[N = n]$$

the *average sample number** (ASN) required by the procedure, and any "good" procedure should have a small ASN for all θ. We call

$$P_\theta = \Pr_\theta[D = H_1] \quad \text{and} \quad Q_\theta = \Pr_\theta[D = H_0]$$

the *power function* and *operating characteristic* (OC) function of the test. If the procedure terminates surely, it is clear that $P_\theta + Q_\theta = 1$ for all θ. The type-I and type-II error probabilities of the test are defined, respectively, as $\alpha(\theta) = P_\theta$ for $\theta \in \omega_0$ and $\beta(\theta) = Q_\theta$ for $\theta \in \omega_1$. Any "good" test should have small values of $\alpha(\theta)$ and $\beta(\theta)$. Criteria for good tests or notions of optimal tests may vary from one problem to another, but they are usually laid down in terms of the distributions of N, the ASN, the error probabilities, and a fixed cost of experimentation. The concepts and definitions above are applicable when the X_i are not i.i.d. but $(X_1, \ldots, X_n)$ possess some joint density or mass function for every n and θ.

Bahadur [5] proved a fundamental principle for reducing the data to construct sequential tests. Suppose S_n, a function of $(X_1, \ldots, X_n)$, insufficient for the joint distributions (i.e., as θ varies) of $(X_1, \ldots, X_n)$ for each n and S_n is also *transitive* (i.e., S_{n+1} is expressible only in terms of S_n and X_{n+1} for each n). Then for every test based on the X_i there exists a test based on the S_n such that the two tests have the same power function and distribution of N. This *principle of sufficiency* simplifies the search for good tests under any reasonable criteria in that one need restrict attention only to values of S_n. But the principle does not describe the actual construction of good tests; this will now be addressed.

SPRT

Suppose ω_0 and ω_1 consist of single points, θ_0 and θ_1, respectively (i.e., H_0 and H_1 are *simple* hypotheses). A *sequential probability ratio test* (SPRT) defines the stopping variable as

$$N = \text{first } n \geqslant 1 \text{ such that } \lambda_n \leqslant A \text{ or } \lambda_n \geqslant B$$

$$= \infty \quad \text{if } A < \lambda_n < B \text{ for all } n \geqslant 1,$$

where A and B are constants satisfying $0 \leqslant A < B \leqslant \infty$ and λ_n is the likelihood ratio

$$\lambda_n = \prod_{i=1}^{n} \{f_1(X_i)/f_0(X_i)\}.$$

The decision function of the SPRT is defined as $D = H_0$ if $\lambda_n \leqslant A$ and $D = H_1$ if $\lambda_n \geqslant B$ when $N = n$ (the decision is indeterminate when $N = \infty$). The description holds when the X_i are not i.i.d., in which case one writes λ_n as the likelihood ratio of the joint distributions of $(X_1, \ldots, X_n)$ under θ_1 and θ_0. The notion of the SPRT is due to Wald [56] and its motivation seems to have come from the success of the Neyman–Pearson theory* in fixed-sample contexts.

A detailed account of the theory and applications of SPRT can be found in refs. 26 and 57. We describe here some of the salient features. The constants A and B satisfy the inequalities

$$A \geqslant \beta(\theta_1)/\{1 - \alpha(\theta_0)\},$$

$$B \leqslant \{1 - \beta(\theta_1)\}/\alpha(\theta_0),$$

and they become approximate equalities if A and B are only slightly overshot by λ_n when sampling stops. Consequently, if the *criteria* for testing $H_0 : \theta = \theta_0$ against $H_1 : \theta = \theta_1$ are that the error probabilities of a test be no more than specified α and β, then an SPRT with $A = \beta/(1 - \alpha)$ and $B = (1 - \beta)/\alpha$

will achieve them closely. Moreover, if the X_i are i.i.d., the SPRT terminates surely, the moments of N are finite, and one can find simple approximations (known as *Wald approximations*) for its OC and ASN functions when $\alpha + \beta < 1$. The most compelling reason for using an SPRT is its *optimal property*, which states If the X_i are i.i.d., then the SPRT minimizes the ASN at $\theta = \theta_0$ and $\theta = \theta$ among all tests whose error probabilities are at most equal to those ot the SPRT. A rigorous proof of this can be found in ref. 12.

In many practical situations θ is often a real-valued parameter and one may be interested in testing the *composite* hypothesis $H_0 : \theta \leqslant \theta_0$ against the composite alternative $H_1 : \theta \geqslant \theta_1$ under the criteria that $\alpha(\theta) \leqslant \alpha$ for all $\theta \leqslant \theta_0$ and $\beta(\theta) \leqslant \beta$ for all $\theta \geqslant \theta_1$. If the joint distribution of $(X_1, \ldots, X_n)$ possesses a monotone likelihood ratio* for every n, then it can be shown [26, p. 100] that the SPRT above provides an appropriate test. However, one cannot assert any optimal property in this case when $\theta \neq \theta_0, \theta_1$. Nevertheless, it often turns out that, for all θ, the ASN of the SPRT is less than the sample size one would need to carry out the corresponding fixed-sample test.

We now illustrate the SPRT for testing H_0: $\mu \leqslant \mu_0$ against $H_1 : \mu \geqslant \mu_1$ when the X_i are i.i.d. normal variables with unknown mean μ $(\equiv \theta)$ and known variance σ^2 (see Example 2). One observes $X_1, X_2, \ldots$ successively. The continuation region $A < \lambda_n < B$ can be equivalently expressed by saying, continue sampling as long as

$$\frac{\sigma^2}{\mu_1 - \mu_0}\log\frac{\beta}{1-\alpha} < n\left(\overline{X}_n - \frac{\mu_0 + \mu_1}{2}\right) < \frac{\sigma^2}{\mu_1 - \mu_0}\log\frac{1-\beta}{\alpha},$$

where $\overline{X}_n$ is the mean of $X_1, \ldots, X_n$. One accepts H_0 if the lower inequality is violated for any n and rejects H_0 if the upper inequality is violated for any n, and sampling stops in either case. One can perform the SPRT graphically by taking $n = 1, 2, \ldots$ along with abscissa, plotting observed values of $\overline{X}_1, \overline{X}_2, \ldots$ along the ordinate, and noting that the continuation region lies between two parallel lines. The Wald approximations for the OC and ASN functions of the SPRT are (see ref. 26, p. 126 and FUNDAMENTAL IDENTITY OF SEQUENTIAL ANALYSIS for details)

$$\begin{aligned} Q_\mu &= (B^h - 1)/(B^h - A^h) \quad \text{if } \mu \neq (\mu_0 + \mu_1)/2 \\ &= (\log B)/(\log B - \log A) \quad \text{if } \mu = (\mu_0 + \mu_1)/2, \\ E_\mu[N] &= \{Q_\mu \log A + (1 - Q_\mu)\log B\}/k \quad \text{if } \mu \neq (\mu_0 + \mu_1)/2 \\ &= (-\log A)(\log B)\sigma^2/(\mu_1 - \mu_0)^2 \quad \text{if } \mu = (\mu_0 + \mu_1)/2, \end{aligned}$$

where

$$A = \frac{\beta}{1-\alpha}, \qquad B = \frac{1-\beta}{\alpha},$$

$$h = (\mu_0 + \mu_1 - 2\mu)/(\mu_1 - \mu_0),$$

$$k = -h(\mu_1 - \mu_0)^2/(2\sigma^2).$$

When α and β lie between 0.005 and 0.1, the sample size m of the fixed-sample test could lie anywhere between 1.67 $E_0[N]$ and 3.29 $E_0[N]$. Moreover, if $\alpha = \beta > 0.01$, then m exceeds $E_\mu[N]$ for all values of μ [26, p. 141]. Note that in this example it is possible (see ref. 26, p. 155) to get exact values of Q_μ and $E_\mu[N]$.

Certain drawbacks of the SPRT and the invariant SPRT below will be pointed out later, to motivate other sequential tests.

Invariant SPRT

There are many situations where θ is a vector-valued parameter with unknown components [e.g., X_i has a normal distribution with $\theta = (\mu, \sigma)$, or a k-variate normal distribution with $\theta = (\mu_1, \ldots, \mu_k)$ and known covariance matrix]. The hypotheses of interest in these situations are typically *composite* about θ (i.e., ω_0 and ω_1 contain more than one point) and the SPRT described above cannot handle them because λ_n may now depend on the unspecified θ. For some of these problems one can develop a new SPRT that uses a derived sequence of variables, V_n say, instead of the X_i directly. A special argument to construct V_n logically is known as the *principle of invariance* and the resulting SPRT based on $V_1, V_2, \ldots$ is called an *invariant SPRT*. Cox [16] first described such tests and they were later given rigorous justification by Hall et al. [29].

Suppose we introduce a group of transformations on the space of $(X_1, \ldots, X_n)$ for each n, and denote the transformed variables by $(Y_1, \ldots, Y_n)$ (e.g., $Y_i = cX_i$ for constants $c > 0$, or $Y_i = X_i + c$ for constants $-\infty < c < \infty$). If it so happens that the joint distribution of $(Y_1, \ldots, Y_n)$ under $\theta = \theta'$ is the same as that of $(X_1, \ldots, X_n)$ under $\theta = \theta''$ for every $\theta', \theta'' \in \theta$ and the statements about H_0 and H_1 do not change, then we say the problem remains invariant under the transformations (see, for instance, the example at the end of this section). The principle of invariance then stipulates that good statistical tests (i.e., N and D) should themselves be invariant under (i.e., not affected by) the same transformations (*see* INVARIANCE CONCEPTS IN STATISTICS). This leads to the conclusion that all tests should be based on a special class of (possibly vector) functions U_n, called *maximal invariants*, for $n = 1, 2, \ldots$. The function $U_n = U_n(X_1, \ldots, X_n)$ satisfies the conditions that $U_n(X_1, \ldots, X_n) = U_n(Y_1, \ldots, Y_n)$ and, if $U_n(x_1', \ldots, x_n') = U_n(x_1'', \ldots, x_n'')$ for two sets of values of the X_i, then the x_i'' must be transformed versions of the x_i'. The distribution of u_n depends on θ only through a (possibly vector) function $\psi(\theta)$, whose form comes out automatically from these arguments, such that H_0 and H_1 can be restated purely in terms of $\psi(\theta)$. By Bahadur's result mentioned earlier, we can then assert that, if V_n is sufficient for the joint distribution (as ψ varies) of $(U_1, \ldots, U_n)$, then all invariant tests of H_0 against H_1 should be based on $V_1, V_2, V_3, \ldots$. We call V_n for $n \geqslant 1$ an *invariantly sufficient* sequence, and Stein (see ref. 29) gave a method of finding V_n from the sufficient statistics. An invariant SPRT of $H_0 : \psi(\theta) = \psi_0$ against $H_1 : \psi(\theta) = \psi_1$ is an SPRT described above, but uses $V_1, \ldots, V_n$ at stage n instead of $X_1, \ldots, X_n$. V_n itself is sufficient for the joint distributions of $(V_1, \ldots, V_n)$ and, consequently, the λ_n of the invariant SPRT is just the likelihood ratio of the distributions of V_n under ψ_1 and ψ_0. Here, again, if $\psi(\theta)$ turns out to be real-valued and the distribution of V_n possesses a monotone likelihood ratio, the same invariant SPRT applies to test $H_0 : \psi(\theta) \leqslant \psi_0$ against $H_1 : \psi(\theta) \geqslant \psi_1$ with maximum error probabilities α and β.

Numerous applications of the invariant SPRT can be found in refs. 26 and 29. Since, unlike the X_i, the V_n are usually nonindependent, the sure termination, the optimal property, and the Wald approximations of the SPRT do not carry over to an invariant SPRT. Wijsman and others (see ref. 66 for references) proved under various conditions that the invariant SPRT terminates surely. Lai [39] showed that if $\alpha(\theta) + \beta(\theta)$ approaches zero in a certain way, then the ASN of all invariant tests satisfying $\alpha(\theta) \leqslant \alpha$ and $\beta(\theta) \leqslant \beta$ is indeed minimized by the invariant SPRT at $\psi(\theta) = \psi_0$ and $\psi(\theta) = \psi_1$. Berk [10] and Lai and Siegmund [40] gave approximations for the ASN of the invariant SPRT.

As an example [26, p. 300], suppose we want to test $H_0 : \mu \leqslant \psi_0\sigma$ against $H_1 : \mu \geqslant \psi_1\sigma$ in a normal distribution with $\alpha(\mu, \sigma) \leqslant \alpha$ and $\beta(\mu, \sigma) \leqslant \beta$. The transformations $Y_i = cX_i$ for $c > 0$ and $i \geqslant 1$ leave the problem invariant. A maximal invariant at stage n is the vector $U_n = (X_1/|X_n|, \ldots, X_n/|X_n|)$, whose distribution depends only on $\psi(\mu, \sigma) = \mu/\sigma$. An invariantly sufficient statistic is

$$V_n = \bar{X}_n\sqrt{n} \Big/ \sqrt{\left(\bar{X}_n^2 + S_n^2\right)},$$

where S_n^2 is the sample variance (with divisor n) of $X_1, \ldots, X_n$. The invariant SPRT then takes the form: Accept H_0 if $V_n \leqslant \underline{v}_n$, reject H_0 if $V_n \geqslant \bar{v}_n$, and continue if $\underline{v}_n < V_n < \bar{v}_n$, for $n = 2, 3, \ldots$, where $\underline{v}_n$ and $\bar{v}_n$ are solved from $g(\underline{v}_n) = \log\{\beta/(1 - \alpha)\}$ and

$g(\bar{v}_n) = \log\{1 - \beta)/\alpha)$, with

$$g(v) = -\tfrac{1}{2}(\psi_1^2 - \psi_0^2)(n - v^2) + \log\{I(v, \psi_1)/I(v, \psi_0)\},$$

$$I(v, \psi) = \int_0^{\infty} s^{n-1}\exp\{-\tfrac{1}{2}(s - \psi v)^2\}\, ds.$$

One can prepare tables of $\underline{v}_n$ and $\bar{v}_n$ by numerical methods before the test is carried out. This is the so-called *sequential t-test*.

Other Sequential Tests

The SPRT and the invariant SPRT have three major drawbacks. First, although they terminate surely, the actual sample size in a specified experiment may be unacceptably large. In many practical situations one deals with a limited amount of data or time. Second, the optimal property of the two tests does not shed any light on the behavior of the ASN when $\theta \neq \theta_0, \theta_1$ [or $\psi(\theta) \neq \psi_0, \psi_1$]. There may indeed be other sequential procedures under which the ASN is lower in some desirable range between θ_0 and θ_1 (or ψ_0 and ψ_1). Third, the tests do not help at all in problems where h_1 is simply the negation of h_0, for then there is no θ_1 or ψ_1 to work with for constructing λ_n. Several alternative tests have been suggested to rectify these drawbacks.

Weiss introduced (see ref. 37) the notion of a *generalized sequential probability ratio test** (GSPRT), which is structurally the same as the SPRT except that the bounds A and B are now replaced by a sequence of constants A_n and B_n, $n \geqslant 1$. Since the bounds of a GSPRT vary, one has greater flexibility in designing the test to satisfy different criteria that may be desirable under varying circumstances. For testing $H_0 : \theta = \theta_0$ against $H_1 : \theta = \theta_1$ when the X_i are i.i.d. and θ is real-valued, Kiefer and Weiss [37] and Weiss [62] showed that, among all tests satisfying $\alpha(\theta_0) \leqslant \alpha$ and $\beta(\theta_1) \leqslant \beta$, a GSPRT minimizes the maximum (over θ) ASN. Eisenberg [21] and Huffman [32] have provided approximations for the A_n and B_n of such GSPRT. Similar tests with an upper bound on N have been proposed by Anderson [1], Armitage [4], Lorden [42], and others (see ref. 26, p. 228) and their efficiency was investigated by Berk and Brown [11]. Some general properties of the GSPRT are discussed by Kiefer and Weiss [37] and Eisenberg et al. [23].

Another class of sequential tests, known as *repeated significance tests**, has been developed by Siegmund [51, p. 71], Sen [47, p. 243; 49], and Woodroofe [68, p. 71] for testing $H_0 : \theta \in \omega_0$ at level α. They are based on the premise that if a good fixed-sample test with m observations were carried out, then the decision to accept or reject h_0 may become obvious with k ($< m$) observations and therefore the procedure should have been terminated at that point. A repeated significance test thus modifies a fixed-sample test to maintain about the same power for a smaller sample size, especially when the true value of θ is far away from ω_0. As an example, suppose the i.i.d. X_i have a one-parameter exponential family of distributions $f_\theta(x) = h(x)\exp\{\theta t(x) - \gamma(\theta)\}$ and one wants to test $H_0 : \theta = \theta_0$ at level α against $H_1 : \theta \neq \theta_0$. A fixed-sample likelihood ratio test observes $X_1, \ldots, X_m$, computes the mean $\bar{t}_m$, and rejects H_0 if $M\phi(\bar{t}_m) > c$, where $\phi(t)$ is the maximum (over θ) of $\{(\theta - \theta_0)t - \gamma(\theta) + \gamma(\theta_0)\}$ and c is a constant satisfying $\alpha(\theta_0) = \alpha$. The repeated significance test of H_0 starts with $r \geqslant 1$ observations, assumes that at most R values can be observed, and uses the stopping variable

$$N = \text{first } n \geqslant r \text{ such that } n\phi(\bar{t}_n) > c$$
$$= R \quad \text{if } n\phi(\bar{t}_n) \leqslant c \text{ for all } n \leqslant R.$$

The decision is to reject H_0 if $n\phi(\bar{t}_n) > c$ for some $n < R$ or if $R\phi(\bar{t}_R) > b$ and to accept H_0 otherwise, where c and b are two constants. Woodroofe [68, p. 74] gives approximations for the power function and the ASN of the test, using which one can choose c and b such that $\alpha(\theta_0) = \alpha$ and some other desirable criterion holds.

Two-stage test procedures, introduced by Dodge and Romig [19] and Stein [53] in different contexts, make a compromise between the ease of a fixed-sample tests and the high economy of truly sequential tests (e.g., SPRT). References for these tests can be found in a survey by Hewett and Spurrier [30].

Tests with power one, introduced by Robbins [45], handle hypotheses of the type $H_0 : \theta \in \omega_0$ at level α and, by nature, can only reject H_0 when sampling stops. They have the property that $\alpha(\theta) \leqslant \alpha$ for all $\theta \in \omega_0$ and $P_\theta = 1$ for all $\theta \in \theta - \omega_0$. Their practicality is rather limited because sampling can continue indefinitely when H_0 is true. A detailed account of these tests is given by Lai [38].

Many of the testing problems described so far have counterparts in a Bayesian context, nonparametric framework or continuous time stochastic processes*. The Bayesian context introduces a prior probablity distribution on the parameter space; the sequential treatment is described in refs. 9, 13, 15, 58, 63, and 64, where further references are given. In a nonparametric framework, no specific form of the distribution $f_\theta(x)$ is assumed and θ may refer to a location or scale parameter of possible distributions. Numerous hypothesis testing problems in such situations can be solved by sequential procedures and an extensive treatment of such procedures can be found in refs. 47–49. In a continuous time stochastic process (e.g., a Poisson process* with intensity θ), one observes a time-dependent process $X(t)$ continuously in time t and poses a hypothesis testing problem for parameters involved in the distribution of $X(t)$. Sequential procedures for such problems were first discussed by Dvoretzky et al. [20] and their theory and applications are described in refs. 26, 33, 34, and 50.

Multihypotheses Problems

Sobel and Wald [52], Lorden [41], and others (see ref. 26, p. 255 for references) have described sequential methods to discriminate among three or more hypotheses. Most of these methods essentially combine the SPRT for pairs of hypotheses. They are intuitively appealing and workable solutions, but have no known optimum property. We describe the method of Sobel and Wald.

Suppose the X_i are i.i.d. according to the exponential family $f_\theta(x) = h(x)\exp\{\theta t(x) - \gamma(\theta)\}$. The problem is to discriminate among the three hypotheses $H_0 : \theta = \theta_0$, $H_1 : \theta = \theta_1$, and $H_2 : \theta = \theta_2$, such that $\alpha_i(\theta_i) \leqslant \alpha_i$ for $i = 0, 1, 2$, where $\alpha_i(\theta) = \Pr_\theta[\text{a test rejects } H_i]$. Suppose we observe $X_1, X_2, \ldots,$ and carry out an SPRT of H_0 against H_1 and a second SPRT of H_1 against H_2 simultaneously. Then the Sobel–Wald test accepts H_0 if the first SPRT accepts H_0, accepts H_1 if both SPRTs accept H_1, accepts H_2 if the second SPRT accepts H_2, and continues sampling otherwise. The Wald approximations for the OC of the two SPRTs can be used to find the $\alpha_i(\theta)$ approximately. This, in turn, enables us to choose (A_1, B_1) and (A_2, B_2) for the two SPRTs such that $\alpha_i(\theta_i) \leqslant \alpha_i$ for $i = 0, 1, 2$. Moreover, since the stopping variable of the Sobel–Wald test is the same as the larger stopping variable of the two SPRTs, we can also get a Wald approximation for the ASN of the Sobel–Wald procedure (see ref. 26, p. 259 for further details).

LITERATURE

Most standard texts in statistics include a chapter or two on sequential analysis, especially in the context of hypothesis testing. Further details on the topics covered in this survey as well as new topics can be found in books by Armitage [4], Chernoff [13], Ghosh [26], Sen [47, 49], Siegmund [51], Wald [57], Wetherill [63], and Woodroofe [68], which are devoted exclusively to sequential analysis. Chow et al. [14] and Shiryayev [50] discuss sequential procedures primarily as applications of probability theory. Wald [58], Ferguson [25], and Berger [9] have given elegant expositions of decision-theoretic problems in both fixed-sample and sequential frameworks. Bechhofer et al. [8] and Gupta and Panchapakesan [27] cover sequential techniques, among others, for a wide variety of screening and ranking problems. Wasan [61] and Nevelson and Hasminskii [43] discuss various methods of stochastic approximation*. Finally, Whittle [65] gives an account of numerous problems that are intrinsically sequential but not of the hypothesis testing or estimation variety. Survey articles and bibliographies in sequential analysis are given in Armitage [3], Jackson [35], Johnson [36], and Schmitz [46]. The survey by Hewett and Spurrier [30] is restricted to two-stage procedures, while Federer's [24] survey on selection and allocation problems includes sequential methods. Darling [18] and Wallis [60] give an historical account of sequential analysis. Most journals in statistics and probability include research articles in or related to sequential analysis. One of them, *Sequential Analysis*, is devoted exclusively to theoretical and applied aspects of the area.

References

[1] Anderson, T. W. (1960). *Ann. Math. Statist.*, **31**, 165–197.

[2] Anscombe, F. J. (1953). *J. R. Statist. Soc. B*, **15**, 1–29.

[3] Armitage, P. (1975). *Sequential Medical Trials*. Wiley, New York.

[4] Armitage, P. (1978). *International Encyclopedia of Statistics*. Macmillan, New York, pp. 937–942.

[5] Bahadur, R. R. (1954). *Ann. Math. Statist.*, **25**, 423–462.

[6] Barnard, G. A. (1944). *Statist. Meth. & Qual. Cont. Rep. No. QC / R / 7*, British Ministry of Supply, London, England.

[7] Bartky, W. (1943). *Ann. Math. Statist.*, **14**, 363–377.

[8] Bechhofer, R. E., Kiefer, J., and Sobel, M. (1968). *Sequential Identification and Ranking Procedures*. Univ. Chicago Press, Chicago, IL.

[9] Berger, J. O. (1980). *Statistical Decision Theory*. Wiley, New York.

[10] Berk, R. H. (1973). *Ann. Statist.*, **1**, 1126–1138.

[11] Berk, R. H. and Brown, L. D. (1978). *Ann. Statist.*, **6**, 567–581.

[12] Burkholder, D. L. and Wijsman, R. A. (1963). *Ann. Math. Statist.*, **34**, 1–17.

[13] Chernoff, H. (1972). *Sequential Analysis and Optimal Design*. SIAM, Philadelphia, PA.

[14] Chow, Y. S., Robbins, H., and Siegmund, D. (1971). *Great Expectations*. Houghton Mifflin, Boston, MA.

[15] Cohen, A. and Samuel-Cahn, E. (1982). *Sequential Anal.*, **1**, 89–100.

[16] Cox, D. R. (1952). *Proc. Camb. Philos. Soc.*, **48**, 290–299.

[17] Dantzig, G. B. (1940). *Ann. Math. Statist.*, **11**, 186–192.

[18] Darling, D. A. (1976). *History of Statistics and Probability*. Dekker, New York, pp. 369–375.

[19] Dodge, H. F. and Romig, H. G. (1929). *Bell Syst. Tech. J.*, **8**, 613–631.

[20] Dvoretzky, A., Kiefer, J., and Wolfowitz, J. (1953). *Ann. Math. Statist.*, **24**, 254–264.

[21] Eisenberg, B. (1982). *Sequential Anal.*, **1**, 81–88.

[22] Eisenberg, G. and Ghosh, B. K. (1980). *Ann. Statist.*, **8**, 1123–1123–1131.

[23] Eisenberg, B., Ghosh, B. K., and Simons, G. (1976). *Ann. Statist.*, **4**, 237–251.

[24] Federer, W. (1963). *Biometrics*, **19**, 53–587.

[25] Ferguson, T. S. (1967). *Mathematical Statistics*. Academic, New York.

[26] Ghosh, B. K. (1970). *Sequential Tests of Statistical Hypotheses*. Addison-Wesley, Reading MA.

[27] Gupta, S. S. and Panchapakesan, S. (1979). *Multiple Decision Procedures*. Wiley, New York.

[28] Haldane, J. B. S. (1945). *Biometrika*, **33**, 222–225.

[29] Hall, W. J., Wijsman, R. A., and Ghosh, J. K. (1965). *Ann. Math. Statist.*, **36**, 575–614.

[30] Hewett, J. E. and Spurrier, J. D. (1983). *Commun. Statist. A*, **12**, 2307–2425.

[31] Hotelling, H. (1941). *Ann. Math. Statist.*, **12**, 20–45.

[32] Huffman, M. D. (1983). *Ann. Statist.*, **11**, 306–316.

[33] Irle, A. and Schmitz, N. (1982). *Modern Applied Mathematics*. North-Holland, New York, pp. 623–653.

[34] Irle, A. and Schmitz, N. (1984). *Math. Operat. Statist.*, **15**, 91–104.

[35] Jackson, J. E. (1960). *J. Amer. Statist. Ass.*, **55**, 561–580.

[36] Johnson, N. L. (1961). *J. R. Statist. Soc. A*, **124**, 372–411.

[37] Kiefer, J. and Weiss, L. (1957). *Ann. Math. Statist.*, **28**, 57–74.

[38] Lai, T. L. (1977). *Ann. Statist.*, **5**, 866–880.

[39] Lai, T. L. (1981). *Ann. Statist.*, **9**, 318–333.

[40] Lai, T. L. and Siegmund, D. (1979). *Ann. Statist.*, **7**, 60–76.

[41] Lorden, G. (1972). *Ann. Math. Statist.*, **43**, 1412–1427.

[42] Lorden, G. (1976). *Ann. Statist.*, **4**, 281–291.

[43] Nevelson, M. B. and Hasminskii, R. Z. (1973). *Stochastic Approximation and Recursive Estimation*. Amer. Math. Soc., Providence, RI.

[44] Neyman, J. (1934). *J. R. Statist. Soc.*, **97**, 558–625.

[45] Robbins, H. (1970). *Ann. Math. Statist.*, **41**, 1397–1409.

[46] Schmitz, N. (1984). *Medizinische Informatik und Statistik*. Springer-Verlag, New York, pp. 94–114.

[47] Sen, P. K. (1981). *Sequential Nonparametrics*. Wiley, New York.

[48] Sen, P. K. (1984). *Sequential Anal.*, **3**, 191–211.

[49] Sen, P. K. (1985). *Theory and Applications of Sequential Nonparametrics*. SIAM, Philadelphia, PA.

[50] Shiryayev, A. N. (1978). *Optimal Stopping Rules*. Springer-Verlag, New York.

[51] Siegmund, D. (1985). *Sequential Analysis*. Springer-Verlag, New York.

[52] Sobel, M. and Wald, A. (1949). *Ann. Math. Statist.*, **20**, 502–522.

[53] Stein, C. (1945). *Ann. Math. Statist.*, **16**, 243–258.

[54] Stein, C. and Wald, A. (1947). *Ann. Math. Statist.*, **18**, 427–433.

[55] Thompson, W. R. (1933). *Biometrika*, **25**, 285–294.

[56] Wald, A. (1943). *Statistics Research Group Report No. 75*, Columbia Univ., New York.

[57] Wald, A. (1947). *Sequential Analysis*. Wiley, New York.

[58] Wald, A. (1950). *Statistical Decision Functions*. Wiley, New York.

[59] Wald, A. and Wolfowitz, J. (1948). *Ann. Math. Statist.*, **19**, 326–339.

[60] Wallis, A. W. (1980). *J. Amer. Statist. Ass.*, **75**, 320–334.

[61] Wasan, M. T. (1969). *Stochastic Approximation*. Cambridge Univ. Press, New York.

[62] Weiss, L. (1962). *J. Amer. Statist. Ass.*, **57**, 551–566.

[63] Wetherill, G. B. (1975). *Sequential Methods in Statistics*. Wiley, New York.

[64] Wetherill, G. B. and Köllerström, J. (1983). *Sequential Anal.*, **2**, 225–242.

[65] Whittle, P. (1982, 1983). *Optimization Over Time*, Vols. I and II. Wiley, New York.

[66] Wijsman, R. A. (1979). *Developments in Statistics*, Vol. 2. Academic, New York, pp. 235–314.

[67] Wolfowitz, J. (1947). *Ann. Math. Statist.*, **18**, 215–230.

[68] Woodroofe, M. (1982). *Nonlinear Renewal Thoery in Sequential Analysis*. SIAM, Philadelphia, PA.

(AVERAGE SAMPLE NUMBER (ASN)
FUNDAMENTAL IDENTITY OF SEQUENTIAL ANALYSIS
DECISION THEORY
ESTIMATION, POINT
FIXED-WIDTH AND BOUNDED-LENGTH CONFIDENCE INTERVALS
HYPOTHESIS TESTING
OPTIMAL STOPPING RULES
RANDOM SUM DISTRIBUTIONS
STOCHASTIC APPROXIMATION
WALD'S EQUATION
WALD'S IDENTITY
SEQUENTIAL (various entries))

B. K. GHOSH

SEQUENTIAL ANALYSIS

HISTORICAL BACKGROUND AND ORGANIZATION

Sequential Analysis was founded (as *Series C* of *Communications in Statistics**) in 1981 by Professors Bhaskar K. Ghosh of Lehigh University and Pranab K. Sen of the University of North Carolina, Chapel Hill. The first issue was published in the Spring of 1982, having been preceded by a Special Issue on Sequential Analysis that appeared in *Communications in Statistics, Theory and Methods, Series A* (Volume 10, No. 21) in the Summer of 1981. During the first two years of operation (that is, 1982 and 1983), this journal was listed as *Series C* of *Communications in Statistics*. However, since the beginning of 1984, it has operated with the title *Sequential Analysis*. It is a quarterly journal, published by Marcel Dekker, Inc., New York

AIMS AND SCOPE

This journal aims to publish papers on the theoretical, methodological, and practical aspects of sequential analysis*. The objective of each article should be to contribute to the understanding of sequential procedures in some area of probability and statistics including hypothesis testing, theory of (point as well as interval) estimation, decision theory*, ranking and selection* of populations, experimental designs, analysis of variance* and covariance, regression* and correlation* analysis, Bayesian* analysis, reliability* and quality control*, sample surveys, and time-series* analysis. The level of mathematical sophistication is flexible, as long as the originality of the work and its relation to some aspects of sequential analysis are apparent. General mathematical theory directed towards this basic goal is also welcome. All papers are refereed. Long papers having substantial contributions need not be condensed excessively for brevity and economy of space.

PUBLISHED RECORDS

This journal has already completed five years of operation; Volumes 1 through 5 are all published. The published articles closely adhere to the general objectives as described above. Contents of the first issue, Volume 1, No. 1 (1982), are summarized as follows:

> Editorial
>
> "Stochastic approximation on a discrete set and the multi-armed bandit problem" by V. Dupač and U. Herkenrath
>
> "*M*-estimators and *L*-estimators of location: uniform itegrability and asymptotic risk-efficient sequential versions" by J. Jurečková and P. K. Sen
>
> "A sequential likelihood ratio test for general hypotheses" by S. I. Bangdiwala
>
> "The asymptotic solution of the Kiefer–Weiss problem" by B. Eisenberg

Generally, reviews of submitted manuscripts are completed within six months of the date of submission, and, following acceptance, the articles appear within four months. There is more room for application-oriented manuscripts.

P. K. SEN

SEQUENTIAL CHI-SQUARED TEST

This test is used to test hypotheses on the expected value vector $\boldsymbol{\xi}$ of a p-variate multinormal* population with *known* variance–covariance matrix $\boldsymbol{\Sigma}$, based on successive values $\mathbf{X}_1, \mathbf{X}_2, \ldots, \mathbf{X}_n, \ldots$ in a sequence chosen randomly from the population.

For testing the hypothesis $H_0 : \boldsymbol{\xi} = \boldsymbol{\xi}_0$ against the alternative $H_1 : \boldsymbol{\xi} = \boldsymbol{\xi}_1$ with

$$\lambda^2 = (\boldsymbol{\xi}_1 - \boldsymbol{\xi}_0)'\Sigma^{-1}(\boldsymbol{\xi}_1 - \boldsymbol{\xi}_0),$$

the sequential probability ratio test* likelihood ratio statistic is

$$L(\lambda|\mathbf{X}) = \exp\left(-\tfrac{1}{2}n\lambda^2\right) {}_0F_1\left(\tfrac{1}{2}p; \tfrac{1}{4}n\lambda^2 Z_n^2\right),$$

where $Z_n^2 = n(\bar{\mathbf{X}}_n - \boldsymbol{\xi}_0)'\boldsymbol{\Sigma}^{-1}(\bar{\mathbf{X}} - \boldsymbol{\xi}_0)$ with $\bar{\mathbf{X}}_n$ the arithmetic mean of the first $n_{x\text{'s}}$ and ${}_0F_1(a; y) = \sum_{j=0}^{\infty}\{y^j/(j!a^{[j]})\}$ with $a^{[j]} = a(a+1)\cdots(a+j-1)$. The continuation region is

$$\frac{\alpha_1}{1-\alpha_0} < L(\lambda|\mathbf{X}) < \frac{1-\alpha_1}{\alpha_0},$$

where α_j is the nominal probability of rejecting H_j when it is valid ($j = 1, 2$). If

$$L(\lambda|\mathbf{X}) \geqslant (1-\alpha_1)/\alpha_0, \quad H_0 \text{ is rejected.}$$

If

$$L(\lambda|\mathbf{X}) \leqslant \alpha_1/(1-\alpha_0), \quad H_1 \text{ is rejected.}$$

The test is equivalent to

$$\text{rejecting } H_0 \text{ if } Z_n^2 \geqslant \bar{\xi}_n,$$

$$\text{rejecting } H_1 \text{ if } Z_n^2 \leqslant \underline{\xi}_n,$$

and continuing to sample otherwise.

Tables of $\bar{\xi}_n$ and $\underline{\xi}_n$ were constructed by Freund and Jackson [1] and have been reproduced by Kres [2]. The fact that the values of $\bar{\xi}_n$ and $\underline{\xi}_n$ depend only on $n\lambda^2$ is helpful in constructing and extending such tables. If $\boldsymbol{\Sigma}$ is not known, the sequential T^2 test* should be used.

References

[1] Freund, R. J. and Jackson, J. E. (1960). Tables to Facilitate Multivariate Sequential Testing from Means, *Tech. Rep. 12*, Dept. of Statistics, Virginia Polytechnical Institute, VA.

[2] Kres, H. (1983). *Statistical Tables for Multivariate Analysis*. Springer, New York.

(MULTIVARIATE ANALYSIS
SEQUENTIAL ANALYSIS
SEQUENTIAL T^2-TEST)

SEQUENTIAL ESTIMATION

Sequential estimation refers to any estimation technique for which the total number of observations used is not a degenerate random variable, i.e., a random variable whose cumulative distribution function assigns probability 1 to a single point. In some problems, sequential estimation must be used because no procedure using a preassigned nonrandom sample size (also denoted by "fixed sample size") can achieve the desired objective. In other problems, there may exist a procedure using a preassigned nonrandom sample size, but a sequential estimation procedure may be better in some way. There is a large body of literature on this subject, and it is growing rapidly. In this article, we will survey some representative examples of sequential estimation procedures.

A CONFIDENCE INTERVAL OF BOUNDED LENGTH FOR A NORMAL MEAN WHEN THE VARIANCE IS UNKNOWN

Suppose $X_1, X_2, \ldots$ are independent and identically distributed (i.i.d.) scalar random variables, each with a normal distribution with unknown mean μ and unknown variance σ^2. We are given a positive value L and a value β in the open interval $(0, 1)$, and the problem is to construct a confidence interval for μ of confidence coefficient at least β and length no greater than L. Dantzig [13] showed that this cannot be done with a preassigned nonrandom sample size. Stein [17] showed that the following sequential estimation procedure achieves the goal. Fix a positive integer n_1 with $n_1 > 1$. Let T_{n_1-1} denote a random variable with a t-distribution with $n_1 - 1$ degrees of freedom. Define the positive value z by the equation $\beta = P[-L/(2\sqrt{z}) < T_{n_1-1} < L/(2\sqrt{z})]$. Observe $X_1, \ldots, X_{n_1}$, and compute

$$S^2 \equiv \frac{1}{n_1-1}\left[\sum_{i=1}^{n_1} X_i^2 - \frac{1}{n_1}\left(\sum_{i=1}^{n_1} X_i\right)^2\right].$$

Define N_2 as $\max\{n_1, [S^2/z] + 1\} - n_1$, where $[S^2/z]$ denotes the largest integer not greater than S^2/z. If $N_2 > 0$, observe $X_{n_1+1}, \ldots, X_{n_1+N_2}$. The confidence interval for μ is then $\bar{X} \pm L/2$, where $\bar{X}$ denotes the arithmetic mean of all $n_1 + N_2$ observations. *See also* FIXED-WIDTH AND BOUNDED-LENGTH CONFIDENCE INTERVALS.

If $\sigma^2 > n_1 z$, then $E\{n_1 + N_2\}$ is approximately σ^2/z, which, as Stein points out, is close to the fixed sample size that would be required to construct the confidence interval if σ^2 were known.

MULTISTAGE PROCEDURES

Stein's sequential procedure is a *two-stage procedure*. For any preassigned positive finite integer k, a k-stage procedure is defined as follows. A positive integral-valued random variable N_1 is observed and N_1 observations are taken: this represents the first stage of sampling. Then N_2 additional observations are taken, where N_2 is a prespecified function of N_1 and the values of the first N_1 observations, N_2 taking nonnegative integral values: this is the second stage of sampling. Continuing this way, at the ith stage of sampling we take N_i additional observations, where N_i is a prespecified function of $N_1, N_2, \ldots, N_{i-1}$ and the values of the first $N_1 + N_2 + \cdots + N_{i-1}$ observations. Sampling ends at the kth stage, where we take N_k additional observations, N_k being a prespecified function of $N_1, N_2, \ldots, N_{k-1}$ and the values of the $N_1 + N_2 + \cdots + N_{k-1}$ observations already taken. The estimator is a prespecified function of $N_1, N_2, \ldots, N_k$ and the values of the $N_1 + N_2 + \cdots + N_k$ observations.

According to this definition, a one-stage procedure with N_1 a degenerate random variable gives the preassigned nonrandom sample size case: N_1 is the nonrandom total sample size.

Any sequential procedure that is not a k-stage procedure for a finite value of k will be called a "fully sequential procedure."

A CONFIDENCE INTERVAL OF BOUNDED LENGTH FOR THE MEAN OF A NORMAL DISTRIBUTION WITH KNOWN VARIANCE

Suppose that $X_1, X_2, \ldots$ are i.i.d. scalar random variables, each with a normal distribution with unknown mean μ and known variance σ^2. We are given a positive value L and a value β in the open interval $(0, 1)$, and the problem is to construct a confidence interval for μ of confidence coefficient at least β and length not greater than L. Stein and Wald [18] showed that the following procedure minimizes the quantity $\max_\mu E_\mu$ {total number of observations} among all procedures achieving the goal. For any nonrandom positive integer m, let $\bar{X}(m)$ denote $(X_1 + \cdots + X_m)/m$, and let $h(m)$ denote $P[-L/2 \leqslant \bar{X}(m) - \mu \leqslant L/2]$. Let m^* denote the smallest integer m such that $h(m^*) \geqslant \beta$. If $h(m^*) = \beta$, observe $X_1, \ldots, X_{m^*}$, and $\bar{X}(m^*) \pm L/2$ is the confidence interval. If $h(m^*) > \beta$, there is a value c in $(0, 1)$ such that $ch(m^*) + (1 - c)h(m^* - 1)$ is equal to β. In this case, the number of observations taken is a random variable with possible values $m^* - 1, m^*$, with respective probabilities $1 - c, c$. Let $\bar{X}^*$ denote the arithmetic mean of the observations taken. Then the confidence interval is $\bar{X}^* \pm L/2$.

Wolfowitz [29] showed that this type of procedure minimizes the maximum expected sample size (*see* AVERAGE SAMPLE NUMBER) when we replace interval estimation by point estimation with various loss functions.

Note that this procedure is either a preassigned nonrandom sample size case, or very close to it. Yet for testing that $\mu = \mu_0$ against $\mu = \mu_1$, a fully sequential test (the sequential probability ratio test) achieves considerable savings compared to the test of the same level of significance and power that uses a preassigned nonrandom sample size. Why this major difference in optimal sampling procedures? For testing the hypothesis, we have two values of μ fixed in advance: μ_0 and μ_1. Observations far above μ_0 enable us to stop early and decide that μ_1 is the value; observations far below μ_1 enable us to stop early and decide that μ_0 is the value. But for the estimation case, we have no such benchmarks fixed in advance, and no such opportunity for stopping early. An interesting contrast to this normal case is given by Wald [20]. If $X_1, X_2, \ldots$ are i.i.d., each uniform over the range $(\theta - \frac{1}{2}, \theta + \frac{1}{2})$, with θ unknown, a fully sequential scheme is optimal for point estimation with squared error loss, because a sample range close to 1 enables us to stop early.

An asymptotically efficient confidence interval for the mean of a general distribution was developed by Chow and Robbins [10]. For concepts and discussion *see* FIXED-WIDTH AND BOUNDED-LENGTH CONFIDENCE INTERVALS.

A CONFIDENCE INTERVAL OF BOUNDED LENGTH FOR A QUANTILE OF A UNIMODAL DISTRIBUTION

Suppose $X_1, X_2, \ldots$ are i.i.d. continuous scalar random variables, and all that is known about their common probability density function $f(x)$ is that it is unimodal: that is, there exists a value x_0 such that $f(x)$ is a nondecreasing function for $x \leqslant x_0$ and a nonincreasing function for $x \geqslant x_0$. Positive values L, β, q are specified, with $\beta < 1$ and $q < 1$. The problem is to construct a confidence interval for the qth quantile of the distribution with confidence coefficient at least β and length no greater than L. No one-stage procedure can achieve this, because such a procedure would cover the special case of estimating a normal mean with unknown variance, which, as stated above, is impossible for a one-stage procedure. The following two-stage procedure solves this problem (Weiss [23]). For any values α, γ both in the open interval $(0, 1)$, define $N(\alpha, \gamma)$ as the smallest positive integer n satisfying the inequality

$$\frac{n!}{([nq]!)(n-[nq]-1)!} \times \int_{\max(0,\, q-\gamma)}^{\min(1,\, q+\gamma)} y^{[nq]}(1-y)^{n-[nq]-1} dy \geqslant \alpha.$$

Choose values α, w, r in $(0, 1)$ with $\alpha w = \beta$ and $r > \max(q, 1-q)$. Observe $X_1, \ldots, X_{n_1}$, where n_1 is the smallest positive integer satisfying the inequality $n_1 r^{n_1 - 1} - (n_1 - 1)r^{n_1} \leqslant 1 - w$. Define S_1, S_2 as the smallest and largest, respectively, of the values $X_1, \ldots, X_{n_1}$. Set γ equal to

$$\min\left[r-q,\, r-(1-q),\, \frac{r-(1-q)}{S_2 - S_1}\left(\frac{L}{2}\right),\, \frac{r-q}{S_2-S_1}\left(\frac{L}{2}\right)\right].$$

Then observe $X_{n_1+1}, \ldots, X_{n_1+N_2}$, where N_2 is the smallest integer that is greater than $N(\alpha, \gamma)$ and such that $N_2 q$ is not an integer. Let Z denote that qth sample quantile of the observations $X_{n_1+1}, \ldots, X_{n_1+N_2}$. Then the confidence interval is $Z \pm L/2$.

Blum and Rosenblatt [6] generalized this result by allowing the unknown common distribution to belong to a class of distributions given as follows. The only thing known about the common cumulative distribution function (CDF) $F(x)$ is that there are values c, d, with $0 < c < 1 < d$, and a unimodal CDF $U(x)$ such that $cP_U(B) \leqslant P_F(B) \leqslant dP_U(B)$ for every Borel set B, where $P_U(B), P_F(B)$ denote the probabilities assigned to B by $U(x), F(x)$, respectively.

ASYMPTOTICALLY MINIMAX PROCEDURES FOR POINT ESTIMATORS

$X_1, X_2, \ldots$ are i.i.d., with common probability density function (PDF) $f(x; \theta)$, θ being an unknown parameter to be estimated. If we take a total of N observations, and then estimate θ by the value D, then our loss is $cN + (D-\theta)^2$, where c is a given positive value. For any estimation procedure T, let $r(\theta, T, c)$ denote the expected loss when T is used and θ is the parameter value. If $\{T_c\}$ is a family of estimation procedures, T_c being a procedure used for the value of c specified by the subscript, then the family is *asymptotically minimax* if

$$\lim_{c \to 0} \frac{\sup_\theta r(\theta, T_c, c)}{\inf_T \sup_\theta r(\theta, T, c)} = 1.$$

Let $d(\theta)$ denote $E\{([\partial \log f(X_1; \theta)]/\partial\theta)^2\}$, and define d_0 as $\inf_\theta d(\theta)$. Under certain regularity conditions, Wald [21] gave two asymptotically minimax families:

Family I: Define n_c as the smallest integer at least as large as $(cd_0)^{-1/2}$, observe $X_1, \ldots, X_{n_c}$, and estimate θ by $\hat{\theta}(n_c)$, the maximum likelihood estimator of θ based on $X_1, \ldots, X_{n_c}$.

Family II: Let $\hat{\theta}(n)$ denote the maximum likelihood estimator of θ based on

$X_1, \ldots, X_n$. Define N as the smallest positive integer n for which

$$\frac{1}{nd(\hat{\theta}(n))} - \frac{1}{(n+1)d(\hat{\theta}(n+1))} \leqslant c,$$

and estimate θ by $\hat{\theta}(N)$.

Although both of these families are asymptotically minimax, for small positive c the procedure from Family I is preferable. The word "asymptotic" is used because as $c \to 0$, the required sample size approaches infinity.

Weiss and Wolfowitz [24] pointed out that Wald's results can be extended to some cases where the regularity conditions of [21] do not hold. Anscombe [3, 4] gives results similar to those of Wald, and some numerical illustrations.

A LOWER BOUND FOR THE VARIANCE OF A SEQUENTIAL ESTIMATOR

$X_1, X_2, \ldots$ are i.i.d. If these random variables are continuous, $f(x; \theta)$ represents the common PDF, where θ is an unknown scalar parameter to be estimated. If the random variables are discrete, $f(x; \theta)$ represents $P(X_1 = x)$ when θ is the parameter. Let N denote the total number of observations that will be taken by a sequential estimation procedure. Let $T(X_1, \ldots, X_N)$ denote a point estimator of θ based on $X_1, \ldots, X_N$, and let $b(\theta)$ denote the bias of $T(X_1, \ldots, X_N)$:

$$b(\theta) = E\{T(X_1, \ldots, X_N)\} - \theta.$$

Under certain regularity conditions, Wolfowitz [28] showed that

$$\operatorname{var}(T(X_1, \ldots, X_N)) \geqslant \left(1 + \frac{db(\theta)}{d\theta}\right)^2 \times \left[E\{N\}E\left\{\left(\frac{\partial}{\partial\theta}\log f(X_1; \theta)\right)^2\right\}\right]^{-1}$$

This generalizes the familiar Cramér–Rao lower bound* for fixed sample size procedures. Wolfowitz also gives results for the case where θ is a vector of unknown parameters, generalizing the ellipsoid of concentration of Cramér [12].

ESTIMATORS BASED ON SEQUENTIAL TESTS

For fixed sample size procedures, the technique of constructing a confidence set using a test of a hypothesis, and the reverse construction of a test of a hypothesis based on a confidence set, are familiar. Wijsman [26] constructs confidence sets based on sequential tests. For given small positive values, α, β, he constructs a confidence set that contains the true parameter point with probability at least $1 - \alpha$ and one or more specified false parameter points with probability at most β, by using a family of sequential tests. In particular, he discusses the case where $X_1, X_2, \ldots$ are i.i.d. normal, with unknown mean μ and known variance. The problem is to construct an upper confidence interval for μ (that is, the upper limit is $+\infty$) of confidence coefficient $1 - \alpha$ and with probability at most β of containing $\mu - \delta(\mu)$, where $\delta(\mu) > 0$ for all μ and $\delta(\mu) \to 0$ as $\mu \to -\infty$. No fixed sample size procedure can achieve this. Wijsman constructs such intervals based on both sequential probability ratio tests and generalized sequential probability ratio tests*, and compares them by a Monte Carlo study. The confidence interval based on the generalized sequential probability ratio test is preferable, having a smaller expected number of observations.

Wolfowitz [27] describes the construction of an unbiased estimator of a Bernoulli parameter based on sequential tests of the parameter.

A BOUNDED CONFIDENCE INTERVAL FOR THE VALUE OF A DENSITY FUNCTION AT A POINT

$X_1, X_2, \ldots$ are i.i.d. with unknown PDF $f(x)$. Values β, L, and t are given, with $L > 0$ and β in the open interval $(0, 1)$. The problem is to construct a confidence interval for

$f(t)$ with confidence coefficient at least β and length no greater than L. [It is assumed that $f'(t) \neq 0$ and that $f''(x)$ is continuous at $x = t$.] This cannot be accomplished by a fixed sample size procedure. Stute [19] accomplishes this approximately by a fully sequential procedure for L close to zero.

ESTIMATING THE MEAN VECTOR OF A MULTIVARIATE NORMAL DISTRIBUTION

Suppose $\mathbf{X}_1, \mathbf{X}_2, \ldots$ are i.i.d. p-dimensional column vectors, each with a joint normal distribution with unknown mean vector $\boldsymbol{\mu}$ and covariance matrix $\sigma^2\boldsymbol{\Sigma}$, where σ^2 is an unknown positive scalar and $\boldsymbol{\Sigma}$ is a known positive definite matrix. The problem is to estimate the vector $\boldsymbol{\mu}$. If the vector $\hat{\boldsymbol{\mu}}$ is the estimate of $\boldsymbol{\mu}$, and N is the number of X's observed, the total loss is $A[(\hat{\boldsymbol{\mu}} - \boldsymbol{\mu})'\boldsymbol{\Sigma}^{-1}(\hat{\boldsymbol{\mu}} - \boldsymbol{\mu})]^{\alpha/2} + N$, where A and α are positive constants. Wang [22] suggested the following sequential estimation procedure. For each integer $n > 1$, define $\overline{\mathbf{X}}_n$ as $\Sigma_{i=1}^{n}\mathbf{X}_i/n$, and define

$$S_n^2 = \frac{1}{p(n-1)} \times \sum_{i=1}^{n} (\mathbf{X}_i - \overline{\mathbf{X}}_n)'\boldsymbol{\Sigma}^{-1}(\mathbf{X}_i - \overline{\mathbf{X}}_n).$$

Define N as the smallest integer $n > 1$ such that $n^{1+\alpha/2}$ is at least equal to

$$A\alpha S_n^{\alpha} 2^{(\alpha/2)-1}\Gamma\{\tfrac{1}{2}(p+\alpha)\}/\Gamma(\tfrac{1}{2}p),$$

and use $\overline{\mathbf{X}}_N$ as the estimate of the vector $\boldsymbol{\mu}$. Wang compares this procedure to the best fixed sample size procedure, and shows that as p increases, the sequential procedure gives a much smaller expected loss than the loss given by the fixed sample size procedure.

ESTIMATING THE MEAN OF A LOGNORMAL DISTRIBUTION

Suppose $X_1, X_2, \ldots$ are i.i.d. random variables, each with a lognormal distribution*: the common density is $(\sigma\sqrt{2\pi})^{-1} \exp\{-[\log x - \mu]^2/(2\sigma^2)\}$ for $x > 0$. μ, σ are unknown parameters, with σ positive. $E\{X_n\} = \exp\{\mu + \frac{1}{2}\sigma^2\} \equiv m$, say. The problem is to construct an estimator $\hat{m}$ of m such that $P[|\hat{m} - m| \leqslant \delta m] \geqslant \gamma$, where δ, γ are given positive values. Nagao [15] constructs a sequential estimator $\hat{m}$ that achieves this, and derives the distribution of the total number of observations used.

ADAPTIVE PROCEDURES

A sequential procedure is called "adaptive" if what is done at a given stage depends particularly heavily on the observations taken in previous stages. An example of such a procedure is given by Weiss and Wolfowitz [25]. $X_1, X_2, \ldots$ are i.i.d. each with PDF $g(x - \theta)$, where θ is an unknown parameter to be estimated by a fixed-length confidence interval with a given confidence coefficient. The function $g(\cdot)$ is known to be symmetric about zero and to satisfy certain mild regularity conditions, but is otherwise unknown. Weiss and Wolfowitz constructed the desired confidence interval by a sequential procedure that estimates the unknown function $g(\cdot)$ at each stage: this is the "adaptive" aspect of the procedure (*see* ADAPTIVE METHODS).

STOCHASTIC APPROXIMATION

This subject was introduced by Robbins and Monro [16], who discussed the following problem. For each given value of the scalar quantity x, we can observe the random variable $Y(x)$. $E\{Y(x)\} = M(x)$, where $M(x)$ is an incompletely known function of x. Given a value α, the problem is to estimate the value of x, say θ, for which $M(\theta) = \alpha$. Robbins and Monro gave a sequential procedure for doing this. Their procedure has been generalized to other problems, for example to the problem of estimating the value that maximizes a regression function. Albert and Gardner [1] give a description of some of the generalizations (*see* STOCHASTIC APPROXIMATION).

NONATTAINABILITY OF CERTAIN GOALS

Suppose that $X_1, X_2, \ldots$ are i.i.d. and all that is known about their common distribution is that the mean exists. Bahadur and Savage [5] showed that it is impossible to construct a confidence interval for the mean of no more than some prescribed finite length and at least some prescribed confidence coefficient. This is easy to see intuitively, because a slight change in the extreme tail of the distribution can change the mean greatly while having a very small effect on what we observe. Blum and Rosenblatt [8] showed that even if the class of distributions is made much smaller than the class of distributions possessing a mean, it is still impossible to construct the confidence interval. Blum and Rosenblatt [7] showed that for each integer $k \geqslant 2$, there is an estimation problem that can be solved by a k-stage procedure but not by a procedure using fewer than k stages.

TIME SERIES

In the problems described above, the observations were mutually independent. Blum and Rosenblatt [9] constructed fixed-length confidence intervals for parameters of a discrete m-dependent stationary Gaussian process* assuming that m is known. If m is unknown, it is impossible to construct such intervals. Zielinski [30] also discusses such confidence intervals.

BAYESIAN SEQUENTIAL PROCEDURES

The preceding problems are non-Bayesian, in that the unknown parameters were not considered random variables. However, some of the results above were derived by using an a priori distribution for the unknown parameters: see Wolfowitz [29] for an example.

In the Bayesian approach, the unknown parameter to be estimated is itself considered a random variable with a known (a priori) distribution. Given this a priori distribution, the theoretical construction of sequential Bayes procedures has been thoroughly described: see Wald [20] and Chow et al. [11]. However, the actual computation of such Bayes procedures is sometimes extremely difficult, and even impossible for practical purposes, and therefore heuristic approximations to Bayes procedures have been developed. An example of such a heuristic procedure is given in Alvo [2], where an ad hoc stopping rule is proposed and studied.

HISTORICAL NOTE

The year 1945 marks the starting point of sequential estimation, with the publication of two unrelated papers. One was the paper by Stein [17] described above. The other was a paper by Haldane [14], who suggested estimating the unknown parameter of a Bernoulli distribution (say the unknown probability of getting a head when tossing a coin) by tossing the coin until a preassigned number (m, say) of heads is observed. Let N denote the total number of tosses required. Haldane showed that $(m - 1)/(N - 1)$ is an unbiased estimator of the parameter, and gave an approximate formula for the variance of this estimator.

References

[1] Albert, A. E. and Gardner, L. A., Jr. (1967). *Stochastic Approximation and Nonlinear Regression*. M. I. T. Press, Cambridge, MA.

[2] Alvo, M. (1977). *Ann. Statist.*, **5**, 955–968.

[3] Anscombe, F. J. (1952). *Proc. Camb. Philos. Soc.*, **48**, 600–607.

[4] Anscombe, F. J. (1953). *J. R. Statist. Soc. B*, **15**, 1–29.

[5] Bahadur, R. R. and Savage, L. J. (1956). *Ann. Math. Statist.*, **27**, 1115–1122.

[6] Blum, J. R. and Rosenblatt, J. (1963). *Ann. Inst. Statist. Math.*, **15**, 45–50.

[7] Blum, J. R. and Rosenblatt, J. (1963). *Ann. Math. Statist.*, **34**, 1452–1458.

[8] Blum, J. R. and Rosenblatt, J. (1956). *Ann. Inst. Statist. Math.*, **18**, 351–355.

[9] Blum, J. R. and Rosenblatt, J. (1969). *Ann. Math. Statist.*, **40**, 1021–1032.

[10] Chow, Y. S. and Robbins, H. (1965). *Ann. Math. Statist.*, **36**, 457–462.

[11] Chow, Y. S., Robbins, H., and Siegmund, D. (1972). *Great Expectations: The Theory of Optimal Stopping*. Houghton-Mifflin, New York.

[12] Cramér, H. (1946). *Mathematical Methods of Statistics*. Princeton University Press, Princeton, NJ.

[13] Dantzig, G. B. (1940). *Ann. Math. Statist.*, **11**, 186–192.

[14] Haldane, J. B. S. (1945). *Biometrika*, **33**, 222–225.

[15] Nagao, H. (1980). *Ann. Inst. Statist. Math.*, **32**, 369–375.

[16] Robbins, H. and Monro, S. (1951). *Ann. Math. Statist.*, **22**, 400–407.

[17] Stein, C. (1945). *Ann. Math. Statist.*, **16**, 243–258.

[18] Stein, C. and Wald, A. (1947). *Ann. Math. Statist.*, **18**, 427–433.

[19] Stute, W. (1983). *Zeit. Wahrsch. verw. Geb.*, **62**, 113–123.

[20] Wald, A. (1950). *Statistical Decision Functions*. Wiley, New York.

[21] Wald, A. (1951). *Proc. Second Berkeley Symp. on Math. Statist. Prob.* University of California Press, Berkeley, CA. pp. 1–11.

[22] Wang, Y. H. (1980). *J. Amer. Statist. Assoc.*, **75**, 977–983.

[23] Weiss, L. (1960). *Naval Res. Logist. Quart.*, **7**, 251–256.

[24] Weiss, L. and Wolfowitz, J. (1969). *Proc. Int. Symp. Prob. Theory*. Springer, New York, pp. 232–256.

[25] Weiss, L. and Wolfowitz, J. (1972). *Zeit. Wahrsch. verw. Geb.*, **24**, 203–209.

[26] Wijsman, R. (1981). *Commun. Statist. A*, **10**, 2137–2147.

[27] Wolfowitz, J. (1946). *Ann. Math. Statist.*, **17**, 489–493.

[28] Wolfowitz, J. (1947). *Ann. Math. Statist.*, **18**, 215–230.

[29] Wolfowitz, J. (1950). *Ann. Math. Statist.*, **21**, 218–230.

[30] Zielinski, R. (1982). *Zastos. Matem.*, **17**, 277–280.

Bibliography

Govindarajulu, Z. (1975). *Sequential Statistical Procedures*. Academic, New York. (This book contains a large section on sequential estimation.)

Wetherill, G. B. (1966). *Sequential Methods in Statistics*. Methuen, London, England. (A largely nonmathematical description, including chapters on estimation and stochastic approximation.)

(ADAPTIVE METHODS
ESTIMATION, POINT
FIXED-WIDTH AND BOUNDED-LENGTH CONFIDENCE INTERVALS
SEQUENTIAL ANALYSIS
SEQUENTIAL ESTIMATION OF THE MEAN, FINITE POPULATIONS
SEQUENTIAL RANK ESTIMATORS
STOCHASTIC APPROXIMATION)

L. WEISS

SEQUENTIAL ESTIMATION OF THE MEAN IN FINITE POPULATIONS

INTRODUCTION

Two procedures for sequential estimation* that are of considerable practical interest have been published. These are (i) Stein's double-sampling* method (*see* FIXED-WIDTH AND BOUNDED-LENGTH CONFIDENCE INTERVALS) for estimating the mean of a normal population by a confidence interval of given width and coefficient, and (ii) Haldane's inverse binomial sampling [6] when it is desired to estimate the proportion p of members of a rare item of a given population (*see* INVERSE SAMPLING). Both Stein's as well as Haldane's method were developed for infinite populations. Fixed-width and bounded-length confidence intervals do not cover finite populations either. We will briefly review here the case of finite populations when the sample observations are not independent.

When n_1/N is negligible, where n_1 is the initial sample size and N is the size of the finite population, the half-width interval for the population mean $\overline{Y}_p$ is given by $t(\alpha, n_1 - 1)s_1/\sqrt{n_1}$, where s_1^2 is the sample variance of the first sample only, and $t(\alpha, \nu)$ is the

(100α)th percentile of the t-distribution* with ν degrees of freedom. If this quantity is $\leqslant d$, the desired half-width, the sample is already sufficiently large. If the quantity exceeds d, we take additional observations so that the total sample size is

$$n^* = \max\left\{n_1, \left[\frac{s_1^2 t^2(\frac{1}{2}\alpha, n_1 - 1)}{d^2}\right] + 1\right\}, \quad (1)$$

where $[x]$ is the greatest integer less than x. Then, if $\bar{y}$ is the mean of the whole sample,

$$P\{|\bar{y} - \bar{Y}_p| \geqslant d\} \leqslant \alpha.$$

If the finite population correction* must be applied, we replace n^* by $n^*/(1 + n^*/N)$.

For estimating a population variance for sample size determinations, a sample may be taken in two steps. The first will be a simple random sample (SRS) of n_1 units out of a population of N units, i.e., every one of the $\binom{N}{n_1}$ distinct samples has an equal chance of being drawn. This process is equivalent to selecting the units at random and without replacement from the finite population. From this, estimate σ^2 by S_1^2 and the required n will be obtained. In the sequel we will define the population variance as

$$\sigma^2 = \sum_{i=1}^{N} (y_i - \bar{Y}_p)^2/(N - 1).$$

Cox [5], following Stein's work, shows how to compute n so that the final estimate $\bar{y}$ or p will have a preassigned variance V or a preassigned coefficient of variation* (CV). The first sample is assumed large enough to neglect terms of order $1/n_1^2$.

Cochran [4] provides values for n when we are required to estimate the population mean $\bar{Y}_p$ or proportion P with a given CV $(= \sqrt{c})$ or variance V assuming $n_1 \leqslant n$, the size of the final sample. Thus, for normal y and given c, additional units will be taken to make the final sample size

$$n = \frac{s_1^2}{c\bar{y}_1^2}\left[1 + 8c + \frac{s_1^2}{n_1\bar{y}_1^2} + \frac{2}{n_1}\right]. \quad (2)$$

The mean $\bar{y}_p$ based on the final sample will be slightly biased and is replaced by $\bar{y}_p(1 - 2c)$.

For estimating the population mean with given V, the total sample size is given by

$$n = \frac{s_1^2}{V}\left[1 + \frac{2}{n_1}\right]. \quad (3)$$

The effect of not knowing σ is to increase the average size in (2) by the factor $[1 + 8c + s_1^2/(n_1\bar{y}_1^2) + 2/n_1]$ and in (3) by $[1 + 2/n_1]$.

Following the approach to asymptotics in finite populations, Ghosh [7] proposes sequential point estimators of the means of U-statistics* in SRS from a finite population. He considers a sequence $\{\pi_{N_k}, k = 1, 2, \ldots\}$ of finite populations, where at the kth stage of the experiment the population consists of a fixed set of N_k symbols, say, $1, 2, \ldots, N_k$. The symbols may denote, for instance, the list of households in a certain community or the list of ports in a certain region at which landings from commercial catches during a season may be regarded as forming successive samples. If an SRS S_{k,n_k} of size n_k units is selected, then each of the

$$\begin{bmatrix} N_k \\ n_k \end{bmatrix}$$

subsets of size n_k has a probability

$$\begin{bmatrix} N_k \\ n_k \end{bmatrix}^{-1}$$

of being selected. The sample mean

$$\eta_k[S_{k,n_k}] = n_k^{-1}\textstyle\sum_{i \in S_{k,n_k}} y_{ki} \quad (4)$$

is an unbiased estimate of the population mean

$$\bar{Y}_k = N_k^{-1}\sum_{i=1}^{N_k} y_{ki}. \quad (5)$$

Assuming the loss function to be squared error plus cost of size n, we have expected loss for a given sample size

$$L[\mathbf{y}_k, \pi_k(S_k, n_k)] = \sigma_k^2[n_k^{-1} - N_k^{-1}] + C_k N_k, \quad (6)$$

where C_k (> 0) is the cost per unit at the

kth stage,

$$\mathbf{y}_k = [y_{k1}, \ldots, y_{kN_k}],$$

$$\sigma_k^2 = N_k^{-1} \sum_{i=1}^{N_k} [y_{ki} - \overline{Y}_k]^2.$$

Regarding n_k as a continuous variable and minimizing expected loss with respect to n_k,

$$n_k{}^* = \sigma_k C_k^{-1/2}. \tag{7}$$

A more general form of the result is given in ref. 12.

Substituting the value $n_k{}^*$ from (7) in (6) we have

$$L[\mathbf{y}_k, \pi_k(S_k, n_k{}^*)] = 2C_k n_k{}^* - \sigma_k N_k^{-1}. \tag{8}$$

Thus, similar to the infinite population case, the loss in sampling from finite populations depends on σ_k, and therefore cannot be minimized simultaneously for all σ_k unless σ_k is known. A sequential procedure is proposed in ref. 7 for which the stopping time $T = T_k$ is the smallest positive integer n ($\geqslant 2$) for which

$$n \geqslant C_k^{-1/2}[\hat{\sigma}_{kn} + n^{\overline{Y}_0}] \tag{9}$$

for some $\overline{Y}_0$, where

$$\hat{\sigma}_{kn}^2 = (n-1)^{-1} \sum\nolimits_{i \in S_{k,n}} [y_{ki} - \eta_k(S_{k,n})]^2. \tag{10}$$

Estimate $\overline{Y}_k$ by $\eta_k(S_{k,T})$.

Sequential procedures of the above type were proposed in ref. 10 for estimating the normal mean for infinite populations. The results and their extensions are given in FIXED-WIDTH AND BOUNDED-LENGTH CONFIDENCE INTERVALS.

Sequential confidence intervals for the mean of a subpopulation from a finite population of known size were considered in ref. 2. Assume that at each stage k there are N_k units having values $x_{k1}, x_{k2}, \ldots, x_{kN_k}$, of which M_k (unknown) are of type I (say), and $N_k - M_k$ are of type II. Let $\xi_{k1}, \xi_{k2}, \ldots, \xi_{kN_k}$ be a random permutation of $x_{k1}, \ldots, x_{kN_k}$ and $\eta_{k1}, \ldots \eta_{kM_k}$ be the type I values arranged by this permutation. Define

$$\mu_k = M_k^{-1} \sum_1^{M_k} \eta_{ki},$$
$$\sigma_k^2 = M_k^{-1} \sum_1^{M_k} (\eta_{ki} - \mu_k)^2. \tag{11}$$

A sequential fixed-width ($2d_k$) confidence interval I_{km} for μ_k is

$$I_{km} = \left[m^{-1} \sum_1^m \eta_{ki} - d_k, m^{-1} \sum_1^m \eta_{ki} + d_k\right], \tag{12}$$

where exactly m units of type I are obtained in a SRS of n units chosen from N_k units.

Assuming $d_k \to 0$ as $k \to \infty$, stopping times $M(k)$ are obtained that satisfy, for a given α,

$$\lim_{k \to \infty} P\{\mu_k \in I_{kM(k)}\} = 1 - \alpha. \tag{13}$$

An illustration is given of a bank that wants to open a branch in the suburbs and would, therefore, like to first obtain a confidence interval for the average deposits of its customers who live within a certain distance of the proposed branch.

We will finally consider the case of sequential sampling* [9], where one is interested in obtaining the current estimates of the mean measurement (say weight) on a characteristic of a rare animal using the capture-mark-recapture (CMR) technique and inverse sampling* from a finite population (*see* INVERSE SAMPLING). This problem is of real interest, since the size N of the population is unknown. A sample (of size n_1) is selected on the first occasion; these are marked and released. The members in the sample are then allowed to mix and on the next occasion, the second sample (of size n) is continued without replacement (WOR) until a prescribed number m of marked animals ($m < n_1$) have been recovered. We will assume that the population is closed. In this setup, generally, m is fixed and n ($\geqslant m$) is a positive integer-valued random variable.

This method provides an unbiased estimator of N. Sampling is done without replacement until m marked animals as well as the $n - m$ unmarked animals are selected to provide the m marked ones, which are measured and transferred back to the population. A second marking is done on the m marked animals, to ensure selection of m distinct animals without removing these from the population.

Let N be the size of the unknown population (assumed to be the same on both the occasions), let $Y_1^{(2)}, \dots, Y_N^{(2)}$ be the population values on the second occasion. In this case, n is a random variable and its probability law is given by the negative hypergeometric distribution*:

$$P_{N, n_1}(n|m) = m\begin{bmatrix} n_1 \\ m \end{bmatrix}\begin{bmatrix} N - n_1 \\ n - m \end{bmatrix}\Big/\left\{n\binom{N}{n}\right\}, \quad (14)$$

for $n = m, m + 1, \dots, N - n_1 + m$. An estimate of the population mean $\overline{Y}^{(2)} = N^{-1}\Sigma_{i=1}^{N} Y_i^{(2)}$ is given by

$$\bar{y}^{(2)} = n^{-1} \sum_{i=1}^{n} y_i^{(2)}. \quad (15)$$

Let

$$n_1 = \alpha N, \qquad m = \beta n_1 = \alpha\beta N,$$
$$0 < \alpha < 1 \quad \text{and} \quad 0 < \beta < 1, \quad (16)$$

where β, m, and n_1 are given and α and N are unknown quantities. We note that

$$E(n) \sim N[\beta(1 - \alpha) + \alpha\beta] = N\beta = n^* \quad \text{(say)}, \quad (17)$$

and the probability law in (14) may be used to verify that

$$n/n^* \to 1 \quad \text{in probability as } m \text{ increases}. \quad (18)$$

Given (18), one may approximate the large sample distribution of $N^{1/2}(\bar{y}^{(2)} - \overline{Y}^{(2)})$ by that of a similar statistic computed for the sample size n^*, provided the classical condition of uniform continuity [1] is satisfied for finite population sampling*. This follows from Theorem 3.3.3 of ref. 10, so that the asymptotic variance of $N^{1/2}(\bar{y}^{(2)} - \overline{Y}^{(2)})$ is given by

$$S_{22}\{(N - n^*)N/[n(N - 1)]\}$$
$$\sim S_{22}\beta^{-1}(1 - \beta), \quad (19)$$

where S_{22} is the variance of $Y_i^{(2)}$. Hence, the variance of $\bar{y}^{(2)}$ is given by

$$\{(N - n^*)/[n(N - 1)]\}S_{22}, \quad (20)$$

where an estimator of N as in ref. 3 is given by

$$\hat{N} = \{n(n_1 + 1)/m\} - 1. \quad (21)$$

References

[1] Anscombe, F. J. (1952). *Proc. Camb. Philos. Soc.*, **48**, 600–607.

[2] Carroll, R. J. (1978). *J. Amer. Statist. Ass.*, **73**, 408–413.

[3] Chapman, D. G. (1952). *Biometrics*, **8**, 286–306.

[4] Cochran, W. G. (1977). *Sampling Techniques*. Wiley, New York.

[5] Cox, D. R. (1952). *Biometrika*, **39**, 217–227,

[6] Haldane, J. B. S. (1945). *Biometrika*, **33**, 222–225.

[7] Ghosh, M. (1981). *Commun. Statist. A*, **10**, 2215–2229.

[8] Robbins, H. (1959). *Probability and Statistics* (H. Cramér. Volume). Almqvist and Wiksell, Upsala, Sweden, pp. 235–245.

[9] Sen, A. R. and Sen, P. K. (1986). In *Applied Probability, Stochastic Processes and Sampling Theory*, **1**, I. B. MacNeill and G. J. Unphrey, eds. Reidel: Boston.

[10] Sen, P. K. (1981). *Sequential Nonparametrics: Invariance Principles and Statistical Inference*. Wiley, New York.

[11] Stein, C. (1945). *Ann. Math. Statist.*, **16**, 235–245.

[12] Yates, F. (1960). *Sampling Methods for Censuses and Surveys*, 3rd ed. Charles Griffin and Co., London, England.

(FIXED-WIDTH AND BOUNDED-LENGTH CONFIDENCE INTERVALS
INVERSE SAMPLING
SEQUENTIAL ESTIMATION
SEQUENTIAL SAMPLING)

A. R. SEN

SEQUENTIAL PROBABILITY RATIO TEST *See* SEQUENTIAL ANALYSIS

SEQUENTIAL PROCEDURES, ADAPTIVE

A sequential procedure consists of a sampling rule, a stopping rule, and a terminal decision rule. The sampling rule specifies the number of observations to be taken at each stage, the stopping rule specifies whether to stop or continue sampling, and the terminal decision rule specifies the decision to be taken when sampling has terminated. A sequential procedure is said to be *adaptive* (or *data-dependent*) if the sampling rule depends on the observations already obtained. In traditional sequential analysis involving a single population, observations are taken one-at-a-time or in one, two, or more stages, the number of observations at each stage being nonrandom; only the stopping rule depends on the data. Of course, the stopping decision can be regarded as a special case of the sampling decision and thus all of sequential analysis* can be regarded as adaptive. For one-population problems however, the term "adaptive sequential procedures" as used in this article refers to those procedures in which the number of observations to be taken at any stage is a function of the data accumulated (for example, Stein's two-stage procedure [22]).

The sampling problem becomes more interesting when two or more populations are involved, and there is the additional aspect of deciding which population (or populations) to sample from at each stage; see Robbins [17]. One most interesting application of adaptive sampling arises in connection with clinical trials*, where ethical reasons might prompt the assignment of fewer patients to treatments that seem inferior. For example, if one is comparing treatments A and B and there is an indication that treatment A is better than treatment B, one might wish to bias the trial so that the next patient receives treatment A with higher probability than treatment B.

Quite often the use of adaptive sequential procedures results in considerable savings in the expected total number of observations in comparison with competing single-stage procedures. Examples of this phenomenon are given in the next two sections. We first describe an adaptive sequential procedure for a two-population hypothesis testing problem. Next, we describe some of the adaptive methods used in the context of selection problems involving two or more populations. Finally we describe, briefly, some situations where the experimenter cannot achieve the desired objective using a single-stage procedure and hence, resorts to sequential procedures that use adaptive sampling.

TWO-POPULATION PROBLEMS

Adaptive sampling is used very often in two-armed bandit problems (*see* ONE- AND TWO-ARMED BANDIT PROBLEMS). Consider a slot machine with two arms A and B having unknown probabilities of success P_A and P_B, respectively. On each trial the player chooses one of the two arms to pull. One objective might be to devise a strategy that maximizes the expected total number of successes in n trials. A commonly used adaptive rule is the *play-the-winner** rule (PWR): One of the two arms is chosen at random to begin the game. If a pull on any arm results in a success, the same arm is chosen for the next trial; otherwise the other arm is chosen. There is a vast literature on two-armed bandit problems and related sampling rules; see Berry [4], for example.

Flehinger and Louis [6] studied some data-dependent allocation rules with Wald-type termination rules for testing which of two treatments π_1 and π_2 is superior (i.e., has the larger mean). The two populations π_1 and π_2 are assumed to be normal with unknown means μ_1 and μ_2, respectively, and a common known variance σ^2. The experiment is said to be at stage (m, n) when m observations on π_1 and n observations on π_2 have been taken. Let $\overline{X}_m$ and $\overline{Y}_n$ denote the sample means of π_1 and π_2, respectively, at stage (m, n). Consider the following sequential test for $H_1 : \Delta = -\Delta^*$ vs. $H_2 : \Delta = \Delta^*$, where $\Delta^* > 0$ is specified. Let $Z_{m,n} = [mn/(m+n)](\bar{X}_m - \bar{Y}_n)$ and $L_{m,n} = \exp\{2\Delta^* Z_{m,n}\}$ and let $B > 1$ be a specified constant.

If $L_{m,n} \leqslant B^{-1}$, stop and choose H_1;

if $L_{m,n} \geqslant B$, stop and choose H_2;

if $B^{-1} < L_{m,n} < B$, continue sampling.

Flehinger et al. [7] showed that the error probabilities for this test are approximately independent of the sampling rule as long as attention is restricted to sampling rules that terminate with probability 1 and depend on the observations only though their differences. This result is very interesting since within this class of sampling rules one can look for data-dependent rules that are optimal in some sense. One such nearly "optimal" procedure is described below.

Louis [15] considers the loss function $E\{N\} + (\gamma - 1)E\{N_{(1)}\}$, where N is the total number of observations, $N_{(1)}$ is the number of observations from the inferior population, and $\gamma \geqslant 1$ is the relative cost of taking an observation from the inferior as opposed to the superior population. Define

$$q_\gamma(L_{m,n}) = \frac{(\gamma L_{m,n} + 1)^{1/2}}{(\gamma L_{m,n} + 1)^{1/2} + (\gamma + L_{m,n})^{1/2}}.$$

At stage (m, n) the next observation is assigned to π_1 if $[m/(m+n)] < q_\gamma(L_{m,n})$, to π_2 if $[m/(m+n)] > q_\gamma(L_{m,n})$, and to π_1 with probability $q_\gamma(L_{m,n})$ if $[m/(m+n)] = q_\gamma(L_{m,n})$. This adaptive rule is nearly optimal for the problem of minimizing the loss function given above. See Robbins and Siegmund [18] for additional results. Hayre [10] describes "nearly optimal procedure" for a related hypothesis testing* problem. Hayre and Turnbull [11] propose a class of sequential procedures for selecting the better of two treatments; these procedures use adaptive sampling and have probability of correct selection approximately independent of the sampling rule.

MULTIPOPULATION PROBLEMS

When more than two populations are involved, the problem becomes even more complex. We shall describe some adaptive sequential procedures in the context of selection problems. First consider the case of k normal populations $\pi_1, \pi_2, \ldots, \pi_k$ with means $\mu_1, \mu_2, \ldots, \mu_k$ and common variance σ^2. Let $\mu_{[1]} \leqslant \mu_{[2]} \leqslant \cdots \leqslant \mu_{[k]}$ denote the ordered means. In the indifference-zone formulation of the problem (*see* SELECTION PROCEDURES) we consider the goal of selecting the population associated with $\mu_{[k]}$ in such a way as to guarantee that

$$\Pr[\text{CS}] \geqslant P^*$$

whenever

$$\mu_{[k]} - \mu_{[k-1]} \leqslant \Delta^*, \tag{1}$$

where P^* $(1/k < P^* < 1)$ and Δ^* $(\Delta^* > 0)$ are specified by the experimenter. Here CS (correct selection) denotes the selection of the population associated with $\mu_{[k]}$. Tamhane and Bechhofer [23, 24] described a two-stage procedure for this problem that uses the data from the first stage to eliminate some populations. Paulson [16] proposed sequential procedures both when σ^2 is known and unknown. We describe the procedure for the case of common known variance σ^2.

Paulson's procedure permits permanent elimination of populations and thus the sampling rule (which specifies which populations to sample from at each stage) can be regarded as adaptive. Let X_{ij} denote the jth observation from population π_i. Define

$$a_\lambda = [\sigma^2/(\Delta^* - \lambda)] \times \ln((k-1)/(1-P^*))$$

and let W_λ be the largest integer less than a_λ/λ. The experiment is started by taking one observation from each population. Any population π_i for which

$$X_{i1} < \max_{1 \leqslant r \leqslant k} X_{r1} - a_\lambda - \lambda$$

is eliminated. If all populations but one are eliminated, the experiment is stopped and the remaining population is selected as the best. At stage m $(m = 2, 3, \ldots, W_\lambda)$ one observation is taken from each population not eliminated after stage $m - 1$; further, any population for which

$$\sum_{j=1}^{m} X_{ij} < \max\left(\sum_{j=1}^{m} X_{rj} \right) - a_\lambda + m\lambda \quad (2)$$

is eliminated. In (2) the maximum is taken over all the populations not eliminated after stage $m - 1$. If there is only one population left after stage m, the experiment is stopped and this population is selected as best; otherwise the experiment proceeds to stage $m + 1$. If more than one population remains after stage W_λ, the experiment is terminated at stage $W_\lambda + 1$ by selecting the population with maximum $\Sigma_{j=1}^{W_\lambda+1} X_{rj}$.

The above procedure guarantees the probability requirement (1) for all $0 < \lambda < \Delta^*$. Although the optimum choice of λ is not known, preliminary calculations indicate that the choice $\lambda = \Delta^*/4$ leads to considerable savings compared to the fixed sample size procedure. Turnbull et al. [25] propose some adaptive sequential procedures for the above problem and compare them with Paulson's. Other multistage selection procedures are described in detail in Gupta and Panchapakesan [9].

Quite often the responses to treatments can be classified simply as success or failure. Suppose we have k Bernoulli populations, $\pi_1, \pi_2, \ldots, \pi_k$, with corresponding probabilities of success, $p_1, p_2, \ldots, p_k$. Let $p_{[1]} \leqslant \cdots \leqslant p_{[k]}$ denote the ordered success probabilities. Several adaptive sequential procedures have been proposed for the problem of selecting the population associated with $p_{[k]}$, subject to the probability requirement

$$\Pr[\text{CS}] \geqslant P^*$$

whenever

$$p_{[k]} - p_{[k-1]} \geqslant \Delta^*, \quad (3)$$

where $1/k < P^* < 1$ and $\Delta^* > 0$ are specified by the experimenter. The papers by Hoel and Sobel [12], Sobel and Weiss [21], and Hoel et al. [13] survey some of the vast literature in this area. A more recent bibliography is given by Bechhofer and Kulkarni [2]. The most commonly used adaptive sampling rule is a generalized version of the two-population play-the-winner rule (PWR) described for the two-armed bandit problem. In the generalized PWR the populations are arranged in random order; population π_j is sampled until a failure is observed and then π_{j+1} is sampled ($j = 1, 2, \ldots, k$); π_{k+1} is identified with π_1. PWR is simple to use and is intuitively appealing; see Zelen [26].

A class of procedures recently proposed by Bechhofer and Kulkarni [2] for the problem of selecting the population associated with $p_{[k]}$ is now described. $n \geqslant 1$ is a fixed integer (the choice of n is not important here) and specifies the maximum number of observations to be taken from any population. At stage m (i.e., when a total of m observations have been taken) let $n_{i,m}$ denote the number of observations that have been taken from π_i and $z_{i,m}$ the number of successes obtained from π_i. The stopping rule $\mathscr{S}^*$ says: Stop at the first stage m at which there exists π_i for which

$$z_{i,m} \geqslant z_{j,m} + n - n_{j,m} \quad \text{for all} \quad j \neq i. \quad (4)$$

At termination, the population π_i that satisfies (4) is chosen as the best, ties being broken at random. The terminal decision rule is denoted by $\mathscr{T}^*$. Let $\mathscr{R}$ denote any sampling rule that takes no more than n observations from any one population. All procedures $(\mathscr{R}, \mathscr{S}^*, \mathscr{T}^*)$ achieve the same probability of correct selection uniformly in $(p_1, p_2, \ldots, p_k)$. Thus, one can use different criteria such as minimizing the expected total number of observations or the expected number of observations from the inferior populations (π_i's with smaller p-values), to choose sampling rules that adapt to the current information from the data. Dynamic programming* techniques were used by Bechhofer and Kulkarni [2] to determine optimal sampling rules for these objectives for $k = 2$ populations. More general results for $k \geqslant 2$ are described in Kulkarni and Jennison [14]. The sampling rule $\mathscr{R}^*$ described below, was found to have very desirable properties. In particular, the adaptive procedure $(\mathscr{R}^*, \mathscr{S}^*, \mathscr{T}^*)$ minimizes the expected total number of observations among all procedures of the form $(\mathscr{R}, \mathscr{S}^*, \mathscr{T}^*)$ if $p_{[1]} + (k-1)^{-1}\Sigma_{i=2}^{k} p_{[i]} \geqslant 1$. The sampling rule $\mathscr{R}^*$ is as follows: At stage m take the next observation from the population with the minimum number of failures; ties are broken by selecting, among the tied populations, the one with the maximum number of successes; further ties are broken at random. An attractive feature of the sequential procedure $(\mathscr{R}^*, \mathscr{S}^*, \mathscr{T}^*)$, and in fact of all the procedures $(\mathscr{R}, \mathscr{S}^*, \mathscr{T}^*)$, is that they are *closed*; i.e., the total number of observations cannot exceed a fixed known number (in this case, $kn - 1$) for all parameter values.

We shall illustrate the advantage of using adaptive sequential procedures instead of fixed sample-size procedures by comparing the performance of $(\mathscr{R}^*, \mathscr{S}^*, \mathscr{T}^*)$ with that of the corresponding single-stage procedure, denoted by (R_{SS}, T_{SS}) in ref. 2. For given n, the single-stage procedure achieves the same probability of correct selection as $(\mathscr{R}^*, \mathscr{S}^*, \mathscr{T}^*)$ does, uniformly in $(p_1, p_2, \ldots, p_k)$, if we take n observations from each of the k populations and select the population with maximum successes as best.

In Table 1 $p_{[1]} \leqslant p_{[2]} \leqslant p_{[3]}$ denote the ordered p-values and $E\{N_{(i)}\}$ denotes the expected number of observations from the population associated with $p_{[i]}$ if $(\mathscr{R}^*, \mathscr{S}^*, \mathscr{T}^*)$ is used. The values of $E\{N_{(i)}\}$ should be compared with n, which is the number of observations taken from each population by the corresponding single-stage procedure. It can be readily seen that considerable savings is achieved by using the adaptive procedure $(\mathscr{R}^*, \mathscr{S}^*, \mathscr{T}^*)$ instead of (R_{SS}, T_{SS}). Further, if $(\mathscr{R}^*, \mathscr{S}^*, \mathscr{T}^*)$ is used, $E\{N_{(1)}\} \leqslant E\{N_{(2)}\} \leqslant E\{N_{(3)}\}$, a desirable property for any procedure used in clinical trials where one would like to minimize the number of patients subjected to inferior treatments. To compare the expected total number of observations, note that $E\{N\} = \Sigma_{i=1}^{3} E\{N_{(i)}\}$. For instance, when $n = 40$ and $(p_{[1]}, p_{[2]}, p_{[3]}) = (0.4, 0.6, 0.8)$, (R_{SS}, T_{SS}) requires 120 observations while $E\{N\}$ using $(\mathscr{R}^*, \mathscr{S}^*, \mathscr{T}^*)$ is 73.069. More detailed tables are available in ref. 3.

ADAPTIVE SAMPLING IN STEIN-TYPE PROBLEMS

In all the procedures described above, the principal reason for using adaptive sampling was because it reduced the expected total number of observations or the expected number of observations from the inferior populations, etc. There are some problems, however, which preclude solution by single-stage (or fixed sample size) procedures, necessitating the use of adaptive methods. One of the earliest examples is that of Stein [22], described in FIXED-WIDTH AND BOUNDED-LENGTH CONFIDENCE INTERVALS. Stein proposed an adaptive two-stage procedure in which the size of the sample in the second stage depends on the data obtained in the first stage.

We describe a two-population problem that cannot be solved using a single-stage procedure. Let π_1 and π_2 be two normal

Table 1

n	$(p_{[1]}, p_{[2]}, p_{[3]}) = (0.5, 0.6, 0.7)$			$(p_{[1]}, p_{[2]}, p_{[3]}) = (0.4, 0.6, 0.8)$		
	$E\{N_{(1)}\}$	$E\{N_{(2)}\}$	$E\{N_{(3)}\}$	$E\{N_{(1)}\}$	$E\{N_{(2)}\}$	$E\{N_{(3)}\}$
10	5.136	6.409	8.354	3.258	4.959	9.178
20	11.038	13.843	18.359	6.535	9.916	19.526
30	17.094	21.421	28.495	9.907	14.931	29.725
40	23.178	29.025	38.648	13.277	19.953	39.839

populations with unknown means μ_1 and μ_2, and unknown variances σ_1^2 and σ_2^2, respectively. The objective is to determine a confidence interval I of width d with coverage probability $\geqslant \alpha$ for the parameter $\Delta = \mu_1 - \mu_2$, where $0 < d < \infty$ and $0 < \alpha < 1$ are prespecified constants. It is not possible to determine the required confidence interval using a single-stage procedure and hence adaptive methods are used. Several Stein-type two-stage procedures have been proposed (Ghosh [8]). Robbins et al. [19] proposed sequential procedures using three more or less equivalent stopping rules and an adaptive sampling rule.

To start the experiment $n_0 \geqslant 2$ observations are taken from each population. Thereafter at each stage, the population from which the next observation is to be taken depends on the current estimates of the two variables. Let $X_1, X_2, \ldots$ denote the observations from Π_1 and $Y_1, Y_2, \ldots$ the observations from Π_2. Let

$$S_{i1}^2 = (i-1)^{-1} \sum_{k=1}^{i} (X_k - \overline{X}_i)^2,$$

$$S_{j2}^2 = (j-1)^{-1} \sum_{k=1}^{j} (Y_k - \overline{Y}_j)^2$$

be the usual estimates of σ_1^2 and σ_2^2, respectively, where $\overline{X}_i = \sum_{k=1}^{i} X_k/i$ and $\overline{Y}_j = \sum_{k=1}^{j} Y_k/j$ are the respective sample means. If at any stage i observations on Π_1 and j observations on Π_2 have been taken, with $n = i + j \geqslant 2n_0$, then the next observation is taken from Π_1 or Π_2 according as $i/j \leqslant S_{i1}/S_{j2}$ or $i/j > S_{i1}/S_{j2}$. In this procedure the sampling rule tends to sample from the population for which the current estimated variance is larger. To describe one of the stopping rules, let $\{a_n\}$ be a given sequence of positive constants such that $a_n \to a$ as $n \to \infty$, and let $b_n = (2a_n/d)^2$. Stop with the first $n \geqslant 2n_0$ such that if i observations on Π_1 and j observations on Π_2 have been taken, with $i + j = n$, then

$$n \geqslant b_n (S_{i1} + S_{j2})^2.$$

The required confidence interval is given [19] by

$$I = \left[\overline{X}_i - \overline{Y}_j - d, \overline{X}_i - \overline{Y}_j + d\right].$$

For generalized Stein-type procedures in the context of selection problems see refs. 1 and 9.

The field of adaptive sequential procedures is rich and varied. There seems to be great potential for beneficial applications in the area of clinical trials*. Unfortunately, these procedures are rarely used in practice. A critical evaluation of the situation is given by Simon [20]. Recently, Edwards and Hsu [5] advocated the use of adaptive methods in clinical trials to allow for early termination of the experiment, in particular, early elimination of very bad treatments. Their paper is one of a series of papers covering sequential elimination procedures, two-stage procedures, etc., prepared under a project sponsored by the National Cancer Institute.

References

[1] Bechhofer, R. E., Dunnett, C. W., and Sobel, M. (1954). *Biometrika*, **41**, 170–176.

[2] Bechhofer, R. E. and Kulkarni, R. V. (1982). *Statistical Decision Theory and Related Topics III*, Vol. 1, pp. 61–108. (Closed adaptive sequential procedures for Bernoulli populations.)

[3] Bechhofer, R. E. and Kulkarni, R. V. (1982). *Commun. Statist. Seq. Anal.*, **1**. (Performance characteristics and tables for procedures in [2] above.)

[4] Berry, D. A. (1972). *Ann. Math. Statist.*, **43**, 871–897.

[5] Edwards, D. and Hsu, J. C. (1983). *Commun. Statist. A*, **12**, 1135–1145.

[6] Flehinger, B. J. and Louis, T. A. (1972). *Proc. Sixth Berkeley Symp. Math. Statist. Prob.*, Vol. **4**. University of California Press, Berkeley, CA, pp. 43–52. (Wald SPRT-type tests with data dependent sampling.)

[7] Flehinger, B. J., Louis, T. A., Robbins, H., and Singer, B. H. (1972). *Proc. Nat. Acad. Sci. USA*, **69**, 2993–2994.

[8] Ghosh, B. K. (1975). *J. Amer. Statist. Ass.*, **70**, 457–462. (Stein-type two-stage procedures for two normal populations.)

[9] Gupta, S. S. and Panchapakesan, S. (1979). *Multiple Decision Procedures: Theory and Methodology*

of Selecting and Ranking Populations. Wiley, New York. (Contains several adaptive multistage and sequential procedures in the context of ranking and selection.)

[10] Hayre, L. S. (1979). *Biometrika*, **66**, 465–474. (Adaptive "nearly optimal" procedure for three-hypotheses test of two normal populations.)

[11] Hayre, L. S. and Turnbull, B. W. (1981), *Commun. Statist. A*, **10**, 2339–2360.

[12] Hoel, D. and Sobel, M. (1972). *Proc. Sixth Berkeley Symp. Math. Statist. Prob.*, Vol. 69. (Compares several adaptive procedures for Bernoulli populations.)

[13] Hoel, D., Sobel, M., and Weiss, G. H. (1975). *Perspectives in Biometry*, R. M. Elashoff ed., Academic, New York, pp. 29–61. (A good survey of the application of adaptive sampling to clinical trials and the problems associated with it.)

[14] Kulkarni, R. V. and Jennison, C. (1983). *Tech. Rep. 600*, School of O.R.21.E., Cornell University, Ithaca, NY. (Optimal properties of the Bechhofer–Kulkarni Bernoulli selection procedure.)

[15] Louis, T. A. (1975). *Biometrika*, **62**, 359–369. ("Optimal" data-dependent sampling for Gaussian populations.)

[16] Paulson, E. (1964). *Ann. Math. Statist.*, **35**, 174–180. (Elimination-type closed sequential procedure for the selection of normal populations.)

[17] Robbins, H. (1952). *Bull. Amer. Math. Soc.*, **58**, 529–532. (One of the earliest papers to refer to adaptive sampling.)

[18] Robbins, H. and Siegmund, D. O. (1974). *J. Amer. Statist. Ass.*, **69**, 132–139.

[19] Robbins, H., Simons, G., and Starr, N. (1967). *Ann. Math. Statist.*, **38**, 1384–1391.

[20] Simon, R. (1977). *Biometrics*, **33**, 743–749.

[21] Sobel, M. and Weiss, G. H. (1972). *Ann. Math. Statist.*, **43**, 1808–1826. (Reviews procedures using play-the-winner sampling rules for $k \geqslant 3$ populations.)

[22] Stein, C. (1945). *Ann. Math. Statist.*, **16**, 243–258.

[23] Tamhane, A. C. and Bechhofer, R. E. (1977). *Commun. Statist. A*, **6**, 1003–1033. (Two-stage procedure with elimination in the second stage based on data obtained in the first stage.)

[24] Tamhane, A. C. and Bechhofer, R. E. (1979). *Commun. Statist. A*, **8**, 337–358.

[25] Turnbull, B. W., Kaspi, H., and Smith, R. L. (1978). *J. Statist. Comput. Simul.*, **7**, 133–150. (Describes and compares several adaptive sequential procedures for selecting among normal populations.)

[26] Zelen, M. (1969). *J. Amer. Statist. Ass.*, **64**, 131–146.

Acknowledgment

The author is grateful to Professors Robert Bechhofer and Ajit Tamhane for helpful suggestions.

(ADAPTIVE METHODS
CLINICAL TRIALS
DOUBLE SAMPLING
FIXED-WIDTH AND BOUNDED-LENGTH CONFIDENCE INTERVALS
ONE- AND TWO-ARMED BANDIT PROBLEMS
PLAY-THE-WINNER RULES
SELECTION PROCEDURES
SEQUENTIAL ANALYSIS
SEQUENTIAL ESTIMATION)

R. KULKARNI

SEQUENTIAL RANK ESTIMATORS

There are two basic reasons why sequential methods (*see* SEQUENTIAL ANALYSIS) are used in statistics. First, to reduce the sample size on an average as compared to the corresponding fixed sample procedure that meets the same error requirements. Wald's sequential probability ratio test (*see* SEQUENTIAL ANALYSIS) is a classic example of this. Second, to solve certain problems that cannot be solved by using a predetermined sample size. *Sequential rank* (*R*-) *estimators* have been developed mainly to address the second problem.

The R-estimators in the one-sample location problem are obtained by equating simple signed linear rank statistics to centers of their distributions. Such linear rank statistics are usually the ones used in the corresponding hypothesis testing* problem (*see* RANKING PROCEDURES). Let $X_1, \dots, X_n$ be i.i.d. (independent and identically distributed) with $P_\theta(X_1 \leqslant x) = F(x - \theta)$, F being an unknown continuous distribution function symmetric about zero, and θ the unknown location parameter. Let R_{ni}^+ denote the rank of $|X_i|$ among $|X_1|, \ldots, |X_n|$, $i = 1, \ldots, n$, and for every $n \geqslant 1$; let $a_n^+(1) \leqslant \cdots \leqslant a_n^+(n)$ denote the scores generated by a score function $\phi^+ : [0,1] \to I$, some interval in the real line in the following way:

$$a_n^+(i) = E\phi^+(U_{ni}), \text{ or } \phi^+(i/(n+1)), \quad 1 \leqslant i \leqslant n,$$

where $U_{n1} \leqslant \cdots \leqslant U_{nm}$ are the ordered random variables in a random sample of size n from the uniform $[0,1]$ distribution, and $\phi^+(u) = \phi((1+u)/2), 0 < u < 1$, ϕ being a nondecreasing skew-symmetric function. The simple signed linear rank statistic for testing $H_0: \theta = 0$ against one- or two-sided alternatives is given by

$$S_n = \sum_{i+1}^{n} \operatorname{sgn} X_i a_n^+(R_{ni}^+),$$

where $\operatorname{sgn} x = 1, 0$, or -1 according as x is greater than, equal to, or less than 0. If we replace $\mathbf{X}_n = (X_1, \ldots, X_n)$ by $\mathbf{X}_n - a\mathbf{1}_n$, where a is real and $\mathbf{1}_n = (1, \ldots, 1)$, and recompute the signed linear rank statistic, the same is denoted by $S_n(a)$; $S_n(a)$ is a step function in a. It is proved in Hodges and Lehmann [6] and Sen [8] that $S_n(a)$ is nonincreasing in a. Then the R-estimator of θ is defined by

$$\hat{\theta}_n^{(R)} = \tfrac{1}{2}(\sup\{a: S_n(a) > 0\} + \inf\{a: S_n(a) < 0\}).$$

In order to find a confidence interval for θ based on signed linear rank statistics, first for every n ($\geqslant 2$) and every $\alpha \in (0,1)$, find an $\alpha_n \leqslant \alpha$ and $S_{n,\alpha}$ such that

$$P_{\theta=0}(-S_{n,\alpha} \leqslant S_n \leqslant S_{n,\alpha}) = 1 - \alpha_n \geqslant 1 - \alpha,$$

where $\alpha_n \to \alpha$ as $n \to \infty$. Define

$$\hat{\theta}_{L,n}^{(R)} = \sup\{a: S_n(a) > S_{n,\alpha}\},$$
$$\hat{\theta}_{U,n}^{(R)} = \inf\{a: S_n(a) < -S_{n,\alpha}\}.$$

Then, $(\hat{\theta}_{L,n}^{(R)}, \hat{\theta}_{U,n}^{(R)})$ is a confidence interval for θ with confidence coefficient $1 - \alpha_n$.

Two important special cases are when $\phi(u) = u$ (the Wilcoxon score) and $\phi(u) = \Phi^{-1}(u)$ (the normal score). In the former case $\hat{\theta}_n^{(R)} = \operatorname{med}_{1 \leqslant i \leqslant j \leqslant n}(X_i + X_j)/2$, the celebrated *Hodges–Lehmann estimator*.

In order to motivate the sequential point and interval estimation procedures, a few asymptotic results available in Hodges and Lehmann [6] and Sen [9] are needed (*see also* SEQUENTIAL ESTIMATION). Assume that F is absolutely continuous with PDF f having finite Fisher information. Define

$$\psi(u) = -f'(F^{-1}(u))/f(F^{-1}(u)), \quad 0 < u < 1.$$

Then, $\sqrt{n}(\hat{\theta}_n^{(R)} - \theta)$ converges in distribution to $N(0, \sigma_R^2)$, where

$$\sigma_R^{-2} = \left(\int_0^1 \phi(u)\psi(u)\,du\right)^2 \Big/ \int_0^1 \phi^2(u)\,du.$$

Under certain additional conditions, it is proved by Sen [10] that $nE_\theta(\hat{\theta}_n^{(R)} - \theta)^2 \to \sigma_R^2$ as $n \to \infty$.

SEQUENTIAL POINT ESTIMATION

In estimating θ by $\hat{\theta}_n^{(R)}$ suppose the loss incurred is weighted squared error plus cost, i.e.,

$$L(\theta, \hat{\theta}_n^{(R)}) = A(\hat{\theta}_n^{(R)} - \theta)^2 + cn,$$

where A (> 0) is the known weight, and c (> 0) is the known cost per unit sample. Since $E_\theta(\hat{\theta}_n^{(R)} - \theta)^2$ behaves asymptotically like $\sigma_R^2 n^{-1}$, and since $A\sigma_R^2 n^{-1} + cn$ is minimized at $n = n^* = (A/c)^{1/2}\sigma_R$ (which for simplicity is assumed to be an integer), the "optimal" sample size if σ_R were known would be n^* and the corresponding risk would be $2cn^*$. However, since σ_R is unknown, there is no fixed sample size that minimizes the risk simultaneously for all F. Accordingly, for estimating θ, one proposes the following stopping rule:

$$N = N_c = \inf\{n \geqslant n_o, (A/c)^{1/2}(\hat{\sigma}_{n,R} + n^{-h})\},$$

where $n_0 \geqslant 2$, h is a number in $(0,1)$, and

$$\hat{\sigma}_{n,R}^2 = \frac{n^{-1}\sum_{i=1}^{n}\{a_n^+(i)\}^2}{\left[2S_{n,\alpha}n^{-1}\left(\hat{\theta}_{U,n}^{(R)} - \hat{\theta}_{L,n}^{(R)}\right)^{-1}\right]^2}$$

is a consistent estimator of σ_R^2 (cf. Sen [9]; Sen and Ghosh [11]).

The above sequential procedure satisfies the following properties (cf. Sen [10]):

(i) $N/n^* \to 1$ a.s. as $c \to 0$;

(ii) $E(N)/n^* \to 1$ as $c \to 0$;

(iii) $EL(\hat{\theta}_N, \theta)/(2cn^*) \to 1$ as $c \to 0$;

(iv) $\sqrt{n}(\hat{\theta}_N^{(R)} - \theta) \xrightarrow{L} N(0, \sigma_R^2)$ as $c \to 0$.

Property **(iii)** is the *first-order risk efficiency* of the proposed sequential procedure. It says that the sequential procedure is asymptotically risk equivalent to the corresponding optimal fixed sample procedure if σ_R^2 were known (see Starr [12]). The normal theory analog of the above point estimation procedure was proposed by Robbins [7].

BOUNDED LENGTH CONFIDENCE INTERVALS

Next, we consider the bounded length confidence interval problem based on R-estimators*. Chow and Robbins [1] considered estimation of the population mean based on the sample mean (*see also* FIXED WIDTH AND BOUNDED LENGTH CONFIDENCE INTERVALS). In our case, we want to propose a confidence interval $[\hat{\theta}_{L,n}, \hat{\theta}_{U,n}]$ for θ with preassigned confidence coefficient $1 - \alpha$ and $\delta_{n,R} = \hat{\theta}_{U,n} - \hat{\theta}_{L,n} \leqslant 2d$ for some preassigned $d > 0$. This problem cannot be solved by employing a fixed sample size (cf. Dantzig [2]), and some sequential procedure must be used. In this case, the stopping rule is defined by

$$N = N_d(R) = \min\{n \geqslant 2 : \delta_n^{(R)} \leqslant 2d\}.$$

The proposed sequential confidence interval for θ is then $[\hat{\theta}_{L,n}^{(R)}, \hat{\theta}_{U,n}^{(R)}]$.

The above problem was first considered by Geertsema [3] for sign and signed rank statistics, and in a more general setup by Sen and Ghosh [11]. The stopping time N satisfies the following properties:

(i′) N is a nonincreasing function of d;

(ii′) $N < \infty$ a.s.;

(iii′) $E(N) < \infty$;

(iv′) $\lim_{d\to 0} N = \infty$ a.s.; $\lim_{d \to 0} E(N) = \infty$;

(v′) $N/n(d) \to 1$ a.s. as $d \to 0$;

(vi′) $\lim_{d\to 0} E(N)/n(d) = 1$;

(vii′) $\lim_{d\to 0} P_\theta(\hat{\theta}_{L,N}^{(R)} \leqslant \theta \leqslant \hat{\theta}_{U,n}^{(R)}) = 1 - \alpha$.

In the above

$$n(d) = A_\phi^2 \tau_{\alpha/2}^2 d^{-2} \left(\int_0^1 \phi(u)\psi(u)\, du\right)^{-2},$$

$\tau_{\alpha/2}$ being the upper $100\alpha/2\%$ point of the $N(0,1)$ distribution.

Note that **(vii′)** ensures that the coverage probability of θ by the proposed sequential confidence interval is asymptotically equal to the preassigned $1 - \alpha$ (*asymptotic consistency*), while **(vi′)** ensures the asymptotic equivalence of the average sample size of the sequential procedure with the sample size one would need if σ_R were known, and a fixed sample procedure were used (*asymptotic efficiency*).

Ghosh and Sen [4] have also obtained a bounded length confidence interval for the regression coefficient in a simple linear regression model based on R-estimators. Consider the model $X_i = \beta_0 + \beta c_i + e_i$, $i \geqslant 1$, where β_0 and β are unknown parameters, the c_i's are known regression constants, and the e_i's are i.i.d. variables with an unknown continuous distribution function F defined on the real line. Our goal is to provide a confidence interval for β with confidence coefficient $1 - \alpha$ $(0 < \alpha < 1)$, and the length bounded by $2d$ (> 0). With this end, for every $n \geqslant 1$, consider a simple linear rank statistic

$$L_n = L_n(\mathbf{X}_n) = \sum_{i=1}^{n} (c_i - \bar{c}_n) a_n(R_{ni}),$$

$$\bar{c}_n = n^{-1} \sum_{i=1}^{n} c_i.$$

In the above R_{ni} is the rank of X_i among $X_1, \ldots, X_n$ for $i = 1, \ldots, n$, and the scores $a_n(i) = E\phi(U_{ni})$ or $\phi(i/(n+1))$ (ϕ nondecreasing), where $U_{n1} \leqslant \cdots \leqslant U_{nn}$ are once again the order statistics in a random sample of size n from the uniform $[0, 1]$ distribution.

We denote by $L_n(b)$ the statistic resulting after replacing $\mathbf{X}_n$ by $\mathbf{X}_n - b\mathbf{c}_n$, b real, $\mathbf{c}_n = (c_1, \dots, c_n)$ in $L_n(\mathbf{X}_n)$. Note that $L_n(b)$ is decreasing in b, and $L_n(\beta)$ has the same distribution as of $L_n(0)$ when $H_0: \beta = 0$ holds. Also, under H_0, $EL_n(0) = 0$ and $V(L_n(0)) = C_n^2 A_n^2$, where

$$C_n^2 = \sum_{i=1}^{n} (c_i - \bar{c}_n)^2,$$

$$A_n^2 = (n-1)^{-1}\left\{\sum_{i=1}^{n} a_n^2(i) - \left(n^{-1}\sum_{i=1}^{n} a_n(i)\right)^2\right\}.$$

Further, under H_0, $L_n(0)$ has a distribution independent of F. As such, it is possible to identify two values $L_{n,\alpha}^{(1)}$ and $L_{n,\alpha}^{(2)}$, such that for every $\alpha \in (0,1)$ and $n \geqslant 2$,

$$P_{\beta=0}\left(L_{n,\alpha}^{(1)} \leqslant L_n(0) \leqslant L_{n,\alpha}^{(2)}\right) = 1 - \alpha_n \geqslant 1 - \alpha,$$

where α_n does not depend on F, $\alpha_n \to \alpha$ as $n \to \infty$, and $C_n^{-1}A_n^{-1}L_{n,\alpha}^{(i)} \to (-1)^i \tau_{\alpha/2}$ as $n \to \infty$. Thus, if we define

$$\hat{\beta}_{L,n}^{(R)} = \sup\{b : L_n(b) > L_{n,\alpha}^{(2)}\},$$

$$\hat{\beta}_{U,n}^{(R)} = \inf\{b : L_n(b) < L_{n,\alpha}^{(1)}\},$$

then $[\hat{\beta}_{L,n}^{(R)}, \hat{\beta}_{U,n}^{(R)}]$ provides a distribution-free confidence interval for β with confidence coefficient $1 - \alpha_n$. However, we want to find a sample size n for which the length $\delta_n^{(R)} = \hat{\beta}_{U,n}^{(R)} - \hat{\beta}_{L,n}^{(R)}$ of the confidence interval is less than or equal to $2d$ as well. Once again, no fixed sample size achieves the objective, and the proposed stopping rule is $N = \min\{n \geqslant 2 : \delta_n^{(R)} \leqslant 2d\}$.

Define now

$$Q(x) = (n + 1 - x)C_n^2 + (x - n)C_{n+1}^2,$$

$$n \leqslant x \leqslant n + 1,\ n \geqslant 0,\ C_0^2 = C_1^2 = 0.$$

Assume that $Q(x)$ is increasing in x and $\lim_{n\to\infty} Q(na_n)/Q(n) = S(a)$ exists for all a_n satisfying $\lim_{n\to\infty} a_n = a$. It is also assumed that $S(a)$ increases in a with $S(1) = 1$. Then N satisfies all the properties similar to **(i′)–(vii′)** associated with the bounded length confidence interval procedure in the one-sample problem with the following change in the definition of n_d. Here

$$n_d = Q^{-1}\left(A_\phi^2 \tau_{\alpha/2}^2 d^{-2}\left(\int_0^1 \phi(u)\psi(u)\, du\right)^{-2}\right),$$

$$Q^{-1}(y) = \inf\{x : Q(x) \geqslant y\}.$$

Gleser [5] considered estimation of β based on the least-squares estimator.

References

[1] Chow, Y. S. and Robbins, H. (1965). *Ann. Math. Statist.*, **36**, 457–462.

[2] Dantzig, G. B. (1940). *Ann. Math. Statist.*, **11**, 186–192.

[3] Geertsema, J. C. (1970). *Ann. Math. Statist.*, **41**, 1016–1026.

[4] Ghosh, M. and Sen, P. K. (1972). *Sankhyā A*, **34**, 33–52.

[5] Gleser, L. J. (1965). *Ann. Math. Statist.*, **36**, 463–467.

[6] Hodges, J. L., Jr. and Lehmann, E. L. (1963). *Ann. Math. Statist.*, **34**, 598–611.

[7] Robbins, H. (1959). *Probability and Statistics*, (H. Cramér Volume). Almqvist and Wiksells, Uppsala, Sweden, pp. 235–245.

[8] Sen, P. K. (1963). *Biometrics*, **19**, 532–552.

[9] Sen, P. K. (1966). *Ann. Math. Statist.*, **37**, 1759–1770.

[10] Sen, P. K. (1980). *Sankhyā A*, **42**, 201–220.

[11] Sen, P. K. and Ghosh, M. (1971). *Ann. Math. Statist.*, **42**, 189–203.

[12] Starr, N. (1966). *Ann. Math. Statist.*, **37**, 1173–1185.

(FIXED-WIDTH AND BOUNDED-LENGTH CONFIDENCE INTERVALS
RANKING PROCEDURES
SEQUENTIAL ESTIMATION)

MALAY GHOSH

SEQUENTIAL SAMPLING

A sampling method in which the units are drawn one by one in a sequence without prior fixing of the total number of observations and the results of drawing at any stage decide whether to terminate sampling or not, is termed sequential sampling. An essential feature of a sequential procedure, as distinguished from fixed sample size procedures, is that because the number of observations required for a sequential method depends on the outcome of the observations it is, therefore, a random variate. A typical example of sequential sampling is inverse sampling* for estimating a proportion in the case of a rare attribute (Haldane [11]). In this case sampling of units is continued until a predetermined number of units having the specified attribute are obtained. Early mention of the idea of sequential sampling was indirectly made by Sukhatme [21], while discussing the problem of optimum sampling under the conditions where stratum sizes were known but the standard deviations were unknown. The word "sequential" was introduced by Wald [23] to describe a procedure for testing hypotheses in which there was no prior fixation of the number of observations.

Sequential sampling has been used both for testing of hypotheses and estimation of parameters. For testing of hypotheses, sequential analysis*, as a well-defined and well-established sector of statistical theory, has developed from the work of Wald*; the sequential probability ratio test (SPRT) has proved of historic importance in the Neyman–Pearson* approach.

The theory of sequential estimation* does not differ greatly in principle from fixed sample size estimation, but the theoretical details are somewhat complex. In sequential methods, an estimator is derived that minimizes a loss function subject to one or more restrictions. Inverse sampling is one of the simplest methods used for sequential estimation. The sampling process is inverted by fixing the number of units in the sample with some specified characteristic(s) rather than fixing the sample size. In other words, the procedure is to continue sampling until a specified number of units with given characteristics are included in the sample. Haldane [11], Tweedie [22], and others introduced this method for the estimation of binomial proportions. Chapman [7] has outlined a sequential procedure for the estimation of the population size with a preassigned degree of precision, which is not possible with any fixed size procedure.

A sampling procedure for inverse sampling with unequal probabilities was considered by Pathak [17]. In this procedure a population containing N units is sampled with replacement with probabilities P_i ($i = 1, \dots, N$), $\Sigma_{i=1}^{N} P_i = 1$. Selection is terminated at the $(r+1)$st draw, when the sample first contains $(n+1)$ different units. The last unit is rejected, and the recorded sample consists of the n different units selected. If an observed sample of r units is $s = (u_1, u_2, \dots, u_r)$, where $u_1, u_2, \dots, u_r$ are, respectively, the 1st, 2nd, ..., rth sample units, then given r, $u_1, u_2, \dots, u_r$ are interchangeable in the sense that their joint distribution is invariant under any permutation of $u_1, u_2, \dots, u_r$. Another instance is found in the context of unequal probability sampling with inclusion probabilities proportional to size (Hanurav [12]).

One can specify stopping rules and examine which stopping rule will satisfy the requirements regarding cost and variance constraints. Once an estimator is decided upon, its sampling variance can be determined, and a sample estimate of this can be associated with each point estimate as a measure of loss. Thus, both the estimate and its loss will depend upon the stopping rules. Anscombe [1] has examined sequential estimation of the mean with these aspects and showed that in large samples it will indeed yield estimates with the desired degree of precision. Barnard et al. [6] have criticised this procedure and argued for direct study of the likelihood as a procedure for estimation. They say that, if n observations with a sample mean $\bar{X}$ are known to be drawn from a normal distribution with variance σ^2, then the likelihood function of the unknown mean θ is proportional to $\exp[-n(\bar{X}-\theta)^2/(2\sigma^2)]$, which is a function of θ, conditional on $\bar{X}$ and n, and independent of the stopping rules. From this point of view, no special methods of estimation are required for sequential sampling. Of course, methods of selecting suitable stopping rules are needed, but these in no way affect the estimation method once the sample is obtained. Thus, one can conclude that confidence intervals* depend upon the stopping rules, whereas interval estimates based on a likelihood approach do not. These points are argued at length in Anscombe [2] and Armitage [5]. Consequently, it is difficult to fix the sample size and the sampling rate, if their determining factors are subject to considerable uncertainties. In many situations some of the determining factors, such as unit variances, unit cost, and population size, may not be known. Sometimes, vagueness about the nonresponse rates and noncoverage introduce further uncertainties. Kish [16] has suggested that flexibility in sampling rates can be attained with a well-designed supplemental sample. The initial sampling rate should be a sensible minimum, based on reasonable extreme expectations, e.g., of lowest estimates of the unit variance and highest estimates of unit cost, population size, and response and coverage rates. Improved estimates of these factors are computed from the results of initial sample. If the recomputed desired sampling rate considerably exceeds the initial rate, the needed supplement may be obtained. However, the method of supplementing should be carefully planned to facilitate the field work.

New approaches to sequential sampling have been attempted by Stenger [20] and Singh [19] using procedures similar to Godambe [10]. Suboptimal sequential procedures for medical trials are evaluated by Petkau [18] and one of these, a Wald-type sequential procedure, is found to be surprisingly efficient. Armitage has published an account of some considerations involved in sequential medical trials (*see* CLINICAL TRIALS) and describes a number of applications in detail. Recently, Chaudhary and Singh [8] have given an interesting approach for the estimation of population size by proposing a technique in sample form.

Sequential sampling is in common use in quality control*, especially in acceptance sampling*. It has been widely used in lot-quality inspection from its very beginning and has not been replaced by any other technique so far. It has applications in screening the efficacy of drugs. The name "up-and-down" given to a sequantial method of approach to bioassays* and quantal response* is a further important offshoot of sequential method applications. Other applications and illustrations are also available in the book by Wetherill; these include estimation of points on regression function curves. In scientific research, there always exists the possibility that the investigator may change his mind in the light of further information gathered in the investigation itself or more generally from other sources as well. The sequential approach is an appropriate technique for minimizing the expected loss as the results become more precise.

The reviews of works up to 1960 on sequential sampling and analysis done by Jackson [14] and Johnson [15] are useful for earlier developments. Wald [24] remains an important source book on sequential analysis containing a good exposition of SPRT. Wetherill [26] and Ghosh [9] have written useful books giving critical discussion on sequential methods. Armitage [5] has described illustrations and applications in medical research.

References

[1] Anscombe, F. J. (1953). *J. R. Statist. Soc. B*, **15**, 1–29. (Contains eight pages of discussion.)

[2] Anscombe, F. J. (1963). *J. Amer. Statist. Ass.*, **56**, 365–383.

[3] Armitage, P. (1950). *J. R. Statist. Soc. B*, **12**, 137–144.

[4] Armitage, P. (1957). *Biometrika*, **44**, 9–26.

[5] Armitage, P. (1975). *Sequential Medical Trials*, 2nd ed. Wiley, New York.

[6] Barnard, G. A., Jenkins, G. M., and Winsten, C. B. (1962). *J. R. Statist. Soc. A*, **125**, 321–372.

[7] Chapman, D. G. (1952). *Biometrics*, **8**, 286–306.

[8] Chaudhary, F. S. and Singh, D. (1984). *Indian J. Agric. Statist.*, **36**, 1–9.

[9] Ghosh, B. K. (1971). *Sequential Tests of Statistical Hypothesis*. Addison-Wesley, Reading, MA.

[10] Godambe, V. P. (1955). *J. R. Statist. Soc. B*, **17**, 269–278.

[11] Haldane, J. B. S. (1945). *Nature*, **155**, 49–50.

[12] Hanurav, T. V. (1962). *Sankhyā A*, **24**, 421–428.

[13] Hill, I. D. (1962). DEF-131. *J. R. Statist. Soc. A*, **125**, 31–87.

[14] Jackson, J. E. (1960). *J. Amer. Statist. Ass.*, **55**, 561–580.

[15] Johnson, N. L. (1961). *J. R. Statist. Soc. A*, **124**, 372–411.

[16] Kish, L. (1965). *Survey Sampling*. Wiley, New York, pp. 277–278.

[17] Pathak, P. K. (1964). *Biometrika*, **51**, 185–193.

[18] Petkau, A. J. (1978). *J. Amer. Statist. Ass.*, **71**, 328–338.

[19] Singh, F. (1977). A Sequential Approach to Sample Surveys. Thesis, Meerut, India.

[20] Stenger, H. (1977). *Sankhyā C*, **39**, 10–20.

[21] Sukhatme, P. V. (1935). *Suppl. J. R. Statist. Soc.*, **2**, 253–268.

[22] Tweedie, M. C. K. (1945). *Nature*, **155**, 453.

[23] Wald, A. (1945). *J. Amer. Statist. Ass.*, **40**, 277–306.

[24] Wald, A. (1947). *Sequential Analysis*, Wiley, New York.

[25] Wald, A. and Wolfowitz, J. (1948). *Ann. Math. Statist.*, **19**, 326–339.

[26] Wetherill, G. B. (1975). *Sequential Methods in Statistics*, 2nd ed. Wiley, New York.

(INVERSE SAMPLING
SAMPLING PLANS
SEQUENTIAL ANALYSIS
SEQUENTIAL ESTIMATION)

D. SINGH

SEQUENTIAL *t*-TEST *See* SEQUENTIAL ANALYSIS

SEQUENTIAL T^2 TEST

The sequential T^2 test is an extension of the sequential t-test to multivariate data developed by Jackson and Bradley [2, 3]. It is used to test the hypothesis (H_0) that the expected value $\boldsymbol{\xi}$ of a p-variate multinormal distribution* equals a specified value $\boldsymbol{\xi}_0$, based on a random sample of size n $\mathbf{X}_1, \mathbf{X}_2, \ldots, \mathbf{X}_n$.

The test is in the form of a sequential probability ratio test with likelihood ratio

$$N_n(T_n) = e^{-n\lambda^2/2} \times {}_1F_1\left(\tfrac{1}{2}n; \tfrac{1}{2}p; \tfrac{1}{2}n\lambda^2 T_n^2 (n-1+T_n^2)^{-1}\right),$$

where $\lambda^2 = (\boldsymbol{\xi}_1 - \boldsymbol{\xi}_0)\boldsymbol{\Sigma}^{-1}(\boldsymbol{\xi}_1 - \boldsymbol{\xi}_0)$ (with $\boldsymbol{\Sigma}$ denoting the variance–covariance matrix of $\mathbf{X}$) corresponds to the specified alternative hypothesis ($H_1 : \boldsymbol{\xi} = \boldsymbol{\xi}_1$),

$${}_1F_1(a; b; x) = \sum_{j=0}^{\infty} \frac{a^{[j]}}{b^{[j]}} \cdot \frac{x^i}{j},$$

$$\text{with } a^{[j]} = a(a+1)\cdots(a+j-1),$$

is the confluent hypergeometric function*, and

$$T_n^2 = n(\overline{\mathbf{X}} - \boldsymbol{\xi}_0)'\mathbf{S}^{-1}(\overline{\mathbf{X}} - \boldsymbol{\xi}_0),$$

with $\overline{\mathbf{X}}$ the sample arithmetic mean vector and $\mathbf{S}$ the sample variance–covariance matrix.

Sampling continues as long as

$$\frac{\alpha_1}{1-\alpha_0} < L_n(T_n) < \frac{1-\alpha_1}{\alpha_0},$$

where α_j is the nominal probability of "rejecting H_j" when H_j is valid ($j = 1, 2$):

$$H_0 \text{ is rejected if } \quad L_n(T_n) \geqslant \frac{1-\alpha_1}{\alpha_0};$$

$$H_1 \text{ is rejected if } \quad L_n(T_n) \leqslant \frac{\alpha_1}{1-\alpha_0}.$$

Tables of upper and lower bounds of T_n^2 for the continuation region were computed by Freund and Jackson [1] and have been reproduced by Kres [4]. The tables cover values of n from 1 to 75, with $p = 2(1)9$ and $\lambda^2 = 0.5(0.5)3(1)6, 10$.

If the variance–covariance matrix $\boldsymbol{\Sigma}$ is known, the sequential chi-squared test* can be used.

References

[1] Freund, R. J. and Jackson, J. E. (1960). Tables to Facilitate Multivariate Sequential Testing of Means. *Tech. Rep. No. 12*, Dept. of Statistics, Virginia Polytechnical Institute, Blacksburg, VA.

[2] Jackson, J. E. and Bradley, R. A. (1959). Multivariate Sequential Procedure for Testing Means. *Tech. Rep. No. 10*, Dept. of Statistics, Virginia Polytechnical Institute, Blacksburg, VA.

[3] Jackson, J. E. and Bradley, R. A. (1961). *Ann. Math. Statist.*, **32**, 1063–1077.

[4] Kres, H. (1983). *Statistical Tables for Multivariate Analysis*. Springer-Verlag, New York.

(HOTELLING'S T^2
MULTIVARIATE ANALYSIS
SEQUENTIAL ANALYSIS
SEQUENTIAL CHI-SQUARED TEST)

SERIAL CORRELATION

A time series* is a sequence of observations, ordered in time (or in one-dimensional space—for instance, the thickness of textile yarn at points along its length). Usually, one only considers a set of values obtained at discrete equispaced instants (or distances), the number of such points being called the "length" of the series. Thus the noon temperatures at Greenwich, from 1 November 1975 to 31 October 1985 inclusive, would yield a time series of length 3653.

Most statistical methodology is concerned with independent sets of observations. Lack of independence is typically considered highly undesirable, and one of the objects of good experimentation is to eliminate dependence. However, with time-series analysis, we are concerned with data that develop through time, and where each observation may depend, to a degree, on earlier observations. It is, in fact, this dependence that is of interest and importance.

Time series may thus possess a "memory" of the past, in the sense that the later values of a series reflect, to some extent, the previous ones. Conversely, this memory implies that the series also possesses a degree of "foresight," since what has already happened will be expected to partially affect the future outcomes. For instance, if the appearance of a time series is very smooth, then a high current value, say, will imply that the next one is likely to be high also.

What the analyst needs to do is to squeeze out as much information as possible from the observed associations between observations. The most important sample statistics used for this are the calculated "autocorrelations" between values at various distances of separation, which measure the corresponding associations. Thus, the association between adjacent values is measured by the first autocorrelation; that between values, separated by a single observation, by the second autocorrelation, and so on.

This sample (auto)correlation between observations on a time (or space) series is called *serial correlation*. For the concept to be meaningful, it is assumed that the series is observed at equidistant points in time (or space) and is a realisation from a second-order stationary process (weak stationarity), defined below. Then, given a time series $\{z_1, \ldots, z_n\}$ of integral length $n \geqslant 2$ and any integer k such that $0 \leqslant k < n$, we have the lag-k serial correlation as the sampled (product-moment-type) correlation between z_i and z_{i-k} for $i = 1 + k, \ldots, n$. Thus, the lag-k serial correlation is based on the $n - k$ pairs of observations $(z_{1+k}, z_1), \ldots, (z_n, z_{n-k})$; that is, on all those pairs that are separated by exactly $k - 1$ consecutive observations from the series.

The lead-k serial correlation is similarly defined in terms of z_i and z_{i+k}, for $i = 1, \ldots, n - k$. Since, clearly, exactly the same pairs are involved, it is immaterial (for univariate series) whether we talk of lags or leads (negative lags). So we can refer to both measures as kth (order) serial correlation, where, conventionally, k is usually considered to be nonnegative and is the fixed "distance" between the elements of each pair being considered in the corresponding sample quantification. Although the idea is framed in terms of time series observed at discrete (equispaced) instants, an analogous discussion can be constructed for series that follow a continuous trace.

In detail, there are several alternative definitions for serial correlation. Before we look at some of the more common ones, we should rigorously define (and briefly discuss) the population or parent measure, the *autocorrelation* of the underlying (stationary) time process, $\{Z_i : i = -\infty, \dots, \infty\}$ say, the infinitely long stochastic process considered as generating the observed series over its short (finite) time span.

First, however, we need to define the kth order *autocovariance* for such a process. This is defined by $\gamma_k = E[(Z_i - \mu)(Z_{i-k} - \mu)]$, for $k = \dots, -1, 0, 1, \dots$, where E denotes the expectation operator and μ is the process mean $E[Z_i]$ that is constant for a stationary process; clearly, $\gamma_{-k} = \gamma_k$. [A process $\{Z_i\}$ is stationary if all its joint probability density functions, $f(Z_{t+i_1}, \dots, Z_{t+i_p})$ for any choice of $i_1, \dots, i_p$ and p, are all defined and independent of t. For a Gaussian process*, having the property that all such functions are multivariate normal, it is sufficient just to ensure the conditions for $p = 1$ and 2 —weak stationarity.] The kth order autocorrelation ρ_k for the process is then defined as $\rho_k = \gamma_k/\gamma_0$, for $k = \dots, -1, 0, 1, \dots$, with $\rho_{-k} = \rho_k$.

It is easily seen that the sequence of Laurent or Toeplitz* autocorrelation matrices $\{\mathbf{P}_k : k = 0, 1, \dots\}$, namely,

$$1, \begin{pmatrix} 1 & \rho_1 \\ \rho_1 & 1 \end{pmatrix}, \dots,$$

$$\begin{pmatrix} 1 & \rho_1 & \rho_2 & \cdots & \rho_{k-1} \\ \rho_1 & 1 & \rho_1 & & \rho_{k-2} \\ \vdots & & & & \vdots \\ \rho_{k-1} & \rho_{k-2} & \rho_{k-3} & \cdots & 1 \end{pmatrix}, \dots,$$

are all positive definitive. [Consider $Y = \lambda_1 Z_1 + \cdots + \lambda_k Z_k$ ($= \boldsymbol{\lambda}^T \mathbf{Z}$, say) with any $\lambda_1, \dots, \lambda_k$ that are not all zero. Then var(Y), which is necessarily positive for any properly stochastic process* $\{Z_1\}$, is equal to $\boldsymbol{\lambda}^T \mathbf{P}_k \boldsymbol{\lambda}$. So $\boldsymbol{\lambda}^T \mathbf{P}_k \boldsymbol{\lambda}$ is positive for all nontrivial $\boldsymbol{\lambda}$, which implies that $\mathbf{P}_k$ is positive definitive.]

Thus there are many constraints on what values a set of autocorrelations may take, because always they must satisfy the determinantal inequalities, $\{|\mathbf{P}_k| > 0 : k = 1, \dots\}$. For instance, for $k = 1$ and 2, these reduce to $-1 < \rho_1 < 1$ and $\rho_2 > 2\rho_1^2 - 1$, respectively, which provide stationarity conditions.

Associated with the autocorrelations are the partial autocorrelations, defined by $\pi_k = |\mathbf{P}_k^*|/|\mathbf{P}_k|$, where $\mathbf{P}_k^*$ is $\mathbf{P}_k$ with every (r, k)th element replaced by ρ_r. Alternatively, one can consider π_k to be the conditional correlation between Z_i and Z_{i-k}, given all the intermediate Z's, namely, $Z_{i-1}, \dots, Z_{i-k+1}$.

We now return to the finite sample. One measure of serial correlation is provided by the product-moment correlation between successive observations of the series:

$$r_{k,1}^{(n)} = \frac{\Sigma(z_i - \bar{z}_n)(z_{i-k} - \bar{z}_{n-k})}{\left[\Sigma(z_i - \bar{z}_n)^2 \Sigma(z_{i-k} - \bar{z}_{n-k})^2\right]^{1/2}}, \tag{1}$$

where the summations run from $i = 1 + k$ to n and $\bar{z}_{n-j} = (z_{1+k-j} + \cdots + z_{n-j})/(n-k)$, for $j = 0$ or k. Here, and elsewhere, we employ the superscript n to emphasise the length of the series with which we are dealing.

For both computational and theoretical convenience, the means of the last $n - k$ and first $n - k$ observations, $\bar{z}_n$ and $\bar{z}_{n-k}$, respectively, are usually replaced in (1) by the mean for the whole series, $\bar{z} = (z_1 + \cdots + z_n)/n$; we may also replace the corresponding separate variances (mean-square deviations about the mean) in the denominator of (1) by the variance for the whole series. This then gives

$$r_{k,2}^{(n)} = \frac{n \sum_{i=1+k}^{n} (z_i - \bar{z})(z_{i-k} - \bar{z})}{(n-k) \sum_{i=1}^{n} (z_i - \bar{z})^2}. \tag{2}$$

An alternative to (2) is

$$r_{k,3}^{(n)} = \frac{\sum_{i=1+k}^{n} (z_i - \bar{z})(z_{i-k} - \bar{z})}{\sum_{i=1}^{n} (z_i - \bar{z})^2}, \tag{3}$$

which has the advantages of (a) (possibly) tending to have a smaller mean-square error than (2) and (b) not allowing values greater than 1 in magnitude, as does (2). Definition (3) is, in fact, the one now almost always used in data analysis.

Again for theoretical and computational convenience, the series is sometimes extended artificially by terms $z_{n+1}, \ldots, z_{2n-1}$, such that $z_{n+i} = z_i$ for $i = 1, \ldots, n-1$. The product moment in the numerator of (3) can then always be summed over n terms, giving the "circular" serial correlation

$$r_{k,4}^{(n)} = \frac{\sum_{i=1}^{n} (z_i - \bar{z})(z_{i-k} - \bar{z})}{\sum_{i=1}^{n} (z_i - \bar{z})^2}. \quad (4)$$

Evidently, unless k is small compared to n, this artifice risks giving misleading results. But for $k \ll n$, the device can provide useful simplification in distribution theory, and was used frequently in classical work.

We now discuss why (3) is to be preferred to (1). This is because, although (1) gives a very reasonable estimate when considered in isolation from the estimated autocorrelations at other lags, it does not yield satisfactory results when a set of estimates is required for the first m autocorrelations $\{\rho_1, \ldots, \rho_m : m < n\}$, which is the normal requirement in practice. The positive definiteness of the sampled Toeplitz matrices* $\{\hat{\mathbf{P}}_k : k = 2, \ldots, m\}$ is no longer assured; this violation of the stationarity constraints can lead to acute problems—as, for instance, when computing the sample spectral density (the Fourier cosine transform of the serial correlations), negative values can be obtained. In what follows, all $r_k^{(n)}$ refer to definition (3), and we drop the suffix 3.

The sequence of serial correlations $\{r_0^{(n)}(\equiv 1), r_1^{(n)}, \ldots, r_{n-1}^{(n)}\}$ is termed the correlogram* or *serial correlation function* (SCF) and it usefully reflects all the internal dependencies of the series. Replacing the ρ's by $r^{(n)}$'s in the definition of π_k gives the sampled partial correlation, or partial serial correlation, p_k. We will term the partial correlogram $\{p_0(\equiv 1), p_1(\equiv r_1^{(n)}), p_2, \ldots, p_{n-1}\}$ the *partial serial correlation function* (PSCF). The parent sequences, $\{\rho_0, \rho_1, \ldots\}$ and $\{\pi_0, \pi_1, \ldots\}$, are referred to as the *autocorrelation* and *partial autocorrelation functions* (ACF and PACF, respectively).

Inspection of the serial correlations and partial serials often allows one to identify plausible explanatory models for the series under consideration, provided that there exists a stationary linear process (of form $Z_i = \sum_{j=0}^{\infty} \psi_j A_{i-j}$, with $\sum_0^{\infty} |\psi_j|$ finite and all the A's independently and identically distributed with finite mean and variance) that gives a reasonably close approximate representation for the underlying mechanism that generated the series. This "identification" is based on the matching of observed serials and partials with the corresponding theoretical patterns for the ACF and PACF, for members of a wide-ranging family of parent linear processes. Once a plausible model has been identified, its parameters can be efficiently estimated (frequently using the $r_k^{(n)}$ to obtain starting values), and then the resulting fit is critically assessed. Any significant shortcomings noted at this "verification" stage are rectified by repeating the cycle of identification, estimation, and verification (using an improved identification resulting from consideration of the earlier fitted model's observed inadequacies), and so on, until a satisfactory final fitted model obtains. This is now a fairly well-developed methodology (see, for instance, the BOX–JENKINS MODEL).

Certain nonstationary situations can also be catered for by first making an appropriate transformation to induce stationarity. Thus, given a realisation from a random walk* process $\{W_i\}$ (defined by $W_i = W_{i-1} + A_i$, with $A_i \sim IN(0, \sigma^2)$ say), we can produce a stationary series $\{z_i\}$ from the $\{w_i\}$ by defining $\{z_i\}$ as the first differences of $\{w_i\}$, i.e., $z_i \equiv w_i - w_{i-1}$. Then $\{z_i\}$ is generated by the process $\{Z_i = A_i\}$, which is clearly stationary. In such a case, the serials and partials of $\{w_i\}$ would indicate nothing except that we were probably dealing with a nonstationarity that could be removed by differencing. Thus, we take differences and look at the SCF and PSCF for the derived $\{z_i\}$. These should indicate no significant correlation structure for $\{Z_i\}$; so, reversing the differencing, we would deduce that $\{W_i\}$ following a random walk was a plausible model.

Unfortunately, time series (even when they contain several hundreds of observations) are frequently anything but effectively "long" (by virtue of their internal serial dependence). Thus, for series encountered in practice (with n often less than 50, in Economics say), asymptotic results risk being very misleading and can provide very poor approximations to the finite sample behaviour, for what are effectively generally very short realisations.

Now, for almost any stationary process that would arise in practice (strictly, an "ergodic" one), ρ_k can be considered as the limit in probability as $n \to \infty$ of $r_k^{(n)}$, the "kth serial correlation for an infinitely long series realisation." Thus, what the standard identification procedure does is to try to compare small sample $r_k^{(n)}$ with an asymptotic parent quantity; and this can be a frustrating task. It is, therefore, suggested that one consider, instead, a comparison of $r_k^{(n)}$ with some parent quantity for *sample realisations of length n* generated by the process. For instance, rather than match up $r_k^{(n)}$ with the ρ_k, for some appropriate process, one could consider matching it with $E[r_k^{(n)}]$ from some model. This suggestion does not provide much practical gain for thoroughly stable series (stationary ones that are not too different from a purely random sequence; that is, those with only weak serial dependence structure). This is because (a) the asymptotic results are not so misleading then (if you like, a given n is effectively less short for a stable series) and (b) any small advantage is reduced by the large sampling errors of the individual $r_k^{(n)}$. (It is not much help to a marksman offsetting correctly for wind, if he waggles his rifle when firing.)

When we consider series that are approaching nonstationarity (or, indeed, certain nonstationary cases), the disparity between ρ_k and $E[r_k^{(n)}]$ grows, while the sampling variability of $r_k^{(n)}$ decreases. Then, trying to compare $r_k^{(n)}$ with ρ_k becomes hopeless, whereas comparison with $E[r_k^{(n)}]$ can be highly successful. For instance, frequently it is impossible for an $r_k^{(n)}$ to attain the corresponding ρ_k value of the underlying process, or even come close to doing so. This becomes even more true when one considers a whole set of $r_k^{(n)}$. (As an example, a simple process gives $\rho_k \equiv 0.999^k$, with $\rho_{25} = 0.975$ and $\sum_{k=1}^{99}\rho_k = 94.2$. However, for any series of length 100, necessarily $r_{25}^{(100)} < 0.866$ and $\sum_{k=1}^{99} r_k^{(100)} = -\frac{1}{2}$.)

Satisfactory study of the sampling distribution and moments for the serial correlations of linear processes is therefore an important goal. To date, little progress has been made at producing exact formulae by analytic means. What can be done, however, is to (a) produce approximate formulae or (b) get results as close as one likes to being exact, by either numerical integration* or by sufficiently replicated simulation. With the speed of modern computers, both the approaches of (b) are now viable, whilst (a) offers less costly insight.

Reference

[1] Bartlett, M. S. (1946). *J. R. Statist. Soc. B*, **8**, 27–41; Correction (1948), **10**, 1.

Bibliography

Anderson, O. D. (1975). *Time Series Analysis and Forecasting: The Box–Jenkins Approach*, Butterworths, London, England. (An introductory treatment with many worked examples of modelling through serial correlation.)

Anderson, O. D. (1976). *Math. Sci.*, **1**, 27–41. (A simple discussion explaining the purposes and methods of time-series analysis, with an emphasis on ideas.)

Anderson, O. D. (1976). *Statist. News*, **32**, 14–20. (A formula-free introduction to time-series forecasting for nonspecialists.)

Anderson, O. D. (1977). *The Statistician*, **25**, 285–303. (An updated view of Box–Jenkins at an elementary level.)

Anderson, O. D., ed. (1979). *Forecasting*. North-Holland, Amsterdam, the Netherlands. (G. M. Jenkins authoritatively discusses practical experience of advanced modelling and forecasting through serial correlation, pp. 43–166.)

Anderson, O. D. (1981). In *Time Series Analysis*. O. D. Anderson, ed. North-Holland, Amsterdam, The Netherlands, pp. 3–26. (Discusses the covariance structure of serial correlations.)

Box, G. E. P. and Jenkins, G. M. (1970). *Time Series Analysis: Forecasting and Control*, 1st ed. Holden-Day, San Francisco, CA. (2nd ed., 1976). (The main reference for modelling through serial correlation.)

Granger, C. W. J. and Newbold, P. (1977). *Forecasting Economic Time Series*. Academic, New York. [A well-respected alternative to Box and Jenkins (1970).]

Jenkins, G. M. and Watts, D. G. (1968). *Spectral Analysis and its Applications*. Holden-Day, San Francisco, CA. (The standard reference for frequency domain; discusses properties of serial correlation estimators.)

Kendall, M. G., Stuart, A, and Ord, J. K. (1983). *The Advanced Theory of Statistics*, 4th ed.,Vol. 3. Griffin, London, England. (The authority on methodology, with very useful discussion of serial correlation.)

Acknowledgments

I would like to thank the following nine people who made suggestions for the improvement of this article: Neville Davies, Jan de Gooijer, Jim Durbin, Norman Johnson, Rául Mentz, Roch Roy, Torsten Söderström, Petre Stoica, and (especially) Morris Walker.

Finally, no discussion on serial correlation would be complete without reference to the classical and seminal paper by Bartlett [1], which treated the asymptotic properties of serial correlations.

(BOX–JENKINS MODEL
CORRELOGRAM
SERIAL DEPENDENCE
TIME SERIES)

O. D. ANDERSON

SERIAL COVARIANCE *See* SERIAL DEPENDENCE

SERIAL DEPENDENCE

For a time-series $\{z_1, \ldots, z_n\}$ of length n, that is not completely random, there will be various dependencies between the ordered observations, by which the past history of the series is, to some extent, reflected in the later observations. (Or, indeed, from which dependencies, partial information can be gleaned on earlier values of the series, given the more recent ones.)

This "serial dependence" between the observations of a series is measured in two related ways—by *serial covariance* and *serial correlation*—that can be considered sample quantities for the finite series realisation, derived from the parent measures of *autocovariance* and *autocorrelation* (*see* SERIAL CORRELATION), the kth order sample and population quantities being denoted by $r_k^{(n)}$ and ρ_k, respectively. Serial covariance and autocovariance of order k, will be written as $c_k^{(n)}$ and γ_k, respectively. We will only entertain one definition for $c_k^{(n)}$, namely:

$$c_k^{(n)} = (1/n) \sum_{i=1+k}^{n} (z_i - \bar{z})(z_{i-k} - \bar{z}),$$

$$k = 0, 1, \ldots, n-1,$$

where $\bar{z} = (z_1 + \cdots + z_n)/n$ is the mean of the realised series. Because $r_k^{(n)} = c_k^{(n)}/c_0^{(n)}$, this then yields the generally preferred measure for serial correlation. The parent quantities are related by $\rho_k = \gamma_k/\gamma_0$, where $\gamma_k = E[(Z_i - \mu)(Z_{i-k} - \mu)]$, for $k = 0, 1, \ldots$. Here $\{Z_i\}$ is taken to be the infinitely long stationary stochastic process*, with mean value μ, that generates the finite series realisation $\{z_1, \ldots, z_n\}$.

[A process $\{Z_i\}$ is stationary if all its joint probability density functions $f(Z_{t+i_1}, \ldots, Z_{t+i_p})$, for any choice of $i_1, \ldots, i_p$ and p, are all defined and independent of t. For a Gaussian process*, having the property that all such functions are multivariate normal, it is sufficient just to ensure the conditions for $p = 1$ and 2, i.e., weak stationarity.]

Define the backshift operator B such that for any function $f(\cdot)$ and all integers i and j, $B^j f(i) \equiv f(i - j)$. We restrict the process $\{Z_i\}$ to the general class of stationary linear processes, with theoretical autocovariances denoted by $\gamma_0, \gamma_1, \ldots$; we also consider certain nonstationary processes $\{W_i\}$. The $\{W_i\}$ are those linear processes that can be reduced to $\{Z_i\}$ by means of a simplifying operator $S_d(B) \equiv 1 + s_1 B + \cdots + s_d B^d$ according to $S_d(B) W_i = Z_i$, where $S_d(\zeta)$, the corresponding dth degree polynomial in the complex variable ζ, has all its zeros precisely on the unit circle, and d is any positive integer. Then, for any stationary length-n series realisation from $\{Z_i\}$ [$S_d(B) = 1$, say], the following holds for $0 \leqslant k \leqslant n-1$:

$$E[c_k^{(n)}]$$

$$= \frac{1}{n^3}\left\{ n(n-k)(n\gamma_k - \gamma_0) + 2n\sum_{j=1}^{k-1}(k-j)\gamma_j \right.$$

$$\left. -2n\sum_{j=1}^{n-k-1}(n-k-j)\gamma_j - 2k\sum_{j=1}^{n-1}(n-j)\gamma_j \right\}. \quad (1)$$

Result (1) was deduced by Anderson [1]; see Anderson and de Gooijer [5] for expressions for $\text{var}[c_k^{(n)}]$ and $\text{cov}[c_k^{(n)}, c_0^{(n)}]$. A general formula for $\text{cov}[c_k^{(n)}, c_h^{(n)}]$, $0 \leqslant h \leqslant k \leqslant n-1$, that includes the latter as special cases, was obtained in Anderson and de Gooijer [4].

For a nonstationary $\{W_i\}$ with $S_d(B) = (1-B)$, Anderson [1] found an expression for $E[c_k^{(n)}(w)]$ for the same range of k [where $c_k^{(n)}(w)$ is used to emphasize that the serial covariances of the nonstationary w-series are being considered]; Anderson and de Gooijer [5] showed how to write down the corresponding $\text{var}[c_k^{(n)}]$ and $\text{cov}[c_k^{(n)}, c_0^{(n)}]$, which are lengthy expressions. Anderson and de Gooijer [4] also gave the general formula for the $\text{cov}[c_k^{(n)}, c_h^{(n)}]$. This type of nonstationarity due to the $(1-B)$ behaves anomalously [within the $S_d(B)$ class] and may be considered as the weakest form of nonstationarity.

For $\{W_i\}$ with other $S_d(B)$, we need to consider the ratios

$$E_k^{(n)} = E\left[c_k^{(n)}\right]/E\left[c_0^{(n)}\right],$$

$$V_k^{(n)} = \text{var}\left[c_k^{(n)}\right]/E^2\left[c_k^{(n)}\right],$$

$$C_{k,h}^{(n)} = \text{cov}\left[c_k^{(n)}, c_h^{(n)}\right]/\left\{E\left[c_k^{(n)}\right]E\left[c_h^{(n)}\right]\right\},$$

rather than $E[c_k^{(n)}]$, $\text{var}[c_k^{(n)}]$, and $\text{cov}[c_k^{(n)}, c_h^{(n)}]$.

Then, for $S_d(B) = (1+B)$ (Anderson [2]), we have for instance that

$$E_k^{(n)} = (-1)^k\left(1 - \frac{k}{n}\right) + O(n^{-2}),$$

$$V_k^{(n)} = 2 + O(n^{-2}), \qquad k \ll n,$$

$$C_{k,\,0}^{(n)} = 2.$$

So, writing

$$E\left[r_k^{(n)}\right] = E_k^{(n)}\left(1 - C_{k,0}^{(n)} + V_0^{(n)}\right) + O(n^{-3/2}),$$

$$\text{var}\left[r_k^{(n)}\right] = E_k^{(n)2}\left\{V_k^{(n)} - 2C_{k,0}^{(n)} + V_0^{(n)}\right\} + O(n^{-3/2}),$$

we find that (Anderson [3]) to $O(n^{-3/2})$ and for all k,

$$E\left[r_k^{(n)}\right] \simeq E_k^{(n)}, \qquad \text{var}\left[r_k^{(n)}\right] \simeq 0, \quad (2)$$

which clearly indicate that $r_k^{(n)}$, from any given realisation, is likely to be very closely characterised by $E_k^{(n)} \simeq (-1)^k(1-k/n)$, whereas one might interpret the parent ρ_k to be $(-1)^k$ (using a limiting argument).

For $S_d(B) = (1-B)^2$, Anderson [1] shows that

$$E_k^{(n)} = \frac{(n-k)(n^2 - 2nk - 2k^2 - 1)}{n(n^2-1)},$$

while for $S_d(B) = (1 - 2B\cos\omega + B^2)$with $0 < \omega \leqslant \pi$,

$$E_k^{(n)} = \left(1 - \frac{k}{n}\right)\cos k\omega + O(n^{-2}).$$

Again we find that, for all these quadratic nonstationarities, (2) holds; so $E_k^{(n)}$ again provides a very good characterisation of an actual sampled $r_k^{(n)}$.

One second degree nonstationarity is not covered by the above, however. This is $S_d(B) = (1-B^2)$ and it provides an interesting example. One might think that the associated serial correlation structure would behave analogously to that for $(1-B)$, but with pure seasonality* of period 2. However, $(1-B^2)$ does in fact give rise to behaviour virtually identical to that of $(1+B)$, because this is the nonstationary factor in $(1-B^2)$ that dominates the other, the weak $(1-B)$.

More complicated $S_d(B)$ formed from powers and products of those already discussed can also be dealt with similarly. Again, $E_k^{(n)}$ always provides a far superior characterisation for $r_k^{(n)}$ than does the parent correlation (obtained from a limiting argument). It is worth noting that, taking limits as $n \to \infty$, in the various formulae for the cases $S_d(B) = 1$ and $(1 - B)$, yields confirmation of asymptotic results achieved by other authors, for instance, Bartlett [6] and Roy and Lefrançois [7].

References

[1] Anderson, O. D. (1979). *Sankhyā B*, **41**, 177–195.

[2] Anderson, O. D. (1979). *Cahiers du CERO*, **21**, 221–237.

[3] Anderson, O. D. (1982). In *Applied Time Series Analysis*, O. D. Anderson and M. R. Perryman, eds. North-Holland, Amsterdam, The Netherlands, pp. 5–14.

[4] Anderson, O. D. and de Gooijer, J. G. (1982). In *Time Series Analysis: Theory and Practice*, Vol. 1, O. D. Anderson, ed. North-Holland, Amsterdam, The Netherlands, pp. 7–22.

[5] Anderson, O. D. and de Gooijer, J. G. (1983). *Sankhyā B*, **45**, 249–256.

[6] Bartlett, M. S. (1946). *J. R. Statist. Soc. B*, **8**, 27–41; Correction (1948), **10**, 1.

[7] Roy, R. and Lefrançois, P. (1981). *Ann. Sci. Math. Québec*, **5**, 87–95.

Bibliography

Anderson, O. D. (1980). *J. Oper. Res. Soc.*, **31**, 905–917. (Surveys serial dependence properties of linear processes.)

Acknowledgments

Neville Davies, Jan de Gooijer, Norman Johnson, Roch Roy, Torsten Söderström and Petre Stoica provided comments which improved this article.

(SERIAL CORRELATION
TIME SERIES)

O. D. ANDERSON

SERIAL TESTS OF RANDOMNESS

These are tests that compare the distribution of observed numbers of *sequences* (pairs, triplets, etc.) of various kinds in a series with what would be expected if the series were random. They are usually based on chi-squared* statistics. Papers by Good [1, 2] and Knuth [3] contain interesting discussions of distributions of such statistics.

References

[1] Good, I. J. (1953), *Proc. Camb. Philos. Soc.*, **49**, 276–284.

[2] Good, I. J. (1957), *Ann. Math. Statist.*, **28**, 262–264.

[3] Knuth, D. E. (1968), *The Art of Computer Programming*, Vol. 2. Addison-Wesley, Reading, MA.

(RANDOMNESS, TESTS FOR
RUNS)

SERIAL VARIATION *See* SERIAL CORRELATION; SUCCESSIVE DIFFERENCES

SERIATION

Seriation theory (and practice) starts with Petrie [9], who wished to place 900 predynastic Egyptian graves in approximately the correct serial order, using only the incidence statistics of 800 different varieties of pottery found in the graves. (See ref. 6 for details of that classical investigation.) A problem of this size is perhaps unusual; it is more common to be asked to seriate, say 100 graves, and it is reasonable to seriate larger collections in overlapping subcollections, and then to fit the bits together, though the question of achieving an optimal scheduling of such an operation has been almost totally neglected and urgently calls for intensive study. The later contributors to the field have spent too much time working on algorithms that are fine for graves in their tens, but almost useless for graves in their hundreds.

The data will in the simplest case be summarised in a 0-1 matrix $\mathbf{A}$ having m rows (graves) and n columns (varieties), 0 denoting absence and 1 presence, but instead of such *incidence* matrices we may have *abundance* matrices with nonnegative entries a_{ij} denoting the number of representatives of the jth variety in the ith grave. For the time being we shall speak of the incidence problem, which is typical and simpler. Ideally one would like to achieve a row permutation that would produce, *simultaneously in each column*, a bunching together of the 1's in a single clump. A 0-1 matrix, so sorted, is called a P matrix (pre-P if it is merely known to possess such a row sorting). It cannot be assumed that a sorting of the Petrie-type automatically gives one a *temporal* seriation (up to complete reversal); it may instead reflect sociological or geographical structure, and there may be more than one relevant dimension. When we have an abundance matrix **A**, then the aim is to achieve unimodality* in each column, and one speaks of Q and pre-Q matrices in a natural generalisation of the earlier definitions. A useful theorem [7] tells one that the row permutations Π that achieve a Q-sorting of a pre-Q matrix are *exactly* those that give, to the symmetric diagonally dominant matrix **S** defined by

$$S_{ij} = \sum_h w_h (a_{ih} \wedge a_{jh}),$$

a unimodal pattern in each row (and so also in each column), the action of Π on **S** being $\mathbf{S} \to \Pi \mathbf{S} \Pi^T$. Here the w's are positive column weights, and are arbitrary, and $\wedge$ denotes "take the minimum."

One successful seriation technique [8] assumes that some approximate form of this theorem holds for a "nearly" pre-Q matrix, and then uses **S** as a similarity matrix for rows, in the context of nonmetric multidimensional scaling* (MDS) with a suitable treatment of ties. [As usual one starts with a random (say three-dimensional) map of rows, performs the scaling, selects the two top principal axes, scales again in two dimensions, and so on down to a one-dimensional arrangement.] It is found [8] that this can reproduce rather closely the sorting achieved by traditional (e.g., stylistic) archaeological methods, with the expenditure of considerably less time. Despite this, the results of such an automatic seriation should always be regarded as being at the most a provisional solution, approximate but *objective*, to be followed by a more accurate, *subjective*, hand-sorting. What is known as the secondary treatment of ties should be avoided; it produces an irritating and unnecessary "horseshoe" effect. An intermediate tertiary treatment, primary on, say, the two least-similar tie-blocks and secondary on the rest, can give very good results. Further improvements can sometimes be obtained by using the fact [7] that all the matrices $\mathbf{S}^{(r)}$ defined by

$$\mathbf{S}^{(0)} = \mathbf{S}, \qquad \mathbf{S}_{ij}^{(r+1)} = \sum_h w_h \left\{ S_{ih}^{(r)} \wedge S_{jh}^{(r)} \right\},$$

also enjoy the property of **S** guaranteed by the theorem above. Finally it should be noted that seriations derived from such MDS methods must be judged by results, not by the stress values.

Quite another possibility is to use **A** as the input matrix for a correspondence analysis*. It was Hill [4] who recognised that seriations could effectively be achieved in this way, and this method has the distinguishing feature that it simultaneously sorts rows *and columns*, producing configurations for each in one and the same two-dimensional map. A peculiarity of this method is that in many cases truly pre-Q matrices generate parabolic plots, and while the seriation can then be obtained by projecting onto the tangent at the vertex of the parabola (which is parallel to the top nontrivial principal axis) one feels that valuable information held in the arms of the parabola may have been lost. (This has nothing to do with the avoidable "horseshoe" effect noted above.)

Both methods (which can, if desired, be combined) are excellent in indicating clearly the two groups of early and late graves, and once these are known then the intermediate graves can of course be inserted in various ways. (The identification of the late as opposed to the early end must however be based on other criteria.)

One can ask, "When is a matrix nearly pre-Q?" No good answer is known, though Laxton [5] shows that the problem is at least connected with the number of triplets of rows and triplets of columns that produce 3×3 submatrices that are bad, in the way that

$$\begin{matrix} 0 & 1 & 1 \\ 1 & 0 & 1 \\ 1 & 1 & 0 \end{matrix}$$

is obviously bad. One would dearly like to know the distribution of the number of such bad submatrices in a random $m \times n$ 0-1 matrix with given row and column sums, and to see a study of the extent to which it is those matrices that achieve a good score in this sense which can be more successfully put into approximate Q form.

One would also welcome an extension of the usual MDS algorithm that would accom-

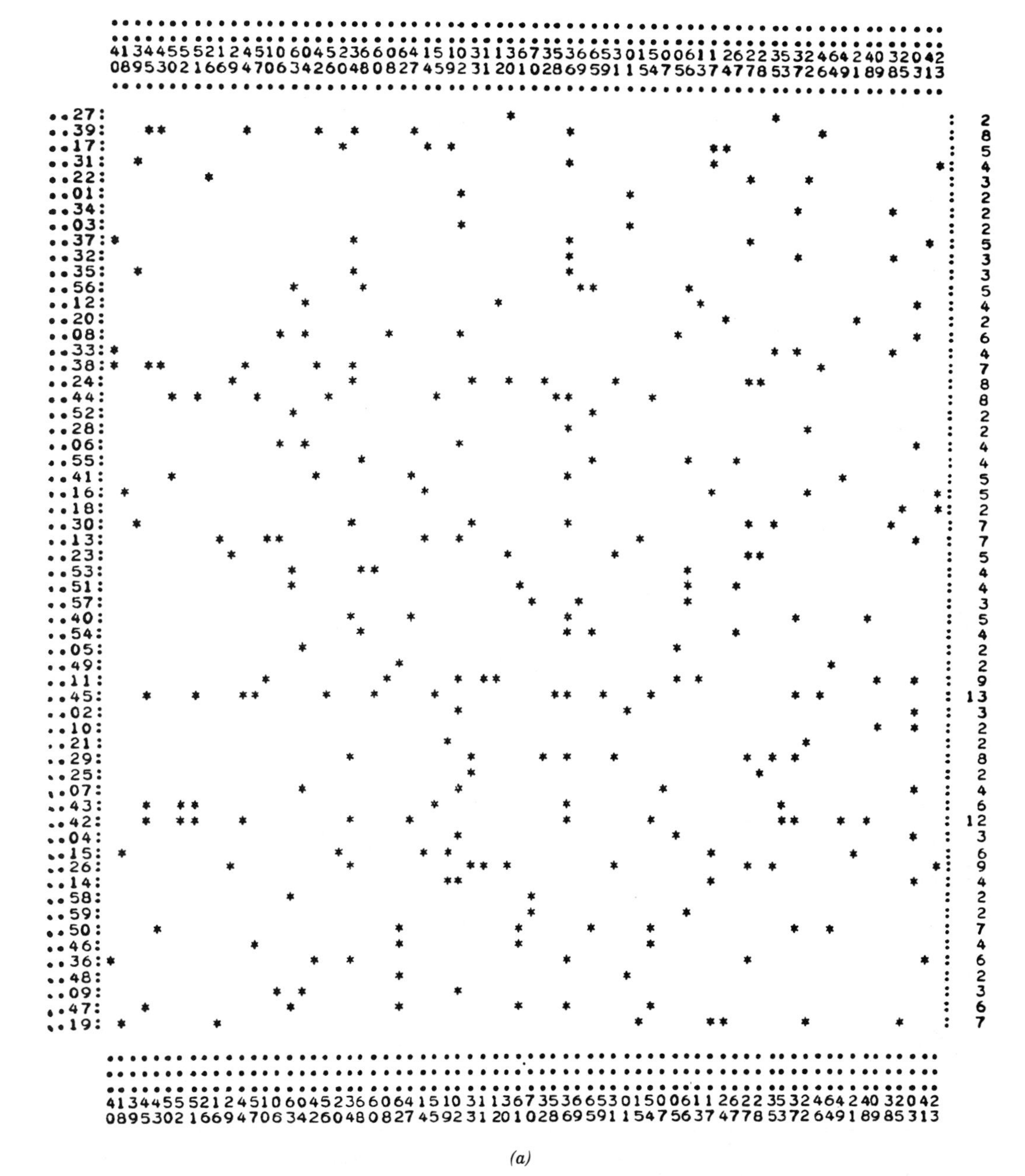

(a)

Figure 1 A 59 × 70 incidence matrix (a) in the raw state, (b) as sorted by the method of ref. 4, and (c) as sorted by the method of ref. 8. (Here * = 1 and space = 0.)

modate more realistically the real order-theoretic structure of the seriation problem; the existing one does so only in a very approximate sense (Sibson [10]). Another open problem is how to handle the seriation of abundance matrices, when the entries are subject to independent statistical errors, in a way that would take into account the fact that, for the elements of **S**, the relation [s_{ij} is not significantly different from s_{kl}] is not transitive. This means producing a useable isotone regression for an unpleasant partial-order relation that is capable of being run in acceptable times for realistically large matrices.

The reader will have noticed that the seriation problem in its original form is discrete in character and that in fact it belongs

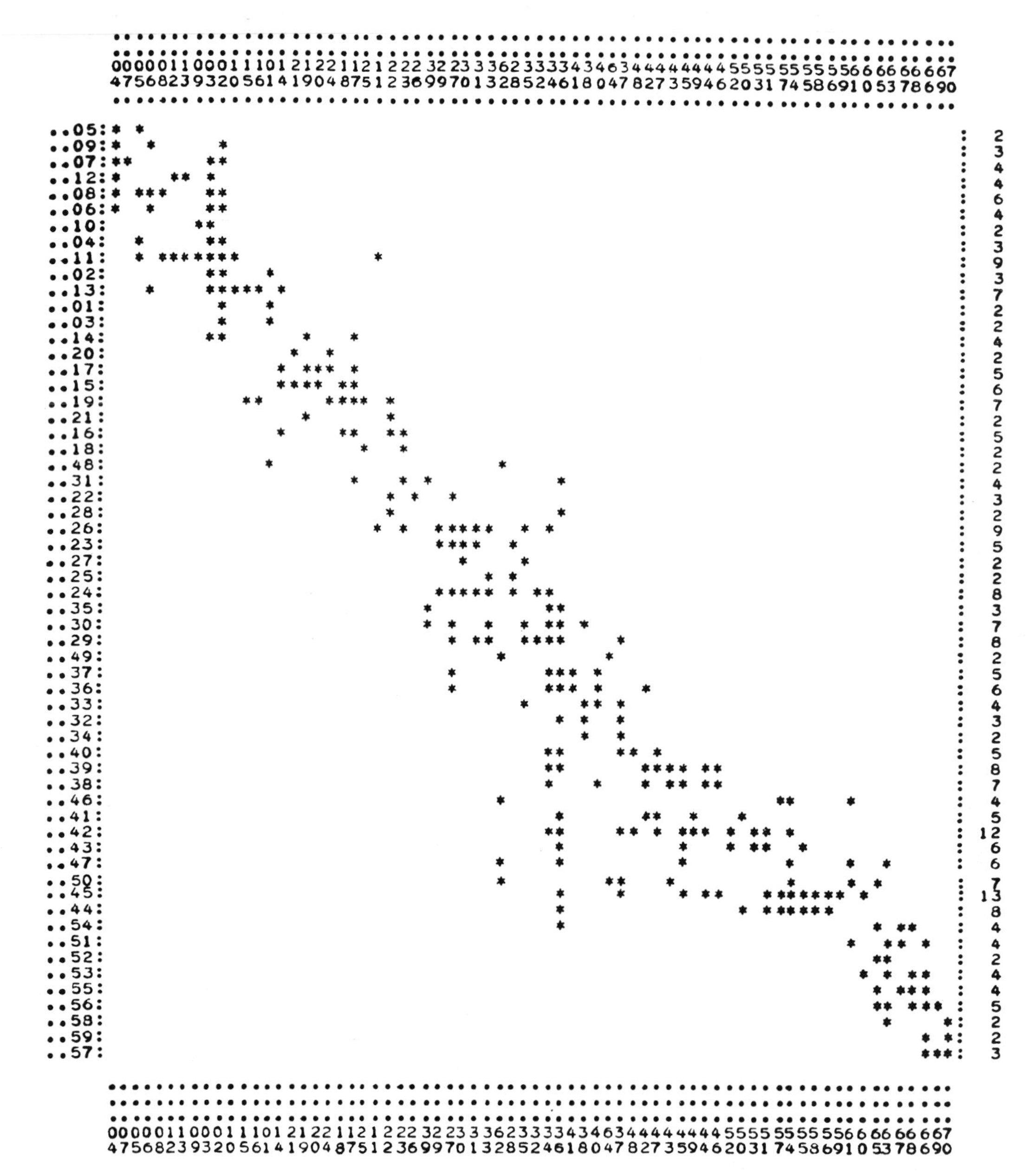

(b)

Figure 1 (*Continued*)

to graph theory* (Fulkerson and Gross [3]; Wilkinson [11]), but that nevertheless both techniques used here to solve it employ continuous variables, this being the way in which the essential compromises are effected when resolving incompatible discrete constraints. There may be other ways of achieving this. Fuzzy sets*?

In Fig. 1 we show (a) a raw 0-1 data matrix (59 rows = graves, 70 columns = varieties of ornament on objects found therein), (b) the row-*and-column* sorting provided by correspondence analysis, and (c) a row sorting provided by the method of ref. 8 (tertiary scaling in three dimensions, followed by tertiary scaling in two dimensions,

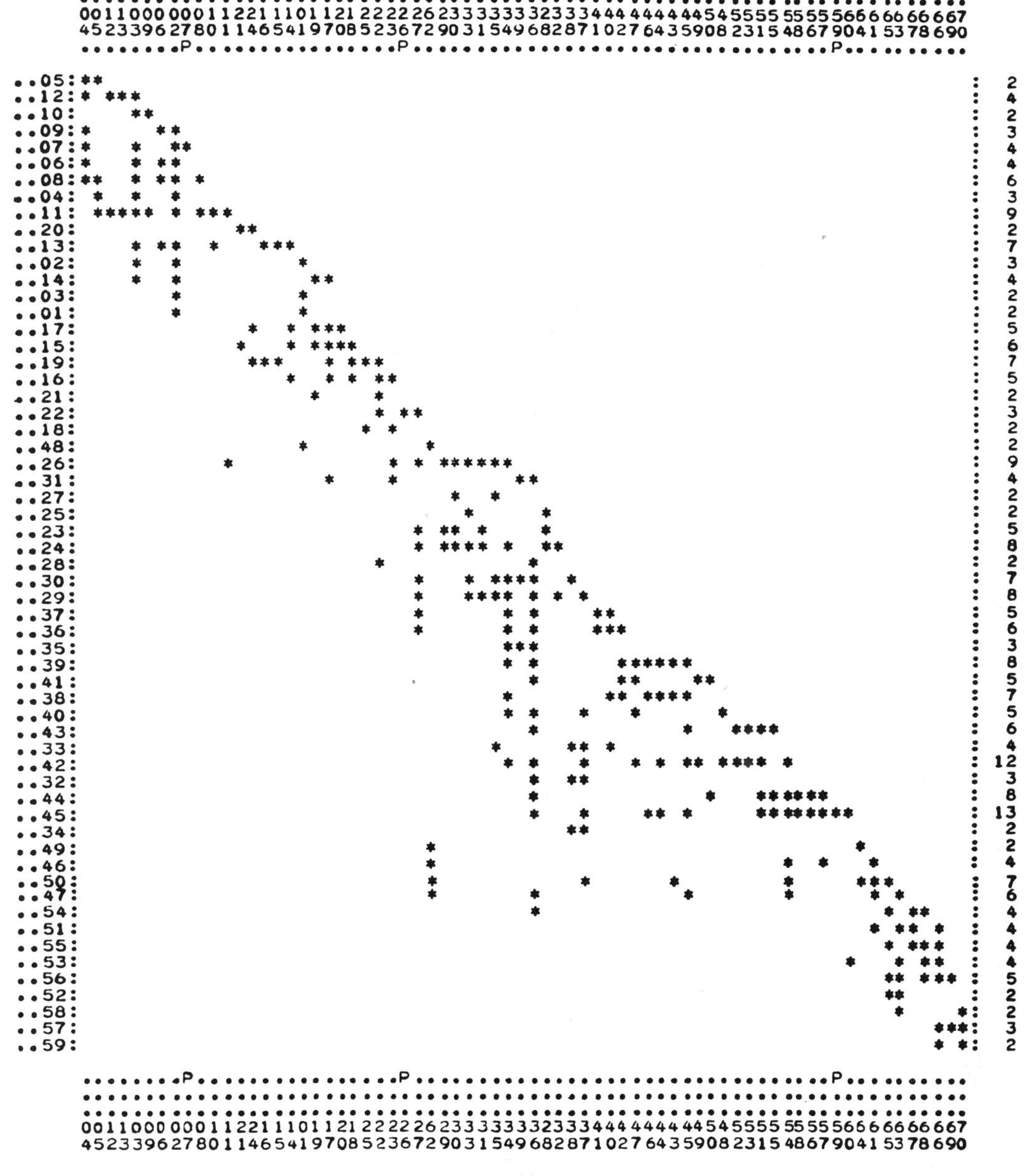

(c)

Figure 1 (*Continued*)

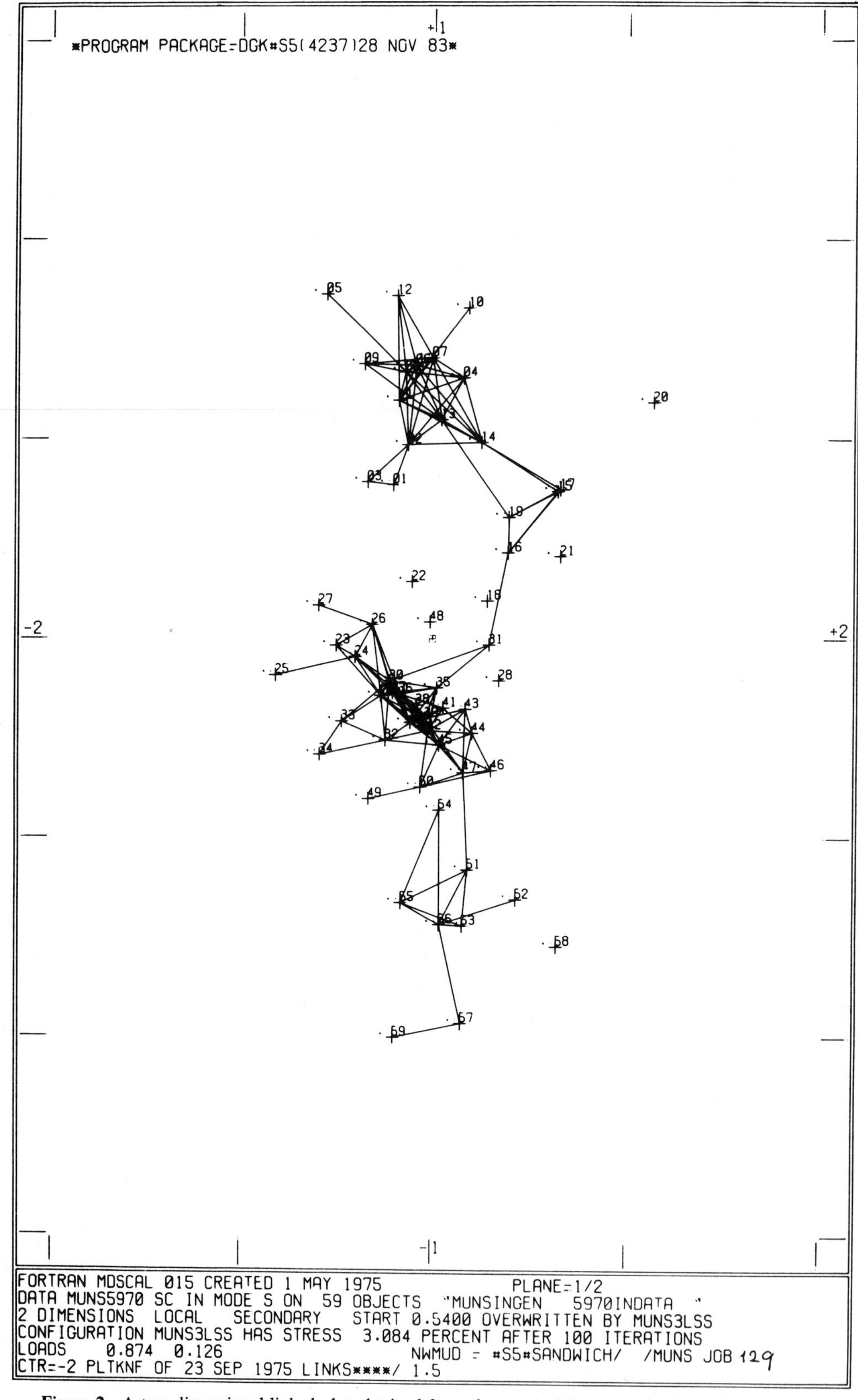

Figure 2 A two-dimensional linked plot obtained from the raw incidence matrix shown in Fig. 1(a) by the method of ref. 8. A pair of graves is shown as linked when the graves have at least two varieties in common. The integers labelling the points (= rows = graves) indicate the serial ordering arrived at by F. R. Hodson using traditional archaeological methods. It is thus possible to compare Hodson's order with that suggested by the MDS linked plot.

followed by primary scaling in one dimension). The column sorting in (c) is cosmetic only. A convenient "figure of demerit" is G = (sum over the columns of the lengths of all gaps between 1's)/(total number of 1's). For these three illustrations we have

$$G(a) = 7.360, \qquad G(b) = 0.978,$$
$$G(c) = 1.103.$$

Traditional sorting by the archaeologist F. R. Hodson of the same material, starting as from (a), yielded $G = 0.901$. Obviously the use of (b) or (c) would have been a useful preliminary to this, yielding a result of at least comparable quality, which could then have been used as a basis for further improvements by other methods.

The reader should notice carefully that algorithms designed to yield a seriation alone, perhaps accompanied by a figure of demerit, which it is the task of the algorithm to minimise, are not enough. It is essential that they be supplemented by additional output, usually graphic in character, which allows the data to tell us how happy or otherwise it feels about being squeezed into a linear shape, and to declare other relevant structures present that we may not have anticipated. The unique advantages of the methods of refs. 4 and 8 are that they do meet this requirement, and that they display the original information in linked plots, showing each strongly associated pair of graves as a pair of points in two dimensions linked by a line segment. The declared seriation, to be acceptable, must adequately summarise such linked plots. (For an example, see Fig. 2.)

References

[1] Benzer, S. (1959, 1961). On the topology (topography) of the genetic fine structure. *Proc. Nat. Acad. Sci. USA*, **45**, 1607–1620; **47**, 403–415. (These papers describe a nonstatistical form of the seriation problem that arises in molecular biology.)

[2] Bølviken, E. et al. (1982), Correspondence analysis: an alternative to principal components. *World Archaeology*, **14**, 41–54.

[3] Fulkerson, D. R. and Gross, O. A. (1965). Incidence matrices and interval graphs. *Pacific J. Math.*, **15**, 835–855. (Here the associated graph-theoretic problem is solved.)

[4] Hill, M. O. (1974). Correspondence analysis: a neglected multivariate method. *Appl. Statist.*, **23**, 340–354. (See also B. F. Schriever (1983) *Int. Statist. Rev.*, **51**, 225–238 for more recent developments.)

[5] Laxton, R. R. (1976). A measure of pre-*Q*-ness with applications to archaeology. *J. Archaeol. Sci*, **3**, 43–54.

[6] Kendall, D. G. (1963), A statistical approach to Flinders Petrie's sequence dating. *Bull. Int. Statist. Inst.*, **40**, 657–680.

[7] Kendall, D. G. (1971). Abundance matrices and seriation in archaeology, *Zeit. Wahrsch. verw. Geb.*, **17**, 104–112.

[8] Kendall, D. G. (1971). In *Mathematics in the Archaeological and Historical Sciences*, F. R. Hodson, D. G. Kendall, and P. Taŭtu, eds. Edinburgh University Press, Edinburgh, Scotland, pp. 215–252. (This article contains a case study.)

[9] Petrie, W. M. F. (1899). Sequences in prehistoric remains. *J. Anthropol. Inst.*, **29**, 295–301.

[10] Sibson, R. (1972). Order invariant methods for data analysis. *J. R. Statist. Soc. B*, **34**, 311–338. (See also Sibson's paper "Some thoughts on sequencing methods" in the volume cited in ref. 8, pp. 263–266.)

[11] Wilkinson, E. M. (1974). Techniques of data analysis: Seriation theory. *Technische Naturwiss. Beiträge Feldarchäologie*, **5**, 1–142.

Bibliography

Some two-dimensional mapping generalisations of seriation theory are discussed in Kendall (1975). The reader may also wish to consult the extensive OR/psychology literature, concerned for the most part with algorithms for the seriation of small collections of up to, say, 15 items. We recommend the following articles for further reading.

Adelson, R. M. et al. (1976). A dynamic programming formulation with diverse applications. *Oper. Res. Quart.*, **27**, 119–121.

Hubert, L. J. (1974). Problems of seriation using a subject by item response matrix. *Psych. Bull.*, **81**, 976–983.

Hubert, L. J. (1974). Some applications of graph theory and related non-metric techniques to problems of approximate seriation: The case of symmetric proximity measures. *Brit. J. Math. Statist. Psych.*, **27**, 133–153.

Hubert, L. J. and Golledge, R. G. (1981). Matrix reorganisation and dynamic programming: Applications to paired comparisons and unidimensional seriation. *Psychometrika*, **46**, 429–441.

Kendall, D. G. (1975). The recovery of structure from fragmentary information. *Philos. Trans. R. Soc. Lond. A*, **279**, 547–582.

(ARCHAEOLOGY, STATISTICS IN
CORRESPONDENCE ANALYSIS
MULTIDIMENSIONAL SCALING)

D. G. KENDALL

SERRA'S CALCULUS

Serra [1] developed a set theoretic calculus useful for statistical and probabilistic analysis of the highly structured data of images arising in fields using remote sensing and microscopy. Serra's calculus was originally developed for problems encountered in the study of porous media, permeability, petrography, and other areas connected with mines and metallurgy. See also Ripley [2] for a lucid survey and extensive bibliography.

The sets involved are sets A of points $\mathbf{x}$ (binary observations on an infinite square lattice in the plane) visualized as *black* points in a black–white "image." The object is to measure the shape of the observed image $A \cap W$, where W is the window in the plane within which the images are observed. A family $\mathscr{T}$ of *test sets* T is introduced that typically comprises finite unions of (open) discs. The operator $T + \mathbf{x}$ denotes translation by $\mathbf{x}$. The operations utilized by Serra [3] (see also Matheron [1]) are:

1. Reflection of A in the origin:

$$\check{A} = \{-\mathbf{x} | \mathbf{x} \in A\}.$$

2. Minkowski addition and subtraction of two sets A and B:

$$A \oplus B = \{\mathbf{x} + \mathbf{y} | \mathbf{x} \in A, \mathbf{y} \in B\};$$

$$A \ominus B = (A^c \oplus B)^c,$$

where $A^c = \bar{A}$ is the complement of A (the set of *white* points in our black–white image).

3. Transformations

$$A \to A \ominus \check{T} \qquad (\text{the } erosion \text{ of } A \text{ by } T)$$

and

$$A \to A \oplus T \qquad (\text{the } dilatation \text{ of } A \text{ by } \check{T}).$$

4. Serra's *opening* of A by T, defined as

$$(A \ominus \check{T}) \oplus T.$$

5. Serra's *closing* of A by T, defined (by duality) as

$$\{(A^c \ominus \check{T}) \oplus T\}^c = (A \oplus \check{T}) \ominus T.$$

Note that an opening is an erosion followed by a dilatation, while a closing is a dilatation followed by an erosion.

T-opened and T-closed sets can be regarded as *smoothed* versions of A. The importance of these concepts is in their use for estimation of (a) $P(T \subset A)$ (here T is fixed, A is random); (b) $P(A \cap T \neq \varnothing) = 1 - P(T \subset A^c)$; (c) $P(\mathbf{x} \in (A \text{ opened by } T))$; and (d) $P(\mathbf{x} \in (A \text{ closed by } T))$. These estimators, denoted by e_T, d_T, o_T, and c_T, respectively, are

$$e_T = \frac{\text{area}\{A \cap W \ominus \check{T}\}}{\text{area}\{W \ominus \check{T}\}},$$

$$d_T = \frac{\text{area}\{(A \oplus \check{T}) \cap (W \ominus \check{T})\}}{\text{area}\{W \ominus \check{T}\}},$$

$$o_T = \frac{\text{area}[\{(A \cap W) \ominus \check{T}\} \oplus T] \cap V}{\text{area } V},$$

$$c_T = \frac{\text{area}[\{(A \cap W) \oplus \check{T}\} \ominus T] \cap V}{\text{area } V.},$$

where $V = W \ominus (T \oplus \check{T})$ is the maximal set such that the opening and closing of A by T are known within V from $A \cap W$. (By duality this must be the same set for *both* operations.) These estimators are plotted against characteristics of T (such as the radius of disc) and the curves e, d, o, and c can be used to extract most of the geometric information available in $A \cap W$.

See, e.g., Ripley [2] and the discussion following that paper for graphical-visual examples.

References

[1] Matheron, G. (1975). *Random Sets and Integral Geometry*. Wiley, New York.

[2] Ripley, B. D. (1986). *Canad. J. Statist.*, **14**, 83–102 (discussion, pp. 102–111).

[3] Serra, J. (1982). *Image Analysis and Mathematical Morphology*, Academic, New York.

(RANDOM FIELDS
RANDOM SETS OF POINTS
STEREOLOGY)

SERVICEABILITY

A term used in reliability* theory to denote the ease with which a system can be repaired.

SERVICE FACTOR; SERVICE PERIOD; SERVICE TIME *See* QUEUEING THEORY

SETWISE DEPENDENCE

In modeling and analyzing multivariate data, interest may focus on relationships among sets of random variables (i.e., random vectors) regardless of the dependence within each set. This study divides roughly into two parts; concepts of setwise dependence and measures of setwise dependence. Setwise dependence concepts, as introduced by Chhetry et al. [1], describe probabilistic and structural properties of relationships among sets of variables, while measures of setwise dependence, e.g., Hotelling's canonical correlation coefficient (*see* CANONICAL ANALYSIS), numerically quantify the strength of the relationship between two (or possibly more) sets of variables. These two facets extend to random vectors some of the ideas, respectively, in DEPENDENCE, CONCEPTS OF and DEPENDENCE, MEASURES AND INDICES OF.

CONCEPTS

A collection $\mathbf{X}_1, \ldots, \mathbf{X}_k$ of random vectors with respective dimensions $p_1, \ldots, p_k$ are *setwise positive upper orthant dependent* (SPUOD) if for all vectors $\mathbf{x}_1, \ldots, \mathbf{x}_k$ of respective dimensions $p_1, \ldots, p_k$,

$$P\left[\bigcap_{j=1}^{k} (\mathbf{X}_j > \mathbf{x}_j)\right] \geqslant \prod_{j=1}^{k} P[\mathbf{X}_j > \mathbf{x}_j] \quad (1)$$

and *setwise positive lower orthant dependent* (SPLOD) if inequality (1) holds with each event $\mathbf{X}_j > \mathbf{x}_j$ replaced by $\mathbf{X}_j \leqslant \mathbf{x}_j$. The random vectors $\mathbf{X}_1, \ldots, \mathbf{X}_k$ are *setwise positive upper set dependent* (SPUSD) if

$$P\left[\bigcap_{j=1}^{k} (\mathbf{X}_j \in U_j)\right] \geqslant \prod_{j=1}^{k} P[\mathbf{X}_j \in U_j] \quad (2)$$

holds for all possible upper sets $U_1, \ldots, U_k$ of appropriate dimensions. (An upper set is a set U for which $\mathbf{x} \in U$ and $\mathbf{y} \geqslant \mathbf{x}$ implies $\mathbf{y} \in U$; equivalently, an upper set is a set whose indicator function is increasing.) Similarly, $\mathbf{X}_1, \ldots, \mathbf{X}_k$ are setwise positive lower set dependent (SPLSD) if (2) holds with upper sets replaced by lower sets (complements of upper sets). When $k = 2$, SPUSD and SPLSD are equivalent, although SPUOD and SPLOD are not. Another concept is *setwise association* (SA), which requires that the k random variables $f_1(\mathbf{X}_1), \ldots, f_k(\mathbf{X}_k)$ be associated for all collections $f_1, \ldots, f_k$ of nondecreasing functions.

If $p_1 = \cdots = p_k = 1$, then SPUOD, SPLOD, and SA reduce, respectively, to positive upper orthant dependence (PUOD), positive lower orthant dependence (PLOD), and association; similarly SPUSD becomes PUOD and SPLSD becomes PLOD. The concepts PUOD and PLOD are discussed in INEQUALITIES ON DISTRIBUTIONS: BIVARIATE AND MULTIVARIATE. Association is discussed in DEPENDENCE, CONCEPTS OF.

Moreover, if the k random vectors are concatenated to form a set of Σp_j random variables, the SPUOD of the original collection of k random vectors neither implies nor is implied by the PUOD of the Σp_j random variables. On the other hand, the association of the Σp_j random variables implies that $\mathbf{X}_1, \dots, \mathbf{X}_k$ are SA, but not conversely.

It is easy to verify that SA ⇒ SPUSD ⇒ SPUOD and that SA ⇒ SPLSD ⇒ SPLOD, and that these implications are strict, If $\mathbf{X}_1, \dots, \mathbf{X}_k$ have a multivariate normal distribution whose covariance matrix $\mathbf{\Sigma}$ is partitioned into a $k \times k$ array of submatrices $\mathbf{\Sigma}_{ij}$, then the concepts SPUOD, SPLOD, SPUSD, and SPLSD are all equivalent to the conditions that $\mathbf{\Sigma}_{ij}$ be a nonnegative matrix whenever $i \neq j$. Note that this condition allows negative correlation between any two random variables within the same vector $\mathbf{X}_j$. Concepts for negative dependence analogous to SPUSD, SPUOD, SPLSD, SPLOD, and SA have also been studied by Chhetry et al. [1].

MEASURES

A number of the measures of setwise dependence between two sets of variables $\{X_{1,1}, \dots, X_{1,p_1}\}$ and $\{X_{2,1}, \dots, X_{2,p_2}\}$ are of the form

$$\delta_{\mathscr{F},\mathscr{G}}(\mathbf{X}_1, \mathbf{X}_2) \equiv \sup_{f \in \mathscr{F},\, g \in \mathscr{G}} \rho\big(f(X_{1,1}, \dots, X_{1,p_1}), g(X_{2,1}, \dots, X_{2,p_2})\big), \quad (3)$$

where ρ is Pearson's correlation* coefficient, and $\mathscr{F}$ and $\mathscr{G}$ are suitable classes of functions mapping $\mathbb{R}^{p_1} \to \mathbb{R}^1$ and $\mathbb{R}^{p_2} \to \mathbb{R}^1$, respectively. Some choices for $\mathscr{F}$ and $\mathscr{G}$ are the class of: (i) linear functions, and then $\delta_{\mathscr{F},\mathscr{G}}$ is Hotelling's [2] canonical correlation* coefficient; (ii) arbitrary functions (with some regularity conditions), and then $\delta_{\mathscr{F},\mathscr{G}}$ is Sarmanov and Zaharov's [6] maximum coefficient of correlation; and (iii) increasing functions, and then $\delta_{\mathscr{F},\mathscr{G}}$ is the multivariate extension of concordant monotone correlation of Kimeldorf et al. [3]. Measures other than ρ in (3) can be employed; for example, a multivariate correlation ratio* can be defined (Sampson [5]). A different approach using information* theory leads to the multivariate logarithmic index, which compares the entropy* in the joint distribution of $\mathbf{X}$ and $\mathbf{Y}$ with the entropy assuming they are independent (Kotz and Soong [4]).

References

[1] Chhetry, D., Kimeldorf, G., and Sampson, A. R. (1983). Concepts of Setwise Dependence. *Tech. Rep. No. 139*, Programs in Mathematical Sciences, University of Texas at Dallas, Richardson, TX.

[2] Hotelling, H. (1936). *Biometrika*, **28**, 321–377.

[3] Kimeldorf, G., May, J., and Sampson, A. R. (1982). *TIMS/Stud. Manage. Sci.*, **19**, 117–130.

[4] Kotz, S. and Soong, C. (1977). On Measures of Dependence. *Tech. Rep.*, Department of Mathematics, Temple University, Philadelphia, PA.

[5] Sampson, A. R. (1984). *Statist. Prob. Lett.*, **2**, 77–81.

[6] Sarmanov, O. V. and Zaharov, V. K. (1960). *Dokl. Akad. Nauk. SSSR*, **130**, 269–271.

Acknowledgments

This work was sponsored by the National Science Foundation under Grant MCS-8301361 (G.K.) and by the Air Force Office of Scientific Research under Contract F49620-82-K-0001 and AFOSR Contract 84-0113 (A.S.). Reproduction in whole or in part is permitted for any purpose of the United States Government.

(ASSOCIATION, MEASURES OF
CANONICAL ANALYSIS
DEPENDENCE, CONCEPTS OF
DEPENDENCE, MEASURES AND INDICES OF
INEQUALITIES ON DISTRIBUTIONS: BIVARIATE AND MULTIVARIATE)

GEORGE KIMELDORF
ALLAN R. SAMPSON

SEXTILE

One of five points on the scale of values of a variable dividing its distribution into six equal parts. The third sextile is the median*.

(QUANTILES)

SHANMUGAN NUMBERS

A generalization of Stirling numbers* of the second kind (SNSK). They were apparently introduced by Shanmugan [1]; it is possible that the concept had been used previously, although the Editors have not been able to locate a source.

For a real valued δ, and nonnegative integers K and n such that $k \geqslant n$, the Shanmugan numbers $S(k, n, \delta)$ are given by the recursive equations

$$\frac{dS(k, n, \delta)}{d\delta} = -kS(k, n-1, \delta),$$

[with $S(k, n, \delta) = 0$ if $k < n$].

It can be shown that

$$S(k, n, \delta) = \sum_{u=n}^{k} \binom{k}{u} (-\delta)^{k-u} S(u, n),$$

where the $S(u, n)$'s are SNSK. So for $\delta = 0$, Shanmugan numbers are SNSK.

Unlike SNSK, which are positive integers, Shanmugan numbers may be integers or fractions. They are positive for $\delta < 0$. Shanmugan numbers are used in defining the *interrupted Poisson distribution*, which arises in models for observations of rare events when the observational apparatus becomes active only when at least one event occurs. See Shanmugan [1, 2] for further details.

References

[1] Shanmugan, R. (1984). *S. Afr. Statist. J.*, **18**, 97–110.

[2] Shanmugan, R. (1984). *Proc. Business Econ. Statist. Sec. Amer. Statist. Ass.*, American Statistical Association, Washington, DC, pp. 612–617.

(POISSON DISTRIBUTION
STIRLING NUMBERS
STUTTERING POISSON DISTRIBUTION)

SHANNON THEOREM *See* INFORMATION AND CODING THEORY

SHANNON–WIENER INDEX *See* DIVERSITY, INDICES OF

SHAPE FACTORS *See* MOMENT RATIOS

SHAPE STATISTICS

At the time of writing, this is a fairly new subject, and a full account of the basic theory [5] has only recently been published. It should be emphasised that (i) we are concerned here with the detailed elaboration of one special part of a broader topic (discussed in SIZE AND SHAPE ANALYSIS), and (ii) there is no connection at all with what topologists have chosen to call "the theory of shape"! We confine ourselves to the two-dimensional case, in which one discusses the shape of a labelled k-ad $(P_1, P_2, \ldots, P_k)$ in R^2. Here *shape is what is left when location, size, and the effects of rotation are ignored.* The form that the theory takes is of course influenced by the way in which this statement is made precise, and in particular by the choice of the quantity to be identified with "size." (For another approach see ref. 1.)

Totally collapsed k-ad's $P_1 = P_2 = \cdots = P_k$ will not be considered. We agree to fix the origin at the centroid G of the k-ad, and to standardise the scale so that $\Sigma GP_j^2 = 1$ (thus removing the effects of size). Each such semi-standardised object (called *pre-shape*) is a point on the sphere S^{2k-3} of radius 1. Finally we identify *pre-shapes* that can be obtained from one another by rotation in R^2, and we topologise this collapsed version of S^{2k-3} with the aid of a so-called "procrustean" metric, which is derived from the natural great-circle metric on S^{2k-3}, and which is simply related to the minimum standardised degree of mismatch between two labelled k-ad's. In this way we obtain the *shape-space* Σ_2^k, the points of which are *shapes*. In the case we are discussing here the shape space is a smooth Riemannian manifold* identical with a version of what is called complex projective space CP^{k-2}; an important feature of it is that the geodesics are closed loops of total length π. In the very important case when $k = 3$ (shapes of triangles!) it is identical with the sphere $S^2(\frac{1}{2})$ of radius $\frac{1}{2}$. The geometry of the shape space is studied by passing the geometrical structure of the sphere S^{2k-3} through the submersive map $S^{2k-3} \to CP^{k-2}$ (under which geodesics behave particularly well).

Statistical studies involving the shapes of triangles have attracted a lot of attention, for example, in connection with the possible existence of "ley lines" ([6], [8]) and the corresponding analysis simplifies considerably when treated from the present point of view, especially as there is usually enough relevant symmetry to allow one to confine attention to one spherical triangle Δ on $S^2(\frac{1}{2})$.

The stochastic theory begins when we give to the vertices P_1, P_2, and P_3 an arbitrary joint distribution, with probability 0 for total coincidence. For example we could take the vertices to be a sample of size 3 from the standard two-dimensional isotropic Gaussian law, and if we do this it turns out that the induced shape measure on $S^2(\frac{1}{2})$ is the uniform measure as usually understood. Thus probability calculations with this model reduce to spherical trigonometry. Any other law for the vertices, which is absolutely continuous, will therefore yield a shape measure that can conveniently be studied via its shape density, i.e., its density relative to the uniform measure that is associated with a Gaussian model.

Such shape measures (and also the empirical shape measures resulting from data analysis or simulation) can conveniently be represented by contours for the density or empirical density, or by scatter plots, on a flat area-preserving projection of an appropriately representative spherical triangle Δ obtained from $S^2(\frac{1}{2})$ by identifying shapes differing only by a reflection or by relabelling. Figure 1 shows the standard projection employed in this case (a doubled projection region is required if reflection effects are to be retained). The figure also shows how 32 typical triangle shapes are located on this, their natural home.

Not many shape measures are known explicitly, but a few have now been studied in detail. These include the shape measure for three points i.i.d. uniform in a circular disk, and that for k points i.i.d. and having a common Gaussian law (not necessarily isotropic). Another example that has been worked out in detail is that for a random Poisson–Delaunay triangle. We recall that any finite set of point in R^2 determines a polygonal tessellation* (that of Voronoi) and a dual triangular tessellation associated with the name of Delaunay (= Delone). The same construction is meaningful for an infinite set of points and so can be applied to the realisation of a two-dimensional Poisson process*. When this is done, the triangular cell containing some preassigned point is called a Poisson–Delaunay triangle. There is an important application for these results because of the interest shown in Delaunay tessellations by geographers concerned with central place theory [7].

There are also some interesting theoretical questions concerned with shape densities, not all at present resolved. Thus if we take a compact convex set K and generate three points i.i.d. uniform within it, can we determine the shape of K from the shape density for such a triad? For important results in this area see Small [9, 10].

Another application of practical importance is associated with tests for collinearity (already briefly referred to above). For k points in two dimensions the relevant statistic turns out to be the minimum geodesic distance from the sample point to the locus in $\Sigma_2^k = CP^{k-2}$ of all collinear shapes [this locus is in fact an embedding in CP^{k-2} of the real projective space RP^{k-2}, and when $k = 3$ it reduces to a certain great circle on $S^2(\frac{1}{2})$.]. For general k the statistical distribution of this minimum geodesic distance is known ([5]) when the vertices are i.i.d. Gaussian (both in the isotropic and the non-isotropic case). Thus we are provided with a natural test statistic (sometimes perhaps preferable to that employed in ref. 6) and it is easy to set up the associated tests and to determine their power, using a Gaussian model. A numerical example will be found in ref. 5 relating to an important question in quasar astronomy.

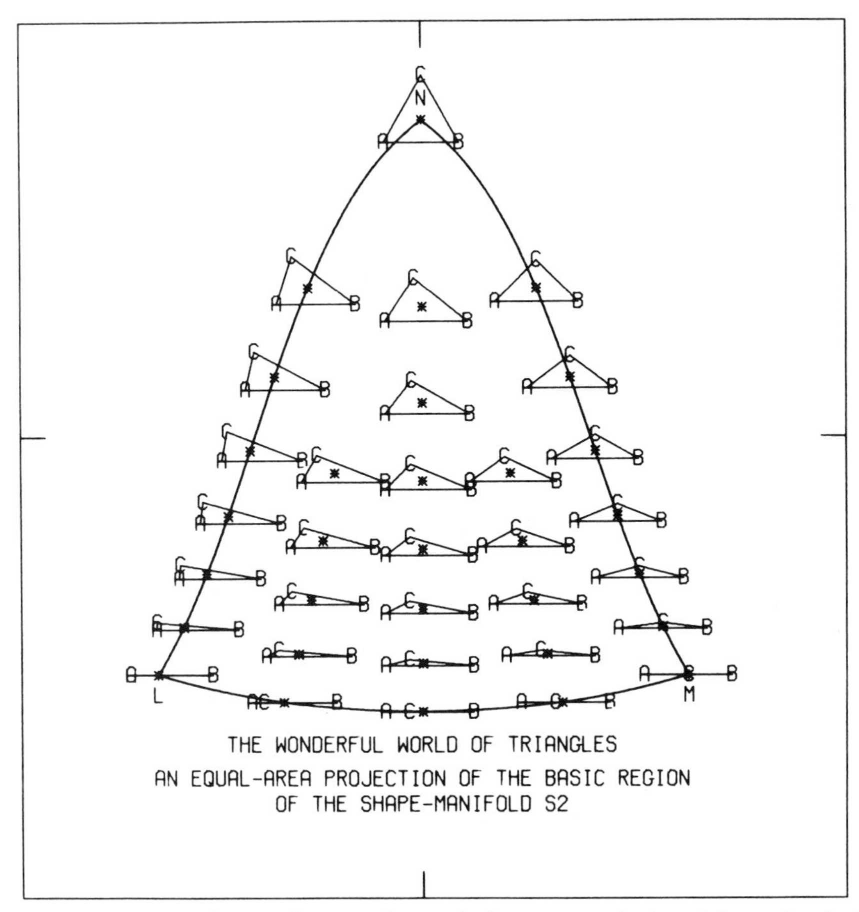

Figure 1 Some triangle shapes at home on the standard area-true projection of the portion Δ of the shape space appropriate when relabelling is unimportant and reflection effects are ignored. (Reproduced from ref. 3.)

If we allow the vertices $P_1, P_2, \ldots, P_k$ to move in R^2, we get a corresponding induced motion of the shape point in the shape space, and in particular, diffusions in R^2 yield time-changed diffusions on CP^{k-2}. These aspects are briefly explored in ref. 2.

References

[1] Ambartzumian, R. V. (1982). In *Statistics in Theory and Practice: Essays in Honour of Bertil Matern*, B. Ranneby, ed. Umeå, Sweden.

[2] Kendall, D. G. (1977). The diffusion of shape. *Adv. Appl. Prob.*, **9**, 428–430.

[3] Kendall, D. G. (1981). In *Interpreting Multivariate Data*, V. Barnett, ed. Wiley, New York, pp. 75–80.

[4] Kendall, D. G. (1983). In *Studies in Probability in Honour of Octav Onicescu*, M. C. Demetrescu and M. Iosifescu, eds. Nagard, Montreal, Canada.

[5] Kendall, D. G. (1984). Shape-manifolds, procrustean metrics, and complex projective spaces. *Bull. Lond. Math. Soc.*, **16**, 81–121. (This gives a comprehensive account of the theory, and some examples.)

[6] Kendall, D. G., and Kendall, W. S. (1980). Alignments in two-dimensional random sets of points. *Adv. Appl. Prob.*, **12**, 380–424.

[7] Mardia, K. V., Edwards, R., and Puri, M. L. (1977). Analysis of central place theory. *Bull. Int. Statist. Inst.*, **47**, 93–110.

[8] Small, C. G. (1982). Random uniform triangles and the alignment problem. *Math. Proc. Camb. Philos. Soc.*, **91**, 315–322.

[9] Small, C. G. (1983). Characterization of distribu-

tions from maximal invariant statistics. *Zeit. Wahrsch. verw. Geb.*, **63**, 517–527.

[10] Small, C. G. (1983). *Ann. Statist.*, **11**, 979–983.

Bibliography

For summaries of the most recent work see:

Kendall, D. G. (1986). *Teor. Veroyat.* **31**, 467–473.

Kendall, D. G. and Le, H.-L., (1986). *Proc. 1st Internat. Congress, Bernoulli Soc.*, Tashkent, USSR.

(MANIFOLDS
SIZE AND SHAPE ANALYSIS)

D. G. KENDALL

SHAPIRO–WILK *W* STATISTICS

Testing for departure from distributional assumptions, particularly from normality, is an important part of statistical practice (*see* DEPARTURES FROM NORMALITY, TESTS FOR). Shapiro and Wilk's W test [9] is one of the most powerful "omnibus" procedures for testing univariate nonnormality [2]. A variant has been developed for the exponential distribution* [10]. There appears to be no reason why the same philosophy cannot be applied to other distributions. It has also been used as part of a test procedure for multivariate nonnormality [6].

The W test is based on the generalised least-squares* regression of ordered sample values on normal scores*, and is computed as follows [9]. Let $\mathbf{M}' = (M_1, \ldots, M_n)$ denote the expected values of standard normal order statistics* for a sample of size n, and $\mathbf{V}$ the corresponding $n \times n$ covariance matrix. Suppose $X = X_1, \ldots, X_n$ is the random sample to be tested, ordered $X_1' < \cdots < X_n'$. Then

$$W = \{\Sigma_{i=1}^n w_i X_i'\}^2 / \Sigma_{i=1}^n (X_i - \bar{X})^2,$$

where

$$\begin{aligned}\mathbf{w}' &= (w_1, \ldots, w_n)\\ &= \mathbf{M}'\mathbf{V}^{-1}[(\mathbf{M}'\mathbf{V}^{-1})(\mathbf{V}^{-1}\mathbf{M})]^{-1/2}\end{aligned}$$

and $\bar{X} = n^{-1}\Sigma_{i=1}^n X_i$. W may be thought of as the squared correlation* coefficient between the ordered sample values (X_i') and the w_i, the latter [7] being approximately proportional to the normal scores M_i. Thus, W is a measure of the straightness of the normal probability plot*, and small values indicate departure from normality. A variant of W called W' [8] is obtained by substituting $M_i/(\mathbf{M}'\mathbf{M})^{1/2}$ for w_i. This is somewhat easier to compute, owing to the existence of a good approximation to the M_i [1, p. 71]. Other weighting systems w_i with differing properties have been used.

COMPUTATION

Shapiro and Wilk [9] give the coefficients w_i and selected percentiles of W for $3 \leqslant n \leqslant 50$. However, Royston [3, 4] has provided an approximation to the null distribution of W and a Fortran algorithm for $n \leqslant 2000$, together with a similar algorithm for W' [5], which simplify the P-value* calculations.

POWER STUDIES

Extensive empirical Monte Carlo* simulation studies (e.g., refs. 2 and 11) have shown that W is powerful against a wide range of alternative distributions. It appears to be especially good against skew or short- or very long-tailed alternatives, even for samples as small as 10 if the departure is strong.

PRACTICAL USES

W has proved to be a very useful routine tool when analysing the residuals* from linear models, by identifying the need for transformation*, the existence of outliers*, and other peculiarities. It should be used in conjunction with a normal plot. Despite the obvious fact that the power to detect nonnormality increases with sample size, models fitted to large data sets (say, $n = 1000$) can still yield residuals with nonsignificant values of W.

References

[1] Blom, G. (1958). *Statistical Estimates and Transformed Beta Variables*. Wiley, New York.

[2] Pearson, E. S., D'Agostino, R. B., and Bowman, K. O. (1975). *Biometrika*, **64**, 231–246. (Contains important information about power of various tests of normality.)

[3] Royston, J. P. (1982). *Appl. Statist.*, **31**, 115–124. (Extends W to $n = 2000$ and gives transformation to normality.)

[4] Royston, J. P. (1982). *Appl. Statist.*, **31**, 176–180. (Fortran algorithms. Vital to applied statisticians.)

[5] Royston, J. P. (1983). *Statistician*, **32**, 297–300. (Algorithm and P value for W'.)

[6] Royston, J. P. (1983). *Appl. Statist.*, **32**, 121–133. (W applied to multivariate distances. Contains some interesting examples.)

[7] Sarhan, A. E. and Greenberg, B. G. (1956). *Ann. Math. Statist.*, **27**, 427–451.

[8] Shapiro, S. S. and Francia, R. S. (1972). *J. Amer. Statist. Ass.*, **67**, 215–216. (Describes W'.)

[9] Shapiro, S. S. and Wilk, M. B. (1965). *Biometrika*, **52**, 591–611. (The essential reference, clear and well written.)

[10] Shapiro, S. S. and Wilk, M. B. (1972). *Technometrics*, **14**, 335–370. (W test for the exponential distribution.)

[11] Shapiro, S. S., Wilk, M. B., and Chen, H. J. (1968). *J. Amer. Statist. Ass.*, **63**, 1343–1372. (Very detailed power study of several tests of nonnormality.)

(DEPARTURES FROM NORMALITY,
TESTS FOR
GOODNESS OF FIT
MULTIVARIATE NORMALITY,
TESTING FOR
OUTLIERS)

J. P. ROYSTON

SHARPE AND LOTKA MODEL *See* MATHEMATICAL THEORY OF POPULATIONS

SHARPENING DATA

It is now generally recognised that graphical and tabular displays of data have different primary strengths. Tables of figures are ideal for record purposes but fairly useless for visual assimilation (for a contrary view, see ref. 1). On the other hand, graphical methods of display, the study of which has enormously increased recently, stimulated in part by developments in computer graphics, are obviously intended to aid effective interpretation and understanding (*see* GRAPHICAL REPRESENTATION, COMPUTER-AIDED). As Mallows and Tukey [4] write:

> The majority of all quite unexpected results are detected from display.

However, graphs of data do not provide a very useful record, particularly when the dimensionality of the data exceeds that of the display.

Once this distinction is established, we can exploit it by deliberately arranging that our graphical displays *record* the data less accurately in order that they can *represent* the data more clearly and with greater impact. Such displays will be valuable both in exploratory work (viewed by the data analyst) and in presentation (to the client). These ideas have perhaps been most strongly advocated and developed by John Tukey. We concentrate here on the implications for scatter plots, displaying the joint distribution of two variables in the data set. (There may only be two variables, or these may be two from many, or two principal components, etc.)

Scatter plots may be *decorated* by various forms of peeling* (see ref. 3) or with smooth middle traces [5, Chap. 8]. More radical possibilities include *agglomeration* and *sharpening* [6, pp. 228–242]. Agglomeration involves replacing the individual data values plotted as identical points in the scatter plot by a variety of symbols representing clusters of different numbers of points, perhaps defined by reference to local density.

Sharpening (and its opposite, blunting) involve instead the deletion, diminution, or displacement of individual data points. The aim is that the contrast between dense and sparse regions of the data set should be increased (or in the case of blunting, decreased). We thus enhance the impact of "typical" points whilst reducing or suppressing that of "unusual" data. Of course, as always, there is no suggestion that such outliers* are henceforth to be forgotten: they should be recorded and investigated, and may in fact hold the true message of the data set; but we want the graphical display to convey the usual, not the discrepant.

Tukey and Tukey [6] suggest two methods for sharpening by deletion. One is to rank

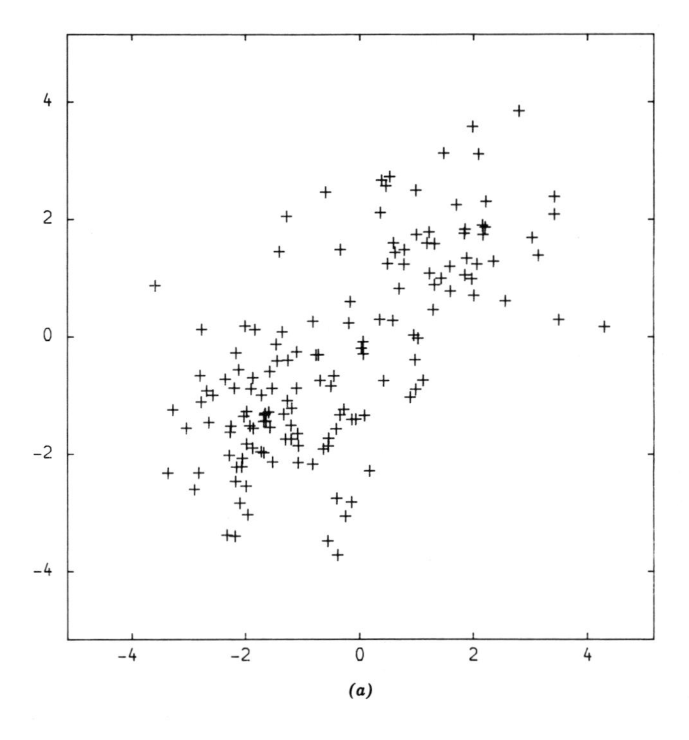

(a)

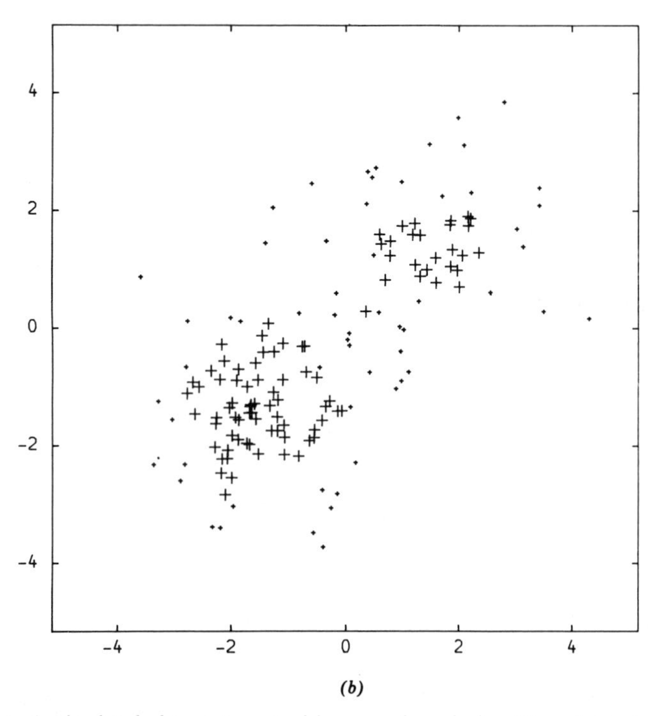

(b)

Figure 1 (a) A simulated data set, comprising samples of size 90 and 60 from two Gaussian distributions with unit dispersion matrices. (b) The same data, but sharpened by plotting with a smaller cross the 40% of data points with the least 8th-nearest-neighbour balloon density estimates.

the data points in decreasing order of "typicality" (they propose using balloon density estimates, but many other possibilities suggest themselves), and then delete points from the end of the list until some criterion is satisfied. The other is to delete points randomly and independently, with survival probabilities dependent on local density. Deleted points may be eliminated altogether, or plotted with smaller or fainter symbols.

Sharpening by displacement [2] proceeds by letting points drift, as if by gravity, towards regions of higher density. Some care is needed to avoid thereby artificially introducing apparent structure into the plot.

Sharpening a scatter plot is not likely to be useful for small data sets—say, of a hundred points or less. Of course, on such a scale the procedures described would in any case behave rather arbitrarily because of poor precision in the density estimates. However, as the size of a data set increases, sharpening may well be a valuable tool. This is especially true when a large number of plots have to be visually digested, perhaps because of high dimensionality in the original data (12 dimensions means 66 pairs of variables). Circumstances in which sharpening might be expected to yield substantial benefit include the presence of two or more overlapping clusters of data points, possibly suggesting a mixture in an underlying population; Fig. 1, and other illustrative examples in ref. 6, demonstrate how much more apparent and separated such clusters are after sharpening the display.

References

[1] Ehrenberg, A. S. C. (1975). *Data Reduction*. Wiley, New York. (A manual on data analysis and presentation. Not always in agreement with ref. 5.)

[2] Friedman, J. H., Tukey, J. W., and Tukey, P. A. (1980). In *Data Analysis and Informatics*, E. Diday et al., eds. North-Holland, Amsterdam, The Netherlands.

[3] Green, P. J. (1981). In *Interpreting Multivariate Data*, V. Barnett, ed. Wiley, Chichester, England, Chap. 1, pp. 3–19. (A review of peeling methods for bivariate data.)

[4] Mallows, C. L. and Tukey, J. W. (1982). In *Some Recent Advances in Statistics*, J. Tiago de Oliveira and B. Epstein, eds. Academic, London, England, Chap. 7, pp. 111–172. (An overview of techniques of data analysis.)

[5] Tukey, J. W. (1977). *Exploratory Data Analysis*. Addison-Wesley, Reading, MA. (The EDA bible.)

[6] Tukey, P. A. and Tukey, J. W. (1981). In *Interpreting Multivariate Data*, V. Barnett, ed. Wiley, Chichester, England, Chap. 11, pp. 215–243. (Part of a wide-ranging and innovative contribution on graphical display of multidimensional data.)

(EXPLORATORY DATA ANALYSIS
GRAPHICAL REPRESENTATION, COMPUTER-AIDED
GRAPHICAL REPRESENTATION OF DATA
MULTIVARIATE GRAPHICS
PEELING DATA)

P. J. Green

SHARP NULL HYPOTHESES

In order to test the statistical significance of a discrepancy between observations and a hypothesis, we must associate a class of probability distributions with the hypothesis and say that one of these distributions governs the observations if the hypothesis is true. Often we can also specify an alternative hypothesis—a further class of probability distributions, one of which governs the observations if the first or "null" hypothesis* is false. The two classes together constitute a statistical model for the observations; we assume these observations are governed by one of the distributions in the model, and the null hypothesis boils down to the further statement that this true distribution is in a certain subset of the model.

The alternative hypothesis is often (but not always—see Cox [3]) broader than the null hypothesis, in the sense that each distribution in the null hypothesis can be approximated by a distribution in the alternative hypothesis. In the parametric case, where the distributions in the statistical model are indexed by a finite-dimensional parameter, this usually means that the null hypothesis corresponds to a lower-dimensional subspace of the model's parameter space. Suppose, for example, that the parameter consists of a pair of means, μ_1 and μ_2, and the null hypothesis is that they are equal. Then the parameter space is a plane and the null hypothesis is a one-dimensional subspace of this plane—the line $\mu_1 = \mu_2$. Any point on the line can be arbitrarily well approximated by a point off the line.

Following Cox [4] (1977), we may divide lower-dimensional null hypotheses into three classes: simplifying hypotheses, dividing hypotheses, and intrinsically plausible hypotheses.

Simplifying hypotheses are considered because they simplify a statistical analysis or its interpretation. Hypotheses such as linearity, normality, equality of variances, and absence of interactions are usually simplifying hypotheses. We do not expect these hypotheses to be exactly true, but we hope they are sufficiently accurate to make simple analyses meaningful, and we test them to check whether there is any evidence against this optimistic assumption.

Dividing hypotheses are considered in order to check whether the observations supply enough evidence to justify more detailed modeling and estimation. The hypothesis that two means μ_1 and μ_2 are equal "divides" the situation where $\mu_1 > \mu_2$ from the situation where $\mu_1 < \mu_2$, and it is reasonable to test this hypothesis before undertaking to estimate the difference between the two means. Similarly, the hypothesis that a certain point process is Poisson, even if it is not plausible on a priori grounds, divides the case of "overdispersion" from the case of "underdispersion," and it is therefore sensible to test for consistency with a Poisson process* before trying to model apparent clustering.

Intrinsically plausible hypotheses are considered because they are implied by a general theory or an a priori argument. In an experiment designed to detect extrasensory perception, for example, we may want to test the hypothesis that a subject's chance p of correctly choosing between two items is equal to $\frac{1}{2}$ because this value is implied by the assumption that extrasensory perception does not exist.

In recent years it has become common to refer to intrinsically plausible lower-dimensional hypotheses as *sharp null hypotheses*. Savage may have coined this term [10]. Fisher*, who introduced the term "null hypothesis" [6, Sec. 8], used it in contexts where alternative hypotheses have not necessarily been formulated, and many later authors have used the term to refer to hypotheses that are not lower dimensional. Thus the adjective "sharp" serves both to indicate that the null hypothesis is embedded in a higher-dimensional model and to suggest that there are theoretical reasons for thinking of it as exactly true.

From a Bayesian point of view, sharp null hypotheses differ from other lower-dimensional null hypotheses in that they are assigned positive prior probability*. A Bayesian analysis of a simplifying or dividing lower-dimensional null hypothesis usually involves assigning zero prior probability and hence zero posterior probability to the hypothesis. (The prior distribution over the finite-dimensional parameter space is defined by a density that assigns zero probability to any lower-dimensional subspace.) The fact that the null hypothesis has zero posterior probability* does not prevent us from giving sensible answers to the questions that are being asked; the real question is whether most of the posterior probability is in a vicinity of the simplifying hypothesis or is to one side or another of the dividing hypothesis. In the case of a sharp null hypothesis, however, we feel there are a priori grounds for thinking the null hypothesis exactly true and so concentrate a portion of the prior probability on it.

Bayesian analyses that concentrate a portion of the prior probability on a lower-dimensional null hypothesis were studied in detail by Jeffreys [7]. As Jeffreys pointed out, such Bayesian "significance tests" sometimes disagree with conventional non-Bayesian significance tests (*see* BAYESIAN INFERENCE). Lindley [8] showed how dramatic this disagreement can be, and it is now often called "Lindley's paradox."

The basic issues can be discussed in terms of a simple example where the model is one dimensional and the null hypothesis is zero dimensional. Suppose X is a random quantity. Its mean μ is unknown, but the probability density of $S = X - \mu$ is known. This density has mean zero; let $f(s)$ denote this density, and let σ^2 denote its variance. There is some reason to think that $\mu = 0$. We observe that $X = x$. The usual two-sided non-Bayesian significance test of the null hypothesis $\mu = 0$ is based on the p value*

$$P = \Pr[X \geqslant |x| \,|\mu = 0] = \int_{|x|}^{\infty} f(s)\,ds + \int_{-\infty}^{-|x|} f(s)\,ds.$$

We may "reject" the hypothesis if P is less than some conventional level, say 0.05. Jeffreys's Bayesian significance test involves assigning prior probability to $\mu = 0$. Suppose we assign prior probability $\frac{1}{2}$ to $\mu = 0$ and distribute the remaining prior probability according to a density $p(\mu)$. Then the posterior probability for $\mu = 0$ will be given by

$$\begin{aligned}
&\Pr[\mu = 0|X = x] \\
&= \lim_{\epsilon \to 0} \{\Pr[\mu = 0|x - \epsilon \leqslant X \leqslant x + \epsilon]\} \\
&= \lim_{\epsilon \to 0} \{(\Pr[\mu = 0] \\
&\qquad \times \Pr[x - \epsilon \leqslant X \leqslant x + \epsilon|\mu = 0]) \\
&\qquad \times (\Pr[\mu = 0]\Pr[x - \epsilon \leqslant X \leqslant x + \epsilon|\mu = 0] \\
&\qquad + \Pr[\mu \neq 0]\Pr[x - \epsilon \leqslant X \leqslant x + \epsilon|\mu \neq 0])^{-1}\} \\
&= \lim_{\epsilon \to 0} \left[\left(\frac{1}{2}\int_{x-\epsilon}^{x+\epsilon} f(s)\,ds\right) \right. \\
&\qquad \times \left(\frac{1}{2}\int_{x-\epsilon}^{x+\epsilon} f(s)\,ds + \frac{1}{2}\int_{-\infty}^{\infty} p(\mu) \right. \\
&\qquad \left.\left. \times \int_{x-\mu-\epsilon}^{x-\mu+\epsilon} f(s)\,ds\,d\mu\right)^{-1}\right] \\
&= \frac{f(x)}{f(x) + E[f(x-\mu)|\mu \neq 0]}. \qquad (1)
\end{aligned}$$

Suppose $|x|$ is several times larger than σ, but $p(\mu)$ is so diffuse that its standard deviation is of a larger order of magnitude than $|x|$. Since $|x|$ is large relative to σ, P will be small, and the non-Bayesian significance test* will reject $\mu = 0$. The quantity $f(x)$ will also be small, but since $p(\mu)$ is diffuse, it will attach only a small probability to the range of values of μ for which $f(x - \mu)$ is substantial, and therefore the average of $f(x - \mu)$ with respect to $p(\mu)$ may be even smaller than $f(x)$. This means the ratio (1) may be close to 1, i.e., the posterior probability of $\mu = 0$ may be very large.

The magnitude of the discrepancy between the two approaches can be striking. Suppose for example, that $f(x)$ and $p(\mu)$ are both normal, $|x| = 2.5\sigma$, and $p(\mu)$ has mean 0 and standard deviation 250σ. Then $P = 0.012$, which suggests strong evidence against $\mu = 0$. But $P[\mu = 0|X = x] = 0.92$.

The cogency of this Bayesian calculation depends, of course, on the evidence for the prior distribution $p(\mu)$. This distribution says that it is very improbable, under the alternative hypothesis $\mu \neq 0$, for μ to fall within a few σ of 0, and the Bayesian calculation is balancing this improbability against the lesser improbability, under the null hypothesis $\mu = 0$, that x should fall as far from 0 as it did. This balancing of improbabilities will be convincing if the improbability $p(\mu)$ assigned to the vicinity of 0 is based on convincing evidence, but not if it is merely the result of our spreading out the probability to reflect our ignorance about μ.

At least two arguments have been advanced for adjusting the non-Bayesian significance test results in the direction of the Bayesian calculation's greater support for the null hypothesis.

1. When most of the possible values of μ are many σ from the null hypothesis, the high power of the non-Bayesian test for these values suggests that greater significance be demanded for rejection of $\mu = 0$ [9].
2. A non-Bayesian p value often needs to be adjusted to account for selection. Selection made possible by optional stopping is of particular interest. If X in

our example has a small variance σ^2 because it is the result of averaging many measurements of μ, then it may fall several σ from μ only because the measurements were continued until this happened [8].

On the other hand, a number of Bayesians have recognized that the Bayesian calculation risks counting mere ignorance about μ as evidence against the alternative hypothesis and have sought to prevent this by limiting the spread of $p(\mu)$ relative to $f(s)$. Jeffreys [7, 3rd ed.] proposed that $p(\mu)$ should be Cauchy*, scaled by the standard deviation of $f(s)$. Zellner and Siow [12] and Bernardo [1] have made similar proposals. These proposals have met some opposition because they violate the principle that $p(\mu)$ should be based on actual knowledge and belief about μ.

Shafer [11] has proposed an alternative approach based on the theory of belief functions* (*see also* NONADDITIVE PROBABILITIES). In this subjective but non-Bayesian approach, the statistical evidence* x and the prior evidence $p(\mu)$ are treated as independent arguments about the value of μ. Each of these arguments is represented by a belief function, and the two belief functions are combined by Dempster's rule (described in BELIEF FUNCTIONS). This approach agrees with the Bayesian calculation if $p(\mu)$ is taken at face value, but comes closer to the conclusions from the usual non-Bayesian test if we are uncertain of the accuracy or relevance of $p(\mu)$ and therefore discount it slightly.

References

[1] Bernardo, J. M. (1980). In [2], pp. 605–618; discussion, pp. 618–647.

[2] Bernardo, J. M., DeGroot, M. H., Lindley, D. V., and Smith, A. F. M., eds. (1980). *Bayesian Statistics: Proceedings of the First International Meeting, Valencia, Spain*. University Press, Valencia, Spain.

[3] Cox, D. R. (1961). In *Proc. 4th Berkeley Symp. Math. Statist.*, Vol. 1. University of California Press, Berkeley, CA, pp. 105–123.

[4] Cox, D. R. (1977). *Scand. J. Statist.*, **4**, 49–70. (An excellent review of the various purposes of significance tests.)

[5] Dickey, J. M. (1976). *J. Amer. Statist. Ass.*, **71**, 680–689. (Treats Jeffreys's Bayesian significance tests as approximations to Bayesian analyses with purely continuous priors.)

[6] Fisher, R. A. (1935). *Design of Experiments*. Oliver and Boyd, Edinburgh, Scotland.

[7] Jeffreys, H. (1938). *Theory of Probability*, 1st ed. Clarendon Press, Oxford, England. [2nd ed. (1948); 3rd ed. (1961); 3rd ed. rev. (1967).]

[8] Lindley, D. V. (1957). *Biometrika*, **44**, 187–192.

[9] Pearson, E. S. (1953). *J. R. Statist. Soc. B*, **15**, 68–69.

[10] Savage, L. J. et al. (1962). *The Foundations of Statistical Inference*, G. A. Barnard and D. R. Cox, eds. Methuen, London, England.

[11] Shafer, G. R. (1982). *J. Amer. Statist. Ass.*, **77**, 325–351. (Includes discussion by six statisticians, a rejoinder, and further references.)

[12] Zellner, A. and Siow, A. (1980). In [2], pp. 585–603; discussion, pp. 618–647.

(BAYESIAN INFERENCE
BELIEF FUNCTIONS
HYPOTHESIS TESTING
NULL HYPOTHESIS
POSTERIOR PROBABILITIES
PRIOR PROBABILITIES
STATISTICAL EVIDENCE)

GLENN SHAFER

SHEFFER POLYNOMIALS

In the statistical sciences, Sheffer polynomials can serve as a mathematical tool to unify and structure certain kinds of problems, like recursions and expansions. There are many other applications that make use of some famous Sheffer polynomials, as can be seen from the list of examples, but they are not discussed here.

HISTORICAL REMARKS

Sheffer's [15] paper "Some properties of polynomial sets of type zero," presented to the American Mathematical Society in 1936, was the first to investigate this subject in some detail. Orthogonality had been studied earlier by Meixner [9]; *see* MEIXNER POLYNOMIALS and ORTHOGONAL EXPANSIONS. A systematic algebraic treatment by Rota et al. [14] followed. This fruitful paper prepared the ground for many generalizations [3, 4, 12, 13].

GENERAL DESCRIPTION

Definition 1. The *shift operator* E^a on the algebra $\mathbb{P}$ of polynomials p is defined by $E^a p(x) = p(x + a)$.

Definition 2. A *delta operator* Q on $\mathbb{P}$ is a linear operator that

(a) is shift-invariant, i.e., $QE^a\, p(x) = E^a Qp(x)$ for all $p \in \mathbb{P}$ and all shift operators E^a;

(b) reduces the degree by 1, i.e., if $\deg(p) = n$, then $\deg(Qp) = n - 1$ for all $n \geqslant 1$, and Q maps only the constant polynomials into zero.

For example, the differential operator $D = d/dx$ is a delta operator.

Definition 3. A *Sheffer sequence* (s_n) for Q is a sequence of polynomials s_n, where $\deg(s_n) = n$, $s_0 \neq 0$ and

$$Qs_n = s_{n-1} \quad \text{for all } n = 1, 2, \ldots. \tag{1}$$

In other words, (s_n) is a solution for the above system of first degree linear operator equations. The delta operator Q and a given sequence of initial values $s_n(v_n)$, $n = 0, 1, \ldots$, uniquely define a Sheffer sequence. The Sheffer sequence with initial values $s_n(0) = \delta_{0,n}$ (= 1 if $n = 0$ and zero otherwise) is called *the basic sequence* for Q. Synonyms for this sequence are Sheffer set, Sheffer polynomials, and polynomials of Sheffer A-type zero. (Sheffer polynomials should not be confused with "Sheffer functions" occurring in logic.) Basic polynomials are also called polynomials of binomial type.

Remark. (1) *shows the way Sheffer* [15] *defined these polynomials. Other authors* [4, 12, 14] *used the relationship* $Qp_n = np_{n-1}$. *Their Sheffer polynomials are larger by a factor of* $n!$.

Examples. Delta operators and Sheffer sequences are modelled after the differential operator D and the sequence $(x^n/n!)$. Sheffer sequences for this operator are called Appell sequences. Hermite* and Bernoulli polynomials* are examples for Appell sequences. The difference operators Δ and ∇ have the basic polynomials $\binom{x}{n}$ and $\binom{x+n-1}{n}$, respectively. Laguerre, Lagrange, Boole, Bell, and Charlier polynomials are all Sheffer polynomials for some delta operators. For example, the Laguerre polynomials

$$L_n^{\alpha}(x) = \sum_{j=0}^{n} (-1)^j \binom{n+\alpha}{n-j} \frac{x^j}{j!}$$

satisfy the functional relation

$$L_{n-1}^{\alpha} = D(L_{n-1}^{\alpha} - L_n^{\alpha}) = -DL_n^{\alpha} + DL_{n-1}^{\alpha}$$

$$= -\sum_{i \geqslant 1} D^i L_n^{\alpha}.$$

Hence, (L_n^{α}) is a Sheffer sequence for the delta operator $Q = -\Sigma_{i \geqslant 1} D^i$, which has the basic sequence (L_n^{-1}). See refs. 12 and 14 for more examples.

Property 1. The following characterisation of a Sheffer sequence (s_n) by generating functions* is often used as the definition:

$$\sum_{n \geqslant 0} s_n(x) t^n = f(t) e^{xu(t)},$$

where $f(t)$ and $u(t)$ are formal power series such that $f(0) \neq 0$, $u(0) = 0$, and u has a nonzero linear term. $u(t)$ has a compositional inverse $q(u)$, i.e., $q(u(t)) = t$ and $q(D) = Q$.

Property 2. Denote the basic sequence for Q by (b_n).

Binomial Theorem

$$s_n(x+y) = \sum_{i=0}^{n} s_i(x) b_{n-i}(y). \quad (2)$$

If E^a is any shift operator, then QE^a is again a delta operator, $(s_n(x - an))$ is a Sheffer sequence for QE^a, and $(xs_n(x - an)/(x - an))$ is the basic sequence for QE^a [14, Proposition 7.4]. Therefore, the representation theorem

$$s_n(x) = \sum_{i=0}^{n} s_i(y - ai) \frac{x - y + an}{x - y + ai} \times b_{n-i}(x - y + ai) \quad (3)$$

follows immediately from (2). It allows one to represent a Sheffer sequence by its initial values $s_i(y - ai)$ $(i = 0, 1, \dots)$ and the basic sequence (b_n).

For more results (umbral calculus, formula of Rodrigues, etc.) see refs. 12 to 14.

RECURSIONS

Wald and Wolfowitz [19] presented a technique to calculate for N independent uniform (0,1) random variables $U_1, \dots, U_N$, the distribution of order statistics* $[0 \leqslant \nu(1) \leqslant \nu(2) \leqslant \cdots]$

$$\Pr[\nu(1) \leqslant U_{1:N}, \dots, \nu(N) \leqslant U_{N:N}] \quad (4)$$

from the following recursion. Let $s_0(x) \equiv 1$ and

$$s_n(x) = \int_{\nu(n)}^{x} \int_{\nu(n-1)}^{u_n} \cdots \int_{\nu(1)}^{u_2} du_1 \cdots \times du_{n-1}\, du_n.$$

Then $Ds_n(x) = s_{n-1}(x)$ and $s_n(\nu(n)) = \delta_{0,n}$. Thus, (s_n) is the Sheffer (Appell) sequence for D with roots in ν. The probability (4) is equal to $N!s_N(1)$. This structure was emphasized again by Whittle [20], when closed forms of (4) had already been worked out for functions ν of the type $\nu(i) = (ic + d)_+ = \max(0, ic + d)$. After the development of the finite operator calculus [14] it became apparent that those closed forms are only special cases of the general representation theorem (3) for Sheffer sequences. The following example, which includes the Kolmogorov–Smirnov test*, demonstrates the method. Let $X_1, \dots, X_N$ be i.i.d. random variables with empirical distribution function $F_N(x)$. The one-sample one-sided Rényi-type distribution*

$$\Pr\left\{ \frac{F_N(x) - cF(x)}{1 - dF(x)} \leqslant s \quad \text{for all } a \leqslant F(x) \leqslant b \right\}$$

can be transformed into the form (4) if we define $\nu(i) = 0$ for $i = 0, \dots, L := \lfloor N(s + a(c - sd)) \rfloor$, $\nu(i) = (s - i/N)/(ds - c)$ for $i = L + 1, \dots, M := \lfloor N(s + b(c - sd)) \rfloor$, and $\nu(i) = M$ for $i > M$ (assuming that $c - sd > 0$ and $d < 1$). Therefore, $s_i(x)$ equals the basic sequence $b_i(x) = x^i/i!$ for all $i = 0, \dots, L$. Now choose y and a in (3) such that $y - ai = (s - i/N)/(ds - c)$. From $s_i(y - ai) = (y - ai)^i/i!$ for $i = 0, \dots, L$ and 0 for $i = L + 1, \dots, M$ one can calculate $s_n(x)$ for all $n \leqslant M$ using (3):

$$s_n(x) = \sum_{i=0}^{L} \left[\left(\frac{s - i/N}{ds - c} \right)^i \times \frac{x - (s - n/N)/(ds - c)}{i!(n-i)!} \times \left(x - \frac{s - i/N}{ds - c} \right)^{n-i-1} \right].$$

But for the remaining degrees $i > M$ we know that $s_i(M) = 0$. Therefore, we choose now $y = M$ and $a = 0$ to calculate $s_n(x)$ for $n > M$, again from (3), using the previously obtained polynomials s_i for $i \leqslant M$. So the desired final answer $N!s_N(1)$ is a double sum if $N > M$. For more details and applications see ref. 10. Similarly, Sheffer polynomials can be used to derive Takács' goodness-of-fit* distribution [17, 11] for the number of crossings between the hypothetical and the empirical distribution function.

The function ν in the above example consists of three affine pieces, i.e., pieces of the form $\nu(i) = iu + v$ for different choices of u and v. The multiplicity of the resulting sum is one less than the minimal number of such affine pieces that are necessary to describe $\nu(i)$ for $i = 0, \dots, N$.

One of the simplest recursions in combinatorial statistics is of the form ($n \geqslant 1$, $k \geqslant 1$)

$$p(n, k) = \xi_k p(n, k-1) + \eta_n p(n-1, k),$$

where the nonzero factor sequences $\xi_0, \xi_1, \dots$ and $\eta_0, \eta_1, \dots$ do not really contribute to the complexity of the problem, because a solution of

$$d(n, k) = d(n, k-1) + d(n-1, k) \quad (5)$$

yields

$$p(n, k) = \left(\prod_{i=1}^{k} \xi_i\right)\left(\prod_{j=1}^{n} \eta_j\right) d(n, k).$$

The difficulties arise only from prescribed initial values $d(n, \nu(n))$, say, where $\nu(0)$, $\nu(1), \dots$ is a sequence of nonnegative integers. In applications, very often $d(n, \nu(n)) = \delta_{0,n}$ and $\nu(n)$ is monotone nondecreasing. If $d(0, \cdot)$ is a nonzero constant, the solution of (5) is the Sheffer sequence for ∇ with initial value sequence $d(n, \nu(n))$. For piecewise affine functions ν, this solution can be calculated from the representation theorem (3). Applications of this method to two-sample Kolmogorov–Smirnov* and Rényi-type distributions* are given in ref. 10. Similar applications occur in the distribution of the number of crossings between two empirical distribution functions [18, 11] (Takács' goodness-of-fit distribution*).

For arbitrary sequences $\nu(0), \nu(1), \dots$ in (4) and (5), only recursive and determinantal solutions are known. In general, the same is true for the two-boundary problems

$$\Pr[\nu(i) \leqslant U_{i:N} \leqslant \mu(i) \quad \text{for all } i = 1, \dots, N] \quad (4')$$

and

$$d(n, k) = \begin{cases} d(n, k-1) + d(n-1, k) \\ \qquad \text{for all } k \leqslant \mu(n), \\ 0 \quad \text{for all } k > \mu(n), \end{cases} \quad (5')$$

with given initial value sequence $d(n, \nu(n))$. The solution can now be expressed by piecewise Sheffer polynomial functions. For applications to Kolmogorov–Smirnov and Rényi-type tests see ref. 10.

ORTHOGONAL SHEFFER POLYNOMIALS

From the general theory of expansions of a measurable function in terms of orthogonal polynomials [16], only expansions using Hermite or Charliér polynomials (Edgeworth's expansion) gained some popularity in statistics [8]. The question as to which Sheffer polynomials are also orthogonal has been answered first by Meixner [9], and was discussed also in refs. 16 and 1. Meixner classified the orthogonal Sheffer polynomials by their defining formal power series $f(t)$ and $u(t)$ (see Property 1). He obtained five classes. Besides some norming factors, the first three classes are the Hermite, Laguerre, and Charlier polynomials. The last two classes are now called the Meixner polynomials* of the first and second kind. Meixner polynomials of the first kind result from

$$f(t) = (1-t)^{-\beta}, \qquad u(t) = \ln\frac{1 - t/c}{1 - t},$$

where $c \neq 0, 1$ and $\beta \neq 0, -1, -2, \dots$. Meixner polynomials of the second kind are obtained from $f(t) = ((1 + \delta t)^2 + t^2)^{-\beta/2}$ and $u(t) = \tan^{-1}[t/(1 + \delta t)]$, where $\beta \neq 0, -1, -2, \dots$. Eagleson [2] showed how to use all these different types of orthogonal Sheffer polynomials for expansions of bivariate distributions. Extending the concept, Freeman [3] was able to find a much larger class of orthogonal generalized Sheffer polynomials.

SERIES EXPANSIONS

Expansions of $f(x+h)$ like the Taylor series $\Sigma_{n\geqslant 0}(h^n/n!)D^nf(x)$ and the difference series $\Sigma_{n\geqslant 0}\binom{h}{n}\Delta^n f(x)$ are only special cases of the class of expansions $\Sigma_{n\geqslant 0}b_n(h)Q^nf(x)$, where (b_n) is the basic sequence for the delta operator Q. Such a series expansion is well defined and terminates if f is a polynomial. For generalizations see ref. 14 (p. 699).

APPROXIMATION OPERATORS

Every Sheffer sequence (s_n) with defining formal power series $f(t)$ and $u(t)$ (see Property 1) operates on the real valued functions F with domain $[0, \infty)$ by

$$T(F, x, a) = \frac{\exp(-xau(1))}{f(1)} \times \sum_{k=0}^{\infty} s_k(ax)F\left(\frac{k}{a}\right), \quad a > 0,$$

whenever the right-hand side is defined. The properties of this approximation operator have been investigated by Ismail [5]. For a connection with exponential operators, see ref. 6 by the same author. If (s_n) is an Appell sequence, the resulting class of operators was studied by Jakimovski and Leviatan [7]. Best known in this class is the Szász operator where $s_n(x) = x^n/n!$.

References

Sheffer polynomials and related calculus are given in refs. 1, 3, 4, 9, and 12–15. The remaining references concern related topics and applications.

[1] Brown, J. W. (1975). On orthogonal Sheffer sequences. *Glasnik Mat.*, **10**, 63–67.

[2] Eagleson, G. K. (1964). *Ann. Math. Statist.*, **35**, 1208–1215. (Applications of orthogonal Sheffer polynomials.)

[3] Freeman, J. M. (1987). *Studies in Appl. Math.*, **77**, No. 2. (Orthogonal generalized Sheffer polynomials.)

[4] Ihrig, E. C. and Ismail, M. E. H. (1981). A q-umbral calculus. *J. Math. Anal. Appl.*, **84**, 178–207. (Generalizes Sheffer sequences.)

[5] Ismail, M. E. H. (1974). On a generalization of Szász operators. *Mathematica (Cluj)*, **16**, 259–267.

[6] Ismail, M. E. H. (1978). Polynomials of binomial type and approximation theory. *J. Approx. Theory*, **23**, 177–186. (Exponential operators.)

[7] Jakimovski, A. and Leviatan, D. (1969). *Mathematica (Cluj)*, **11**, 79–103.

[8] Kendall, M. G. and Stuart, A. (1952). *The Advanced Theory of Statistics*, Vol. I. Griffin, London, England.

[9] Meixner, J. (1934). *J. Lond. Math. Soc.*, **9**, 6–13. (Classifies orthogonal Sheffer polynomials.)

[10] Niederhausen, H. (1981). Sheffer polynomials for computing exact Kolmogorov–Smirnov and Rényi-type distributions. *Ann. Statist.*, **9**, 923–944.

[11] Niederhausen, H. (1986). *J. Statist. Plann. and Inference*, **14**, 95–114. (Survey of applications of Sheffer polynomials to random walks.)

[12] Roman, S. M. and Rota, G.-C. (1978). The umbral calculus. *Adv. Math.*, **27**, 95–188. (Introduces delta functionals. Extensive bibliography.)

[13] Roman, S. (1984). *The Umbral Calculus*. Academic Press, New York.

[14] Rota, G.-C., Kahaner, D., and Odlyzko, A. (1973). Finite operator calculus. *J. Math. Anal. Appl.*, **42**, 684–760. (Basic reference, contains many examples, problems and history.)

[15] Sheffer, I. M. (1939). *Duke Math. J.*, **5**, 590–622. (Origin of the theory of Sheffer sequences.)

[16] Szegö, G. (1939). *Orthogonal Polynomials*, 4th ed. (1975). AMS, Providence, RI.

[17] Takács, L. (1971). *J. Appl. Prob.*, **8**, 321–330. (Exact distribution of the number of crossings. One-sample case.)

[18] Takács, L. (1971). *Ann. Math. Statist.*, **42**, 1157–1166. (Exact distribution of the number of crossings. Two-sample case.)

[19] Wald, A. and Wolfowitz, J. (1939). *Ann. Math. Statist.*, **10**, 105–118. (Recursive approach to distributions of Kolmogorov–Smirnov type.)

[20] Whittle, P. (1960). *Ann. Math. Statist.*, **32**, 499–505. (Appell sequences in Kolmogorov–Smirnov distributions.)

(BELL POLYNOMIALS
BERNOULLI POLYNOMIALS
CHEBYSHEV–HERMITE POLYNOMIALS
LAGRANGE EXPANSIONS
LAGUERRE SERIES

MEIXNER POLYNOMIALS
ORTHOGONAL EXPANSIONS)

H. NIEDERHAUSEN

SHEPPARD CORRECTIONS *See* CORRECTION FOR GROUPING; GROUPED DATA

SHEPPARD'S FORMULA

A formula relating the probability (P) that two bivariate normal* variables both exceed their respective expected values to their correlation coefficient* (ρ). The formula is

$$P = \frac{1}{4} + \frac{1}{2\pi}\sin^{-1}\rho.$$

It has been used to construct an estimator for ρ based on the observed proportion of individuals in a sample for which both variables exceed their respective sample *medians*.

Bibliography

Sheppard, W. F. (1899). *Philos. Trans. R. Soc., London Ser. A.*, **192**, 101–167.

(BIVARIATE NORMAL DISTRIBUTION
CORRELATION
TETRACHORIC CORRELATION
TWO-BY-TWO TABLES)

SHERMAN DISTRIBUTION

A continuous variable X possesses the *Sherman distribution* (Sherman [3]) with parameter n (called degrees of freedom*) if its density is given by

$$f(x) = \begin{cases} \sum_{m=1}^{n} m\, b_m x^{m-1}, & x \in \left[0, \frac{n}{n+1}\right], \\ 0, & \text{otherwise.} \end{cases}$$

The coefficient

$$b_m = \sum_{j=0}^{r} \left[(-1)^{(m+j+1)} \binom{n+1}{j+1} \binom{m+j}{j} \times \binom{n}{m} \left(\frac{n+j}{n+1}\right)^{m-n}\right],$$

where r is a nonnegative integer satisfying

$$\frac{n-r-1}{n+1} \leqslant x < \frac{n-r}{n+1}.$$

The first two moments of this distribution are

$$E(X) = \left(1 - \frac{1}{n+1}\right)^{n+1},$$

$$\operatorname{var}(X) = \frac{2n^{n+2} + n(n-1)^{n+2}}{(n+2)(n+1)^{n+2}} - \left(1 - \frac{1}{n+1}\right)^{2(n+1)}.$$

The genesis of this distribution, useful for testing for randomness* (see, e.g., Bartholomew [1] and Johnson and Kotz [2]), is as follows: Let n points on the interval [0, 1] be randomly chosen. Order these points and let X_i be the length of the ith interval from the left ($i = 1, \ldots, n+1$). The random variable $X = \frac{1}{2}\sum_{j=1}^{n+1}|X_j - 1/(n+1)|$ has a Sherman distribution with n degrees of freedom.

References

[1] Bartholomew, D. J. (1954). *Biometrika*, **41**, 556–559.

[2] Johnson, N. L. and Kotz, S. (1970). *Distributions in Statistics: Continuous Univariate Distributions*, Vol. 2. Wiley, New York, Chap. 33.

[3] Sherman, B. (1950). *Ann. Math. Statist.*, **21**, 339–361.

(RANDOMNESS, TESTS FOR)

SHERMAN TEST STATISTIC *See* SHERMAN DISTRIBUTION

SHEWHART CHARTS *See* CONTROL CHARTS; QUALITY CONTROL, STATISTICAL

SHEWHART PRINCIPLE

This is a basic philosophical principle underlying statistical quality control* methodology (and Shewhart control charts* in particular).

Measured quality of manufactured products is always subject to a certain amount of variation as a result of chance. Some stable system of chance causes is inherent in any particular scheme of production and inspection. The reasons for variation outside this stable pattern may be discovered and corrected (Shewhart [2]).

See also Levey and Jennings [1].

References

[1] Levey, S. and Jennings, E. R. (1980). *Amer. J. Chem. Pathol.*, **20**, 1059–1066.

[2] Shewhart, W. A. (1939). In *Statistical Methods from the Viewpoint of Quality Control*, W. E. Deming, ed. Washington, DC.

(CONTROL CHARTS
QUALITY CONTROL, STATISTICAL)

SHIFT MODEL *See* LOCATION-SCALE PARAMETER

SHIFT, TEST FOR *See* MEAN SLIPPAGE PROBLEM

SHOCK MODELS

The class of shock models arises in reliability theory* when a device is subjected to shocks occurring over time according to some stochastic process*. The device is assumed to have an ability to withstand a random number of these shocks. A common assumption is that the shock process and the random number of shocks survived are independent. If we denote the stochastic process that counts the shocks by $\{S(t): t \geqslant 0\}$ and the random number of shocks survived by M, then the survival function or probability that an item will survive t units of time has the mathematical form

$$\bar{F}(t) = \sum_{k=0}^{\infty} P(S(t) = k) P(M > k).$$

One area that has been investigated extensively is that of class preservation theorems. There are several nonparametric classes of life distributions of interest in reliability theory. These classes represent various models of adverse (beneficial) stochastic aging. Included among these are the increasing (decreasing) failure rate, increasing (decreasing) failure rate average, new better (worse) than used, new better (worse) than used in expectation, and decreasing (increasing) mean residual life classes of life distributions. These are commonly referred to in the literature by the acronyms IFR (DFR), IFRA (DFRA), NBU (NWU), NBUE (NWUE), and DMRL (IMRL). A description of these classes in both the continuous and discrete cases can be found in Esary et al. [17] and in HAZARD RATE AND OTHER CLASSIFICATIONS OF DISTRIBUTIONS; *see also* RELIABILITY THEORY, PROBABILISTIC.

Class preservation theorems deal with conditions under which membership by M to a certain class of discrete distributions implies the membership of the lifetime of the item to the analogous class of continuous distributions. Results of this type have been obtained by a number of authors. In their remarkable 1973 paper Esary, Marshall, and Proschan [17] obtained preservation theorems when $\{S(t): t \geqslant 0\}$ is an ordinary Poisson process*. A-Hameed and Proschan [3, 4] subsequently generalized this to the cases where $\{S(t): t \geqslant 0\}$ is a nonhomogeneous Poisson process or a nonstationary pure birth process. Klefsjö [27, 28] has obtained DMRL and IMRL results under conditions different from A-Hameed and Proschan in the pure birth shock model and has obtained results for another pair of classes of life distributions, called the harmonic new better (worse) than used in expectation (HNBUE and HNWUE) classes when the shock process is either a Poisson process or a nonhomogeneous Poisson process. Thall [41] considers the case in which the shock process is a Poisson cluster process and obtains a certain preservation result in the DFR case. He also provides a counterexample to demonstrate that an IFR preservation result does not hold.

Ohi et al. [34], Ghosh and Ebrahimi [22], and Joe and Proschan [26] have obtained class preservation results assuming that the stochastic process governing the shocks has stationary independent nonnegative increments. Additional assumptions dealing with total positivity* are made in some cases. This was motivated by the fact that the original paper of Esary et al. used properties of total positivity in the case of a shock process that is Poisson to obtain closure results in the IFR, DFR, IFRA, DFRA, IMRL, and DMRL cases. Some authors have provided other proofs of the Esary et al. results using different techniques, e.g., Derman and Smith [13] and Griffith [24].

Block and Savits [8] have obtained a NBUE closure result under conditions more general than Esary et al. and A-Hameed and Proschan; they obtain a NBUE (NWUE) closure result if the shock process $\{S(t): t \geqslant 0\}$ has independent NBUE (NWUE) interarrival times with decreasing (increasing) mean interarrival times. They have also obtained NBU (NWU) preservation results where the shock process satisfies what can be viewed as a certain type of NBU (NWU) condition. They use this to reprove a result of A-Hameed and Proschan, which says that the NBU (NWU) property is preserved when the shock process has independent NBU (NWU) interarrival times with stochastically decreasing (increasing) times between successive shocks.

Klefsjö [28] has obtained analogous results for the HNBUE and NHWUE cases. Neuts and Bhattacharjee [33] have obtained preservation theorems for phase-type distributions*, where the random variable M is of phase-type and the shock process is a phase-type renewal process.

An interesting variation on this general univariate shock model is to hypothesize a cumulative damage mechanism to determine the random number of shocks that an item can survive. Associated with each shock is a random amount of damage (*see* CUMULATIVE DAMAGE MODELS). These damages accumulate in some manner. A common assumption is that the damages are additive, although more general models can be treated. Failure occurs when the accumulated damage exceeds a particular threshold x (fixed or random) that is inherent in the item. Thus for an additive damage model we can describe the random variable M by

$$M = m \Leftrightarrow \sum_{i=1}^{m-1} X_i \leqslant x < \sum_{i=1}^{m} X_i,$$

where $\{X_i\}$ represents the sequence of random damages.

Note that the class of IFRA distributions arises naturally in this context. Esary et al. [17] have proved that if the damages caused by shocks are nonnegative, independent and identically distributed (i.i.d.), accumulate additively, occur at times governed by a Poisson process, and cause failure when a fixed threshold is exceeded, then the lifetime of the item belongs to the IFRA class of life distributions. To emphasize the generality of this model, no assumption is made on the common distribution of the damages other than being the distribution of a nonnegative random variable.

The strong assumption that the damages are i.i.d. can, in fact, be weakened. If independence is retained but successive shocks are assumed to cause stochastically greater damages, then the life distribution of the item will still be IFRA. It is possible to weaken the assumptions even further by allowing dependence, but making technical assumptions that essentially say that an accumulation of damage stochastically lowers resistance to further damage and that for any given level of accumulated damage, later shocks are stochastically more severe.

Returning to the case of i.i.d. damages but permitting the threshold to be random rather than constant, an IFRA distribution is obtained for the lifetime of the item if the threshold is assumed to be IFRA.

A variation of these results, in which the accumulation of damage is more general than the additive damage model, is due to Ross

[36]. The damages are again i.i.d. and nonnegative. The accumulated damage due to the first n shocks is $D(X_1,, \dots, X_n)$, which is assumed to be a symmetric nondecreasing function of the damages. The shock process is assumed to be a nonhomogeneous Poisson process with a star-shaped mean value function. Then without any assumption (other than nonnegativity) on the common distribution of the damages, an IFRA distribution will result for the item.

Another natural physical model for IFRA distributions can be formulated combining results from a paper by Ross et al. [37] with results from Esary et al. [17]. Suppose that the item subjected to a stream of shocks is actually a system of components. Each shock causes the failure of one component and the affected component is equally likely to be any of the components that were still working just prior to the shock. Then for any arbitrary system configuration, the lifetime of the system is IFRA if the shock process is Poisson.

Wear processes* generalize these cumulative damage models. The wear process $\{Z(t) : t \geqslant 0\}$ describes the amount of wear or damage that has accumulated by time t. In the case of the additive damage models previously described, the wear process $Z(t)$ is given simply by $Z(t) = \sum_{i=1}^{S(t)} X_i$, where X_i is the random damage caused by the ith shock.

More generally we can have continuous wear processes. Such models have been treated by Esary et al. [17], A-Hameed [1, 2], Gottlieb [23], and Çinlar [12] among others.

Esary et al. have proven, for example, that if the wear process is a Markov process* with nonnegative increments in which the device is stochastically more prone to additional wear as its age or accumulated wear increase, then the waiting time for the first passage time beyond any fixed threshold is IFRA.

A-Hameed has obtained class preservation results for gamma wear processes with random thresholds. Gottlieb has investigated conditions that result in IFR and IFRA distributions for the item under consideration. Çinlar has studied the model under the assumption that the wear process is a Markov additive process.

Several authors have investigated replacement policies for items subject to failure according to various shock, cumulative damage, or wear process models. Borland and Proschan [10] consider a model in which each shock adds a fixed cost per unit of time to the operation of the system and find an optimal replacement time. Taylor [40] obtains an optimal policy when the item suffers shocks according to a Poisson process, the damages accumulate additively, and the failure probability is an increasing function of the accumulated damage.

A-Hameed and Shimi [5] have examined optimal replacement policies where the cost structure is more general but the shocks occur at discrete time points. Zuckerman [43] removed the restriction that the device can be replaced only at a shock point of time.

A-Hameed [2] has also investigated optimal replacement models for devices subject to a gamma wear process. Feldman [18–20] and Zuckerman [42] have studied replacement policies where the damage process is semi-Markov*. Chikte and Deshmukh [11] consider the joint problem of optimal maintenance and replacement in an additive damage model.

Bivariate and multivariate shock models have been studied by several authors. These have given rise to certain parametric distributions as well as nonparametric classes of life distributions. Marshall and Olkin [30] have studied a particular multivariate exponential distribution*. The bivariate version can be thought of as the joint distribution of a pair of components subjected to three independent Poisson streams of shocks. One stream shocks both components, while the other streams provide shocks to the first component only and the second component only. A component fails when it sustains a shock. This can be extended to a nonfatal shock model in which shocks are not necessarily fatal.

Arnold [6] and Block [7] have investigated nonfatal (or hierarchical) shock models in which a bivariate geometric compounding mechanism is used. This approach unifies a number of the bivariate exponential distributions such as those of Marshall and Olkin [30], Downton [14], Hawkes [25], and Paulson [35].

Marshall and Shaked [31] have studied multivariate cumulative damage models that produce a joint distribution lying in the multivariate increasing failure rate average class of Esary and Marshall [16]. Savits and Shaked [38] show that in a special case the joint distribution lies in the multivariate increasing failure rate average class of Block and Savits [9]. In addition, they consider a shock model for an IFRA process. Various other authors including Griffith [24], Ghosh and Ebrahimi [21, 22], El-Neweihi et al. [15], Klefsjö [29], and Marshall and Shaked [32] have investigated bivariate and multivariate shock models giving rise to various other multivariate nonparametric classes of life distributions of interest in reliability theory. Shaked [39] has surveyed some of these results.

References

[1] A-Hameed, M. S. (1975). *IEEE Trans. Rel.*, **24**, 152–154.

[2] A-Hameed, M. S. (1977). In *The Theory and Applications of Reliability*, Vol. 1, C. P. Tsokos and I. N. Shimi, eds. Academic, New York, pp. 397–412.

[3] A-Hameed, M. S. and Proschan, F. (1973). *Stoch. Processes Appl.*, **1**, 383–404.

[4] A-Hameed, M. S. and Proschan, F. (1975). *J. Appl. Prob.*, **12**, 18–28.

[5] A-Hameed, M. S. and Shimi, I. N. (1978). *J. Appl. Prob.*, **15**, 153–161.

[6] Arnold, B. C. (1975). *J. Appl. Prob.*, **12**, 142–147.

[7] Block, H. W. (1977). In *The Theory and Applications of Reliability*, Vol. 1, C. P. Tsokos and I. N. Shimi, eds. Academic, New York, pp. 349–372.

[8] Block, H. W. and Savits, T. H. (1978). *J. Appl. Prob.*, **15**, 621–628.

[9] Block, H. W. and Savits, T. H. (1980). *Ann. Prob.*, **8**, 793–801.

[10] Borland, P. J. and Proschan, F. (1983). *Operat. Res.*, **31**, 697–704.

[11] Chikte, S. D. and Deshmukh, S. D. (1981). *Naval Res. Logist. Quart.*, **28**, 33–46.

[12] Çinlar, E. (1977). In *The Theory and Applications of Reliability*, Vol. 1, C. P. Tsokos and I. N. Shimi, eds. Academic, New York, pp. 193–214.

[13] Derman, C. and Smith, D. R. (1980). *Naval Res. Logist. Quart.*, **27**, 703–708.

[14] Downton, F. (1970). *J. R. Statist. Soc. B.*, **32**, 408–417.

[15] El-Neweihi, E., Proschan, F., and Sethuraman, J. (1983). *Operat. Res.*, **31**, 177–183.

[16] Esary, J. D. and Marshall, A. W. (1979). *Ann. Prob.*, **7**, 359–370.

[17] Esary, J. D., Marshall, A. W., and Proschan, F. (1973). *Ann. Prob.*, **1**, 627–649.

[18] Feldman, R. M. (1976). *J. Appl. Prob.*, **13**, 108–117.

[19] Feldman, R. M. (1977). In *The Theory and Applications of Reliability*, Vol. 1, C. P. Tsokos and I. N. Shimi, eds. Academic, New York, pp. 215–226.

[20] Feldman, R. M. (1977). *Ann. Prob.*, **5**, 413–429.

[21] Ghosh, M. and Ebrahimi, N. (1981). *Egypt. Statist. J.*, **2**, 36–55.

[22] Ghosh, M. and Ebrahimi, N. (1982). *J. Appl. Prob.*, **19**, 158–166.

[23] Gottlieb, G. (1980). *J. Appl. Prob.*, **17**, 745–752.

[24] Griffith, W. S. (1982). *Naval Res. Logist. Quart.*, **29**, 63–74.

[25] Hawkes, A. G. (1972). *J. R. Statist. Soc. B.*, **34**, 129–131.

[26] Joe, H. and Proschan, F. (1980). Shock Models Arising from Processes with Stationary, Independent, Nonnegative Increments. *Tech. Rep. No. M*557, Department of Statistics, Florida State University, Tallahassee, FL.

[27] Klefsjö, B. (1981). *J. Appl. Prob.*, **18**, 554–560.

[28] Klefsjö, B. (1981). *Scand. J. Statist.*, **8**, 39–47.

[29] Klefsjö, B. (1982). *IAPQR Trans., J. Indian Ass. Productivity, Quality, Rel.*, **7**, 87–96.

[30] Marshall, A. W. and Olkin, I. (1967). *J. Amer. Statist. Ass.*, **62**, 30–44.

[31] Marshall, A. W. and Shaked, M. (1979). *Ann. Prob.*, **7**, 343–358.

[32] Marshall, A. W. and Shaked, M. (1983). *Adv. Appl. Prob.*, **15**, 601–615.

[33] Neuts, M. F. and Bhattacharjee, M. C. (1981). *Naval Res. Logist. Quart.*, **28**, 213–220.

[34] Ohi, F., Kodama, M., and Nishida, T. (1977). *Rep. Stat. Appl. Res.*, **24**, 181–190.

[35] Paulson, A. S. (1973). *Sankhyā A.*, **35**, 69–78.

[36] Ross, S. M. (1981). *Ann. Prob.*, **9**, 896–898.

[37] Ross, S. M., Shashahani, M., and Weiss, G. (1980). *Math. Operat. Res.*, **5**, 358–365.

[38] Savits, T. H. and Shaked, M. (1981). *Stoch. Processes Appl.*, **11**, 273–283.

[39] Shaked, M. (1983). Wear and Damage Processes from Shock Models in Reliability Theory. *Proceedings of the Conference on Stochastic Failure Models*, to appear.

[40] Taylor, H. M. (1975). *Naval Res. Logist. Quart.*, **22**, 1–18.

[41] Thall, P. F. (1981). *J. Appl. Prob.*, **18**, 104–111.

[42] Zuckerman, D. (1978). *J. Appl. Prob.*, **15**, 629–634.

[43] Zuckerman, D. (1980). *Naval Res. Logist. Quart.*, **27**, 521–524.

Bibliography

Birnbaum. Z. W. and Saunders, S. C. (1958). *J. Amer. Statist. Ass.*, **53**, 151–160.

Cox, D. R. (1962). *Renewal Theory*. Methuen, London, England.

Epstein, B. (1958). *Industrial Quality Control*, **15**, 2–7. (His random peaks model is one of the early shock models.)

Gaver, D. P. (1963). *Technometrics*, **5**, 211–226. (One part of the paper describes a model of system life where random shocks cause failure.)

Hill, D. L., Saunders, R., and Laud, P. W. (1980). *Canad. J. Statist.*, **8**, 87–93. (Maximum likelihood estimation of the probability mass function for the random number of shocks that the device can survive.)

Lo, A. Y. (1981). *Scand. J. Statist.*, **8**, 237–242. (Bayesian nonparametric inference for the shock model.)

Mercer, A. and Smith, C. S. (1959). *Biometrika*, **46**, 30–35. (Study of a problem arising from wear of conveyor belting. Fixed and linearly decreasing threshold models and approximations for the moments are discussed.)

Mercer, A. (1961). *J. R. Statist. Soc. B.*, **23**, 368–376. (The probability that a component fails is dependent upon the wear it has received and its age. Replacement strategies are compared.)

Morey, R. C. (1966). *Operat. Res.*, **14**, 902–908. (An early cumulative damage paper.)

Proschan, F. and Sullo, P. (1976). *J. Amer. Statist. Ass.*, **71**, 465–472. (Estimation for the Marshall–Olkin multivariate exponential distribution.)

(CUMULATIVE DAMAGE MODELS
HAZARD RATE AND OTHER
CLASSIFICATIONS OF DISTRIBUTIONS
RELIABILITY, PROBABILISTIC
WEAR PROCESSES)

WILLIAM S. GRIFFITH

SHORACK ESTIMATORS

Suppose independent observations $Y_1, \ldots, Y_n$ all have the same density that is symmetric about θ; that is, $Y_i - \theta$ has density f that is symmetric about 0. Many interesting robust* estimators $\hat{\theta}_n$ of θ are solutions of an equation of the form $0 = \Sigma_1^n \psi((Y_i - \theta)/D_n)$, for some estimator of scale $D_n \to_p$ (some d), where d depends on f. We shall use the robust scaling estimator

$$D_n \equiv median\{|Y_i - \hat{\theta}_n^0| : 1 \leqslant i \leqslant n\},$$

where $\hat{\theta}_n^0$ is a preliminary estimator of scale satisfying $n^{1/2}(\hat{\theta}_n^0 - \theta) = O_p(1)$. Under regularity, the above estimating equation is solved by iteration, and the resulting estimator satisfies

$$n^{1/2}(\hat{\theta}_n - \theta) \underset{d}{\to} N(0, V^2),$$

with

$$V^2 = \frac{d^2 E \psi^2((Y - \theta)d)}{\{E\psi'((Y - \theta)/d)\}^2}.$$

The first iteration toward $\hat{\theta}_n$ is just the mean of the "pseudo-observations"

$$Y_i^* = \hat{\theta}_n^0 + \frac{D_n \psi\left((Y_i - \hat{\theta}_n^0)/D_n\right)}{\Sigma_1^n \psi'\left((Y_j - \hat{\theta}_n^0)/D_n\right)/n}$$

One then supposes that $T_n \equiv n^{1/2}(\hat{\theta}_n - \theta_0)/s_n$, where $s_n^2 \equiv \Sigma_1^n (Y_i^* - \overline{Y}^*)^2/(n-1)$, has approximately a t distribution whose degrees of freedom are one less than the number of distinct pseudovalues. Two especially interesting ψ functions have been proposed. If antisymmetric ψ equals t or k as $0 \leqslant t \leqslant k$ or $k \leqslant t$, then $\hat{\theta}_n$ will be called the *Huber* (k) *estimator*. If $\psi(t)$ equals t, a, $a(a + b + c - t)/c$, or 0 as $0 \leqslant t \leqslant a$, $a \leqslant t \leqslant a + b$, $a + b \leqslant t \leqslant a + b + c$, or

$a + b + c \leqslant t$, then $\hat{\theta}_n$ will be called the *Hampel* (a, b, c) *estimator*.

Suppose we numerically integrate the functions $f(x)$, $xf(x)$, and $x^2 f(x)$ on the intervals $[(i-1)\Delta, i\Delta]$ for $i = 1, 2, \ldots, I$, and store the results. Performing simple algebra on these stored results allows us to compute the asymptotic variance V^2 for any Huber (k) or Hampel (a, b, c) estimator whose parameters a, b, c, and k are all multiples of Δ. Dividing the Cramér–Rao bound* by V^2 then gives us the asymptotic efficiency of our estimator. This procedure was carried out in ref. 3, where the efficiencies of 16 Hubers and 43 Hampels are tabled for each of 20 different densities f. Small-sample robustness of level of significance, of power, and of length of confidence interval of the studentized version of these estimators are also studied there via Monte Carlo methods*. [Note the criticism of ref. 2 (p. 150), and our present definition of s_n.] The upshot of all this is that the studentized version of the estimator Hampel(3, 0, 11) performs very well. [The authors of ref. 1 effectively replace D_N by $D_N^* \equiv D_N/0.6745$ so that $D_N^* \rightarrow_p 1$ if f is a normal density. They also use a, b, c, and k to denote the corner points we denote by a, $a+b$, $a+b+c$, and k. We let Huber*(k) and Hampel*(a, b, c) denote estimators in their scheme of labeling. Then Huber(k) = Huber*$(0.6745k)$ and Hampel(a, b, c) = Hampel*$(0.6745a, 0.6745(a+b), 0.6745(a+b+c))$. Hence Hampel(3, 0, 11) = Hampel*(2.02, 2.02, 9.44).]

Another interesting estimator, to be labeled *adaptive*, is Hampel$(2.25A_n, 0, 7A_n)$, in which A_n denotes the natural estimate of the variance of Huber(1) divided by the natural estimate of the variance of Huber(3). See ref. 3 for its impressive asymptotic efficiencies. It is the estimators Hampel (3, 0, 11) = Hampel*(2, 2, 9.5) and adaptive to which the editors' suggested title for this article could reasonably be applied.

References

[1] Andrews, D., Bickel, P., Hampel, F., Huber, P., Rogers, W., and Tukey, J. (1972). *Robust Estimates of Location*. Princeton University Press, Princeton, NJ.

[2] Huber, P. (1981). *Robust Statistics*. Wiley, New York.

[3] Shorack, G. (1976). *Statist. Neerlandica*, **30**, 119–141.

(ROBUST ESTIMATION)

GALEN R. SHORACK

SHORE APPROXIMATIONS (NORMAL DISTRIBUTION)

Shore [1] derived several approximations for the inverse of the standard normal distribution among which the most accurate is

$$Z = -5.5310\left\{\left[(1-p)/p\right]^{0.1193} - 1\right\}, \quad p \geqslant \tfrac{1}{2}.$$

A simpler form is

$$Z = -0.4115\left[\{(1-p)/p\} + \ln\{(1-p)/p\} - 1\right], \quad p \geqslant \tfrac{1}{2},$$

where $F(Z) = \int_{-\infty}^{Z} (2\pi)^{-1/2} \exp(-u^2/2)\, du = p$ and $Z = F^{-1}(p)$.

These approximations were extended in shore [2], where additional references and results are presented.

References

[1] Shore, H. (1982). *Appl. Statist.*, **31**, 108–114.

[2] Shore, H. (1986). *SIAM J. Statist. Comput.*, **7**, 1–23.

(APPROXIMATIONS TO DISTRIBUTIONS NORMAL DISTRIBUTION)

SHORTCUT METHODS

In earlier usage, this term referred to ways of reducing the amount of computation required in applying statistical procedures. The resulting procedures are usually not "optimum" according to some specified requirements, but should be of reasonably comparable efficiency.

The term "short-cut" was used by Hartley [1, 2] to describe replacement of sums of squares in analysis-of-variance* tests by ranges*. Similar examples are estimation of parameters by order statistics* and certain distribution-free* tests.

The methods are sometimes called "quick and dirty," the last adjective referring to lack of optimality, but not, usually, having a perjorative connotation.

At present the term "short-cut" is commonly applied to any method of quick appraisal—for example by use of simple graphical representation*. This, of course, includes the earlier usage, but is not restricted to it.

References

[1] Hartley, H. O. (1950). *Biometrika*, **37**, 145–148.

[2] Hartley, H. O. (1950). *Biometrika*, **37**, 308–312.

(DISTRIBUTION-FREE METHODS
HARTLEY'S $F_{\max}$ TEST)

SHORTH

Shorth is an adaptive L estimator defined as the sample mean of the "shortest half" of the sample (chosen as the first $l + [n/2]$ order statistics*, $x_{(1)}, \dots, x_{(l+[n/2])}$, where l minimizes $(x_{(l+[n/2])} - x_{(l)})$. See Andrews et al. [1] for details.

Reference

[1] Andrews, D. F., Bickel, P. J., Hampel, F. R., Huber, P. J., Rogers, W. H., and Tukey, J. W. (1972). *Robust Estimates of Location: Survey and Advances*. Princeton University Press, Princeton, NJ.

(L-STATISTICS)

SHOT-NOISE PROCESSES AND DISTRIBUTIONS

Many statistical distributions appear as marginal distributions for shot-noise processes. Here such distributions are called *shot-noise distributions*. For instance, very simple shot-noise processes produce stable, gamma, and negative binomial distributions.

SHOT-NOISE PROCESSES

Assume that an electron that arrives at the anode in a vacuum tube at time u generates a current of magnitude $h(t - u)$ at time t in the circuit. The response function h vanishes here for physical reasons on $(-\infty, 0)$. Under additivity, the total current is given by

$$X(t) = \sum h(t - T_k), \tag{1}$$

where $\{T_k\}_{-\infty}^{\infty}$ denote the successive arrival times of the electrons. These times are random and are most often assumed to stem from a Poisson (point) process* of constant intensity λ on $(-\infty, \infty)$. This realistic assumption makes the analysis of the stochastic process* $X(t)$, $t \in R$, simple. The process $X(t)$ is the classical *shot-noise process* that has a very rich literature; cf., e.g., ref. 17, p. 423, and ref. 18, p. 150. Early works are those by Campbell [5] and Schottky [21]. A thorough mathematical study was made by Rice [20].

The process is strictly stationary with (Campbell's theorem) $E[X(t)] = \lambda \int h(u)\, du$ and

$$\operatorname{cov}[X(t), X(t - \tau)] = \lambda \int h(u) h(u - \tau)\, du.$$

For λ large, $(X(t_1), \dots, X(t_m))$ has approximately a multivariate normal distribution.

There are many stochastic processes appearing in different applied fields (astronomy, biology, hydrology*, insurance mathematics, queuing theory*, etc.) that can be considered as shot-noise processes or simple generalizations thereof. The response is often random, in which case

$$X(t) = \sum h(t - T_k) V_k, \tag{2}$$

where the V_k's are i.i.d. random variables independent of the Poisson process. More generally, one may have

$$X(t) = \sum Z(t, T_k), \qquad (3)$$

where $\{Z(\cdot, u), u \in R\}$ is a family of independent stochastic processes. The interpretation is that $Z(t, u)$ represents the random effect (noise, response) a *possible* event (shot, impulse) at u will have a t. Here t and u may also be points in, e.g., R^n (or R^m and R^n, respectively) and the $\{T_k\}$ may denote points in a spatial Poisson process of constant intensity. Of course, $Z(t, u)$ may vanish for t and u outside certain regions. The process $X(t)$ is stationary if the distribution of $(Z(t_1, u), \ldots, Z(t_m, u))$ depends only on $(t_1 - u, \ldots, t_m - u)$. This happens in particular when $Z(t, u) = h(t - u, V(u))$, where $\{V(u), u \in R\}$ is a family of i.i.d. random variables or vectors.

Some more or less well-known examples are presented below; Example 3 may be new.

Example 1. Insurance Mathematics and Waterflows. Let $\{T_k\}$ be the (Poisson) claim epochs and $\{V_k\}$ the corresponding i.i.d. claim sizes on an insurance company. The discounted value at time t, $X(t)$, of all claims after t may then be represented by (2), where $h(u) = e^{\rho u}\mathbf{1}_{(-\infty,0]}(u)$. Here $\rho > 0$ and $\mathbf{1}_A(\cdot)$ denotes the indicator function of a set A. (The interest is assumed constant and inflation neglected.)

With $h(u) = Ce^{-\rho u}\mathbf{1}_{[0,\infty)}(u)$ (C is a positive constant) and V_k the amount of rain in a rainstorm at T_k, (2) has been used as an approximate model for a steamflow, $X(t)$, at time t; see ref. 25. In this case $X(t)$, $t \in R$, is Markovian and $X(t)$, $t \in Z$, is an autoregressive* process of order 1. By formula (4) below for the characteristic function* for $X(t)$ it can be shown that if the V_k's are exponentially distributed, then $X(t)$ is gamma* distributed (and any value of the shape parameter is possible). This result was perhaps first noticed by Bartlett [1].

Example 2. Busy Lines and Thickness of Yarn. Calls to a telephone exchange with infinitely many lines arrive according to a Poisson process with intensity λ. Let $X(t)$ be the number of busy lines at time t and let $L(u)$ stand for the stochastic length of a call arriving at u; the $L(u)$'s are assumed to be i.i.d. random variables. Then $X(t)$ is given by (3) with $Z(t, u) = 1$ if $L(u) > t - u$, and $Z(t, u) = 0$ otherwise. By simple considerations or from (4) below it follows that $X(t)$ is Poisson distributed with mean $\lambda E[L(u)]$.

If, instead, $L(u)$ denotes the length of a fibre with left endpoint at position u in a thread of yarn, then $X(t) = \Sigma Z(t, T_k)$ can be interpreted as the number of fibres at position t, i.e., the thickness of the thread at t; cf., e.g., ref. 8, pp. 366–368.

Example 3. Generalization of Example 2. Groups of people arrive at a service station with infinitely many servers according to a Poisson process. The number of persons in a group arriving at u is a random variable $N(u)$ and their service times are i.i.d. random variables $L_j(u)$, $j = 1, \ldots, N(u)$. The number of people under service at t is given by $X(t) = \Sigma Z(t, T_k)$, where $Z(t, u) = \sum_{j=1}^{N(u)} \mathbf{1}_{\{L_j(u) > t-u\}}$. Example 2 covers this case if the arrival process is compound Poisson. The process $X(t)$, $t \in R$, is Markovian if the $L_j(u)$'s are exponentially distributed. If, moreover, $N(u)$ is geometrically distributed, then $X(t)$ has a negative binomial distribution* (and any value of the shape parameter is possible). This result seems to be little known; cf. ref. 4.

Example 4. Gravitation Force and Traffic Noise. Point masses (stars) are assumed to be distributed in R^3 according to a spatial Poisson process of constant intensity. Let $V(\underline{u})$ denote the mass of a star at $\underline{u}$; the $V(\underline{u})$'s are assumed to be i.i.d. random variables. Then $X(\underline{t}) = \Sigma Z(\underline{t}, T_k)$, where

$$Z(\underline{t}, \underline{u}) = C(u_1 - t_1)\|\underline{u} - \underline{t}\|^{-3}V(\underline{u})$$

is the total gravitation force in the t_1 direction on a unit mass at $\underline{t}$. (The sum is not absolutely convergent and must be interpreted suitably.) It was shown by the astronomer Holtsmark [14] that $X(t)$ has a symmetric stable distribution* of index $\frac{3}{2}$. A simple derivation is given in ref. 9, pp. 173–174. See also refs. 7 and 12. If the space is R^n and the attractive force is proportional to $\|u - t\|^{-\beta}$, $\beta > n/2$, the index will be n/β. Unsymmetric stable distributions for $X(\underline{t})$ appear if the intensity of the Poisson process is permitted to be direction dependent viewed from $\underline{t}$ ($\underline{t}$ fixed).

The case $n = 1$, $\beta = 2$, leads to an application to traffic noise. Consider a long straight highway with cars with random velocities and positions (at any moment) according to a Poison point process. Then, under very weak assumptions, the intensity (effect) of the traffic noise at an arbitrary point at the road will follow at any time a positive stable distribution with index $\frac{1}{2}$. The case when the noise is registered at a distance from the road is treated in ref. 24. See also, e.g., ref. 16.

For an application of shot-noise processes in photographic science, see ref. 13. The papers of refs. 6, 11, 15, and 19 contain further applications or serve as guides to the more recent literature.

SHOT-NOISE DISTRIBUTIONS

Let $X(t)$ be given by (3). Then

$$\begin{aligned}\varphi_{X(t)}(s) &= E\left[\exp\{isX(t)\}\right]\\ &= \exp\{\lambda\textstyle\int\left(E\left[\exp\{isZ(t, u)\}\right]\right.\\ &\qquad\left.-1\right) du\} \quad (4)\end{aligned}$$

and the multivariate characteristic function of $(X(t_1), \ldots, X(t_m))$ admits an analogous expression from which Campbell's theorem can be derived; cf. ref. 18, pp. 152–155. Formula (4) is a consequence of the fact that the distribution of $X(t)$ equals the limit distribution of $\sum_j Z(t, j/n)\delta_{jn}$ as $n \to \infty$, where $\delta_{jn} = 1$ if there is an event in $((j - 1)/n, j/n]$ and 0 otherwise; the terms of the sum are independent. It is easy to generalize (4) to be valid also when the points $\{T_k\}$ stem from a nonstationary compound spatial Poisson process.

Formula (4) shows that a *shot-noise distribution*, i.e., the marginal distribution of a shot-noise process $X(t)$, is infinitely divisible*. This result follows immediately also from the fact that, for any n, a Poisson (point) process of intensity λ can be considered to be the superposition of n independent Poisson (point) processes of intensity λ/n.

A converse result is that any infinitely divisible distribution without a normal component appears as the marginal distribution of

$$\lim_{T\to\infty}\left[\sum_{|T_k - t| < T} h(t - T_k) - b(T)\right]$$

for appropriate functions h and b. Stable distributions with index α, $0 < \alpha < 2$, are obtained by choosing $h(u) = C\,\mathrm{sgn}(u)|u|^{-1/\alpha}$, where the constant C may have different values for $u > 0$ and $u < 0$. An infinitely divisible distribution on $[0, \infty)$, with left extremity 0 and with characteristic function

$$\varphi(s) = \exp\left\{\int_{(0,\infty)} (e^{isy} - 1) N(dy)\right\},$$

is the distribution of $X(t) = \sum h(t - T_k)$ if h is nonnegative with $h(u) = 0$ for $u < 0$ and related to the Lévy measure $N(dy)$ on $(0, \infty)$ by

$$\int_{(x,\infty)} N(dy) = \lambda\mu\{u; h(u) > x\}, \qquad x > 0,$$

where μ is Lebesgue measure; see, e.g., ref. 3. The sum giving $X(t)$ will converge almost surely. To obtain the gamma distribution, one may choose h to be proportional to the inverse of the function $\int_{(x,\infty)} y^{-1}e^{-y}\,dy$ and a suitable λ. However there is a more explicit representation of the gamma distribution as shot-noise distribution; cf. Example 1.

For the case $X(t) = \sum h(t - T_k)V_k$, with h nonnegative and 0 on $(-\infty, 0)$, some interesting subclasses of the infinitely divisible distributions on $[0, \infty)$ are obtained as classes of shot-noise distributions by restricting the form of h and/or the possible distributions of the nonnegative i.i.d. random variables V_k as noticed in ref. 3.

(a) If $h(u) = Ce^{-\rho u}$, then $X(t)$ has a *self-decomposable distribution** (class L distribution), which means that the Lévy measure has a density $n(y)$ such that $yn(y)$ is nonincreasing. See, e.g., ref. 9, pp. 588–590, for a description of the self-decomposable distributions as limit distributions for normed sums of independent random variables. Any self-decomposable distribution with $\lim_{y \downarrow 0} yn(y) < \infty$ is a shot-noise distribution for an appropriate distribution of the V_k's.

(b) If $h(u) = ce^{-\rho u}$ and the distribution of the V_k's is a mixture of exponential distributions*, then the shot-noise distribution is a *generalized gamma convolution*, i.e., a distribution with characteristic function of the form

$$\varphi(s) = \exp\left\{ ias + \int \log\left(\frac{1}{1 - is/y}\right) U(dy) \right\},$$

where $a \geqslant 0$ and $U(dy)$ is a nonnegative measure on $(0, \infty)$. Moreover, any generalized gamma convolution with $a = 0$ and $\int U(dy) < \infty$ is possible as a shot-noise distribution. The class of generalized gamma convolutions, introduced by Thorin [23], can be described as the class of limits of finite convolutions of gamma distributions. It contains several of the continuous standard distributions on $(0, \infty)$, in particular positive stable distributions and distributions with densities of the form

$$f(x) = Cx^{\beta - 1} \prod_{j=1}^{N} (1 + c_j x)^{-\gamma_j},$$

where the parameters are nonnegative and limits of such distributions; see ref. 2.

(c) When there is no restriction on h but the V_k's are exponentially distributed, the possible shot-noise distributions are precisely the *T_2 distributions** with left extremity 0. The T_2 distributions are defined as the limits of finite convolutions of mixtures of exponential distributions. Every generalized gamma convolution is in the T_2 class.

The distributions encountered in **(a)**–**(c)** above have discrete analogs on Z_+ that are marginal distributions for shot-noise processes of the kind considered in Example 3 with suitable restrictions; e.g., if the $L_j(u)$'s are exponentially distributed, then $X(t)$ has a *discrete self-decomposable distribution*. These distributions were introduced in ref. 22.

Finally, an interesting shot-noise representation of Ferguson's [10] *Dirichlet process* is worth mentioning. A Dirichlet process with parameter space Ω, equipped with a σ field $\mathscr{A}$, is a stochastic process $X(A)$, $A \in \mathscr{A}$, such that, for every partition $A_1, \ldots, A_m$ of Ω, $(X(A_1), \ldots, X(A_m))$ has a Dirichlet distribution* $D(\alpha(A_1), \ldots, \alpha(A_m))$, where α is a finite positive measure on $\mathscr{A}$. Then, as shown in ref. 10, $X(A)$ may be represented as $Y(A)/Y(\Omega)$, with

$$Y(A) = \sum_{k=1}^{\infty} h(T_k) \mathbf{1}_A(W_k),$$

where h is the inverse of the function $\int_{(x,\infty)} y^{-1} e^{-y}\, dy$ and the W_k's are i.i.d. random variables with values in Ω and probability distribution $\alpha(d\omega)/\alpha(\Omega)$. The intensity λ of the Poisson process on $(0, \infty)$ should equal $\alpha(\Omega)$. For disjoint $A_1, \ldots, A_m$, the variables $Y(A_1), \ldots, Y(A_m)$ are independent and gamma distributed. A somewhat more explicit representation is obtained by changing $Y(A)$ to

$$Y(A) = \sum_{k=1}^{\infty} \exp\{-T_k\} V_k \mathbf{1}_A(W_k),$$

where the V_k's are independent and exponentially distributed with mean 1; cf. Example 1.

References

[1] Bartlett, M. S. (1957). *J. R. Statist. Soc. B*, **19**, 220–221.

[2] Bondesson, L. (1979). *Ann. Prob.*, **7**, 965–979.

[3] Bondesson, L. (1982). *Adv. Appl. Prob.*, **14**, 858–869.

[4] Boswell, M. T. and Patil, G. P. (1970). *Random Counts in Scientific Work*, Vol. I, G. P. Patil, ed. Pennsylvania State University Press, University Park, PA, pp. 3–22.

[5] Campbell, N. R. (1909). *Proc. Camb. Philos. Soc. Math. Phys. Sci.*, **15**, 117–136, 310–328.

[6] Chamayou, J. M. F. (1978). *Stoch. Processes Appl.*, **6**, 305–316.

[7] Chandrasekhar, S. (1954). *Selected Papers on Noise and Stochastic Processes*. N. Wax, ed. Dover, New York, pp. 3–91.

[8] Cox, D. R. and Miller, H. D. (1965). *The Theory of Stochastic Processes*. Methuen, London, England.

[9] Feller, W. (1971). *An Introduction to Probability Theory and Its Applications*, 2nd ed., Vol. II. Wiley, New York.

[10] Ferguson, T. (1973). *Ann. Statist.*, **1**, 209–230.

[11] Gilchrist, J. H. and Thomas, J. B. (1975). *Adv. Appl. Prob.*, **7**, 527–541.

[12] Good, I. J. (1961). *J. R. Statist. Soc. B*, **23**, 180–183.

[13] Hamilton, J. F., Lawton, W. H., and Trabka, E. A. (1972). *Stochastic Point Processes: Statistical Analysis, Theory and Applications*, P. A. W. Lewis, ed. Wiley, New York, pp. 818–867.

[14] Holtsmark, J. (1919). *Annalen Physik*, **58**, 577–630.

[15] Lawrance, A. J. and Kottegoda, N. T. (1977). *J. R. Statist. Soc. A*, **140**, 1–31.

[16] Marcus, A. H. (1975). *Adv. Appl. Prob.*, **7**, 593–606.

[17] Moran, P. A. P. (1968). *A Introduction to Probability Theory*. Clarendon, Oxford, England. (Emphasizes the connection between shot-noise processes and infinite divisibility.)

[18] Parzen, E. (1962). *Stochastic Processes*. Holden-Day, San Francisco, CA. (A fairly elementary systematic treatment of shot-noise processes, called filtered Poisson processes in the most general versions; recommended reading.)

[19] Rice, J. (1977). *Adv. Appl. Prob.*, **9**, 553–565.

[20] Rice, S. O. (1954). *Selected Papers on Noise and Stochastic Processes*, N. Wax, ed. Dover, New York, pp. 133–294.

[21] Schottky, W. (1918). *Annalen Physik*, **57**, 541–567.

[22] Steutel, F. W. and van Harn, K. (1979). *Ann. Prob.*, **7**, 93–99.

[23] Thorin, O. (1977). *Scand. Actuarial J.*, **60**, 31–40.

[24] Weiss, G. H. (1970). *Transport. Res.*, **4**, 229–233.

[25] Weiss, G. H. (1977). *Water Resour. Res.*, **13**, 101–108.

(INFINITE DIVISIBILITY
POINT PROCESSES, STATIONARY
POISSON PROCESSES
STABLE DISTRIBUTIONS
STOCHASTIC PROCESSES, POINT)

L. BONDESSON

SHOVELTON'S FORMULA

This is the following quadratic formula, using values of the integral at 11 equally spaced values of the variable:

$$\begin{aligned}\int_a^{a+10h} f(x)\,dx \\ \doteqdot \frac{5h}{126}\big[8\{f(a) + f(a+10h)\} \\ +35\{f(a+h) + f(a+3h) \\ +f(a+7h) + f(a+9h)\} \\ +15\{f(a+2h) + f(a+4h) \\ +f(a+6h) + f(a+8h)\} \\ +36f(a+5h)\big].\end{aligned}$$

It gives the exact value of the integral if $f(x)$ is a polynomial of degree 5 or less.

(NUMERICAL INTEGRATION
SIMPSON'S RULE
THREE-EIGHTS RULE
TRAPEZOIDAL RULE
WEDDLE'S RULE)

SHRINKAGE ESTIMATORS

A shrinkage (shrunken) estimator is an estimator obtained through modification of the usual (maximum likelihood*, minimum variance unbiased*, least squares*, etc.) estimator in order to minimize (maximize) some desirable criterion function (mean square error*, quadratic risk*, bias*, etc.).

Shrinkage estimators of various types are found in the literature, e.g., ordinary shrinkage, preliminary test (shrinkage), Stein-type, ridge regression*, empirical Bayes* estimators, etc. The desire to improve on an existing estimator lies at the root of each of these estimators. Among the earliest contributions in this respect are those of Goodman [6] and

Stein [21]. Thompson [22] was apparently the first to use the term "shrinkage" in connection with estimators that have been modified.

SHRINKAGE IN THE DIRECT SENSE

First, consider the problem of estimating the value of some unknown parameter θ (possibly vector-valued). In practice the experimenter often possesses some knowledge of the experimental conditions, based on acquaintance with the behaviour of the system under consideration or from past experience or from some extraneous source, and is thus in a position to give an educated guess or an initial estimate, say θ_0, of the value of the parameter. In such cases, it may be reasonable to take the usual estimator for θ, say $\hat{\theta}$, and move it closer to (or shrink it toward) this so-called natural origin θ_0 by multiplying the difference $\hat{\theta} - \theta_0$ by a shrinking factor k and adding it to θ_0, i.e.,

$$\hat{\theta}_s = k(\hat{\theta} - \theta_0) + \theta_0$$

$$= k\hat{\theta} + (1 - k)\theta_0, \qquad 0 \leqslant k \leqslant 1.$$

The resulting estimator, though perhaps biased, has a smaller mean square error* (MSE) than $\hat{\theta}$ for θ in some interval around θ_0 (the so-called *effective interval*). It is generally accepted (James and Stein [11]) that minimum MSE is a highly desirable property, and it is therefore used as a criterion to compare different estimators with each other.

In order to explain the construction of shrinkage estimators in this class further, consider $T = a[k\hat{\theta} + (1 - k)\theta_0]$, $0 \leqslant k \leqslant 1$, a and k constants. Different approaches toward a and k lead to a variety of shrinkage estimators that have been proposed for the parameters of normal, gamma, Poisson, and binomial distributions. For illustration we consider estimators for the mean μ of a normal (μ, σ^2) distribution with σ^2 unknown. Thompson [22] considered the case $a = 1$ and determined the best value of k by solving for k in the equation $\partial \text{MSE}(T)/\partial k = 0$. His estimator for μ is then

$$T_T = (\bar{x} - \mu_0)^3 \Big/ \left[(\bar{x} - \mu_0)^2 + s^2/n\right] + \mu_0$$

with $\bar{x}$ the mean and s^2 the variance of a sample of size n from the distribution and μ_0 the guessed value of μ. Pandey [16] used k as a constant specified by the experimenter according to his belief $(1 - k)$ in θ_0 and determined a by solving the equation $\partial \text{MSE}(T)/\partial a = 0$. He thus obtained

$$T_P = \hat{d}^2\bar{x}^3 \big/ \left(\hat{d}^2\bar{x}^2 + k^2 s^2/n\right)$$

with $\hat{d} = k + (1 - k)\mu_0/\bar{x}$.

Both Thompson's and Pandey's estimators suffer from the disadvantage that the constants estimated are functions of $\hat{\theta}$, which makes their estimators heavily dependent on $\hat{\theta}$. The estimator of Lemmer [12], namely

$$T_L = k\bar{x} + (1 - k)\mu_0$$

with $(1 - k)$ proportional to the experimenter's confidence in μ_0, was found to be better than the other two. A slightly more complicated shrinkage estimator was proposed by Mehta and Srinivasan [14], namely

$$T_M = \bar{x} - a(\bar{x} - \mu_0)\exp\left[-nb(\bar{x} - \mu_0)^2/s^2\right]$$

with a and b constants, having the property that its MSE is everywhere bounded, at the expense of a higher MSE than T_L, at least in the effective interval of T_L. If $b = 0$, T_M reduces to T_L. For examples, see Lemmer [13] for a comparison between these estimators and for a discussion of Bayesian shrinkage estimators, that is, estimators of the form

$$T_B = k\hat{\theta} + (1 - k)\hat{\theta}_0$$

with $\hat{\theta}_0$ the Bayes estimator derived from a prior distribution* that places a weight $(1 - b)$ on θ_0 and distributes the rest of the probability mass b according to some probability distribution around θ_0. The idea behind this type of estimator is that one's knowledge of θ can best be expressed by means of a prior distribution of θ around some value θ_0 rather than a one-point distribution in θ_0. The desire to accentuate the value θ_0 is accommodated by placing a weight $(1 - b)$ on the point θ_0. If $b = 0$, T_B reduces to the ordinary shrinkage estimator and if $k = 0$ and $b = 1$, T_B becomes an ordinary Bayes estimator with quadratic loss function.

Instead of guessing the value θ_0 of θ, it may be more feasible to shrink toward an interval, i.e., guess an interval (θ_1, θ_2), that we believe contains θ; cf. Lemmer [12] and Thompson [23].

Shrinkage estimators are also used in the case of preliminary test estimators: Test the hypothesis $H_0 : \theta = \theta_0$ at a specified level of significance. If H_0 is not rejected, use the shrinkage estimator for θ, and if H_0 is rejected, use $\hat{\theta}$; cf. Upadhyaya and Srivastava [24] and Hirano [8].

In Pandey and Singh [17], shrinkage and pretest shrinkage estimators are proposed for the scale parameter θ of an exponential distribution* when the observations become available from life test* experiments. In Singh and Bhatkulikar [20] some pretest shrinkage estimators of the shape parameter of the Weibull distribution* are proposed under censored sampling. Another application of shrinkage estimation is in nonparametric Bayesian analysis* using censored data, where the shrinkage is toward a prior family of exponential survival curves; cf. Rai et al. [18].

In all the shrinkage estimators discussed so far, the ordinary estimator $\hat{\theta}$ is shrunken toward the guessed value θ_0 or an interval (θ_1, θ_2).

SHRINKAGE IN A WIDER SENSE

We now come to estimators that are known under a variety of names, but clearly have the idea of shrinkage built into them.

Consider the multiple linear regression model $\mathbf{Y} = \mathbf{X}\boldsymbol{\beta} + \mathbf{e}$, where $\mathbf{X}(n \times p)$ is of rank p, $\boldsymbol{\beta}(p \times 1)$ is unknown, $E(\mathbf{e}) = \mathbf{0}$, and $E(\mathbf{ee}') = \sigma^2\mathbf{I}_n$. Assume that $\mathbf{X}$ has been standardized so that $\mathbf{S} = \mathbf{X}'\mathbf{X}$ has the form of a correlation matrix. For estimating $\boldsymbol{\beta}$, the technique of shrinking the *least-squares* estimator* $\hat{\boldsymbol{\beta}} = \mathbf{S}^{-1}\mathbf{X}'\mathbf{Y}$ toward a point, or more generally toward a subspace, has received much attention. Two popular approaches are the ridge estimators or Hoerl and Kennard [10] and the Stein-type estimators of James and Stein [11] (*see* JAMES–STEIN ESTIMATORS). Ridge estimators were introduced in order to improve on the unsatisfactory properties of the least-squares estimator when multicollinearity* exists, that is, if $\mathbf{S}$ has eigenvalues $\lambda_1 \geqslant \lambda_2 \geqslant \cdots \geqslant \lambda_p > 0$ with at least one λ_i "close" to 0. Then the MSE of $\hat{\boldsymbol{\beta}}$ will be large, or viewed differently; some of the components of $\hat{\boldsymbol{\beta}}$ can become inaccurate because they have large variances. The *ridge regression estimator*

$$\hat{\boldsymbol{\beta}}^* = \left[\mathbf{X}'\mathbf{X} + k\mathbf{I}_p\right]^{-1}\mathbf{X}'\mathbf{Y}, \qquad k \geqslant 0,$$

has been found to help circumvent many of the difficulties associated with the usual least-squares estimator. Hoerl and Kennard [10] discussed the choice of the constant k by means of the so-called *ridge trace*. Two other estimators of k have been proposed, namely

$$\hat{k}_1 = p\hat{\sigma}^2/\hat{\boldsymbol{\beta}}'\hat{\boldsymbol{\beta}} \text{ and } \hat{k}_2 = p\hat{\sigma}^2/\hat{\boldsymbol{\beta}}'\mathbf{X}'\mathbf{X}\hat{\boldsymbol{\beta}},$$

with $\hat{\sigma}^2$ an estimator of σ^2. Galpin [4] pointed out that the ridge estimators based on $\hat{k}_1$ and $\hat{k}_2$ dominate the least-squares estimator if $\mathbf{X}'\mathbf{X}$ is well conditioned (not multicollinear) and also when $\mathbf{X}'\mathbf{X}$ has at least two very small eigenvalues. Only in the case of exactly one small eigenvalue may the ridge estimators sometimes be better and sometimes worse than the least-squares estimator.

If $\mathbf{S}$ has $p - r$ zero eigenvalues, i.e., $\mathbf{X}'\mathbf{X}$ is of rank r, the umber of input variables can be reduced by $p - r$. Let $\mathbf{T}'\mathbf{S}\mathbf{T} = \Lambda$ with Λ the diagonal matrix of eigenvalues λ_i of $\mathbf{S}$ and $\mathbf{T}$ the orthogonal matrix of eigenvectors, i.e., $\mathbf{T}'\mathbf{T} = \mathbf{I}$. Now the inverse $\mathbf{S}^{-1} = \mathbf{T}\Lambda^{-1}\mathbf{T}'$ does not exist. Partition $\mathbf{T}$ as follows: $\mathbf{T} = (\mathbf{T}_r, \mathbf{T}_{p-r})$ and similarly

$$\Lambda = \begin{bmatrix} \Lambda_r & \mathbf{0} \\ \mathbf{0} & \Lambda_{p-r} \end{bmatrix}.$$

Thus the inverse becomes

$$\mathbf{S}_r^+ = \mathbf{T}_r \Lambda_r^{-1} \mathbf{T}'_r = \sum_{j=1}^{r} \lambda_j^{-1} \mathbf{T}_j \mathbf{T}'_j$$

where $\mathbf{T}_j$ is the eigenvector of $\mathbf{S}$ corresponding to λ_j. Now the *principal components estimator* is given by $\hat{\boldsymbol{\beta}}_r^+ = \mathbf{S}_r^+ \mathbf{X}'\mathbf{Y}$. In many cases $\mathbf{X}$ may be of rank p but have some very small eigenvalues, in which case it may be reasonable to suppose that $\mathbf{X}$ has fractional rank f where $r < f < r + 1$. The *generalized inverse estimator* is then given by

$$\hat{\boldsymbol{\beta}}_f^g = (1 - f + r)\hat{\boldsymbol{\beta}}_r^+ + (f - r)\hat{\boldsymbol{\beta}}_{r+1}^+.$$

For further examples, see Hocking [9], Oman [15], and Farebrother [3].

In *discriminant** and *canonical* variate analysis*, the use of shrinkage estimators, by adding shrinkage constants to the eigenvalues, leads to more stable estimators when the between-groups sum of squares for a particular principal component is small and the corresponding eigenvalue is also small; cf. Campbell [1].

The *James–Stein estimator* of $\boldsymbol{\beta}$ can, under certain conditions, including normality of $\mathbf{e}$, be written as

$$\tilde{\boldsymbol{\beta}} = \left[1 - (p - 2)\sigma^2/(\hat{\boldsymbol{\beta}}'\mathbf{X}'\mathbf{X}\hat{\boldsymbol{\beta}})\right]\hat{\boldsymbol{\beta}},$$

and has the advantage of improving uniformly on the maximum likelihood estimator in terms of total squared error risk; cf. Oman [15] and Rolph [19].

Other related estimators, which will not be discussed here, are *generalized ridge estimators* (Hoerl and Kennard [10]) and *empirical Bayes estimators* cf. Rolph [19] for references and still further types.) *See also* EMPIRICAL BAYES THEORY and Hocking [9] and Goldstein and Brown [5] for the use of shrinkage estimators in prediction.

A number of shrinkage formulas have been proposed to reduce the bias so that the sample squared multiple correlation* coefficient R^2 becomes a less biased estimator of the squared multiple correlation coefficient ρ^2; cf. Carter [2].

In many epidemiological* studies, a single measurement of some variable is made on each member of a population, and all individuals whose values exceed some truncation point are classified as high. The use of an empirical Bayes adjustment formula, whereby extreme values are shrunken toward the population mean, substantially improves the average accuracy of estimation of a subject's true mean and reduces the probability of false positive classification; cf. Harris and Shakarki [7].

References

[1] Campbell, N. A. (1980). *Appl. Statist.*, **29**, 5–14. (Discusses canonical variate analysis and its relationship with generalized ridge and principal component estimators.)

[2] Carter, D. S. (1979). *Educ. Psychol. Meas.*, **39**, 261–266. (A number of bias reduction formulas for the multiple correlation coefficient are discussed.)

[3] Farebrother, R. W. (1978). *J. R. Statist. Soc. B*, **40**, 47–49. (A class of shrinkage estimators is defined, which includes ridge regression, principal component, and minimum conditional mean square error estimators.)

[4] Galpin, J. S. (1980). *Commun. Statist. A*, **9**, 1019–1024. (Discusses the conditions under which certain ridge regression estimators are always better than least-squares estimators.)

[5] Goldstein, M. and Brown, P. J. (1978). *Math. Operat. Statist.*, **9**, 3–7, (Discusses prediction with shrinkage estimators.)

[6] Goodman, L. A. (1953). *Ann. Math. Statist.*, **24**, 114–117. (A simple technique for improving estimators is proposed.)

[7] Harris, E. K. and Shakarki, G. (1979). *J. Chronic Diseases*, **32**, 233–243. (The shrinking of extreme values towards the population mean, in order to reduce the probability of false classification, is discussed.)

[8] Hirano, K. (1977). *Ann. Inst. Statist. Math.*, **29**, 21–34. (Discusses estimation procedures based on preliminary tests, shrinkage, and Akaike's information criterion.)

[9] Hocking, R. R. (1976). *Biometrics*, **32**, 1–49. (Reviews methods of variable selection in linear regression models and discusses various shrinkage estimators. A comprehensive list of 170 references is given.)

[10] Hoerl, A. E. and Kennard, R. W. (1970). *Technometrics*, **12**, 55–67. (A basic paper on ridge regression.)

[11] James, W. and Stein, C. (1961). *Proc. 4th Berkeley Symp. Math. Statist.*, Vol. 1. University of California Press, Berkeley, CA, pp. 361–379. (A basic paper on Stein-type estimators.)

[12] Lemmer, H. H. (1981). *Commun. Statist. A*, **10**, 1017–1027. (Ordinary and interval type shrinkage estimators for the binomial distribution are discussed.)

[13] Lemmer, H. H. (1981). *S. Afr. Statist. J.*, **15**, 57–72. (A comprehensive discussion of various shrinkage estimators, including Bayesian shrinkage estimators, is presented.)

[14] Mehta, J. S. and Srinivasan, R. (1971). *J. Amer. Statist. Ass.*, **66**, 86–90. (A shrinkage estimator with bounded mean square error is proposed and discussed.)

[15] Oman, S. D. (1978). *Commun. Statist. A*, **7**, 517–534. (Compares ridge, principal components, generalized inverse, and Stein estimators from a Bayesian point of view.)

[16] Pandey, B. N. (1979). *Commun. Statist A*, **8**, 359–365. (A shrinkage estimator for the normal population variance is proposed and studied.)

[17] Pandey, B. N. and Singh, P. (1980). *Commun. Statist. A*, **9**, 875–882. (Shrinkage estimators for the scale parameter of the exponential distribution are proposed.)

[18] Rai, K., Susarla, V., and Van Ryzin, J. (1978–1979). *Proc. Social Statist. Sec., Amer. Statist. Ass.*, 96–99. (Treats shrinkage estimation in survival analysis.)

[19] Rolph, J. E. (1976). *Commun. Statist. A*, **5**, 789–802. (The relationship between ridge and Stein estimators is discussed.)

[20] Singh, J. and Bhatkulikar, S. G. (1977). *Sankhyā B*, **39**, 382–393. (Shrinkage estimation in the Weibull distribution is discussed.)

[21] Stein, C. (1956). *Proc 3rd Berkeley Symp. Math. Statist.*, Vol. 1. University of California Press, Berkeley, CA, pp. 197–206. (Discusses the inadmissibility of the usual estimator for the mean.)

[22] Thompson, J. R. (1968). *J. Amer. Statist. Ass.*, **63**, 113–122. (A shrinkage estimator for the mean of a normal distribution is proposed.)

[23] Thompson, J. R. (1968). *J. Amer. Statist. Ass.*, **63**, 953–963. (A shrinkage estimator is proposed where shrinkage is made toward an interval.)

[24] Upadhyaya, L. N. and Srivastava, S. R. (1975). *J. Statist. Res.*, **9**, 67–74. (Preliminary test shrinkage estimators are discussed.)

(EMPIRICAL BAYES THEORY
JAMES–STEIN ESTIMATORS
LEAST SQUARES
MEAN SQUARED ERROR
MULTICOLLINEARITY
RIDGE REGRESSION)

HERMANUS H. LEMMER

SICHEL'S COMPOUND POISSON DISTRIBUTION

This distribution is obtained by ascribing to the expected value parameter θ of a Poisson distribution a mixing distribution with density function

$$f_\theta(t) = \tfrac{1}{2}Ct^{\alpha-1} \times \exp\{-(\theta^{-1}-1)t - \tfrac{1}{4}\alpha^2\theta t^{-1}\}, \tag{1}$$

where

$$C = \frac{1}{2}\frac{\{2\sqrt{(1-\theta)}\}^\gamma}{(\alpha\theta)^\gamma K_\gamma\{\alpha\sqrt{(1-\theta)}\}},$$

$0 < \theta < 1$, $\alpha > 0$, and $K_\gamma(\cdot)$ is a modified Bessel function* of the second kind of order γ. (1) may be regarded as a generalization of the inverse Gaussian distribution*, since the latter is obtained by putting $\gamma = -\frac{1}{2}$.

The distribution has probability function

$$\Pr[X = x] = \frac{(1-\theta)^{\alpha/2}}{K_\gamma\{\alpha\sqrt{(1-\theta)}\}}\frac{(\alpha\theta/2)^x}{x!}K_{x+\gamma}(\alpha), \quad x = 0, 1, \ldots. \tag{2}$$

Its mean and variance are

$$E[X] = \frac{\alpha\theta}{2\sqrt{(1-\theta)}}\frac{K_{\gamma+1}\{\alpha\sqrt{(1-\theta)}\}}{K_\gamma\{\alpha\sqrt{(1-\theta)}\}}$$

and

$$\frac{(\alpha\theta)^2}{4(1-\theta)}\frac{K_{\gamma+2}\{\alpha\sqrt{(1-\theta)}\}}{K_\gamma\{\alpha\sqrt{(1-\theta)}\}} + E[X]\{1 - E[X]\},$$

respectively. The distribution has a long positive tail, but all moments are finite. It was introduced by Sichel [3], developing a suggestion by Good [2], and applied by Sichel [4, 5] to the distribution of sentence length. Sichel [3] gave formulae for maximum likelihood estimators of α and θ, if γ is known. For fits by moments, see Sichel [4]. Further inferential properties of the distribution are described by Aitchison and Lam Yeh [1].

References

[1] Aitchison, A. C. and Lam Yeh (1982). *J. Amer. Statist. Assoc.*, **77**, 153–158.

[2] Good, I. J. (1953). *Biometrika*, **40**, 237–264.

[3] Sichel, H. S. (1971). *Proc. 3rd Symp. Math. Statist.*, N. F. Laubscher, ed. CSIR., Pretoria, South Africa, pp. 51–97.

[4] Sichel, H. S. (1974). *J. R. Statist. Soc. A.*, **137**, 25–34.

[5] Sichel, H. S. (1975). *J. Amer. Statist. Ass.*, **70**, 542–547.

(BESSEL FUNCTIONS
COMPOUND DISTRIBUTION
HYPERBOLIC DISTRIBUTIONS
INVERSE GAUSSIAN DISTRIBUTION
POISSON DISTRIBUTIONS)

SIEGEL–TUKEY TEST

In 1960, Sidney Siegel and John Tukey [7] developed a distribution-free rank-sum test of a difference in scale parameters in two populations (*see* SCALE TESTS). Suppose that independent samples $X_1, \dots, X_m$ and $Y_1, \dots, Y_n$ of observations are taken:

$$X_i = \sigma_1 e_i + \mu, \qquad Y_j = \sigma_2 e_{m+j} + \mu,$$

where $i = 1, \dots, m$ and $j = 1, \dots, n$; $e_1, \dots, e_{m+n}$ are independent random variables from a common continuous distribution with zero median; the X and Y populations have common unknown median μ; σ_1 and σ_2 are scale parameters. To test the null hypothesis

$$H_0 : \sigma_1^2/\sigma_2^2 = 1$$

against $H_1 : \sigma_1^2/\sigma_2^2 \neq 1$, pool the data in increasing order, and replace the combined ordered sample by ranks

$$1\ 4\ 5\ 8\ 9 \dots 7\ 6\ 3\ 2.$$

The test statistic is

$$R = \text{sum of ranks in the smaller group.}$$

If the X group is smaller, i.e., $m \leqslant n$, then under H_0 [3],

$$E(R) = \tfrac{1}{2}m(m + n + 1),$$

$$\operatorname{var}(R) = \tfrac{1}{12}mn(m + n + 1).$$

H_0 is rejected for large values of R. The ranking procedure is close to that used in deriving Ansari–Bradley W-statistics*, and the two tests are asymptotically equivalent [3, 4]. Limiting Pitman efficiencies of R relative to locally most powerful tests and to Mood's dispersion test* relative to R can thus be read from Table 1 of the entry MOOD'S DISPERSION TEST. The Siegel–Tukey statistic R, however, has the same distribution under H_0 as the two-sample Wilcoxon rank-sum statistic. *See* MANN–WHITNEY–WILCOXON STATISTIC, therefore, for discussion of tied data and other matters. Tables of percent points of the Wilcoxon statistic may thus be used; these appear, e.g., in the original source [7] as exact critical values of R for $m \leqslant n = 1(1)20$ and $0.001 \leqslant \alpha \leqslant 0.10$, and in Conover [1, Table A7]. Upper tail probabilities are given in Hollander and Wolfe [5, Table A.5] for $m = 3(1)10$, $n \leqslant m$, and $m = 11(1)20$, $n = 1(1)4$.

For large sample sizes m and n with $m \leqslant n$, the statistic

$$Z = \frac{2R - m(m + n + 1) \pm 1}{\sqrt{\{mn(m + n + 1)/3\}}}$$

has an approximate standard normal distribution under H_0, where the $\pm$ sign in the numerator is chosen to make $|Z|$ smaller. See also Iman [6].

As with the Ansari–Bradley and Mood dispersion tests, this test is applicable to nonnumerical ordinal data*, which can be ranked. None of these three tests depends on any normality assumption, an important consideration in light of the sensitivity of the parametric F-test* to departures from normality*.

The requirement of equal medians in the X and Y populations is a necessity for the

Siegel–Tukey, Ansari–Bradley, and Mood tests. For a discussion of alternative tests that accommodate location differences *see* MOOD'S DISPERSION TEST, Hollander and Wolfe [5, Chap. 5], and HOLLANDER EXTREME TEST. Other distribution-free tests of dispersion appear in the list of related entries. A useful comparative survey and bibliography up to 1976 is by Duran [2].

References

[1] Conover, W. J. (1980). *Practical Nonparametric Statistics*, 2nd ed. Wiley, New York.

[2] Duran, B. S. (1976). *Commun. Statist. A*, **5**, 1287–1312.

[3] Eeden, C. van (1964). *J. Amer. Statist. Ass.*, **59**, 105–119.

[4] Hájek, J. and Šidák, Z. (1967). *Theory of Rank Tests*. Academic, New York. (See p. 126.)

[5] Hollander, M. and Wolfe, D. A. (1973). *Nonparametric Statistical Methods*. Wiley, New York.

[6] Iman, R. L. (1976). *Commun. Statist. A*, **5**, 587–598.

[7] Siegel, S. and Tukey, J. W. (1960). *J. Amer. Statist. Ass.*, **55**, 429–445.

(ANSARI–BRADLEY *W*-STATISTICS
CAPON TEST
DISTRIBUTION-FREE METHODS
F-TESTS
HOLLANDER EXTREME TEST
KLOTZ TEST
MOOD'S DISPERSION TEST
RANK TESTS
SCALE TESTS
ZAREMBA TEST STATISTIC)

CAMPBELL B. READ

SIEVES, METHOD OF

The method of sieves* is a technique of nonparametric estimation in which estimators are restricted by an increasing sequence of subsets of the parameter space (which comprises the *sieve*) with the subsets indexed by the sample size. The need for this technique arises where the parameter space is too large for the existence or consistency of unconstrained maximum likelihood* (ML) or least-squares* estimators. Grenander [11] developed the abstract theory of the method of sieves and provided a wealth of examples.

Geman and Hwang [10] have shown that the method leads to consistent nonparametric estimators in very general settings. In practice, the sieve needs to be carefully chosen to exploit the specific structural properties of the problem. It would be desirable to know how to construct the sieve to yield an optimal rate of convergence of the sieve estimator, but this question has only been studied in some special cases.

In some nonparametric problems, typically where a monotonicity condition holds, the method of maximum likelihood is directly applicable without the need for a sieve. For instance, under monotonicity of the probability density function the ML estimator, based on an independent identically distributed (i.i.d.) sample, exists and is consistent in L^1 norm; see Grenander [11, p. 402]. Similar results have been obtained for monotone failure rate functions [16] and unimodal densities [20]. However, without order restrictions the direct method of maximum likelihood usually fails in nonparametric problems. The method of sieves then presents itself as one of several alternative approaches, others being the method of penalized maximum likelihood*, orthogonal series methods, kernel methods*, spline methods, and the Bayesian approach. These techniques are themselves closely related to the method of sieves; see Grenander [11, p. 7] and Geman and Hwang [10, p. 403]. The distinguishing feature of the method of sieves is its use of an optimization principle subject to constraints that depend on the sample size. The following examples supplement those in the entry METHOD OF SIEVES.

TRANSLATE OF WIENER PROCESS

Let $W(t)$, $t \geqslant 0$, be a standard Wiener process (*see* BROWNIAN MOTION) and α an unknown function of $t \in [0, 1]$. Suppose that n i.i.d. copies X_i, $i = 1, \ldots, n$, of the signal +

noise process,

$$X(t) = \int_0^t \alpha(s)\,ds + W(t), \qquad t \in [0,1], \quad (1)$$

are observed. The parameter space is $L^2[0,1]$, the space of square integrable functions on $[0,1]$. Grenander [11, p. 424] considered a sieve of the form

$$S_{d_n} = \left\{ \alpha(t) : \alpha(t) = \sum_{r=1}^{d_n} \alpha_r \phi_r(t) \right\}, \quad (2)$$

where $(\phi_r, r \geqslant 1)$ is a complete orthonormal sequence in $L^2[0,1]$. The ML estimator contained in S_{d_n} is given by

$$\hat{\alpha}^{(n)}(t) = \sum_{r=1}^{d_n} \hat{\alpha}_r^{(n)} \phi_r(t), \quad (3)$$

where

$$\hat{\alpha}_r^{(n)} = \frac{1}{n} \sum_{i=1}^{n} \int_0^1 \phi_r(t)\,dX_i(t).$$

It can be shown that $\hat{\alpha}^{(n)}$ is consistent in L^2 norm as $n \to \infty$, provided $d_n \uparrow \infty$ and $d_n/n \to 0$; see Nguyen and Pham [19] and McKeague [18]. The estimator (3) was first studied by Ibragimov and Khasminskii [12], who defined it from a point of view suggested by Cencov's [6] method of orthogonal series for density estimation*; within the parameter space of Lipschitz functions of order γ, $0 < \gamma \leqslant 1$, the estimator $\hat{\alpha}^{(n)}$ can be designed to attain the optimal rate of convergence (in the sense of an asymptotic minimax property) over all estimators. The optimal rate of convergence of the mean square error is $O(n^{-2\gamma/(2\gamma+1)})$ and this can be achieved by using the Fourier sieve

$$S_{d_n} = \left\{ \alpha : \alpha(t) = \sum_{r=-d_n}^{d_n} \alpha_r e^{2\pi i r t} \right\}$$

with $d_n = [n^{1/(2\gamma+1)}]$, where [ ] denotes the integer part.

Another sieve for this problem is given by

$$S_{m_n} = \left\{ \alpha \in L^2[0,1] : \sum_{r=1}^{\infty} a_r^2 \langle \alpha, \phi_r \rangle^2 \leqslant m_n \right\},$$

where $\langle\ ,\ \rangle$ denotes the inner product in $L^2[0,1]$ and $\sum_{r \geqslant 1} a_r^{-2} < \infty$. This sieve has been studied by Geman and Hwang [10], and Antoniadis [3] for general Gaussian processes*. Antoniadis showed that this sieve estimator is consistent provided $m_n \uparrow \infty$ and $m_n = O(n^{1-\epsilon})$ for some $\epsilon > 0$. Beder [5] has studied sieves of the form (2) for general Gaussian processes. Other approaches to the problem can be found in [15, 17, 21].

INTENSITY OF A POINT PROCESS

Let $N(t)$, $t \geqslant 0$, be a point process (*see* STOCHASTIC PROCESSES, POINT) with intensity

$$\lambda(t) = \sum_{j=1}^{p} \alpha_j(t) Y_j(t), \quad (4)$$

where $\alpha_1, \ldots, \alpha_p$ are unknown functions and $Y_1, \ldots, Y_p$ are observable covariate processes. Practical examples of this model arise in reliability* and biomedical settings. For instance, suppose that a subject has been exposed to p carcinogens. Let X be the time of the initial detection of cancer. Then a plausible "competing risks model" for the hazard function $\lambda(t)$ of X is given by (4) where $\alpha_1, \ldots, \alpha_p$ represent the changes in the relative hazard rates of the p carcinogens with age and $Y_j(t)$ is the cumulative exposure to the jth carcinogen by age t. The model (4) was introduced by Aalen [1, 2] as an alternative to the proportional-hazard regression model of Cox [7] (*see* COX'S REGRESSION MODEL). Aalen introduced an estimator of the integrated hazard function $\int_0^t \alpha_j(s)\,ds$.

The method of sieves is able to provide estimators of the α_j's themselves. Suppose that n i.i.d. copies of the processes $N(t)$ and $Y_j(t)$ are observed over $[0,1]$. In the case $p = 1$, Karr [13] used the sieve

$$S_{a_n} = \{ \alpha \in L^1[0,1] : \alpha \text{ is absolutely continuous, } a_n \leqslant \alpha \leqslant a_n^{-1} \text{ and } |\alpha'| \leqslant a_n^{-1}\alpha \},$$

and showed that the ML estimator of α_1

restricted by this sieve is strongly consistent in L^1 norm, where $a_n = n^{-(1/4)+\eta}$, with $0 < \eta < \frac{1}{4}$. For models with more than one covariate, McKeague [18] has used the orthogonal series sieve (2) to obtain consistent estimators of $\alpha_1, \ldots, \alpha_p$ for a general semimartingale regression model that contains the point process model (4) and diffusion process * models as special cases.

STATIONARY PROCESSES

Some recent applications of the method of sieves have been motivated by problems in the area of engineering known as *system identification*. A stationary process is observed over a long period of time and the engineer seeks to reconstruct the "black box" that produced the process. In practice this amounts to estimation of a spectral density or a transfer function and similar considerations, which led to the use of the method of sieves for probability density estimation, are involved here.

Chow and Grenander [7] consider estimation of the spectral density of a stationary Gaussian process $\{X_t,\ t = 1, 2, \ldots\}$ with mean zero and covariance $r_t = E(X_s X_{s+t}) = \int_{-\pi}^{\pi} e^{it\lambda} f(\lambda)\, d\lambda$, where f is the spectral density. They employ a sieve of the form

$$S_{\mu_n} = \left\{ f : f = \frac{1}{g} \text{ and } \int_{-\pi}^{\pi} \left(\frac{d}{d\lambda} g \right)^2 d\lambda \leqslant \frac{1}{\mu_n} \right\},$$

where n is the length of observation time. They show that an approximate maximum likelihood estimator of f restricted to S_{μ_n} is strongly consistent in $L^1[-\pi, \pi]$ provided $\mu_n = n^{-(1-\delta)}$, where $0 < \delta < 1$.

Ljung and Yuan [14] consider the problem of estimating the transfer function* of a linear stochastic system given by

$$y(t) = \sum_{k=1}^{\infty} g_k u(t-k) + w(t),$$

$$t = 1, 2, \ldots .$$

Here $u(t)$ and $y(t)$ are the input and output, respectively, at time t, and $\{w(t)\}$ is supposed to be a stationary process*. A reasonable sieve for the transfer function $h(\omega) = \sum_{k=1}^{\infty} g_k e^{-ik\omega}$, $\omega \in [-\pi, \pi]$, is given by

$$S_{d_n} = \left\{ h : h(\omega) = \sum_{k=1}^{d_n} g_k e^{-ik\omega} \right\},$$

where n is the length of observation time of input and output processes. The results of Ljung and Yuan show that the sieve estimator, formed by using the least-squares* estimates of $g_1, \ldots, g_{d_k}$, is uniformly consistent provided $d_n = [n^\alpha]$, $0 < \alpha < \frac{1}{4}$.

Bagchi [4] has used the method of sieves to estimate the distributed delay function α of the linear time-delayed system

$$dX_t = \left[\int_{-b}^{0} \alpha(u) X_{t+u}\, du \right] dt + dW_t,$$

where $\{W_t, -\infty < t < \infty\}$ is a standard Wiener process. The sieve is given by

$$S_{d_T} = \left\{ \alpha \in L^2[-b, 0] : \right.$$

$$\left. \alpha(u) = \sum_{r=1}^{d_T} \alpha_r \phi_r(u) \right\},$$

where $(\phi_r, r \geqslant 1)$ is a complete orthonormal sequence in $L^2[-b, 0]$ and T is the length of observation time of the process X. The ML estimator restricted to S_{d_T} is consistent in L^2 norm provided $d_T \uparrow \infty$ and $d_T^2/T \to 0$ as $T \to \infty$.

References

[1] Aalen, O. O. (1978). *Ann. Statist.*, **6**, 701–726.

[2] Aalen, O. O. (1980). *Lecture Notes in Statistics*, Vol. 2. Springer-Verlag, New York, pp. 1–25.

[3] Antoniadis, A. (1985). Technical Report, Institut de Mathématiques Appliquées de Grenoble, St.-Martin-d'Hères, France.

[4] Bagchi, A. (1985). *Syst. Control Lett.*, **5**, 339–345.

[5] Beder, J. (1987). *Ann. Statist.*, **15**, 59–78.

[6] Cencov, N. N. (1962). *Sov. Math.*, **3**, 1559–1562.

[7] Chow, Y.-S. and Grenander, U. (1985). *Ann. Statist.*, **13**, 998–1010.

[8] Cox, D. R. (1972). *J. R. Statist. Soc. B.*, **34**, 187–229.

[9] Geman, S. (1982). In *Nonparametric Statistical Inference*, B. V. Gnedenko, M. L. Puri, and I. Vincze, eds. North-Holland, Amsterdam, The Netherlands, pp. 231–252.

[10] Geman, S. and Hwang, C.-R. (1982). *Ann. Statist.*, **10**, 401–414.

[11] Grenander, U. (1981). *Abstract Inference*. Wiley, New York.

[12] Ibragimov, I. A. and Khasminskii, R. Z. (1981). *Theory Prob. Appl.*, **25**, 703–719.

[13] Karr, A. F. (1983). *Technical Report No. 46*, Center for Stochastic Processes, Department of Statistics, University of North Carolina, Chapel Hill, NC.

[14] Ljung, L. and Yuan, Z.-D. (1985). *IEEE Trans. Automat. Contr.*, **AC-30**, 514–530.

[15] Mandelbaum, A. (1984). *Ann. Statist.*, **12**, 1448–1466.

[16] Marshall, A. W. and Proschan, F. (1965). *Ann. Math. Statist.*, **36**, 69–77.

[17] McKeague, I. W. (1984). *Ann. Statist.*, **12**, 1310–1323.

[18] McKeague, I. W. (1986). *Ann. Statist*, **14**, 579–589.

[19] Nguyen, H. T. and Pham, T. D. (1982). *SIAM J. Control Optimization*, **20**, 603–611.

[20] Prakasa Rao, B. L. S. (1969). *Sankhyā A*, **31**, 23–36.

[21] Tsirelson, B. S. (1982). *Theory Prob. Appl.*, **27**, 411–418.

[22] Walter, G. G. and Blum, J. R. (1984). *Ann. Statist.*, **12**, 372–379.

Acknowledgment

Research supported by the U.S. Army Research Office uder Grant DAAG 29-82-K-0168.

(DENSITY ESTIMATION
KERNEL ESTIMATORS
MAXIMUM LIKELIHOOD ESTIMATION
METHOD OF SIEVES
PENALIZED MAXIMUM LIKELIHOOD ESTIMATION)

I. W. McKeague

SIGMA-FIELD *See* MEASURE THEORY IN PROBABILITY AND STATISTICS

SIGMA ISOCHRON

This is an obsolete term for 0.07 × (standard deviation). It refers to the division of the range of values

mean ± 3.5 × (standard deviation),

which includes "most" (actually 99.53%) of a normal distribution*, into one hundred equal parts.

SIGMA LATTICE *See* DENSITY ESTIMATION

SIGMA RESTRICTION

This term is sometimes used for a constraint on the values of a number of parameters in the form of a requirement that a linear function of them (usually a sum) shall have a specified value (often zero). For example, in the one-way classification* linear model

$$X_{ti} = \alpha + \lambda_t + Z_{ti},$$
$$t = 1, \ldots, k, \; i = 1, \ldots, n_t,$$

the constraint

$$\sum_{t=1}^{k} n_t \lambda_t = 0$$

is a sigma restriction (or restraint). It is a particular case of an identifiability* restriction.

(CONTRAST)

SIGNAL *See* COMMUNICATION THEORY

SIGNAL DETECTION THEORY *See* PATTERN RECOGNITION

SIGNAL-TO-NOISE RATIO *See* COMMUNICATION THEORY; NOISE

SIGNAL TRACKING *See* FORECASTING

SIGNED-RANK STATISTICS

The classical sign statistic is the precursor of general signed-rank statistics. Whereas a sign statistic is employed for testing a null hypothesis concerning the median* (or a quantile*) of a distribution (without necessarily assuming the symmetry of the distribution), a signed-rank statistic is generally used for testing the hypothesis of symmetry of a distribution (around a specified median) against shift or other alternatives. For a random sample $X_1, \dots, X_n$ of size n from a population with a distribution F, typically, a signed-rank statistic is expressed as

$$T_n = \sum_{i=1}^{n} (\text{sign } X_i) a_n^*(R_i^+), \qquad (1)$$

where $a_n^*(1), \dots, a_n^*(n)$ are suitable *scores*, sign x is equal to $+1$, 0 or -1, according as x is greater than, equal to, or less than 0, and R_i^+ is the rank of $|X_i|$ among $|X_1|, \dots, |X_n|$, for $i = 1, \dots, n$. If F is continuous (almost) everywhere, ties among the observations may be neglected with probability 1, so that $(R_1^+, \dots, R_n^+)$ represents a (random) permutation of $(1, \dots, n)$. Adjustments for ties will be discussed later on. Special cases of T_n having practical importance (and uses) are the following:

(i) Sign Statistic. If in (1) $a_n^*(k) = 1$ for every $k: 1 \leqslant k \leqslant n$, then T_n reduces to the classical sign statistic $\sum_{i=1}^{n} \text{sign } X_i$ [which does not utilize any information contained in the rank vector $\mathbf{R}^+ = (R_1^+, \dots, R_n^+)$ and depends only on the vector $\mathbf{S} = (\text{sign } X_1, \dots, \text{sign } X_n)$ of the signs of the observations].

(ii) The Wilcoxon Signed-Rank Statistic. The term "signed-rank statistic" was coined by Wilcoxon [26], who considered the statistic

$$W_n = \sum_{i=1}^{n} (\text{sign } X_i) R_i^+, \qquad (2)$$

which corresponds to the scores $a_n^*(k) = k$, for $k = 1, \dots, n$. A variant form of W_n, due to Tukey [24], is

$$\sum_{1 \leqslant i \leqslant j \leqslant n} \left(\text{sign}(X_i + X_j)\right). \qquad (3)$$

(iii) One-Sample Normal Scores* Statistic. Let $a_n^*(k)$ be the expected value of the kth order statistic* of a sample of size n from the central chi distribution* with 1 degree of freedom (i.e., the folded normal distribution), for $k = 1, \dots, n$. With these scores, T_n in (1) is termed a *normal scores statistic* (see Fraser [3]). A variant form involves the scores

$$a_n^*(k) = \Phi^{-1}((1 + k/(n+1))/2), \quad k = 1, \dots, n,$$

where $\Phi(\cdot)$ is the standard normal distribution* function.

All these particular cases may be formally defined by letting

$$a_n^*(k) = E\phi^*(U_{nk}) \text{ or } \phi^*\left(\frac{k}{n+1}\right), \quad k = 1, \dots, n; \qquad (4)$$

$$\phi^*(u) = \phi((1+u)/2), \quad 0 < u < 1,$$

where $U_{n1} < \cdots < U_{nn}$ are the ordered random variables of a sample of size n from the uniform $(0, 1)$ distribution, and ϕ is a skew-symmetric function, i.e., $\phi(u) + \phi(1-u) = 0$, for every $u \in (0, 1)$. The sign, Wilcoxon, and the normal scores statistics correspond, respectively, to $\phi(u) = \text{sign}(2u - 1), 2u - 1$, and $\Phi^{-1}(u)$. We refer to DISTRIBUTION-FREE METHODS and RANK TESTS, where the basic ideas have been elaborately discussed at a more elementary level.

Consider the null hypothesis H_0 that the distribution F is symmetric about a specified median θ_0, and, without any loss of generality, we may take $\theta_0 = 0$. Then, under H_0, $\mathbf{S}$ and $\mathbf{R}^+$ are mutually stochastically independent, $\mathbf{S}$ can have 2^n equally likely sign inversions (with the common probability 2^{-n}), and $\mathbf{R}^+$ takes on each permutation of $(1, \dots, n)$ with the common probability $(n!)^{-1}$. Since T_n in (1) depends on the sample observations through $\mathbf{S}$ and $\mathbf{R}^+$ only, it follows that under H_0, T_n is a genuinely distribution-free statistic, and its null distribution is generated by the $2^n(n!)$ equally likely realizations of $(\mathbf{S}, \mathbf{R}^+)$. This procedure enables us to enumerate the exact distribution of T_n under H_0, for specific scores and small values of n. Some tables for these are also available in the literature; see for example, Owen [12]. With increasing n, this task becomes prohibitively laborious, even with the advent of modern computers. However, if we define $T_n^* = \sum_{i=1}^n a_n^*(i)\text{sign } X_i$, then under H_0 T_n and T_n^* both have the same distribution, while for T_n^* having independent summands the classical central limit theorem* holds whenever for n increasing,

$$A_n^2 = n^{-1} \sum_{i=1}^n (a_n^*(i))^2 \quad \text{is finite and}$$

$$\max_{1 \leqslant k \leqslant n} |a_n^*(k)|/n^{1/2} \to 0. \qquad (5)$$

Note that under (4), for square-integrable ϕ, (5) holds. Thus, under H_0,

$$n^{-1/2}A_n^{-1}T_n \quad \text{is asymptotically } N(0,1), \quad (6)$$

and this provides simple large-sample approximations to the critical values of T_n. Note that for $a_n^*(k) = E\phi^*(U_{nk})$, $1 \leqslant k \leqslant n$, on letting $T_n^o = \sum_{i=1}^n \phi(F(X_i))$ and denoting by E_0 the expectation under H_0, $T_n = E_0(T_n^o|\mathbf{S}, \mathbf{R}^+)$. Further, for square-integrable ϕ, denoted by $A^2 = \int_0^1 \phi^2(u)\, du$, we have (cf. Hájek [6])

$$\begin{aligned} n^{-1}E_0(T_n - T_n^o)^2 &= n^{-1}\{E_0(T_n^{o2}) - E_0(T_n^2)\} \\ &= A^2 - A_n^2 \to 0, \qquad (7) \end{aligned}$$

as $n \to \infty$, so that by the Chebyshev inequality*, under H_0, $n^{-1/2}(T_n - T_n^o) \to 0$, in probability. For T_n^o, the classical central limit theorem* holds (when $A^2 < \infty$), and hence, the above projection yields an easy proof for the asymptotic normality of $n^{-1/2}T_n$ (under H_0). Note that unlike T_n^o or T_n^*, T_n in (1) does not have independent summands. Nevertheless, under H_0 some martingale* characterizations of $\{T_n;\, n \geqslant 1\}$ due to Sen and Ghosh [20, 22] enable incorporation of the asymptotic distributional properties of martingales to derive parallel results on $\{T_n\}$; see Sen [18, Chap. 5]. Further, T_n is scale-equivariant, i.e., a positive scalar multiplication of the X_i does not affect the value of T_n; in fact, T_n is invariant under any skew-symmetric monotone transformations on the observations [i.e., $X_i \to Y_i = g(X_i)$, where $g(\cdot)$ is monotone and $g(-x) = -g(x)$, for every x]. Actually, ($\mathbf{S}$, $\mathbf{R}^+$) constitutes the maximal invariant* with respect to this general group of transformations.

LOCALLY MOST POWERFUL (LMP) SIGNED-RANK TEST (STATISTICS)

If the distribution F has an absolutely continuous probability density function f (whose first derivative is denoted by f'), we may define $\phi_f^*(u) = \phi_f((1+u)2)$, $0 < u < 1$, by letting

$$\phi_f(u) = -f'(F^{-1}(u))/f(F^{-1}(u)),$$
$$0 < u < 1. \qquad (8)$$

Also, as in (4), we let $a_{nf}^*(k) = E\phi_f^*(U_{nk})$, $k = 1, \dots, n$, and the corresponding T_n in (1) is denoted by T_{nf}. Then, for testing H_0 against an alternative that F is symmetric about some θ (> 0), T_{nf} is an LMP rank statistic (viz. Hájek and Šidák [7, pp. 73–74]). Note that for (i) normal F, $\phi_f(u) = \Phi^{-1}(u)$, (ii) logistic F, $\phi_f(u) = 2u - 1$, and (iii) double exponential F, $\phi_f(u) = \text{sign}(u - \frac{1}{2})$, so that the normal scores statistic is an LMP rank test statistic for normal densities, the Wilcoxon signed-rank statistic for logistic* densities, and the classical sign statistic for double exponential* densities. Such LMP test statistics have also been worked out for other alternatives.

NONNULL DISTRIBUTION THEORY

When F is not symmetric about 0, $\mathbf{S}$ and $\mathbf{R}^+$ are not necessarily independent, and, moreover, each one of them may have a distribution dependent on the unknown F. Thus, T_n is not distribution-free when H_0 is not true. The exact distribution of T_n depends on the joint distribution of $\mathbf{S}$ and $\mathbf{R}^+$, and thereby may become so much involved, even for small values of n, that one is naturally inclined to adapt suitable approximations. The asymptotic distributional equivalence of T_n and T_n^o (or T_n^*) may not hold when H_0 does not hold. The basic approach is to find out suitable normalizing constants a_n and b_n (> 0), such that $(T_n - a_n)/b_n$ has asymptotically a unit normal distribution, and these a_n, b_n depend on the underlying distribution F in an estimable pattern. In the null case, $a_n = 0$ and $b_n^2 = nA_n^2$. For this problem, the Chernoff–Savage approach [2] is worked out in detail in Puri and Sen [13, Chap. 3]. An alternative approach based on the weak convergence of some empirical processes is due to Pyke and Shorack [14] (*see* PROCESSES, EMPIRICAL). For contiguous (local) alternatives. Hájek [6] has an elegant proof requiring the least restrictive regularity conditions on the score function (ϕ) but a finite Fisher information* on the density function. For such contiguous alternatives, martingale characterizations, considered in detail in Sen [18, Chap. 5] yield deeper results. For some stationary stochastic processes*, for signed-rank statistic, asymptotic results have been studied by Sen and Ghosh [21] and Yoshihara [27], among others. Hušková [9] has considered a more general form of signed-rank statistics, namely,

$$\sum_{i=1}^{n} c_{ni} a_n^*(R_i^+) \operatorname{sign} X_i, \qquad (9)$$

where the c_{ni} are given (regression) constants; he also studied the general distribution theory for both local and general alternatives.

ADJUSTMENT FOR TIES

For continuous F, ties among the observations may be neglected with probability 1. However, in practice, one may have a discrete distribution, or due to the process of rounding up, one may even have data classified into grouped (or ordered) class intervals. Here sign X_i may assume the value 0 with a positive probability, and ties among the $|X_i|$ may occur with a positive probability. In such a case, for tied observations, either one may use the *mid-ranks* (i.e., the average of the ranks for these tied observations) or *average scores* [i.e., the average of the $a_n^*(\cdot)$ corresponding to the tied ranks]. For the Wilcoxon signed-rank* statistic, either process will lead to the same statistic, while for general scores these are not necessarily the same. In the context of grouped data, Ghosh [4] has given useful account of these developments (*see also* RANK TESTS, GROUP). In the tied case, the signed-rank statistics are generally only conditionally distribution-free (under h_0), and the locally optimal scores are defined in terms of ϕ_f in (8) along with the adjustments for the ties.

ALIGNED SIGNED-RANK STATISTICS

In the context of testing for symmetry of a distribution whose median (θ) is unspecified, often an estimator $\tilde{\theta}_n$ of θ is incorporated in aligning the observations to $\hat{X}_i = X_i - \tilde{\theta}_n$, $i = ,\dots, n$, and a signed-rank statistic $\hat{T}_n$, based on these aligned observations [and defined as in (1)], is termed an *aligned signed-rank statistic*. The $\hat{X}_i$ are no longer independent, and in general the distribution of $\hat{T}_n$ depends on the underlying F (through the estimator $\hat{\theta}_n$), so that $\hat{T}_n$ is not necessarily distribution-free (even when F is symmetric about θ). Nevertheless, one may obtain asymptotically distribution-free versions of $\hat{T}_n$ under quite general conditions. Gupta [5] used the Wilcoxon signed-rank statistic for $\hat{T}_n$, where for $\tilde{\theta}_n$ the sample median was used; more general treatment for arbitrary score functions is owing to Sen [17]. If the score function ϕ is nondecreasing and if for every real b, the signed-rank statistic based on the $X_i - b$ [and defined as in (1)] is denoted by $T_n^{(b)}$, then $T_n^{(b)}$ is nonincreasing in b. Since $T_n(\theta)$ has a distribution symmetric about 0 (when F is symmetric about θ), we may virtually equate $T_n(b)$ to 0, and as in Hodges and Lehmann [8] and Sen [16] consider the following estimator:

$$\tilde{\theta}_n = \left(\sup\{b : T_n(b) > 0\} + \inf\{b : T_n(b) < 0\}\right), \qquad (10)$$

which is known as an *R-estimator of location*.

$\tilde{\theta}_n$ is a robust, translation-invariant, scale-equivariant*, symmetric, and consistent estimator of θ. Essentially (10) is based on an alignment principle, and in this context some asymptotic linearity results play a very important role. Basically, for every (fixed) $K: 0 < K < \infty$, as $n \to \infty$,

$$\sup\left\{ n^{-1/2}|T_n(\theta + b) - T_n(\theta) + nb\gamma| : |b| \leqslant n^{-1/2}K \right\} \to 0 \quad (11)$$

(almost surely, due to Sen and Ghosh [20], and weakly, due to van Eeden [25]), where γ is a suitable constant dependent on the distribution F and the score function ϕ. The asymptotic properties of $\hat{\theta}_n$ may most conveniently be studied by using (11). Aligned rank statistics have also been employed in the change point problem (*see* Sen [19] and ROBUST TESTS FOR CHANGE POINT MODELS).

MULTIVARIATE SIGNED-RANK STATISTICS

A natural extension of the hypothesis of symmetry to the multivariate case is the hypothesis of *diagonal symmetry*. A $p(\geqslant 1)$-variate distribution F of a random vector $\mathbf{X}$ is said to be diagonally symmetric about the origin if both $\mathbf{X}$ and $(-1)\mathbf{X}$ have the same distribution F. If $\mathbf{X}_i = (X_{i1}, \dots, X_{ip})'$, $i = 1, \dots, n$, are n independent random vectors drawn from the distribution F, based on the jth coordinate variables $X_{1j}, \dots, X_{nj}$, we may define the signed-rank statistic T_{nj} as in (1), where, to be more flexible, we may also take the scores as $a_{nj}{}^*(i)$, $i = 1, \dots, n$, generated by score functions ϕ_j^*, $j = 1, \dots, p$. Though marginally the T_{nj} are distribution-free under the null hypothesis, the joint distribution of the vector $\mathbf{T}_n = (T_{n1}, \dots, T_{np})'$ generally depends on F even under H_0. To overcome this problem, Sen and Puri [23], following Chatterjee [1], considered the following rank-permutation approach. Let $E_n = (\mathbf{X}_1, \dots, \mathbf{X}_n)$ be the $p \times n$ matrix (sample point), and consider the set of 2^n matrices

$$E_n(\mathbf{j}) = \left((-1)^{j_1}\mathbf{X}_1, \dots, (-1)^{j_n}\mathbf{X}_n\right), \quad \mathbf{j} = (j_i, \dots, j_n), \quad (12)$$

where each j_i can be either 0 or 1. Over the set $\mathscr{E}_n$ of 2^n matrices in (12), the conditional distribution of E_n is uniform (under H_0). With respect to this conditional law (P_n), $\mathbf{T}_n$ has null mean vector and dispersion matrix $\mathbf{V}_n = (V_{njj'})$ defined by

$$V_{njj'} = \sum_{i=1} \operatorname{sign} X_{ij} \operatorname{sign} X_{ij'} \times a_{nj}^*(R_{ij}^+) a_{nj'}^*(R_{ij'}^+) \quad (13)$$

for j, $j' = 1, \dots, p$, where R_{ij}^+ = rank of $|X_{ij}|$ among $|X_{1j}|, \dots, |X_{nj}|$, for $i = 1, \dots, n$ and $j = 1, \dots, p$. Sen and Puri [23] considered the multivariate test statistic

$$L_n = \mathbf{T}_n'\mathbf{V}_n^-\mathbf{T}_n, \quad (14)$$

where $\mathbf{V}_n^-$ is a generalized inverse* of $\mathbf{V}_n$. For small values of n, the conditional distribution of L_n (over the set $\mathscr{E}_n$) may be obtained by direct enumeration, while for large n, this can be approximated by the (central) chi-square distribution* with p degrees of freedom. For nonnull distributional results, see Sen and Puri [23] and Hušková [10].

SEQUENTIAL SIGNED-RANK STATISTICS

If R_{ni}^+ be the rank of $|X_i|$ among $|X_1|, \dots, |X_n|$, for $i = 1, \dots, n$, $n \geqslant 1$, then a variant form of (1) is

$$T_n^S = \sum_{i=1}^{n} (\operatorname{sign} X_i) a_i^*(R_{ii}^+), \quad (15)$$

where under H_0, sign X_i and R_{ii}^+ are mutually independent and for different i, the summands in (15) are also independent. Here ranking is made sequentially and so are the scores $a_i^*(\cdot)$ chosen. Reynolds [15] considered the particular case of Wilcoxon scores and the general case has been studied by Müller-Funk [11]. Though T_n in (1) does not have independent summands, under H_0, it has some nice martingale properties and, further, T_n and T_n^S are asymptotically stochastically equivalent.

References

[1] Chatterjee, S. K. (1966). *Ann. Math. Statist.*, **37**, 1771–1782.

[2] Chernoff, H. and Savage, I. R. (1958). *Ann. Math. Statist.*, **29**, 972–994.

[3] Fraser, D. A. S. (1957). *Ann. Math. Statist.*, **28**, 1040–1043.

[4] Ghosh, M. (1973). *Ann. Inst. Statist. Math.*, **25**, 109–122.

[5] Gupta, M. K. (1967). *Ann. Math. Statist.*, **38**, 849–866.

[6] Hájek, J. (1962). *Ann. Math. Statist.*, **33**, 1124–1147.

[7] Hájek, J. and Šidák, Z. (1967). *Theory of Rank Tests.* Academic, New York.

[8] Hodges, J. L., Jr., and Lehmann, E. L. (1963). *Ann. Math. Statist.*, **34**, 598–611.

[9] Hušková, M. (1970). *Zeit. Wahrsch. verw. Geb.*, **12**, 308–322.

[10] Hušková, M. (1971). *J. Multivariate Anal.*, **1**, 461–484.

[11] Müller-Funk, U. (1983). *Sequential Anal.*, **2**, 123–248.

[12] Owen, D. B. (1962). *Handbook of Statistical Tables.* Addison-Wesley, Reading, MA.

[13] Puri, M. L. and Sen, P. K. (1971). *Nonparametric Methods in Multivariate Analysis.* Wiley, New York.

[14] Pyke, R. and Shorack, G. R. (1968). *Ann. Math. Statist.*, **39**, 1675–1685.

[15] Reynolds, M. R. (1975). *Ann. Statist.*, **3**, 382–400.

[16] Sen, P. K. (1963). *Biometrics*, **19**, 532–552.

[17] Sen, P. K. (1977). *Ann. Statist.*, **5**, 1107–1123.

[18] Sen, P. K. (1981). *Sequential Nonparametrics: Invariance Principles and Statistical Inference.* Wiley, New York.

[19] Sen, P. K. (1985). *Chernoff Festschrift*, M. H. Rizvi et al., eds. Academic, New York, pp. 371–391.

[20] Sen, P. K. and Ghosh, M. (1971). *Ann. Math. Statist.*, **42**, 189–203.

[21] Sen, P. K. and Ghosh, M. (1973). *Sankhyā A*, **35**, 153–172.

[22] Sen, P. K. and Ghosh, M. (1973). *Ann. Statist.*, **1**, 568–576.

[23] Sen, P. K. and Puri, M. L. (1967). *Ann. Math. Statist.*, **38**, 1216–1228.

[24] Tukey, J. W. (1949). *Statistics Research Group Mimeo, Rep. No. 17.* Princeton University, Princeton, NJ.

[25] Van Eeden, C. (1972). *Ann. Math. Statist.*, **43**, 791–802.

[26] Wilcoxon, F. (1945). *Biometrics*, **1**, 80–83.

[27] Yoshihara, K.-I. (1978). *Zeit. Wahrsch. verw. Geb.*, **43**, 101–127.

(DISTRIBUTION-FREE METHODS
NORMAL SCORES TESTS
RANK ORDER STATISTICS
RANK TESTS
SCORE STATISTICS
SIGN TESTS
WILCOXON SIGNED-RANK TEST)

P. K. SEN

SIGNIFICANCE LEVEL *See* HYPOTHESIS TESTING; SIGNIFICANCE TESTS, HISTORY AND LOGIC

SIGNIFICANCE PROBABILITY *See* *P*-VALUES

SIGNIFICANCE TESTS, HISTORY AND LOGIC

Significance tests comprise a wide variety of statistical techniques. Well known examples include Student's *t*-test* for equality of means, the chi-square test* for goodness of fit, and the *F*-test* used in the analysis of variance*. Such tests are used frequently in the social and biological sciences, and to a lesser extent in the physical sciences. Despite their diversity, all significance tests share a common logic.

LOGIC

A satisfactory significance test must (1) stipulate a suitable hypothesis of chance, (2) find a test statistic to rank possible experimental outcomes, (3) determine the level of significance of the experimental outcome from the test statistic probability distribution, and (4) reject or fail to reject the hypothesis of chance. The history of significance tests may be viewed as a concerted effort to construct tests that satisfy these criteria. Much controversy remains today over the degree to which tests can and do satisfy these four criteria.

Stipulating a Chance Hypothesis

Significance tests are used in situations where the phenomena under investigation have a chance component. This chance component may be introduced artificially, as it is in random sampling, or it may occur naturally, as it does in radioactive decay. In either case, the presence of a chance component makes possible the occurrence of unusual and misleading results: 90 heads in 100 flips of a fair coin, for example. Significance tests provide operational bounds on the degree to which such unusual results can be attributed to the working of chance* rather than to some systematic source. In order to provide these bounds, a model is used to describe the operation of chance effects. If chance is the only source for the observed results, then the model must be sufficient by itself to account for the results. Significance tests test the sufficiency of the model in this respect. The hypothesis that the model is sufficient is commonly called the *null* hypothesis*.

Finding a Test Statistic

In order to conduct significance test, an appropriate means of summarizing the data is necessary. Appropriateness makes two demands on this summary statistic: First, it must rank all the possible outcomes with regard to how adequately the chance model accounts for their occurrence. Second, probabilities of test statistic values, conditional on the null hypothesis, must be well defined and must yield the same ordering. The first demand allows evidence to bear on the adequacy of the chance model. The second demand allows results from different significance tests to be compared with each other. The most important consequence of these demands is that experimental results that the model cannot account for well are placed in the "tails" of the probability distribution of the test statistic.

Determining the Level of Significance

Once the trial is performed, the resulting value of the test statistic is calculated. The *level of significance* of the outcome is the probability, on the null hypothesis, of any result *as bad or worse* than the observed (understood in terms of the ranking provided by the test statistic). The smaller this probability, the stronger the evidence against the null hypothesis. "Tail area probabilities" measure the strength of evidence.

Rejecting or Failing to Reject the Null Hypothesis

When the null hypothesis does not adequately account for experimental results it is rejected. Adequacy reflects several features of the test situation, among them the strength of the evidence, measured by the level of significance, against the null hypothesis. However, the adequacy of the null hypothesis also depends on other features of the test situation, such as what alternative hypotheses to the null are available and how fully the phenomena are theoretically understood. A commonly accepted level of significance in the social sciences is 0.05.

HISTORY

The first published use of reasoning that looks like a significance test is commonly attributed to John Arbuthnot (1710). Arbuthnot noted that during those years when birth records were kept in London, male births exceeded female births each year. He calculated that, on the assumption of chance (which for Arbuthnot meant equal chances), it was exceedingly improbable that 82 out of 82 years would be "male years."

Arbuthnot went on to generalize his observation beyond the past 82 years in London to "Ages of Ages, and not only at London, but all over the World." He concluded that chance cannot by itself account for the gender of an infant at birth.

Arbuthnot provides a clear example of the problem of finding a suitably inclusive model for the operation of chance. Nicholas Bernoulli (1713) criticized Arbuthnot for not including binomial* probabilities other than $\frac{1}{2}$ to describe chance. Furthermore, for all its appearance of a significance test, Arbuthnot's test was not really a significance test—Arbuthnot did not reach his conclusion based on the small probability or level of significance of the occurrence. He reached his conclusion on the basis of a simple inductive generalization of the 82 observed years, which resulted in an infinitely small probability.

Other examples of significance tests were rare. One well known exception was J. Mitchell's argument that the stars were not distributed at random. This appeared in 1767.

In 1812, Laplace's *Théorie Analytique des Probabilités* presented a pattern of reasoning that would now be called a significance test. In one of his examples, Laplace* showed that the height of the barometer at 9:00 A.M. and at 4:00 P.M. is systematically different. Laplace used a normal distribution* with a mean of 0 to model the putative effects of chance *differences* in the barometer readings at the two times of the day and estimated the standard deviation. Since his observations were more than 7 standard deviations from the mean, he rejected the hypothesis of chance.

In contrast to the modern significance test, however, Laplace adopted a purely *subjective* interpretation of probability. The shift from subjective* to objective probabilities occurred over the next 60 years as statistical descriptions of nature blossomed, from statistical thermodynamics to Galton's hereditary statistics.

In 1885, F. Y. Edgeworth* presented a detailed theory for testing the adequacy of a statistical description. Edgeworth's description of his procedure sounds remarkably modern:

> In order to determine whether the difference between two proposed Means is or is not accidental, form the probability curve under which the said difference, supposing it were accidental, would range. Consider whether the difference between the observed Means exceeds two or three times the modulus [2.8–4.2 standard deviations] of that curve. If it does, the difference is not accidental.

Later Edgeworth described such a result as "significant." For Edgeworth, the stringency of a test was measured by an equivalent to the standard deviation of the normal curve fit to the data. This was in keeping both with Laplace's treatment and its use in Gauss' theory of error. Levels of significance were established as tail area probabilities by a link to the theory of error (*see* LAWS OF ERROR).

In 1900, Karl Pearson's chi-squared paper addressed the problem of measuring the goodness of fit* between data and distributions other than the normal distribution. This was a problem Pearson* himself had generated with his 1895 paper that generalized the use of the normal distribution to his system of frequency distributions (*see* PEARSON SYSTEM OF DISTRIBUTIONS). Pearson considered a case of N normally distributed correlated variables. He defined a function χ^2 (chi squared) of these variables that increases as values of the variables deviate from *expected* values. Happily, the probability distribution of this function behaved properly with values for poor fit in the tails. Given enough observations per cell, an N-celled multinomial chance setup can be treated as a set of N normally distributed correlated variables. In such a setup, χ^2 equals the familiar sum of normalized squared deviations from expectations:

$$(O - E)^2/E.$$

Broadly inclusive null hypotheses, of necessity, include parameters whose values are not identified by the model. When samples are small, the asymptotic normal properties of these models cannot be used and no determinant probability distribution can be found for the null hypothesis. Pearson's test is a large-sample test because a large number of observations are required to make use of the properties of asymptotic normalcy.

In 1908, writing under the pen name of "Student," W. S. Gosset* proposed the first solution to the problem of small-sample tests. As usual, he modeled chance using the normal distribution. The key to Student's achievement was his ability to determine an exact probability distribution even when the standard deviation had to be approximated by the sample standard deviation. The result was Student's t-distribution* with its characteristic tails fatter than the normal distribution.

In the early 1920s R. A. Fisher* greatly extended Student's result by incorporating his theory of parameter estimation into test construction. By using data to estimate parameter values, the "degrees of freedom*" that the data had to deviate from expectations were reduced. Nonetheless, Fisher showed that in a wide variety of cases a distribution for the test statistic, conditional on the null hypothesis, could still be found when the free parameters were estimated using his method of maximum likelihood*. Fisher's method had some unexpected consequences. In particular, Fisher's technique allowed for cases where one hypothesis i would be rejected but another h would not, even though i is a consequence of h. It has been suggested that "reject i" should be understood as "take i to be unacceptable for further consideration" rather than "take i to be false"; for if i were false, h would have to be false as well.

Starting in 1928, J. Neyman* and E. S. Pearson* developed a testing methodology that frequently has been confused with significance tests, since many specific Neyman–Pearson hypothesis tests virtually coincide with a corresponding significance test. Neyman, however, explicitly denied that any form of statistical inference could produce the logical concepts that significance tests purport to produce. Neyman asked "What justifies the ordering provided by the test statistic?" The answer—that they occur in the tails of the statistic distribution and are, consequently, improbable given the hypothesis—did not satisfy Neyman. He argued that by means of a 1 : 1 transformation, these "improbable" tail values could be made "probable" center-of-the-distribution values. Neyman chose to avoid the problem by requiring tests to be comparative between the null hypothesis and some other well specified alternative (*see* HYPOTHESIS TESTING). To the contrary, it has been argued that Neyman's demand for specific alternative hypotheses was unnecessary and undesirable. In 1979, Seidenfeld noted that Neyman's 1 : 1 transformation requires information not available at the time a significance test is conducted, and in many instances there seems adequate independent justification for the particular test statistic employed.

Another problem has arisen in the recent history of significance tests. The result of a significance test is either *rejection* of the null hypothesis or no rejection of the null hypothesis. The question remains, however, whether any hypotheses can legitimately be *accepted* or provisionally adopted as a result of a significance test. Suppose a test turns up a significant result and the null hypothesis is rejected: Does this justify the acceptance of another hypothesis about the existence of a systematic source for the observations? Strictly speaking, adopting the hypothesis of *some* systematic source for the observations requires that the hypothesis of chance be completely inclusive, a difficult demand to meet. Conversely, suppose a test turns up a strongly nonsignificant result ($p \sim 0.75$): does that justify the acceptance of the null hypothesis? There are strong reasons to suppose not. A nonsignificant result may show that the null hypothesis *could* adequately account for the data, but without additional assumptions, such a result need not show that the null hypothesis is the only possible account for the data. The confusion surrounding this issue can be relieved somewhat by attending to a subtle shift in the interpretation of significance tests. Pearson's 1900 chi-square test uses levels of significance as a *measure* of fit between hypothesis and data. Nonsignificant results may well indicate good fit, but not the truth, of the null hypothesis.

Bibliography and References

Logic

Most elementary statistics texts include some discussion of the logic of significance tests.

Henkel, R. (1976). *Tests of Significance*, Sage Publications, Beverly Hills, CA. (A recent booklet that elaborates on the logic of significance tests.)

Morrison, D. E. and Henkel, R. E., eds. (1970). *The Significance Test Controversy—A Reader*, Aldine Publishing Company, Chicago, IL. (This anthology usefully describes the confusion and controversy surrounding the logic of signficance tests.)

Seidenfeld, T. (1979). *Philosophical Problems of Statistical Inference: Learning from R. A. Fisher*, Chap. 3, Reidel, London, England. (The best recent philosophical examination of the logic of significance tests.)

History

Baird, D. (1983). The Fisher/Pearson chi-squared controversy: A turning point for inductive inference. *Brit. J. Philos. Sci.*, **34**, 109–118. (The controversy surrounding Fisher's introduction of degrees of freedom is discussed.)

Berkson, J. (1942). Tests of significance considered as evidence, *J. Amer. Statist. Ass*, **37**, 325–335. (Argues for taking nonsignificant outcomes as evidence for the null hypothesis.)

Box, J. F. (1978). *R. A. Fisher, The Life of a Scientist.* Wiley, New York. (Biography of Fisher; an enjoyable and informative history of much of the important history connected with Fisher's contributions to statistics.)

Edgeworth, F. Y. (1885). Methods of statistics, *J. R. Statist. Soc.* (jubilee volume), 181–217. (Not reprinted.)

Fisher, R. A. (1959). *Statistical Methods and Scientific Inference.* Hafner, New York. (Fisher defends and elaborates significance testing in his last book on statistical methods.)

Fisher, R. A. (1970). *Collected Papers of R. A. Fisher*, J. H. Bennett, ed. University of Adelaide Press, Adelaide, Australia. (Contains reprints of Fisher's two most important papers on significance tests: his 1922 paper "On the Mathematical Foundations of Theoretical Statistics" and his 1924 paper "On a Distribution Yielding the Error Functions of Several Well-known Statistics." Also included is his 1922 paper "On the Interpretation of Chi-Square for Contingency Tables and the Calculation of *P*," which introduced the concept of degrees of freedom.)

Hacking, I. (1975). *The Emergence of Probability*. Cambridge University Press, Cambridge, England. (Includes a contemporary analysis of Arbuthnot's paper with mention of Nicholas Bernoulli's criticism.)

Hall, P. and Selinger, B. (1986). Statistical significance: balancing evidence against doubt, *Austral. J. Statist.*, **28**, 354–370. (History of development of 5% and 1% conventions, with emphasis on interactions between legal and statistical aspects.)

Kendall, M. G. and Plackett, R. L., eds. (1977). *Studies in the History of Statistics and Probability*. Griffin, London, England. (A useful collection that includes a reprint of Arbuthnot's 1710 paper, "An Argument for Divine Providence, Taken from the Constant Regularity Observed in the Births of Both Sexes.")

Laplace, P. S. de (1812). *Théorie Analytique des Probabilités*. Paris. (Not reprinted.)

Laplace, P. S. de (1951). *A Philosophical Essay on Probabilities*. Dover, New York. (Philosophical introduction to the theory of probability; is reprinted.)

MacKenzie, D. A. (1981). *Statistics in Britain*, 1865–1930. Edinburgh University Press, Edinburgh, Scotland. (One of the few general histories of statistics. An interesting sociological perspective on the history of statistics.)

Mitchell, J. (1767). An inquiry into the probable parallax and magnitude of the fixed stars, from the quantity of light which they afford us, and the particular circumstances of their situation. *Philos. Trans. R. Soc. Lond.*, **87**. (Not reprinted.)

Neyman, J. (1938). Lectures and Conferences on Mathematical Statistics. The Graduate School of the United States Department of Agriculture, Washington, DC. (Some of Neyman's criticisms of significance tests.)

Neyman, J. and Pearson, E. S. (1967). *Joint Statistical Papers*. Cambridge University Press, Cambridge, England. (Reprints of the original sources for Neyman and Pearson's alternative testing methodology.)

Pearson, K. (1948). *Karl Pearson's Early Statistical Papers*, E. S. Pearson, ed. Cambridge University Press, London England. (Pearson's generalized system of frequency distributions in his second "Contribution to the Mathematical Theory of Evolution" and his chi-squared paper, "On the Criterion that a Given System of Deviations from the Probable in the case of a Correlated System of Variables is such that it Can be Reasonably Supposed to Have Arisen from Random Sampling," are reprinted in this excellent collection of his statistical papers.)

Pearson, K. (1978). *The History of Statistics in the 17th and 18th Centuries*, E. S. Pearson, ed. MacMillan, New York. (Thé statistical methods presented in the *Theorie Analytique*, including the barometer example, are discussed at length in these lecture notes.)

Pearson, E. S. and Kendall, M. G., eds. (1970). *Studies in the History of Statistics and Probability*. Griffin, London, England. (An earlier anthology of articles about the history of statistics that includes reprints.)

Pearson, E. S. and Wishart, J., eds. (1942). *Student's Collected Papers*. Cambridge University Press, Cambridge, England. (Gosset's 1908 paper presenting the t-test, "The Probable Error of a Mean," is reprinted in this collection of his papers.)

Walker, H. M. (1929). *Studies in the History of Statistical Method*. Williams and Wilkins, Baltimore, MD. (Interesting both as a history and as an item out of history.)

(CHANCE
FISHER, RONALD AYLMER
HYPOTHESIS TESTING
LOGIC OF STATISTICAL REASONING
NEYMAN, JERZY
PEARSON, EGON SHARPE
PEARSON, KARL
NULL HYPOTHESIS
P-VALUES
STATISTICS, HISTORY OF)

DAVIS BAIRD

SIGNIFICANT FIGURE

This is any digit in a number that contributes to specification of its magnitude, apart from 0's, which solely determine the position of the decimal point.

For example 345 and 0.000345 each have three significant figures; 34.05 and 34.50 each have four significant figures (but 34.5 has only three); 345,000 has three, four, or five significant figures according as it means 3.45×10^5, 3.450×10^5 or 3.4500×10^5, respectively.

(ROUND-OFF ERROR)

SIGN TESTS

The sign test is a simple, quick, and versatile distribution-free test (*see* DISTRIBUTION-FREE METHODS) that was introduced in its basic form very early by Arbuthnot [1]. The primary applications covered here are a special case of the binomial sign test (*see* BINOMIAL TEST), a test and confidence interval for location in a one-sample or paired-sample situation (*see* LOCATION TESTS), a test for trend*, and McNemar's test (*see* MCNEMAR STATISTIC). These procedures are discussed in more detail in many nonparametric text and reference books, including Bradley [5], Conover [9], Daniel [11], Gibbons [18, 19], Hájek [20], Hájek and Šidák [21], Hettmansperger [24], Hollander and Wolfe [26], Lehmann [29], Marascuilo and McSweeney [33], Pratt and Gibbons [38], and Randles and Wolfe [40]. Bradley [6] gives a good summary of these tests and some modifications.

BINOMIAL SIGN TEST

Assume that the data consist of n independent trials of an event that has only two possible outcomes, called success (S) and failure (F) for convenience, and that $\Pr[S] = 1 - \Pr[F] = \theta$ on every trial. The exact distribution of the observed number of successes R is binomial with parameters n and θ, or

$$\Pr[R = r] = \binom{n}{r}\theta^r(1-\theta)^{n-r}. \quad (1)$$

R provides a logical test statistic for the null hypothesis $H_0: \theta = 0.5$, and we expect R to be about $n/2$ if H_0 is true. A large value of R would support the one-sided alternative $A_+: \theta > 0.5$ and hence the appropriate rejection region for a test at nominal level α is $R \geqslant c_\alpha$, where c_α is the smallest integer which satisfies

$$\sum_{r=c_\alpha}^{n} \binom{n}{r}(0.5)^n \leqslant \alpha; \quad (2)$$

the exact level of this test is the value of the left-hand side of (2). If a P-value* is desired for this alternative A_+, its value is $\Pr[R \geqslant r]$ for an observed value r. Similarly, a small value of R supports the alternative $A_-: \theta < 0.5$ and the rejection region is $R \leqslant c'_\alpha$, where c'_α is the largest integer that satisfies

$$\sum_{r=0}^{c'_\alpha} \binom{n}{r}(0.5)^n \leqslant \alpha; \quad (3)$$

the exact level is the value of the left-hand side of (3). The appropriate P-value here is $\Pr[R \leqslant r]$. For a two-sided alternative $A: \theta \neq 0.5$, we use a two-tailed rejection region, each tail of size $\alpha/2$.

Any table of the binomial distribution can be used to find critical values or P-values. Extensive tables are given in Harvard University Computation Laboratory [22] and National Bureau of Standards [36], as well as many other sources.

For n large (at least 20), the distribution of $Z = (R - n/2)/\sqrt{n/4}$ is approximately standard normal. A continuity correction* of ± 0.5 ($+0.5$ if $r < n/2$ and -0.5 if $r > n/2$) can be incorporated in the numerator of Z. A convenient approximation for a two-tailed test at level 0.05 is to reject H_0 when $|r - (n - r)| \geqslant 2\sqrt{n}$ (Duckworth and Wyatt [16]).

The first implicit use of this test in the literature (Arbuthnot [1]) is for a comparison of the number of male and female births in London over a period of $n = 82$ years (*see* SIGNIFICANCE TESTS, HISTORY AND LOGIC). If we define a success as more male than female births in a given year and regard the years as independent trials, we can test $H_0: \theta = 0.5$ versus $A: \theta \neq 0.5$. Since a success occurred in each year, we have $r = 82$; the exact two-tailed P-value is

$$\Pr[R = 0] + \Pr[R = 82] = 4.13 \times 10^{-25}$$

and the normal approximation test statistic is

$$Z = (82 - 41)/\sqrt{82/4} = 9.055.$$

Hence the null hypothesis is rejected at any reasonable level.

The binomial sign test of $H_0: \theta = 0.5$ can be generalized to a test of $H_0: \theta = \theta_0$ using (1) (*see* BINOMIAL TEST).

LOCATION TEST FOR THE MEDIAN (OR MEDIAN DIFFERENCE)

The sign test for the median applies in a situation where the data consist of a random sample $X_1, X_2, \ldots, X_n$ from any infinite population or drawn with replacement from a finite population with median M. The null hypothesis is $H_0: M = M_0$. We assume that the population is continuous at M_0 and the data are measured on at least an ordinal scale relative to M_0. Here we define a success as the event $X_i - M_0 > 0$ so that $\theta = \Pr[X_i - M_0 > 0]$ and the null hypothesis can again be written as $H_0: \theta = 0.5$. If we define R as the number of positive differences among the observed $X_i - M_0$ (the number of observations that exceed M_0), the binomial test applies exactly as before to the respective alternatives $A_+: \Pr[X_i > M_0] > 0.5$ or equivalently $A_+: M > M_0$, $A_-: \Pr[X_i > M_0] < 0.5$ or equivalently $A_-: M < M_0$, and $A: \Pr[X_i > M_0] \neq 0.5$ or equivalently $A: M \neq M_0$.

The assumption that the population is continuous at M_0 guarantees that $\Pr[X_i = M_0] = 0$. Nevertheless, it may happen in practice that $X_i - M_0 = 0$, called a zero, because the measurement is not sufficiently refined. One method of handling zeros is to calculate the test statistic R (or the P-value) at the two possible extreme values, both when all the zeros are counted as positives and when all the zeros are counted as negatives. If both calculations lead to rejection of H_0, or both lead to acceptance, the decision is clear; if these two calculations lead to opposite conclusions, the experimenter should either take more refined measurements or additional data. The other method of handling zeros is to ignore them and reduce n accordingly; this procedure is justified if the distribution of X_i is symmetric or if the population is defined as excluding any $X_i = M_0$. Additional discussion of zeros is given in Hemelrijk [23], Putter [39], and Krauth [28]. Ties, i.e., $X_i = X_j \neq M_0$ for some $i \neq j$, present no problems for the sign test.

In the paired sample situation with $H_0: M_D = M_0$, where M_D denotes the median of the population of differences $D = X - Y$, the exact same procedure can be followed, where R is defined as the number of positive differences among the $X_i - Y_i - M_0$.

Hemelrijk [23] shows that the sign tests with appropriate critical regions are unbiased and consistent against all alternatives in the corresponding direction. The exact power* of the sign test against the alternative $\theta = \theta_1 > 0.5$ is

$$\sum_{r=c_\alpha}^{n} \binom{n}{r} [F(\theta_1)]^r [1 - F(\theta_1)]^{n-r},$$

where F is the CDF of the population X (or population of differences $X - Y$). Comparisons of the power of the sign test with analogous parametric (Student's t*, normal theory) and/or nonparametric (Wilcoxon signed rank*) tests are reported in Walsh [47, 48], Dixon [14], Hodges and Lehmann [25], Blyth [4], Bahadur [2], David and Perez [13], and Gibbons [17]. MacStewart [30] and Dixon and Mood [15] give tables of sample sizes required to achieve a specified power at a given level for various values of θ. Cohen [8, pp. 147–173] gives extensive tables of the power of these sign tests and illustrates their use.

One difficulty with exact power calculations for comparison of two or more nonparametric tests is that the discreteness of the sampling distributions usually makes it impossible to have exactly equal α levels. One solution to this problem is to resort to randomized decision rules, as in Gibbons [17]. Another solution is to make Monte Carlo comparisons of power. Randles and Wolfe [40, p. 116] give some Monte Carlo comparisons of the sign test, Wilcoxon's signed rank test, and Student's t test with $n = 10, 15, 20$ for the uniform, normal, logistic, double exponential, and Cauchy distributions. The sign test is the best test for small amounts of shift in the double exponential and Cauchy distributions, but has the poorest performance in all other cases studied.

The asymptotic relative efficiency* of the sign test relative to Student's t test is at least $1/3$ for any unimodal symmetric distribution, equals 0.637 for the normal, 0.333 for the uniform, 2.0 for the double exponential, and 0.822 for the logistic distributions. These and other efficiencies are summarized in Pratt and Gibbons [38, p. 384].

The location test for the median can be generalized to a test of $H_0 : Q = Q_0$, where Q is a quantile* of any order p; (1) applies with $\theta = \Pr[X_i > Q_0] = 1 - p$ and R is the number of sample observations that exceed Q_0.

CONFIDENCE INTERVAL FOR THE MEDIAN (*see* NONPARAMETRIC CONFIDENCE INTERVALS)

The sign test for the median M (or median difference M_D) has a corresponding procedure for constructing a confidence interval for M (or M_D) with confidence coefficient $1 - \alpha$. The endpoints for a two-sided confidence interval on M are the order statistics $X_{c'_\alpha+1:n}$ and $X_{n-c'_\alpha:n}$, and on M_D are the same order statistics for the differences of pairs $X - Y$. For large samples, the normal approximation (with continuity correction) gives

$$c_\alpha' = 0.5\left(n - 1 - z_{\alpha/2}\sqrt{n}\right)$$

(rounded down to the next smaller integer), where $z_{\alpha/2}$ is the positive standard normal variate that satisfies $\Phi(z_{\alpha/2}) = 1 - \alpha/2$ for Φ the standard normal CDF. These procedures are developed in Thompson [44], Savur [42], and David [12].

SIGN TEST FOR TREND

Cox and Stuart [10] suggest that the sign test be used to test for trend* in location of a set of $(2n)$ time ordered or otherwise sequenced observations on a continuous variate X measured on at least an ordinal scale. (The middle observation is discarded if the total number is odd.) Here we form the n differences $X_{n+i} - X_i$ and R is the number of positive signs among these differences. The null distribution of R for the null hypothesis of a random sequence is as in (1) with

$$\begin{aligned}\theta &= \Pr[X_{n+i} - X_i < 0] \\ &= \Pr[X_{n+i} - X_i > 0] = 0.5\end{aligned}$$

and effective sample size n, and the critical regions specified by (2) and (3) are appropriate for the alternatives A_+: positive trend, A_-: negative trend, and A: trend in unspecified direction. Here we assume that $\Pr[X_{n+i} - X_i = 0] = 0$ so that zeros do not occur in theory; in practice zeros are usually ignored and n is reduced accordingly. A variation of this test uses a sequence of $3n$ observations and forms n differences using only the first third and last third of the sequence, or $X_{2n+i} - X_i$.

Stuart [43] shows that the asymptotic relative efficiencies of these tests for X normally distributed at each point in time are 0.78 and 0.83, compared to the parametric regression coefficient test, and 0.79 and 0.84, respectively, compared to Spearman's or Kendall's nonparametric tests for trend. Mansfield [31] and Olshen [37] give some power and efficiency results for these tests.

These same tests can be modified to test for trend in dispersion by taking the signs of ranges of blocks of observations. Such modifications are discussed in Ury [45] and Rao [41].

These tests for trend can also be applied to test correlation in bivariate continuous variables (X, Y) for sample data measured on at least an ordinal scale. The null hypothesis here is H_0: X and Y are independent, with alternatives A_+: positive correlation, A_-: negative correlation, or A: correlation in unspecified direction. The procedure is to order the X (or Y) values from smallest to largest and compute R as defined above for the arrangement that results for the corresponding Y (or X) values. Ordinarily it does not matter which set is ordered; however, if there are ties, the set with fewer ties should be ordered and arranged so that the test is conservative, i.e., rejection is least probable.

OTHER TESTS

The McNemar [35] test (*see* MCNEMAR STATISTIC) is also frequently called a sign test because it is based on the difference of two frequencies and the test statistic follows the binomial distribution with $\theta = 0.5$. It is applicable in a paired sample situation, where the observations in each sample have only two possible outcomes so that many ties are likely, and we wish to compare the proportions of successes in the two groups. Bennett and Underwood [3] study the power of the McNemar test. Various extensions of this test have been introduced in the literature, including Ury [46], Mantel and Fleiss [32], McKinlay [34], and Cochran's Q test (Cochran [7]).

The two-sample median test is sometimes called a two-sample sign test (*see* BROWN–MOOD MEDIAN TEST) because it is based on the signs of the differences of the sample values and either the median of the pooled samples or some fixed value chosen in advance.

References

[1] Arbuthnot, J. (1710). *Philos. Trans.* **27**, 186–190.

[2] Bahadur, R. R. (1960). In *Contributions to Probability and Statistics, Essays in Honor of Harold Hotelling*, Stanford University Press, Stanford, CA, pp. 79–88.

[3] Bennett, B. M. and Underwood, R. E. (1970). *Biometrics*, **26**, 339–343.

[4] Blyth, C. R. (1958). *Ann. Math. Statist.*, **29**, 898–903.

[5] Bradley, J. V. (1968). *Distribution-Free Statistical Tests*. Prentice-Hall, Englewood Cliffs, NJ. (Elementary; Chapter 7 covers sign tests and references.)

[6] Bradley, J. V. (1969). *J. Quality Tech.*, **1**, 89–101. (Elementary; survey article; extensive references.)

[7] Cochran, W. G. (1950). *Biometrika*, **37**, 256–266.

[8] Cohen, J. (1969). *Statistical Power Analysis for the Behavioral Sciences*. Academic, New York. (Covers power and sample size determination for the sign test.)

[9] Conover, W. J. (1980). *Practical Nonparametric Statistics*. Wiley, New York. (Elementary; Chapter 3 covers sign tests and references.)

[10] Cox, D. R. and Stuart, A. (1955). *Biometrika*, **42**, 80–95.

[11] Daniel, W. W. (1978). *Applied Nonparametric Statistics*. Houghton Mifflin, Boston, MA. (Elementary; Chapter 2 covers sign tests.)

[12] David, H. A. (1981). *Order Statistics*. Wiley, New York.

[13] David, H. A. and Perez, C. A. (1960). *Biometrika* **47**, 297–306.

[14] Dixon, W. J. (1953). *Ann. Math. Statist.*, **24**, 467–473.

[15] Dixon, W. J. and Mood, A. M. (1946). *J. Amer. Statist. Ass.*, **41**, 557–566.

[16] Duckworth, W. E. and Wyatt, J. K. (1958). *Operat. Res. Quart.*, **9**, 218–233.

[17] Gibbons, J. D. (1964). *J. Amer. Statist. Ass.*, **59**, 142–148.

[18] Gibbons, J. D. (1985). *Nonparametric Methods for Quantitative Analysis*. American Sciences Press, Columbus, OH. (Elementary; Chapter 3 covers sign tests; applied approach; many numerical examples.)

[19] Gibbons, J. D. (1985). *Nonparametric Statistical Inference*. Dekker, New York. (Intermediate level; sign tests are covered in Chapters 6, 7, and 14; mostly theory.)

[20] Hájek, J. (1969). *A Course in Nonparametric Statistics*. Holden-Day, San Francisco, CA. (Intermediate level; mostly theory.)

[21] Hájek, J. and Šidák, Z. (1967). *Theory of Rank Tests*. Academic, New York. (Intermediate to advanced; all theory.)

[22] Harvard University Computation Laboratory (1955). *Tables of the Cumulative Binomial Probability Distribution*. Harvard University, Cambridge, MA. (Extensive tables.)

[23] Hemelrijk, J. (1952). *Proc. Kon. Ned. Akad. Wet. A*, **55**, 322–326.

[24] Hettmansperger, T. P. (1984). *Statistical Inference Based on Ranks*. Wiley, New York. (Intermediate sign tests are covered in Chapter 1.)

[25] Hodges, J. L. and Lehmann, E. (1956). *Ann. Math. Statist.*, **27**, 324–335.

[26] Hollander, M. and Wolfe, D. A. (1973). *Nonparametric Statistical Methods*. Wiley, New York. (Elementary; sign tests are covered in Chapters 2 and 3.)

[27] Hwang, T. Y. and Klotz, J. (1970). On the Approach to Limiting Bahadur Efficiency. *Technical Report No. 237*, Dept. of Statistics, University of Wisconsin, Madison, WI.

[28] Krauth, J. (1973). *Ann. Statist.*, **1**, 166–169.

[29] Lehmann, E. L. (1975). *Nonparametrics: Statistical Methods Based on Ranks*. Holden-Day, San Francisco, CA. (Intermediate level; sign tests are covered in Chapters 3 and 4.)

[30] MacStewart, W. (1941). *Ann. Math. Statist.*, **12**, 236–239.

[31] Mansfield, E. (1962). *Technometrics*, **4**, 430–432.

[32] Mantel, N. and Fleiss, J. L. (1975). *Biometrics*, **31**, 727–729.

[33] Marascuilo, L. A. and McSweeney, M. (1977). *Nonparametric and Distributions-Free Methods for the Social Sciences*. Brooks/Cole, Monterey, CA. (Elementary cookbook approach; many numerical examples; Chapter 3 covers sign tests.)

[34] McKinlay, S. M. (1975). *Biometrics*, **31**, 731–735.

[35] McNemar, Q. (1947). *Psychometrika*, **12**, 153–157.

[36] National Bureau of Standards (1949). *Tables of the Binomial Probability Distribution*. U.S. GPO, Washington, DC. (Extensive tables.)

[37] Olshen, R. A. (1967). *Ann. Math. Statist.*, **38**, 1759–1769.

[38] Pratt, J. W. and Gibbons, J. D. (1981). *Concepts of Nonparametric Theory*. Springer-Verlag, New York. (Intermediate to advanced; conceptual approach to theory; sign tests are discussed in Chapters 2, 5, and 8.)

[39] Putter, J. (1955). *Ann. Math. Statist.*, **26**, 368–386.

[40] Randles, R. H. and Wolfe, D. A. (1979). *Introduction to the Theory of Nonparametric Statistics*. Wiley, New York. (Intermediate level; mostly theory.)

[41] Rao, T. S. (1968). *Biometrika*, **55**, 381–386.

[42] Savur, S. R. (1937). *Proc. Indian. Acad. Sci. Ser. A*,**5**, 564–576.

[43] Stuart, A. (1956). *J. Amer. Statist. Ass.*, **51**, 285–287.

[44] Thompson, W. R. (1936). *Ann. Math. Statist.*, **7**, 122–128.

[45] Ury, H. K. (1966). *Biometrika*, **53**, 289–291.

[46] Ury, H. K. (1975). *Biometrics*, **31**, 643–649.

[47] Walsh, J. E. (1946). *Ann. Math. Statist.*, **17**, 358–362.

[48] Walsh, J. E. (1951). *Ann. Math. Statist.*, **22**, 408–417.

(BROWN–MOOD MEDIAN TEST
COCHRAN'S Q-STATISTIC
DISTRIBUTION-FREE METHODS
LOCATION TESTS
MCNEMAR STATISTIC
MEDIAN ESTIMATES AND SIGN TESTS
NONPARAMETRIC CONFIDENCE INTERVALS
QUANTILES
TREND)

JEAN DICKINSON GIBBONS

SIGNUM FUNCTION

This function is defined by

$$\operatorname{sgn}(x) = \begin{cases} 1 & \text{for } x > 0, \\ 0 & \text{for } x = 0, \\ -1 & \text{for } x < 0. \end{cases}$$

It is sometimes written sign(x). It can be

represented as the Dirichlet discontinuous integral

$$\operatorname{sgn}(x) = \frac{1}{\pi}\int_{\infty}^{\infty} \frac{\sin xu}{u}\, du.$$

Also, for any real number x,

$$|x| = \int_0^x \operatorname{sgn}(u)\, du.$$

SIMILARITY MEASURES *See* MEASURES OF SIMILARITY, DISSIMILARITY AND DISTANCE

SIMILAR MATRICES

Two square $m \times m$ matrices, **A** and **B** are said to be *similar* it there exists a nonsingular $m \times m$ matrix **P** such that

$$\mathbf{B} = \mathbf{P}^{-1}\mathbf{A}\mathbf{P}, \quad \text{or equivalently} \quad \mathbf{A} = \mathbf{P}\mathbf{B}\mathbf{P}^{-1}.$$

Similar matrices have the same eigenvalues*, the same trace*, and the same characteristic equation.

SIMILAR REGIONS AND TESTS

A term introduced by Neyman and Pearson [1] in 1933. A similar region with respect to parameter(s) ϕ is a region of the sample space* with probability content that does not depend on the value of ϕ. "Similar" means "similar to the whole sample space" (in respect to the property of constant probability content, which for the whole sample space is equal to 1).

Regions that are similar with respect to nuisance parameters* are useful in the construction of tests of composite hypotheses*. For example, if $X_1, X_2, \ldots, X_n$ are independent variables with a common normal distribution having expected value ξ and standard deviation σ, the statistic

$$T = \sqrt{n}\,(\bar{X} - \xi)/S,$$

where $\bar{X} = n^{-1}\Sigma_{i=1}^{n} X_i$; $S^2 = (n-1)^{-1} \times \Sigma_{i=1}^{n}(X_i - \bar{X})^2$ has a t-distribution* with $(n-1)$ degrees of freedom, whatever the value of σ. The test of the (composite) hypothesis $\xi = \xi_0$ with critical region*

$$\left|\sqrt{n}\,(\bar{X} - \xi_0)/S\right| > K \qquad (1)$$

has a significance level depending on K but not on σ. The region defined by (1) is *similar with respect to* σ. *See also* NEYMAN STRUCTURE, TESTS WITH.

A test with significance level that does not depend on nuisance parameters, such as that defined by (1), is a *similar test*. The adjective "similar" is less relevant in reality to a *test* than it is to a *region*. In the latter case the region *is* similar to the whole sample space in the sense that the probability of a sample point falling in it does not depend on certain parameter values. In the case of a test, however, there need be no other test to which the given one is similar in any sense.

Reference

[1] Neyman, J. and Pearson, E. S. (1933) *Philos. Trans. R. Soc. London, A*, **231**, 289–337.

(HYPOTHESIS TESTING
INFERENCE, STATISTICAL, I, II
NEYMAN STRUCTURE, TESTS WITH
SUFFICIENT STATISTICS)

SIMPLE EXPANSION

In sample surveys it often occurs that the domain of interest is a subpopulation of the population from which the sample is selected. Several situations have been discussed to estimate the total of a quantitative variable over such a subpopulation. The choice of methods depends on available information about the subpopulation.

Cochran [1, pp. 35–38] presented three methods for three different situations. In the first, the total count of units in the subpopulation is known, and an estimate is simply the product of the total count and the mean of the sample units that fall in the subpopulation. In the second, the total of the quantitative variable over the entire population is known. In this case, a ratio estimation* may be employed; the sample gives an estimate of the ratio of the total for units in the subpopulation over the total for units in the entire population. This is multiplied by the known total of the variable to get the desired estimate. The third is the situation where neither the total count nor the total of the quantitative variable is available. An estimate in this case is the product of the sample total over the subpopulation and the reciprocal of the sampling fraction. This is the *simple expansion method*; see Jones and Coopersmith [6] and Perng [8].

It is useful when data are lacking. A mass production of tables is one such situation in which the estimation for each cell deals with an estimation over a subpopulation.

The simple expansion method has been widely discussed. Sukhatme and Sukhatme [9], Cochran [1], and Kish [7], among others, presented the estimate and its variance. Sukhatme and Sukhatme [9, pp. 36–37], gave an example of sample size computation. Kish [7, pp. 434–436] and Cochran [1, p. 38] compared its variance to the variance of an estimate when the subpopulation count is known. Jones and Coopersmith [6] compared the simple expansion estimate with a ratio estimate where the total of the auxiliary variable over the subpopulation is estimated. They showed that the simple expansion estimate would do better if the correlation between the variable, whose total is to be estimated, and the auxiliary variable is low. Perng [8] studied the variance and the coefficient of variation* of the estimate, and the variance of the variance estimate, to see how they depend on what proportion the subpopulation is of the entire population under various circumstances; the simple expansion estimate is well behaved when the coefficient of variation per unit in the subpopulation is not less than one. Both the variance of the estimate and the variance of the variance estimate are essentially proportional to the proportion of the subpopulation. When the coefficient of variation per unit is less than 1, the simple expansion estimate and its variance estimate are much less reliable for certain ranges of the proportion.

References

[1] Cochran, W. G. (1977). *Sampling Techniques*, 3rd ed. Wiley, New York.

[2] Cox, D. R. and Snell, E. J. (1979). *Biometrika*, **66**, 125–132.

[3] Hansen, M. H., Hurwitz, W. N., and Madow, W. G. (1953). *Sample Survey Methods and Theory*, Vol. 1. Wiley, New York.

[4] Hansen, M. H., Hurwitz, W. N., and Madow, W. G. (1953). *Sample Survey Methods and Theory*, Vol. 2. Wiley, New York.

[5] Holt, D. and Smith, T. M. F. (1979). *J. R. Statist. Soc. A*, **142**, 33–46.

[6] Jones, D. H. and Coopersmith, L. (1976). *Commun. Statist. A*, **5**, 251–260.

[7] Kish, L. (1965). *Survey Sampling*. Wiley, New York.

[8] Perng, S. (1982). *1982 ASA Proc., Sec. Sur. Res. Meth.* (pp. 99–104) or *Statist. of Income and Related Administ. Record Res.: 1982* (pp. 143–149). Internal Revenue Service, Washington, DC.

[9] Sukhatme, P. V. and Sukhatme, B. V. (1970). *Sampling Theory of Surveys with Applications*, 2nd ed. Iowa State University Press, Ames, IA.

SHIEN-SEN PERNG

(RATIO ESTIMATORS
SAMPLE SURVEYS)

SIMPLE HYPOTHESIS

This is a hypothesis that completely determines the joint distribution of the random variables in a model. Usually the variables (or random vectors) are independently distributed and the hypothesis then specifies the common parent distribution(s) completely. For example, if the parent is a gamma distribution* with probability density function

$$f(x; \theta, a) = (\Gamma(a))^{-1}\theta^a x^{a-1} e^{-\theta x},$$

$$x > 0,\ a > 0,\ \theta > 0,$$

the hypothesis that asserts that $a = 2, \theta = 1$,

is simple. However, the hypothesis that only asserts that $a = 2$ is composite*.

(COMPOSITE HYPOTHESIS
HYPOTHESIS TESTING)

SIMPLE n-PERSON GAME

A cooperative game* in characteristic function form is called a *simple n-person game* if only *two* payoffs are possible and if it is specified in advance precisely which coalitions can achieve the winning payoff.

(GAME THEORY
POWER INDEX OF A GAME)

SIMPLE RANDOM SAMPLING

A random sample of size n, $\mathbf{S} = (x_1, \ldots, x_n)$, is said to be a *simple random sample* if all possible samples $\mathbf{S}$ are equally probable. Sampling is said to be *with* or *without replacement* according as to whether or not the same member of the population may be selected more than once. If the population is comprised of N members, the probability of $\mathbf{S}$ is $1/N^n$ or $1/\binom{N}{n}$ according as the sampling is with or without replacement. Conversely, if $\mathbf{S}$ has these probabilities, then $\mathbf{S}$ is a simple random sample. The probability that all members are distinct in a simple random sample with replacement is $N!/[(N - n)!N^n]$. A somewhat surprising numerical illustration of this formula is the consequence that among 30 people the probability is about 0.706 that at least two persons have the same birthday.

The probability of inclusion of the ith member of the population in a simple random sample is n/N. In sampling with replacement the inclusions of the ith and jth $(i \neq j)$ members of the population are statistically independent. On the other hand, these events are not independent in sampling without replacement*. In this case, the probability of inclusion of both the ith and jth population members is given by

$$\pi_{i,j} = n(n-1)/[N(N-1)], \; i \neq j.$$

In a population comprised of two types of members, the number of members of one type in a simple random sample has a binomial* or hypergeometric* distribution according as the sampling is with or without replacement. Thus sampling with replacement provides an alternative method to that of Bernoulli trials for the derivation of the binomial distribution [12].

In the study of sampling distributions* in mathematical statistics a random variable X is associated with a probability space*. It is also sometimes helpful to visualize a hypothetical infinite population in which the frequency with which each member appears is in accordance with the distribution function of X ([2, Chap. 25] and [12, p. 22]). In any case, a vector of n independent realizations of X is considered to be a simple random sample.

Simple random sampling with replacement is used in bootstrapping* to estimate nonparametric standard errors.

Simple random sampling without replacement is the fundamental technique used in survey sampling* [1, 13]. It is also a component of many other probability sampling techniques (*see* CLUSTER SAMPLING and STRATIFIED DESIGNS). Historically, nonrandom techniques including purposive sampling and full enumeration were used [9]. However, from the viewpoint of validity and cost effectiveness, probability sampling techniques are now generally considered to be much preferable. In fact, if the finite population is homogeneous and the statistician's resources permit taking a sample size n at most, it may be shown [6] that the best strategy for estimating the population mean is to draw a simple random sample of size n by simple random sampling without replacement and to take the sample mean as the estimate.

A list of all population members is called a *frame*. A simple random sample without replacement is obtained from a frame using a random number table [1, p. 19] or a computer algorithm [7, 10]. Many computer languages and statistical packages have built-in functions or procedures for drawing a simple random sample (for example, the deal function in APL). In actual application, obtaining a suitable frame may present practical difficulties. Numerous illuminating examples of simple random sampling with survey data are presented in refs. 8 and 11.

The purpose of a sample survey is to obtain information about a numerical characteristic of a population at minimum cost. Let u_i $(i = 1, \ldots, N)$ denote the numerical characteristic of interest for the ith member of the population. Then the purpose is often to estimate the population mean

$$\mu = \sum_{i=1}^{N} u_i/N \tag{1}$$

or the population total $N\mu$. Note that the proportion of the population possessing some attribute is just the population mean of the binary variable u_i, which is 1 or 0 according as the ith population member does or does not possess the attribute. The sample mean

$$\bar{x} = \sum_{j=1}^{n} x_j/n \tag{2}$$

is an unbiased* estimate of μ. In sampling without replacement,

$$\begin{aligned} \operatorname{cov}(x_j, x_l) &= \sum_{i \neq h} (u_i - \mu)(u_h - \mu)\pi_{i,h} \\ &= -\sigma^2/N, \qquad j \neq l, \end{aligned} \tag{3}$$

where $\sigma^2 = \Sigma(u_i - \mu)^2/(N-1)$. Hence,

$$\operatorname{var}(\bar{x}) = \frac{\sigma^2}{n}\frac{N-n}{N}. \tag{4}$$

Similarly, if sampling is with replacement,

$$\operatorname{var}(\bar{x}) = \frac{\sigma^2}{n}\frac{N-1}{N}. \tag{5}$$

A powerful general methodology for finding mean values and unbiased estimates of moments and of other symmetric functions is presented in ref. 3. Comparing (4) and (5), we see that sampling without replacement is statistically more efficient. The term $(N - n)/N$ in (4) is referred to as the *finite population correction**; it is often neglected when the sampling fraction $f = n/N$ is less than a tenth [1, p. 25]. The sample mean is admissible [4] in sampling without replacement but inadmissible in sampling with replacement [2, p. 30]. The sample estimate of σ^2 is

$$s^2 = \sum_{j=1}^{n} (x_j - \bar{x})^2/(n-1). \tag{6}$$

Because $\bar{x}$ is approximately normally distributed [2, p. 39], an approximate confidence interval for μ may be calculated by estimating the appropriate standard deviation.

Often multiple numerical characteristics are available. Thus $u_i = (u_{1i}, \ldots, u_{Ki})$ and $x_j = (x_{1j}, \ldots, x_{Kj})$ if there are K characteristics. We will now assume that sampling is without replacement. Then it is somewhat remarkable that the sample mean is admissible for all K [5]. Frequently it is of interest to estimate the population ratio. For example, $R = u_{1.}/u_{2.}$ The sample estimate of R is $\hat{R} = x_{1.}/x_{2.}$ In large samples, $\hat{R}$ is approximately normal with mean R. The estimated variance of $\hat{R}$ is

$$\text{est. var}(\hat{R}) = \frac{(1-f)\sum\left(x_{1j} - \hat{R}x_{2j}\right)^2}{n(n-1)\bar{x}_{2.}^2}. \tag{7}$$

If the population total for the second characteristic, $u_{2.}$, is known, the ratio estimator* of the population total of the first characteristic is $\hat{R}u_{2.}$. Thus, it is possible to estimate the population total without knowing N. Moreover, the ratio estimator has smaller variance than $N\bar{x}_{1.}$ if the correlation between u_{1i} and u_{2i} is large enough [1, p. 157]. A regression estimator* may do even better if the linear regression of u_{1i} on u_{2i} does not pass through the origin.

References

[1] Cochran, W. G. (1977). *Sampling Techniques*. Wiley, New York. (The standard reference for sample survey methodology.)

[2] Cramér, H. (1946). *Mathematical Methods of Statistics*. Princeton University, Princeton, NJ.

[3] Herzel, A. (1982). *Statistica*, **42**, 315–350.
[4] Joshi, V. M. (1968). *Ann. Math. Statist.*, **39**, 606–620.
[5] Joshi, V. M. (1977). *Ann. Statist.*, **5**, 1501–1503.
[6] Joshi, V. M. (1979). *Ann. Statist.*, **7**, 531–536.
[7] McLeod, A. I. and Bellhouse, D. R. (1983). *Appl. Statist.*, **32**, 182–184.
[8] Slonim, M. J. (1960). *Sampling in a Nutshell.* Simon and Schuster, New York. (Humorous and entertaining examples.)
[9] Smith, T. M. F. (1976). *J. R. Statist. Soc. A*, **139**, 183–204.
[10] Vitter, J. S. (1984). *Commun. ACM*, **27**, 703–718.
[11] Wallis, W. A. and Roberts, H. V. (1956). *The Nature of Statistics*. MacMillan, New York. (Contains numerous illuminating examples.)
[12] Whittle, P. (1976). *Probability*. Wiley, New York. (An interesting novel approach in which the basic discrete distributions are derived from sampling.)
[13] Williams, B. (1978). *A Sampler on Sampling*. Wiley, New York. (An excellent nonmathematical introduction to sample survey methodology.)

(CLUSTER SAMPLING
INSPECTION SAMPLING
PROBABILITY PROPORTIONAL TO SIZE SAMPLING
RATIO ESTIMATORS
STRATIFIED DESIGNS
SURVEY SAMPLING
SYSTEMATIC SAMPLING)

A. IAN MCLEOD

SIMPLEX DESIGN *See* MIXTURE EXPERIMENTS

SIMPLEX METHOD *See* NELDER–MEAD SIMPLEX METHOD

SIMPSON INDEX *See* LINEAR PROGRAMMING

SIMPSON'S DISTRIBUTION

This distribution has probability density function

$$f_X(x) = \begin{cases} \dfrac{2}{(b-a)} - \dfrac{2}{(b-a)^2}|a+b-2x| & \text{for } x \in (a,b), \\ 0 & \text{for } x \notin (a,b), \end{cases} \qquad (1)$$

symmetrical about its mean of $(a+b)/2$ (Fig. 1).

The particular case when $a = 0$, $b = 2$, so that the density is given by

$$f_X(x) = \begin{cases} x & \text{for } x \in (0,1), \\ 2-x & \text{for } x \in (1,2) \\ 0 & \text{otherwise,} \end{cases}$$

is sometimes called a *standard* Simpson distribution. The characteristic function* of the distribution (1) is

$$\phi_X(t) = \left[\frac{2}{b-a}\,\frac{e^{itb/2} - e^{ita/2}}{it}\right]^2.$$

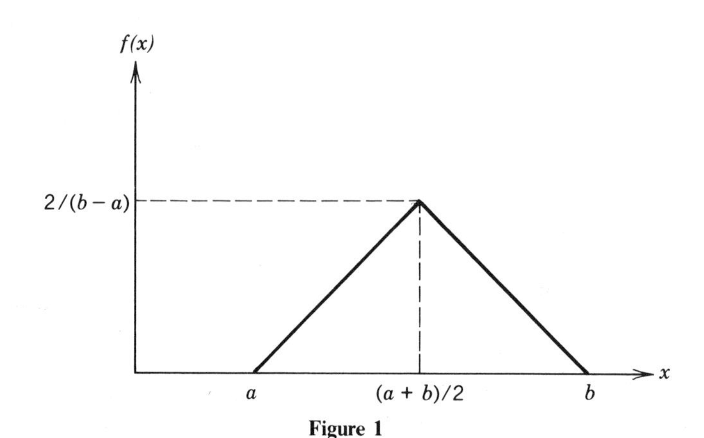

Figure 1

Its moments are

$$E(X^k) = \frac{4}{(b-a)^2(k+1)(k+2)} \times \left[a^{k+2} + b^{k+2} - 2\left(\frac{a+b}{2}\right)^{k+2} \right], \quad k = 1, 2, \ldots.$$

Its variance is $(b-a)^2/24$.

If X_1 and X_2 are independent random variables uniformly distributed on $(a/2, b/2)$, then $X = X_1 + X_2$ has the Simpson distribution (1). This genesis of Simpson's distribution is a useful feature, utilized for Monte Carlo generation of random variables*. The distribution seems to have been suggested originally by Thomas Simpson (1710–1761) in ref. 1; it is a special case of a *triangular distribution* [*see* UNIFORM (RECTANGULAR) DISTRIBUTIONS].

Reference

[1] Simpson, T. (1755). *Philos. Trans. R. Soc. London Ser. A.*, **49**, 82–93.

(GENERATION OF RANDOM VARIABLES
UNIFORM (RECTANGULAR) DISTRIBUTIONS)

SIMPSON'S PARADOX *See* FALLACIES, STATISTICAL

SIMPSON'S RULE

This is the following quadrature formula, using values of the integrand at three equally spaced values of the variable:

$$\int_a^{a+2h} f(x)\,dx \doteqdot \tfrac{1}{3}h\{f(a) + 4f(a+h) + f(a+2h)\}.$$

It gives the exact value of the integral if $f(x)$ is a polynomial of degree 3 or less. If the fourth derivative of $f(x)$ is continuous, the error in the formula is $(1/90)h^5 f^{(4)}(\xi)$, where ξ is some value between a and $a + 2h$.

The formula is sometimes called the *parabolic rule*.

(NUMERICAL INTEGRATION
SHOVELTON'S FORMULA
THREE-EIGHTHS RULE
TRAPEZOIDAL RULE
WEDDLE'S RULE)

SIMULATION MODELS, VALIDATION OF

INTRODUCTION

The validation of a simulation model involves measuring how well the model mimics the system it is intended to represent. Model validation can best be understood within the context of how a simulation study should be conducted. The literature describes the following steps in a well-planned simulation study (see ref. 9, p. 23): (1) system identification, (2) model development, (3) model verification, (4) model validation, and (5) model analysis. The last step refers to the process of indirectly studying the real system by studying the behavior of the simulation model. In this outline, model validation is essential before the model can be used to make inferences regarding the real system. Unfortunately, however, the validation step has been omitted or has been given only token attention in many simulation studies.

DEFINING VALIDITY

A model is said to be *valid* in the strictest sense if the relationships between input values and output values are the same in the model as they are in the real system. This definition requires that a model provide a *perfect representation* of every aspect of the real system in order to be valid.

In practice, however, it is seldom necessary for a simulation model to represent the real system exactly. Simulation models are usually developed to provide insights into particular parts of the system. For example, the main objective of many simulation studies is to predict system responses under certain hypothetical operating conditions. In this setting it is required only that the model provide accurate predictions, without necessarily duplicating every detail of the real system. Hence, the adjective "valid" will refer here to any model that gives an adequate representation of the components of the system that are of interest to the model user.

PRACTICAL PROBLEMS

The practical difficulties in model validation are evident from the expansive literature devoted to the topic over the last 20 years. Among the difficulties are the following (see ref. 5).

LACK OF DATA. The complexity of the system or the fact that the system does not yet exist (as with a proposed new highway system) can make it difficult or even impossible to obtain sufficient real-life data.

LACK OF CONTROL OF THE INPUT VARIABLES. A comparison of the model with the real system is best accomplished when both the model and the system can be observed under several operating conditions chosen by the experimenter. Unfortunately, many real-life systems do not permit such control.

STATISTICAL DIFFICULTIES. The distributional properties of both model- and real-system output are often unknown. In addition, the sequences of output data are frequently known to be autocorrelated and nonstationary. While these problems are not insurmountable, one must have more than a cursory knowledge of statistical techniques for analyzing such data.

STATISTICAL METHODS FOR EVALUATING VALIDITY

Several statistical techniques have been proposed in the literature, which are useful in checking the validity of the output from a simulation model. Most of the procedures suggested are not new, but are specific applications of well-known statistical procedures.

For a fixed set of operating conditions c, we let $R_1, R_2, \ldots, R_m$ and $S_1, S_2, \ldots, S_n$ represent sequences of output observations taken from the real system (R) and the simulation model (S), respectively. Each observation may be univariate or multivariate. Most tests for model validity involve comparisons of properties of the two sequences $\{R\}$ and $\{S\}$. Several of the more often cited procedures are described below. For a detailed description of these procedures, see the appropriate entries in this encyclopedia, and also the references in the bibliography of this article.

TEST OF EQUALITY OF LOCATION AND SCALE PARAMETERS. If the sequences $\{R\}$ and $\{S\}$ are composed of mutually independent random variables, then standard testing procedures can be used to test if the two sequences were generated from probability distributions having equal means and dispersion matrices. Normal theory tests or nonparametric tests can be used, depending on whether or not the underlying distributions are from the normal family (*see* LOCATION TESTS and SCALE TESTS).

GOODNESS-OF-FIT* TESTS. The model user may require that the distributions from which $\{R\}$ and $\{S\}$ are sampled be identical. Several tests for equality of two distributions are available.

PROCEDURES FOR TIME-SERIES OUTPUT. If the output sequences $\{R\}$ and $\{S\}$ are composed of time-series data that are serially dependent and possibly nonstationary, then the data must be analyzed using the techniques of time-series* analysis. Autoregressive models and spectral analysis* have been used to analyze stationary output (see refs. 2, 3, and 6), and ARIMA models could be used for many nonstationary time series.

SENSITIVITY ANALYSIS. The goal of sensitivity analysis is to quantify the effects that individual input variables have on the output of the model and the real system. This is accomplished by imposing slight perturbations on the input variables and observing the change in the output (see refs. 4 and 9, pp. 235–236). One major limitation of this type of analysis is that the model evaluator must be able to exercise very specific control over the values of the input variables in the real system.

THE TURING TEST. The Turing test allows the model developer to test how well a panel of system experts can distinguish between output from the model and output from the real system. See refs. 8 and 10 for a detailed discussion.

Combining the Results From Several Tests

Usually several tests of model validity are conducted on a single model. For example, the model and the system might be compared under several operating conditions $c_1, c_2, \ldots, c_k$, and it may be desirable to combining the results of several hypothesis tests into an overall test for validity. There is an expansive literature on combining results of several tests. For a discussion within the context of model validation, see ref. 7.

Sample Reuse* Methods

Methods such as the bootstrap* and the jackknife* can sometimes be used to estimate the prediction mean squared error for very complicated models whenever theoretical estimates are unavailable; see also ref. 1.

References

[1] Efron, B. (1982). *The Jackknife, the Bootstrap and Other Resampling Plans*. SIAM, Philadelphia, PA. (Good overview of sample reuse methods. Requires an intermediate level of mathematical and statistical expertise.)

[2] Fishman, G. S. and Kiviat, P. J. (1967). *Management Sci.*, **13**, 525–557. (Good introductory discussion of spectral analysis and its applications to simulation.)

[3] Hsu, D. A. and Hunter, J. S. (1977). *Management Sci.*, **24**, 181–190. (A method for testing the equivalence of two stochastic processes. Fairly mathematical, with a good numerical example.)

[4] Miller, D. R. (1974). *J. Theor. Biol.*, **48**, 345–360. (Describes sensitivity analysis for a deterministic model.)

[5] Naylor, T. H. and Finger, J. M. (1967). *Management Sci.*, **14**, B92–B101 (with discussion). (A keynote paper in the philosophy of model validation.)

[6] Naylor, T. H. , Wertz, K., and Wonnacott, T. H. (1971). In *Computer Simulation Experiments with Models of Economic Systems*, T. H. Naylor, ed. Wiley, New York, pp. 247–268. (Introduces spectral analysis and discusses its uses in simulation studies.)

[7] Reynolds, M. R., Jr., and Deaton, M. L. (1982). *Commun. Statist. Simul. Comp.*, **11**, 769–799. (Theoretical and empirical comparison of several validation tests.)

[8] Schruben, L. W. (1980). *Simulation*, **22**, 101–105. (Good discussion of the Turing test, with a suggested method of analysis.)

[9] Shannon, R. E. (1975). *Systems Simulation: The Art and Science*. Prentice-Hall, Englewood Cliffs, NJ. (Outlines several statistical procedures.)

[10] Van Horn, R. L. (1971). *Management Sci.*, **17**, 247–258. (Good overview with several examples and a discussion of the Turing test.)

Bibliography

Balci, O. and Sargent, R. G. (1981). *Commun. ACM*, **24**, 190–197. (A method for maximizing the power of tests used in a validity study, subject to budgetary constraints.)

Balci, O. and Sargent, R. G. (1980). *Newsletter–TIMS College on Sim. Gaming*, **4**, 11–15. (A good bibliography.)

Freedman, H. T. , Harms, T. R., and Koontz, W. L. G. (1979). *Proc. Winter Simulation Conf.*, pp. 167–178. (Example of using a field study to validate a simulation model in the telecommunications industry.)

Law, A. M. and Kelton, W. D. (1982). *Simulation Modeling and Analysis*, McGraw-Hill, New York. (Contains a good summary of validation procedures.)

Mihram, G. A. (1972). *Simulation: Statistical Foundations and Methodology*, Academic, New York. (Gives several normal-theory and nonparametric procedures.)

Mihram, G. A. (1973). *Operat. Res. Quart.*, **23**, 17–29. (Philosophical discussion of verification and validation, with some discussion of statistical procedures.)

Naylor, T. H., Wallace, W. H., and Sasser, W. E. (1967). *J. Amer. Statist. Ass.*, **62**, 1338–1364. (A good validation study of a simulation model of the textile industry, including spectral analysis.)

Sargent, R. G. (1979). *Proc. Winter. Simulation Conf.*, pp. 497–503. (Good overview of model validation and associated terminology. Good bibliography.)

(BOOTSTRAPPING
GOODNESS OF FIT
JACKKNIFE
MONTE CARLO METHODS
SAMPLE REUSE
TIME SERIES)

MICHAEL L. DEATON

SIMULTANEOUS CONFIDENCE INTERVALS

The *confidence coefficient* $(1 - \alpha)$ associated with a rule for constructing a confidence interval or region for a (possibly vector-valued) parameter $\underset{\sim}{\theta}$ is the probability that intervals or regions constructed according to the rule will contain the true value of $\underset{\sim}{\theta}$.

If the region for $\underset{\sim}{\theta}$ based on data values $\underset{\sim}{X}$ is a rectangular parallelopiped made up of intervals $\underline{\theta}_i(\underset{\sim}{X}) < \theta_i < \bar{\theta}_i(\underset{\sim}{X})$ $(i = 1, 2 \ldots)$, these are said to be a set of *simultaneous confidence intervals* for $\theta_1, \theta_2, \ldots$ with *joint confidence coefficient* $(1 - \alpha)$.

The joint confidence coefficient is the probability that *every one* of the intervals $(\underline{\theta}_i,(x), \bar{\theta}_i(x))$ will include the corresponding true value θ_i $(i = 1, 2, \ldots)$. It is not the expected proportion of intervals containing the appropriate true values, which is usually a larger amount, and never less.

Bibliography

Miller, R. G. (1981). *Simultaneous Statistical Inference*, 2nd ed. Springer, New York.

(CONFIDENCE INTERVALS AND REGIONS
MULTIPLE COMPARISONS)

SIMULTANEOUS EQUATION MODELS

See ECONOMETRICS; FIX-POINT METHOD

SIMULTANEOUS INFERENCE

See MULTIPLE COMPARISONS; SIMULTANEOUS CONFIDENCE INTERVALS

SIMULTANEOUS TESTING

INTRODUCTION

Many studies are designed to simultaneously test a number of hypotheses, rather than only one. For example, consider an experiment in which the physiological effects of k different diets are to be compared. A standard initial approach is to test the single hypothesis that all diets have the same effects. However, a decision to reject this hypothesis is rarely sufficient; rather, we are generally interested in more specific information. Thus, for example, we might be interested in testing each of the set of $k(k-1)/2$ hypotheses

$$H_{ij}: \theta_i - \theta_j = 0, \qquad i < j = 1, \ldots, k, \quad (1)$$

where θ_i is some parameter of the distribution of effects of diet i (i.e., in making all *pairwise comparisons* of the diets) or even the infinite set

$$H_i: \sum_j a_{ij}\theta_j = 0 \qquad (2)$$

over all sets a_{ij} such that $\sum_j a_{ij} = 0$ (i.e., in testing that all *contrasts** among the diets equal zero).

There is a considerable body of work devoted to procedures designed for simultaneous testing of a number of hypotheses. For an early history, see Harter [18]. (For related work on ranking and selection of populations *see* RANKING PROCEDURES and SELECTION PROCEDURES.) There have been decision-theoretic approaches, both non-Bayesian (e.g., Lehmann [30]) and Bayesian (e.g., Waller and Duncan [51, 52]). Since often the information required for use of decision-theoretic procedures is not available, the most commonly used methods have been based on extensions of the Neyman–Pearson approach for testing single hypotheses (*see* HYPOTHESIS TESTING): these extensions will be discussed.

THE NEYMAN–PEARSON APPROACH

In the Neyman–Pearson approach for testing a single hypothesis, consideration is limited to tests that bound the probability of rejecting the hypothesis if true (the probability of a *type I error*) below some specified value (the *significance level*). Given that restriction, the aim is to choose a test with a high probability of rejecting the hypothesis if false (*power*). In simultaneous testing of a number of hypotheses, the concepts of type I error, significance level, and power* are generalized to apply to the whole set of tests. The reader should be warned that terminology is not standard, and the same terms are sometimes used for slightly different concepts by different authors.

Given a family of n hypotheses $H_1, \ldots, H_n$, suppose each hypothesis H_i is tested at significance level α_i. A *false rejection* is the rejection of a true hypothesis. *The expected proportion of the total number of tests that result in false rejections* (a) is then bounded above by $\Sigma\alpha_i/n$, *the expected number of false rejections* (b) is bounded above by $\Sigma\alpha_i = n(\Sigma\alpha_i/n)$, and *the probability of one or more false rejections* (c) is bounded above by some value between $\Sigma\alpha_i/n$ and $\Sigma\alpha_i$, depending on the joint distribution of the test statistics. The value (a) is the *per-comparison error rate*, (b) the *per-family error rate*, and (c) the *familywise error rate*. The definition of error rate (c) applies also to infinite families. Since the family often consists of all hypotheses tested in an experiment, (b) and (c) are sometimes called the *per-experiment* and *experimentwise error rates*, respectively.

If the decisions resulting from the individual hypothesis tests are unrelated, the bound (a) may be the only one of interest. However, if the outcome of all tests is to be considered as a whole, bounds on (b) and/or (c) may be desired. If each hypothesis is tested individually at a typical level of α, the least upper bounds on (b) and (c) will usually then be unacceptably high.

Most of the work in multiple comparisons* has been based on a bound on (c) rather than on (b); a bound of α on (c) allows the simple interpretation that with some acceptably high probability $1 - \alpha$, no incorrect rejections will occur, and the interpretation is applicable to infinite sets of hypotheses such as the set (2). Spjøtvoll [48, 49] however, argues for a bound on (b). Since (c) is always less than or equal to (b), a bound of α on the latter also bounds the former at α. On the other hand, with a bound of α on (c) rather than on (b), more powerful procedures are possible. The issues are illustrated by the comparison of two closely related procedures below.

THE BONFERRONI AND SEQUENTIALLY REJECTIVE BONFERRONI PROCEDURES

One simple way of controlling (b) is to test each hypothesis H_i at some level α_i such that $\Sigma\alpha_i$ equals the desired α. This is the *Bonferroni procedure*, usually applied by setting $\alpha_i = \alpha/n$ for all i (*see* BONFERRONI INEQUALITIES AND INTERVALS). If the type I error probabilities can be arbitrarily close to α_i simultaneously for all hypotheses, the least upper bound on (b) is exactly α. Given a bound on (b) for the set of hypotheses (1) with normally distributed means having equal variance, Spjøtvoll [48] proved optimal minimum and average power properties for the Bonferroni t-tests (*see* MULTIPLE COMPARISONS) with $\alpha_i = \alpha/n$ for all i.

The sequentially rejective Bonferroni procedure (SRBP) bounds the error rate (c) at some specified value α. It was proposed by Holm [25], who also proposed more general sequentially rejective procedures (Holm [24]); special cases of them were recommended earlier by Hartley [19], Naik [36], and Larzelere and Mulaik [29]. Sequentially rejective procedures can be considered special cases of the closed testing procedures described by Marcus et al. [34]. In the SRBP, as in the Bonferroni procedure, there are specified test statistics T_i and significance levels α_i for each H_i such that $\Sigma\alpha_i = \alpha$. Let the T_i be defined so that large values lead to rejection, and let p_i be the *significance probability* of T_i, i.e., the supremum of the probability that $T_i \geqslant t_i$ when H_i is true, where t_i is the observed value of T_i. Let $p_{(i)}$ be the ordered values of p_i, $p_{(1)} \leqslant p_{(2)} \leqslant \cdots \leqslant p_{(n)}$, with arbitrary ordering in case of ties, and let $H_{(i)}$ be the correspondingly ordered hypotheses. For simplicity, we will consider only the case $\alpha_i = \alpha/n$ for all i. In the Bonferroni procedure, each hypothesis H_i is rejected if and only if $p_i \leqslant \alpha/n$. In the SRBP, hypothesis $H_{(1)}$, corresponding to the "most significant" T_i, is rejected if and only if $p_{(1)} \leqslant \alpha/n$. If $H_{(1)}$ is accepted, all hypotheses are accepted; if rejected, then $H_{(2)}$ is rejected if $p_{(2)} \leqslant \alpha/(n-1)$. Given any $H_{(j)}$, $j = 1, \ldots, n$, it is rejected if and only if $H_{(1)}, \ldots, H_{(j-1)}$ have been rejected *and* $p_{(j)} \leqslant \alpha/(n-j+1)$.

Note that any hypothesis that is rejected by the Bonferroni procedure, which bounds (b) below α, is also rejected by the SRBP, which bounds (c) below α, and the latter often results in a number of additional rejections [at the price, of course, of a greater value of (b)]. These two procedures are examples of single-stage and multistage procedures, respectively, as described below.

SINGLE-STAGE AND MULTISTAGE TESTING PROCEDURES

Simultaneous testing procedures with upper bounds on (c) may be divided into single-stage and multistage (or stagewise) procedures. Assume that to each hypothesis H_i there corresponds a test statistic T_i that would be used if it were the only hypothesis being tested. In single-stage procedures, the decision with respect to H_i depends only on the value of T_i, with critical values adjusted to achieve the appropriate overall error control. Two subclasses of single-stage procedures, union–intersection* procedures and simultaneous test procedures, will be considered in the next section. In multistage procedures, the decision with respect to any hypothesis depends on other aspects of the situation, possibly on the values of the test statistics corresponding to other hypotheses; possibly on the values of other relevant statistics. Many multistage procedures have a more complicated structure than that of the SRBP described above, such as the methods described in MULTIPLE RANGE AND ASSOCIATED TEST PROCEDURES, and the subset and partition methods described in Shaffer [45].

All procedures to be discussed here are based on fixed sample sizes. For sequential two- and three-stage procedures, where stages refer to taking additional (random) numbers of observations, see Hochberg and Lachenbruch [20], and Hochberg and Marcus [21].

Some single-stage procedures, particularly some nonparametric ones, are based on test statistics T_i for which it is known that $\Pr\{T_i \leqslant c_i \text{ for all } i\} \geqslant 1-\alpha$ only *under the assumption that all n hypotheses are true* (this intersection of all hypotheses will be called the *overall null hypothesis*). A hypothesis H_i is then rejected if $T_i > c_i$. While such a procedure has an upper bound α on (c) under the overall null hypothesis, (c) may exceed α when some hypotheses are false; α is therefore not an upper bound on (c) (see, e.g., Petrondas and Gabriel [38] and Fligner [12]). For a related problem in clustering of means, see Carmer and Lin [8]. Some multistage procedures that control the error rate (c) under the overall null hypothesis also suffer from the same defect; see Spjøtvoll [49] and Ramsey [39]. Since simultaneous testing is useful primarily when the overall null hypothesis is not true, these procedures should be used with caution, after further investigation of their properties under various combinations of true and false hypotheses. Other procedures, such as the Bonferroni procedure and the SRBP described above, have the stated bounds on (c) under all possible combinations of true and false hypotheses.

The Bonferroni procedure and the SRBP can be used for simultaneous tests of any *finite* set of hypotheses, and can sometimes even be extended to permit tests of infinite sets without increasing (c): see, e.g., Richmond [40]. They have the virtue of simplicity, but can generally be improved, among procedures that bound (c), both because their least upper bound on (c) is usually strictly less than α and because tests of more complicated structure may achieve greater power. Given any specific situation, then, special more powerful procedures can be devised by taking into account both the joint distribution of relevant test statistics and logical relations among the hypotheses, while maintaining an upper bound α on (c). As an example of the former, if the test statistics are independent, the Bonferroni procedure can be used with a set of significance levels α_i' such that $1 - \prod(1 - \alpha_i') = \alpha$ instead of a set α_i such that $\sum\alpha_i = \alpha$; the set α_i' can be chosen so that $\alpha_i' > \alpha_i$ for all i, thus giving greater power. The SRBP can be modified similarly. In many situations, known inequalities permit the use of the levels α_i' even without independence (see, e.g., Miller [35]). For other modifications of the SRBP based on relations among test statistics, see Holm [24].

There are obviously many logical relations among the set of hypotheses (1); e.g., H_{12} true and H_{13} true imply H_{23} true. See Shaffer [46] for direct modifications of the SRBP when such relationships exist.

HIERARCHICAL RELATIONS AMONG HYPOTHESES

Sometimes logical relations exist because some of the n hypotheses in the set being tested are equivalent to intersections of others. For example, the overall null hypothesis, which is the intersection of all other hypotheses in the set, may actually be a member of the set. In these cases, the Bonferroni procedure and its simple modifications are generally not advisable for the whole set. Alternative procedures will be indicated below.

Following Gabriel [16], those hypotheses that are not equivalent to intersections of any others will be called *minimal hypotheses*. Let I be the set of indices $1, 2, \ldots, n$, and let K be the set of indices of the minimal hypotheses $K \subset I$. Then any nonminimal hypothesis H_j is equivalent to the intersection of a subset $J \subset K$ of minimal hypotheses. Given such a nominimal hypothesis H_j, it seems essential for logical consistency that the rejection of any H_i, $i \in J$, should imply the rejection of H_j. Furthermore, it seems desirable, for ease in interpretation, that rejection of H_j should imply rejection of some H_i, $i \in J$. These two properties are denoted *coherence* and *consonance*, respectively, by Gabriel [16]. While all simultaneous testing procedures in use are coherent, not all are consonant.

Procedures that are both coherent and consonant result when the union–intersection principle of Roy [41, 42] is applied. Assume, for each minimal hypothesis H_i, $i \in K$, an associated test statistic T_i with critical value c_i and significance level α_i. Using the union–intersection principle, the rejection region for the nonminimal hypothesis H_j is the union of the rejection regions for $\{H_i, i \in J\}$, and the acceptance region for H_j is the intersection of the acceptance regions for $\{H_i;\ i \in J\}$. (In other words, H_j is rejected if any H_i, $i \in J$, is rejected, and is accepted if all H_i, $i \in J$, are accepted. Stated in this form, the principle can be applied in multistage procedures as well.) By putting various conditions on the T_i, c_i, and α_i [which of course are adjusted to control the error rate (c)], different types of union–intersection procedures (UIPs) are obtained.

The simultaneous test procedures (STPs) are an extensive subclass of single-stage procedures introduced by Gabriel [16] to be applied when some of the hypotheses being tested are nonminimal. Different types of

STPs, conditions necessary for them to be coherent and/or consonant, and their relationship to likelihood ratio tests* and UIPs are explored in Gabriel [15, 16].

AREAS OF APPLICATION

Probably the most intensively investigated area of application has been that of comparing distributions or populations with respect to some specified characteristics. For a general discussion of the set of hypotheses (1) and other sets involving symmetric treatment of the populations, see Shaffer [44, 45]; some optimality results for a class of procedures that bound (c) are given in Lehmann and Shaffer [31]. For treatment of the special case of comparison of normally distributed means, *see* the methods and references in MULTIPLE COMPARISONS; MULTIPLE RANGE AND ASSOCIATED TEST PROCEDURES; and BONFERRONI t-STATISTIC; methods for comparing means that are robust against nonnormality are investigated by Dunnett [10]. For simultaneous tests for values of specified subsets of linear combinations of normally distributed means, see Dunnett [9], Hochberg and Rodriguez [22], and Richmond [40]. For tests of effects in factorial analysis of variance* designs see Spjøtvoll [49], Hartley [19], Krishnaiah and Yochmowitz [27], and Johnson [26]; the latter two deal exclusively with interaction* effects. References and methods for distribution-free* comparisons of populations, for comparisons of proportions, and for testing values of regression coefficients can be found in MULTIPLE COMPARISONS. Among other multiple comparison problems that have been investigated are the comparison of effects in analysis of covariance* (Bryant and Paulson [7], Bryant and Bruvold [6], and Hochberg and Varon-Salomon [23]), the comparison of multinomial populations and contingency tables* (Gabriel [13], Bjornstad [1], and GOODMAN Y^2), tests on values of correlation coefficients (Larzelere and Mulaik [29] and Eagleson [11]), and the comparison of mean vectors and covariance matrices of multivariate normal distributions (MULTIVARIATE MULTIPLE COMPARISONS, Gabriel [14], and Krishnaiah et al. [28]).

AN EXAMPLE

Hartley [19] discussed the analysis of a 2^3 factorial experiment* laid out as an 8×8 Latin square*, reported by Goulden [17], which compared the effects of three fertilizers on wheat yields; Table 1 is adapted from Table 1 of Hartley. (It is assumed that the standard analysis of variance assumptions are satisfied.) If the SRBP is applied, treating the seven hypotheses as a family for which the error rate (c) is to be controlled at 0.05, the smallest significance probability is compared with 0.05/7; if that effect is significant, the next smallest is compared with

Table 1 Analysis of Variance of Effects in a 2^3 Factorial Experiment

Effect	Difference Between Levels	Mean Square	Degrees of Freedom	F-ratio	Significance Probability
N	−29.6	13.7	1	6.26	0.02
P	176.8	488.4	1	223.0	< 0.00001
K	−9.8	1.5	1	0.68	0.41
NP	15.2	3.6	1	1.6	0.21
NK	−8.2	1.0	1	0.46	0.50
PK	−15.8	3.9	1	1.8	0.19
NPK	10.2	1.6	1	0.73	0.40
Error	···	2.19	42	···	···

0.05/6, etc. A slightly less conservative procedure, justified by the use of Šidák's [47] inequality and the results in Holm [24], is to compare the significance probabilities with $1 - 0.95^{(1/7)}$, $1 - 0.95^{(1/6)}$, etc.; Hartley's results [19] indicate that these latter values are very close to the largest that can be used for overall level 0.05. Using this procedure, only the main effect* of P is significant. Hartley recommended this special case of the SRBP.

In factorial experiments of this kind, the conventional procedure is to test each effect at some standard level α, thus implicitly treating each as a separate family. The error rate (c) for the whole experiment can then be much greater than α; in this 2^3 example, if all hypotheses are true and α for each comparison is 0.05, the error rate is close to $1 - (0.95)^7 = 0.30$. If this conventional procedure is followed here, so that each significance probability in Table 1 is compared with 0.05, the effects for both P and N are significant, while if the SRBP is used, as above, only the P effect is significant. As Hartley points out, it seems likely that the N effect is not real, since the observed effect is in the opposite direction to what would be expected. Thus, the use of the SRBP, controlling error rate (c) for the whole experiment, apparently results in the avoidance of a type I error in this example.

DIRECTIONAL DECISIONS

Returning to the set of hypotheses (1), note that rejection of a hypothesis H_{ij} permits only the conclusion $\theta_i - \theta_j \neq 0$, but in most pairwise comparison situations, and in many other multiple testing situations, the sign, or direction, of a nonzero difference or nonzero parameter value is of major importance. Often a direction for the difference (or sign of a nonzero value) is asserted based on an estimated direction; however, there is then the risk of making an error not allowed for or controlled in the previous analysis and possibly more serious: Rejecting the hypothesis of equality but choosing the wrong sign or direction. Such directional errors have been called *type III errors*. Directional conclusions can be encompassed within the multiple testing framework by redefining the hypotheses in an appropriate manner; for example, the set of $k(k-1)/2$ hypotheses (1) could be replaced by the $k(k-1)$ hypotheses

$$H_{ij}: \theta_i - \theta_j \leqslant 0, \qquad i \neq j = 1, \ldots, k. \quad (3)$$

Directional errors are type I errors in tests of the set of hypotheses (3). For discussions of methods and issues in directional testing, see Peritz [37], Marcus et al. [34], Holm [25], and Shaffer [43].

Sometimes there is an interest only in one direction, e.g., in testing either H_{ij} *or* H_{ji} for some or all pairs $i \neq j$ in (3), or it may be known a priori that some differences or parameter values, if nonzero, must be positive (or negative). See Marcus and Peritz [33], Spjøtvoll [50], and Holm [25] for procedures appropriate in some situations of this type.

Sometimes (i) it is reasonable to assume a priori that all differences or values are nonzero or (ii) there may be no concern about what decision is made if any differences or values are zero. In these cases, directional errors are the only ones to be controlled. For procedures that bound the probability of one or more directional errors in (i), see Bofinger [2]; in (i) or (ii), see Bohrer [3], Bohrer et al. [5] and Bohrer and Schervish [4]. (See Lewis [32] for a partially Bayesian approach in (i), with a bound on the expected number of errors.)

References

[1] Bjornstad, J. F. (1982). *Commun. Statist. A*, **11**, 673–686.

[2] Bofinger, E. (1985). *J. Amer. Statist. Ass.*, **80**, 433–437.

[3] Bohrer, R. (1979). *J. Amer. Statist. Ass.*, **74**, 432–437.

[4] Bohrer, R. and Schervish, M. (1980). *Proc. Nat. Acad. Sci.*, **77**, 52–56.

[5] Bohrer, R., Chow, W., Faith, R., Joshi, V. M., and Wu, C. F. (1981). *J. Amer. Statist. Ass.*, **76**, 119–124.

[6] Bryant, J. L. and Bruvold, N. T. (1980). *J. Amer. Statist. Ass.*, **75**, 874–880.

[7] Bryant, J. L. and Paulson, A. S. (1976). *Biometrika*, **63**, 631–638.

[8] Carmer, S. G. and Lin, W. T. (1983). *Comm. Statist. B*, **12**, 451–466.

[9] Dunnett, C. W. (1955). *J. Amer. Statist. Ass.*, **50**, 1096–1121.

[10] Dunnett, C. W. (1982). *Comm. Statist. A*, **11**, 2611–2629.

[11] Eagleson, G. K. (1983). *Aust. J. Statist.*, **25**, 256–263.

[12] Fligner, M. A. (1984). *J. Amer. Statist. Ass.*, **79**, 208–211.

[13] Gabriel, K. R. (1966). *J. Amer. Statist. Ass.*, **61**, 1081–1096.

[14] Gabriel, K. R. (1968). *Biometrika*, **55**, 489–504.

[15] Gabriel, K. R. (1968). In *Essays in Probability and Statistics*, R. C. Bose et al., eds. University of North Carolina Press, Chapel Hill, NC.

[16] Gabriel, K. R. (1969). *Ann. Math. Statist.*, **40**, 224–250.

[17] Goulden, C. H. (1936). *Methods of Statistical Analysis*. Burgess, Minneapolis, MN.

[18] Harter, H. L. (1980). In *Handbook of Statistics*, Vol. 1, P., R. Krishnaiah, ed. North-Holland, Amsterdam, The Netherlands.

[19] Hartley, H. O. (1955). *Comm. Pure Appl. Math.*, **8**, 47–72.

[20] Hochberg, Y. and Lachenbruch, P. A. (1976). *Comm. Statist. A*, **5**, 1447–1454.

[21] Hochberg, Y. and Marcus, R. (1981). *Comm. Statist. A*, **10**, 597–612.

[22] Hochberg, Y. and Rodriguez, G. (1977). *J. Amer. Statist. Ass.*, **72**, 220–225.

[23] Hochberg, Y. and Varon-Salomon, Y. (1984). *J. Amer. Statist. Ass.*, **79**, 863–866.

[24] Holm, S. (1977). *Statistical Research Report No. 1977-1*, University of Umea, Umea, Sweden.

[25] Holm, S. (1979). *Scand. J. Statist.*, **6**, 65–70.

[26] Johnson, D. E. (1976). *Biometrics*, **32**, 929–934.

[27] Krishnaiah, P. R. and Yochmowitz, M. C. (1980). In *Handbook of Statistics*, Vol. 1, P. R. Krishnaiah, ed. North-Holland, Amsterdam, The Netherlands.

[28] Krishnaiah, P. R., Mudholkar, G. S., and Subbaiah, P. (1980). In *Handbook of Statistics*, Vol. 1, P. R. Krishnaiah, ed. North-Holland, Amsterdam, The Netherlands.

[29] Larzelere, R. E. and Mulaik, S. A. (1977). *Psychol. Bull.*, **84**, 557–569.

[30] Lehmann, E. L. (1957). *Ann. Math. Statist.*, **28**, 1–25, 547–572.

[31] Lehmann, E. L. and Shaffer, J. P. (1979). *Ann. Statist.*, **7**, 25–45.

[32] Lewis, C. (1981). *VVS Bull.*, **14**, 38–41.

[33] Marcus, R. and Peritz, E. (1976). *J. R. Statist. Soc. Ser. B*, **38**, 157–165.

[34] Marcus, R., Peritz, E., and Gabriel, K. R. (1976). *Biometrika*, **63**, 655–660.

[35] Miller, R. G., Jr. (1981). *Simultaneous Statistical Inference*, 2nd ed. Springer, New York.

[36] Naik, U. D. (1975). *Commun. Statist.*, **4**, 519–535.

[37] Peritz, E. (1965). *Biometrics*, **21**, 337–344.

[38] Petrondas, D. and Gabriel, K. R. (1983). *J. Amer. Statist. Ass.*, **78**, 949–957.

[39] Ramsey, P. H. (1981). *Psychol. Bull.*, **90**, 352–366.

[40] Richmond, J. (1982). *J. Amer. Statist. Ass.*, **77**, 455–460.

[41] Roy, S. N. (1953). *Ann. Math. Statist.*, **24**, 220–238.

[42] Roy, S. N. (1957). *Some Aspects of Multivariate Analysis*. Wiley, New York.

[43] Shaffer, J. P. (1980). *Ann. Statist.*, **8**, 1342–1347.

[44] Shaffer, J. P. (1981). *J. Amer. Statist. Ass.*, **76**, 395–401.

[45] Shaffer, J. P. (1984). In *Proc. Seventh Conf. Probability Theory*, M. Iosifescu, ed. Editura Academiei Republicii Socialiste Romania, Bucharest, Romania.

[46] Shaffer, J. P. (1986). *J. Amer. Statist. Ass.*, **81**, 826–831.

[47] Šidák, Z. (1967). *J. Amer. Statist. Ass.*, **62**, 626–633.

[48] Spjøtvoll, E. (1972). *Ann. Math. Statist.*, **43**, 398–411.

[49] Spjøtvoll, E. (1974). *Scand. J. Statist.*, **1**, 97–114.

[50] Spjøtvoll, E. (1977). *Biometrika*, **64**, 327–334.

[51] Waller, R. A. and Duncan, D. B. (1969). *J. Amer. Statist. Ass.*, **64**, 1484–1503.

[52] Waller, R. A. and Duncan, D. B. (1974). *Ann. Inst. Statist. Math.*, **26**, 247–264.

(BONFERRONI INEQUALITIES AND INTERVALS
HYPOTHESIS TESTING
MULTIPLE COMPARISONS
MULTIPLE DECISION THEORY
MULTIPLE RANGE AND ASSOCIATED TEST PROCEDURES
MULTIVARIATE MULTIPLE COMPARISONS
PAIRED COMPARISONS
RANKING PROCEDURES
SCHEFFÉ'S SIMULTANEOUS COMPARISON PROCEDURE
SELECTION PROCEDURES
UNION–INTERSECTION PRINCIPLE)

JULIET P. SHAFFER

SINGH–MADDALA DISTRIBUTION

This class of distributions was suggested by Singh and Maddala [3] for use in descriptive models of distribution of income. The CDF is of the form

$$F(x) = 1 - (1 + a_1 x^{a_2})^{-a_3},$$
$$x \geqslant 0;\ a_1, a_2, a_3 \geqslant 0.$$

For $a_3 = 1$ it is known as a *Fisk distribution*. The distribution is in fact a Burr type XII distribution* and has Pareto distributions* as limiting cases. It often seems to give a better fit to income data than does a gamma distribution*. Geneses are described in ref. 3; further information can be obtained from refs. 1 and 2 and INCOME DISTRIBUTIONS.

References

[1] Cronin, D. C. (1979). *Econometrica*, **47**, 773–774.

[2] McDonald, J. B. (1984). *Econometrica*, **52**, 647–663.

[3] Singh, S. K. and Maddala, G. S. (1976). *Econometrica*, **44**, 963–970.

(INCOME DISTRIBUTIONS
PARETO DISTRIBUTION)

SINGLE-HUMPED CURVE *See* UNIMODALITY

SINGLE LINKAGE CLUSTERING METHOD *See* HIERARCHICAL CLUSTER ANALYSIS

SINGLE-PEAKEDNESS AND MEDIAN VOTERS

Majority rule is commonly used when collective (or group) decisions have to be made. However, as is widely known, the rule has a potential flaw: For a given set of alternatives, it may be the case that, for each alternative, there is some other alternative that is preferred by a majority of the eligible voters. For an example where each alternative can be beaten under majority rule, *see* VOTING PARADOX. Recognition of this problem with majority rule has led researchers to focus on (i) estimating the likelihood of there being no unbeaten alternative (discussed in detail in VOTING PARADOX) and (ii) identifying distributions of voter preferences for which there is an unbeaten alternative. The best-known distributions of voter preferences with unbeaten alternatives are ones where the voters have what are known as *single-peaked preferences*. The distinguishing feature of these preferences was described in Black [3, p. 24], where this special term was first introduced. His description was specifically for a "committee considering different possible sizes of a numerical quantity and choosing one size in preference to the others." The description was for any particular member:

> Once he had arrived at his view of the optimum size, the farther any proposal departed from it on the one side or the other, the less he would favor it. The valuations carried out by the member would then take the form of points on a single-peaked or ∩-shaped curve.

Interestingly, Black [3] established that (for any such distribution of voter preferences) if a voter's most preferred alternative (or "optimum") is (in a certain specific sense) a median for the corresponding distribution of the voters' most preferred points (or "optima"), then that alternative is unbeaten under majority rule. Precursors who had informally described the link between single-peakedness and "median voters" prior to Black [3] include Galton* [8] and Hotelling [9]. For further information on the history of the recognition of this link and related ideas, see Black [4].

MAJORITY RULE AND MEDIAN VOTERS

Two examples will be used. In order to keep the subsequent discussion as clear as possible, these simple examples have been tailored to illustrate the basic concepts and theorems that are covered in this entry. Both of the examples will have the following common features: A committee is considering the budget of a particular government agency. The set of alternatives available to the committee is $\{r, s, t\}$, where r denotes retrenchment (with large budget cuts), s denotes the same budget as at present, and t denotes a tremendous increase in the agency's budget.

Example 1. There are three individuals, $i = 1, 2, 3$, on the committee. Their preferences are rP_1sP_1t, sP_2tP_2r, and tP_3sP_3r, respectively (where xP_iyP_iz means: i prefers x to y, prefers y to z, and prefers x to z).

Example 2. There are four individuals, $i = 1, 2, 3, 4$, on the committee. $i = 1, 2, 3$ have the same preferences as in Example 1 and $i = 4$ has rP_4sP_4t.

X will denote a set of alternatives. $N = \{1, \ldots, n\}$ will be an index set for the finite set of individuals who get to vote on the alternatives. For each $i \in N$, R_i will be a preference ordering on X (i.e, a complete reflexive transitive binary preference relation on X—*see* UTILITY THEORY). xR_iy means "x is at least as good as y, for i." xP_iy means "i prefers x to y" (i.e., xR_iy holds, but yR_ix does not). These features are all clearly present in the two examples.

Definition. $x \in X$ "gets a majority over" $y \in X$ if and only if $|\{i \in N : xP_iy\}| > |\{i \in N : yP_ix\}|$, where $|A|$ denotes the number of elements in (or "cardinality of") the finite set A.

In Example 1: s gets a majority over both r and t. In Example 2: no alternative gets a majority over every other alternative—but, at the same time, there is also no alternative that gets a majority over either r or s.

Following Black [3, pp. 24–25]: "We shall refer to the motion corresponding to the peak of any curve—the most-preferred motion for the member concerned—as his optimum." Stating this in terms of an individual's preference relation R_i:

Definition. $x_i \in X$ is an *optimum* for $i \in N$ if and only if x_iP_iy, $\forall y \in X - \{x_i\}$.

As in Black [3], in what follows it will be assumed that each individual in N has an optimum. In the examples, the committee members' optima are $x_1 = x_4 = r$, $x_2 = s$, and $x_3 = t$.

Definition. The *distribution of voter optima* is the discrete probability distribution $P(\cdot)$ on X that satisfies

$$P(x) = |\{i \in N : x_i = x\}|/N, \forall x \in X.$$

In Example 1, the distribution of voter optima is $P(r) = P(s) = P(t) = \frac{1}{3}$. In Example 2, it is $P(r) = \frac{1}{2}$, $P(s) = P(t) = \frac{1}{4}$.

The symbol $\leqslant_o$ will denote a linear order on X (i.e., a complete, transitive, antisymmetric binary relation on X—see, for instance, Rubin [11] or Denzau and Parks [5]). The interpretation of $x \leqslant_o y$ will be "x is *either* to the left of y *or* in the same place as y." $x <_o y$ will mean "x is to the left of y." The linear order $\leqslant_o$ gives us a way of lining up the alternatives from left to right. The most familiar example of a linear order is, of course, the relation "less than or equal to" for the real numbers.

Definition. $m \in X$ *is a median for the distribution of voter optima with respect to* $\leqslant_o$ if and only if $P(\{x \in X : x \leqslant_o m\}) \geqslant \frac{1}{2}$ *and* $P(\{x \in X : m \leqslant_o x\}) \geqslant \frac{1}{2}$.

For the examples, let the linear order be r is left of s, s is left of t, and (hence) r is left of t (i.e., $r <_o s$, $s <_o t$, and $r <_o t$). In Example 1, there is a unique median $m = s$ (the unique alternative that beats every other alternative). In Example 2, there are two medians $m' = s$ and $m'' = r$ (the two unbeaten alternatives). A voter whose optimum is a median for the distribution of voter optima (with respect to a given linear order) is, accordingly, called a "median voter" (with respect to the given linear order). In Example 1, $i = 2$ is the unique median voter. In Example 2, every individual except $i = 3$ is a median voter.

SINGLE-PEAKEDNESS AND MEDIAN VOTER THEOREMS

The preferences of the four individuals used in the two examples in this entry can be represented by the following utility functions (*see* UTILITY THEORY):

$$u_1(r) = 3, \quad u_1(s) = 2, \quad u_1(t) = 1,$$
$$u_2(s) = 3, \quad u_2(t) = 2, \quad u_2(r) = 1,$$
$$u_3(t) = 3, \quad u_3(s) = 2, \quad u_3(r) = 1,$$
$$u_4(r) = 3, \quad u_4(s) = 2, \quad u_4(t) = 1,$$

(where the subscript denotes the individual whose preferences are being represented). The utility function for any given individual $i \in \{1, 2, 3, 4\}$ can be graphed as follows. First, place r, s, and t on the horizontal axis with r left of s and s left of t (corresponding to the linear order being used for the examples). Second, letting the vertical axis be measured in utility units, place dots at the three points $(r, u_i(r))$, $(s, u_i(s))$, and $(t, u_i(t))$. Third, draw lines connecting $(r, u_i(r))$ with $(s, u_i(s))$ and connecting $(s, u_i(s))$ with $(t, u_i(t))$. When this is done, it becomes clear that, for each individual, the three dots "take the form of isolated points on single-peaked curves" (Black [3, p. 24]). More specifically, if you start at the leftmost alternative (r) and move to the right, you can easily see that:

(i) for $i = 1$ and $i = 4$, the curve that you have drawn consistently goes down [from $(r, u_i(r)) = (r, 3)$ to $(s, u_i(s)) = (s, 2)$ and then to $(t, u_i(t)) = (t, 1)$];

(ii) for $i = 3$, the curve that you have drawn consistently goes up [from $(r, u_3(r)) = (r, 1)$ to $(s, u_3(s)) = (s, 2)$ and then to $(t, u_3(t)) = (t, 3)$];

(iii) for $i = 2$, the curve that you have drawn first goes up [from $(r, u_2(r)) = (r, 2)$ to $(s, u_2(s)) = (s, 3)$] and then goes down [to $(t, u_2(t)) = (t, 1)$]. Thus each curve is "one which changes its direction at most once, from up to down" (Black [4, p. 7]).

When each voter has an ideal point, this statement implies that each voter has a single "peak" for his preference ordering (occurring, in each case, at the voter's optimum). In the examples, this is easily seen by looking at the graphs of the voter's utility functions.

In the preceding paragraphs, the preferences of the four individuals in the examples in the entry have been used to illustrate the simple geometric idea behind the concept of "single-peaked" preferences. Assuming that each voter i has an optimum x_i, the concept itself can be stated precisely as follows (following Denzau and Parks [5], for instance):

Definition. R_i is "single-peaked with respect to $\leqslant_o$" if and only if, for each pair $y, z \in X$, (i) $[y <_o z \,\&\, z \leqslant_o x_i] \Rightarrow [zP_iy]$ and (ii) $[x_i \leqslant_o y \,\&\, y <_o z] \Rightarrow [yP_iz]$.

The following two "median voter theorems" were established in Black [3, pp. 27 and 28, respectively]:

Theorem 1. Suppose that X is finite, each voter has an optimum, each voter's preference ordering is single-peaked with respect to $\leqslant_o$, and *n is odd*. Then the voter optimum x_i gets a majority over every other alternative in X if and only if X_i is a median for the distribution of voter optima (with respect to $\leqslant_o$).

When the premise for Theorem 1 is satisfied, the discrete distribution of voter optima necessarily has a unique median. Therefore Theorem 1 implies that, when its premise is satisfied, there exists a voter optimum that beats every other alternative. Theorem 1 and this implication are both illustrated by Example 1 (where $n = 3$, $i = 2$ is the median voter and $x_2 = s$ gets a majority over r and t).

Theorem 2. Suppose that X is finite, each voter has an optimum, and each voter's preference ordering is single-peaked with respect to $\leqslant_o$, and *n is even*. Then there is no alternative in X that gets a majority over the voter optimum x_i if and only if X_i is a median for the distribution of voter optima (with respect to $\leqslant_o$).

When the premise for Theorem 2 is satisfied, the discrete distribution of voter optima has at least one voter optimum that is a median and no more than two distinct voter optima that are medians. Therefore, when its premise is satisfied, Theorem 2 implies that there is at least one voter optimum that is unbeaten and no more than two distinct voter optima that are unbeaten. Theorem 2 and this implication are illustrated by Example 2 (where $n = 4$, voters 1, 2, and 4 are the median voters and the optima that are unbeaten under majority rule are $x_2 = s$ and $x_1 = x_4 = r$). Black [3, p. 28] also pointed out that, *if* the premise of Theorem 2 is satisfied *and* there are two distinct optima that are unbeaten *and* there is one individual in N (such as a committee chairman) who can break ties, *then* the unbeaten optimum that is closer to that individual's optimum will be the final decision.

The two median voter theorems have been generalized and extended in Black [4], Fishburn [7], Denzau and Parks [5], Enelow and Hinich [6], and elsewhere. They have also been applied in a large number of studies of the public sector (see the discussion plus the references cited in Atkinson and Stiglitz [2, Chap. 10])— fulfilling Black's prediction that his theorems would "provide the basis for a theory of the equilibrium distribution of taxation or of public expenditure" (Black [3, p. 23]). For further information about single-peakedness and its implications, the reader should also see Arrow [1], Mueller [10], Sen [12], and/or White [13].

References

[1] Arrow, K. J. (1951). *Social Choice and Individual Values*, 1st ed. Wiley, New York (2nd ed., 1963.) (Especially Chapter VII, which provided one of the first formal analyses of single-peakedness and its implications.)

[2] Atkinson, A. and Stiglitz, J. (1980). *Lectures on Public Economics*. McGraw-Hill, New York. (Especially Chapter 10, which surveys applications of median voter theorems in analyses of the public sector.)

[3] Black, D. (1948). *J. Polit. Economy*, **56**, 23–34. [Reprinted in Arrow, K. and Scitovsky, T., eds. (1969). *Readings in Welfare Economics*. Irwin, Homewood, IL.] (*The* pioneering formal analysis of the link between single-peakedness and median voters.)

[4] Black, D. (1958). *The Theory of Committees and Elections*. Cambridge University Press, Cambridge, England. (Especially Chapters I–V, which elaborate at length on the ideas first developed in Black [3], and p. 188, where Galton [8] is discussed.)

[5] Denzau, A. and Parks, R. (1975). *Econometrica*, **43**, 853–866.

[6] Enelow, J. and Hinich, M. (1984). *The Spatial Theory of Voting*. Cambridge University Press, Cambridge, England. (Especially Chapter 2, which provides an intermediate-level textbook treatment of the link between single-peakedness and median voters.)

[7] Fishburn, P. (1973). *The Theory of Social Choice*. Princeton University Press, Princeton, NJ. (Especially Chapter 9, which provides a formal analysis of single-peakedness and its implications.)

[8] Galton, F. (1907). *Nature*, **75**, 414.

[9] Hotelling, H. (1929). *Econ. J*, **39**, 41–57. (Especially pp. 54–55.)

[10] Mueller, D. (1979). *Public Choice*. Cambridge University Press, Cambridge, England. (Especially Chapters 3, 6, and 10, which survey research on single-peakedness and its implications.)

[11] Rubin, J. (1967). *Set Theory for the Mathematician*. Holden-Day, San Francisco, CA.

[12] Sen, A. (1970). *Collective Choice and Social Welfare*. Holden-Day, San Francisco, CA. (Especially Chapter 10, which provides an informal discussion of single-peakedness and related restrictions on preferences.)

[13] White, D. (1976). *Fundamentals of Decision Theory*. North-Holland, Amsterdam, The Netherlands. (Especially Chapter 9, which provides an intermediate-level textbook treatment of single-peakedness and related concepts.)

(DECISION THEORY
ELECTION FORECASTING

UTILITY THEORY
VOTING PARADOX)

PETER J. COUGHLIN

SINGLE SAMPLING *See* SAMPLING PLANS

SINGLE-SERVER QUEUE *See* QUEUEING THEORY

SINGULAR DISTRIBUTION *See* DEGENERATE DISTRIBUTION

SINGULAR MATRIX

A square matrix is called *singular* if its determinant is zero. An alternative (and equivalent) definition is that its rank is less than its dimension. A singular matrix does not have a regular inverse. Singular matrices play important roles in the theory of linear models and in multivariate analysis* and distribution theory.

(GENERALIZED INVERSES)

SINGULAR NORMAL DISTRIBUTIONS *See* DEGENERATE DISTRIBUTION; MULTINORMAL DISTRIBUTIONS

SINGULAR-VALUE DECOMPOSITION

If $\mathbf{A}$ is a $n \times p$ matrix ($n \geq p$), then it is possible to express it in the form

$$\mathbf{A} = \mathbf{U}\mathbf{D}_\mu\mathbf{V}',$$

where $\mathbf{U}$ and $\mathbf{V}$ are each orthogonal matrices (satisfying $\mathbf{U}'\mathbf{U} = \mathbf{V}'\mathbf{V} = \mathbf{I}_p$), and $\mathbf{D}_\mu = \text{diag}(\mu_1, \dots, \mu_p)$ is a diagonal matrix. The diagonal elements $\mu_1, \dots, \mu_p$ are called the *singular values* of $\mathbf{A}$.

Note that $\mathbf{U}'\mathbf{A}\mathbf{V} = \mathbf{D}_\mu$, so that a bilinear form $\Sigma_{i=1}^n \Sigma_{j=1}^p a_{ij} X_i Y_j = \mathbf{X}'\mathbf{A}\mathbf{Y}$ in mutually independent standard normal variables $\mathbf{X}' = (X_1, \dots, X_n)$ and $\mathbf{Y}' = (Y_1, \dots, Y_p)$ can be transformed into the canonical form

$$\mathbf{W}'\mathbf{D}_\mu\mathbf{Z} = \sum_{k=1}^{p} \mu_h W_h Z_h,$$

where $\mathbf{W}' = \mathbf{X}'\mathbf{U}$ and $\mathbf{Z} = \mathbf{V}'\mathbf{Y}$ are the $2p$ independent standard normal variables $(W_1, \dots, W_p)'$ and $(Z_1, \dots, Z_p)'$.

This decomposition is used in regression diagnostics* (see ref. 1) to measure (multi-) collinearity* among predicting variables in multiple linear regression*.

If $\mathbf{A}$ is a real symmetric square ($n = p$) matrix, it is possible to have $\mathbf{U} = \mathbf{V}$, giving a *spectral decomposition**.

Reference

[1] Belsley, D. A., Kuh, E., and Welsch, R. E. (1980). *Regression Diagnostics*. Wiley, New York.

(LINEAR ALGEBRA, COMPUTATIONAL
MULTICOLLINEARITY
MULTIVARIATE ANALYSIS
QUADRATIC FORMS
REGRESSION (various entries)
SPECTRAL DECOMPOSITION)

SINGULAR WEIGHING DESIGNS *See* WEIGHING DESIGNS

SINUSOIDAL LIMIT THEOREM

The oscillations observed in economic, meteorological, geophysical, and many other physical time series* have puzzled and intrigued statisticians for many years. In trying to model such phenomena, a problem of interest is to measure the degree to which an oscillatory time series is close to a pure sinusoid or to a superposition of several sinusoids. An early result along these lines is the sinusoidal limit theorem (SLT) of Slutsky* [7], which provides conditions under which a second-order stationary time series $\{Z_t\}$, $t = \cdots, -1, 0, 1, \dots,$ tends toward a sinusoid or sine function of the form

$$A \cos \omega t + B \sin \omega t, \qquad (1)$$

where ω is a fixed frequency. This result of

Slutsky should not be confused with a number of other results associated with his name and that deal with a variety of topics.

Slutsky's (SLT) is inspired by an interesting fact concerning the second difference of a sinusoid of the form (1). Define the difference operator ∇ by $\nabla Z_t \equiv Z_t - Z_{t-1}$ and $\nabla^2 Z_t \equiv \nabla(\nabla Z_t)$. If $\{Z_t\}$ is a second-order stationary time series of the form (1), then $\nabla^2 Z_t$ is perfectly negatively correlated with Z_{t-1}. The SLT essentially states that the converse in some sense holds as well. This is made precise in the following theorem [7].

Theorem (SLT). Let $Z_1, Z_2 \ldots$ be a second order stationary time series such that $EZ_t = 0$ and

$$EZ_t^2 = \sigma^2 = h(n),$$

$$\frac{EZ_t Z_{t+k}}{EZ_t^2} = \rho_k = \phi(k, n),$$

where n is a parameter specifying the sequence as a whole, and $h(n)$ and $\phi(k, n)$ are independent of t. If

$$|\rho_1| \leqslant \text{constant} < 1 \quad \text{as } n \to \infty,$$

and the correlation coefficient η_1 between $\nabla^2 Z_t$ and Z_{t-1} is such that

$$\eta_1 \to -1 \quad \text{as } n \to \infty,$$

then the finite section $(Z_t, \ldots, Z_{t+s})$ converges in probability to a sinusoid of the form (1) with frequency ω determined from the equation $\rho_1 = \cos(\omega)$.

Some authors refer to a corollary of this theorem, also due to Slutsky [7], as the SLT. Let $\mathscr{B}$ be the shift operator defined by $\mathscr{B}Z_t \equiv Z_{t-1}$.

Corollary. Let

$$\{X_t\}, t = \cdots, -1, 0, 1, \ldots,$$

be a random sequence such that $EX_t = 0$, $EX_t^2 = \sigma_x^2$, and $EX_t X_s = 0$ for $t \neq s$. Define a second-order stationary sequence by

$$Z_t = (1 - \mathscr{B})^m (1 + \mathscr{B})^n X_t,$$

and let $m, n \to \infty$ such that $m/n =$ constant. Then the finite section $(Z_t, \ldots, Z_{t+s})$ converges in probability to a sinusoid of the form (1) with frequency

$$\omega = \cos^{-1}\left(\frac{1 - m/n}{1 + m/n}\right).$$

Note that $(1 - \mathscr{B})$ stands for a difference and $(1 + \mathscr{B})$ stands for a sum of two consecutive values X_t and X_{t-1}. Thus the corollary says that repeated summation followed by repeated differencing of "random causes" produces sinusoidal processes. This is an example of the so-called Slutsky–Yule effect to be discussed below.

Slutsky's original results are discussed and extended in Romanovsky [5, 6], Wold [8], and Moran [3, 4]. These authors put linear restrictions on the correlation function that force the corresponding time series to follow difference equations* whose solutions are sums of several sinusoids.

In the Gaussian case the SLT can be stated in terms of the expected number of zero crossings in the series and in its first difference. Let D_1 be the number of zero crossings observed in the graph of $Z_1, \ldots, Z_N$ and let D_2 be the number of zero crossings observed in the graph of $\nabla Z_1, \ldots, \nabla Z_N$. Then the quantities $ED_j/(N-1)$, $j = 1, 2$, are independent of the series size N. The following theorem is proved in Kedem [1].

Theorem (SLT). Let $\{Z_t\}$ be a zero mean stationary Gaussian sequence and assume $0 < ED_1$. If

$$\frac{ED_1}{N-1} = \frac{ED_2}{N-1},$$

then the vector $(Z_t, \ldots, Z_{t+s})$ lies with probability 1 on a sinusoid of the form (1) with frequency $\omega = \pi ED_1/(N-1)$.

The converse of this version of the SLT is also true. Since, except for end-effects, D_2 is the number of peaks and troughs observed in the graph of $Z_1, \ldots, Z_N$, a Gaussian time series tends to obey the SLT when the difference between the number of zero crossings and the number of peaks and troughs tends to zero. In general however $ED_1 \leqslant ED_2$. For examples see refs. 1, and 2.

SLUTSKY-YULE EFFECT

The above corollary is an example of a linear operation applied repeatedly to a random sequence that produces regular oscillations. This effect, or tendency towards sinusoidal fluctuations resulting from linear operations, was observed independently by Yule [9, 10] and by Slutsky [7]. Both authors were interested in modeling periodic phenomena in physical and economic time series, and their studies show that periodicities observed in such time series may be due to linear operations applied to "random causes," or to chaotically random elements with no definite structure, and not to any strictly periodic mechanism. It is important to note though that the Slutsky–Yule effect is concerned with the *rise* in regularity as expressed by a more pronounced cyclical oscillation; completely random sequences also display fluctuations but of a lesser regularity. In a language more akin to the present day state of the art, the Slutsky–Yule effect is the effect produced by a time invariant linear filter when applied to a stationary time series. In this case there is a change in the spectral weight given to the frequencies present in the spectrum. A repeated application of the filter may cause the spectral mass to concentrate at a few frequencies only, in which case the original series converges to a sum of several sinusoids.

The investigations of Yule gave rise to the important class of *autoregressive processes*, and those of Slutsky led to another equally important class, *moving averages**. In both cases a stream of random elements is operated on by a linear system.

References

[1] Kedem, B. (1984). *Ann. Statist.*, **12**, 665–674.

[2] Kedem, B. (1986). *Amer. Math. Monthly*, **93**, 430–440 (Here and in ref. 1, a discussion of the SLT and its application in the Gaussian case.)

[3] Moran, P. A. P. (1949). *Biometrika*, **26**, 63–70. (Gives a spectral-theoretic proof of the SLT.)

[4] Moran, P. A. P. (1950). *Proc. Camb. Philos. Soc.*, **46**, 272–280. (A discussion of repeated linear operations to random sequences.)

[5] Romanovsky, V. I. (1932). *Rend. Circolo. Mat. Palermo*, **56**, 82–111. (A detailed proof of the SLT under a somewhat modified condition, and a generalization of the SLT.)

[6] Romanovsky, V. I. (1933). *Rend. Circolo. Mat. Palermo*, **57**, 130–136.

[7] Slutsky, E. E. (1927). *Problems of Economic Conditions*, **3**, No. 1. [In Russian; English transl. *Econometrica*, **5**, 105–146 (1937)]. (The original SLT is stated and proved. A pioneering paper that discusses moving averages and in particular those obtained by repeated summation and differencing of random sequences.)

[8] Wold, H. (1938). *A Study in the Analysis of Stationary Time Series*. Almqvist and Wiksells, Uppsala, Sweden. (Discusses singular processes in general.)

[9] Yule, G. U. (1921). *J. R. Statist. Soc.*, **84**, 497–526.

[10] Yule, G. U. (1926). *J.R. Statist. Soc.*, **89**, 1–69. (Here and in ref. 9, a discussion of repeated differences of harmonic functions and of random series.)

(LIMIT THEOREMS
TIME SERIES)

BENJAMIN KEDEM

SIZE AND SHAPE ANALYSIS

The terms "size" and "shape" have been used in a variety of contexts (often biological) in statistical and quantitative analyses of data; an indication of some of this variety will be given later. Here the approach will be of broad applicability in the probability modelling and statistical analysis of positive random variables.

The thrust of size and shape analysis is to consider models and methods for the analysis of random proportions or ratios. If proportions are regarded as fixed values, the concepts considered here are less relevant. However, when the proportions or ratios are random quantities, the concepts that follow assume importance. The results are applicable to any vector of positive random variables.

In many scientific studies, the data are random proportions or ratios that represent dimensionless quantities in a physical sense. When these data represent the same physical dimensions (say lengths, counts) and are expressed in the same units (say, millimeters, dozens), their ratios are dimensionless numbers, which can be viewed as expressing "shape" in some general sense.

Thus given any two similar plane triangles of different size, any ratio of the sides of one triangle will be the same as the corresponding ratio in the other, reflecting their similarity. They have the same "shape." Now let (A, B, C) be random variables representing the respective lengths of corresponding sides of random triangles. Then random ratios like A/C and B/C represent the "shape" of the triangle, while the length of one of the sides, say A, might be chosen to represent "size."

Alternatively, suppose one has measurements of the length, width, and height of the shell, or carapace, of a number of turtles of the same species. Let the random vector (L, W, H) represent these measurements. For some species of turtles, recently hatched individuals are nearly round when viewed from above so that the ratio of length to width L/W is nearly 1. Subsequently length grows faster than width, so that the ratio L/W is greater in larger turtles (with L, W, and H all large) [36]. Thus there is a positive association of the random "shape" variable L/W with each of the random "size" variables L, W, and H, respectively. In biology the study of such associations of dimensionless shape variables, like L/W, with variables containing scale or size information is a part of the field of allometry. For references see refs. 20–22, 24, 28, 33, 36, 38, and 49 and ALLOMETRY.

As a further illustration, let $(T_1, \dots, T_k)$ be the intervals of time between the occurrence of successive events in a Poisson process*. Then $T_1, \dots, T_k$ are mutually independent exponential variables with a common scale parameter, say λ. Let $S_k = T_1 + \cdots + T_k$. Then S_k is statistically independent of the shape vector of proportions $(T_1/S_k, \dots, T_k/S_k)$, reflected in the fact that S_k is a sufficient statistic* for the scale parameter λ, which itself contains all size information. For examples of studies of statistics sufficient for scale or size under various models, see refs. 17, 25, and 59. The basic concepts that follow are developed in refs. 38, 40, 41, and 45.

SHAPE VECTORS AND SIZE VARIABLES

A *positive* vector in k-dimensional Euclidean space is one whose coordinates are all positive. The set of all such positive vectors is denoted by R_+^k; for example, R_+^1 is the set of positive real numbers. For each positive vector, a pair of associated variables is defined: a scalar "size variable," which contains all size or scale information, and a "shape vector," which contains all shape or ratio information. A *size variable* is a scalar-valued function G from R_+^k to R_+^1 with the homogeneity property that $G(\alpha\mathbf{x}) = \alpha G(\mathbf{x})$ for all $\mathbf{x} \in R_+^k$ and all $\alpha > 0$. (Thus if each coordinate of the vector $\mathbf{x}$ is expressed in millimeters, then the size variable G is expressed in millimeters.) A *shape vector* is a vector-valued function $\mathbf{z}$ from R_+^k to R_+^k, defined using a given size variable G by $\mathbf{z}(\mathbf{x}) = \mathbf{x}/G(\mathbf{x})$ for all $\mathbf{x} \in R_+^k$.

Examples of size variables (in notation used throughout) are:

the ordinary sum $\quad S_k = \sum_{i=1}^{k} x_i;$

the geometric mean* $\quad M_k = \left(\prod_{i=1}^{k} x_i\right)^{1/k};$

the distance $\quad D_k = \left(\sum_{i=1}^{k} x_i^2\right)^{1};$

the maximum coordinate $\quad \max_k = \max(x_1, \dots, x_k);$

the last coordinate $\quad C_k = x_k.$

Corresponding to each size variable is a shape vector. Thus $\mathbf{x}/S_k$ is the usual vector of proportions, while $\mathbf{x}/D_k$ is the vector of direction cosines. Other shape vectors, such as $\mathbf{x}/M_k$, have no particular name, but the name *ratio vector* will be adopted for a shape vector where $\mathbf{x}$ is divided by a single coordinate, e.g. $\mathbf{x}/x_k$.

SIZE VARIABLES COMPRISE A BROAD CLASS

The image of any size variable is R^1_+, the positive real numbers. The image of a shape vector is never all of R^k_+, but is that subset of positive vectors constrained so that the defining size is 1, because by homogeneity, $G(\mathbf{z}) = G(\mathbf{x}/G(\mathbf{x})) = G(\mathbf{x})/G(\mathbf{x}) = 1$. This subset of vectors of size 1 can be thought of as the unit sphere of the size variable G, e.g., the surface of the positive simplex if $G = S_k$. A size variable can be defined by arbitrarily determining a unit sphere, that is, by selecting a point on each positively directed ray from the origin, and assigning size 1 to that point. By homogeneity then, the size variable is defined everywhere in R^k_+. While any size variable must be continuous along a ray because of homogeneity, size variables need not be continuous functions on R^k_+. The class of size variables is a very broad one that includes nonmeasurable functions unsuitable for defining random variables. Excluding these, the class remains very broad. Thus there is a one–one correspondence of values of a given shape vector with the positive rays from the origin. Simply put, a shape vector names the rays, while a size variable reveals where a given vector terminates on its particular ray.

REGULAR SEQUENCES OF SIZE VARIABLES

Functions can be defined that relate shape and size in a lower-dimensional space with shape and size, respectively, in a higher-dimensional space. To define a "related pair" of size variables consider a k-dimensional positive vector $\mathbf{x}$, and append to it a new measurement x_{k+1}. Let G_k be a size variable defined using $\mathbf{x}$ alone (i.e., from R^k_+); let G_{k+1} be defined using $\mathbf{x}'_{k+1} = (\mathbf{x}'; x_{k+1})$, so that G_{k+1} is defined from R^{k+1}_+. If knowing only G_k and G_{k+1}, one also can know the pair G_k, x_{k+1}, and conversely, knowing G_k and x_{k+1}, one can also know G_k and G_{k+1}, then G_k and G_{k+1} are said to be *related*. The idea is that all scale information needed to capture G_{k+1} is encapsulated in G_k and x_{k+1}, and correspondingly, one can recapture x_{k+1} from G_k and G_{k+1} alone. For example, S_k and S_{k+1} are related, but $\max_k$ and $\max_{k+1}$ are not (x_{k+1} cannot be recaptured from $\max_k$ and $\max_{k+1}$). Also C_k, C_{k+1}, M_k, M_{k+1}, and D_k, D_{k+1} are all related pairs, but the pair M_k, S_{k+1} is not.

One consequence of having a related pair, G_k, G_{k+1}, is that knowledge of k-dimensional shape and of the ratio G_{k+1}/G_k gives one knowledge of $(k+1)$-dimensional shape without knowing any scale information.

A sequence of size variables $G_1, \ldots, G_{k+1}$ is *regular* if each contiguous pair G_j, G_{j+1} is a related pair. (Here G_1 is defined from R^1_+ based on the first coordinate of $\mathbf{x}$; G_2 from R^2_+ based on the first two coordinates of $\mathbf{x}$; and so forth.)

SIZE RATIOS OF A REGULAR SEQUENCE REPRESENT SHAPE

When $(G_1, \ldots, G_{k+1})$ is a regular sequence of size variables, there is an inverse function by which the vector $\mathbf{x}$ may be known from $(G_1, \ldots, G_{k+1})$. The vector of size ratios $(G_2/G_1, \ldots, G_{k+1}/G_k)$ is also in one–one correspondence with the $(k+1)$-dimensional rays. Therefore the size ratios of a regular sequence represent shape as well as any shape vector [40].

Any data vector $(x_1, \ldots, x_{k+1})$ can be represented by the size ratios with size appended $(G_2/G_1, \ldots, G_{k+1}/G_k, G_{k+1})$. An advantage of this is that when the latter is partitioned as two vectors, the first $(G_2/G_1, \ldots, G_r/G_{r-1})$ represents r-dimensional shape while involving none of the coordinates from $r+1$ to k. In contrast, the second vector $(G_{r+1}/G_r, \ldots, G_{k+1}/G_k, G_{k+1})$ represents equally well either the higher-dimensional size variables

$(G_r, \ldots, G_{k+1})$ or r-dimensional size and the higher coordinates $(G_r, x_{r+1}, \ldots, x_{k+1})$. These representations hold when the size sequence is regular, but need not be valid when it is not.

INDEPENDENCE OF SHAPE AND SIZE; ISOMETRY AND NEUTRALITY

Let $\mathbf{X}_k$ be a k-dimensional positive random vector. It follows from the preceding discussion that if some random shape vector, say $\mathbf{Z} = \mathbf{X}_k/G(\mathbf{X}_k)$, is statistically independent of a random variable, say H, then every random shape vector [including, say, $\mathbf{U} = \mathbf{X}_k/D_k(\mathbf{X}_k)$] is independent of H. Thus it is possible to speak unambiguously of the independence of shape and any random H. (This implies that the probability distribution of the rays does not change given different values of H.) When H is a size variable, so that shape is independent of size, then one has *isometry with respect to* H.

Because the regular size ratios of $\mathbf{X}_k$ may be used to represent the rays as well as any shape vector, isometry with respect to H occurs if and only if the size ratio vector for every regular size sequence is independent of H. Also, because the lower size ratios represent lower-dimensional shape (e.g., G_2/G_1, two-dimensional shape) then H independent of k-dimensional shape implies H independent of $(k-1)$-dimensional shape, and hence H must thereby be independent of the shape defined by any subset of the coordinates of $\mathbf{X}_k$, without regard to their order within $\mathbf{X}_k$.

Consider $\mathbf{X}_{k+1}$ of $k+1$ coordinates and denote its first r coordinates by $\mathbf{X}_r$. Then the size ratios $(G_2/G_1, \ldots, G_r/G_{r-1})$ represent the r-dimensional shape of $\mathbf{X}_r$. The condition that r-dimensional shape be independent of the vector of higher-dimensional size variables $(G_r, \ldots, G_{k+1})$ is *isometry with respect to* $G_r, \ldots, G_{k+1}$. The weaker condition that r-dimensional shape be independent of the higher-dimensional size ratios $(G_{r+1}/G_r, \ldots, G_{k+1}/G_k)$ is *neutrality with respect to* $G_r, \ldots, G_{k+1}$. Isometry with respect to $G_r, \ldots, G_{k+1}$ implies neutrality with respect to $G_r, \ldots, G_{k+1}$.

Complete isometry with respect to $G_2, \ldots, G_{k+1}$ occurs when r-dimensional shape is independent of the vector of higher size variables $(G_r, \ldots, G_{k+1})$, for $r = 2, \ldots, k+1$. Complete isometry is equivalent to the mutual independence of the k size ratios and size: $G_2/G_1, \ldots, G_{k+1}/G_k, G_{k+1}$. Correspondingly *complete neutrality* is equivalent to the mutual independence of the size ratios $G_2/G_1, \ldots, G_{k+1}/G_k$. The concept of neutrality is scale-free, and applicable to shape vectors alone. Nonetheless it is a concept applicable to any positive random vector, including those containing scale information. Thus the size ratios are the same, whether determined directly from $\mathbf{X}_{k+1}$ or from any of its shape vectors.

A pivotal result, (given in ref. 38, with proof completed in ref. 45) can be simply stated: The shape of $\mathbf{X}_k$ can be statistically independent of at most one size variable $G(\mathbf{X}_k)$. Thus if some shape vector $\mathbf{Z}(\mathbf{X}_k)$ is independent of a size variable $G(\mathbf{X}_k)$, then shape cannot be independent of any other size variable $H(\mathbf{X}_k)$. The only exception to this sweeping statement is virtually no exception at all; namely, shape is also independent of any other size variable that, with probability 1, is a scalar multiple of $G(\mathbf{X}_k)$.

A comparable result holds for related size ratios. As before, let $\mathbf{X}_k$ denote the initial k coordinates of $\mathbf{X}_{k+1}$, and suppose that G_k, G_{k+1} and H_k, H_{k+1} are related pairs. If k-dimensional shape is independent of the ratio $G_{k+1}(\mathbf{X}_{k+1})/G_k(\mathbf{X}_k)$, then it cannot be independent of the ratio $H_{k+1}(\mathbf{X}_{k+1})/H_k(\mathbf{X}_k)$ [40]. Again the only exception is when G_k is a scalar multiple of H_k with probability 1.

ISOMETRY AND NEUTRALITY CHARACTERIZE DISTRIBUTIONS

Specification of a size variable or of a regular sequence of size variables is thus necessarily important in questions of isometry or neutrality, in striking contrast to the representation of shape, which, for questions of independence, may be specified by any shape vector or vector of "regular size ratios"; that is, size ratios from a regular sequence.

Because the independence of shape and size is such a strong condition, it is not surprising that the additional condition of mutual independence of the coordinate variables themselves results in a specific distributional form. Consider the characterization of the gamma distribution* given by Lukacs [34]: Suppose (1) that X_1 and X_2 are independent positive (nondegenerate) random variables. Suppose also (2) that X_1/X_2 is independent of $X_1 + X_2$; then X_1 and X_2 each have a gamma distribution with the same scale parameter λ. Condition (2) above is simply the independence of two-dimensional shape [say $(X_1/X_2, 1)$] and size $S_2 = X_1 + X_2$.

The result proved by Lukacs for two dimensions holds also in k dimensions. Thus (1) the mutual independence of positive and nondegenerate $X_1, \ldots, X_k$ along with (2) the independence of k-dimensional shape and S_k implies that each X_i has a gamma distribution with the same scale parameter [37]. If one has mutually independent, not necessarily identically distributed, positive variables, and shape independent of M_k, then each coordinate variable must follow a lognormal distribution*. If size is D_k, then each coordinate must have a generalized gamma distribution. For these and other results see Mosimann [38] and James [25]; see also Ferguson [17]. James identified the class of size variables for which both (1) shape may be independent of size, and simultaneously (2) the original coordinates may be mutually independent.

Neutrality properties can also be used to characterize distributions. Thus suppose $\mathbf{X}_{k+1}$ is constrained so that $S_{k+1}(\mathbf{X}_{k+1}) = 1$ with probability 1, and that $\mathbf{X}_{k+1}$ is completely neutral with respect to $S_2, \ldots, S_{k+1}$. Then consider a vector whose coordinate variables are simply those of $\mathbf{X}_{k+1}$ in reverse order. If the latter is completely neutral with respect to $S_2, \ldots, S_{k+1}$, then $\mathbf{X}_{k+1}$ must follow a Dirichlet distribution*, the characterizations of which by Darroch and Ratcliffe [14], Fabius [15], and James and Mosimann [27] all concern neutrality properties with respect to the additive sequence $S_1, \ldots, S_{k+1}$ for $\mathbf{X}_{k+1}$ along with at least one of its permutations. A subfamily of the multivariate lognormal family is similarly characterized by neutrality properties with respect to the multiplicative sequence $M_1, \ldots, M_{k+1}$ [27, 41].

A concept closely related to additive neutrality is that of F-independence; see Darroch and James [12]. A recent discussion of additive neutrality is given in ref. 26.

Briefly then, whether the positive measurements are independent or not, there are no probability distributions on R_+^k for which two truly distinct size variables are both independent of shape. The specification of a size variable is crucial; even when working with scale-free shape vectors alone, and forming concepts of independence based on size ratios alone, the choice of a size sequence is crucial. Independence of the rays in the lower-dimensional space, and a size ratio that determines the rays in the next higher-dimensional space, can only occur with respect to at most one related pair of size variables.

DIVERSITY OF HYPOTHESES FOR ISOMETRY (OR NEUTRALITY)

Let $\mathbf{X}_k$ be a k-dimensional positive random vector, and $\mathbf{Y}_k$ the random vector whose coordinates are, respectively, logarithms of the coordinates of $\mathbf{X}_k$. Suppose that $\mathbf{Y}_k$ follows a multivariate normal distribution with covariance matrix Σ and mean vector μ. Then $\mathbf{X}_k$ follows a multivariate lognormal distribution with parameters Σ and μ, and hypotheses of isometry can be discussed through examination of Σ and its submatrices. Let Σ_r denote the covariance matrix of the initial r coordinates of $\mathbf{Y}_k$, so that Σ_r comprises the upper left $r \times r$ elements of Σ. The following statements are all equivalent:

(M1) r-dimensional shape, say $(X_1/X_r, \ldots, X_{r-1}/X_r, 1)$, is independent of the r-dimensional geometric mean $M_r = (X_1 \cdots X_r)^{1/r}$.

(M2) $(Y_1 - Y_r, \ldots, Y_{r-1} - Y_r)$ is independent of $T_r = Y_1 + \cdots + Y_r$.

(M3) $\text{cov}(Y_i - Y_r, T_r) = 0$, $i = 1$ to $r - 1$.

(M4) $\text{cov}(Y_i, T_r) = \text{cov}(Y_r, T_r)$, $i = 1$ to $r - 1$.

(M5) The column totals of Σ_r are the same.

(M6) $(1, \ldots, 1)$ is an eigenvector of Σ_r.

Statement **(M5)** shows that the independence of r shape and the geometric mean M_r is equivalent to the columns of Σ_r having the same column totals. Since the column totals of Σ_r may be the same without those of Σ_{r+1} being the same, r-dimensional shape may be independent of M_r without affecting the independence or lack thereof of $(r+1)$-dimensional shape and M_{r+1}. Thus complete isometry with respect to $M_2, \ldots, M_k$ occurs if the column totals of Σ_r are the same for $r = 2, \ldots, k$.

The hypotheses above are for the specific order represented by $\mathbf{X}_k$. If the coordinates of $\mathbf{X}_k$ are permuted then isometry with respect to $M_2, \ldots, M_k$ represents different hypotheses. The situation is complex indeed.

If instead of multiplicative isometry with respect to $M_2, \ldots, M_k$ we consider coordinate isometry (with respect to $C_2, \ldots, C_k$), then the following statements are equivalent:

(C1) r-dimensional shape, say $(X_1/X_r, \ldots, X_{r-1}/X_r, 1)$, is independent of the rth coordinate $C_r = X_r$.

(C2) $(Y_1 - Y_r, \ldots, Y_{r-1} - Y_r)$ is independent of Y_r.

(C3) $\text{cov}(Y_i - Y_r, Y_r) = 0$, $i = 1, \ldots, r - 1$.

(C4) $\text{cov}(Y_i, Y_r) = \text{var}(Y_r)$, $i = 1$ to $r - 1$.

As a consequence, if for any pair (X_i, X_j), $\text{cov}(Y_i, Y_j) = \text{var}(Y_i)$, then two-dimensional shape $(X_j/X_i, 1)$ is independent of X_i. [If both **(M1)** and **(C1)** are true, then the linear combination

$$Y_r - \{1/(r-1)\}(Y_1 + \cdots + Y_{r-1})$$

has variance zero, so that X_r/M_r is degenerate as required.]

It is thus clear that isometry and neutrality hypotheses for loglinear size variables are readily framed within the multivariate lognormal context. For example the appropriate multiple correlation* coefficient can be used to test such hypotheses, as well as to measure the degree of association* of size with shape [40, 43, 46].

Unfortunately neither additive isometry nor neutrality (with respect to $S_2, \ldots, S_k$) can occur within the multivariate lognormal family, so that such hypotheses are not readily studied in this framework [41]. A related result is that for every nonsingular member of the multivariate lognormal family, there exists a unique loglinear size variable that is independent of shape [52].

Aitchison [2] has studied the (closely related) additional concept of "subcompositional independence" by assumption of multivariate lognormality for a variety of log ratios, which can be expressed as log ratios of additive size variables. The term *additive logistic distribution* is the distribution of proportions $\mathbf{X}_k/S_k$ when the ratios $\mathbf{X}_k/X_k$ follow a multivariate lognormal distribution, while the *multiplicative logistic distribution* is the distribution of proportions when logs of functions of additive size ratios of the permuted vector $(X_k, \ldots, X_2, X_1)$ follow a multivariate normal distribution.

PRACTICAL CONSIDERATIONS

In any statistical distribution, k-dimensional shape can be independent of at most one size variable or related size ratio, but in the sample, shape may appear to be independent of more than one size variable. Thus suppose that all positive probability occurs only on a single ray from the origin in R_{+}^{k}. Then shape is constant, and thereby independent of all size variables. (In fact, any size variable must be a scalar multiple of any other with probability 1.) If now we perturb the situation by letting the positive probability occur in a long very narrow ellipse, say, with the ray as its major axis, then shape will be "nearly" independent of a variety of size variables, and there would be little prospect of detection of the nonindependence in moderate-sized samples. In the sample, shape could appear to be independent of more than one size variable.

Tests of independence in practice are often tests of zero covariance. Thus while shape cannot be independent of S_k in the multivariate lognormal family, if the logs $Y_1, \ldots, Y_k$ are independently and identically distributed, each $P_i = X_i/S_k$ is uncorrelated with S_k.

Such a distinction is unnecessary when a multivariate lognormal assumption is made for various size ratios of **X** or for **X** itself. The most successful analyses involving size and shape are those in which the range of methods based on an underlying multivariate or univariate normal distribution can be applied; see Aitchison [1, 2] and Aitchison and Shen [4].

Prior to 1970 there was no emphasis on the need to specify a particular size variable for study. Jolicoeur [28] proposed that the major axis of the ellipse of concentration of the log measurements be used to represent the allometric relationship (cf. Gould [20, 21], Huxley [24], Reeve and Huxley [49], Griffiths and Sandland [22], and Jolicoeur [29]). He defines isometry to be the condition that this axis be the equiangular line. From statement (**M6**), this is completely consistent with the results here, but it does implicitly select M_k, the geometric mean of the measurements, as size. Hopkins [23] and Sprent [57, 58] present models in which the observed covariance matrix is the sum of a "model" or structural matrix plus an error matrix. Again the definitions here are applicable, but to the structural portion of the model. The presence of an error matrix often destroys the ability to test certain isometric hypotheses, but Mosimann et al. [46] were able to study some isometric hypotheses in the presence of error. (See also Jolicoeur and Heusner [30].)

THE CHOICE OF A SIZE VARIABLE

The choice of a size variable should be specific for a given study, and will ideally be based on underlying processes thought to generate the random quantities under analysis. In studies of random proportions or ratios it seems particularly important to think about the processes that generate the measurements being studied. Methods for studying such measurements will likely differ, based on the underlying model, which may dictate which size variables are appropriate. The examples that follow are illustrative of the possibilities, and are by no means definitive.

Let the random vector (L, W, H) represent the length, width, and height, respectively, of the carapace, or shell, of turtles of a given species. The three turtle measurements are perpendicular to each other and thus $M_3 = (LWH)^{(1/3)}$ may be a reasonable measurement of "volume" (how reasonable, for three distinct species, was studied empirically by Mosimann [36]). The hypothesis that three-dimensional shape is independent of volume M_3 is distinct from the hypothesis that two-dimensional shape $(L/W, 1)$ is independent of "transverse sectional area" $M_2 = (LW)^{(1/2)}$. Rearranging the order of the measurements changes the two-dimensional space represented by the first two coordinates. Thus for (W, H, L), the size variable M_2 becomes $(WH)^{(1/2)}$, or "cross-sectional area." It makes a difference. In the map turtle, two-dimensional shape $(W/H, 1)$ is isometric with respect to cross-sectional area, $(WH)^{(1/2)}$, but, since length grows faster than height or width, two-dimensional shape

$(W/L, 1)$ is not isometric with respect to $(LW)^{(1/2)}$ [36; 43, p. 447].

On the other hand, in the study of proportions of various mid-line measurements of scales (or scutes) of turtles [35; 11, p. 204], all measurements are taken along a single physical dimension, and a natural size variable is the sum S_k.

In another context, let $\mathbf{T}' = (T_1, \dots, T_k)$ represent the time intervals between successive events in a Poisson process*. Here there is a single time dimension, and S_k is a natural size variable. (The T_i's are independent exponential* random variables and exhibit complete isometry with respect to $S_2, \dots, S_k$ for every permutation of $\mathbf{T}$.)

In a study of geographic variation in blackbirds, Mosimann and James [43] used standard measurements available from museum "study-skins" to examine geographic variation in bill and wing–tail proportions and size. Various size variables were studied, and a coordinate size variable, wing length, exhibited the highest between/within ratio of variance components. One advantage of considering a separate definition of shape vector was that it was possible to relate shape to a measurement of size (weight) apart from the linear measurements themselves. Jungers [32], in studying limb proportions of primates, gives reasons for choosing weight as size. In the same volume, Ford and Corruccini [18] discuss practical difficulties that arise with this choice. Baron and Jolicoeur [5] use body weight as a size variable against which the relative volumes of various portions of the brain are studied in comparing species of bats and insectivorous mammals.

Veitch [60] thoughtfully uses a multivariate linear model to study changes in the proportions of claws and appendages of two species of fiddler crabs. The size variable chosen is the geometric mean of carapace length and breadth. Reyment [50] compared relations of three-dimensional shape with the geometric mean of height, breadth, and diameter, as well as diameter alone, in three species of fossil ammonites. Fordham and Bell [19] consider the behaviour of the geometric mean size variable with certain principal components in an artificial data set. Sampson and Siegel [52] define a "residual" size variable, with an interesting interpretation of the allometry* of antlers in the Irish elk.

Seebeck [53, 54] has used linear models in the analysis of size and shape. His work has focused on growth and body composition in cattle and other livestock. In ref. 53, a variety of covariance analyses of log shape show that, when size is taken into account, faster growing animals have a lower yield of carcass. Seebeck's studies exploit, in an interesting way, the use of loglinear size variables embedded in experimental designs permitting the application of the general linear model*.

Darroch and Mosimann [13] present canonical discriminant (and principal component) analyses for log shape that are invariant when applied to any log shape vector formed from a loglinear size variable. The component analyses are the same as those presented for composition data by Aitchison [3], based on a different argument.

The choice of a suitable size variable for chemical composition data seems important. For such data both Connor and Mosimann [11] and Aitchison [2] implicitly chose the sum S_k. Whether such a choice is a suitable one may vary, depending on the processes generating the data. A point worth noting is that unequal changes of the scale of measurement affect additive isometry and neutrality. Thus if chemical compositions by weight are additively neutral (or isometric), then the comparable compositions by volume cannot be so [45].

Unequal changes of scale are important in other analyses involving proportions. Differential production of pollen by different kinds of plants may affect interpretation of fossil pollen diagrams [16, 39, 42]. In such studies a weighted sum (with positive weights adding to 1) may be as appropriate for size as the sum S_k.

Finally, investigators in different scientific fields may habitually normalize their data in different ways; for example, as proportions

$\mathbf{X}/S_k$ or ratios $\mathbf{X}/X_k$. It would be unwise to assume without reflection that the particular size used for the normalization is relevant. The underlying processes may be such that any of a variety of size variables, and their related size ratios, could be the random quantities pertinent to the scientific investigation.

RELATIONSHIPS TO OTHER CONCEPTS OF SIZE AND SHAPE

The terms *size* and *shape* are often used in a distinct, but closely related, sense in the statistical literature. If in discriminant or principal component analysis one linear combination of the coordinate variables is identified as a size variable, then other variables orthogonal to size may be designated as shape variables. By construction the shape variables are uncorrelated with size, and cannot be used to study the association of size with shape (Rao [48], Jolicoeur and Mosimann [31], many examples in Reyment et al. [51], and Burnaby [10]). In a typical situation the investigator wishes to classify organisms into one of several populations or species, and to classify small and large individuals from the same species correctly. Hence the shape variables are chosen for the purpose of classification, and should not depend on size. The need for such variables stems from the lack of isometry in the first place; although useful for classification, they generally contain a mixture of scale and scale-free information.

Penrose [47] and Spielman [56] have partitioned the squared distances represented by Pearson's coefficient of racial likeness and Mahalanobis' D^2 into size and shape components, respectively. Mosimann and Malley [44] show that Penrose's partition is based on the size variable S_k, while that of Spielman is based on D_k. In both cases the shape component may reflect scale information [56, 44].

Reyment et al. [51] present a particularly informative discussion of a host of applications of multivariate methods in morphometry, including uses of the concept of size and shape variables. The work of Blum [6], Siegel and Benson [55], and much of the work of Bookstein and collaborators [7–9] is relevant to matching points and choosing measurements that are most suitable for description of shape as an outline. Important points relevant to pattern recognition* are revealed in such studies, particularly with respect to meaningful measurements to describe the deformation of one outline to another. Whatever measurements best define such deformation, if the choice results in a vector of positive random variables, whose coordinates are potentially expressible in the same scale, then the concepts of size and shape presented here will be relevant. In this article, the vector of measurements is taken as given, and may or may not be associated with outlines in the physical world. The methods are applicable to discrete and continuous measurements.

In the entry SHAPE STATISTICS, D. G. Kendall describes models for generating random polygonal figures in the plane. Consider triangles; after translation, scaling, and rotation, Kendall defines a space in which each triangle shape corresponds to a single point. He treates the cases of labeled and unlabeled vertices. Labeled vertices are identified externally to the analysis so that permutations of the labels are not allowed. Unlabeled vertices are not so identified and, in effect, all permutations of "labels" are permitted. In a typical application with triangles in morphometrics, the vertices and the distances between them are labeled externally to the analysis and therefore may not lose their identities under permutations. For example, suppose three points of an animal are observed from the lateral aspect (tip of snout, posterior of eye, rear of jaw) with the corresponding distances (snout-jaw, jaw-eye, eye-snout). The distances $(3, 4, 5)$ and $(5, 3, 4)$ have distinct appearances and distinct shape vectors, but would correspond to the same point in a shape space for unlabeled vertices.

Models may be constructed with labeled vertices whose placement is restricted so that permutations are not relevant. (In such cases shape vectors correspond to shapes of both labeled and unlabeled triangles.) An example of unlabeled vertices is that of triangles formed by the positions of three similar stones in a field. Kendall's methods represent a major development for understanding randomly generated polygonal shapes.

References

[1] Aitchison, J. (1981). In *Statistical Distributions in Scientific Work*, Vol. 4, C. Taillie, G. P. Patil, and B. A. Baldessari, eds. Reidel, Dordrecht, The Netherlands, pp. 147–156.

[2] Aitchison, J. (1982). *J. R. Statist. Soc. Ser. B*, **44**, 139–177.

[3] Aitchison, J. (1983). *Biometrika*, **70**, 57–65.

[4] Aitchison, J. and Shen, S. M. (1980). *Biometrika*, **67**, 261–272.

[5] Baron, G. and Jolicoeur, P. (1980). *Evolution*, **34**, 386–393.

[6] Blum, H. (1973). *J. Theor. Biol.*, **38**, 205–287.

[7] Bookstein, F. L. (1978). The Measurement of Biological Shape and Shape Change. *Lect. Notes in Biomath.*, **24**. Springer-Verlag, Berlin, Germany.

[8] Bookstein, F. L. (1984). *J. Theor. Biol.*, **107**, 475–520.

[9] Bookstein, F., Chernoff, B., Elder, R., Humphries, J., Smith, G., and Strauss, R. (1985). *Morphometrics in Evolutionary Biology*, Special Publication 15. The Academy of Natural Sciences of Philadelphia, Philadelphia, PA.

[10] Burnaby, T. P. (1966). *Biometrics*, **22**, 96–110.

[11] Connor, R. J. and Mosimann, J. E. (1969). *J. Amer. Statist. Ass.*, **64**, 194–206.

[12] Darroch, J. N. and James, I. R. (1974). *J. R. Statist. Soc. Ser. B*, **36**, 467–483.

[13] Darroch, J. N. and Mosimann, J. E. (1985). *Biometrika*, **72**, 241–252.

[14] Darroch, J. N. and Ratcliff, D. (1971). *J. Amer. Statist. Ass.*, **66**, 641–643.

[15] Fabius, J. (1973). *Ann. Statist.*, **1**, 583–587; Correction, **9**, 234 (1981).

[16] Fagerlind, F. (1952). *Botaniska Notiser*, **2**, 185–224.

[17] Ferguson, T. S. (1962). *Ann. Statist.*, **33**, 986–1001.

[18] Ford, S. M. and Corruccini, R. S. (1985). In *Size and Scaling in Primate Biology*, W. L. Jungers, ed. pp. 401–435.

[19] Fordham, B. G. and Bell, G. D. (1978). *Math. Geol.*, **10**, 111–139.

[20] Gould, S. J. (1966). *Biol. Rev.*, **41**, 587–640.

[21] Gould, S. J. (1977). *Ontogeny and Phylogeny*. The Belknap Press of Harvard University Press, Cambridge, MA.

[22] Griffiths, D. A. and Sandland, R. L. (1982). *Growth*, **46**, 1–11.

[23] Hopkins, J. W. (1966). *Biometrics*, **22**, 747–760.

[24] Huxley, J. S. (1932). *Problems of Relative Growth*. The Dial Press, New York.

[25] James, I. R. (1979). *Ann. Statist.*, **7**, 869–881.

[26] James, I. R. (1981). In *Statistical Distributions in Scientific Work*, Vol. 4, C. Taillie, G. P. Patil, and B. A. Baldessari, eds. Reidel, Dordrecht, The Netherlands, pp. 125–136.

[27] James, I. R. and Mosimann, J. E. (1980). *Ann. Statist.*, **8**, 183–189.

[28] Jolicoeur, P. (1963). *Biometrics*, **19**, 197–499.

[29] Jolicoeur, P. (1984). *Biometrics*, **40**, 685–690.

[30] Jolicoeur, P. and Heusner, A. A. (1971). *Biometrics*, **27**, 841–855.

[31] Jolicoeur, P. and Mosimann, J. E. (1960). *Growth*, **24**, 339–354.

[32] Jungers, W. L. (1985). In *Size and Scaling in Primate Biology*. W. L. Jungers, ed. pp. 401–435.

[33] Kapur, B. D. and Patil, G. P. (1979). *J. Indian Statist. Ass.*, **17**, 69–82.

[34] Lukacs, E. (1955). *Ann. Math. Statist.*, **26**, 319–324.

[35] Mosimann, J. E. (1956). *Miscellaneous Publications of the Museum of Zoology*, No. 97. University of Michigan, Ann Arbor, MI, pp. 1–43.

[36] Mosimann, J. E. (1958). *Rev. Canad. Biol.*, **17**, 137–228.

[37] Mosimann, J. E. (1962). *Biometrika*, **49**, 66–82.

[38] Mosimann, J. E. (1970). *J. Amer. Statist. Ass.*, **65**, 930–945.

[39] Mosimann, J. E. (1970). In *Random Counts in Scientific Work*, Vol. 3, G. P. Patil, ed. The Pennsylvania State University Press, University Park, PA, pp. 1–30.

[40] Mosimann, J. E. (1975). In *Statistical Distributions in Scientific Work*, Vol. 2, G. P. Patil, S. Kotz, and J. K. Ord, eds. Reidel, Dordrecht, The Netherlands, pp. 187–217.

[41] Mosimann, J. E. (1975). In *Statistical Distributions in Scientific Work*, Vol. 2, G. P. Patil, S. Kotz, and J. K. Ord, eds. Reidel, Dordrecht, The Netherlands, pp. 219–239.

[42] Mosimann, J. E. and Greenstreet, R. L. (1971). In *Statistical Ecology*, Vol. 1, G. P. Patil, E. C. Pielou, and W. E. Waters, eds. The Pennsylvania State University Press, University Park, PA, pp. 23–58.

[43] Mosimann, J. E. and James, F. C. (1979). *Evolution*, **33**, 444–459.

[44] Mosimann, J. E. and Malley, J. D. (1979). In *Statistical Ecology*, Vol. 7, L. Orloci, C. R. Rao, and W. M. Stiteler, eds. International Cooperative Publishing House, Fairland, MD, pp. 175–189.

[45] Mosimann, J. E. and Malley, J. D. (1981). In *Statistical Distributions in Scientific Work*, Vol. 4, C. Taillie, G. P. Patil, and B. A. Baldessari, eds. Reidel, Dordrecht, The Netherlands, pp. 137–145.

[46] Mosimann, J. E., Malley, J. D., Cheever, A. W., and Clark, C. B. (1978). *Biometrics*, **34**, 341–356.

[47] Penrose, L. S. (1954). *Ann. Eugen.*, **18**, 337–343.

[48] Rao, C. R. (1964). *Sanhkyā A*, **26**, 329–358.

[49] Reeve, E. C. R. and Huxley, J. S. (1945). In *Essays on Growth and Form*, W. E. Le Gros Clark and P. B. Medawar, eds. Oxford University Press, Oxford, England, pp. 121–156.

[50] Reyment, R. A. (1982). *Stockholm Contrib. Geol.*, **37**, 201–214.

[51] Reyment, R. A., Blackith, R. E., and Campbell, N. A. (1984). *Multivariate Morphometrics*, 2nd ed. Academic, New York.

[52] Sampson, P. D. and Siegel, A. F. (1985). *J. Amer. Statist. Ass.*, **80**, 910–914.

[53] Seebeck, R. M. (1983). *Animal Production*, **37**, 53–66.

[54] Seebeck, R. M. (1983). *Animal Production*, **37**, 321–328.

[55] Siegel, A. F. and Benson, R. H. (1982). *Biometrics*, **38**, 341–350.

[56] Spielman, R. S. (1973). *Amer. Naturalist*, **107**, 694–708.

[57] Sprent, P. (1968). *Biometrics*, **24**, 639–656.

[58] Sprent, P. (1972). *Biometrics*, **28**, 23–27.

[59] Tate, R. F. (1959). *Ann. Math. Statist.*, **30**, 341–366.

[60] Veitch, L. G. (1978). *Math. Scientist*, **3**, 35–45.

(ALLOMETRY
SPATIAL DATA ANALYSIS
SHAPE STATISTICS
STEREOLOGY)

JAMES E. MOSIMANN

SKANDINAVISK AKTUARIETIDSKRIFT *See* SCANDINAVIAN JOURNAL OF STATISTICS

SKEW-NORMAL DISTRIBUTIONS

These distributions are given by the density

$$f(z;\lambda) = 2\phi(z)\Phi(\lambda z), \; -\infty < z < \infty, \tag{1}$$

where ϕ and Φ are the standard normal density and distribution functions, respectively. For $\lambda = 0$, (1) reduces to the standard normal density and as $\lambda \to \infty$, $f(z,\lambda)$ approaches the half-normal* density; it is strongly unimodal for all λ.

This density was introduced by O'Hagan and Leonard [2] as a prior density for estimating normal location parameters and was investigated in detail by Azzalini (1985), who uses the notation SN(λ) and proposes various generalizations as well as some multivariate extensions. (It is interesting to note that if a random variable Z possesses either a standard normal or a SN(λ) density, then Z^2 is distributed as a chi square* with 1 degree of freedom.)

References

[1] Azzalini, A. (1985). *Scand. J. Statist.*, **12**, 171–718. (Extension in *Statistica*, **46** (1986), 199–208.)

[2] O'Hagan, A. and Leonard, T. (1976). *Biometrika*, **63**, 201–202.

(FOLDED DISTRIBUTIONS)

SKEW REGRESSION *See* NONLINEAR REGRESSION

SKEW-SYMMETRIC MATRIX

A square $m \times m$ matrix $\mathbf{A}$ is skew-symmetric if

$$\mathbf{A} = -\mathbf{A}',$$

where $\mathbf{A}'$ is the transpose of $\mathbf{A}$. The diagonal elements of a skew-symmetric matrix are zero, and its rank* is an even number.

Let $\mathbf{C}$ be an $m \times m$ matrix. Then the quadratic form* $\mathbf{x}'\mathbf{C}\mathbf{x} \equiv 0$ (i.e., for all $m \times 1$

real-valued vectors $\boldsymbol{x}$) if and only if $\mathbf{C}$ is skew-symmetric [1].

Reference

[1] Graybill, F. A. (1969). *Introduction to Matrices with Applications in Statistics*. Wadsworth, Belmont, CA, p. 336.

SKIP-LOT SAMPLING PLANS

These are acceptance sampling* plans of a type introduced by Dodge [1]. The basic idea is that when an acceptable level of quality is established, only a fraction (f, say) of lots are subjected to inspection*. Establishment of an acceptable level of quality is achieved by acceptance of a specified number (m, say) of consecutive lots. The quantities f and m are the *parameters* of the plan. Reduced inspection is continued until a lot is rejected. Inspection of every lot is then resumed.

An attractive feature of skip-lot sampling plans is the reduction in the expected amount of inspection. This has to be set against the possibility of missing occasional lots of poor quality. There is a full discussion of theoretical and practical aspects of skip-lot sampling in Stephens [3].

There are, of course, many possible variants of the basic plan described above. Two or more fractions ($f_1 > f_2 \cdots$) for reduced inspection may be used, for example, with a lower fraction being attained after satisfactory performance for a period with a higher fraction. (It is desirable to return immediately to sampling every lot if a lot is rejected, whatever the fraction of lots currently being inspected.) Also, provision may be made for taking a very small sample (even of size 1!) rather than completely omitting inspection of any lot (see, e.g., Parker and Kessler [2]).

References

[1] Dodge, H. F. (1955). *Ind. Qual. Control*, **11** (5), 3–5.

[2] Parker, R. D. and Kessler, L. (1981). *J. Qual. Technol.*, **13**, 31–35.

[3] Stephens, K. S. (1982). How to Perform Skip-Lot and Chain Sampling, *ASQC Basic References in Quality Control: Statistical Techniques*, Vol. 4, American Society for Quality Control, Milwaukee, WI.

(ACCEPTANCE SAMPLING
INSPECTION SAMPLING
QUALITY CONTROL, STATISTICAL)

SKIPPING

This notion is owing to Tukey [1] and is related to formation of adaptive L estimators* (linear order statistics estimates) of location. Let h_1 and h_2 be the hinges (the first and third sample quartiles, respectively). Consider quantities of the form

$$c_1 = h_1 + \eta(h_2 - h_1),$$

$$c_2 = h_1 - \eta(h_2 - h_1),$$

defined for a prescribed η (typically 1, 1.5, or 2). The *skipping* process involves deletion of observations in the tails of the sample [outside the interval (c_1, c_2)] before calculating the trimean* of the *retained* observations. *Iterative skipping* consists of repeating the process with recalculated hinges at each stage until the retained data set remains constant. *Multiple skipping* consists of repeating the skipping process applied to the *retained* data set with difference choices of η at each stage. See Andrews et al. [1] for some numerical examples.

References

[1] Andrews, D. F., Bickel, P. J., Hampel, F. R., Huber, P. J., Rogers, W. H., and Tukey, J. W. (1972). *Robust Estimates of Location: Survey and Advances*. Princeton University Press, Princeton, NJ.

[2] Tukey, J. W. (1977). *Exploratory Data Analysis*. Addison-Wesley, Reading, MA.

(EXPLORATORY DATA ANALYSIS
FIVE-NUMBER SUMMARIES
SLICING)

SKITOVITCH–DARMOIS THEOREM *See* DARMOIS–SKITOVITCH THEOREM

SKOROHOD CONSTRUCTION *See* HUNGARIAN CONSTRUCTIONS OF EMPIRICAL PROCESSES; PROCESSES, EMPIRICAL

SKOROHOD EMBEDDINGS

Let $\mathscr{L}_t$ be the sigma-field generated by Brownian motion* up to time t. To embed a given process $\{X(t), t \geqslant 0\}$ in Brownian motion means to construct a standard Brownian motion $\{B(t), t \geqslant 0\}$ and a nondecreasing family of $\mathscr{L}_t$-measurable stopping times T_t such that the processes $\{B(T_t), t \geqslant 0\}$ and $\{X(t), t \geqslant 0\}$ have the same (probability) laws. For the partial sum process of independent Bernoulli trials, a first result in this direction was proved by Knight [6, p. 226, Remark (a)]. In the general case of sums or independent identically distributed random variables with mean zero and finite variance such an embedding theorem for the partial sum process was first proved by Skorohod [12]; hence the term Skorohod embedding. Skorohod's theorem had a profound impact on the development of limit theorems* in probability theory. Very readable presentations are given in Freedman [4] and Breiman [1]. Soon after the publication of Skorohod's book, it was realized by Freedman and Strassen that the independence and identical distribution could be replaced by a martingale difference structure (Strassen [14]). Other constructions of the stopping times were given by Dubins [3], Root [10] and others. The latest versions of the Skorohod embedding theorem for discrete parameter martingales* are due to Hall and Heyde [5] and Scott and Huggins [11]. The Hall and Heyde version is as follows.

Let $\{X_n, \mathscr{F}_n, n \geqslant 1\}$ be a square-integrable martingale difference sequence. Then there exists a probability space supporting a standard Brownian motion $\{B(t), t \geqslant 0\}$ and a sequence $\{t_n, n \geqslant 1\}$ of nonnegative random variables with the following properties. Set

$$T_n := \sum_{j=1}^{n} t_j, \qquad n \geqslant 1, T_0 = 0,$$

$$Y_n := B(T_n) - B(T_{n-1}), n \geqslant 1,$$

and let $\mathscr{G}_n$ be the sigma-field generated by $Y_1, \ldots, Y_n$ and $B(t)$ for $0 \leqslant t \leqslant T_n$. Then

(i) $\{X_n, n \geqslant 1\}$ and $\{Y_n, n \geqslant 1\}$ have the same laws;

(ii) T_n is $\mathscr{G}_n$-measurable;

(iii) $E(t_n|\mathscr{G}_{n-1}) = E(Y_n^2|\mathscr{G}_{n-1})$, $n \geqslant 1$ a.s.;

(iv) for each real $r \geqslant 1$,

$$E(t_n^r|\mathscr{G}_{n-1}) \leqslant C_r E(|Y_n|^{2r}|\mathscr{G}_{n-1}) = C_r E(|Y_n|^{2r}|Y_1, \ldots, Y_{n-1}) \quad \text{a.s.},$$

where

$$C_r = 2(8/\pi^2)^{r-1}\Gamma(r+1).$$

The Scott and Huggins version leaves the constants C_r undetermined, but replaces the sigma-fields $\mathscr{G}_{n-1}$ on the right-hand sides of (iii) and (iv) by $\mathscr{F}_{n-1}$, a definite advantage in some applications.

To demonstrate the power and utility of such results, we return to the original Skorohod embedding theorem, which treats the partial sum process of independent identically distributed random variables with mean 0 and variance 1. Then $\{t_n, n \geqslant 1\}$ is also a sequence of independent identically distributed random variables with $Et_n = 1, n \geqslant 1$. Hence by the Kolmogorov strong law of large numbers $n^{-1}T_n \to 1$ a.s. This relation fed into $S_n = B(T_n)$ yields $S_n - B(n) = ((n \log\log n)^{1/2})$ a.s., and this is Strassen's [13] almost sure invariance principle* for the law of the iterated logarithm*.

For continuous parameter martingales, Skorohod embeddings were established independently by Dambis [2] and Kunita and Watanabe [8]. The restriction of Kunita and Watanabe that the martingale have no intervals of constancy was later removed by Knight [7]. Monroe [9] identified all processes that can be embedded in Brownian motion as the class of local martingales.

Although special $\mathbb{R}^d$-valued martingales have been considered in Kunita and Watanabe [8] and Knight [7], the Skorohod embedding theorem applies in essence only to real-valued processes.

References

[1] Breiman, L. (1968). *Probability*. Addison-Wesley, Reading, MA.

[2] Dambis, K. E. (1965). On the decomposition of continuous sub-martingales. *Theory Prob. Appl.*, **10**, 401–410

[3] Dubins, L. (1968). On a theorem of Skorohod. *Ann. Math. Statist.*, **39**, 2094–2097.

[4] Freedman, D. (1971). *Brownian Motion and Diffusion*. Holden-Day, San Francisco, CA.

[5] Hall, O. and Heyde, C. C. (1980). *Martingale Limit Theory and its Applications*. Academic Press, New York.

[6] Knight, F. B. (1962). On the random walk and Brownian motion. *Trans. AMS*, **103**, 218–228.

[7] Knight, F. B. (1971). A Reduction of Continuous Square-Integrable Martingales to Brownian Motion. *Lecture Notes in Math.*, **190**, Springer-Verlag, New York, pp. 19–31.

[8] Kunita, H. and Watanabe, S. (1967). On square integrable martingales. *Nagoya Math. J.*, **30**, 209–245.

[9] Monroe, I. (1978). Processes that can be embedded in Brownian motion. *Ann. Prob.*, **6**, 42–56.

[10] Root, D. H. (1969). The existence of certain stopping times on Brownian motion. *Ann. Math. Statist.*, **40**, 715–718.

[11] Scott, D. J. and Huggins, R. M. (1983). On the embedding of processes in Brownian motion and the law of the iterated logorithm for reverse martingales. *Bull. Aust. Math. Soc.*, **27**, 443–459.

[12] Skorohod, A. V. (1965). *Studies in the Theory of Random Processes*. Addison-Wesley, Reading, MA.

[13] Strassen, V. (1964). An almost sure invariance principle for the law of the iterated logorithm. *Zeito Wahrsch. verw. Geb.*, **3**, 211–226.

[14] Strassen, V. (1967). Almost sure behaviour of sums of independent random variables and martingales. *Proc. Fifth Berkeley Symp. Math. Statist. Prob.*, Vol. II. University of California Press, Berkeley, CA, Part 1, pp. 315–343.

(BROWNIAN MOTION
INVARIANCE PRINCIPLES AND
FUNCTIONAL LIMIT THEOREMS
MARTINGALES)

Walter Philipp

SLANTED PROBABILITY

An obsolete term for non-additive probability* (see, e.g., Fellner [1]).

Reference

[1] Fellner, W. J. (1965). *Probability and Profit*. Irwin, Homewood, IL.

SLASH DISTRIBUTION

The slash distribution is the distribution of the ratio of a Gaussian (normal) random variable to an independent uniform random variable. Its density is symmetric and can be expressed in terms of the standard Gaussian density, $\phi(x)$, as

$$f(x) = [1 - \exp\{-x^2/2\}]/(\sqrt{2\pi}x^2)$$

$$= \begin{cases} [\phi(0) - \phi(x)]/x^2, & x \neq 0 \\ \phi(0)/2, & x = 0. \end{cases}$$

The slash is used primarily in simulation* studies (e.g., ref. 1), where, like the Gaussian, it represents an extreme situation in which to compare the performance of statistical procedures. Typically, a simulation study is designed to evaluate the performance of a procedure assuming various underlying distributions. The Gaussian represents one extreme, as real data often arise from distributions having straggling tails [4, p. 23]. The slash represents the opposite extreme, as its tails, similiar to those of the Cauchy, are much heavier than those likely to arise in practice. Thus, a statistical procedure that performs well on these two extremes is likely to perform well on a wide variety of distri-

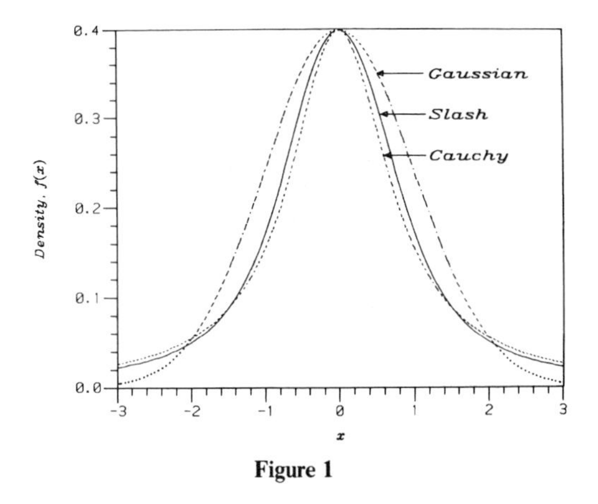

Figure 1

butions with varying degrees of tail behavior. Sometimes the Cauchy* is used as an extreme. However, its sharp central peak violates Winsor's principle (most distributions from large samples are reasonably Gaussian in the middle [4, p. 12]), making the Cauchy even less realistic in practice than the slash (Figure 1).

Representation of the slash as a ratio (Gaussian/independent positive random variable) permits the application of a Monte Carlo swindle*, such as that described in ref. 2, for more efficient estimation of characteristics of a random variable (variances, percent points, etc.). A location and scale form of the slash may be defined by using a Gaussian density with mean μ and variance σ^2 in the numerator. Additional characteristics of the slash distribution (e.g., representing function, percent points) may be found in ref. 5, and maximum likelihood* estimates for location and scale parameters are derived in ref. 3.

References

[1] Andrews, D. F., Bickel, P. J., Hampel, F. R., Huber, P. J., Rogers, W. H. and Tukey, J. W. (1972). *Robust Estimates of Location: Survey and Advances*. Princeton University Press, Princeton, NJ.

[2] Gross, A. M. (1973). *Appl. Statist.*, **22**, 347–353.

[3] Kafadar, K. (1982). *J. Amer. Statist. Ass.*, **77**, 416–424.

[4] Mosteller, F. and Tukey, J. W. (1977). *Data Analysis and Regression: A Second Course in Statistics*. Addison-Wesley, Reading, MA.

[5] Rogers, W. H., and Tukey, J. W. (1972). *Statist. Neerlandica*, **26**, 213–226.

Acknowledgment

The author wishes to thank Alan M. Gross for his comments on an earlier version of this article.

(CAUCHY DISTRIBUTION
MONTE CARLO SWINDLE
NORMAL DISTRIBUTION
UNIFORM (RECTANGULAR)
DISTRIBUTIONS)

KAREN KAFADAR

SLEPIAN PROCESS

The name *Slepian process* is sometimes applied to the stationary zero-mean Gaussian process* with triangular autocorrelation function. That is, if $\{X(t), -\infty < t < \infty\}$ is the standardized Slepian process, its autocorrelation function is

$$\begin{aligned} R(\tau) &\triangleq EX(t)X(t+\tau) \\ &= \max(0, 1 - |\tau|). \end{aligned}$$

This process was first studied by Slepian [4], who used its "peculiar Markoff-like property," later identified as the reciprocal property by Jamison [3], to derive an explicit expression for its first passage time density on [0, 1]. As the Slepian process is one of the few non-Markov processes for which any explicit first passage time density function is known, many generalizations and extensions of Slepian's process have been studied in the context of first passage time or level-crossing problems, including, for example, a process with a sawtooth autocorrelation function and a random field* with an autocorrelation function given by the product of triangles. See Blake and Lindsey [2] and Abrahams [1] for surveys of the level-crossing problem including discussions of the Slepian process and its extensions.

References

[1] Abrahams, J. (1986). In *Communications and Networks: A Survey of Recent Advances*. I. F. Blake and H. V. Poor, eds. Springer, New York.

[2] Blake, I. F. and Lindsey, W. C. (1973). *IEEE Trans. Inf. Theory*, IT-19, 295–315.

[3] Jamison, B. (1970). *Ann. Math. Statist.*, **41**, 1624–1630.

[4] Slepian, D. (1961). *Ann. Math. Statist.*, **32**, 610–612.

(GAUSSIAN PROCESSES)

JULIA ABRAHAMS

SLEPIAN'S INEQUALITY

Let $\mathbf{X}$ have a multinormal distribution* with $E[\mathbf{X}] = \mathbf{0}$, unit variances, and correlation matrix $\boldsymbol{\Sigma}$. Let $\mathbf{R} = (\rho_{ij})$ and $\mathbf{T} = (\tau_{ij})$ be two positive semidefinite correlation matrices. If $\rho_{ij} \geqslant \tau_{ij}$ for all i, j, then

$$P_{\boldsymbol{\Sigma}=\mathbf{R}}\left[\bigcap_{i=1}^{k} (X_i \leqslant a_i)\right] \geqslant P_{\boldsymbol{\Sigma}=\mathbf{T}}\left[\bigcap_{i=1}^{k} (X_i \leqslant a_i)\right]$$

for all $\mathbf{a} = (a_1, \ldots, a_k)'$. The inequality is *strict* if $\mathbf{R}, \mathbf{T}$ are positive definite *and* if $\rho_{ij} > \tau_{ij}$ holds for *some* i, j. As the individual correlation coefficients increase, both upper and lower orthants receive more probability mass.

Using the above probability inequality it was shown by Joag-Dev et al. [1] that, if the entries of $\boldsymbol{\Sigma}$ are nonnegative, then the vector $\mathbf{X}$ is associated. This was a simplification of the proof given by Pitt [2].

For applications, see Tong [4].

References

[1] Joag-Dev, K., Perlman, M. D., and Pitt, L. D. (1983). *Ann. Prob.*, **11**, 451–455.

[2] Pitt, L. D. (1982). *Ann. Prob.*, **10**, 496–499.

[3] Slepian, D. (1962). *Bell Syst. Tech. J.*, **41**, 463–501.

[4] Tong, Y. L. (1980). *Probability Inequalities in Multivariate Distributions*. Academic, New York.

(DEPENDENCE,CONCEPTS OF
INEQUALITIES ON DISTRIBUTIONS)

SLICING

Slicing is a term used in exploratory data analysis* for dividing the data points into a series of non-overlapping groups (slices). This is the first step in determining median ad quartiles (the so-called "median" and "hinge" tracing).

(SKIPPING)

SLIPPAGE PARAMETER

A parameter that measures the shift of one distribution in a family relative to another. If the CDFs of X and Y are $F(x - \theta)$ and $F(y)$, respectively, then θ is a slippage parameter. (*See also* LOCATION AND SCALE PARAMETERS.)

(MEAN SLIPPAGE PROBLEMS)

SLIPPAGE TESTS *See* MEAN SLIPPAGE PROBLEMS

SLOPE, ESTIMATION OF *See* REGRESSION (various entries)

SLUTSKY (SLUTSKII), EVGENII EVGENIEVICH

Born: April 7, 1880, in Yaroslavl province, Russia.

Died: March 10, 1948 in Moscow, USSR.

Contributed to: mathematical economics, econometrics, mathematical statistics.

Slutsky's (we use the commonly accepted transliteration of his name) academic inclination on entering the Physics–Mathematics Faculty of Kiev University in his father's native Ukraine in 1899 was toward physics, and he regarded mathematics as merely a tool. Caught up in the wave of revolutionary fervour current among students in the Russian empire of the time, he was finally expelled in 1902 and forbidden to enter any Russian tertiary institution. Subsidised and encouraged by a grandmother, he spent a period over the years 1902–1905 at the Polytechnic Institute in Munich, ostensibly studying mechanical engineering, for which he showed no aptitude, but in fact deepening his knowledge of economics, until the events of 1905 made it possible for him to begin studies in political economy at Kiev University. After another turbulent period, he ultimately completed his university studies in this area in 1911, with a gold medal for a study entitled "The Theory of Limiting Utility," in which he applied mathematical methods to economic problems. This probably led to the writing of the now famous, though long overlooked, paper [8] on the theory of consumer behaviour, in which ideas of F. Y. Edgeworth* and V. Pareto* are developed.

His interest in statistics, and the theory of probability as a theoretical basis for it, was stimulated by a personally presented copy of the elementary book [5] by the eminent physiologist A. V. Leontovich, which exposited techniques of Gauss* and Pearson*. This led Slutsky to produce the book [6], said to be, for its time, a significant contribution to statistical literature. As a result, in 1913 Slutsky was appointed lecturer at the Kiev Commercial Institute where he worked, rising to the rank of professor, until 1926 when he left for Moscow.

In training, continuing interests, and early career, Slutsky's development will be seen to parallel closely that of his equally eminent countryman A. A. Chuprov*; it is therefore not surprising that the two men established close academic contact that continued until emigré Chuprov's untimely death in 1926. In Moscow Slutsky decided to pursue theoretical problems of statistics (although, as will be seen, the direction of these investigations was influenced by his interests in economics and geophysics), working at least until 1931 at the Koniunkturnyi Institute (an institute for the study of business cycles) and the Central Statistical Office. From 1931 to 1934 he worked at the Central Institute of Meteorology, from 1934 at Moscow State University, and from 1938 until his death at the Steklov Mathematical Institute of the USSR Academy of Sciences.

One of Slutsky's first papers in statistics [7] relates to fitting a function $f(x;\beta_1,\ldots,\beta_r)$ of one variable x (the β's being parameters) when there are repeated readings on response Y for each value of x considered, the system being normal and possibly heteroscedastic. If there are n_n in responses for the value x_i, and their average is $\bar{Y}_i$, $i = 1,\ldots,N$, then

$$\bar{Y}_i = f(x_i;\beta_1,\ldots,\beta_r) + \epsilon_i, \qquad i = 1,\ldots,N,$$

where the ϵ_i are independent, $\epsilon_i \sim N(0,\sigma_i^2/n_i)$, and σ_i^2 is the variance of the normal response corresponding to setting x_i. Slutsky proposes to estimate $\boldsymbol{\beta}$ by minimizing

$$\chi^2 = \sum_{i=1}^{n} \left[n_i \{ \bar{Y}_i - f(x_i;\beta_1,\ldots,\beta_r) \}^2 / \sigma_i^2 \right],$$

which he recognizes as a chi-square variable, so we may (if we assume σ_i^2 known) regard this procedure as an early instance of minimum chi-square* estimation. [It is also obviously maximum likelihood estimation* and (without the normality assumption) an instance of weighted least squares*.]

Slutsky's views on the abstract formalization of the probability calculus [9] do not refer to any specific axiomatic system, but are of general philosophical kind as to what features a rigorous mathematical formalization of it should contain. In particular, no subjective elements should enter. There are no references to other authors, but it is likely [10] Slutsky was aware of the attempts at formalization by Bernstein* and von Mises*.

The paper in ref. 10 has had fundamental influence in elucidating the notion of convergence* in probability. In this Slutsky was anticipated by Cantelli [2], but in some respects he went further, formalizing the notion of "stochastic asymptote," which generalizes that of "stochastic limit" (if for a sequence of random variables $\{X_n\}$, $n \geqslant 1$, $X_n - EX_n \xrightarrow{p} 0$, EX_n is said to be the "stochastic asymptote" of X_n) and establishing a form of what is now known as Slutsky's Theorem. The paper also contains, as does subsequent work, results on the weak law of large numbers (from which the convergence in probability* notions derive); in this direction the influence of Chuprov, and interaction with M. Watanabe, is apparent.

A number of Slutsky's papers treat stationary sequences and have had significant influence on the development of time-series* analysis. These contributions were stimulated by manifestations and investigations of periodicity. The most famous is ref. 11, in which the *Slutsky effect* is demonstrated: essentially that repeated filtering of even a purely random sequence of readings may in the limit produce a purely periodic sequence. More generally this paper made manifest that observed quasiperiodicity may simply be a result of statistical stationarity rather than the result of real periodic effects. Slutsky was aware of the work of G. U. Yule in a similar direction.

The notion of stochastic limit also led him to study random functions more generally (in particular to develop the notions of stochastic continuity, differentiability, and integrability) and he may thus be regarded, with A. I. Khinchin, as one of the founders of the theory of stationary random processes. Slutsky's approach to this work was centered on the consideration of moments of the random function (under the influence of Markov* and Chuprov, it would seem) up to fixed order, which notion more recently occurs in the guise of second-order (wide-sense) stationary stochastic processes*. His best known work in the area is, perhaps, ref. 12.

References

[1] Allen, R. G. D. (1950). The work of Eugen Slutsky. *Econometrica*, **18**, 209–216. (By an eminent mathematical economist who rediscovered ref. 8 in 1936. Includes a bibliography, pp. 214–216.)

[2] Cantelli, F. P. (1916). La tendenza ad un limite nel senzo del calcolo delle probabilità. *Rend. Circolo. Mat. Palermo*, **16**, 191–201.

[3] Gnedenko, B. V. (1960). Evgenii Evgenievich Slutskii. In *E. E. Slutsky*, *Izbrannie Trudy*. Izd. ANSSSR, Moscow, USSR, pp. 5–11. (in Russian.) (Biographical sketch and survey of his work.)

[4] Kolmogorov, A. N. (1948). Evgenii Evgenievich Slutskii (1880–1948). *Uspekhi Mat. Nauk*, **3**, 143–151 (in Russian). (Obituary.)

[5] Leontovich, A. (1911). *Elementarnoe Posobie k Primeneniu Metodov Gaussa i Pearsona pri Otsenke Oshibok v Statistike i Biologii*. Kiev, Ukraine, USSR.

[6] Slutsky, E. E. (1912). *Posobie k Izucheniu Nekotorikh Vazhneishikh Metodov Sovremennoy Statistiki*. Kiev, Ukraine, USSR.

[7] Slutsky, E. E. (1914). On the criterion of goodness of fit of the regression lines and on the best method of fitting them to data. *J. R. Statist. Soc.*, **77**, 78–84. (Reprinted in Russian in ref. 13, pp. 12–17; commentary by N. V. Smirnov on p. 283.)

[8] Slutsky, E. E. (1915). Sulla teoria del bilancio del consumatore. *Giornale degli Economisti*, *Ser.* 3, **51**, 1–26. [In Russian: *Ekon.-mat. Metody*, **1** (1963)]

[9] Slutsky, E. E. (1922). On the problem of the logical foundations of the probability calculus. *Vestnik Statist.* **12**, 13–21. (In Russian.) (A version with some changes and corrections in text appeared in 1925, in a collection of essays in statistics in memory of N. A. Kablukov. Reprinted in ref. 13, pp. 18–24.)

[10] Slutsky, E. E. (1925). Über stochastische Asymptoten und Grenzwerte. *Metron*, **5**, 3–89. (In Russian in ref. 13, as pp. 25–90.)

[11] Slutsky, E. E. (1937). The summation of random causes as the source of cyclic processes. *Econometrica*, **5**, 105–146. [An earlier version appeared in Russian in *Voprosy Koniunktury* **3**, 34–64 (1927). Reprinted in ref. 13, pp. 99–132.]

[12] Slutsky, E. E. (1938). Sur les fonctions aléatoires presque périodiques et sur la décomposition des fonctions aléatoires stationnaires en composantes. *Actual. Sci. Ind. Paris*, **738**, 33–55. (In Russian in ref. 13, pp. 252–268.)

[13] Slutsky, E. E. (1960). *Izbrannie Trudy*. Izd. ANSSSR, Moscow, USSR. (Selected works. Contains commentaries by N. V. Smirnov, B. V. Gnedenko, and A. M. Yaglom on aspects of Slutsky's work, a photograph, and a complete bibliography.)

(CHUPROV, ALEXANDER
ALEXANDROVICH

CONVERGENCE OF SEQUENCES OF RANDOM VARIABLES
ECONOMETRICS
LAWS OF LARGE NUMBERS
STOCHASTIC PROCESSES
TIME SERIES)

E. SENETA

SLUTSKY EFFECT *See* SINUSOIDAL LIMIT THEOREM

SLUTSKY–FRÉCHET THEOREM

The following theorem, which is useful in asymptotic probability theory, was proved by Slutsky [4] in a slightly less general form and generalized by Fréchet [2]. See also Cramér [1] and Serfling [3].

"If the sequence of random variables $\{X_n\}$ converges in probability (or almost certainly) to a random variable X_n, then so does $f(X_n)$ to $f(X)$ for any continuous function $f(\cdot)$."

References

[1] Cramér, H. (1946). *Mathematical Methods in Statistics*. Almqvist & Wiksell, Stockholm, Sweden; Princeton University Press, Princeton, NJ.

[2] Fréchet, M. (1937). *Recherches Théoriques Modernes sur la Théorie des Probabilités*, 1er livre, Gauthier-Villars, Paris, France.

[3] Serfling, R. (1980). *Approximation Theorems of Mathematical Statistics*. Wiley, New York.

[4] Slutsky, E. (1925). *Metron*, **5**, 3–89.

SMARTINGALE *See* MARTINGALES

SMEAR-AND-SWEEP

INTRODUCTION

The smear-and-sweep procedure was developed in the *National Halothane Study* (Bunker et al. [1]) to analyze the effect on the postoperative death rate of patients receiving anesthetic agents such as ether and halothane during general surgery. Because the choice of agent was associated with several variables such as age, sex, and type of operation that are also known to influence death rates, failure to adjust for these *confounding* variables* could have led to severe bias in the estimation of effects of anesthetic on death rate (*see* RATES and VITAL STATISTICS). A full contingency table* approach was impossible because it would have contained millions of cells, a great many of which would have been empty. The halothane statisticians referred to this as a problem of "sparse" contingency tables. Consequently, smear-and-sweep, together with other techniques, was used to summarize the effects of the confounding variables in a single index with the objective of reducing bias.

DESCRIPTION

Smear-and-sweep is an iterative procedure with each iteration consisting of two steps. It is started by "smearing" the data into a two-way classification based on two of the confounding variables. In the next step, the resulting cells are "swept" into categories according to their ordering on the criterion variable (death rates). The groups of cells make up the sweep index, which is then used with another of the confounding variables to produce another two-way classification and a new sweep index. The procedure is repeated until all of the confounding variables have been entered into the index. The final index should represent the effects of the confounding variables on the criterion variable. If successful, smear-and-sweep should yield less bias than techniques that do not attempt to account for the effects of the confounding variables.

Example. Consider the hypothetical problem (motivated by the National Halothane Study) of analyzing the effect of the anesthetic (ether and halothane) on death

Table 1 Smear of Death Rates[a] by Age by Sex[b]

Sex	Age 20–29	30–44	45–59	60–69
Male	90	109	118	128
	(I)	(II)	(III)	(III)
Female	68	85	101	112
	(I)	(I)	(II)	(III)

[a]Entries in Arabic numerals are death rates (the mean death rate in that cell expressed in number of deaths per 100,000 patients).
[b]Roman numerals indicate the results of sweeping into a three-category sweep index.

Table 2 Smear of Death Rates by Operation and by Sweep[a]

Operation	Sweep (From Previous Iteration) I	II	III
A	61	78	97
	(I)	(I′)	(II′)
B	81	102	123
	(I′)	(II′)	(III′)
C	103	120	138
	(II′)	(III′)	(III′)

[a]I′ refers to the category of the index being created at this iteration.

rates (the criterion variable). The confounding variables are age, sex, type of operation, hospital where the operation is performed, length of operation, and the patient's previous health status. The first iteration is performed by selecting two variables to produce a table of means (the smear). In this example, age and sex were selected as the starting variables. It is not known how to choose the variables to be entered first; consequently this choice is arbitrary. Table 1 illustrates the eight cells of death rates (these are commonly reported in deaths per 100,000 operations) by age and sex.

The "sweeping" is performed by creating new categories based on similar death rates. The Roman numerals show the regrouping into a three-level sweep variable (actually, a composite of effects). In the example, the three cells grouped into category I of the sweep have death rates of 68, 85, and 90 per 100,000 operations. At this point, the user would create a sweep variable in the data set using Table 1 to score the index value for each individual observation. For example, for the observations for which age is in the 20–29 range and for which sex is male, the sweep variable would score a I. Similarly the cells with death rates from 101 to 109 have been grouped into sweep category II; the observations that fall into those cells would then be scored as II's. At this point, the sweep index should reflect the effects of age and sex on death rate.

The second iteration is performed by selecting another variable to include in the sweep composite. Again the selection is entirely up to the user's judgment; here, operation type has been selected. Table 2, illustrates the smear step with the death rates recorded by sweep and operation type. Again the cells are classified into three similar groups, which are indicated by the numerals I′, II′, and III′. As in the first iteration, the data would be scored according to Table 2, thus creating a new sweep variable. At this point the sweep index should reflect the effects of age, sex, and operation on death rates.

This process would be repeated as in the second iteration for each variable until all variables have been included in the sweep composite. Finally, a two-way table of means (Table 3) based on the sweep variable and the anesthetic would be used to analyze the effect of anesthetic. From the data in this table, one could compute for each anesthetic an overall death rate that is "adjusted" for

Table 3 Final Two-Way Table of Means: Death Rates Tabled on Sweep and Anesthetic[a]

Anesthetic	Sweep (From Final Iteration) I^f	II^f	III^f
Ether	71	101	119
Halothane	84	103	131

[a]I^f represents category I of the final sweep index.

the sweep index. To the degree that the sweep index represents the effects of the confounding variables on death rate, this procedure accounts for their effects. Space limitations preclude showing a complete analysis of Table 3; however, one could use a procedure such as ANOVA (*see* ANALYSIS OF VARIANCE).

CAVEATS

The user of smear-and-sweep should be aware that the procedure is relatively new and many of its qualities are not yet known. It has been shown that in some circumstances (Scott [2]) it could lead to increased bias. It is not known how frequent such occurrences are. Such issues as order of entry and criteria for sweeping are not yet settled.

References

[1] Bunker, J. P., Forrest, W. H., Jr., Mosteller, F., and Vandam, L. D., eds. (1969), *The National Halothane Study*. (This is an extensive technical study of the relation of death rates to anesthetics.)

[2] Scott, R. C. (1978), *J. Amer. Statist. Ass.*, **73**, 714–18. (This is a study of a situation in which smear-and-sweep fails to accomplish the goal of lessening bias.)

(ANALYSIS OF VARIANCE
CONFOUNDING
RATES
STRATIFIED DESIGNS)

R. C. SCOTT

SMELTING

"Smelting" refers to a class of techniques introduced by Tukey [1] for smoothing by excluding data arising in regression problems [series of pairs of values (u_i, v_i), where v is trying to "guide" u]. The quantitative nature of the series is used to indicate which (u, v) pairs to keep and which to set aside. A detailed discussion of the technique with a numerical example is given in Tukey [1].

Reference

[1] Tukey, J. W. (1982). The use of smelting in guiding re-expression. In *Modern Data Analysis*, R. L. Launer and A. F. Siegel, eds. Academic, New York, pp. 83–102.

(EXPLORATORY DATA ANALYSIS
OUTLIERS)

SMIRNOV TESTS *See* KOLMOGOROV–SMIRNOV TESTS

SMITH'S TEST OF INDEPENDENCE IN CONTINGENCY TABLES

Given an $r \times c$ contingency table* with N_{ij} denoting the observed frequency in the cell defined by the ith row and the jth column, the classical chi-squared statistic for testing independence is

$$X^2 = \sum_{i=1}^{r} \sum_{j=1}^{c} \left(N_{ij} - N_{i\cdot}N_{\cdot j}/N\right)^2 \times \left(N_{i\cdot}N_{\cdot j}/N\right)^{-1},$$

where $N_{i\cdot} = \Sigma_{i=1}^{r} N_{ij}$, $N_{\cdot j} = \Sigma_{i=1}^{c} N_{ij}$, and $N = \Sigma_{i=j}^{r} \Sigma_{j=1}^{c} N_{ij}$. Smith [1, 2] proposed using the statistic

$$\Psi = \sum_{i=1}^{r} \sum_{j=1}^{c} \left(N_{ij} - N_{i\cdot}N_{\cdot j}/N\right)^2 N_{ij}^{-1}$$

in place of X^2. An appropriate null distribution (for N large) is obtained [1] by taking $\sqrt{\Psi}$ to have a normal distribution with expected value and standard deviation $(\gamma^2 - \delta)^{1/4}$ and $\{\lambda - \sqrt{(\lambda^2 - \delta)}\}^{1/2}$, respectively, where

$$\gamma = (c-1)\left(N^2 - \sum_{i=1}^{k} N_{i\cdot}^2\right)N^{-2},$$

$$\delta = (c-1)\left[\left(N^2 - \sum_{i=1}^{k} N_{i\cdot}^2\right)\left(\sum_{i=1}^{k} N_{i\cdot}^2 - N\right)\right.$$
$$\left. +2\left\{\left(\sum_{i=1}^{k} N_{i\cdot}^2\right)^2 - N\sum_{i=1}^{k} N_{i\cdot}^2\right\}\right]N^{-4}.$$

This test is planned to be especially sensitive to deviations from independence of more or less equal size. It is not powerful with respect to alternative hypotheses specifying a few large deviations and many small ones.

For 2×2 tables the test is identical with Pearson's chi-squared test.

References

[1] Smith, C. A. B. (1951). *Ann. Eugen.*, **16**, 16–25.
[2] Smith, C. A. B. (1952). *Ann. Eugen.*, **17**, 35–36.

(CHI-SQUARE TESTS
CONTINGENCY TABLES)

SMOOTH

A term used in exploratory data analysis* to denote the underlying simplified structure or "pattern(s)" in a given set of data points.

The term "smooth" is also used with other connotations. *See*, for example, GRADUATION and SMOOTHNESS PRIORS.

(ROUGH)

SMOOTHING *See* GRADUATION; INTERPOLATION; MOVING AVERAGES; SUMMATION $[n]$

SMOOTHING, KERNEL METHOD *See* DENSITY ESTIMATION

SMOOTHNESS PRIORS

The term *smoothness priors* is very likely due to Shiller [27], who modeled the linear distributed lag (impulse response) relationship between the regularly spaced discrete time-series* inputs and outputs of economic series under difference equation* constraints on the solution (*see* LAG MODELS, DISTRIBUTED). Akaike [1] completed the analysis initiated by Shiller. The origin of the problem solved by the Shiller–Akaike smoothness priors is in a problem posed by Whittaker in 1923 [33]: Let

$$y(n) = f(n) + e(n), \qquad n = 1, \dots, N \tag{1}$$

with the $e(n)$ i.i.d. $N(0, \sigma^2)$, σ^2 unknown and $f(\cdot)$ an unknown "smooth" function. The problem is to estimate $\{f(n),\ n = 1, \dots, N\}$.

Whittaker suggested that the solution balance a tradeoff between infidelity to the data and infidelity to a kth order difference equation* constraint. For a fixed value of k and λ, the solution achieves the minimization of

$$\sum_{n=1}^{N} [y(n) - f(n)]^2 + \lambda^2 \sum_{n=1}^{N} [\nabla^k f(n)]^2 \tag{2}$$

where ∇ is a backward difference* operator $[\nabla f(x) = f(x) - f(x-1)]$. The first term in (2) is a measure of the infidelity of the solution to the data. The second is a measure of the infidelity of the solution to the smoothness constraint. The difference equation constraints are

$$\nabla f(n) : f(n) = f(n-1) + w(n);$$
$$\nabla^2 f(n) : f(n) = 2f(n-1) - f(n-2) + w(n);$$

etc., where $w(n)$ is a zero mean i.i.d. sequence with variance τ^2 (assumed Gaussian here for convenience).

The properties of the solution to (2) are clear. If $\lambda = 0$, $f(n) = y(n)$, and the solution is a replica of the observations. The sum of squares of errors is zero and the solution is uninteresting. As λ becomes increasingly large, the smoothness constraint dominates and the solution satisfies a kth order least-squares constraint. For very large λ and $k = 1$, the solution is the mean of the data; for $k = 2$ it is a straight line, and so on for increasing k. Whittaker left the choice of the smoothness tradeoff parameter λ to the investigator.

Shiller and Akaike interpreted this problem from a Bayesian* point of view. In that context, λ is a hyperparameter (Lindley and Smith [24]). Shiller determined the hyperparameter in an ad hoc manner. Akaike provide a complete solution of the problem, showed a method for computing the likelihood of the hyperparameter, interpreted the likelihood as a measure of goodness of fit* of the model, and showed applications of that procedure to a variety of interesting linear model data analysis problems.

There are numerous connections in the literature to the Whittaker problem. Under the assumptions that $f(\cdot) = f(x)$ has at least m continuous derivatives on $[0, 1]$ and that the observations $y(x_i)$, $i = 1, \ldots, N$, are at discrete but not necessarily equally spaced points in that interval, the constraints can be expressed as $\{\int_0^1 |f^{(m)}(x)|^2\, dx\}$; the solution is a spline of degree $2m - 1$ with knots at the x_i, $i = 1, \ldots, N$. This approach yields elegant convergence proofs as $N \to \infty$, a maximum likelihood goodness-of-fit statistic for the choice of λ and confidence interval* estimates and bounds for the solution. Wahba et al. exploited the cubic splines ($m = 2$) case, used an $O(N^3)$ generalized cross-validation computation to select λ, and developed relationships between the smoothing problem and problems of statistical regularization, and with problems of stochastic modeling [29, 30]. With considerable success she extended the problem to smoothing on the plane and on the sphere [31]. Solutions to those problems involve elliptical partial differential equations.

In applications to regression* with observations $\{(y_i, x_i), i = 1, \ldots, N\}$, ordering the independent variables x_i with a Bayesian interpretation of the $f(x_i)$ as random variables allows flexible state-space modeling (Kalman filter time-series methods) to bear on the subject. The Kalman filter* yields an $O(N)$ computational solution for the likelihood of the smoothness tradeoff parameters. Sallas and Harville [26], Wecker and Ansley, [32], and Kitagawa and Gersch [20, 21], respectively, employed this state-space modeling approach in the mixed model in ANOVA*, in the random effects model in nonparametric regression, and in nonstationary time-series modeling. Correspondingly, the vast literature on random and changing coefficient models, particularly as emphasized in the econometrics* time-series literature, is also relevant [6].

Also closely related are the bump hunting, penalized likelihood* methods introduced by Good [13] and the Bayesian and empirical Bayesian* analyses of the linear model by Box [4] and by Morris [25]. In what follows we present a simple treatment of the Shiller–Akaike smoothness priors and the smoothness priors state-space modeling Kalman filter methodology. We mention, but do not emphasize, objections to this Bayesian modeling that are critical of the seemingly arbitrary assumptions on the prior distribution. Kiefer's comment, "...an estimator should not be chosen on the basis of Bayes risk unless the prior law is firmly believed to be a physical truth," summarizes one objectionist point of view [19]. The literature is rich with others [7, 9, 28]. Parametric modeling is almost always done without an objective confirmation of the model structure. A rational objective in statistical modeling is the construction of a predictive distribution. In that case, modeling the distribution of future observations as a function of present data justifies probabilistic reasoning, including Bayesian modeling, only if the final model performance is satisfactory.

SMOOTHNESS PRIORS IN THE LINEAR MODEL

Consider the linear model

$$\mathbf{y} = \mathbf{X}\boldsymbol{\theta} + \mathbf{e}, \tag{3}$$

with $\mathbf{e} \sim N(\mathbf{0}, \sigma^2\mathbf{I})$ and $\boldsymbol{\theta}$ an unknown coefficient vector. Given the observations y_i, $i =$

$1, \ldots, N$, the maximum likelihood* estimator of $\boldsymbol{\theta}$ achieves the minimization of

$$\sum_{i=1}^{N} (y_i - \mathbf{x}_i'\boldsymbol{\theta})^2 = \|\mathbf{y} - \mathbf{X}\boldsymbol{\theta}\|^2, \qquad (4)$$

where $\mathbf{y} = (y_1, \ldots, y_N)'$, $\mathbf{x} = (x_1, \ldots, x_N)'$, and $\mathbf{A}'$ denotes the transpose of $\mathbf{A}$. This estimate is unstable or arbitrary when the number of parameters is comparable to or larger than N.

If $\boldsymbol{\theta}$ is thought to approximately satisfy the linear constraint $\mathbf{D}\boldsymbol{\theta} = \mathbf{0}$, then it is reasonable to estimate $\boldsymbol{\theta}$ by minimizing

$$\|\mathbf{y} - \mathbf{X}\boldsymbol{\theta}\|^2 + \lambda^2 \|\mathbf{D}\boldsymbol{\theta}\|^2. \qquad (5)$$

For the Whittaker smoothing problem cast into this linear model framework, $\mathbf{X}$ is the $N \times N$ identity matrix $\mathbf{I}$; we let $\mathbf{D} = \mathbf{D}_k$ be an $N \times N$ matrix expression of the kth difference equation constraint, and $\boldsymbol{\theta} = (f(1), \ldots, f(N))'$. For example,

$$\mathbf{D}_1 = \begin{bmatrix} \alpha & \cdot & \cdot & \cdot & \cdots & 0 \\ -1 & 1 & \cdot & \cdot & & \cdot \\ \cdot & & -1 & 1 & & \cdot \\ \cdot & & \cdot & \cdot & \ddots & \\ \cdot & & & & \cdot & \cdot \\ 0 & \cdot & \cdot & \cdot & -1 & 1 \end{bmatrix},$$

$$\mathbf{D}_2 = \begin{bmatrix} \alpha & & & \cdot & \cdot & 0 \\ -\beta & \beta & & & & \cdot \\ 1 & -2 & 1 & & & \cdot \\ & \cdot & \cdot & \cdot & & \cdot \\ & & \cdot & \cdot & \cdot & \cdot \\ 0 & \cdot & \cdot & 1 & -2 & 1 \end{bmatrix} \qquad (6)$$

for $k = 1$, $k = 2$, etc.; $\boldsymbol{\alpha}$ and $\boldsymbol{\beta}$ are chosen to satisfy initial conditions on the solution of the difference equation. If $\mathbf{D}_k$, the constraint matrix, and λ, the smoothness tradeoff parameter, are known, the solution of (5) is explicitly that of a least-squares problem. That is, $\boldsymbol{\theta}$ minimizes the Euclidean norm

$$\left\| \begin{pmatrix} \mathbf{y} \\ \mathbf{0} \end{pmatrix} - \begin{pmatrix} \mathbf{I} \\ \lambda \mathbf{D}_k \end{pmatrix} (\boldsymbol{\theta}) \right\|^2. \qquad (7)$$

For a Bayesian interpretation of this problem, multiply by $\sigma^2/2$ under the temporary assumption that σ^2 is known. Then the minimization of (5) is equivalent to the maximization of

$$\exp\left\{ -\frac{1}{2\sigma^2} \|\mathbf{y} - \mathbf{X}\boldsymbol{\theta}\|^2 \right\} \exp\left\{ -\frac{\lambda^2}{2\sigma^2} \|\mathbf{D}\boldsymbol{\theta}\|^2 \right\}. \qquad (8)$$

In this form (8) is proportional to the posterior distribution for $\boldsymbol{\theta}$,

$$\pi(\boldsymbol{\theta}|\lambda^2, \sigma^2, y) = p(\mathbf{y}|\sigma^2, \boldsymbol{\theta})\pi(\boldsymbol{\theta}|\lambda^2, \sigma^2, \mathbf{D}), \qquad (9)$$

where $p(\mathbf{y}|\lambda^2, \boldsymbol{\theta})$ is a conditional data distribution and $\pi(\boldsymbol{\theta}|\lambda^2, \sigma^2, \mathbf{D})$ is a prior on the solution $\boldsymbol{\theta}$. Akaike considered the marginal likelihood (*see* PSEUDO-LIKELIHOOD) for the unknown tradeoff parameter,

$$L(\lambda^2, \sigma^2|\mathbf{y}) = \int p(y|\sigma^2, \boldsymbol{\theta}) \times \pi(\boldsymbol{\theta}|\lambda^2, \sigma^2, \mathbf{D})\, d\boldsymbol{\theta}. \qquad (10)$$

Direct integration of (10) yields the explicit formula for L,

$$L(\lambda^2, \sigma^2|\mathbf{y}) = (2\pi\sigma^2)^{-N/2} |\lambda^2 \mathbf{D}'\mathbf{D}|^{1/2} |\lambda^2 \mathbf{D}'\mathbf{D} + \mathbf{X}'\mathbf{X}|^{-1/2} \times \exp\left(-\frac{1}{2\sigma^2} \mathrm{SSE}(\lambda^2) \right), \qquad (11)$$

where SSE (λ^2, σ^2) is the residual sum of squares from (5) and $\hat{\sigma}^2 = \mathrm{SSE}\ (\lambda^2)/N$. Akaike's procedure is to select a model that maximizes the marginal likelihood (L). Good [12] called this the *type II maximum likelihood method*. The maximization of (10) with respect to λ^2 is a nonlinear optimization problem. The computational problem can be imbedded into an EM algorithm or other equivalent way of maximizing the likelihood, but that is computationally expensive. In the vicinity of the maximized likelihood, the likelihood is a rather flat function of the hyperparameters. So for practical applications, it is usually satisfactory to consider a finite set of possible values of λ, solve the corresponding least-squares* problem, and choose the solution for which L is maximized.

Equation (10) is known as the *predictive distribution* in Bayes and empirical Bayes modeling (Box [4] and Morris [25]). The smoothness priors model solution is the mean of the (normal) predictive distribution.

The smoothness priors method is extensively applicable in this least-squares computational formulation. Akaike [1, 2] showed applications to the distributed lag estimation problem, to seasonal adjustment of time series, and to polynomial regression*. It has been applied to ridge regression* [7], to demography* [15], to the analysis of earth tides [16], and to dose–response curve estimation [17]. The modeling in ref. 16 includes fixed* effects regression and an autoregressive model effect; that in ref. 17 is related to logistic modeling. We used this formulation to fit long autoregressive models to relatively short data span stationary time-series* data for spectral estimation [22]. The computational complexity of the modeling shown in this section is $O(N^3)$. Next we discuss the expression of smoothness priors constraints in a state-space model form that yields the likelihood of the hyperparameters in a Kalman filter with a $O(N)$ computational complexity solution.

SMOOTHNESS PRIORS IN STATE SPACE

An irregularly spaced data regression version of (1) is

$$y_i = f(x_i) + e_i, \qquad i = 1, \dots, N, \quad (12)$$

where (y_i, x_i), $i = 1, \dots, N$, are the observed data, $\{e_i;\ i = 1, \dots, N\}$ is an independent unobserved sequence from $e \sim N(0, \sigma^2)$, and again $f(\cdot)$ is "smooth." Wahba [30], introduced the integrated Wiener process*, a stochastically perturbed m-degree polynomial model for $f(x)$,

$$f(x) = \sum_{k=i}^{m-1} \alpha_k \frac{(x-a)^k}{k!}$$

$$+ \xi^{1/2}\sigma^2 \int_\alpha^h \frac{(x-h)^{m-1}}{(m-1)!}\, dW(h), \quad (13)$$

where $W(h)$ is a Wiener process, and showed a correspondence to smoothing polynomials. If $\boldsymbol{\alpha} = (\alpha_0, \dots, \alpha_{m-1})'$ has a diffuse prior distribution, then the conditional expectation of $f(x)$ given the data is the smoothing polynomial spline discussed earlier. Wecker and Ansley [32] implemented a state-space version of the model in (12) and (13) in nonparametric regression modeling. Incidentally, in 1880, Thiele proposed and solved the problem in (12) and (13) with $m = 1$ (the random walk* case), and estimated the regression function with a recursive computational Kalman filter-like procedure. (That remarkable work is discussed in detail by Lauritzen [23].) In what follows we emphasize the state-space smoothness priors for the regularly spaced data case (1), instead of (12), and show applications to nonstationary modeling in the mean time series.

Consider a generalization of the problem in (1) to the case

$$y(n) = t(n) + s(n) + v(n) + d(n) + e(n), \qquad n = 1, \dots, N. \quad (14)$$

In (14) $t(n)$, $s(n)$, $v(n)$, and $d(n)$ are, respectively, the local polynomial trend, seasonal globally stochastic perturbation, and "trading day" effects that are frequently associated with economic time series. In (14) $e(n)$ is an observation error sequence.

The State-Space Models

The generic state-space or signal model for the regularly spaced observations $\mathbf{y}(n)$, $n = 1, \dots, N$, is

$$\begin{aligned} \mathbf{x}(n) &= \mathbf{F}(n)\mathbf{x}(n-1) + \mathbf{G}(n), \\ \mathbf{y}(n) &= \mathbf{H}(n)\mathbf{x}(n) + \mathbf{e}(n), \end{aligned} \quad (15)$$

where $\mathbf{F}$, $\mathbf{G}$, and $\mathbf{H}$ are $M \times M$, $M \times L$, and $1 \times M$ matrices, respectively, and it is assumed that $\mathbf{w}(n) \sim N(\mathbf{0}, \mathbf{Q}(n))$ and $\mathbf{e}(n) \sim N(\mathbf{0}, \mathbf{R}(n))$. $\mathbf{x}(n)$ is the state vector at time n and $\mathbf{y}(n)$ is the observation at time n. For any particular model of the time series, the matrices $\mathbf{F}$, $\mathbf{G}$, and $\mathbf{H}$ are known, and the observations are generated recursively from an initial state that is assumed to be normally distributed with mean $\mathbf{x}(\mathbf{0})$ and covariance matrix $\mathbf{V}(\mathbf{0})$.

In particular, a state-space model for the time series $\mathbf{y}(1), \dots, \mathbf{y}(N)$ that includes the effects of components indicated in (14) is written in the schematic form:

$$\mathbf{x}(n) = \begin{bmatrix} F_1 & 0 & 0 & 0 \\ 0 & F_2 & 0 & 0 \\ 0 & 0 & F_3 & 0 \\ 0 & 0 & 0 & F_4 \end{bmatrix} \mathbf{x}(n-1) + \begin{bmatrix} G_1 & 0 & 0 & 0 \\ 0 & G_2 & 0 & 0 \\ 0 & 0 & G_3 & 0 \\ 0 & 0 & 0 & G_4 \end{bmatrix} \mathbf{w}(n), \tag{16}$$

$$\mathbf{y}(n) = [\mathbf{H}_1\ \mathbf{H}_2\ \mathbf{H}_3\ \mathbf{H}_4(n)]\mathbf{x}(n) + \mathbf{e}(n).$$

In (16) the overall state-space model $(\mathbf{F}, \mathbf{G}, \mathbf{H})$ is constructed by the component models $(\mathbf{F}_j, \mathbf{G}_j, \mathbf{H}_j)$, $j = 1, \dots, 4$, respectively representing, the polynomial trend, a stationary autoregressive time series, the seasonal, and the trading day effects component models. The component models satisfy particular difference equation constraints on the components. By the orthogonality of the representation in (16), $(2^4 - 1)$ alternative model classes of trend* and seasonality* may be constructed from combinations of $(\mathbf{F}_j, \mathbf{G}_j, \mathbf{H}_j)$, $j = 1, \dots, 4$. Kitagawa and Gersch [20] consider that situation. The key computation there and indeed in any model selection procedure is the computation of the likelihood for the hyperparameters of the model.

Given the observations $\mathbf{y}(1), \dots, \mathbf{y}(N)$ and the initial conditions $\mathbf{x}(\mathbf{0})$ and $\mathbf{V}(\mathbf{0}|\mathbf{0})$, the distributions of the one-step-ahead predictor and filter are obtained from the Kalman filter algorithm [3, 8, 14, 18].

TIME UPDATE (PREDICTION).

$$\begin{aligned} \mathbf{x}(n|n-1) &= \mathbf{F}(n)\mathbf{x}(n-1|n-1), \\ \mathbf{V}(n|n-1) &= \mathbf{F}(n)\mathbf{V}(n-1|n-1)\mathbf{F}(n) \\ &\quad + \mathbf{G}(n)\mathbf{Q}(n)\mathbf{G}(n). \end{aligned} \tag{17}$$

OBSERVATION UPDATE (FILTERING).

$$\begin{aligned} \mathbf{K}(n) &= \mathbf{V}(n|n-1)\mathbf{H}(n) \\ &\quad \times (\mathbf{H}(n)\mathbf{V}(n|n-1)\mathbf{H}(n) + \mathbf{R}(n))^{-1}, \\ \mathbf{x}(n|n) &= \mathbf{x}(n|n-1) + \mathbf{K}(n)(\mathbf{y}(n) \\ &\quad - \mathbf{H}(n)\mathbf{x}(n|n-1)), \\ \mathbf{V}(n|n) &= (\mathbf{I} - \mathbf{K}(n)\mathbf{H}(n))\mathbf{V}(n|n-1). \end{aligned} \tag{18}$$

In (17) and (18), $\mathbf{x}(n|n-1)$ and $\mathbf{V}(n|n-1)$ are the estimates of the expected value of the state and the estimate of the state covariance at time n, given the data $\mathbf{y}(1), \dots, \mathbf{y}(n-1)$, respectively. The smoothed value of the state $\mathbf{x}(n)$ given all of the data $\mathbf{y}(1), \dots, \mathbf{y}(N)$ is obtained by the fixed interval smoothing algorithm:

$$\begin{aligned} \mathbf{A}(n) &= \mathbf{V}(n|n)\mathbf{F}(n)\mathbf{V}(n+1|N)^{-1}, \\ \mathbf{x}(n|N) &= \mathbf{x}(n|n) + \mathbf{A}(n) \\ &\quad \times [\mathbf{x}(n+1|N) - \mathbf{x}(n+1|n)], \\ \mathbf{V}(n|N) &= \mathbf{V}(n|n) + \mathbf{A}(n) \\ &\quad \times [\mathbf{V}(n+1|N) \\ &\quad - \mathbf{V}(n+1|n)]\mathbf{A}(n)'. \end{aligned} \tag{19}$$

The state-space representation and the Kalman filter yield an orthogonal decomposition of the data with a corresponding efficient algorithm for the computation of the likelihood of time-series models. That procedure obtains

$$f(\mathbf{y}(1), \dots, \mathbf{y}(N)) = \prod_{n=1}^{N} f(\mathbf{y}(n)|\mathbf{y}(1), \dots, \mathbf{y}(n-1)), \tag{20}$$

with

$$\begin{aligned} f(\mathbf{y}(n)|\mathbf{y}(1), \dots, \mathbf{y}(n-1)) &= (2\pi v(n))^{-1/2} \\ &\quad \times \exp\{-[\mathbf{y}(n) - \mathbf{H}(n)\mathbf{x}(n|n-1)]^2 \\ &\quad /[2\mathbf{V}(n)]\}, \end{aligned} \tag{21}$$

where $v(n)$ is the variance of $y(n|n-1)$. From (21) the log likelihood of the model is obtained by

$$l = -\tfrac{1}{2}\left\{ N \log 2\pi + \sum_{n=1}^{N} \log \mathbf{V}(n) + \sum_{n=1}^{N} (\mathbf{y}(n) - \mathbf{H}(n)\mathbf{x}(n|n-1))^2/[2\mathbf{V}(n)] \right\}. \tag{22}$$

In terms of the simplest trend* estimation problem, as in (1) [with the notation $t(n) = f(n)$], the state-space model yields the joint density function of the state and the observations $f(\mathbf{y}(n), t(n))$. Then

$$f(\mathbf{y}(n)|\mathbf{y}(1),\ldots,\mathbf{y}(n-1)) = \int f(\mathbf{y}(n)t(n)|\mathbf{y}(1),\ldots,\mathbf{y}(n-1))\,dt(n) = \int f(\mathbf{y}(n)|t(n))f(\mathbf{t}(n)|\mathbf{y}(1),\ldots,\mathbf{y}(n-1))\,dt(n). \quad (23)$$

Thus, as in (10), the state-space Kalman filter yields the marginal likelihood of the smoothness priors model.

Consider the simplest trend and seasonal component model. A state-space signal model for a simple trend–simple seasonal model is

$$\mathbf{x}(n) = \begin{bmatrix} t(n) \\ t(n-1) \\ s(n) \\ s(n-1) \\ s(n-2) \end{bmatrix} = \begin{bmatrix} 2 & -1 & 0 & 0 & 0 \\ 1 & 0 & 0 & 0 & 0 \\ 0 & 0 & -1 & -1 & -1 \\ 0 & 0 & 1 & 0 & 0 \\ 0 & 0 & 0 & 1 & 0 \end{bmatrix} \mathbf{x}(n-1) + \begin{bmatrix} 1 & 0 \\ 0 & 0 \\ 0 & 1 \\ 0 & 0 \\ 0 & 0 \end{bmatrix} \mathbf{w}(n), \quad (24)$$

$$\mathbf{y}(n) = [1 \quad 0 \quad 1 \quad 0 \quad 1]\mathbf{x}(n) + \mathbf{e}(n),$$

$$\begin{pmatrix} \mathbf{w}(n) \\ \mathbf{e}(n) \end{pmatrix} \sim N\left(\begin{pmatrix} 0 \\ 0 \\ 0 \end{pmatrix}, \begin{pmatrix} \tau_1^2 & 0 & 0 \\ 0 & \tau_3^2 & 0 \\ 0 & 0 & \sigma^2 \end{pmatrix} \right).$$

In (24), the trend $t(n)$ satisfies a second-order difference equation perturbed by a normal zero mean uncorrelated sequence with variance τ_1^2. Also in (24) the seasonal component satisfies

$$\sum_{i=0}^{L-1} s(n-i) = w_3(n), \qquad w_3(n) \sim N(0, \tau_3^2).$$

This smoothness priors problem corresponds to the maximization of

$$\times \exp\left\{ -\frac{1}{2\sigma^2} \sum_{n=1}^{N} [y(n) - t(n) - s(n)]^2 \right\}$$
$$\times \exp\left\{ \frac{-1}{2\sigma^2} \sum_{n=1}^{N} [\nabla^k t(n)] \right\}$$
$$\times \exp\left\{ -\frac{\lambda^2}{2\sigma^2} \sum_{n=1}^{N} \left[\sum_{i=0}^{L-1} s(n-i) \right]^2 \right\}. \quad (25)$$

Equation (25) clarifies the role of the hyperparameters τ_1^2 and τ_3^2 as measures of the uncertainty of belief in the priors. The ratios σ^2/τ_j^2, $j = 1$ or 3, can be interpreted as signal-to-noise ratios.

As before, we advocate a discrete likelihood hyperparameter space computational strategy and a search in that space to determine the approximate likelihood maximizing hyperparameters. That strategy retains the $O(N)$ computational complexity of the Kalman filter.

The results of illustrative computations on Bureau of the Census* North Central U.S. housing starts data 1969–1979, $N = 132$, are shown in Fig. 1A–D. As anticipated, small values of the hyperparameters yield relatively wiggly trends (seasonals). Relatively large values of the hyperparameters yield relatively smooth trends (seasonals).

The smoothness priors methodology shown here is an alternative to the Box–Jenkins* methodology for the modeling and prediction of time series with trends and seasonalities. More complete treatments of the modeling and prediction of such time series are in refs. 10 and 20. There we demonstrate that the model that is best for one-step-ahead prediction is not necessarily the same model that is best for k-step-ahead prediction. This is contrary to the conventional practice in the Box–Jenkins ARIMA modeling approach. Other smoothness priors

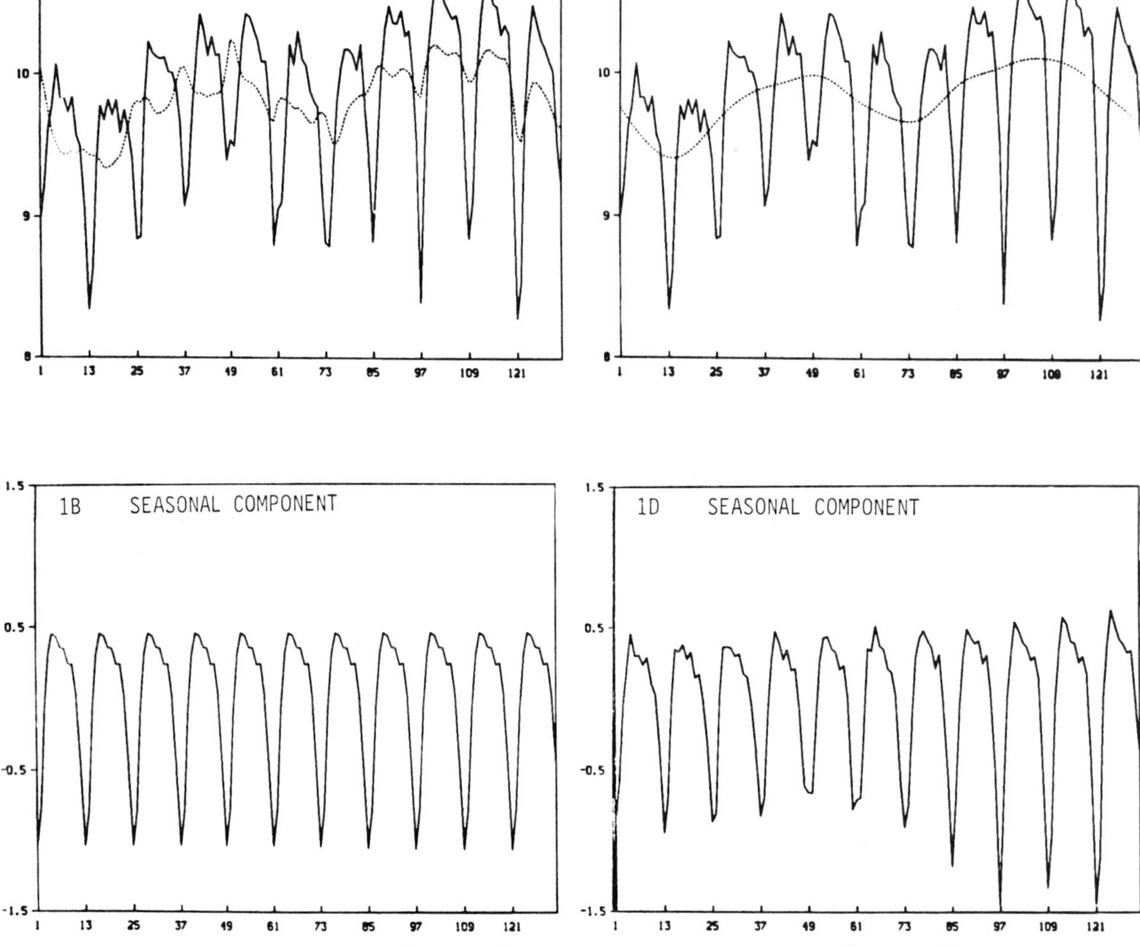

Figure 1 (A) Original and trend, $\tau_1^2 = 1.0$, $\tau_3^2 = 16.0$. (B) Seasonal component, $\tau_1^2 = 1.0$, $\tau_3^2 = 16.0$. (C) Original and trend, $\tau_1^2 = 16.0$, $\tau_3^2 = 1.0$. (D) Seasonal component, $\tau_1^2 = 16.00$, $\tau_3^2 = 1.00$.

treatments of series with seasonalities are in refs. 2 and 11.

The Box–Jenkins method and the aforementioned modeling procedure treat time series that are nonstationary in the mean. Another topic of interest in time series is the modeling of time series that are nonstationary in the covariance. A state-space smoothness priors time-varying AR coefficient model for the modeling of such series is in ref. 21. Applications of that model to the modeling of seismic data and to the estimation of instantaneous power spectrum density are shown in ref. 21.

Numerous theoretical questions and other engineering and statistical data analysis applications of the smoothness priors state-space modeling methodology remain to be investigated.

References

[1] Akaike, H. (1980). In *Bayesian Statistics*, J. M. Bernado, M. H. DeGroot, D. V. Lindley, and

A. F. M. Smith, eds. University Press, Valencia, Spain, pp. 141–166. (A likelihood of the Bayes model solution to Shiller's problem and interesting linear model applications.)

[2] Akaike, H. (1980). *J. Time Series Anal.*, **1**, 1–14. (Seasonal adjustment by smoothness priors.)

[3] Anderson, B. D. O. and Moore, J. B. (1979). *Optimal Filtering*. Prentice-Hall, Englewood Cliffs, NJ. (A sound engineering treatment of Kalman filtering.)

[4] Box, G. E. P. (1980). *J. R. Statist. Soc. Ser. A*, **143**, 383–430. (Bayes inference and the role of the predictive distribution.)

[5] Box, G. E. P. and Jenkins, G. M. (1970). *Time Series Analysis, Forecasting and Control*. Holden-Day, San Francisco, CA. (A standard reference on parametric modeling of time series, particularly those with trends and seasonality.)

[6] Chow, G. C. (1983). *Handbook of Econometrics*, Vol. 2, In Z. Griliches and M. Intriligator, eds. North-Holland, Amsterdam, The Netherlands, Chap. 21. (A review of random coefficient modeling.)

[7] Draper, N. R. and Van Nostrand, R. C. (1978). *Technometrics*, **21**, 451–466. (A review of biased estimation; criticism of unsubstantiated use of priors.)

[8] Duncan, D. and Horn, S. (1972). *J. Amer. Statist. Ass.*, **67**, 815–821. (The first treatment of Kalman filtering in the statistics literature.)

[9] Fomby, T. B. (1979). *Int. Econ. Rev.*, **20**, 203–215. (A mean square error anti "arbitrary priors treatment" of Shiller's model.)

[10] Gersch, W. and Kitagawa G. (1983). *J. Bus. Econ. Statist.*, **1**, 253–264. (State-space smoothness priors prediction of seasonal time series.)

[11] Gersovitz, M. and MacKinnon, J. G. (1978). *J. Amer. Statist. Ass.*, **73**, 264–273. (Smoothness priors application to seasonal data.)

[12] Good, I. J. (1965). *The Estimation of Probabilities*. MIT Press, Cambridge, MA. (Perhaps the first development of the role of the predictive distribution, here called the type II ML procedure.)

[13] Good, I. J. and Gaskins, R. A. (1980). *J. Amer. Statist. Ass.*, **75**, 42–73. (Bump hunting and the penalized likelihood method and references.)

[14] Harrison, P. J. and Stevens, C. F. (1976). *J. R. Statist. Soc. Ser. B*, **38**, 205–248. (A statistics literature treatment of Kalman filtering.)

[15] Hickman, J. C. and Miller, R. B., (1981). *Scand. Actuarial J.*, **64**, 129–150. (An application of smoothness priors in demography.)

[16] Ishiguro, M., Akaike, H., Doe, M., and Nakai, S. (1981). *Proc. 9th Int. Conf on Earth Tides*. (An application of smoothness priors that includes regression on fixed components.)

[17] Ishiguro, M. and Skarmoto, Y. (1983). *Ann. Inst. Statist. Math.*, Tokyo, **35B**, 115–137. (A smoothness priors logistic model analysis.)

[18] Kalman, R. E. (1960). *Trans. ASME Ser. D Basic Eng.*, **82**, 35–45. (The original Kalman paper.)

[19] Kiefer, J. (1978). *J. R. Statist. Soc. Ser. B*, **40**, 34. ("An estimator should not be chosen on the basis of the Bayes rule unless the prior law is firmly believed to be a physical truth.") In discussion of O'Hagan, A., *J. R. Statist. Soc. Ser. B*, **40**, (1978) 1–24.

[20] Kitagawa, G. and Gersch, W. (1984). *J. Amer. Statist. Ass.*, **79**, 378–389. (Smoothness priors modeling of time series with trends and seasonalities, an alternative to Box–Jenkins ARIMA modeling.)

[21] Kitagawa, G. and Gersch, W. (1985). *IEEE Trans. Automat. Control*, **AC-30**, 48–56. (A smoothness priors time-varying autoregressive coefficient modeling.)

[22] Kitagawa, G. and Gersch, W. (1985). *IEEE Trans. Automat. Control*, **AC-30**, 57–65. (A smoothness priors long autoregressive model.)

[23] Lauritzen, S. L. (1981). *Int. Statist. Rev.*, **49**, 319–331. [A review of the remarkable smoothness priors recursive computational paper by Thiele (1880).]

[24] Lindley, D. V. and Smith, A. F. M. (1972). *J. R. Statist. Soc. Ser. B.*, **34**, 1–41. (Hierarchical Bayesian models; the role of the hyperparameter.)

[25] Morris, C. N. (1983). *J. Amer. Statist. Ass.*, **78**, 47–165. (A review of parametric empirical Bayes inference.)

[26] Sallas, W. M. and Harville, D. A. (1981). *J. Amer. Statist. Ass.*, **76**, 860–869. (Kalman filter recursive computations in the mixed effects ANOVA model.)

[27] Shiller, R. (1973). *Econometrica*, **41**, 775–778. (The first use of the term smoothness priors, distributed lag model.)

[28] Thurston, S. S. and Swamy, P. A. V. B. (1980). *Special Studies Paper No. 142*. Federal Reserve Board, Washington, DC. (A non-Bayesian smoothness priors distributed lag model.)

[29] Wahba, G. (1977). In *Application of Statistics*, P. R. Krishnaiah, ed. North-Holland, Amsterdam, The Netherlands, pp. 507–524. (A first survey of smoothing problems via the method of generalized cross-validation.)

[30] Wahba, G. (1978). *J. R. Statist. Soc. Ser. B*, **40**, 364–372. (The relationship between spline smoothing and stochastic processes.)

[31] Wahba, G. (1982). In *Proc. Signal Processing in the Ocean Environment Workshop*, E. J. Wegman and J. Smith, eds. Dekker, New York. (A survey of thin plate spline smoothing problem solutions on the plane and on the sphere.)

[32] Wecker, W. E. and Ansley, C. F. (1983). *J. Amer. Statist. Ass.*, **78**, 81–89. (A state-space smoothing problem of treatment of nonparametric regression.)

[33] Whittaker, E. T. (1923). *Proc. Edinburgh Math. Soc.*, **41**, 63–75. (The original smoothing problem paper.)

(GRADUATION
KALMAN FILTERING
LAG MODELS, DISTRIBUTED
SEASONALITY
SPLINE FUNCTIONS
TIME SERIES
WHITTAKER–HENDERSON GRADUATION)

W. GERSCH

SNEDECOR DISTRIBUTION *See* F-DISTRIBUTION

SNEDECOR, GEORGE WADDEL

Born: October 20, 1881, in Memphis, Tennessee.

Died: February 15, 1974, in Amherst, Massachusetts.

Contributed to: analysis of variance* and covariance*, applied sampling, data analysis, design of experiments*, statistical methods, world-wide use of statistical methods.

Through his many years of personal experiences in statistical consulting* with research workers, particularly in the biological and agricultural sciences, his equally many years of teaching statistical methods using real experimental data, usually drawn from such consultations, and through the seven editions of *Statistical Methods*, George Waddel Snedecor is among the greatest pioneers in improving the quality of scientific methods insofar as it concerns the use of analytical statistical methodologies. These applications using real data were based primarily on new statistical methodologies developed by R. A. Fisher* and other English statisticians.

The Snedecor family traces its genealogy from a Dutch settler in New Amsterdam (New York) in 1639, with descendants moving south after the Revolutionary War, reaching Alabama in 1818 to enter an aristocratic plantation period, and later becoming involved with law, education, and dedicated service. The oldest of eight children, George Snedecor was born in Tennessee but grew up mainly in smaller towns and rural areas of Florida and Alabama; his lawyer father became a minister and finally an educator working with young blacks.

Snedecor earned two degrees in mathematics and physics in Alabama and Michigan and had taught for eight years before coming to Iowa State College (now University) in 1913 as an assistant professor of mathematics. Quickly promoted, he began teaching the first formal statistics courses in 1915.

In 1924 Snedecor assisted with the historically important Saturday seminars conducted by Henry A. Wallace, during the spring quarter, on statistics, including machine calculation of correlation coefficients*, partial correlation*, and calculation of regression* lines. Wallace and C. F. Sarle borrowed some card-handling equipment from an insurance company in Des Moines, Iowa, to bring to Ames on certain Saturdays to illustrate their use. Thus business machines (including IBM punch-card tabulation equipment) were used at this early time for research computations. These seminars led to the ISU bulletin "Correlation and Machine Calculations" by Snedecor and Wallace (1925), which attained world-wide distribution. See Jay L. Lush, "Early Statistics at Iowa State University," in Bancroft [1]; also see ref. 3.

In 1927 Snedecor and A. E. Brandt became directors of a newly created Mathematics Statistical Service in the Department of Mathematics to provide a campus-wide statistical consulting and computation service. It was the forerunner of the Statistical Laboratory, organized in 1933, under the ISU President's Office, with Snedecor as the first director.

While special courses in statistics were taught in several departments, degree programs were with Mathematics, and in 1931 the first degree in statistics at ISU, an M.S., was awarded to Snedecor's student Gertrude M. Cox.

Snedecor was instrumental in bringing R. A. Fisher to Ames as a visiting professor twice, in 1931 and 1936. Since the combination of the service functions of the Statistical Laboratory and the degree programs in statistics through Mathematics (both provided primarily by part-time service from the same faculty members) was unique among universities, many outstanding statisticians visited ISU, some to lecture and some just to observe.

In order to strengthen the graduate program in statistics at the Ph.D. level, Snedecor brought C. P. Winsor from Harvard as a faculty member in 1938. W. G. Cochran came from Rothamsted in 1938 as a visiting professor and continued as a regular professor in 1939. In recognition of the Statistical Laboratory's leadership role, contractual and cooperative projects were initiated in the 1940s by the U.S. Weather Bureau and the Bureau of the Census*.

Snedecor became both a dedicated research worker and teacher and an able administrator, yet retained a personal humility. Tall, lean, and vigorous, with a direct, unpretentious, patient, and kind manner, Snedecor was warmly regarded by colleagues and students.

In recognition of the importance of his contributions, he was given many honors and awards by ISU, other universities (U.S. and foreign), and the statistical societies. At ISU: establishment of the Snedecor Ph.D. Student Award, Faculty Citation, honorary D.Sc., dedication of Snedecor Hall, and designation as Professor Emeritus. At other universities: honorary D.Sc., visiting lecturer and/or professor appointments in the 1950s and 1960s at North Carolina state universities, Virginia Polytechnic Institute and State University, University of Florida, Alabama Polytechnic Institute, and, for the Rockefeller Foundation, the University at São Paulo, Brazil. The statistical societies: President of the American Statistical Association*, honorary membership in the Royal Statistical Society*, Samuel S. Wilks Memorial Medal award, establishment of a Snedecor Award by the American Statistical Association and Iowa State University for the best publication in biometrics.

Snedecor's world-wide leadership among research workers has been based on the seven editions of *Statistical Methods*, beginning with his first in 1937 and extending to the last, in 1980, which was coauthored with Cochran. Note that the early editions' full titles were *Statistical Methods Applied to Experiments in Biology and Agriculture*. As of April 1984, over 207,000 copies of the seven editions had been published in English. In addition, these editions had been translated and published abroad in nine languages, including Spanish, French, Roumanian, Japanese, and Hindi. There were over 3,000 entries in the 1981 volume of *Science Citation Index* for Snedecor's *Statistical Methods*, establishing it as among the most cited publications (see David [6]). Snedecor also wrote numerous papers, including noteworthy expository works on data analysis, statistical methods, design of experiments, sampling, analysis of variance and covariance, biometry, and scientific method.

Snedecor retired as director of the Statistical Laboratory in 1947; however, he remained active both at ISU and on off-campus visiting appointments until 1958. During this period he advised on the establishment, in 1947, of a separate Department of Statistics at Iowa State, initiated and prepared a text for a pioneering, introductory course in statistics [8, 9], and conducted a special seminar, often using queries submitted to *Biometrics**, for graduate students majoring in statistics. After 1958 Snedecor spent another five years as consultant at the U.S. Naval

Electronics Laboratory in San Diego—to complete an exceptionally lengthy and vigorous career.

At the age of 92, George Snedecor died. Brief but insightful appraisals [2, 5, 7] of his character and contributions are given by Cox and Homeyer and in obituaries in statistical journals. Apart from his books on statistical methods, his great contribution was his vision in propagating the role of statistics in quantitative studies.

The life of G. W. Snedecor is encapsulated by the statement of W. G. Cochran [4]:

> Our profession owes Snedecor a great debt for his vision in foreseeing the contributions that statistical methods can make in quantitative studies, for his book made these methods available to workers with little mathematical training, and his administrative skill in building a major training center and in attracting leaders like Fisher, Yates, Mahalanobis, Kendall and Neyman to Ames as visitors.

References

[1] Bancroft, T. A., ed. (assisted by S. A. Brown) (1972). *Statistical Papers in Honor of George W. Snedecor.* Iowa State University Press, Ames, IA.

[2] Bancroft, T. A. (1974). *Amer. Statist.*, **28**, 108–109.

[3] Bancroft, T. A. (1982). *Iowa State J. Res.*, **17**, 3–10.

[4] Cochran, W. G. (1974). *J. R. Statist. Soc. Ser. A*, **137**, 456–457.

[5] Cox, G. M. and Homeyer, P. G. (1975). *Biometrics*, **31**, 265–301. (Contains a bibliography.)

[6] David, H. A. (1984). In *Statistics: An Appraisal*, H. A. David and H. T. David, eds. Iowa State University Press, Ames, IA, pp. 3–18.

[7] Kempthorne, O. (1974). *Int. Statist. Rev.*, **42**, 319–321. (Is followed by a bibliography, pp. 321–323.)

[8] Snedecor, G. W. (1948). *J. Amer. Statist. Ass.*, **43**, 53–60.

[9] Snedecor, G. W. (1950). *Everyday Statistics—Facts and Fallacies*, 1st ed. William C. Brown Co., Dubuque, IA. [Limited 1st ed.; limited 2nd ed. (1951).]

(AGRICULTURE, STATISTICS IN)

T. A. BANCROFT

SNOWBALL SAMPLING—I

INTRODUCTION: NETWORK SAMPLING

Social scientists, biologists, and systems analysts conceive a population as being an organic system characterized by important relations of interaction and structural position of the elementary units that comprise it. These units can be cells, business firms, individuals, or social groups. In network sampling these relationships are frequently more important for the subsequent analysis than the elements they connect. The selection of the sample is necessarily affected by the network of relationships that exists in the population (cf. Stephan [14]).

Network methods are obviously distinct from methods such as survey sampling*, which are designed to provide information about attributes of aggregated units. Conventional sample survey procedures can yield only limited information about aspects of respondents' egocentric networks.

Pioneering work in a sociological context was done by Moreno in the 1930s (see, e.g., Moreno and Jennings [12]), when existing mathematical and statistical tools were inadequate to deal with sociometric data describing social configurations. We now have at our disposal graph-theoretic concepts for organizing relational data and for expressing theoretical ideas about social structure (see, e.g., Harary et al. [9]). The book by Frank and subsequent work by him [6, 7] represent an attempt at a systematic treatment of statistical and inferential aspects of graph sampling.

All methods of obtaining network data, regardless of the procedure, trace paths (chains) from one individual to another on the basis of relationships between them. When making inferences based on chains, chain length is a quantity of particular interest, as it is often related to important variables, such as the quality of jobs found through contacts, access to elected representatives, and the modification of information by word of mouth transmission [5]. Connectedness, network density, and the prevalence of symmetric relationships are other topics that may be addressed with the aid of network sampling. The aim is, of course, to describe a structure represented by a network in terms of a few, essential parameters.

SNOWBALL SAMPLING

Snowball sampling, by some authors called *chain referral sampling*, is a method that has been widely used in qualitative sociological research. Coleman (as quoted by Biernacki and Waldorf [2] has even argued that it is a method uniquely designed for sociological research as it allows for the sampling of natural interactional units. By definition, a chain referral sample, or snowball sample, is created through a series of referrals that are made within a circle of people who know one another.

A survey of snowball type thus proceeds from an initial sample, in which information is obtained from each element about other elements to which it is connected in the system. The next step is to add to the first sample some or all of these related elements, acquiring data from them and also information about still other individuals to whom they are connected. Thus, step-by-step, the sampler proceeds from a starting set of elements to a larger set connected with them by one or more links of relationships. The number of waves or stages, the number of nominations per individual, can be varied at will by the researcher.

Probabilistic and statistical properties of this type of sampling has been examined in detail in an early paper by Goodman [8]. Snowball sampling, unlike other network methods, permits loops in which a person named in a later wave in turn names someone from an earlier wave. The occurrence of loops of various lengths in a population is often of substantial interest; for example, one might often want to know the frequencies of reciprocated choices in various friendship networks. Estimates using loops of various lengths have been developed by Goodman [8], who finds that snowball sampling is particularly efficient for this purpose.

In order to illustrate these ideas, we need some notation. Let s be the number of waves in a snowball design, not counting the initial sample, and k the number of names, constant through the procedure. Consider the case $s = k = 1$, i.e., an initial sample and a first wave and only one name is mentioned. Let the population size be N and the number of mutual relationships be N_{11}. Thus $2N_{11}$ individuals in the population would, if asked, name each other. We are interested in estimating the parameter N_{11} on the basis of a snowball sample of type $s = k = 1$. Suppose now, as Goodman does, that the initial sample is a Bernoulli sample with selection parameter p, i.e. each of the N individuals has an equal probability p of being included in the initial sample. Clearly, the expected sample size is equal to Np.

Let Y denote the number of individuals in the initial sample who enter mutual relationships with individuals either in the initial sample or individuals who are included in the first wave. The random variable Y now has a binomial distribution and the expected value

$$E(Y) = 2N_{11}p.$$

Hence we can use an observed value of Y, say y, to form a moment estimate of the parameter N_{11},

$$N_{11} = y/(2p).$$

Several other unbiased estimates can be formed, based on snowball data. See Goodman [8], who also extends the results to more general parameters and snowball designs.

SNOWBALL SAMPLING—RANDOM CHOICES

An interesting special case is that in which nominations are made at random: each individual in the initial sample mentions k individuals chosen at random, and similarly in later waves. The individuals included at each stage are thus paired off with k individuals chosen at random from the population, forming random cliques of size $k + 1$.

Numerical Example. Moreno and Jennings [12] compared a group of 26 fictitious individuals making three random choices each with a population of New York school girls. The girls belonged to groups of size 26 and each girl made three choices of table partner within her group. Under randomness, the expected number of members never chosen —the number of "isolates"—is clearly

$$N\left[\binom{N-2}{3}\bigg/\binom{N-1}{3}\right]^{N-1}$$
$$= 26\left(\frac{22}{25}\right)^{25} = 1.06.$$

The actual frequency distribution, based on the school girl data, showed the number of isolates to be 250% greater. Similarly, the number of mutual choices in the actual configurations was greater (by 213%) than was expected under randomness*, the latter figure computed as

$$N\left(\frac{N-1}{2}\right)\left(\frac{3}{N-1}\right)^2 = \frac{26(25)}{2}\left(\frac{3}{25}\right)^2$$
$$= 4.68.$$

In his classic text on mathematical sociology, Coleman [3] writes

> Often in measures of characteristics of networks, for example, we need a model to show what the network would look like if all connections were random. It is the departure from randomness, the residue remaining after application of the model, which is of interest.

Thus one possible use of the randomness model is as a baseline scheme, or null hypothesis*, against which actual outcomes of snowball, or chain referral processes, can be compared. In his study of draft resisters, Useem [15] found the ratio between new and old nominations for each wave and compared these figures to the ones obtained for a baseline random net, in which choices were made at random. People on later waves may be systematically different from choices on earlier waves, having different patterns of relationships or different biases in choosing which relationship to report. The amount of deviation from what is expected under random choices can be taken as a measure of the "tightness" of the social space—the degree to which friends of friends are also friends and the degree to which friendship is symmetric.

Consider now a snowball sampling design with $s = 1$, k arbitrary and random choices. We are interested in the total number of interviews we can expect with this design. Given the size of the initial sample n, we obtain

$$E\left(\begin{matrix}\text{number of}\\ \text{interviews}\end{matrix}\,\bigg|\, n\right)$$
$$= N - (N-n)\left(\frac{N-1-k}{N-1}\right)^n.$$

If the initial sample is a Bernoulli sample with selection parameter p, then, as the population size N tends to infinity, we get

$$E\left(\begin{matrix}\text{fraction}\\ \text{interviewed}\end{matrix}\right) \to 1-(1-p)\exp(-kp).$$

More generally, for an s-stage k-name random choice scheme the fraction involved will converge in probability to Q_s, where

$$Q_s = 1-(1-p)\exp(-kQ_{s-1}),$$
$$Q_0 = Np.$$

Hence for large s, an approximation to Q_s is obtained by finding the appropriate root of the transcendental equation [8]:

$$Y = 1-(1-p)\exp(-k\gamma),$$
$$\gamma = \lim_{s\to\infty} Q_s.$$

Thus, as an example, if $p = 0.1$ and $k = 2$, then, ultimately, 82.8% of the population will have been contacted in the process. The reader will notice the apparent similarity with simple stochastic models of contagion and diffusion phenomena, such as the spread of a rumor through a population (see, e.g., Berg [1]).

Next set $k = 1$, i.e., only one name is mentioned, let s be arbitrary, and the choices made at random. We then have a situation analogous to that described by Harris [10] under the heading random mappings, which in turn can be subsumed under the more general title random graphs*. A random mapping space is a triplet $(\mathbf{X}, \mathbf{T}, \mathbf{P})$, where $\mathbf{X}$ is a set of cardinality N, $\mathbf{T}$ a set of mappings of $\mathbf{X}$ into itself, and, finally, $\mathbf{P}$ is a family of probability measures. Harris posed the question: How many distinct elements of $\mathbf{X}$ are contained in the set of successors, or images, of a given element $x \in \mathbf{X}$, when a randomly chosen mapping $T \in \mathbf{T}$ is iterated s times? The similarity with problems in snowball sampling with random referrals is obvious.

The random graphs alluded to here arise as the result of a tracing procedure, developing sequentially in the course of a stochastic process*. In this respect they differ somewhat from the classical concept of a random graph, as introduced by Erdös and Rényi [4]. See the review article by Karónski [11], which contains an extensive bibliography.

SNOWBALL SAMPLING IN APPLICATIONS

In a snowball sampling scheme, the obvious basis for inference to individuals is the initial sample, which ideally is a random sample of individuals and can be analyzed accordingly. Frequently, however, the researcher wants to make use of individual data from all the people interviewed, not just from the initial sample. This is the case when the chain referral technique has been used to ensure an adequate number of interviews from a relatively inaccessible population, or when it is required to estimate the prevalence of a rare attribute.

To begin with the latter case, a sampling design specifically intended as a means for estimating the frequency of a rare attribute is multiplicity sampling, as suggested by Sirken [13]. It may be viewed as a special case of a two-wave snowball design. In a household survey* with multiplicity, sample households report information about their own residents as well as about persons who live elsewhere, such as neighbors or relatives, as specified by a multiplicity rule adopted in the survey. Statistical inference here is relatively straightforward, as demostrated by Sirken.

Snowball sampling or chain referral sampling is often resorted to in sociological research when the focus of study is on a sensitive issue. Finding ex-heroin addicts and starting referral chains among them, as reported by Biernacki and Waldorf [2], is one example of this; Useem's study of draft resisters [15] is another. In the absence of a sampling frame* covering the population, insiders' knowledge is often required to locate people for study and to start referral chains.

Monitoring the quality of data being collected is certainly not a problem unique to snowball sampling. However, certain nonsampling error problems emerge when the method is used for the present purpose [2]. Verifications of eligibility, as well as of the accounts provided by respondents, tend to become problematic as the sources used to initiate referrals become more distant and knowledge of the sources less personal. Another problem is an ethical one: respondents might feel that information they give about themselves is not adequately protected in a chain referral process.

An important issue that must be addressed concerns the generality of the data on individuals provided by the snowball method. As a rule, a snowball sample will be strongly biased toward inclusion of those who have many interrelationships with, or are coupled to, a large number of other individuals. In the absence of knowledge of individual inclusion probabilities in different waves of the snowball sample, unbiased estimation is not possible.

We now turn to inference about population structure, as described by parameters such as average chain length, frequency of mutual relationships, network density, and the like. Here also the applicability of snowball sampling is limited by the likely presence of substantial nonsampling errors. In an excellent review article, Erickson [5] mentions a number of such problems.

First there is an obvious need for fairly cooperative subjects, as well as for unambiguous relational questions. Secondly, because of the possible confusion between chaining processes and structural effects, the snowball technique is difficult to apply to weak relationships, or to ties of which the respondent may have a great many. Chaining processes are present whenever the respondent has some choice in how the chains are constructed. Ideally, no chaining processes should occur in snowball sampling, because each respondent makes nominations according to criteria specified by the investigator. Furthermore, it is impractical to use more than a few waves in a snowball design, lest nonresponse become severe. Consequently, the overall structure of large networks is difficult to assess through snowball sampling. There is again the ethical issue of relational questions being seen as a threat to the privacy of the respondent or the people he or she mentions, especially if the respondent knows that these people will be interviewed in their turn.

References

[1] Berg, S. (1983). *J. Appl. Prob.*, **20**, 31–46.

[2] Biernacki, P. and Waldorf, D., (1981). *Sociological Meth. Res.*, **10**, 1414-1463.

[3] Coleman, J. S. (1964). *Introduction to Mathematical Sociology*. Free Press, Glencoe, IL and Collier Macmillan, London, England.

[4] Erdös, P. and Rényi, A. (1961). *Publ. Math. Inst. Hungarian Acad. Sci.*, **5**, 17–61.

[5] Erickson, B. H. (1979). In *Sociological Methodolgy*, K. F. Schluessler, ed. Jossey-Bass, SanFrancisco, CA pp. 276–302.

[6] Frank, O. (1971). *Statistical Inference in Graphs*. Foa Repro, Stockholm, Sweden.

[7] Frank, O. (1981). In *Sociological Methodology*, S. Leinhardt, ed. Jossey-Bass, San Francisco, CA, pp. 110–155.

[8] Goodman, L. A. (1961). *Ann. Math. Statist.*, **32**, 148–170.

[9] Harary, F., Norman, R. Z., and Cartwright, D. (1966). *Structural Models: An Introduction to the Theory of Directed Graphs*. Wiley, New York.

[10] Harris, B. (1960). *Ann. Math. Statist.*, **31**, 1045–1062.

[11] Karoński, M. (1982). *Graph Theory*, **6**, 359–389.

[12] Moreno, J. L. and Jennings, H. H. (1938). *Sociometry*, **2**, 342–374.

[13] Sirken, M. G. (1970). *J. Amer. Statist. Ass.*, **65**, 280–294.

[14] Stephan, F. F. (1969). In *New Developments in Survey Sampling*, N. L. Johnson and H. Smith Jr., eds. Wiley Interscience, New York, pp. 81–104.

[15] Useem, M. (1973). *Conscription, Protest, and Social Conflict*. Wiley, New York.

Bibliography

Frank, O. (1977). *J. Statist. Plann. Inf.*, **1**, 235–264. (Discusses survey sampling in graphs, e.g., the estimation of a total from a snowball sample.)

Granowetter, M. (1976). *Amer. J. Sociology*, **81**. 1267–1303, (Deals with network sampling in a sociological context. The article is followed by a discussion.)

Rapoport, A. (1980). *Social Networks*, **2**, 1–18. (Approaches the use of networks as structural models from a probabilistic viewpoint.)

Wasserman, S. (1980). *J. Amer. Statist. Ass.*, **75**, 280–294. (Presents a methodolgy for studying social networks based on stochastic processes.)

Finally, as an introduction to the topic discussed in the article, there is the Sage series book:

Knoke, D. and Kuklinski, J. H. (1982). *Network Analysis*. Sage University Paper.

(RANDOM GRAPHS
SOCIAL NETWORK ANALYSIS
SOCIOMETRY
SURVEY SAMPLING)

SVEN BERG

SNOWBALL SAMPLING—II

In snowball (or refutational) sampling for sampling *rare* populations, one creates a frame of members of that population. The approach is to identify a few members, to ask each of them to identify other members, to contact those so identified and ask them to identify others, and so on. When the frame has been compiled a probability sample can then be drawn from it. The crucial issue of this type of snowballing is the completeness of the frame.

An alternative, more common approach—without construction of the frame—is simply to continue the snowballing process until a "sufficient" number of members of the rare population has been found for the survey. See Biernacki and Waldorf [1] for a detailed review of problems and techniques of snowballing sampling, and Welch [4] and Snow et al. [3] for applications.

Kalton and Anderson [2] present a lucid and comprehensive review of various procedures for sampling rare populations.

References

[1] Biernacki, P. and Waldorf, D. (1981). *Sociological Meth. Res.*, **10**, 141–163.

[2] Kalton, G. and Anderson, D. W. (1986). *J. R. Statist. Soc. Ser. A*, **149**, 65–82.

[3] Snow, R. E., Hutcheson, J. D., and Prather, J. E. (1981). *Proc. Sec. Survey Res. Meth. Amer. Statist. Ass.*, pp. 101–109.

[4] Welch, S. (1975). *Publ. Opinion Quart.*, **39**, 237–245.

(PROBABILITY PROPORTIONAL TO SIZE SAMPLING
SNOWBALL SAMPLING—I
SURVEY SAMPLING)

SNOWFLAKES

A *snowflake* is a graphical technique for displaying sets of multivariate data (*see* MULTIVARIATE GRAPHICS). For n variates, a snowflake is constructed by plotting the magnitude of each variate along equiangular rays originating from the same point. Each multivariate observation constitutes one snowflake and snowflakes are often displayed side-by-side for quick visual comparisons. (See Fig. 1.)

The term *snowflake* was coined in 1972 by Herman and Montroll [4], who used the method to portray the development of countries based on the fraction of the labor force in six different economic sectors. Siegel et al. [11] used the method in 1971 to portray different types of shock in humans. Friedman et al. [2] used the same technique in 1972, referring to them as *circle diagrams*. Snowflakes have also been called *circular profiles* (Mezzich and Worthington [10]), *polygons* (Jacob [5]), *Kiviat figures* (Kolence and Kiviat [6]), and, most commonly, *stars*.

Computer construction of snowflakes can be accomplished using the TROLL (time-shared reactive on-line laboratory) system of Welsh [12] or a SAS procedure STARS developed by Gugel [3]. Options in these packages include connecting the endpoints of each ray to form polygons, displaying any or all of each ray, labelling each ray, labelling each snowflake, and displaying reference circles. These options are used to suit individual preferences. Other computer packages can be modified to construct snowflakes.

Negativity causes special problems with snowflakes. The TROLL package automatically scales negative values so that the minimum equals zero. The circle diagrams of Siegel et al. [11] and Friedman et al. [2] avoid the negativity problem by plotting 6 + the standardized value obtained by subtracting the mean value across observations and dividing by the standard deviation across observations. Reference circles indicate the number of standard deviations an observation is from the mean. As the added constant is decreased, the visual difference between snowflakes is emphasized but the probability of observing a negative number is increased. The constant 5 seems to work well in practice.

There are two primary uses for snowflakes: pattern recognition* and magnitude comparisons. The primary purpose of the snowflakes of Herman and Montroll [4], Siegel et al. [11], Friedman et al. [2], and

McDonald and Ayers [9] is pattern recognition. However the snowflakes of Lorenzen and McDonald [8] are used for magnitude comparisons.

When used for pattern recognition, a competing graphical technique is the Chernoff face [1], which maps a multivariate observation into a cartoon-like face. Chernoff faces* are of special interest since humans possess a special innate ability to visually process faces. A study conducted by Jacob [5] suggests that Chernoff faces are more effective than snowflakes in certain multivariate applications like clustering. However, Chernoff faces are more difficult to set up than snowflakes and variable assignments to the facial features can have a significant effect on the visual impact. Snowflakes have a natural ordering and can be used to display correlation within observations as well as patterns across observations. Snowflakes are easily overlayed to provide direct comparisons of multivariate data while overlays of Chernoff faces tend to be confusing.

The snowflakes of Lorenzen and McDonald [8] are used to compare the fatality rates of various states over the years 1967 to 1975 (see Fig. 1). Here, a "small" snow-

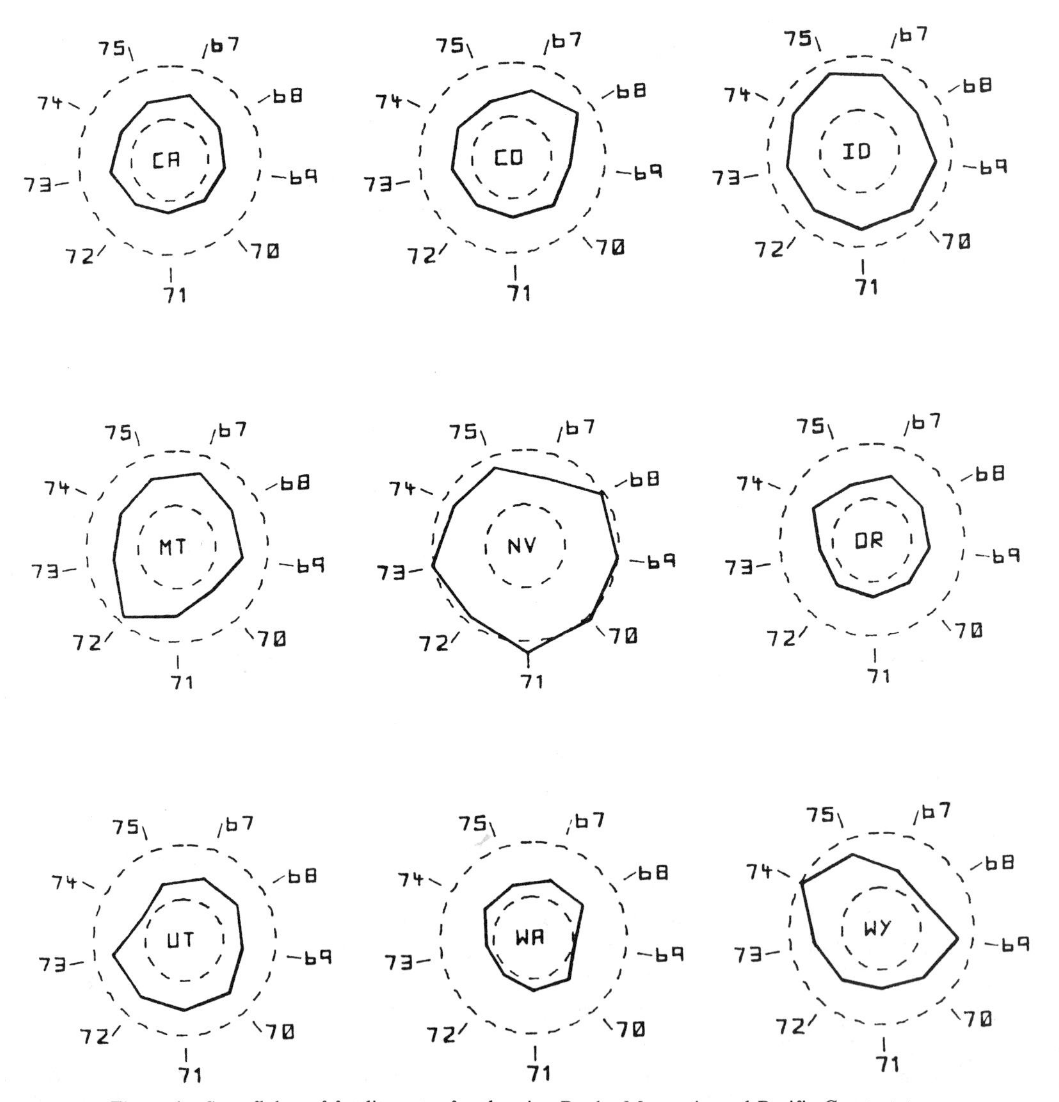

Figure 1 Snowflakes of fatality rates for the nine Rocky Mountain and Pacific Coast states.

flake (e.g., Washington) indicates a state with a relatively low fatality rate while a "large" snowflake (e.g., Nevada) indicates a state with an overall high fatality rate. Chernoff faces cannot be used to portray similar information in a meaningful fashion.

In a technical report, Lorenzen [7] considered area as a measure of size of a snowflake. It is assumed that the snowflake was constructed by using 5 + the standardized value obtained by subtracting the mean within years and dividing by a pooled estimate of the standard deviation. The distribution of the area is derived and, under an assumption of normality, a table of approximate percentiles is given. Note that, since the area is not invariant to permutations of the variates, some natural ordering should exist.

References

[1] Chernoff, H. (1973). *J. Amer. Statist. Ass.*, **68**, 361–368.

[2] Friedman, H. P., Farrel, E. J., Goldwyn, R. M., Miller, M., and Siegel, J. H. (1972). *Proc. Computer Sci. Statist. Sixth Annual Symp. Interface*, pp. 56–59.

[3] Gugel, H. W. (1985). *Proc. Tenth Annual SAS Users Group Int. Conf.*, pp. 253–258.

[4] Herman, R. and Montroll, E. W. (1972). *Proc. Nat. Acad. Science USA*, **69**, 3019–3023.

[5] Jacob, R. J. K. (1978). In *Graphical Representation of Multivariate Data*, P. C. C. Wang, ed. Academic, New York, pp. 143–168.

[6] Kolence, K. W. and Kiviat, P. J. (1973), *ACM SIGMETRICS Performance Evaluation Rev.*, **2**, 2–12.

[7] Lorenzen, T. J. (1980). The Distribution of the Area in a Snowflake. *Report No. GMR-3330*, Mathematics Dept., General Motors Research Laboratories, Warren, MI.

[8] Lorenzen, T. J. and McDonald, G. C. (1984). In *Design of Experiments Ranking and Selection*, T. J. Santner and A. C. Tamhane, eds. Dekker, New York, pp. 143–163.

[9] McDonald, G. C. and Ayers, J. A. (1978). In *Graphical Representation of Multivariate Data*, P. C. C. Wang, ed. Academic, New York, pp. 183–197.

[10] Mezzich, J. E. and Worthington, D. R. L. (1982). In *Graphical Representation of Multivariate Data*, P. C. C. Wang, ed. Academic, New York, pp. 123–141.

[11] Siegel, J. H., Goldwyn, R. M., and Friedman, H. P. (1971). *Surgery*, **70**, 232–245.

[12] Welsch, R. E. (1976). *Computers and Graphics*, **2**, 31–37.

(CHERNOFF FACES
GRAPHICAL REPRESENTATION OF DATA
MULTIVARIATE GRAPHICS)

T. J. LORENZEN

SOBOLEV SPACES

A Sobolev space is a function space with high order continuity properties. In order to clearly understand the relationship of Sobolev spaces to other spaces of continuous functions consider a domain $\Omega \subset R^n$. For any nonnegative integer, $C^m(\Omega)$ i defined as the vector space of all functions f such that all their partial derivatives $D^j f$, $j \leqslant m$, are continuous on Ω. $C^0(\Omega)$ is abbreviated as $C(\Omega)$ and $C^\infty(\Omega) = \bigcap_{m=0}^{\infty} C^m(\Omega)$. If Ω is open, $f \in C^m(\Omega)$ need not be bounded. If, however, Ω is closed and $D^j f$ is bounded and uniformly continuous for $j = 0, 1, \ldots, m$, then $C^m(\Omega)$ is a Banach space with norm given by

$$\|f\| = \max_{0 \leqslant j \leqslant m} \sup_{x \in \Omega} |D^j f(x)|,$$

where $D = d/dx$ is the differential operator.

If we now let p be any positive real number, then $L^p(\Omega)$ is the class of measurable functions f defined on Ω such that

$$\int_\Omega |f(x)|^p \, dx < \infty.$$

For $1 \leqslant p < \infty$, $L^p(\Omega)$ is a Banach space. Indeed if the inner product is defined as

$$\langle f, g \rangle = \int_\Omega f(x) g(x) \, dx,$$

then $L^2(\Omega)$ is a Hilbert space, which is a fact frequently used in second order statistical inference. If Ω is compact then $C(\Omega)$ is dense in $L^2(\Omega), 1 \leqslant p < \infty$.

Thus any p integrable function over a compact space can be approximated by a continuous function over the same compact space. We may further refine the notion of approximation with the Sobolev space. Let the functional $\|\cdot\|_{m,p}$ be defined as

$$\|f\|_{m,p} = \left\{ \sum_{j=0}^{m} \|D^j f\|_p^p \right\}^{1/p} \quad \text{if } 1 \leqslant p < \infty,$$

$$\|f\|_{m,\infty} = \max_{0 \leqslant j \leqslant m} \|D^j f\|_\infty,$$

where m is a nonnegative integer, $1 \leqslant p < \infty$, and $\|\cdot\|_p$ is the usual L^p norm. The Sobolev space is then defined as

$$W^{m,p}(\Omega) = \text{completion of}$$
$$\left\{ f \in C^m(\Omega) : \|f\|_{m,p} < \infty \right\}.$$

The spaces $W^{m,p}(\Omega)$ were introduced by Sobolev [9, 10] with a number of related spaces investigated by other authors, notably Morrey [7] and Deny and Lyons [2]. A variety of symbols is used for these spaces in the various literatures, including $W^{m,p}$, $H^{m,p}$, $P^{m,p}$, and L_m^p. Several other names were also used before these spaces became commonly known as Sobolev spaces. Perhaps the most common alternative name was Beppo Levi spaces.

In recent functional analysis literature, numerous generalizations and extensions have been made, largely originating in the Soviet Union. Notable among these extensions are

1. Arbitrary real values of m interpreted as fractional differentiation. See Lions [5, 6].
2. Weighted spaces with weight functions in the L^p norms.
3. Sobolev spaces with different orders of differentiation in distinct coordinate directions.
4. Orlicz–Sobolev spaces based on generalization of L^p spaces known as *Orlicz spaces*. See Krasnolsel'skii and Rutickii [4] and Donaldson and Trudinger [3].

It is relatively easy to demonstrate that $W^{m,p}(\Omega)$ is a Banach space. In addition, a wide variety of approximation and imbedding results are available. These, however, must be formulated with considerably more care and functional analytic machinery than space allows. Adams [1] provides thorough exposition of the basic theory of Sobolev spaces in his Chapters 3 to 6. Perhaps the key result is the Sobolev imbedding theorem which in essence says that $W^{m,p}(\Omega)$ is imbedded in $L^q(\Omega)$ for $p \leqslant q \leqslant np/(n - mp)$, where $n < mp$ and $\Omega \subset R^n$.

The principal applications of Sobolev spaces are to the solution of ordinary or partial differential equations and to approximation theory. In both of these settings it is assumed that the solution is a function on Ω with a suitable number of derivatives. Thus a Sobolev space is the natural representation class for the set of candidate solutions. In the approximation theory context, a problem of interest is to interpolate a set of points with a smooth function f minimizing some norm such as $\int_\Omega (Lf(x))^2\,dx$, where L is an mth order differential operator. In this case the solution to this optimization problem is the so-called interpolating L spline. See Schultz and Varga [8]. In the case that $L = D^2$, the solution is the well-known cubic interpolating spline*.

It is via splines that Sobolev spaces enter the realm of statistical theory. The generalized optimization problem,

$$\text{minimize} \left[\sum_{j=1}^{n} (y_j - f(t_j))^2 + \lambda \int_R (D^2 f(t))^2\,dt \right]$$

for $f \in W^{3,2}(R)$, provides the solution to the nonparametric regression problem

$$y_j = f(t_j) + \epsilon_j,$$

ϵ_j uncorrelated, 0 mean, constant variance. Indeed, the solution is a smoothing cubic spline. In this problem, λ is the so-called smoothing parameter, which may be determined by cross validation (Wahba [11]) or other techniques. A general survey treatment of splines in statistics is found in Wegman and Wright [12], which discusses many variants of this basic optimization problem.

References

[1] Adams, R.A. (1975). *Sobolev Spaces*, Academic, New York. (This book is an excellent monograph on the basic theory of Sobolev spaces including generalizations in several directions. Chapters 3 through 6 contain the basic exposition of ordinary Sobolev spaces. A function theory background is required.)

[2] Deny, J. and Lions, J. L. (1955). *Ann. Inst. Fourier (Grenoble)*, **5**, 305–377. (A general treatment in French of Sobolev spaces under the name of Beppo Levi spaces.)

[3] Donaldson, T. K. and Trudinger, N. S. (1971). *J. Funct. Anal.* **8**, 52–75. (A discussion of the basic theory of Orlicz–Sobolev spaces with imbedding theorems.)

[4] Krasnosel'skii, M. A. and Rutickii, Ya. B. (1961). *Convex Functions and Orlicz Spaces*. Groningen, The Netherlands. Noordhoff. (A monograph on Orlicz spaces, a generalization of L^p spaces.

[5] Lions, J. L. (1961). *Math. Scand.*, **9**, 147–177. (This and next paper are the basic works on the generalization of Sobolev spaces for m real value, i.e., fractional order spaces. Both in French.)

[6] Lions, J. L. (1963). *Math. Ann.*, **151**, 41–56.

[7] Morrey, C. B. (1940). *Duke J. Math.*, **6**, 187–215.

[8] Schultz, M. and Varga, R. (1967). *Numer. Math.*, **10**, 345–369.

[9] Sobolev, S. L. (1938). *Mat. Sb.*, **46**, 471–496. (This paper and next monograph are the two that introduced the notion of Sobolev spaces. Both in Russian.)

[10] Sobolev, S. L. (1950). *Applications of Functional Analysis in Mathematical Physics*, Leningrad, USSR. [English translation: *Amer. Math. Soc. Transl., Math. Mono.* **7**, (1963).]

[11] Wahba, G. (1976). In *Applications of Statistics*, P. R. Krishnaiah, ed. North-Holland, Amsterdam, The Netherlands, pp. 507–524.

[12] Wegman, E. J. and Wright, I. W. (1983). *J. Amer. Statist. Ass.*, **78**, 351–365. (A basic discussion of splines in statistics—Sobolev spaces play an integral role in formulating the optimization problem for which splines are the solution.)

(CURVE FITTING
DENSITY ESTIMATION
GRADUATION
INTERPOLATION
SPLINE FUNCTIONS)

EDWARD J. WEGMAN

SOCIAL NETWORK ANALYSIS

Social network analysis is the modern day extension of the structural analysis of sociometric data (*see* SOCIOMETRY). Data that are collected by sociometric tests or similar procedures that record attractions and rejections can be represented graphically (by sociograms) or by matrices. The complexity of the choices made can vary from simple choices of others to weighted or positive and negative choices, and there can be a number of criteria for choices, involving different aspects of task or social–emotional behavior. With large group size or with complex data collection, analytic procedures quickly become difficult to interpret and standards of comparisons of structures become an unspecified area.

Early procedures for the analysis of networks were mechanical, direct operations on the matrices, in large part because computing power that is currently available simply was unheard of at that time. Thus, a procedure advanced by Forsyth and Katz [8] required rearrangement of rows and columns to permit examination of structures. The principle was extended by Beum and Brundage [4], who suggested the analytic principle of maximizing the choices on the main diagonal, which was then carried out as an iterative procedure. The diagonalizations were not restricted to discrete choices, but could accommodate any weighted values, and thus resembled the procedures of cluster analysis. Bock and Husain [5] moved the analytic development of the rearrangement procedures exactly in this direction, and then brought to the analysis specific questions about the boundaries of subgroups. Unfortunately, the advance of procedure was limited by the analytic technique, and so criteria for comparison of subgroup structures were not actually established. Since algorithms exist for the rearrangement of matrices, these types of structural analyses are commonly available, but it is clear that once applied the rearranged matrices still require an interpretation that is not standardized, i.e., intuitive interpretation. Similar more formalized procedures have been tried extensively, including component analysis* and factor analysis*, but interpretation after the fact has not encouraged broad use of the procedures.

Matrix multiplication and other similar procedures were initiated at about the same time by Luce and Perry [15] and Katz [12]. Raising the matrix of choices or nonchoices (values of 0 and 1) to a power of 2 (squared matrix) leads to the identification of mutual pairs, and summary statistics such as the number of mutual pairs for each person or the group are readily derived in the procedure. Raising the matrix to a higher power permits identification of cliques based on mutual choices, and requiring that all members in the clique choose each other. This relatively direct and simple procedure has been too limiting to be useful because clique definitions have been more loosely considered, emphasizing that in the relatively arbitrary choice behavior of subgroup members, some choices may not be mutual or some paths of choice between some of the subgroup members may be altogether vacant. Luce [14] noted how the procedures could be specified so that looser definitions of cliques could be used, and this additional modification essentially set the course of what has become social network analysis.

While historians may find different circumstances to suggest the reason for the growth of interest in social network analysis in the 1970s, the availability of large computing facilities and myriad programs of algorithms that could be adapted to the analytic problems obviously must be recognized. The objectives of analysis from the early period persist in the distinction made by Burt [6] of methods to delineate relations or group structures, methods to delineate the positions occupied by persons, and alternate ways of describing the attributes associated with a specific person as defined by the choice structure. To the extent that specific patterns can be identified, patterns occupied by persons may be compared.

Alba's review of network analysis [1] particularly relates more recent developments of analytic techniques to their antecedents in the 1940s and 1950s. Alba emphasizes the importance of graph theory* applications to the development, including his own work (Alba [2]). Additional developments include the work of Schwartz [16] on block model construction, which emphasizes the areas of high density of choice and those of high density of nonchoices, using procedures familiar in matrix analysis (component analysis). While publication on the technology of social network analysis continues to accumulate (e.g., Batchelder and Lefebre [3], Freeman [9], Holland and Leinhardt [10, 11], Knoke and Kuklinski [13], Burt and Minor [7]), the frequency of application in actual research is quite rare. The social network analysis procedures do not generate easily interpretable results, and there is frequently an impression that intuitive analyses or more rudimentary procedures are as effective as the technological refinements. Additionally, structural hypotheses about comparisons of groups and research questions that are amenable to exploration through network analysis are not easily formulated.

References

[1] Alba, R. D. (1981). From small groups to social networks. *Amer. Behav. Sci.*, **24**, 681–694.

[2] Alba, R. D. (1973). A graph-theoretic definition of a sociometric clique. *J. Math. Soc.*, **3**, 113–126.

[3] Batchelder, W. H. and Lefebre, V. A. (1982). A mathematical analysis of a natural class of partitions of a graph. *J. Math. Psychol.*, **26**, 124–148.

[4] Beum, C. O. and Brundage, E. G. (1950). A method for analyzing the sociomatrix. *Sociometry*, **13**, 141–145.

[5] Bock, R. D. and Husain, S. Z. (1950). An adaptation of Holzinger's *B*-coefficients for the analysis of sociometric data. *Sociometry*, **13**, 146–153.

[6] Burt, R. S. (1978). Cohesion versus structural equivalence as a basis for network subgroups. *Sociological Meth. Res.*, **7**, 189–212.

[7] Burt R. S. and Minor, M. J. (1983). *Applied Network Analysis*. Sage, Beverly Hills, CA.

[8] Forsyth, E. and Katz, L. (1946). A matrix approach to the analysis of sociometric data: Preliminary report. *Sociometry*, **9**, 340–347.

[9] Freeman, L. C. (1980). *Q*-analysis and the structure of friendship networks. *Ind. J. Man-Machine Studies*, **12**, 367–378.

[10] Holland, P. W. and Leinhardt, S. (1977). A dynamic model for social networks. *J. Math. Soc.*, **5**, 5–20.

[11] Holland, P. W. and Leinhardt, S. (1973). The structural implications of measurement error in sociometry. *J. Math. Soc.*, **3**, 85–111.

[12] Katz, L. (1947). On the matrix analysis of sociometric data. *Sociometry*, **10**, 233–241.

[13] Knoke, D. and Kuklinski, J. H. (1983). *Network Analysis*. Sage, Beverly Hills, CA.

[14] Luce, D. R. (1950). Connectivity and generalized cliques in sociometric group structure. *Psychometrika*, **15**, 169–190.

[15] Luce, D. R. and Perry, A. D. (1949). A method of matrix analysis of group structure. *Psychometrika*, **14**, 95–116.

[16] Schwartz, J. E. (1977). An examination of Concor and related methods for blocking sociometric data. In *Sociological Methodology*, D. R. Heise, ed. Jossey-Bass, San Francisco, CA.

(SOCIOLOGY, STATISTICS IN SOCIOMETRY)

EDGAR F. BORGATTA
KYLE KERCHER

SOCIAL SECURITY STATISTICS

The primary role of a social security program is to provide economic protection to individuals when events occur that result in an individual's loss of income from possible employment. In accomplishing this purpose, the system develops a vast amount of statistical data. Much of this data is needed for the administration of the program and for planning for future changes in it. Still other data are valuable for general demographic, economic, and social analyses.

SCOPE OF SOCIAL SECURITY PROGRAMS

The International Labor Office has defined "social security" as having nine branches. Many systems integrate several branches in the operation of a single program. These branches are old-age retirement pensions, invalidity (or long-term disability) pensions, survivor pensions, unemployment benefits, sickness (or short-term disability) cash benefits, medical care, maternity benefits (cash and medical care), work-connected accident and disease benefits (cash and medical care), and family (or child) allowances [4].

Social security programs often cover the entire population of a nation, or else the entire working population (including retired former workers) and their dependents. In some countries—especially economically developing ones—only select groups of workers are covered; either those in certain geographic areas or those in certain types of employment.

Governmentally administered social security programs may take one of three forms. Many countries have two or even all three forms. The most common form is social insurance, under which the benefit amounts are precisely defined (usually based on prior earnings) and are payable as a right if specified insured-status conditions are met and the insured risk occurs. Social (or public) assistance provides benefits based on the needs or means of the eligible person. Demogrant programs pay flat benefit amounts to all eligibles who meet the demographic requirements, regardless of need or any insured-status conditions.

Mandatory–employer plans are required by governmental law or regulation, but are entirely administered by employers. Subsidized–voluntary plans are administered by private organizations or by governmental agencies, with people being encouraged to participate by the presence of a subsidy from governmental funds.

Social security programs are financed in several different ways. Payroll taxes paid by employers and employees (not necessarily in equal proportions, and in some instances only by employers) and by self-employed persons are usually used in social insurance systems. Taxes of a uniform amount are sometimes used in demogrant programs, although more often the financing is through general revenues. Social assistance programs are financed through general revenues. Employer-mandated plans are financed by employers (possibly with some employee contributions), while subsidized-voluntary plans are paid for by premiums from the participants (plus the government subsidy).

The various types of social security programs in other countries are summarized in *Social Security Programs Throughout the World* [6].

Other countries frequently publish recurring statistical reports on the operations of their social security programs. On the international front, the diverse nature of social security programs (as compared to, say, vital statistics*) makes publication of comparable data very difficult. However, the International Labor Office periodically compiles data on the financial transactions of social security systems throughout the world [1].

Administrative procedures often hinder or prevent the tabulation of the pertinent data, so that subsequent analysis is not possible. It almost goes without saying that the actual operation of the system—the collection of contributions and the payment of benefits—must take precedence over the collection of statistical data.

Frequently, social security institutions are so heavily pressed in maintaining their vast operations that no time and resources are available to "mine" the statistical wealth that they possess. For example, their electronic data processing equipment might be so overburdened with day-to-day operations that it is never (or only rarely) available for statistical and research purposes (and, even then, such work has a low time priority, and the resulting data may be quite delayed in becoming available).

A vast statistical system is sometimes developed at the inception of a program. However, over the years, significant parts of it are dropped. Also, many statistics that are obtained are often not available until several years after the event, thus making them less valuable.

Certainly, this has been the case with the U.S. Old-Age, Survivors, and Disability Insurance (OASDI) program. In the early years of operation (the late 1930s and the 1940s), vast tabulations of data on both covered workers and beneficiaries were planned for and were begun, but many of these plans have fallen by the wayside. For example, for some years, an extensive amount of data on covered workers, including earnings histories, was published, but such publication has ceased [5]. The same was also true for the Medicare program, which began operations in 1966. For several years, an extensive series of statistical reports on all aspects of the program was issued. Now, this has tapered off, and the latest ones (which are only a portion of what was originally prepared) are for 1974 and 1975 [2].

U.S. DATA POTENTIALLY AVAILABLE

Social Security data are basically available from two different sources—accounting procedures and reporting procedures. Quite naturally, any program involving the receipt and disbursement of money must maintain adequate accounting records, so that proper fiscal control is present. As a result, such data are almost always readily available. They show the income to the system by type (such as payroll contributions, payments from the government, and income from any invested assets) and the disbursements by type (such as administrative expenses and benefit payments, the latter often being subdivided by category—old-age pensions, widow pensions, etc.).

The operation of the U.S. Old-Age, Survivors, and Disability Insurance systems makes a large amount of data with regard to beneficiaries potentially available (both for new awards and for beneficiaries on the roll).

This program is financed by contributions from covered persons and employers on the basis of earnings. Consequently, a wealth of information can be obtained as to wages and self-employment income. Not only is it possible to have earnings data for a particular period (such as a year), subdivided by such variables as age, sex, industry, geographical areas, etc., but also lifetime work histories can be traced through time.

PUBLICATIONS OF U.S. PROGRAM DATA

Relatively current data are available each month (with more detail quarterly) in the *Social Security Bulletin*. The data available on a monthly basis include the fiscal operations of the trust funds and the number of beneficiaries and average benefit amounts by type of benefit for both new awards and those on the roll. The number of covered workers and their total earnings (both taxable and total) are available on a quarterly basis, as also is more detailed information on beneficiaries (such as data by states, benefits being withheld because of employment, and the extent of early retirement with reduced benefit amounts).

Furthermore, the *Annual Statistical Supplement to the Social Security Bulletin* gives a vast amount of data, especially historical series, with much detail by age, sex, and race, for both covered workers and beneficiaries.

With the split-off of the Medicare program from the jurisdiction of the Social Security Administration to the Health Care Financing Administration (both in the Department of Health and Human Services, formerly the Department of Health, Education, and Welfare), data on this program is available in HCFA's *Review* and its *Financing Notes*. Due to the nature of medical-service benefits, Medicare data are more difficult to gather for the periodic cash-benefits OASDI program. However, data on the fiscal operations of the Medicare trust funds and on aggregate operations (by numbers of bills, amounts reimbursed, and days of hospital care, where applicable) for broad categories of services (e.g., hospital, skilled nursing facilities, home health, physicians, outpatient, and independent laboratories), subdivided between aged and disabled beneficiaries, are relatively currently available. But detailed data by such elements as type of illness, duration of services, age, and sex are either unavailable or published with a considerable time lag.

The Social Security Administration also carries out longitudinal* studies of beneficiaries on a small-sample basis over extended periods of time, such as for new retirement beneficiaries and for disabled-worker beneficiaries. Valuable statistical data are also contained in the three annual reports of the Boards of Trustees of the OASDI and Medicare trust funds and in various research reports of the Social Security Administration and the Health Care Financing Administration.

Data on the various public assistance programs in the United States are published monthly in the *Social Security Bulletin*. Such monthly data consist of numbers of persons and amounts of expenditures, by broad categories by states. More detailed data are available in special reports.

The unemployment insurance program in the United States consists of separate systems in each state. The operations are summarized for each year, for the nation as a whole and for each state separately [9]. The data available include average monthly employment, total taxable wages, average weekly wages, trust fund operations, average benefit cost relative to payroll, average contribution rates, reserve ratios, number of first payments, number exhausting benefit rights, average duration of benefits, and average weekly benefits. Detailed data by such elements as age, sex, and occupation or industry are not available.

COMPARING AND LINKING PROGRAM DATA WITH SURVEY DATA

Because program data are collected for the purpose of operating the OASDI system, they tend to be quite complete and accurate

—perhaps more so than is information collected in survey interviews [8]. For example, an individual's age must usually be determined with considerable accuracy when benefits are claimed. Often, such determination is largely based on the reported (or even proven) age of the person at the time of enrollment in the program many years previously. Thus, a specific age is recorded in the records and is maintained over the years. On the other hand, in survey interviews, a particular person might very well not give consistent responses over the years.

Also, earnings data is likely to be more accurately recorded for OASDI purposes than in survey interviews. This is because of the tax aspects of the program and the possibility of verification through detailed inspection of the employer's records and comparison with other tax information available in the government. On the other hand, when survey data are collected, an interviewer may not be given correct information to a query on income (either because of lack of knowledge or poor memory of the respondent—or even intentional misreporting, which carries no penalties).

Also, gaps in data due to nonresponse are less likely in OASDI data than in data obtained by interviews, because of the continuing nature of the person's contact with the system (especially so as to beneficiaries). For example, in the case of deaths, data obtained from the required vital statistics registration procedures may be incomplete because of failure to report in some cases and also may be inaccurate as to the details (such as age) in other cases [3]. On the other hand, OASDI data as to deaths of wage-earners or beneficiaries is more complete (because of the financial-transactions aspects) and accurate (because the demographic details had been recorded many years ago).

On the other hand, because OASDI data are collected for program operational purposes, some types of information are not collected, even though they would be extremely useful for research and policy planning purposes. For example, OASDI earnings records are a valuable data source for analyzing lifetime work and earnings patterns. However, because earnings are recorded and taxes are paid on an individual basis for program purposes, information is not collected on the marital or family status of individual workers. If such data are needed for research or planning, they must come from other sources. Also, because OASDI taxes are paid only on earnings up to a specified maximum, individual earnings above this level are not reported. If total earnings are needed for research or planning, income from earnings for those who earn more than the taxed amount must be estimated or obtained from other sources.

Finally, while benefit data are quite complete and accurate, they too are limited to the requirements of program operations. Information on marital or family characteristics of retirees is obtained when they file for benefits. If, however, these characteristics do not affect the actual or potential benefits payable in the individual case, that information has not in the past been entered into the computerized record system. Also, because an individual's old-age benefit amount is usually not affected by the other retirement income that he or she receives, that information is not collected. Other data sources, such as periodic surveys, are used to estimate the total income of beneficiaries and the extent to which OASDI benefits are supplemented by private pensions, income from assets, or other types of retirement resources.

In some instances, OASDI record data is linked with information from special-purpose surveys or with income tax returns so as to merge the unique advantages of each data source. Care is taken to protect the confidentiality of individual information. Data linkages can be time-consuming and, as a result, may not be as up-to-date as one might wish. Nonetheless, they are a valuable resource for some types of policy analysis. Examples of research done with the OASDI program and interview data are compiled in a research report, "Policy Analysis with Social Security Research Files" [7].

SUMMARY

Social security systems, by their very operation, develop a vast amount of relatively accurate data, which can serve valuable purposes in many social and economic areas outside of themselves. For example, this can be the case in connection with studies of national income and of employment, unemployment, and underemployment. Similarly, much valuable national demographic data can best be obtained from the social security records. In some cases, social security data are supplemented by, or merged with, data obtained from survey interviews so as to utilize the unique advantages of each data source.

References

[1] International Labor Office (1985). The Cost of Social Security—Ninth International Inquiry, 1978–80. International Labor Office, Geneva, Switzerland.

[2] Health Care Financing Administration (1978). Medicare, 1973–74, Section 1.2, Summary and Medicare, 1975, Section 2, Enrollment. Health Care Financing Administration, Washington, DC. (The latest.)

[3] Myers, R. J. (1940). Errors and bias in the reporting of ages in census data. *Trans. Actuarial Soc. Amer.*, **41**, 395–415.

[4] Myers, R. J. (1951). New International Convention on Social Security. *Social Security Bulletin*, October.

[5] Social Security Administration (1968). Workers under Social Security. Social Security Administration, Washington, DC. (The latest and the last.)

[6] Social Security Administration (1977). Social Security Programs Throughout the World, 1977. *Research Report No. 50*, Social Security Administration, Washington, DC.

[7] Social Security Administration (1978). Policy Analysis with Social Security Research Files. *Research Report No. 52*, Social Security Administration, Washington, DC.

[8] Spiegelman, M. (1968). *Introduction to Demography*. Harvard University Press, Cambridge, MA.

[9] U.S. Department of Labor (1978). *Handbook of Unemployment Insurance Financial Data, 1938–76*. Employment and Training Administration, U.S. Department of Labor, Washington, D.C. (With annual up-dating inserts.)

(BUREAU OF LABOR STATISTICS
DEMOGRAPHY
FEDERAL STATISTICS
SOCIAL STATISTICS)

ROBERT J. MYERS

SOCIAL STATISTICS

Social statistics is concerned with the collective aspects of human society. Many of the pioneers of modern statistics were social statisticians, including John Graunt (1620–1674), Sir William Petty (1623–1687), Sir John Sinclair (1754–1835), Adolphe Quetelet* (1796–1874), William Farr (1807–1883), and Florence Nightingale* (1820–1910). The Statistical Society of London, later to become the Royal Statistical Society*, was founded in 1834 with the object of procuring, arranging, and publishing "Facts calculated to illustrate the Condition and Prospects of Society," limited as far as possible to "facts which can be stated numerically and arranged in tables." Its original motto, later deleted, "to be threshed by others" shows that the founders drew a sharp distinction between the collection and the analysis of data. Although such separation would, today, be considered unnecessary and undesirable, the collection of accurate and timely social data is a vital part of the work of government and United Nations statistical agencies. They, together with other organizations, have greatly benefitted from the information-processing capabilities of modern computers. There is also a longstanding tradition of private statistical studies on the social condition of the people, going back to such classics as Charles Booth's *Life and Labour of the People of London* published between 1889 and 1903. Further historical material in this vein is given in Kendall [30], Bulmer [11], and Marsh [36]. A history of social statistics in Britain from 1660 to 1830 and an account of the early statistical societies and government statistics is given by Cullen [15].

Modern social statistics has developed rapidly under the twin influences of computers and theoretical work in modelling and inference. In addition to their traditional role of quantifying such matters as poverty, crime, welfare, health, education, and employment, social statisticians have increasingly turned to modelling the random processes that arise in the detailed study of phenomena such as absenteeism, labour turnover, and recidivism. The storage and processing facilities of large computers have led, among other things, to the establishment of data archives with the opportunities that they offer for secondary analyses and the consequent strengthening of the theoretical basis of the subject. On the methodological side, the annual publication *Sociological Methodology* (Jossey-Bass) includes much relevant material, as do many of the Sage University *Paper Series in Quantitive Applications in the Social Sciences* and the eight volumes published in the series *Progress in Mathematical Social Science* (Elsevier). Although there is no journal of social statistics, the field is well catered to by the main statistical journals as well as some of the specialist journals in the neighbouring fields of psychology, sociology, education, and economics. The American Statistical Association* publishes an annual *Proceedings on Social Statistics*. Bartholomew [4] provides a review of the methodology of social statistics with extensive references and reviews [6] some recent developments. There are numerous elementary texts on statistics for social scientists, but many are little more than routine treatments with social examples. A notable exception is Blalock's *Social Statistics* [10], which is firmly rooted in the social sciences, and is sound statistically. Two other recommended texts are Fuller and Lury [19] and Loether and McTavish [34]. On the practical side there is a growing use of *social indicators*. [A social indicator is a statistic that "is interpreted as measuring progress or retrogression towards or away from an accepted goal" (Hauser [24]).] They take their place beside the more familiar economic indicators as measures of social well-being. Starting with *Social Trends*, published annually in the U.K. since 1970, there are now at least 30 similar publications throughout the world (see Horn [26]).

CENSUSES AND SURVEYS

Censuses, in which data are obtained from every member of the population, are long established. Population censuses go back to 1790 in the United States and 1801 in Britain. However, the usefulness of information based on random samples was recognized by Arthur Bowley in the early years of this century and used by him in his study of poverty published in 1915 as *Livelihood and Poverty*. With the development of the theory of sampling from finite populations* stemming from Neyman [38], the sample survey has come to be the principal tool of data gathering. Simple random samples are rarely used nowadays and a substantial literature on complex designs intended to give maximum precision at minimum cost has grown. The theory of inference from such complex samples has lagged behind practice but considerable strides have been made to close the gap (see, for example, Kish and Frankel [32], Bebbington and Smith [9], Holt et al. [25], Nathan and Holt [37], and, for further references, Kalton [29]). A simple expedient is to determine what is known as the *design effect*, the factor by which the simple random sampling* variance of an estimator must be multiplied to yield the true variance. An alternative approach to the analysis of survey data is the so-called *model-based treatment* as discussed in Ericson [16], Smith [46], and Cassel et al. [13], for example. This brings finite population inference within the general framework of statistical inference by supposing that the population itself is a sample from a superpopulation. The merits of the two approaches from a practical point of view are considered by Kalton [29].

Uncertainty in surveys* arises not only from the act of sampling but also from response errors. These may arise, either from failure to contact the respondent, or from the inaccurate recording of information. A review of methods and problems is given in O'Muircheartaigh [39].

There are many well-established texts on survey design and analysis, including Kish [31], Konijn [33], Cochran [14], and Yates [52]. For a practical guide to data collection* in developing countries see Casley and Lury [12].

METHODOLOGY OF SOCIAL STATISTICS

The methods available to analyse social data are strongly influenced by the practical and, sometimes, ethical restrictions imposed on their collection. It is rarely possible to experiment in a social context so we have to make do with observational data (but see Fienberg et al. [17] for examples of social experiments). This makes it inevitable that important factors will often be confounded, making causal inference virtually impossible. Social data are particularly prone to missing values and incompleteness due to limitations on what can be observed. For example, social processes develop in time so that full observation over a lengthy period is necessary to obtain the full picture. Yet, for practical reasons, it may only be possible to take cross-sectional views at widely spaced intervals of time. If, for example, one wishes to estimate the distribution of the duration of some socially interesting phenomenon, the data at hand are liable to be truncated* or censored*.

Where the process of change is of the essence, as in studies of child development or changes in attitudes, it is sometimes possible to collect longitudinal data* by observing particular individuals over an extended period of time. This has many advantages but it is costly and sometimes means that the results are of historical interest only by the time they are obtained. Goldstein [23] provides a useful guide to problems in this field.

Another characteristic of social data that has strongly influenced the methodology of social statistics is that they are frequently both categorical and multivariate. Even when it is possible to measure variables in units of time or money, it is common to find that they have highly skewed distributions that are not well suited to the normal theory methods of the text books. Traditional multivariate analysis*, as exemplified in Anderson [1], is heavily concentrated on inference about the multivariate normal distribution and therefore has relatively little to offer. The emphasis in social research has been on exploratory techniques making minimal distributional assumptions and capable of using categorical*, as well as metrical data. These include cluster analysis, multidimensional scaling*, and principal components*. A good account of the use of such methods in survey analysis is in O'Muircheartaigh and Payne [40]. Log-linear models and methods of latent structure analysis* are likewise particularly suited to social data. Many of the standard statistical methods, including those based on the general linear model*, find many applications in social statistics though the distributional assumptions are often suspect. For this reason distribution-free methods* have an important role to play.

MODELLING SOCIAL PROCESSES

Social systems involve a high degree of uncertainty, stemming both from the unpredictability of human behaviour and the randomness of the economic and social environment. Dynamic models for such systems therefore need to be stochastic and the theory of stochastic processes* thus provides the tools for studying social change. Some of the work in this field has been highly practical, aimed, for example, at predicting wastage, recruitment, and career prospects for manpower planners. Other developments have been directed at gaining insight into social mechanisms. For example, studies of competing social groups aim to determine the conditions under which a viable equilibrium may be established or where one group may eventually eliminate its rivals. The former approach is illustrated in Bartholomew and Forbes [8] and the latter in Bartholomew [5]; both books contain extensive bibliographies. The statistical analysis of data arising from such stochastic processes is relatively undeveloped and its literature is widely scattered. A brief introduction is given in Bartholomew [2] and examples of recent progress will be found in Ginsberg [20–22], Singer and Cohen [43], Singer and Spilerman [44, 45], Plewis [41], and Tuma et al. [49]. There has been work on the stochastic analysis of patterns of interpersonal relations, where members of a population are classified according to whether or not a link of some kind exists between them. Volume 5, Number 1, of the *Journal of Mathematical Sociology* (1977) was devoted to this topic; see also Wasserman [51], Sorensen and Hallinan [48], and Frank [18].

MEASUREMENT (OR SCALING)

Measurement is basic to science, yet in the social field it presents peculiar difficulties. Many of the quantities that occur in social science discourse are not susceptible to direct measurement. Cost of living, general intelligence, and quality of life are examples. The traditional way of dealing with this problem has been to measure a number of indicator variables that are supposed to be correlated with the quantity in question. An index is then constructed from them by some form of averaging. Principal components provide one technique for constructing linear combinations of indicators that may serve this purpose. Other methods of scaling are in wide use (see, for example, van der Ven [50]), but in recent work the trend has been toward using models in which the quantity to be measured is represented by one or more latent variables. Factor analysis* and latent trait analysis are both based on such models. The linear factor model has been incorporated, as the measurement component, in a widely used model for linear structural relations implemented in the LISREL* program as described in Jöreskog and Sorböm [28] and Jöreskog [27].

If the indicator variables are categorical, as in much work in educational testing, the treatment is similar, originating in Birnbaum's contribution to Lord and Novick [35]. A general class of models was proposed [3] and reviewed [7] by Bartholomew.

A somewhat different class of measurement problems arises in the study of social processes, illustrated by the example of social mobility. Individuals or families move between occupational or social classes; in some societies this happens more frequently than in others. A single measure of mobility is required so that comparative judgments can be made about the degree of mobility in different societies or in the same society at different times. Various ways of summarizing flow information have been proposed, but the position is greatly clarified if we start with a model of the mobility process. The parameters of that model can then be used to suggest and construct an appropriate measure. For example, if mobility is adequately described by a Markov chain the problem is one of mapping the set of transition matrices onto some convenient interval of the real line. This approach is used in Shorrocks [42] and Sommers and Conlisk [47]. See also Bartholomew [5].

THE FUTURE

It is probable that there will be a growing demand from governments and society in general to be more adequately informed about "the Conditions and Prospects of Society." There will be an increasing need for international cooperation, and the example of the World Fertility Survey shows what can be achieved. Sponsored by the International Statistical Institute*, this was probably the largest and most complex exercise undertaken in social statistics. It required questionnaires to be administered in many languages in very varied cultural settings. In addition to the vast store of information on fertility* that it provided, it has contributed significantly to the methodology of social research. Sir Maurice Kendall, its director, was awarded a United Nations peace medal in recognition of his work.

A broad ranging survey of Social Statistics in the year 2000 was made by Hauser [24], who saw computers* playing a key role in providing comprehensive data banks for households and individuals. This poses formidable problems in reconciling the demands for privacy and confidentiality with the legitimate needs of research. Computers will obviously play a major role both in collection and storage and in analysis. The ability of quite small computers to present data graphically in a sophisticated manner will stimulate the search for better ways of exploring data structures. The recognition that an adequate model is a prerequisite for sound analysis will provide a further impetus for new theoretical developments.

References

[1] Anderson, T. W. (1985). *An Introduction to Multivariate Analysis*, Wiley, New York. 2nd. ed.

[2] Bartholomew, D. J. (1977). In *The Analysis of Survey Data*, Vols. 1 and 2, C. A. O'Muircheartaigh and C. Payne, eds. Wiley, Chichester, England, pp. 145–174. (Deals with inference about stochastic processes in a social context.)

[3] Bartholomew, D. J. (1980). *J. R. Statist. Soc. Ser. B*, **42**, 293–321. (Introduces a family of models for factor analysis of categorical data. There is a discussion and reply.)

[4] Bartholomew, D. J. (1981). *Mathematical Methods in Social Science*. Wiley, Chichester, England. (The first guide book in Wiley's *Handbook of Applicable Mathematics*. It reviews much of the methodology of social statistics.)

[5] Bartholomew, D. J. (1982). *Stochastic Models for Social Processes*, 3rd ed. Wiley, Chichester, England. (Contains over 700 references to work in this field.)

[6] Bartholomew, D. J. (1983). *Int. Statist. Rev.*, **51**, 1–9. (A review of recent developments in social statistics.)

[7] Bartholomew, D. J. (1983). *J. Econometrics*, **22**, 229–243.

[8] Bartholomew, D. J. and Forbes, A. F. (1979). *Statistical Techniques for Manpower Planning*. Wiley, Chichester, England. (A manual with many worked examples and several computer programs.)

[9] Bebbington, A. C. and Smith, T. M. F. (1977). In *The Analysis of Survey Data*, C. A. O'Muircheartaigh and C. Payne, eds. Wiley, New York, pp. 175–192.

[10] Blalock, H. (1981). *Social Statistics*, rev. 2nd ed. McGraw-Hill, New York. (A well-tried text that has served students well over many years but with very little material on multivariate methods.)

[11] Bulmer, M. (1982). *The Uses of Social Research*. Allen and Unwin, London, England.

[12] Casley, D. J. and Lury, D. A. (1981). *Data Collection in Developing Countries*. Oxford University Press, Oxford, England.

[13] Cassel, C. M., Wretman, J. H., and Sarndal, C. E. (1977). *Foundations of Inference in Survey Sampling*. Wiley, New York.

[14] Cochran, W. G. (1977). *Sampling Techniques*, 3rd ed. Wiley, New York. (A widely used text with a practical slant.)

[15] Cullen, M. J. (1975). *The Statistical Movement in Early Victorian Britain*. The Harvester Press Ltd., Barnes and Noble Books, New York. (Based on a thesis, this book makes extensive use of primary sources.)

[16] Ericson, W. A. (1969). *J. R. Statist. Soc. Ser. B*, **31**, 195–234. (Describes a Bayesian approach to inference in sampling finite populations.)

[17] Fienberg, S. E., Singer, B., and Tanur, J. M. (1984). Large scale social experimentation in the U.S.A. In *Centennial Volume of the International Statistical Institute*, to appear.

[18] Frank, O. (1981). In *Sociological Methodology 1981*, S. Leinhardt, ed. Jossey-Bass, San Francisco, CA. (A survey of statistical methods in graph analysis.)

[19] Fuller, M. and Lury, D. A. (1977). *Statistics Workbook for Social Science Students*. Philip Alan, Deddington, Oxford, England. (An introductory book for beginners with minimal mathematical knowledge. The exposition includes many examples.)

[20] Ginsberg, R. B. (1978). *Environ. Plann. A*, **10**, 667–679.

[21] Ginsberg, R. B. (1979). *Stochastic Models of Migration: Sweden 1961–1975*. North-Holland, Amsterdam, The Netherlands.

[22] Ginsberg, R. B. (1979). *Environ. Plann. A*, **11**, 1387–1404.

[23] Goldstein, H. (1979). *The Design and Analysis of Longitudinal Studies: Their Role in the Measurement of Change*. Academic, London, England.

[24] Hauser, P. M. (1977). Social Statistics in 2000. *ASA Proc. Social Statis.*, pp. 46–52.

[25] Holt, D., Smith, T. M. F., and Winter, P. D. (1980). *J. R. Statist. Soc. Ser. A*, **143**, 474–487. (Discusses regression analysis from complex surveys.)

[26] Horn, R. V. (1978). *Aust. J. Statist.*, **20**, 143–152.

[27] Jöreskog, K. G. (1977). In *Applications of Statistics*, P. R. Krishnaiah, ed. North-Holland, Amsterdam, The Netherlands, pp. 265–287. (Theory and applications of structural equation models in the social sciences.)

[28] Jöreskog, K. G. and Sorböm, D. (1977). In *Latent Variables in Sociometric Models*. North-Holland, Amsterdam, The Netherlands, pp. 285–325. (A fundamental paper describing the LISREL model and its applications.)

[29] Kalton, G. (1983). *Int. Statist. Rev.*, **51**, 175–188.

[30] Kendall, M. G. (1972). In *Man and the Social Sciences*, W. Robson, ed. Allen and Unwin, London, England, pp. 131–147. (A masterly survey of the origins and development of social statistics with prospects for the future.)

[31] Kish, L. (1965). *Survey Sampling*. Wiley, New York. (A classic in its field. Still worth reading in spite of its age.)

[32] Kish, L. and Frankel, M. R. (1974). *J. R. Statist. Soc. Ser. A*, **126**, 557–565. (A pioneering paper on inference from complex sample designs.)

[33] Konijn, H. S. (1973). *Statistical Theory of Sample Survey Design and Analysis*. North-Holland, Amsterdam, The Netherlands.

[34] Loether, H. J. and McTavish, D. G. (1974). *Descriptive Statistics for Sociologists, An Introduction* and *Inferential Statistics for Sociologists, An Introduction*, Vols. 1 and 2. Allyn and Bacon, Boston, MA. (A good introductory treatment with a strong orientation to sociology.)

[35] Lord, F. M. and Novick, M. R. (1968). *Statistical Theories of Mental Test Scores*. Addison-Wesley, Reading, MA. (The best starting point for reading in this field.)

[36] Marsh, C. (1982). *The Survey Method*. Allen and Unwin, London, England.

[37] Nathan, G. and Holt, D. (1980). *J. R. Statist. Soc. Ser. B*, **42**, 377–386.

[38] Neyman, J. (1934). *J. R. Statist. Soc. Ser. A*, **97**, 558–625. (The classic paper on inference in samples from finite populations.)

[39] O'Muircheartaigh, C. A. (1977). In *The Analysis of Survey Data*, Vol. 2, C. A. O'Muircheartaigh and C. Payne, eds. Wiley, Chichester, England, pp. 193–239. (A useful summary of recent work on response errors.)

[40] O'Muircheartaigh, C. A. and Payne, C., eds. (1977). *The Analysis of Survey Data*, Vols. 1 and 2. Wiley, Chichester, England. (Covers both exploratory and model-based methods and gives many examples.)

[41] Plewis, I. (1981). *J. Educ. Statist.*, **6**, 237–255.

[42] Shorrocks, A. F. (1978). *Econometrica*, **46**, 1013–1024. (Should be read in conjunction with Sommers and Conlisk [47].)

[43] Singer, B. and Cohen, J. E. (1980). *J. Math. Biosci.*, **49**, 273–305.

[44] Singer, B. and Spilerman, S. (1975). *Bull. Int. Statist. Inst.*, **46**, 681–697.

[45] Singer, B. and Spilerman, S. (1976). *Ann. Econ. Social Meas.*, **5**, 447–474. (References 43, 44, and 45 are a major contribution to the statistical analysis of stochastic social processes.)

[46] Smith, T. M. F. (1976). *J. R. Statist. Soc. Ser. A*, **139**, 183–204. [A review paper on the foundations of survey sampling (with discussion). Strongly recommended.]

[47] Sommers, P. M. and Conlisk, J. (1979). *J. Math. Sociol.*, **6**, 169–234. (Should be read in conjunction with Shorrocks [42].)

[48] Sorensen, A. B. and Hallinan, M. T. (1977). In Mathematical Models of Sociology, *Sociological Review Monograph 24*, P. Krishnan, ed. University of Keele, Keele, United Kingdom.

[49] Tuma, N. B., Hannan, M. T., and Groenveld, L. P. (1979). *Amer. J. Sociol.*, **84**, 820–854. (Dynamic analysis of event histories.)

[50] van der Ven (1980). *Introduction to Scaling*. Wiley, Chichester, England. (An elementary treatment of many methods including those based on latent variable models.)

[51] Wasserman, S. S. (1978). *Adv. Appl. Prob.*, **10**, 803–818.

[52] Yates, F. (1980). *Sampling Methods for Censuses and Surveys*. 4th ed. Griffin, High Wycombe, England. (First edition appeared in 1949. Based on much practical experience but now somewhat dated.)

(BUREAU OF THE CENSUS
CENSUS
DEMOGRAPHY
INDEX NUMBERS
SOCIOLOGY, STATISTICS IN
SURVEY SAMPLING)

D. J. BARTHOLOMEW

SOCIOLOGY, STATISTICS IN

Those who have written histories of sociology have variously begun their accounts with ancient civilizations, with the Greeks, or with nineteenth century social thought. But most scholars agree that as a specialized, empirical social science, sociology is a twentieth century product, especially if demography* and associated topics such as vital statistics* and census* taking are removed from the discussion.

The development of sociology as a social science is coincident with a shift in emphasis from speculative theorizing to pure fact finding and then to a program involving, to some degree, the cross-fertilization of theory and research. The pre-twentieth century sociologist was basically a social philosopher engaged in purely speculative theorizing. This is not to say that there was no empirical research prior to the turn of the twentieth century. In fact there was, as exemplified in the works of Charles Booth and others (see Lécuyez and Oberschall [31], but there is little doubt that empiricism became a dominant feature of sociology only after the turn of the century.

The trend toward empiricism was most manifest in the United States, where the movement was led by W. I. Thomas, Florian Znaniecki, Robert E. Park, F. S. Chapin, Franklin Giddings, W. F. Ogburn, and Howard W. Odum, among others. Not every one of these men was engaged in data collection* and analysis, but all of them encouraged the empirical approach to the study of social problems. In the 1920s and 1930s a large number of empirical studies were completed, most of them linked to bringing about social amelioration of one kind or another.

After about 1940, however, some sociologists began to feel that empiricism in sociology had gone too far. Pure fact finding, with little investment in theory building, was felt to be as unproductive as pure arm-chair theorizing with no supporting empirical investigations (Merton [32]). Those who subscribed to this view increasingly began to shift their focus from social problems to sociological concerns, that is, to theoretical and methodological issues.

Coincident with the changes outlined above, major shifts in the materials (data) and methods of sociology occurred, and with those shifts a controversy emerged as to whether the adoption of statistical orientation in sociological investigations is profitable for the development of sociology.

MATERIALS OF SOCIOLOGY

Existing Data

Early sociologists used information available from census reports, administrative records, ethnographic sketches, and the like in their investigations. The main analytical strategy was comparison of cases on a nonquantitative basis. The works of Herbert Spencer (1820–1903), Ferdinand Tönnies (1855–1936), and William Graham Sumner (1840–1910) illustrate this strategy. The earlier works of Émile Durkheim (1858–1917) and most of Max Weber's (1864–1920) works also belong to this category.

Some early sociologists did demonstrate their interest in and their capability of analyzing quantitative information available in the sources mentioned above (Durkheim's *Suicide* [14] in 1897, is an example). With the emergence of data archives, however, and especially with the increasing access to public data tapes in recent decades, secondary data analysis (that is, analysis of existing data) has become very common among sociologists. The methods applicable to such analyses resemble those applicable to one's own data. But secondary data are beset with certain problems of their own (Hyman [22]). Two such problems are (1) the information available in the existing data may not be exactly the kind of information needed for the chosen investigation, but only a more or less useful approximation to it and (2) data from different sources may not be comparable because of differences in concepts and definitions used in their collection. There are no easy solutions to problems such as these. The user should recognize them and take them into account when analyzing and interpreting the data. One thing that makes secondary data analysis highly attractive, despite such problems, is that such data are cheap and less time consuming to obtain.

New Data

Although the practice of collecting new data tailored to specific needs obtained from very early times, the procedures for the collection and processing of data and related steps in research began to be codified only in the twentieth century. Two modes of data collection may be distinguished: *participant observation* and *surveys**.

In participant observation, the researcher or a team of researchers goes into a community to take temporary residence there and to observe, interview, and prepare field notes, based on which reports are written on the social life of the community. Quantitative analysis is a subordinate mode in the preparation of such reports. Studies conducted on the basis of participant observation have the advantage that they give an "inside" view of the workings of the life in the community studied. They have the disadvantage, however, that the patterns identified may be of questionable validity, reliability, and representativeness (Friedrichs and Lüdtke [15]).

Survey research differs from studies based on participant observation in that surveys tend to focus on a narrower segment of life (for example, voting behavior), use probability sampling to select units for observation (measurement), and employ structured questionnaires or interview schedules. In the analysis of survey data, quantitative methods are emphasized. Survey research techniques began to be developed in the 1920s, first in connection with market research, and later with reference to opinion polling. Various fields such as health and vital statistics, attitude and knowledge studies, communication research, mobility studies, and many others use survey research today. Different aspects of survey methodology have been codified, so as to make communication easier. An extensive literature is available on sampling, questionnaire construction, coding, interviewing, and analytical techniques (Sonquist and Dunkelberg [40]).

Experiments

In some areas of sociology, such as small group studies, laboratory experiments are practicable. Social psychologists have used experiments to study decision making, problem solving, and other topics. Laboratory experiments have been criticized for their lack of generalizability, a drawback mainly due to the artificiality of the laboratory setting and the nonrepresentativeness of the experimental units (subjects) commonly used. By relaxing experimental control, it may be possible to approximate reality in the laboratory, but then the problem is that it would be difficult to determine the causes of the observed effects (see comments on quasiexperiments below).

METHODS

Collectivities as Units of Analysis

One of the interesting recent developments in sociological methodology has been the increasing attention given to empirical studies of collectivities such as colleges, communities, and corporations. The information on collectivities taken for analysis may be of the *global* type, such as the volume of export or annual budget, which is defined without reference to the properties of micro units that comprise the collectivities, or it may be of the *analytical* type, such as the mean or variance of a trait of constituent units (Lazarsfeld and Menzel [28]). When working with analytical data, problems of aggregation* and disaggregation arise. A very extensive literature on these problems has appeared in econometric* publications. Sociologists also have made some contribution to the literature (Hannan [18]). Recent attempts at codification of methods of collection and analysis of data on collectivities (macro data) include those of Pennings [35] and Price [37].

Contextual Analysis

An important application of macro data occurs in what has become known as *contextual analysis* (Boyd and Iversen [9]). This involves the use of macro variables (global or analytical) in micro equations to account for the variation in, or to explain the relationship among, micro units. The simplest form of contextual analysis is an examination of the relationship between individual attitude or behavior and the attributes of the collectivity of which the individuals in question are members. For example, one may examine whether the tendency of college students to cheat on tests is a function of college size. In more complex analyses, one examines the nature of the relationships among individual characteristics, when various attributes of collectivities are introduced as controls. Thus, for example, one may investigate whether the nature of the association between an individual's self-esteem and religious background varies according to whether the individual belongs to a religious minority in his or her neighborhood.

Causal Models

Another recent development is the emergence of interest in the construction of multiequation stochastic models in which each equation represents a causal relationship rather than a mere empirical association. Models of this kind have been variously referred to in the literature as causal models, path models, or structural equation models; *see also* CAUSATION and PATH ANALYSIS. Causal modeling has a long history in econometrics centering on simultaneous equation systems and in psychology* under factor analysis* or measurement models.

Sociologists have applied causal models in their research with some frequency since the middle of the 1960s. The formal statistical framework underlying such models is rather simple (see, for example, Jöreskog [23]).

Let $\mathbf{F}' = (F_1, F_2, \ldots, F_m)$ be a vector of "dependent" (endogenous) variables and $\mathbf{G}' = (G_1, G_2, \ldots, G_n)$ a vector of "independent" (exogenous) variables. Suppose $\mathbf{F}$ and $\mathbf{G}$ are linked by the linear system of relations

$$\mathbf{BF} = \mathbf{\Gamma G} + \mathbf{H}, \tag{1}$$

where $\mathbf{B}$ and $\mathbf{\Gamma}$ are parameter (coefficient) matrices and $\mathbf{H}' = (H_1, H_2, \ldots, H_m)$ a random vector of residuals (errors in equations, random disturbances). With no loss of generality it may be assumed that $\mathbf{F}$, $\mathbf{G}$, and $\mathbf{H}$ have zero expectations. It is common to assume that $\mathbf{B}$ is nonsingular and that $\mathbf{G}$ and $\mathbf{H}$ are uncorrelated.

Now suppose that $\mathbf{F}$ and $\mathbf{G}$ cannot be directly measured. (They may, for example, be composed of elements standing for unobservable variables such as "cohesion," "permanent income," and "envy.") Suppose $\mathbf{Y}' = (Y_1, Y_2, \ldots, Y_p)$ and $\mathbf{X}' = (X_1, X_2, \ldots, X_q)$ are vectors of indicators (proxies) of $\mathbf{F}'$ and $\mathbf{G}'$, respectively, such that $\mathbf{Y}$ and $\mathbf{X}$ can be directly measured and that

$$\mathbf{Y} = \mathbf{\Lambda}_y \mathbf{F} + \mathbf{U}, \tag{2}$$

$$\mathbf{X} = \mathbf{\Lambda}_x \mathbf{G} + \mathbf{V}, \tag{3}$$

where $\mathbf{U}$ and $\mathbf{V}$ are measurement errors* associated with $\mathbf{Y}$ and $\mathbf{X}$, respectively. Assume that $\mathbf{U}$ and $\mathbf{V}$ have zero expectations and that $\mathbf{U}$ is uncorrelated with $\mathbf{F}$ and $\mathbf{V}$ with $\mathbf{G}$.

Let $\mathbf{\Phi}$, $\mathbf{\Psi}$, $\mathbf{\Theta}_u$, and $\mathbf{\Theta}_v$ be the dispersion matrices of $\mathbf{G}$, $\mathbf{H}$, $\mathbf{U}$, and $\mathbf{V}$, respectively. Then the elements of $\mathbf{\Sigma}$, the dispersion matrix of the observed indicators, are functions of the elements of the parameter matrices $\mathbf{\Lambda}_y$, $\mathbf{\Lambda}_x$, $\mathbf{B}$, $\mathbf{\Gamma}$, $\mathbf{\Phi}$, $\mathbf{\Psi}$, $\mathbf{\Theta}_u$, and $\mathbf{\Theta}_v$. In particular applications, some of the elements of the parameter matrices are assigned specific values, such as zero or one, some are constrained to be equal to known combinations of other unknown parameters, and the rest are subjected to no constraints at all. Thus, in any particular case, there are fixed, constrained, and free parameters.

It should be noted that many structures of the parameter matrices may give rise to the same dispersion matrix $\boldsymbol{\Sigma}$. If the value of a parameter is the same in all parameter structures that give rise to the same $\boldsymbol{\Sigma}$, consistent estimation of the parameter is possible, otherwise not.

To estimate the parameters in a model, one fits $\boldsymbol{\Sigma}$ to the sample dispersion matrix $\mathbf{S}$. Maximum likelihood* methods can be applied for the purpose (Jöreskog and Sörbom [24]). Goodness of fit* of a given model can be tested, in large samples, using the likelihood ratio* technique.

Equations (2) and (3) can be combined to yield

$$\mathbf{Z} = \boldsymbol{\Lambda}\mathbf{M} + \mathbf{E}, \tag{4}$$

where

$$\mathbf{Z} = \begin{bmatrix} \mathbf{Y} \\ \mathbf{X} \end{bmatrix}, \quad \boldsymbol{\Lambda} = \begin{bmatrix} \boldsymbol{\Lambda}_y & \mathbf{0} \\ \mathbf{0} & \boldsymbol{\Lambda}_x \end{bmatrix},$$

$$\mathbf{M} = \begin{bmatrix} \mathbf{F} \\ \mathbf{G} \end{bmatrix}, \quad \text{and} \quad \mathbf{E} = \begin{bmatrix} \mathbf{U} \\ \mathbf{V} \end{bmatrix}.$$

Equation (4) shows that the measurement model represented by (2) and (3) can be viewed as a restricted factor analysis model in which the factors F and G are constrained to satisfy a linear system of the form (1), and in which certain zero factor loadings are specified in advance. Note, however, that the general model outlined above does not require that m be less than p, that n be less than q, and that the dispersion matrices of the measurement errors be diagonal, as in the usual factor analysis model.

The mathematical principles of path analysis* are the same as those of the general model presented above, although, historically, path analysis (and the same may be said of factor analysis) has taken the correlation matrix as a point of departure, the more current view being that it is better to work with unstandardized regression coefficients because they show more autonomy (invariance over different populations).

It may be noted before changing the subject that multivariate analysis* with latent (unmeasured) variables is, to some extent, a controversial topic. Latent variables are hypothetical constructs introduced by the analyst for the purpose of understanding a research area; in general there exists no operational method for directly measuring these constructs. Since the latent variables are abstractions, precise inferences as to their substantive meaning may be problematical, and they are subject to certain problems of indeterminacy (Bentler [3]).

Quasiexperiments

Yet another development in sociological methodology has been an increase in the application of experimental modes of analysis and interpretation to bodies of data generated by quasiexperiments such as social intervention or action programs where complete experimental control may not be possible. The most recent attempt at cataloging the pitfalls of such applications is that of Cook and Campbell [10]. These authors list 35 problems or pitfalls that the analyst should worry about when dealing with quasi-experimental data. They use the phrase *threats to validity* for such pitfalls. These cover problems related to drawing valid statistical inferences from the available data, those related to leaping to theoretical constructs from actual measurements, those of generalizing the results of the analysis to the desired population, and those of imputing causal forces to chosen treatments (intervention strategies).

Elaboration Technique

In the area of categorical data* analysis, three developments are worth mentioning: elaboration technique, latent structure* models, and panel analysis, all of which were initially developed by Paul F. Lazarsfeld and his colleagues. The elaboration technique is essentially percentage analysis of contingency tables*. It starts with the cross-tabulation of two variables, and then introduces additional variables into the analysis to define subclasses in each, of which the relationship between the two original variables is examined separately. Initially, the intent of the technique was to clarify the causal linkages between variables, but, gradually, the focus shifted to the study of *conditional relations*, that is, relations between variables in subclasses of populations (Rosenberg [38]). Log-linear and other modern methods of analysis of categorical data (see, for example, Bishop et al. [5] and Goodman [7] seem to have replaced the elaboration technique.

Latent Structure Models

These models deal with a situation in which observable polytomous variables are related to an unobservable (latent) polytomous variable, such that within each class of the latent variable, the observable variables are mutually independent (Lazarsfeld [25], Anderson [1], Lazarsfeld and Henry [19]). The simplest latent structure model is the latent dichotomy model in which the latent variable is a dichotomy. Even in the case of the latent dichotomy models, there are unresolved problems; for example, sometimes the two latent classes identified may not have any reasonable interpretation. In many ways, including this last, the latent structure model is analogous to the factor analysis model. Goodman and his colleagues have discussed latent structure models and modifications thereof using the concepts of independence and quasiindependence in contingency tables (Goodman [16] and Haberman [20]). A major criticism of the earlier literature on the subject is that the method of estimation emphasized in that literature is generally cumbersome and inefficient. A computer program called MLLSA (maximum likelihood latent structure analysis) written by Clifford Clogg is now available for fitting latent structure models.

Panel Analysis

Panel studies*, in which the same sample of units is observed two or more times in sequence, have become increasingly common in social research. The first major sociological study using a panel design was one by Lazarsfeld [27] on the decision making of voters during the 1940 Presidential election campaign. In that study, a sample of voters was interviewed, once each month from May to October. The purpose of the study was to identify factors that produced changes in voting intentions during an election campaign.

The starting point of panel analysis, as Lazarsfeld conceived it, is the turnover table, showing a categorical variable (for example, voting intention) at time 1 cross-classified with itself at time 2. Turnover tables are often formed separately for individuals in subclasses obtained by stratifying the sample on the basis of various characteristics such as sex, religious affiliation, and race, in order to see whether such factors (sometimes called *qualifiers*) affect the turnover tendency. A discrete-state discrete-time Markov chain can serve as a model for panel data* (Lazarsfeld [26]). Turnover tables can also be analyzed as contingency tables (see, e.g., Bishop et al. [5]).

In some panel surveys the responses may be quantitative. An example is the answer to the question: How many hours did you spend last week watching television? A possible model for panel data on such variables is a univariate or multivariate autoregressive process (Anderson [2]).

The general structural equation model outlined earlier can also be applied to the analysis of panel data. In such applications, it is important to seriously consider the possibility that measurement errors associated with observations obtained on successive occasions using the same measurement procedures are likely to be correlated (Jöreskog [23] touches this issue; Hannan and Young

[19] address this and a few other issues not well understood by users of panel models).

OTHER DEVELOPMENTS

Computers have contributed enormously to the development of quantitative methodology in sociology during the past three decades. The implementation of various statistical techniques in widely available package programs such as SPSS (Nie et al. [34]), SAS (SAS Institute [39]), and BMDP (Dixon et al. [12]) has led to the frequent use as well as abuse of statistical methods in social research.

Another development worth mentioning is that in 1961 the Section on Methodology was established within the American Sociological Association and in 1968 the first issue of the Association-sponsored annual *Sociological Methodology* appeared. This annual continues to provide a forum for interpretation of sound statistical practices and illustrative application or novel adaptation of statistical techniques to the solution of selected substantive problems. Among the topics covered in recent issues of *Sociological Methodology* are, to mention just a few, cohort analysis*, contingency table and categorical data analysis, event histories (survival time analysis), social mobility matrices, with special emphasis on clustering on the main diagonal, and analysis of time series*.

CRITICISMS

There are many sociologists who seriously doubt whether statistical methodology can be of any substantial help in developing sociology. Some skeptics believe that adherence to methodological rigors in the conduct of inquiry stifles one's imagination, dampens one's creativity, and causes one to abandon a potentially productive line of investigation. This group prefers a setup akin to a child at play, with no constraints whatsoever as to the design, measurement, or analysis (see, e.g., Phillips [36]). It remains to be demonstrated, however, that the playful and free-floating approach to inquiry advocated by this group promotes significant discoveries as the group claims.

A position overlapping with the one just described is that social reality is not amenable to statistical treatment in terms of fixed precise concepts and categories. The reasoning is that social conduct and group processes are so fluid that they cannot be studied in terms of preconceived notions. The only viable option, it is contended, is to use an empirical approach letting facts speak for themselves—facts, it may be noted, that are collected under broad guidelines that suggest where to look, but not what precisely to look for (see, e.g., Blumer [8]). The dominant methodological position in sociology, however, is that, while it may be useful and often desirable in the initial stages of an inquiry to employ an exploratory approach, unbridled by fixed precise concepts and categories, a time comes when the exploratory phase should be replaced by a confirmatory phase (see the debate between Blumer and Huber on the matter: Blumer [8] and Huber [21]).

Yet another position worth mentioning is that statisical methods are unlikely to make any significant contribution to the development of sociology because not only is there a widespread tendency among sociologists to use statistical methods indiscriminately, but the availability of quantitative methods is frequently allowed to dictate the choice of problems for investigation, so much so that trivial problems are treated with utmost refinement while complex problems of central concern are left unattended (Coser [11, p. 692] and Bierstedt [4, p. 5]). No doubt, statistical methods have been misused by sociologists [leading quantitative methodologists are irritated by this tendency (see, e.g., Duncan [13, p. 150])]. But one cannot judge the potential contribution of a tool by examining the frequency of its abuse. Moreover, the chances are that users of statistical methods in sociology would become progressively more self-disciplined as time goes by if the following are indicative of anything: (1) Increased attention is now being paid to properly training sociologists in the use of statistical methods, through, for example, placing progressively greater emphasis on quantitative training in graduate curricula and holding frequent methodology workshops on university campuses or in connection with national or regional meetings of professional associations. (2) Excellent books on statistical methods, written especially for sociologists are now available (e.g., Blalock [6] and Mueller et al. [33]). (3) Publications such as the American Sociological Association's annual *Sociological Methodology* illustrate sound statistical practices and call attention to pitfalls.

As for the other criticism, namely that complex problems are left unattended to, the implication is that such problems are not amenable to statistical treatment. But the fact of the matter is that the real stumbling block in modeling complex social phenomena is the lack of clarity in the concepts used in the substantive literature and the absence of appropriate "correspondence rules" linking unmeasured concepts with their measurable counterparts. The situation calls for increasing efforts on the part of substantive-area specialists to removing this stumbling block. Recent publications such as Blalock and Wilkin's [7] *Intergroup Processes* indicate that complex social phenomena can be analyzed systematically, applying scientifically oriented methodological approaches. Only time will tell whether other investigators will follow suit with similar exercises so as to help build a solid cumulative knowledge base for sociology.

References

[1] Anderson, T. W. (1959). In *Probability and Statistics*. U. Grenander, ed. Wiley, New York, pp. 9–38.

[2] Anderson, T. W. (1979). In *Qualitative and Quantitative Social Research: Papers in Honor of Paul F. Lazarsfeld*, R. K. Merton et al., eds. Free Press, New York, pp. 82–97.

[3] Bentler, P. M. (1980). Multivariate analysis with latent variables: causal modeling. *Ann. Rev. Psychol.* **31**, 419–456.

[4] Bierstedt, R. (1960). Sociology and humane learning. *Amer. Sociol. Rev.* **25**, 3–9.

[5] Bishop, Y. M. M., Fienberg, S. E., and Holland, P. W. (1975). *Discrete Multivariate Analysis: Theory and Practice*. MIT Press, Cambridge, MA.

[6] Blalock, H. M., Jr. (1979). *Social Statistics*. Rev. 2nd ed. McGraw-Hill, New York. (Contains references to a number of other introductory statistics textbooks used by sociologists.)

[7] Blalock, H. M., Jr. and Wilkin, P. H. (1979). *Intergroup Processes: A Micro-Macro Perspective*. Free Press, New York.

[8] Blumer, H. (1973). A note on symbolic interactionism. *Amer. Sociol. Rev.* **38**, 797–798.

[9] Boyd, L. and Iversen, G. (1979). *Contextual Analysis: Concepts and Statistical Techniques*. Wadsworth, Belmont, CA.

[10] Cook, T. D. and Campbell, D. T. (1976). In *Handbook of Industrial and Organizational Psychology*, M. D. Dunnette, ed. Rand McNally, Chicago, IL, pp. 223–326.

[11] Coser, L. A. (1975). Presidential address: two methods in search of a substance. *Amer. Sociol. Rev.*, **40**, 691–700.

[12] Dixon, W. J. (1981). *BMDP Statistical Software, 1981*. University of California Press, Berkeley, CA.

[13] Duncan, O. D. (1975). *Introduction to Structural Equation Models*. Academic, New York.

[14] Durkheim, E. [(1897)] (1951). *Suicide: A Study in Sociology*. Free Press, Glencoe, IL. (First published in French.)

[15] Friedrichs, J. and Lüdtke, H. (1975). *Participant Observation: Theory and Practice*. Heath, Lexington, MA. (Contains an extensive bibliography.)

[16] Goodman, L. A. (1975). A new model for scaling response patterns: an application of the quasi-independence concept. *J. Amer. Statist. Ass.*, **70**, 755–768.

[17] Goodman, L. A. (1978). *Analyzing Qualitative/Categorical Data: Log-Linear Models and Latent Structure Analysis*, J. Magidson, ed. Abt Books, Cambridge, MA.

[18] Hannan, M. T. (1971). *Aggregation and Disaggregation in Sociology*. Heath, Lexington, MA.

[19] Hannan, M. T. and Young, A. A. (1977). In *Sociological Methodology* 1977, D. R. Heise, ed. Jossey-Bass, San Francisco, CA, pp. 52–83.

[20] Haberman, S. J. (1979). *Analysis of Qualitative Data*, Vol. 2, Academic, New York.

[21] Huber, J. (1973). Reply to Blumer: but who will scrutinize the scrutinizers? *Amer. Sociol. Rev.* **38**, 798–800.

[22] Hyman, H. H. (1972). *Secondary Analysis of Surveys: Principles, Procedures, and Potentialities*. Wiley, New York.

[23] Jöreskog, K. G. (1977). In *Applications of Statistics*. P. R. Krishnaiah, ed. North-Holland, Amsterdam, The Netherlands, pp. 265–287.

[24] Jöreskog, K. G. and Sörbom, D. (1978). *LISREL IV User's Guide*. National Education Research, Chicago, IL.

[25] Lazarsfeld, P. F. (1950). In *Measurement and Prediction*, S. A. Stouffer et al., eds. Princeton University Press, Princeton, NJ, pp. 362–412.

[26] Lazarsfeld, P. F., ed. (1954). *Mathematical Thinking in the Social Sciences*. Free Press, Glencoe, IL. (Reprinted in 1969 by Russel and Russel, New York.)

[27] Lazarsfeld, P. F., Berelson, B., and Gaudet, H. (1960). *The People's Choice: How the Voter Makes Up His Mind in a Presidential Campaign*, 2nd ed. Columbia University Press, New York.

[28] Lazarsfeld, P. F. and Menzel, H. (1961). In *Complex Organization: A Sociological Reader*, A. Etzioni, ed. Holt, Rinehart, and Winston, New York, pp. 422–440.

[29] Lazarsfeld, P. F. and Henry, N. W. (1968). *Latent Structure Analysis*. Houghton-Mifflin, Boston, MA.

[30] Lazarsfeld, P. F., Pasanella, A. K., and Rosenberg, M., eds. (1972). *Continuities in the Language of Social Research*. Free Press, New York.

[31] Lécuyez, B. and Oberschall, A. R. (1968). In *International Encyclopedia of the Social Sciences* Vol. **15**, D. L. Sills, ed. MacMillan and Free Press, New York, pp. 36–52.

[32] Merton, R. K. (1968). *Social Theory and Social Structure*. Free Press, New York.

[33] Mueller, J. H., Schuessler, K. F., and Costner, H. L. (1977). *Statistical Reasoning in Sociology*, 3rd ed. Houghton-Mifflin, Boston, MA.

[34] Nie, N. H. et al. (1975). *SPSS: Statistical Package for the Social Sciences*. McGraw-Hill, New York.

[35] Pennings, J. (1973). Measures of organizational structure: a methodological note. *Amer. J. Sociol.*, **79**, 687–704.

[36] Phillips, D. L. (1973). *Abandoning Method*. Jossey-Bass, San Francisco, CA.

[37] Price, J. L. (1972). *Handbook of Organizational Measurement*. Heath, Lexington, MA.

[38] Rosenberg, M. (1968). *The Logic of Survey Analysis*. Basic Books, New York.

[39] SAS Institute (1979). *SAS User's Guide, 1979 Edition*. SAS Institute, Raleigh, NC.

[40] Sonquist, J. A. and Dunkelberg, W. C. (1977). *Survey and Opinion Research: Procedures for Processing and Analysis*. Prentice-Hall, Englewood Cliffs, NJ.

(DEMOGRAPHY
ECONOMETRICS
INDEX NUMBERS
LATENT STRUCTURE ANALYSIS
PANEL DATA
PATH ANALYSIS
SOCIAL STATISTICS)

N. Krishnan Namboodiri

SOCIOMETRY

The term sociometry has had a number of meanings, and in the early use of the term there was some debate as to what it should mean. Probably the person most closely identified with the development of the field was J. L. Moreno [14], whose vision for the term was inclusive. Sociometry from his point of view would have included all of the measurement of social behavior and social phenomena, and, in a sense, the term could have been co-extensive with sociology * and social psychology. While a number of sociologists supported this inclusive use of the term (Chapin [6] and Lundberg, [13]), as a practical matter it came to be more associated with the *sociometric test* and the more expansive applications of sociometric testing. In one sense, the final rejection of the more inclusive term is marked by the replacement, in 1978, of the name for the journal *Sociometry* by the title *The Social Psychology Quarterly*. This reflected a move by the American Sociological Association to make the journal title inclusive, giving more or less formal recognition to the notion that, in practice, sociometry defined a much narrower field of interest.

Sociometric procedures are closely identified with Moreno, but reference to his original work reflects that his collaborator was

Helen Hall Jennings, who probably was responsible for the development of techniques and procedures. Jennings subsequently published a book, *Leadership and Isolation* [9], which provided systematic work on sociometric measurement and formalized some of the procedures focusing on specific analysis of structures. An outline of some of the early procedures is implicit in our discussion of the sociometric test and sociometry in general.

SOCIOMETRIC TEST

The sociometric test is the systematic gathering of information based on a specific criterion question about the relationships between individual members of a group. While the nature of a group is not defined in any of the early writings, the notion is that any aggregate may be studied, whether informal or formal, for which some encompassing concept exists with implicit or explicit boundaries. The sociometric test gathers the choices people make of each other within the group based on a specific criterion. Information is gathered from each member of the group, so that presumably a network of choices can be described that represents a form of structure of the group.

In the original formulation, Moreno did not place any restrictions on the number of persons within the group who could be chosen, and the form of the question could be of choice or rejection. In practice, however, most sociometric measurements have been concerned with attraction or positive choice rather than rejection. In general, choices are requested in an open-ended way, so that they correspond to a concept of recall rather than recognition. Choices are limited, thus, in any larger group, to persons who are known to the chooser, and the concept of subgroups implicitly comes into operation quickly as group size increases.

The general form of a sociometric question is "With whom would you like to do this activity again?" The activity or criterion should be specific, and each sociometric question becomes a separate basis for the analysis of the sociometric structure. The issue of the appropriate selection of activities, therefore, is critical from a social psychological point of view.

With regard to how choices are made, considerable variation can occur, depending upon the objectives of the application of the sociometric method. For example, in addition to open-ended choices, it is possible to restrict choices to a single choice, to the top three choices, to request rankings, and other alternatives. Some of the data collection techniques provide more information than others, but may be more difficult to handle in a statistical analytic sense. Most of the accumulated experience suggests that the original unrestricted choice procedure probably provides information that is as effective in the analysis of structures as any other subsequently suggested procedure.

GRAPHIC REPRESENTATIONS

Part of the appeal of the sociometric procedure has been associated with graphic representations, in particular with *sociograms* (see Fig. 1). In the diagrammatic representation of structures there has been the opportunity for the development of great versatility for description. For example, characteristics of the persons in the group can be classified by different designations, such as circles for girls and triangles for boys. Such classificatory variations, obviously, can be extended to any social category. A choice can be designated by an arrow from one person to another, and mutual choices can be designated in any number of ways. For example, a plain line between two persons may indicate mutual choice, or arrows in both directions, or a single line with arrow points at each end.

When a person is taken as the reference point for choices made and choices received, this representation is identified as the *social atom*. Examination of how individuals fit in a structure leads to a number of concepts.

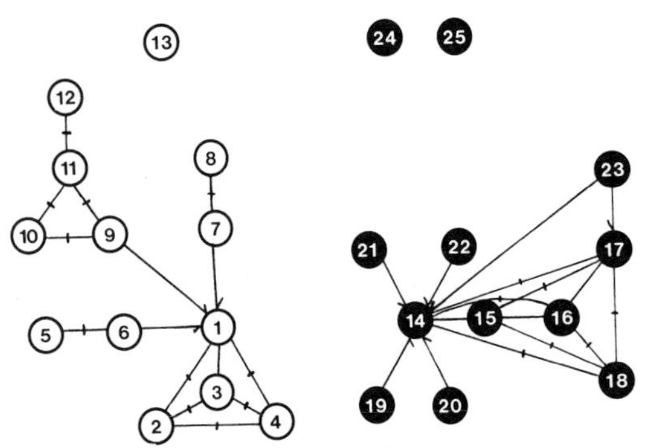

Figure 1 An illustrative sociogram, minimizing crossing lines. The sociometric question, asked of a second grade class, is: With whom do you want to work tomorrow? Boys are denoted by open circles and girls by solid circles. Mutual choices are with line crosses, single direction choices with points. Tight cliques are 1, 2, 3, 4; 9, 10, 11; and 14, 15, 16, 17, 18. Mutual choice dyads are 5, 6; and 7, 8. Girl 14 is a "star" or popular leader. Boy 1 is a "powerful" leader. Girls 19, 20, 21, 22, and 23 are unchosen. Boys 13 and girls 24 and 25 are isolates. Boys and girls self-segregate.

For example, if the person receives no choices and makes no choices, the person then is identified as an *isolate*. By contrast, a person who receives and makes a great many choices could be identified as a *leader*, but what type of leader would depend upon the criterion question used. Thus if a social activity were the basis of the sociometric test, the person might be a *popular leader*. Such a person may also be described as *overchosen*, which suggests the notion of a normative basis and expectation of choice behavior.

Moving to a more interpretive notion associated with the sociogram, a person may be a leader without being overchosen. In particular, if a person receives only a few choices, but they are from persons who are overchosen, that person may be in a critical position. If the criterion is one of leadership with regard to a task, for example, then a common designation for such a person would be a *power figure*. Other concepts have been suggested such as the *star*, and appropriate language has been plentiful in describing structures. If a dual sociometric question is used involving both choices and rejections, obviously additional distinctions can be made. For example, there will be isolates who are ignored or invisible, and then there may be isolates who are rejected persons.

Structures of two or more persons increase the complexity of graphic representations. If two persons choose each other, they may be designated as a pair, three persons as a triangle, and, in general, any mutually selective group may be designated as a *clique*. The more neutral term "subgroup" is used but is more general. Linkages may be described that are *chains* as well as cliques.

The appeal of sociograms has been tempered by several criticisms. Although one may view sociograms and get a "feel" for the structure of a group, it is difficult to appraise the meaning of the structure with any accuracy. Standards of comparison are at best vague and intuitive. Attempts at standardization that might make comparison more feasible are noted as early as Northway's suggestion [16] for the use of concentric circles to create a "target" sociogram. However, such a structure is difficult to interpret for larger groups where there are subgroup structures. Borgatta [4] proposed a criterion of minimizing the number of crossed lines in choices made, a procedure which when used manually requires many iterations, but provides a more readable sociogram. Such a procedure is intrinsically programmable for computer application.

QUANTITATIVE APPROACHES

Quantitative approaches have emphasized the manipulation of square matrices of choices (cf. Forsyth and Katz [8], Beum and Brundage [2], Bock and Husain [3], Luce and Perry [12], and Festinger [7]; *see also* SOCIAL NETWORK ANALYSIS). Much of the thinking about sociometric analysis was done three decades or more ago during the period when *small group research* was a vigorous and intensive area of development in social psychology. In the more recent period, two directions of development may be identified, dealing with the characteristics of persons as contrasted to the characteristics of the group structure. The development of the latter has centered in the area of *social network analysis*. In general, this has been the movement of the procedures noted for matrix analysis into the world of modern computer applications. While intrinsically attractive, such applications have not at this point proved to be of great interest to the social sciences. The emphasis on the characteristics of individuals in many ways blended with the development of aspects of personality research concerned with behavioral descriptions.

With regard to the development of interest in the characteristics of individuals, the general direction has been one of elaboration of the issues of the criteria used in sociometric tests. As is emphasized in the work of Jennings [9, 10], the distinction between the psyche group (socio-emotional criteria) and the socio group (task-oriented criteria) has been emphasized by many theoretical orientations. These developments have in large part been associated with behavioral ratings, but also in the attribution of more qualitative concepts, such as leaders vs. followers. With regard to behavioral description, the development of summary statistics on individuals for data collected by the sociometric test becomes one of many ways of collecting data by ratings and observations. So, in a tradition of small group research (Bales [1]), sociometrically identified task leaders and socio-emotional leaders are analyzed in comparison to the characteristics of group participation in social interaction. This has been generalized in sociological theory more generally as the distinction between instrumental and expressive roles, a major distinction in Parsonian theory. The models of interpersonal dimensions of personality developed by Leary and his associates [11] centers on this basic distinction as well. Factor analytic analyses of peer rating and ranking criteria have confirmed the centrality of the two concepts (Borgatta [5]) for the behavioral aspects of description of personality.

References

[1] Bales, R. F. (1950). *Interaction Process Analysis: A Method for the Study of Small Groups*. Addison-Wesley, Reading, MA.

[2] Beum, C. O. and Brundage, E. G. (1950). A method for analyzing the sociomatrix. *Sociometry*, **13**, 141–145.

[3] Bock, R. D. and Husain, S. Z. (1950). An adaptation of Holzinger's *B*-coefficients for the analysis of sociometric data. *Sociometry*, **13**, 146–153.

[4] Borgatta, E. F. (1951). A diagnostic note on the construction of sociograms and action diagrams. *Group Psychotherapy*, **3**, 300–308.

[5] Borgatta, E. F. (1964). The structure of personality characteristics. *Behav. Sci.*, **9**, 8–17.

[6] Chapin, F. S. (1940). Trends in sociometrics and critique. *Sociometry*, **3**, 245–262.

[7] Festinger, L. (1949). The analysis of sociograms using matrix algebra. *Hum. Rel.*, **2**, 153–158.

[8] Forsyth, E. and Katz, L. (1946). A matrix approach to the analysis of sociometric data: Preliminary report. *Sociometry*, **9**, 340–347.

[9] Jennings, H. H. (1943). *Leadership and Isolation*. Longmans, Green, New York.

[10] Jennings, H. H. (1947). Sociometric differentiation of the psychegroup and sociogroup. *Sociometry*, **10**, 71–79.

[11] Leary, T. (1957). *Interpersonal Dimensions of Personality*. Ronald Press, New York.

[12] Luce, D. R. and Perry, A. D. (1949). A method of matrix analysis of group structure. *Psychometrika*, **14**, 95–116.

[13] Lundberg, G. A. (1943). Discussion of sociometry. *Sociometry*, **6**, 219–220.

[14] Moreno, J. L. (1934). Who shall survive? *Nervous and Mental Disease Monograph No.* 58, Washington, DC.

[15] Norman, W. T. (1963). Toward an adequate taxonomy of personality attributes: Replicated factor structure in peer nomination personality ratings. *J. Abnorm. Soc. Psychol.*, **66**, 574–583.

[16] Northway, M. L. (1940). A method for depicting social relationships obtained by sociometric testing. *Sociometry*, **3**, 144–150.

(SOCIAL NETWORK ANALYSIS
SOCIAL STATISTICS
SOCIOLOGY, STATISTICS IN)

EDGAR F. BORGATTA
KYLE KERCHER

SOFT MODELING *See* PARTIAL LEAST SQUARES

SOFTWARE RELIABILITY

INTRODUCTION

As a result of rapid technological advances in microelectronics, microprocessors, and computer science, there has been growing concern that system reliability problems have made a transition from hardware to software. Current estimates show that software costs amount to about 60–80% of total systems costs. These and other considerations have given birth to the subject of *software engineering*, which concerns itself with study of the various aspects and phases of the life cycle of a piece of computer software. Fundamental aspects are a specification of the computing environment, a definition of what is meant by a software error, the sources and classification of software errors, and finally, the assessment of software quality. *Software reliability* is a measure of software quality; it is the probability of failure-free operation of the software in a specified environment for a specified period of time. By "time" we mean "execution time," whose magnitude (seconds, microseconds, nanoseconds, etc.) would depend on the complexity of the task undertaken and the sophistication of the hardware system used. Kline [10] compares and contrasts hardware and software reliability problems with respect to their defining concepts, engineering aspects, management, and maintenance. Of interest to statisticians are issues pertaining to the measurement and the quantification of software reliability, the assessment of reliability growth, and the analysis of software failure data. To facilitate these, several models for describing the stochastic behavior of software running times have been proposed in the literature, and are discussed next. A good model can aid program developers in estimating the time and effort required to achieve a desired level of reliability, and if acceptance testing is to take place, a valid model can assist in decision making.

MODELS

The models we will describe stand out as representatives of different schools of thought; they have been motivated by different considerations of the software failure process and can be unified and comprehensively viewed. In order to achieve this unification, it is easiest if we start with the following view of the software failure process, and then introduce the shock model* proposed by Langberg and Singpurwalla [13].

The software segment of a computer system involves instructions or codes used to program the hardware system. Let N^* be the total number of distinct "input types" to the software system; N^* is assumed large, conceptually infinite. By an input type we mean a specific type of job, dataset, or function that the software system is required to undertake. Let $N \leqslant N^*$ be the number of input types that result in the inability of the software system to perform its desired function; N is assumed unknown. Such input types lead to software failures, which are due either to errors in the logic of the instructions, to errors in the coding of the instructions, or to an input that is incompatible with the design of the system. We assume that the processing of an input (if successful) is instantaneous, and also allow for the possibility that the same type of input can arrive at the software system over and over again.

Whenever a software failure occurs, two types of action are taken by the user, depending upon the nature of the application:

1. A diagnosis is made of the cause of failure, and either

   (i) error(s) in either the logic or the coding, or both, are corrected so that N is reduced by at least one; or

   (ii) if it is determined that the input causing failure is incompatible with the software design, then that is eliminated from further consideration by an appropriate modification of the software.

2. The computer system is reinitialized, without any modifications or corrections to it, and is allowed to continue operation. Modifications to the software are eventually made as a *group*, rather than the individual modifications described under **1** above.

Much of the current literature on software reliability assumes that it is action **1** above that is undertaken. The only exception is the work of Crow and Singpurwalla [2], which focuses on the case of action **2**, and in so doing argues that software failures could occur in clusters or bunches. The authors propose an empirically developed Fourier series model to describe the software failure process and demonstrate its usefulness by considering some real life software failure data.

In what follows we assume that action **1** is taken at the times of failure.

A Shock Model* for Software Failures

Suppose that the inputs arrive at the software system according to the postulates of a Poisson process* with an intensity function ω. Then given N^* and N, the probability that the software encounters no failures in time $[0, t)$ is

$$P\{T \geqslant t|N, N^*, \omega\} = \bar{F}(t|N, N^*, \omega) = \sum_{j=0}^{\infty}\left(\frac{e^{-\omega t}(\omega t)^j}{j!}\right) \times\left(\frac{N^* - N}{N^*}\right)^j. \quad (1)$$

We justify (1) by noting that the first factor inside the summation sign denotes the probability of j inputs (shocks) in time $[0, t)$, and the second, the probability that none of the j inputs leads to a failure of the software. The above model, which can be conveniently extended and generalized to other arrival schemes, was proposed by Langberg and Singpurwalla [13]. Under the error correction policy of action 1, T_i, the time between the $(i-1)$st and ith failures, $i \leqslant N$, has the survival function, given N and N^*,

$$\begin{aligned} P\{T_i \geqslant t|N, N^*, \omega\} &= \bar{F}_i(t|N, N^*, \omega) \\ &= \exp(-\omega(N - i + 1)t/N^*), \quad t \geqslant 0. \quad (2)\end{aligned}$$

The Model by Jelinski and Moranda

One of the oldest, and certainly the most often referenced, models for describing software failures is the model proposed by Jelinski and Moranda [8], henceforth called J-M. Attention is focused on N, the input types that lead to software failures; these inputs are viewed as bugs or faults, and the following assumptions are made:

(i) The failure rate of the software at any point in time is proportional to the residual number of faults in the program; the program begins life with N faults.

(ii) Each of the N faults contributes an equal amount, say Λ (unknown), to the failure rate.

Then, given N and Λ, $T_1, T_2, \ldots, T_N$, the times between successive failures of the program are independently distributed with density functions

$$f(t_i|N, \Lambda) = \Lambda(N - i + 1) \times \exp\{-\Lambda(N - i + 1)t_i\} \quad (3)$$

and survival function

$$\bar{F}_i(t_i|N, \Lambda) = \exp\{-\Lambda(N - i + 1)t_i\}. \quad (4)$$

A Modified Version of the J-M Model

Arguing that the different portions of the program code are exercised with varying frequencies and that there is some uncertainty in the fault removal process, Littlewood and Verrall [14], denoted L-V, retain the assumption of J-M that the T_i's are conditionally independent with densities

$$f(t_i|\Lambda_i) = \Lambda_i \exp(-\Lambda_i t_i),$$

but require that the Λ_i's also be conditionally independent [given $\psi(i)$ and α] with densities

$$f(\lambda_i|\psi(i),\alpha) = \frac{\psi(i)\{\psi(i)\lambda_i\}^{\alpha-1}}{\Gamma(\alpha)} \cdot e^{-\psi(i)\lambda_i}$$

To describe reliability growth that is expected to occur under action **1**, $\psi(i)$ is taken to be $\beta_0 + \beta_1 i$, an increasing function of i. This choice of $\psi(i)$ ensures that the sequence $\{\Lambda_i\}$ stochastically increases in i.

A Nonhomogeneous Poisson Process Model

Goel and Okumoto [5], denoted G-O, take what *appears* to be a distinct departure from the bug counting scenario of J-M and L-V, and consider $M(t)$, the cumulative number of software failures in time $[0, t)$. They model the counting process $\{M(t); t \geqslant 0\}$ as a nonhomogeneous Poisson process* with a mean value function $a(1 - e^{-bt})$, where a is the expected number of faults to be eventually detected and b is a constant of proportionality. Kyparisis and Singpurwalla [12] propose a nonhomogeneous Poisson process with a Weibull intensity function (the Weibull process*) to assess software reliability growth, and apply their model to some software failure data.

Other Models

The preceding models do not account for the time needed to correct detected errors, and moreover, assume that the error correction process (debugging) is perfect. Furthermore, the interplay between the hardware and the software errors in the evaluation of the operational performance of a system is an important part of the reliability evaluation process.

Claudio and Kasperson [1] and Kremer [11] use a birth-and-death process* to model the discovery and correction of errors, the latter also considering the possibility of imperfect correction. Goel and Soenjoto [6] use a Markov model for studying the availability of a hardware–software system subject to failures and imperfect maintenance.

MODEL UNIFICATION

The shock model for software failures is a *central one*, in the sense that the other models of the previous section are derivatives of it. This plus the fact that the shock model can be extended in several possible directions makes it, conceptually, fundamental for software reliability.

To begin the unification process, we note that the J-M model is a special case of the shock model if we set $\Lambda = \omega/N^*$. in so doing we must recognize that even though the mathematical forms of the J-M model and the shock model (with $\omega/N^* = \Lambda$) are alike, their motivations are different, and their parameters have different interpretations.

Since the mathematical form of the J-M model is the one that is familiar to those working in software reliability we shall use the form (3).

Recall that $M(t)$ is the number of times that the software fails in $[0, t)$. Langberg and Singpurwalla [13] take a Bayesian* approach and show that if the prior distribution of N is a Poisson* with a mean θ, and if Λ is degenerate at λ, then $\{M(t), t \geqslant 0\}$ is a nonhomogeneous Poisson process with $EM(t) = \theta(1 - e^{-\lambda t})$, precisely the model considered by G-O. On the other hand, if the probability mass function of N were assumed to be such that all its mass is concentrated at some known N, and Λ assumed to have a gamma density function, then the model L-V would result. In view of the above, the models considered by Goel and Okumoto, and Littlewood and Verrall, are special cases of the model by Jelinski and Moranda (itself a special case of the shock model), when specific prior distributions* are assumed for its parameters.

The two key features of the model by L-V are that the times between failures, $T_1, T_2, \ldots,$ have a decreasing failure rate (*see* RELIABILITY, PROBABILISTIC) and that the sequence $\{T_i\}$ is stochastically increasing in i. This latter property is a consequence of a specific choice of $\psi(i)$. Langberg and Singpurwalla [13] also prove that the above two features are a natural consequence of assigning *any* prior distributions to the two parameters of the J-M model.

Stefanski [22, 23] has also consolidated the models due to J-M, L-V, and G-O by using an order statistic property of renewal* processes. It is because of the above arguments that several other models proposed in the literature (see Shooman [20] and Schick and Wolverton [19]), which are related to the three representative models cited above, are not elaborated upon here.

STATISTICAL INFERENCE FOR SOFTWARE RELIABILITY MODELS

Some practical difficulties with a serious implementation of the three representative models have been posed by the problem of inference for their parameters. The typical approach has been the method of maximum likelihood* (ML), or a combination of ML and Bayes (see, for example, L-V). In all these cases, the effect of the stopping rule, important in software reliability testing (see Langberg and Singpurwalla [13]) has been overlooked. Forman and Singpurwalla [3] show that the ML estimator of the parameter N of the J-M model can be highly misleading and often nonsensical. Meinhold and Singpurwalla [16] give an explanation of why this happens, and argue that this could be true of any model describing reliability growth. Analogous difficulties, though in contexts different from those above, have been reported by Johnson [9], Sanathanan [18], and Marcus and Blumenthal [15].

Langberg and Singpurwalla [13] argue in favor of a Bayesian approach for the parameters of the J-M model and obtain expressions for their posterior distributions*. These are used by Meinhold and Singpurwalla [16] for analyzing some real life software data.

A related issue, namely, the optimal time for which to test the software before releasing it, has been addressed by Forman and Singpurwalla [4] and Okumoto and Goel [17]. An empirical Bayes* approach to assess software reliability growth or decay has been proposed and applied by Horigome et al. [7]. Singpurwalla and Soyer [21] have considered ramifications of the random coefficient autoregressive process, and have used these for analyzing some real life software failure data.

References

[1] Claudio, L. F. and Kaspersen, D. L. (*1981*). *Proc. 1981 Army Numerical Anal. Comput. Conf.*, pp. 41–58.

[2] Crow, L. H. and Singpurwalla, N. D. (1984). *IEEE Trans. Rel.*, **R-33**, 176–183.

[3] Forman, E. H. and Singpurwalla, N. D. (1977). *J. Amer. Statist. Ass.*, **72**, 750–757.

[4] Forman, E. H. and Singpurwalla, N. D. (1979). *IEEE Trans. Rel.*, **R-28**, 250–253.

[5] Goel, A. L. and Okumoto, K. (1979). *IEEE Trans. Rel.*, **R-28**, 206–211.

[6] Goel, A. L. and Soenjoto, J. (1981). *IEEE Trans. Rel.*, **R-30**, 232–239.

[7] Horigome, M., Singpurwalla, N. D., and Soyer, R. (1984). In *Comput. Sci. Statist*: *Proc. 16th Symp. on the Interface*, L. Billard, ed. North-Holland, Amsterdam, The Netherlands, pp. 47–56.

[8] Jelinski, Z. and Moranda, P. B. (1972). In *Statistical Computer Performance Evaluation*, W. Freiberger, ed. Academic, New York, pp. 485–502.

[9] Johnson, N. D. (1962). *Technometrics*, **4**, 59–67.

[10] Kline, M. B. (1980). *1980 Proc. Ann. Reliability Maintainability Symp.*, pp. 179–185.

[11] Kremer, W. (1983). *IEEE Trans. Rel.*, **R-32**, 37–46.

[12] Kyparisis, J. and Singpurwalla, N. D. (1984). In *Comput. Sci. Statist*: *Proc. 16th Symp. on the Interface*, L. Billard, ed. North-Holland, Amsterdam, The Netherlands, pp. 57–64.

[13] Langberg, N. and Singpurwalla, N. D. (1985). *SIAM J. Sci. Statist. Comp.*, **6**, 78–790.

[14] Littlewood, B. and Verrall, J. L. (1973). *Record IEEE Symp. Comp. Software Reliability*, pp. 70–77.

[15] Marcus, R. and Blumenthal, S. (1974). *Technometrics*, **16**, 229–234.

[16] Meinhold, R. J. and Singpurwalla, N. D. (1983). *The Statistician* (*Lond.*), **32**, 168–173.

[17] Okumoto, K. and Goel, A. L. (1980). *J. Syst. Software*, **1**, 315–318.

[18] Sanathanan, L. P. (1972). *Ann. Math. Statist.*, **43**, 142–152.

[19] Schick, G. J. and Wolverton, R. W. (1973). In *Proc. Operations Res.* Physica-Verlag, Wurzburg-Wien, Germany, pp. 395–422.

[20] Shooman, M. L. (1972). In *Statistical Computer Performance Evaluation*, W. Freiberger, ed. Academic, New York, pp. 485–502.

[21] Singpurwalla, N. D. and Soyer, R. (1985). *IEEE Trans. Software Eng.*, **SE-11**, 1456–1464.

[22] Stefanski, L. A. (1981). A Review of Software Reliability. *Technical Report*, Mathematics Division, U.S. Army Research Office, Research Triangle Park, NC.

[23] Stefanski, L. A. (1982). *Proc. 27th Conf. on Design of Experiments Army Research and Testing*, pp. 101–118.

Acknowledgment

Work supported by the Army Research Office under Grant No. DAAG-29-83-K-0013, and by the Office of Naval Research under Contract No. N00014-77-C-0263, Project NR 042-372.

(COMPUTERS AND STATISTICS
RELIABILITY, PROBABILISTIC
SHOCK MODELS
STATISTICAL SOFTWARE)

NOZER D. SINGPURWALLA

SOFTWARE, STATISTICAL *See* STATISTICAL SOFTWARE

SOI BULLETIN *See* *STATISTICS OF INCOME (SOI) BULLETIN*

SOJOURN TIME

The sojourn time of a Markov process* at a stable state is roughtly the duration of its temporary stay at that state, starting at the moment it enters that state and ending when it leaves that state to enter a new state. To define it appropriately, consider a Markov process $X(t, \ )$ with standard transition function $P(t)$ and a countable state space E. Define the random variable W_t by $W_t = \inf\{s > 0: X_{s+t} \neq X_t\}$. Then for $u \geqslant 0$ (see ref. 1),

$$\Pr(W_t > u | X_t = i) = \exp\{-\beta(i)u\},$$

where $\beta(i)$ is a number of $[0, \infty]$ and the right-hand expression above is zero for all $u \geqslant 0$ when $\beta(i) = \infty$. Let us assume that for all i in E, $0 < \beta(i) < \infty$; that is, all states are stable. Define the random variables (T_n) as

$$T_0 = 0 \quad \text{and} \quad T_{n+1} = \inf\{t > T_n: X_t \neq X_{T_n}\}.$$

If $X_{T_n} = i$, then $T_{n+1} - T_n$ is the *sojourn time* in i. The sequence $Y_n \equiv X_{T_n}$ is a Markov chain on E, almost surely, $0 < T_{n+1} - T_n < \infty$ for each n, and

$$\Pr(T_{n+1} - T_n > u | Y_n = i, Y_{n+1} = j) = \exp\{-\beta(i)u\}.$$

Thus, the sojourn time has an exponential distribution* with a parameter that depends on the state in which the sojourn passes and not on the next state.

The concept of the random variables T_n and in particular, the sojourn time, is useful in various contexts. For example, the transition function $P(t)$ can be described in terms of the parameter $\beta(i)$ in the distribution of the sojourn time and the transition matrix $\mathbf{Q}$ of the Markov chain (Y_n). Assuming that the map $t \to X_t$ is almost surely right continuous and that almost surely, $\sup_n T_n = \infty$ [this is true, for example, when $\sup_{i \in E} \beta(i) < \infty$],

$$P(t) = \exp\{t\mathbf{A}\} \quad \left[\equiv \sum_{n=0}^{\infty} (t^n/n!)\mathbf{A}^n\right],$$

where the matrix $\mathbf{A}$ is given by

$$A_{ij} = -\beta(i), \qquad i = j;$$
$$= \beta(i)Q_{ij}, \qquad i \neq j.$$

There are various other applications of sojourn time. See, for instance, Kelly and Pollett [2].

References

[1] Çinlar, E. (1975). *Introduction to Stochastic Processes.* Prentice-Hall, Englewood Cliffs, NJ.

[2] Kelly, F. P. and Pollett, P. K. (1983). *Adv. Appl. Prob.*, **15**, 638–658.

(MARKOV PROCESSES)

ARUNAVA MUKHERJEA

SOLUTION MATRIX

This is another name for a generalized inverse* of a matrix.

SOMERS' *d*

These indices are asymmetric ordinal measures of association for grouped data, closely related to Wilson's e^*. For a $k \times k$ contingency table* based on two polytomous variables X and Y, they are defined [2] by the formulas

$$d_{YX} = \frac{C - D}{C + D + T_Y}, \quad d_{XY} = \frac{C - D}{C + D + T_X},$$

where C is the number of concordant pairs, D is the number of discordant pairs, and $T_X(T_Y)$ is the number of pairs tied on X (Y) but not on $Y(X)$.

Further details are given in ref. 1.

References

[1] Blalock, H. M. (1979). *Social Statistics*, 2nd ed. McGraw-Hill, New York.

[2] Somers, R. H. (1962). *Amer. Sociol. Rev.*, **27**, 799–811.

(ASSOCIATION, MEASURES OF
WILSON'S *e*)

SOMERVILLE'S MULTIPLE RANGE SUBSET SELECTION PROCEDURE

Given independent samples each of size n from k populations, the population with the largest mean will be called the *best*. Using the sample data, we would like to select a subset of populations that includes the best one with a prescribed probability P^*. The basic procedure for doing this, among others, was proposed by Gupta [1] (*see also* RANKING PROCEDURES and SELECTION PROCEDURES).

In the case where the ith population is $N(\mu_i, \sigma^2)$ with known σ^2 and $\mu = (\mu_1, \ldots, \mu_k)$ unknown, Somerville [3] proposed the following procedure:

Let $d_0 = 0$ and let $d_1, d_2, \ldots$ be an increasing sequence of numbers depending on P^* (but not on k). Then, if the range of all k sample means is less than $d_{k-1}\sigma/n^{1/2}$, include all k populations in the selected subset; otherwise, eliminate the population with the smallest sample mean and calculate the range of the remaining sample means. If this is less than $d_{k-2}\sigma/n^{1/2}$, include the populations corresponding to the $k - 1$ largest sample means; otherwise, eliminate the population with the second smallest sample mean, and so forth. Somerville [3] gives tables for $d_1, d_2, \ldots$ and reports numerical and Monte Carlo* results that show that his procedure compares favorably with Gupta's in terms of expected subset sizes.

Detailed analysis of this procedure for the case of three populations is given by Du Preez et al. [2].

References

[1] Gupta, S. S. (1965), *Technometrics*, **7**, 225–245.

[2] du Preez, J. P., Swanepoel, J. W. H., Venter, J. H., and Somerville, P. N. (1985). *S. Afr. Statist. J.*, **19**, 45–72.

[3] Somerville, P. N. (1984), *J. Statist. Comp. Simul.*, **19**, 215–226.

(RANKING PROCEDURES
SELECTION PROCEDURES)

SOR METHOD *See* GAUSS–SEIDEL ITERATION

SOUTH AFRICAN STATISTICAL JOURNAL

The *South African Statistical Journal* was founded in 1967 by the South African Statistical Association. From the outset it has appeared twice a year, two issues constituting a volume. The size of the *Journal* has grown from 80 pages in Volume 1 (1967) and 60 pages in Volume 2 (1968) to 194 pages in Volume 15 (1981) and 164 pages in Volume 16 (1982). The first editor was D. J. Stoker, who served from 1967–1968. The current editorial address is Dr. T. de Wet, *South African Statistical Journal*, Institute for Maritime Technology, P. O. Box 181, Simon's Town, 7995, Republic of South Africa.

The main aim of the *Journal* is to publish original research in the general area of mathematical statistics and probability theory. Papers have traditionally been fairly theoretical, although applied articles are welcome. Authoritative review articles on topics of general interest that are not readily accessible in a coherent form will also be considered for publication. Abstracts of papers presented at the annual conference of the Association are published in the *Journal*, as are summaries of theses obtained at South African universities. An international perspective and coverage is intended and contributions from nonmembers of the South African Statistical Association have always been welcomed.

All papers are refereed, by at least one referee and in many cases referees outside South Africa are used. A large percentage of submissions come from outside South Africa; this was approximately 35% during 1983. The average time between submission and editorial decision is six months. Approximately 55% of the papers submitted are rejected. The contents of a recent issue (Vol. 17, No. 2, 1983) are as follows:

"Counting processes within Markov renewal processes" by Y. P. Gupta and Sunita Gupta.

"Recent views in the foundational controversy in statistics" by J. C. Geertsema.

"Model selection for future data in the case of multivariate regression analysis" by J. A. van der Merwe, P. C. N. Groenewald, D. J. de Waal and C. A. van der Merwe.

"Union–intersection tests for sphericity" by C. J. Lombard.

T. DE WET

SPACE, PROBABILITY *See* PROBABILITY SPACES, METRICS AND DISTANCES ON

SPACE, SAMPLE *See* SAMPLE SPACE

SPACE, SEPARABLE *See* SEPARABLE SPACE

SPACINGS

Let $X_1, \ldots, X_n$ be a random sample of size n from a distribution with cumulative distribution function F, and $X_{(1)} \leqslant X_{(2)} \leqslant \cdots \leqslant X_{(n)}$ be the order statistics* of the sample. The successive differences

$$D_i = X_{(i)} - X_{(i-1)} \qquad (1)$$

are the *spacing* or *first-order gaps* of the sample. Depending on whether F has an unbounded support*, a support bounded on the left, or a support bounded on the left and right, the definition applies for $i = 2, \ldots, n$, for $i = 1, \ldots, n$ or for $i = 1, 2, \ldots, n+1$, with D_1 and D_{n+1} suitably defined in terms of the bounds on the support. Certain goodness-of-fit* tests based upon spacings are discussed later.

In a different context (which we do not consider here) the term "spacing" denotes a set of numbers $u_1 < u_2 < \cdots < u_n$, all in the interval $(0, 1)$, that relate to certain quantile estimators; *see* OPTIMAL SPACING PROBLEMS.

The most informative review of the properties of spacings is the 1965 survey by Pyke [12]; see also Pyke [13]. Many of the properties given here appear in the first of these two surveys, along with references to key sources. Pyke includes discussions of spacings based on uniform, exponential and general parent populations, tests of hypotheses, and limit theorems, and he provides a substantial bibliography.

UNIFORM SPACINGS

When F is continuous, the variate $U_i = F(X_i)$ has a uniform distribution* on $(0, 1)$ (*see* PROBABILITY INTEGRAL TRANSFORMA-

TION). The corresponding order statistics are

$$U_{(1)} \leqslant U_{(2)} \leqslant \cdots \leqslant U_{(n)}.$$

Without loss of generality tests of $H_0: F(x) = F_0(x)$, based on spacings of a sample, can therefore frequently be specified in terms of F_0 uniform on (0, 1). Then

$$D_i = U_{(i)} - U_{(i-1)}, \qquad i = 2, \ldots, n,$$
$$D_1 = U_{(1)} = F(X_{(1)}), \qquad (2a)$$
$$D_{n+1} = 1 - U_{(n)} = 1 - F(X_{(n)}).$$

Wilks [14, pp. 235–243] calls the statistics

$$D_i' = F(X_{(i)}) - F(X_{(i-1)}),$$
$$i = 1, \ldots, n+1, \quad (2b)$$

defined in this way, the *coverages* of the random sample $X_1, \ldots, X_n$ drawn from F.

The uniform spacings (2) have the following properties:

(a) $D_1, D_2, \ldots, D_{n+1}$ are exchangeable random variables, each with a beta $(1, n)$ distribution, having probability density function (PDF)

$$f(u) = n(1-u)^{n-1}, \qquad 0 < u < 1. \quad (3)$$

Huang et al. [9] give a related characterization of the uniform distribution. The mean and variance of each D_i are $1/(n+1)$ and $n/[(n+1)^2(n+2)]$, respectively.

(b) The exchangeability* of the spacings implies that the joint distribution of (D_i, D_j), $i \neq j$, is the same as that of (D_1, D_2). The joint PDF is

$$f(u, v) = n(n-1)(1-u-v)^{n-2},$$
$$u \geqslant 0,\ v \geqslant 0,\ u + v \leqslant 1,$$

with corresponding CDF

$$F(u, v) = 1 - \{(1-u)^n + (1-v)^n - (1-u-v)^n\}.$$

The covariance and correlation between D_i and D_j, $i \neq j$, are

$$\operatorname{cov}(D_i, D_j) = -1\big/\big[(n+1)^2(n+2)\big];$$
$$\rho(D_i, D_j) = -1/n.$$

(c) The joint PDF of $D_1, D_2, \ldots, D_{n+1}$ is

$$f(d_1, d_2, \ldots, d_{n+1}) = n!,$$
$$d_i \geqslant 0\ (i = 1, \ldots, n+1),$$
$$\sum_{i=1}^{n+1} d_i = 1,$$

noting that the distribution is nonsingular only if it is restricted to the hyperplane $\sum_{i=1}^{n+1} d_i = 1$ of Euclidean $(n+1)$-space. The joint distribution of any k of the coverages defined by (2b) is a k-dimensional Dirichlet distribution* [15, p. 238], $1 \leqslant k \leqslant n$.

(d) Let $D_{(1)} \leqslant D_{(2)} \leqslant \cdots \leqslant D_{(n+1)}$ be the *ordered* spacings of the sample from the uniform distribution, so that $D_1, D_2, \ldots, D_{n+1}$ are ranked from lowest to highest. Then the CDF of $D_{(n-j)}$ is given by [7]

$$\Pr(D_{(n-j)} \leqslant x) = \sum_{r=0}^{j} \binom{n}{r} \sum_{s=0}^{n-r} (-1)^s \times \binom{n-r}{s} [1 - (r+s)x]_+^{n-1},$$
$$0 < x < 1,$$

where $a_+ = \max(a, 0)$. This result can be traced back to Whitworth [14] in 1897.

(e)

$$E(nD_{(i)}) = \sum_{r=0}^{i-1} (n-r)^{-1}.$$

As n increases [7],

$$E(nD_{(n)}) = \log n + \gamma + o(1),$$

where γ is Euler's constant*, 0.577216....

(f) Let $D_{(n)}' = nD_{(n)} - \log n$. Then as $n \to \infty$ [8]

$$\Pr(D_{(n)}' \leqslant x) \to \exp(-e^{-x}),$$

and the moment generating function* of $D_{(n)}'$ converges to $\Gamma(1-t)$. Holst [7] also gives asymptotic distributions as $n \to \infty$ for $D_{(n-j)}$ when j is fixed, i.e., for other extreme values.

(g) Holst [7] also gives conditions on constants $a_1, \ldots, a_n$, such that $\sum_{i=1}^{n} a_i [nD_{(i)} - E(nD_{(i)})]$ is asymptotically normal. These include the asymptotic distributions of $D_{(i)}$ when $i/n \to b$ $(b \neq 0, b \neq 1)$ as $n \to \infty$.

Further properties of uniform spacings, based on constructions, are given in **(i)–(k)** of the next section.

SPACINGS CONSTRUCTED FROM EXPONENTIAL VARIABLES

Let F be the negative exponential distribution* with PDF

$$f(x) = \lambda \exp(-\lambda x), \qquad x > 0;\ \lambda > 0. \tag{4}$$

(h) If (1) is defined for $i = 2, \ldots, n$, and $D_1 = X_{(1)}$, then $D_1, \ldots, D_n$ are independent exponential random variables; D_i has PDF

$$f_i(x) = \lambda(n - i + 1)\exp\{-\lambda(n - i + 1)x\}, \qquad x > 0. \tag{5}$$

The *normalized spacings* $\{\lambda(n - i + 1)D_i\}$, $i = 1, \ldots, n$, are thus independent identically distributed exponential variables with mean equal to unity. Ahsanullah [1] gives a related characterization of (4).

(i) When $X_1, \ldots, X_{n+1}$ is a random sample from (4), let

$$S = X_1 + \cdots + X_{n+1},$$
$$D_i = X_i/S, \qquad i = 1, \ldots, n + 1.$$

Then $(D_1', \ldots, D_{n+1}')$ is distributed as the set of $n + 1$ spacings determined by n independent U(0, 1) random variables. Thus an *ordering* of uniform spacings can be represented as a *normalized ordering* of exponential variables $X_1, \ldots, X_{n+1}$.

(j) With the notation of **(i)**, the conditional distribution of $X_1, \ldots, X_{n+1}$, given $S = 1$, is that of $n + 1$ uniform spacings $(D_1', \ldots, D_{n+1}')$ defined as in **(i)**.

(k) Let $\{N(t) : t \geqslant 0\}$ be a Poisson process* with parameter λ and $T_1 \leqslant T_2 \leqslant \cdots$ the successive times of occurrence of events. Then the independent exponential variables $X_1, X_2, \cdots$ with common PDF (4) can be represented as interarrival times

$$X_1 = T_1, \qquad X_i = T_i - T_{i-1}, \qquad i = 2, 3, \ldots,$$

and the conditional distribution of $(X_1/t, X_2/t, \ldots, X_{n+1}/t)$, given $N(t) = n$, where $X_{N(t)+1} = t - T_{N(t)}$, is that of $n + 1$ uniform spacings $(D_1', \ldots, D_{n+1}')$ as defined in **(i)**.

TESTS

For the null hypothesis $H_0: F(x) = F_0(x)$ against the alternative $H_a: F(x) \neq F_0(x)$, where F_0 is a specified continuous CDF, many test statistics have either been based on the ordered uniform spacings $D_{(1)}, \ldots, D_{(n+1)}$ of property **(d)**, or on sums of the form

$$G_n = \sum_{i=1}^{n} g_n(D_i). \tag{6}$$

Various choices of $g_n(\cdot)$ were unified by Darling [5], who reviewed these and developed a characteristic function* for G_n for quite arbitrary choices of g_n. In his review, Pyke [12] surveyed these choices briefly, including

$$g_n(x) = x^r\ (r > 0), \{x - 1/(n + 1)\}^2,$$
$$|x - 1/(n + 1)|, \log x, \text{ and } (1/x),$$

as well as references for limit distributions for G_n. Koziol [10] used Hájek's projection* method to derive many of these asymptotic results [e.g., the asymptotic normality* of $g_n(x) = x^r$ for $r > 0$, $r \neq 1$]. For further details of the case when $g_n(x) = x^2$, *see* GREENWOOD'S STATISTIC.

Pyke [12] pointed out that the models for working with spacings could be approached in terms of "order statistics," "point processes," or "renewal processes," as appropriate. Under H_0 the theory for these models is identical, but different approaches emerge when the behavior of the test statistics under H_a is under scrutiny.

*m*th ORDER SPACINGS

Instead of basing goodness-of-fit tests on spacings (1), more powerful test statistics can be constructed from *m*th-*order spacings* or mth-*order gaps*,

$$D_i^{(m)} = X_{(i)} - X_{(i-m)}. \qquad (7)$$

When these are based on a uniform distribution on (0, 1), each $D_i^{(m)}$ has a beta distribution with PDF [15, p. 238]

$$\frac{1}{B(m, n-m+1)} x^{m-1}(1-x)^{n-m}, \quad 0 < x < 1.$$

Cressie [2, 3] and Holst [6] derived properties of a test of uniformity based on the test statistic

$$L_n = \sum_{i=0}^{R-(m-1)} \log\left(D_i^{(m)}\right),$$

that performs well against alternatives with peaks or bumps; a simulation showed $m = 3$ to be a recommended choice of gap order for $25 \leqslant n \leqslant 100$ [3]. See these sources for further references.

With Pitman asymptotic relative efficiency (ARE) as a criterion (*see* PITMAN EFFICIENCY), the test statistic

$$S_n = \sum_{i=0}^{n-(m-1)} \left(nD_i^{(m)}\right)^2$$

is optimal [4], although the ARE of S_n relative to L_n tends to 1 as m increases. Under H_0 and certain alternatives, L_n and S_n are asymptotically normal.

There is an interesting property of scan statistics* that is related to mth order spacings. Let

$$N(x, h) = \sum_{i=1}^{n} I\{X_{(i)} \in (x, x+h]\},$$

where $0 \leqslant x \leqslant 1-h$, h is fixed, $0 < h < 1$, and I is the indicator function. The scan statistic is defined as [11]

$$N(h) = \sup_x \{N(x, h); 0 \leqslant x \leqslant 1-h\}.$$

Then

$$\Pr(N(h) > m) = \Pr\left(\min\{D_i^{(m)}; i = 0, 1, \ldots, n+1-m\} \leqslant h\right).$$

References

[1] Ahsanullah, M. (1978). *J. Appl. Prob.*, **15**, 650–653.

[2] Cressie, N. (1976). *Biometrika*, **63**, 343–355.

[3] Cressie, N. (1978). *Biometrika*, **65**, 214–218.

[4] Cressie, N. (1979). *Biometrika*, **66**, 619–627.

[5] Darling, D. A. (1953). *Ann. Math. Statist.*, **24**, 239–253.

[6] Holst, L. (1979). *Ann. Prob.*, **7**, 1066–1072.

[7] Holst, L. (1980). *J. Appl. Prob.*, **17**, 623–634.

[8] Holst, L. (1981). *Ann. Prob.*, **9**, 648–655.

[9] Huang, J. S., Arnold, B. C., and Ghosh, M. (1979). *Sankhyā B*, **41**, 109–115.

[10] Koziol, J. A. (1977). *Zeit. Wahrsch. verw. Geb.*, **50**, 55–62.

[11] Naus, J. I. (1966). *J. Amer. Statist. Ass.*, **61**, 1191–1199.

[12] Pyke, R. (1965). *J. R. Statist. Soc. B*, **27**, 395–436. (This is the primary source for references on spacings prior to 1965, and is followed on pp. 436–449 by an informative discussion.)

[13] Pyke, R. (1972), *Proc. Sixth Berkeley Symp. Math. Statist. Prob.*, Vol. 1, University of California Press, Berkeley, CA, pp. 417–427.

[14] Whitworth, W. A. (1897). *Choice and Chance*. Cambridge University Press, Cambridge, England. (See Problem 667.)

[15] Wilks, S. S. (1962). *Mathematical Statistics*. Wiley, New York.

Bibliography

Blumenthal, S. (1966). *Ann. Math. Statist.*, **37**, 904–924, 925–939. (Properties are studied of tests of a two-sample problem and of goodness of fit, both in the presence of nuisance location and scale parameters, and where test statistics are based on sample spacings.)

Finch, S. J. (1977). *J. Amer. Statist. Ass.*, **72**, 387–392. (The author presents a robust test of symmetry, with a test statistic constructed from spacings.)

(EXPONENTIAL DISTRIBUTION
GAPPING
ORDER STATISTICS
SCAN STATISTICS
UNIFORM DISTRIBUTIONS)

CAMPBELL B. READ

SPACINGS, MAXIMUM PRODUCT OF

See MAXIMUM PRODUCT OF SPACINGS (MPS) ESTIMATION

SPAN TESTS

These are tests for trend* in expected values of a sequence of independent random variables $X_1, X_2, \ldots,$ with constant standard deviation σ. If σ is known, the test criterion, based on a "span" of n values $X_1, X_2, \ldots, X_n$ is

$$T = \sigma^{-1} \max_{1 \leqslant r < n} \left| \sum_{j=1}^{r} \left(X_j - \bar{X} \right) \right|,$$

where $\bar{X} = n^{-1}\Sigma_{j=1}^{n} X_j$. Large values of T are regarded as significant.

Bissell [1] has provided tables of estimated upper significance limits for T (on the assumption of constant expected value) that are appropriate when the distribution of each X is normal.

If σ is not known, it is replaced by an estimate. Williamson [2] has provided tables of estimated upper significance limits of T^* and T^{**}, statistics obtained by replacing σ in the formula for T by

$$\sigma^* = \left\{ \frac{1}{2(n-1)} \sum_{j=1}^{n-1} \left(X_{j+1} - X_j \right)^2 \right\}^{1/2}$$

and

$$\sigma^{**} = \frac{8}{9(n-1)} \sum_{j=1}^{n-1} |X_{j+1} - X_j|,$$

respectively (*see* SUCCESSIVE DIFFERENCES).

The use of σ^* or σ^{**} in preference to the more usual

$$\hat{\sigma} = \left\{ \frac{1}{n-1} \sum_{j=1} \left(X_j - \bar{X} \right)^2 \right\}^{1/2}$$

is recommended because if a trend exists it will tend to bias σ^* and σ^{**} upward less than $\hat{\sigma}$. The tests are especially relevant to the interpretation of cumulative sum* (CUSUM) control charts* on the mean.

References

[1] Bissell, A. F. (1984). *An Introduction to Cusum Charts*. Institute of Statistics, Bury St. Edmunds, England.

[2] Williamson, R. J. (1985). *The Statistician (Lond.)*, **34**, 345–356.

(CUMULATIVE SUM CONTROL CHARTS
QUALITY CONTROL, STATISTICAL
SUCCESSIVE DIFFERENCES
TREND TESTS)

SPATIAL DATA ANALYSIS

The study of spatial patterns and processes by statistical means has a long history in applications such as forestry* [7] and there has been more recent extensive interest from fields such as ecology, astronomy, geology* and mining, geography*, and archaeology*. Statisticians were minimally involved until the 1970s, when a number of computer-based methods revolutionized the subject and are gradually having a profound impact on practice. This revolution has included moves away from simple questions such as "is this pattern of trees random, regular or clustered" toward more complete summaries of the data and the fitting of stochastic models to data (*see* SPATIAL PROCESSES).

Two of the characteristics of spatial data analysis are the lack of ordering of the data (which usually entails heavy computations) and the presence of nonnegligible edge effects, unfortunately often ignored. The most natural division of the subject is by the type of data. Here we consider *spatial series*, the spatial analogs of time-series* analysis, *smoothing and interpolation** for continuous surfaces, *spatial point patterns*, and *image analysis*, dealing with patterns made up from points and more complex shapes, respectively.

Most books on the subject are based within specific applications. At the time of writing the only general reference was [11]; it is planned to periodically review subsequent work in the *International Statistical Review**.

SPATIAL SERIES

Spatial series are multidimensional analogs of time series. Data on two-dimensional lattices occur almost exclusively from sampling or experiments; for example, systematic sampling* (*see* SPATIAL SAMPLING), agricultural field trials, and satellite images. The main interest has been in defining models; key references are refs. 2, 6, 9, and 13. Consider random variables (X_{ij}), where i and j denote spatial coordinates. A typical *simultaneous* spatial autoregression is defined by

$$X_{ij} = \alpha(S_{i+l,j} + X_{i-l,j}) + \beta(X_{i,j-1}) + \epsilon_{ij},$$
$$\epsilon_{ij} \text{ independent } N(0, \sigma^2),$$

whereas a typical *conditional* autoregression might be

$$\begin{aligned} E(X_{ij}|X_{kl}, kl \neq ij) &= \alpha(X_{i+1,j} + X_{i-1,j}) \\ &\quad + \beta(X_{i,j+1} + X_{i,j-1}), \\ \mathrm{var}(X_{ij}X_{kl}, kl \neq ij) &= \kappa. \end{aligned}$$

These are different processes, with different correlation structures, and each has some of the properties of a time-series autoregression. There is also a significant difference between the process defined on the observed sites and that defined throughout the lattice and restricted to the observed sites, due to edge effects. The references describe a number of methods of estimation for these processes. An important application has been to modelling residual fertility in field trials; see refs. [1 and 14].

These models can be generalized to irregular lattices (such as systems of counties) if they are given a notion of neighbor. Geographers work with a weighting matrix (w_{ij}) that gives the proximity of site i to site j. Then a simultaneous autoregression is $\mathbf{X} = \mathbf{WX} + \boldsymbol{\epsilon}$ for $\mathbf{X} = (X_i)$, and $\boldsymbol{\epsilon} = (\epsilon_i)$ independent $N(0, \sigma^2)$ random variables. The geographical literature is dominated by "tests of spatial autocorrelation," for example

$$I = \frac{n \sum_{i \neq j} w_{ij}(x_i - \bar{x})(x_j - \bar{x})}{\Sigma(x_k - \bar{x})^2 \sum_{i \neq j} w_{ij}}$$

and their distribution, analogs to the Durbin–Watson test*. (*See also* GEOGRAPHY, STATISTICS IN.)

SMOOTHING AND INTERPOLATION

Most contouring packages for spatial data use ad hoc methods, based on distances between points. For example, with data $(Z_1, \ldots, Z_n)$ at sites $(\mathbf{x}_1, \ldots, \mathbf{x}_n)$, one might estimate the surface $Z(\mathbf{x})$ at $\mathbf{x}$ by

$$\hat{Z}(\mathbf{x}) = \Sigma w_i Z_i / \Sigma w_i, \qquad w_i = d(\mathbf{x}, \mathbf{x}_i)^{-2}.$$

This, more sophisticated variants, and methods based on spatial analogs of splines are discussed in ref. 11, pp. 36–44. Another idea is *trend surfaces*, polynomial regression in two or more dimensions [11, pp. 29–36].

All these methods have serious drawbacks; one way to assess them is to use a spatial process model for $Z(\;)$. Then one could use standard methods to find the conditional distribution of $Z(\mathbf{x})$ given the data $Z_1, \ldots, Z_n$. For a Gaussian random process $(Z(\mathbf{x}), Z_1, \ldots, Z_n)$ has a multivariate normal distribution, and the conditional distribution of $Z(\mathbf{x})$ is normal with known mean and variance. These remarks are the basis of an often-discovered method that was most developed in mining by Matheron and his colleagues and is now referred to as *kriging**. An extension in which $Z(\mathbf{x}) = f(\mathbf{x}, \boldsymbol{\beta}) + \eta(\mathbf{x})$, where $f(\mathbf{x}, \boldsymbol{\beta})$ is a parametric surface such as a polynomial trend surface, and η is a Gaussian random process, is known as "universal kriging." An account in statistical language is in ref. 11, pp. 44–74; ref. 8 gives a good description of mining practice.

The drawback of the kriging approach is the large number of assumptions made. Transformations to normality have been considered, but this is difficult for correlated random variables. The major problem of identifying and estimating the covariance function of the Gaussian process* has received little attention in the past but is a topic of current research.

POINT PATTERNS

There are many examples of spatial patterns of objects that can be regarded as points, for example trees, market towns, galaxies, and pebbles on a sand beach. Automated techniques are increasingly making available a complete map of the objects within a small region. Until recently this was often too laborious to produce, and special sampling methods were devised to obtain more limited information. These are still needed, for example in forestry where mapping is uneconomic.

Field methods

Two basic techniques of sampling are used; counting within small squares (quadrat sampling*) and measuring distances to the nearest object (distance sampling or nearest-neighbour methods*). Both can aim either to estimate the intensity λ (number of objects per unit area) or to test for randomness. The latter is a significance test of a spatial Poisson process* and usually provides an indication of "random, regular, or clustered" (*see also* ECOLOGICAL STATISTICS).

Traditionally, quadrat sampling has been analysed by regarding the counts as independent samples from a discrete distribution. It usually provides good estimates of intensity but little indication of the spatial pattern of points. (Methods based on systematic quadrat grids are discussed under Image Analysis below.)

Distance methods depend on the observation that λ should be proportional to d^{-2}, where d is the distance from any point to the nearest tree. For a Poisson process $E(d^2) = 1/(\pi\lambda)$, but the constant of proportionality depends quite strongly on the pattern. Many of the methods depend on the observation that, as the pattern becomes regular, distances from object to object increase, and those from point to object decrease, so some combination might be "robust" to the type of pattern. See ref. 11, pp. 131–139.

A specialized form of distance sampling, known as line transect sampling*, is used to assess the intensity of a population of a mobile population of animals. See ref. 11, pp. 139–142.

Mapped Patterns

The Clark–Evans test [3], based on the average distance from each object to its nearest neighbour, has been popular. Its commonly quoted distribution illustrates two common fallacies by ignoring edge effects and dependence between the distances. Better approximations are given in ref. 11, p. 153. However, much more information is available from a mapped point pattern. The two main approaches use the distributions of all nearest-neighbour distances and of all distances between pairs of objects or produce more extensive data summaries and means of identifying and estimating models. For nearest-neighbour methods see ref. 5. Second-moment methods use plots of functions such as

$$\hat{K}(t) = A\Sigma k(x, y)/n^2,$$

where the sum is over all those pairs of n points at most distance t apart. Here $k(x, y)$ is one of several possible devices to correct for edge effects. From such plots one can identify regularity or clustering at each scale t. It is also possible to derive $E\hat{K}(t)$ for many models and so build a theoretical understanding of the behaviour of $\hat{K}$. For a Poisson process* $E\hat{K}(t) \approx \lambda\pi t^2$, and it transpires that working with $L(t) = \sqrt{\{\hat{K}(t)/\pi\}}$ has both linearizing and variance-stabilizing effects. Full details are given in ref. 11, pp. 158–190.

IMAGE ANALYSIS

Image analysis is used here to cover methods for the analysis of complex patterns, such as microscopic sections of rock or tissue, remotely sensed images, and patterns of communities of vegetation. Again, there is a split into simple field methods and those that use digitized images. This is an area of spatial data analysis that to date has received scant attention.

Plant ecologists have developed a family of methods based on systematic grids of quadrats, either say a 16 × 16 layout or a line of 128 to 512 abutting quadrats. An observation is taken within each quadrat, possibly presence/absence of a species, a count of stems, or a measure of abundance. The original analysis, owing to Greig–Smith, was a nested analysis of variance* based on successive binary division of the grid. This and later variations aim to detect block sizes of blocks of quadrats with significantly high (or low) variation and so ascribe "scales" to the pattern. Critical surveys are given in refs. 10 and 11, pp. 108–129.

Automated image analysers and remote sensing devices produce a digitized version of an image, usually on a square grid with 512 × 512 or 1024 × 1024 pixels. Each pixel* may be a 0 or 1, to represent presence or absence of rock, say, or a gray level of colour. There is little *statistical* theory for such images, although ref. 12 does develop probability models and computational considerations. Random sets* as developed by Matheron and Kendall are one type of model on which one could base a statistical analysis.

The spatial series models mentioned above are also beginning to be used as models for images in the pattern recognition*/image processing literature. These can only adequately model the local characteristics of the digitized image, but this is sufficient to be helpful, for example, in classifying the pixels of a Landsat photograph of the earth's surface into various agricultural uses.

References

[1] Bartlett, M. S. (1978). *J. R. Statist. Soc. Ser. B*, **40**, 147–174.

[2] Besag, J. (1974). *J. R. Statist. Soc. Ser. B*, **36**, 142–236.

[3] Clark, P. J. and Evans, F. C. (1954). *Ecology*, **35**, 445–453.

[4] Cliff, A. D. and Ord, J. K. (1981). *Spatial Processes. Models and Applications.* Pion, London. (Details the practice of geographers. Principally on tests for spatial autocorrelation.)

[5] Diggle, P. J. (1979). *Biometrics*, **35**, 87–101.

[6] Guyon, X. (1982). *Biometrika*, **69**, 95–105.

[7] Holgate, P. (1972). In *Stochastic Point Processes*, P. A. W. Lewis, ed. Wiley, New York, pp. 122–135.

[8] Journel, A. and Huijbregts, C. J. (1978). *Mining Geostatistics*. Academic, London, England, (Best source for kriging practice.)

[9] Ord, K. (1975). *J. Amer. Statist. Ass.*, **70**, 120–126.

[10] Ripley, B. D. (1978). *J. Ecology*, **66**, 965–981.

[11] Ripley, B. D. (1981). *Spatial Statistics*. Wiley, New York. (Most comprehensive reference available to date.)

[12] Serra, J. (1982). *Mathematical Morphology and Image Analysis*. Academic, London, England. (Idiosyncratic account. More probability and philosophy than data analysis.)

[13] Whittle, P. (1954). *Biometrika* **41**, 434–449.

[14] Wilkinson, G. N., Eckert, S. R., Hancock, T. W., and Mayo, O. (1983). *J. R. Statist. Soc. Ser. B*, **45**, 151–211.

Bibliography

Besag, J. (1986). *J. R. Statist. Soc. Ser. B*, **48**, 259–302. (An account of recent spatial statistics in image analysis.)

Diggle, P. J. (1983). *Statistical Analysis of Spatial Point Patterns*. Academic, London, England. (Aimed at biometricians. Little mathematics and many examples.)

Hodder, I. and Orton, C. (1976). *Spatial Analysis in Archaeology*. Cambridge University Press, London, England. (Many interesting case studies with simple methods.)

Pielou, E. C. (1977). *Mathematical Ecology*. Wiley, New York. (Compendium of ecologists' methods, right and wrong.)

Ripley, B. D. (1977). *J. R. Statist. Soc. Ser. B*, **39**, 172–212. (Mathematical account of spatial point processes, with discussion.)

Ripley, B. D. (1984). Spatial statistics–developments 1980–3. *Int. Statist. Rev.*, **52**, 141–150. (First of a projected series of review papers.)

Ripley, B. D. (1986). *Canad. J. Statist.*, **14**, 83–111. (Survey of more recent work in image analysis.)

Rogers, A. (1974). *Statistical Analysis of Spatial Dispersion. The Quadrat Method*. Pion, London, England. (Comprehensive account of obsolete method.)

(GEOGRAPHY, STATISTICS IN
GEOSTATISTICS
KRIGING
NEAREST-NEIGHBOR METHODS
QUADRAT SAMPLING
SPATIAL PROCESSES
SPATIAL SAMPLING
STEREOLOGY)

B. D. RIPLEY

SPATIAL DISTRIBUTIONS *See* SPATIAL PROCESSES

SPATIAL MEDIAN

The spatial median (also called *mediancentre*, see Gower [3]) is the multivariate location measure that minimizes the sum of absolute distances to observation points, in direct analogy to the ordinary, univariate sample median. Gower [3] defined it and discussed some of the properties that are now listed below, for the *bivariate* case only; see also Brown [1]. Broadly speaking, its properties resemble those of the univariate median (*see* MEDIAN ESTIMATES and SIGN TESTS), except for its efficiency performance, which is somewhat better.

COMPUTATION

Let observations be (x_i, y_i), for $i = 1, 2, \ldots, n$, and let $\boldsymbol{\mu} = (\mu_1, \mu_2)$ be a trial median value. Let θ_i be the angle of the line from $\boldsymbol{\mu}$ to (x_i, y_i) with an arbitrary zero direction, and let $C = \Sigma_i \cos\theta_i$ and $S = \Sigma_i \sin\theta_i$. Then the spatial median $\hat{\boldsymbol{\mu}}$ satisfies *either*

$$C = S = 0, \tag{1}$$

in the case where $\hat{\boldsymbol{\mu}}$ is *not* one of the observations, *or*

$$\{C^2 + S^2\}^{1/2} \leqslant \text{multiplicity of observation}(x_j, y_j), \tag{2}$$

where $\hat{\boldsymbol{\mu}} =$ the observation(x_j, y_j), which is then omitted from the sums yielding C and S.

There are various ways of iteratively calculating $\hat{\boldsymbol{\mu}}$. Gower [3] outlines a first-order method. Another possibility is to use the Newton–Raphson method* as follows. Let r_i be the distance from $\boldsymbol{\mu}$ to (x_i, y_i). The estimate $\hat{\boldsymbol{\mu}}$ minimizes $\Sigma_i r_i$, whose second derivative matrix is

$$\Gamma = \begin{pmatrix} -\Sigma r_i^{-1}\cos^2\theta_i & \Sigma r_i^{-1}\cos\theta_i \sin\theta_i \\ \Sigma r_i^{-1}\cos\theta_i \sin\theta_i & \Sigma r_i^{-1}\sin^2\theta_i \end{pmatrix}.$$

Let $\boldsymbol{\mu}_n$ be the nth iterate for $\hat{\boldsymbol{\mu}}$, with corresponding values C_n, S_n, and Γ_n of C, S, and Γ. Then

$$\boldsymbol{\mu}_{n+1} = \boldsymbol{\mu}_n - \Gamma_n^{-1}\begin{pmatrix} C_n \\ S_n \end{pmatrix}.$$

However, because of the nonsmooth nature of the surface $(\boldsymbol{\mu}, \Sigma r_i)$ near the observation points, it is wise to include an "escape hatch" in the Newton–Raphson iteration and only move from $\boldsymbol{\mu}_n$ to $\boldsymbol{\mu}_{n+1}$ when, say Σr_i, or perhaps $C_n^2 + S_n^2$, is thereby reduced; otherwise move back from the suggested $\boldsymbol{\mu}_{n+1}$ toward $\boldsymbol{\mu}_n$ until a point is found yielding such an improvement. Such a point does exist because of the convexity of Σr_i as a function of $\boldsymbol{\mu}$. A further modification is that a check of which of the conditions (1) or (2) holds must be built in. However, after including this and the "escape hatch," the procedure is guaranteed to converge.

ROBUSTNESS

The estimating equations $C = S = 0$ show that observation points can be moved arbitrarily along rays connecting $\hat{\boldsymbol{\mu}}$ to each (x_i, y_i) (thus preserving each θ_i) without changing $\hat{\boldsymbol{\mu}}$. In this sense the spatial median $\hat{\boldsymbol{\mu}}$ is robust against large outliers*, and in a manner exactly analogous to the similar property of the univariate median.

EFFICIENCY

The efficiency properties of the univariate median are sometimes unwisely labelled as "poor" because its efficiency relative to the sample mean is $2/\pi = 0.637$ for normal data. The corresponding efficiency of the spatial median relative to the multivariate sample mean actually increases as k, the number of dimensions, increases. For spherical normal data, it is $\pi/4 = 0.785$ for $k = 2$ and $\to 1$ as $k \to \infty$.

OTHER PROPERTIES

There are other aspects of the spatial median of possible potential interest yet to be investigated. Their companion tests are "angle tests" (see Brown, [1]), which become the multivariate analog of univariate sign tests*; the test statistics are null versions of C and S, emphasizing the polar coordinate/angular aspect of the spatial median.

The spatial median version described here is invariant to rotation of coordinate axes, but not (as some applications demand) to arbitrary changes of scale along individual axes. One way of ensuring the latter invariance is to choose axis scales so that $\hat{\mu}$ satisfies $\Sigma\cos^2\theta_i = \frac{1}{2} = \Sigma\sin^2\theta_i$; details are given in Brown [2].

The spatial median may have a role in developing other rank-analog multivariate procedures. For instance a multivariate Hodges–Lehmann estimate* appears to be the spatial median of all pairwise averages of observations.

References

[1] Brown, B. M. (1983). *J. R. Statist. Soc. Ser. B*, **45**, 25–30.

[2] Brown, B. M. (1984). *The Standardized Spatial Median*. Technical Report, University of Tasmania, Hobart, Australia.

[3] Gower, J. C. (1974). *Appl. Statist.*, **23**, 466–470.

(DIRECTIONAL DATA ANALYSIS
MEAN, MEDIAN, MODE AND SKEWNESS
REGRESSION QUANTILE
SPHERICAL MEDIAN)

B. M. BROWN

SPATIAL PROCESSES

Let $Y(\mathbf{x})$ describe a random variable at each of a set of possible sites $\mathbf{x}$. For example, Y might denote height above sea level or the diameter at breast height of a tree. In the first case the process is defined for all possible values of $\mathbf{x}$ and Y is said to be a *continuous* spatial process. In the second case, Y is defined only at the finite set of sites $\mathbf{x}$ at which trees are to be found and Y describes a *discrete* spatial process. The same distinction is used in time series*; indeed, many similarities exist between the two fields. Y may denote either a continuous or a discrete random variable. Usually $\mathbf{x}$ describes two-dimensional space, although the extension to higher dimensions is often straightforward.

When the location of entities is of primary importance, the set of observed entities describe a spatial point process. Much of the theory of point processes* applies here and interest often focuses upon distances between individuals; *see* NEAREST NEIGHBOR METHODS.

Spatial processes have found application in many disciplines. For example, processes defined on regular lattices are used in the Ising model for magnetism [16] and in digital image processing [7] (*see* LATTICE SYSTEMS). Sampling at regular or irregular patterns of sites is carried out extensively in the study of hydrological [18], geological [12], and ecological* processes [10, 13] (*see* HYDROLOGY, STOCHASTIC and GEOSTATISTICS). Also, geography* [8] might claim the study of spatial processes as one of its primary activities. The growth of regional econometric* modeling [24] has stimulated interest among economists and social scientists. Recently, considerable interest has also been shown in the use of spatial models to analyze agricultural field trials [35].

The study of spatial processes has evolved in a rather sporadic fashion. Some of the basic results on spatial point processes* were known by the 1900s and, in 1914, Student [32] suggested the use of polynomial regression for modeling spatial trends. The basic notion of a Markov random field* derived from Ising's work in the 1920s on ferromagnetism; see ref. 16. The proposal to use observations in adjacent plots as covariates in the analysis of agricultural field trials

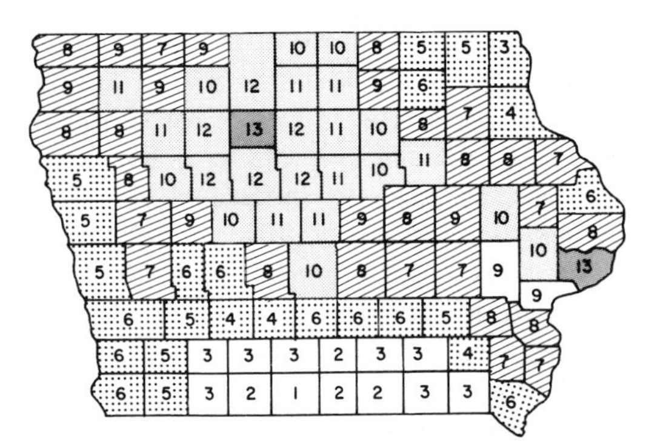

Figure 1

surfaced briefly in the late 1930s, but lay dormant for many years until revived recently; see Bartlett [3].

The modern development of spatial modeling can be traced to Whittle's 1954 paper [34] and the work by Matheron and his co-workers [12, 23].

An example of spatial data is given in Fig. 1, which displays the spatial pattern, by county, of agricultural land prices in Iowa in 1978. That is, each value has been averaged spatially within the county. Such spatial aggregation* is fairly common, but may complicate the modeling process. The similarities between neighboring counties are evident from the data. The figure illustrates a *mapped* pattern where the relative locations of all sites are known; data may also be "sparsely" sampled [13, p. 4] when a set of small areas, sometimes known as *quadrats*, are selected; the relative locations of these sample areas are unknown.

In this discussion we shall concentrate upon random variables $Y(\mathbf{x})$ defined at specified sites $\mathbf{x}$. Spatial point processes will not be considered; *see* NEAREST-NEIGHBOR METHODS and SPATIAL DATA ANALYSIS.

STATIONARITY

A process is said to be *weakly* spatially stationary (or wide-sense or covariance stationary) if

$$E[Y(\mathbf{x})] = \mu \quad \text{for all } \mathbf{x},$$
$$V[Y(\mathbf{x})] = \sigma^2, \quad \text{for all } \mathbf{x},$$

and

$$\operatorname{cov}[Y(\mathbf{x}), Y(\mathbf{x}')] = \sigma^2 \rho(\mathbf{s}),$$
$$\mathbf{s} = |\mathbf{x} - \mathbf{x}'|;$$

that is, the covariances depend only on the relative positions of the two sites. If the process is only defined at certain sites then stationarity may still be assumed, conditional upon the set of sites. As for time series, the process is said to be *strictly* stationary if the joint distribution of $\{Y(\mathbf{x}_1), \ldots, Y(\mathbf{x}_k)\}$ depends upon their relative, but not their absolute positions, for all $\mathbf{x}$ and k.

If the process is direction-invariant, so that $\operatorname{cov}[Y(\mathbf{x}), Y(\mathbf{x}')] = \sigma^2\rho(d)$, where $d = d(\mathbf{x}, \mathbf{x}')$ denotes the distance between the two sites, the process is said to be *isotropic*. A process may be isotropic but nonstationary, as occurs when

$$E[Y(\mathbf{x}) - Y(\mathbf{x}')] = 0,$$
$$V[Y(\mathbf{x}) - Y(\mathbf{x}')] = \gamma(d),$$

but $\lim_{d \downarrow 0} \gamma(d) > 0$. The function $\gamma(d)$ is known as the *variogram*, and the $Y(\mathbf{x})$ are known as regionalized variables. The theory of *regionalized variables* has been developed by Matheron [23] and his co-workers and this theory forms the basis of the method of spatial interpolation known as kriging* (*see also* GEOSTATISTICS).

When the process is weakly stationary, its spectral density function (*see* SPECTRAL ANALYSIS) is given by

$$f(\omega_1, \omega_2) = f(\boldsymbol{\omega})$$
$$= \int \exp(is_1\omega_1 + is_2\omega_2)\rho(\mathbf{s})\, d\mathbf{s},$$

where $\mathbf{x} - \mathbf{x}' = \mathbf{s}$. For a full development of the spectral theory of spatial processes, see Vanmarcke [33].

ANALYSIS OF TRENDS

When trends exist in the study area, the simplest approach to modeling is to use polynomial regression (*see* REGRESSION POLYNOMIALS). Thus, if a quadratic surface is used,

$$Y(x_1, x_2) = \beta_{00} + \beta_{10}x_1 + \beta_{01}x_2 + \beta_{20}x_1^2 + \beta_{11}x_1x_2 + \beta_{02}x_2^2 + \epsilon(x_1, x_2),$$

which may be fitted by least squares* if it is assumed that the errors are independent and identically distributed. Interestingly, this approach was first suggested by Student in 1914 [32], but was not heavily used until the 1960s; it is now known as *trend surface analysis*. For further references *see* KRIGING.

When a spatial point process is modeled as a nonhomogeneous Poisson process* with intensity $\lambda(\mathbf{x})$ at $\mathbf{x}$, either $\lambda(\mathbf{x})$ or $\ln \lambda(\mathbf{x})$ may be modeled as a polynomial in $\mathbf{x}$; see refs. 17 and 8, pp. 111–112.

TESTS FOR SPATIAL AUTOCORRELATION

A natural first step in modeling is to check whether the data exhibit any spatial dependence. In general, this involves testing the null hypothesis

$$H_0 : \rho(\mathbf{s}) = 0 \quad \text{for all } \mathbf{s} \neq \mathbf{0}.$$

The test statistics employed are usually of the form either

$$\frac{\Sigma\Sigma w_{ij}z_iz_j}{g(\mathbf{z})} \quad \text{or} \quad \frac{\Sigma w_{ij}(z_i - z_j)^2}{g(\mathbf{z})}, \tag{1}$$

where the $\{w_{ij}\}$ are a set of nonnegative weights specified in light of the alternative hypothesis, $g(\mathbf{z})$ is a symmetric function of the z's used to make the statistics scale invariant, and the z_i are monotone transforms of the y_i. When

$$z_i = \begin{cases} 1 & \text{if } y_i \geqslant y_0, \\ 0 & \text{otherwise,} \end{cases}$$

we have the joint count statistics; see ref. 8, Chap. 1 and 2. Test statistics based upon $z_i = y_i - \bar{y}$ and for regression residuals are described in ref. 8, Chaps. 1, 2, and 8; *see also* CLIFF–ORD STATISTIC. These tests consider only the null hypothesis of no spatial autocorrelation. A test has been developed [31] for $H_0 : \rho = \rho_0 \neq 0$. Rather than use a single function, we may define a spatial correlogram* by

$$r(\mathbf{s}) = \frac{(n-1)^{-1}\mathbf{z}'\mathbf{W}_s\mathbf{z}}{\{\text{var}(\mathbf{z})\text{var}(\mathbf{z}_s^*)\}^{1/2}}, \tag{2}$$

where $\mathbf{z}_s^* = \mathbf{W}_s\mathbf{z}$ and $\mathbf{W}_s$ is a spatial adjacency weighting matrix. For example, we may take the (i, j)th element $w_{ij}(\mathbf{s}) = 1$ if $|\mathbf{x}_i - \mathbf{x}_j| = \mathbf{s}$, and 0 otherwise. Guyon [14] shows that corrections for edge effects, implicit in (2), should be included in the spatial autocorrelation coefficients, even though they are commonly ignored in time series. For irregularly spaced data, some grouping of observations is necessary to produce reasonable, if somewhat biased, estimators [17]. For tests of the spatial correlogram, see ref. 25.

For point patterns, the null hypothesis of a spatially independent stationary scheme is expressed by the statement that the underlying stochastic mechanism is a Poisson process. The Poisson assumption implies that counts in nonoverlapping areas are independently Poisson* distributed. Particularly in the ecological literature, tests of the Poisson assumption are known as tests of *randomness**.

Tests based on areal counts are of the form (1), but a considerable number of tests have been developed that use measurements from randomly selected points (or events) to nearest neighbors* (events); see Cormack [9, p. 161] for a detailed listing. For sparsely sampled populations, satisfactory distributional results are available [9], but for mapped data sets Monte Carlo methods* must often be used; see Diggle [13, Chap. 2].

In recent years new methods of spatial data analysis* have been developed, notably by Ripley [30]. These techniques provide a basis for testing not only the Poisson process but also more complex spatial models. Again, the general approach is to use Monte Carlo tests.

MODELS IN THE SPATIAL DOMAIN

Models for spatial processes may be specified in either the spatial domain or the frequency domain. As for time series, a formal equivalence between these classes exists through the Fourier transform, although the nature of the process and estimation problems may make one approach more useful in any given problem. In the frequency domain, the estimation procedures are essentially the same as those used in the study of time series; *see* SPECTRAL ANALYSIS. An interesting application of the spectral approach to spatial data appears in ref. 29.

In the spatial domain, the process might be specified by its covariance structure or the variogram. Given that the population autocorrelations are a function of p parameters, $\boldsymbol{\theta}$ say, estimators for $\hat{\boldsymbol{\theta}}$ and hence $\hat{\rho}(s)$, $s = 1, 2, \ldots,$ may be derived; see refs. 30, pp. 54–74, and 19 for further details.

The other approach to modeling is to use an autoregressive structure. In his seminal paper on the subject, Whittle [34] considered the *simultaneous* scheme

$$Y_i = \sum_{j \neq 1} g_{ij} Y_j + \epsilon_i, \tag{3}$$

where the $\{\epsilon_i\}$ are uncorrelated error terms with zero means and variances σ_i^2. When the $\{g_{ij}\}$ are linear functions of unknown parameters $\boldsymbol{\theta}$, Whittle showed that the least-squares estimator derived from (3) was inconsistent for $\boldsymbol{\theta}$ because of the dependence between Y_j and ϵ_i.

An alternative scheme is the *conditional* model developed in ref. 4. Assume that

$$\begin{aligned} E\left[Y_i | Y_j = y_j,\ j \neq i\right] &= \Sigma g_{ij} y_j, \\ V\left[Y_i | Y_j = y_j,\ j \neq i\right] &= \sigma_i^2. \end{aligned} \tag{4}$$

These conditions imply multivariate normality* (MVN), provided the process is truly multilateral ($g_{ij} \neq 0$ and $g_{ji} \neq 0$ for some i, j). It follows that

$$\mathbf{Y} \sim \text{MVN}\left(\mathbf{0}, \boldsymbol{\Sigma}(\mathbf{I} - \mathbf{G}')^{-1}\right), \tag{5}$$

where $\boldsymbol{\Sigma} = \text{diag}(\sigma_i^2)$ and $\mathbf{G} = \{g_{ij}\}$. This contrasts with the simultaneous scheme for which

$$\mathbf{Y} \sim \text{MVN}\left(\mathbf{0}, (\mathbf{I} - \mathbf{G})^{-1}\boldsymbol{\Sigma}(\mathbf{I} - \mathbf{G}')^{-1}\right).$$

The conditional scheme, also known as the *autonormal*, describes a Markov random field*. Besag [4] exploits this structure, via the Hammersley–Clifford theorem*, to generate conditional schemes for nonnormal processes. For a development of spatial moving average* models, see ref. 14.

```
X • X • X • X •
• X • X • X • X
X • X • X • X •
• X • X • X • X
```

Figure 2

Estimation

Whittle [34] developed an approximate large-sample procedure for maximum likelihood* estimation for the first-order autoregressive scheme. When $\mathbf{G} = \theta\mathbf{W}$ and $\mathbf{W}$ is known, the exact ML estimators may be obtained [26]. This approach is extended in ref. 6 to cover $p = 2$ parameters, but computational difficulties grow rapidly as p increases. Martin [20] avoids these difficulties by considering models for which

$$\mathbf{V}(\mathbf{Y}) = \boldsymbol{\Sigma}\left\{\prod_{i=1}^{p}(\mathbf{I} - \theta_i \mathbf{W}_i)\right\}^{-1},$$

with the $\mathbf{W}_i$ known.

The computational problems associated with likelihood* procedures have led to the development of less efficient but more tractable methods. Besag [4] developed the method of coding that codes sites such that the sets of variates $\mathbf{Y}_r (r \in A_r)$, where $A_r \cap A_s = \phi$, $r \neq s$, satisfy the assumption that $\mathbf{Y}_r | \mathbf{Y}_s$, $s \neq r$, are conditionally independent. The coding for the first-order scheme is illustrated in Fig. 2. Although these estimators are rather inefficient for larger values of θ [5], they are becoming popular in the analysis of large data sets, in such areas as image processing [11]. Estimation methods for incomplete data* are discussed in ref. 21.

Nonnormal schemes

This section has concentrated upon normal processes; various nonnormal schemes are described in ref. 4. In particular, when the random variables are binary, the conditional distributions are autologistic in form:

$$P(Y_r = y_r | y_s, s \neq r) = \frac{\exp(y_r u_r)}{1 + \exp(u_r)},$$

where $u_r = \alpha_r + \sum_{s \neq r} \beta_{rs} y_s$. This model also describes a Markov random field*.

ANALYSIS OF FIELD TRIALS

The methods used in the analysis of field trials (*see* AGRICULTURE, STATISTICS IN) continue to be based largely on the Fisherian approach of randomization*. However, there has been a recent resurgence of interest in the use of neighboring plots as covariates. Starting with the usual model

$$\mathbf{Y} = \mathbf{D}\boldsymbol{\beta} + \boldsymbol{\delta},$$

where $\mathbf{D}$ is the design matrix* for the experiment and $\boldsymbol{\beta}$ denotes the unknown parameters, we add the spatial autoregressive model for the error terms:

$$\boldsymbol{\delta} = \theta \mathbf{W} \boldsymbol{\delta} + \boldsymbol{\epsilon}.$$

For details of the estimation procedure and its performance relative to the usual estimators, see Bartlett [3]. Other approaches that incorporate the spatial component into the analysis of designed experiments are described in Wilkinson et al. [35].

Similar arguments can be brought to bear to argue for systematic, rather than purely random, sampling in a spatial context (Matérn [22]).

SPATIAL TIME SERIES

The Box–Jenkins* approach to modeling time series may be extended to consider spatiotemporal autoregressive integrated moving-average* models of the form

$$Z_x(t) = \sum_{i>0} \sum_{k} \phi_{ki} Z_{x+k}(t-i) + \sum_{i>0} \sum_{k} \theta_{ki} \epsilon_{x-k}(t-i) + \epsilon_x(t), \tag{6}$$

where $Z_x(t)$ denotes the random variable for site $\mathbf{x}$ at time t, possibly after differencing to induce stationarity. The $\epsilon_x(t)$ have zero means and are uncorrelated across both time and space; further, $E[Z_x(t)\epsilon_{x+k}(t+i)] = 0$ for all k and $i > 0$. The $Z_x(t)$ are then assumed to be stationary both in time and in space. This is the approach adopted by Aroian and his co-workers; see ref. 2 and earlier references quoted therein. Pfeifer and Deutsch [27, 28] use a somewhat different formulation, replacing the inner summations in (6) by $\phi_{ki}\mathbf{W}_k\mathbf{Z}(t-i)$ and $\theta_{ki}\mathbf{W}_k\boldsymbol{\epsilon}_x(t-i)$, respectively. These two models may be made equivalent, but typically the treatment of boundary effects is different.

The sample autocorrelation and partial autocorrelation functions may be used to identify the model; ref. 27 gives conditional maximum likelihood estimators for (6). Since $Z_x(t)$ depends only on past values of X and ϵ, the problems that arise in the purely spatial case are no longer present. When current values of Z_{x+k} are allowed on the right-hand side of (6), the numerical problems are severe, although these may be alleviated to some extent; see ref. 1.

When the number of sites is small or spatial stationarity cannot be assumed, multiple time-series* or econometric* models may be employed.

References

[1] Ali, M. M. (1979). *Biometrika*, **66**, 513–518.

[2] Aroian, L. A. (1985). In *Time Series Analysis: Theory and Practice*, Vol. 6, O. D. Anderson, J. K. Ord, and E. A. Robinson, eds. North-Holland, Amsterdam, The Netherlands, pp. 241–261.

[3] Bartlett, M. S. (1978). *J. R. Statist. Soc. Ser. B*, **40**, 147–174.

[4] Besag, J. E. (1974). *J. R. Statist. Soc. Ser. B*, **36**, 192–236.

[5] Besag, J. E. and Moran, P. A. P. (1975). *Biometrika*, **62**, 555–562.

[6] Brandsma, A. S. and Ketellapper, R. H. (1979). *Environ. Plann. A*, **11**, 51–58.

[7] Chellappa, R. and Kashyap, R. L. (1983). *IEEE Trans. Inf. Theory*, **IT-29**, 60–72.

[8] Cliff, A. D. and Ord, J. K. (1981). *Spatial Processes: Models and Applications*. Pion, London, England.

[9] Cormack, R. M. (1979). In *Spatial and Temporal Analysis in Ecology*. International Cooperative Publishing House, Fairland, MD, pp. 151–212.

[10] Cormack, R. M. and Ord, J. K., eds. (1979). *Spatial and Temporal Analysis in Ecology*. International Cooperative Publishing House, Fairland, MD

[11] Cross, G. R. and Jain, A. K. (1983). *IEEE Trans.*, **PAMI-5**, 25–39.

[12] David, M. (1977). *Geostatistical Ore Reserve Estimation*. Elsevier, Amsterdam, The Netherlands.

[13] Diggle, P. J. (1983). *Statistical Analysis of Spatial Point Patterns*. Academic, London, England.

[14] Guyon, X. (1982). *Biometrika*, **69**, 95–105.

[15] Haining, R. P. (1979). *Geogr. Anal.*, **11**, 45–64.

[16] Kindermann, R. and Snell, L. J. (1980). *Markov Random Fields and Their Applications*. American Mathematical Society, Providence, RI.

[17] Kooijman, S. A. L. M. (1976). *Ann. Syst. Res.*, **5**, 113–132.

[18] Lloyd, E. H., O'Donnell, T., and Wilkinson, J. C., eds. (1979). *The Mathematics of Hydrology and Water Resources*. Academic, London, England.

[19] Mardia, K. V. and Marshall, R. J. (1984). *Biometrika*, **71**, 135–146.

[20] Martin, R. J. (1979). *Biometrika*, **66**, 209–217.

[21] Martin, R. J. (1984). *Commun. Statist. A*, **13**, 1275–1288.

[22] Matérn, B. (1960). Spatial variation. *Medd. Statens Skogsforsk.*, **49**, 1–148.

[23] Matheron, G. (1971). *The Theory of Regionalised Variables*. Centre de Morphologie Mathematique, Fontainebleau, Paris, France.

[24] Nijkamp, P. (1979). *Multidimensional Spatial Data and Decision Analysis*. Wiley, New York.

[25[Oden, N. L. (1984). *Geogr. Anal.*, **16**, 1–16.

[26] Ord, J. K. (1975). *J. Amer. Statist. Ass.*, **70**, 120–126.

[27] Pfeifer, P. E. and Deutsch, S. J. (1980). *Technometrics*, **22**, 35–47.

[28] Pfeifer, P. E. and Deutsch, S. J. (1980). *Commun. Statist. B*, **9**, 533–549, 551–562.

[29] Renshaw, E. and Ford, E. D. (1983). *Appl. Statist.*, **32**, 51–63.

[30] Ripley, B. D. (1981). *Spatial Statistics*. Wiley, New York.

[31] Singh, B. B. and Shukla, G. K. (1983). *Biometrika*, **70**, 523–527.

[32] Student (1914). *Biometrika*, **10**, 179–180.

[33] Vanmarcke, E. (1983). *Random Fields: Analysis and Synthesis*. MIT Press, Cambridge, MA.

[34] Whittle, P. (1954). *Biometrika*, **41**, 434–449.

[35] Wilkinson, G. N., Eckert, S. R., Hancock, T. W. and Mayo, O. (1983). *J. R. Statist. Soc. B*, **45**, 151–211.

Bibliography

Anderson, O. D., Ord, J. K., and Robinson, E. A., eds. (1985). *Time Series Analysis: Theory and Practice, Vol. 6*. North-Holland, Amsterdam, The Netherlands. (A volume of recent papers on spatial time series.)

Bartlett, M. S. (1975). *The Statistical Analysis of Spatial Pattern*. Chapman and Hall, Andover, Hampshire, England. (A concise description of the theory and practice of spatial modeling.)

Bennett, R. J. (1979). *Spatial Time Series: Analysis, Forecasting and Control*. Pion, London, England. (Wide-ranging coverage of time-series methods useful in the analysis of spatiotemporal data.)

Cliff, A. D., Haggett, P., and Ord, J. K. (1986). *Spatial Aspects of Influenza Epidemics*. Pion, London, England. (Use of spatial time series in modeling the spread of a disease.)

Cliff, A. D. and Ord, J. K. (1981). See ref. 8. (Describes tests and estimation procedures for spatial models and their applications.)

Diggle, P. J. (1983). See ref. 13. (Uses a minimum of mathematics and provides many examples.)

Matérn, B. (1960). See ref. 22. (A classic in the development of stochastic spatial models.)

Ripley, B. D. (1981). See ref. 30. (Comprehensive coverage of spatial modeling and spatial data analysis.)

Vanmarcke, E. (1983). See ref. 33. (Develops the theory of random processes, with particular emphasis on the use of local averages. Also discusses distributions of extrema in several dimensions.)

Whittle, P. (1954). See ref. 34. (The seminal paper on spatial models.)

(AGRICULTURE, STATISTICS IN
AUTOREGRESSIVE INTEGRATED MOVING AVERAGE MODELS
CLIFF–ORD STATISTIC
ECOLOGICAL STATISTICS
GEOGRAPHY, STATISTICS IN
GEOLOGY, STATISTICS IN
GEOSTATISTICS
HYDROLOGY, STOCHASTIC
KRIGING

MARKOV RANDOM FIELDS
MULTIPLE TIME SERIES
NEAREST-NEIGHBOR METHODS
RESPONSE SURFACE METHODS
SPATIAL DATA ANALYSIS
SPATIAL SAMPLING
SPECTRAL ANALYSIS
STOCHASTIC PROCESSES, POINT)

KEITH ORD

SPATIAL SAMPLING

Spatial sampling is that area of survey sampling* concerned with sampling in two dimensions; for example, the sampling of fields, groups of contiguous quadrats, or other planar areas. The area of application of these techniques is very wide and includes fields such as archaeology*, ecology, geography*, geology*, cartography, and forestry*.

One approach to spatial sampling is through a population of MN units, usually points or quadrats, arranged in M rows and N columns. The sampling designs to choose mn units fall into three distinct types: designs in which the sample units are aligned in both the row and column directions; designs in which the sample units are aligned in one direction only, say the rows, and unaligned in the column direction; and designs in which the sample units are unaligned in both directions. For designs that have the sample units aligned in both directions, the number of sample elements in any row of the population will be 0 or n and the number of sampled elements in any column of the population will be 0 or m. For designs that have sample units aligned in the rows and unaligned in the columns, the number of sample elements in any row of the population will be 0 or n and the number of sample elements in any column will be at most m. With the exception of simple random sampling* without replacement of mn units from the MN in the population, designs that have sample units unaligned in both directions are characterized by having at most n sample elements in any row and at most m elements in any column of the population.

Figure 1 illustrates three examples of these types of designs. In case (a), a simple random sample of row positions is taken. The intersection of row and column choices defines the sample. In case (b), a stratified sample of rows is taken. The row strata are

(a) Simple random sampling, aligned in both directions.

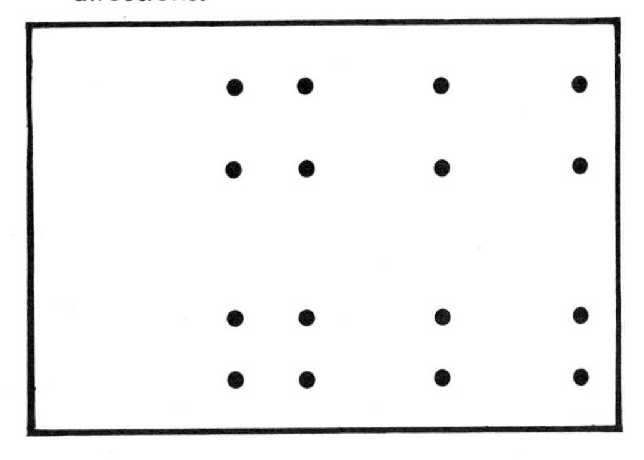

(b) Stratified sampling, aligned in the rows, unaligned in the columns.

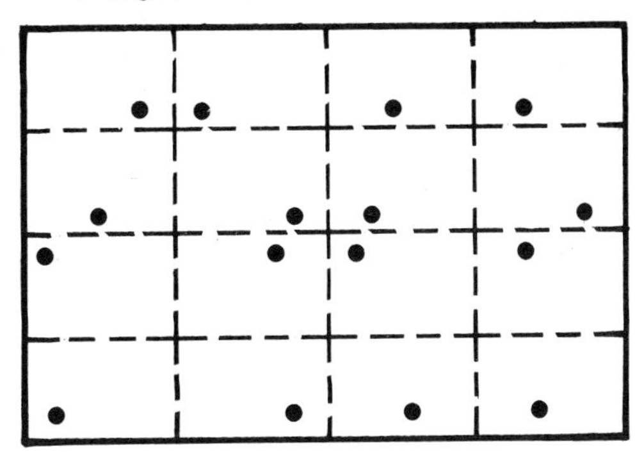

(c) Systematic sampling, unaligned in both directions.

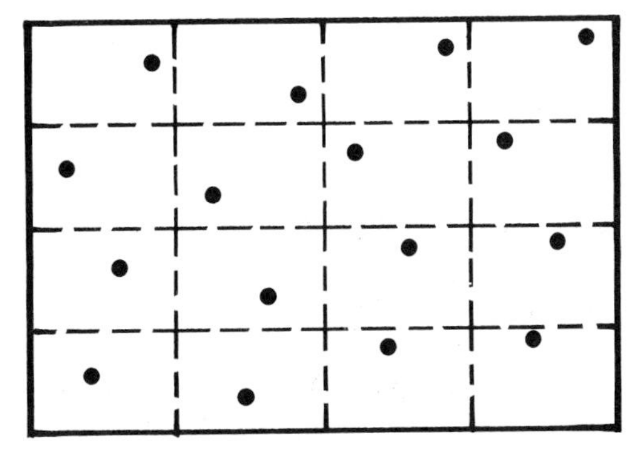

Figure 1

demarcated by the horizontal dotted lines. For each chosen row, an independent stratified sample of columns is chosen with the strata defined by the vertical dotted lines. Without going into the details of the sampling design in case (c), note that in any row stratum the points are equally spaced in the columns and for any column stratum the points are equally spaced in the rows. Method (c) in Fig. 1, systematic sampling* unaligned in both directions, is known in the geographical literature as *stratified systematic unaligned sampling*, a term first used by Berry [3]. The most commonly used designs are unaligned simple random sampling, unaligned stratified sampling, aligned systematic sampling, and stratified systematic unaligned sampling; in each case alignment or nonalignment of the sample units is in both directions. Further illustrations of these may be found in Quenouille [10], the originator of these sampling methods, Berry and Baker [4], Koop [8], and Ripley [11].

For any of these two-dimensional sampling designs based upon simple random, stratified, or systematic sampling in each direction, Bellhouse [2] has given the finite population variance of the sample mean. For the first three commonly used designs, the finite population variances of the sample mean may be obtained as straightforward applications of one-dimensional results. The variance for the last design, stratified systematic unaligned sampling, is

$$\sum_i \sum_v \frac{(\bar{y}_{i\ldots v} - \bar{y}_{\ldots v})^2}{n^2 k} + \sum_u \sum_j \frac{(\bar{y}_{.uj.} - \bar{y}_{.u..})^2}{m^2 l}$$

$$+ \sum_i \sum_u \sum_j \sum_v \frac{(y_{iujv} - \bar{y}_{iu.v} - \bar{y}_{.ujv} + \bar{y}_{.u.v})^2}{m^2 n^2 kl},$$

where y_{iujv} is the measurement in the ith row of the uth row stratum and the jth column of the vth column stratum ($i = 1, \ldots, k$; $u = 1, \ldots, m$; $j = 1, \ldots, l$; $v = 1, \ldots, n$; $M = mk$; $N = n$). The dot notation is used to determine the appropriate mean values. If two subscripts are used to determine the location of the y measurement, for example y_{rc} is the measurement in the rth row and cth column of the population, then y_{iujs} is equivalent to y_{rc} when $r = i + (u - 1)k$ and $c = j + (v - 1)l$.

Koop [8] has given the finite population variance of the estimate of a cover-type area on a map. For example, if A_α is the area of interest to be estimated on a map of total area A, the variance of the estimate of A_α (A times the proportion of sample points falling in the area of interest) based on unaligned simple random sampling is

$$A_\alpha(A - A_\alpha)/(mn) = V,$$

where mn is the total sample size. Letting $A_{ij\alpha}$ be the coverage area of interest in each of the mn population strata ($i = 1, \ldots, m$; $j = 1, \ldots, n$), the variance based on unaligned stratified sampling is

$$V - \sum_{i=1}^{m} \sum_{j=1}^{n} (mnA_{ij\alpha} - A_\alpha)^2/(mn)^2.$$

A second approach to spatial sampling is in a more general population structure. In this case, the spatial population is composed of a number of nonoverlapping domains that are congruent by translation. Without imposing any more structure on the population, three sampling schemes can be considered: random, stratified, and systematic sampling. In random sampling, sample points are chosen with uniform probability over the union of the domains. To obtain a stratified sample, points are independently chosen with uniform probability in each domain. The systematic sample is obtained by choosing a point at random in one domain and then a point in each of the remaining domains by a translation that establishes the congruence of the domains. This approach was taken by Zubrzycki [12], and, with the addition of more structure to the population, also by Ripley [11].

In many situations, it is reasonable to assume a superpopulation model on the measurements in the spatial population. Upon letting x and y be measurements at two points in the planar population, the model most frequently used is

$$E(y) = E(x) = \mu;$$
$$\text{var}(y) = \text{var}(x) = \sigma^2; \qquad (1)$$
$$\text{cov}(x, y) = \sigma^2\rho(u, v),$$

where the correlation function ρ depends on u, the difference in the latitude, and v, the difference in the longitude, for the points in the plane associated with the measurements x and y. A further assumption often made is that

$$\rho(u, v) = \rho(d), \qquad (2)$$

where $d = (u^2 + v^2)^{1/2}$ is the Euclidean distance between the points.

Model (1) was used by Das [6] and model (1) with $\rho(u, v) = \rho_1(u)\rho_2(v)$ was used by Quenouille [10]. Quenouille showed that unaligned designs are more efficient than aligned sampling designs and that within each type of alignment, a systematic sampling scheme was most efficient, followed by stratified and then random sampling. Efficiency was defined as inverse to the finite population variance of the sample mean averaged over the superpopulation model. Das also made efficiency comparisons but the results were restricted to three sampling designs only. Bellhouse [1] has shown, for correlation functions in model (1) satisfying the restriction $\Delta_u^2\Delta_v^2(u, v) \geqslant 0$, that within each type of alignment, a systematic sampling design is the most efficient design in a wide class of sampling designs. The operators Δ_u^2 and Δ_v^2 are the second finite difference* operators acting on u and v, respectively; for example,

$$\Delta_u^2\rho(u, v) = (u + 2, v) - 2\rho(u + 1, v) + \rho(u, v).$$

An example of the optimality result is shown numerically in Table 1. This shows the percentage gain in efficiency of unaligned systematic sampling over unaligned stratified sampling, that is, $100(V_2/V_1 - 1)$, where V_2 is the finite population variance averaged over model (1) for unaligned stratified sampling and V_1 is the same for unaligned systematic sampling. A correlation structure $\rho(u, v) = \theta^u\phi^v$, for various values of θ and ϕ, is used. The results are adapted from Quenouille [10, Table 3]. Then as the gains increase, the stronger the correlation. Table 2 shows the effect of alignment of the sample units. In this table, the percentage gains in efficiency of unaligned systematic over aligned systematic sampling are presented. The numbers are obtained by a manipulation of the results presented in Quenouille [10, Table 3]. Substantial gains in precision are made, especially at high levels of correlation.

Table 1 Percent Gains in Efficiency—Systematic Over Stratified (Unaligned)

$\phi \backslash \theta$	0.1	0.3	0.5	0.7	0.9
0.1	21	28	31	33	33
0.3		41	49	53	55
0.5			59	66	70
0.7				75	82
0.9					92

Model (2) was studied by Zubrzycki [12] and Hájek [7]. In this setup, Zubrzycki found that stratified sampling was more efficient than simple random sampling. Hájek and Zubrzycki found that there is no simple relationship between systematic and stratified sampling as there is in the Quenouille [10] case. Dalenius et al. [5] investigated the relationship further by obtaining optimal sampling schemes under model (2). The schemes consist of various triangular, rectangular, and

Table 2 Percent Gains in Precision—Unaligned Over Aligned Systematic Sampling

$\phi \backslash \theta$	0.1	0.3	0.5	0.7	0.9
0.1	24	68	165	416	1735
0.3		89	162	386	1669
0.5			203	380	1588
0.7				467	1492
0.9					1806

hexagonal lattices, the choice depending on the range of sampling point densities. Using an exponential correlation structure, Matérn [9] earlier investigated the problem of the choice of sample points under model (2). For systematic sampling, he found triangular lattices to be more efficient than square lattices, although the gains in efficiency were slight. For stratified random sampling, he found the opposite to be true: squares were better than triangles, and also hexagonal lattices were more efficient than either square or triangular lattices. In an empirical study of these sampling designs, he found that in systematic sampling, the square lattice design was usually slightly more efficient than the triangular design, the opposite of what the theory had predicted. Matérn (9) also investigated approximate variance estimation techniques, based on squared differences in the measurements, for systematic sampling. He found that the methods were biased and that there was a considerable amount of overestimation of the variance.

References

[1] Bellhouse, D. R. (1977). Some optimal designs for sampling in two dimensions. *Biometrika*, **64**, 605–611.

[2] Bellhouse, D. R. (1981). Spatial sampling in the presence of a trend. *J. Statist. Plann. Inf.*, **5**, 365–375.

[3] Berry, B. J. L. (1962). *Sampling, Coding, and Storing Flood Plain Data*. Agriculture Handbook No. 237, U.S. Dept. of Agriculture, Washington, DC.

[4] Berry, B. J. L. and Baker, A. M. (1968). Geographical sampling. In *Spatial Analysis*, B. J. L. Berry and D. F. Marble, eds. Prentice-Hall, Englewood Cliffs, NJ, pp. 91–100.

[5] Dalenius, T., Hájek, J., and Zubrzycki, S. (1961). On plane sampling and related geometrical problems. In *Proc. Fourth Berkeley Symp. Math. Statist. Probab.*, Vol. 1, J. Neyman, ed. University of California Press, Berkeley, CA, pp. 125–150.

[6] Das, A. C. (1950). Two-dimensional systematic sampling and the associated stratified and random sampling. *Sankhyā*, **10**, 95–108.

[7] Hájek, J. (1961). Concerning relative accuracy of stratified and systematic sampling in a plane. *Colloq. Math.*, **8**, 133–134.

[8] Koop, J. C. (1976). Systematic sampling of two-dimensional surfaces and related problems. *Technical Report*, Research Triangle Institute, NC.

[9] Matérn, B. (1960). Spatial variation: Stochastic models and their application to some problems in forest surveys and other sampling investigations. *Meddelanden fran Statens Skogsfor.*, **49**, 1–144.

[10] Quenouille, M. H. (1949). Problems in plane sampling. *Ann. Math. Statist.*, **20**, 355–375.

[11] Ripley, B. D. (1981). *Spatial Statistics*. Wiley, New York.

[12] Zubrzycki, S. (1958). Remarks on random, stratified and systematic sampling in a plane. *Colloq. Math.*, **6**, 251–264.

Bibliography

Bellhouse, D. R. (1980). Sampling studies in archaeology. *Archaeometry*, **22**, 123–132.

Holmes, J. H. (1970). The theory of plane sampling and its application in geographic research. *Econ. Geogr.*, **46**, 379–392.

(ECOLOGICAL STATISTICS
GEOGRAPHY, STATISTICS IN
GEOLOGY, STATISTICS IN
HYDROLOGY, STOCHASTIC
KRIGING
SPATIAL DATA ANALYSIS
SPATIAL PROCESSES
STEREOLOGY
STRATIFIED DESIGNS
SURVEY SAMPLING)

D. R. BELLHOUSE

SPEARMAN–BROWN FORMULA *See* GROUP TESTING

SPEARMAN ESTIMATORS *See* KARBER METHOD

SPEARMAN RANK CORRELATION COEFFICIENT

The psychologist Charles Spearman introduced his correlation of ranks procedure in 1904 [13, 14]. His rank correlation coefficient r_s, often called *Spearman's rho*, is thus one of the oldest rank statistics, and clearly the oldest still in common use (*see also* RANKING PROCEDURES).

Of the three algebraically equivalent forms of r_s below, the first is the oldest and shows it to be the familiar sample correlation coefficient r (*see* CORRELATION) computed on ranks in place of data. Specifically, given a sample of paired data, $(X_1, Y_1), \ldots, (X_n, Y_n)$, and ranking the X's (separately the Y's) with $R_i = \text{rank } X_i$ ($S_i = \text{rank } Y_i$), then

$$r_s = \frac{\Sigma_{i=1}^{N}(R_i - \bar{R})(S_i - \bar{S})}{\sqrt{\Sigma_i(R_i - \bar{R})^2 \cdot \Sigma_i(S_i - \bar{S})^2}}$$

$$= \frac{\{\Sigma_i R_i S_i - N(N+1)^2/2\}}{\{N(N^2-1)/12\}}$$

$$= \frac{1 - 6\Sigma_i(R_i - S_i)^2}{\{N(N^2-1)\}},$$

where $\bar{R} = \Sigma_i R_i/N = (N+1)/2 = \bar{S}$.

The two succeeding forms are computationally more convenient, with the final one generally preferred in practice. All three appear commonly in the literature.

Early distribution tables were for the quantity $\Sigma_i(R_i - S_i)^2$, but the most complete table currently is for r_s, compiled by Zar [15].

In the presence of ties among X's or among Y's the standard procedure is to use midranks*. If $u_1, u_2, \ldots$ represent the sizes of tied X groups, $v_1, v_2, \ldots$ the tied Y groups, and $U = \Sigma(u_i^3 - u_i)$, $V = \Sigma(v_j^3 - v_j)$, then the appropriate form is

$$r_s = \frac{N(N^2-1) - 6\Sigma_i(R_i - S_i)^2 - 6(U+V)}{[\{N(N^2-1) - U\}\{N(N^2-1) - V\}]^{1/2}}.$$

The purpose of a correlation coefficient is to characterize some aspect of the relationship between the two variates in a bivariate population. Although many others have been proposed and studied, in practice only three are commonly used: the parametric sample correlation coefficient r and the two nonparametric versions r_s, and $\hat{\tau}$, the Kendall rank correlation coefficient or Kendall's tau*. Each measures a different, specific form of dependency or association*, and therefore comparisons can be misleading.

In a careful definition, correlation* refers specifically to a measure of linear relationship between X and Y, whereas the two rank versions are more general, being measures of monotone relationships. That is easily seen from the fact that ranks are unaffected by any monotone, strictly increasing transformation of the ranked variable(s). Modern terminology as in ref. 1, therefore, refers to r_s and $\hat{\tau}$ as *measures of association* rather than correlation.

Differences among the three are perhaps most easily seen by examining their expected values*. In the interest of simplicity, only the asymptotic results are given below. For a more detailed discussion see ref. 1. If (X_1, Y_1), (X_2, Y_2), and (X_3, Y_3) are independent pairs from a continuous bivariate population, then

$$\lim_{N\to\infty} E[r_s] = \text{corr}[F_X(X), F_Y(Y)]$$

$$= 3\{2\Pr[(X_2 - X_1) \times (Y_3 - Y_1) > 0] - 1\} = \rho_s,$$

$$E[\hat{\tau}] = 2\Pr[(X_2 - X_1) \times (Y_2 - Y_1) > 0] - 1 = \tau,$$

and

$$\lim_{N\to\infty} E[r] = \text{corr}[X, Y] = \rho,$$

where ρ_s is called the *grade correlation*, τ is Kendall's tau, and ρ is the Pearson product moment correlation coefficient*.

Although these expectations all have the range $[-1, 1]$, have the value zero when X and Y are independent, and values $+1(-1)$ for positive (negative) linear relationships, they are otherwise quite different functions. As a result, comparisons among the three estimates are not particularly relevant.

Another way to interpret and relate the rank correlations is as functions of inversions to achieve identical ordering of the two variables (see [5, 6]).

Each of the correlation coefficients can be used to construct a test, usually described as a test of (in)dependence, although association would be more accurate, since there exist highly dependent bivariate distributions where all three of the measures have expected values of zero (e.g., uniform on a disk).

A distribution-free* test of association is obtained for sample sizes up to $N = 100$ by referring r_s to Zar's table and rejecting for large values. For larger samples and/or in the presence of frequent ties, an approximate test proposed by Olds [10] is given by referring

$$Z_r = r_s\sqrt{N-1}$$

to a standard normal table. Other approximate tests are proposed in refs. 4 and 11.

Distribution theory of Spearman's rho under dependence is very complicated and not completely solved. Recent advances have been made in refs. 2 and 8, the latter work also including extensions to partial and multiple correlation*.

Example. Consider a sample of $N = 5$ pairs ordered by ascending y values

y:	5	7	8	11	14
x:	41	17	12	26	19

Replacing the observations by their ranks yields

S:	1	2	3	4	5
R:	5	2	1	4	3
$(R-S)$:	4	0	−2	0	−2

and

$$r_s = 1 - \frac{6}{5(25-1)} \times \left\{4^2 + 0^2 + (-2)^2 + 0^2 + (-2)^2\right\} = -0.2,$$

a slight indication of negative association.

From Zar's table, the two-sided p-value* for $r_s = -0.2$, $N = 5$, exceeds 0.5, strongly indicating that random chance can account for the observed association.

EFFICIENCIES

Efficiency comparisons among tests based on the correlation coefficients are not as straightforward a matter as they are for tests for location shift. One must first choose a model for a specific form of dependency or association. There are many possible forms, and efficiencies vary accordingly.

If Pitman asymptotic relative efficiency* is considered, then $\text{ARE}(r_s, \hat{\tau}) = 1$ for all distributions. For one form of alternative, Konijn [7] found the following results:

Distribution	$\text{ARE}(r_s, r)$
Normal	0.912
Uniform	1
Parabolic	0.857
Double Exponential	1.266

Lehmann [9] gives the range $0.746 \leqslant \text{ARE}(r_s, r) \leqslant \infty$ for all distributions.

OTHER APPLICATIONS

Because the null distributions of r_s and $\hat{\tau}$ are unchanged under random permutations among pairs, e.g., reordering by the values of X, say, the tests based on the two rank coefficients can be applied to problems of monotone trend or regression. (Cf. refs. 3, 9, and 12, and THEIL TEST.)

Tests based on r_s and $\hat{\tau}$ have also been developed for ordered alternatives in one- and two-way layouts. *See* JONCKHEERE TEST and PAGE TEST (both FOR ORDERED ALTERNATIVES).

References

[1] Gibbons, J. D. (1971). *Nonparametric Statistical Inference*. McGraw-Hill, New York, Chap. 12. (Good development of distribution theory and moments for r_s and $\hat{\tau}$ plus some efficiencies.)

[2] Henze, F. H.-H. (1979). *J. Amer. Statist. Ass.*, **74**, 459–464. (Nonnull distribution of r_s based on bipermutations, plus extensions of partial and multiple correlations*.)

[3] Hollander, M. and Wolfe, D. A. (1973). *Nonparametric Statistical Methods*. Wiley, New York. (Strictly a methods book which has good references and tables. Chapters 8 and 9 use $\hat{\tau}$ for association and regression.)

[4] Iman, R. L. and Conover, W. J. (1978). *Commun. Statist. B*, **7**, 269–282. (Approximate critical values with and without ties, including tables for $n = 3(1)30$, $\alpha = 0.10, 0.05, 0.025, 0.01, 0.005, 0.001$.)

[5] Kendall, M. G. (1948). *Rank Correlation Methods*, 4th ed. Griffin, London, England. (Early thorough discussion of $\hat{\tau}$ and r_s. No modern distribution theory or efficiencies but otherwise well worth reading.)

[6] Kendall, M. G. and Stuart, A. (1973). *The Advanced Theory of Statistics*, 3rd ed., Vol. 2. Hafner, New York, Chap. 31. (Brief summary of ref. [5] plus some efficiencies.)

[7] Konijn, H. S. (1956). *Ann. Math. Statist.*, **27**, 300–323; Errata, **29**, 935 (1958).

[8] Kraemer, H. Ch. (1974). *J. Amer. Statist. Ass.*, **69**, 114–117. (Nonnull distribution theory of r_s.)

[9] Lehmann, E. L. (1975). *Nonparametrics: Statistical Methods Based on Ranks*. Holden-Day, San Francisco, CA. (Introductory theory and methods. Chapter 7 develops r_s for trend, paired association, and association in contingency tables*.)

[10] Olds, E. G. (1949). *Ann. Math. Statist.*, **20**, 117–118. (Proposes the normal approximate test for r_s.)

[11] Pitman, E. J. G. (1937). *J. R. Statist. Soc. Suppl.*, **4**, 225–232. (Proposes an approximate t-test for r_s.)

[12] Randles, R. H. and Wolfe, D. A. (1979). *Introduction to the Theory of Nonparametric Statistics*. Wiley, New York. (Intermediate level theory of rank statistics developed from U-statistics*. Chapter 12 uses r_s for association and regression.)

[13] Spearman, C. (1904). *Amer. J. Psych.*, **15**, 88.

[14] Spearman, C. (1906). *Brit. J. Psych.*, **2**, 89.

[15] Zar, J. H. (1972). *J. Amer. Statist. Ass.*, **67**, 578. (Most complete table available for r_s.)

Bibliography

The following articles represent additional recent work related to Spearman's rho.

Ghosh, M. (1975). *Ann. Inst. Statist. Math. Tokyo*, **27**, 57–68. (Develops a sequential procedure based on r_s for bounded length confidence intervals in simple regression.)

Schulman, R. S. (1979). *Amer. Statist.*, **33**, 77–78. (A geometrical model and interpretation of r_s and $\hat{\tau}$ with a method for sequential computation.)

Shirahota, S. (1981). *Biometrika*, **68**, 451–456. (A modified version of r_s when X and Y have identical marginal distributions and may be permuted within pairs.)

Simon, A. (1978). *J. Amer. Statist. Ass.*, **73**, 545–551. (Efficacies* of rank correlation and other measures of association applied to contingency tables* with ordered categories.)

(ASSOCIATION, MEASURES OF
CORRELATION
DISTRIBUTION-FREE METHODS
JONCKHEERE TEST FOR ORDERED ALTERNATIVES
KENDALL'S TAU
PAGE TEST FOR ORDERED ALTERNATIVES
RANK TESTS)

W. PIRIE

SPECIES, RICHNESS OF *See* DIVERSITY INDICES

SPECIFICATION, PREDICTOR

The least-squares* (LS) principle was all-pervading in the statistical sciences in the nineteenth century and well into the twentieth. In the 1940s LS was pushed aside in favour of the maximum likelihood* (ML) principle, and the ensuing ML mainstream of contemporary statistics soon developed an elaborate framework for scientific model building, including ML designs for (i) model specification, (ii) model estimation, (iii) hypothesis testing*, and (iv) assessment of standard errors (SEs). LS counterparts to (i)–(iv) have emerged, here denoted (i*)–(iv*).

Predictor specification, or briefly Presp, introduced by Wold [39–41, 44, and 48–51], provides a general rationale for (i*) LS specification and (ii*) LS estimation, and thereby also for the application of (iii*), the cross-validation test for predictive relevance by Stone [34] and Geisser [10], and (iv*) the assessment of SEs by Tukey's jackknife* [36]. Presp marks the comeback of least squares [51], in particular as the Stone–Geisser test and Tukey's jackknife apply in the general LS context of (i*)–(ii*).

ML modeling assumes that the observables are subject to measurement errors* and are jointly ruled by a specified multivariate distribution that is subject to independent observations (the iid assumption). Presp is a version of LS that models the data as observed, the modeling is distribution-free, except for Presp, and the observations are independence-free. Thanks to the generality of its basic assumptions, LS is of broad scope in theory and practice, is flexible and easy in the statistical implementation, and is easy and speedy on the computer.

ML is parameter-oriented; LS is prediction-oriented. Interweaving LS with ML assumptions, ML gives optimal parameter accuracy and LS gives optimal prediction accuracy. There is a choice; in general you cannot have both kinds of accuracy [5; 28, p. 221; 49].

SCIENTIFIC MODEL BUILDING BY MEANS OF PRESP

We shall consider two types of models [47–50]: path models with manifest (directly observed) variables (briefly MVP) models and path models with latent (indirectly observed) variables (LVP) models. All through, the models are linear in the parameters, and the MVs under analysis are assumed to have finite variance.

To begin with the case of *simple regression*, the model is

$$y_n = \alpha + \beta x_n + \epsilon_n, \qquad n = 1, N, \quad \text{(1a)}$$

and is subject to Presp, i.e.,

$$E(y|x) = \alpha + \beta x. \quad \text{(1b)}$$

(NOTE: In this entry the notation $n = 1, N$ means $n = 1, \ldots, N$ throughout.) In words, the conditional expectation $E(y|x)$ equals the systematic part of y as given on the right in (1a).

Comments. Let $\varphi(x)$ and $\varphi(x, y)$ be the frequency functions of the random variables x and (x, y). Then

$$\varphi(x, y) = \varphi(x)\varphi(y|x), \quad \text{(1c)}$$

where $\varphi(y|x)$ is the conditional frequency function of y for given x. Assuming Presp (1b), the distribution of $\epsilon_n = y_n - \alpha - \beta x_n$ may depend on x, and (x_n, y_n) in general will not be independent observations of (x, y).

For an example where x_t varies with t in the irregular and nonrandom fashion that is typical of nonexperimental time series*, suppose

$$x_t = \Sigma_j\left[w_j \sin(q_j t)\right], \qquad j = 1, J, \quad \text{(1d)}$$

where w_j and q_j are constants, and the q_j's are incommensurable. As shown by Allais [1], the distribution of x_t over x is asymptotically normal if J is large and the w_j's are small.

Three fundamental implications of (1a, b) will be noted.

1. $E(\epsilon) = 0$. In fact, $E(\epsilon|x) = 0 = \int E(\epsilon|x)\,d\Phi(x) = E(\epsilon)$, where $\Phi(x)$ denotes the distribution function of x.
2. $E(\epsilon x) \equiv r(\epsilon, x) = 0$. In fact, $E(\epsilon x|x) = xE(\epsilon|x) = 0$, which gives $E(\epsilon x) = \int E(\epsilon x|x)\,d\Phi(x) = 0$.
3. Forming the OLS (ordinary least squares) regression of y and x, say,

$$y_n = a + bx_n + e_n, \quad \text{(2a)}$$

where

$$b = \sum_1^N [y_n(x_n - \bar{x})] / \sum_1^N (x_n - \bar{x})^2; \quad \text{(2b)}$$

$$a = \bar{y} - b\bar{x}, \quad \text{(2c)}$$

the regression coefficients a, b will on mild supplementary conditions be consistent estimates of α, β, i.e.,

$$\operatorname{prob} \lim_{N\to\infty} a = \alpha, \quad \text{(2d)}$$

$$\operatorname{prob} \lim_{N\to\infty} b = \beta. \quad \text{(2e)}$$

The following condition is sufficient for consistency:

$$r(\epsilon_t, \epsilon_{t+k}) = 0, \quad k = K+1, K+2, \ldots, \quad (3a)$$

for some finite $K < \infty$. More generally, the following condition is sufficient:

$$\sum_k |r(\epsilon_n, \epsilon_{n+k})| < \infty; \quad k = 1, 2, \ldots. \quad (3b)$$

We can now spell out the operative use of Presp, and thereby its general rationale: *Presp is imposed on relations that the investigator wants to use for prediction, be it in theoretical or estimated form, and Presp provides the ensuing predictions.*

Thus when x is known in model (1a) the theoretical prediction of y is given by (1b), and ϵ_n is the prediction error. The empirical prediction is obtained from (2a):

$$\text{pred } y_n = a + bx_n \quad (4)$$

with prediction error e_n. Thanks to (2d, e) the prediction (4) is consistent, and among linear models (1a) the predictions (1b) and (4) have minimum variance.

In the subsequent discussion the argument (1)–(4) is of general scope, and carries over under the requisite adaptation. Without loss of generality the observables are assumed to be measured as deviations from the mean, in (2a) giving $\bar{x} = \bar{y} = 0$, and all location parameters will be ignored. Throughout, it is an immediate matter to carry over formula (2c).

MULTIPLE REGRESSION

Using H explanatory variables, the model reads

$$y_n = \beta_1 x_{1n} + \cdots + \beta_H x_{Hn} + \epsilon_n, \quad n = 1, N, \quad (5a)$$

and Presp takes the form

$$E(y|x_1, \ldots, x_H) = \beta_1 x_1 + \cdots + \beta_H x_H + \epsilon. \quad (5b)$$

See refs. 39 and 53 for examples.

The parity principle, a characteristic and fundamental property of model building based on Presp, is illustrated by the multiple regression (5): The prediction error ϵ_n is uncorrelated over n with each predictor variable x_h, the estimation of the unknown parameters thus being performed in terms of just as many zero correlations between the prediction error and the predictors. In this sense, Presp modeling is a full information* method.

CAUSAL CHAIN SYSTEMS

We consider a causal chain [39] (cf. [4]) with M endogenous and K exogenous variables, and write its structural form (SF)

$$y_{mt} = \sum_{h=1}^{M-1} \beta_{mh} y_{ht} + \sum_{k=1}^{K} \gamma_{mk} x_{kt} + \delta_{mt}, \quad t = 1, T, \quad (6a)$$

where $\beta_{mh} = 0$ for $h \geqslant m$, and t runs over time or a cross section. The reduced form (RF) of the system is obtained by iterative substitution of $y_1, \ldots, y_{m-1}$ into y_m, say,

$$y_{mt} = \sum_{k=1}^{K} \omega_{mk} x_{kt} + \epsilon_{mt}, \quad t = 1, T. \quad (6b)$$

In (6a), $\beta_{mh} = 0$ if y_{ht} does not occur in the mth relation, and similarly for γ_{mk} in (6a) and ω_{mk} in (6b).

To make Presp possible for all relations both in SF and RF we assume

$$E(\epsilon_m | x'_1, \ldots, x'_{K(m)}) = 0, \quad m = 1, M, \quad (7)$$

where x'_k $[k = 1, K(m)]$ denotes those exogenous variables that occur in the m first relations of SF. Assumption (7) implies Presp on the one hand for SF,

$$E(y_m | y''_1, \ldots, y''_{m-1}; x''_1, \ldots, x''_K) = \sum_{h=1}^{m-1} \beta_{mh} y_h + \sum_{k=1}^{K} \gamma_{mk} x_k, \quad (8a)$$

where $y''_1, \ldots, y''_{m-1}; x''_1, \ldots, x''_K$ denote the predictor variables in the mth relation of SF. On the other hand, (7) implies Presp for RF,

$$E(y_m|x'_1, \ldots, x'_{K(m)}) = \sum_{k=1}^{K(m)} \omega_{mk} x_k. \quad (8b)$$

CLASSICAL INTERDEPENDENT (ID) SYSTEMS

Classical ID systems [4, 11, 35, 47] are models with SF (6a), where the first sum runs from $m = 1$ to $m = M$ with

$$\beta_{mm} = 0. \quad (9a)$$

The RF is formally the same as (6b); all exogenous variables x_k are uncorrelated with all residuals δ_m in SF, or equivalently in RF, in symbols,

$$r(x_k, \epsilon_m) = r(x_k, \delta_m) \quad (9b)$$

$$= 0, \qquad k = 1, K;\ m = 1, M. \quad (9c)$$

OLS estimation of RF (6b) is consistent under mild supplementary conditions. As applied to the SF (6a) of classical ID systems, however, OLS is not consistent [11].

In classical ID systems the Presp assumption (7) of RF does *not* extend to SF (6b) [42, 43], that is,

$$E(y_m|y'_1, \ldots, y'_M;\ x'_1, \ldots, x'_K) \neq \sum_{h=1}^{M} \beta_{mh} y_h + \sum_{k=1}^{K} \gamma_{mk} x_k. \quad (9d)$$

Comments. In classical ID systems the solving of the SF for the endogenous variables cannot be performed by mere substitutions; it involves the reversion of at least one SF relation. Conditional expectations are not reversible and this is why (9d) differs from (8a).

The two-stage least squares* (TSLS) estimation of classical ID systems [3, 35] uses OLS regression in the first stage to estimate the RF (8b), where the ensuing systematic part of each endogenous variable, say y_{ht}^{TSLS}, is transferred to the SF (8a) to replace its endogenous predictor variables; the SF is then estimated in the second stage by OLS.

REFORMULATED ID (REID) SYSTEMS AND GENERAL ID (GEID) SYSTEMS

REID and GEID [26, 42, 43, 47] are models with formally the same RF (6b) as in causal chain systems and classical ID systems, with SF (6a) reformulated as

$$y_{mt} = \sum_{h=1}^{M} \beta_{mh} \eta_{ht}^* + \sum_{k=1}^{K} \gamma_{mk} x_{kt} + \delta_{mt} \quad (10a)$$

$$= \eta_{mt}^* + \epsilon_{mt}, \quad (10b)$$

where $\beta_{mm} = 0$ as before, the parameters β_{mh}, γ_{mk} are formally the same as in the SF (9a), and the endogenous predictors in (6a) are replaced by their expected values η_{mt}^* as given either by the SF (10a, b) after dropping the residuals ϵ_{mt} or, equivalently, by the RF,

$$y_{mt} = \eta_{mt}^* + \epsilon_{mt} \quad (11a)$$

$$= \sum_{k=1}^{K} \omega_{mk} x_{kt} + \epsilon_{mt}. \quad (11b)$$

The salient point of the reformulation is that SF (10) and RF (11) have the same residuals ϵ_{mt}.

REID SYSTEMS. REID and classical ID systems assume the same zero correlations (9b, c) between residuals and exogenous variables. REID systems allow Presp both in SF and RF, giving for SF,

$$E(y_m|\eta_1^{\prime *}, \ldots, \eta_M^{\prime *};\ x'_1, \ldots, x'_K) = \sum_{h=1}^{M} \beta_{mh} \eta_m^* + \sum_{k=1}^{K} \gamma_{mk} x_k, \quad (11c)$$

and again (7) for RF.

GEID SYSTEMS. GEID systems adopt Presp relations (10a) for the reformulated SF, but not for the RF, thus cancelling (7). Hence for each SF residual ϵ_m the parity principle implies the zero correlations

$$r(\eta_h^*, \epsilon_m) = 0, \qquad h = 1, M, \quad (12a)$$

$$r(x_k, \epsilon_m) = 0, \qquad k = 1, K. \quad (12b)$$

The zero correlations in (9c) and (12a, b) are equal in number in the special case of the ID systems just identified. Otherwise there are more correlations in (9c) than in (12a, b): the GEID system is then more general than the REID system.

FIX-POINT* (FP) ESTIMATION OF REID AND GEID SYSTEMS

OLS estimation applies to the reformulated SF relations (9a) by virtue of (12a, b), but there is the snag that in the product terms $\beta_{mh}\eta_{mt}^*$ both factors are unkown. FP estimation [7, 20, 42, 43, 47] solves the problem by an iterative LS procedure, say with steps $s = 1, 2, \ldots$. Letting $y_{mt}^{(s)}$ denote the proxy estimate of η_{mt} in step s, the limiting proxy as $s \to \infty$, say y_{mt}^*, is the FP estimate of η_{mt}^*,

$$\lim_{s \to \infty} y_m^{(s)} = y_{mt}^* = \text{est } \eta_{mt}^*,$$

$$m = 1, M; \; t = 1, T. \quad (13)$$

Using vector and matrix notation the FP algorithm in step s gives (14a) for the SF, and (14b) for the RF:

$$Y_t = Y_t^{(s)} + E_t^{(s)}$$

$$= B^{(s)}Y_t^{(s-1)} + G^{(s)}X_t + E_t^{(s)} \quad (14a)$$

$$= [I - B^{(s)}]^{-1}G^{(s)}X_t + E_t^{(s)}. \quad (14b)$$

The FP algorithm stays in the SF (14a) in line with the parity principle, but unlike the TSLS method. Alternatively, and with numerically the same results except for rounding errors, FP in the sth step alternates between (14a) and (14b).

Start, $s = 1$: Starting values $y_{nt}^{(0)}$ are needed. The choice of starting values is largely arbitrary; customary choices are

$$y_{mt}^{(0)} = y_{mt}, \quad (15a)$$

$$y_{mt}^{(0)} = 0, \quad (15b)$$

$$y_{mt}^{(0)} = y_{mt}^{\text{TSLS}}. \quad (15c)$$

General Step from $s - 1$ to s: Using $Y^{(s-1)}$ computed in step $s - 1$, OLS regression applied to the SF relations (14a) gives $B^{(s)}$ and $G^{(s)}$. Next, dropping the residuals $E_t^{(s)}$ in (14a) gives $Y^{(s)}$; alternatively, the RF (14b) gives $Y^{(s)}$ in terms of $B^{(s)}$ and $G^{(s)}$.

Stopping Rule: The FG algorithm stops by a conventional rule, say at step s, as soon as

$$\left|\left(p^{(s-1)} - p^{(s)}\right)/p^{(s)}\right| \leqslant 10^{-5} \quad (16a)$$

for all parameters p in B and G, giving

$$Y_t^* = Y_t^{(s)}. \quad (16b)$$

Comments.

In REID and GEID models the FP algorithm gives a correct estimate of the standard error.

The FP algorithm is the only known method for estimation of general GEID systems. Thanks to the parity principle the FP algorithm is a full information method. The FIML method is too complicated to be available beyond quite small models; cf. ref. 20.

Applications of FP estimation to ever larger GEID systems have been reported, and the performance of FP turns out to be satisfactory with respect to important aspects, including prediction accuracy, convergence, and uniqueness.

Until recently the FP developments were mainly a Swedish affair; cf. ref. 47. In 1980 a Polish GEID system with 25 SF relations was estimated using FP and FIMD (FIML with diagonal error variance matrix), giving somewhat larger R^2 for FP than for FIMD [30]. This result inspired simulation experiments on FP and FIMD [5]. When Presp is combined with FIMD assumptions it turns out that FIMD estimation gives somewhat more accurate parameter estimates than FP, whereas FP gives somewhat more accurate predictions than FIMD.

As to uniqueness of FP estimation, Bodin [7] has explored the simulation* experiments in ref. 26 with regard to multiplicity of FP estimation. The FP estimates are unique in all of these many hundreds of GEID systems.

Because the FP algorithm is based solely on the Presp assumptions on the model, it is questionable whether general theorems can be proved on such desirable properties as convergence of the FP algorithm and uniqueness of the FP estimates; cf. ref. 21.

RATIONAL EXPECTATIONS (RE) MODELS

In the RE models the rational expectations are conditional expectations formed in accordance with the model [27]. General RE models may involve conditional expectations of present, past, and future endogenous variables; hence GEID systems are the special case of RE models that involve expectations of the present, but not of the past or the future. Lösch [22] has extended the FP method to the estimation of general RE models. ML estimation of RE models has been reported only for quite small models [22], but the scope of RE models is extended by a quantum jump by Lösch's FP algorithms. Excellent results are reported for RE models with more than 24 structural relations and 30 variables [31].

LVP MODELS

We now come to LVP models with LVs indirectly observed by multiple manifest variables, called *indicators*. This is the territory of partial least squares* (PLS) estimation [48].

Principal Components

This topic is covered in refs. 18, 43, 45, 48, 55, and 56. Let

$$[x_{hn}] = X, \qquad h = 1, H;\ n = 1, N, \quad (17)$$

denote raw data on N observational units (cases) observed by H manifest indicators, the subscript n referring to time or a cross section.

The *first principal component*, denoted ξ_1, is an LV designed to predict the manifest indicators x_h (*see also* COMPONENTS ANALYSIS):

$$x_{hn} = \pi_{h1}\xi_{1n} + \epsilon_{1hn}, \quad (18a)$$

$$E(x_h|\xi_1) = \pi_{h1}\xi_1, \quad (18b)$$

where, in accordance with (1)–(4),

$$E(\epsilon_{1h}) = 0, \quad (18c)$$

$$r(\epsilon_{1h}, \xi_1) = 0, \qquad h = 1, H. \quad (18d)$$

The parameter π_{h1} $(h = 1, H)$ is called the (first) *loading* of x_h.

In estimated form (18a) reads:

$$x_{hn} = p_{h1}\hat{\xi}_{1n} + e_{1hn}, \quad (19a)$$

$$\hat{\xi}_{1n} = f_1 \sum_{h=1}^{H} p_{h1}x_{hn}. \quad (19b)$$

Since both factors in $\pi_{h1}\xi_{1n}$ are unknown their scales are ambiguous. For *standardization of scale unambiguity* (SSU), PLS modeling sets the variance of the LVs equal to unity; that is, in the present case,

$$\text{var}(\xi_1) = \text{var}(\hat{\xi}_1) \quad (19c)$$

$$= 1, \quad (19d)$$

where (19d) determines the scalar factor f_1 in (19b).

PLS estimation estimates first the LV ξ_1 and then the parameters π_{h1}. The estimation of ξ_1 is iterative, each step alternating between (19a) and (19b). As in (13) the proxy estimates in step s are denoted by superscript (s).

Start, $s = 1$: Again the start is largely arbitrary, say,

$$\hat{\xi}_{1n}^{(1)} = x_{1n}, \qquad n = 1, N. \quad (20)$$

General step from $s - 1$ to s: First, using $\hat{\xi}_{1n}^{(s-1)}$ computed in step $s - 1$, the simple OLS regressions with fixed h $(= 1, H)$

$$x_{hn} = p_{h1}^{(s)}\hat{\xi}_{1n}^{(s-1)} + e_{1hn}^{(s)} \quad (21a)$$

give the H loading proxies $p_{h1}^{(s)}$. Second,

weighting the indicators x_{hn} by $p_{h1}^{(s)}$ gives $\hat{\xi}_{1_H}^{(s)}$:

$$\hat{\xi}_{1n}^{(s)} = f_1^{(s)} \sum_{h=1}^{H} p_{h1}^{(s)} x_{hn}, \qquad n = 1, N, \tag{21b}$$

with

$$f_1^{(s)} = \pm N^{1/2} \left\{ \sum_{n=1}^{N} \left[\sum_{h=1}^{H} p_{h1}^{(s)} x_{hn} \right]^2 \right\}^{-1/2}. \tag{21c}$$

Stopping rule: A conventional rule (16a) applied to the loadings $p_{h1}^{(s)}$ $(h = 1, H)$ concludes the PLS algorithm (21a–c).

When forming the model (18a, b) the investigator should specify the expected sign of each loading π_{h1} $(h = 1, H)$. Then the $\pm$ sign in (21c) is chosen so as to agree with the majority of the expected signs.

The PLS algorithm (20)–(21) is numerically equivalent to the algebraic method in terms of eigenvalues and eigenvectors, and is always convergent except when the two largest eigenvalues are equal [43].

Principal Components of Higher Orders

The model for the second principal component is

$$x_{hn} = \pi_{h1}\xi_{1n} + \pi_{h2}\xi_{2n} + \epsilon_{2hn}, \tag{22a}$$

$$E(x_h|\xi_1, \xi_2) = \pi_{h1}\xi_1 + \pi_{h2}\xi_2, \tag{22b}$$

with

$$E(\epsilon_{2h}) = 0, \tag{23a}$$

$$r(\epsilon_{2h}, \xi_1) = r(\epsilon_{2h}, \xi_2) = 0, \tag{23b}$$

$$r(\xi_1, \xi_2) = 0. \tag{23c}$$

The PLS estimation algorithm remains the same (20)–(21) as for the first component, and uses e_{1hn} as data input instead of x_{hn}. The third principal component is estimated with e_{2hn} as data input, and so on. See refs. 48 and 49 for further information.

FACTOR MODELS

The general (one-) factor model (Spearman [33]) is formally the first principal component model (18a) with (18c, d), whereas (18b) is replaced by (24a):

$$r(\epsilon_{1h}, \epsilon_{1k}) = 0, \tag{24a}$$

$$r(\epsilon_{2h}, \epsilon_{2k}) = 0, \qquad h, k = 1, H;\ h \neq k. \tag{24b}$$

In words, ξ_1 is the general factor and ϵ_{1h} is the specific factor of the hth indicator ($H = 1, H$). The argument carries over to the multiple factor models of L. L. Thurstone (*see* FACTOR ANALYSIS). Thus the two-factor model is (22a) with (23a–c) and (24a, b).

The stringent assumptions of type (24a, b) make it difficult to estimate factor models. The first operative method to estimate them, an ML method, is due to Jöreskog [13]. Until then principal components in terms of eigenvalues and eigenvectors were often used as an approximation to factor models.

CANONICAL CORRELATIONS

Given two blocks of variables observed over N cases, say,

$$x_{hn}, \qquad h = 1, H,\ n = 1, N \tag{25a}$$

$$y_{kn}, \qquad k = 1, K,\ n = 1, N, \tag{25b}$$

and letting x_w, y_w denote weighted aggregates of the two blocks,

$$x_{wn} = \sum_h w_{1h} x_{hn}, \tag{25c}$$

$$y_{wn} = \sum_k w_{2k} y_{kn}, \tag{25d}$$

the *first canonical correlation* as defined by Hotelling [12] is the maximum R_1 for the correlation of x_w and y_w:

$$R_1 = \max r(x_w, y_w), \tag{25e}$$

Hotelling calculates the maximizing weights w_{1h}, w_{2h}, in R_1 by an algebraic method [12]; *see* CANONICAL ANALYSIS.

Canonical correlations can be interpreted in terms of PLS estimation of a path model with two LVs, say ξ, η [18, 19, 42, 48]. The PLS model is defined by two types of relations, both subject to Presp:

Inner Relations: Between the latent variables,

$$\eta_n = \beta_{21}\xi_n + u_n, \tag{26a}$$

$$E(\eta|\xi) = \beta_{21}\xi. \tag{26b}$$

Outer Relations: Between the manifest variables and the LVs,

$$x_{hn} = \pi_{1h}\xi_n + \epsilon_{1h}, \tag{27a}$$

$$E(x_h|\xi) = \pi_{1h}\xi, \tag{27b}$$

$$y_{kn} = \pi_{2k}\eta_n + \epsilon_{2k}, \tag{28a}$$

$$E(y_k|\eta) = \pi_{2n}\eta. \tag{28b}$$

In the PLS model the aggregates (25c, d) are estimates of the two LVs:

$$\hat{\xi}_n = \text{est } \xi_n = f_1 \sum_h (w_{1h} x_{hn}), \tag{29a}$$

$$\hat{\eta}_n = f_2 \sum_k (w_{2k} y_{kn}), \tag{29b}$$

where f_1, f_2 are scalars that give $\hat{\xi}_n$ and $\hat{\eta}_n$ unit variance; cf. (19c, d). Auxiliary parameters of the model, the weights w_{1h}, w_{2k} are determined by the *weight relations* defined by

$$\hat{\eta}_n = \sum_h (w_{1h} x_{hn}) + d_{1n}, \tag{30a}$$

$$\hat{\xi}_n = \sum_k (w_{2k} y_{kn}) + d_{2n}. \tag{30b}$$

The PLS estimation of the LVs is iterative. Alternating between (29a, b) and (30a, b), each step has four substeps. First, using weight proxies w_{1h} computed in the previous step, the aggregate (29a) gives the LV proxies $\hat{\xi}_n$; second, the multiple OLS regression (30b) gives the proxies w_{2k}; third, the aggregate (29b) gives the LV proxies $\hat{\eta}_n$; fourth, the OLS regression (30a) gives weight proxies w_{1h} as the start for the next step. The starting weights w_{1h} are largely arbitrary, say $w_{1h} = 1$; $h = 1, H$. A conventional stopping rule (16a) concludes the PLS estimation of ξ_n, η_n.

Using the estimates $\hat{\xi}_n$, $\hat{\eta}_n$, the inner and outer relations are estimated by OLS regressions, giving

$$\hat{\eta}_n = b_{21}\hat{\xi}_n + u_n, \tag{31}$$

$$x_{hn} = p_{1h}\hat{\xi}_n + e_{1hn}, \tag{32a}$$

$$y_{kn} = p_{2k}\hat{\eta}_n + e_{2kn}. \tag{32b}$$

The correlation of the PLS estimates $\hat{\xi}_n$, $\hat{\eta}_n$ is numerically the same as Hotelling's first canonical correlation, and is furthermore the PLS estimate of β_{21}:

$$R_1 = r(\hat{\xi}_n, \hat{\eta}_n) = N^{-1} \sum_n (\hat{\xi}_n, \hat{\eta}_n)$$

$$= \text{est } \beta_{21} = b_{21}. \tag{33}$$

As always in LS modeling it is an immediate matter to estimate the location parameters. Thus for (29a) and (31),

$$\text{mean}(\hat{\xi}_n) = N^{-1} \sum_n (\hat{\xi}_n)$$

$$= N^{-1} f_1 \sum_n (w_{1h} \bar{x}_n), \tag{34a}$$

$$b_{20} = \text{mean}(\hat{\eta}_n) - b_{21}\text{mean}(\hat{\xi}_n). \tag{34b}$$

Canonical Correlations of Higher Order

Hotelling [12] obtains canonical correlations of second and higher orders in terms of algebraic eigenvalues and eigenvectors. His weighted aggregates (25c, d) of higher order are uncorrelated with the aggregates of lower order.

Applying the PLS algorithm (29)–(34) to the residuals e_{1hn}, e_{2kn} of the outer relations (32a, b), the ensuing correlation of type (33) is numerically the same as Hotelling's canonical correlation of second order. The extension to canonical correlations of third and higher orders is straightforward. See also refs. 18, 42, and 55.

LINEAR PATH MODELS WITH LATENT VARIABLES INDIRECTLY OBSERVED BY MULTIPLE MANIFEST INDICATORS

Briefly, LVP models in sociology were introduced in the mid-1960s [8]. They combine econometrics (path models with manifest variables) and psychometrics (modeling with LVs indirectly observed by multiple manifest indicators). Figure 1 shows an LVP model illustrated by what is called the *arrow scheme*.

The LVP models posed entirely new problems of statistical estimation. Jöreskog [14] launched the first general algorithm for estimation of LVP models, an ML approach called LISREL, a landmark achievement. Seeing the LISREL* arrow schemes gave the author the thought that his iterative LS estimation of principal components models with one or two LVs [43] might be extended to LVP models with three or more LVs. Work on this idea from 1973–1977 led to the partial least squares (PLS) algorithm for estimation of LVP models [45, 48]. In partial least squares the PLS algorithm is set forth with emphasis on its key features. We note the following points, without entering into technical details.

PLS provides prediction based on inner and/or outer relations, whereas LISREL is parameter oriented and estimates parameters that model the covariance matrix of the observables, called the *structure* of the data.

If the model does not involve two or more time points, the PLS algorithm gives the (classical) standard error. If the model involves two or more time points, the PLS algorithm gives an underestimate of the standard error. (See [5] for some examples.)

There is a trade-off between the estimation of parameters and LVs. LISREL provides consistent estimation of the parameters, but no prediction or estimation of the LVs. PLS provides estimation of both parameters and LVs. The PLS estimation is *consistent at large*, i.e., is consistent in the limit as the number of indicators of each LV increases indefinitely.

THE BROAD SCOPE OF PRESP

As emphasized in the introduction, modeling on the basis of Presp is not hampered by the stringent ML assumptions that the observations are jointly ruled by a specified multivariate distribution, and that the data are independent observations of this distribution. Presp is of broad scope in three respects: data input, theoretical model, and operative purpose, as will now be illustrated with reference to published applications to real-world models and data. For brevity the illustrations are selective, and the examples from the literature mainly take up applied work on large complex systems, which is the forte of PLS.

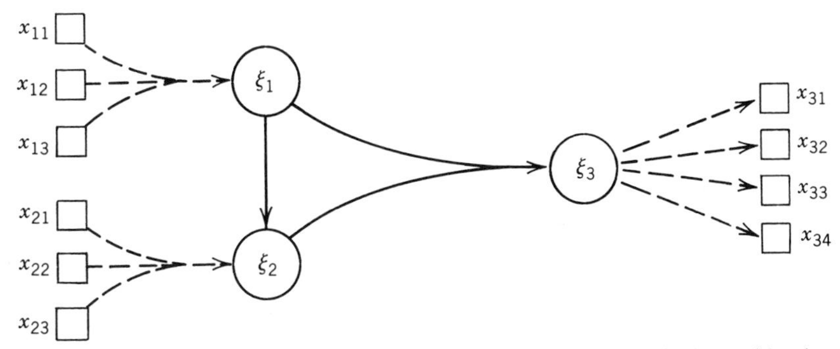

Figure 1 Arrow scheme for an LVP model with three latent variables, each observed by three or four manifest indicators.

Different Types of Data Modeled by Means of PLS

Cardinal data, i.e., measurable variables. Estimation by OLS [53]; by FP [5a, 22, 26, 31, 47]; by PLS [16, 32, 38, 52]. Also see ref. 15, Vol. 2.

Ordinal data*.

Categorical data* and contingency tables*. Estimation by PLS [6, 52].

Cardinal, ordinal, and categorical data often occur in one and the same PLS model [9, 19, 23, 29, 38, 56].

The data may be reproducible or nonreproducible. The data in all the given references are nonreproducible. For reproducible data, cardinal or categorical, see ref. 56.

Different Types of Model Estimated on the Basis of Presp

Path models with manifest variables. FP estimation [5a, 47].

Path models with latent variables. PLS estimation [23, 29, 32].

Rational expectations (RE) models. FP estimation [22, 31].

Nonlinear LVP models estimated by PLS [48, 54].

LVP models with LVs in higher dimensions. Estimation by PLS [2, 18, 48, 56].

LVP models for three-way data matrices [17, 18].

Different Purposes of Modeling

Classification. Principal components models estimated by PLS [55, 56].

Prediction. Estimation by FP [42, 43, 47].

Causal interpretation of predictive models [44, 52].

Controlled experiments. PLS modeling [56].

Field experiments. PLS modeling [56].

MODEL EVALUATION

The Stone–Geisser (SG) test (iii*) for predictive relevance evaluates the model as a whole, whereas Tukey's jackknife (iv*) measures the accuracy of specific parameter estimates. The SG test criterion Q^2 and Tukey's jackknife suit Presp modeling as hand in glove. Relative to the basic ideas of Tukey [36], Stone [34], and Geisser [10], the FP and PLS applications mark a shift both in scope and purpose. To specify, with reference to FIX-POINT METHOD* and PARTIAL LEAST SQUARES*: deleting (blindfolding) the N observations one by one, Stone and Tukey reestimate the model N times on the basis of $N - 1$ observations; the N sets of reestimated parameters give new estimates of parameters (Stone) and of standard errors for the estimated parameters (Tukey), new estimates which are more accurate on the assumption of independent observations, [25]. Geisser allows the blindfolding to be performed group-wise, say G observations, simultaneously. The grouping reduces the computations by a factor.

In the context of PLS modeling the SG blindfolding was extended from a parametric test to give a holistic test for predictive relevance [51, 55, 56]. The ensuing test criterion, denoted Q^2, is an R^2 that, thanks to the blindfolding, is evaluated without loss of degrees of freedom.

Presp modeling is distribution-free in the exogenous variables. Real-world data often are interdependent, and then the jackknife applies and remains consistent, whereas the classical standard errors are biased downward [5, 5a].

CONCLUSIONS

By the late 1950s it was widely understood that in the passage from small and simple to large and complex models ML modeling soon becomes intractable. This insight was reinforced by the advent and the rapid diffusion of the computer. Hence it is no wonder that LS approaches that break away from the

ML mainstream entered almost simultaneously; among those are the jackknife, the fix-point method, the Stone–Geisser parametric test with its predictive twist, partial least squares and the SG test criterion Q^2.

Scientific models are more or less successful approximations; they are never exactly true. The ML null hypothesis is that the model is true; this hypothesis is wrongly posed inasmuch as the ML test rejects the model sooner or later as the sample size N increases indefinitely. In contrast, the test criterion Q^2 poses the question whether the model is predictive. The answer is yes or no, depending on whether $Q^2 > 0$ or $Q^2 < 0$. As N increases a positive Q^2 usually stabilizes at some level $0 < Q^2 < 1$. A resulting degree of predictive relevance $Q^2 > 0$ may or may not be satisfactory to the investigator. Since the test did not ask for the best model, let alone the true model, the door is left open for improving the model. The criterion Q^2 is in line with the fact that model building in practice is a trial and error procedure, a dialogue with the computer.

The standard ML test is asymptotic for $N \to \infty$. In contrast, the Q^2 test applies for any finite N, even for quite small N; cf. refs. 49, p. 369, 51 and 52. The model is sometimes assumed to change, to take different forms for consecutive periods of time. The test criterion Q^2 then applies to the different models separately, whereas the standard ML test does not apply [51, 52]. Still more broadly, the Q^2 test applies whether or not the model is estimated by a consistent method. The crucial requisite is that the data possess a sufficient degree of statistical regularity and homogeneity.

References

[1] Allais, M. (1983). *C. R. Acad. Sci. Paris Ser. I*, **296**, 829–832.

[2] Apel, H. and Wold, H. (1982). In *Systems under Indirect Observation: Causality, Structure, Prediction*, Vol. 2, K. G. Jöreskog and H. Wold, eds. North-Holland, Amsterdam, The Netherlands, pp. 209–249.

[3] Basmann, R. L. (1957). *Econometrica*, **25**, 77–83.

[4] Bentzel, R. and Wold, H. (1946). *Skand. Aktuarietidskr.*, **29**, 95–114.

[5] Bergström, R. (1986). Jackknifing of some econometric estimators. *Research Rep. 86-4*, University of Uppsala, Dept. Statistics, Uppsala, Sweden.

[5a] Bergström, R. and Wold, H. (1983). *The Fix-Point Approach to Interdependent Systems*, Vol. 22, G. Tintner, H. Strecker, and E. Féron, eds. Vandenhoeck and Ruprecht, Göttingen, Federal Republic of Germany.

[6] Bertholet, J.-L. and Wold, H. (1980). In *Measuring the Unmeasurable*, P. Nijkamp, H. Leitner, and N. Wrigley, eds. Martinus Nijhoff, Dordrecht, The Netherlands, pp. 253–286.

[7] Bodin, L. (1974). *Recursive Fix-Point Estimation: Theory and Applications*. Published Ph.D. Dissertation, Uppsala University, Uppsala, Sweden.

[8] Duncan, O. D. (1966). *Amer. J. Sociol.*, **72**, 1–16.

[9] Falter, J. F. et al. (1983). *Kölner Zeit. Soziologie Sozialpsychologie*, **35**, 525–554.

[10] Geisser, S. (1974). Discussion contribution to M. Stone (1974). *J. R. Statist. Soc. Ser. B*, **36**, 141.

[11] Haavelmo, T. (1942). *Econometrica*, **11**, 1–12.

[12] Hotelling, H. (1936). *Biometrika*, **28**, 321–377.

[13] Jöreskog, K. G. (1967). *Psychometrika*, **32**, 443–482.

[14] Jöreskog, K. G. (1970). *Psychometrika*, **36**, 409–426.

[15] Jöreskog, K. G. and Wold, H., eds. (1982). *Systems under Indirect Observation: Causality, Structure, Prediction*, Vols. 1 and 2. North-Holland, Amsterdam, The Netherlands.

[16] Knepel, H. (1981). *Sozialökonomische Indikatormodelle zur Arbeitsmarktanalyse*. Campus, Frankfurt, Germany.

[17] Lohmöller, J.-B. (1981). *LVPLS Program Manual, Version 1.6: Latent Variables Path Analysis with Partial Least Squares Estimation*. Hochschule der Bundeswehr, München, Bavaria, Federal Republic of Germany. [2nd ed., Zentralarchiv für empirische Sozialforschung, Koln, Federal Republic of Germany (1984).]

[18] Lohmöller, J.-B. (1986). *Path Models with Latent Variables and Partial Least Squares (PLS) Estimation*. Physika, Stuttgart, Federal Republic of Germany.

[19] Lohmöller, J.-B. and Wold, H. (1984). In *Cultural Indicators: An International Symposium*, G. Melischek et al., eds. Austrian Academy of Sciences, Vienna, Austria, pp. 501–519.

[20] Lyttkens, E. (1973). *J. R. Statist. Soc. Ser. A*, **136**, 353–394.

[21] Lösch, M. (1980). *Identifikations- und Schätzprobleme linearer ökonometrischer Modelle*. Athenäum, Königstein/Ts.

[22] Lösch, M. (1984). *Fixpunkt-Schätzverfahren für Modelle mit rationalen Erwrtungen*. Athenäum, Köningstein/Ts.

[23] Meissner, W. and Uhle-Fassing, M. (1983). *Weiche Modelle und iterative Schätzung. Eine Anwendung auf Probleme der Neuen Politischen Ökonomie*. Campus, Frankfurt, Federal Republic of Germany.

[24] Mensch, G. O., Kaasch, K., and Wold, H. (1986). In *Proc. IIASA Conf. on Long Term Fluctuations in Economic Growth: Their Causes and Consequences*.

[25] Miller, R. G. (1974). *Biometrika*, **61**, 1–15.

[26] Mosbaek, E. J. and Wold, H. (1970). *Interdependent Systems: Structure and Estimation*. North-Holland, Amsterdam, The Netherlands. (Includes contributions by E. Lyttkens, A. Ågren and L. Bodin.)

[27] Muth, J. F. (1961). *Econometrica*, **29**, 315–335.

[28] Nijkamp, P., Leitner, H., and Wrigley, N., eds. (1985). *Measuring the Unmeasurable*. Martinus Nijhoff, Dordrecht, The Netherlands.

[29] Noonan, R. and Wold, H. (1983). *Evaluating School Systems Using Partial Least Squares*. Pergamon, Oxford, England.

[30] Romański, J. and Welfe, W. (1980). On forecasting efficiency of different estimation methods for interdependent systems. *4th World Congr. Econometric Soc.*

[31] Schips, B. (1986). In *Theoretical Empiricism. A General Rationale of Scientific Model-Building*, H. Wold, ed. Paragon House, Washington, DC.

[32] Schneewind, K. A., Beckman, M., and Engfer, A. (1983). *Eltern und Kindern*. Kohlhammer, Stuttgart, Federal Republic of Germany.

[33] Spearman, C. (1904). *Amer. J. Psychol.*, **15**, 72–101.

[34] Stone, M. (1974). *J. R. Statist. Soc. Ser. B*, **36**, 111–147.

[35] Theil, H. (1953). *Mimeo paper*. Central Planning Bureau, The Hague, The Netherlands.

[36] Tukey, J. W. (1958). *Ann. Math. Statist.*, **29**, 614.

[37] Tinbergen, J. (1937). *An Econometric Approach to Business Cycle Problems*. Hermann, Paris, France.

[38] Wilkenfeld, J., Hopple, G. W., Rossa, P. J., and Andriole, S. J. (1982). *Interstate Behavior Analysis*. Sage, Beverly Hills, CA.

[39] Wold, H. (1959). In *Probability and Statistics: The Harald Cramér Volume*, U. Grenander, ed. Almqvist and Wiksell, Uppsala, Sweden, pp. 355–434.

[40] Wold, H. (1961). *Proc. Fourth Berkeley Symp. Math. Statist. Prob.*, Vol. 1. University of California Press, Berkeley, CA, pp. 719–761.

[41] Wold, H. (1963). *Sankhyā A*, **25**, 211–240.

[42] Wold, H. (1965). *Ark. Mat.*, **6**, 209–240.

[43] Wold, H. (1966). In *Research Papers in Statistics: Festschrift for J. Neyman*, F. N. David, ed. Wiley, New York, pp. 411–444.

[44] Wold, H. (1969). *Synthese*, **20**, 427–482.

[45] Wold, H. (1975). In *Perspectives in Probability and Statistics: Papers in Honour of M. S. Bartlett*, J. Gani, ed. Academic, London, pp. 117–142.

[46] Wold, H. (1980). *Math. Statist., Banach Center Publ. (Warsaw)*, **6**, 333–346.

[47] Wold, H. (1980). *The Fix-Point Approach to Interdependent Systems*. North-Holland, Amsterdam, The Netherlands.

[48] Wold, H. (1982). In *Systems under Indirect Observation: Causality, Structure, Prediction*, Vol. 2, K. G. Jöreskog and H. Wold, eds. North-Holland, Amsterdam, The Netherlands, pp. 1–54.

[49] Wold, H. (1983). In *Applied Time Series Analysis of Economic Data*, A. Zellner, ed. U.S. Department of Commerce, Bureau of the Census, Washington, DC, pp. 363–371.

[50] Wold, H. (1985). In *Measuring the Unmeasurable*, P. Nijkamp, H. Leinter, and N. Wrigley, eds. Martinus Nijhoff, Dordrecht, The Netherlands, pp. 211–251.

[51] Wold, H. (1985). In *Contributed Papers, I–II, Centenary Session of International Statistical Institute (ISI)*. ISI, Amsterdam, The Netherlands, pp. 19–20.

[52] Wold, H., ed. (1987). *Theoretical Empiricism. A General Rationale of Scientific Model-Building*. Paragon House, Washington, DC.

[53] Wold, H. and Juréen, L. (1952). *Demand Analysis. A Study in Econometrics*. Geber, Uppsala, Sweden. [3rd reprint edition, Greenwood Press, Westport, CT (1982).]

[54] Wold, H. and Mensch, G. O. (1983). *Working Paper No. WSCM 83-017*, Case Western Reserve University, Cleveland, OH.

[55] Wold, S. (1978). *Technometrics*, **20**, 397–405.

[56] Wold, S. et al. (1983). In *Food Research and Data Analysis*, H. Martins and H. Russwurm, Jr., eds. Applied Science, London, England, pp. 147–188.

Bibliography

See the following paper, as well as the references just cited, for further information on

Presp (predictor specification) from the general point of view of scientific model-building.

Wold, H. (1982). In *The Making of Statisticians*, J. Gani, ed. Springer, New York, pp. 190–212.

(ECONOMETRICS
FIX-POINT METHOD
FULL-INFORMATION ESTIMATORS
LEAST SQUARES
MAXIMUM LIKELIHOOD ESTIMATION
PARTIAL LEAST SQUARES
SAMPLE REUSE)

HERMAN WOLD

SPECIFICITY

Specificity has two meanings:

1. In factor analysis*, the proportion of the variance of observed values of a variable due to factors specific to that variable, and not to common factors. It can be regarded as an index of "uniqueness."
2. A property of a diagnostic test *(see* SENSITIVITY AND SPECIFICITY).

SPECTRAL ANALYSIS

Spectral analysis has its origin in the notion that an emitted signal may be partitioned into a set of components, each of which is a pure wave motion. For example, a ray of white light passed through a prism divides into distinct bands of color. Each color is a ray with a different *wavelength* (length of a single complete cycle), which may also be described by its *frequency* (the number of cycles per unit length). We shall use the notion of frequency hereafter, since frequency $\propto 1/\text{wavelength}$.

The most natural mathematical representation of such a phenomenon is the sine wave, so that a wave with a frequency of λ cycles per unit time may be represented by

$$x(t) = A\cos\omega t + B\sin\omega t$$
$$= S\cos(\omega t - \phi), \tag{1}$$

where $S^2 = A^2 + B^2$ and $\omega = 2\pi\lambda$; ω is the *angular frequency*. S represents the amplitude of the sine wave, and $\phi = \tan^{-1}(B/A)$ represents the *phase* or displacement from the origin, since $x(t) = S$ when $\omega t = \phi$.

If $\{x(t)\}$ represents a (single) realization of a time series* over the period $[0, T]$, we may represent the series by a weighted average of sine waves. This was the motivation for Shuster's development [14] in 1898 of the periodogram*. In turn, this led to the development of tests for the existence of strict periodicities in the data (*see* PERIODOGRAM ANALYSIS; PRIESTLEY'S TEST FOR HARMONIC COMPONENTS). For further discussion of the historical developments, see ref. 3, pp. 2–7.

Since the aim of this approach is to represent the time series by a set of sine waves of different frequencies, such methods are described as *frequency domain* analyses. However, rather than presuppose the existence of a finite set of exact periodicities, spectral analysis allows all possible frequencies to be represented. If the time series is discrete (recorded at times $t = 1, 2, \ldots, T$), then $0 \leqslant \omega \leqslant \pi$, since any wave with frequency $\omega > \pi$ is totally indistinguishable from one with frequency $\omega^* = \omega - k\pi$, where k is the largest integer such that ω^* remains nonnegative. π is known as the *Nyquist frequency**, since it represents the highest frequency that can be detected in the series. For example, if a daily cycle is to be detected, setting $\omega = \pi$ or $\lambda = \frac{1}{2}$ implies that at least two observations per day must be taken.

When the times series is continuous (measured for all $0 \leqslant t \leqslant T$), $0 \leqslant \omega < \infty$. In the next section, which deals with mathematical developments, we assume that the series is defined in discrete time. However, analogous results are obtained for continuous time upon replacing the summations over time by integrals and extending the range of ω to $[0, \infty)$.

THEORETICAL PROPERTIES OF THE SPECTRUM

We now consider a random process rather than a single realization. Let $X(t)$ denote a time series that is weakly stationary; that is, for all s and t,

$$E[X(t)] = \mu,$$
$$V[X(t)] = \sigma_x^2,$$
$$\operatorname{cov}[X(t), X(t-s)] = \gamma(s) = \sigma_x^2 \rho(s), \tag{2}$$

where $\rho(s) = \rho(-s)$. Since we are interested only in the covariance structure, we may put $\mu = 0$. Hence $X(t)$ may be represented in the frequency domain by

$$X(t) = \int_0^{\pi} [A(\omega)\cos \omega t + B(\omega)\sin \omega t]\, d\omega, \tag{3}$$

where $A(\omega)$ and $B(\omega)$ are orthogonal random processes. Although $X(t)$ is a real-valued process, there are considerable theoretical benefits to be gained from reformulating (3) as

$$X(t) = \int_0^{\pi} G_T(\omega) e^{i\omega t}\, d\omega, \tag{4}$$

where $e^{i\omega t} = \cos \omega t + i \sin \omega t$ and $G_T(\omega)$ is a complex-valued process. One such benefit from (4) is that $X(t)$ is the Fourier transform of $G_T(\omega)$ (*see* INTEGRAL TRANSFORMS), so that

$$G_T(\omega) = \pi^{-1} \sum_{i=1}^{T} e^{-i\omega t} X(t). \tag{5}$$

By extension from (1), we can see that $|G_T(\omega)|^2$ is the squared amplitude for the wave with angular frequency ω. For any interval $(\omega_1, \omega_2]$, the total power of the signal may be defined as

$$h(\omega_1, \omega_2) = \int_{\omega_1}^{\omega_2} h(\omega)\, d\omega, \tag{6}$$

where $h(\omega) = \lim_{T\to\infty}[E|G_T(\omega)|^2/T]$; see ref. 13, pp. 206–210. $h(\omega_1, \omega_2)$ is the *band spectrum* for the frequency band $(\omega_1, \omega_2]$ and $h(\omega)$ is the *spectral density function*. It follows from Parseval's identity that

$$\sigma_x^2 = \int_0^{\pi} h(\omega)\, d\omega; \tag{7}$$

that is, the total power in the spectrum is equal to the variance of the process. Using (7), the *normalized* spectral density function $f(\omega) = h(\omega)/\sigma_x^2$ is seen to behave like a probability density function, hence the name.

It follows from (2), (4), and (5) that

$$\rho(s) = \int_0^{\pi} \cos \omega s f(\omega)\, d\omega \tag{8}$$

and

$$f(\omega) = \pi^{-1}\left[1 + 2\sum \rho(s)\cos \omega s\right]. \tag{9}$$

That is, the autocorrelation function and the normalized spectral density function are a Fourier transform pair. Since $f(\omega)$ and $\rho(s)$ are so related, they contain equivalent information about the underlying process. However, in any particular case, one may be easier to interpret than the other.

It should be noted that many texts define $f(\omega)$ for $-\pi \leqslant \omega \leqslant \pi$. The results are unchanged apart from constant factors, and that approach has some theoretical attractions. However, the specification on $[0, \pi]$ is easier to interpret.

STATISTICAL INFERENCE

Given the relationships (8) and (9) and the knowledge that the sample autocorrelations, or serial correlations*,

$$r(s) = \frac{\Sigma[x(t) - \bar{x}][x(t-s) - \bar{x}]}{\Sigma[x(t) - \bar{x}]^2}, \quad s = 1, 2, \ldots, \tag{10}$$

are consistent estimators for $\rho(s)$, it would appear that the Fourier transform of the $\{r(s)\}$ should produce consistent estimators for $h(\omega)$. Unfortunately, this is not so; see ref. 10, pp. 584–585. The lack of consistency may lead to a large number of spurious peaks with no real meaning (*see* PERIODOGRAM ANALYSIS for an example).

To overcome this difficulty, we may smooth the estimators to induce consistency. The general form of such estimators is

$$\hat{f}_q(\omega) = \pi^{-1}\left[1 + 2\sum_{s=1}^{q} \alpha_s r(s)\cos \omega s\right], \quad (11)$$

where q is a suitably chosen truncation point and the $\{\alpha_s\}$ are said to form a *lag window*. For details and examples, see ref. 10, pp. 590–594. If we take the Fourier transform of the $\{\alpha_s\}$, we obtain the *spectral window*, which enables us to produce consistent estimates of $f(\omega)$ by smoothing the periodogram rather than by using (11). Estimation of the spectral density using (10) and (11) used to be the standard approach until the rediscovery of the fast Fourier transformation (FFT); see ref. 5 for an historical account.

The Fast Fourier Transform

Assume that the time series contains $T = rs$ terms. If T does not factor, we add zeros to the ends of the mean-adjusted series, $u(t) = x(t) - \bar{x}$. If we write

$$t = rt_1 + t_2,$$

it follows that

$$\sum_{t=0}^{T-1} u(t)e^{i\omega t} = \sum_{t_2=1}^{r-1} e^{i\omega t_2} \sum_{t_1=0}^{s-1} u(t)e^{i\omega t_1}.$$

Hence the spectrum can be computed as the result of $T(s + \frac{1}{2}r)$ arithmetic operations rather than $\frac{1}{2}T^2$. When $T = 2^p$, the total effort reduces to the order of $T \log_2 T$ operations; see ref. 13, pp. 575–577, or ref. 10, pp. 595–597. For long series, it is much quicker to use the FFT; indeed, it may be quicker to compute the sample spectrum and invert that to obtain the serial correlations. The major statistical packages allow either approach.

Other proposed estimators include fitting a high-order autoregressive scheme [12], and robust methods [11].

CROSS-SPECTRA

Consider a pair of series $X_1(t)$ and $X_2(t)$. Their individual spectra are defined as in (9), while the cross-spectrum is

$$f_{12}(\omega) = \pi^{-1}\sum_{s} e^{-i\omega s}\rho_{12}(s), \quad (12)$$

where $\rho_{12}(s) = \mathrm{corr}[X_1(t), X_2(t-s)]$. As before, f_{12} and ρ_{12} form a Fourier transform pair. We may express f_{12} in the forms

$$\begin{aligned} f_{12}(\omega) &= c(\omega) + iq(\omega) \\ &= \alpha(\omega)\exp[i\psi(\omega)], \end{aligned} \quad (13)$$

where $c(\omega)$ is the *co-spectrum*, $q(\omega)$ is the *quadrature spectrum*, $\alpha(\omega)$ is the *cross-amplitude spectrum*, and $\psi(\omega)$ is the *phase spectrum*. The phase spectrum enables us to determine the extent to which $X_1(t)$ leads or lags $X_2(t)$ at each frequency ω. The coherence

$$c(\omega) = |\alpha(\omega)|^2/[f_1(\omega)f_2(\omega)] \quad (14)$$

may be interpreted like a squared correlation coefficient between the signals at frequency ω, and the *transfer function**

$$T_{12}(\omega) = f_{12}(\omega)/f_2(\omega) \quad (15)$$

may be interpreted as the regression coefficient for $X_1(t)$ on $X_2(t)$ at frequency ω. $|T_{12}(\omega)|$ is known as the *gain*. For further details, see refs. 10, pp. 662–668, and 13, pp. 655–676.

The estimation of cross-spectra proceeds as in the univariate case, using the same windows to produce consistent estimates. The bias in the estimates may be reduced substantially if the two series are *aligned* so that, when the maximum of the cross-correlation function occurs at lag t_0, we may use $X_2^*(t) = X_2(t - t_0)$; see ref. 13, pp. 692–718.

ADDITIONAL TOPICS

Prediction

Given the values $X_1, \ldots, X_T$, the best m-step ahead predictor for X_{T+m} is its conditional expectation, given $(X_1, \ldots, X_T)$; *see* TIME SERIES. In the absence of specific distributional assumptions, we use the best linear predictor

$$\tilde{X}_{T+m} = a_0 X_T + a_1 X_{T-1} + \cdots + a_{T-1} X_1.$$

The optimal weights $\{a_j\}$ may be derived from the spectral density to yield the Wiener–Kolmogorov filter; see refs. 13, pp. 727–761, and 15.

Evolutionary Spectra

When a series is trend free, but has a spectrum that is changing slowly over time, we define

$$X(t) = \int_0^{\pi} G(\omega, t) e^{i\omega t}\, d\omega; \qquad (16)$$

$G(\omega, t)$ is the *evolutionary spectrum*. For further details on theoretical properties and estimation procedures, see ref. 13, pp. 821–855.

Band-Pass Filters

If we were to eliminate all frequencies except those in the interval $[\omega_1, \omega_2]$, and then examine the modified $X^*(t)$ series defined by

$$X^*(t) = (\text{const.}) \int_{\omega_1}^{\omega_2} e^{-i\omega t} G_T(\omega)\, d\omega,$$

we would have reconstructed $X^*(t)$ by means of a *band-pass filter*; that is, one that "passes" only those frequencies in a certain interval. When $\omega_1 = 0$, we speak of a *low-pass filter* and when $\omega_2 = \pi$, of a *high-pass filter*. Perfect filters cannot be constructed, but it is easy to construct filters that favor different ranges of frequencies. For example, the first difference $X(t) - X(t-1)$ is a high-pass filter that attenuates the spectrum for low frequencies. *Complex demodulation*, a form of local harmonic analysis, involves using observations in the neighborhood of t to examine local changes in amplitude and phase at a particular frequency ω. For further details, see refs. 13, pp. 848–855, and 3, pp. 118–145.

References

[1] Anderson, T. W. (1971). *The Statistical Analysis of Time Series*. Wiley, New York. (A comprehensive account of univariate time and frequency domain analysis.)

[2] Blackman, R. B. and Tukey, J. W. (1959). *The Measurement of Power Spectra*. Dover, New York. (A classic text in the development of spectral analysis.)

[3] Bloomfield, P. (1976). *Fourier Analysis of Time Series: An Introduction*. Wiley, New York. (A clear introduction at an intermediate level.)

[4] Brillinger, D. R. (1975). *Time Series: Data Analysis and Theory*. Holt, Rinehart and Winston, New York. (Good coverage of frequency domain analysis, both theoretical and applied.)

[5] Cooley, J. W., Lewis, P. A. W., and Welch, P. D. (1967). *IEEE Trans. Aud. Electr.*, **AU-15**, 76–79.

[6] Fuller, W. A. (1976). *Introduction to Statistical Time Series*. Wiley, New York. (Coverage of theory at an advanced level.)

[7] Granger, C. W. J. and Hatanaka, M. (1964). *Spectral Analysis of Economic Time Series*. Princeton University Press, Princeton, NJ. (Coverage at an intermediate level with several economic applications.)

[8] Hannan, E. J. (1970). *Multiple Time Series*. Wiley, New York. (Treats theory of multiple time series at an advanced level.)

[9] Jenkins, G. M. and Watts, D. G. (1968). *Spectral Analysis and its Applications*. Holden-Day, San Francisco, CA. (A clear introduction at an intermediate level.)

[10] Kendall, M. G., Stuart, A., and Ord, J. K. (1983). *The Advanced Theory of Statistics*, 4th ed., Vol. 3. Griffin, London, England. (Discusses univariate and multiple time series in both the time and frequency domains.)

[11] Kleiner, B., Martin, R. D., and Thomson, D. J. (1979). *J. R. Statist. Soc. Ser. B*, **41**, 313–351.

[12] Parzen, E. (1969). In *Multivariate Analysis*, Vol. 2, P. R. Krishnaiah, ed. Academic, New York.

[13] Priestley, M. B. (1981). *Spectral Analysis and Time Series*, Vols. 1 and 2. Academic, London, England. (A comprehensive account of the theory of spectral analysis.)

[14] Shuster, A. (1898). *Terr. Mag. Atmos. Electr.*, **3**, 13–41.

[15] Whittle, P. (1963). *Prediction and Regulation*. English University Press, London, England. (A classic account of the theory of prediction.)

(MULTIPLE TIME SERIES
NYQUIST FREQUENCY
PERIODOGRAM ANALYSIS
SEASONALITY
TIME SERIES)

J. K. Ord

SPECTRAL DECOMPOSITION

The *spectral decomposition* of a real symmetric $m \times m$ matrix $\mathbf{A}$ is

$$\mathbf{A} = \mathbf{P}\mathbf{D}_\lambda\mathbf{P}'$$

$$= \lambda_1\mathbf{P}_1\mathbf{P}_1' + \lambda_2\mathbf{P}_2\mathbf{P}_2' + \cdots + \lambda_m\mathbf{P}_m\mathbf{P}_m',$$

with $\mathbf{D}_\lambda = \text{diag}(\lambda_1, \ldots, \lambda_m)$, where λ_j is the jth eigenvalue* (characteristic root) of $\mathbf{A}$ and $\mathbf{P}_j$ is the corresponding normalized eigenvector; $\mathbf{P} = (\mathbf{P}_1, \ldots, \mathbf{P}_m)$ is an orthogonal matrix*.

An important application of spectral decomposition is in the theory of quadratic forms*. If $\mathbf{U}' = (U_1, \ldots, U_m)$, then

$$\mathbf{U}'\mathbf{A}\mathbf{U} = \sum_{i=1}^{m}\sum_{j=1}^{m} a_{ij}U_iU_j$$

$$= \mathbf{U}'\mathbf{P}\mathbf{D}_\lambda\mathbf{P}'\mathbf{U}$$

$$= \mathbf{W}'\mathbf{D}_\lambda\mathbf{W} = \sum_{j=1}^{m}\lambda_j W_j^2,$$

where $\mathbf{W} = \mathbf{P}'\mathbf{U}$ is an orthogonal transform of $\mathbf{U}$. If $U_1, \ldots, U_m$ are mutually independent standard normal variables, so are $W_1, \ldots, W_m$.

For further details, see, e.g., Rao [1, Section 1c.3].

Reference

[1] Rao, C. R. (1973). *Linear Statistical Inference and Applications*, 2nd ed. Wiley, New York.

(LINEAR ALGEBRA, COMPUTATIONAL
SINGULAR-VALUE DECOMPOSITION)

SPECTRAL DENSITY; SPECTRAL DISTRIBUTION: SPECTRAL ESTIMATION; SPECTRAL FUNCTION *See* SPECTRAL ANALYSIS

SPECTRAL WINDOW *See* KERNEL ESTIMATORS

SPENCER'S GRADUATION FORMULAE

There are two of these formulae—one based on 15 and one on 21 consecutive values of a series, centered at the point for which the graduated value is being calculated. In these formulae the summation operator $[n]$ is used; *see* SUMMATION $[n]$.

15-point formula: Graduated value is

$$u_x' = \frac{1}{320}[4]^2[5]\{1 + 6[3] - 3[5]\}u_x.$$

21-point formula: Graduated value is

$$u_x' = \frac{1}{350}[5]^2[7]\{1 + [3] + [5] - [7]\}u_x.$$

The 15-point formula has been used less commonly than the 21-point formula.

(GRADUATION)

SPHERICAL DISTRIBUTIONS *See* DIRECTIONAL DISTRIBUTIONS

SPHERICAL MEDIAN

The concept of the median as a measure of the centre of a distribution can usefully be applied to unimodal distributions of three-dimensional unit vectors. Let $\mathbf{X}$ be a random vector from such a distribution, and $\mathbf{a}$ an arbitrary fixed unit vector, and define $\theta_\mathbf{a} = \cos^{-1}(\mathbf{X}'\mathbf{a})$, the angular deviation of $\mathbf{X}$ from $\mathbf{a}$. The *spherical median direction* is the direction $\mathbf{a}_0$ that minimises $E\theta_\mathbf{a}$. By way of comparison, the spherical *mean* direction is that $\mathbf{a}$-direction, which minimises $E(1 - \cos\theta_\mathbf{a})$; *see* DIRECTIONAL DISTRIBUTIONS.

Given a random sample of unit vectors $\mathbf{X}_1, \ldots, \mathbf{X}_n$, the *sample spherical median* is then defined as the direction $\hat{\mathbf{a}}_0$ that minimises $\Sigma \cos^{-1}(\mathbf{X}_i'\mathbf{a})$. It has the following robustness* property, analogous to one enjoyed by the median of linear data and the mediancentre of spatial data [1, 3]: If $\hat{\mathbf{a}}_0$ is taken, for convenience, to be the direction of the North Pole, then it depends only on the longitudes of the data points relative to it, and not on their latitudes (subject to the obvious restriction that the majority of data points lie in the same hemisphere, otherwise the median direction would be $-\hat{\mathbf{a}}_0$). Large-sample theory for the spherical median direction is given in Fisher [2]. The notion generalises in straightforward fashion to higher-dimensional spheres.

For a distribution of three-dimensional unit axes, or undirected lines, the spherical median axis can be defined analogously. Let $M(\mathbf{X}, \mathbf{a}) = \pi/2 - |\pi/2 - \cos^{-1}(\mathbf{X}'\mathbf{a})|$, the smaller of the two angles between a random unit axis $\mathbf{X}$ and the axis $\mathbf{a}$. If $\mathbf{X}$ has a bipolar distribution, the *spherical median axis* is that value of $\mathbf{a}$ for which $EM(\mathbf{X}, \mathbf{a})$ is minimum; if $\mathbf{X}$ has a great-circle distribution, the *median polar axis* is the value of $\mathbf{a}$ maximising $EM(\mathbf{X}, \mathbf{a})$.

References

[1] Brown, B. M. (1983). Statistical uses of the spatial median. *J. R. Statist. Soc. Ser. B*, **45**, 25–30.

[2] Fisher, N. I. (1985). Spherical medians. *J. R. Statist. Soc. Ser. B*, **47**, 342–348.

[3] Gower, J. C. (1974). The mediancentre. *Appl. Statist.* **23**, 466–470.

(DIRECTIONAL DISTRIBUTIONS
SPATIAL MEDIAN)

N. I. FISHER

SPHERICITY, TESTS OF

Traditional univariate analyses of variance* for repeated measures* (or mixed effect) designs are extensively used in empirical research [20]. When all subjects (or sampling units) belong to the same population, the k repeated measures may be represented by the vector $\mathbf{X}$ with mean $\mu(\mathbf{X})$ and covariance matrix $\Sigma(\mathbf{X})$. A comparison of the k components of this mean vector may be carried out via a traditional F test* as long as $\mathbf{X}$ follows a nonsingular k-variate normal distribution with some restriction on the covariance matrix $\Sigma(\mathbf{X})$. Let $\mathbf{C}$ denote an appropriate orthogonal contrast matrix and let $\mathbf{Y} = \mathbf{CX}$. Then the F test is valid as long as the covariance of $\mathbf{Y}$ satisfies the condition $\Sigma(\mathbf{Y}) = \mathbf{C}\Sigma(\mathbf{X})\mathbf{C}' = \sigma^2\mathbf{I}_p$, where $\mathbf{I}_p$ is the identity matrix of order $p = k - 1$ [5, 19].

When $\Sigma(\mathbf{Y}) = \sigma^2\mathbf{I}_p$, the covariance matrix is said to satisfy the *sphericity condition*. When the vector $\mathbf{Y}$ follows a p-variate normal distribution*, sphericity is equivalent to the requirement that the p components of $\mathbf{Y}$ are independent and equally variable (e.g., having the same variance σ^2). A test for sphericity is a test of the null hypothesis $H_0 : \Sigma(\mathbf{Y}) = \sigma^2\mathbf{I}_p$ against the alternative $H_1 : \Sigma(\mathbf{Y}) \neq \sigma^2\mathbf{I}_p$. Let $\Sigma(\mathbf{Y})$ be estimated by the sample covariance matrix $\mathbf{S}$ based on a random sample of n vectors. The likelihood ratio test* statistic for H_0 against H_1 is of the form $W = |\mathbf{S}|/|\operatorname{trace} \mathbf{S}/p|^p$ [13].

The exact sampling distribution of W under H_0 was derived by Mauchly [13] for $p = 2$; by Consul [2] for $p = 3, 4, 6$; by Consul [3], Mathai and Rathie [12], and Mathai [11] in the general case; and more recently by John [9] for $p = 3$. The results of Consul are in series of G-functions; those by Mathai and by Mathai and Rathie are in series form. The exact nonnull distribution of W was first derived by Pillai and Nagarsenker [18] and later independently by Khatri and Srivastava [10]. Critical values for W are provided by Nagarsenker and Pillai [16, 17].

Approximate large-sample critical values for W may be obtained by using Box's series [1, p. 263]. From these series, $-n\rho \log W$ may be approximated by a chi-square distribution* with f degrees of freedom, where

$$\rho = 1 - (2p^2 + p - 2)/(6pn)$$

and

$$f = [p(p+1) - 1]/2.$$

The results provided in Nagarsenker and

Pillai [17] indicate that Box's series approximation is reasonably good even for small p and moderate n. In practical applications, it seems safe to use this approximation when n is at least twice as large as p.

The Mauchly criterion W has been extended to cover the case of more than one covariance matrix. In repeated measures designs involving q k-variate normal vectors $\mathbf{X}_g$, each being distributed with mean vector $\boldsymbol{\mu}_g$ and covariance matrix $\boldsymbol{\Sigma}_g$, $g = 1, \ldots, q$, the validity of the traditional F tests regarding the $\boldsymbol{\mu}_g$'s requires that the vectors $\mathbf{Y}_g = \mathbf{C}\mathbf{X}_g$ share the common covariance matrix $\sigma^2\mathbf{I}_p$ [4, 5, 15]. Here $\mathbf{C}$ is an appropriate contrast matrix and $p = k - 1$. The q-distribution version of W is the likelihood ratio (LR) test statistic for the null hypotheses

$$H_0 : \boldsymbol{\Sigma}_1(\mathbf{Y}_1) = \cdots = \boldsymbol{\Sigma}_q(\mathbf{Y}_q) = \sigma^2\mathbf{I}_p$$

against the alternative H_1 in which at least one of these equalities does not hold. As before, let the matrix $\boldsymbol{\Sigma}_g$ be estimated by the sample covariance matrix $\mathbf{S}_g$ $(= \mathbf{A}_g/N_g)$ based on a random sample of N_g vectors. The LR test is based on the ratio

$$\lambda = H \prod_{g=1}^{q} |\mathbf{A}_g|^{N_g/2} \Big/ \left\{ \operatorname{tr}(\mathbf{A}/p)^{Np/2} \right\},$$

where $H = N^{Np/2}/\prod_{g=1}^{q} N_g^{N_g p/2}$, $N = \sum_{g=1}^{q} N_g$, and $\mathbf{A} = \sum_{g=1}^{q} \mathbf{A}_g$. When the sample sizes N_g are sufficiently large, $-2\ln\lambda$ is approximately distributed as chi-square with $[qp(p+1)/2] - 1$ degrees of freedom. Mendoza [14] proposed the use of the modified LR criterion λ^* obtained by replacing each N_g in the expression of λ by the corresponding degrees of freedom $n_g = N_g - 1$. Let

$$\phi = 1 - \frac{\left[np^2(p+1)(2p+1) - 2np^2\right] \times \left(\sum_{g=1}^{q}(1/n_g) - 1\right) - 4}{6np\left[qp(p+1-2)\right]}.$$

Then $-2\phi\log\lambda^*$ is approximately distributed as a chi-square with $[qp(p+1)/2] - 1$ degrees of freedom.

Being a test for a covariance matrix, the Mauchly criterion W (and, in general, a Mendoza λ^*) may be suspected of being oversensitive to departure from normality*. This is known to occur for various tests of variance homogeneity such as the Bartlett's test* and two of its competitors, Hartley's $F_{\max}$* and the Cochran criterion. Huynh and Mandeville [8] simulate the behavior of the type I error associated with the use of W under a number of nonnormality conditions. For distributions with tails lighter than that of a normal distribution (light-tailed distributions), W errs on the conservative side in terms of type I error, i.e., from these distributions the actual type I error tends to be less than the nominal level posted for the case of normal distributions. Huynh and Mandeville explain this finding by noting that W is an increasing function of the variability in the eigenvalues* of the sample covariance matrix $\mathbf{S}$. A light-tailed distribution is usually more dense near the mean (or median) than is a normal distribution. Thus, the eigenvalues for a sample covariance matrix $\mathbf{S}$ based on a light-tailed distribution would be expected to display less variability than those derived from a covariance matrix based on a normal distribution. Thus, the critical value under a light-tailed distribution would be expected to be smaller than the one established under normality. The simulation also points out the fallibility of the Mauchly criterion in the case of heavy-tailed distributions*. Departures from normality result in substantial increases in the actual type I error beyond the nominal level posted under normality.

Since sphericity is required for the validity of traditional F tests in repeated measures design, the Mauchly criterion W may be used to assess the tenability of this condition. The simulation conducted by Huynh and Mandeville indicates that such preliminary testing should not be performed if there are reasons to suspect that the various distributions have tails heavier than a normal distribution. In any case, the use of W may not be necessary in many instances. Huynh and Feldt [7] indicate that some departure from sphericity may not substantially change the nature of the traditional F tests. In studies of approximate tests for repeated measures designs, Huynh [4] and Huynh and Feldt [6] also note that an adjustment in the degrees of freedom would be sufficient to account for most examples of departure of the covariance matrix from sphericity.

References

[1] Anderson, T. W. (1960). *An Introduction to Multivariate Statistical Analysis*. Wiley, New York.

[2] Consul, P. C. (1967). *Ann. Math. Statist.*, **38**, 1170–1174.

[3] Consul, P. C. (1969). In *Multivariate Analysis II*. Academic, New York.

[4] Huynh, H. (1978). *Psychometrika*, **43**, 161–175.

[5] Huynh, H. and Feldt, L. S. (1970). *J. Amer. Statist. Ass.*, **65**, 1582–1589.

[6] Huynh, H. and Feldt, L. S. (1976). *J. Educ. Statist.*, **1**, 69–82.

[7] Huynh, H. and Feldt, L. S. (1980). *Commun. Statist. A*, **9**, 61–74.

[8] Huynh, H. and Mandeville, G. K. (1979). *Psychol. Bull.*, **86**, 964–973.

[9] John, S. (1972). *Biometrika*, **59**, 169–173.

[10] Khatri, C. G. and Srivastava, M. S. (1971). *Sankhyā A*, **33**, 201–206.

[11] Mathai, A. M. (1970). *Applications of Generalized Special Functions*. McGill University, Montreal, Quebec, Canada.

[12] Mathai, A. M. and Rathie, P. N. (1970). *J. Statist. Res.*, **4**, 140–159.

[13] Mauchly, J. W. (1940). *Ann. Math. Statist.*, **11**, 204–209.

[14] Mendoza, J. L. (1980). *Psychometrika*, **45**, 495–498.

[15] Mendoza, J. L., Toothaker, L. E., and Cain, B. R. (1976). *J. Amer. Statist. Ass.*, **71**, 992–993.

[16] Nagarsenker, B. N. and Pillai, K. C. S. (1972). The Distribution of the Sphericity Test Criterion. *NTIS Report No. AD754-232*, Washington, DC.

[17] Nagarsenker, B. N. and Pillai, K. C. S. (1973). *J. Multivariate Anal.*, **3**, 226–235.

[18] Pillai, K. C. S. and Nagarsenker, B. N. (1971). *Ann. Math. Statist.*, **42**, 764–767.

[19] Rouanet, H. and Lepine, D. (1970). *Brit. J. Math. Statist. Psychol.*, **23**, 147–163.

[20] Winer, B. J. (1971). *Statistical Principles in Experimental Design*, 2nd ed. McGraw-Hill, New York.

(BARTLETT'S TEST OF HOMOGENEITY OF VARIANCES
HARTLEY'S F_{max} TEST
HETEROSCEDASTICITY
HOMOGENEITY AND TESTS OF HOMOGENEITY
MULTIVARIATE ANALYSIS)

HUYNH HUYNH

SPITZER'S IDENTITY

This identity is used in fluctuation and queueing theory* (see, e.g., Feller [1]) in order to study the distributions of sums of independent identically distributed random variables $X_1, X_2, \dots$.

Defining

$$S_n = \sum_{i=1}^{n} X_i \quad \text{with } S_0 = 0,$$

$$S_n^* = \max_{0 \leqslant k \leqslant n} S_k \quad \text{and} \quad S_n^+ = \max(0, S_n),$$

Spitzer's identity is [3, 4]

$$\sum_{n=0}^{\infty} z^n E\left[\exp(itS_n^*)\right] = \exp\left\{ \sum_{n=1}^{\infty} z^n E\left[\exp(itS_n^+)\right] \right\},$$

$$|z| < 1,\ t \text{ real}.$$

An elementary probabilistic proof was given by Heinrich [2].

References

[1] Feller, W. (1966). *An Introduction to Probability Theory and its Applications*, Vol. 2. Wiley, New York.

[2] Heinrich, L. (1985). *Statistics*, **16**, 249–252.

[3] Spitzer, F. (1956). *Trans. Amer. Math. Soc.*, **82**, 323–339.

[4] Spitzer, F. (1960). *Trans. Amer. Math. Soc.*, **94**, 150–169.

(RANDOM WALKS)

SPJØTVOLL–STOLINE COMPARISONS *See* MULTIPLE COMPARISONS

SPLINE FUNCTIONS *See Supplement*

SPLIT-PLOT DESIGNS *See* REPEATED MEASUREMENTS

SPORTS, SCORING SYSTEMS IN

Familiar sports such as tennis, table tennis, and volleyball comprise a sequence of contested *points*, each of which is won by one or the other of the players (or teams)—A and B, say. *Scoring systems* serve to regulate play and determine an eventual match winner; as will presently be seen, they may also be regarded as sequential* statistical tests.

Arlott [2] is a good reference for all aspects, including scoring systems, of popular and obscure sports. The historical origins of scoring systems are very much shrouded in mystery—only limited information may be found in the Arlott volume and general encyclopedias. Generally speaking, they have been regarded as "facts of life," of minor importance relative to the more tangible aspects of a sport. Changes have only been made when there were pressing reasons to do so. For example, "sudden death" tiebreaker games were introduced into tennis circa 1970 to decide sets reaching six games all. Hitherto, play had continued until one player (or doubles team) established a lead of two games. One set in a top class men's doubles match in 1967 was decided at 49–47 games!

Sporting competition between three or more players (or teams) is discussed in SPORTS, STATISTICS IN and PAIRED COMPARISONS.

The space of possible scoring systems is immense. Any sequential prescription for the next point in terms of the results of the past points (e.g., A wins, B wins, or play continues...) determines a possible scoring system; in practice it must have a simple, easily memorizable specification.

UNIPOINTS

The simplest stochastic model supposes the results of the points are independent and identically distributed, with

$$\begin{aligned}\Pr[A \text{ wins each point}] &= p = 1 - q\\ &= 1 - \Pr[B \text{ wins each point}].\end{aligned}$$

Then, if $p > \frac{1}{2} > q$, A is the *better player* (and B the poorer). For any scoring system, elementary probability (tree diagrams, conditional expectations, etc.) permits calculations or computation of the basic quantities

$$\begin{aligned}P(p) &= \Pr[A \text{ wins match}]\\ &= 1 - \Pr[B \text{ wins match}]\\ &= 1 - Q(p),\\ \mu(p) &= E[\text{point duration of match}].\end{aligned}$$

Various authors have calculated P and μ for various sports; an example with good references is Phillips [11]. As a simple example,

$$\begin{aligned}&\Pr[A \text{ wins a tennis game}]\\ &\quad = p^4(1 - 16q^4)/(p^4 - q^4),\end{aligned}$$

is a compact form due to Kemeny and Snell [5, Sec. 7.2]. Thus *the better player doesn't always win*!

A natural requirement, reflecting the notion of fairness, is that scoring systems be *symmetric* in A, B: reversing the result of every point leads to the same outcome, but with $A \leftrightarrow B$. Then

$$\begin{aligned}P(q) &= Q(p) \quad (\Rightarrow P(\tfrac{1}{2}) = \tfrac{1}{2}),\\ \mu(q) &= \mu(p).\end{aligned}$$

A further natural requirement is that $P(p)$ increases continuously from $\frac{1}{2}$ to 1 as p increases from $\frac{1}{2}$ to 1. Call such scoring systems *uniformats*. The similar scoring systems of volleyball, badminton, and squash rackets (international rules), in which a player scores a won point *only when serving*, are not uniformats, essentially because of the advantage of serving first (Watson [16]). A surprising result in this connection is due to Kingston (see Anderson [1]).

In a pioneering paper, Mosteller [9] applied the unipoints model to the 44 best-of-seven World Series baseball competitions played between 1905 and 1951 (each match equal to one point). He concluded that the American League had the better team about

75% of the time, and that the average value of p was about 0.65. In such competition, $\lambda \equiv \Pr[6 \text{ point duration} | 6 \text{ or } 7 \text{ point duration}] = 1 - 2pq$, which increases from $\frac{1}{2}$ to 1 as p increases from $\frac{1}{2}$ to 1. Groeneveld and Meeden [4] considered long term records of best-of-seven competition in (1) baseball, (2) basketball, and (3) ice hockey, finding $\hat{\lambda}_1 = 0.34$, $\hat{\lambda}_2 = 0.41$, and $\hat{\lambda}_3 = 0.57$, respectively. To account for the unduly low values of $\hat{\lambda}_1$ and $\hat{\lambda}_2$, they allowed λ to be a free parameter, and then were able to satisfactorily fit the data. Specific suggestions were offered as to why λ_1 and λ_2 might be less than $\frac{1}{2}$. Their work is based on the premise that p is constant from year to year, whereas in reality it would vary, as supposed by Mosteller. See also Gibbons et al. [3].

Regarding p as a statistical unknown, uniformats fit into the framework of sequential analysis* as follows. Consider, for $\frac{1}{2} < p' < 1$, the simple hypotheses*

$$H_0: p = p', \qquad q = q',$$

$$H_1: p = q', \qquad q = p'.$$

A *uniformat* is a sequential test of H_0 vs. H_1, with

type I error: B wins when A is the better player;

type II error: A wins when B is the better player.

In fact,

$$\alpha = \Pr[\text{type I error}] = Q(p') = P(q')$$

$$= \Pr[\text{type II error}] = \beta.$$

The symmetry of A, B and H_0, H_1, and the consequent equality of α, β, suggest the term "symmetric sequential analysis." If uniformats U_1, U_2 have $P_1(p) = P_2(p)$, then U_1 is $\mu_2(p)/\mu_1(p)$ times as efficient as U_2 at p. Wald and Wolfowitz [15] showed that the SPRT uniformats $\{W_n\}$ $(n = 1, 2, \dots)$, where W_n = (first to establish a lead of n points) is symmetrical gambler's ruin, have optimal efficiency. Hence, defining the efficiency of each W_n as 1 and extrapolating W_n to all positive n, the *efficiency* of $U(P, \mu)$ at p is

$$\rho_U(p) = \frac{(P - Q)\ln(P/Q)}{\mu(p - q)\ln(p/q)}.$$

As an example (Miles [7]), the efficiency of (first to n points) = (best of $2n - 1$ points) tends to $-p \ln(4pq)/\{(p - q)\ln(p/q)\}$ as $n \to \infty$ $(\frac{1}{2} < p < 1)$. The $\{W_n\}$ are exceptionally efficient relative to most other uniformats, but have rather high point duration variances, which may prove somewhat inconvenient in practice. However, the latter may be significantly reduced by truncation* (first to either establish a lead of n points or to score m points—see Wald, [14, Sect. 3.8]) whilst retaining high efficiency.

BIPOINTS AND BIFORMATS

Unipoints is inappropriate as a model for top class men's (and especially doubles) tennis, since typically each player (team) wins well over half of the points he (it) serves. This suggests the more general bipoints model: the results of points are mutually independent, and A [B] wins each a point [b point] with probability p_A [p_B]. Then A [B] is the better player if $p_A > p_B$ [$p_A < p_B$]. The p_A, p_B parameter space is the unit square, with diagonals $p_A = p_B$ ("equality" diagonal) and $p_A + p_B = 1$ ("unipoints" diagonal). The specification of a biformat, relative to that of a uniformat, is more complex, not only because the alternation of a and b points must be determined, but also because P, μ becomes P^a, μ^a and P^b, μ^b depending upon the type of the initial point. Beyond symmetry analogous to that of uniformats, biformats require $P^a = P^b$, $\mu^a = \mu^b$. In the scoring systems of tennis and table tennis, $P^a = P^b$ and μ^a very nearly equals μ^b.

The symmetric hypotheses

$$H_0: p_A = p'_A, \qquad p_B = p'_B,$$

$$H_1: p_A = p'_B, \qquad p_B = p'_A,$$

imply that $\alpha = \beta$ also for biformats. Wald

[14, Chap. 6] showed that biformats of the class $\{W_{2n}\}$ $(n = 1, 2, \ldots)$, applied to the tiebreaker point sequence *abbaa* ..., are very efficient. Standardizing by taking their efficiencies to be 1, the *efficiency* of a general biformat $V(P, \mu)$ at (p_A, p_B) is

$$\rho_V(p_A, p_B) = \frac{2(P - Q)\ln(P/Q)}{\mu(p_A - p_B)\ln[p_A q_B/(p_B q_A)]}.$$

An answer to the question of optimal biformat efficiency has recently been given. Pollard [12] has shown that, for $0 < p < 1$, the maximal limiting value of $\rho(p + t, p - t)$ as $t \to 0$ is 1; and that biformats with this property form a class $\{W_{i,j}\}$, for each of which "winner" and "duration" are independent, with

$$P_{i,j}/Q_{i,j} = (p_A/p_B)^i (q_B/q_A)^j \quad (i, j = 1, 2, \ldots).$$

Miles subsequently derived the corresponding duration probability generating functions

$$g_{i,j}(s) = \frac{(p_A^i q_B^j + p_B^i q_A^j)s^{i+j}}{\{1 - h_+(s)\}^i h_-(s)^j + \{1 - h_-(s)\}^i h_+(s)^j}$$

where

$$h_\pm(s) = \tfrac{1}{2}\Big[1 + (q_A q_B - p_A p_B)s^2 \pm \{1 - 2(q_A q_B + p_A p_B)s^2(q_A q_B - p_A p_B)^2 s^4\}^{1/2}\Big].$$

The case $i = j = n$ corresponds to Wald's biformats and, roughly speaking, the cases $i = 1$ and $j = 1$ correspond to W_n applied to "play the loser" and "play the winner" point sequences (Sobel and Weiss [13], Miles [7]). Among $\{W_{i,j}\}$, maximal efficiencies (off the equality and unipoints diagonals) only marginally exceed 1 for most (p_A, p_B) values (Miles [7, Table 3]).

Taking $p_A = 0.7$ and $p_B = 0.6$ as typical values in top men's tennis, without tiebreakers, $\rho = 0.55$ (three sets) and 0.51 (five sets). This is less efficient than the corresponding women's play, where for typical values $p_A = 0.6$ and $p_B = 0.5$, $\rho = 0.67$ (three sets). The following describe changes relative to these as base values. The effect of using tiebreakers has been to *reduce* these efficiencies by about 8% (men) and 4% (women). On the other hand, by the entropy*-increasing device of starting every game at the score 0–15, the men's efficiencies would each be increased by 11%, or with tiebreakers 7%. The gains in starting every game at 0–30 would be even greater: 15% and 12%, respectively! The corresponding percentages for women's play are 2, −2; −2 and −5 (Miles [6, 7]).

IMPORTANCE OF POINTS

The importance of a given point π in a match has been formalized by Morris [8] as

$$\begin{aligned} I(\pi) &= \Pr[X \text{ wins match} | X \text{ wins } \pi] \\ &\quad - \Pr[X \text{ wins match} | X \text{ loses } \pi], \\ X &= A \text{ or } B. \end{aligned}$$

He showed that it is multiplicative under nesting; thus in tennis

$$\begin{aligned} I(\text{point/match}) &= I(\text{point/game}) \\ &\quad \times I(\text{game/set}) \\ &\quad \times I(\text{set/match}). \end{aligned}$$

Actually, efficiency is also multiplicative under nesting. He also investigated the increased probability of winning a match for a player who is able to raise his game on more important points.

OTHER SPORTS

The realm of scoring systems extends further, for example to fixed time period competition, as opposed to the point sequence competition so far considered. Mosteller [10] investigated the aggregate of individual match scores in U.S. collegiate football for the 1967 and 1968 seasons, with fascinating results.

References

[1] Anderson, C. L. (1977). Note on the advantage of first serve. *J. Combin. Theory A*, **23**, 363.

[2] Arlott, J., ed. (1975). *The Oxford Companion to Sports and Games*. Oxford University Press, London, England.

[3] Gibbons, J. D., Olkin, I., and Sobel, M. (1978). Baseball competitions—are enough games played? *Amer. Statist.*, **32**, 89–95.

[4] Groeneveld, R. A. and Meeden, G. (1975). Seven game series in sports. *Math. Mag.*, **48**, 187–192.

[5] Kemeny, J. G. and Snell, J. L. (1960). *Finite Markov Chains*. Van Nostrand, Princeton, NJ.

[6] Miles, R. E. (1982). Why tennis scoring should be changed. *Tennis Australia–Asia–The Pacific*, **April**, pp. 54–55.

[7] Miles, R. E. (1984). Symmetric sequential analysis: the efficiencies of sports scoring systems (with particular reference to those of tennis). *J. R. Statist. Soc. Ser. B*, **46**, 93–108.

[8] Morris, C. (1977). The most important points in tennis. In *Studies in Management Science and Systems*, Vol. 5, S. P. Ladany and R. E. Machol, eds. North-Holland, Amsterdam, The Netherlands, pp. 131–140.

[9] Mosteller, F. (1952). The World Series competition. *J. Amer. Statist. Ass.*, **47**, 355–380.

[10] Mosteller, F. (1970). Collegiate football scores, U.S.A. *J. Amer. Statist. Ass.*, **65**, 35–48.

[11] Phillips, M. J. (1978). Sums of random variables having the modified geometric distribution with application to two-person games. *Adv. Appl. Prob.*, **10**, 647–665.

[12] Pollard, G. H. (1986). A stochastic analysis of scoring systems. Ph.D. thesis, Australian National University, Canberra.

[13] Sobel, M. and Weiss, G. H. (1970). Play-the-winner sampling for selecting the better of two binomial populations. *Biometrika*, **57**, 357–365.

[14] Wald, A. (1947). *Sequential Analysis*. Wiley, New York.

[15] Wald, A. and Wolfowitz, J. (1948). Optimum character of the sequential probability ratio test. *Ann. Math. Statist.*, **19**, 326–339.

[16] Watson, D. (1970). On scoring in games. *Math. Gazette*, **54**, 110–113.

(KNOCK-OUT TOURNAMENTS
SPORTS, STATISTICS IN)

R. E. MILES

SPORTS, STATISTICS IN

Theoretical and quantitative studies of sports outside the sports pages of newspapers and magazines are scattered throughout the literature (but see refs. 18 and 19); they involve game theory*, mathematical modeling, operations research*, and combinatorics* in addition to statistics and applied probability. The main topics of interest are tournaments, strategy of play, and scoring systems (*see* SPORT, SCORING SYSTEMS IN).

TOURNAMENTS

The principal types of tournaments are round robin (RR) and knock-out* (KO). In the former, each team in a league plays every other team the same number of times. In the latter, in each round only the winner of each game goes on to the next round, and teams are knocked out until only two teams remain. Thus a KO has $n-1$ and a RR, minimally, has $n(n-1)/2$ games. For a meaningful analysis of tournaments, assumptions are required about the preference system for the set of teams or players $x_1, \ldots, x_n$. If $p_{ij} = \Pr(x_i \text{ wins against } x_j)$, *stochastic transitivity* (ST) holds if for every three players x_i, x_j, and x_k,

$$p_{ij} \geqslant \tfrac{1}{2},\ p_{jk} \geqslant \tfrac{1}{2} \rightarrow p_{ik} \geqslant \tfrac{1}{2}.$$

Stong stochastic transitivity (SST) is more stringent:

$$p_{ij} \geqslant \tfrac{1}{2},\ p_{jk} \geqslant \tfrac{1}{2} \rightarrow p_{ik} \geqslant \max(p_{ij}, p_{jk}).$$

Much of the research makes the latter assumption, but either assumption leads to a clear-cut ranking of all n players. Other rankings are possible, such as that determined by the average probabilities

$$p_i = (n-1)^{-1} \sum_{j \neq i} p_{ij}, \qquad i = 1, \ldots, n.$$

See David [8, pp. 12–13] for further discussion.

Round robins can be conveniently studied as balanced designs for paired comparisons*; see David [8, Chaps. 5 and 6] and Gibbons et al. [10, Chap. 8]. If each team has an effect on its opponents that may carry over to the next match, a changeover design* is appropriate [24]. Tournaments as a whole and RRs in particular can be studied from a graph-theoretic approach; see the book by Moon [21], which investigates structural and combinatorial properties.

A *deterministic tournament* (DT) is one for which $p_{ij} = 1$ or 0, for all $i \neq j$. Although they rarely occur in the real world, study of DTs throws light on the reliability of tournaments. In his essay "Lawn Tennis Tournaments," Lewis Carroll [4; 8, p. 85] had noticed how unreliable deterministic KO tournaments might be in identifying the second-best player, although they correctly identify the best. In fact, the second best player may then have only slightly more than an even chance of being runner-up [25]. In order to correctly identify the second-best player in a DT, Carroll proposed that no player be eliminated until two others are shown to be better. In a *double elimination tournament* players are not eliminated until they have lost two games [8, p. 84; 9]. "A swifter, if less sporting way, would be to arrange a Knock-out tournament between all the players defeated by the champion" [8, p. 85]. Katz [16] developed a sequential procedure for a DT in which the best m out of n players are to be ranked and in which the number of games is to be kept, in a sense, to a minimum.

Since DTs are the exception and not the rule in the real world, some research has focused on the *effectiveness* of tournament designs in selecting the "best" of n players; David [8] measured effectiveness by Pr("best" player wins), using a suitable model. See Maurer [20] for further discussion, some results, and other references. Hwang [14] shows that traditional methods of seeding are not necessarily effective, and suggests a seeding strategy for SST schemes.

MISCELLANEOUS

The occurrence of infrequent events involving groups of players in a game sometimes gives rise to the negative binomial distribution* [22]. This does not seem to happen if the sole performances of individual players are involved. For example, the distribution describes the number of pass-moves in soccer, seasonal collegiate football scores, runs per half-inning in baseball, goals per match by National Hockey League teams, and the combined scores of particular batsmen-pairs in cricket.

Scoring systems are discussed in SPORT, SCORING SYSTEMS IN. In tournaments the aim is to assign "optimum" scores that rank teams in order of ability. Daniels [7] introduces the concept of fairness, e.g., in resolving ties when total scores in RRs are used to rank teams, and he proposes some fair scoring methods. On the other hand, the national champion collegiate football team in the United States is decided annually by opinion polls. This method, which is not objective, arises partly because the teams play an incomplete tournament. Jech [15] develops an objective method of ranking teams when there are enough results to make meaningful comparisons. The issue of fairness arises in games incorporating an inherent advantage to one contestant, such as service in tennis. Kingston [17] shows that the probability of winning in such competitions is independent of whether the two players alternate this advantage or whether the winner of part of the contest goes on to have the same advantage in the next part.

The aim of the drafting system by professional sports associations is to balance team strengths. A number of drafting schemes are discussed from a statistical and game-theoretic standpoint by Brams and Straffin [2] and by Price and Rao [23]. Pricing strategies are discussed by Heilman and Wendling [13].

A number of statistical studies have focused on particular sports. Two collections of articles worth noting are *Optimal Strategies in Sports* (OSSp) [18] and *Management Science in Sports* (MSSp) [19]. The former is less technical than most references, and deals largely with baseball and football, although other sports are also featured. Individual sports are also the subject of articles in the second, more theoretical collection. OSSp also contains a bibliography of statistics in sports.

BASEBALL

In OSSp there are discussions of strategy (also in MSSp), batting performance, valuation of players, numbers of games, adjusting standings for team strength, baseball as a game of chance* and as a Markov process*, distribution of runs, and of batting order. Gibbons et al. [11] investigated the number of games played in a World Series, and in each league before 1969 (a RR with 18 games per pair of teams). They concluded that, in order to have a reasonable level of confidence that the best team wins, seven games is highly inadequate for the World Series and the total of 162 games per league is barely sufficient. See Bennett and Flueck [1] for an evaluation of major league baseball offensive performance models.

AMERICAN FOOTBALL

In OSSp, analyses of field position, collegiate football scores and Ivy League game scores appear, and also a discussion of extra-point strategy. In MSSp a method of ranking teams is applied to football, the ability of pro football handicappers is found to predict a median value very well, and professional football scores are analyzed. Harville [12] used mixed linear models to predict National Football League (NFL) outcomes, with an average absolute error of 10.68 for 1320 games, as compared with 10.49 for bookmakers. In 1970, the NFL merged with the American Football League, along with a statistically significant drop of nearly six in the average number of points scored per game [3].

OTHER SPORTS

In OSSp performance evaluation in professional basketball is discussed, as are playoff structures in the National Hockey League, cricket, the most important points in tennis, the handicap system in golf, the long jump (also in MSSp), assignments of swimmers and runners to relay teams, arrangements in rowing eights, and competitive running. In MSSp, basketball scheduling, optimal weightlifting, serving strategies in tennis, goalie strategy in ice hockey, and scoring in cross-country competitions are featured. Croucher has analyzed the finals of the Wimbledon tennis championships [6] and Rugby League Football scores [5].

References

[1] Bennett, J. M. and Flueck, J. A. (1983). *Amer. Statist.*, **37**, 76–82.

[2] Brams, S. J. and Straffin, P. D., Jr. (1979). *Amer. Math. Monthly*, **86**, 80–88.

[3] Canavos, G. C. and Koutrouvelis, I. A. (1979). *Proc. Amer. Statist. Ass., Bus. Econ. Statist. Sec.*, pp. 407–410.

[4] Carroll, L. (1936). In *The Complete Works of Lewis Carroll*. Modern Library, New York, pp. 1201–1211.

[5] Croucher, J. S. (1979). In *Interactive Statistics*, D. R. McNeil, ed. North-Holland, Amsterdam, The Netherlands, pp. 137–145.

[6] Croucher, J. S. (1981). *Teaching Statist.*, **3**, 72–75.

[7] Daniels, H. E. (1969). *Biometrika*, **56**, 295–299.

[8] David, H. A. (1963). *The Method of Paired Comparisons*. Griffin, London, England.

[9] Fox, M. (1973). *Amer. Statist.*, **27**, 90–91.

[10] Gibbons, J. D., Olkin, I., and Sobel, M. (1977). *Selecting and Ordering Populations: A New Statistical Methodology*. Wiley, New York.

[11] Gibbons, J. D., Olkin, I., and Sobel, M. (1978). *Amer. Statist.*, **32**, 89–95.

[12] Harville, D. (1980). *J. Amer. Statist. Ass.*, **75**, 516–524.

[13] Heilman, R. L. and Wendling, W. R. (1976). In *Management Science in Sports*, R. E. Machol, S. P. Ladany, and D. G. Morrison, eds. North-Holland, Amsterdam, The Netherlands, pp. 91–100.

[14] Hwang, F. K. (1982). *Amer. Math. Monthly*, **89**, 235–239.

[15] Jech, T. (1983). *Amer. Math. Monthly*, **90**, 246–266.

[16] Katz, L. (1977). *J. Amer. Statist. Ass.*, **72**, 841–844.

[17] Kingston, J. G. (1976). *J. Comb. Theory*, **20**, 357–362.

[18] Ladany, S. P. and Machol, R. E., eds. (1977). *Optimal Strategies in Sports*. North-Holland, Amsterdam, The Netherlands.

[19] Machol, R. E., Ladany, S. P., and Morrison, D. G., eds. (1976). *Management Science in Sports*. North-Holland, Amsterdam, The Netherlands.

[20] Maurer, W. (1975). *Amer. Statist.*, **3**, 717–727.

[21] Moon, J. W. (1968). *Topics on Tournaments*. Holt, Rinehart, and Winston, New York.

[22] Pollard, R., Benjamin, B., and Reep, C. (1977). In *Optimal Strategies in Sports*, S. P. Ladany and R. E. Machol, eds. North-Holland, Amsterdam, The Netherlands, pp. 188–195.

[23] Price, B. and Rao, A. G. (1976). In *Management Science in Sports*, R. E. Machol, S. P. Ladany, and D. G. Morrison, eds. North-Holland, Amsterdam, the Netherlands, pp. 79–90.

[24] Russell, K. G. (1980). *Biometrika*, **67**, 127–131.

[25] Wiorkowski, J. J. (1972). *Amer. Statist.*, **26**(3), 28–30.

Bibliography

Hooke, R. (1972). In *Statistics: A Guide to the Unknown*, J. M. Tanur et al., eds. Holden-Day, San Francisco, CA., pp. 244–252.

(GAME THEORY
KNOCK-OUT TOURNAMENTS
PAIRED COMPARISONS
RANDOM GRAPHS
RANKING AND SELECTION
SPORT, SCORING SYSTEMS IN)

CAMPBELL B. READ

S.P.R.T. *See* SEQUENTIAL ANALYSIS

SPURIOUS CORRELATION

The term *spurious correlation*, as used by Pearson [14], referred to the case where two ratios containing a common component were found to have a nonzero correlation even though the variables making up the ratios were random. However, others such as Simon [16] used the term spurious correlation to refer to the more general problem of a misleading correlation between two variables resulting from some third causal factor. Kendall and Buckland [8] called this an *illusory correlation*

Pearson's original formulation is largely of historical import. His approximate formula for the correlation between ratios derived from random variables (w, x, y, and z) is

$$r_{(y/z)(x/w)} = \frac{r_{yx}V_yV_x - r_{yw}V_yV_w - r_{xz}V_xV_z + r_{zw}V_zV_w}{\sqrt{V_y^2 + V_z^2 - 2r_{yz}V_yV_z} \times \sqrt{V_x^2 + V_w^2 - 2r_{xw}V_xV_w}},$$

where V is the coefficient of variation* (standard deviation/mean).

Though numerous researchers have tried to use his approximate formula for determining a spurious correlation when using ratios, his assumptions have been shown to be faulty due to his ignoring the higher-order terms in his derivation. The higher-order terms are quite important (Kunreuther [10] and Kuh and Meyer [9]). An interesting historical note is Pearson's own work (Pearson et al. [15]), where he attempted to use his spurious correlation formula and concluded that cancer and diabetes were causally related. Neyman [13] attempted to combine Pearson's treatment of spurious correlation with the more common concept of a third misleading causal variable.

To answer the question of just how important spurious correlation is requires a careful concern for model specification. Researchers (Bollen and Ward [3], Kasarda and Nolan [7], and MacMillan and Daft [12]) have agreed that correlating theoretically meaningful ratio variables should not be considered spurious. Indeed Snedecor [17] noted that the term "spurious" was rather derogatory and could needlessly lead researchers astray. Logan [11], however, strongly questions the correlation of ratios, warning that "artifactual" or spurious correlation may be present, and recommends using partial correlations*.

Reformulating the spurious correlation argument to one of a specification problem provides a different and clearer perspective. Aitchison [1] offers a comprehensive treatment of relating ratios giving numerous examples, though only one is from the social sciences (household expenditure budgets). He presents specific distributions (transformed normal) for relating ratios. Viewing spurious correlations of ratios as a specification problem, Kuh and Meyer [9] note that a common denominator among ratios could be considered weighted least squares with the inverse denominator variable being the weight. Belsley [2] shows that relating ratio variables can be considered a special case of generalized least squares* with the common denominator being a correction for heteroscedastic disturbances (see Johnston [6, Chap. 7, especially pp. 214–221]). Thus, when relating ratios in a properly specified model, and provided the other assumptions of least-squares analysis are met, linear unbiased coefficients are attained.

While the regression coefficients are unbiased measures of relationship, measures of association* and correlation are much more difficult to interpret. An important analogy here is the work done with multiple correlations* in using generalized least squares (Buse [4]). Additionally, ratios are directly analogous to regression through the origin (no constant term), where there exists difficulty in finding a meaningful measure of correlation (Casella [5]). Ultimately, spurious correlation is a problem of properly specifying the model on the one hand, and on the other, finding the right interpretation for the correlation coefficient.

References

[1] Aitchison, J. (1982). *J. R. Statist. Soc. Ser. B*, **44**, 139–177. (Highly recommended to those with a statistical background who wish to build models using proportions of some whole.)

[2] Belsley, D. A. (1972). *Econometrica*, **40**, 923–927. (A thorough treatment of spurious correlation as a specification problem.)

[3] Bollen, K. A.and Ward, S. (1979). *Sociol. Meth. Res.*, **7**, 431–450.

[4] Buse, A. (1973). *Amer. Statist.*, **27**, 106–108.

[5] Casella, G. (1983). *Amer. Statist.*, **37**, 147–152.

[6] Johnston, J. (1972). *Econometric Methods*, 2nd ed. McGraw-Hill, New York. (Clearly presents generalized least squares and discussed its role when using ratio variables.)

[7] Kasarda, J. D. and Nolan, P. D. (1979). *Social Forces*, **58**, 212–227.

[8] Kendall, M. G. and Buckland, W. R. (1982). *A Dictionary of Statistical Terms*, 4th ed. Longman Group Ltd., New York.

[9] Kuh, E. and Meyer, J. R. (1955). *Econometrica*, **23**, 400–416. (A comprehensive review of Pearson's 1897 work, showing it to be of only historical interest.)

[10] Kunreuther, H. (1966). *Econometrica*, **34**, 232–234.

[11] Logan, C. H. (1982). *Social Forces*, **60**, 791–810. (Definitive statement of those who hold that spurious correlation remains a substantive issue.)

[12] MacMillan, A. and Daft, R. L. (1980). *Social Forces*, **58**, 1109–1128.

[13] Neyman, J. (1952). *Lectures and Conferences on Mathematical Statistics and Probability*, 2nd ed. U.S. Department of Agriculture, Washington, DC. (Neyman's treatment of spurious correlation remains fresh and insightful decades later, filled with irony and perception.)

[14] Pearson, K. (1897). *Proc. R. Soc. Lond.* **60**, 489–497. (Sadly, this effort is of only historical interest, the formulas being useless for real data sets.)

[15] Pearson, K., Lee, A., and Elderton, E. M. (1910). *J. R. Statist. Soc.*, **73**, 534–539. (Perhaps a lesson to all who use Pearson's spurious correlation formulas and come up with wrong results.)

[16] Simon, H. A. (1954). *J. Amer. Statist. Ass.*, **49**, 467–479. (Remains a valid statement of how two variables can have a spurious correlation due to some third causal factor working upon both variables.)

[17] Snedecor, G. W. (1946). *Statistical Methods*, 4th ed. Iowa State College Press, Ames, IA.

(CORRELATION
MULTIPLE CORRELATION
PARTIAL CORRELATION)

JAMES E. PRATHER

SQUEEZE SIGNIFICANCE TESTS

These are tests proposed by Burch and Parsons [1] for hypotheses specifying the form of distribution function (CDF) of a random variable as $F((y - \eta)/\theta)$, where η and θ are unknown, but $F(\cdot)$ is known.

On probability paper corresponding to the CDF $F(\cdot)$, two plots are made, one joining points with coordinates $(y_r, r/(n+1))$ and the other joining points with coordinates $(y_r, (r - \frac{1}{2})/n)$, where $y_1 \leqslant \cdots \leqslant y_n$ are the order statistics from a random sample of size n. The two closest parallel straight lines containing all these points are drawn. The vertical distance between these two lines is the test statistic.

Distributional results so far available are based on Monte Carlo (simulation*) studies.

Reference

[1] Burch, C. R. and Parsons, I. T. (1976). *Appl. Statist.*, **25**, 287–291.

(EDF STATISTICS
KOLMOGOROV–SMIRNOV STATISTICS)

SSR$_2$ (STRICTLY SIGN REVERSE REGULAR FUNCTIONS OF ORDER 2)

A function $p(x, y)$ is *sign reverse regular* of order $k(\text{SRR}_k)$ if for every $x_1 < x_2 < \cdots < x_m$, $y_1 < y_2 < \cdots < y_m$, and $1 < m < k$ [13, 31],

$$(-1)^{[m(m-1)]/2} p\begin{pmatrix} x_1, x_2, \ldots, x_m \\ y_1, y_2, \ldots, y_m \end{pmatrix} \geqslant 0, \quad (1)$$

where

$$p\begin{pmatrix} x_1, x_2, \ldots, x_m \\ y_1, y_2, \ldots, y_m \end{pmatrix} = \begin{vmatrix} p(x_1, y_1) & \cdots & p(x_1, y_m) \\ \vdots & & \vdots \\ p(x_m, y_1) & \cdots & p(x_m, y_m) \end{vmatrix}. \quad (2)$$

Here $x_i \in X$ and $y_i \in Y$, where X and Y are linearly ordered one-dimensional sets, typically representing intervals of the real line or countable sets of discrete values on the real line, such as the set of all integers. If strict inequality holds in (1), the function $p(x, y)$ is said to be *strictly sign reverse regular* of order k (SSR_k) [17]. These functions are closely related to totally positive (TP) functions (*see* TOTAL POSITIVITY). For example, a function $p(x, y)$ is SRR_k if $p(x, -y)$ is TP_k in $x \in X$ and $y \in Y$.

A relationship between a SRR_2 function and a Pólya type II frequency* function (PF_2) is that if $f(x)$ is a PF_2, then $f(x+y)$ is a SRR_2 for $x > 0$ and $y > 0$.

Let $\mu_s = \int x^s f(x)\,dx$ be the sth moment of a density $f(x)$ for $s > -1$ and let

$$r_s = \mu_s / \Gamma(s+1)$$

be called the sth *normalized moment*. If f is a PF_k density of a nonnegative random variable, such as the life length of a unit, with normalized moment r_s $(s > -1)$, then [17] r_{s+t} is SSR_k in $s > -\frac{1}{2}$ and $t > -\frac{1}{2}$.

A similar result holds for random variables on the nonnegative integers with the binomial* moments $B_m = \Sigma p_k \binom{k}{m}$. If $\{p_k\}$ is a PF_k sequence, then B_{m+n} is SSR_k in $m = 0, 1, 2, \ldots$ and $n = 0, 1, 2, \ldots$.

The function $h(u) = u^r$ if $u > 0$ and 0 otherwise, generates a PF_∞ function whenever r is a positive integer. This follows from Schoenberg's [30] fundamental representation theorem concerning the Laplace transform of Pólya frequency functions. Let $f(t)$ be a PF_2 density, which is continuous throughout, except for at most a single discontinuity of the first kind. Define

$$g(x) = \int f(u) h(u - x)\,du, \qquad r = 1.$$

Then $g(x+y)$ is SRR_2. The following identity (Pólya and Szegö [26, p. 48, Prob. 68]) is very much used to establish several moment inequalities involving SRR_k functions. If $r(x, w) = \int p(x, t) q(t, w)\,d\sigma(t)$, then

$$r\begin{pmatrix} x_1, x_2, \ldots, x_k \\ w_1, w_2, \ldots, w_k \end{pmatrix} = \iint_{t_1 < t_2 < \cdots < t_k} \cdots \int p\begin{pmatrix} x_1, x_2, \ldots, x_k \\ t_1, t_2, \ldots, t_k \end{pmatrix} \times q\begin{pmatrix} t_1, t_2, \ldots, t_k \\ w_1, w_2, \ldots, w_k \end{pmatrix} \times d\sigma(t_1)\,d\sigma(t_2) \ldots d\sigma(t_k)$$

in the notation of (2). Karlin and Proschan

[18] proved that $f(0) \leqslant 1/\mu_1$, where $\mu_1 = \int_0^\infty t f(t)\, dt$, with equality if and only if $f(t) = (1/\mu_1)e^{-t/\mu_1}$, $t \geqslant 0$. With $r = 2$ in the definition of $h(u)$, the function $g(x + y)$ becomes SRR$_3$, provided f is PF$_3$. Then $f'(0) \leqslant 2/(2\mu_1^2 - \mu_2)$, provided $f(0) = 0$, where $\mu_i = \int_0^\infty t^i f(t)\, dt$; also $2\mu_1^2 > \mu_2$. Several other moment inequalities are proved in refs. 14–16.

Schoenberg [30] has made an extensive study of the variation diminishing property (VDP) of TP$_k$ kernels, where the variables x and y occur in translation form. Karlin [13, 16, 19] extended their applications in characterizing best statistical procedures for a variety of decision problems. More recently, Brown et al. [5] adopt a more direct approach and take the variation reducing (VR) property as basic, and give criteria for a family of distributions to have the VR. Their method emphasizes the importance of the VR property to statistics. Karlin's [13, Sec. 5.3.1] theorem asserts the equivalence of STP$_2$ and SVR$_2$, which is equivalent to the strict monotone likelihood ratio* (MLR). More generally, Brown et al. [5] have recently proved the equivalence of STP$_r$ and SVR$_r$. These results are, of course, related to SSR functions.

The SVR$_r$ property has been applied in many statistical contexts since Karlin's work. Examples are the study of the combination of independent one-sided test statistics (van Zwet and Oosterhoff [32]), scale families of symmetric and one-sided stable densities (Kanter [12]), the structure of optional screening and classification* procedures (Marshall and Olkin [24]), qualitative properties of the power function of the F test* (Farrell [8]), inequalities on the multinormal* integral useful in optimal design* (Rinott and Santner [28]), comparison of large deviation rates for differing distribution functions (Lynch [23]), and theory of hypothesis tests* (Cohen [6] and Meeden [25]).

Karlin and Rinott [19] surveyed and obtained total positivity properties of absolute value multinormal variables with applications to simultaneous confidence interval* estimates and related multivariate probabilistic inequalities. Many of these results are directly related to SSR$_k$ functions. These results use the concept of (multivariate) MTP$_2$ functions and extend the results of Kemperman [21], Rinott [27], and Lorentz [22]. The theorems of Sarkar [29], Barlow and Proschan [4], and Fortuin et al. [9] on the multivariate Tchebycheff rearrangement inequality are generalized. The connections between MTP$_2$ and other notions of positive quadrant dependence [1, 7, 10, 31] are extended.

Let $\mathbf{X} \sim N(\mathbf{0}, \boldsymbol{\Sigma})$; then the principal theorem of Karlin and Rinott [19] is that a necessary and sufficient condition for the density of $(|x_1|, \ldots, |x_n|)$ to be MTP$_2$ is that there exists a diagonal matrix $\mathbf{D}$ with elements ± 1 such that the off-diagonal elements of $-\mathbf{D}\boldsymbol{\Sigma}^{-1}\mathbf{D}$ are all nonnegative. It was applied to obtain positive dependence and associated inequalities for the multinormal, multivariate t*, and Wishart distributions*. These inequalities yield bounds for multivariate confidence set probabilities.

References

[1] Abdel-Hameed, M. and Sampson, A. R. (1978). *Ann. Statist.*, **6**, 1360–1368.

[2] Barlow, R. E., Marshall, A. W., and Proschan, F. (1963). *Ann. Math. Statist.*, **34**, 375–389.

[3] Barlow, R. E. and Proschan, F. (1965). *Mathematical Theory of Reliability*. Wiley, New York.

[4] Barlow, R. E. and Proschan, F. (1975). *Statistical Analysis of Reliability and Life Testing*. Holt, Rinehart, and Winston, New York.

[5] Brown, L. J., Johnstone, I. M., and MacGibbon, K. B. (1981). *J. Amer. Statist. Ass.*, **76**, 824–832.

[6] Cohen, A. (1965). *Ann. Math. Statist.*, **36**, 1185–1206.

[7] Esary, J. D., Proschan, F., and Walkup, D. W. (1967). *Ann. Math. Statist.*, **38**, 1466–1474.

[8] Farrell, R. H. (1968). *Ann. Math. Statist.*, **39**, 1978–1994.

[9] Fortuin, C. M., Kastelyn, P. W., and Ginibre, J. (1971). *Commun. Math. Phys.*, **22**, 89–103.

[10] Jogdeo, K. (1977). *Ann. Statist.*, **5**, 495–504.

[11] Johnson, N. L. and Kotz, S. (1970). *Continuous Univariate Distributions*, Vol. 1. Wiley, New York.

[12] Kanter, M. (1975). *Ann. Prob.*, **3**, 697–707.

[13] Karlin, S. (1968). *Total Positivity*, Vol. I. Stanford University Press, Stanford, CA.

[14] Karlin, S., Proschan, F., and Barlow, R. E. (1961). *Pacific J. Math.*, **11**, 1023–1033.

[15] Karlin, S. (1956). *Proc. Third Berkeley Symp. Math. Statist. Prob.*, Vol. 1. University of California Press, Berkeley CA, pp. 115–128.

[16] Karlin, S. (1957). *Ann. Math. Statist.*, **28**, 281–308.

[17] Karlin, S. (1964). *Trans. Amer. Math. Soc.*, **111**, 33–107.

[18] Karlin, S. and Proschan, F. (1960). *Ann. Math. Statist.*, **31**, 721–736.

[19] Karlin, S. and Rinott, Y. (1981). *Ann. Statist.*, **9**, 1035–1049.

[20] Karlin, S. and Rubin, H. (1956). *Ann. Math. Statist.*, **27**, 272–299.

[21] Kemperman, J. H. B. (1977). *Indag. Math*, **39**, 313–331.

[22] Lorentz, G. G. (1953). *Amer. Math. Monthly*, **60**, 176–179.

[23] Lynch, J. (1979). *Ann. Prob.*, **7**, 96–108.

[24] Marshall, A. W. and Olkin, I. (1968). *J. R. Statist. Soc. Ser. B*, **30**, 407–435.

[25] Meeden, G. (1971). *Ann. Math. Statist.*, **42**, 1452–1454.

[26] Pólya, G. and Szegö, G. (1925). *Aufgaben und Lehrsatze aus der Analysis*, Vol. I. Springer, Berlin, Germany.

[27] Rinott, Y. (1973). *Israel J. Math.*, **15**, 60–77.

[28] Rinott, Y. and Santner, T. T. (1977). *Ann. Statist.* **5**, 1228–1234.

[29] Sarkar, T. K. (1969). *Tech. Report No. 124*, Department of Operations Research and Statistics, Stanford University, Stanford, CA.

[30] Schoenberg, I. J. (1951). *J. d'Analyse Math., Jerusalem*, **1**, 33.

[31] Shaked, M. (1975). Ph.D. dissertation, University of Rochester, Rochester, NY.

[32] van Zwet, W. R. and Osterhoff (1967). *Ann. Math. Statist.*, **38**, 659–680.

(PÓLYA TYPE 2 FREQUENCY
DISTRIBUTIONS
STP_2
TOTAL POSITIVITY)

B. RAJA RAO

SSQCOR *See* GENERALIZED CANONICAL VARIABLES

STABLE DISTRIBUTIONS

Let $X, X_1, \ldots$ be independent identically distributed random variables. The distribution of X is *stable* (in the broad sense) if it is not concentrated at one point and if for each n there exist constants $a_n > 0$ and b_n such that

$$\frac{X_1 + X_2 + \cdots + X_n}{a_n} - b_n$$

has the same distribution as X. The random variable X is *stable* if it has a stable distribution. If the above holds with $b_n = 0$ for all n then the distribution of X is *stable in the strict sense*. For every stable distribution we have $a_n = n^{1/\alpha}$ for some *characteristic exponent* α with $0 < \alpha \leqslant 2$. For example, the family of Gaussian distributions is the unique family of distributions that are stable with $\alpha = 2$. The Cauchy distributions* are stable with $\alpha = 1$.

Cauchy studied the symmetric stable distributions around 1850. The general stable distributions were discovered by Lévy in 1919 when someone made the erroneous claim that the Gaussian distributions were the only stable distributions. Lévy showed that the characteristic function* of a stable distribution has the canonical representation

$$\phi(t) = \exp(\psi(t)), \qquad -\infty < t < \infty,$$

where

$$\psi(t) = i\mu t - c|t|^{\alpha}\{1 + i\beta(t/|t|)\omega(t, \alpha)\}.$$

Here μ is a centering constant, c is a positive scaling constant, $|\beta| \leqslant 1$, and

$$\omega(t, \alpha) = \begin{cases} \tan(\pi\alpha/2), & \alpha \neq 1, \\ (2/\pi)\log|t|, & \alpha = 1. \end{cases}$$

Thus, ignoring scaling and centering constants, the stable distributions form a two-parameter family indexed by $-1 \leqslant \beta \leqslant 1$, $0 < \alpha \leqslant 2$; β is called the *skew parameter*. It follows from the characteristic function representation that a stable distribution is symmetric if and only if $\beta = 0$ and $\mu = 0$. If X is Gaussian with mean μ, the distribution of $X - \mu$ is strictly stable. Similarly, if X is stable with exponent $\alpha \neq 1$ and the above characteristic function, then $X - \mu$ is strictly stable. (In the case $\alpha = 1$, X is strictly stable if and only if $\beta = 0$.)

All stable random variables are of continuous type. In most cases ($\alpha = \frac{1}{2}, 1, 2$ provide exceptions) it appears impossible to express the densities of stable laws in closed form. However, series expansions for the densities do exist (see refs. 15 and 9).

If a stable distribution is not completely asymmetric (if $|\beta| < 1$), then its density $p(x)$ has two "fat tails"; As $x \to \infty$,

$$p(x) \sim Ax^{-(\alpha+1)}, \qquad p(-x) \sim Bx^{-(\alpha+1)},$$

for some positive constants A and B. It is these fat tails that account for the "outliers"* in data following a stable distribution.

If $|\beta| = 1$, then the density $p(x)$ has only one fat tail. The other tail vanishes if $\alpha < 1$ and goes to zero faster than the tails of the Gaussian densities if $1 \leqslant \alpha < 2$.

We encounter the stable laws as a solution to an important subproblem of the central limit problem. Let $X_1, X_2, \ldots$ be independent identically distributed (i.i.d.) random variables. If there exist $a_n > 0$ and b_n such that

$$\frac{X_1 + X_2 + \cdots + X_n}{a_n} - b_n$$

converges in distribution* to a nondegenerate random variable X, then X_1 is in the *domain of attraction* of X. Determining what random variables are attracted to what limiting random variables and what choices of a_n, b_n work constitutes the domain of attraction* problem. The importance of the stable distributions is that they constitute exactly the class of possible limit distributions for this problem. (They form a subclass of the infinitely divisible* distributions, which constitute the class of possible limit distributions for a related problem.) This fact has important implications for applied work; *see* DOMAIN OF ATTRACTION.

Without the benefit of Lévy's analysis, Holtsmark [11] showed by direct computation that you can use certain stable laws in modeling electrical fields induced at a fixed point in space due to charged particles in random motion. We can simplify the argument as follows:

Let θ be a fixed point in three-dimensional space, and for each number $R > 0$ let B_R denote the ball centered at θ with a radius R. Consider a homogeneous gas of identical particles uniformly distributed in B_R. We shall consider three cases:

1. The particles are charged ions.
2. The particles are dipoles.
3. The particles are quadrupoles.

Let $X^{(R)}$ denote the electrical field in a given direction induced at θ. (Below, by "field" is meant the field in the given direction.) Since the particles are in constant movement, we are looking for a probabilistic description of $X^{(R)}$, rather than a deterministic one. $X^{(R)}$ is the sum of the fields at θ induced by the individual particles. The electrical field at θ induced by a single particle at a distance r from θ equals

$$\omega_p r^{-p},$$

where $p = 2$, 3, and 4 in cases 1, 2, and 3, respectively, and ω_p does not change if we move the particle in a straight line towards θ. It follows that if we move each particle in B_R into B_1 by reducing its distance from θ by a factor of R, then the field increases at θ by a factor of R^p. The same effect could have been achieved by increasing the particle density in B_1 by a factor of R^3 and removing all particles outside B_1. Hence ($\stackrel{d}{=}$ denoting equal in distribution)

$$R^p X^{(R)} \stackrel{d}{=} X_1^{(1)} + X_2^{(1)} + \cdots + X_{R^3}^{(1)}, \quad (1)$$

where $X_k^{(1)}$, $k = 1, 2, \ldots$, are independent and have the same distribution as $X^{(1)}$. (Here we are assuming the density of particles is the *same* for $X^{(1)}$ and $X^{(R)}$ and are taking R such that R^3 is an integer.) If we put

$n = R^3$ we get

$$X^{(R)} \stackrel{d}{=} \left(X_1^{(1)} + X_2^{(1)} + \cdots + X_n^{(1)}\right)/n^{p/3}.$$

If we let the radius R go to infinity, keeping the particle density constant, then the limit distribution

$$X^{(\infty)} = \lim_{R \to \infty} X^{(R)}$$

must be stable with index $\alpha = 3/p$. (Here "lim" refers to limit in distribution.)

There is a vigorous literature on the modeling of random economic phenomena by stable laws. Unlike the above example from physics, the justification depends heavily on data-based (i.e., statistical) conclusions, such as that the tails of the distributions generating the data have a particular shape or that the distributions generating the data have infinite variances. The stable models obviously include the traditional Gaussian models. So statistical methods developed for stable models also apply to Gaussian models. The new feature is that second-order methods used to make inferences about means, trends, etc., such as least-squares forecasting*, often lead to poor inferences. More robust methods have to be applied.

The pioneering article putting forth the viewpoint that stable laws are appropriate distributions to use in the modeling of certain economic phenomena is that of Mandelbrot [12]. A very readable article is that of Fama [5]. This viewpoint is grounded in the empirical claim that economics (in particular, finance) abounds with "fat tailed" data, that is, data sets that statistical analysis suggests were sampled from infinite variance populations. For example, Mandelbrot [12] did a thorough analysis of the fluctuation of cotton prices and concluded that the statistical properties of the data set were inconsistent with the variance being finite. For more discussion and other examples, see refs. 6, 13, and 16.

Three somewhat different arguments have been presented to reach the conclusion that the data are well modeled by a nonnormal stable law. Let $X_1, X_2, \ldots, X_n$ represent a set of economic data. In many cases, it is reasonable to argue that the X_i are approximately independent and approximately identically distributed. Hence, one assumes that $X_1, X_2, \ldots, X_n$ is a sequence of (i.i.d.) random variables.

The first argument [1] is based upon the empirical claim that

$$P[X_1 > x] \sim C/x^\alpha, P[X_1 < -x] \sim D/x^\alpha \tag{1}$$

($\sim$ denotes proportional to) as $x \to \infty$ for some $0 < \alpha < 2$, $\max(C, D) > 0$. Then it follows inescapably from the domain of attraction theorem that $\sum_{i=1}^n X_i$ is approximately stable with index $\alpha < 2$ when n is "large." (Unfortunately, n may have to be *very* large.) Sometimes in this argument (1) is replaced by the weaker statement that the variance of X_i is infinite, thus producing a persuasive but not tight argument for stability.

The second argument, which supports the conclusion that X_1 *itself* is stable nonnormal, is based upon the presentation of empirical evidence to support the claim that

$$\sigma^2(X_1) = \infty, \tag{2}$$

and

$$S_m = X_1 + X_2 + \cdots + X_m \text{ and } X_1, \tag{3}$$

properly normalized and centered, have the same distribution. This is the viewpoint in Fama [6], where α is actually estimated in each case for a large number of stocks in the New York stock exchange and $m = 4$ is used to investigate (3). Clearly it is delicate to justify (3) statistically.

The third argument [1] is really a variation of the second. It *postulates* that X_1 results from an additive process of the form

$$X_1 = \sum_{m=1}^{M} X_{1m},$$

where $\{X_{1m}\}$ are i.i.d. and M is large. Then

it is very plausible that X_1 is approximately stable when one takes into consideration the empirically argued fact that $\sigma^2(X_1) = \infty$.

Because of the established tradition of using stable laws in modeling, several authors have addressed the estimation and hypothesis testing problems that naturally arise in these models. (See, for example, refs. 3, 7, and 8.) This is an interesting challenge since closed form expressions for the stable law densities are usually not available. Some people consider the stable models controversial. In the first place, most supporting arguments depend heavily on empirically based conclusions about the shapes of distributions. But you need a large amount of data in order to be confident about the correctness of such decisions (e.g., see ref. 2, pp. 202–206). A particularly troubling concern is the observation of Du Mouchel [3], relying on a result of Cramér, that convergence to a nonnormal stable law can be very slow compared to convergence to a normal law, particularly if α is close to 2. Hence, in many situations where the model might apply, there is likely not enough of a time span to justify using the model. Moreover, in a recent paper, Du Mouchel [4] points out that slow convergence to the limit law can happen with economic data *even* when the limit law is normal (concluding that 200–300 observations may not even suffice to reliably conclude that the limit law of partial sums of i.i.d. economic variables is not normal). Finally, if, as several authors postulate, the appropriate model is that of contamination of finite variance data with either large variance or infinite variance outliers, then the limiting distribution resulting from summing these data will be a mixture, either of two normals or of a normal likely with a stable law. Thus, the question is often more complex than merely deciding whether the limiting distribution is normal or nonnormal stable. Time will of course clarify just how beneficial the use of stable laws in economic modeling will be.

Good references for the theory of stable laws are Feller [9] and Gnedenko and Kolmogorov [10]. Du Mouchel [4] is quite informative in its discussion of the use of stable laws in economic modeling.

References

[1] Bartels, R. (1977). On the use of limit theorem arguments in economic statistics. *Amer. Statist.*, **31**, 85–87.

[2] Breiman, L. (1973). *Statistics: With a View Toward Applications*. Houghton Mifflin, Boston, MA.

[3] Du Mouchel, W. (1973). Stable distributions in statistical inference: 2. Information from stably distributed samples. *J. Amer. Statist. Ass.*, **70**, 386–393.

[4] Du Mouchel, W. (1980). Stable Distributions and Infinite Variance Models in Statistics. *MIT Technical Report No. 16*, Department of Mathematics, MIT, Cambridge, MA.

[5] Fama, E. F. (1963). Mandelbrot and the stable Paretian hypothesis. *J. Bus. Univ. Chicago*, **36**, 420–429.

[6] Fama, E. F. (1965). The behavior of stockmarket prices. *J. Bus. Univ. Chicago*, **38**, 34–105.

[7] Fama, E. and Roll, R. (1968). Some properties of symmetric stable distributions, *J. Amer. Statist. Ass.*, **63**, 817–836.

[8] Fama, E. and Roll, R. (1971). Parameter estimates for symmetric stable distributions, *J. Amer. Statist. Ass.*, **66**, 331–338.

[9] Feller, W. (1966). *An Introduction to Probability Theory and its Applications*, Vol. 2. Wiley, New York.

[10] Gnedenko, B. V. and Kolmogorov, A. N. (1954). *Limit Distributions for Sums of Independent Random Variables*. Addison-Wesley, Reading, MA.

[11] Holtsmark, J. (1919). Über die Vebreitering von Spektrallinier. *Ann. Physik*, **58**, 577–630.

[12] Mandelbrot, B. (1963). The variation of certain speculative prices. *J. Bus. Univ. Chicago*, **36**, 394–419.

[13] Mandelbrot, B. (1963). New methods in statistical economics. *J. Political Econ.*, **71**, 421–440.

[14] Parzen, E. (1962). *Stochastic Processes*. Holden-Day, San Francisco, CA.

[15] Skorokhod, A. V. (1961). Asymptotic formulas for stable distribution laws. *Select. Trans. Math. Statist. Prob.*, **1**, 157–161.

[16] Steindl, J. (1965). *Random Processes and the Growth of Firms*. Griffin, London, England.

(DOMAIN OF ATTRACTION

(HEAVY-TAILED DISTRIBUTIONS
INFINITE DIVISIBILITY)

DITLEV MONRAD
WILLIAM STOUT

STABLE ESTIMATION

The concept of stable estimation arises when use is made of Bayesian* methods of estimation of parameters, or of their distributions. We here express the concept informally and therefore intelligibly. Suppose (i) that outside some region R of parameter space the likelihood is small compared with its mode, which of course occurs at some point within R; (ii) that the prior* for the parameters does not vary much (proportionately) within R and is not relatively large anywhere outside R; (iii) that the prior is "proper," that is, it integrates or sums to 1; (iv) that the parameter space is bounded, though this last condition is not always important. Then the Bayesian estimates will be much the same as if the prior had been assumed to be uniform (the so-called "Bayes postulate"). The estimates are then said to be *stable*.

A nice example is given by Edwards et al. [1]. You feel feverish and take your temperature. If the reading is 102.3° you accept an interval estimate resembling (102.25°, 103.35°) because the conditions mentioned are satisfied. But if the reading is 109.2° or 84.3° you do not accept it because condition (ii) is not satisfied. In such circumstances everybody would take a prior into account at least implicitly, even those who had never heard of the Reverend Thomas Bayes*. In this example (when the thermometer reading is unreasonable), the likelihood is large in a region of parameter space where the prior is small. In such cases the conditions of stable estimation do not apply, and, if "you" (as a Bayesian) are unprepared to revise your judgement of the prior distribution, your results will conflict with those of a strict non-Bayesian (if there are any). The revision of a prior after seeing data is undogmatic but of course has its dangers.

Reference

[1] Edwards, W., Lindman, H., and Savage, L. J. *Psychol. Rev.*, **70**, 193–242. [Reprinted in Kadane (1984). Contains a very readable introductory account of stable estimation.]

Bibliography

Berger, J. O. (1984). In *Robustness of Bayesian Analysis*. North-Holland, Amsterdam, The Netherlands, pp. 63–144, with discussion. [In this paper, and in Berger (1985, pp. 223–226), the author develops the concept of stable estimation in considerable detail in the context of Bayesian robustness.]

Berger, J. O. (1985). *Statistical Decision Theory and Bayesian Analysis*, 2nd ed. Springer-Verlag, New York.

Good, I. J. (1950). *Probability and the Weighing of Evidence*. Griffin, London, England. [Page 55 gives a succinct account of the idea of stable estimation in a form like that described by Savage (1962). The principle was not given a name here.]

Jeffreys, H. (1939). *Theory of Probability*, 1st ed. Clarendon Press, Oxford, England. (On page 146, he says "An accurate statement of the prior probability is not necessary in a pure problem of estimation when the number of observations is large.")

Jeffreys, H. (1948). *Theory of Probability*, 2nd ed. Clarendon Press, Oxford, England. (On page 121, he says in effect that variations in the prior can be ignored in a range where the likelihood is sharply peaked.)

Kadane, J. B. (1984). *Robustness of Bayesian Analysis*. North-Holland, Amsterdam, The Netherlands. (All the articles in this book bear to some extent on the topic of stable estimation.)

Kadane, J. B. and Chuang, D. T. *Ann. Statist.*, **6**, 1095–1110. [Reprinted in Kadane (1984). Discusses how the *utility* of the estimates change when the prior is varied.]

Savage, L. J. (1962). In *The Foundations of Statistical Inference*, G. A. Barnard and D. R. Cox, eds. Methuen, London, England, pp. 9–35. (Gives a typically homely example concerning the weighing of a potato. Calls stable estimation "precise measurement.")

(BAYESIAN INFERENCE
PRIOR DISTRIBUTIONS)

I. J. GOOD

STABLE POPULATION *See* DEMOGRAPHY; LIFE TABLES

STAGEWISE REGRESSION *See* STEPWISE REGRESSION

STAIRCASE METHOD (UP-AND-DOWN)

Early use of this experimental method occurred in the field of explosives testing in the 1940s. Later developments considered modifications of the basic method to be described here. For small or moderate sample sizes, little is gained by these modifications.

The response is all-or-none (examples, explosion or death). A traditional method for performing an analysis of a sensitivity experiment is to test a prescribed number of subjects at each of several fixed stimulus levels. In the up-and-down method the dose levels are determined sequentially; under most circumstances the accuracy of the usual method can be obtained with less experimentation. In explosives testing the subjects are primer caps or other detonators and the stimulus is a particular amount of force or impact. In the biological application, the stimulus is dose and the subjects are animals or in vitro preparations. Up-and-down testing finds extensive application across the entire spectrum of the social, life, physical, and engineering sciences; we shall use the biological application in the remainder of this section.

The improved estimates [3] are available directly from tables. One merely records the sequence of outcomes (○ for alive and × for dead) and reads the estimate from Table 1. For example, suppose a series of concentrations of 1, 2, 4, 8, and 16% was used for a drug known to require analysis as log dose. Suppose testing began at 8% or log dose = 0.903, and proceeded with results as shown in Fig. 1; that is, tests were performed at 0.903, 1.204, 0.903, 0.602, 0.903, 0.602 on successive animals. The estimate is 0.602 + 0.831(0.301) = 0.852 from the rule (final test level) + (value from table)(difference between dose levels) = LD_{50} ("lethal dose 50"–the dose producing response in 50% of subjects).

The following steps are used in an experiment using the up-and-down procedure:

1. A series of test levels is chosen with equal spacing between doses (equal spacing on the appropriate scale, usually log dose). This spacing is chosen approximately equal to the standard deviation σ of the distribution of threshold response levels. More simply stated, there should be two or three levels at which some animals respond and some do not. It is not too important if the interval is actually incorrect by as much as 50%.
2. A series of trials is carried out using this rule: Increase the dose following a negative response and decrease the dose following a positive response. The first test should be performed at a level as near as possible to the LD_{50}.
3. The number N' is the total number of tests performed in each series. Testing continues until a chosen "nominal" sample size N is reached. N is the total number of trials reduced by one less than the number of like responses at the beginning of the series. The example in Fig. 1 is $N = 6$. For the series ○○○ × × ○ × ○, $N' = 8$ and $N = 6$.
4. The resulting configuration of responses for each series is referred to the table of maximum likelihood* solutions (Table 1) for the LD_{50}, and $X_f + kd$ is computed, where X_f is the last dose administered, k is the tabular value, and d is the interval between doses. Table 1 lists all solutions for all N' and for $N \leqslant 6$. If the series begins with more than four like responses (i.e., $N' - N \geqslant 3$) the entry in the final column of Table 1 can be used (except for five tabular entries where an additional increment in the third decimal place is indicated).
5. The estimates have homogeneous variance when tests are performed for a fixed nominal sample size and thus are easily used in further analyses to investigate the dependence of these LD_{50}'s on related experimental factors. For N

Table 1. Maximum likelihood estimates of LD_{50}.[a]

N	Second Part of Series	k for Test Series Whose First Part is ○	○○	○○○	○○○○		Standard Error of LD_{50}
2	×	−0.500	−0.388	−0.378	−0.377	○	0.88σ
3	×○	0.842	0.890	0.894	0.894	○×	0.76σ
	××	−0.178	0.000	0.026	0.028	○○	
4	×○○	0.299	0.314	0.315	0.315	○××	0.67σ
	×○×	−0.500	−0.439	−0.432	−0.432	○×○	
	××○	1.000	1.122	1.139	1.140	○○×	
	×××	0.194	0.449	0.500	0.506	○○○	
5	×○○○	−0.157	−0.154	−0.154	−0.154	○×××	0.61σ
	×○○×	−0.878	−0.861	−0.860	−0.860	○××○	
	×○×○	0.701	0.737	0.741	0.741	○×○×	
	×○××	0.084	0.169	0.181	0.182	○×○○	
	××○○	0.305	0.372	0.380	0.381	○○××	
	××○×	−0.305	−0.169	−0.144	−0.142	○○×○	
	×××○	1.288	1.500	1.544	1.549	○○○×	
	××××	0.555	0.897	0.985	1.000^{+1}	○○○○	
6	×○○○○	−0.547	−0.547	−0.547	−0.547	○××××	0.56σ
	×○○○×	−1.250	−1.247	−1.246	−1.246	○×××○	
	×○○×○	0.372	0.380	0.381	0.381	○××○×	
	×○○××	−0.169	−0.144	−0.142	−0.142	○××○○	
	×○×○○	0.022	0.039	0.040	0.040	○×○××	
	×○×○×	−0.500	−0.458	−0.453	−0.453	○×○×○	
	×○××○	1.169	1.237	1.247	1.248	○×○○×	
	×○×××	0.611	0.732	0.756	0.758	○×○○○	
	××○○○	−0.296	−0.266	−0.263	−0.263	○○×××	
	××○○×	−0.831	−0.763	−0.753	−0.752	○○××○	
	××○×○	0.831	0.935	0.952	0.954	○○×○×	
	××○××	0.296	0.463	0.500	0.504^{+1}	○○×○○	
	×××○○	0.500	0.648	0.678	0.681	○○○××	
	×××○×	−0.043	0.187	0.244	0.252^{+1}	○○○×○	
	××××○	1.603	1.917	2.000	2.014^{+1}	○○○○×	
	×××××	0.893	1.329	1.465	1.496^{+1}	○○○○○	
		×	××	×××	××××	Second Part of Series	
		$-k$ for Series Whose First Part is					

[a]Values of k for estimating LD_{50} from up-and-down sequence of trials of nominal length N. The estimate of LD_{50} is $x_f + kd$, where x_f is the final test level and d is the interval between dose levels. If the table is entered from the foot, the sign of k is to be reversed.

> 6, additional tables are included in Dixon [3], but it is usually preferable to use $N \leqslant 6$ and associate the separate short runs with design variables (either covariates or time); i.e., separate the components of variation due to the design variables of an experiment by using factorial designs, Latin squares*, etc.

Most other estimating procedures for this case, probit analysis, logit analysis, etc. (see e.g., ref. 1), as well as up-and-down procedures [4, 2], result in estimates with widely varying standard error, making it difficult to use these estimates in comparative analyses. Furthermore, the estimates obtained are in general very inefficient.

Log Dose	Results of Tests					
1.204		×				
0.903	O		×	×		
0.602			O		O	
0.301						
0						

Figure 1 Example of test series. For this series O × × O × O the estimate of LD_{50} is 0.602 + 0.831(0.301) = 0.852.

The estimates provided in Table 1 are maximum likelihood estimates for each possible configuration of responses, assuming a normal cumulative distribution of response thresholds. See ref. 5 for alternate assumptions.

An experiment performed by Dr. Donald J. Jenden, Department of Pharmacology, UCLA, illustrates the use of the up-and-down method for a study of the dependence of time to death on the dose of ryanodine. Since several series of tests could be in progress at the same time, additional animals could be treated without waiting for the outcome of each separate trial. Also, it was not necessary to use the same test levels for all series. However, previous experimentation had indicated a standard deviation near 1.0, so a common dose interval $d = 1$ was used for log dose for all trials. Ryanodine was administered intravenously to male mice in volume 0.1 to 0.2 milliliters in saline. The end point was considered to be the time of last visible movement. Four cutoff points with equal spacing in log time, 64, 96, 144,

Table 2 Analysis of ryanodine experiment.

Weight (in grams)		Time in Seconds to Cutoff Point			
		64	96	144	216
	Tests	OOO × × O ×	OOOO × × O ×	OO × × OO	× × O × × O
18–20	x_f	2.000	−0.107	−3.213	−7.213
	k	−0.144	−0.142	0.372	0.861
	Est.	1.856	−0.249	−2.841	−6.352
	Tests	OOO × O × ×	O × × × ×	× O × × ×	× × O × O ×
21–23	x_f	4.00	−1.107	−5.213	−6.213
	k	0.181	0.555	0.157	−0.737
	Est.	4.181	−0.552	−5.056	−6.950
	Tests	× × × O × × O	× × O × × O	O × × O ×	O × O × ×
24–26	x_f	1.000	−3.107	−4.213	−6.213
	k	0.860	0.861	−0.305	0.084
	Est.	1.860	−2.246	−4.518	−6.129

Effect	SS
L_t	139.086
Q_t	1.286
C_t	0.010
L_w	1.485
Q_w	0.145
L_tL_w	0.021
Q_tL_w	1.902
C_tL_w	0.014
L_tQ_w	4.159
Q_tQ_w	0.877
C_tQ_w	0.338

Analysis of Variance

Source	SS	df	MS
L_t	139.086	1	139.086
L_w	1.485	1	1.485
Remainder	8.752	9	0.972

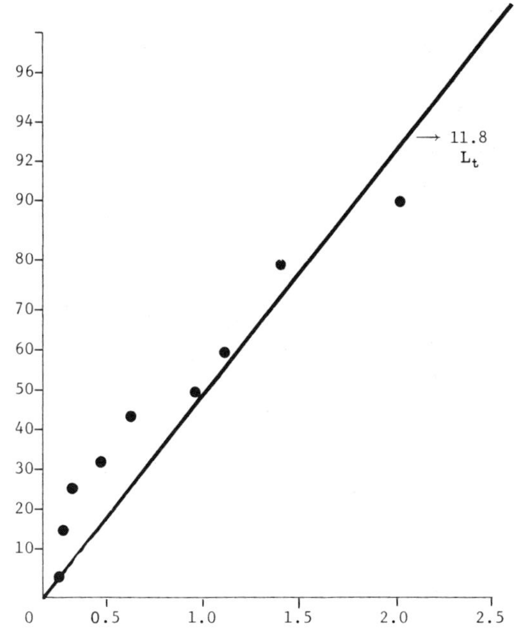

Figure 2 Half-normal plot for ryanodine experiment.

and 216 seconds, were chosen for observing the status of the animal. Dosage was computed by body weight, and a randomized block* experiment with body weight as the blocking variable was used to obtain information on the validity of the use of the body weight basis for dosage. The results of the tests are given in Table 2. The analysis indicates a time–dose dependence that is linear between log dose and log time and is established with an efficient use of animals and with a small standard error. See Dixon [3] for a discussion of the indications in this experiment for improved dosage intervals.

The analysis of variance* was completed using single degrees of freedom for the time and weight variables. The weight variable has three levels requiring two degrees of freedom, coded $-1, 0, 1$ and $1, -2.1$, labeled L_w and Q_w for linear and quadratic effects. The time variable has four levels equally spaced in log time and one coded $-3, -1, 1, 3$ for L_t; $1, -1, -1, 1$ for Q_t; and $-1, 3, -3, 1$ for C_t, representing three degrees of freedom for linear, quadratic, and cubic effects in log time.

Figure 2 shows that half-normal plot* for eleven single degrees of freedom.

We see a uniform size to the sum of squares for all terms except linear "time," i.e., log response is linearly related to log time.

References

[1] Berkson, J. (1965). *Proc. 3rd Berkeley Symp. Math. Statist. Probab.*, Vol. 1, J. Neyman, ed. University of California Press, Berkeley, CA. (A general discussion of a problem with fixed sample sizes at each level, p. 11.)

[2] Brownlee, K. A., Hodges, J. L., and Rosenblatt, M. (1953). *J. Amer. Statist. Ass.*, **48**, 262–277.

[3] Dixon, W. J. (1965). *J. Amer. Statist. Ass.*, **60**, 967–978. (More complete presentation of the staircase analysis.)

[4] Dixon, W. J. and Mood, A. M. (1948). *J. Amer. Statist. Ass.*, **43**, 109–126. (The early large-sample results for the staircase design.)

[5] Little, R. E. (1974). *J. Amer. Statist. Ass.*, **69**, 202–206. (A comparison of certain up-and-down estimates.)

(BIOASSAY, STATISTICAL METHODS IN QUANTAL RESPONSE ANALYSIS STOCHASTIC APPROXIMATION)

W. J. DIXON

STANDARD DEVIATION

This is the most commonly used measure of spread or variability. It defines an average size of the deviation of a single observation of a random variable X from its expected value $E[X]$. The population *variance** σ^2 is defined by

$$\sigma^2 = E\left[(X - E[X])^2\right] = \mu_2(X)$$

or equivalently

$$\sigma^2 = E[X^2] - (E[X])^2.$$

The *standard deviation* σ is the positive square root of σ^2. It describes how values clump around the mean; for a normal distribution* 68% of values lie within $\pm\sigma$ of the mean, and 95% within $\pm 2\sigma$ of the mean.

In order to calculate σ it is necessary to know the distribution function, but in practical work it is more common to estimate σ from a random sample $x_1, x_2, \ldots, x_n$. The usual estimate is the sample standard deviation S, where

$$S = \sqrt{\frac{\sum_{i=1}^{n}(x_i - \bar{x})^2}{n-1}},$$

$$\bar{x} = n^{-1}\sum_{i=1}^{n} x_i. \qquad (1)$$

If the divisor n is used, this defines the standard deviation of the sample itself, denoted by S_n.

S_n is the maximum likelihood* estimator of σ when the population mean is unknown, which is generally the case. S^2 is an unbiased estimator of σ^2. However, neither S nor S_n is an unbiased estimator of σ; for this S must be scaled by a factor that depends on n and the distribution. In the case of a normal distribution $C_n S$ is unbiased, where

$$C_n = \Gamma\left(\frac{n-1}{2}\right)\sqrt{\frac{n-1}{2}}\bigg/\Gamma\left(\frac{n}{2}\right).$$

For $n > 10$, C_n can be approximated by $1 + \frac{1}{4}(n-1)^{-1}$. Other unbiased linear estimates, based on an ordered sample, have been suggested (*see* L-ESTIMATORS).

The standard deviation gives an indication of the difference that could reasonably be expected between the observed and expected values of a random variable. It is therefore used as the scale of measurement in significance tests* and confidence intervals*. Other measures of spread such as range* and mean deviation* are used in descriptive statistics, but play little part in the important area of significance tests. They are related by

$$\text{mean deviation} \leqslant S \leqslant \frac{\text{range}}{2}\sqrt{\frac{n}{n-1}}. \qquad (2)$$

Sometimes it is required to estimate σ from a number of different samples from populations with the same standard deviation. In this case a pooled estimate is used. If S_i is obtained from the ith sample size n_i, $i = 1, \ldots, k$, then (*see* ANOVA)

$$S = \sqrt{\frac{(n_1-1)S_1^2 + (n_2-1)S_2^2 + \cdots + (n_k-1)S_k^2}{(n_1-1) + (n_2-1)\cdots(n_k-1)}}.$$

HISTORICAL DEVELOPMENT

The term *standard deviation* was introduced by Karl Pearson* in a lecture to the Royal Society in 1893. His work marked the culmination of ideas from the previous two centuries, and formed the basis for much of the subsequent development, particularly the mathematical theory of estimation* and hypothesis testing*.

Original ideas on variability were founded on the study of Laws of Error*. This was primarily the concern of the eighteenth century astronomers who wanted to combine experimental measurements in order to determine the position of heavenly bodies. They were interested in deducing the distribution of errors, assessing the likelihood of different-sized errors, and determining the maximum possible error. Notable contributors to the work were Galileo, de Moivre*, and Bernoulli*. At the turn of the century the principle of summing squares of errors was well established, and the method of least squares* was gaining acceptance. In this context several people derived the normal law. Over the period 1810–1825 Gauss* made a very significant contribution to this work and gained a large following. He derived the normal law, using h [$= 1/(\sigma\sqrt{2})$], a measure of *precision*, as a parameter. He looked at the variability of the mean $\bar{x}$., and effectively stated that the standard deviation of $\bar{x}$. from a sample of size n is $\sigma/\sqrt{n}$. He later adopted the term probable error* and gave the probable error of a moment. Gauss also used the mean deviation. He published tables to show the probabilities of an observation deviating from the mean by more then certain amounts. He proved inequalities related to $\Pr[|X - E[X]| \geqslant \lambda\sigma]$ for certain types of distribution.

By the middle of the nineteenth century there were several measures of dispersion in use. Airy used the *modulus* C ($= \sigma\sqrt{2}$), the *mean error* (C/π), the *mean square* $C^2/2$ ($= \sigma^2$), the *error of mean square* ($= \sigma$), and *probable error**. Then Chebyshev* published his inequalities, which are more general than those of Gauss. The ideas of summing squares were also being developed in the context of regression* and correlation*, but were not combined with those of the astronomers until later. The normal distribution had been thought to apply only to experimental error. This was changed in the 1880s when Galton* fitted genetics* data to the normal distribution, and Edgeworth observed the central limit theorem*. Galton extended his work on the normal distribution to many observed deviations, dispersions and differences, and developed work on regression, correlation, and bivariate normal* distribution.

Pearson worked on moments and probable errors, making analogies with moments in statics. He derived the standard deviation of several statistics, introducing the χ^2-test* statistic. At the start of the twentieth century Gosset* (Student) formalized much of the standard deviation work. He explained the need to distinguish between σ and S, and derived the t-test. By this time the normal distribution and standard deviation were more commonly accepted. In the 1920s Fisher* started to produce a formal mathematical basis for estimation.

COMPUTATION

An alternative version of the formula for S is

$$S = \sqrt{\frac{\sum_{i=1}^{n} x_i^2 - \left(\sum_{i=1}^{n} x_i\right)^2/n}{n-1}}. \quad (3)$$

(1) and (3) are mathematically equivalent, but (3) should not be used on a finite-arithmetic machine, as it can produce highly inaccurate results. Whenever possible data should be scaled, and (2) can be used as an elementary check on a computed value.

Computational problems arise if the standard deviation is much smaller than the mean. In using a computer it may be impractical to use (1) owing to restrictions on memory size. In this case an updating algorithm should be used. The accuracy of the result depends on the relative uncertainty in the data, the precision of the computer, and the size of the standard deviation in comparison with the mean. West's algorithm is reasonably accurate. It uses the relations

$$M_N = M_{N-1} + (x_N - M_{N-1})/N,$$

$$T_N = T_{N-1} + \frac{(N-1)}{N}(x_N - M_{N-1})^2,$$

where

$$M_N = \frac{1}{N}\sum_{i=1}^{N} x_i, \qquad T_N = \sum_{i=1}^{N} x_i^2 - NM_N^2.$$

If the data are grouped in a frequency table, then the formula needs adjustment. If f_i is the number of observations in the ith interval and x_i is the mid-point of the interval then

$$S = \sqrt{\frac{\sum_{i=1}^{n} x_i^2 f_i - \left(\sum_{i=1}^{n} x_i f_i\right)^2/N}{N-1}},$$

$$N = \sum_{i=1}^{n} f_i.$$

Grouping data and using the mid-points causes errors in the calculation of moments. If the original readings are fairly accurate and the distribution tails off at each end, then it is appropriate to use Sheppard's correction (*see* CORRECTION FOR GROUPING and GROUPED DATA).

Bibliography

Cooke, D., Craven, A. H., and Clarke, G. M. (1982). *Basic Statistical Computing*. Edward Arnold, London, England. (Gives a very readable account of techniques used in programming statistical problems.)

Ehrenberg, A. S. C. (1982). *A Primer in Data Reduction*. Wiley, Chichester, England. (This is an excellent introductory text.)

Kendall, M. G. and Stuart, A. (1977). *The Advanced Theory of Statistics*, Vol. 1. Griffin, London, England. (This is a more formal reference text.)

Pearson, E. S. and Kendall, M. G. (1970). *Studies in the History of Statistics and Probability*. Griffin, London, England. (This is a detailed history of statistical ideas with a very full list of references.)

The following papers discuss estimation of the standard deviation.

Bolch, B. W. (1968). More on unbiased estimation of the standard deviation. *Amer. Statist.*, **22**, 27.

Cureton, E. E. (1968). Unbiased estimation of the standard deviation. *Amer. Statist.*, **22**, 22.

Downton, F. (1966). Linear estimates with polynomial coefficients. *Biometrika*, **53**, 129–141.

Holtzman, W. H. (1950). The unbiased estimate of the population variance and standard deviation. *Amer. J. Psychol.*, **63**, 615–617.

Jarrett, R. F. (1968). A minor exercise in history. *Amer. Statist.*, **22**, 25–26.

Markowitch, E. (1968). Minimum mean-square error estimation of the standard deviation of the normal distribution. *Amer. Statist.*, **22**, 26.

Pearson, K. (1931). *Biometrika*, **23**, 416–418. (Historical note on the distribution of the standard deviations of samples of any size drawn from an indefinitely large normal parent population.)

The most significant original works of the pioneers are:

Airy, G. B. (1861). *On the Algebraical and Numerical Theory of Errors of Observations and the Combinations of Observations*. London.

Edgeworth, F. Y. (1883). The methods of least squares. *Philos. Mag., 5th Series*, **16**, 360–375.

Edgeworth, F. Y. (1885). Methods of statistics. *J. R. Statist. Soc.*, Jubilee volume, 181–217.

Edgeworth, F. Y. (1886). On the determination of the modulus of errors. *Philos. Mag. 5th Series*, **21**, 500–507.

Edgeworth, F. Y. (1908). On the probable errors of frequency constants. *J. R. Statist. Soc.*, **71**, 381–397, 499–512, 652–678.

Galton, F. (1886). Family likeness in structure. *Proc. R. Soc. Lond.*, **40**, 42–73.

Galton, F. (1888). Co-relations and their measurement, chiefly from anthropometric data. *Proc. R. Soc. Lond.*, **45**, 135–145.

Galton, F. (1889). *Natural Inheritance*. Macmillan, London, England.

Gauss, C. F. (1809). *Theoria Motus Corporum Coelestium*. This is available in Borsch, A. and Simon, P. (1887, 1964). *Abhandlungen zur Methode der kleinsten Quadrate von Carl Friedrich Gauss*. Wurzburg, Federal Republic of Germany. It is summarized in Whittaker, E. T. and Robinson, G. (1924, 1926). *The Calculus of Observations*. Blackie, London, England.

Gauss, C. F. (1821). *Theoria Combinatioris Erroribus Minimis Obnoxiae*.

Gosset, W. S. (Student) (1908). The probable error of a mean. *Biometrika*, **6**, 1–25.

Gosset, W. S. (Student) (1908). The probable error of a correlation coefficient. *Biometrika*, **6**, 302–310.

Gosset, W. S. (Student) (1917). Tables for estimating the probability that the mean of a unique sample of observations lies between $-\infty$ and any given distance of the mean of the population from which the sample is drawn. *Biometrika*, **11**, 179, 414–417.

Pearson, K. (1893). (For a report on Pearson's paper to the Royal Society, see *J. R. Statist. Soc.*, **56**, 676.)

Pearson, K. (1894). Contributions to the mathematical theory of evolution. *Philos. Trans. R. Soc. A*, **185**, 71–110.

Pearson, K. (1900). On the criterion that a given system of deviations from the probable in the case of a correlated system of variables is such that it can reasonably be supposed to have arisen from random sampling. *Philos. Mag., 5th Series*, **50**, 157–175.

Pearson, K. (1903). The law of ancestral heredity. *Biometrika*, **2**, 211–229.

Pearson, K. and Filon, L. N. G. (1898). Mathematical contributions to the theory of evolution. IV: On the probable errors of frequency constants and on the influence of random selection on variation and correlation. *Philos. Trans. R. Soc. A*, **191**, 229–311.

For a discussion of computational procedures see:

Chan, T. F. and Lewis, G. (1979). Computing standard deviations: Accuracy. *Commun. A.C.M.*, **22**, 526–531.

(CORRECTION FOR GROUPING
INTERQUARTILE RANGE
LAWS OF ERROR
LEAST SQUARES

MEAN DEVIATION
NORMAL DISTRIBUTION
PROBABLE ERROR
RANGES
VARIABILITY
VARIANCE)

ANNA HART

STANDARD ERROR

This term is often used interchangeably with standard deviation*. Although there is no established "correct" usage, *standard error* often refers to an *estimate* of standard deviation. Thus, the standard deviation of the arithmetic mean $\bar{X} = n^{-1}\Sigma_{i=1}^{n} X_i$ of n mutually uncorrelated random variables $X_1, \ldots, X_n$ with common standard deviation σ, is $\sigma/\sqrt{n}$, while the estimator $S/\sqrt{n}$ with

$$S^2 = n^{-1}\sum_{i=1}^{n}(X_i - \bar{X})^2$$

[or $(n-1)^{-1}\Sigma_{i=1}^{n}(X_i - \bar{X})^2$] may be referred to as the "standard error."

STANDARDIZATION

Standardization is anticipatory data analysis*. It has to do with describing similarities and dissimilarities between data sets in such a way that certain possibly confounding reasons for them are already taken into account. *See also* DEMOGRAPHY and RATES, STANDARDIZED.

Let Ω be a finite population and $(x(\omega), y(\omega))$ be a bivariate response for each ω in Ω, with y a variable of interest for comparison with other populations and x a confounding covariate that takes the values $1, 2, \ldots, K$. Let Ω_k be the subpopulation for which $x = k$. We suppose $y = 1$ or 0 but the argument applies generally, for instance when y is vector-valued. For $F \subseteq \Omega$, (F) is its size, $y(F)$ is the sum of the y values over F, and $\bar{y}(F) = y(F)/(F)$ is their mean. For definiteness, we interpret $x(\omega)$ as the age group to which ω belongs at the beginning of a given period, $y(\omega)$ being 1 or 0 according as ω is alive or dead at the end of it. Thus $\bar{y}(\Omega)$ is the *crude death rate* in Ω and $\bar{y}(\Omega_k)$ is the *age-specific death rate* in age group k. Consider also another finite population Δ, disjoint from Ω, with analogous responses $(x(\delta), y(\delta))$.

Comparing mortality in Ω and Δ through their total deaths $y(\Omega)$ and $y(\Delta)$ does not take account of their relative sizes. To do so we compare the total deaths $N\bar{y}(\Omega)$ and $N\bar{y}(\Delta)$ in two hypothetical populations that have the same size N but are otherwise like Ω and Δ in their respective mortality experiences. Their ratio $\bar{y}(\Omega)/\bar{y}(\Delta)$ involves only the original crude death rates; the choice of standardizing size N has no bearing on the comparison. However, this comparison can be misleading. Since

$$\bar{y}(\Omega) = \sum_{k=1}^{K}(\bar{\Omega}_k)\bar{y}(\Omega_k),$$

$$\bar{y}(\Delta) = \sum_{k=1}^{K}(\bar{\Delta}_k)\bar{y}(\Delta_k),$$

where $(\bar{\Omega}_k) = (\Omega_k)/(\Omega)$ and $(\bar{\Delta}_k) = (\Delta_k)/(\Delta)$, we can have

$$\bar{y}(\Omega) \neq \bar{y}(\Delta) \tag{1}$$

even though

$$\bar{y}(\Omega_k) = \bar{y}(\Delta_k), \qquad k = 1, 2, \ldots, K. \tag{2}$$

For example, we cannot use (1) to argue that unequal crude death rates in two occupations result from corresponding inequalities in working conditions. The corresponding age-specific death rates might be the same and (1) simply result from different age compositions for the two occupations. It is implicit here that if (2) does hold, then we regard Ω and Δ as experiencing the same mortality relative to the covariate x.

Given the covariate x, comparison of mortality reduces to K pair-wise comparisons between corresponding subpopulations Ω_k and Δ_k. With size standardization as above, each of them may be effected through the ratio of its subpopulation crude death rates. Since these crude rates are age-specific death rates of Ω and Δ, the component pair-wise comparisons give the K age-specific mortality ratios

$$r_k = \bar{y}(\Omega_k)/\bar{y}(\Delta_k), \qquad k = 1, 2, \ldots, K. \tag{3}$$

These may be regarded as arising from the comparison of two hypothetical populations with the same age compositions but otherwise like Ω and Δ in their respective age-specific mortality experiences. Following Yule [6], we see the practical side of standardization as the problem of finding a satisfactory summary of the K ratios in (3). In particular, Ω and Δ should play symmetric roles. It should not matter whether one works with the r_k or their reciprocals.

One way of doing this was proposed recently in Finch [3]. The $(r_k - 1)/(r_k + 1)$ are regarded as comprising a noisy constant in the sense of Finch [2] and their mean μ is calculated. The quantity $\lambda = (1 + \mu)/(1 - \mu)$ is then used to summarize the K ratios in (3). In words, the age-specific death rates in Ω are, in summary, $100\lambda\%$ those of Δ. Equivalently, the age-specific death rates in Δ are, in summary, $100\lambda^{-1}\%$ of those in Ω.

Historically, however, the practical side of standardization has been dealt with by introducing a population S with a so-called standard age composition (S_k), $1 \leqslant k \leqslant K$. The standardized death rate in Ω is then defined to be

$$\bar{y}^S(\Omega) = \sum_{k=1}^{K} (\bar{S}_k)\bar{y}(\Omega_k), \tag{4}$$

viz. the crude death rate in a population with the standard age composition but the age-specific death rates of Ω. The standardized death rate in Δ is defined in a similar way. Mortality in Ω and Δ is then compared through the ratio

$$r^S = \bar{y}^S(\Omega)/\bar{y}^S(\Delta) \tag{5}$$

of these standardized death rates. Difficulties with this approach, in particular its dependence on an arbitrarily chosen standard, are discussed by Yerushalmy [5]. It seems to be based on a misunderstanding. The rate $\bar{y}^S(\Omega)$ is a particular type of summary of the age-specific death rates of Ω and similarly for $\bar{y}^S(\Delta)$. The ratio (5) is a comparison of summaries of the age-specific death rates of Ω and Δ, whereas what is needed is a summary of the K pair-wise comparisons of corresponding age-specific death rates in (3). As Yule [6] said, standardized rates are superfluous.

Both Yule and Yerushalmy had misgivings about usual standardization procedures but did not break away from them. Yule felt one should take account of the widths of the age groups. But this would seem to be unnecessary because, although the number of ages in age group k might vary with k, each pair-wise comparison (3) is based on two age groups with the same width. Yerushalmy constructed a standardized comparative index through individual comparisons with a standard population. Similarly Kitagawa [4] distinguished between aggregative and relative standardized indices but did not pursue these as a contrast between a comparison of summaries and a summary of comparisons. He saw them as alternative ways of constructing standardized indices when one of the populations in question was the standard one.

The rates in (4) are said to be *directly standardized* to distinguish them from the so-called *indirectly standardized rates* that are used when one knows the crude death rates of Ω and Δ but not their age-specific death rates. There are then no pair-wise comparisons (3) and hence no question of summarizing them. In these circumstances it seems natural to proceed by comparing summaries. The first use of a standard rate occurred in 1855 in Farr's 16th Annual Report of the Registrar General and was an instance of indirect standardization [1]. Direct standardization came later and this may be why it has been seen as a variant of indirect standardization and, like it, involved with the comparison of summaries. Indirect standardization is discussed in Kitagawa [4] and Yerushalmy [5].

References

[1] Benjamin, B. (1968). *Health and Vital Statistics*. Allen and Unwin, London, England.

[2] Finch, P. D. (1980). *Biometrika*, **67**, 539–550.

[3] Finch, P. D. (1983). *Biometrics*, **39**, 275–279.

[4] Kitagawa, E. M. (1964). *Demography*, **1**, 296–315. (A scholarly discussion of the principles and practice of standardization.)

[5] Yerushalmy, J. (1951). *Amer. J. Public Health*, **41**, 907–922. (A penetrating and thoughtful discussion of the practical dangers and pitfalls of conventional standardization procedures.)

[6] Yule, G. U. (1934). *J. R. Statist. Soc.*, **97**, 1–84. (Mainly of historical interest now, but well worth serious study.)

(ACTUARIAL STATISTICS—LIFE
DEMOGRAPHY
RATES, STANDARDIZED
SIMPSON'S PARADOX
VITAL STATISTICS)

PETER D. FINCH

STANDARDIZATION, INTERNATIONAL *See* INTERNATIONAL STANDARDIZATION: APPLICATION OF STATISTICS

STANDARDIZED DISTRIBUTION

If X is a random variable with mean μ and standard deviation σ, then the transformed variable

$$Z = (X - \mu)/\sigma$$

is said to have a *standardized distribution* or to be in *standard measure*, having mean 0 and variance 1. Then X and Z have the same shape factors (*see* MOMENT RATIOS), and the cumulants* κ_r of X and Z are related by

$$\kappa_r(Z) = \sigma^{-r}\kappa_r(X), \qquad r \geqslant 2.$$

Standardization is useful for comparing distributions differing in their means and variances. The correlation* coefficient between two random variables X and Y is the covariance between them when they are expressed in standard measure. It is also customary to state the central limit theorem* in terms of convergence in distribution to the standard normal distribution, a device that avoids the introduction of means and variances that become infinitely large or infinitesimally small in the limit.

(MOMENT RATIOS
STANDARD DEVIATION)

STANDARDIZED MORTALITY RATIO *See* DEMOGRAPHY; RATES, STANDARDIZED

STANDARDS *See* INTERNATIONAL STANDARDIZATION: APPLICATION OF STATISTICS; NATIONAL BUREAU OF STANDARDS

STANINE SCALE

A contraction of *standard nine-point scale*, referring to a nine-point scale of grouped and normalized standard scores taking values $1, 2, \ldots, 9$ and having mean value 5 and standard deviation 2. This scale is used in educational* and psychological testing* methodology.

(NORMALIZED *T* SCORES)

STARS *See* SNOWFLAKES

STAR SLOPES

Suppose that data $(x_1, y_1), \ldots, (x_n, y_n)$ are fitted to a simple linear regression model. By drawing a line from each data point to a central value and by measuring the slope of these lines, one can construct a sample of *star slopes*. The median of these slopes is the *median star*, and has been proposed [1] as a simple but resistant estimator of the overall slope.

Reference

[1] Simon, S. D. (1986). *Abstracts, Joint Statistical Meetings*. American Statistical Association, Washington, D.C., p. 205.

(LINEAR REGRESSION
RESISTANT TECHNIQUES)

START

In exploratory data analysis*, this refers to a constant added to observed values before applying a logarithmic or square root transformation. A recommended value is $\frac{1}{6}$ (see ref. 1).

Reference

[1] Tukey, J. W. (1977). *Exploratory Data Analysis*, Addison-Wesley, Reading, MA.

STATES, IN MARKOV CHAINS *See* MARKOV PROCESSES

STATIONARY DISTRIBUTION

This is the probability distribution of a homogeneous Markov* chain that is independent of "time." More precisely, let $x(t)$ be a homogeneous Markov chain with the set of states $\mathscr{S}$ and transition probabilities $p_{ij}(t) = P(x(t) = j|x(0) = i)$. The stationary distribution is a set of numbers $\{\pi_j, j \in \mathscr{S}\}$ such that $\pi_j \geqslant 0$, $\Sigma_{j \in \mathscr{S}} \pi_j = 1$, and

$$\sum_{i \in \mathscr{S}} \pi_i p_{ij}(t) = \pi_j, \qquad j \in \mathscr{S}, t > 0.$$

See MARKOV PROCESSES for more details and bibliography.

STATIONARY POPULATION *See* DEMOGRAPHY; LIFE TABLES

STATIONARY PROCESSES

In the theory of stochastic processes* one frequently encounters processes $X(t)$ that have equilibrium distributions as $t \to \infty$, in the sense that these do not depend on initial conditions. The process is thus stationary in time. A stationary process has the important property that the joint distribution of $X(t)$, $X(t + a)$ is a function of time only through the interval of time a from t to $t + a$, and does not depend on t.

Formally, the process $X(t)$ is *stationary in the strict sense* if $X(t_1), X(t_2), \dots, X(t_k)$ and $X(t_1 + a), X(t_2 + a), \dots, X(t_k + a)$ have the same joint distribution for every $t_1, t_2, \dots, t_k, a$, where $k = 2, 3, \dots$. The definition holds in discrete or continuous time.

There are practical data-analytic difficulties in working with this definition, and a simpler approach requires stationarity through the moments only. A process is *stationary of order h* if all moments up to order h have the stationarity property. For second-order stationarity (which suffices for much work in practice) this means that

$$E(X(t)) = \mu,$$

$$\operatorname{cov}(X(t), X(t + a)) = \gamma(a),$$

say, so that the mean and autocovariance function are free of t.

An example of a second-order stationary process that is also strictly stationary is a Gaussian process* [for which the joint distribution of $X(t_1), X(t_2), \dots, X(t_k)$ is multivariate normal] in which $E(X(t_i)) = \mu$, $\operatorname{cov}(X(t_i), X(t_j)) = \gamma|t_i - t_j|$. For further details see ref. 1, Chap. 7, and STOCHASTIC PROCESSES.

Reference

[1] Cox, D. R. and Miller, H. D. (1965). *The Theory of Stochastic Processes*. Methuen, London, England.

STATIONARY TIME SERIES *See* TIME SERIES

STATISTIC

Generally, this term is used to denote a quantity derived from observed values—for example, an index number*. It is usually not something observed *directly*, but rather calculated from directly observed values. (This is true even of order statistics* that, although they take values that are directly observed, need all available variables for their definition.)

In mathematical statistics the term can be applied to any mathematical function of random variables, such as, for example, an arithmetic mean ($\bar{X} = n^{-1}\Sigma_{i=1}^{n} X_i$). The term is also applied to a particular realized (observed) value of such a function—there is no distinguishing term available, as there is, for example with "estimate" and "estimator." A statistic can also be a vector, such as $(\bar{X}, S)$ with $S^2 = (n-1)^{-1}\Sigma_{i=1}^{n}(X_i - \bar{X})^2$.

A more abstract definition is as follows: Let a random variable **X** take values in a sample space $(\mathscr{X}, \mathscr{B}, \underline{P}^X)$. Then any σ-measurable mapping $T(\cdot)$ of the space $\mathscr{X}$ into a measurable space* $(\mathscr{Y}, \mathscr{A})$ is called a *statistic*. The *probability distribution* of T is defined by

$$\begin{aligned} P^T(\mathscr{B}) &= \Pr[T(x) \in \mathscr{B}] \\ &= \Pr[X \in T^{-1}(\mathscr{B})] \\ &= P^X(T^{-1}(\mathscr{B})) \quad \text{for all } \mathscr{B} \in \mathscr{A}. \end{aligned}$$

In applications, a statistic does not involve unknown (and hence unobservable) parameters. $\Sigma_{i=1}^{n}(X_i - \bar{X})^2$ is a statistic, but $\Sigma_{i=1}^{n}(X_i - \mu)^2$ is not, if μ is an unknown population parameter.

STATISTICA

The journal *Statistica* was founded in 1941 by Paolo Fortunati, who was the editor until 1980. Since 1981, the Director has been Italo Scardovi. The journal is owned by the University of Bologna. The administration and editorial board are located in the Department of Statistical Sciences (Via Belle Arti 41, 40126 Bologna, Italy). *Statistica* comes out every three months, in issues of about 150–200 pages. It is devoted to theoretical and applied problems of statistics, statistical analysis in the most varied fields of scientific research, and review and information on relevant publications. The journal represents Italian statistics in the world; its international standing is witnessed by its circulation in over 100 countries including some 20 important international organizations, in addition to over 130 exchanges with the most reputable scientific periodicals.

In an editorial in its first issue, *Statistica* offered its pages to the contributions of scholarly minded people "...who believe in quantitative analysis and the logical conception of collective phenomenon (...) as the basis of progress of natural and social disciplines... ."

Statistica is open to critical debate by Italian and foreign authors in the objective, wide ranging, and free spirit of research. As a consequence, it has become increasingly directed toward epistemological and logical developments of statistical method and to the deepening of quantitative analytical techniques for natural and social phenomena, without bias against any specific scientific trends and without a priori opposition to any one school of thought.

In this way *Statistica* also strives to encourage statisticians not to remain strangers to the conceptual debates taking place among scientists, philosophers, and probabilists on the meaning and value of statistics as an instrument of scientific research and as an inductive method within a phenomenology that is no longer deterministic or limited to "universal" statements.

The partial list of contents of a recent issue [Vol. XLII (1982)] indicates some of the present major interests of the journal:

"Chance, statistics, sampling" by L. Kish.

"Le utilizzazioni scientifiche del censimento demografico" by B. Colombo.

"Paolo Fortunati, uno statistico tra vecchio e nuovo" by V. Castellano.

"On mean values and unbiased estimators in simple random sampling" by A. Herzel.

"Sulle curve logonormali di ordine r quali famiglie di distribuzioni di errori di proporzione" by S. Vianelli.

"Generazione di simboli binari pseudocasuali median te polinomi primitivi" by A. Rizzi.

"Effetti della variazione di morbosità sulla mortalità e sull'ammontare e la struttura di una popolazione" by A. Golini and V. Egidi.

"La legge di Benford-Furlan come legge statistica" by E. Regazzini.

"Inferenza predittiva e fatti" by D. Costantini and P. Monari.

"Regression with stable errors: an empirical characteristic function approach" by I. A. Koutrouvelis.

"Computing methods for restricted estimation in linear models" by C. Corradi.

"La stima puntuale dei parametri nel modello input-output" by A. Gardini.

"Maximum likelihood estimation of the parameters of the Makeham survival function" by I. D. Patel and A. V. Gajjar.

"Necessità del caso" by I. Scardovi.

Papers are accepted in Italian, English, French, German, and Spanish. Papers should be submitted for possible publication to the Editor. Instructions to authors for preparation of the manuscripts are given at the end of each issue of *Statistica*.

All papers are refereed. Three months is the expected refereeing time.

I. SCARDOVI

STATISTICAL ABSTRACT OF THE UNITED STATES

The *Statistical Abstract of the United States*, published since 1878, is the standard summary of statistics on the social, political, and economic organization of the United States. It is designed to serve as a convenient volume for statistical reference and as a guide to other statistical publications and sources. The latter function is served by the introductory text to each section, the source note appearing below each table, and an appendix which comprises the Guide to Sources of Statistics and the Guide to State Statistical Abstracts.

The most recent edition, as of the time of this writing, the 105th Annual Edition (1985), was published by the U.S. Bureau of the Census* in December of 1986, comprises 1008 pages, and was compiled by a team from the Bureau headed by G. W. King, Chief, Statistical Compendia Staff.

The *Abstract* is sold by the Superintendent of Documents, U.S. Government Printing Office, Washington, DC 20402, or any U.S. Department of Commerce district office.

(BUREAU OF LABOR STATISTICS
BUREAU OF THE CENSUS
DEMOGRAPHY
FEDERAL STATISTICS)

STATISTICAL CATASTROPHE THEORY

Elementary catastrophe theory* is concerned with the qualitative behavior of the equilibria of dynamic models that are nonlinear in their variables. In the typical formulation we have a system with state variable y:

$$\dot{y} = h(\mathbf{x}, y), \tag{1}$$

where $\mathbf{x} \in R^k$ is a vector of exogenous variables, possibly under experimental control, h is a function $h: R^{k+1} \to R$, and $\dot{y}$ stands for dy/dt. We will consider only one-dimensional y, though catastrophe theory* extends easily to multiple dimensions. The equilibria of this system are the points $(\mathbf{x}, y)$ such that $\dot{y} = 0$. A qualitative change is said to have occurred if, as a result of a change in the exogenous variable, the *number* of equilibria changes.

To fix ideas, consider the system $\dot{y} = 2x + 3y - y^3$. If $|x| > 1$, the system has only one equilibrium point, whereas if $|x| < 1$, there are three. The points $x = 1$ and $x = -1$ are the catastrophe points for this system. (See Fig. 1.)

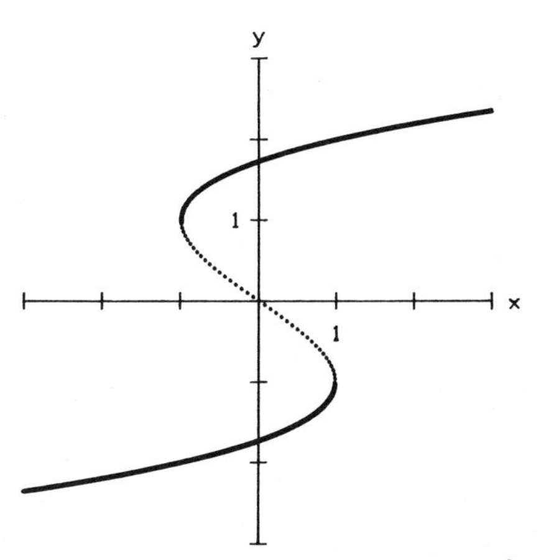

Figure 1 The equilibria of $\dot{y} = 2x + 3y - y^3$. The unstable portion is indicated with a dotted line.

The asymptotic states of a general dynamical system may be stable points, limit cycles, strange attractors (chaos), etc., but in the context of one-dimensional systems only stable points arise. An equilibrium point y_0 is asymptotically stable (an *attractor*) if a sufficiently small neighborhood of y_0 can be found within which the state variable moves asymptotically toward y_0. The point y_0 is a *repellor* if the state variable permanently departs every sufficiently small neighborhood of y_0 (unless, of course, its initial state is y_0). The attractors in Fig. 1 are drawn with a solid line, the repellors with a dashed line. Catastrophe points separate the attractors from the repellors. In an experiment in which an exogenous variable x is slowly varied from -2 to $+2$, observations of y would exhibit a slow decrease until just after $x = 1$, whereupon a rapid decrease (catastrophe) would be seen. Reversing the experiment does not reverse the observations: the return catastrophe does not occur until just after $x = -1$. This *hysteresis phenomenon* is characteristic of most catastrophe models.

In general the function h in (1) will not be a convenient polynomial. However, it is catastrophe theory's greatest strength to be able to tell us precisely which polynomial approximation to h will provide the same *qualitative* behavior with the minimum number of terms. These results can be illustrated by the following heuristic argument. Consider the case of a system known a priori to have at most three equilibria (two stable, one unstable). The fact that there are at most three equilibria suggests the use of a polynomial approximation to h of degree 3:

$$h(\mathbf{x}, y) = a(\mathbf{x}) + b(\mathbf{x})y + c(\mathbf{x})y^2 + d(\mathbf{x})y^3.$$

Now the problem is to specify the functions $a(\mathbf{x}), \dots, d(\mathbf{x})$. To do this, recall the method for discriminating between cubic polynomials that have three real roots and those that have only one. The cubic equation $a + by + cy^2 + dy^3 = 0$ can be converted to the form $\alpha + \beta z - z^3 = 0$ with the transformation

$$z = \sqrt[3]{d}\,[y - c/(3d)].$$

Cardan's discriminant for the cubic is now

$$\delta = (\alpha/2)^2 - (\beta/3)^3.$$

If $\delta > 0$ there are three real roots, while $\delta < 0$ implies just one. Because the discriminant depends only on α and β, there is no loss of *qualitative* generality entailed by the assumption that $c(\mathbf{x})$ and $d(\mathbf{x})$ are constants (i.e., do not depend on $\mathbf{x}$). This suggests that the two-dimensional control space spanned by α and β is the *minimum* necessary to reproduce all the possible qualitative behaviors of a system with cubic dynamics. Figure 2 shows a graph of the equilibria of $\dot{y} = \alpha + \beta y - y^3$ as a function of the parameters α and β. This is the famous *cusp catastrophe* model. Typically α and β depend nondegenerately on $\mathbf{x}$ around the cusp point (where $\alpha = \beta = 0$), so that α and β can be used as the essential control factors that determine the qualitative behavior of the system. The residual coordinates of $\mathbf{x}$ then have no importance qualitatively, although they may still be correlated with y.

Catastrophe theory has four major points of contact with contemporary statistical theory. These are in completely different areas: time-series* analysis, exponential families*, response surface analysis*, and classification of densities. In each case the connection is made by introducing suitable probabilistic elements into the deterministic models of catastrophe theory.

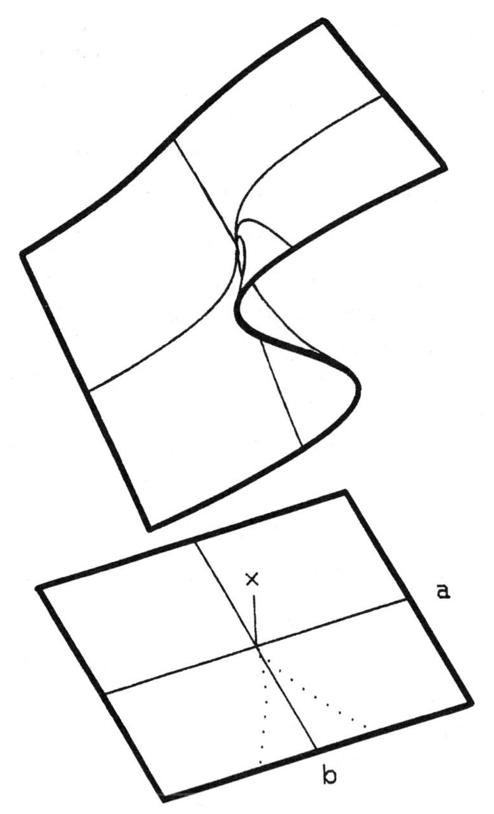

Figure 2 The equilibria of the cusp catastrophe model $\dot{y} = a + by - y^3$.

TIME-SERIES ANALYSIS

The cusp model described in the previous section can be adapted to the time-series framework by specifying its discrete-time analog, the stochastic difference equation

$$\Delta Y_t = a + bY_{t-1} + cY_{t-1}^2 + dY_{t-1}^3 + U_t, \quad (2)$$

where $\Delta Y_t = Y_t - Y_{t-1}$ and U_t is normally distributed and independent of Y_t. Note that this is autoregressive of order 1 and polynomial of degree 3. More complex cusp models could be specified, with a higher order of autoregression for instance, but this is the model that more nearly captures the essence of the continuous-time version.

As before, exogenous control variables can be introduced into the model by allowing a and b to be specified functions of these exogenous variables. For example, to implement the model whose equilibria are shown in Fig. 1, we simply replace the parameter a in the above model with

$$a(X) = a_0 + a_1 X_{t-1}.$$

To span the two-dimensional control space of Fig. 2 with linear functions of the exogenous variables, it is necessary to have at least two exogenous variables, say X_1 and X_2:

$$a(X_1, X_2) = a_0 + a_1 X_{1,t-1} + a_2 X_{2,t-1},$$
$$b(X_1, X_2) = b_0 + b_1 X_{1,t-1} + b_2 X_{2,t-1}.$$

When so specified, the complete model is linear in its parameters, and parameter estimation can be accomplished using standard techniques for polynomial regression*.

EXPONENTIAL FAMILIES

It is reasonable to wonder whether the simple discrete-time model of the previous section might lead to a continuous-time model by an appropriate limiting process as the increment of time (Δt) goes to zero. It turns out that the relevant theory does exist in the literature on stochastic differential equations* (Soong [3]), and that the models so produced have some fascinating and nontrivial properties. These lead ultimately to a large class of multimodal exponential families of great potential usefulness to statistics.

Let $\Delta Y_t = Y_{t+\Delta t} - Y_t$ and consider the discrete-time dynamic model

$$\Delta Y_t / \Delta t = h(X, Y_t) + U_t,$$

in which Δt is a given increment of time. The limit $\Delta t \to 0$ can be taken if we assume that $U_t = \Delta W_t / \Delta t$, where W_t is a standard Wiener process (the increments $\Delta W_t = W_{t+\Delta t} - W_t$ of a Wiener process are independent and normally distributed). The limit as $\Delta t \to 0$ is usually written as

$$dY_t = h(X, Y_t)\,dt + dW_t,$$

due to the nondifferentiability in mean square of W_t. A slightly more general stochastic differential equation (SDE) is

$$dY_t = h(X, Y_t)\,dt + \sigma(Y_t)\,dW_t. \quad (3)$$

This allows a kind of heteroscedasticity to

appear, because the function $\sigma^2(y)$ determines the variance of the random input; $\sigma^2(y)$ is usually called the *variance function* for the SDE (3), while the function $h(x, y)$ is called the *drift function*.

The probability density function (PDF) $f(y_t, t|x, y_0)$ of Y_t evolves with time. If we assume that all sample trajectories start from an initial position y_0 at time $t = 0$, then the initial PDF is a delta function located at this initial position. As t increases the PDF broadens out and evolves towards some stationary form, which is possibly degenerate (e.g., uniform over the entire real line). In most cases of interest, however, the stationary limiting PDF is nondegenerate. In any case the evolution of this PDF is governed by the forward Kolmogorov equation:

$$\partial_t f = -\partial_y(hf) + \partial_y^2(\sigma^2 f)/2. \quad (4)$$

The stationary density may be obtained by setting $\partial_t f = 0$ and integrating with respect to y. With the appropriate boundary conditions this integration yields

$$\partial_y(\sigma^2 f) = 2hf.$$

We now expand the derivative and divide by $\sigma^2 f$:

$$\partial_y f/f = (2h - \partial_y \sigma^2)/\sigma^2. \quad (5)$$

The solution to this differential equation is a conditional PDF $f(y|x)$, which is no longer dependent upon time or initial conditions. Its relevance depends upon the assumption, not always realistic, that convergence to the stationary form is more rapid than changes in x.

The differential equation (5) for the stationary density is remarkably similar to the equation defining the venerable Pearson system* of probability densities (Ord [2]):

$$f'(y)/f(y) = (a - y)/(b + cy + dy^2). \quad (6)$$

The Pearson equation (6) coincides with (5) when

$$\sigma^2(y) = b + cy + dy^2,$$

$$h(x, y) = 0.5(a + c) + (d - 0.5)y.$$

Thus a Pearson probability density can originate as the stationary density of a stochastic differential equation with a linear drift function and a quadratic (or lower) variance function.

The various densities of the Pearson system (normal*, gamma*, beta*, Student's t*, etc.) are classified according to the form taken by the variance function. By contrast, the various models of catastrophe theory are classified according to the maximum number of equilibria they can have. If we rewrite equation (5) as

$$f'(y)/f(y) = -g(y)/\sigma^2(y), \quad (7)$$

where the function $g(y)$ is defined as

$$g(y) = -2h(y) + \sigma^{2\prime}(y),$$

then the Pearson classification applies to the denominator of (7), while the catastrophe classification scheme can be applied to the numerator, i.e., to the function g. [The argument x in $h(x, y)$ has been suppressed in the above formulae for notational convenience. Of course, when h depends on x then so also does g.]

Recall that the equilibria of the system (1) are the points y such that $h(y) = 0$. If the variance function is a constant [e.g., $\sigma^2(y) = 1$], then these are also the points for which $g(y) = 0$, and therefore $f'(y) = 0$. Thus in this case the equilibria of the deterministic system coincide with the modes and antimodes of the stationary density of the stochastic system. The modes correspond to the stable equilibria, and the antimodes to the unstable equilibria.

When the variance function is not a constant the modes and antimodes are perturbed away from the stable and unstable equilibria, and the story becomes more interesting. Consider first the case in which $h(y)$ is linear, i.e., the original Pearson system. Then the normal density occurs when $\sigma^2(y) = 1$, the gamma density when $\sigma^2(y) = y$, the beta density when $\sigma^2(y) = y(1-y)$, and the inverse gamma density when $\sigma^2(y) = y^2$. For our purposes, we can consider these the principal Pearson types. Given any one of them, we can construct a set of catastrophe densities by letting the degree (d) of the function g increase beyond 1. These catastrophe types have the following names (d in parentheses): Fold (2), Cusp (3), Swallowtail (4), Butterfly (5), and Wigwam (6). Within each type the maximum total of modes and antimodes that a representative density can exhibit is given by the degree of g. In view of its importance, the function g has been called the *shape polynomial* for a density of this type (Cobb et al. [1]).

The general form for the densities that satisfy (7) is

$$f(y) = \eta \exp\left[-\int^y \{g(s)/\sigma^2(s)\}\, ds\right]. \quad (8)$$

The integral in this expression is indefinite, and the constant η is chosen so that each density integrates to 1 over its domain. If $\sigma^2(y)$ is one of the above principal types and $g(y)$ is a polynomial, then these densities are all exponential families parametrized by the coefficients of $g(y)$. Let us call then type N, G, B, or I, if the form of $\sigma^2(y)$ is 1, y, $y(1-y)$, or y^2, respectively. Thus the type G Fold density is

$$f(y) = \eta \exp\left[-\int^y \{(a + bs + cs^2)/s\}\, ds\right]$$
$$= \eta y^{-a} \exp\left[-by - cy^2/2\right].$$

These are the four principal Cusp densities:

$$\text{N:} \quad f(y) = \eta \exp\left[\alpha y + \beta y^2 + \gamma y^3 + \delta y^4\right], \quad -\infty < y < \infty,$$

$$\text{G:} \quad f(y) = \eta y^{\alpha} \exp\left[\beta y + \gamma y^2 + \delta y^3\right], \quad 0 < y < \infty,$$

$$\text{I:} \quad f(y) = \eta y^{\alpha} \exp\left[\beta/y + \gamma y + \delta y^2\right], \quad 0 < y < \infty,$$

$$\text{B:} \quad f(y) = \eta y^{\alpha}(1-y)^{\beta} \exp\left[\gamma y + \delta y^2\right], \quad 0 < y < 1.$$

These densities belong to multiparameter exponential families, and each can be either unimodal or bimodal. A graph of the modes and antimodes as a function of the first two coefficients of $g(y)$ for each of these densities would look exactly like Fig. 2.

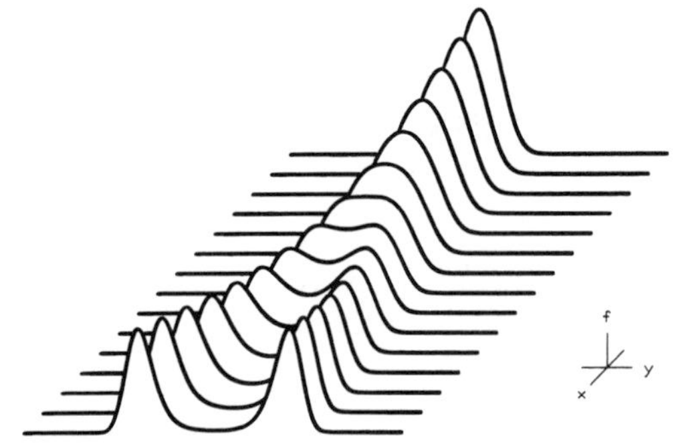

Figure 3 The symmetrical type N cusp densities, with $f(y|x) = \eta \exp[xy^2/2 - y^4/4]$.

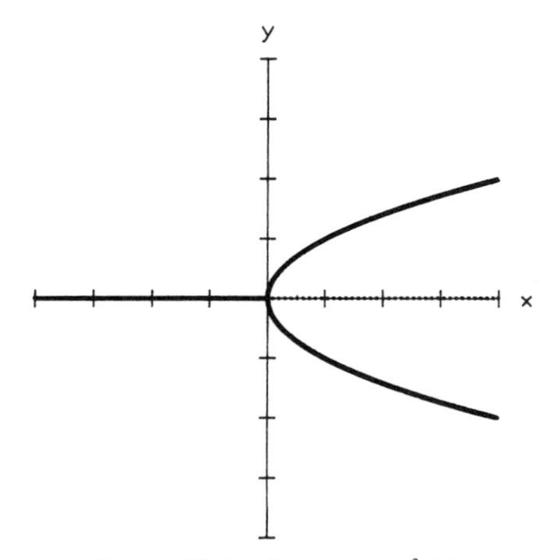

Figure 4 The equilibria of $\dot{y} = xy - y^3$. These are the modes and antimodes of the densities in Fig. 3.

When the function g depends upon an exogenous variable x, we obtain conditional PDFs in place of the above. For example, when $\sigma^2(y) = 1$ and $g(x, y) = xy - y^3$, we obtain

$$f(y|x) = \eta(x)\exp\left[xy^2/2 - y^4/4\right].$$

A sequence of graphs of this PDF for various values of x is shown in Fig. 3, with the corresponding graph of the equilibria of the deterministic system $\dot{y} = xy - y^3$ shown in Fig. 4.

Virtually all the catastrophe densities derivable from (7) are multiparameter exponential families. Standard results indicate that maximum likelihood estimators exist for each. Further, since each such density is a generalization of one of the principal Pearson types, it is possible to construct tests of hypotheses concerning the degree of g using the asymptotic normality* of the maximum likelihood* estimators. It is therefore possible to distinguish between the various catastrophe models using standard statistical techniques.

RESPONSE SURFACE ANALYSIS

Response surface* analysis is concerned with models in which the regression function $E[Y|x_1, \ldots, x_k]$ is specified as a quadratic form* in $(x_1 - \lambda_1), \ldots, (x_k - \lambda_k)$. This type of model is appropriate when it is desired to find the precise vector of values for $\mathbf{x} = (x_1, \ldots, x_k)$ such that the regression function attains its maximum (or minimum) value. The statistical task is to estimate the optimal setting vector $\mathbf{x} = \boldsymbol{\lambda}$ from an independent set of observations of $(Y, \mathbf{x})$.

A response surface like that in Fig. 2 permits both the hysteresis effect seen in Fig. 1 and the bifurcation effect seen in Fig. 4, neither of which are possible in the usual (quadratic) formulation of response surface models. Estimation of the system parameters proceeds by an extension of the previous methods. Let the function g of equation (7) have the form

$$g(y, \mathbf{x}) = a(\mathbf{x}) + b(\mathbf{x})y + cy^2 + dy^3.$$

The predictions for Y given $\mathbf{x}$ are the values of Y such that $a(y, \mathbf{x}) = 0$. In the case of multiple predictions, modes and antimodes alternate. The PDF of Y given $\mathbf{x}$ is

$$f(y|\mathbf{x}) = \eta(\mathbf{x})\exp\left[-\int^{y} \{g(s, \mathbf{x})/\sigma^2(s)\}\, ds\right].$$

Now the maximum likelihood methods mentioned in the previous section can be employed to estimate the coefficients of g and to test their values against common null hypotheses. For example, the null hypothesis that $b(\mathbf{x}) =$ constant and $c = d = 0$ is equivalent to the statement that the regression function of Y given $\mathbf{x}$ is planar with Pearson-distributed residuals. The further restriction $\sigma^2(y) = 1$ reduces this to the usual multiple regression model with normally distributed residuals. This, in brief, describes the use of the Cusp catastrophe models in response surface analysis.

EQUIVALENCE OF DENSITIES

Catastrophe theory proper is primarily a set of concepts and theorems by which an equivalence relation can be defined on families of smooth functions. Each polynomial catastrophe model represents an equivalence class of families. A connection between catastrophe theory and statistics can be made by an attempt to apply the above ideas directly to families of probability density functions. Unfortunately, this attempt seems to fail almost at once, because of the requirement that densities integrate to 1. To see why, it is necessary to explore the equivalence relation used by catastrophe theory.

Two functions f and g are *source-equivalent* if there exists a constant γ and diffeomorphism ρ such that $g = \gamma + f \circ \rho$. A *diffeomorphism* is an invertible function with derivatives of all orders and an inverse with derivatives of all orders. (Arctan and exp are diffeomorphisms, but x^3 is not if the domain includes zero.)

The problem with applying source-equivalence to PDFs lies in the interpretation of the transformation ρ. Ordinarily, if one random variable is a transformation of another, say $Y = \rho(X)$, where Y has PDF f and X has PDF g, then the relationship between their densities is:

$$g = J_\rho f \circ \rho,$$

where J_ρ stands for the Jacobian* determinant of ρ. It is the function $J_\rho(x)$ that creates the problem.

A resolution of this problem is possible if we define two random variables to be *density-equivalent* if there exists a constant γ and diffeomorphism ρ such that $g = \gamma f \circ \rho$. Under this definition it is not true that $Y = \rho(X)$, but it is true that their densities have many qualitative features in common. Among other things, it is possible to prove that the modes of f are preserved under ρ. This definition of equivalence is easily extended to families, after which it may be seen that the Cauchy, normal, gamma, and beta densities all fall into one equivalence class, while the four cusp densities and various two-population mixture* densities fall into a second class, and so on. The double exponential*, $\exp(-|x|)$, falls into a residual class of densities that are not classifiable by these methods (because they violate the necessary smoothness conditions). Thus the outlook is good for an adaptation of catastrophe theory that will provide a useful classification of probability density functions.

References

[1] Cobb, L., Koppstein, P., and Chen, N. B. (1983). Estimation and moment recursion relations for multimodal exponential distributions. *J. Amer. Statist. Ass.*, **78**, 124–130.

[2] Ord, J. K. (1972). *Families of Frequency Distributions*. Hafner, New York.

[3] Soong, T. T. (1973). *Random Differential Equations in Science and Engineering*. Academic, New York.

Bibliography

Cobb, L. and Watson, W. B. (1980). Statistical catastrophe theory: An overview. *Math. Modelling*, **1**, 311–317.

Cobb, L. and Zacks, S. (1985). Applications of catastrophe theory for statistical modeling in the biosciences, *J. Amer. Statist. Soc.*, **80**, 793–802.

Poston, T. and Stewart, I. (1978). *Catastrophe Theory and Its Applications*. Pitman, London, England.

Thom, R. (1975). *Structural Stability and Morphogenesis*. Benjamin, New York. (Translated from French by D. H. Fowler.)

Zeeman, E. C. (1977). *Catastrophe Theory: Selected Papers*. Addison-Wesley, Reading, MA.

(CATASTROPHE THEORY
EXPONENTIAL FAMILIES
FREQUENCY CURVES, SYSTEMS OF
RESPONSE SURFACE DESIGNS
STOCHASTIC DIFFERENTIAL EQUATIONS
TIME SERIES)

LOREN COBB

STATISTICAL COMPUTATION AND SIMULATION, JOURNAL OF

The *Journal of Statistical Computation and Simulation* began publication in January 1972 in order to accommodate the rapidly growing but difficult-to-publish field of statistical computation and simulation. The journal started publishing with one volume per year, then expanded to two volumes and at present has a publish-as-required policy that prevents backlog and publication delays. Dr. Richard G. Krutchkoff, of Virginia Polytechnic Institute and State University, was the founding editor and is still editor of the journal.

The *Journal* is truly an international one, having authors and subscribers from more than 30 countries including Third World and Soviet bloc countries. All articles, however, are published in English.

The *Journal* publishes significant and original work in areas of statistics that are related to, or dependent upon, the computer. Fields covered included computer algorithms related to probability or statistics, studies in statistical inference by means of simulation techniques, and implementation of interactive statistical systems. There is a little used section for tutorials on topics of interest to our readers, but which might not be totally original. There is a camera-ready section called "Comments, Conjectures and Conclusions," edited by I. J. Good, which includes brief articles of general interest. The *Journal* usually ends with news, notices, and abstracts.

Manuscripts intended for consideration should be sent in triplicate to Dr. R. G.

Krutchkoff, Department of Statistics, Virginia Polytechnic Institute and State University, Blacksburg, Virginia 24061.

R. G. Krutchkoff

STATISTICAL CONFIDENTIALITY

The grist of statistical data products is data on individual subjects (persons, households, businesses, etc.). The data producer typically receives such data under an obligation to treat them as *confidential*, i.e., not to be divulged to a third party. A social contract is formed, whereby the subject relinquishes some informational privacy; in return the producer agrees to guard against disclosure to others of the subject's data in identifiable form. The assurance of confidentiality requires that the data be held securely, with access limited and strictly controlled, that enforceable sanctions against unauthorized release be established and made widely known, and that derived data products are secure from *statistical disclosure*. Such disclosure could result from the release of microdata containing sufficient information to identify a respondent and thereby divulge additional data present on the microrecord, from the release of small frequency counts from which precise information on individuals could be inferred, or from the release of aggregates of magnitude data from which a narrow estimate of individual subject data could be derived. *Statistical confidentiality* is preserved when the risk of statistical disclosure is sufficiently reduced. Reference 12 provided an overview to approaches to preserving statistical confidentiality in the U.S. federal statistical system in the 1970s. Reference 4 described the approach taken by the U.S. Census Bureau in 1985.

Techniques for preserving statistical confidentiality vary according to the release format for the data product and factors affecting respondent identifiability, such as sampling rates and the presence of identifying information available to a third party. Such techniques *mask* true data values sufficiently to ensure that attempts at identification are thwarted or, where reidentification cannot be prevented, that for identifiable subjects unequivocal attribution of characteristics or narrow estimation of magnitude data cannot be made. Several techniques are available, although not in widespread use.

Reference 5 synopsized the state-of-the-art in confidentiality preserving methodology in 1979. Reference 11 described the confidentiality problem in microdata in relation to modern record linkage technology, and demonstrated the vulnerability of microdata masking methods in current use to attacks via linkage. References 10, 9, and 7, and 3 introduced different methods applicable to frequency count tables, respectively, random perturbation, random rounding, and controlled rounding of counts. Reference 1 described how cell suppression can preserve confidentiality in tables of aggregate magnitude data. Reference 2 dealt with defining disclosure in statistical tables. Reference 6 provided an overview and references to the confidentiality problem in statistical databases open to public use. An area less developed in the literature is that of assessing and controlling the distorting effects of masking on subsequent analyses of the data. This is the "statistical" side of the statistical confidentiality problem.

The salient ethical issues arising from confidentiality concerns are the responsibility of the data producer to develop and apply confidentiality preserving methods to released data (discussed above), the collection and use of the data *for statistical purposes only* [12], and the issue of how to achieve *informed consent* when data may be shared, combined, or reanalyzed for future statistical purposes [8]. An emerging ethical concern is the responsibility of the data producer to *subgroups* that become clearly identified by the released statistics. If such subgroups could be singled out for increased scrutiny from official sources or targeted for receipt of large quantities of junk mail based on market studies, should the producer censor the data before release [4]? Another important component is the relationship between the pledge or assurance of confidentiality and the willingness of subjects to respond truthfully to statistical inquiries [4].

References

[1] Cox, L. (1980). Suppression methodology and statistical disclosure control. *J. Amer. Statist. Ass.*, **75**, 377–385.

[2] Cox, L. (1981). Linear sensitivity measures in statistical disclosure control. *J. Statist. Plann. Inf.*, **5**, 153–164.

[3] Cox, L. and Ernst, L. (1982). Controlled rounding. *INFOR—Canad. J. Oper. Res. Inf. Process.*, **20**, 423–432.

[4] Cox, L., Johnson, B., McDonald, S-K., Nelson, D., and Vasquez, V. (1985). Confidentiality issues at the Census Bureau. *Proc. First Annual Census Bureau Res. Conf.*, pp. 199–218.

[5] Cox, L. and Sande, G. (1979). Techniques for preserving statistical confidentiality. *Bull. Int. Statist. Inst., Proc. 42nd Sess.*, pp. 499–512.

[6] Denning, D. (1982). *Cryptography and Data Security*, Addison-Wesley, Reading, MA.

[7] Fellegi, I. (1975). Controlled random rounding. *Survey Meth.*, **1**, 123–135.

[8] Gastwirth, J. (1986). Comment. *J. Amer. Statist. Ass.*, **81**, 43–47.

[9] Nargundkar, M. and Saveland, W. (1972). Random rounding to prevent statistical disclosure. *Proc. Social Statist. Sec., Amer. Statist. Ass.*, 1972, 382–385.

[10] Newman, D. (1975). Techniques for ensuring the confidentiality of census information in Great Britain. *Bull. Int. Statist. Inst., Proc. 40th Sess.*

[11] Paass, G. (1985). Disclosure risk and disclosure avoidance for microdata. Presented at the IAS-SIST Conference on Public Access to Public Data, Amsterdam.

[12] U.S. Department of Commerce, (1978). *Statistical Policy Working Paper 2: Report on Statistical Disclosure and Disclosure-Avoidance Techniques*. U.S. Government Printing Office, Washington, DC.

(PRINCIPLES OF PROFESSIONAL STATISTICAL PRACTICE)

LAWRENCE H. COX

STATISTICAL CONTROL *See* CONTROL CHARTS

STATISTICAL CURVATURE

For statistical inference*, statisticians often use a parametrized statistical model of the form $M = \{p(x, \theta)\}$, where $p(x, \theta)$ is a probability density function specified by a (vector) parameter θ of a vector random variable x with respect to some measure on the sample space. A model M occupies only a part of the function space $F = \{p(x)\}$ of all the density functions $p(x)$ of regular distributions, and M forms an n-dimensional manifold* in F under certain regularity conditions, where n is the number of dimensions of θ. Here the following questions naturally arise: What are the geometric characteristics of M in F? Is it flat or curved? If curved, what role does the curvature play in statistical inference?

It was Efron [6] who first introduced statistical curvature and pointed out its importance in statistics. However, in order to answer the above questions completely, it is necessary to introduce differential-geometrical notions, such as a Riemannian metric and an affine connection. A Riemannian metric was first introduced in statistics by Rao [12], and affine connections by Chentsov [5]. A differential-geometric framework was given for statistical inference by Amari [1], where α-affine connections and α curvatures were introduced; statistical α curvatures play a fundamental role in the higher-order asymptotic properties of statistical inference [1–4] (see also [7–11]). It is expected that the geometrical approach will grow to provide an indispensible theoretical method of statistics, which not only is useful for higher-order asymptotic theory but is applicable to wider realms of statistics [3, 11].

STATISTICAL CURVATURE

This is introduced here following Efron [6] in an intuitive way. Let $M = \{p(x, \theta)\}$ be a scalar parameter model of the form

$$p(x,\theta) = g(x)\exp\left[\sum_i \xi^i(\theta)x_i - \psi\{\xi(\theta)\}\right],$$

where $x = (x_i)$ and $\xi = (\xi^i)$, $i = 1,\ldots,n$. This is called an $(n,1)$-*curved exponential family**, because M is represented by a curve $\xi = \xi(\theta)$ in the n-dimensional natural parameter space $S = \{\xi\}$ of the exponential family of distributions. The curvature of M represents the deviation of M from exponentiality. The tangent vector of the curve at θ is a vector $\dot{\xi}(\theta)$, where the overdot denotes $d/d\theta$. The curvature is measured by the changing rate in the direction of the tangent vector $\dot{\xi}(\theta)$ as θ changes. Let Δ be the angle between two tangent vectors $\dot{\xi}(\theta)$ and $\dot{\xi}(\theta + d\theta)$, calculated from

$$\cos\Delta = \frac{\dot{\xi}(\theta)\cdot\dot{\xi}(\theta+d\theta)}{|\dot{\xi}(\theta)||\dot{\xi}(\theta+d\theta)|},$$

by the use of the inner product of two vectors. Then, the *curvature* γ is given by $\gamma = \Delta/d\theta$ in the limit as $d\theta \to 0$. The inner product of two vectors $\dot{\xi}(\theta)$ and $\dot{\xi}(\theta + d\theta)$ is defined by the quadratic form

$$\sum g_{ij}(\theta)\dot{\xi}^i(\theta)\dot{\xi}^j(\theta + d\theta),$$

where g_{ij} is the Fisher information* matrix at $\xi = \xi(\theta)$. Calculations yield the explicit form of the statistical curvature γ,

$$\gamma^2 = \left(|\dot{\xi}|^2|\ddot{\xi}|^2 - |\dot{\xi}\cdot\ddot{\xi}|^2\right)/|\dot{\xi}|^3.$$

For a general scalar parameter statistical model $M = \{p(x,\theta)\}$, the random variable $\dot{l}(x,\theta)$, where $l(x,\theta) = \log p(x,\theta)$, can be regarded as the tangent vector of the curve M in F. Hence, the statistical curvature is defined by replacing $\dot{\xi}$ by $\dot{l}$ and $\ddot{\xi}$ by $\ddot{l}$ in the above, where the inner product of $a(x)$ and $b(x)$ is defined by their covariance with respect to $p(x,\theta)$.

α CURVATURE OF STATISTICAL MODELS

The idea of statistical curvature can be extended to yield the α curvature in a general

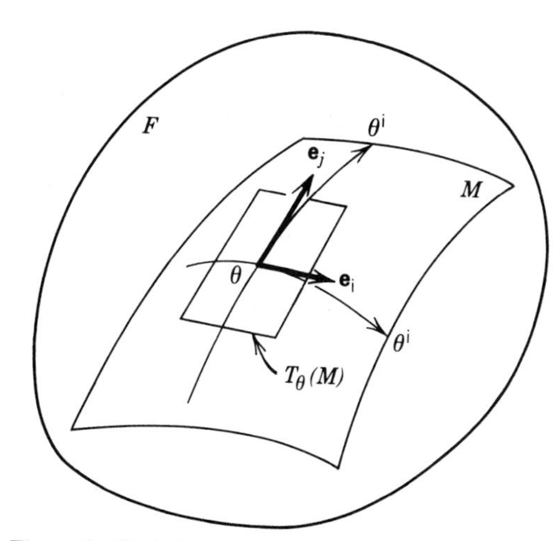

Figure 1 Statistical model M, tangent space $T_\theta(M)$, and basis vectors $\mathbf{e}_i$.

statistical model as follows. Let $M = \{p(x,\theta)\}$ be a regular statistical model imbedded in the set $F = \{p(x)\}$ of mutually absolutely continuous distributions. A point $p(x,\theta)$ of M is specified by the vector parameter $\theta = (\theta^i) = (\theta^1,\ldots,\theta^n)$. Hence, M can be regarded as an n-dimensional manifold, and θ plays the role of a coordinate system. For $l(x) = \log p(x)$, its small deviation $\delta l(x) = \delta p(x)/p(x)$ satisfies $E_p[\delta l(x)] = 0$, where E_p is the expectation with respect to $p(x)$. Therefore, we may regard the linear space of the random variables $T_p = \{r(x)|E_p[r(x)] = 0\}$ as the tangent space at $p(x)$ of F (see Dawid's discussion in [6]). The tangent space $T_\theta(M)$ at $p(x,\theta)$ or shortly at point θ of M is a vector space spanned by n random variables $\mathbf{e}_i = \partial_i l(x,\theta)$, $i = 1,\ldots,n$, where $\partial_i = \partial/\partial\theta^i$ is the partial derivative with respect to the ith component of the vector parameter $\theta = (\theta^i)$. Hence $\{\mathbf{e}_i\}$ defines a basis for the vector space $T_\theta(M)$ that is a subspace of T_p, $p = p(x,\theta)$. The vector $\mathbf{e}_i = \partial_i l$, $i = 1,\ldots,n$, is the tangent vector of the coordinate curve θ^i in the direction of increasing the ith coordinate θ^i (Fig. 1). The inner product is naturally introduced in T_p by

$$\langle s,t\rangle = E[s(x)t(x)],$$
$$s = s(x),\ t = t(x) \in T_p.$$

Two vectors s and t are orthogonal when $\langle s,t\rangle = 0$. The inner product of the basis vectors $\mathbf{e}_i$ and $\mathbf{e}_j$ of $T_\theta(M)$ is

$$g_{ij}(\theta) = \langle \mathbf{e}_i, \mathbf{e}_j\rangle = E\left[\partial_i l(x,\theta)\partial_j l(x,\theta)\right],$$

which defines a *Riemannian metric* in M. This g_{ij} is the Fisher information matrix.

The curvature of M in F can be obtained by knowing how the directions of the tangent space $T_\theta(M)$ or its basis $\{\mathbf{e}_i\}$ change as the point θ moves. The random variable $\partial_i\,\partial_j l(x,\theta)$ might be expected to represent the rate of change in $\mathbf{e}_j(\theta)$ when θ moves in the direction of $\mathbf{e}_i$. However, this does not belong to $T_{p(x,\theta)}$ since $E[\partial_i\,\partial_j l] = -g_{ij}(\theta) \neq 0$. We modify it as

$$l_{ij}^{(\alpha)}(x,\theta) = \partial_i\partial_j l + \frac{1-\alpha}{2}\partial_i l\,\partial_j l + \frac{1+\alpha}{2}g_{ij},$$

where α is a real parameter and $E[l_{ij}^{(\alpha)}] = 0$. The quantity $l_{ij}^{(\alpha)}$, called the *α-covariant derivative* of $\mathbf{e}_j$ in the direction of $\mathbf{e}_i$ in F, represents how the tangent vector $\mathbf{e}_i$ at θ changes intrinsically as θ moves in the direction of $\mathbf{e}_i$ in the sense of the α-affine connection.

The vector $l_{ij}^{(\alpha)} \in T_p$ can be decomposed into two parts; one, denoted by $H_{ij}^{(\alpha)}(x,\theta)$, is orthogonal to the tangent space $T_\theta(M)$. This is the *α-curvature tensor* of M in F, because it shows how the directions of the tangent space $T_\theta(M)$ change as θ moves in M. The other, denoted by $\nabla_i^{(\alpha)}\mathbf{e}_j$, is called the *$\alpha$-covariant derivative* of $\mathbf{e}_j$ in the direction $\mathbf{e}_i$ in M. It is the projection of $l_{ij}^{(\alpha)}$ onto $T_\theta(M)$. It represents how the coordinate system θ is curved in M. By using the components of the α-affine connection of M defined by

$$\Gamma_{ijk}^{(\alpha)} = \left\langle l_{ij}^{(\alpha)}, \mathbf{e}_k\right\rangle = E\left[(\partial_k l)l_{ij}^{(\alpha)}\right],$$

they are explicitly given as

$$\nabla_i^{(\alpha)}\mathbf{e}_j = \sum_{k,m}\Gamma_{ijk}^{(\alpha)}g^{km}\mathbf{e}_m,$$

$$H_{ij}^{(\alpha)} = l_{ij}^{(\alpha)} - \Gamma_i^{(\alpha)}\mathbf{e}_j,$$

where g^{km} is the inverse matrix of g_{mk}. Intuitively, the one-curvature ($\alpha = 1$) represents the deviation from the exponential family, because it vanishes for an exponential family; the -1-curvature ($\alpha = -1$) represents the deviation from the mixture family, because it vanishes for a mixture family.

The square of the α-curvature is given by

$$R_{ij}^{(\alpha)} = \sum_{k,m} g^{km}\left\langle H_{ik}^{(\alpha)}(x), H_{jm}^{(\alpha)}(x)\right\rangle.$$

This is a nonnegative definite matrix. The statistical curvature γ^2 defined by Efron [6] in a one-dimensional model M is $\gamma^2 = \Sigma g^{ij}R_{ij}^{(1)}$, which is a scalar.

STATISTICAL ROLE OF CURVATURE

An *(m,n)-curved exponential family*, $M = \{p(x,\theta)\}$ is an n-dimensional statistical model imbedded in an m-dimensional exponential family $S = \{q(x,\xi)\}$ by $p(x,\theta) = q[x,\xi(\theta)]$, i.e., $\xi = \xi(\theta)$. The statistical role of the α curvature will be explained in a curved exponential family, although similar discussions hold in the general case by treating the fibre bundle of local exponential families*. Given a large number N of independent observations $x_1,\ldots,x_N$ from $p(x,\theta) \in M$, their arithmetic mean $\bar{x} = (1/N)\Sigma x_i$ defines a distribution $q(x,\bar{\xi})$ in S such that the expectation of x under $q(x,\bar{\xi})$ is equal to $\bar{x}$, i.e., the expectation parameter $\bar{\eta}$ of the distribution $q(x,\bar{\xi})$ in the exponential family S is equal to $\bar{x}$.

Since the sufficient statistic* $\bar{x}$ defines a point $\bar{\eta}$ in S, an estimator $\hat{\theta} = \hat{\theta}(\bar{\eta})$ is a mapping from S to M, where the sample space is identified with S by using the expectation parameter η. Let $A(\theta)$ be the inverse image of the mapping $\hat{\theta}$,

$$A(\theta) = \left\{\eta \in S|\hat{\theta}(\eta) = \theta\right\},$$

which forms an $(m-n)$-dimensional submanifold in S (Fig. 2). The properties of an estimator are evaluated by the geometrical characteristics of the associated $A(\theta)$'s. First,

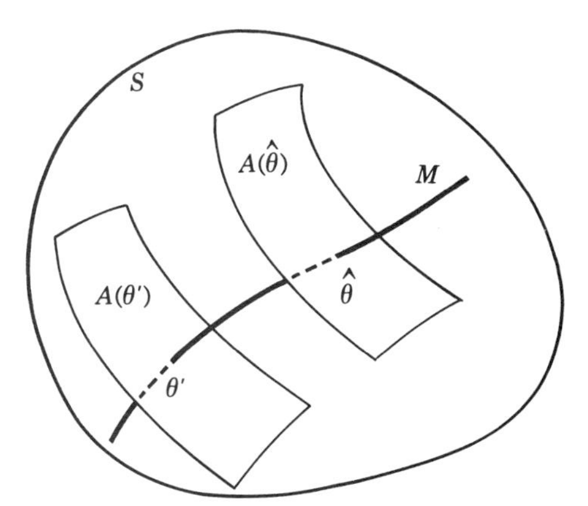

Figure 2 Submanifold $A(\theta) = \hat{\theta}^{-1}(\theta)$ associated with estimator $\hat{\theta}$.

it is consistent iff $\theta \in A(\theta)$ for all $\theta \in M$. A consistent estimator is efficient iff the tangent space of $A(\theta)$ is orthogonal to $T_\theta(M)$ at θ for all $\theta \in M$. In order to compare the higher-order efficiencies of efficient estimators, we modify an efficient estimator $\hat{\theta}$ into $\hat{\theta}^* = \hat{\theta} - b(\hat{\theta})$, where $b(\theta) = E[\hat{\theta}] - \theta$ is the bias of the estimator $\hat{\theta}$ when the true parameter is θ. The covariance matrix of $\hat{\theta}^*$ is expanded in the power series

$$\operatorname{cov}[\hat{\theta}^{*^i}, \hat{\theta}^{*^j}] = (1/N)g_1^{ij} + (1/N^2)g_2^{ij} + O(1/N^3),$$

where the first-order term g_1^{ij} coincides with the inverse g^{ij} of the Fisher information matrix (Cramer–Rao theorem). The second-order term g_2^{ij} or its covariant expression

$$g_{2ij} = \Sigma g_2^{mk} g_{ki} g_{mj}$$

is decomposed into the sum of three nonnegative matrices [1, 3]:

$$g_{2ij} = R_{ij}^{(1)}(M) + \tfrac{1}{2}R_{ij}^{(-1)}(A) + \tfrac{1}{2}\Gamma_{ij}^{(-1)}.$$

The first term is the square of the 1-curvature of the model M. The second is a half of the square of the -1-curvature of the submanifold $A(\theta)$, which depends on the estimator $\hat{\theta}$. The third term is a half of the square of the -1-connection

$$\Gamma_{ij}^{(-1)} = \Sigma \Gamma_{sti}^{(-1)} \Gamma_{krj}^{(-1)} g^{sk} g^{rt}.$$

The second is the only term depending on the estimator $\hat{\theta}$. Hence, an efficient estimator is higher-order efficient when it has -1 flat $A(\theta)$, i.e., when its -1-curvature vanishes. The maximum likelihood* estimator has -1 flat $A(\theta)$ and hence is higher-order efficient.

Let M be a scalar parametric model, and consider the problem of testing a null hypothesis $H_0 : \theta = \theta_0$ against the others, $H_1 : \theta > \theta_0$ in the one-sided case and $H_1 : \theta \neq \theta_0$ in the two-sided case. A test T is determined by the critical region R, which is a subset of S such that H_0 is rejected when $\bar{\eta} = \bar{x}$ falls in R. R is bounded by two $(n-1)$-dimensional submanifolds $A(\theta_-)$ and $A(\theta_+)$ (cf. Fig. 2, where $\theta_- = -\infty$ in the one-sided case). A test T is asymptotically (first-order) uniformly efficient when $A(\theta_-)$ and $A(\theta_+)$ are asymptotically orthogonal to M. However, in order to get higher-order efficiency, it is necessary to incline $A(\theta_\pm)$ slightly depending on the 1-curvature of M. Various widely used efficient tests have different inclinations of $A(\theta_\pm)$, so that they have different higher-order characteristics. Let $\Delta P_T(t, \alpha)$ be the deficiency of the power of a test T of significance level α at $\theta_t = \theta_0 + t/\sqrt{Ng}$,

$$\Delta P_T(t, \alpha) = \lim_{N \to \infty} \{N P_{\text{env}}(\theta_t, \alpha) - N P_T(\theta_t, \alpha)\},$$

where $P_T(\theta, \alpha)$ and $P_{\text{env}}(\theta, \alpha)$ are, respec-

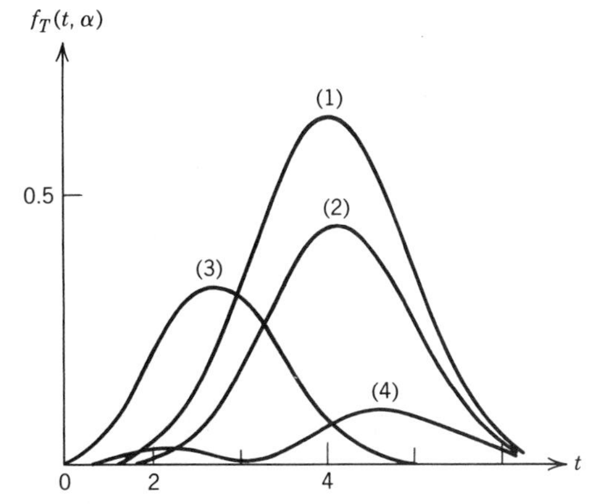

Figure 3 Deficiency of various unbiased tests ($\alpha = 0.05$); (1) Rao test (efficient-score test); (2) locally most powerful test; (3) Wald test (test based on the MLE); (4) likelihood-ratio test.

tively, the power functions of test T and the envelope function, and g is the Fisher information. Then there exists a universal function $f_T(t, \alpha)$ for test T such that $\Delta P_T(t, \alpha) = f_T(t, \alpha)\gamma^2$, i.e., the deficiency depends on the model M only through the statistical curvature γ. Figure 3 shows $f_T(t, \alpha)$ for various two-sided unbiased tests [3, 9]. The likelihood-ratio test has a good performance for a wide range of θ_t.

The curvatures play an important role also in analyzing interval estimation, conditional inference*, information and ancillarity, etc. [3].

References

[1] Amari, S. (1982). *Ann. Statist.*, **10**, 357–387.

[2] Amari, S. (1982). *Biometrika*, **69**, 1–17. [Correction (1983), **70**, 303.]

[3] Amari, S. (1985). Differential geometry of statistics. *Springer Lecture Notes in Statistics*, **28**.

[4] Amari, S. and Kumon, M. (1983). *Ann. Inst. Statist. Math.*, **35**, 1–24.

[5] Chentsov, N. N. (1982). *Statistical Decision Rules and Optimal Conclusions*. American Mathematical Society, Providence, RI. (English translation of the Russian book published in 1972, Nauka, Moscow.)

[6] Efron, B. (1975). *Ann. Statist.*, **3**, 1189–1242.

[7] Efron, B. (1978). *Ann. Statist.*, **6**, 362–376.

[8] Efron, B. and Hinkley, D. V. (1978). *Ann. Statist.*, **6**, 362–376.

[9] Kumon, M. and Amari, S. (1983). *Proc. R. Soc. Lond. A*, **387**, 429–458.

[10] Madsen, T. L. (1979). *Research Report No. 79-1*, Statistics Research Unit, Danish Council.

[11] Nagaoka, H. and Amari, S. (1982). *Technical Report, MEJR 82-7*, University of Tokyo, Japan.

[12] Rao, C. R. (1945). *Bull. Calcutta Math. Soc.*, **37**, 81–91.

(EFFICIENCY, SECOND ORDER
FISHER INFORMATION
PROBABILITY SPACES, METRICS AND DISTANCES ON)

S. AMARI

STATISTICAL DIFFERENTIALS, METHOD OF

This is a method of approximating the expected value of a function of random variables. It consists essentially of expanding the function as a Taylor* series about the expected values of the random variables and then taking expected values, term by term, using only a few terms in the expansion.

Formally, with $E[X_j] = \xi_j$ $(j = 1, \ldots, m)$,

$$f(X_1, \ldots, X_m) = f(\xi_1, \ldots, \xi_m) + \sum_{j=1}^{m} (X_j - \xi_j)\frac{\partial f}{\partial \xi_j} + \frac{1}{2!}\left(\sum_{j=1}^{m} (X_j - \xi_j)\frac{\partial}{\partial \xi_j}\right)^2 f + \cdots, \tag{1}$$

where

$$\frac{\partial f}{\partial \xi_j} = \left.\frac{\partial f(X_1, \ldots, X_m)}{\partial X_j}\right|_{\mathbf{x}=\boldsymbol{\xi}}$$

and

$$\left(\sum_{j=1}^{m} (X_j - \xi_j)\frac{\partial}{\partial \xi_j}\right)^r$$

is to be expanded as an operator, so that the last term in (1) is

$$\frac{1}{2}\left[\sum_{j=1}^{m} (X_j - \xi_j)^2 \frac{\partial^2 f}{\partial \xi_j^2} + 2\sum_{i<j}^{m-1} \sum_{i<j}^{m} (X_i - \xi_i)(X_j - \xi_j)\frac{\partial^2 f}{\partial \xi_i \partial \xi_j}\right].$$

Taking expected values we obtain, stopping with the terms shown in (1),

$$E[f(X_1, \ldots, X_m)] \doteqdot f(\xi_1, \ldots, \xi_m) + \frac{1}{2}\left\{\sum_{j=1}^{m} \mathrm{var}(X_j)\frac{\partial^2 f}{\partial \xi_j^2} + 2\sum_{i<j}^{m-1} \sum_{i<j}^{m} \mathrm{cov}(X_i, X_j)\frac{\partial^2 f}{\partial \xi_i \partial \xi_j}\right\}$$

(note that $E[X_j - \xi_j] = 0$). Of course,

higher-order terms can be included as desired, though the expressions rapidly become more elaborate.

The deviations of the random variables from their expected values $(X_j - \xi_j)$ are called *statistical differentials*. They are often denoted by the compound symbol (δX_j), giving rise to the alternative name of *delta method* for the way of obtaining these approximations.

It should be noted that the approach is purely formal. There is no guarantee that the sum of expected values of terms on the right-hand side of (1), even if taken to infinity, will equal the expected value of $f(X_1, \dots, X_m)$. Indeed for some values of the X's the series may not even converge. The smaller the deviations $(X_j - \xi_j)$ the better the approximation to be expected, broadly speaking.

In the special case of a single variable X with expected value $E[X] = \xi$ and central moments* $\{\mu_r\}$ (with $\mu_2 = \sigma^2$):

$$E[f(X)] \doteqdot f(\xi) + \tfrac{1}{2}\sigma^2 f''(\xi) + \tfrac{1}{6}\mu_3 f'''(\xi) + \tfrac{1}{24}\mu_4 f''''(\xi) + \cdots .$$

Further, we obtain

$$\begin{aligned}\operatorname{var}(f(X)) &\doteqdot E\left[\left\{(x-\xi)f'(\xi) + \tfrac{1}{2}\{(x-\xi)^2\}f''(\xi)\right\}^2\right] \\ &\doteqdot \sigma^2[f'(\xi)]^2 + \mu_3 f'(\xi)f''(\xi)\end{aligned}$$

to terms of third order in the statistical differentials. Retaining only the first term of (2) gives

$$\operatorname{var}(f(X)) \doteqdot \sigma^2[f'(\xi)]^2,$$

which is commonly used as a quick approximation of $\operatorname{var}(f(X))$. This formula has been used to derive many (approximately) variance-equalizing transformations, from the condition

$$\sigma^2[f'(\xi)]^2 = \text{constant},$$

where σ is a function of ξ.

Another special case of common application is approximating the variance of a ratio X_1/X_2. Using the statistical differential notation,

$$\begin{aligned}\frac{X_1}{X_2} &= \frac{\xi_1}{\xi_2}\,\frac{1 + (\delta X_1)/\xi_1}{1 + (\delta X_2)/\xi_2} \\ &= \frac{\xi_1}{\xi_2}\left\{1 + \frac{(\delta X_1)}{\xi_1}\right\} \\ &\quad \times \left\{1 - \frac{(\delta X_2)}{\xi_2} + \frac{(\delta X_2)^2}{\xi_2^2} - \cdots\right\},\end{aligned}$$

where, retaining terms up to second order in statistical differentials,

$$E\left[\frac{X_1}{X_2}\right] \doteqdot \frac{\xi_1}{\xi_2}\left\{1 - \frac{\operatorname{cov}(X_1, X_2)}{\xi_1\xi_2} + \frac{\operatorname{var}(X_2)}{\xi_2^2}\right\}.$$

The formula

$$\operatorname{var}\left(\frac{X_1}{X_2}\right) \doteqdot \left(\frac{\xi_1}{\xi_2}\right)^2 \times \left\{\frac{\operatorname{var}(X_1)}{\xi_1^2} - \frac{2\operatorname{cov}(X_1, X_2)}{\xi_1\xi_2} + \frac{\operatorname{var}(X_2)}{\xi_2^2}\right\}$$

can be obtained in a similar way.

For a formal discussion of regularity conditions, rates of convergence, and so on, see Chapter 14 of Bishop et al. [1].

References

[1] Bishop, Y. M. M., Fienberg, S. E., and Holland, P. W. (1975). *Discrete Multivariate Analysis: Theory and Practice*. MIT Press, Cambridge, MA.

(ASYMPTOTIC EXPANSION
EQUALIZATION OF VARIANCE)

STATISTICAL EDUCATION

INTRODUCTION

For purposes of this article, statistics will be broadly viewed as the collection, display, summarization, and interpretation of data, and the related development of probabilistic models for real phenomena. Statistical education will be divided into the training of professional statisticians (those who receive degrees in statistics at either the graduate or undergraduate level) and the educating of users of statistics. The latter can, and does, take place in precollege settings, college and university settings, and in various continuing education settings for employees of government and industry.

Since statistical methods have a wide range of uses, education in these methods takes place in a variety of settings across many disciplines. This article will concentrate on statistical education provided by professional statisticians or others well trained in some aspect of the discipline.

HISTORICAL DEVELOPMENT

Statistics is a relatively new field, coming into its own as a discipline only since the early 1900s. Although some statistical methods date from the time of Gauss* and Laplace*, modern statistics grew out of the work of applied scientists in the early twentieth century. Thus, early research and education in these methods was conducted by specialists in fields such as the biological sciences, economics, the social sciences, and demography. The growing field of statistics began to attract its own specialists, including some from mathematics, in the 1920s and 1930s, and began to emerge as a separate discipline. This early development took place first in England and Western Europe and later in the United States, India, and Australia.

The first formal academic programs in statistics in the United States were developed in the late 1930s and early 1940s at Iowa State, North Carolina State, Virginia Polytechnic Institute, and George Washington University. The influence of the biological sciences and agriculture was significant in those early programs, and remains so even today in many statistics programs.

World War II caused a tremendous surge in statistical research and education because of the need to solve quantitative problems quickly and efficiently (*see* MILITARY STATISTICS). Areas such as design of experiments*, decision theory*, sample survey* methods, quality control*, and operations research* came to the fore as important fields for practical problem solving. The mathematical theory underlying statistical methods began to receive attention by a much broader group of mathematicians.

Shortly after World War II, many programs in statistics began in the United States and throughout the world. A number of these resulted in departments of statistics, but most became attached to existing programs in mathematics, agriculture*, biological sciences, psychology*, engineering*, business, and others. This splitting of statistics in many directions caused a lack of cohesiveness in statistical education that still exists. Also, most of these programs trained statisticians at the graduate level, with most students coming from undergraduate backgrounds in agriculture, the sciences, or mathematics. Most undergraduate degree programs in statistics date from the mid 1960s, or later.

STATISTICAL EDUCATION PROGRAMS IN COLLEGES AND UNIVERSITIES

The American Statistical Association* lists approximately 240 degree-granting programs in statistics in the United States (see ref. 3). Canada lists approximately 25 such programs, and Great Britain has a similar number. Good educational programs in statistics are found in most of Europe, India, Australia, Japan, and South Africa, with a scattering of programs throughout the remainder of the world.

Among the 240 U.S. programs, 86 give only a master's and Ph.D degree, 64 give bachelor's, master's and Ph.D. degrees, 42 give bachelor's and master's degrees, 27 give only master's degrees, and 21 give only bachelor's degrees. The emphasis is clearly seen to be on graduate education. Interestingly, the sizes of the three programs tend to be nearly equal in terms of students majoring in statistics. A study for the 1981–1982 academic year shows the median number of students to be near 10 for bachelor's, master's, and doctoral programs (see ref. 2). The U.S. programs grant approximately 200 Ph.D's, 700 master's degrees, and 400 bachelor's degrees in statistics per year.

Degrees with concentration in statistics are granted by a variety of programs in mathematics, the natural and social sciences, business, and education, in addition to departments of statistics. Of the 240 U.S. programs, only 52 are in departments of statistics and an additional 30 are in departments of biostatistics or biometry. Another 32 are in departments with statistics in the title, most commonly linked with mathematics. Others are mainly in departments of quantitative business methods, psychology, educational testing and measurement, industrial engineering, and computer science.

EDUCATION OF PROFESSIONAL STATISTICIANS

Since statistics has traditionally been emphasized only in graduate study, most professional statisticians have either a master's or doctoral degree, or both, in the subject. A typical master's program in statistics requires a good working knowledge of calculus and matrix or linear algebra as a prerequisite. Required courses in statistics generally include some probability theory, mathematical statistics (estimation and hypothesis testing), design and analysis of experiments, and regression theory and methods. The program may contain courses in multivariate techniques, nonparametric techniques, sampling theory and methods, stochastic processes, and time-series analysis. Programs specializing in biostatistics* would require courses in that subject, perhaps including epidemiology, statistical genetics*, and biomathematics. Since most master's-level statisticians work on applications in government, business, or industry, at least one course in statistical computing is recommended.

Doctoral programs often have core requirements similar to a strong master's program. In addition, mathematics at the level of real analysis or functional analysis is often required for advanced work in probability theory and theory of statistical inference*. Advanced courses may be available in any of the specific areas mentioned under the discussion of master's programs.

Many graduate degrees with a concentration in statistics are offered in quantitative departments other than mathematics or statistics. Study of psychometrics generally includes much work in design of experiments, multivariate methods, and nonparametric methods. Econometrics* includes work in regression modeling and time-series* analysis. Statistics taught in an engineering environment usually emphasizes stochastic processes and probability theory, as well as quality assurance techniques. The social sciences rely heavily upon techniques of categorical data* analysis, in addition to regression, in research and teaching.

Professional statisticians with only a bachelor's degree are becoming more in evidence as more high quality bachelor's programs are being developed. Such programs usually emphasize a mix of mathematics, computer science, and statistics. The statistical component is, typically, methodologically oriented so that graduates are well versed in common statistical methods and the related computing techniques. There still are mixed opinions in the statistical profession as to the value of a bachelor's degree in statistics, many arguing that statistics programs should only grant degrees at the graduate level after a strong background in mathematics is achieved.

STATISTICAL EDUCATION FOR NONSTATISTICIANS

Students in many disciplines are required to have some knowledge of statistical techniques and, as a result, many academic statisticians spend much of their time teaching introductory courses to students from a wide variety of areas. At the graduate level such courses emphasize design of experiments, analysis of data, and regression* techniques, since these methods are widely used in research. At the undergraduate level such courses include an introduction to probability, usually including simple combinatorial rules, and the basic notions of estimation* and hypothesis testing*. Undergraduate courses for students in the sciences and engineering could include a more mathematical orientation to probability and basic statistical theory.

The depth of courses for nonstatisticians varies widely, at both the graduate and undergraduate levels. Some are the equivalent of courses for statistics majors while others merely show students how to use computer software packages to analyze data, with little regard for the understanding of the principles of statistics.

STATISTICAL EDUCATION IN INDUSTRY

Industry now realizes that sound statistical principles must be followed in the design phase as well as the production phase of manufactured items if optimal productivity is to be achieved. Thus, many major industries have massive in-house education programs to teach workers the basic ideas of data collection*, display, summarization, and interpretation. These educational programs may begin at the level of defining proportions and averages, constructing histograms* and other graphical* displays, and showing how control charts* work. More advanced programs may give instruction in regression techniques, designs and analysis of experiments, stochastic processes*, and time-series analysis. Most industries employ some professional statisticians to conduct their training programs, but many rely heavily upon hiring academic statisticians as temporary consultants.

Japan has long been the world's leader in the use of statistics in industry, and has extensive training programs for workers at all levels. Industries in Canada, England, South Africa, and the United States are now beginning programs to train workers in quality control* techniques and in other areas of statistics (design of experiments, for example) that are important for improving productivity.

STATISTICAL EDUCATION AT THE PRECOLLEGE LEVEL

There is now much activity and interest in the United States, Canada, Western Europe, and other countries in the area of statistical education at the precollege level. Statistics at this level has long been ignored, or thought impractical. However, the current demands of industry for workers with statistical skills, coupled with the computer revolution and the need to revitalize mathematics and science education, has generated a strong movement toward including statistics in general mathematics and science education. Statistical education at the schools level usually involves hands-on experiences with real data and simulations, with a strong emphasis on descriptive and graphical techniques. Of course, the computer is used extensively in many programs.

England and Sweden have led the rest of the world in developing statistical curricula for the schools, England now having a Centre for Statistical Education. These countries have extensive programs and materials for both the primary and secondary grades. Students in primary grades tend to work with real data in tables, graphs and charts, and with simple summary statistics, while the secondary students study more sophisticated summarizing statistics along with more formal notions of probability and inference. Statistics is taught in science and social science classes, as well as in mathematics. Australia and Japan have similar programs for students of all ages.

Other countries (Canada, West Germany) have ambitious programs in statistics and probability only at the secondary level, but are now beginning to expand the teaching of statistics at the lower levels. The United States has only recently begun efforts to make statistics and probability a requirement in mathematics and science education, but these efforts vary from state to state.

Some countries (South Africa, France) have little in the way of formal training in statistics at the school level, but are planning to move in that direction. Teaching statistics at this level is a relatively new phenomenon, and one that now requires much study as to what topics should be taught and how they can effectively be presented. See ref. 1 for more details on the subject.

References

[1] Barnett, V., ed. (1982). *Teaching Statistics in Schools Throughout the World*. International Statistical Institute, Voorburg, The Netherlands.

[2] Bisgyer, E. M., ed. (1983). *Amstat News*, **92**, 19.

[3] Bisgyer, E. M., ed. (1983). *Amstat News*, **99**, 33–44.

Bibliography

Bradley, R. A. (1982). The future of statistics as a discipline. *J. Amer. Statist. Ass.*. **77**, 1–10.

Committee on Training Statisticians for Government (1982). *Amer. Statist.*, **36**, 69–81.

Committee on Training Statisticians for Industry (1980). *Amer. Statist.*, **34**, 65–75.

Jordan, E. W. and Stroup, D. F. (1984). The image of statistics. *Collegiate News Views*, **37**, 11–13.

Minton, P. D. (1983). The visibility of statistics as a discipline. *Amer. Statist.*, **37**, 284–289.

Råde, L., ed. (1982). *First Int. Conf. Teaching Statist.*, final report.

Tanur, J. M. (1982). Quality of statistical education: should ASA assist or assess? *Amer. Statist.*, **36**, 90–102.

RICHARD L. SCHEAFFER

STATISTICAL EQUILIBRIUM *See* ERGODIC THEOREMS

STATISTICAL EVIDENCE

Evidence E concerning a hypothesis H is the data on which may be based the acceptability or probability of H. (See the *American Heritage Dictionary*.) In non-Bayesian statistics it is in some sense the "acceptability" of H, and in Bayesian* statistics it is the "epistemic" probability of H, that is affected by the evidence. [By epistemic probability is meant either logical or subjective* (personal) probability, or some compromise between the logical and the subjective.] In common sense, and in Bayesian statistics, evidence can support or undermine a hypothesis. Some non-Bayesians consider that in most circumstances a hypothesis cannot be supported but can often be undermined by evidence, namely by the smallness of the P value (tail-area probability) of a statistic. Most of this article will be concerned with evidence from a Bayesian point of view, but tail-area probabilities will also be discussed briefly.

It is natural, in accordance with the normal usage of the English language, to think of *weight of evidence* as a quantitative or semiquantitative measure of *how much* the evidence E supports or undermines H. For example, H might denote the proposition that asserts the guilt of an accused person. The merit of this example is its familiarity. Let us denote by $W(H:E)$ the weight of evidence in favor of H provided by E. If we want to specify some background or given knowledge, G, we can use the notation $W(H:E|G)$, the *weight of evidence concerning H provided by E given G*. Let us consider how $W(H:E|G)$ should be expressed in terms of epistemic probability. When discussing this question we shall take G for granted and write $W(H:E)$ for the sake of brevity. We shall also assume, for the most part, that the epistemic probabilities occuring in the discussion have sharp values satisfying the usual axioms*, although this assumption is seldom more than a good approximation at any rate when the probabilities are subjective. When the probabilities are not sharp then neither is the weight of evidence.

It is difficult to see how $W(H:E)$ can depend on anything other than (i) the probability of the evidence given that the accused is guilty (or, more generally, given H), and (ii) the probability given that he is innocent (or, more generally, given $\overline{H}$ the negation of H). In other words there must be a function f, of two variables, such that

$$W(H: E) = f[P(E|H), P(E|\overline{H})].$$

The probabilities are epistemic unless H and $\overline{H}$ are both "simple statistical hypotheses." Now the final (posterior) probability of guilt must surely depend only on the initial (prior) probability and on the weight of evidence. That is, there should be a function g such that

$$P(H|E) = g[W(H: E), P(H)].$$

From these two equations it can be deduced [7] that $W(H: E)$ must be some function of $P(E|H)/P(E|\overline{H})$. By using the product axiom of probability four times, or by using Bayes' theorem, we have

$$\frac{P(E|H)}{P(E|\overline{H})} = \frac{P(E \& H)P(\overline{H})}{P(H)P(E \& \overline{H})} = \frac{P(H|E)P(E)P(\overline{H})}{P(H)P(\overline{H}|E)P(E)} = \frac{O(H|E)}{O(H)}, \qquad (1)$$

the ratio of the final to the initial odds of H, where the odds corresponding to a probability p are defined as $p/(1-p)$. [The simple theorem (1) was stated in ref. 29, p. 387.] Accordingly $P(E|H)/P(E|\overline{H})$ or $O(H|E)/O(H)$ is often called the (Bayes) *factor in favor of H provided by E* and may be denoted by $F(H: E)$. This terminology was suggested in 1941 by A. M. Turing for a cryptanalytic attack, known as Banburismus, against the Enigma enciphering machine. The Bayes factor is called K by Jeffreys [21, 22, p. 53], and more recently a few writers have called it the "posterior odds ratio" although "odds ratio" would be preferable. When H and $\overline{H}$ are both simple statistical hypotheses, and only in this case, the Bayes factor is equal to the simple likelihood ratio (called by Wald [28] the "probability ratio"). Its use for hypothesis testing* in this case is justified by the non-Bayesian by means of the Neyman–Pearson lemma*. The Bayesian justification is equation (1). The Bayes factor has the multiplicative property

$$F[H:(E \& E')] = F(H:E) \times F(H:E'|E). \qquad (2)$$

It is natural to use the logarithm of the Bayes factor (the log-factor) as the technical definition of weight of evidence, for it has the additive property

$$W(H:(E \& E')|G) = W(H:E|G) + W(H:E'|E \& G). \qquad (3)$$

In words, the weight of evidence provided by E & E' is equal to that provided by E plus that provided by E' when E has already been taken into account (given G all along). This additive property corresponds to putting weights on the scales of Themis, the goddess of justice. The positive weights go on one scale and negative ones (or rather their absolute values) go on the other scale. A zero weight of evidence corresponds to a Bayes factor of 1; that is, it makes no change in the odds (or probability) of H.

Weight of evidence was called "support" by Jeffreys [20], but he dropped the expression in ref. 21 because there he nearly always assumed that the initial probability of H was $\frac{1}{2}$. This assumption of Jeffreys' might be a reasonable approximation in many ordinary statistical applications but in legal and diagnostic contexts it would be a travesty of justice or medicine. Apart from its use by Jeffreys, the concept of weight of evidence or its expected value is heavily used, for example, in refs. 3, 24, and 27.

Weight of evidence encapsulates the concept of degree of corroboration or confirmation of concern to philosophers of science, although some philosophers, following Humpty Dumpty, misuse the word "confirmation" to mean $P(H|E)$. Popper [26, p. 394] said "I regard the doctrine that the *degree of corroboration or acceptability cannot be a probability* as one of the most interesting findings of the philosophy of knowledge." Some philosophers were still using such inappropriate measures as $P(H|E) - P(E)$ and $P(H|E)/P(H)$ as late as 1982. If either of these measures were used, a long enough sequence of independent but equal supporting weights of evidence would make the probability of H exceed unity. In other words such a sequence would be impossible under either of these definitions. This conclusion does not depend on an assumption of additivity or multiplicativity.

The expression "weight of evidence" was used by C. S. Peirce in 1878 in a technical sense that can be seen to be a special case of the sense defined here, although he was not sympathetic to the Bayesian point of view. For details see ref. 17.

Turing pointed out the convenience of having a name for the unit in terms of which weight of evidence is measured. If the base of logarithms is 10 he suggested the name *ban*, and *deciban* for one-tenth of a ban. The deciban is analogous to the decibel in acoustics. For base e the name he suggested was *natural ban*. These units were used, for example, in the cryptanalytic application in 1941, which was a very early form of sequential analysis*.

For deciding in advance whether Banburismus was likely to be successful, Turing computed the expected weight of evidence, given H, which he pointed out has the desirable property of being nonnegative. In other words, for the discrete case,

$$\sum p_i \log(p_i/q_i) \geqslant 0$$

when $p_i > 0$, $q_i > 0$, and $\Sigma p_i = \Sigma q_i = 1$. This inequality also occurs in Gibbs [2, p. 136]. A more general inequality is given in ref. 16. In 1941, Turing also proved that if $W(H:E)$ has a normal distribution when H is true, then

$$\text{var}[W(H:E)|H] = 2\mathscr{E}[W(H:E)|H]$$

when natural bans are used. This result was later published in connection with radar [25]. The result, expressed in decibans, is $\sigma \approx 3\sqrt{\mu}$, where σ is the standard deviation and μ is the mean of the weight of evidence. For example, if the median Bayes factor is 4000, then $\mu = 36$ and $\sigma = 18$, so there is a disturbingly large chance, about 1/40, that the weight of evidence will be negative, and an equal chance that it will exceed 72 decibans (a factor of 16,000,000).

An extension of this result to the case where the weight of evidence is only approximately normally distributed near its mean was treated in ref. 5 in connection with false alarms in signal detection.

Other relationships between moments and cumulants of weight of evidence can be deduced from the formal identity $\phi(t + \sqrt{-1}) = \bar{\phi}(t)$, where ϕ and $\bar{\phi}$ denote the characteristic functions of $W(H:E)$ given H and $\bar{H}$, respectively [15, 18].

Expected weight of evidence is in effect a generalization of entropy* as understood in Shannon's theory of communication, and Shannon's coding theorems can be expressed with intuitive advantage in terms of expected weight of evidence [18]. Expected weight of evidence is a more fundamental concept than entropy (see, for example, ref. 16). Other names for it are "relative entropy," "cross-entropy," "dinegentropy," and "discrimination information" (*see* KULLBACK INFORMATION). This last name is used in ref. 23 for hypothesis testing. For further discussion of relationships between expected weight of evidence and maximum entropy, see refs. 14, and 16. The technique of maximizing entropy for the selection of prior distributions* was introduced by Jaynes [19], and was adapted to the formulation of hypotheses in ref. 6.

Weight of evidence can be regarded as a quasiutility and when this is done one can interpret Jeffreys' invariant prior, the principle of maximum entropy, and that of minimum expected weight of evidence, as minimax* procedures [1, 8].

For relationships between weight of evidence and errors of the first and second kinds, see refs. 9 and 17 and for the simple relationship to sufficiency see, for example, ref. 16.

It is often much easier to judge a weight of evidence or a Bayes factor than to judge $P(E|H)$ or $P(E|\bar{H})$, because these probabilities are liable to be exceedingly small. Just as for other quantities, such as length or weight, the judgment of ratios can be easier than that of absolute values.

One method of weighing evidence is to use part of the evidence to judge that the log-odds of H are not far from zero, and to use the remainder for estimating the further weight of evidence. It is to be hoped that this is often roughly the situation when a scientist proposes a null hypothesis for testing. His progress will be faster if his hypotheses are neither far-fetched nor too obviously (approximately) true and this perhaps provides some justification for the assumption made by Jeffreys [21] that one can often take $O(H) = 1$ (at least as an approximation).

When probabilities are not assumed to have sharp values, but are regarded only as interval-valued, then weights of evidence and Bayes factors are also only interval-valued; that is, they have upper and lower values. They can then be used as part of the input to a black box [3, 4]. A scientific theory should combine a black box with grey matter. Shafer (*see* BELIEF FUNCTIONS) defines degrees of belief* as lower probabilities, but the present writer thinks upper and lower probabilities have equal claims to be called degrees of belief, and some intermediate value has a better claim. At any rate the main difficulty in Bayesian* statistics, and therefore also in the use of weights of evidence, is that epistemic probabilities can often not be assigned precise values, and this is why some statisticians avoid Bayesian methods.

For more detailed discussion of weight of evidence, with numerous references, see refs. 3, 13, and 16.

In non-Bayesian statistics the evidence against H is often judged in terms of a tail-area probability. It is, however, possible to give examples where a tail-area probability of say 5% is conclusive evidence *in favor* of the null hypothesis! A simple and natural example is quoted in ref. 16. It depends on having a large sample. Another "paradox" arises if optional stopping is permitted. For a discussion and history of this paradox see ref. 11. For further discussion of tail probabilities*, including references to the relationship between them and Bayes factors, see ref. 10.

Because tail-area probabilities are often used, it is worth considering how to improve their use. Two proposals, both of which are examples of a Bayes/non-Bayes compromise, are (i) to regard a tail-area probability P that occurs for sample size N as equivalent to a "standardized" one of $P(N/100)^{1/2}$ for sample size 100 when both P and $P(N/100)^{1/2}$ are small [12]; and (ii) to make sure that a use of tail-area probabilities is not in serious conflict with your mature judgement of the Bayes factor or weight of evidence against the null hypothesis [16].

References

The parenthetic comments mention only what is relevant to the present encyclopedia article.

[1] Bernardo, J. (1979). *J. R. Statist. Soc. B*, **41**, 113–147 (with discussion). (A systematic application of minimizing expected weight of evidence for the determination of standard prior distributions.)

[2] Gibbs, J. W. (1902). *Elementary Principles in Statistical Mechanics*. Constable, London, England. [Reprint, Dover, New York (1960).] (Applies the algebraic expression for expected weight of evidence, but usually for entropy, to statistical mechanics but not to ordinary statistics.)

[3] Good, I. J. (1950). *Probability and the Weighing of Evidence*. Charles Griffin, London, England. (Applications of weight of evidence to subjectivistic or personalistic Bayesian statistics, involving Turing's concept of the deciban.)

[4] Good, I. J. (1960). In *Logic, Methodology, and Philosophy of Science*. Nagel, Suppes, and Tarski, eds. Stanford University Press, Stanford, CA, pp. 319–329. (Reprinted in ref. 13, pp. 73–82.) (Describes a black box theory of partially ordered, or upper and lower, probability, and permits judgments of inequalities between weights of evidence as input to "deluxe" black boxes.)

[5] Good, I. J. (1961). In *Information Theory, Fourth London Symposium*, Colin Cherry, ed. Butterworths, London, England, pp. 125–136. (Gener-

alizes the theorem, relating variance of weight of evidence to its expectation, to the case where weight of evidence is only approximately normal near its mean, with application to false-alarm probabilities.)

[6] Good, I. J. (1963). *Ann. Math. Statist.*, **34**, 911–934. (Reinterprets maximum entropy as a method for formulating hypotheses instead of for determining priors, and shows that it leads to log linear models for multidimensional contingency tables.)

[7] Good, I. J. (1968). *Brit. J. Philos. Sci.*, **19**, 123–143. (Demonstrates the uniqueness of the explicatum for weight of evidence based on reasonable desiderata.)

[8] Good, I. J. (1969). In *Multivariate Analysis* II, P. R. Krishnaiah, ed. Academic, New York, pp. 183–203. (Discusses weight of evidence as a quasiutility, and points out that this leads to a minimax interpretation of the principles of maximum entropy, minimum expected weight of evidence, and of Jeffreys' invariant prior.)

[9] Good, I. J. (1980). *J. Statist. Comput. Simul.*, **10**, 315–316. (Shows a relationship between expected weight of evidence and errors of the first and second kinds, with a medical application.)

[10] Good, I. J. (1981). In *Philosophy in Economics*, J. C. Pitt, ed., Reidel, Dordrecht, The Netherlands, pp. 149–174. (Reprinted in ref. 13, pp. 129–148.) (Discusses the logic and history of hypothesis testing and relates testing to weight of evidence.)

[11] Good, I. J. (1982). *J. Amer. Statist. Ass.* **77**, 342–344. (Discusses the "paradox" of optional stopping, and its history.)

[12] Good, I. J. (1982). *J. Statist. Comput. Simul.*, **16**, 65–66. (Suggests that tail-area probabilities should be standardized for sample size if they are to be interpreted evidentially, or as "distances" from the null hypothesis.)

[13] Good, I. J. (1983). *Good Thinking: The Foundations of Probability and its Applications*. University of Minnesota Press, Minneapolis, MN. (Contains 23 of the author's more philosophical papers many of which relate to weight of evidence.)

[14] Good, I. J. (1983). *J. Amer. Statist. Ass.*, **78**, 987–989. (A critique of maximum entropy.)

[15] Good, I. J. (1983). *J. Statist. Comput. Simul.*, **17**, 315–319; **18**, 85. (Discusses moments and cumulants of weights of evidence.)

[16] Good, I. J. (1983). In *Second Valencia International Meeting on Bayesian Statistics*, University of Valencia Press, Valencia, Spain. (A survey of weights of evidence. Gives an example to show that a tail-area probability of 5% can be overwhelming evidence *in favor* of the null hypothesis.)

[17] Good, I. J. (1983). *J. Statist. Comput. Simul.*, **18**, 71–74. (Gives a Bayesian interpretation of Neyman–Pearson "hypothesis determination." Also mentions that C. S. Peirce, in spite of being against Bayesianism, used the expression "weight of evidence" in 1878 in the sense of the present article, but only in a very special case.)

[18] Good, I. J. and Toulmin, G. H. (1968). *J. Inst. Math. Appl.*, **4**, 94–105. (Shows that Shannon's coding theorems can be intuitively well grasped in terms of expected weight of evidence instead of entropy.)

[19] Jaynes, E. T. (1957). *Phys. Rev.*, **106**, 620–630. (Suggested the maximization of entropy for the choice of priors.)

[20] Jeffreys, H. (1936). *Proc. Camb. Philos. Soc.*, **32**, 416–445. (Suggested the name "support" for weight of evidence.)

[21] Jeffreys, H. (1939). *Theory of Probability*. Oxford University Press, London, England. (Dropped the term "support" because the emphasis was on hypotheses with initial odds of unity.)

[22] Jeffreys, H. (1957). *Scientific Inference*, 2nd ed. Cambridge University Press, Cambridge, England. (Applies weight of evidence to the philosophy of scientific inference.)

[23] Kullback, S. (1959). *Information Theory and Statistics*. Wiley, New York. (Bases hypothesis testing on the minimization of expected weight of evidence, by analogy with the Neyman–Pearson–Wilks likelihood ratio, i.e., ratio of maximum likelihoods.)

[24] Mosteller, F. and Wallace, D. L. (1964). *Inference and Disputed Authorship*. Addison-Wesley, Reading, MA. (Uses weight of evidence, though not with that name, for discriminating between two authorships. A careful scholarly analysis.)

[25] Peterson, W. W., Birdsall, T. G., and Fox, W. C. (1954). *Trans. Inst. Radio Eng.*, **PGIT-4**, 171–212. (An early publication and rediscovery of the theorem relating the variance to the expectation of weight of evidence, a theorem discovered in 1941 by Turing but not published by him.)

[26] Popper, K. (1959). *The Logic of Scientific Discovery*. Hutchinson, London, England. (Emphasizes that corroboration cannot be a probability, but does not give the explicatum weight of evidence as used here.)

[27] Tribus, M. (1969). *Rational Descriptions, Decisions and Design*. Pergamon Press, New York. (Makes extensive use of weight of evidence for statistical inference.)

[28] Wald, A. (1947). *Sequential Analysis*. Wiley, New York. [Dover reprint, New York (1973).] (Uses the Bayes factor, called the "probability ratio," or simple likelihood ratio, for sequential analysis.

Independent of the work of A. M. Turing and of G. A. Barnard.)

[29] Wrinch, D. and Jeffreys, H. (1921). *Philos. Mag., Ser. 6*, **42**, 369–390. [Mentions equation (1) of the present paper, and shows its relevance for the philosophy of science.]

(BAYESIAN INFERENCE
BELIEF FUNCTIONS
DEGREES OF BELIEF
HYPOTHESIS TESTING
INFERENCE, STATISTICAL: I, II
LIKELIHOOD PRINCIPLE, THE
SUBJECTIVE PROBABILITIES)

I. J. Good

STATISTICAL FUNCTIONALS

Let $X_1, X_2, \dots, X_n$ be a random sample from a population with cumulative distribution function (CDF) F and let F_n be the corresponding empirical CDF (*see* EDF STATISTICS). If a statistic $T_n = T_n(X_1, \dots, X_n)$ can be written as a functional T of the empirical CDF F_n, $T_n = T(F_n)$, where T does not depend on n, then T will be called a *statistical functional*. The domain of definition of the statistical functional T is assumed to contain the empirical CDFs F_n for all $n \geqslant 1$ as well as the population CDF F. The parameter to be estimated in estimation problems is $T(F)$.

Von Mises [7, 8] introduced statistical functionals and proposed that a form of Taylor expansion* be used to approximate a given statistic $T(F_n)$. Currently, statistical functionals are used in the theory of robust procedures where complicated statistics can often be represented in concise functional form. The following are some examples of common types.

Example 1. A *linear statistical functional* is of the form

$$T(F_n) = \int g(x)\, dF_n(x) = \frac{1}{n} \sum_{i=1}^{n} g(X_i),$$

where g is a real valued function. For $g(x) = x$, $T(F_n)$ is the sample mean.

Example 2. For $0 < q < 1$, let

$$T(F_n) = F_n^{-1}(q) = \inf\{x : F_n(x) \geqslant q\}.$$

$T(F_n)$ is a qth *sample quantile** and it estimates the population quantile $T(F) = F^{-1}(q)$.

Example 3. Let ψ be a real valued function. The implicitly defined statistical functional $T(F_n) = \theta$, where θ is a root of

$$\int (x - \theta)\, dF_n(x) = 0,$$

is called an *M-estimator** of location. When $\psi(x) = -(\partial/\partial x)\log f(x)$, where $f = F'$ is the density of F, then $T(F_n)$ is the maximum likelihood* estimator of $T(F)$.

Von Mises [8] used Taylor expansions to approximate a given statistical functional $T(F_n)$ in order to analyze its asymptotic behavior. To calculate the terms of such an expansion, von Mises introduced a derivative, usually called the *Gateaux derivative* or sometimes the "von Mises derivative." This is similar to, but not exactly the same as, the Gateaux derivative that appears in the mathematical literature (see Yamamuro [11] or Fernholz [3]).

Definition 1. Let T be a statistical functional defined on a convex set of functions containing a fixed CDF F and the empirical CDFs F_n for all $n \geqslant 1$. For any G in the domain of T, the *Gateaux derivative* T_F' of T at F is defined by

$$\begin{aligned} T_F'(G - F) &= \frac{d}{d\epsilon} T((1-\epsilon)F + \epsilon G)|_{\epsilon=0} \\ &= \lim_{\epsilon \to 0} \frac{T((1-\epsilon)F + \epsilon G) - T(F)}{\epsilon} \qquad (1) \end{aligned}$$

if there exists a measurable function ϕ_F, independent of G, such that

$$T_F'(G - F) = \int \phi_F(x)\, d(G - F)(x). \qquad (2)$$

The function ϕ_F in equation (2) is unique up to an additive constant, and is usually normalized so that $\int \phi_F(x)\, dF(x) = 0$. $\phi_F(x)$ can be calculated from equation (1) by setting $G = \delta_x$, the CDF of the point mass 1 at x. Hampel [5] observed that for large n, $\phi_F(x)$ measures the effect on T_n of a single additional observation with value x. He called this function the *influence curve* or *influence function** of T at F and introduced notation of the form $\mathrm{IC}(x; F, T) = \phi_F(x)$.

Example 4. For the M-estimators of Example 3, the Gateaux derivative is

$$T'_F(G - F) = \int \phi_F(x) d(G - F)(x),$$

where

$$\phi_F(x) = \frac{\psi(x - T(F))}{\int \psi'(x - T)(F))\, dF(x)}.$$

VON MISES EXPANSIONS

If a functional T is Gateaux differentiable at F, then $T((1 - \epsilon)F + \epsilon G)$ can be expressed as a Taylor expansion in ϵ about the point $\epsilon = 0$. If we let $\epsilon = 1$, this provides a first-order expansion for T of the form

$$T(G) = T(F) + T'_F(G - F) + \mathrm{Rem}(G - F),$$

where $\mathrm{Rem}(G - F)$ represents the remainder term. When $G = F_n$, we have

$$\begin{aligned} T(F_n) &= T(F) + T'_F(F_n - F) + \mathrm{Rem}(F_n - F) \\ &= T(F) + \int \phi_F(x)\, dF_n(x) \\ &\quad + \mathrm{Rem}(F_n - F), \end{aligned} \tag{3}$$

since $\int \phi_F(x)\, dF(x) = 0$. An expansion of the form (3) is called a first-order *von Mises expansion** of T at F. A statistical functional that has a von Mises expansion is often called a *von Mises functional.*

Higher-order Gateaux derivatives for statistical functionals were defined by von Mises in a similar manner by

$$\begin{aligned} T_F^{(k)}(G - F) &= \frac{d^k}{d\epsilon^k} T((1 - \epsilon)F + G)|_{\epsilon=0} \\ &= \int \cdots \int \phi_k(x_1, \ldots, x_k) d(G - F)(x_1) \cdots d(G - F)(x_k) \end{aligned}$$

for some measurable function ϕ_k defined on $\mathbb{R}^k$, where $k \geqslant 1$. The corresponding k-th order von Mises expansion for T at F is given by

$$\begin{aligned} T(F_n) &= T(F) + T'_F(F_n - F) \\ &\quad + \cdots + T_F^{(k)}(F_n - F)\frac{1}{k!} \\ &\quad + \mathrm{Rem}_k(F_n - F). \end{aligned} \tag{4}$$

In this expansion, the first nonvanishing term after $T(F)$ is a form of U statistic*, and under appropriate conditions provides the asymptotic distribution of $n^{k/2}(T(F_n) - T(F))$; see von Mises [8], Filippova [4], Reeds [9], and Serfling [10].

ASYMPTOTIC NORMALITY*

When a statistical functional T has a von Mises expansion with nonvanishing linear term, then from (3) we have

$$\sqrt{n}(T(F_n) - T(F)) = \frac{1}{\sqrt{n}} \sum_{i=1}^{n} \phi_F(X_i) + \sqrt{n}\, \mathrm{Rem}(F_n - F). \tag{5}$$

If

$$0 < E(\phi_F(X_1))^2 = \sigma^2 < \infty \tag{6}$$

and

$$\sqrt{n}\, \mathrm{Rem}(F_n - F) \xrightarrow{P} 0 \tag{7}$$

as $n \to \infty$, where $\xrightarrow{P}$ denotes convergence in probability, then the central limit theorem* and Slutsky's* lemma imply that

$$\sqrt{n}(T(F_n) - T(F)) \xrightarrow{\mathscr{D}} N(0, \sigma^2)$$

as $n \to \infty$, where $\xrightarrow{\mathscr{D}}$ denotes convergence in distribution or law. Condition (6) can be verified immediately for any statistical functional that has an influence curve. Condition (7) is more difficult to establish and indeed is not satisfied by all statistical functionals that are Gateaux differentiable (see Fernholz [3]). To satisfy condition (7), von Mises assumed that the functional T was twice Gateaux differentiable with a second-order von Mises expansion of the form (4). Under these assumptions, if the second derivative satisfies certain uniformity conditions, it can be shown that (7) holds. However, the requirement of twice Gateaux differentiability is unnecessarily restrictive and has since been replaced by stronger forms of derivative, the Fréchet derivative and the Hadamard (or compact) derivative, for which single differentiation suffices.

THE FRÉCHET DERIVATIVE

On a vector space with norm $\| \ \|$, the *Fréchet derivative* of a functional T is the linear functional T_F' that satisfies

$$\lim_{G \to F} \frac{T(G - F) - T(F) + T_F'(G - F)}{\|G - F\|} = 0. \tag{8}$$

Huber [6] and Serfling [10] have defined modified versions of the Fréchet derivative for which a statistical functional T need not be extended to a vector space, and for which the derivative T_F' has a representation of the form (2). For these modified versions, Fréchet differentiability implies Gateaux differentiability, and asymptotic normality results for a statistical functional T can be obtained if the norm on the domain of T satisfies

$$\|F_n - F\| = O_p(n^{1/2}). \tag{9}$$

In this case

$$\sqrt{n}\,(T(F_n) - T(F)) = \sqrt{n}\,T_F'(F_n - F) + \sqrt{n}\,o_p(\|F_n - F\|),$$

so condition (7) is valid and asymptotic normality follows. Condition (9) is satisfied, for example, when

$$\|F_n - F\| = \sup_{x \in \mathbb{R}} |F_n(x) - F(x)|,$$

because of the well known properties of the Kolmogorov–Smirnov test statistic. Using this approach, Boos [1] and Boos and Serfling [2] proved the asymptotic normality of M- and L-estimators*, as well as the law of the iterated logarithm* for these statistics.

Huber [6] further modified the Fréchet derivative to apply when the domain of T is the space $\mathscr{M}_{\mathbb{R}}$ of probability measures on $\mathbb{R}$ and the norm is replaced by a metric that generates the weak topology on $\mathscr{M}_{\mathbb{R}}$. The latter is the weakest topology on $\mathscr{M}_{\mathbb{R}}$ for which all transformations of the form

$$F \to \int \phi(x)\,dF(x)$$

are continuous, where ϕ is a bounded continuous function on $\mathbb{R}$. Huber used this generalized form of Fréchet derivative not only to prove the asymptotic normality of statistical functionals, but also to investigate their robustness* properties. Huber characterized the weak topology on $\mathscr{M}_{\mathbb{R}}$ as natural for statistics, and observed that a reasonable requirement for the robustness of a statistical functional is that it be continuous in this topology. For details see Huber [6].

The use of the Fréchet derivative or any of its generalizations presents a difficulty, because Fréchet differentiability is such a restrictive condition that many statistical functionals are simply not Fréchet differentiable. A classical statistic, such as the sample median, provides an example (see Fernholz [3]). For this reason the weaker Hadamard (or compact) derivative is preferable.

THE HADAMARD DERIVATIVE

Let T be a statistical functional and suppose that it has been extended to an open subset $\mathscr{A}$ of a topological vector space V. T is *Hadamard differentiable* at a CDF $F \in \mathscr{A}$ if there exists a continuous linear functional $T_F' : V \to \mathbb{R}$ such that for any compact $K \subset V$,

$$\lim_{t \to 0} \frac{T(F + tH) - T(F) - T_F'(tH)}{t} = 0 \tag{10}$$

uniformly for $H \in K$. T_F' is called the Hadamard (or compact) *derivative* of T at F. If we define

$$\text{Rem}(tH) = T(F + tH) - T(F) - T_F'(tH),$$

then (10) can be expressed as

$$\lim_{t \to 0} \frac{\text{Rem}(tH)}{t} = 0 \tag{11}$$

uniformly for $H \in K$.

For a normed vector space V, condition (8) in the definition of the Fréchet derivative is equivalent to (11) holding uniformly for $H \in B$, where B is any bounded subset of V. Since compact sets are bounded, it follows that Fréchet differentiability implies Hadamard differentiability. Gateaux differentiability is roughly equivalent to the condition that (11) holds for each $H \in V$ (assuming that certain weak continuity properties are satisfied), so Hadamard differentiability implies Gateaux differentiability. See Yamamuro [11], Reeds [9], and Fernholz [3] for details.

Now let $H = \sqrt{n}(F_n - F)$ and $t = n^{-1/2}$. If T is a Hadamard differentiable statistical functional with derivative T_F' at F that satisfies a condition of the form (2), then a von Mises expansion as in (3) can be obtained, and asymptotic normality will follow if

$$\sqrt{n}\,\text{Rem}(F_n - F) \xrightarrow{P} 0.$$

This will hold if the family of measures induced on V by $\{\sqrt{n}(F_n - F),\ N \geqslant 1\}$ is *tight*, which means that for any $\epsilon > 0$, there exists a compact set $K \subset V$ such that for all $n \geqslant 1$, $P\{\sqrt{n}(F_n - F) \in V\} > 1 - \epsilon$.

This approach was first used by Reeds [9], who proposed that for a given statistical functional T, a topological vector space V be constructed in such a way that the extension of T to V be Hadamard differentiable and the family of measures $\{\sqrt{n}(F_n - F),\ n \geqslant 1\}$ be tight, in which case asymptotic normality would be proved.

Fernholz [3] proposed a unified approach to the analysis of the asymptotic properties of a statistical functional using the Hadamard derivative for functionals defined on the space $D[0, 1]$ of functions on $[0, 1]$ that are right continuous with left limits. She used the fact that for a given CDF F, a statistical functional T induces a functional τ on $D[0, 1]$ by the relation

$$\tau(G) = T(G \circ F)$$

for $G \in D[0, 1]$. For a sample $X_1, \ldots, X_n$ from F, it follows that

$$\tau(U_n) = T(U_n \circ F) = T(F_n),$$

where U_n is the empirical CDF corresponding to $F(X_1), \ldots, F(X_n)$. If U denotes the uniform CDF on $[0, 1]$, then

$$\tau(U) = T(F);$$

under appropriate regularity conditions the Hadamard differentiability of τ at U implies the asymptotic normality of $\sqrt{n}(T(F_n) - T(F))$. The advantage of this approach is that a functional τ on $D[0, 1]$ can often be expressed as a composition of simpler transformations each of which is Hadamard differentiable, and hence τ itself will be Hadamard differentiable. Thus a calculus of standard Hadamard differentiable transformations can be developed and used routinely for the factorization of diverse types of statistical functionals.

References

[1] Boos, D. D. (1979). *Ann. Statist.*, **7**, 955–959.

[2] Boos, D. D. and Serfling, R. J. (1980). *Ann. Statist.*, **8**, 618–624.

[3] Fernholz, L. T. (1983). von Mises Calculus for Statistical Functionals. *Lecture Notes in Statistics*, No. 19. Springer-Verlag, New York.
[4] Filippova, A. A. (1962). *Theory Prob. Appl.*, **7**, 24–57.
[5] Hampel, F. R. (1974). *J. Amer. Statist. Ass.*, **69**, 383–393.
[6] Huber, P. J. (1981). *Robust Statistics*. Wiley, New York.
[7] von Mises, R. (1937). Sur Les Functions Statistiques. In *Conférence de la Réunion Internationale des Mathematiciens*. Gauthier-Villars, Paris, France.
[8] von Mises, R. (1947). *Ann. Math. Statist.*, **18**, 309–348.
[9] Reeds, J. A. (1976). On the Definition of von Mises Functionals. Ph.D. dissertation, Harvard University, Cambridge, MA.
[10] Serfling, R. J. (1980). *Approximation Theorems of Mathematical Statistics*. Wiley, New York.
[11] Yamamuro, S. (1974). Differential Calculus in Topological Linear Spaces. *Lecture Notes in Mathematics*, No. 374. Springer-Verlag, Berlin, Germany.

(ASYMPTOTIC NORMALITY
EDF STATISTICS
M-ESTIMATORS
VON MISES EXPANSIONS)

LUISA TURRIN FERNHOLZ

STATISTICAL HYPOTHESIS *See* HYPOTHESIS TESTING

STATISTICAL INDEPENDENCE

If A_1 and A_2 are two events, one says that "A_1 is *independent of* A_2" if $\Pr(A_1) = \Pr(A_1|A_2)$, the conditional probability* of A_1, given A_2. If in addition $\Pr(A_1) > 0$ and $\Pr(A_2) > 0$, then A_2 is also independent of A_1, and one then says that "A_1 and A_2 are *independent*." The case in which either or both of $\Pr(A_1)$ and $\Pr(A_2)$ are zero is covered by defining A_1 and A_2 to be independent whenever

$$\Pr(A_1 \cap A_2) = \Pr(A_1) \cdot \Pr(A_2). \quad (1)$$

Events $A_1, A_2, \ldots, A_n$ are *mutually independent* if for any subset $A^{(1)}, \ldots, A^{(k)}$ of them,

$$\Pr[A^{(1)} \cap A^{(2)} \cap \cdots \cap A^{(k)}] = \Pr(A^{(1)}) \cdot \Pr(A^{(2)}) \cdots \Pr(A^{(k)}). \quad (2)$$

Pairwise independence* does not imply mutual independence. Customarily $A_1, \ldots, A_n$ are events in a sigma algebra over a sample space* on which a probability measure is defined.

The random variables $X_1, X_2, \ldots, X_n$ are said to be *mutually independent* if

$$\Pr\left[\bigcap_{i=1}^{n} (X_i \in B_i)\right] = \prod_{i=1}^{n} \Pr(X_i \in B_i) \quad (3)$$

for any Borel sets $B_1, \ldots, B_n$ on the real line. Suppose that $X_1, \ldots, X_n$ have joint cumulative distribution function (CDF) F, and respective marginal CDFs $F_1, \ldots, F_n$. Then $X_1, \ldots, X_n$ are mutually independent if and only if [3]

$$F(x_1, x_2, \ldots, x_n) = F_1(x_1) \cdot F_2(x_2) \cdots F_n(x_n) \quad (4)$$

for any real-valued $x_1, \ldots, x_n$. Equation (4) is a special case of (3), with B_i as the interval $(-\infty, x_i]$, $i = 1, \ldots, n$. When (4) holds, one also talks of the *statistical independence* or *stochastic independence* of the variables $X_1, \ldots, X_n$, to distinguish this property from other mathematical forms of independence (e.g., linear independence of rows in an orthogonal matrix or the distribution of a statistic being functionally independent of one of the parameters in the parent distribution of the sample on which the statistic is based). When no ambiguity can arise, $X_1, \ldots, X_n$ are simply said to be independent.

If $X_1, \ldots, X_n$ have discrete or continuous distributions with joint density function $f(x_1, \ldots, x_n)$ and marginal densities $f_1(x_1), \ldots, f_n(x_n)$, the necessary and sufficient condition for independence (4) is equivalent to

$$f(x_1, x_2, \ldots, x_n) = f_1(x_1)\, f_2(x_2) \cdots f_n(x_n). \quad (5)$$

The independence of $X_1, \ldots, X_n$ is also characterized by the property

$$\phi(t_1, t_2, \ldots, t_n) = \phi_1(t_1)\phi_2(t_2) \cdots \phi_n(t_n), \quad (6)$$

where $\phi(\ )$ is the joint and $\phi_i(t_i)$ $(i = 1, \ldots, n)$ are the marginal characteristic functions* of the variables [1, Sec. 6.6].

If (4) holds and the expected values* in the following equation exist, then

$$E(X_1 X_2 \cdots X_n) = E(X_1) E(X_2) \cdots E(X_n). \quad (7)$$

The converse is false; when (7) holds, $X_1, \ldots, X_n$ are said to be *uncorrelated*.

Long before a formal set of axioms of probability* had been laid out, the notion of independence was implicitly utilized. Cardano used the multiplication rule (1) in computing the odds for throwing an ace twice in rolling three dice in two series; see Maistrov [2, p. 23], who also states that Pascal*, Fermat*, and Huygens* "were familiar with notions such as dependence and independence of events" [2, p. 39].

Stochastic independence is occasionally confused with mutual exclusiveness. In general [unless $P(A)$ or $P(B) = 0$] mutually exclusive events* are stochastically dependent.

References

[1] Chung, K. L. (1974). *A Course in Probability Theory*, 2nd ed. Academic, New York. (See Secs. 3.3 and 6.6.)

[2] Maistrov, L. E. (1974). *Probability Theory—A Historical Sketch*, translated from the 1967 Russian edition by S. Kotz. Academic, New York.

[3] Wilks, S. S. (1962). *Mathematical Statistics*. Wiley, New York. (See Secs. 2.4 and 2.6.)

(CONDITIONAL PROBABILITY AND EXPECTATION
PAIRWISE INDEPENDENCE
PROBABILITY THEORY (OUTLINE)
SEMI-INDEPENDENCE)

STATISTICAL INFERENCE *See* INFERENCE, STATISTICAL

STATISTICAL JOURNAL OF THE U.N. ECONOMIC COMMISSION FOR EUROPE

The Statistical Journal of the UN Economic Commission for Europe was founded in 1982. It publishes studies relating to the work programme of the Conference of European Statisticians in four issues comprising one volume a year. The Conference is composed of the executive heads of national statistical offices of Europe (including the USSR) and North America. Members of the Conference meet annually in plenary to direct their work. The Conference seeks to intensify such cooperation in six main directions. First, enhanced international comparability of statistics presupposes harmonization of their underlying concepts and classification. Consequently, the Conference strives to adopt compromise recommendations with regard to concepts and classifications, which are based on a review of national practices. Second, the Conference stimulates research and international exchange of experience regarding statistical methods applied or likely to be applied in statistical production processes. Third, the development of software for all purpose of automated data processing in statistical services looms large in the work undertaken. Fourth, statistical analyses are undertaken in cooperation between two or more member countries with a view to learning lessons about international statistical comparability. Fifth, how best to organize national information systems and how to determine the place of statistical services therein is increasingly being discussed, although the Conference certainly does not envisage any sort of international harmonization in this regard. And finally, the Conference attaches great importance to statistical cooperation in these matters between countries of the East and the West.

The Statistical Journal primarily reflects the preoccupations of the Conference, although articles may be submitted from authors outside the ECE region. Articles that have appeared so far under the heading of "descriptions of national practices" concentrate either on inadequately documented practices or on practices that are likely to respond to a high level of professional interest. They include articles on enterprise statistics in countries with centrally planned economies; on measuring flows in Finnish manpower statistics; on the French natural patrimony accounts; on the Norwegian system of resource accounts; on Danish experience with a register-based statistical system; on USSR experience in training of staff of statistical offices in the use of electronic data processing equipment; on the French business statistics system; on elaboration of prices with hand-held computers in the Netherlands; on recent developments in Hungarian national accounts; on gross flow estimates in Swedish labour force surveys; on software architecture of the statistical information system of the German Democratic Republic; on production and productivity indexes in the Federal Republic of Germany; and on the use of industrial statistics to estimate the generation of recycled and waste residuals in Finland.

National illustrations can also be found in other articles, focusing to a greater extent on discussion of broad conceptual approaches to well-defined problems. Such articles include a comparison of retail price indices with and without indirect taxes; on the adjustment of consumer price* indices for changes in the rates of indirect taxes and subsidies; on health accounts; on problems and techniques of harmonizing manpower statistics from different sources; on the stress-response environmental statistical system; on aims and methods of statistics on the natural environment; on the possible impact of national accounts and balances on environment statistics; on statistics on the production and use of high and low temperature geothermal energy; on the use of the International Classification of Impairments, Disabilities and Handicaps in household surveys; on major issues of the System of National Accounts; on historical and methodological aspects of commodity price indices; on considerations of price impacts and statistics in energy demand modelling; on the conceptual development of energy statistics in terms of useful energy; on coordination of statistics on households and family; on estimating after-tax income using survey and administrative data; on the use of population censuses as multisubject data bases; on the integration of economic and social statistics; on methods of international comparisons; and on industrial water use statistics.

Articles on methodological issues typically treat problems of practical relevance to the work of statistical offices. Past contributions relate to nonresponse problems; to seasonal adjustment by frequency determined filter procedures; to population and sampling approaches in synoptic monitoring of water quality and water resources; and to coverage and content errors in censuses. Several articles dealing with automated data processing (ADP) were on the borderline between methodological issues and ADP development. Such contributions notably include papers on semantic structures of statistical meta-data; on the meta information system and the conceptual level of statistical data modelling; on automated cell suppression to preserve confidentiality of business statistics; on data structuring in a general statistical data bank system; on relational modelling; on the Statistical Computing Project; on user-friendliness; and on statistical information systems design.

Statistical analyses are found in articles on the Austria–Poland comparison of prices and GDP levels; on migration between Canada and the United Kingdom; on the European Comparison Project; on a comparison of purchasing power parities and real economic aggregates in 15 African countries; and on an international comparison of industrial pollution control costs. More general questions in the area of statistical analyses such as the attitude of statistical offices towards studies appeared under the heading of organizational problems, which, however, was largely dominated by questions related to management of registers by statistical offices; to the maintenance of confidentiality of data; to the preservation of high quality of statistics; to planning and coordination problems for statistical activities; to the use of computers in statistical services; and, last but not least, to the optimal integration of automated data processing into the work of statistical offices.

Submissions have to be written in the English language. They should be addressed to the Editor-in-Charge, ECE Statistical Division, Palais des Nations, CH-1211 Geneva 10, Switzerland. All articles are refereed prior to acceptance. The *Statistical Journal* can be ordered at Elsevier Science Publishers, P. O. Box 1991, 1000 BZ Amsterdam, The Netherlands *or*, for customers in North America, Elsevier Science Publishing Company Inc., Journal Information Center, P. O. Box 1663, Grand Central Station, New York, NY 10163, USA.

A. KAHNERT

STATISTICALLY EQUIVALENT BLOCKS *See* NONPARAMETRIC DISCRIMINATION

STATISTICAL MATCHING

Statistical matching procedures have been developed to permit analysis of data from two or more samples of cases from a single population. Existing statistically matched files, and the procedures used to create them, are described in Radner et al. [4]; a typical match brings together administrative data (for example, Social Security earnings data) and survey data (for example, data about other sources of income from a current population survey). In contrast to exact matching procedures, whereby the record for a particular case from one source is linked to the record for the same case from a second source using identifiers such as Social Security numbers, in statistical matching each record from one data source is matched with a record from the second source that generally does *not* represent the *same* case but only a "similar" case.

A statistical matching process depends on there being a set of common characteristics, referred to as X variables, measured for cases in both input files. Let X_i be the vector of these variables for record i on file A (the "base file"), and let X_j be the corresponding vector for record j on file B (the "supplemental file"). The remaining variables in each of the files will be referred to as Y_i variables in file A and as Z_j variables in file B. The matching process requires the definition of a distance function D_{ij} in terms of the X variables, which permits the similarity of any pair of cases to be assessed (*see* MEASURES OF SIMILARITY, DISSIMILARITY AND DISTANCE).

Two basic types of statistical matching procedures have been used. An *unconstrained match* (e.g., Okner [2]) is one in which no restrictions are placed on the number of A records to which the values of the Z_j variables of a case in file B may be imputed. In a *constrained match* (e.g., Barr and Turner [1]), restrictions are imposed that result in the transfer of the multivariate distribution of the Z variables as observed in file B to the matched file, but at the cost of less closely matched pairs and more computer time for implementation as compared to an unconstrained match.

The relationships among the X and Y variables as observed in file A and among the X and Z variables as observed in file B provide only broad constraints on the possible values of the relationships among the Y and Z variables. The assumption inherent in statistical matching is that the conditional distribution of Y given X is independent of the conditional distribution of Z given X. This point was made by Sims [6] and repeatedly by others since then.

The conditional independence assumption is a strong one for which little justification has generally been offered. If information about the relationship of a Y–Z pair is available from another source, it can be incorporated into the matching process (Paass [3]); but this does not change the fact that statistical matching does not generate new information about the conditional relationship of the Y–Z pair. The best that can be hoped for in a statistical match is that no information is lost, and in general even this goal is not achieved. Simulations of a variety of statistical matching procedures (summarized in Rodgers [5]) confirm that estimates based on statistically matched files must be regarded with skepticism.

References

[1] Barr, R. S. and Turner, J. S. (1978). In *1978 Compendium of Tax Research*. Office of the Treasury, U.S. Government Printing Office, Washington, DC, pp. 131–149.

[2] Okner, B. A. (1972). *Ann. Econ. Social Measurement*, **1**, 325–342.

[3] Paass, G. (1985). In *Microanalytic Simulation Models to Support Social and Financial Policy*, Orcutt, Merz, and Quinke, eds. North-Holland, Amsterdam, The Netherlands, in press.

[4] Radner, D. B., Allen, R., Gonzalez, M. E., Jabine, T. B., and Muller, H. J. (1980). Report on Exact and Statistical Matching Techniques. *Statistical Policy Working Paper No.* 5, U.S. Department of Commerce. U.S. Government Printing Office, Washington, DC.

[5] Rodgers, W. L. (1984). *J. Bus. Econ. Statist.*, **2**, 91–102.

[6] Sims, C. A. (1972). *Ann. Econ. Social Measurement*, **1**, 343–345.

(MATCHED PAIRS
RECORD LINKAGE AND MATCHING SYSTEMS)

WILLARD L. RODGERS

STATISTICAL MECHANICS *See* STOCHASTIC MECHANICS

STATISTICAL NEWS

Statistical News was first published by Her Majesty's Stationery Office (HMSO) for the Central Statistical Office in May 1968. Since then it has been published quarterly in February, May, August, and November of each year.

Early issues contained approximately 38 pages, while current ones are around 60 pages. The original aim was to provide a comprehensive account of developments in British official statistics and to help all those who use, or who would like to use, official statistics. Recently the scope has been widened, to allow the inclusion of articles on matters related to British official statistics, such as international statistical organisations, computer matters, and data protection. The subtitle on the cover "Developments in British Official Statistics" has now been dropped.

Each of the more recent issues contains at least four articles and often more, dealing with a subject in depth. In addition, shorter notes give news of the latest developments in many fields. The cumulative index going back over previous issues included in the November edition of *Statistical News* each year provides a permanent and comprehensive guide to developments in all areas of U.K. official statistics and other regular features deal with recently available statistical series and publications. Important appointments and changes in the Government Statistical Service are noted. While most articles are submitted by members of the Government Statistical Service, there is no objection to the publication of articles from sources outside the official service. All manuscripts are reviewed by persons selected by the Editor as competent in the subject matter. The contents page of a recent issue (August 1985) gives a good indication of the subject matter covered in articles:

"The Home Office Statistical Department" by Rita Maurice. The sixth in a series of articles on the structure and functions of the Government Statistical Service.

"The General Household Survey: its policy uses by Government" by Maureen Bennett. The article gives a broad idea of the uses made by Government departments of GHS data.

"The development of a new register of businesses" by John Perry. Progress in the development of a central register since 1975.

"Statistics of domestic proceedings in magistrates' courts, England and Wales, 1983" by Lawrence Davidoff. The development and refinement of collecting magistrates' courts statistics.

"United Kingdom National Accounts —sources and methods" by Tim Jones. New edition—an essential companion to the annual CSO Blue Book.

"Schools annual statistics competition."

The editor's address is: The Editor, Statistical News, Room 74A/3, Central Statistical Office, Great George Street, London, SW1P 3AQ, England. The USA agents for the supply of Statistical News are: Bernan Associates Inc., 9730-E, George Palm Highway, Lanham, Maryland 20706.

I. M. McKinney

STATISTICAL OFFICE OF THE UNITED NATIONS *See* UNITED NATIONS, STATISTICAL OFFICE OF THE

STATISTICAL PACKAGES *See* STATISTICAL SOFTWARE

STATISTICAL PHYSICS

There are a number of entries on various aspects of the use of statistical theory in physics. Major aspects include:

(i) Distributions of particles in space (*see* MAXWELL DISTRIBUTION; GIBBS DISTRIBUTIONS).

(ii) Distributions of states of particles [e.g., the two QUANTUM MECHANICS entries and QUANTUM PHYSICS AND FUNCTIONAL INTEGRATION; also FERMI-DIRAC STATISTICS (WITH MAXWELL–BOLTZMANN AND BOSE–EINSTEIN STATISTICS].

(iii) Diffusion theory (e.g., DIFFUSION PROCESSES).

(iv) Thermodynamics (*see* MAXWELL, JAMES CLERK and BOLTZMANN, LUDWIG EDWARD).

Overall treatments of statistical physics are available in the books in refs. 1–3, among many others.

References

[1] Isihara, A. (1971). *Statistical Physics*. Academic, New York.

[2] Mandl, F. (1971). *Statistical Physics*. Wiley, New York.

[3] Vasilyev, (1983). *Statistical Physics*. Mir, Moscow, USSR. (English translation. A compact and lucid treatment, not requiring advanced mathematics.)

(BOLTZMANN, LUDWIG EDWARD
DIFFUSION PROCESSES
FERMI–DIRAC STATISTICS (WITH MAXWELL–BOLTZMANN AND BOSE–EINSTEIN STATISTICS)
GIBBS DISTRIBUTIONS
LATTICE SYSTEMS
MAXWELL DISTRIBUTION
MAXWELL, JAMES CLERK
NYQUIST FREQUENCY
ORNSTEIN–UHLENBECK PROCESS
PHASE PROBLEM OF X-RAY CRYSTALLOGRAPHY, PROBABILISTIC METHODS IN
QUANTUM MECHANICS AND PROBABILITY
QUANTUM MECHANICS: STATISTICAL INTERPRETATION
QUANTUM PHYSICS AND FUNCTIONAL INTEGRATION
RAYLEIGH DISTRIBUTION
SHOT-NOISE PROCESSES AND DISTRIBUTIONS
STATISTICS IN CRYSTALLOGRAPHY
STOCHASTIC DIFFERENTIAL EQUATIONS
STOCHASTIC MECHANICS)

STATISTICAL QUALITY CONTROL *See* QUALITY CONTROL, STATISTICAL

STATISTICAL SCIENCE

The first issue of *Statistical Science*, a review journal published by the Institute of Mathematical Statistics* (IMS), appeared in February, 1986. As stated in an editorial printed at the beginning of that inaugural issue:

> A central purpose of *Statistical Science* is to convey the richness, breadth, and unity of the field by presenting the full range of contemporary statistical thought at a modest technical level accessible to the wide community of practitioners, teachers, researchers, and students of statistics and probability.
>
> The scientific journals of the IMS, *The Annals of Statistics* and *The Annals of Probability*, together with their predecessor, *The Annals of Mathematical Statistics*, have distinguished themselves as the premier research journals of their type in the world. With the publication of *Statistical Science*, the IMS introduces a journal that is designed to complement the *Annals* by coordinating, integrating, and explicating important current statistical research. In line with this goal, the journal will publish articles describing influential new methodological and theoretical topics, reviews of substantive areas of scientific research with promising statistical applications, evaluations of research papers and books in diverse branches of statistics and probability, and discussions of classic articles with commentary on their impact on contemporary thought and practice. The journal will also publish articles on the history of statistics and probability, articles on teaching and educational programs, and interviews with distinguished statisticians and probabilists.

Statistical Science is a quarterly publication, appearing in February, May, August, and November of each year. The original conception of the journal was by Prof. Ingram Olkin of Stanford University, and it was largely through his efforts during 1982–1984 that the journal was approved and established by the IMS. Other members of the IMS organization who contributed in important ways to bringing *Statistical Science* into existence were Prem K. Goel as Managing Editor, Jose L. Gonzalez as Business Manager, and Bruce E. Trumbo as Treasurer. The first editorial board comprised an Executive Editor, Morris H. DeGroot of Carnegie-Mellon University, and three Editors, David R. Brillinger of the University of California, Berkeley, J. A. Hartigan of Yale University, and Paul Switzer of Stanford University.

A feeling for the nature of *Statistical Science* can be obtained by considering the contents of the first two issues. The first issue contained an article criticizing the use of regression models to adjust the Census for the number of people who were not counted; a survey of bootstrap methods for standard errors, confidence intervals, and other measures of statistical accuracy; a study of the status of the central limit theorem around 1935; a review of the collected works of George E. P. Box; a review of methods for combining different expert opinions that are expressed as probability distributions; an essay describing the contents of the first issue of *The Annals of Mathematical Statistics*; and interviews with T. W. Anderson and David Blackwell.

The second issue (May, 1986) contained a discussion of the statistical applications of Poisson's work; an article on morphometrics studying size and shape spaces in two dimensions; a description of computing environments for data analysis; a review of the contributions of Herbert Robbins to mathematical statistics; two essays on the origins and early days of the IMS; and an interview with Erich L. Lehmann.

A characteristic of many of the articles in *Statistical Science* is that they present highly personal views of various areas of statistics and probability by some of the leading researchers of our day. Several of the articles are controversial and are followed by comments from discussants who present somewhat divergent views. Through this editorial approach, contemporary statistics and probability are seen to be lively and rapidly developing fields of research.

MORRIS H. DEGROOT

STATISTICAL SELF-SIMILARITY *See* SELF-SIMILAR PROCESSES

STATISTICAL SOFTWARE

INTRODUCTION

Statistical methods are historically bound up with computational algorithms. The earliest efforts at simplifying computational tasks associated with statistical analysis procedures were, of course, the construction and use of tables. With the invention of the stored program computer in the mid to late 1940s, the possibility of permanent statistical coding became a reality. The style of the 1950s, however, did not emphasize the use of computing by researchers, who were generally not expert programmers. The prevalent code in this era was machine language or the only slightly higher-level language of assembly language. These languages required substantial coding skill for efficient use, and since such machine specific languages inhibited portability, it was typically left to each programmer to devise his or her own statistical code. By the late 1950s and early 1960s, the relatively high-level language Fortran became more prevalent on such computers as IBM 1620, 7090, and 7094. Early IBM scientific software for these machines included simple statistical routines, the most notorious of which was a random number generator called RANDU. This routine had the distinctly unpleasant property of generating coplanar vectors when contiguous numbers were taken as order pairs, triples, etc.

With the relative portability offered by languages such as Fortran (and, of course, Algol and PL/1), coordinated packages of statistical software became feasible. The systematic development of statistical software took place during the 1960s, BMD and SPSS being two of the earliest efforts, followed soon by SAS. During the 1970s this style of coordinated integrated statistical software package continued to mature. Since 1980, however, the increasing prevalance of mini- and microcomputer resources has returned dedicated computer resources to the hands of individuals rather than requiring the shared use of batch mode or time-share facilities on a mainframe. A major difference between the dedicated computer user of the 1950s and that of the 1980s is frequently the programming skill level of the user. In the 1950s, the user of necessity was an expert programmer. This is not so in the 1980s. In particular, the prevalance of the Apple and IBM PCs and their look-alikes has created a large demand for software of all types, including statistical software (see Foxwell [9] and Carpenter et al. [5]). It is probably safe to describe the state of statistical software as confusing, with many new products available, but without sufficient time passing to sort out the features and quality.

Since statistical software is the vehicle by which products from the statistical research community reach the consumer market, its ultimate quality is clearly of signal importance to the discipline as a whole. Software has been developed with various degrees of formality, perhaps initially with very informal, one-shot, throw-away software. However, with increasing complexity of computing hardware and increased sophistication of statistical methodology, it is no longer desirable or economical to treat software in this way.

In this article, we discuss several formally documented software packages or systems, including some of a more experimental nature. Many resources are available describing software. Several of the more popular software systems have individual entries in the encyclopedia. Perhaps the most extended review of statistical software is the treatment by Francis [10]. This is a key work for anyone interested in using statistical packages extensively. A work with a slightly different intent is Hayes [17], who investigates the characteristics of some 213 different generic statistical software packages (GSSP) with the intent of uncovering statistical topics that have not been effectively coded in software. In addition she indicates languages and machines on which the software is implemented and gives a nice summary of resources, including points of contact.

We shall discuss statistical software in a variety of forms, including subroutine libraries, software systems, languages, and software/hardware combinations. Subroutine libraries typically are collections of subroutines that are installed on a mainframe system and may be accessed by subroutine calls in much the same way as system subroutines such as SIN, COS, ABS, SORT. Typically there is no driver program that must be user created and there may not be a common data format for all routines. Examples of the libraries are those of NAG and IMSL discussed later.

The software systems are rather more complete packages, typically having a common data structure and minimal programming requirements. Typically, an operating system call is sufficient to initiate the program. Examples of software systems include SAS, BMDP, P-STAT, and S. If the software system is at the one extreme essentially eliminating all user programming, languages oriented towards statistical usage are at the other. The prime example is APL, which is essentially a language oriented toward array manipulation. APL is discussed later.

Finally, of a more recent and more experimental character are the software/hardware combinations. These are now feasible because of the relative economy of workstation-type computers and raster-scan high resolution graphics. The archetype of this class is PRIM-9 with several successors including PRIM-H and ORION-1.

Hayes [17] offers a taxonomy of statistical software that is convenient to use here. She distinguishes between advanced and basic (A & B), the former refering to methodology that would normally be taught to students at the graduate level, while the latter normally would be taught at the introductory level. In addition she defines 16 topical areas, summarized in Fig. A. Figure A, in fact, gives a useful perspective on statistical software as a whole.

THE LIBRARIES

The NAG Library of Algorithms [25] consists of about 600 subroutines that were programmed in three of the languages (Algol 60, Algol 68, and ANSI Fortran) most commonly used in engineering and scientific computing. The most recent Mark 9 Fortran version (9th ed. of the NAG Library) is composed of 468 algorithms that operate in 13 hardware families from minicomputers (DEC VAX 11) to supercomputers (CRAY-1) with on-line documentation at the disposal of international subscribers. It is meant for mathematical and statistical computation in research, development, and production as well as teaching. It has highly advanced algorithms dealing with asymptotic results, simulation, regression, spline estimation, optimization techniques, time series, and graphical analysis. In addition, it also offers basic statistical techniques for complex surveys, multivariate analysis, and nonparametrics.

NAG Inc. also distributes GENSTAT (a generalized statistical program) prepared by the Statistics Department, Rothamsted Experimental Station, U.K. and the GLIM* system (generalized linear models) developed by the Working Party on Statistical Computing of the Royal Statistical Society*.

The 1981 survey of software users by Datapro Research Corporation [6], indicates that NAG's Mark IV edition appears highly reliable (rated at 3.6 out of a possible score of 4.0), with an overall satisfaction of 3.1 rating among sample respondents. The same report earmarked Mark IV as one of the 19 top-rated packages across all nine software quality attributes assessed.

In an effort to provide quality methodology, the NAG staff, since its inception in 1970 in England, has attempted to adhere to criteria such as accuracy, numerical stability, robustness, speed, and utility.

For more information contact Numerical Algorithms Group Ltd., NAG Central Office, 7 Banbury Road, Oxford, OX2 6NN, U.K. or Numerical Algorithms Group Inc., 1250 Grace Court, Downers Grove, IL 60516.

The IMSL (International Mathematics and Statistics Library [21]) is a collection of 535 Fortran-based subroutines designed and tested on 18 types of hardware currently utilized in 45 countries. Datapro's survey revealed that the IMSL ranked 13th among the 417 packages sampled in 1981, based on customer satisfaction. This finding is compatible with Hayes [17], which indicates that the IMSL addressed all but the quality assurance topic shown in Fig. A. In fact, the statistical subroutines marketed by IMSL are at the highly advanced level in 13 of the 15 clusters.

Edition 9 of the IMSL fielded 40 new subroutines that were designed to eliminate shortcomings of earlier editions and to test some of the latest statistical developments. Recent subroutines include those for life table analysis; boundary problems in nonlinear systems, techniques for generalized eigenvalue problems, de Boor's bicubic spline estimators, and utility functions for plotting and printing of complex vectors and matrices.

In addition to IMSL's software packages, it distributes subroutines developed elsewhere, such as the B-Spline, Rosepack, and ACM Algorithms. For more detailed information contact IMSL Inc., NBC Building, 7500 Bellaire Boulevard, Houston, TX 77036.

SOFTWARE SYSTEMS

BMDP statistical software [3, 4] written in Fortran is enhanced by its tailor-made BMDP control language and has been greatly improved over the past 25 years at UCLA by experienced statisticians and computer scientists. BMDP subroutines have been tested by batch and interactively on 22 major hardware clusters, including minicomputers. Datapro's survey reported 3.4 for reliability and 3.3 for consumer satisfaction of BMDP products. On-site tutorials are conducted in the U.S. and Europe at basic, intermediate, and advanced levels to familiarize new users and to assist perennial clients in keeping abreast of the latest advanced BMDP offerings.

BMDP is well known for its highly advanced algorithms in 10 of the 16 categories examined in Fig. A. Emphasis is on sample surveys, graphical analyses, multivariate techniques, nonparametrics, linear and nonlinear regression, and time series. Since BMDP programs evolved from medical and educational research as well as business and government operations, the techniques validated over the years tend to be concentrated on traditional statistical methods peculiar to the above-mentioned disciplines. Hence, there has been little development of subroutines utilizing Bayesian methods, order preserving procedures, quality assurance, sequential analysis, and spline estimation.

An excellent BMDP feature is its flexible data handling capabilities, permitting sizable numbers of variables and cases built into the system, thus accommodating large-scale and complex problems. Easy interface between BMDP and other systems that are summarized here and elsewhere (i.e., SAS, P-STAT, and MINITAB*, has extended additional computational power to BMDP clients. For further details write to BMDP Statistical Software, Department of Biomathematics, University of California, Los Angeles, CA 90024.

P-STAT is a conversational computing system (with batch mode edition) that originated in Princeton University in 1962, but is now independently managed by its developers under P-STAT Inc. [26]. It has been tested on 11 of the most widely used hardware systems and is highly recommended for big minicomputers. P-STAT pioneered in areas of programming language, conversational, and data management features that enhance its routines for ANOVA and MANOVA involving complex experimental designs and for graphical analysis, nonparametrics, and exploratory data analytical techniques. For more information contact P-STAT Inc., P.O. Box AH, Princeton, NJ 08540.

The Statistical Analysis System (SAS) provides a range of capabilities in the library system that includes an SAS basic and macro language applicable to mainframes, microcomputers, and minicomputers. It is known for interfacing with other statistical software systems such as the IMSL, particularly in the area of multivariate statistical techniques ranging from regression and analysis of variance for linear models to time-series analysis and forecasting. The SAS graphics procedures were developed for a variety of colored scatterplots with repeated values with mapping and cartographic applications. For more information contact SAS Institute Inc., Box 8000, Cary, NC 27511.

The Statistical Package for the Social Sciences (SPSS) by Nie and Hull is one of the commonly used general purpose software packages for both the mainframe and microcomputer systems. It features interactive capabilities for handling large data bases. Its algorithms have been tested for the computation of statistical problems from simple descriptive statistics to the more complex multivariate and nonparametric techniques. In addition to the Report Writer, the graphic facilities extends from contour plots to dendograms. For detailed information contact SPSS Inc., 444 North Michigan Avenue, Chicago, IL 60611.

The S system is an expression language for simple and advanced analyses of databases developed at Bell Laboratories under the leadership of Becker and Chambers [2]. This high-level programming language was developed and tested, during the 1970s, on the UNIX operating system. Its special features include computations in terms of uniform data structures as a whole; the S macroprocessor; multivariate analytical methods; high-level graphical functions; and advanced graphical algorithms utilizing subroutines with or without stand-alone tools. For details concerning documentation and latest developments on S contact AT & T Bell Laboratories, Computing Information Service, 600 Mountain Avenue, Murray Hill, NJ 07974.

LANGUAGES

The quest for a highly flexible representational language grew out of the desire to have a general representational system for complex software and hardware systems. Iverson [22] introduced a system later known as A Programming Language (APL) as a concise, interactive language that exhibits all the earmarks of a general purpose language that new users can effectively use within a span of a few hours. Several features, notably, (a) APL's capability to deal with scalars, vectors, matrices, and multidimensional arrays (both numerical and character arrays) without any loops and (b) APL's generality of definitions, nonexistence of arbitrary features, as well as easy, consistent rules governing its syntax reduce the vocabulary and taxonomy to be learned and make it an attractive language to users with statistical applications and limited programming skills. Statistical inferences using graphical analysis are commonly effective in APL since character arrays and numerical arrays are easily manipulated in APL. APL language employed as a conversational computing system uses statistical techniques with a latitude of flexibility not common to other general purpose languages. Notable textbooks on this language recommended to statisticians are Gilman and Rose [15] as well as Ramsey and Musgrave [27]. Illustrative APL applications in research studies and teaching of probability and statistics are summed up in Heilberger's compendium [18]. The text by Anscombe [1] is an excellent treatment of APL and its use as a statistics tool.

We indicate selected statistical software libraries in APL to convey some of APL's versatility. (a) IBM's selection [19] of scientific computing routines include its *APL Advanced Statistical Library* package. This interactive system covers topics on graphical analysis, multivariate analysis, time-series, and stochastic processes, simulation, and modeling, as well as nonparametrics. For more information contact IBM Corporation, 1133 Westchester Avenue, White Plains, NY 10604. (b) STATAPL is the British version of a statistical software library in APL featured in the *International Directory of Software* [20]. It consists of four packages with about 300 programs addressing basic statistical methods, analysis of variance, regression, and time-series analysis as well as simulation. For details contact the developers Gamma Fars, Bath, U.K. (c) STATPACK1 and STATPACK2 are Canadian products devised by Surillie [28] for elementary statistical inference, regression analysis, analysis of variance of rectangular arrays based on complex factorial experiments, regression analysis, and time-series studies. For more information contact Department of Computer Science, University of Alberta, Edmonton, Alberta, Canada. (d) APL Library of Mathematical Software was devised at the public library of Brown University's Computer Center under the direction of Grenander [16]. The library was meant to be a repository for programs required to pursue computational probability research. Hence, algorithms at advanced levels are available in asymptotic theory, graphical analysis, simulation and modeling spline estimation, time series, and stochastics and nonparametrics. For detailed information contact Ulf Grenander, Division of Applied Mathematics, Box F, Brown University, Providence, RI 02912.

SOFTWARE / HARDWARE COMBINATIONS

The revolutionary impact of computational graphics as an alternative methodology to classical statistical analysis has been intensified by the availability of the current generation computer system (Yasaki [31]). In fact, the ideas and techniques found in Tukey's [29] exploratory data analysis, which includes both graphical and nongraphical techniques were considerably extended through his *Projection, Rotation, Isolation, and Masking (PRIM) System* and later developments of it including the ORION-1 system.

PRIM-9 was the first interactive computer graphics statistical system that demonstrated the feasibility of employing kinematic displays to achieve spatial impressions with three-dimensional scatterplots (Tukey et al. [30]). PRIM-9 was created at the Stanford Linear Accelerator Center (SLAC) in 1972, where several million dollars of IBM machines used for graphical analysis were available. The prohibitive costs associated with PRIM-9 made it only a demonstration system and not feasible for wide use among statisticians. With cheaper hardware, more capable and less costly versions could be pursued. In 1978, PRIM-S, a Swiss version, was developed and tested on a DEC-10 and Evans and Sutherland Picture System 2 (PS2) by Peter Huber and his students in Zurich. In 1979, SLAC began development of PRIM-79, while PRIM-H was begun at Harvard, both of which proved to be superior to their predecessors in statistical power and cost efficiency (Donoho [8]). PRIM-H has all the features of PRIM-9 but is not limited to nine dimensions. It uses a VAX-11/780 with PS2 interfacing with Interactive Statistical Package (ISP), an interactive system operating on alphanumeric arrays. The latest edition of PRIM-H uses Friedman and Tukey's [14] projection pursuit techniques for reducing dimensionality. Most recently, Huber has been pursuing developments on Apollo microcomputers using its high resolution graphics. For more information contact Peter J. Huber, Department of Statistics, Harvard University, Cambridge, MA 02138.

ORION-1 is a SLAC developed workstation system particularly designed for applying interactive kinematic graphics (McDonald [24]). ORION-1 processes large, multidimensional data sets (e.g., 1,000 observations in 20 dimensions) that can be simultaneously represented in a single picture through two methods: (a) comparison of several scatterplots interactively colored to indicate relationships among their contents; (b) application of projection pursuit techniques (Friedman and Stuetzle [13]) to highlight and compress structures peculiar to the many-dimensional data into lower-dimensional pictures. Software packages for projection pursuit regression, classification, and density estimation have been developed. Two films (Friedman et al. [11, 12]) for loan and other information are available from Jerry Friedman, Bin-88 Computation Research Group, SLAC, P.O. Box 4349, Stanford, CA 94305.

(*Added in Proof*) The evolution of the graphics-based exploratory techniques has been rapid. The prototype systems described as ORION-1 and PRIM-H have been commercialized as a $100 program for the Apple Macintosh known as MacSpin, written by David and Andrew Donohoe and available through Wadsworth Advanced Books and Software based in Monterey, CA. The

ORION-1 system could be best described as an experimental workstation that was extremely difficult to program, and hence software was extensible only with great difficulty. Current thinking of many focuses on increasingly sophisticated workstation environments such as the Apollo 3000, SUN 3, MicroVAX, and Symbolics 3600 series. Incorporating UNIX and high resolution graphic capabilities, these recent products allow duplication of ORION-1 capabilities with relative ease. The Symbolics 3600 is a LISP machine. A number of influential people have argued that the LISP language is ideal for the exploratory data analysis environment. There is much feeling that object oriented programming is also fruitful for statistical computing.

OTHER LITERATURE

In addition to the (mainly commercially) available software, there has been increasing emphasis on the publication of algorithms and related material on statistical computing. Several journals are note-worthy resources for much material.

1. *SIAM Journal of Scientific and Statistical Computing*, published by Society of Industrial and Applied Mathematics. Editor: Professor William Gear, SIAM Publications, Box 7541, Philadelphia, PA 19101.
2. *Statistical Software Newsletter*, published by International Association for Statistical Computing. Editors: Professor K. Berk, Department of Mathematics, Illinois State University 313 Stevenson Hall, Normal, IL 61761 and Dr. W. J. Dixon, BMDP Statistical Software, 1964 Westwood Boulevard, Suite 202, Los Angeles, CA 90025.
3. *American Statistician* (*Statistical Computing Section*), published by American Statistical Association. Editor: Professor Charles E. Gates, Institute of Statistics, Texas A & M University, College Station, TX 77843.
4. *Journal of Statistical Computing and Simulation*, published by Gordon and Breach, Inc. Editor: Dr. Richard Krutchkoff, Department of Statistics, Virginia Polytechnic Institute and State University, Blacksburg, VA 24061.
5. *Communications in Statistics—Simulation and Computation*, published by Marcel Dekker, Inc. Editor-in-Chief: Professor Don Owen, Department of Statistics, Southern Methodist University, Dallas, TX 25275.
6. *Computational Statistics and Data Analysis*, published by North-Holland, Inc. Editor: Professor Stanley P. Azen, School of Medicine, University of Southern California, Los Angeles, CA 90033.

SUMMARY

This article is a very compact survey of several packages not covered elsewhere in this encyclopedia. Particularly noteworthy software not discussed here includes LINPACK and MINITAB*. See specific articles in this encyclopedia or Francis [10] or Hayes [17] for more specific information on these systems.

As we write this, we are on the threshold of a computational revolution in two extremes. At the low end of the scale, inexpensive microprocessors on a single chip are making the personal computer ubiquitous. As these are networked, we shall see an extremely powerful distributed computing resource that will make mainframes obsolete for many purposes, perhaps including statistical computing. At the other end of the scale are the supercomputers employing highly parallel architectures. These computing giants will revolutionize the type of thing we are able to compute as well as the scale of problems we can address. While statistical computing will continue to be an important facet of the statistical enterprise, the present revolution will no doubt reshape our thinking about the type and scale of problems we will address in the future. It is thus our expectation that the character of statistical software will be changing very significantly in the next decade.

References

[1] Anscombe, F. (1981). *Computing in Statistical Science through APL*. Springer-Verlag, New York. (This is a major work documenting the use of APL in statistical computing. APL is a vector-array oriented language with a particularly simple syntax that makes it ideal for matrix oriented computations. The Anscombe volume is the definitive work on this topic.)

[2] Becker, R. A. and Chambers, J. M. (1981). *S-a Language and System for Data Analysis*. Bell Laboratories, Murray Hill, NJ.

[3] *BMDP Statistical Software, Communications* (1981). The University of California Press, Los Angeles, CA.

[4] *BMDP Tutorial Bulletin* (1981). The University of California Press, Los Angeles, CA.

[5] Carpenter, J., Deloria, D., and Morganstein, D. (1984). *Byte*, **9**, 234–251. [This is an article on statistical software for microcomputers. It contains a review of some 25 packages. A review of seven major statistical software packages (SPSS, BMDP, etc.) is expected in the spring of 1985 by the same authors.]

[6] Datapro Research Corporation (1982). *User Ratings of Proprietary Software*. Datapro Research Corporation, NJ. (This document attempts a quantitative assessment of some of the major statistical packages.)

[7] Dixon, W. J. (1981). *BMDP Statistical Software*. University of California Press, Los Angeles, CA.

[8] Donoho, D., Huber, P. J., and Thoma, H. M. (1981). *Computer Science and Statistics: Proc. 13th Symposium on the Interface*, W. F. Eddy, ed. Springer-Verlag, New York.

[9] Foxwell, H. J., ed. (1984). *Capital PC Monitor*, **6**, No. 3. (This is a special issue on statistical software for the IBM PC published by the Capital PC User Group, Inc. P.O. Box 3189, Gaithersburg, MD 20878. This issue contains a list of 65 statistical software packages available for the IBM PC.)

[10] Francis, I. (1981). *Statistical Software: A Comparative Review*, North Holland, New York. (The most extended review of statistical software, this book documents some 60 packages with specific detail.)

[11] Friedman, J. H., McDonald, J. A., and Stuetzle, W. (1982). *Exploring Data with the Orion I Workstation* (16 mm sound film). SLAC, Stanford, CA.

[12] Friedman, J. H., McDonald, J. A., and Stuetzle, W. (1982). *Projection Pursuit Regression with the Orion I Workstation* (16 mm sound film). SLAC, Stanford, CA.

[13] Friedman, J. H. and Stuetzle, W. (1981). *J. Amer. Statist. Ass.*, **76**, 817–823.

[14] Friedman, J. H. and Tukey, J. W. (1974). *IEEE Trans. Comput.*, **C23**, 881–890.

[15] Gilman, L. and Rose, A. J. (1976). *APL: An Interactive Approach*. Wiley, New York.

[16] Grenander, U. (1981). *APL Library of Mathematical Software Technical Manual*. Department of Applied Mathematics, Brown University, Providence, RI.

[17] Hayes, A. R. (1982). *Statistical Software: A Survey and Critique of Its Development*. Office of Naval Research, Arlington, VA. (This paper is an analytic survey of some 213 statistical software packages available in 1982. This paper treats individual packages with less detail than Francis [10], but is somewhat more exhaustive in scope. It identifies sources for all of the software packages covered.)

[18] Heilberger, R. M. (1979). *ACM*, **9**, 134–137.

[19] *APL Advanced Statistical Library: Program Description/Operations Manual*, (1978). International Business Machines Corporation, New York.

[20] (1981). *International Directory of Software*. CUYB Publications, Gamma Fars, Bath, England.

[21] *IMSL Library Information* (1981). IMSL, Houston, TX.

[22] Iverson, K. E. (1962). *A Programming Language*. Wiley, New York. (This book describes the fundamentals of APL. Iverson is the inventor of APL.)

[23] Management Science Associates, Inc. (1980). *User's Guide to BMDP on the DEC VAX-11/780 Computer*. Management Science Associates, Arlington, VA.

[24] McDonald, J. A. (1983). *Naval Research Reviews*, Office of Naval Research, Arlington, VA.

[25] Numerical Algorithms Group, Ltd. (1981). *The NAG Library Service*. Numerical Algorithms Group, Ltd., Oxford, UK.

[26] P-STAT Inc. (1978). *User's Manual*. P-STAT Inc., Princeton, NJ.

[27] Ramsey, J. B. and Musgrave, G. L. (1981). *APL-STAT: A Do-It-Yourself Guide to Computational Statistics Using APL*. Lifetime Learning Publications, Belmont, CA.

[28] Surillie, K. W. (1968/1969). STATPACK 1 and 2: An APL Statistical Package. *Publications Nos. 9 and 17*. Department of Computing Science, University of Alberta, Edmonton, Alberta, Canada.

[29] Tukey, J. W. (1977). *Exploratory Data Analysis*, Addison-Wesley, Reading, MA.

[30] Tukey, J. W., Friedman, J. H., and Fisherkeller, M. A. (1975). *Proc. 4th International Congress for Stereology*.

[31] Yasaki, E. K. (1982). *Datamation*, **28**, 110–115.

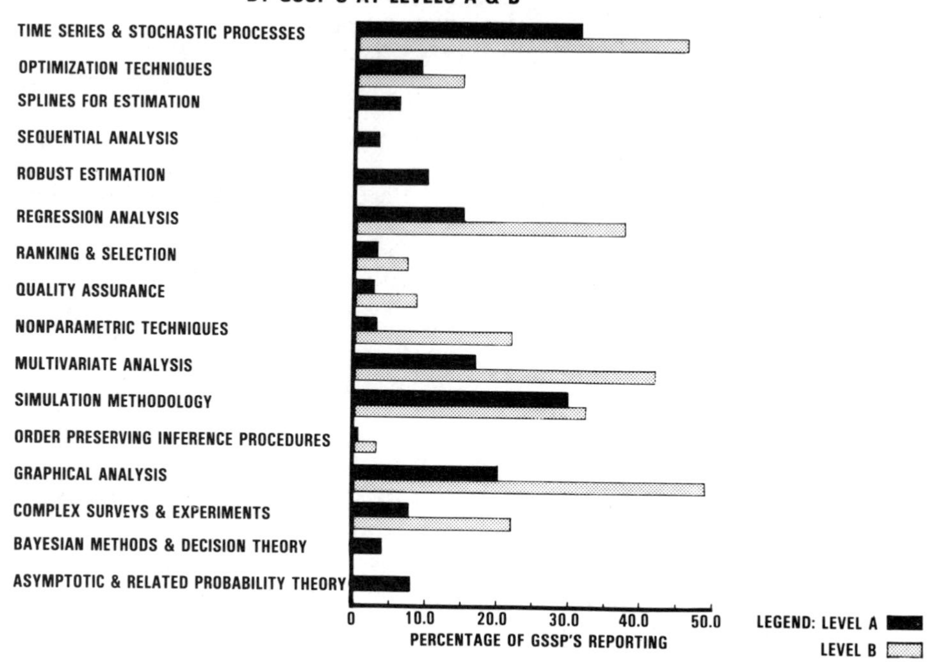

Figure A

(COMPUTERS AND STATISTICS
GLIM
GRAPHICAL REPRESENTATION, COMPUTER-AIDED
MINITAB
SOFTWARE RELIABILITY)

EDWARD J. WEGMAN
ANNIE R. HAYES

STATISTICAL THEORY AND METHOD ABTRACTS

This journal is published by the International Statistical Institute*. Volume 1 appeared in 1959–1960; subsequent volumes have appeared annually in four parts. The increase in the numbers of abstracts is indicated in the following table:

Volume No.	Year	Number of Abstracts
6	1965	1254
11	1970	1456
16	1975	2160
21	1980	2726
26	1985	3803

The abstracts are in English. The journal abstracts articles that report advances in statistical theory and techniques. It does not abstract papers reporting application of established standard statistical techniques. For the following journals the abstracting is "virtually complete":

Adv. Appl. Prob.; *Ann. Inst. Statist. Math.*; *Ann. Prob.*; *Ann. Statist.*; *Appl. Statist.*; *Austral. J. Statist.*; *Biometrics*; *Biometrika*; *Bull. Inf. Cybernet.*; *Calcutta Statist. Ass. Bull.*; *Canad. J. Statist.*; *Commun. Statist.-Sequent. Anal.* (now *Sequent. Anal.*); *Commun. Statist.-Simul. Comp.*; *Commun. Statist.-Theor. Math.*; *Comp. Statist. Data Anal.*; *Int. Statist. Rev.*; *J. Amer. Statist. Ass.* (*Section Theory and Methods*); *J. Appl. Prob.*; *J. Econometrics*; *J. Indian Statist. Ass.*; *J. Japan Statist. Soc.*; *J. Multivariate Anal.*; *J. R. Statist. Soc. B*; *J. Statist. Comp. Simul.*; *J. Statist. Planning Infer.*; *J. Time Ser. Anal.*; *Metrika*; *Metron*; *S. Afr. Statist. J.*; *Sankhyā* (*A and B*); *Scand. J. Statist.*; *Statist. and Dec.*; *Statist. Neerl.*;

Statist. Prob. Letters; *Stochastics*; *Stoch. Processes Appl.*; *Theory Prob. Appl.*; *Technometrics*; *Zeit. Wahrscheinl.-theorie*

Abstracts of appropriate articles in many other journals are also included.

Books are not abstracted (except in Vol. 26); the reader is referred to *Short Book Reviews*, also published by the International Statistical Institute. However, abstracts are included for "monographs in a series, which are not proper textbooks but have the character of research papers which are too extensive to be published in a journal."

Originally, long abstracts (of about 400 words) were aimed at. However, pressure on space has led to a gradual shortening of abstracts, and even, in some cases giving only authors name(s), title, and journal reference. Each of the four issues per year contains its own author index. Abstracts are arranged in groups according to main topics. A new system of classification was introduced in Vol. 26.

A consolidated index arranged in alphabetical order by authors' names is issued, in an Index Supplement, at the end of each year.

Since 1965 (starting with Vol. 6), Dr. J. W. E. Vos (Technische Hoogschool, Delft, Netherlands) has been General Editor. The only previous General Editor was Dr. W. R. Buckland.

The business address is Publications Department, International Statistical Institute, P. O. Box 950, 2270 AZ Voorburg, Netherlands.

(INTERNATIONAL STATISTICAL INSTITUTE)

STATISTICA NEERLANDICA

Statistica Neerlandica was founded in 1946 by the Vereniging voor Statistiek (VVS, the Netherlands Society for Statistics, Biometrics, Econometrics, and Operational Research). The first, introductory, paper by chairman Van Ettinger of the VVS was titled "Statistics at the service of recovery" (the war had just ended), and perhaps set the applied tone of the journal.

In the first few years, *Statistica* (*Neerlandica* was added later to avoid confusion with the earlier Italian journal *Statistica**) appeared bimonthly, and after 1951 quarterly. The number of pages per year has varied and now fluctuates around 225. Among the first editors were the applied statisticians A. R. van den Burg, J. Sittig, and H. C. Hamaker. By contrast, many of the later editors have been mathematical statisticians, or even probabilists; among them were J. Hemelrijk, W. R. van Zwet, and C. J. Scheffer. This contrast has led to a good balance between theory and application.

Gradually, *Statistica Neerlandica* has become more international; almost all papers are now in English.

In 1975 the journal's editorial policy was first stated explicitly; it calls for expository papers (in Dutch, English, German, or French) in probability, statistics and operations research and their applications, with emphasis on clarity, applicability, and accessibility for the general reader rather than on formal novelty and mathematical detail. All papers are refereed, usually by two referees.

Along with papers, the journal publishes notes and occasional book reviews (many book reviews now appear in *VVS Bulletin*); it also features information on recent Ph.D. theses in the Netherlands on probability and statistics, and a problem section.

The contents of Vol. 36, No. 3 (1982) are:

"Asymptotic confidence intervals for the effect of closing an area upon the number of birds' nests" by P. M. de Jong.

"Modelling association football scores" by M. J. Maher.

"Distribution free tests for comparing several treatments with a control" by M. A. Fligner and D. A. Wolfe.

"Minimax character of the two-sample chi-squared test" by H. Luschgy.

"Asymptotic normality of vacancies on a line" by J. Th. Runnenburg.

Notes about five recent Ph.D. theses and the problem section complete the issue (a total of 60 pages).

The current editorial address is: Dr. T. A. B. Snÿders, Faculty of Actuarial Science & Econometrics, University of Groningen, P. O. Box 800, 9700 AV Groningen, The Netherlands.

F. W. Steutel

STATISTICIAN, THE

The Statistician first appeared in 1950 as *The Incorporated Statistician*, being the journal of the newly formed Association of Incorporated Statisticians Limited. In 1962 the Association changed its name to the Institute of Statisticians and the journal became *The Statistician*, with the alternative title *Journal of the Institute of Statisticians*. The journal from its inception was published in four issues per volume (now five), each volume appearing annually. The size of the journal has grown from 62 pages in Vol. 1 (1950) to 371 pages in Vol. 35 (1986). The current (1987) editor is Dr. R. R. Harris and the editorial address is School of Mathematics and Statistics, Lancashire Polytechnic, Preston, PRJ 2TQ, U.K. All papers submitted are independently refereed and the mean times between submission, editorial decision, and publication are approximately four and seven months, respectively.

The journal publishes original articles, reviews and case studies from all areas of statistical application. The emphasis is on communication, both between members of the statistical profession and between the profession and other interested parties. This editorial policy has changed little since the inception of the journal; *The Statistician* has always carried critical Book Reviews and has correspondence pages. Until the Institute began publication of a monthly newsletter in 1974 (which in 1982 became the monthly publication *The Professional Statistician*), *The Statistician* carried the Institute's notices together with information about the work of the Institute's committees as well as the results of the Institute's worldwide examinations.

One issue each year (usually 1) contains the invited papers from the Institute's Annual International Conference (held in July); themes in recent years have been Statistics in Medicine, Practical Bayesian Statistics, Energy Statistics, and Statistics in Health and Survey Sampling. Although Members and Fellows of the Institute receive the journal as of right, the subscription list includes many nonmembers around the world as well as the libraries of a large number of institutions. The international flavour of the authorship has increased in recent years, and like the readership of the journal, is by no means restricted to members of the Institute. To indicate the type of papers that appear in *The Statistician* the following are the contents of issues Nos. 2 and 3 of Vol. 32 (1983), neither of which are conference issues.

"Bayes' estimation in sampling for auditing" by J. J. A. Moors.

"Analysis of clay strata orientations" by R. J. Gadsden and G. K. Kanji.

"A simple method for evaluating the Shapiro–Francia W' test for non-normality" by J. P. Royston.

"Wider application of survival analysis: an evaluation of an unemployment benefit procedure" by G. H. D. Royston.

"Measurement in medicine: the analysis of method comparison studies" by D. G. Altman and J. M. Bland.

"Observed betting tendencies and suggested betting strategies for European football pools" by Raymond T. Stefani.

"Rank analysis of covariance: alternative approaches" by Andrew Lawson.

"Bayesian estimation of the mean of a normal distribution when the coefficient of variation is known" by S. K. Sinha.

"A simple method for monitoring routine statistics" by M. J. R. Healy.

"Asymptotic variance formulae for some measures used as prognostic indices" by John O'Quigley.

"Assessing goodness of fit of ranking models to data" by Ayala Cohen and C. L. Mallows.

"The monstrous regiment of mathematicians" by S. C. Pearce.

"On the use of minimum chi-square estimation" by R. R. Harris and G. K. Kanji.

"On an exploratory analysis of contingency tables" by Ron S. Kenett.

"A graphical identification procedure for growth curves" by D. I. Holmes.

"Sequential detection of unusual points in regression" by S. R. Paul.

"Further investigation of robustness of power in the analysis of variance" by G. K. Kanji and C. K. Liu.

Book Reviews

Obituary

Volume Index

G. K. Kanji
R. R. Harris

STATISTICS, ANCILLARY *See* ANCILLARY STATISTICS

STATISTICS AND ARTIFICIAL INTELLIGENCE *See Supplement*

STATISTICS AND DECISIONS

HISTORICAL BACKGROUND AND ORGANIZATION

Statistics and Decisions is an international journal for decision theoretically oriented mathematical statistics and stochastic models. Largely owing to the pioneering efforts of Professor D. Plachky (of the University of Münster, W. Germany) and the financial undertakings of the Akademische Verlagsgesellschaft, Wiesbaden, W. Germany, the journal was founded in 1981, and its first issue was published in the fall of 1982. The journal is planned to have four issues per volume for each calendar year, although the first volume has five numbers and the second one has an additional supplement.

The journal currently has three coordinating editors: Professors E. J. Dudewicz (Syracuse University, New York), D. Plachky (Münster), and P. K. Sen (University of North Carolina, Chapel Hill); an editorial board of more than 30 eminent statisticians from all corners of the globe (with more representation from Central Europe); and an advisory board of other prominent statisticians (all actively participating in the reviews and editorial decisions on the submitted manuscripts). In 1983, the financial undertakings were taken over by the R. Oldenbourg Verlag, München, W. Germany; also, Professor W. Wertz (Technical University, Vienna, Austria) joined the editorial board as the Book Review Editor.

AIMS AND SCOPE

Statistics and Decisions intends to provide an international forum for the discussion of theoretical and applied aspects of decision theoretically oriented mathematical statistics. Topics covered include classical and multiple statistical decision procedures (relating to estimation, testing, ranking, and selection), asymptotic and nonparametric statistical procedures, sequential analysis (parametric, nonparametric as well as asymptotic), and abstract and applied statistical inference for stochastic processes. Research as well as expository papers of fundamental mathematical importance giving impetus to further developments in decision theoretically oriented mathematical statistics and its applications also fall within the jurisdiction of this journal. The major objectives are the following:

(i) To encourage the trend of growing interaction between problems in mathematical statistics (as well as in stochastic models and methods) and current research in information theory, probability theory, game theory, control theory, approximation theory, functional analysis, stochastic automata, and other areas of mathematics useful for modelling, analysis and (optimal) solution of decision theoretically oriented statistical problems.

(ii) To encourage the further development of statistical decision theory in all its aspects, and to nurture applications in diverse fields (viz. clinical trials, management problems) based on sound mathematical theory.

All papers are refereed; excessively long papers are not encouraged. Manuscripts for publication should be submitted to one of the three coordinating editors (depending on the area of the work).

PUBLISHED RECORDS

During 1982–1986, 18 numbers of the journal have been published. For the calendar year 1987, four issues for Vol. 5 are already in press. The published articles represent a very broad spectrum of decision theoretic stochastic methods and models and these are very much in line with the general objectives of the journal stressed earlier. Traditionally, in continental Europe, developments in statistics have taken place in a somewhat different setup than in the United States, England, Australia, and India. This journal attempts to retain this continental flavour to a greater extent, nurturing at the same time a meaningful blending of the two setups for a relatively unified approach to the mathematical theory. Toward this objective, the contents of the first issue [Vol. 1, No. 1 (1982)], listed below, provide a clear picture:

"Third order efficiency of the maximum likelihood estimator in the multinomial distribution" by C. R. Rao, B. K. Sinha, and K. Subramanyam.

"On asymptotic deficiency of estimators in pooled samples in the presence of nuisance parameters" by M. Akahira and K. Takeuchi.

"Remarks on linear parametric inference for stochastic processes" by B. L. S. Prakasa Rao.

"A joint study of the Kolmogorov–Smirnov and the Eicker–Jeschke statistics" by P. Révész.

"Consistency of maximum penalized likelihood density estimators based on initial estimators" by R.-D. Reiss.

"Asymptotic properties of likelihood ratio tests based on conditional specification" by P. K. Sen.

Generally, the reviewing process for submitted manuscripts is quick and articles accepted for publication appear within a reasonable time period. There is good scope for some application oriented theoretical papers in this journal, and the editors welcome such submissions.

P. K. Sen

STATISTICS AND PROBABILITY LETTERS

Statistics and Probability Letters is an international journal that first appeared in July 1982. It is published by North-Holland Publishing Company, Amsterdam, The Netherlands. Richard A. Johnson (Madison) was appointed to serve as the journal's first Editor and he assembled the international editorial board consisting of 22 members.

The current editorial address for the submission of papers is Richard A. Johnson,

Editor, *Statistics and Probability Letters*, Department of Statistics, University of Wisconsin, 1210 West Dayton Street, Madison, Wisconsin 53706.

Statistics and Probability Letters takes a highly innovative approach to the publication of research findings in statistics and probability. It features (i) concise articles, (ii) rapid publication, and (iii) broad coverage of the statistics and probability literature. Articles are limited to six journal pages in double column format (13 double spaced typed pages) including references and figures. Accepted manuscripts are published within about three to four months of submission.

Letters journals are a proven medium of communication in the physical sciences and, in recent years, they have been introduced to computer science, economics, operations research, and other fields.

Statistics and Probability Letters is a refereed journal. Apart from the six-page limitation, manuscripts are judged primarily on originality and quality. The subject matter can be theory, methodology, empirical studies, or applications. The major emphasis will be on new theorems, novel statistical methods, and probabilistic or statistical models that the author wants to disseminate very rapidly. We will also publish descriptions of applications and case studies that demonstrate a novel use of existing techniques or have interesting innovative ideas about data collection or description, modeling or inference.

Statistics and Probability Letters is published in volumes consisting of six issues. These are nearly equally spaced within the year in order to avoid publication delays once a paper is accepted. The first volume contained 53 papers and totaled 135 pages in length. The international nature of the journal is reflected in the fact that, of the 108 papers published in the first 10 issues, the 152 authors came from 19 countries.

The extent of coverage and diversity of topics is exemplified by a list of selected papers published in the first 10 issues.

"On the fallacy of the likelihood principle" by H. Akaike.

"A note on Hájek projections and the influence curve" by W. C. Parr.

"The hat matrix for smoothing splines" by R. L. Eubank.

"Strong approximation of certain stopped sums" by L. Horvath.

"On Kersting's proof of the central limit theorem" by G. Ch. Pflug.

"Maximum entropy interpretation of autoregressive spectral densities" by E. Parzen.

"A note on the 'underadjustment phenomenon'" by H. Robbins and B. Levin.

"On the exponential rate of convergence of the least square estimator in the nonlinear regression model with Gaussian errors" by B. L. S. Prakasa Rao.

"Density estimation based on line transect samples" by P. V. Rao.

"A multivariate correlation ratio" by A. R. Sampson.

"Poisson on the Poisson distribution" by S. M. Stigler.

R. A. JOHNSON

STATISTICS AND RELIGIOUS STUDIES

The relationship between statistics and religious studies appears in three main areas: (1) the use of statistical methodology to analyze religious literature (particularly, authorship studies), (2) the use of statistics to study sociological and historical aspects of religious movements, and (3) the appearance of statistical and probabilistic ideas in religious writings and their use in theological investigation.

A major application of statistics to religious studies occurs in the study of religious literature (*see* STATISTICS IN LITERATURE). Compilation of data from religious writings can be dated to the work of the Masoretes,

Jewish scribes who worked during the sixth through tenth centuries and were concerned with accurate preservation of the Hebrew Bible. Lists of word counts, letter counts, and the occurrence of certain words were made, but apparently no statistical analysis was performed on such data until quite recently.

Although study of biblical material dates back 2000 years or more, modern biblical scholarship is generally dated to the mid-nineteenth century (e.g., see ref. 6). Following initial suggestions of Astruc (in the eighteenth century), Wellhausen and others developed the "documentary hypothesis," a thesis holding that several narrative strands can be discerned in the early Hebrew Bible (particularly, the Pentateuch and early historical books), and that these strands could be attributed to different sources. Similar hypotheses concerning sources have also been developed for the New Testament. However, the use of formal statistical analysis to explicate such hypotheses did not occur until the mid-twentieth century. Following Wake [35] (see also ref. 8), Morton and his colleagues applied statistical methods to the Pauline Epistles [19, 20]. He measured such variables as sentence length and the frequency of occurrence of specific words, and compared the writings traditionally attributed to Paul with each other and with other classical Greek writings. He found that Romans, I and II Corinthians, and Galatians formed a relatively coherent group, but that other Epistles differed from these by significantly more than would be expected by examining the variation within single Greek authors. On this basis, Morton claimed that the other Epistles were not written by Paul. Since this work, a number of authors [2, 5, 11, 21, 32, and 33] have analyzed New Testament materials (particularly, Matthew, Mark, Luke, and Acts) using measures of vocabulary agreement between different books (although generally with little formal statistical inference). At the same time, Radday [29–31], Adams [1], Bee [3], Houk [12, 13], and others have applied statistical methods to authorship problems in the Hebrew Bible.

However, these statistical applications to both the Hebrew and the Greek Bible involve a serious problem: for biblical literature there is no large body of writings by known authors. Essentially all biblical authorship problems concern partitioning writings traditionally attributed to a single author among several unknown authors. Morton tried to solve this problem by considering known classical Greek authors; but his work is still somewhat problematic because statistical measures of an author's style may not be similar for Paul and for classical Greek authors (particularly since the authors he chose spanned a period of over 500 years). This problem is especially critical for the Hebrew Bible. Portnoy and Petersen [26] strongly criticize Radday et al. [30] for interpreting statistically significant differences between values of certain measures of linguistic behavior as differences between authors. Similarly, in several studies (see ref. 3 for references), Bee has suggested using statistics based on verb counts to analyze whether Hebrew passages derive from written or oral material; and he has also applied these methods to source criticism, again finding multiple sources in the material studied. As noted by Weitzman [36], however, any conclusions suffer from inadequate sampling theory, the use of possibly inappropriate statistical models (and the consequent use of critical values determined in a very ad hoc manner), and again from lack of evidence that the measures reasonably reflect source differences.

Conversely, the lack of significant statistical variation within a literary unit *cannot* be taken as evidence of single authorship. For example, Radday et al. [29] attack the documentary hypothesis by considering a very large number of statistical measures and showing that differences within two of the strands are larger than differences between them. These results are criticized by Portnoy and Petersen [27], who describe problematic features both of statistical methodology and biblical scholarship. Similarly, Adams [1] measured the frequency of certain Hebrew prefixes and compared two sections of the book of Isaiah with each other and with samples from other prophetic books. He found that the two sections of Isaiah (generally considered by scholars to have different authors) are more similar in terms of Mahalanobis* distance (*see* DISCRIMINANT ANALYSIS) than any other pair of samples. He then concluded unique authorship for the book of Isaiah. Clearly the appropriate conclusion is that prefix frequencies do not indicate statistical differences between parts of Isaiah—proof of unique authorship is simply beyond statistical verification, particularly since other measures show substantial variation between the sections (see ref. 31).

The second area of relationship between statistics and religious studies concerns application of statistical methodology in sociological, historical, or archeological investigations in religious studies. Three journals [37–39], are dedicated to such investigations, although similar studies appear in a wide variety of sources. A number of references can be found in Greeley [9]. Since the statistical methodologies and formulations of problems are generally those of social statistics, the reader is referred to entries under SOCIOLOGY, STATISTICS IN; STATISTICS IN HISTORICAL STUDIES, and related entries.

The third area of relationship between statistics and religion concerns the appearance of statistical and probabilistic ideas in religious writings (*see* STATISTICS, HISTORY OF). These are mainly of historical interest to statisticians, but probabilistic and statistical ideas concerning "randomness" and "inference" seem to have profound implications in wide areas of philosophical and theological thought.

One example of such references would be the discussions of censuses in the Hebrew Bible (see Madansky [18]). Other examples are presented in Hasofer [10] and Rabinovitch [28], who describe some occurrences of probabilistic or statistical reasoning in ancient Jewish writings. These are often similar to the following example of "significance testing" given by Høyrup [14]: in the late sixth century, Gregory of Tours described an attack on a monastery by a group of Frankish raiders. When the raiders tried to cross a river after the attack, their boat capsized and only one raider survived—a man who had rebuked the others for attacking the monastery. Gregory wrote: "If anyone thinks that this happened by chance, let him consider the fact that one innocent man was saved among so many who perished."

Another example not reported in these references is a very early example of a comparison of estimators presented in the Mishnah [7] (writings of Rabbis compiled in the early third century). The Rabbis held that "an egg's bulk" of unclean food would contaminate clean food. In providing an operational definition of "an egg's bulk," the majority prescribed the volume of a "middle sized egg" (the median?). But one Rabbi suggested submerging the largest and smallest eggs and halving the volume of displaced liquid. This estimator was rejected, though the reason given was not based on statistical theory: "but who can tell which is the largest and which is the smallest? Rather all depends on the observer's estimate."

More substantive application of probabilistic or statistical ideas to philosophical or theological thought can be dated to the seventeenth and eighteenth centuries when philosophers like Pascal* and others considered questions concerning causation*, divine provenance, determinism, and so forth. For example, Pascal posed the following "game": one may either believe or disbelieve in God's existence. Since the loss incurred if one disbelieves in error (eternal damnation) is so much greater that that incurred if one believes in error, it is irrational to disbelieve (Pascal actually assumed that God's existence and nonexistence were equally likely events). A modern application of game-theoretic arguments to theology is given in Brams [4]. An excellent modern exposition of the application of probabilistic and statistical reasoning to theological concerns is given by Bartholomew [1a] (see also ref. 34).

A more historical discussion of such probabilistic and statistical arguments in theology can be found in Kruskal [16] and in

Pearson [25]. Other similar material is irregularly presented in *Biometrika* and collected in refs. 24 and 15. With the modern development of statistical mechanics and quantum theory, probabilistic and statistical ideas have become more critical to philosophical and theological inquiry, particularly in questions concerning cosmology (see refs. 1a, 22, 23, or 17 for examples of such applications).

References

[1] Adams, L. F. and Reucher, A. C. (1973). The popular critical view of the Isaiah problem in light of statistical style analysis. *Computer Stud.*, **4**, 149–157.

[1a] Bartholomew, D. J. (1984). *God of Chance*. SCM Press, London, England.

[2] Beck, R. E. (1977). Common authorship of Luke and Acts. *New Testament Stud.*, **23**, 346–352.

[3] Bee, R. E. (1983). Statistics and source criticism. *Vetus Testamentum*, **33**, 483–488.

[4] Brams, S. J. (1983). *Superior Beings: If They Exist, How Should We Know*. Springer-Verlag, New York.

[5] Carlston, C. E. and Norlin, D. (1971). One more —statistics and *Q*. *Harvard Theol. Rev.*, **64**, 59–78.

[6] Eissfeldt, O. (1965). *The Old Testament: An Introduction*. Harper and Row, New York.

[7] Epstein, I., translator (1948). *The Babylonian Talmud*. Soncino Press, London, England.

[8] Grayston, K. and Herdan, G. (1958). The authorship of the Pastorals in light of statistical linguistics. *New Testament Stud.*, **6**, 1–15.

[9] Greeley, A. M. (1982). *Religion: A Secular Theory*. The Free Press, Macmillan, New York.

[10] Hasofer, A. M. (1966). Random mechanisms in Talmudic literature. *Biometrika*, **54**, 316–321.

[11] Honoré, A. M. (1968). A statistical study of the synoptic problem. *Novum Testamentum*, **10**, 95–147.

[12] Houk, C. B. (1983). A syllable–word structure analysis of Genesis 12–23. *The Word Shall Go Forth*, Meyers and O'Connor, eds. ASOR Philadelphia, PA, pp. 191–201.

[13] Houk, C. B. (1981). A statistical linguistic study of Ezekiel 1:4–3:11. *Zeit. für Alttestamentliche Wissenschaft*, **93**, 76–84.

[14] Høyrup, J. (1983). Sixth century intuitive probability: the statistical significance of a miracle. *Historia Math.*, **10**, 80–83.

[15] Kendall, M. G. and Plackett, R. L., eds. (1977). *Studies in the History of Statistics and Probability*, Vol. II. Macmillan, New York.

[16] Kruskal, W. (1982). Miracles and statistics: the casual assumption of independence. *J. Amer. Statist. Ass.*, to be published.

[17] Leslie, J. (1982). III. Anthropic Principle, world ensemble, design. *Amer. Philos. Quart.*, **19**, 141–151.

[18] Madansky, A. (1984). *On Biblical Censuses*. To be published.

[19] Michaelson, S. and Morton, A. Q. (1972). Last words: a test of authorship of Greek writers. *New Testament Stud.*, **18**, 192–208.

[20] Morton, A. Q. and McLeman, J. (1966). *Paul, the Man and the Myth: A Study in the Authorship of Greek Prose*. Harper and Row, New York.

[21] O'Rourke, J. J. (1974). Some observations on the synoptic problem and the use of statistical procedures. *Novum Testamentum*, **16**, 272–277.

[22] Peacocke, A. R., ed. (1981). *The Sciences and Theology in the 20th Century*. University of Notre Dame Press, Notre Dame, IN.

[23] Peacocke, A. R. (1979). *Creation and the World of Science*. Clarendon Press, Oxford, England.

[24] Pearson, E. S. and Kendall, M. G., eds. (1970). *Studies in the History of Statistics and Probability*, Vol. 1. Hafner, Darien, CT.

[25] Pearson, K. (1978). *The History of Statistics in the 17th and 18th Centuries*, E. S. Pearson, ed. Griffin, London, England.

[26] Portnoy, S. and Petersen, D. (1984). Biblical texts and statistical analysis: Zechariah and beyond. *J. Biblical Lit.*, **103**, 11–21.

[27] Portnoy, S. and Petersen, D. (1984). Genesis Wellhausen and the computer: a response. *Zeit. für Alttestamentliche Wissenschaft*, **96**, 421–426.

[28] Rabinovitch, N. L. (1973). *Probability and Statistical Inference in Ancient and Medieval Jewish Literature*. University of Toronto Press, Toronto, Canada.

[29] Radday, Y., Shore, H., Pollatschek, M., and Wickmann, D. (1982). Genesis, Wellhausen and the computer. *Zeit. für Alttestamentliche Wissenschaft*, **94**, 467–81.

[30] Radday, Y. and Wickman, D. (1975). The Unity of Zechariah in the light of statistical linguistics, *Zeit. für Alttestamentliche Wissenschaft*, **87**, 30–55.

[31] Radday, Y. (1973). *The Unity of Isaiah in the Light of Statistical Linguistics*. Verlag Gerstenberg, Hildesheim, W. Germany.

[32] Rosche, T. (1960). The words of Jesus and the future of the *Q* hypothesis. *J. Biblical Lit.*, **79**, 210–212.

[33] Spottorno, V. (1982). The relative pronoun in the New Testament: some critical remarks. *New Testament Stud.*, **28**, 132–141.
[34] Swinburne, R. (1979). *The Existence of God.* Clarendon, Oxford, England.
[35] Wake, W. C. (1957). Sentence length distributions of Greek authors. *J. R. Statist. Soc. Ser. A.*, **120**, 331–346.
[36] Weitzman, M. P. (1981). Verb frequency and source criticism. *Vetus Testamentum*, **31**, 451–471.
[37] *Journal for the Scientific Study of Religion*, Society for the Scientific Study of Religion, Inc., Storres, CT.
[38] *Review of Religious Research*, Religious Research Association, New York, NY.
[39] *Sociological Analysis*, Association for the Sociology of Religion, Storrs, CT.

(LITERATURE AND STATISTICS
STATISTICS IN HISTORICAL STUDIES)

STEPHEN PORTNOY

STATISTICS: AN OVERVIEW

Contents

INTRODUCTION

The term "statistics" is used in different ways. "Accident statistics" or "sales statistics" refer to numerical data in these areas. (The corresponding theoretical terminology defines statistics to be any functions of observable random variables.) Common problems encountered in the work with such data and those collected by scientists, engineers, government officials, lawyers, doctors, etc., have led to the development of general methods and principles concerning the collection, presentation, and analysis of data. The term "statistics" also denotes the discipline concerned with such methods.

The present article considers statistics broadly as the field comprising all of the above concerns. It might be described as the enterprise dealing with the collection of data sets, and extraction and presentation of the information they contain.

Comprehensive surveys of the present state of the field of statistics are provided by the nine volumes of this encyclopedia and the two volumes of the *International Encyclopedia of Statistics.* (Reference to articles in the first of these will be indicated by an asterisk (*) and in the latter by a dagger (†).) However, like any other scientific disciplines, statistics is an ongoing, ever changing endeavor. This is obvious for the collection of data relating to constantly changing populations. Regarding published guides to the enormous volume of government statistics are provided by the *American Statistical Index, The Statistical Abstract of the United States**, and the *Statistical Reference Index* (for the U.S.), and by the *Index to International Statistics*. Many other types of data are covered by *Statistical Sources: A Subject Guide to Data on Industrial, Business, Social, Educational, Financial, and Other Topics for the U.S. and Internationally*, which is also periodically updated.

On the methodological side, new methods and formulations are constantly being developed, and new types of application come into view and in turn give rise to new problems. This flow of research is disseminated through a multitude of journals being published around the world. A listing of annual contributions (more than 8000 in 1984) is available in the *Current Index to Statistics**:

Applications, Methods, and Theory, published since 1975. Much of the material prior to 1970 is listed in the five volume *Index to Statistics and Probability* (Ross and Tukey, eds. R and D Press).

What establishes statistics as a discipline is that the same kind of data requiring the same kind of concepts and methods turn up in many different fields. On the other hand, special areas of application also may require some specialization and adaptation to particular needs. Separate entries in this encyclopedia outline how statistics is used in actuarial work, agriculture, animal science, anthropology, archaeology, auditing, crystallography, demography, dentistry, ecology, econometrics, education, election forecasting and projection, engineering, epidemiology, finance, fisheries research, gambling, genetics, geography, geology, historical studies, human genetics, hydrology, industry, law, linguistics, literature, manpower planning, management science, marketing, medical diagnosis, meteorology, military science, nuclear material safeguards, ophthalmology, pharmaceutics, physics, political science, psychology, psychophysics, quality control, religious studies, sociology, survey sampling, taxonomy, and weather modification.

DATA INTERPRETATION

Statistical Methodology

It is a central fact, underlying essentially all statistical thinking, that an actual data set typically is only one of many possible such sets that might have been obtained under the given circumstances. (For a possible exception, see Diaconis [11].) Measurements vary when they are repeated. A store inventory, besides depending on the day on which it is taken, will be affected by bookkeeping and counting errors. In a survey, different households would be obtained if a new sample were drawn; even if the same households are visited on another occasion, different members may be at home and provide different answers; and, finally, even the same family member may answer the same questions differently on another day. As a consequence, interpretation of a data set depends not only on the actual data but also on what (if anything) is assumed about the possible alternative observations that might have been obtained instead. The following sections consider three categories of such assumptions, and briefly indicate the kinds of statistical procedures that can be based on each.

Data Analysis

It is rare that a data set is studied without any preconceived notions. Consider, however, the idealized approach of pure data analysis in which the data are considered on their own terms. The statistical methods developed on this basis have as their primary aim (i) exploration of the data to uncover the features of principal interest, and (ii) presentation of the data in a manner that will bring out and highlight these features. The set of techniques dealing with (i) and (ii) are called *exploratory data analysis** and *descriptive statistics*†, respectively. Both employ a great variety of numerical and graphical *† techniques.

The simplest, most basic data analytic methods concern a single batch of numbers, for example, the first-year sales of the 12 novels of a successful author, 40 measurements of corrosion taken at different locations on a copper plate, or the ages of 350,000 cancer patients listed in a tumor registry. Histograms*, stem and leaf displays, and one-dimensional scatter plots* are some of the many ways of presenting the numbers of a batch. From these one can get an impression of where the data are centered (the general level of their values), and how spread out they are. Observations that lie far from the bulk of the data, so-called *outliers**†, may correspond to errors or exceptional cases and may deserve special attention. The display may exhibit unusual features such as bimodality, perhaps suggesting the possibility of a mixture of two batches, each unimodal but with different *modes*. If there is marked asymmetry, the possibility arises of making a transformation* (such as taking logarithms) to obtain a more symmetric data set.

Instead of a fairly detailed display of the data, authors frequently present only one or two summary statistics; for example, the mean* or median to indicate the general level of the numbers, and a measure of their variability such as the standard deviation*, median absolute deviation, or interquartile range*. A compromise is the *five number summary**, consisting of the smallest, largest, and median observation, and the first and third quartiles. These may be displayed graphically as a *box plot* (*see* NOTCHED BOX AND WHISKER PLOTS*).

Additional information concerning the numbers of a batch may be available and important. If the order in which the 12 novels appeared is known, the data may show that the sales steadily increased, or that they increased up to a certain point and then leveled off, and so on, thus providing an indication of the author's changing reputation and success.

Example 1. Corrosion Data. The exploratory use to which even very simple data can be put is illustrated by 40 corrosion readings taken at random over a metal plate (Campbell [6], Wolfowitz [36]). No particular pattern emerged when the observations were plotted according to their position on the plate. However, when they were plotted in the order in which they were taken, a bunching together in runs* of high and low observations suggested (as it turned out, correctly) a malfunctioning of the delicate measuring device.

Much of data analysis deals with batches of more complex units such as pairs of numbers, more general vectors, matrices, or curves, and it need not be restricted to a single batch. With multivariate data, one may be interested in the separate features of the different variables, in relationships among the variables or sets of variables, or in aspects of the overall pattern of the multidimensional data set. The development of graphical (including, in particular, computer-displayed) and numerical methods for these purposes is a very active area of statistical research (*see* MULTIVARIATE GRAPHICS*). It includes, among others, such approaches as cluster* analysis, and multivariate classification†, multidimensional scaling*, pattern recognition*, and factor analysis*†. When several batches are being considered simultaneously, comparisons of the batches will tend to be of primary interest.

Statistical Analysis Based on Probability Models

The data-analytic approach indicated in the preceding section can provide clarification of the phenomena represented by both simple and very complex data sets, and can lead to important new insights and hypotheses. In its simplest forms such as numerical summaries and histograms, it is the statistical presentation most frequently encountered by the general public in newspaper articles and magazine reports. (Through misleading use, it also lends itself to much mischief. See for example, Huff [23].) However, this approach lacks what is often an essential requirement of the resulting inferences: Because of the fact, mentioned earlier, that the same phenomena might have led to different observations, it is impossible to assess the reliability of the conclusions.

Such an assessment requires knowledge concerning the alternative data sets that might have been observed in the given situation instead of the set that was observed. The crucial step underlying modern theories of statistical inference and decision making is to consider the observed data as realizations of random variables. The possible values of these variables are governed by probability distributions specifying the probabilities of observing the various possible data sets. These distributions are not assumed to be known, but only to belong to a postulated family of possible distributions.

The lack of complete specification represents the unknown aspect of the situation for the clarification of which the data were collected.

For a single batch $X_1, \ldots, X_n$, for example a set of n measurements of some quantity, investigators often assume that the n observations are independent, and that each has the same probability distribution. If this common distribution is denoted by F, the model is completed by specifying a family $\mathbb{F}$ to which F is assumed to belong. This may for example be the family of all possible distributions F (i.e., no further assumptions are made) or the family of all distributions that are symmetric with respect to a specified point of symmetry. Such broad families are called *nonparametric*† in distinction to parametric families where F is known except for the values of some parameters, for example, the family of all normal, Poisson, or Weibull distributions*. Once the model is specified, one can now ask, and answer, the type of question concerning a single batch considered in the preceding section, with more precision. In particular, the conclusions no longer refer to this particular data set, but to the underlying process that produced it.

Suppose for example that the aspect of interest is the overall level previously represented by the average of the observations. Suppose that $\mathbb{F}$ is the family of normal distributions with mean ξ and unit variance. Then ξ is the average value, not of these particular n measurements but rather of all potential measurements, each weighted according to its probability. The average $\bar{X} = (X_1 + \cdots + X_n)/n$, which before was a descriptive measure of the general level of the batch, now becomes an estimator of the unknown ξ: for example, the true value of the quantity being measured or the average value of the characteristic (e.g., height, age, or income) in the population from which the X's were obtained as a sample.

The model assumptions make it possible to get an idea of the accuracy of an estimator such as $\bar{X}$, for example in terms of the expected closeness to the true value ξ. One commonly used measure of this closeness is the expected squared error, which for the case of n independent measurements with variance 1 equals

$$E(\bar{X} - \xi)^2 = 1/n.$$

This formula shows for instance how the accuracy improves with n, and enables one to determine the sample size n required to achieve a given accuracy.

This approach also provides a basis for comparing the accuracy of competing estimators. For the median $\tilde{X}$ of the X's for example, one finds that approximately (if n is not too small) $E(\tilde{X} - \xi)^2 = 1.57/n$, more than 50% larger than the corresponding value $1/n$ for $\bar{X}$. The median is thus considerably less accurate than the mean. This conclusion depends strongly on the assumption that the X's are normally distributed. For other distributions the result may be just the reverse. (See for example the section on robustness in ESTIMATION*.)

Another way of describing the accuracy of $\bar{X}$ is obtained by noting that

$$P\left[|\bar{X} - \xi| \leqslant 1.96/\sqrt{n}\right] = 0.95, \qquad (1)$$

so that with probability 0.95 the estimator $\bar{X}$ will differ from the true value ξ by less than $1.96/\sqrt{n}$. The statement (1) can be paraphrased by saying that the random interval $(\bar{X} - 1.96/\sqrt{n}, \bar{X} + 1.96/\sqrt{n})$ will contain the unknown ξ with probability 0.95. Random intervals that cover an unknown parameter ξ with probability greater than or equal to some prescribed value γ are called *confidence intervals**† at confidence level γ.

Point estimation and estimation by confidence intervals provide two of the classical approaches to statistical inference. The third is *hypothesis testing**†.

Example 2. Extrasensory Perception. Suppose the claim of a subject A to have extrasensory perception (ESP) is to be tested by tossing a coin 100 times at a location invisible to A, and recording A's perception (heads or tails) for each toss. Suppose A obtains the correct result on 54 of the tosses.

This clearly is not an indication of a strong ability at ESP. However, even a very slight ability would be of extraordinary interest. Is there support for such a finding, or is the result compatible with purely random guessing? Under the null hypothesis* of pure guessing, the probability of calling a toss correctly is $\frac{1}{2}$ for each toss, and the probability of getting 54 or more right is then $\frac{1}{4}$. The result is therefore not particularly surprising under H, and a case for A's ability to do better than chance has not been made. Had A correctly identified 65 tosses rather than 54, the conclusion would be quite different. The probability of 65 or more correct calls is only 0.002 when H is true. In the light of so extreme a result, one would have to give serious attention to A's claim.

Hypothesis testing and point and interval estimation* are all used extensively in applications, with each being more useful and popular in some areas than in others. The theory of these three methodologies is concerned with the performance of proposed procedures (including the determination of sample size to achieve a desired performance), the comparison of different procedures, and the determination of optimal ones. An important consideration is the robustness* of a given procedure under violation of the model assumptions. If the procedures are very sensitive to these assumptions, one may want to study the problem in a nonparametric setting of the kind described for a single batch at the beginning of this subsection. Nonparametric (and semiparametric*) models are particularly important for the analysis of large data sets, which has become more practicable as a result of increased computer capabilities and availability.

A unified framework for the three areas is provided by Wald's decision theory*. This very general theoretical approach deals with the choice of one of a set of possible decisions d on the basis of observations x_i. Let the chosen d be denoted by $\delta(\mathbf{x})$. The observed value $\mathbf{x}$ is assumed to be the realization of a random quantity $\mathbf{X}$ with probability density $p_\theta(\mathbf{x})$, θ unknown. A loss function $L(\theta, d)$ measures the loss resulting from decision d when θ is the true parameter value, and the performance of a decision procedure δ is measured by its *risk function*

$$R(\theta, \delta) = E_\theta[L(\theta, \delta(\mathbf{X})],$$

the average loss incurred by δ when θ is true. A principal concern of the theory is the determination of a δ for which the risk function is as small as possible in some suitable sense. Another problem is the characterization of all admissible* procedures, i.e., all procedures whose risk cannot be uniformly improved.

Decision theory can also be made to encompass sequential analysis*, the design of experiments* (by letting the loss function take account of the cost of observations), and the choice of model (by imposing a penalty that increases with the complexity of the model). However, it is an abstract approach that has been useful primarily for exhibiting general relationships rather than for its impact on specific methods. An important consequence of Wald's theory has been its liberating effect on the formal consideration of new types of situations, among them multiple comparisons* and other simultaneous inference* procedures.

Exploration vs. Verification

To illustrate both the relation and the difference between the approaches described in the preceding two sections, consider once more the ESP example.

Example 2. Extrasensory Perception (Continued). When looking over the results of the 100 tosses, suppose the experimenter notices that of the 54 successes, 33 occurred during the last and only 21 during the first 50 tosses. The probability of 33 or more successes in 50 tosses with success probability $p = \frac{1}{2}$ is only about 0.01. One might be tempted to explain away the poor performance in the first half of the experiment by the theory that it requires some warming up before the ability hits its stride, and to declare the success of the second half significant. However, such a conclusion would not be justified on the basis of this analysis since the calculation does not take account of the fact that the particular test (restricting oneself to the second half) was not originally planned but was suggested by the data.

Suppose the situation had been reversed, that there had been 33 successes in the first and 21 in the second half. A possible explanation: the exercise of ESP requires great concentration, and after 50 attempts the subject is likely to get tired. Similar explanations could have been found if success had been unusually high in the middle 50, or on the last 25, or the first 25, and so on. Instead of high concentration of successes in a particular segment of the sequence, other patterns might have struck an observer: for example, a gradual rise in the frequency of success, or a cyclic pattern of successes and failures, and so on. For each, an explanation could have been found.

This is the problem of *multiplicity*. Every set of observations—even a completely random one— will show some special features, and explanations can usually be found post facto to account for them. Unfettered examination of many different aspects of a data set is legitimate, and in fact a primary purpose, of exploratory data analysis. However, the results will then tend to look more impressive than they really are since attention is likely to focus on the extremes of a large number of possibilities. To legitimize a theory suggested by the data one must test it, for example, on a separate part of the data not used at the exploratory stage or from new data specially obtained for this purpose. (A somewhat more limiting alternative to such a two-stage procedure is to formulate a number of possible theories to be considered before any observations are taken. The simultaneous testing of such a number of possibilities provides a legitimate calculation for the probability of the most striking of the associated results.)

The two stages, exploration of the data leading to the formulation of a hypothesis, followed by an independent test of this hypothesis, constitute the basic pattern of scientific progress as described by scientists (see for example, Feynman [13, Chap. 7]) and discussed by philosophers of science. Before considering a third aspect in the next section, let us briefly mention another distinction, which relates to the purpose of a statistical investigation. This is the difference between inference* and decision making. It may be illustrated by the problem faced by a doctor who wants to arrive at a diagnosis (inference) but must also select a treatment (decision).

Example 3. Medical Diagnosis*. Suppose there are k possible conditions (diagnoses) $\theta_1, \ldots, \theta_k$ that might have led to the observed symptoms and test results $\mathbf{X}$ (the data), including for example measurements of temperature, blood pressure, and so on. Under condition θ, the observations $\mathbf{X}$ have a distribution $p_\theta(\mathbf{x})$. The problem of diagnosis is thus the statistical problem of using the observed value of x to determine the correct value of θ. (A standard procedure is to select the value $\hat{\theta}$ of θ that maximizes $p_\theta(\mathbf{x})$ for the given observation $\mathbf{x}$, the so-called *maximum likelihood estimate**† of θ.) The associated decision procedure might be to select the treatment that would be most appropriate for the chosen θ. The situation is however more complicated since the choice of treatment must also take account of the severity of the consequences in case of an incorrect diagnosis resulting in a nonoptimal treatment, the so-called loss function. The distinction between inference and decision making is reviewed in Barnett [2]. For a discussion of computer-based medical diagnosis and treatment choice, see Shortliffe [30] and Spiegelhalter and Knill-Jones [33].

Bayesian Inference* and Decision Making

Example 3. Medical Diagnosis (Continued). Suppose a patient P being tested for the conditions $\theta_1, \ldots, \theta_k$ of the last example is told that the tests point to $\hat{\theta}$ ($= \theta_1$, say) as the most likely cause of the symptoms. Not unnaturally, P wants to know just how likely it is that θ_1 is in fact the true cause. The doctor has to admit that the term "most likely" was used imprecisely; that $\hat{\theta} = \theta_1$ is the condition that assigns the highest probability to the observed test results, not necessarily the most likely of the conditions $\theta_1, \ldots, \theta_k$, and that in fact no probability can be assigned to these conditions.

Actually, in this example it may be possible to make such an assignment. Suppose that π_i is the incidence of condition θ_i in the population of sufferers from the given symptoms, and hence is the probability that a patient drawn at random from the population of such sufferers has condition θ_i. The condition of such a patient is then a random variable Θ, with prior probability $P(\Theta = \theta_i) = \pi_i$ before the tests are taken, and posterior probability $P(\Theta = \theta_i | \mathbf{x})$ in the light of the test results $\mathbf{x}$. The latter probability can be calculated from the π_i and the $p_{\theta_i}(\mathbf{x})$ by Bayes' theorem*.

On being presented with this probability, P may however still not be satisfied but complain to the doctor: "You have treated me for 20 years, you know my complete medical history, life style, and habits. In the light of all this additional information, what is the probability of suffering from θ_i not for a random patient but for me personally?" Unfortunately, the interpretation of probability as frequency, which was tacitly assumed up to now, and which in particular applied to the prior probabilities π_i of the preceding paragraph, precludes assigning probabilities (other than 0 or 1) to unique events such as this particular patient's suffering from condition θ_i.

This difficulty is met head-on by the Bayesian approach according to which π_i can be chosen to represent the physician's probability that condition θ_i obtains for this particular patient. The meaning of probability is, however, different from the earlier one. Probability is no longer a frequency but the degree of belief* attached to the event in question. (If the event is repetitive, this probability typically draws close to the observed frequency as the number of cases gets large.)

In general, the Bayesian approach assigns a *prior probability distribution* to a parameter θ before the observations $\mathbf{X}$ are taken. Once the values $\mathbf{x}$ of $\mathbf{X}$ are available, the prior distribution of θ is updated to the posterior (i.e., conditional) distribution of θ given $\mathbf{X} = \mathbf{x}$, which shows how the prior beliefs regarding the chances of different θ values have changed in the light of the data.

In the decision theoretic terminology introduced earlier, the relevant assessment of the performance of a procedure δ from a Bayesian point of view is not the risk function $R(\Theta, \delta)$ but rather the posterior expected loss

$$r(\mathbf{x}, \delta) = E[L(\theta, \delta(\mathbf{x}) | \mathbf{x}],$$

calculated according to the conditional distribution of θ given $\mathbf{x}$.

Ideally a Bayesian has a comprehensive, consistent view of the world with a probability attached to every unknown fact. These probabilities must satisfy an appealing set of axioms (the axioms of *coherence**), and must be updated as new information becomes available. For an individual's response to the world (or even a specific problem), a chief difficulty in implementing this program is the determination of the prior distribution*. A considerable literature deals with methods for eliciting a person's degrees of belief* with respect to a given situation. [There is also an objective Bayesian school in which the prior distribution represents "total lack of information." This will not be discussed here, but *see* STATISTICAL INFERENCE—I (subsection on Logical Bayesianism) and INVARIANT PRIOR DISTRIBUTIONS.]

Ideally, each person has a single correct personal probability regarding any unknown event. However, in practice, these probabilities "can never be quantified or elicited exactly (i.e., without error) especially in a finite amount of time" (Berger [3, p. 64]). This has led to the suggestion of a robust Bayesian viewpoint according to which one "should strive for Bayesian behavior which is satisfactory for all prior distributions which remain plausible after the prior elicitation process has been terminated" (Berger [3]).

A second difficulty arises in situations in which the decision or opinion concerns not a single person, but represents a joint problem for a group or occurs in the public domain, as in the case for example in the publication of the analysis of a scientific investigation. Some aspects of this problem will be considered in the next section.

The Bayesian / Frequentist Controversy

The mutual criticism of Bayesians and Frequentists has given rise to a lively (and sometimes acrimonious) debate, which has helped to clarify a number of basic statistical issues. (There are many different variants of Bayesians and Frequentists, not all of which will agree with the positions ascribed to these approaches here.) One of the central concerns is that of subjective versus objective data evaluation in scientific inference and reporting. Fisher*, Neyman*, and Pearson* developed their frequentist theories in a deliberate effort to free statistics from the Bayesian dependence on a prior distribution, and this aspect has continued as the central frequentist objection. The Bayesian response to this criticism is twofold.

On the one hand, it is pointed out that frequentist analysis involves similar types of specification. There is the choice of model and loss function, both of which must be chosen in the light of previous experience and involve judgments that are likely to vary from one person to another. In addition, there is the problem of selecting a frame of reference that forms the basis of the frequency calculations. In assessing the incidence of the conditions $\theta_1, \dots, \theta_k$ of the preceding section, for example, for what population should this be calculated: the population of the world, or the country, or the city in which the person lives; should the comparison be restricted to patients of the same age, sex, etc.? This is the problem of conditional inference*, which so far has found no satisfactory frequentist solution. (For the Bayesian, the problem does not arise in this form since the probabilities will always refer to this particular patient, but the same issue arises when one must decide how to weigh the experience with other patients in forming an opinion about this one.)

As a more positive response, there have been recent Bayesian suggestions (for example, Dickey [12], Smith [32]), that in scientific inference the analysis should be reported under a variety of different priors that—it is hoped—will include the opinions of the readers. The view that "any approach to scientific inference which seeks to legitimize *an* answer to complex uncertainty is a totalitarian parody of a would-be rational human learning process" (Smith [33]) considerably narrows the gulf between the two approaches.

The reason for this narrowing can be found in the combination of two facts. The first is a basic result of Wald's (frequentist) decision theory to the effect that every admissible procedure is a Bayes solution or a limit of Bayes solutions. Secondly, frequentists tend not to believe in a unique correct approach, and may therefore try a number of different solutions corresponding to different optimality principles, robustness properties, and so on. If these lead to similar conclusions, any one of them can be adopted. Otherwise, a careful examination of the differences may clarify the reason for the discrepancies and point to one as the most appropriate. Lacking such a resolution, one may instead prefer to report a number of different procedures.

The Bayesian and frequency approaches lead to different ways of assessing the performance of a decision procedure. From a strict Bayesian point of view, only the posterior distribution of θ given $\mathbf{x}$, and the posterior expected loss $r(\mathbf{x}, \delta)$, are relevant, while frequentists measure the performance of a procedure by its risk function. However, on this issue also, an accommodation to statistical practice and the need for communication has narrowed the gap by generating a Bayesian interest in risk functions.

Thus Rubin [29, p. 116] writes: "Frequency calculations that investigate the operating characteristic* of Bayesian procedures are relevant and justifiable for a Bayesian when investigating or recommending procedures for general consumption." Similar considerations can be found in Berger [4].

In the other direction, the frequentist approach has been strongly influenced by Bayesian ideas, in particular, by recognizing that it is natural and useful to consider the prior distribution leading to a proposed admissible procedure. An example in which such a Bayesian interpretation has made an important contribution to a theory developed in a decision theoretic framework is that of Stein estimation.

DATA ACQUISITION

Measuring Single Units

The first part of this entry dealt with the interpretation of data once they have been collected. In this and the following sections, we consider some of the processes that produce data, and the statistical problems arising at this earlier stage. (An extensive discussion of various types of data is provided in Hoaglin et al. [20].)

The basic data units are the numbers, symbols, words, or other entries making up a data set; for example, measurements of the height of a person or plant, or of the weight of a wagonload of fruit or a minute amount of some chemical; barometer or temperature readings; or the scores of a student on an intelligence or aptitude test. Alternatively, the observations could be the information provided by a person answering a questionnaire or interviewer such as family size, last year's income, or religious and political affiliation.

A concern for data quality, for their reliability and validity, is an important task preceding the collection of data. Efforts must be made to eliminate bias*, reduce variability, and eliminate sources of error. In constructing a questionnaire, great care is required to avoid ambiguities. Checks on the reliability of responses can often be built into the data set, for example by asking for the same information in a number of different ways or in different contexts.

Another aspect of data improvement arises after the data have been obtained. Data editing* (or "cleaning") involves the deletion or modification of entries that do not appear to be in consonance with the rest of the data and that sometimes represent obvious errors (for example, in a series of monthly measurements of a baby's head circumference when one month's measurement is smaller than the preceding ones). A variety of statistical methods have been developed for this purpose. In addition, robust statistical procedures are available that satisfactorily control the effect of outlying observations (see for example Hoaglin et al. [21] and Hampel [17]). On the other hand, data cleaning by inspection, without clearly stated rules, runs the risk of introducing a subjective element. (For example, just which observations are to be singled out for such treatment?) Even with good rules, it may remove valuable evidence and destroy the basis for probability calculations. For this reason, it is typically better to consider the editing of data as part of the statistical analysis and, while perhaps indicating definite or suspected errors, not to change the original data.

The improvement of data quality mentioned above is not the only aspect of data collection to be considered before obtaining the observations for an investigation. Two questions, in particular, that are of crucial importance and that will be discussed in the next sections are how many and what kind of observations are required.

Assessing Population Characteristics

The target of most investigations is not a single unit but a population of such units. A set of professors, apples, days, mice, or light bulbs is typically examined not because of interest in these particular specimens but in order to reach conclusions about the populations they represent. (It is this aspect and the associated quantification that has earned statistics its reputation of being antihumanistic.)

Efforts to obtain information about every member of the target population through a census*† go back to at least the third millenium B.C. in Babylonia, China, and Egypt. (For a nontechnical history of the census, see Alterman [1].) The purpose was usually to provide a basis for taxation or proscription. Today, population censuses seeking a great variety of information are carried out, many of them at regular intervals, in nearly all countries of the world.

However, taking a complete inventory is not the only way, and often not the best way, to obtain accurate statistical information about a population. The same end can usually be achieved more economically by collecting information only from the members of a sample, taken from the population by a suitable sampling* method. The much smaller size of the sampling operation tends to make this procedure not only cheaper but also more accurate since it permits better control of the whole process and hence the quality of the resulting data. On the other hand, a census has the advantage—not shared by any sample—of providing information even for very small subpopulations.

Suppose a population Π consists of N units, each of which has a value v of some characteristic of interest, such as the age or income in the case of a human population, the number or weight of the apples on each of the N trees in an orchard, or the length of life or brightness of the lightbulbs in a shipment received from the factory. To obtain an estimate of the average v value $\bar{v} = (v_1 + \cdots + v_N)/N$, a sample is taken from the population. In the early days of sampling, investigators usually relied on judgment samples, in which judgment is used to obtain a sample "representative"* of the population. It is now realized that such sampling tends to lead to biases, and does not provide a basis for calculating accuracy or the sample size needed to achieve a desired accuracy. The methods used instead are probability sampling schemes* that select the units to be included in the sample according to stated probabilities. The simplest such scheme is *simple random sampling**, according to which n units are chosen in such a way that every possible sample of size n has the same probability of being drawn. The average of the sample v values is then the natural estimate of $\bar{v}$.

Better accuracy can often be attained, and the needed sample size and resulting cost therefore reduced, by dividing the population to be sampled into strata within which the v values are more homogeneous than in the population as a whole but which differ widely among each other. (The population of school children of a city might for example be stratified by school, grade, and gender.) A stratified* sample of size n is then obtained by drawing a simple random sample of size n_i from the ith stratum for each i, where $\Sigma n_i = n$.

In a different direction, the cost of sampling can often be reduced by combining units into clusters* (for example, all the apartments in an apartment house, all houses in a city block, or all patients in a hospital ward), and obtaining the required information for each member of a sampled cluster. The two approaches can be combined into stratified cluster sampling, and many other designs are possible. For references to the extensive and sophisticated methodology available, see the articles on sampling*† and sample surveys*†.

Sample surveys to obtain information about a population are in widespread use and have become familiar through election polls, market surveys, and surveys of television viewers to establish ratings. However, they are not very well understood. In particular, it appears puzzling how it is possible to obtain an accurate estimate of the opinions or intentions of many millions from information concerning just a few thousand.

Example 4. An Election Poll. To get some insight into this question, let us simplify the situation, and suppose that a population Π consists of a large number N of voters, each of whom supports either candidate A or B. A simple random sample of n voters is drawn from Π, and the preference of each member of the sample is ascertained. If X is the number of voters in the sample favoring A, then X/n, the proportion of A supporters in the sample, is the natural estimate of the proportion p of A supporters in the population.

The question at issue is how large a sample is required for X/n to achieve a prescribed accuracy as an estimate of p, for example, for the standard deviation (SD) of X/n to satisfy

$$\mathrm{SD}(X/n) \leqslant 0.01. \qquad (2)$$

It is often felt intuitively that the required sample size n should be roughly proportional to the population size N. However, it turns out that this intuition is misleading and that in fact, for large populations, the required n is essentially independent of the value of N.

To see this, suppose for a moment that the sampling is done "with replacement," i.e., that the members are drawn successively at random, with each—after giving the required information—being put back into the population before the next member is drawn, again at random. This method is slightly less efficient than the original simple random sampling (and therefore requires a larger sample size), because it allows the same unit to be drawn more than once. It is introduced here because it is particularly simple to analyze. Sampling with replacement is characterized by two properties: (i) on each draw, the probability of obtaining an A supporter is p; (ii) the results of the n draws are independent in the sense that the probability of getting an A supporter on any given draw does not depend on the results of the earlier draws.

Let us now compare this situation with a quite different one. Suppose a coin with probability p of falling heads when spun on its edge (this probability may be far from $\frac{1}{2}$), is spun n times and that X is the number of times it falls heads. Then (i) the probability of heads is p on each spin, and (ii) the results of the n spins are independent. The standard deviation of X/n is therefore the same for this coin problem as in the election sampling with replacement. The number n required to reduce this standard deviation to 0.01 is therefore also the same in both cases. Note however that the coin problem involves only n and p; no population is involved. Therefore the required n cannot depend on N.

This argument depends of course crucially on the assumption that the sampling was random. It is the randomness that insures that with high probability the sample contains approximately the same proportion of A supporters as the population. In addition, it was tacitly assumed that the preference of each voter can be ascertained without error. In practice, the possibility of "response error" can rarely be ruled out.

Data From Experiments

A study is called an *experiment* if its data are produced by an intervention (i.e., do not occur naturally) for the purpose of gathering information. It is a *comparative experiment* if its purpose is to compare several ways of doing something (e.g., different teaching methods, medical treatments, fertilizers, and so on) rather than to determine some absolute value.

Example 5. Weather Modification*. Consider a company's claim to be able to increase precipitation by seeding the clouds of suitable storms. Here the comparison is between seeding and not seeding. How can one obtain data to test the claim and to provide an estimate of the amount of increase?

As one possibility, the company might seed all suitable storms in the given location, and compare the resulting rainfall with that during the corresponding period in the preceding years. Of course, if the results are very striking (for example, if an enormous downpour occurs immediately following each seeding) this may settle the issue. However, typically the results are less clear. Suppose, for example, that the total rainfall matches, or even slightly exceeds, that of the wettest of the last five years. This may be the result of the seeding; or it may just be the consequence of an exceptionally wet year.

This difficulty is inherent in studies in which there is no randomness in the assignment to the experimental units of the conditions or treatments being compared. A better basis for the establishment of a causal relationship—in this case that the increased rainfall is due to the seeding—is obtained if storms are compared within the same season, and if they are assigned to the two treatments (seeding and not seeding) according to a random mechanism. This can be done in a variety of ways, corresponding to different *experimental designs**.

As a simple possibility, suppose that the experiment is to extend to the first 20 storms that are suitable for seeding. Of these, 10 will be seeded and 10 not, the latter providing the *controls*. According to one design (*complete* randomization), 10 of the numbers $1, \ldots, 20$ are selected at random; the storms bearing these numbers will be seeded, the remaining 10 will not. An alternative design (*paired comparisons**) pairs the storms $(1, 2)$, $(3, 4), \ldots, (19, 20)$, and within each pair assigns at random (e.g., by tossing a coin) one storm to seeding, the other to control. This second design will be particularly effective if storms occurring close together in time are more likely to be similar in strength than storms separated by longer time intervals.

It is interesting to note the close relationship of these designs to the sampling schemes of the preceding section. In the first design, the storms assigned to seeding are a simple random sample of the 20 available storms; in the second case, they constitute a stratified sample, with samples of size 1 from each of 10 strata of size 2.

Unfortunately, random assignment of the conditions being compared is not always possible. Consider for example a study that reports that married men tend to live longer than unmarried ones. It is tempting to draw the conclusion that marriage prolongs life, perhaps by providing a more regulated life style. However, such a conclusion is not justified. Married and unmarried men constitute groups that differ in many ways. The latter for instance includes men with health problems that preclude marriage and that also tend to shorten life. It is thus not clear whether the observed effect is the result of the difference in marital status or of conditions leading to this difference. All that can be safely concluded is that marriage is associated with longer life expectancy. This may be enough for an insurance company but does not answer the sociological or public health question regarding the effect of marriage. Quite generally, observational studies* in which subjects have not been assigned to the conditions being compared (marital status, smoking habits, religious beliefs, etc.) can establish associations (such as that between marital status and longevity) but have great difficulty in validating causal relationships. To establish causation*† is a cherished goal of statistical methodology but tends to be rather elusive. (For some further discussion of this point, see for example Mosteller and Tukey [27, pp. 260–261].)

Example 6. A Headache Remedy. Suppose that you wake up with a headache, take a headache remedy, and an hour later find that the pain is gone. Based on this isolated instance (which, in words attributed to R. A. Fisher, is an experience rather than an experiment), it is clearly not possible to conclude that the result is due to the remedy. The cause might instead have been another intervening event such as breakfast, or possibly the headache had run its course and would have dissipated in any case.

The attribution of the cure to the medication becomes much stronger if the incident is not isolated. If you have suffered from headaches before, took the medicine sometimes soon after onset, at other times only after having waited in vain for the pain to subside on its own; if on different occasions you took the tablets at different times of day, and if under all these different circumstances the headache disappeared shortly after treatment and never (or only very rarely) before, the robustness of the effect over many different conditions would tend to carry conviction where a single instance would not. The finding could be strengthened further if your own experience could be merged with that of others. (However, even if the evidence for a treatment effect is convincing, observations of this kind cannot determine whether the ingredients of the medication are responsible for the improvement, or whether it is a *placebo effect*, i.e., the patient's belief in the effectiveness of the treatment, which relieves the pain.)

To see how to systematize the informal reasoning of the preceding paragraph, consider another example.

Example 7. New Traffic Signs. Suppose that four-way stop signs have been installed at a dangerous intersection, and that four accidents had occurred in the month preceding the installment of the signs but only one in the month following it. These facts by themselves provide little basis for an inference. Suppose for example that the monthly accidental frequencies constitute a purely random sequence of which the value 4 was a chance high, which however caused the installment of the signs. Then even without this intervention, a decrease could have been expected in the succeeding month.

A more meaningful comparison can be obtained if the accident statistics are available not only for one month before and after, but for several months in both directions. One is then dealing with an *interrupted time series**, for which it is possible to compare the observations before with those after the "interruption" (the installment of the signs). Even the nonrandom choice of the time of intervention would then have only a relatively minor effect.

While the interrupted time series can establish that the accident rate is lower after installment than it was before, it cannot establish the new signs as the cause of the decrease since other changes might have occurred at the same time. To mention only one possibility, the community may have been affected by editorials published at the times of the third and fourth accidents. Some control—although not as firm as that resulting from randomization—can be obtained by studying the accident rates during the same months also at some other intersections, a *multiple time-series* design*. If the other series do not show a corresponding decrease, this will clearly strengthen the causal argument.

Quasi-experiments† (a better term might be "quasi-controlled experiments") such as the interrupted time series and multiple time series described above, which try to identify and control the most plausible alternatives to the treatment being the cause of the change, are treated by Cook and Campbell [9] and Cochran [8]; for an elementary discussion see Campbell [7]. They can greatly strengthen the causal attribution although they cannot be expected to be as convincing as a controlled randomized experiment.

Serial Data

Most of the data considered so far were assumed to be collected on a one-shot basis. Often, they are obtained instead consecutively over a period of time. Such data are particularly useful in assessing changes over time. (Are winters getting colder? Is the birthrate in the U.S. falling? Are an author's novels becoming more popular?)

An important application is provided by statistical *quality control**. An established production process is monitored by taking observations of the quality of the product at regular intervals. The process is said to be in control if the successive observations are independent and have the same distribution. A *control chart** on which the successive observations are plotted provides signals when the process appears to be going out of control. A similar approach is used in monitoring the cardiogram of a heart patient in intensive care, where the data consist of a continuous graph instead of a discrete sequence of points. Analogously, seismographs provide continuous data for the monitoring of seismic activity while persons watching their weight through regular weighings provide another illustration of the discrete case.

In these applications, a single process is followed to check that no changes are occurring and to alert us when they are. A different situation arises when one is interested in assessing the changes occurring in a population as part of a developing pattern or as the result of some intervention. For example, to study the pattern of growth (the *growth curve**) of children, plants, or institutions, or the effect of several different programs (say, of different rehabilitation programs on juvenile offenders), one observes each member of a sample from the population over a period of time. In such *longitudinal studies** the observations of the same units at different times often are dependent. Probability models for sequences of dependent observations are considerably more complicated than those for sequences of i.i.d. variables. Such models and their statistical analysis are treated in the theory of *time series** or more generally of *stochastic processes**.

Observations arising serially, for instance on patients coming to a clinic, successive books by an author, or stockmarket performance in successive time periods, provide an opportunity to economize by letting the size of the study be determined by the data (according to a clearly specified rule) instead of fixing it in advance. Suppose for example that a shipment of goods is being sampled to determine whether its overall quality meets certain specifications. Items are being drawn at random, and examined, one by one. It may then be possible to stop early when the quality of the initial observations is either very satisfactory or very unsatisfactory, while one may wish to take a larger sample when the initial observations are mixed. The working out of economical stopping rules providing the desired statistical information, and the analysis of the resulting data, is the problem treated in *sequential analysis**.

Designing Experiments

The first part of this article was concerned with the interpretation of data once they have been obtained, and the preceding sections with various processes used to acquire the data. However, preceding even this stage, it is necessary to plan how many and what kind of data will be needed to answer the questions under consideration.

As an illustration, consider once more the ESP study of Example 2, based on 100 tosses of a coin. Suppose the investigator had decided, before the experiment was carried out, to give serious attention to the possibility of ESP provided the number of subject A's correct calls would be at least 65. As was pointed out in Example 2, under the hypothesis of pure guessing the probability of 65 or more answers is 0.002. There is thus little danger of paying attention to the claim if it has no validity.

However, is this study giving subject A a fair shake? What is the probability of getting 65 or more of the tosses right if A really does possess the claimed ability? The answer depends of course on the extent of the ability, which can be measured by the probability of calling a toss correctly. Here are some values, computed under the simplifying assumptions that the 100 calls are independent, and that the probability p of a correct call is the same for each toss.

p	0.55	0.6	0.7	0.75
P(no. correct calls $\geqslant 65$) = power	0.027	0.179	0.884	0.990

The probability in this table measures the power* of the test of the hypothesis $H: p = \frac{1}{2}$, i.e., the probability of following up A's claim against various alternative values of p. It shows that the power is quite satisfactory when p is 0.7 or more, but not, for example, when $p = 0.6$. In this case, despite the large discrepancy between pure guessing and an ability corresponding to $p = 0.6$, the probability is less than 20% that serious attention would be given to the claim. To get higher power would require a larger number of tosses. For example, to increase the power against $p = 0.6$ to 0.9, while keeping the probability of following up the claim when $p = 0.5$ at 0.002, would require a sample of about 425 tosses instead of the previous 100. A statement of the goals to be achieved, e.g., the required power of a test or accuracy of an estimate, and determination of the sample size needed for this purpose, are crucial aspects of planning a study.

Taking a fixed number of observations (100 or 425, etc.) is not necessarily the best design. Suppose for example that A calls all of the first 25 tosses correctly. This may be enough to provide the desired evidence of A's ability (or suggest that something is wrong with the experiment). Thus, a sequential stopping rule of the type indicated at the end of the last section might be more efficient.

Consider next the problem of determining an optimal design when different types of observations are involved. Suppose we are concerned with the comparison of two treatments, and that the total number of observations is fixed at $N = 2k$. Then it will typically be best to assign k subjects to each treatment, since this will tend to maximize the power of the tests and minimize the variance of the estimate of the difference. This may of course not be the case if observations on one of the treatments have smaller variance than on the other.

Finding the optimum assignment becomes much more difficult when observations are taken sequentially, and a decision is required at each stage which treatment to apply next. What is the best (or even a good) rule depends strongly on the purpose of the procedure. If the only concern is to decide whether the treatments are equally effective and, if not, which is better and by how much, the treatments should be assigned in a balanced way, i.e., so that each is received by the same number of subjects. However, if one is comparing two treatments of a serious medical condition, an additional consideration arises: a desire to minimize the number of patients in the study who receive the inferior treatment. A sequential procedure that has been suggested for such a case is the "play the winner rule" in which the next patient receives the treatment that at that point looks better. (For details, see for example, Flehinger and Louis [14] and Siegmund [31].) It should be noted however, that such an unbalanced procedure will require more observations than the best balanced rule, and hence will take longer to lead to termination of the study and hence to a recommendation. Thus, the inferior treatment will be assigned to fewer patients within the experimental group but to a larger number outside of it.

Comparative experiments typically involve more than the study of just one difference. Consider an experiment to investigate the effect of a number of factors (to which experimental units can be assigned at random) on some observable response such as rainfall, crop yield, or length of life. Two possible designs are one-at-a-time experiments, in which each factor is studied in a separate experiment with all other factors held constant, and *factorial experiments**, which provide a joint study of the effects of all factors simultaneously. The latter type of design has two important advantages.

Two or more effects may interact in the sense that the effect of one factor depends on the level of the other. In a study of the effect of textbooks and instructors on the performance of students, for example, it may turn out that a textbook that works very well for one instructor is quite uncongenial to another. The existence and size of such interactions* are most easily investigated by studying the various factors simultaneously. When no interactions are present, joint experimentation has the advantage of requiring far fewer observations to estimate or test the various effects than would be needed if each factor were studied separately. (For further discussion of these and related issues, see for example, Cox [10] and Box et al. [5].)

A factorial experiment is concerned with the study of a number of factors, each of which can occur at several levels. A study of cancer treatments, for example, might involve different types of surgery, a number of surgeons, and several kinds of postoperative therapy (radiation, chemotherapy, etc.). The factorial experiment is said to be *complete* if one or more observations are obtained at all possible combinations of levels. If the number of such combinations is too large, the complete design may be replaced by a *fractional factorial design**, in which only certain combinations of levels are observed. The theory of experimental design* is concerned with the search for good (or possibly optimal) designs able to provide the basis for a suitable analysis of the resulting data.

Unless the experimental units (the patients, students, agricultural plot, etc.) are fairly homogeneous, it may be advisable to divide them into more homogeneous strata, as was discussed earlier for the sampling of populations. In the present context, such stratification is called *blocking**. The strata or blocks then play a role that in some respects is similar to that of an additional factor.

Much of the theory of factorial designs was originated by R. A. Fisher* in the 1920s, in the context of agricultural field trials, where it continues to play a central role; in addition, its uses have since expanded to many other areas of application.

CONCLUSION

Statistics has been discussed in this article as dealing with the collection and interpretation of *data* to obtain information. We have been particularly concerned with the uncertainty caused by the fact that the observations might have taken on other values, and with the resulting variability of the data. An alternative view of statistics is obtained by noting that uncertainty attaches not only to data, but also to many other other "chancy" events such as length of life, the occurrence of accidents, the quality of a manufactured article, or the fall of a die. Correspondingly, statistics has also been defined as the subject concerned with understanding, controlling, and reducing *uncertainty*. Actually, there is little difference between these two descriptions: Data involve uncertainty, and the study of any particular uncertainty requires the collection of appropriate data. It is a matter of emphasis.

Table 1 Adapted from Meier [24]

	Total Number	Number Confirmed Polio
Vaccinated	200,745	57
Placebo	201,229	142

Earlier sections have given a general indication of the pervasiveness and impact of statistical considerations relating to both data and uncertainty. To illustrate the power and wide range of the statistical approach we shall in this concluding section take a brief look at three specific studies.

Example 8. Polio Vaccine Trial. To test whether a proposed polio vaccine (the Salk vaccine) was effective, a large scale study was carried out in 1954. The most useful part of the study was a completely randomized trial in which about half of approximately 400,000 school children were assigned at random to receive the vaccine, with the other half receiving a placebo (injection of an ineffective salt solution). The study was double blind, i.e., neither the children nor the physicians making the diagnosis knew to which of the two groups any given child belonged. The results in Table 1 show that vaccination has cut the rate of polio by about 60%. The probability of that marked a decrease under the hypothesis that the vaccine has no effect is about $1/10^7$, much too small to reasonably attribute the effect to pure chance. [Note that this example is of the same type as the hypothetical Example 2 (ESP).]

Example 9. Medical Progress. In the preceding example we were concerned with the effectiveness of a single medical innovation —the Salk vaccine. The study reported in the present example (Gilbert et al. [16]) addresses a much broader question: What can be said about the effectiveness of present-day medical (or rather more specifically, surgical and anesthetic) innovations as a whole? Such an investigation of the combined implications of a whole area of studies is called a *meta-analysis*. (For a recent account of meta-analysis with many references to the literature, see Hedges and Olkin [19].)

A first step toward such an analysis is to decide what data to collect, in particular, which studies to include in the investigation. Ideally, one would like to have a list of all relevant studies for the period in question. For medical research, an approximation to this ideal is available in MEDLARS (National Library of Medicine's MEDical Literature And Retrieval System), which provides exhaustive coverage of the world's medical literature since 1964. From MEDLARS, the authors obtained a set of 107 papers ("the sample") evaluating the success of specific innovative surgical and anesthetic treatments, and used these to assess the effectiveness of such treatments as a whole. The authors view this set of papers as a sample from the flow of such research studies, and therefore believe that the results should give a realistic idea of what to expect from future innovations, at least for the short term.

The sample contained a total of 48 comparisons of a new treatment with a standard (control). In 36 of the cases, the assignment to treatment and control was randomized, but not in the remaining 12. The results are summarized in Table 2.

A striking feature of the results for the randomized trials is that only 5 out of 36 or about 14% of the innovations were highly preferred. This suggests, the authors point out, that medical science is sufficiently well established so that substantial improvements over the standard treatments are difficult to achieve yet that it is not so settled that only a major theoretical advance will lead to any substantial further improvements.

A more sanguine assessment is obtained from the nonrandomized studies, where 5 out of 12 or about 42% of the innovations were highly preferred. Since randomization provides a surer foundation, the results in the first row may be deemed to be more reliable than those in the second. However, such a judgment overlooks the fact that the comparison of randomized with nonrandomized studies is itself not randomized, i.e., the assignment of studies to these two types was not, and was in fact far from, random. As a result, the two types of studies may not be directly comparable. For example, a surgeon who is strongly convinced of the great superiority of the new treatment, might for ethical or other reasons not be willing to apply the standard treatment, and would therefore have to look for controls from an earlier period or from other surgeons, while no such qualms might arise for a less dramatic innovation where a randomized study would then be acceptable. Many other explanations could be imagined, and much more information would be needed before one could attempt to assign a cause.

The authors go beyond the frequencies displayed in Table 2 to estimate the size of the treatment effects. For this purpose, they utilize a sophisticated technique, empirical Bayes*, which is particularly suited for meta-analysis.

Example 10. Literary Detection. The Federalist papers are an historically important set of 85 short political essays published mostly anonymously in 1787–1788 by Alexander Hamilton, James Madison, and

Table 2 Adapted from Gilbert et al. [16]

	Innovation Highly Preferred	Innovation Preferred	About Equal	Standard Preferred	Standard Highly Preferred	Total
Randomized	5	7	14	6	4	36
Nonrandomized	5	2	3	1	1	12

Table 3 Adapted from Mosteller and Wallace [28]

	Word Frequency of "on" in 200-Word Blocks							Total Number of Blocks
	0	1	2	3	4	5	6	
H	145	67	27	7	1	—	—	247
M	63	80	55	32	20	8	4	262

John Jay. The authorship of most of the papers was eventually established but that of 12 of them has remained in dispute between Madison (M) and Hamilton (H), with recent historical research leaning towards, but not clearly deciding in favor of, Madison.

An effort to resolve the doubt by statistical methods was undertaken by Mosteller and Wallace [28]. The basic idea of such literary detection is to find aspects of the writing styles of the two (or more) authors in question that have good ability to discriminate between the various possibilities. This was particularly difficult in the present case because the two authors have very similar styles. However, by comparing known texts of M and H, Mosteller and Wallace were able to identify a number of words that were used with much higher frequency by one of the authors than the other. As an illustration Table 3 shows the frequency of occurrence of the word "on" in blocks of about 200 words from texts by the two authors. (For example, the number of blocks in which the word "on" occurred exactly twice was 27 for the 247 H blocks, but 55 for the 262 M blocks.) The rate of occurrence of the word "on" per thousand words of text was 3.38 for H and 7.57 for M.

The present problem is similar to the diagnostic problem of Example 3, with θ taking on the two values H and M. (The statistical methodology dealing with the attribution of an item to one of two or more classes, a patient to a disease, or a piece of writing to an author, is called classification*† or discrimination.) Mosteller and Wallace's first task was to decide on observations X that might provide them with the evidence needed to settle the question. They selected a total of 30 words (including the word "on") with high discriminating ability, and used as their observations X the frequencies with which these words occurred in a given text of disputed authorship.

Next the distribution $p_\theta(x)$ for the frequency of a given word had to be specified for each author with the help of the known H and M texts. Bayesian analyses for each of the 30 words based on the chosen family of distributions were combined into overall odds for M and H. For all but two of the papers, these overwhelmingly (several hundred million to one) favored Madison, given any reasonable prior odds for M and H. For each of these papers, the analysis leaves little room for doubt. In the remaining two cases Madison is also favored, but the odds are more modest, and the statistical attribution therefore less certain.

Statistical considerations arise in nearly all fields of human endeavor. Many of the associated activities are carried out by large numbers of full or part time professional statisticians. Some of the demands and satisfactions of a statistical career are described by Healy [18] in the pamphlet *Careers in Statistics* (rev. 3rd ed., 1980) published by the American Statistical Association*, and in PRINCIPLES OF PROFESSIONAL STATISTICAL PRACTICE. More detailed descriptions of the work and requirements of statisticians at various levels in government and industry can be found in articles on "Preparing statisticians for careers in federal government and in industry" (*Amer. Statist.*, **36**, 69–89, (1982); **34**, 65–80, (1980) and in Moser [25]).

The basic training in statistics, as in most other fields, occurs at universities and colleges where Departments of Statistics offer both undergraduate and graduate degrees in statistics. (At some institutions, the principal statistics courses are instead provided by the mathematics department.) In addition, there may be degrees in biostatistics*. Statistics courses and quantitative programs with a strong statistical component may also be offered in operations research, business schools, demography, economics, education, psychology, and sociology. In addition to courses and programs preparing for a profession in statistics, Statistics Departments also provide "service courses" both at introductory and advanced levels to students in other fields who may need to use statistical methods in their work.

Another need of statistical education is filled by courses in statistical concepts as part of a general education. Acquiring the ability to think in statistical terms is of great importance, even for persons without a quantitative bent, in view of the pervasive occurrence of statistical ideas in newspapers and magazines, and in the terminology we use to describe and discuss the world around us. What do we mean by saying that women tend to live longer than men, or that cancer patients survive longer today than 10 years ago? Does it mean they live longer on the average, that the median length of their life is longer, or that they have a better chance to survive to any given age? And is the longer survival of persons diagnosed to have cancer primarily due to the availability of more effective treatments, including earlier diagnosis? In fact, is there even an increase in length of survival, or is the apparent increase just a statistical consequence of earlier diagnosis? If the disease is diagnosed a year earlier, the survival after diagnosis has increased by a year even if nothing else has changed.

To consider another example, what is meant by the "rate of unemployment"? Roughly speaking, it is the proportion among the people wishing to be employed who are not. But do the official rates include those past seekers for jobs who have given up in despair? A change in the figure for unemployment may be the result of a small change in the definition, or of some sociological change such as the increased entry of women into the work force.

Many of the considerations involved in such issues are statistical. Fortunately, the basic ideas needed for general discussion and comprehension can be communicated at a fairly nontechnical level, as is done, for example, in Tanur et al. [35], Mosteller et al. [26], and Freedman et al. [15]. Books such as these, and the increasing availability of first courses in probability and statistics in high schools, should have the effect of gradually raising the general level of statistical literacy.

References

[1] Alterman, H. (1969). *Counting People*. Harcourt, Brace and World, New York.

[2] Barnett, V. (1982). *Comparative Statistical Inference*, 2nd ed. Wiley, New York.

[3] Berger, J. (1984). In *Robustness of Bayesian Analysis*, J. Kadane, ed. North-Holland, Amsterdam, The Netherlands.

[4] Berger, J. (1985). *Statistical Decision Theory and Bayesian Analysis*, 2nd ed. Springer, New York.

[5] Box, G. E. P., Hunter, W. G., and Hunter, J. S. (1978). *Statistics for Experimenters*. Wiley, New York.

[6] Campbell, W. E. (1942). *Monograph No. B-1350*, Bell Telephone System Technical Publication.

[7] Campbell, D. T. (1978). In *Statistics: A Guide to the Unknown*, 2nd ed., J. Tanur et al., eds. Wadsworth, Belmont, CA.

[8] Cochran, W. G. (1983). *Planning and Analysis of Observational Studies*. Wiley, New York.

[9] Cook, T. D. and Campbell, D. T. (1979). *Quasi-Experimentation*. Rand McNally, Chicago, IL.

[10] Cox, D. R. (1958). *Planning of Experiments*. Wiley, New York.

[11] Diaconis, P. (1985). In *Exploring Data Tables, Trends, and Shapes*, D. C. Hoaglin, F. Mosteller, and J. W. Tukey, eds. Wiley, New York.

[12] Dickey, J. M. (1973). *J. R. Statist. Soc. Ser. B*, **35**, 285–305.

[13] Feynman, R. (1965). *The Character of Physical Law*. The British Broadcasting Corp., London, England.

[14] Flehinger, B. J. and Louis, T. A. (1972). *Biometrika*, **58**, 419–426.

[15] Freedman, D., Pisani, R., and Purves, R. (1978). *Statistics*. Norton, New York.

[16] Gilbert, J. P., McPeek, B., and Mosteller, F. (1977). In *Costs, Risks and Benefits of Surgery*, Bunker et al., eds. Oxford University Press, New York.

[17] Hampel, F. R., Ronchetti, E. M., Rousseeuw, P. J., and Stahel, W. A. (1986). *Robust Statistics*. Wiley, New York.

[18] Healy, M. J. R. (1973). *J. R. Statist. Soc. Ser. A*, **136**, 71–74.

[19] Hedges, L. V. and Olkin, I. (1985). *Statistical Methods for Meta-Analysis*. Academic, Orlando, FL.

[20] Hoaglin, D. C., Light, R., McPeek, B., Mosteller, F., and Stoto, M. A. (1982). *Data for Decisions*. Abt Associates, Cambridge, MA.

[21] Hoaglin, D. C., Mosteller, F., and Tukey, J. W. (1983). *Understanding Robust and Exploratory Data Analysis*. Wiley, New York.

[22] Hoaglin, D. C., Mosteller, F., and Tukey, J. W. (1985). *Exploring Data Tables, Trends, and Shapes*, Wiley, New York.

[23] Huff, D. (1954). *How to Lie with Statistics*. Norton, New York.

[24] Meier, P. (1978). In *Statistics: A Guide to the Unknown*, 2nd ed., J. Tanur et al., eds. Wadsworth, Belmont, CA.

[25] Moser, C. A. (1973). *J. R. Statist. Soc. Ser. A*, **136**, 75–88.

[26] Mosteller, F., Kruskal, W. H., Link, R. F., Pieters, R. S., and Rising, G. R., eds. (1973). *Statistics by Example*, 4 vols. Addison-Wesley, Reading, MA.

[27] Mosteller, F. and Tukey, J. W. (1977). *Data Analysis and Regression*. Addison-Wesley, Reading, MA.

[28] Mosteller, F. and Wallace, D. L. (1984). *Applied Bayesian and Classical Inference*. Springer, New York.

[29] Rubin, D. B. (1984). *Ann. Statist.*, **12**, 1151–1172.

[30] Shortliffe, E. H. (1976). *Computer-based Medical Consultations, MYCIN*. Elsevier, New York.

[31] Siegmund, D. (1985). *Sequential Analysis*. Springer, New York.

[32] Smith, A. F. M. (1984). *J. R. Statist. Soc. Ser. A*, **147**, 245–259.

[33] Spiegelhalter, D. J. and Knill-Jones, R. P. (1984). *J. R. Statist. Soc. Ser. A*, **147**, 35–77.

[34] Stigler, S. M. (1986). *The History of Statistics*, Harvard University Press, Cambridge, MA.

[35] Tanur, J., Mosteller, F., Kruskal, W. H., Link, R. F., Pieters, R. S., Rising, G. R., and Lehmann, E. L., eds. (1978). *Statistics: A Guide to the Unknown*, 2nd ed. Wadsworth, Belmont, CA.

[36] Wolfowitz, J. (1943). *Ann. Math. Statist.*, **14**, 280–288.

Acknowledgments

I should like to thank David Cox, Persi Diaconis, David Freedman, William Kruskal, Frederick Mosteller, Juliet Shaffer, and the editors for reading a draft of this article, and providing me with comments and suggestions which resulted in many improvements.

E. L. LEHMANN

STATISTICS CANADA

Statistics Canada is the nation's central statistical agency, mandated by the Statistics Act of 1971 to "collect, compile, analyse, abstract and publish statistical information relating to the commercial, industrial, financial, social, economic and general activities and condition of the people." The Act also made Statistics Canada responsible for collaborating with departments of government in the collection, compilation, and publication of statistical information, and for promoting and developing integrated social and economic statistics for Canada and each of the provinces.

The Agency was first established in 1918 as the Dominion Bureau of Statistics under the original Statistics Act. The Act was revised in 1971, primarily to clarify the earlier legislation. Statistics Canada's mandate has thus remained basically unchanged for more than 65 years, a testimony to the visionary architecture of the designers of the original legislation. The agency has grown in size, complexity, and scope, reflecting comparable social and economic changes and concomitantly increasing requirements for statistical information as a basis for decision making. Statistics Canada is now the core of Canada's national statistical system. It is one of the largest centralized statistical organizations in the western world, with an operating budget of $190 million in 1983–1984 and it employs over 650 statisticians, economists, sociologists, and other professionals, along with technical, administrative, and clerical staffs of over 3,700.

Today, Statistics Canada's economic statistics provide descriptions and analyses of the national economy, prices, the labour market, international transactions, and business and public finance. They also encompass all Canadian industry—manufacturing, merchandising and services, transportation, communications, construction, wholesale and retail trade, and governments.

Social and socio-economic statistics monitor the health and social well being of Canadians, their educational status and activities, labour force experiences, personal finances, cultural and educational activities, and justice systems.

The Statistics Act enjoins Statistics Canada to conduct the nation's Census of Population and Housing and a Census of Agriculture. The Census of Population and Housing is a national inventory of key socio-economic changes, with coverage extending to every person in the country. Taken every five years, it provides basic demographic measures such as geographic population distributions and family characteristics, as well as information on language, education, ethnicity, income, industries, occupations, and housing. The Census of Agriculture, conducted at the same time as the Census of Population and Housing, covers all Canadian farmers, producing information on topics such as land use, crops, livestock, and farm machinery.

As a common service organization, Statistics Canada carries out surveys funded by other federal departments and agencies, provides other consultative and advisory services, and collaborates with them in the development and dissemination of statistical information. Statistical focal points, appointed by each provincial and territorial government, cooperate with Statistics Canada's staff in many aspects of statistical collection, production, and dissemination.

Statistics Canada participates in the programs of the United Nations and its agencies, the Organization for Economic Cooperation and Development, the Commonwealth, and other international organizations. Much of its work in the international statistical arena relates to the development of common standards and frameworks for internationally comparable statistics and the advancement of statistical methodologies. The agency also maintains extensive bilateral relations with statistical agencies of other countries. These contacts provide mutual benefits in the further improvements of statistical methodology and techniques.

The agency's regional offices in principal cities play an important role in both data collection and dissemination. Household and business surveys are conducted by field employees of nine such offices and regional advisory services staff assist clients in the location, interpretation, and use of statistical data and analyses. Advisory services centres are located in St. John's, Halifax, Montréal, Ottawa, Sturgeon Falls, Toronto, Winnipeg, Regina, Edmonton, and Vancouver.

Statistics Canada makes intensive efforts to minimize the burden on survey respondents. Research and development in the use of administrative records, improved sampling techniques, and questionnaire design are resulting in a marked reduction of response burden, particularly for small businesses, while maintaining or enhancing the quality of data. At the same time, Statistics Canada is committed to maintaining, without exception, the confidentiality of respondents. The Statistics Act provides unconditional and very strong guarantees of confidentiality, ensuring that individual returns to Statistics Canada are used only for statistical purposes and seen only by employees of the agency, who are sworn to secrecy.

INFORMATION DISSEMINATION

Statistics Canada's information is presented in a variety of media suitable to the different needs of users. These include hardcopy publications, microforms, computer tapes, and a large electronic dissemination system.

The agency prepares and issues more than 500 publications a year. Among these are: the *Statistics Canada Daily*, the vehicle of first release for agency data; the monthly *Canadian Statistical Review* and *Current*

Economic Analysis; the periodical *Survey Methodology*; compendia, such as the *Canada Year Book* and *Canada Handbook*; monthly and annual statistical summaries; and research works on key social and economic issues. Many Statistics Canada publications, such as the *International Balance of Payments*, the *Gross National Product*, and *External Trade*, provide data essential to international comparisons and analyses. Single copies of the *Statistics Canada Catalogue*, which lists all agency publications, can be ordered free of charge from the regional reference centres or from Statistics Canada, Central Inquiries, R.H. Coats Building, Ottawa, Ontario K1A 0T6, Canada.

The past decade has been marked by a rapid increase in the agency's use of electronic communications. Most important is CANSIM, the Canadian SocioEconomic Information Management System, Statistics Canada's machine-readable data base accessible to users by means of computer terminal. It has three modules: a very large number of economic time series, cross-classified social statistics, and census summary tapes.

Statistics Canada has also developed an innovative combination of statistics and graphics called TELICHART, which represents a bridge between TELIDON videotex display technology and the CANSIM data bases. TELICHART provides dynamic, rapid graphic displays of economic and social statistics on terminals capable of receiving TELIDON signals.

Statistics Canada's programs describe most aspects of the nation's social and economic life, and also monitor its natural environment and social institutions. Some areas of statistical activity are listed below.

Balance of Payments: Current and capital account transactions between Canada and other countries, and year-end claims and liabilities of nonresidents.

Business: A wide range of data on finances, production, sales, inventories and employment, salaries and wages, industries, and commodities.

Energy: The supply and demand for all types of fuels, household and industrial consumption, shipments, and price levels.

External Trade: The type, quantity, and value of exports and imports by country of destination or origin.

Gross National Product: Estimates of the total value of production and consumption of goods and services by Canadian residents at market prices and inflation-adjusted dollars.

Labour Force: All important aspects of the labour force related activities of Canadians, including levels and duration of periods of employment and unemployment; employee levels, hours and earnings for business, institutions and governments; industries and occupations; the labour market; and place of work for major urban areas.

Manufacturing Industries: Inventories, shipments, orders, capacity, utilization, and materials used for all types of manufacturing.

Prices: Information on price changes within various levels of the economy, including consumer prices, construction, farming, and manufacturing.

Health, Education, and Justice: Statistics on the operations of hospitals, morbidity and mortality, teachers, students, educational finance, crime, police, the courts, and prisons.

L. ROWEBOTTOM

STATISTICS, HISTORY OF

ORIGINS AND DISTRIBUTIONS

Statistics arose as a study of the description of states, especially their economic and demographic aspects. An excellent history of these aspects is available in Westergaard's text [22], largely written before the expansion of modern mathematical statistics. It appears that much of refs. 22 and 17 would now be judged to belong to the history of economics. If modern statistics is to be characterized after ref. 8 as the study of populations, variation, and the reduction of data, some themes can be picked out. The first classic was the skillful analysis [12] in which John Graunt (1620–1674) pointed out the value of the reduction of several, large confused volumes into a few well chosen (perspicacious) tables. He noticed a statistical regularity in the number of deaths per year in a great city in nonepidemic times; further the sex ratio of births was about 1.08, 14 males to every 13 females born. Urban death rates were found to be greater than rural. By a consideration of the deaths by known causes and without a knowledge of the ages at death, Graunt was able to estimate the loss of life between birth and 6 years of age. His estimate of the maternal mortality rate as 1.5% seems reasonable. He was thus able to draw important conclusions from unpromising material. He also gave a new life table* (*see also* BIOSTATISTICS).

Edmond Halley (1656–1742), using data from Breslau with ages at death recorded, was able to improve on Graunt's life table and to introduce the notion of a death rate. Leonhard Euler (1707–1783) had the notion of the stationary and stable life table population. Pension and life assurance schemes were early discussed by Johan de Witt* (1625–1672) and others. The significance of the life table theory for population dynamics of human and other aggregates was studied by Thomas Robert Malthus* (1766–1834), Alfred James Lotka (1880–1949), Ronald Aylmer Fisher* (1890–1962), and William Feller (1906–1970) among others [21], with increasing mathematical sophistication.

William Petty (1623–1687), Graunt's contemporary and friend, saw the need to extend demographic knowledge of the human population by the creation of a central statistical department; specifically, data were to be collected by parishes by means of an enumeration schedule in which were to be recorded details of age, sex, marital status, etc.; there were to be statistics of births, deaths, marriages, revenue, education, and trade [13].

With the examination of time series*, such as of deaths, random jumps and falls were noted; Graunt thought of them in mechanical terms and related them to the jerky motion of clocks; but such irregular variations also affected the time series of gambling and astronomy. So the next advance was the development of the notions of random errors and error distribution (*see* LAWS OF ERROR).

Gambling* yielded the first models of chance* events, with equal probabilities assigned to the sides of a coin or to the faces of a die, leading to evaluations of the probabilities of the number of heads after tosses of a true coin or the total of points after throwing a true die. Some generality was achieved when Abraham de Moivre* (1667–1754) derived an approximation to the terms of the binomial distribution*, which enabled each probability to be equated with an area under the normal curve—an early version of the central limit theorem*. Pierre Simon Laplace* (1749–1827) derived similar asymptotic formulas for the proportion of male births. Jakob Bernoulli* (1654–1705), with his weak law of large numbers*, gave support to the use of means calculated from a large sample. Thomas Simpson (1710–1761) calculated the exact distribution of sums of identically distributed random variables (*see* LAWS OF ERROR*), giving similar support.

In astronomy, estimates, in some sense "the best," were made of unknown parameters, usually the positions of stars or the motions of planets, from observations which were necessarily discordant because of errors in measurement whenever the observations exceeded the parameters in number; among others, Roger Cotes (1682–1716), Thomas Bayes (1702–1761), Euler, Johann Tobias Mayer (1723–1762), Rudger Josif Boškovič* (1711–1787), Laplace, and Adrien Marie Legendre (1752–1833) struggled with the problem, substantially solved by Carl

Friedrich Gauss* (1777–1855). John Michell (?1724–1793) proved the existence of double stars by a statistical argument.

It would be false to believe, however, that modern statistical theory developed out of the needs of actuarial* science, demography*, and astronomy; rather it developed from the needs of psychology, medicine, anthropometry, genetics*, and agriculture*.

Almost all empirical distributions up to 1830 were one dimensional of error or of a qualitative (nonnumerical) variable. After 1830, empirical distributions of metrically valued variables, such as height or weight, were popularized by the astronomer and sociologist Lambert Adolphe Jacques Quetelet* (1796–1874), who made much use of the theoretical binomial and normal* distributions in his biometrical studies. Later the Poisson distribution* was found to be relevant to official statistics by Ladislaus von Bortkiewicz* (1868–1931), who reported on the incidence of injuries from horse kicks in Prussian regiments. The same distribution appeared in the work of Ernst Abbe* (1840–1905) on the counting of red blood cells. It has since been much applied in counting experiments, e.g., scintillation counting.

In biology, statistical methodology enabled Johann Gregor Mendel (1822–1884) to recognize the existence of certain major genetic factors, appearing at levels 0, 1, and 2, of which 0 (double recessive) could be distinguished from levels 1 or 2 (dominant). He was able to determine the results of mating between individuals of like or unlike levels and to suggest a model whereby certain biological events were equivalent to the tossing of a coin; he was able to assign probabilities to the results of arbitrary matings and to verify his hypotheses by experiment [1].

Although economics failed to yield problems that required more than elementary theory for solution, there was interest early in medical statistics. Philippe Pinel (1745–1826) and Pierre Charles Alexandre Louis (1787–1872) began the difficult task of establishing a classification of disease; these workers kept accurate and thorough records of all cases and were able to give statistics relating to prognosis. Louis was able to refute claims in favor of extensive blood-letting by follow-up* methods. Three of his many "pupils" may be mentioned; Jules Gavarret (1809–1890) wrote a textbook on medical statistics that contains many uses of the test of the difference between two proportions according to the theory of Siméon-Denis Poisson* (1781–1840); Oliver Wendell Holmes (1809–1894) and his unknown mathematical consultant gave an interesting analysis of sequences of cases of childbed fever, proving that it was infective, which was superior to any similar study of the nineteenth century; William Farr (1807–1883) established new traditions in official vital statistics*.

A more direct impetus came from genetics* (actually eugenics). Francis Galton* (1822–1911), studying the correlations between the weights of parent and offspring peas in 1886, found linear regression* of Y on a normal variable X and lines of equal probability on approximately similar and similarly situated ellipses, from which James Douglas Hamilton Dickson (1849–1931) deduced the joint normal form with density proportional to $\exp(-\frac{1}{2}\mathbf{x}^T\mathbf{A}\mathbf{x})$; in standard notation, $\mathbf{x}^T\mathbf{A}\mathbf{x}$ would become $x^2 + (y - \rho x)^2/(1 - \rho^2)$. ρ was identified as the slope of the regression line of Y on X. Now the multivariate normal distribution had often appeared in the literature; bivariate and trivariate normal distributions were known to Laplace. The joint distribution can be constructed by linear transformation of mutually independent normal variables, e.g., as by Giovanni Antonio Amedeo Plana (1781–1864), and conversely it can be resolved into a product of mutually independent normal variables as by Auguste Bravais (1811–1863) and Irénée-Jules Bienaymé* (1796–1878). Isaac Todhunter (1820–1884), writing on the theory of least squares, derived a multivariate joint normal distribution of general form, constant $\cdot \exp(-\frac{1}{2}\mathbf{x}^T\mathbf{A}\mathbf{x})$; Arthur Cayley (1821–1895) reduced $\mathbf{x}^T\mathbf{A}\mathbf{x}$ to a sum of squares and so determined the value of the constant. None of these authors was interested in the nondiagonal elements of $\mathbf{A}^{-1} = \mathbf{V}$, where $\mathbf{V}$ is the covariance matrix. Galton [11] was later to write that "these errors or deviations were the very things I wanted to preserve and know about."

The normal distribution* plays a dominant role in theoretical statistics. There are many characterizations to show why this is so [15]; in general, if a model contains several nontrivial properties of the normal distribution it often implies them all.

In 1895, Karl Pearson* (1857–1936) recognized the need for more theoretical distributions and obtained the density functions as the solution of a differential equation (*see* PEARSON SYSTEM); like some other statisticians, Andreĭ Andreevič Markov* (1856–1922) was reluctant to use the Pearson system* of distributions because there was no evident model generating them [18], even though he obtained a Pearsonian Γ curve as a limiting distribution—see ref. 18, pp. 53–54. Markov further identified the Pearson χ^2 as the sample size multiplied by the coefficient of dispersion of Wilhelm Hector Richard Albrecht Lexis* (1837–1914). Walter Frank Raphael Weldon (1860–1906) obtained joint distributions in binomial variables by taking sums of independent elementary binomial variables with some held in common. Many authors, e.g., Alexander Craig Aitken (1895–1967), have taken part in developing this idea; but many other methods have been used to obtain joint distributions [16]. After the failure of Karl Pearson's methods to generate more joint distributions, Sergeĭ Natanovič Bernstein* (1880–1968) [2] suggested that a more productive approach might be in the field of stochastic processes*.

KARL PEARSON ERA, 1890–1920

The principal achievements of the English biometric school* up to 1920 were (i) collection and reduction of much empirical data; (ii) definition of the joint normal distributions, with partial, multiple and total coefficients of correlation ρ, and also the joint distribution of errors of estimation; (iii) χ^2 test of goodness of fit*, comparing the observed with the theoretical distributions, including the introduction of conditional Poisson variables by Herbert Edward Soper (1865–1930); (iv) analysis of contingency tables*, especially by χ^2; (v) estimation of ρ by maximum likelihood* when the marginal distributions were sufficiently finely partitioned; (vi) estimation of ρ when the marginal distributions were not fully defined; (vii) description of a system of curves with a unified system of estimation of the parameters, the method of moments*; (viii) applications of normal theory to the genetical selection problem; (ix) some steps toward a general theory of dependence [19]; (x) some steps toward a theory of estimation and tests of the accuracy of the estimates; (xi) construction of appropriate tables [20].

At this time, Félix Édouard Justin Émile Borel (1871–1956), Maurice Fréchet (1878–1973), and Jules Henri Poincaré (1854–1912) in France, and Aleksandr Aleksandrovič Čuprov* (1874–1926), A. A. Markov, and Vsevolod Ivanovič Romanovsky (1879–1954) in Russia, were making contributions, in particular adding rigor to the mathematical treatment of data.

R. A. FISHER ERA, 1921–1936

All Pearsonian methods were applicable to large samples, in which close estimates of the variance could be made. With small samples in practical applications, this was no longer so. William Sealy Gosset* (1876–1937), therefore, devised a test, which, after a transformation suggested by R. A. Fisher, is now known as Student's t-test*.

Fisher opened a new era with four memorable articles: the exact distribution of the estimate of the coefficient of correlation [4]; reconciliation of Mendelian and biometrical approaches to genetics* [5]; correct interpretation of contingency tables [6]; general theory of estimation and inference [7]. After

1920, at Rothamsted Experimental Station, Fisher developed the theories of the analysis of variance* and the analysis and design of experiments* [8, 9] of very general applicability. Fisher had great mathematical power, especially in combinatorics*, and was able to attract other mathematicians as assistants. His field of application was fortunately chosen; the results had immediate application with assessable economic value; realistic simplifying hypotheses, such as the normality of the errors and their mutual independence, could be made; there were few entrenched doctrines to combat; the price of experiment was low; there were no ethical problems. Great steps were taken in developing the branches of activity of the Pearson school mentioned above. In (iii) and (iv) many outstanding problems were solved; correct numbers of degrees of freedom were assigned; as Eisenhart's biography [3] makes clear, K. Pearson had already gone some way towards this goal; in (vii) Fisher was able to devise more efficient methods of estimation; he rejected the fitting of distributions by the method of moments; in (x) Fisher made many advances; in (xi) Fisher and Yates published [10].

Fisher's assumption on the distribution of the errors and their mutual independence enabled him to retain the mutual independence of linear and quadratic forms* by orthogonal transformations and so to justify the use of the t- and F-tests*. Fisher saw that agricultural experiments permitted far more complex design. Thus the effects in a two-way layout could be analysed into parts due to geographical factors (rows and columns) and treatments. Each effect could be analysed separately. This could be extended by the use of $n \times n$ Latin squares*, with treatments applied to the plots corresponding to the various letters of the Latin square*; the procedure could be carried out with Graeco-Latin squares*. Problems of design, missing values, nonorthogonality, etc., were treated by Fisher and his assistants and colleagues, among whom may be mentioned Maurice Stevenson Bartlett (1910–), William Gemmell Cochran (1909–1980), especially influential later in America, David John Finney (1917–), Joseph Oscar Irwin (1898–1982), Kenneth Mather (1911–), and Frank Yates (1902–).

NEYMAN–PEARSON ERA, 1937–1949

Jerzy Neyman* (1894–1981) and Egon Sharpe Pearson* (1895–1980), in a series of remarkable papers, clarified the theory of inference, in particular, the rationale of significance tests, the justifications of which had often been criticised. Early significance tests* had been of the difference between binomial variables or between means and they had been extended to the multivariate case by K. Pearson with the χ^2 test and by R. A. Fisher with the F-test generalizing Student's t-test. Neyman and E. S. Pearson saw that such tests, to be effective, must consider the alternative hypotheses against which the null hypothesis was tested. They set out the two kinds of error in such testing and thus arrived at their fundamental lemma, the likelihood ratio test* and the notion of power*; they incidentally validated most of the uses of the common tests of significance*; they also introduced confidence* limits*; but their system was never recognized by Fisher. The consequences of the work of Neyman and Pearson have been worked out by many authors, especially American.

THE MODERN PERIOD, 1950–

Statistics has become increasingly mathematized. Measure theory* is required for general descriptions of distribution and inference theories; Fourier analysis is the natural tool for the study of waves; group and number theories are applied to inference in the analysis of variance and in the algebra of design with their symmetries and special configurations such as the Graeco-Latin squares and Steiner triples. The same combinatorial theory can be applied to coding theory and finite geometries. Consequently statistical mathematics becomes part of pure mathematics studied for its usefulness in the various applications. As the common statistical tests have been thoroughly researched and are applied by practitioners often employed full time in a single applied field, a gap has arisen between researchers and practitioners; but this can be observed in many other fields of endeavour.

Electronic computers have brought about great changes. Data, such as the height of water level in oceanography, flux of electromagnetic energy (especially radio), state of industrial processes and biological states, can be collected with their aid, collections quite impractical or impossible otherwise. There have been great savings in manpower, especially in those areas where repeated computations are made on the same input, for example, in the computation of the coefficients of correlation and other test statistics in multivariate analysis. The savings are even greater now that computer packages are available for all the common tests, especially those of the analysis of variance. Rapid computation has made possible the use of matching and permutation tests*. Significance levels can be computed when the distributions are not defined by a closed analytical formula; at other times the machine can compute the probability of every event or approximate to the significance levels by Monte Carlo methods*. Ease of computation with the use of packages has sometimes led to conclusions made without proper consideration of the statistical problems arising, especially that of multiple comparisons*.

The role of models in statistical and scientific work has now become generally recognized; the choice of model is to some extent aesthetic and arbitrary, although founded on experience and knowledge of the field of application; but once the choice is made and the model determined, all the inferences must be made with the aid of mathematics without the introduction of further hypotheses or principles. The system of inference used is also to some extent arbitrary; alternative systems have been much studied. Fiducial inference* no longer plays an important role. Bayesian* methods have become more widely used after their partial eclipse in the Fisher era. Information theory* has been introduced. Most statistical inference seems still to be carried out in a manner consistent with Neyman–Pearson theory using tests introduced in the K. Pearson and Fisher eras.

The extended powers and interests of the modern state require the collection of more data at moderate cost. Anders Nicolai Kiaer* (1838–1919), ahead of his time, suggested that probability sampling should supplement census* methods. Such sampling became standard practice in India and elsewhere after its introduction by Prasanta Chandra Mahalanobis* (1893–1972). Quality control* methods, popularized by Walter Andrew Shewhart (1891–1967) in industry, had similar aims.

Many new branches or specializations and applications have been developed: decision theory, time series, multivariate analysis, econometrics, games theory, clinical trials, nonparametric inference, sequential analysis, mathematical taxonomy, and reliability. Mathematical statistics and its applications continue to develop and expand.

References

[1] Bennett, J. H., ed. (1965). *Experiments in Plant Hybridisation*: *Gregor Mendel*. Oliver and Boyd, Edinburgh, Scotland.

[2] Bernstein, S. N. (1932). *Verh. Math. Kongr. Zürich*, **1**, 288–309.

[3] Eisenhart, C. (1974). In *Dictionary of Scientific Biography*, Vol. 10, C. C. Gillispie, ed. Scribners, New York, pp. 447–473.

[4] Fisher, R. A. (1915). *Biometrika*, **10**, 507–521.

[5] Fisher, R. A. (1918). *Trans. R. Soc. Edinburgh*, **52**, 399–433.

[6] Fisher, R. A. (1922). *Philos. Trans. R. Soc. Lond. A*, **122**, 309–368.

[7] Fisher, R. A. (1922). *J. R. Statist. Soc.*, **85**, 87–94.

[8] Fisher, R. A. (1925). *Statistical Methods for Research Workers*. Oliver and Boyd, Edinburgh, Scotland. [13th ed. (1958).]

[9] Fisher, R. A. (1935). *The Design of Experiments*. Oliver and Boyd, London, England. [8th ed. (1966).]

[10] Fisher, R. A. and Yates, F. (1938). *Statistical Tables for Biological, Agricultural and Medical Research*. Oliver and Boyd, Edinburgh, Scotland. [6th ed. (1963).]

[11] Galton, F. (1908). *Memories of My Life*. Methuen, London, England.

[12] Graunt, J. (1662). *Natural and Political Observations Made Upon the Bills of Mortality*, reprint (1939), W. F. Willcox, ed. Johns Hopkins University Press, Baltimore, MD.

[13] Greenwood, M. (1948). *Medical Statistics from Graunt to Farr*. Cambridge University Press, Cambridge, England. (The Fitzpatrick Lectures for the years 1941 and 1943.)

[14] Heyde, C. C. and Seneta, E. (1977). *I. J. Bienaymé: Statistical Theory Anticipated*. Springer, New York.

[15] Kagan, A. M., Linnik, Yu. V., and Rao, C. R. (1973). *Characterization Problems in Mathematical Statistics*, B. Ramachandran, transl. Wiley, New York.

[16] Lancaster, H. O. (1972/1983). *Math. Chronicle (New Zealand)*, **2**, 1–16; **12**, 55–92.

[17] Meitzen, A. (1891). *History, Theory and Technique of Statistics*, R. P. Faulkner, transl. Kraus Reprint Co., New York. [From the 1886 German original. Reprinted (1970) from the supplement to the *Annals of the American Academy of Political and Social Science*, March–May, No. 2 (1891).]

[18] Ondar, K. O., ed. (1981). *The Correspondence between A. A. Markov and A. A. Chuprov on the Theory of Probability and Mathematical Statistics*, C. Stein and M. Stein, transl. Springer, Berlin, Germany.

[19] Pearson, K. (1904). *On the Theory of Contingency and its Relation to Association and Normal Correlation*. Drapers' Company Research Memoirs, Biometric Series I.

[20] Pearson, K., ed. (1914/1931). *Tables for Statisticians and Biometricians, Parts I and II*. Cambridge University Press, Cambridge, England.

[21] Smith, D. P. and Keyfitz, N. (1977). *Mathematical Demography*. Springer, New York.

[22] Westergaard, H. (1932). *Contribution to the History of Statistics*, reprinted (1970). Mouton, Paris, France.

Bibliography

There is no standard history of statistics. The actuarial and demographic traditions are given in refs. 17, 21, and 22.

Cramér, H. (1946, 1963). *Mathematical Methods of Statistics*. Princeton University Press, Princeton, NJ. (A text of mathematical statistics presented as a branch of mathematics.)

Gillispie, C. C., ed. (1970/1980). *Dictionary of Scientific Biography*, Vols. I–XVI. Scribners, New York. (Contains many fine biographies of statisticians.)

Kendall, M. G. and Doig, A. (1962, 1965, 1968). *Bibliography of Statistical Literature*. Oliver and Boyd, Edinburgh, Scotland. [Gives an unclassified list of references of articles of mathematical statistical interest. (1962) 1950–1958; (1965) 1940–1949; (1968) pre-1940 with supplements to the volumes for 1940–1949 and 1950–1958.]

Kendall, M. G. and Plackett, R. L. (1977). *Historical Studies in Probability and Statistics*. Griffin, London, England. (Contains many fine biographies and some histories of special topics.)

Kendall, M. G. and Stuart, A. (1973, 1977, 1983). *The Advanced Theory of Statistics*, 4th ed. Vols. 1–3. Griffin, London, England. [Vol. 1 (1977): Distribution theory; Vol. 2 (1973): Inference and relationship; Vol. 3 (1983): Design and analysis and time-series (J. K. Ord, coauthor). Eclectic, free of polemics and with many historical references in the text and exercises. Volume 1 is the best introduction to mathematical statistics for mathematicians.]

Lancaster, H. O. (1968). *Bibliography of Statistical Bibliographies*. Oliver and Boyd, Edinburgh, Scotland. (Gives lists of (a) biographies (b) bibliographies, and (c) national bibliographies of scientists and topics of interest to statisticians.)

Lancaster, H. O. (1969). *The Chi-squared Distribution*. Wiley, New York. (Gives the history of chi-squared, contingency tables and dependence.)

Lancaster, H. O. (1980). In *Developments in Statistics*, Vol. 3, P. R. Krishnaiah, ed. Academic, New York, pp. 99–157. (Many historical references.)

Lancaster, H. O. (1982). *Int. Statist. Rev.*, **50**, 195–217. (Indexes the biographies in the *Bibliography of Statistical Bibliographies* series with about 500 authors listed, including all authors cited above.)

MacKenzie, D. A. (1981). *Statistics in Britain 1865–1930*. Edinburgh University Press, Edinburgh, Scotland.

Meyer, A. (1874). *Calcul des Probabilités*, 2nd ed., Vol. 4. Soc. Royale des Sciences de Liège, Liège, France, pp. 1–458. (A text of contemporary mathematical statistics.)

Pearson, E. S. and Kendall, M. G., eds. (1970). *Studies in the History of Statistics and Probability*. Griffin, London, England. (Contains biographies and histories of special topics.)

Pearson, K. (1978). *The History of Statistics in the 17th and 18th Centuries Against the Changing Background of Intellectual, Scientific and Religious Thought: Lectures by Karl Pearson, 1921–1933*, E. S. Pearson, ed. Griffin, London, England. (Covers the period from political arithmetic up to the death of Laplace.)

Stigler, S. (1986). *The History of Statistics: The Measurement of Uncertainty Before 1900.* Harvard University Press, Cambridge, MA. (A clear and thorough exposition for students of mathematics and statistics; comprehensible and interesting to all general readers. Part 1 deals with the development of discrete probability, the method of least squares, and probability curves. Parts 2 and 3 cover the synthesis of least squares with the law of errors in its application to the social sciences.)

Tankard, J. W. Jr. (1984) *The Statistical Pioneers.* Shankman, Cambridge, MA. (Comparative biographies of F. Galton, K. Pearson, W. Gosset, and R. A. Fisher, with numerous references.)

Yule, G. U. (1910). *An Introduction to the Theory of Statistics.* Griffin, London, England. [9th ed. (1929). A well known and used elementary text. It does not contain the usual mathematical errors of such texts.]

Editors' Note. Each of the following related entries contains a substantial amount of historical material, but readers can find relevant information in many other entries, too numerous to list here.

(AGRICULTURE, STATISTICS IN
CHANCE
COMBINATORICS
CORRELATION
DEMOGRAPHY
ENGLISH BIOMETRIC SCHOOL
FREQUENCY CURVES, SYSTEMS OF
GAMBLING, STATISTICS IN
HYPOTHESIS TESTING
INFERENCE, STATISTICAL
LAWS OF ERROR
LEAST SQUARES
MILITARY STATISTICS
PHYSICS, STATISTICS IN (EARLY HISTORY)
PROBABILITY, HISTORY OF
QUALITY CONTROL, STATISTICAL
SIGNIFICANCE TESTS, HISTORY AND LOGIC
STATISTICS IN PSYCHOLOGY
Various biographical entries)

H. O. LANCASTER

STATISTICS IN ANIMAL SCIENCE

INTRODUCTION AND HISTORY

The introduction of modern statistical methods has been a slower process in animal science than in agriculture*. In designing experiments there have been limitations in the maintenance costs and duration of some of the experiments. Sampling methods for estimating animal numbers have been difficult to adopt, because animals, being mobile, may be counted repeatedly.

The first application of statistical methods to animal populations was in 1890 when Weldon [44] showed that the distributions of different measurements (ratios to total length) made on four local races of shrimp (*Crangon vulgaris*) closely followed the normal law. Weldon [45] found that the frontal breadths of Naples crabs formed into asymmetric curves with a double hump. It was this problem of dissection of a frequency distribution into two normal components that led to Karl Pearson's [27] first statistical memoir. The capture–mark–recapture method was first used by Petersen [28] in his studies of the European plaice and the earliest beginnings in the quantitative study of bird populations were made by Jones [21].

Some developments in the planning and analysis of animal experiments and in sampling methods for estimating animal numbers will be reviewed in the following sections.

DESIGN OF EXPERIMENTS

Choice of Experimental Unit

In feeding trials the experimental unit may be a pen. Often one pen of animals is fed on each treatment. The animals are assigned at random to each pen and the treatments are assigned at random to the pens. With a single pen of animals receiving a treatment, the effect of factors such as breed and age cannot be separated from treatment differences to provide a valid estimate of error [19, 23].

Change-over Designs

One way of controlling the variation among animals in experiments is to adopt change-over designs* in which different treatments are tested on the same animal, such that each animal receives in sequence two or more treatments in successive experimental periods. The analysis of switchback trials for more than two treatments is dealt with in ref. 24. Incomplete block designs*, where the number of treatments exceeds the number of experimental units per block, are given in refs. 25 and 26.

Complete Block Designs

Randomized block designs (RBD) and Latin squares are common in animal experiments. A series of Latin squares is recommended for use in large experiments on rodents [39]. Alternatively, analysis of covariance* has been used in which a concomitant observation is taken for eliminating the effect of variations arising out of the peculiarities of individual animals. Thus, to test the effect of treatments on milk production of cows, the yield in the previous lactation is used as a concomitant variate to reduce the error in the experimental yield. The method assumes that (i) the regression of y on the concomitant variate x is significant and (ii) x does not vary significantly among treatments.

Scientists often object to use of RBDs or Latin squares*, which make dosing and sampling procedures difficult to operate in practice. In such cases, a standard order of groups may be used within each block, e.g., use of a standard Latin square within a battery of cages in rodent studies.

Split-Plot Designs*

Incomplete blocks or fractional replication designs may be used to eliminate the influence of litter differences from treatment effects when the number of treatments is large. An experiment on mice in which larger units are split into smaller ones is described in ref. 8 to provide increased precision on the more interesting comparisons.

Repeated Measurements* Experiments

A RBD in which the experimental units are animals that receive each treatment in turn is a repeated measurement experiment. This is generally analyzed as a split-plot design, which assumes equal correlation for every pair of subplot treatments. Multivariate methods that do not require this assumption are given in ref. 5.

Regression*

A common problem in animal studies is the estimation of x from the regression of y on x when y is easy to measure but x is difficult and expensive. Thus y and x may represent the length and age of a fish and we may want to estimate the age composition to predict the status of the stock in future years. Other examples are given in ref. 16. A polynomial regression* of average daily gains of each of several Holstein calves on time was fitted in ref. 1 and analysis of variance* (ANOVA) done on the regression coefficients. Nonlinear regressions for describing weight–age relationship in cattle are compared in ref. 3. Fitzhugh [14] reviewed analysis of growth curves* and strategies for altering their shape.

ANOVA Models

Sometimes observations may be lost due to death or premature drying of lactation. The estimation of variance components* in mixed models with unbalanced data are dealt with in ref. 17 and illustrations from animal experiments are given in ref. 2. Analysis of balanced experiments for linear and normal models is discussed in ref. 18.

Transformations

Often the data violate basic assumptions of the methods of analysis. For violations due to heterogeneity, transformation of the individual observations into another scale may reduce heterogeneity, increase precision in treatment comparisons, and reduce nonadditivity [41]. Illustrations from animal experiments are given in ref. 16.

In management trials with pigs a score of x (out of 100) is assigned to each pig for treatment comparisons. Treatment effects will not be additive and the transformation $\log[(x + \frac{1}{2})/(100 + \frac{1}{2} - x)]$ given in ref. 8 would rectify this.

Categorical Data

Log-linear and generalized least-squares approaches have been used for categorical data* on the incidence of pregnancy in ewes and of buller steers [33]. Log-linear models for capture–recapture* data from closed populations are mainly discussed in refs. 6 and 34; extensions to open populations are in ref. 7.

ESTIMATION OF ANIMAL NUMBERS OR DENSITY

Two important reasons for using sampling methods for estimation of animal numbers or density are (1) limitation of funds, time, and resources and (2) lesser disturbance of the population and its environment than by total count operations. Some important methods for estimating animal numbers will be summarized in the following sections.

Strips, Transects, or Quadrats

A common method is to count the number within a long rectangle of known area called a *transect*, a short rectangle called a *strip*, or a square called a *quadrat*. The average density per unit area is estimated by taking total counts over randomly chosen areas. The total population is estimated by multiplying average density by the total area of the population. Quadrat sampling* is preferred for some big game species and its size and shape will depend upon the habitat, abundance, and mobility of the species. When a population is randomly distributed, the number(s) of quadrats to be sampled in an area is given by $s = S/(1 + NC^2)$, where S is the total number of quadrats, N is the size of the population being sampled, and C is the coefficient of variation* of $\hat{N}$ [34].

When the population density varies over large areas, stratified sampling with optimum allocation* is recommended. Siniff and Skoog [40] allocated a sampling effort for six caribou strata on the basis of preliminary population estimates of big game in aerial surveys of Alaska caribou conducted in 22,000 square miles of the Nelchina area. The method was based on rectangular sample plots 4 square miles in area on which observers attempted to count all animals. Two important advantages of the method were (i) reduction of the observer bias in sighting by using large transects and (ii) the use of an efficient design through optimum allocation of effort based on available information on caribou distribution. Sampling effort allocated to strata reduced the variance by more than half over simple random sampling*. Other examples are given in refs. 13 and 36. Where it is difficult or time consuming to count all animals in all the sampled quadrats, two-phase sampling using ratio or regression methods may be adopted. Somewhat similar to quadrat sampling in birds and mammals is the use of sampling in estimating fish population in a section of a stream. Sample areas are selected and fish captured by seining or electrical shocking. The areas are screened to prevent the dispersal of the fish while they are being caught. Stratification in weirs or seines [20] is common to the Atlantic herring or sardine fishing off the New England and New Brunswick coasts.

STRIP SAMPLING. Parallel lines one strip width ($2W$, say) apart determine the population of strips. All the n animals observed within the sampled strips are counted. The estimate of total number of animals (N) is given by $A \cdot n/(2LW)$, where L and A represent the length and population area of the strip. Strip sampling involves less risk of repeated counting and is generally recommended in aerial surveys [42, 43].

LINE TRANSECT SAMPLING. An observer walks a fixed distance L along a transect or set of transects and records for each of the n sighted animals, its right angle distance y_i ($i = 1, 2, \ldots, n$) from the transect or its distance r_i from the observer or both. The technique is most useful when the animals move so fast that they can only be seen when they are flushed into the open. Various parametric estimators have been suggested in refs. 15 and CAPTURE–RECAPTURE METHODS. Nonparametric estimators are given in refs. 4 and 10. References 15 and 34 deal with consequences for departures from the assumptions. All the estimates are of the form $A \cdot n/(2LW)$, where W is some measure of one-half the "effective width" of the strip covered by the observer as he moves down the transect. The transect method can also be used in aerial surveys. The procedure is the same as for walking or driving and is likewise subject to error. *See also* LINE INTERCEPT SAMPLING and LINE INTERSECT SAMPLING.

Capture–Mark–Recapture (CMR) Methods

A number of M animals from a closed population are caught, marked, and released. On a second occasion, a sample of n individuals are captured. If m is the number of marked animals in the sample, a biased estimate of N and its variance are

$$\hat{N} = \frac{n}{m}M,$$

$$v(\hat{N}) = \frac{\hat{N}^2(\hat{N} - m)(\hat{N} - n)}{Mn(\hat{N} - 1)}.$$

This was first given in ref. 28 using tagged plaice. Because the coefficient of variation of $\hat{N}$ is approximately given by $1/m^{1/2}$, it follows that for $\hat{N}$ to be efficient, we should have sufficient recaptures in the second sample. For closed populations $\hat{N}$ appears to be the most useful estimate of N if the basic assumptions [34] underlying the method are met. CAPTURE–RECAPTURE METHODS contains recent developments in estimating N for "closed" and "open" populations when marking is carried over a period of time. A capture–recapture design robust to unequal probability of capture is given in ref. 29. The linear logistic binary regression model has been used in ref. 30 to relate the probability of capture to auxiliary variables for closed populations. With tag losses due to nonreporting or misreporting of tag returns by hunters, $\hat{N}$ will be an overestimate of N [35]. A treatment of response and nonresponse errors in Canadian waterfowl surveys is given in refs. 37 and 38; response errors were found larger than the sum of sampling and nonresponse errors.

CHANGE-IN-RATIO AND CATCH-EFFORT METHODS. The change-in-ratio method estimates the population size by removing individual animals if the change in ratio of some attribute of the animal, e.g., age or sex composition, is known. For a closed population the maximum likelihood estimator (MLE) of the population total N_t ($t = 1, 2$), based on samples n_t at the beginning and end of the "harvested period," is given by

$$\hat{N}_t = \left(R_m - R\hat{P}_t\right)/\left(\hat{P}_1 - \hat{P}_2\right),$$

where R_m and R_f ($R = R_m + R_f$) are, respectively, the males and females caught during the harvested period $\hat{P}_t = m_t/n_t$ ($t = 1, 2$), where m_t is the number of males at time t. The method assumes (i) a closed population, (ii) $\hat{P}_t = P =$ const for all t, and (iii) R_m and R_f are known exactly. A detailed discussion when the assumptions are violated is given in ref. 34. In fisheries attempts have been made to estimate populations using harvest and age structure data [32].

In the catch-effort method, one unit of sampling effort is assumed to catch a fixed proportion of the population. It is shown in refs. 9 and 22 that

$$E(C_t|k_t) = K(N - k_t),$$

where C_t is the catch per unit effort in the tth period, k_t is the cumulative catch through time period $(t - 1)$, K is the catchability coefficient, and N is the initial population size. C_t plotted against k_t provides estimates of the intercept KN and the slope K, whence N can be estimated.

Indices

Indices are estimates of animal population derived from counts of animal signs, e.g., pellet group, roadside count of breeding birds, nest counts, etc. The investigator records the presence or absence of a species in a quadrat and the percentage of quadrats in which the species is observed, giving an indication of its relative importance. Stratified sample surveys are conducted annually in U.S.A. and Canada for measuring changes in abundance of nongame breeding birds during the breeding season [12, 31].

Big game populations are estimated by counting pellet groups in sample plots or transects. A method for calibrating an index by using removal data is given in ref. 11.

SUMMARY

A basic problem in experimental design with animals is substantial variation among animals owing to too few animals per unit. Hence, the need for concomitant variables* and their relationship with the observations for reducing the effect of variations due to individual animals in a unit. The alternative approach, to reduce this variation by picking experimental animals from a uniform population is not satisfactory, since the results may not necessarily apply to populations with much inherent variability.

Animal scientists are often not convinced of the importance of good experimentation, so that laboratory experiments are not necessarily the best to detect small differences. Closer cooperation is needed between the animal scientist and the statistician to yield the best results. Where the underlying assumptions are false, it is important to investigate whether this would invalidate the conclusions. Animal scientists should be alert to the severity of bias in the use of regression for calibration of one variable as an indicator for another when the problem is an inverse one and the relation between the variables is not very close.

In experiments with repeated observations, with successive observations correlated on the same animal, the animal scientist should consult the statistician in the use of the most appropriate method of analysis.

Sampling techniques for estimating the size of animal populations are indeed difficult to implement since animals often hide from us or move so fast that the same animal may be counted more than once. The choice of method would depend on the nature of the population, its distribution, and the method of sampling. Where possible, the design should be flexible enough to enable the use of more than one method of estimation.

In the past, the Petersen method has been used with too few captures and recaptures, leading to estimates with low precision. As far as possible, more animals should be marked and recaptured to ensure higher precision. Planning for a known precision has been difficult owing to lack of simple expressions for errors. Where the total sample can be split into a number of interpenetrating subsamples*, separate estimates of population size can be formed, resulting in simpler expressions for overall error. This point needs examination for population estimates.

The CMR method assumes that both mortality and recruitment are negligible during the period of data collection*. Another assumption underlying the method is that marked and unmarked animals have the same probability of being caught in the second sample. When these assumptions are violated, it is useful to compare the results with other estimating procedures and, if possible, test on a known population. The latter is recommended for use in animal populations when the methods are likely to vary widely in accuracy.

Nonresponse and response errors often form a high proportion of the total error in estimation of animal numbers or their density. As an example of such errors, visibility bias of the observer in aerial surveys, which results in a proportion of the animal being overlooked, may be cited. In such cases, the population total or its density should be estimated by different groups of experienced observers, either through the use of interpenetrating subsamples or bias corrected by use of alternative methods, e.g., air–ground comparisons.

References

[1] Allen, O. B., Burton, J. H., and Holt, J. D. (1983). *J. Animal Sci.*, **55**, 765–770.

[2] Amble, V. M. (1975). *Statistical Methods in Animal Sciences.* Indian Society of Agricultural Statistics, New Delhi, India.

[3] Brown, J. E., Fitzhugh, H. A., and Cartwright, T. C. (1976). *J. Animal Sci.*, **42**, 810–818.

[4] Burnham, K. P. and Anderson, D. R. (1976). *Biometrics*, **32**, 325–336.

[5] Cole, J. W. L. and Grizzle, J. E. (1966). *Biometrics*, **22**, 810–828.

[6] Cormack, R. M. (1979). In *Sampling Biological Populations*, R. M. Cormack, G. P. Patil, and D. S. Robson, eds. Satellite Program in Statistical Ecology, International Co-operative Publishing House, Fairland, MD, pp. 217–255.

[7] Cormack, R. M. (1984). *Proc. XIIth International Biom. Conf.*, pp. 177–186.

[8] Cox, D. R. (1958). *Planning of Experiments.* Wiley, New York.

[9] De Lury, D. B. (1947). *Biometrics*, **3**, 145–167.

[10] Eberhardt, L. L. (1978). *J. Wildl. Manag.*, **42**, 1–31.

[11] Eberhardt, L. L. (1982). *J. Wildl. Manag.*, **46**, 734–740.

[12] Erskine, A. J. (1973). *Canad. Wildl. Serv. Prog. Note*, **32**, 1–15.

[13] Evans, C. D., Troyer, W. A., and Lensink, C. J. (1966). *J. Wildl. Manag.*, **30**, 767–776.

[14] Fitzhugh, H. A. (1976). *J. Animal Sci.*, **42**, 1036–1051.

[15] Gates, C. E. (1969). *Biometrics*, **25**, 317–328.

[16] Gill, J. L. (1981). *J. Dairy Sci.*, **64**, 1494–1519.

[17] Henderson, C. R. (1953). *Biometrics*, **9**, 226–252.

[18] Henderson, C. R. (1969). In *Techniques and Procedures in Animal Science Research.* American Society for Animal Sciences, pp. 1–35.

[19] Homeyer, P. G. (1954). *Statistics and Mathematics in Biology.* O. Kempthorne et al., eds. Iowa State College Press, Ames, IA, pp. 399–406.

[20] Johnson, W. H. (1940). *J. Fish. Res. Board Canad.*, **4**, 349–354.

[21] Jones, L. (1898). *Wils. Orn. Bull.*, **18**, 5–9.

[22] Leslie, P. H. and Davis, D. H. S. (1939). *J. Animal Ecol.*, **8**, 94–113.

[23] Lucas, H. L. (1948). *Proc. Auburn Conference on Applied Statistics*, p. 77.

[24] Lucas, H. L. (1956). *J. Dairy Sci.*, **39**, 146–154.

[25] Patterson, H. D. (1952). *Biometrika*, **39**, 32–48.

[26] Patterson, H. D., and Lucas, H. L. (1962). *Tech. Bull. No. 147*, N.C. Agric. Exper. Stn.

[27] Pearson K. (1894). *Philos. Trans. R. Soc. Lond.*, *A*, **185**, 71–110.

[28] Petersen, C. E. J. (1896). *Rep. Dan. Biol. Stat.*, **6**, 1–48.

[29] Pollock, K. H. and Otto, M. C. (1983). *Biometrics*, **39**, 1035–1049.

[30] Pollock, K. H., Hines, J. E., and Nichols, J. D. (1984). *Biometrics*, **40**, 329–340.

[31] Robbins, C. S. and Vanvelzen, W. T. (1967). *Spec. Sci. Rep. Wildl. No. 102*, U.S. Fish Wildl. Serv.

[32] Ricker, W. E. (1958). *Bull. Fish. Board Canad.*, **119**, 1–300.

[33] Rutledge, J. J. and Gunsett, F. C. (1982). *J. Animal Sci.*, **54**, 1072–1078.

[34] Seber, G. A. F. (1980). *The Estimation of Animal Abundance and Related Parameters*, 2nd ed. Griffin, London, England.

[35] Seber, G. A. F. and Felton, R. (1981). *Biometrika*, **68**, 211–219.

[36] Sen, A. R. (1970). *Biometrics*, **26**, 315–326.

[37] Sen, A. R. (1972). *J. Wildl. Manag.*, **36**, 951–954.

[38] Sen, A. R. (1973). *J. Wildl. Manag.*, **37**, 485–491.

[39] Shirley, E. (1981). *Bias*, **8**, 82–90.

[40] Siniff, D. B. and Skoog, R. O. (1964). *J. Wildl. Manag.*, **28**, 391–401.

[41] Tukey, J. W. (1949). *Biometrics*, **5**, 232–242.

[42] Watson, R. M., Parker, I. S. C., and Allan, T. (1969). *E. Afr. Wildl. J.*, **7**, 11–26.

[43] Watson, R. M., Graham, A. D., and Parker, I. S. C. (1969). *E. Afr. Wildl. J.*, **7**, 43–59.

[44] Weldon, W. F. R. (1890). *Proc. R. Soc. Lond.*, **47**, 445–453.

[45] Weldon, W. F. R. (1893). *Proc. R. Soc. Lond.*, **54**, 318–329.

Bibliography (Editors' note)

Seber, G. A. F. (1986). *Biometrics*, **42**, 267–292. (Reviews animal abundance estimation techniques developed between 1979 and 1985, some of which supplant previous methods. This is a very thorough survey, with over 300 references.)

(CAPTURE–RECAPTURE METHODS
ECOLOGICAL STATISTICS
LINE TRANSECT SAMPLING
QUADRAT SAMPLING
WILDLIFE SAMPLING)

A. R. SEN

STATISTICS IN AUDITING

Auditing is a specialized area of accounting involving examination by independent auditors of the operation of an organization's system of internal control and of the financial statements in order to express an opinion on the financial statements prepared by the management of the organization. Internal auditors, who are members of the organization, also review the operation of the system of internal control to see if it is operating effectively.

Uses of statistical methods in auditing are a relatively recent phenomenon and are still evolving. Three major current uses are sampling of transactions to study the operation of the internal control system, sampling of accounts to study the correctness of recorded account balances, and regression analysis for analytical review.

STUDY OF INTERNAL CONTROL SYSTEM

Transactions, such as payments of bills received, are sampled to make inferences about the effectiveness of the internal control system. Usually with this type of sampling, auditors are concerned with a qualitative characteristic, namely, whether the internal control was operating improperly for the transaction. Inferences are made about the process proportion of transactions that are handled improperly by the internal control system. Typically, auditors desire assurance that the process proportion is reasonably small. Random samples of transactions are utilized and evaluated by standard statistical procedures based on the binomial* distribution or the Poisson* approximation. No special statistical problems are encountered in this application. Consequently, little research is currently being done in this area. A good summary is contained in Roberts [8].

STUDY OF ACCOUNT BALANCES

Accounts, such as accounts receivable and inventory, often consist of thousands of individual line items that may have a total value of millions of dollars. For such large accounts, it is not economical for the auditor to audit every line item. Consequently, auditors frequently select a sample of line items and audit these, on the basis of which inferences are made about the total error amount in the population. Thus, the characteristic of interest here is quantitative. Denoting the total amount recorded by the company for the account by Y and the total amount that the auditor would establish as correct if each line item in the account were audited by X, the total error amount in the account is $E = Y - X$.

Since the auditor usually knows the amounts recorded by the company for each line item in the population (called the *book amounts*), this information can be utilized in estimating the total error amount E. One estimator that incorporates this supplementary information is the difference estimator, which for simple random sampling of line items is

$$\hat{E} = \frac{N}{n} \sum_{i=1}^{n} d_i = N\bar{d},$$

where

$$d_i = y_i - x_i$$

is the difference between the book and audit amounts for the ith sample line item, $\bar{d}$ is the mean difference per sample line item, and N is the number of line items in the population.

Other estimators that incorporate the information on book amounts are the ratio and regression estimators. Each of these can be used with simple or stratified random sampling of line items. Still another means of incorporating the information about book amounts is through sampling with probability proportional to book amount and then utilizing an unbiased estimator.

Each of these procedures has some serious limitations in auditing applications. When the sample contains no errors, the estimated standard error of the estimator equals zero, as can be seen from the estimated variance of the difference estimator with simple random sampling of line items:

$$s^2(\hat{E}) = N^2 \frac{N-n}{Nn(n-1)} \sum_{i=1}^{n} (d_i - \bar{d})^2.$$

An estimated standard error of zero suggests perfect precision, which would be an unwarranted conclusion.

A second limitation of these supplementary information procedures is that a number of simulation* studies have found that large-sample confidence limits based on the normal distribution are frequently not applicable for sample sizes used in auditing. Neter and Loebbecke [5] studied four actual accounting populations and from these constructed a number of study populations with varying error rates. They utilized sample sizes of 100 and 200 with unstratified, stratified, and PPS sampling*. They found for the supplementary information procedures that the coverages for the large-sample upper confidence bounds for the total error amount (i.e., the proportion of times the bound is correct in repeated sampling) were frequently substantially below the nominal confidence level.

The search for alternative inference procedures that do not depend on large-sample theory has frequently involved monetary unit sampling, which involves the selection of individual monetary units from the population. Since the auditor cannot audit a single monetary unit but only the line item to which the unit pertains, any error found is then prorated to the monetary units belonging to the line item. The prorated errors are called *taints* in auditing, and are denoted by $t_k = d_k/y_k$, where $d_k \neq 0$. An important case in auditing is when the errors in the population are all overstatement errors and the taints are restricted to positive values not exceeding 1.0. For this case, a conservative confidence bound for the total error amount can be constructed by assuming that all overstatement taints are at the maximum value of 1.0. The problem then becomes one of obtaining an upper confidence bound for the population proportion of monetary units that contain an error. Multiplying this bound by Y, the total number of monetary book units in the population, yields a conservative upper confidence bound for the total error amount E.

A difficulty with this conservative bound is that it is usually not tight enough. Stringer [9] developed a heuristic for reducing this bound when all observed taints are not at the maximum value of 1.0:

$$Yp_u(1-\alpha; n, m) - Y \sum_{k=1}^{m} [p_u(1-\alpha; n, k) - p_u(1-\alpha; n, k-1)](1-t_k),$$

where m is the observed number of errors in the sample of n monetary units, $t_1 \geqslant t_2 \geqslant \cdots \geqslant t_m$ are the observed taints, and $p_u(1-\alpha; n, m)$ is the $1-\alpha$ upper bound for a binomial proportion when m errors are observed in a sample of n.

Simulation studies (e.g., Reneau [7]) have shown that this Stringer bound has coverages always exceeding the nominal level and often close to 100%; also that the Stringer bound is not very tight and may involve substantial risks of making incorrect decisions when the bound is used for testing purposes. Leslie et al. [3] developed another heuristic bound, called a *cell bound*, that tends to be tighter than the Stringer bound and has good coverage characteristics for many accounting populations. The cell bound is based on cell sampling, where the frame of monetary units is divided into strata or cells of equal numbers of monetary units and one monetary unit is selected at random from each cell, the cell selections being made independently.

Fienberg et al. [2] developed a bound based on the multinomial distribution* by viewing monetary unit sampling in discretized form. The multinomial classes represent the different possible taints rounded to a specified degree. The procedure involves obtaining a joint confidence region for the multinomial parameters and then maximizing a linear function of the parameters, representing the total error amount in the population, over the confidence region. Because of the complexities of computation, Fienberg et al. utilized only a partial ordering of the sample outcomes for developing the joint confidence region. However, simulation studies (e.g., Plante et al. [6]) have shown that coverages for the multinomial bound are near to or above the nominal level for a variety of populations, particularly if cell sampling is employed.

Several approaches based on Bayesian* methodology have been proposed when monetary unit sampling is employed. Cox and Snell [1] proposed a Bayesian bound for which they assume that the population error rate for monetary units and the population mean taint for monetary units in error are independent parameters. These and other assumptions lead to a posterior distribution* of the total error amount that is a simple multiple of the F distribution*. Hence Bayesian bounds are very easy to obtain by this method. Neter and Godfrey [4] studied the behavior of the Cox and Snell bound in repeated sampling from a given population for sample size 100, and found that conservative prior parameter values exist so that the Cox and Snell bound has coverages near or above the Bayesian probability level for a wide range of populations. However, research in progress suggests that robustness may be a function of sample size, and not include all possible populations. Another recent proposal by Tsui et al. [11] is to combine the multinomial sampling model with a Dirichlet* prior distribution* to obtain Bayesian bounds for the total error amount.

ANALYTICAL REVIEW

Analytical review procedures are utilized by auditors to make internal and external consistency comparisons, such as comparing the current financial information with comparable information for preceding periods, or comparing operating data for the firm with data for the industry. Ratios have frequently been used in making these comparisons, such as comparing the ratio of cost of sales to sales for the current period with corresponding ratios for prior periods. Use of ratios for making comparisons assumes that the relationship between the two variables is linear through the origin. Often, this relationship may not be of this form. Also, use of ratio analysis does not permit the study of the simultaneous relationship of one variable with several others, such as when it is desired to examine the relationship of the revenue of a public utility to kilowatt hours, rate per kilowatt hour, and seasonal effects.

Regression analysis is now being used by some auditors for certain analytical review procedures. The data often are time series*, as when current periods are to be compared to prior periods. Sometimes, the data are cross-sectional, as when data for several hundred retail outlets of a firm are studied to identify ones that are outliers*. No special problems in applying regression analysis for analytical review have been encountered. The regression models employed have tended to be relatively simple ones. In using regression models, auditors are particularly concerned about identifying outliers that are worthy of further investigation. Much of the research to date has been concerned with developing rules that relate the identification of outliers with the amount of subsequent audit work that should be performed. Stringer and

Stewart [10] have provided a comprehensive summary of the use of regression methods in auditing for analytical review.

References

[1] Cox, D. R. and Snell, E. J. (1979). *Biometrika*, **66**, 125–132.

[2] Fienberg, S., Neter, J., and Leitch, R. A. (1977). *J. Amer. Statist. Ass.*, **72**, 295–302.

[3] Leslie, D. A., Teitlebaum, A. D., and Anderson, R. J. (1979). *Dollar-Unit Sampling*. Copp, Clark, Pitman, Toronto, Canada. (This book provides a comprehensive description of the Stringer and cell bounds and their uses in auditing.)

[4] Neter, J. and Godfrey, J. (1985). *J. R. Statist. Soc. C*, (*Appl. Statist.*), **34**, 157–168.

[5] Neter, J. and Loebbecke, J. K. (1975). *Behavior of Major Statistical Estimators in Sampling Accounting Populations*. American Institute of Certified Public Accountants, New York.

[6] Plante, R., Neter, J., and Leitch, R. A. (1985). *Auditing*, **5**, 40–56.

[7] Reneau, J. H. (1978). *Accounting Rev.*, **53**, 669–680.

[8] Roberts, D. M. (1978). *Statistical Auditing*. American Institute of Certified Public Accountants, New York.

[9] Stringer, K. W. (1963). *Proc. Bus. Econ. Statist. Sec., 1963*. American Statistical Association, pp. 405–411.

[10] Stringer, K. W. and Stewart, T. R. (1985). *Statistical Techniques for Analytical Review in Auditing*. Deloitte, Haskins, and Sells, New York.

[11] Tsui, K. W., Matsumura, E. M., and Tsui, K. L. (1985). *Accounting Rev.* **60**, 76–96.

(CONTROL CHARTS
FINANCE, STATISTICS IN
QUALITY CONTROL, STATISTICAL)

JOHN NETER

STATISTICS IN CRYSTALLOGRAPHY

The crystalline state in everyday experience is exemplified by substances such as salt, sugar, and precious stones such as diamonds and emeralds [1]. Crystals are generally bounded by flat surfaces that are an external manifestation of inner order usually consisting of translationally invariant unit cells. The unit cell is the smallest structure (not necessarily unique) containing all of the atomic species comprising the molecule. The science of crystallography attempts to deduce information related to the configuration of atoms within a unit cell from a knowledge of patterns of diffracted x-ray radiation that have been scattered from the crystal at different orientations [8, 20, 22, 24]. A typical pattern for one type of experiment is shown in Fig. 1. Most of our current knowledge of the configuration of proteins is derived from crystallographic analyses. Any molecules that can be crystallized can be analyzed by crystallographic techniques.

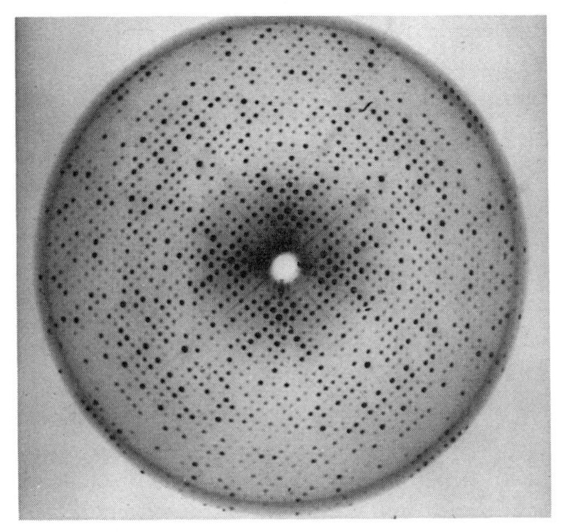

Figure 1 Typical scattering data found in an investigation of the structure of γ-chymotrypsin in ref. 3. The figure was kindly provided by Dr. G. H. Cohen. These data are typical of a large number of experiments, although other types of data may also be available.

Although it might appear that the analysis of crystallographic data should not pose any but the most superficial statistical problems, probability and statistics lie at the heart of most modern techniques of structure determination. This article contains a description of only the simplest of these, to illustrate the basic ideas. A number of crystallographic texts can be consulted for extensions and details [5, 6, 8, 20, 24]. The origin of the diffraction pattern shown in Fig. 1 can be understood by considering the

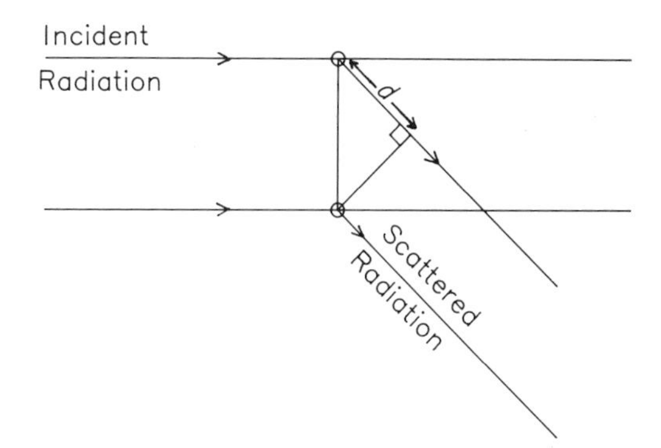

Figure 2 A schematic diagram of incident radiation scattered by a pair of atoms. The path difference to a detector located at a considerable distance from the atoms is denoted by d. Since the radiation can be regarded as a wave, d is the path difference between radiation scattered by the two atoms. Depending on the wavelength of the radiation this difference can cause a reinforcement or cancellation of the two waves from the atoms leading, in general, to a pattern exemplified by that in Fig. 1.

diagram of the scattering of radiation from two atoms as shown in Fig. 2. There is a difference in path length to the detector of the radiation scattered from the two atoms. Since radiation can be pictured as a wave, this difference d can correspond to a reinforcement or cancellation of the contributions from individual atoms. The general pattern of Fig. 1 is the resultant of scattering from all of the atoms. The orientation of the radiation with respect to the unit cell is generally expressed in terms of *Miller indices* $\mathbf{h} = (h, k, l)$, which are triads of integers. These specify the orientation of a plane with respect to the three axes of the unit cell. If the lengths of these axes are (a, b, c), then the plane identified by the set of Miller indices $\mathbf{h}$ cuts the unit cell at $(a/h, b/k, c/l)$.

Much information about molecular structure is available in terms of the space group to which the structure belongs [1, 24]. The space group specifies the kinds of internal symmetries satisfied by the atomic positions. An example is space group $P\bar{1}$, which indicates the presence of a center of symmetry such that for an atom located at $\mathbf{r}_j$ there is an identical atom at $-\mathbf{r}_j$. There are 230 distinct space groups [1, 7, 24]. Crystallographic techniques were the first available to identify the space group dependence of molecules. Since one can associate an intensity and phase with radiation scattered from an atom, one can represent the contribution from each atom as a complex number, which we write as $f_j \exp(2\pi i \mathbf{h} \cdot \mathbf{r}_j)$ where f_j is the scattering factor that is known once the chemical species is known. The $\mathbf{r}_j$ are generally unknown. The f_j can also be related to the electron density, and is calculated from quantum-mechanical models of the electron density. The scattering factors are tabulated for all atomic species [6]. Hence the total scattered radiation can be represented as a sum

$$F(\mathbf{h}) = A + iB = \sum_j f_j \exp(i\theta_j), \quad (1)$$

where the sum is over all atoms and $\theta_j = 2\pi \mathbf{h} \cdot \mathbf{r}_j$. The $F(\mathbf{h})$ are also Fourier components of the electron density of the molecule. Knowing the space group, parameters (and atomic positions) are estimated by least squares*.

The fundamental problem of crystallography is that only the intensity $|F(\mathbf{h})|^2$ is measurable while the phase must be inferred (*see* PHASE PROBLEM OF X-RAY CRYSTALLOGRAPHY, PROBABILISTIC METHODS IN) from the data. This inference is heavily dependent on the theory of random walks*. This can be established from a theorem by Weyl [21] that states that when x, y, z are rationally independent (there exist no integers m_1, m_2, m_3 such that $m_1x + m_2y + m_3z =$ integer) then the fractional parts of the set $hx + ky + lz$ are asymptotically distributed modulus 1 uniformly as h, k, and l range independently over the integers. Thus if all of the atoms occupy general positions in the unit cell (in some cases atoms can occupy special positions that have rational coordinates [6]) and intensities are measured at a large number of values of $\mathbf{h}$, the angles θ_j in (1) can be assumed to be uniformly distributed in $(0, 2\pi)$. Hence $F(\mathbf{h})$ can be regarded as the vector resultant of a two-dimensional random walk* of a type whose study was first suggested by Karl Pearson* [9, 11].

To see how this can be used to obtain structural information consider the problem of distinguishing between crystals belonging to space groups $P\bar{1}$ (center of symmetry) and those belonging to P1 (absence of a center of symmetry) using intensity statistics [20, 22]. Since, in $P\bar{1}$, whenever a θ_j appears in (1), a value $-\theta_j$ must also occur by the assumed symmetry, it follows that $B = 0$ so that $F(\mathbf{h})$ can be represented as a symmetric random walk in one dimension. If there are sufficient number of atoms in the unit cell so that the central limit theorem* can be invoked, one can conclude that the probability density function (PDF) of

$$E(\mathbf{h}) = F(\mathbf{h}) \Big/ \left[E\{|F^2(\mathbf{h})|\}^{1/2} \right]$$

in space group $P\bar{1}$ is

$$p(E) = (2/\pi)^{1/2} \exp(-E^2/2). \quad (2)$$

Notice that $E(\mathbf{h})$ is standard crystallographic notation for the normalized structure factor and should not be confused with an expectation operator. In the absence of centrosymmetry, the parameter B in (1) is not identically zero, so that $F^2 = A^2 + B^2$. Again, assuming that there are a sufficient number of atoms to justify use of the normal approximation, one finds

$$p(|E|) = 2|E| \exp(-E^2). \quad (3)$$

The two PDFs in (2) and (3) are qualitatively different, and there are usually enough measured values of the $|F(\mathbf{h})|$ for different $\mathbf{h}$ to allow one to distinguish between the presence or absence of a center of symmetry. This is the simplest distinction that can be made using intensity statistics. Many more complicated extensions have been analyzed in the crystallographic literature. For example, there may be one or more outstandingly large values of f_j in (1), in which case the resulting terms must be handled separately from those whose PDF is approximated by the normal [19, 20]. There may also be auxiliary symmetries. For example, if in addition to the center of symmetry defined earlier there is a center of symmetry at $\mathbf{d}$, so that an atom at $\mathbf{r}_j$ generates three other symmetrically situated atoms, PDF's different from those in (2) and (3) can arise [12, 17]. Different space groups give rise to different PDFs of intensity [20].

While intensity statistics were extensively used in the early 1950s, direct computer methods for phase determination are used in modern crystallography. Although these were originally suggested in the 1950s, their use generally requires a considerable amount of computing. Direct methods attempt to infer information about phases from measured intensities [3, 6, 10]. These generally require the calculation and use of multivariate PDFs.

For example, one of the earliest techniques analyzed allows one to calculate the probability that the phase of $E(2\mathbf{h})$ is positive, given measured values of $|E(\mathbf{h})|$ and $|E(2\mathbf{h})|$ [3]. In space group $P\bar{1}$ one can use a normal approximation to derive the much used formula

$$p_+ = \frac{1}{2}\left[1 + \tanh\left\{\frac{\sigma_3}{2\sigma_2^{3/2}}|E(2\mathbf{h})|(E^2(\mathbf{h}) - 1)\right\}\right], \quad (4)$$

in which the σ_m are defined by

$$\sigma_m = \sum_j f_j^m. \quad (5)$$

Equation (4) is an approximation whose range of validity has only recently been subjected to scrutiny [18]. Equation (4) is generally used to make a tentative identification of phase, provided that $p_+ > 0.95$. Many variations of direct methods are discussed in greater detail in ref. 4 and in PHASE PROBLEM IN X-RAY CRYSTALLOGRAPHY, PROBABILISTIC METHODS IN.

All of the methods discussed so far depend on the central limit theorem*, possibly corrected by Edgeworth or Gram–Charlier series (*see* CORNISH–FISHER AND EDGEWORTH EXPANSIONS) for heterogeneous f_j's [6, 10, 14, 15]. Some recent work has concentrated on obtaining very accurate results by expressing the relevant PDFs in terms of characteristic functions* [9, 16–18]. Since maximum intensities are bounded, the Fourier integrals can be expressed as Fourier series whose coefficients are just the characteristic functions* evaluated at appropriate arguments. These calculations circumvent difficulties associated with atomic heterogeneities and may be used in place of Edgeworth expansions. An alternative approach to the derivation of crystallographic PDFs is based on the maximum entropy* method [2], which is currently enjoying some popularity in crystallography as well as in other fields of spectroscopy. Finally, although the analysis of crystallographic data presupposes periodic repetitions of a unit cell, recent experiments have raised the possibility of nonperiodic space filling structures that may represent a new state of matter [13]. This possibility will inevitably raise new statistical problems.

References

[1] Bragg, L. (1955). *The Crystalline State. A General Survey*. Bell, London, England. (In spite of its age, this is a still valuable account of the characterization of crystals.)

[2] Bricogne, G. (1984). *Acta Crystall. A*, **40**, 410–445. (An exposition of the maximum entropy method as used for deriving approximations to PDFs.)

[3] Cochran, W. and Woolfson, M. M. (1955). *Acta Crystall.*, **8**, 1–12.

[4] Cohen, G. H., Silverton, J. E., and Davies, D. R. (1981). *J. Mol. Biol.*, **148**, 449–479.

[5] Giacovazzo, C. (1980). *Direct Methods in Crystallography*. Academic, New York. (A fairly exhaustive monograph on the very commonly used direct methods. This is mainly a theoretical treatise.)

[6] Hauptman, H. and Karle, J. (1953). *Solution of the Phase Problem I. The Centrosymmetric Crystal.* Polycrystal Book Service, Pittsburgh, PA. (One of the earliest expositions of direct methods. Since this monograph was written before computers were generally available, direct methods were considered of theoretical interest only.)

[7] Henry, N. F. M. and Lonsdale, K., eds. (1965). *International Tables for X-ray Crystallography*, Vols. I–IV. (A complete listing of properties of all 230 symmetry groups in addition to a discussion of the fundamentals of crystalline symmetry. Further tables of parameters needed for structure determination.)

[8] James, R. W. (1958). *The Optical Principles of the Diffraction of X-rays*. Bell, London, England. (An account of the physical principles underlying crystallography.)

[9] Kiefer, J. E. and Weiss, G. H. (1984). In *AIP Proceedings 109, Random Walks and Their Applications in the Physical and Biological Sciences*. American Institute of Physics, New York, pp. 11–32 (A review of many applications and approximations of the Pearson random walk.)

[10] Klug, A. (1958). *Acta Crystall.*, **11**, 515–543. (A clear exposition of central limit theorem approximations for joint PDFs of structure factors.)

[11] Pearson, K. (1905). *Nature*, **72**, 294.

[12] Rogers, D. and Wilson, A. J. C. (1953). *Acta Crystall.*, **6**, 439–449.

[13] Schechtman, D., Blech, I., Gratias, D., and Cahn, J. W. (1984). *Phys. Rev. Lett.*, **53**, 1951–1956.

[14] Shmueli, U. (1979). *Acta Crystall. A*, **35**, 282–286.

[15] Shmueli, U. and Wilson, A. J. C. (1983). *Acta Crystall. A*, **39**, 225–233.

[16] Shmueli, U., Weiss, G. H., Kiefer, J. E. and Wilson, A. J. C. (1984). *Acta Crystall. A*, **40**, 651–660.

[17] Shmueli, U., Weiss, G. H., and Kiefer, J. E. (1985). *Acta Crystall. A*, **41**, 55–59.

[18] Shmueli, U. and Weiss, G. H. (1985). *Acta Crystall. A*, **41**, 401–408.

[19] Sim, G. A. (1959). *Acta Crystall.*, **12**, 813–814.

[20] Srinivisan, R. and Parthasarathy, S. (1976). *Some Statistical Applications in X-ray Crystallography*. Pergamon Press, London, England, (A comprehensive discussion of techniques that use intensity statistics.)

[21] Weyl, H. (1916). *Math. Annalen*, **77**, 313–352.

[22] Wilson, A. J. C. (1949). *Acta Crystall.*, **2**, 318–323. (The earliest discussion of intensity statistics based on the normal approximation.)

[23] Wilson, A. J. C. (1980). *Acta Crystall. A*, **36**, 945–946.

[24] Woolfson, M. M. (1970). *An Introduction to X-ray Crystallography*. Cambridge University Press, Cambridge, England. (An elementary introduction to the geometric characterization of crystals together with an elementary account of crystallographic techniques.)

(PHASE PROBLEM OF X-RAY CRYSTALLOGRAPHY, PROBABILISTIC METHODS IN
RANDOM WALKS)

GEORGE H. WEISS

STATISTICS IN DENTISTRY

In order to give their patients the best available treatment, dentists need to be able to interpret the statistics that are used in evaluating new and possibly improved treatments appearing in dental journals. About one-fifth of such articles in the 1981 issues of the *Journal of the American Dental Association*, which has the widest circulation of any dental journal, used inferential statistics to test hypotheses about their assertions. The *Journal of Dental Research* has a smaller circulation, but its research orientation may make it even more important to a dentist searching for improved treatments; approximately half of the articles published in 1981 used inferential statistics to substantiate their analysis.

Most of the statistical methods used by these and other journals can be included in a fairly short list: paired and unpaired t tests; one-way analysis of variance*; correlation* and regression*; chi-square*. Other statistical methods, such as multiple range* tests, two-way analysis of variance, and nonparametric analyses (other than chi-square) appear occasionally in the literature. We have listed references to articles containing such infrequently used statistics at the beginning of our bibliography.

Beertsen and Everts used the paired t test [1]. It revealed that freezing the periodontal ligament of one central incisor decreased that ligament's fibroblastic nuclei compared to the ligament of the paired incisor on the other side of the jaw. Although before and after comparisons make use of the paired t test in dentistry as well as in other disciplines, pairing opposite sides of the mandible makes the paired t test particularly suitable to many dental experiments.

Dentists, like other biological scientists, use the unpaired t test to compare a difference between the means of two independent groups. For example, Ranney et al. used that test to show that subjects with severe periodontal disease had a significantly different mean level of IgG immunoglobin in their saliva than normal subjects [4].

Greenberg et al. used one-way analysis of variance to compare the mean upper body strengths of groups with different mandibular positions [3]. The F statistic obtained from that experiment failed to show that the groups differed significantly. Although such a negative result is not conclusive (perhaps a larger number of subjects could reveal some significant difference among groups), their article did show that their experiment did not support any relationships between mandibular position and upper body strength.

Linear regression* predicts a dependent variable's value from the values of an independent variable. Correlation then estimates the strength of association* between those two variables. These two methods are commonly applied to predict caries activity from a variable suspected of influencing the incidence (e.g., fluoride concentration in water and lacobacillus count on tooth surfaces) and to predict behavior from an attitudinal survey. An example of the former is Crossner's finding that regression can predict caries activity from salivary lactobacillus counts, and that the correlation is significant [2].

The *chi-square* method is commonly used by dentists when they compare the effects of different kinds of injections, commonly classified as satisfactory or unsatisfactory, or into less than four categories of satisfaction. They then count the number in each category for each kind of injection. A typical example is Walton and Abbott's demonstration that back pressure significantly increased the proportion of satisfactory anesthesia in periodontal ligament injections [5].

Before 1980, dentists needed to understand a fairly restricted list of statistical analyses in order to evaluate evidence for better treatments. That limited list does not now suffice for all articles. The Duncan multiple range test to determine pairwise differences in means for more than two groups, replacing the t test as it should, appears in almost every recent issue of the *Journal of Dental Research*. Nonparametric statistics like the Mann–Whitney* U, Wilcoxon's signed rank, and Kendall's correlation coefficient are beginning to appear more often in dental journals.

As dental researchers continue to expand their ability to analyze data and refine their ability to choose the correct tests, the clinicians will have to improve their ability to understand the increasingly complex statistics that appear in those articles. Only by so doing can they hope to keep abreast of current knowledge in the field.

References

[1] Beertsen, W. and Everts, V. (1981). *J. Periodontal Res.*, **16**, 524–541.

[2] Crossner C.-G. (1981). *Community Dentistry and Oral Epidemiol.*, **9**, 182–189.

[3] Greenberg, M. S., Cohen, S. G., Springer, P., Kotwick, J. E., and Vegso, J. J. (1981). *J. Amer. Dental Ass.*, **103**, 576–579.

[4] Ranney, R. R., Ruddy, S., Tew, J. G., Welshimer, H. J., Palcanis, K. G., and Segreti, A. (1981). *J. Periodontal Res.*, **16**, 390–402.

[5] Walton, R. E. and Abbott, B. J. (1981). *J. Amer. Dental Ass.*, **103**, 571–575.

Bibliography

Examples of some infrequently used statistical methods in the dental journals.

Duncan's Multiple Range Test

Re, G. J. and Norling, B. K. (1981). *J. Dental Res.*, **60**, 805–808.

Kendall's Partial Correlation

Demirjian, A. and Levesque, G.-Y. (1980). *J. Dental Res.*, **59**, 1110–1122.

Mann–Whitney U-Test

Korberly, B. H., Shreiber, G. F., Kilkuts, A., Orkland, R. K., and Segal, H. (1980). *J. Amer. Dental Ass.*, **100**, 39–42.

Student–Newman–Keuls Comparison

Re, G. J., Draheim, R. N. and Norling, B. K. (1981). *J. Amer. Dental Ass.*, **103**, 580–583.

Two-Way Analysis of Variance

Going, R. E., Hsu, S. C., Pollack, R. L., and Haugh, L. D. (1980). *J. Amer. Dental Ass.*, **100**, 27–33.

Wilcoxon's test

Ooshima, T., Sobue, S., Hamada, S., and Kotani, S. (1981). *J. Dental Res.*, **60**, 855–859.

Textbooks on dental statistics:

Chilton, N. W. (1982). *Design and Analysis in Dental and Oral Research*, 2nd ed. Praeger, New York. (This is the most comprehensive text written on design and analysis in dental research.)

Darby, M. L. and Boven, D. M. (1980). *Research Methods for Oral Health Professionals—An Introduction*. Mosby, St. Louis, MO. (This book has one chapter on statistics. Each statistical test is described in half to one page and its use and limitations are explained in simple terms.)

Miller, S. L. (1981). *Introductory Statistis for Dentistry and Medicine*. Reston, Reston, VA. (This text contains experimental designs and uses dental and medical examples in its examples and exercises. Some dental articles are reprinted in this book.)

Weinberg, R. and Cheuk, S. L. (1980). *Introduction to Dental Statistics*. Noyes Medical Publications, Park Ridge, NJ. (This book is written like an arithmetic text with step by step calculation on each statistical method followed by exercises of more dental examples.)

Dental articles with a fine reference source:

Canonical and Cluster Analysis

Roth, H. D. and Koritzer, R. (1972). *Georgetown Dental J*, **38**, 27–30.

Ridit Analysis

Fleiss, J. L., Chilton, N. W., and Wallenstein, S. (1979). *J. Dental Res.*, **58**, 2080–2084.

Sequential Analysis

Smith, V. O. and O'Mullane, D. M. (1977). *J. Dental Res. Special Issue C*, **56**, 112–115.

Statistics Books in Medical, Social, and Behavioral Sciences

Wagner, A. G. (1973). *J. Prosthetic Dentistry*, **30**, 446–453.

(BIOSTATISTICS
CLINICAL TRIALS
MEDICAL DIAGNOSIS, STATISTICS IN)

SHU L. CHEUK
ROGER WEINBERG

STATISTICS IN FORESTRY

ORIGIN AND HISTORY

Statistics in forestry has its origin with the need to measure the forest estate, not just in terms of the area occupied by forest, but also the volume of utilizable lumber contained therein and projections of what this volume would be in future years, particularly in response to alternative management practices. While the censusing of wooded lands has, no doubt, been practised for several centuries, forest management as a science appears to have emerged around the middle of the nineteenth century, particularly in Central Europe.

In the assessment of the forest estate there are essentially two components: first, *mensuration*, the determination of the volume of wood from readily measurable characteristics such as the tree diameter at an accessible height, bark thickness, and total height; second, the *method of sampling* individual trees or plots of trees as representative of the forest of interest. In earlier days, tree volume was usually related to diameter, or more particularly to cross sectional (basal) area at breast height (i.e., 4.5 foot or 1.3 meter above ground level) by harmonizing a set of freehand curves from various tree heights, the data being obtained by climbing or felling, and sectionally measuring individual trees. Volume projection or yield was likewise obtained by harmonizing curves visually fitted to volume–age data, sometimes from permanent plots so that the growth of individual trees could be followed, sometimes from trees of different ages measured at a single point in time, the two approaches in general not being equivalent. Sampling was usually by means of plots of the order of 1/10 to 1/4 acre, rarely selected at random and not infrequently laid out systematically to facilitate mapping.

It was not until the mid 1930s that statistical methods noticeably became incorporated into these operations, the basic tools being least-squares* fits of multiple and polynomial regressions*, simple random sampling*, the construction of confidence intervals*, and, to some extent, the analysis of variance*. Perhaps the most influential North American pioneer in the application of statistical methods to forestry was F. X. Schumacher of Duke University, who received much of his inspiration from his contact with Sir Ronald Fisher*. The text "Sampling Methods in Forestry and Range Management" by Schumacher and Chapman [20], while unquestionably dated, makes a number of points that are regrettably too often overlooked in more modern texts. Other pioneering forest statisticians in the United States were E. Osbourne, who succeeded Schumacher at Duke, T. Evans of the U. S. Forest Service, and C. A. Bickford, who concluded his career at the State University of New York, Syracuse.

Development was not restricted to North America; prominent early workers in the field included M. Prodan (Germany), L. Strand (Norway), B. Matérn (Sweden), J. N. R. Jeffers (United Kingdom), P. Schmid-Haas (Switzerland), P. Arbonnier, R. Tomassone (France), V. J. Chacko (India), G. MacIntyre (Australia), and K. Kinashi (Japan). (Kinashi spent a year with Schumacher at Duke shortly after World War II.) Some of these people were primarily statisticians who developed an interest in forestry problems; others were foresters who later acquired a facility in statistics. (One even claimed that he went into forestry to avoid mathematics.)

Recognition that the role that statistics could play in forestry was not confined to inventory and mensurational problems came about slowly as reflected in the history of the Advisory Group of Forest Statisticians of the International Union of Forest Research Organizations (IUFRO). This group had its inaugural meeting at the British Forestry Commission's Research Station at Alice Holt Lodge in 1963, and was formalized within IUFRO as a component of Section 25, Study of Growth and Yield and Forest Management. In 1971, pressure to have the broader function of the group acknowledged resulted in the formation within Division 6, General Subjects, of a separate, distinct subject group, S6.02, Statistical Methods, Mathematics and Computers, with a mandate to interact with subject groups in any division, and to promote the proper use of statistical and mathematical methods and computers wherever it appeared relevant and advantageous to do so.

Leaders of the group have been, in order, B. Matérn (Sweden), J. N. R. Jeffers (United Kingdom), W. G. Warren (Canada), and currently G. Z. Gertner (U.S.A.). The group participates in the IUFRO World Congresses, held at approximately 5-year intervals, and joins with other subject groups in jointly sponsoring meetings. Its newsletter was first distributed in 1968 to approximately 150 individuals; the current mailing list numbers close to 700 in more than 50 countries.

CONTRIBUTIONS

Sampling

While it may not be possible to attribute the derivation of statistical theory in itself to the community of forest statisticians, they have been in the forefront of the operational development of certain statistical methods. The relascope, introduced by Bitterlich [2], and its derivatives, in particular the use of the optical wedge or prism sampling, are specializations of frameless sampling with probability proportional to size*. The Bitterlich principle revolutionized forest inventory since the fundamental measurement of stand basal area could be estimated in a 360° sweep about sample points simply by applying a constant factor to the count of trees that appear to have their diameters greater than a specified critical value. Related to this is line intersect sampling*, introduced by Warren and Olsen [27] for the assessment of logging residue. Since the volume of such residue is relatively small, but the amount contained in conventional plots is highly variable and time consuming to find and measure, a more economical method of sampling was required. Line intersect sampling, which can be looked on as an extension of Buffon's needle* problem, provided this and is again a variant on sampling with probability proportional to size without a frame. DeVries [5] gives a detailed account of line intersect sampling and its relationship to other methods.

Transect* sampling has the same attraction as line intersect sampling in that observations are made continually as one traverses the area of interest, in contrast to plot methods, where much time and effort can be expended in travelling from plot to plot, especially in rugged and/or densely forested regions. While users of transect sampling have sometimes overlooked the spatial correlation that undoubtedly exists, other workers have attempted to exploit the characteristics of this approach, for example the run lengths of diseased and healthy trees along narrow belt transects in estimating, not just the proportion of diseased trees, but also the dispersal of disease in terms of the number and size of diseased patches and the intensity of disease within the patch, since the disease is often truly contagious (Pielou [16, 17]; see also Vithayasai [24], Chacko and Negi [4], and Warren [26]). Accordingly some forest statisticians have shown interest in methods for what can be called "areal inference from linear samples." Others have found use for nearest neighbor* distances (e.g., Batcheler and Hodder [1]).

Related to sampling with probability proportional to size, and with an impact in forestry that once seemed to rival that of Bitterlich sampling, is 3P sampling or sampling with probability proportional to prediction, suggested by Grosenbaugh [9]. 3P sampling is in fact a specialization of Poisson sampling (Hájek [10]). While remarkably efficient for certain purposes, its reputation suffered through its indiscriminate use, particularly in situations where it could have no advantage over, and would likely be inferior to, alternative more standard techniques. It can, however, be used effectively as one stage in a process that also involves the more conventional methods.

Notwithstanding this interest in sophisticated techniques, forest inventory and the measurement of growth and yield have been and remain based very much on conventional sample plots. These can be placed in one of two categories, temporary and permanent. The establishment of permanent plots is relatively expensive: since they may have to last for several decades, their boundaries have to be well demarcated and the individual trees identified. The cost factor, in general, limits their use for inventory purposes. On the other hand, growth or increment can be estimated much more precisely from permanent plots. These considerations led Ware and Cunia [25] into detailing the theory of sampling with partial replacement as applied to forestry. While, in general, on the first sampling occasion most plots are temporary, some plots are designated to be maintained until the second occasion at which time further temporary plots are added. This process may, of course, be repeated indefinitely. Ware and Cunia examined the optimum balance of temporary and permanent plots in relation to the costs, the required precision and the variability commonly met with in forest measurement.

Other sampling concepts that have been carried across to forestry include: multiphase sampling*, particularly with respect to LANDSAT imagery, high- and low-level aerial photography, and ground control (e.g., Frayer [7]), which requires the development of explicit formulae for the standard errors of the estimated means and totals; empirical Bayes* estimation (e.g., Ek and Issos [6], Burk and Ek [3]), since one often has available the results of previous inventories of forests of similar age and composition; and, recently, an interest in model-based (purposive) sampling (Wood et al. [29]).

Mensuration, Growth and Yield

While a proportion of the wood extracted from a forest goes for pulp (and thence paper) or fuel, much is still in the form of sawlogs for use in construction. Thus it is not simply the volume that is of concern but the disposition of that volume in terms of the log-size distribution, i.e., the combination of diameter and length, which can be derived, of course, from the tree profiles once a rule for choosing log lengths has been given. Tree profiles are described by "taper" equations that express the diameter at any specified height as a function, principally, of the diameter at breast height and total height. The volume up to a specified height is also commonly expressed as a function of the breast-height diameter and the total height. This volume should also be obtainable as an integral of the square of the taper function, but the value so computed differs in general from that obtained directly from a volume equation. This incompatibility stems largely from a reliance on ordinary least-squares methods of fitting and the failure to recognize that if the conditions for best linear unbiased estimation hold for the taper equation they cannot hold for the volume equation, and vice versa. Only recently have there been attempts at developing compatible systems of taper and volume equations that come to grips with the real problems, which include correlated errors due to a lack of independence of measurements made on a single tree, whereas measurements made on different trees are clearly independent.

A natural extension brought about by the computer age is the simulation* of forest growth. At opposite ends of the spectrum are stand models and individual-tree models. The former simply project the total volume or basal area of the stand as a whole; the latter build these up as the sum of individually "grown" trees, taking into account spatial relationships and individual tree mortality due to competition (or other causes) and, in its most sophisticated form, incorporate stochastic or random elements. The principal objective of such simulations is to evaluate the effect of various management policies. Clearly the individual-tree models contain much more information concerning the resultant "crop," in particular the size class distribution of the "logs" that, of course, affects the dollar value and thus determines the economics of the chosen regime. Individual-tree model simulations are, clearly, much more demanding of computer storage and execution time.

There has therefore been much interest in stand models from which individual-tree information, in the sense of the distribution of tree diameter, for example, can be inferred. The parameters of such distributions are related to such factors as age, total basal area, and site index. This presupposes knowledge of the appropriate distributional form. While early work was based on the normal distribution, interest is now focussed more on the Weibull* and Johnson* S_B distributions. Although no biophysical model on which to base a preference for any one form has been suggested to date, such as weakest-link theory in the case of the Weibull, there are, however, bivariate and trivariate extensions to the S_B distribution that give acceptable fits to diameter, height, and volume data jointly (Schreuder and Hafley [18], Schreuder et al. [19]). The Weibull appears to permit no such satisfactory extension (Hafley and Schreuder [10]).

Recent attempts at breaking away from the conventional distribution fitting and regression approach are in favor of describing the stand by a system of stochastic differential equations* (e.g., Garcia [8], Sloboda [21]) with promising results.

Silviculture, Forest Establishment and Protection, and Related Areas

It is impossible to give a comprehensive account of all the uses of statistics in forestry; an attempt is made, therefore, to highlight some of the more distinctive applications.

With respect to experimental design, the best maxim appears to be "keep it simple." Since the time taken for a forest to reach maturity is often of the order of 70–80 years or more, and rarely less than 25–30 years, forestry experiments are commonly long term, of necessity, with a very good chance therefore that some experimental units will experience extraneous damage due to such things as animal browsing, insect infestations, fire, snow breakage, or even unauthorized felling for firewood or Christmas trees. Accordingly, for field experiments, it would seem wise to employ designs no more complex than a randomized block. For example, the almost inevitable loss of one or more plots that would occur under actual forest conditions would make impossible the analysis, as planned, of a fractional factorial design*. Of course, there is scope and indeed necessity for the more complex designs in, say, forest nurseries, and within the laboratory, where conditions can be monitored and controlled.

In field experiments a vexing question has long been the choice of plot size. By virtue of the size of trees, experimental plots of but a few trees are necessarily of greater area than the plots customarily used for agricultural crops. Accordingly there is far more difficulty in maintaining reasonable within-block homogeneity of natural growing conditions. This has led some workers to suggest the use of single-tree plots. While from the purely design point of view this may appear to have an advantage, with the possible exception of the first few years a major factor in a tree's growth is the competition from its neighbors. In practice, a treatment, say fertilization, would be applied to all the trees in a stand; thus trees would be competing against similarly treated trees. The response may well differ then from that indicated by single-tree plot experiments where the trees would be competing against differently treated neighbors. A common practice therefore has been to make plots as small as possible, perhaps 16–25 trees, but with a buffer region of 1–2 rows of similarly treated individuals.

An additional problem arises in the case of provenance trials, where seedlings obtained from a range of source locations of a single tree species are being tested for performance over a range of sites where that species does not occur naturally, e.g., seed collected in Europe to aid reforestation somewhere in the Southern Hemisphere. Conventionally randomized block designs have been used, but often the experimental area available at one site is insufficient to accommodate all provenances and/or there are insufficient seedlings of some provenances to permit their being represented at all sites of interest. Such trials are then not simply unbalanced in design but also involve missing cells, and since the possibility of site by provenance interaction cannot be dismissed, their interpretation by standard analysis of variance procedures is by no means easy.

In other respects statistics in forest genetics* and tree breeding parallel the work elsewhere, e.g., the estimation of hereditability and genetic gain. One interesting problem lies in the design of seed orchards, where a spatial arrangement is desired such that the probability of all possible crosses of the clones employed is constant. Langner and Stern [14] also consider the design problem of combining a clonal trial with a seed orchard.

Strangely, multivariate methods have not been extensively applied. Their primary use to date has been in the ordination- and classification-type problems of forest ecology. Recently, however, some forest statisticians have attempted to stimulate the use of multivariate analysis of variance*, etc., as the appropriate tool for interpreting forestry experiments (e.g., van Laar [23]). There has been, perhaps, even less utilization of stochastic processes* per se, although Suzuki [22] has treated forest succession as such.

A natural use of sequential methods arises in the assessment of the incidence of forest disease or insect infestations that, at times, have the potential to cause significant or even total tree mortality. Morris [15] and Waters [28] saw sequential sampling* as a means of determining whether insect infestations had attained levels that would necessitate control, usually by chemical spraying, and of monitoring the success of such treatment.

Forest Products

Since it deals with the end product, forest products research should not be divorced from forestry; however, it introduces some statistical problems of a different kind. Whereas in conventional forestry the interest is usually in averages or totals, the integrity of a wooden structure is not dependent on the member of average strength but on the weakest element, particularly if placed in a critical position. Interest is thus in the statistics of (low) extremes and, because lumber strength can be accurately determined only from destructive testing, inferences must be based on samples of relatively small size. For many years estimation of low-order fractiles was based on the testing of small dimension specimens of clear wood (often with many specimens cut from a single tree but relatively few trees actually sampled), and on the application of normal distribution theory coupled with empirical correction factors. While it was recognized that the strength of structural lumber is not governed by the strength of whatever clear wood it may contain, but on the incidence, magnitude, and location (center or edge) of knots and deviations in the grain angle, there was until recently little extensive testing of lumber as it would be used.

While the normality assumption may be adequate for the distribution of clear-wood strength it is in general not viable for that of structural lumber, which commonly although not universally exhibits a degree of positive skewness. Various distributional forms have been assumed including the lognormal*, which, however, tends to have greater skewness than indicated by modulus-of-rupture data, and the three-parameter Weibull. The latter can be justified through weakest-link theory (which also explains the size effect, i.e., the reduction in strength with increase in board size) and appears to be negligibly inferior in tracking data to the more flexible, but somewhat arbitrary, Pearson* type III and Johnson S_B distributions*. That the Weibull should have a nonzero location parameter, i.e., a strictly positive minimum strength, seems reasonable on the grounds that lumber of zero or nearly zero strength, even if it existed, would never survive long enough to be used or tested. Thus statisticians involved with forest products research have been faced with the problems of small sample inference of low-order fractiles from a nonnormal universe (see, e.g., Johnson [12], Johnson and Haskell [13]).

One should not give the impression that forest products are restricted to wood engineering. Other areas include wood preservation, with the need to determine the minimum effective level of the preservative, wood identification (i.e., classification problems), wood chemistry, and wood machining, the last of which is a fruitful area for the application of response-surface* designs. One challenging problem is in the prediction of the properties of a wood product, e.g., the bursting strength of paper, from measures of wood morphology such as cell wall thickness and fiber length, and the determination of how such properties can be influenced by silvicultural practices.

THE CHALLENGE OF THE FUTURE

Tree growth is the product of a multitude of environmental factors. In many tree species the year-by-year record of growth is embodied in the sequence of ring widths (and, perhaps, densities) that can be measured on small cylindrical cores drawn from representative trees. Superimposed upon a natural trend, rings often increasing in width until a maximum is reached and then decreasing, are fluctuations due to climatic variation (temperature and precipitation patterns), depressions due to disease or insect infestations, accelerated growth due to the death or physical removal of a neighboring tree (thereby reducing competition and/or increasing available growing space), and, of current concern, the effects of atmospheric deposition (acid rain). How to quantify the effect of these various factors on tree growth and, in particular, to determine precisely the effect on the forest of industrial pollution is, possibly, the most challenging problem facing forest statisticians in the 1980s.

All in all, forestry is a fruitful area for the application of statistics. Some three decades ago, the director of a forestry-related research establishment remarked "We have no use for statistics; our data are too variable." In the intervening years, forestry populations have become no less variable; fortunately there has been a growing appreciation of the role that statistics can and should be playing.

References

[1] Batcheler, C. L. and Hodder, R. A. C. (1975). *N. Z. J. Forestry Sci.*, **5**, 3–17.

[2] Bitterlich, W. (1948). *Allg. Forst. Holzwirtsch. Zeit.*, **59**, 4–5.

[3] Burk, T. E. and Ek, A. R. (1982). *Forest Sci.*, **28**, 753–771.

[4] Chacko, V. J. and Negi, G. S. (1965). *Proc. Second Conf. Advisory Group of Forest Statisticians, IUFRO. Rapp. Uppsats. Inst. Skoglig. Mat. Statist.*, Vol. 9. Skoghogskolan, Stockholm, Sweden, pp. 45–66.

[5] DeVries, P. G. (1978). *Statistical Ecology: Sampling Biological Populations*, Vol. 5, R. C. Cormack, G. P. Patil, and D. S. Robson, eds. International Co-operative Publishing House, Fairland MD, pp. 1–70.

[6] Ek, A. R. and Issos, J. N. (1978). *Proc. Subject Groups S4.02 and S4.04, IUFRO*. Institutul de Cercetari si Amenajari Silvice, Bucharest, Romania, pp. 34–35.

[7] Frayer, W. E. (1981). *Proc. XVII World Congress, IUFRO, Kyoto, Japan*, Interdivisional, pp. 172–176.

[8] Garcia, O. (1983). *Biometrics*, **39**, 1059–1072.

[9] Grosenbaugh, L. R. (1964). *Proc. Soc. Amer. Forestry*. Boston, MA, pp. 36–42.

[10] Hafley, W. L. and Schreuder, H. T. (1976). *Proc. XVI World Congress, IUFRO, Oslo, Norway*, Division 6, pp. 104–114.

[11] Hájek, J. (1957). *Bull. Int. Statist. Inst.*, **36**, 127–133.

[12] Johnson, R. A. (1980). *Forestry Prod. J.*, **30**, 14–22.

[13] Johnson, R. A. and Haskell, J. H. (1983). *Canad. J. Statist.*, **11**, 155–167.

[14] Langner, W. and Stern, K. (1955). *Zeit. Forstgenet.*, **4**, 81–88.

[15] Morris, R. F. (1954). *Canad. J. Zool.*, **32**, 302–313.

[16] Pielou, E. C. (1963). *Biometrics*, **19**, 603–614.

[17] Pielou, E. C. (1965). *Forest Sci.*, **11**, 18–26.

[18] Schreuder, H. T. and Hafley, W. L. (1977). *Biometrics*, **33**, 471–478.

[19] Schreuder, H. T., Bhattacharyya, H. T., and McClure, J. P. (1982). *Biometrics*, **38**, 137–142.

[20] Schumacher, F. X. and Chapman, R. A. (1942). Sampling Methods in Forestry and Range Management. *Bulletin No. 7*, Duke Univ. School of Forestry, Durham, NC.

[21] Sloboda, B. (1978). *Proc. Fifth Conference Subject Group S6.02, IUFRO. Mitt. Forstl. Versuchs Forstungsanstalt (Baden-Wurtemburg)*, **91**, 133–146.

[22] Suzuki, T. (1971). *Mitt. Forstl. Bundes Versuchsanst (Wien)*, **91**, 69–88.

[23] van Laar, A. (1986). *Proc. XVIII World Congress, IUFRO, Ljubljana, Yugoslavia*, Division 6, in press.

[24] Vithayasai, C. (1972). *Forest Sci.*, **18**, 87–92.

[25] Ware, K. D. and Cunia, T. (1962). *Forest Sci. Monograph No. 3.*

[26] Warren, W. G. (1970). *Proc. Third Conf., Advisory Group of Forest Statisticians, IUFRO. Publ. No. 72-2*, Institute National de Recherche Agronomique, Paris, France, pp. 261–275.

[27] Warren, W. G. and Olsen, P. F. (1964). *Forest Sci.*, **10**, 267–276.

[28] Waters, W. E. (1955). *Forest Sci.*, **1**, 68–79.

[29] Wood, G. B., Schreuder, H. T., and Brink, G. E. (1985). *Canad. J. Forestry Res.*, **15**, 83–86.

(AGRICULTURE, STATISTICS IN
ECOLOGICAL STATISTICS
GEOGRAPHY, STATISTICS IN
LINE INTERCEPT SAMPLING
LINE INTERSECT SAMPLING
TRANSECT METHODS)

WILLIAM G. WARREN

STATISTICS IN HISTORICAL STUDIES

Statistics are, on the one hand, and statistics is, on the other hand, important to historical studies. In the plural, in the sense of numbers or data subject to analysis, statistics underlie or adorn much current historical writing, part of a broader movement of "quantitative history" whose spread and persistance mark the 1960s, 1970s, and 1980s. In the singular, in the sense of a conceptual discipline constructing inferences under uncertainty, statistics has begun to add history to the roster of fields like genetics, economics, health sciences, thermodynamics, and agriculture from which its challenging problems are drawn.

Numbers enter significantly into accounts of the past at least as early as Homer's catalogue of ships in the *Iliad*, Moses' poll of fighting men in the *Book of Numbers*, and Thucydides' sober recital of Athenian resources in Book Two of *The Peloponnesian War*. Livy cites Roman census totals, which still impinge on modern debates in Brunt [3]. But the sceptical scrutiny of historical numbers is modern, Hume's 1752 essay "On the Populousness of Ancient Nations" contrasts received numbers to expose their inconsistencies, conducting what might now be called an analysis of outliers* guided by informal models of the demography* and economics of slavery and the relation of national wealth to political structure. Macaulay's famous "Third Chapter" of 1849, an account of English life and society at the accession of James II, recognizes the superiority of several independent measures over a single one when it takes as its starting point three concordant estimates of total English population. One of these estimates is that made in 1696 by the great initiator of national statistics, Gregory King. Statistical history and statistics arise together. Indeed, Karl Pearson [25] in his *History of Statistics* sees them sharing the same precise birthday, the 1662 publication of John Graunt's *Natural and Political Observations ... upon the Bills of Mortality*.

It is no accident that all these examples involve demographic statistics, for historical demography has proved to be the most soundly quantifiable aspect of the study of the past. Hume's other preoccupations, the economics of slavery, national wealth, and political systems adumbrate the other main themes on which statistics and historical studies meet.

This article treats in turn research in historical demography and social structure, in economic history, and in political science history. Each of the three sections, while mentioning works notable for their use of statistics (plural), dwells mainly on a few examples involving conceptual contributions from statistics (singular).

Some historical work, like Kendall's grave site seriation [18] or Mosteller and Wallace's ascription of authorship to disputed Federalist papers [24] employ the most advanced statistical techniques extant. Kendall, for instance, constructs a matrix of dissimilarities between archaeological grave sites based on numbers of artifact types in common. He then represents sites via multidimensional scaling as points in a two-dimensional space and interprets horseshoe-shaped clusters as a sinuous time line that can be unwound to order sites chronologically. Multidimensional scaling*, however complex and experimental in 1970, proved suited to the task.

But often, advanced techniques outrun the quality of historical data, and the more basic statistical methods presented in Floud [7], *An Introduction to Quantitative Methods for Historians*, are those that lead to important insights, demanding for their application to problematic data sets historical sensitivity and finesse. At its limits, statistical analysis merges into a broader field of formal analysis, surveyed by Herlihy [12], and it connects to historical schools like the "New Social History" surveyed by Tilly [31] or the cliometric movement surveyed in Kousser [20]. Reviews of data and analyses, nation by nation, are to be found in Lorwin and Price [22]. The *Journal of Interdisciplinary History*, *Historical Methods*, and *Social Science History*, along with the specialized journals of economic history, are rallying points for statistically informed studies.

Demography, the first area under discussion, permeates statistical history, since birth, death, marriage*, and age structure interactions are often the aspect of an historical context that can be pinned down most firmly from surviving data, and since disagreements about what is being measured are not so rife as elsewhere. Outstanding use has been made of parish registers, local church records of baptisms, marriages, and burials, through the technique of "family reconstitution." Though marriage records usually omit ages, the ages can be found by linking the record to the baptism records with the partners' names, and ages at childbirths and death by further linkage to baptisms and burials. However, migration* out of a parish and thus out of observation was common and unrecorded. Most "reconstituted" families are missing some dates, and so the numbers at risk of being observed to die, say, before the age of five or give birth between 20 and 25—the denominators of the demographic rates—are elusive.

It was Louis Henry who around 1955 shifted emphasis away from the genealogical aim of all possible true linkages from parish records to the statistical aim of controlling bias, even at the cost of discarding data. Henry developed rules for reserving certain linked dates for the sole purpose of attesting a family's presence at risk of events in the parish, removing those dates from demographic calculations to reduce sample-inclusion bias. Henry (see e.g., ref. 11) in France, Wrigley (e.g., ref. 35) in England, and many colleagues have applied family reconstitution to questions of the prominence of crisis mortality, the existence of family limitation practices in preindustrial societies, and the leverage of marriage age as a demographic regulator of growth, shaping a rural demographic history of pre-census times.

Sampling of parishes has made possible the transition from local instances to a national picture in France (cf. INED [16]). In England, reconstitutable parishes are too sparse and survival bias is severe. In a monumental work, Wrigley and Schofield [36] have subjected monthly totals of baptisms and burials from 404 parishes to a battery of corrections and multipliers to generate aggregate series of births and deaths over time from 1541 to 1871. Survival bias is reduced by devising sampling weights based on parish size. Populations at risk to use as denominators of rates are obtained by forecasting age structure backward in time from the earliest censuses. This "back projection" is far from elementary because of the all-important, unobserved component of net international migration. Among their findings, Wrigley and Schofield assert the preeminence of rising fertility* over falling mortality as a cause of accelerating growth rates in early industrial England.

Demographic studies naturally broaden into studies of historical social structure. Tables and graphs enhance Stone's [29] recasting of the venerable controversy over the rise of the English gentry before the English Civil War. Comprehensive projects on whole industrializing communities are reported for Hamilton, Ontario, around 1850 in Katz [17] and for Philadelphia after 1850 in Hershberg [14]. Thernstrom (e.g., ref. 30) has pioneered the use of repeated city directories to trace geographical and occupational mobility in the American northeast of the 1800s.

Among the richest compilations for any era is the Florentine tax census or catasto of 1427. It has been analyzed and made accessible for computer analysis by Herlihy and Klapisch-Zuber [13]. Florence continues to stimulate the blending of humanistic with quantitative concerns in Cohn [5] and Weissman [34].

Numerical data always relate only to a limited subset of the whole range of subjects of interest to historians. For instance, numerical data exist on the prevalence of households containing grandparents or other kin, but not, generally, on the strength of affective ties between relatives nor on frequency of social contact (besides legal proceedings) outside the household. Those few parts of the historical picture that can be pinned down statistically play a special role vis a vis the other parts, of equal or greater inherent interest, for which, at best, the relationship between evidence and conclusions is not a matter of formalized systematic scrutiny, and for which, at worst, it is a matter of impressionistic inference from anecdotes. The settling of a few key issues statistically can supply fixed points, which then constrain the scope within which conjecture operates, as it constructs a full picture. This view, of course, rejects the popular caricature of the statistical historian seeking a number for every phenomenon and filling articles with endless tables. Conceptual contributions of the discipline of statistics to history are concentrated around issues, and the paradigm is one of hypothesis testing* rather than estimation*.

The kin composition of preindustrial English and European households is one issue on which statistical thinking has focussed. Peter Laslett, whose 1965 book *The World We Have Lost* crystallized interest in historical sociology and who with Wrigley and Schofield founded the Cambridge Group for the History of Population and Social Structure, called into question in the 1960s the standard view that preindustrial households contained large numbers of kin and that the nuclear family only came to the fore with industrialization. Whereas others had selected communities for study on the basis of family forms that interested them, Laslett and his colleagues set about determining the distribution of communities according to the prevalance of household types. Eventually, they assembled a set of 64 English pre-census listings of communities by household membership that, if not strictly random, were at least selected without regard to household characteristics, as in Wachter ([33, Chap. 5]). These data for England showed a paucity of multigenerational households and coresiding kin. Skeptics, however, argued that the limited portion of the life cycle during which a household could contain, say, grandparents, under late marriage and severe mortality conditions, might produce small numbers of complex households in cross-sectional data in spite of social practices favoring the inclusion of kin.

In the late 1970s simulation*, that is, computer microsimulation, was used to test the theory that ascribed the observed paucity of complex households in England before 1800 to the constraints of demographic rates. Different sets of demographic rates spanning a range suggested by the reconstitution studies were paired with three alternative detailed models of household formation, and the cross-sectional consequences of this life cycle behavior were observed in different random realizations under replicable conditions. The results, presented in Wachter [33], contradict the theory and show little leverage over household composition from demographic rates.

The name "experimental history" has been claimed for computer simulation studies. They provide a way to operationalize the *ceteris paribus* that modifies most explanatory sentences in historical writing. Would unobserved behavior in accordance with a particular account, holding all else constant, have the observed consequences that is adduced to explain? Such tests are particularly important if data pertain to traditional village-size populations, in which random effects are prominent. Since semiclosed small populations cannot be regarded as independent cases like samples from large populations, standard formulas for variances and other indices of randomness do not apply. Analytic solutions from stochastic demography* are as yet available mainly for unrealistically stylized cases, so that simulation offers the promising approach.

In economic history, even more so than in historical demography, statistics, as numbers, are ubiquitous. Aggregate estimates for western economies, including time series indices of prices, real wages, and business activity, now stretch back more than a century and a half. Claiming a mid-1800s precursor in James Thorwold Rogers, a tradition of ingenious and painstaking compilation has flourished continuously since the time of Mitchell in the 1920s. The advantage that this shared empirical base gives to debate is well illustrated by the comprehensive *Economic History of Britain Since 1700* edited by Floud and McCloskey [8], where references abound. As Morgenstern [23] taught, problems of definitional change, hidden bias, and data contamination require hardy vigilance. A piquant example from the history of the female labor force is given by Smuts [28].

Conceptual interchange between statistics and economic history is typified by research on business cycles*. Aggregate historical series pertaining to periodic or irregular fluctuations in business or trade have been a chief proving ground for statistical time-series* methodology, first for spectral analysis* and later for ARIMA. The 1920s and 1930s saw the empirical teamwork of the National Bureau for Economic Research culminate in the definition of leading and lagging indicators for present-day economic forecasting. With a longer time horizon, Ronald Lee in Wrigley and Schofield ([36]) has illuminated interdependencies in fluctuations over three centuries of English grain prices, births, and deaths.

The best-known part of the whole movement of quantitative history, or "Cliometrics", has been work on American slavery, inaugurated by a paper of Conrad and Meyer in 1958 that challenged views that slavery was or became unprofitable, adducing an econometric calculation of a positive rate of return on investment in slaves. More statistical investigations followed on the drawing by Parker and Gallman of random samples from the 1860 manuscript censuses of population, agriculture, and slaves for 11 southern states. Bateman and Foust drew similar samples for the North, and Ransom and Sutch drew them for 1880, enabling comparisons to be made.

In 1974 Fogel and Engerman [9] published *Time on the Cross*, the most widely read and debated of all works by quantitative historians. This book ventures a synthetic picture of the economics of American slavery. In it, antebellum plantation owners and slaves are seen in the image of the "rational economic man," slave agriculture is profitable, efficient, and competitive, slave family stability is encouraged, positive incentives are present and net expropriation of earnings is low. Placed within this picture, each separate technical issue takes on a definite polarity. Furthermore, something like a canonical set of particular calculations emerges, whose reworking would bear directly on and span the larger picture.

One by one the Fogel and Engerman calculations have been taken up, among others by David et al. [6], whose *Reckoning with Slavery* argues the preponderance of Fogel and Engerman's inferences to be unfounded. Passionate discussion spread far beyond the technical controversies, as could be expected from the sensitive subject of slavery and the popular audience that Fogel and Engerman attracted. While complex statistical methods are not involved, most questions turn partly on the interpretation of graphs, the suitability of samples, and the bounding of cumulative biases, making the debate statistical in a fundamental sense.

The value of sizable samples from historical records drawn with regard to statistical sampling theory is clear in many areas besides slavery. Jones [15] has collected and analysed a cluster sample by counties of probate inventories for the American colonies in 1774, giving a picture of aggregate and per capita wealth, its size, composition, and unequal distribution in *Wealth of a Nation to Be*. In studies of this type, it is usually necessary to impute missing values, often by linear regression equations, or to adjust for biases suspected in certain types of records, by regressions with dummy variables for record types. An alternative to such highly parametric analysis is represented by Singer [27], who applies graphical techniques of exploratory data analysis* to probate inventories.

Statistical contributions to political history are largely to the history of political expression by constituencies or other groups, what might be called political science history. The statistical history of voting in America is particularly active and extensive; in terms of method it coalesces with contemporary voting statistics* and election forecasting*. Seminal studies before 1961 relied entirely on percentages and cross tabulations, including Key's work on the concept of critical elections, rare events in which longstanding political allegiances give way to realignments, and Benson's [2] new emphasis on persistent ethnocultural and religious affiliations in place of transient economic partitions. Work of the 1960s and 1970s summarized in the introduction to Silbey et al. [26] shifted to correlational methods. Recently, statistical "unfolding methods" related to multidimensional scaling have been tried by Calhoun [4]. For legislative rather than constituency voting, Aydelotte [1] has pursued Guttman scaling. For all the electoral research, the Survey Research Center at Michigan has been an important catalyst, and the archiving and dissemination of data bases through the Inter-University Consortium for Political and Social Research have accelerated cumulative research.

The methodological cause célèbre of voting studies has been the problem of ecological correlation. Whatever the true individual-level correlation between variables, the sample correlation between subgroup averages can rise as high as ever if individuals are allocated nonrandomly to subgroups. Republicans might comprise a mere 51% of all Protestants compared to 50% of all Catholics regardless of precinct, but if all Catholics were allocated to some precincts and all Protestants to others, a precinct by precinct graph of proportion Republican by proportion Protestant, having only two distinct points, would show a sample correlation of 1.0. Goodman's [10] solution to work through regression slopes rather than correlations has been brought by Kousser into general use [20]. Interactions still cause problems, however. If a high proportion Catholic reduces the percentage Republican among Protestants in the precincts, only individual-level data can reveal individual-level associations.

Dissent and expression outside electoral channels are more difficult subjects; the systematic treatment of the unsystematic makes unconventional demands. Charles Tilly and his collaborators have gradually developed methods for coding incidents culled especially from newspaper reports. The focus on collective violence, often emanating from riots, demonstrations, and strikes, has broadened to the study of "contentious gatherings." Coding criteria have evolved, sufficiently specific to allow replication and cross-period and cross-national comparison, but giving scope to the self-expressions of the historical participants. The analysis of small incidents in large numbers reveals regularities of which the large incidents, few in number, that traditionally make their way into textbooks give little hint. Behind the issues and stated aims of protest, changes in the repertory from which protestors took the modes of action of which they availed themselves have come to attention. *The Rebellious Century* by Tilly et al. [32] treats France, Italy, and Germany from 1830 to 1930. A more intensive study of contentious gatherings in Great Britain in 1832 is underway, with independent replication for Ireland. The Tillys' work has inspired a growing literature on strikes, on labor action, on dissent, and even on crime, some of it rural, much of it with close ties to the movement of New Urban History.

While advanced statistical technique has only a small role in the Tillys' work, and while tables and graphs of numbers take second place to historical commentary and argument in ordinary prose, this research exemplifies the involvement of statistics in historical studies. Its subject matter consists in regularities that emerge out of randomness only when large numbers of independent instances are brought together. Its methodological concerns, with sampling, measurement artifacts, outliers, and biasses, recapitulate the concerns of all statistics, but within the more difficult arena of the past, where experimental treatments and controls are out of grasp.

References

[1] Aydelotte, W., ed. (1977). *The History of Parliamentary Behavior.* Princeton University Press, Princeton, NJ, pp. 3–27, 225–246.

[2] Benson, L. (1961). *The Concept of Jacksonian Democracy.* Princeton University Press, Princeton, NJ.

[3] Brunt, P. A. (1971). *Italian Manpower 225 B.C.–A.D. 14.* Clarendon Press, Oxford, England.

[4] Calhoun, D. (1981). From collinearity to structure. *Historical Meth.*, **14**, 107–121.

[5] Cohn, S. (1980). *Laboring Classes in Renaissance Florence.* Academic, New York.

[6] David, P. A., Gutman, H. G., Sutch, R., Termin, P., and Wright, G. (1976). *Reckoning with Slavery.* Oxford University Press, Oxford, England. (A stimulating debate, worth close study.)

[7] Floud, R. (1979). *An Introduction to Quantitative Methods for Historians*, 2nd ed. Methuen, London, England. (An excellent methodological exposition requiring little prior knowledge.)

[8] Floud, R. and McCloskey, D. (1981). *The Economic History of Britain Since 1700.* Cambridge University Press, Cambridge, England.

[9] Fogel, R. and Engerman, S. (1974). *Time on the Cross.* Little Brown, Boston. (The best-known work of quantitative history, it continues to deserve attention.)

[10] Goodman, L. (1959). Some alternatives to ecological correlation. *Amer. J. Sociol.*, **64**, 610–625.

[11] Henry, L. (1980). *Techniques d'Analyse en Demographie Historique.* Editions de l'Institut National d'Etudes Demographiques, Paris, France. (A formal handbook.)

[12] Herlihy, D. (1981). Numerical and formal analysis in European history. *J. Interdisciplinary History*, **12**, 115–135. (Provides a valuable overview of recent work.)

[13] Herlihy, D. and Klapisch-Zuber, C. (1978). *Les Toscans et leurs Familles.* Fondation Nationale des Sciences Politiques, Paris, France.

[14] Hershberg, T. ed. (1981). *Philadelphia: Work, Space, Family, and Group Experience in the Nineteenth Century*, Oxford University Press, Oxford, England.

[15] Jones, A. H. (1980). *Wealth of a Nation to Be.* Columbia University Press, New York. (A readable application of sampling.)

[16] INED (1975). *Population, Numero Special*, **30**.

[17] Katz, M. (1975). *The People of Hamilton, Canada West.* Harvard University Press, Cambridge, MA.

[18] Kendall, D. (1971). In *Mathematics in the Archeological and Historical Sciences*, Kendall, Hodson, and Tautu, eds. Edinburgh University Press, Edinburgh, Scotland. (The whole volume repays perusal.)

[19] Kousser, J. M. (1974). *The Shaping of Southern Politics.* Yale University Press, New Haven, CT.

[20] Kousser, J. M. (1983). *History in Our Time.* Chapter 17, to appear.

[21] Laslett, P. (1971). *The World We Have Lost.* 2nd ed. Methuen, London, England. (A classic.)

[22] Lorwin, V. and Price, J. M. (1972). *The Dimensions of the Past.* Yale University Press, New Haven, CT.

[23] Morgenstern, O. (1950). *On the Accuracy of Economic Observations.* Princeton University Press, Princeton, NJ.

[24] Mosteller, F. and Wallace, D. (1964). *Inference and Disputed Authorship: The Federalist.* Addison-Wesley, Reading, MA.

[25] Pearson, K. (1978). *The History of Statistics in the 17th and 18th Centuries.* Griffin, High Wycombe, England.

[26] Silbey, J., Bogue, A., and Flanigan, W. (1978). *The History of American Electoral Behavior.* Princeton University Press, Princeton, NJ.

[27] Singer, B. (1976). Exploratory strategies and graphical displays. *J. Interdisciplinary History*, **7**, 57–70.

[28] Smuts, R. W. (1960). The female labor force: A case study of the interpretation of historical statistics. *J. Amer. Statist. Ass.*, **55**, 71–79.

[29] Stone, L. (1965). *The Crisis of the Aristocracy.* Clarendon Press, Oxford, England.

[30] Thernstrom, S. (1973). *The Other Bostonians.* Harvard University Press, Cambridge, MA.

[31] Tilly, C. (1983). The old new social history and the new old social history. *Working Paper 218*, Center Res. Social Organiz., University of Michigan.

[32] Tilly, C., Tilly, R., and Tilly, L. (1975). *The Rebellious Century*. Harvard University Press, Cambridge, MA.

[33] Wachter, K. W., Hammel, E. A., and Laslett, P. (1978). *Statistical Studies of Historical Social Structure*. Academic, New York.

[34] Weissman, R. (1982). *Ritual Brotherhood in Renaissance Florence*. Academic, New York.

[35] Wrigley, E. A. (1965). Family limitation in pre-industrial England. *Econ. History Rev. 2nd Ser.*, **19**, 82–109.

[36] Wrigley, E. A. and Schofield, R. (1981). *The Population History of England 1541–1871*. Harvard University Press, Cambridge, MA.

(ARCHAEOLOGY, STATISTICS IN
DEMOGRAPHY
EXPLORATORY DATA ANALYSIS
MARRIAGE
POLITICAL SCIENCE, STATISTICS IN
PROBABILITY, HISTORY OF)

KENNETH W. WACHTER

STATISTICS IN MEDICINE

Statistics in Medicine is an international journal that was first published in 1982. Each volume comprises a year's publication made up of six issues, each of approximately 100 pages.

The aim of the journal is to enhance communication between statisticians, clinicians, and medical researchers. To this end, the journal publishes papers on practical applications of statistics and other quantitative methods to medicine and its allied sciences. Specific areas include: clinical trials and surveys; methods of quantification in medicine; descriptive and analytic epidemiology; mathematical modelling and evaluation of health services; laboratory quality control; statistical methods in laboratory experiments; methods, mathematical models, and evaluation of screening for disease; statistical methods in diagnosis; management of medical data and statistical analysis by computer; mathematical models of biological processes; teaching statistics; training medical statisticians and statistical consulting.

Examples of applications of statistics in specific projects, articles explaining new statistical methods, and reviews of general topics are published. The journal emphasizes the relevance of numerical techniques and aims to communicate statistical and quantitative ideas in a medical context. The main criteria for publication are therefore appropriateness of the statistical methods to the particular medical problem and clarity of exposition.

The editors of the journal are T. Colton (Boston) and L. S. Freedman and A. L. Johnson (Cambridge). The journal operates jointly from two editorial offices: one at Boston University School of Public Health, 80 East Concord Street, Boston, Massachusetts 02118, U.S.A.; the other at Medical Research Council Centre, Hills Road, Cambridge CB2 2QH, U.K. Each office has an Editorial Board of approximately 20 members. Authors from North or South America should submit their articles to the U.S.A. office, while the remainder of the world should communicate with the U.K. office.

Contributions to the journal may be in the form of articles or letters to the Editors. Book reviews and announcements of forthcoming meetings are published. All manuscripts are refereed, normally by both a statistician and a medical scientist.

The contents of a recent issue (Vol. 2, No. 1) included:

"Epidemiology of prenatal infections: An extension of the congenital rubella model" by E. G. Knox.

"Relative risks of diseases in the presence of incomplete penetrance and sporadics" by P. P. Majumder, R. Chakraborty and K. M. Weiss.

"Reducing classification errors in cohort studies: The approach and a practical application" by K. F. Schultz et al.

"An analysis of repeated biopsies following cardiac transplantation" by D. J. Spiegelhalter and P. G. I. Stovin.

"Unbiased assessment of treatment effects on disease recurrence and survival in clinical trials" by R. Kay and M. Schumacher.

"Log-linear model selection in a rural dental health study" by J. Oler and J. M. Bentley.

"Bias correction in maximum likelihood logistic regression" by R. L. Schaefer.

"The association between HLA antigens and the presence of certain diseases" by J. R. Green et al.

"On the standard error of the probability of a particular diagnosis" by D. Machin et al.

"Discriminant analysis based on multivariate response curves: A descriptive approach to dynamic allocation" by A. Albert.

The journal is published by John Wiley and Sons in Chichester, Sussex, England.

L. S. FREEDMAN

STATISTICS, OPTIMIZATION IN *See* OPTIMIZATION IN STATISTICS; OPTIMIZATION, STATISTICS IN

STATISTICS IN PSYCHOLOGY

Psychology and statistics can well claim to be sister sciences. Both lay claim to Francis Galton* (1822–1911) as one of its forefathers. Karl Pearson* (1857–1936), in his biography of Galton, places considerable emphasis on Galton's psychological research, and *Biometrika*, which he and Galton founded, was intended to publish the results of mathematical research in biology and psychology. Galton has been described as primarily responsible for launching the psychological testing* movement, and may well be regarded as the father of quantitative psychology.

Statistical methods were used early in the history of psychology. Walker [26], in her classic history of statistical methods, notes that the psychologist Ebbinghaus (1850–1909) in 1885 gave the probable error* of every mean, while Fechner (1801–1887), the "father of the new Psychology" invented the median independently of Galton, pointing out that it minimizes the absolute deviations. She credits the psychologist Cattell (1860–1944) as having taught the second college course in statistics in 1887. Spearman (1863–1945) invented rank correlation* in 1904 and urged psychologists to become familiar with the new methods of correlation and regression*, writing [22, p. 73]

> One after another, laborious series of experiments are executed and published with the purpose of demonstrating some connection between two events, wherein the otherwise learned psychologist reveals that his art of proving and measuring correspondence has not advanced beyond that of lay persons.

He noted that

> Psychologists, with scarcely an exception, never seem to have become acquainted with the brilliant work being carried on since 1886 by the Galton–Pearson school. The consequence has been that they do not even attain to the first fundamental requisite of correlation, namely, a precise quantitative expression.

This deficiency was soon remedied as psychologists quickly came to appreciate the utility of these methods. Brown [3] published a text in 1911 "primarily for the professed psychologist" that he hoped would "also prove of use to the educationalist." This text combined two chapters on psychophysics* with four on correlation, three of the latter being his doctoral thesis, *The Use of the Theory of Correlation in Psychology*.

It was primarily the educational psychologists who took up the new statistical methods. For many years educational statistics and psychological statistics have been virtually synonymous. The *Journal of Educational Psychology*, founded in 1910, specifically included experimental pedagogy, child physiology and hygiene, and educational statistics* in its charter. For years it emphasized the applications of correlation and regression in psychology. Walker [26] was to write in 1931 that

> The influence of Pearson upon educational statistics in America has been so profound that it is all but impossible to overstate the extent to which he is responsible for modern statistical theory.

Twenty years later such a statement could no longer be made, as the influence of R. A. Fisher* became pervasive in psychology. Thus in 1940, McNemar [16] wrote that

> Eventually, the analysis of variance will come into use in psychological research; an expository paper thereon would not be without value.

Only 10 years later Webb and Lemmon wrote

> In problems involving more than two conditions... the increasing use of the analysis of variance has been apparent. That this statistical technique has become a typical method of analysis in the problems of psychology and education is evident by its inclusion in most of the recent statistics texts in this field.

There have always been critics of the uses of statistics of psychology, particularly among psychologists. Brown [3] in 1911 warned that

> the theory of correlation is likely to be found especially valuable in educational psychology, if only its presuppositions are clearly grasped and the requisite precautions in its application exercised.

He criticized Spearman's use of small samples and his adjustment of correlations based on the unwarranted assumption of uncorrelated errors. In a second edition revised by Thompson [4], Spearman's theory of a single common factor was attacked through use of what we would today call a Monte Carlo* study, with a population containing no single common factor. These authors were themselves savagely attacked by Yule [27], who wrote:

> *Essentials*—a few experimental methods, rather difficult to interpret: ... *Mental*? well it is no good attempting to pretend that many of the data (or any of the data?) can be termed mental. ... and *measurement*! O dear! isn't it almost an insult to the word to term some of these numerical data measurements? ... I have begun to visualize the title of the book as "The essentials(?) of mental(?) measurement(?).

He goes on to say that statistical methods

> should be regarded as ancillary, not essential. ... statistical methods are only necessary in so far as experiment fails to attain its ideal, the ideal of only permitting one causal circumstance to vary at a time. ... Statistical methods are not only ancillary; they are, to the experimenter, a warning of failure.

This dictum of varying only one factor at a time, which was the dictum of psychophysics, was of course demolished by Fisher (1890–1962), whose first publication on analysis of variance* was still a year away. The American tradition of multiple factor analysis* was initiated by L. L. Thurstone (1887–1955).

Psychologists have often been criticized for their emphasis on tests of significance*, but it was Fisher who was responsible for this change in direction toward statistical testing and designed experiments. In the first edition of *Statistical Methods for Research Workers* [6] he noted:

> Little experience is sufficient to show that the traditional machinery of statistical processes is wholly unsuited to the needs of practical research. Not only does it take a cannon to shoot a sparrow, but it misses the sparrow!

Although correlational methods were well treated in texts in psychological statistics through the 1930s, there was little treatment of Fisherian methods prior to Snedecor's publication of *Statistical Methods* [20] in 1937. Sorenson's 1936 *Statistics for Students of Psychology and Education* [21] devoted 11 pages to the reliability of differences between means.

A series of papers from 1931 through 1933 in the *Journal of Educational Psychology* illustrates the sorry state of affairs at that time. It also illustrates the sometimes unfortunate collaboration of psychologists and statisticians. Lindquist [15] writing on "The Significance of a Difference Between Matched Groups" pointed out that such experiments were universally analyzed incorrectly in education and psychology. He sought out graduate student S. S. Wilks, who published an accompanying paper [28] entitled "The Standard Error of the Means of Matched Samples." In a remarkably awkward derivation, Wilks obtained as his standard error what is, in fact, the standard error of the mean of deviations from the regression line, i.e.,

$$\sigma_i = \sqrt{\left(1 - r_i^2\right)\sigma_{Y_i}^2 / n_i}\,.$$

Lindquist then computes

$$\sigma_d^2 = \sigma_1^2 + \sigma_2^2.$$

In a followup article, Ezekiel [5] writing on "Student's Method for Measuring the Significance of a Difference Between Matched Groups" stated:

> Over twenty years ago, an eminent English statistician "Student" (he writes under this pseudonym) considered the same type of problem which Lindquist attacks, and arrived at parallel conclusions.

He proceeded to give the correct solution based on differences of the matched observations, noting only that the Wilks–Lindquist formula "much under estimates the significance of the results, as judged by 'Student's' method." Lindquist then replied that

> the method suggested by Ezekiel has long been known to workers in the field of educational statistics. It was discarded by Wilks and the writer, however, because it is not valid in the situation for which Wilks' formula is intended.

In the same year Peters and Van Voorhis [18] presented "A New Proof and Corrected Formulae for the Standard Error of a Mean and of a Standard Deviation," noting that their derivation follows a proof given by Lindquist; "But we believe Lindquist makes an incorrect application of his conclusion." It is interesting that Hotelling's original paper on principal components* [11] was published in the same year in the same journal at the initiative of a psychologist, Truman Kelley, the author of a significant text in psychological statistics [13].

The new journal *Psychometrika** was founded in 1936, emphasizing articles on "the development of quantitative rationale for the solution of psychological problems," but specifically included statistical techniques. Its first few years included articles by Hotelling [12] on principal components, by Travers [23] on the discriminant function*, by Kurtz [14] on the use of the Doolittle method (later renamed the square root method) in multiple correlation*, as well as articles on the effect of heterogeneity of variance on F-tests* in ANOVA*. Other early issues treated nonorthogonal ANOVA [24] and t-tests with unequal variances [19]. The fiftieth anniversary issue of *Psychometrika* in 1986 surveys the major areas covered by the journal over this period, including test theory, factor analysis*, structural modeling, statistics and data analysis, and scaling. Currently at least a dozen journals deal, at least in part, with applications of statistics in psychology, and it is no longer feasible for the *Annual Review of Psychology* to include a chapter covering statistical methods for the past year. Every department of psychology in the United States currently requires advanced training in statistics for its graduate students, and applied statistics is a recognized subspecialty in many graduate departments. The emphasis nowadays is on Fisherian methods, as treated by such psychological statistics textbook authors as Hays [10] and Winer [29]. Psychologists were among the first to adapt multivariate methods* to computers and to make use of these methods in research (Bock [1, 2], Hall and Cramer [9]). Psychologists have also played a leading role in the use of Monte Carlo methods in statistics. Statisticians have on occasion rediscovered methods long used by psychologists such as the minimum normit method of Berkson, which has been continuously used in psychophysics for over 70 years under the name of the "least-squares solution with Mueller–Urban weights." Similarly, methods of stepwise regression* have long been used by psychologists. Psychologists today are continuing the tradition of Galton in their use of statistical methods.

References

[1] Bock, R. D. (1964). Programming univariate and multivariate analysis of variance. *Technometrics*, **6**, 95–117.

[2] Bock, R. D. (1975). *Multivariate Statistical Methods in Behavioral Research*. McGraw Hill, New York.

[3] Brown, W. (1911). *Essentials of Mental Measurement*. Cambridge University Press, Cambridge, England.

[4] Brown, W. and Thompson, G. (1921). *Essentials of Mental Measurement*. Cambridge University Press, Cambridge, England.

[5] Ezekiel, M. (1932). Student's method for measuring the significance of a difference between matched groups. *J. Educ. Psychol.*, **23**, 446–450.

[6] Fisher, R. A. (1925). *Statistical Methods for Research Workers*. Oliver and Boyd, Edinburgh, Scotland.

[7] Garrett, H. (1943). The discriminant function and its use in psychology. *Psychometrika*, **8**, 65–79.

[8] Godard, R. H. and Lindquist, E. F. (1940). An empirical study of the effect of heterogeneous within-group variance upon certain *F*-tests of significance in analysis of variance. *Psychometrika*, **5**, 263–274.

[9] Hall, C. E. and Cramer, E. M. (1963). A general purpose program to compute multivariate analysis of variance on an IBM 7090 computer. Biometric Laboratory The George Washington University, Washington, DC.

[10] Hays, W. L. (1972). *Statistics of the Social Sciences*. Holt, Rinehart, and Winston, New York.

[11] Hotelling, H. (1933). Analysis of a complex of statistical variables into principal components. *J. Educ. Psychol.*, **24**, 417–520.

[12] Hotelling, H. (1936). Simplified calculation of principal components. *Psychometrika*, **1**, 27–36.

[13] Kelley, T. L. (1923). *Statistical Methods*. Macmillan, New York.

[14] Kurtz, A. (1936). The use of the Doolittle method in obtaining related multiple correlation coefficients. *Psychometrika*, **1**, 45–52.

[15] Lindquist, E. F. (1931). The significance of a difference between "matched" groups. *J. Educ. Psychol.*, **22**, 197–204.

[16] McNemar, Q. (1940). Sampling in psychological research. *Psychol. Bull.*, **37**, 331–365.

[17] Pearson, K. (1914, 1924, 1930), *The Life, Letters, and Labours of Francis Galton*, Vols. 1–3. Cambridge University Press, Cambridge, England.

[18] Peters, C. C. and Van Voorhis, W. R. (1983). A new proof and corrected formulae for the standard error of a mean and of a standard deviation. *J. Educ. Psychol.*, **24**, 620–633.

[19] Satterthwaite, F. (1941). Synthesis of variance. *Psychometrika*, **6**, 309–316.

[20] Snedecor, G. W. (1937). *Statistical Methods*. Collegiate Press, Ames, IA.

[21] Sorenson, H. (1936). *Statistics for Students of Psychology and Education*. McGraw-Hill, New York.

[22] Spearman, C. (1904). The proof and measurement of association between two things. *Amer. J. Psychol.*, **15**, 72–101.

[23] Travers, R. M. W. (1939). The use of a discriminant function in the treatment of psychological group differences. *Psychometrika*, **4**, 25–32.

[24] Tsao, F. (1942). Tests of statistical hypotheses in the case of unequal or disproportionate numbers of observations in the subclasses. *Psychometrika*, **7**, 195–212.

[25] Urban, F. M. (1908). *The Application of Statistical Methods to Problems of Psychophysics*. Psychological Clinic Press, Philadelphia, PA.

[26] Walker, H. M. (1929). *Studies in the History of Statistical Methods*. Williams and Wilkins, Baltimore, MD.

[27] Yule, G. U. (1921). Critical notice. *Brit. J. Psychol.*, **12**, 100–107.

[28] Wilks, S. S. (1931). The standard error of the means of "matched samples," *J. Educ. Psychol.*, **22**, 205–208.

[29] Winer, B. J. (1971). *Statistical Principles in Experimental Design*. McGraw-Hill, New York.

(COMPONENT ANALYSIS
DISCRIMINANT ANALYSIS
EDUCATIONAL STATISTICS
FACTOR ANALYSIS
GROUP TESTING
MEASUREMENT THEORY
MULTIDIMENSIONAL SCALING
MULTIVARIATE ANALYSIS
PSYCHOLOGICAL DECISION MAKING
PSYCHOLOGICAL SCALING
PSYCHOLOGICAL TESTING THEORY
PSYCHOPHYSICS, STATISTICAL METHODS IN
RASCH MODEL, THE
SPEARMAN, RANK CORRELATION COEFFICIENT
THURSTONE'S THEORY OF COMPARATIVE JUDGMENTS)

ELLIOT M. CRAMER

STATISTICS METADATA

Statistics metadata contain some numerical functions (could be vector valued) of observed data, which can be used to compute other statistics. Thus they may contain partially computed sums of observations, partial sums of squares of observations, partial sums of products of observations, etc. They also contain the rules for processing these objects, which are derived data. Some database management researchers would like to label statistical metadata as *abstract data type objects*, but this is inaccurate because of the unique generic nature of statistics metadata and the associate processing rules.

Consider the random variables X_1 = age, X_2 = height, and X_3 = weight of individuals. If we are interested in the average height of individuals whose age is 20 years, then we have to select the records of all individuals whose age is 20 years. This set will be an aggregation of individuals of all weights, so long as their ages are 20 years. If we had constructed some statistics metadata for the heights of the individuals then a desirable property for these statistics metadata, for expediting computations, should be that *they be additive under aggregation* of the values of the weights. This additive property would enable the derivation of the statistics metadata needed for computation of the average height to be much faster than if the average height had to be calculated from the observed values of the heights. If the average heights of all individuals were stored as the statistics metadata for every pair of values of age and weight, they would not have the additive property needed to compute the average height of individuals whose ages are 20 years. The same problem would arise if we were interested in the standard deviation of the heights of individuals whose ages are 20 years, and the statistics metadata were standard deviations of heights for every combination of values of age and weight. If instead, we had stored the sum of the heights, the sum of squares of the heights, the number of individuals for each pair of values of age and weight, and the rules for computing the means and standard deviations under aggregation, then the average height and standard deviation for individuals whose ages are 20 years could be calculated very rapidly (from these statistics metadata).

Most of the statistical metadata provided by different database management systems (DBMS), contain rules for processing raw data. As the volume of raw data is very large in statistical databases, providing only the rules for processing raw data is not sufficient to allow real time statistical analysis. *It is important that statistical metadata contain some complex data types and the rules for processing them*. These new objects are referred to as *statistics metadata*. The rules for processing the complex data objects in the statistics metadata are different from those for processing or representing or retrieving (information) of raw data. The statistics metadata contain both computation rules for the complex data objects and logical manipulation rules for these objects.

The author (Ghosh [8]) has provided some methods for the constructions of certain complex data objects needed for statistics metadata, which have additive properties under consolidation of categories and statistical consolidation of tables. Partial sums, partial sums of squares, sums of products, frequency distributions, etc., all preserve additive properties under consolidation of categories and consolidation of statistical tables. These complex data objects are components of our statistics metadata.

Researchers in database management have felt for a long time that the data collected from the real world and stored in the computer are not adequate for complex manipulation of information and extraction of all the implicit knowledge in the database. Thus they have invented different *semantic models* to represent the semantics of data, and *physical models* to represent storage of data (Date [4], Ghosh [6], Ullman [15]). Most of these models were invented to facilitate processing of logical information contained in the data. Statisticians on the other hand, have been concerned with the processing of statistical information contents of data (e.g., averages, standard deviations, etc.), for more than a century (Kendall and Stuart [11]). In the last few years statisticians and database management researchers have been working together to combine logical and statistical processing of data (Wong [16, 17], Hammond and McCarthy [9], Annual Conferences on Computer Science and Statistics).

The concept of *metadata*, which has been around for many years (Teitel [14]) received new life in the hands of statistical database management system (SDBMS) researchers (McCarthy [13]). Many early researchers in information retrieval and information management systems (Baxendale [2]) have argued in detail that derived or descriptive data should be treated as metadata. Many derived or descriptive data objects like *data dictionaries* (components of database management systems containing descriptions of data) have been used in DBMS. Similarly *data directories* (lists of databases in a DBMS) have been used in DBMS for many years (Lefkovits [12] and Allen et al. [1]) without being referred to as metadata (i.e., data derived from other data). Many researchers in SDBMS have rediscovered that metadata can play a vital role not only in data description but also in enhancing statistical processing of the data (Bishop and Freedman [3]). Researchers have a good idea about some of the properties of metadata, but no formal definition has been agreed upon up to now, because it is a new subject and researchers are not sure that all the necessary properties have been identified.

McCarthy's paper [13] provided an excellent overview of metadata, including statistical metadata, as it is known to the researchers in DBMS. According to McCarthy, metadata are data about data, i.e., systematic descriptive information about data content and organization that can be retrieved, manipulated, and displayed in various ways. His definition of statistical metadata includes default specifications at the database level, error expressions, weighting expression, aggregate or disaggregate expressions, suppression expression, etc. Most of the researchers of DBMS agree with these definitions and at the same time complain that statistical processing of very large statistical databases is extremely time consuming.

According to DBMS researchers metadata have two broad kinds of properties, one based on usage (functional properties), the other based on structural and descriptive properties of data (operational properties). The functional properties are storage and retrieval of data, presentation and display of data, analysis of data, and physical manipulation of the data files. The operational properties of metadata consist of information that is most commonly included in various data dictionaries and directories, e.g., data structures, long names, abbreviated names, titles, column headers, footnotes, data types, summary statistics, etc. Most of these types of metadata have played some role in SDBMS but have not succeeded in reducing the time needed for the common types of statistical analysis. Some attempts have been made in this area (Ghosh [7, 8]). Most statistical estimates of parameters are mathematical transformations of large sets of observed data. If the data are stored as raw data, then calculation of the estimates of the parameters can be a very time consuming effort. This is a major complaint of many statisticians using a DBMS. The commercially available SDBMS do not solve this problem.

Some of the advantages of the use of metadata have been shown in manufacturing test environments (Ghosh [7]) and in an interactive information display system (Ghosh [8]). Occasionally certain types of computational logic may be included as a part of statistics metadata. When computational logic is included it becomes modular (i.e., is a self-contained object) and can also be transported (i.e., can be executed in different computers without changes). Certain types of *statistics metadata* can be used to calculate estimates of parameters and perform statistical data analysis (including analysis of variance, analysis of data from an experimental design, and multiple regression and correlation) much faster than conventional statistical formulas.

The *granularity of statistics metadata* plays a very important role in SDBMS. In general, the complex data objects are values of set functions of raw data, i.e., many pieces of raw data are mapped into one value. Thus identities of real world objects associated with the raw data are lost. The operations associated with the statistics metadata allow the creation of more complex aggregate data objects but the disaggregation of metadata back to raw data is not possible. The purpose of creating complex data objects in the statistics metadata is to reduce computation time. The granularity selection (i.e., sizes of the raw data sets that are used to create metadata) is a trade-off between (i) the query creation capability of the SDBMS and (ii) the desired real time response of query transactions.

In a multiuser database it is important that instances of data segments are locked during dynamic updating so that the responses provided by the system to different users are consistent. In most databases, limited access privileges (who can see what data, who can modify what data, etc.) are provided to different users. The granularity of the locking is determined by the nature of the updating transaction, which varies from one transaction to another. The granularity of the access privileges or the sensitivity of the data segments are relatively static and are determined from security considerations.

As the complex data objects in the statistics metadata are usually set functions of raw data, they do not compromise any security at the individual data level. Sometimes it is possible to direct different statistical queries at the same database and penetrate some individual data security (Denning and Schlorer [5]). *A set of access privileges on complex data objects provides a higher level of security than the same access privileges on the raw data.*

If transactions are directed at the statistics metadata only, then very often it is possible to achieve desired security by restricting the execution authority of the rules in the statistics metadata rather than by restricting access privileges to complex data objects. Suppose that the system manager would like to restrict the disclosure of salary information of individuals and at the same time to permit statistical queries directed at atomic categories. If the number of individuals in a category is 2, then by requesting the mean salary and the standard deviation of salaries, it is possible to deduce the salaries of the individuals. Suppose that the rules of the statistics metadata are modified to include the following rule: "If the number of individuals in an atomic category is equal to n, then do not disclose to the same userid (i.e., the name of the user authorized to use the computer) more than $n - 1$ statistics calculated from the same statistics attribute". This would prevent access to individual salary information by a single computer user. Under this rule, it would still be possible for two userids to deduce the individual salary information. This can also be prevented by tightening the rule, by never disclosing to the user community more than $n - 1$ statistics on the same statistics attribute until the data have been updated. This security rule can even be relaxed by restricting the constraints to $n - 1$ independent statistics, but it would be very time consuming to enforce this rule when the computer is executing a program.

References

[1] Allen, F. W., Loomis, E. S., and Mannino, M. V. (1982). The integrated dictionary/directory system. *ACM Computing Surv.* **14**, 245–286.

[2] Baxendale, P. (1966). In *Annual Review of Information Science and Technology*, Vol. 1, C. A. Cuadra, ed. Wiley, New York.

[3] Bishop, Y. M. and Freedman, S. R. (1983). In *Proc. Second Int. Workshop Statist. Database Manag.*, pp. 230–234.

[4] Date, C. J. (1977). *An Introduction to Database Systems*. Addison-Wesley, Reading, MA.

[5] Denning, D. E. and Schlorer, J. (1983). Inference controls for statistics databases. *IEEE Computer*, July.

[6] Ghosh, S. P. (1986). *Data Base Organization for Data Management*. 2nd ed. Academic, New York, Chap. 1.

[7] Ghosh, S. P. (1985). An application of statistical databases in manufacturing testing. *IEEE Trans. Software Eng.*, **SE-11**, 591–598. [Also published as a IBM Research Report No. R 4055 (1983).]

[8] Ghosh, S. P. (1984). Statistical Relational Tables for Statistical Database Management. *IEEE Trans. Soft. Engin.*, **SE-12**, 1106–1116.

[9] Hammond, R. and McCarthy, J. L., eds. (1983). *Proc. Second Int. Workshop Statist. Database Manag.*

[10] Kendall, M. G. and Stuart, A. (1958). *The Advanced Theory of Statistics*, Vol. 1. Griffin, London, England, Chaps. 1–6.

[11] Kendall, M. G. and Stuart, A. (1961). *The Advanced Theory of Statistics*, Vol. 2. Hafner, New York, Chaps. 17–20.

[12] Lefkovits, H. (1977). *Data Dictionary Systems*. QED Information Sciences, Wellesley, MA.

[13] McCarthy, J. L. (1982). *Proc. Eighth Int. Conf. Very Large Databases*, pp. 234–243.

[14] Teitel, R. F. (1977). *Proc. Computer Sci. Statist.: Tenth Ann. Symp. Interface*, pp. 165–177.

[15] Ullman, J. D. (1982). *Principles of Database Systems*. Computer Science Press, Potomac, MD.

[16] Wong, H. K. T., ed. (1982). *Proc. First LBL Workshop Statist. Database Manag.*

[17] Wong, H. K. T., ed. (1982). *A LBL Perspective on Statistical Database Management*. University of California Press, Berkeley, CA.

(COMPUTERS AND STATISTICS
STATISTICAL SOFTWARE)

SAKTI P. GHOSH

STATISTICS OF INCOME *See* INCOME DISTRIBUTIONS

STATISTICS OF INCOME (SOI) BULLETIN

The *SOI Bulletin* provides the earliest published annual financial statistics from the various types of tax and information returns filed with the U.S. Internal Revenue Service. The Bulletin also includes information from periodic or special analytical studies of particular interest to tax administrators. In addition, historical data from 1970 to the present are provided for selected types of taxpayers, as well as on tax rates for individuals and gross internal revenue collections. The *SOI Bulletin* started publication in 1981 and is published quarterly. It is available from the Superintendent of Documents, U.S. Government Printing Office, Washington, DC 20402, U.S.A. Subscription price (1985): $20.00 domestic, $25.00 foreign; single copy price: $5.50 and $6.88, respectively.

The table of contents of Vol. 4, issue 4 (April 1985) is:

"Individual income tax rates, 1982" by Dan Holik.

"Taxpayers classified by sex" by Patricia A. Crabbe and Elizabeth L. Gross.

"Fiduciary income tax returns, 1982" by Gary J. Estep.

"Environmental taxes: Superfund and hazardous waste, 1981–83" by Rashida Belal.

"Crude oil windfall profit tax, second quarter 1984" by Ed Chung.

Selected statistical series (1970–1985).

Appendix—General description of SOI sample procedures and data limitations.

STATISTICS OF SHAPE *See* SHAPE STATISTICS

STATISTICS, OPTIMIZATION IN *See* OPTIMIZATION IN STATISTICS

STATISTICS, SUFFICIENT *See* SUFFICIENT STATISTICS

STATISTICS SWEDEN

Statistics Sweden is now the officially adopted English name of *Statistiska Centralbyrån*, abbreviated SCB. Statistics Sweden is the central statistical office in Sweden and is responsible for most of Swedish official statistics.

Swedish population statistics have deep historical roots. They have a solid basis in the continuous population registration, which has been done for a very long time by the State Church. In 1686 this registration became stipulated by law for the whole country. Regular Swedish population statistics began in 1749, partly under the auspices of the Royal Academy of Sciences, and soon a Royal Commission on Tabulation was permanently established. The statistics consisted of sets of population tables for the whole country (*Tabellverket*), regularly compiled from reports based on the population registration. The concern of the state authorities for further development of official statistics led to the start of a national statistical office, *Statistiska Centralbyrån*, i.e., Statistics Sweden in 1858. The responsibilities of the office included both population statistics and other important branches of official statistics, such as statistics on agriculture, education, and finances. During the last few decades Statistics Sweden has developed rapidly. Sample surveys in a modern sense became substantially important around 1950. From around 1960 the electronic computer quickly got a key role in the routines of the office. A considerable expansion followed.

Today the work of Statistics Sweden covers most subject matters of Swedish official statistics. It thus includes a wide range of areas such as demographic data, living conditions, education, justice, housing, employment, price indexes, national accounts, trade, enterprises, finances, agriculture, and environment. Data collection* is made by various means, comprising mail questionnaires, interviews, administrative records, and even direct physical measurements. The number of employees is presently (in 1985) about 1800, including about 200 interviewers. The office is located at two sites, the cities of Stockholm and Örebro. It is financed mainly by government appropriations, but to a notable extent also by income from commissioned work.

The publication activities in Swedish official statistics were integrated into a systematic programme at the start of Statistics Sweden in 1858. Today's programme contains a variety of publications with widely different scopes. It ranges from series of detailed specific reports to comprehensive overviews. Public statistical detabases are also provided. Recently developed new publications are to a great extent informative overviews of a popular and readable kind. Statistics Sweden also publishes an international scientific journal, the *Journal of Official Statistics*, which since 1985 replaced the former *Statistisk Tidskrift* (Statistical Review). The mailing address of the Stockholm site of Statistics Sweden is Statistics Sweden, S-115 81 Stockholm, Sweden.

The use of large computer-based registers is a particular feature of the work of Statistics Sweden. The office thus maintains the Register of the Total Population, which is updated weekly on the basis of the general population registration. This register is essential in work on statistics on individuals. Demographic statistics are obtained from it, and it is highly useful as a frame for censuses and surveys. Similarly there is the Central

Register of Enterprises, updated monthly, and the Agricultural Enterprise Register. The latter is updated yearly and in itself contains much of the data needed for basic agricultural statistics. Another feature of the work consists in a rather extensive use of administrative records as a data source. For instance, taxation data are used as a basis for statistics on income and employment.

Data quality is a matter of great concern. Essential efforts are made to keep both sampling errors and various kinds of nonsampling errors on an acceptably low level. But the quality standards encompass more than reliability. Thus relevance, comparability, and timeliness of statistics are also important aims for quality efforts. It is the policy of Statistics Sweden to try to supply users of statistics with adequate descriptions of the quality. Furthermore, the protection of privacy and confidentiality is imperative.

Research and development activities have always had a strong position at Statistics Sweden. Present efforts include extension of subject matters covered, improvements in presentation of statistics, and development of tools in the fields of statistical science and computer science.

(*STATISTISK TIDSKRIFT*)

M. RIBE

STATISTISCHE GESELLSCHAFT, DEUTSCHE

The Deutsche Statistische Gesellschaft [German Statistical Society (DStG)] founded in 1911, is among the large academic corporations in Germany. The purpose of the DStG is the promotion of the exchange of knowledge in both theoretical and applied statistics among statisticians and those involved with statistics in universities, industry, statistical offices, and statistical agencies or institutions, both in Germany and abroad.

The DStG is governed by an elected board including the president, the treasurer, two vice-presidents, and the editor of the journal *Allgemeines Statistisches Archiv*. Membership at the end of 1986 was about 680 individual and 40 institutional members. The society itself is affiliated with the International Statistical Institute*. The current address of the society is Deutsche Statistische Gesellschaft, Mertonstr. 17, 6000 Frankfurt 11, West Germany.

There are six sections within the DStG. They cover the main activities of the society. The sections meet at least once a year to facilitate exchange of ideas, knowledge, and experience in a group of experts. The sections are

New Statistical Methods,

Business and Market Statistics,

Regional Statistics,

Empirical Economic Research and Applied Econometrics,

Technical Statistics,

Statistical Education.

The annual meeting of the society generally is held as the "Statistical Week" in conjunction with the annual meeting of the Association of German Municipal Statisticians.

Annual short courses in statistical methodology are part of the continuing education for statisticians in government offices, industry, and professional institutions. They provide the opportunity to explore status and recent developments of the statistical praxis in one field of application.

Publications include the quarterly *Allgemeines Statistisches Archiv*, which covers a broad spectrum of articles in theoretical and applied statistics. The journal presents recent developments and review articles. One issue each year is devoted to the publication of the presentations given at the plenary session of the annual meeting. In addition, special issues of the journal cover the proceedings of short courses, section meetings, or topics of special interest. The spectrum may be exemplified by the table of contents of a recent issue:

"Concentration measures as instruments of evaluation of monopolistic and oligopolistic structures" by W. Deffaa, Stuttgart-Hohenheim.

"A regression model with mixed additive–multiplicative seasonal variation in time series" by H. Gollnick, Hamburg.

"Iterated Aitken estimators" by K. H. Jöckel, Dortmund.

"A plea for pseudo-Bayes estimators in distributed-lag models" by G. Ronning, Konstanz.

"Official statistics in the People's Republic of China (organisation and offer of services)" by R. Cremer, Darmstadt.

"Report on the meeting of the section on Empirical Economic Research and Applied Econometrics" by J. Frohn, Bielefeld.

"Report on the meeting of the section on New Statistical Methods" by S. Heiler, Dortmund.

Personal News

Book Reviews

The current editor is H. Rinne. The editorial address is Allgemeines Statistisches Archiv, Statistik und Ökonometrie, Universität Giessen, Licher Strasse 64, 6300 Giessen, West Germany.

CHRISTIAN LÖSCHCKE

STATISTISCHE HEFTE / STATISTICAL PAPERS

Statistical Papers was founded by G. Menges in 1960. In the beginning it appeared only once a year, a single issue constituting a volume. The size of the journal has grown steadily and since 1967 it has appeared four times a year. Except for the years 1966–1975, when R. Wagenführ served as co-editor, G. Menges was the only editor until his death (January 10, 1983). An advisory board consisting of some 20 scientists from Germany, France, Switzerland, Austria, and Czechoslovakia was established in 1966. Since August 1983 the journal has been published by Springer Publishing Company. The Advisory Board has been enlarged by statisticians from the United States, Canada, Sweden, and the Netherlands [T. S. Ferguson (Los Angeles); J. Kmenta (Ann Arbor); P. de Jong (Vancouver); C. R. Rao (Pittsburgh); R. Bartels (Sydney); D. A. S. Fraser (Toronto); D. A. Sprott (Waterloo); F. C. Palm (Maastricht); H. Wold (Uppsala)]. The Editorial Board now consists of four members: G. Bamberg (Augsburg); W. Janko (Wien); H. Schneeweiss (München); H. J. Skala (Paderborn). The current editorial address is H. Schneeweiss, Universität München, Akademiestrasse 1, D-8000 München 40, West Germany.

The aims and scope of the journal are formulated as follows: "The journal *Statistische Hefte/Statistical Papers* addresses itself to all persons and organizations that have to deal with statistical methods in their own field of work. It attempts to provide a forum for the presentation and critical assessment of statistical methods, in particular for the discussion of their methodological foundations as well as their potential applications. Methods that have broad applications will be preferred. However, special attention is given to those statistical methods which are relevant to the economic and social sciences. The journal publishes original papers written in German or English, i.e., research papers, short notes, survey articles, reports on statistical software, and, occasionally, book reviews." The journal also provides a comprehensive Calendar of Conferences. Furthermore, there is a new section dealing with algorithms, highlighting the importance of computers in theoretical and applied statistics. A contribution to this section may take the form of an algorithm, an algorithmic comparison, a certification, or a translation (meeting certain criteria).

The first volumes were characterized by a distinct majority of submissions from Germany; since then a shift towards submissions from outside Germany has taken place. There is now a strong tendency toward articles written in English. As of January 1988, only papers written in English will be considered for publication. The contents of a recent issue [Vol. 24, No. 1 (1983)] included:

> "On models and theories of inference; structural or pivotal analysis" by D. Brenner, D. A. S. Fraser, and G. Monette.
>
> "Offene versus vollständige Modellbildung in der Ökonometrie" by R. Pauly.
>
> "Adaptive verteilungsfreie Tests" by H. Büning.
>
> "A note on the true factorial price index" by B. M. Balk.
>
> "Local consistency of risk preferences" by K. Spremann and G. Bamberg.
>
> "A remark to G. Singh's paper: A note on inclusion probability proportional to size sampling" by S. Gabler.

All submitted papers are refereed by at least one referee.

G. BAMBERG

STATISTISK TIDSKRIFT

Statistisk Tidskrift (*Statistical Review*) is a journal published until recently by Statistics Sweden*. In 1985 it was replaced by the *Journal of Official Statistics*, which is also presented below.

Statistisk Tidskrift first appeared in 1860. Originally it served as a publication for statistical data, and from 1871 it contained a Statistical Yearbook of Sweden. Statistics Sweden had been established in 1858. Swedish official statistics are much older than the statistical agency. Articles in the *Statistisk Tidskrift* in the early years often presented plans for various developments. The first series of the journal lasted until 1913; during this period there were also many articles that analyzed and interpreted statistical data. Other articles dealt with statistics from earlier times, or discussed statistical activities in other countries or general methodology.

In 1914 the *Statistical Yearbook* became a separate publication. Other publications were considered suitable for the presentation of short-term statistics and for articles on statistical issues. Thus it was decided to discontinue the *Statistisk Tidskrift*. Later on its reintroduction was suggested, and the second series started in 1952. An English name, *Statistical Review*, was added and English summaries of the articles were provided. The articles, occasionally in English, were mostly expository and dealt with methods or with the organization of Swedish official statistics, or gave analyses of specific data. In 1963 a new publication series was created for the presentation of statistical data, and therefore the *Statistisk Tidskrift* no longer had to serve that purpose. So the journal went into its third series with extended space for articles. Through many expository articles it can be seen how new possibilities, by such means as electronic computers, modern survey theory, and large computer-based registers, helped to build a statistical office in its modern form.

Articles published during this last series gradually assumed a more universal character. The authors are to a great extent from outside Statistics Sweden, and articles in English are far from uncommon. There is a recognition of fundamental statistical questions, which are common for different countries. Thus it seemed natural to turn the *Statistisk Tidskrift* into an international journal. In 1985 it was replaced by the new *Journal of Official Statistics*, abbreviated *JOS*, also published by Statistics Sweden.

JOS deals with methods useful in the various steps of preparation of statistics; also with questions concerning the use of statistics, methods of analysis, and policies for statistical work. The language is English. The bulk of each issue consists of a section for refereed articles on methodology or policy. There are also sections for Miscellanea, Letters to the Editor, and Book Reviews. Each year, one of the four issues is devoted to a special topic. In 1985 the topic was questionnaire design; in 1986, population censuses*; and in 1987, nonsampling errors. The contents of the 1985 special issue are:

"Questionnaire design activities in government statistics offices" by S. Sudman and C. D. Cowan.

"Some important issues in questionnaire development" by R. Platek.

"Laboratory and field response research studies for the 1980 census of population in the United States" by N. D. Rothwell.

"Testing assessing question quality" by M. Thorslund and B. Wärneryd.

"Informal testing as a means of questionnaire development" by D. D. Nelson.

"Flow charts—A tool for developing and understanding survey questionnaires" by T. B. Jabine.

"Questionnaire design with computer assisted telephone interviewing" by C. C. House.

"A comparative study of field and office coding" by M. Collins and G. Courtenay.

"Questionnaire measurement of drinking behaviour in sample surveys" by J. C. Duffy.

"Telescoping—The skeleton in the recent reading closet" by V. Appel.

Special Notes.

Book Reviews.

The contents of Vol. 2, No. 3 are:

"The codification of statistical ethics" by R. Jowell.

"Record linkage for statistical purposes: Methodological issues" by T. B. Jabine and F. J. Scheuren.

"Sampling frames for agriculture in the United States" by R. Fecso, R. D. Tortora, and F. A. Vogel.

"Applying a British class scheme to Swedish data" by O. Lundberg.

"Handling wave response in panel surveys" by G. Kalton.

Miscellanea.

Special News.

Book Reviews.

The address for submission of manuscripts and for other communications is Journal of Official Statistics, Statistics Sweden, S-115 81 Stockholm, Sweden.

JOS will serve as an international forum for methodology and policy matters in official statistics. The intended readers are those who work at either statistical agencies, universities, or private organizations, and must deal with problems that concern official statistics. A purpose of *JOS* is to communicate valuable contributions of the kind that very often remain unpublished and not widely circulated. *JOS* will have a wider methodological scope than most other statistical journals, since official statistics may benefit from many disciplines, such as statistics, economics, computer science, and social science. Articles and other materials in *JOS* may either focus on specific topics and applications, or present general overviews and discussions.

(STATISTICS SWEDEN)

L. Lyberg
M. Ribe

STEEL STATISTICS

Steel statistics are two types of distribution-free* multiple comparisons* techniques. The many–one Steel statistics are methods for determining which of k treatment population location parameters differ significantly from a control population location parameter. The Steel k-sample statistics are methods for determining which pairs of k treatment population location parameters differ significantly from each other. In each case, they are multiple comparison procedures in that for a test at level α, $1 - \alpha$ is the probability that all statements of significance are correct under the null hypothesis* (*see* SIMULTANEOUS TESTING and MULTIPLE COMPARISONS). Steel statistics are based on either the signs or rank sums of appropriate differences of observations. Each test procedure has a corresponding confidence interval procedure for estimating the appropriate differences between location parameters.

The Steel statistics are rarely covered in books on nonparametric or distribution-free statistics. Lehmann [15] discusses both the Steel many–one rank sum statistic (pp. 228–229) and the k-sample rank sum statistic (pp. 238–244) and Kleijnen does also [13, pp. 338–342 and pp. 544–545, respectively], but other books either do not mention them or give only a reference. They are presented fully in Miller [18], which also covers many other simultaneous inference* techniques. See also Gabriel [9].

NOTATION

In the many–one situation, let F_0 be a control population and $F_1, F_2, \ldots, F_k$ be a set of k treatment populations. The general null hypothesis is that all distributions are the same, $H_0: F_0(x) = F_1(x) = \cdots = F_k(x)$ for all x. Since the procedures are especially sensitive to location differences, H_0 is frequently rewritten in terms of location parameters as

$$H_0': M_0 = M_1 = \cdots = M_k.$$

The alternative may be one-sided or two-sided, as

$$A_-: M_i < M_0 \quad \text{for at least one } i,$$

$$A_+: M_i > M_0 \quad \text{for at least one } i,$$

$$A: M_i \neq M_0 \quad \text{for at least one } i.$$

A random sample of n observations is taken from each of the k treatment populations, and X_{ij} denotes the jth observation from the ith population for $j = 1, 2, \ldots, n$ and $i = 1, 2, \ldots, k$. Similarly, X_{0j} denotes the jth observation from the control population for $j = 1, 2, \ldots, m$.

In the k-sample situation, the notation is identical except that there is no control population; the null hypothesis is $H_0': M_1 = M_2 = \cdots = M_k$ with alternative A: equality does not hold (always two-sided).

MANY–ONE STEEL STATISTICS BASED ON SIGNS

Assume that $m = n$ and that the observations occur in n blocks of size $k + 1$, a randomized block design. Let $D_{ij} = X_{ij} - X_{0j}$ for $j = 1, 2, \ldots, n$ be the differences between the ith treatment and the control. A special case of the null hypothesis is

$$H_0: \Pr[D_{ij} < 0] = \Pr[D_{ij} > 0] = 0.5$$

or equivalently H_0: median of $D_{ij} = 0$, for $i = 1, \ldots, k$, $j = 1, \ldots, n$. Under H_0, for any i we expect $n/2$ of the D_{ij} to be positive and $n/2$ to be negative for each population. Let S_i^+ (S_i^-) denote the number of positive (negative) differences among the D_{ij} for $j = 1, 2, \ldots, n$; the one-sided Steel test statistic is

$$S^+ = \max(S_1^+, S_2^+, \ldots, S_k^+)$$

or

$$S^- = \max(S_1^-, S_2^-, \ldots, S_k^-),$$

which are identically distributed. A large value of S^+ supports the alternative A_+ and a large value of S^- supports the alternative A_-. The two-sided Steel test statistic is $S = \max\max_i (S_i^+, S_i^-)$. Approximate critical values of these test statistics are given as a function of k and n in Table VI of Miller [18] for $\alpha = 0.01$ and $\alpha = 0.05$, adapted from Steel [25], where the null distributions of S_i^+ and S^+ are derived. Rhyne and Steel [22] extend these tables for larger n and $\alpha = 0.10$.

If ties occur between treatment and control observations so that $D_{ij} = 0$ for some i, j, it is recommended that + and − signs be assigned randomly to those zeros.

Table 1

Patient	Control	Method 1	2	3	D_{1j}	D_{2j}	D_{3j}
1	260	240	270	200	−20	10	−60
2	300	290	290	240	−10	−10	−60
3	290	320	320	240	30	30	−50
4	250	240	270	210	−10	20	−40
5	270	250	260	190	−20	−10	−80
6	180	220	230	160	40	50	−20
7	200	190	210	140	−10	10	−60
8	220	230	250	180	10	30	−40
9	410	420	430	270	10	20	−140
10	310	300	320	200	−10	10	−110
S^+					4	8	0
S^-					6	2	10

In an estimation context, these same tables can be used to develop joint level $1 - \alpha$ confidence intervals for the differences of the median of each treatment $M_1, M_2, \dots, M_k$, and the control median M_0. Suppose the two-sided critical value at level a for the given k, n is s. The confidence interval end points for $M_i - M_0$ are the order statistics $D_{n-s+1:n}$ and $D_{s:n}$ for the differences $D_{i1}, D_{i2}, \dots, D_{in}$, which must be calculated for each of $i = 1, 2, \dots, k$. One-sided confidence intervals are given similarly using one-tailed critical values.

Rhyne and Steel [22] give an example that compares three methods of measuring serum cholesterol with a standard method for a common blood sample in 10 patients. The data [22, p. 401] are given in Table 1.

For a two-tailed test with $\alpha = 0.05$, $k = 3$, and $n = 10$, Table VI of Miller [28] gives the critical value $s = 10$, and our test statistic here is $\max\max_i(S_i^+, S_i^-) = 10$ and thus only method 3 is found to be significantly different from the control.

Now we obtain joint confidence limits for the differences between the median of each method and the control method. Since $s = 10$, the confidence interval end points for each difference $M_i - M_0$ are the first $(n - s + 1 = 1)$ and tenth order statistics of the respective differences $D_{i1}, D_{i2}, \dots, D_{i10}$. The results here are:

$$-20 \leqslant M_1 - M_0 \leqslant 40$$

$$-10 \leqslant M_2 - M_0 \leqslant 50$$

$$-140 \leqslant M_3 - M_0 \leqslant -20.$$

These numbers are the 95% joint two-sided confidence limits. Since the $M_3 - M_1$ confidence interval end points are both negative, we can also conclude that method 3 gives significantly lower measurements than the control method.

Note that this procedure of testing and estimation is applicable only in a randomized complete block design with a common number of observations in each treatment and the control. Thus it is a direct competitor of the Friedman test procedure (*see* DISTRIBUTION-FREE METHODS and FRIEDMAN'S CHI-SQUARE TEST), except that the latter procedure is usually applied to a comparison of treatments among themselves without a control. A large-sample approximate many–one Friedman test procedure is given in Steel [25, p. 774], but corresponding estimation procedures have apparently not been developed. Power* comparisons of the tests have not resulted in a clear recommendation for use, but the Steel procedures are easier to apply.

Signed rank* procedures have been proposed for the treatment vs. control comparisons, primarily by Nemenyi [20], and Hollander [12], but they are not distribution-free. Robson [23] shows how Dunnett's many–one procedure [4, 5] for the completely randomized design can be adapted for use in the balanced incomplete block design, but the procedure assumes normal distributions and common variance.

MANY–ONE STEEL TESTS BASED ON RANK SUMS

Assume now that the $m + kn$ observations are mutually independent so that we have a completely randomized design. For each set of treatment observations separately, pool them with the control observations and order them from smallest to largest and rank them from 1 to $m + n = N$ while keeping track of which are the treatment observations. The rank sum of the treatment observations for the ith treatment is

$$R_i = \sum_{j=1}^{n} \text{rank } X_{ij}, \quad \text{for } i = 1, 2, \ldots, k,$$

and the Steel test statistic is

$$R = \max(R_1, R_2, \ldots, R_k).$$

A large value of R supports the alternative A_+ and a large value of

$$R' = \max(R_1', R_2', \ldots, R_k'),$$

where $R_i' = n(N + 1) - R_i$, supports A_-. Each treatment population with R exceeding the critical value is inferred to be significantly different from the control population in the appropriate direction. The two-sided test statistic is $R^* = \max \max_i(R_i, R_i')$. Approximate critical values of these test statistics are given in Table VIII of Miller [18] as a function of k and $m = n$ for $\alpha = 0.01$ and $\alpha = 0.05$, adapted from Steel [26], where the null distributions of R, R', and R^* are derived.

If ties occur between the treatment and control observations, the average rank procedure is recommended, where all tied observations are assigned the rank they would have if they were not tied.

In an estimation context, these same tables can be used to develop joint level $1 - \alpha$ confidence intervals for the differences of the median of each treatment and the control. Suppose the two-sided critical value for the given m, n, k at level α is s. Compute the mn differences $D_{i,jl} = X_{ij} - X_{0l}$ for $j = 1, \ldots, n$, $l = 1, \ldots, m$ in the ith sample and arrange them from smallest to largest. The confidence interval end points for $M_i - M_0$ are the order statistics*

$$D_{mn+[n(n+1)/2]-s+1:mn}, \qquad D_{s-[n(n+1)/2]:mn}$$

among $(D_{i,11}, \ldots, D_{i,mn})$. One-sided confidence intervals are given similarly using one-tailed critical values.

Steel [26] gives an example that compares IQ scores of female white children classified as premature, anoxic, or Rh negative at birth with a control of normal children at birth, six in each group. The desired alternative is that the treatment IQ medians are lower than the control IQ median. The data [26, p. 563] are given in Table 2. The corresponding

Table 2

Normal	Premature	Anoxic	Rh Negative
103 (4,5,3)	92 (2)	119 (10)	89 (2)
106 (5,6,4)	119 (9)	89 (2)	114 (7)
111 (6,7,5)	114 (7.5)	100 (4)	132 (11)
114 (7.5,9,7)	94 (3)	86 (1)	125 (10)
122 (10,11,9)	131 (11)	112 (8)	114 (7)
136 (12,12,12)	86 (1)	97 (3)	86 (1)
R_i	33.5	28	38
R_i'	44.5	50	40

Table 3

X_{ij}	$X_{ij} - X_{01}$	$X_{ij} - X_{02}$	$X_{ij} - X_{03}$	$X_{ij} - X_{04}$	$X_{ij} - X_{05}$	$X_{ij} - X_{06}$
86	−17	−20	−25	−28	−36	−50
192	−11	−14	−19	−22	−30	−44
94	−9	−12	−17	−20	−28	−42
114	11	8	3	0	−8	−22
119	16	13	8	5	−3	−17
131	28	25	20	17	9	−5

ranks of each treatment when pooled with the control are shown in parentheses.

For a one-tailed test with $\alpha = 0.05$, $k = 3$, and $m = n = 6$, Table VIII of Miller [18] gives the critical value $s = 53$, and our test statistic against A_- is 50. We conclude that none of the treatment groups has a lower median IQ than the normal group.

Now we obtain joint one-sided confidence limits of the differences $M_i - M_0$. The confidence interval is

$$M_i - M_0 \geqslant D_{mn+[n(n+1)/2]-s+1:mn},$$

which here is the fifth smallest difference among $(D_{i,11}, \ldots, D_{i,mn})$. We illustrate for $i = 1$ only. The $mn = 36$ differences for $X_{ij} - X_{0l}$ are in Table 3. The one-sided confidence intervals are $M_1 - M_0 \geqslant -30$, $M_2 - M_0 \geqslant -36$, and $M_3 - M_0 \geqslant -28$, with joint confidence 0.95.

This Steel test is the nonparametric analog of the Dunnett many–one t test, which is based on the assumption of normal distributions with common but unknown variance, and also provides simultaneous estimation techniques through confidence intervals. The relative performance of these procedures parallels the performance of the Mann–Whitney–Wilcoxon* test relative to the two-sample normal theory t test in that the Steel test is almost as good for normal distributions and better for many other distributions (see Sherman [24] and Hodges and Lehmann [11]). Another parametric procedure that assumes normal distributions but unknown and unequal variances and provides confidence intervals is given in Dudewicz et al. [2]. However, it is a two-stage procedure and as such is not comparable to the Steel test.

The nonparametric competitor to the Steel test is that proposed by Dunn [3]. The Dunn procedure ranks all the observations in a single array, while Steel ranks separately each set of treatment observations with the control. The computation for Dunn may take less effort because only one ranking is required, but it is questionable whether the ranks of observations from treatment j should affect a conclusion about differences between treatment i and the control. The Steel procedure uses information about only treatment i and the control for the inference about $M_i - M_0$. Fligner [8] shows that only Steel controls the maximum type I error rate. The Dunn and Steel procedures have the same power against Dunnett's test under normality, but Steel has greater power generally. All three procedures are conservative.

A general extension of this Steel procedure that can be used for all contrast comparisons and unequal sample sizes is given in Fligner [8]. The method is approximate and requires numerical integration* but a computer program is available; a simple but less accurate approximation is also given.

STEEL k-SAMPLE RANK SUM STATISTICS

The Steel k-sample rank sum statistics provide a method for comparing all pairs of a set of k treatments with each other in a completely randomized design, as opposed to the situation in the preceding section where we are comparing each treatment with

Table 4

T_1	T_2	T_3	T_1 vs. T_2		T_1 vs. T_3		T_2 vs. T_3	
1.8	3.2	1.1	3	5.5	2	1	7.5	22
11.0	1.2	3.1	8	2	8	6	3	6
2.7	3.2	2.4	4	5.5	5	4	7.5	5
8.5	0.7	2.2	7	1	7	3	1	4
Rank Sums			22	14	22	14	19	17
R^*			22		22		19	

a control. The primary references are Steel [27, 28] and Dwass [6].

The test procedure is to pool the treatment observations for each possible pair of treatments and compute the rank sums for each pair. Let R_{hi} denote the smaller rank sum for the (h, i) treatment combination and $R'_{hi} = n(2n+1) - R_{hi}$. There are $\binom{k}{2}$ values for R_{hi}. The two-sided Steel test statistic is

$$R^* = \max_{1 \leqslant h,\, i \leqslant k} (R_{hi}, R'_{hi})$$

and large values of R^* support rejection of $H'_0 : M_1 = M_2 = \cdots = M_k$. Critical values of R^* are given in Table IX of Miller [18], adapted from Steel [27, 28] and based on the large-sample approximation.

Confidence intervals* for $M_h - M_i$ can be constructed in a manner that parallels that for the Steel many–one rank sum procedure. We compute the n^2 differences, $D_{hi,\,jl} = X_{hj} - X_{il}$ for $j, l = 1, 2, \ldots, n$, of the observations on treatments h and i and pool them in an array. The $1 - \alpha$ confidence interval end points are the order statistics

$$D_{n^2 + [n(n+1)/2] - s + 1 \,:\, n^2}, \qquad D_{s - [n(n+1)/2] \,:\, n^2}$$

of $(D_{hi,11}, \ldots, D_{hi,\,nn})$, where s is the corresponding critical value of the test statistic R^* for a two-tailed test at level α.

Steel [27] gives an example with measurements on the speed of flow of air in soil in units of square microns for three different soil treatments, four observations on each treatment, in a completely randomized design. The data and corresponding ranks are given in Table 4. The P-value* is 0.625 and no treatment is judged significantly different from any other treatment.

The approximation to the critical value of R^* suggested by Steel [28] is

$$n(2n+1)/2 + q^{\alpha}_{k,\infty} \sqrt{n^2(2n+1)/24}\,,$$

where $q^{\alpha}_{k,\infty}$ is the upper $100\alpha\%$ point of the studentized range* distribution with parameters k and ∞. Gabriel and Lachenbruch [9] give a Monte Carlo study of the adequacy of this approximation and find it to be very conservative, the bias increasing with k for fixed n; they recommend its use only for k small and n large. By contrast, they find the chi-square approximation to the Kruskal–Wallis test* statistic, a nonparametric competitor, to be only slightly conservative for small n and quite adequate otherwise. These conclusions are reinforced by Lin and Haseman [16].

Morley [19] reports a simulation study of the power of the Steel k-sample rank sum statistic, the Kruskal–Wallis statistic, and the normal theory studentized range* statistic for equal sample sizes, six different distributions, and thirteen configurations of means, for $k = 4$, 7, and 10. The Steel statistic is found to be nearly always more powerful than the Kruskal–Wallis statistic when using the criterion of all-pairs power. Lin and Haseman [16] give Monte Carlo comparisons that suggest the opposite conclusion, but this Monte Carlo power study uses the criterion of power* based on individual comparisons rather than experimentwise comparisons and hence may be misleading because of differences in type I error probabilities. See Elinot and Gabriel [7] for some discussion and Carmer and Swanson [1] for a related study.

The primary nonparametric competitors to this Steel test include those proposed by Kruskal and Wallis [14], Nemenyi [20], Dunn [3], Tobach et al. [29], and McDonald and Thompson [17]. Of these, only the Steel procedures use information only about treatments h and i when making inferences about $M_h - M_i$. Lehmann [15] recommends the pairwise ranking over the joint ranking, although the former can lead to inconsistencies in rare circumstances (see Gabriel [8]). Further, only the Steel tests have easy corresponding confidence intervals for purposes of estimation. These other tests are more frequently used, however, probably because they can be applied with unequal sample sizes.

References

[1] Carmer, S. G. and Swanson, M. R. (1973). *J. Amer. Statist. Ass.*, **68**, 66–74.

[2] Dudewicz, E. J., Ramberg, J. S., and Chen, H. J. (1975). *Biom. Zeit.*, **17**, 13–26.

[3] Dunn, O. J. (1964). *Technometrics*, **6**, 241–252.

[4] Dunnett, C. W. (1955). *J. Amer. Statist. Ass.*, **50**, 1096–1121.

[5] Dunnett, C. W. (1964). *Biometrics*, **20**, 482–491.

[6] Dwass, M. (1960). In *Contributions to Probability and Statistics*, I. Olkin et al., eds. Stanford University Press, Stanford, CA, pp. 198–202.

[7] Elinot, I. and Gabriel, K. R. (1975). *J. Amer. Statist. Ass.*, **70**, 574–583.

[8] Fligner, M. A. (1984). *J. Amer. Statist. Ass.*, **79**, 208–211.

[9] Gabriel, K. R. (1969). *Ann. Math. Statist.*, **40**, 224–250.

[10] Gabriel, K. R. and Lachenbruch, P. A. (1969). *Biometrics*, **25**, 593–596.

[11] Hodges, J. L., Jr. and Lehmann, E. L. (1956). *Ann. Math. Statist.*, **27**, 324–335.

[12] Hollander, M. (1966). *Ann. Math. Statist.*, **37**, 735–738.

[13] Kleijnen, J. P. C. (1975). *Statistical Techniques in Simulation, Part II*. Marcel Dekker, New York.

[14] Kruskal, W. H. and Wallis, W. A. (1952). *J. Amer. Statist. Ass.*, **47**, 583–621; errata, **48**, 907–911.

[15] Lehmann, E. L. (1975). *Nonparametric Statistical Methods Based on Ranks*. Holden-Day, San Francisco, CA. (Intermediate; Steel tests are covered on pp. 228–229 and 238–244.)

[16] Lin, F. A. and Haseman, J. K. (1978). *Commun. Statist. B*, **7**, 117–128.

[17] McDonald, B. J. and Thompson, W. A. (1967). *Biometrika*, **54**, 487–498.

[18] Miller, R. G., Jr. (1981). *Simultaneous Statistical Inference*., 2nd ed. Springer, New York (Intermediate; mostly theory; Steel tests are covered in Secs. 1, 3, and 4 of Chap. 4.)

[19] Morley, C. L. (1982). *Austral. J. Statist.*, **24**, 201–210.

[20] Nemenyi, P. (1963). Distribution-Free Multiple Comparisons. Ph.D. dissertation, Princeton University, Princeton, NJ.

[21] O'Neill, R. and Wetherill, G. B. (1971). *J. R. Statist. Soc. B*, **33**, 218–250.

[22] Rhyne, A. L. and Steel, R. G. D. (1965). *Technometrics*, **7**, 293–306.

[23] Robson, D. S. (1961). *Technometrics*, **3**, 103–105.

[24] Sherman, E. (1965). *Technometrics*, **7**, 255–256.

[25] Steel, R. G. D. (1959). *J. Amer. Statist. Ass.*, **54**, 767–775.

[26] Steel, R. G. D. (1959). *Biometrics*, **15**, 560–572.

[27] Steel, R. G. D. (1960). *Technometrics*, **2**, 197–207.

[28] Steel, R. G. D. (1961). *Biometrics*, **17**, 539–552.

[29] Tobach, E., Smith, M., Rose, G., and Richter, D. (1967). *Technometrics*, **9**, 561–568.

(DISTRIBUTION-FREE METHODS
FRIEDMAN'S CHI-SQUARE TEST
MULTIPLE COMPARISONS
SIMULTANEOUS TESTING)

JEAN DICKINSON GIBBONS

STEEPEST DESCENT, METHOD OF

See SADDLE POINT APPROXIMATIONS

STEIN EFFECT, THE

Stein [30] established the remarkable fact that (in frequentist decision-theoretic terms) the usual estimator of a p-variate normal mean, namely the vector of sample means, could be improved upon for $p \geqslant 3$. This was particularly surprising since Blyth [11] had shown that the sample mean could not be improved upon for $p = 1$. Hence, "optimal" estimators in "independent" problems were not necessarily optimal if the problems were considered simultaneously. This has become known as the *Stein effect*, and it exists in a very wide variety of situations.

PRELIMINARIES

Suppose independent random variables X_i, $i = 1, \ldots, p$, having (for simplicity) continuous or discrete densities $f_i(x_i|\theta_i)$ on R^1, are to be observed. It is desired to estimate the unknown θ_i, with presumed losses in estimating θ_i by a_i of $L_i(\theta_i, a_i)$, for $i = 1, \ldots, p$. Letting $\mathbf{X} = (X_1, \ldots, X_p)$, an estimator $\delta_i(\mathbf{X})$ of θ_i is to be evaluated by its risk function (or expected loss)

$$R_i(\theta_i, \delta_i) = E_{\boldsymbol{\theta}}[L_i(\theta_i, \delta_i(\mathbf{X}))].$$

If it is desired to simultaneously estimate all coordinates, i.e., to estimate $\boldsymbol{\theta} = (\theta_1, \ldots, \theta_p)$ by $\boldsymbol{\delta}(\mathbf{X}) = (\delta_1(\mathbf{X}), \ldots, \delta_p(\mathbf{X}))$, the loss will be presumed to be $L(\boldsymbol{\theta}, \boldsymbol{\delta}) = \sum_{i=1}^{p} L_i(\theta_i, \delta_i)$ and hence the risk function will be

$$R(\boldsymbol{\theta}, \boldsymbol{\delta}) = E_{\boldsymbol{\theta}} L(\boldsymbol{\theta}, \boldsymbol{\delta}(\mathbf{X})) = \sum_{i=1}^{p} R_i(\theta_i, \delta_i).$$

An estimator $\boldsymbol{\delta}$ is *inadmissible* if there exists a $\boldsymbol{\delta}^*$ with $R(\boldsymbol{\theta}, \boldsymbol{\delta}^*) \leqslant R(\boldsymbol{\theta}, \boldsymbol{\delta})$, where the inequality is strict for some $\boldsymbol{\theta}$. Otherwise $\boldsymbol{\delta}$ is *admissible* (*see* ADMISSIBILITY).

It is frequently the case that "standard" estimators $\delta_i^0(\mathbf{X})$ (usually depending only on X_i) are admissible in the individual problems of estimating the θ_i based on the risks R_i, but that $\boldsymbol{\delta}^0(\mathbf{X}) = (\delta_1^0(\mathbf{X}), \ldots, \delta_p^0(\mathbf{X}))$ is inadmissible in estimating $\boldsymbol{\theta}$ using the overall risk R, even though the X_i are independent and the θ_i need not be related in any way. This surprising phenomenon, that one can combine unrelated problems to get improved estimators, is what is commonly called the Stein effect.

The original example of this is that in which the X_i are normal $\mathcal{N}(\theta_i, 1)$, $L_i(\theta_i, a_i) = (\theta_i - a_i)^2$, and $\delta_i^0(\mathbf{X}) = X_i$. Then each δ_i^0 is admissible for estimating δ_i alone under R_i (Blyth [11]), but $\boldsymbol{\delta}^0(\mathbf{X}) = \mathbf{X}$ has everywhere larger risk R than the James–Stein estimator*

$$\boldsymbol{\delta}^{\text{J-S}}(\mathbf{X}) = \left\{1 - (p-2) \Big/ \sum_{i=1}^{p} X_i^2 \right\} \mathbf{X}$$

for $p \geqslant 3$ (see James and Stein [25] and Stein [31]).

EXAMPLES

There is a huge literature concerning the Stein effect. In this section the literature is categorized, and a few simple examples are given.

The General Linear Model*

The Stein effect has been found to hold in essentially any normal estimation problem involving the estimation of at least a three-dimensional mean. Independence of coordinates is not needed, and hence the results apply to quite general estimation in the normal linear model; for instance, the ordinary least-squares* estimator of three or more regression coefficients* in a linear regression with normal error structure can be improved upon via the Stein effect. Most of the results that have been obtained are for known error covariance matrix and quadratic loss [i.e., $L(\boldsymbol{\theta}, \boldsymbol{\delta}) = (\theta - \delta)^t \mathbf{Q} (\theta - \delta)$, $\mathbf{Q}$ a $(p \times p)$ positive definite matrix]. See Berger [5; 7, pp. 360–369], Bock [26], Stein [31], Strawderman [32], and Zidek [33] for some such results and other references.

Results for unknown error covariance matrix can be found (with other references) in Gleser [21], and results and references for other losses [of the form $L(\boldsymbol{\theta} - \boldsymbol{\delta})$] are in Berger [3] and Brandwein and Strawderman [12]. The Stein effect has also been shown to exist in the normal problem with the "control" loss $L(\boldsymbol{\theta}, \boldsymbol{\delta}) = (\boldsymbol{\theta}^t\boldsymbol{\delta} - 1)^2$ (cf. Berger et al. [8]), and in estimating the mean function of a Gaussian process* (cf. Berger and Wolpert [9]). Finally, the Stein effect extends to the existence of improved confidence intervals and confidence ellipsoids for a normal mean vector in three or more dimensions (cf. Hwang and Casella [24]).

Continuous Exponential Families

A variety of instances of the Stein effect has been observed in other continuous exponential families*. General references are Berger [4] and Hudson [22]. As a typical example, suppose

$$f_i(x_i|\theta_i) = \theta_i^{\alpha_i} x_i^{(\alpha_i - 1)} e^{-x_i\theta_i}/\Gamma(\alpha_i),$$
$$x_i > 0,\ \alpha_i > 2,$$

and $L_i(\theta_i, a_i) = (\theta_i - a_i)^2$. Thus it is desired to estimate gamma scale parameters under squared error loss. The usual (best invariant) estimator of θ_i is $\delta_i^0(X) = (\alpha_i - 2)/X_i$, and it is admissible under the coordinatewise risk R_i. However (Berger [4]), $\boldsymbol{\delta}^*(\mathbf{X}) = (\delta_1^*(\mathbf{X}), \dots, \delta_p^*(\mathbf{X}))$, where

$$\delta_i^*(\mathbf{X}) = \frac{\alpha_i - 2}{X_i}\left\{1 + \frac{cX_i^2}{(\alpha_i - 2)\left(b + \Sigma_{j=1}^p X_j^2\right)}\right\},$$

has $R(\boldsymbol{\theta}, \boldsymbol{\delta}^*) < R(\boldsymbol{\theta}, \boldsymbol{\delta}^0)$ for all $\boldsymbol{\theta}$ if $b \geqslant 0$, $0 < c < 4(p - 1)$, and $p \geqslant 2$; thus a Stein effect is present.

Several interesting features of this example deserve mention and are typical of nonnormal estimation problems (and even some normal problems, as shown in Brown [15]). First, only *two* dimensions are needed for the Stein effect to manifest itself. Second, the improved estimator $\boldsymbol{\delta}^*$ does not "shrink" $\boldsymbol{\delta}^0$ toward a point, as is typical in normal situations ($\boldsymbol{\delta}^{\text{J-S}}$ shrinks $\boldsymbol{\delta}^0$ toward zero), but expands $\boldsymbol{\delta}^0$. Third, the distributions of the X_i need not be the same for the Stein effect to be present; indeed (Berger [4]) the effect exists even if normal random variables are combined with gamma random variables.

Another important area is that of normal covariance matrix estimation (cf. Efron and Morris [18] and Haff [21]).

Discrete Exponential Families

For general results on the Stein effect in discrete expotential families and other references see Hwang [23] and Ghosh et al. [19]. As an example (from Clevenson and Zidek [16]), suppose

$$f_i(x_i|\theta_i) = \theta_i^{x_i} e^{-\theta_i}/x_i!, \qquad x_i = 0, 1, 2, \dots,$$

and $L_i(\theta_i, a_i) = (\theta_i - a_i)^2/\theta_i$. Thus it is desired to estimate Poisson means under a weighted squared error loss. The usual estimator $\delta_i^0(\mathbf{X}) = X_i$ is admissible under R_i, but the estimator

$$\boldsymbol{\delta}^*(\mathbf{X}) = \left(1 - \frac{c\Sigma_{i=1}^p X_i}{p - 1 + \Sigma_{i=1}^p X_i}\right)\mathbf{X}$$

has $R(\boldsymbol{\theta}, \boldsymbol{\delta}^*) < R(\boldsymbol{\theta}, \boldsymbol{\delta}^0)$ for all $\boldsymbol{\theta}$ if $0 < c < 2(p - 1)$ and $p \geqslant 2$.

Location Families

There is substantial literature on the Stein effect when $\mathbf{X}$ has a location density, i.e., of the form $f(\mathbf{x} - \boldsymbol{\theta})$. The most practical results include those in Berger [1a] and Brandwein and Strawderman [12], who develop estimators similar to $\boldsymbol{\delta}^{\text{J-S}}$, which improve on the best invariant estimator $\boldsymbol{\delta}^0$ under squared error loss when f is spherically symmetric about $\boldsymbol{\theta}$ (or more generally an elliptic distribution of the form

$$f\{(\mathbf{x} - \boldsymbol{\theta})^t \boldsymbol{\Sigma}^{-1}(\mathbf{x} - \boldsymbol{\theta})\}.$$

Also of practical interest are the results in Shinozaki [29], which concern a variety of cases involving independent X_i (such as independent uniform and t distributions).

There is also a substantial *theoretical* literature for location families, in which the Stein effect is shown to *exist*, but explicit improved estimators are not exhibited. For essentially any three- or higher-dimensional location family and loss of the form $L(\boldsymbol{\theta} - \boldsymbol{\delta})$, the Stein effect exists (Brown [13]). It exists in a variety of location problems with nuisance parameters, the needed dimension depending on the number of nuisance parameters (Berger [2]). Since a log transform converts a scale parameter problem into a location parameter problem, all these results also apply to the existence of the Stein effect in scale problems.

Two final articles deserve mention. In Blackwell [10], there is the first actual example of a Stein effect, although it was for a highly artificial discrete location distribution. Brown [14] is noteworthy for its general heuristic development of the Stein effect.

THE STEIN EFFECT AND BAYESIAN ANALYSIS

There are intimate relationships between the Stein effect and Bayesian* analysis. The first observed relationship (cf. Lindley and Smith [27] and Efron and Morris [17]) was that Stein effect estimators (such as $\boldsymbol{\delta}^{\text{J-S}}$) are often very similar to hierarchial Bayes and empirical Bayes estimators, developed for situations in which the θ_i are thought to be related in some fashion (*see* EMPIRICAL BAYES THEORY). Furthermore, the greatest gains in risk from the Stein effect seem to occur in precisely these hierarchical Bayes or empirical Bayes settings (cf. Morris [28]).

The above point is related to the general fact, discussed in Berger [5; 7, pp. 364–369], that to take practical advantage of the Stein effect, prior information about the θ_i must be taken into account. The reason for this is that any improved Stein effect estimator $\boldsymbol{\delta}^*$ will have risk *significantly* smaller than that of the standard estimator $\boldsymbol{\delta}^0$, only in a relatively small region or subspace of the parameter space. Thus it is important to choose a Stein effect estimator tailored to do well in the region or subspace where $\boldsymbol{\theta}$ is a priori suspected to lie.

Although closely related to Bayesian ideas in most applications, it is important to realize that the Stein effect is a distinct phenomenon. This is so because the θ_i need not be related in any way or even be the same type of parameter for the Stein effect to hold (as mentioned in Sec. 2). Furthermore, it will not generally be true that hierarchial Bayes or empirical Bayes estimators dominate standard estimators in terms of risk. Of course, risk dominance is not of direct interest to Bayesians because it is a frequentist concept (involving averaging over $\mathbf{X}$), but, nevertheless, the Stein effect can be of substantial aid in attaining Bayesian robustness. For example (Berger [6]), if the X_i are $\mathcal{N}(\theta_i, \sigma^2)$, $L(\theta_i, a_i) = (\theta_i - a_i)^2$, $p \geqslant 3$, and the θ_i are thought (a priori) to be $\mathcal{N}(\mu_i, \tau^2)$, then use of the Stein effect estimator [defining $\boldsymbol{\mu} = (\mu_1, \ldots, \mu_p)$]

$$\boldsymbol{\delta}^*(\mathbf{X}) = \mathbf{X} - (\mathbf{X} - \boldsymbol{\mu})\min\left\{\frac{\sigma^2}{(\sigma^2 + \tau^2)}, \frac{2(p-2)\sigma^2}{\sum_{i=1}^{p}(X_i - \mu_i)^2}\right\}$$

is virtually as good as use of the conjugate prior Bayes estimator if the θ_i really are $\mathcal{N}(\mu_i, \tau^2)$, and is *never* worse than use of the standard estimator $\boldsymbol{\delta}^0(\mathbf{X}) = \mathbf{X}$ no matter how much the specification of the prior distribution is in error. (In contrast, the conjugate prior Bayes estimator can be very bad in case of prior misspecification.) This phenomenon, of having an estimator with essentially optimum Bayesian performance and that is, at the same time, extremely robust with respect to possible misspecification of the prior distribution, seems to be due to the Stein effect. If, however, it is certain that the θ_i are a priori independent, then the Stein effect does not seem to be necessary or helpful in obtaining Bayesian robustness (Angers and Berger [1]).

References

[1] Angers, J. F. and Berger, J. (1986). *Commun. Statist.–Theor. Meth.*, **15**, 200–220.
[1a] Berger, J. (1975). *Ann. Statist.*, **3**, 1318–1328.
[2] Berger, J. (1976). *Ann. Statist.*, **4**, 1065–1076.
[3] Berger, J. (1978). *J. Multivariate Anal.*, **8**, 173–180.
[4] Berger, J. (1980). *Ann. Statist.*, **8**, 545–571.
[5] Berger, J. (1982). *Ann. Statist.*, **10**, 81–92.

[6] Berger, J. (1982). *J. Amer. Statist. Ass.*, **77**, 358–368.

[7] Berger, J. (1985). *Statistical Decision Theory and Bayesian Analysis*, 2nd ed. Springer-Verlag, New York.

[8] Berger, J., Berliner, L. M., and Zaman, A. (1982). *Ann. Statist.*, **10**, 838–856.

[9] Berger, J. and Wolpert, R. (1983). *J. Multivariate Anal.*, **13**, 401–424.

[10] Blackwell, D. (1951). *Ann. Math. Statist.*, **22**, 393–399.

[11] Blyth, C. R. (1951). *Ann. Math. Statist.*, **22**, 22–42.

[12] Brandwein, A. and Strawderman, W. (1980). *Ann. Statist.*, **8**, 279–284.

[13] Brown, L. D. (1966). *Ann. Math. Statist.*, **37**, 1087–1137.

[14] Brown, L. D. (1979). *Ann. Statist.*, **7**, 960–994.

[15] Brown, L. D. (1980). *Ann. Statist.*, **8**, 572–585.

[16] Clevenson, M. and Zidek, J. (1975). *J. Amer. Statist. Ass.*, **70**, 698–705.

[17] Efron, B. and Morris, C. (1973). *J. Amer. Statist. Ass.*, **68**, 117–130.

[18] Efron, B. and Morris, C. (1976). *Ann. Statist.*, **4**, 22–32.

[19] Ghosh, M., Hwang, J., and Tsui, K. (1983). *Ann. Statist.*, **11**, 351–376.

[20] Gleser, L. (1979). *Ann. Statist.*, **7**, 838–846.

[21] Haff, L. R. (1980). *Ann. Statist.*, **8**, 586–597.

[22] Hudson, H. M. (1978). *Ann. Statist.*, **6**, 473–484.

[23] Hwang, J. T. (1982). *Ann. Statist.*, **10**, 857–867.

[24] Hwang, J. T. and Casella, G. (1982). *Ann. Statist.*, **10**, 868–881.

[25] James, W. and Stein, C. (1960). In *Proc. Fourth Berkeley Symp. Math. Statist. Prob.*, Vol. 1. University of California Press, Berkeley, CA, pp. 361–379.

[26] Judge, G. and Bock, M. E. (1978). *Statistical Implications of Pre-Test and Stein-Rule Estimators in Econometrics*. North-Holland, Amsterdam, The Netherlands.

[27] Lindley, D. V. and Smith, A. F. M. (1972). *J. R. Statist. Soc. Ser. B*, **34**, 1–41.

[28] Morris, C. (1983). *J. Amer. Statist. Ass.*, **78**, 47–65.

[29] Shinozaki, N. (1984). *Ann. Statist.*, **12**, 322–335.

[30] Stein, C. (1956). In *Proc. Third Berkeley Symp. Math. Statist. Prob.*, Vol. 1. University of California Press, Berkeley, CA, pp. 197–206.

[31] Stein, C. (1981). *Ann. Statist.*, **9**, 1135–1151.

[32] Strawderman, W. E. (1978). *J. Amer. Statist. Ass.*, **73**, 623–627.

[33] Zidek, J. (1978). *Ann. Statist.*, **6**, 769–782.

(DECISION THEORY
EMPIRICAL BAYES THEORY
JAMES–STEIN ESTIMATORS
MINIMAX ESTIMATION
RIDGE REGRESSION)

JAMES O. BERGER

STEIN ESTIMATOR *See* EMPIRICAL BAYES THEORY

STEM-AND-LEAF DISPLAY

A stem-and-leaf display (Fig. 1) is a graphical display of data that provides information about certain characteristics in a batch of numbers. It is one of the most familiar tools in exploratory data analysis* and hence its origin is associated most commonly with Tukey [2]. The concept is based on the histo-

A. The data (in units of gram/gram):

11.7, 13.3, 13.4, 13.5, 13.6, 13.7, 14.0, 14.1, 14.2,
14.3, · · · , 16.5, 16.8, 16.8, 16.8, 17.1, 18.0, 18.3.

B. Stem-and-leaf display (one line per stem):

```
11|7
12|
13|34567
14|01233334445566777888889999
15|0001222334455666677777788899
16|0001111133355888
17|1
18|03
```

C. Stem-and-leaf display (two lines per stem):

```
11 · |7
12 * |
12 · |
13 * |34
13 · |567
14 * |0123333444
14 · |556677788889999
15 * |00012223344        ← median (39th value)
15 · |5566667777788899           = 15.3
16 * |00011111333
16 · |55888
17 * |1
17 · |
18 * |03
```

Figure 1 Stem-and-leaf display of 77 male hemoglobin values.

gram*, which dates back to the eighteenth century.

The basic idea is to provide distributional information similar to that from a histogram*, but in addition to retain the numerical information from which it is constructed. Whereas a histogram indicates merely the count of the values in a given interval, a stem-and-leaf display retains the significant digits of the values themselves. Thus, the number of such data values in a given interval is the same as that provided by the corresponding histogram, but the data values themselves may be read off from the display.

To construct a stem-and-leaf display from a batch of numbers, each number is "split" into its "stem," or major part of the number, and its "leaf," or minor part. For example, 4.2 or 42 or 420 could be split as "4" for the stem and "2" for the leaf. Then the stems are listed vertically, one line per stem (generally smallest to largest), and the leaves are written alongside. For example, if a batch includes the numbers 4.2, 4.6, 4.8, 4.3, 4.1, these may be displayed on a single line as 4|26831. Notice that there is no need to repeat the stem in every case since it is always the same for all numbers on the line. For ease in calculating statistics from the display (e.g., median, hinges, extremes), the leaves are usually sorted within the line, e.g., 4|12368. Figure 1 shows a batch of numbers and the corresponding stem-and-leaf display.

The stem-and-leaf display is an extremely effective tool because:

1. it offers a qualitative view of the batch (symmetry, spread, concentrations, gaps, outliers*);
2. it retains the numerical information, which can aid in identifying a particular data value with its source;
3. it results in a sorted batch, from which certain order statistics* (e.g., median, hinges, extremes) can be calculated easily. These values may be used to construct other displays such as a box-and-whisker plot.

A. The data (in microvolts):

Cell A: 8189.4, 8189.0, 8188.7, 8188.8, 8188.8, 8188.9, 8189.0, 8189.2, 8189.2, 8189.2, 8189.4, 8189.3

Cell B: 8188.8, 8190.0, 8190.0, 8190.2, 8190.2, 8190.3, 8190.2, 8190.6, 8190.6, 8190.6, 8190.7, 8190.7

B. The data, minus 8189.0 microvolts:

Cell A: 0.4, 0.0, −0.3, −0.2, −0.2, −0.1, 0.0, 0.2, 0.2, 0.2, 0.4, 0.3

Cell B: − 0.2, 1.0, 1.0, 1.2, 1.2, 1.3, 1.2, 1.6, 1.6, 1.6, 1.7, 1.7

C. Stem-and-leaf display:

Cell A		Cell B
322	−T	2
1	−O	
00	O	
33222	T	
4	F	
	S	
	*	
	10	00
	T	2223
	F	
	S	66677

Figure 2 Stem-and-leaf display of NBS measurements on two standard cells.

Sometimes, there may be too many numbers having the same stem to show them all on one line. The ten leaves corresponding to the one stem may be subdivided into two lines, as in 4*|01234 and 4 · |56789, or five lines, as in 4O|01, 4T|23, 4F|45, 4S|67, and 4*|89. The letters O (0, one), T (two, three), F (four, five), and S (six, seven), and the symbol * (eight, nine) remind the user which digits are placed on the lines. Part C of Fig. 1 shows an expanded display using two lines per stem. Emerson and Hoaglin [1, p. 11] recommend the value $L = 10 \log_{10} n$ for the total number of lines in a stem-and-leaf display of n values.

Comparing samples from two populations can be done effectively using a back-to-back stem-and-leaf display, where both samples share the same set of stems, and the leaves from the two samples are placed on either side. Figure 2 illustrates such a comparison for two sets of measurements obtained on two standard cells at the National Bureau of Standards in March 1979. On the average, Cell A contains 8189.1 microvolts, whereas

Cell B contains 8190.2 microvolts. In addition, the display highlights the unusually low value on Cell B, far from the other values; in fact, it was the first measurement in the measurement process.

Emerson and Hoaglin [1] discuss additional variations to the basic display and its relation to the histogram and provide numerous examples. Basic and Fortran programs for its construction are available in ref. 3.

References

[1] Emerson, J. D. and Hoaglin, D. C. (1983). In *Understanding Robust and Exploratory Data Analysis*, D. C. Hoaglin, F. Mosteller, and J. W. Tukey, eds. Wiley, New York.

[2] Tukey, J. W. (1977). *Exploratory Data Analysis*. Addison-Wesley, Reading, MA.

[3] Velleman, P. F. and Hoaglin, D. C. (1982). *Applications, Basics, and Computing of Exploratory Data Analysis*. Duxbury Press, New York.

(EXPLORATORY DATA ANALYSIS
GRAPHICAL REPRESENTATION OF DATA
HISTOGRAMS
NOTCHED BOX-AND-WHISKER PLOT)

KAREN KAFADAR

STEP

In exploratory data analysis*, this term is used to denote 1.5 times the interquartile range* or $1.5 \times H$ spread; see ref. 1 and FIVE-NUMBER SUMMARIES.

Reference

[1] Tukey, J. W. (1977). *Exploratory Data Analysis*. Addison-Wesley, Reading, MA.

(EXPLORATORY DATA ANALYSIS)

STEP-DOWN *See* ELIMINATION OF VARIABLES

STEP FUNCTION

Generally, a step function is a function $g(x)$ with derivative equal to zero except at an enumerable number of values of x. In statistics and probability, the most frequently encountered step function is the cumulative distribution function* (CDF) of a random variable X taking integer values $0, 1, 2, \ldots$. If

$$\Pr[X = j] = p_j, \qquad j = 0, 1, 2, \ldots,$$

then the CDF is

$$F_x(x) = \begin{cases} 0 & \text{for } x < 0, \\ p_0 & \text{for } 0 \leqslant x < 1, \\ p_0 + p_1 & \text{for } 1 \leqslant x < 2, \\ p_0 + p_1 + \cdots + p_{j-1} & \\ \text{for } j - 1 \leqslant x < j;\ j = 3, 4, \ldots & \end{cases}$$

(CUMULATIVE DISTRIBUTION FUNCTION
DISCRETE DISTRIBUTIONS)

STEP DISTRIBUTION APPROXIMATION, VON MISES'

The von Mises approximation theorem [3] states: "Let $M_0(= 1), M_1, \ldots, M_{2n-1}$ be the moments of order zero to $(2n - 1)$ of a distribution $F(x)$ with support* containing at least n points. Then there is a unique n-step distribution $V_n(x)$ that has these moments. The n steps are in the *interior* of the smallest interval containing all points of increase of $F(x)$. Either

$$F(x) = V_n(x)$$

or $F(x)$ crosses each step of $V_n(x)$."

von Mises [3] developed an algorithm for computing the appropriate values of x and corresponding step size for $V_n(x)$ (see, e.g., Springer [1]). Thompson and Palicio [2] apply this method to approximate a posterior distribution of system availability.

References

[1] Springer, M. E. (1979). *The Algebra of Random Variables*. Wiley, New York, pp. 269–270.

[2] Thompson, W. E. and Palicio, P. A. (1975). *IEEE Trans. Rel.*, **R-24**, 118–120.

[3] von Mises, R. (1964). *A Mathematical Theory of Probability and Statistics*. Academic, New York, pp. 384–400.

(n-POINT METHOD)

STEP-SIZE CLUSTERING

This is a method of applying hierarchical clustering procedures. As described by Johnson [1] and Sokal and Sneath [3], it consists of examining the difference in fusion values between successive hierarchy levels. A large difference suggests that the data were overclustered (i.e., insufficiently subdivided) at the first level. The maximum difference serves as an indication of the optimal number of clusters. A recent Monte Carlo* study [2] indicates that the step-size rule is very simple to apply, as compared to the likelihood* criterion proposed by Wolfe [4] for testing the hypothesis of k clusters *versus* $(k-1)$ clusters. It also showed that under certain conditions there is usually little difference in performance of the two methods. In particular the step-size gives good results at the level of two clusters.

References

[1] Johnson, S. C. (1967). *Psychometrika*, **32**, 241–254.

[2] Milligan, G. W. and Cooper, M. C. (1985). *Psychometrika*, **50**, 159–179.

[3] Sokal, R. R. and Sneath, P. H. A. (1963). *Principles of Numerical Taxonomy*. Freeman, San Francisco, CA.

[4] Wolfe, J. H. (1970). *Multivariate Behav. Res.*, **5**, 329–350.

(HIERARCHICAL CLUSTER ANALYSIS)

STEP-UP *See* ELIMINATION OF VARIABLES

STEPWISE REGRESSION

The stepwise regression method as it is presented in standard statistical computing packages, e.g., SAS, BMDP, SPSS, etc. (*see* STATISTICAL SOFTWARE), is a sequential procedure for entering and deleting variables in building a linear regression* model. Originally formulated by Efroymson in 1960 [2], the stepwise procedure became the most popular method for applied statisticians in the 1960s and 1970s. Draper and Smith, in the first edition of *Applied Regression Analysis* [2], recommended it to users as a useful beginning in the model building process. Subsequently many other techniques have been developed; see Hocking [4]. In the 1980s, stepwise regression is viewed as one of many techniques. It is usually applied after extensive analysis on the raw data is done and in conjunction with such techniques as "best" subsets regression, etc. Draper and Smith [1] still recommend it as a useful tool for the applied statistician.

The statistical problem for which stepwise regression was created is: Given a set of n observations on each of k x variables and a response variable Y, where $n \gg k$, choose a subset of p x variables that "best" predicts (explains) the response Y. A linear additive model is used:

$$Y_{ij} = \beta_0 + \sum_{j=1}^{p} \beta_j x_{ij} + \epsilon_{ij}.$$

Least squares* is used to estimate the unknown β's at each step of the selection process. This is a univariate statistical procedure in that the only random variable is Y and the x's are treated as nonrandom.

ILLUSTRATION

The shortened version of the Hald data [3] with $n = 13$ observations on four x-variables and a response Y and the steps in the stepwise procedure are shown in Table 1 (references to parts of Table 1 are shown by numbers in parentheses in the following paragraphs).

The procedure starts by calculating the correlation matrix (1) for all the variables in the data set. In order to begin the stepping procedure, criteria for entering a variable and for removing a variable must be established. There are two methods presently used. SAS and SPSS require probability levels for entering and removing variables, while BMDP requires arbitrary F values. F, in this context, refers to the usual kind of F statistic, i.e., the ratio of two variances (*see* F-TESTS). In the illustration arbitrary F criteria are used: F to-Enter = 4.00, and F-to-Remove = 3.90.

Step 0. The intercept b_0 is entered into the model (2). This is always done if the model includes a constant. The format now shows two columns: Variables in Equation and Variables Not in Equation (3). In the first column is shown the variable entered, its coefficient, and its F-to-Remove, while in the second column are shown the variables not entered, their partial correlations* with the response, and their F's-to-Enter.

Step 1. The procedure examines the Variables Not in Equation in Step 0. It selects that x variable that has the largest correlation with Y. This variable will also have the largest F-to-Enter. If the value of this F-to-Enter statistic is larger than the arbitrary criterion F-to-Enter, the variable is entered into the linear model. Here, variable 4 is entered with an F-to-Enter of 22.80 (4), $F = 22.80 > F = 4.00$. After the ANOVA table is printed out, the Variables in Equation and Not in Equation are shown once again. Inspecting the Variables Not in Equation at this step one sees that variable 1 has the largest F-to-Enter: $F = 108.22$ (5). Variable 1 is the variable with the highest partial correlation with the response variable Y; i.e., after Y and all the variables not in the equation are adjusted for x_4, Y and x_1 are the most highly correlated.

Step 2. Variable 1 is entered. After the ANOVA is printed, the Variables in Equation and Variables Not in Equation are shown. At this point the procedure examines those Variables in Equation to determine the contribution of each of them as if they had been the last one to be added; that is, their F's-to-Remove (6). Since both of the variables have F's-to-Remove that exceed the criterion $F = 3.90$, none are removed. This means that each of the variables contribute significantly in explaining the response Y when added last in this two variable model. Next the Variables Not in Equation are examined (7). Variable 2 has the largest F-to-Enter: $F = 5.03$ and it is greater than the criterion $F = 4.00$.

Step 3. Variable 2 is entered. After the ANOVA is shown, the Variables in Equation are checked for their F's-to-Remove (8). Variable 4 has $F = 1.86$, which is less than $F = 3.90$. This means that variable 4 does not explain enough of the variation in Y after variables 1 and 2 are entered into the model first.

Step 4. Variable 4 is removed. The procedure now examines the Variables Not in Equation for any that have F's-to-Enter that exceed $F = 4.00$. There are none. It then examines the Variables in Equation against the F-to-Remove criterion. All of them exceed $F = 3.90$. Thus, the procedure stops, and chooses the model:

$$\hat{Y}_i = 52.58 + 1.47x_{1i} + 0.66x_{2i}.$$

Comment. This procedure does not examine all possible subsets of the k x-variables. Thus, there is no guarantee that it will produce the uniquely "best" subset regression model, regardless of the criteria used for "best." However, it has been a very useful tool for the model builder and will continue to be.

References

[1] Draper, N. R. and Smith, H. (1966). *Applied Regression Analysis*. Wiley, New York. [Revised version (1981).]

Table 1 Hald Data

$X1$	$X2$	$X3$	$X4$	Y
7	26	6	60	78.5
1	29	15	52	74.3
11	56	8	20	104.3
11	31	8	47	87.6
7	52	6	33	95.9
11	55	9	22	109.2
3	71	17	6	102.7
1	31	22	44	72.5
2	54	18	22	93.1
21	47	4	26	115.9
1	40	23	34	83.8
11	66	9	12	113.3
10	68	8	12	109.4

Variable	Mean	Standard Deviation	Coefficient Of Variation
$X1$	7.46	5.88	0.7884
$X2$	48.15	15.56	0.3232
$X3$	11.77	6.41	0.5442
$X4$	30.00	16.74	0.5579
Y	95.42	15.04	0.1577

Correlation Matrix (1)

	$X1$	$X2$	$X3$	$X4$	Y
$X1$	1.0000				
$X2$	0.2286	1.0000			
$X3$	−0.8241	−0.1392	1.0000		
$X4$	−0.2454	−0.9730	0.0295	1.0000	
Y	0.7307	0.8163	−0.5347	−0.8213	1.0000

Step No. 0. Variable Entered: Intercept (2)

Analysis of Variance

	Sum of Squares	d.f.	Mean Square
Residual	2715.7666	12	226.3139

(3) Variables in Equation			Variables Not in Equation		
Variable	Coefficient	F to Remove	Variable	Partial Corr.	F to Enter
Intercept	95.4230		$X1$	0.7307	12.60
			$X2$	0.8163	21.96
			$X3$	−0.5347	4.40
			$X4$	−0.8213	22.80
					(4)

Step No. 1. Variable Entered: $X4$

Analysis of Variance

	Sum of Squares	d.f.	Mean Square	F Ratio
Regression	1831.8994	1	1831.8994	22.80
Residual	883.8669	11	80.3515	

Table 1 *(Continued)*

Variables in Equation			Variables Not in Equation		
Variable	Coefficient	F to Remove	Variable	Partial Corr.	F to Enter
Intercept	117.5679				(5)
$X4$	−0.7382	22.80	$X1$	0.9568	108.22
			$X2$	0.1302	0.17
			$X3$	−0.8951	40.29

Step No. 2. Variable Entered: $X1$

Analysis of Variance

	Sum of Squares	d.f.	Mean Square	F Ratio
Regression	2641.0032	2	1320.5016	176.62
Residual	74.7634	10	7.4763	

Variables in Equation			Variables Not in Equation		
Variable	Coefficient	(6) F to Remove	Variable	Partial Corr.	F to Enter
Intercept	103.0973				(7)
$X1$	1.4400	108.22	$X2$	0.5986	5.03
$X4$	−0.6140	159.29	$X3$	−0.5657	4.24

Step No. 3. Variable Entered: $X2$

Analysis of Variance

	Sum of Squares	d.f.	Mean Square	F Ratio
Regression	2667.7920	3	889.2639	166.83
Residual	47.9744	9	5.3305	

Variables in Equation			Variables Not in Equation		
Variable	Coefficient	F to Remove	Variable	Partial Corr.	F to Enter
Intercept	71.6482				
$X1$	1.4519	154.00	$X3$	0.0477	0.02
$X2$	0.4161	5.03			
$X4$	−0.2365	1.86			
		(8)			

Step No. 4. Variable Removed: $X4$

Analysis of Variance

	Sum of Squares	d.f.	Mean Square	F Ratio
Regression	2657.8606	2	1328.9303	229.50
Residual	57.9059	10	5.7906	

Variables in Equation			Variables Not in Equation		
Variable	Coefficient	F to Remove	Variable	Partial Corr.	F to Enter
Intercept	52.5773				
$X1$	1.4683	146.52	$X3$	0.4113	1.83
$X2$	0.6623	208.58	$X4$	−0.4141	1.86

F-levels (4.000, 3.900) or tolerance insufficient for further stepping

[2] Efroymson, M. A. (1960). In *Mathematical Methods for Digital Computers*, A. Ralston and H. S. Wilf, eds. Wiley, New York.

[3] Hald, A. (1952). *Statistical Theory with Engineering Applications*. Wiley, New York.

[4] Hocking, R. R. (1983). Developments in linear regression methodology: 1959–1982. *Technometrics*, **25**, 219–230.

(LINEAR REGRESSION
REGRESSION COEFFICIENTS
REGRESSION VARIABLES, SELECTION OF)

HARRY SMITH, JR.

STEREOGRAM

A three-dimensional generalization of a histogram* used to represent bivariate distributions, a stereogram consists of a set of rectangular parallelepipeds erected on bases formed by cells of a two-dimensional rectangular grid, representing groups of values of the two variables (X, Y). For the parallelepiped based on the cell

$$(x_i < X < x_{i+1}) \cap (y_j < Y < y_{j+1}),$$

the volume should be proportional to P_{ij}, the proportion of (X, Y) values satisfying these inequalities. That is, the height should be proportional to

$$P_{ij}\{(x_{i+1} - x_i)(y_{j+1} - y_j)\}^{-1}.$$

Stereograms provide an excellent way of representing bivariate distributions, but their regular use became practicable only with the development of computer graphics. Previously, scatter diagrams were most commonly used. For an illustration, *see* CORRELATION, Fig. 5.

(GRAPHICAL REPRESENTATION, COMPUTER-AIDED
HISTOGRAMS
MULTIVARIATE GRAPHICS)

STEREOLOGY

Stereology (Greek *stereos* meaning solid, stiff) is the science of estimating geometric aspects of the spatial structure of solid materials, by means of random planar sections of the material, random orthogonal projections through the material, and similar geometric probes. The statistician will recognize the main form of stereology—plane section stereology—as *geometric* sampling theory, in the sense that the population is continuous and geometric, and the random samples are correspondingly geometric. Thus the geometric equivalents of standard sampling theory techniques such as simple random sampling*, systematic sampling*, weighted sampling*, multistage sampling, and ratio estimation* prove basic. Since sections and projections are usually random relative to the material, the results of geometric probability* (Kendall and Moran [6]; Solomon [21]) and integral geometry (Santaló [19]) are also basic.

Serious problems not encountered in sampling theory arise. In sampling theory the population is well-defined, even if difficult to sample in ideal fashion. However, in stereology the very entity of the population is far less certain, e.g., an ore body yet to be mined or the *future* product of an industrial process. Thus population specification and the resulting sampling plan are usually ad hoc matters of critical importance, at the heart of stereology (Miles [10]).

For a solid specimen $X \subset \mathbb{R}^3$ comprising two phases (i.e., labelled regions) α, β separated by interface surface(s) $\mathscr{S}$, write V_V for the volume fraction (α volume divided by total volume) and S_V for the surface density (interface area/total volume ratio). Similarly for specimens with embedded curve(s) $\mathscr{C}$, write L_V for the curve length/total volume ratio. Corresponding ratios A_A, B_A, N_A may be measured (or estimated!) in planar sections $X \cap T$ [A is the area of $X \cap T$ or $\alpha \cap T$, B is the length of $\mathscr{S} \cap T$, and $N = \text{card}(\mathscr{C} \cap T)$]. Historically the first stereological result $A_A = V_V$ was stated by Delesse [3]. That is, in some properly defined sense (see below), A_A is an unbiased estimator of V_V:

$$E[A_A] = V_V.$$

Subsequently, in the first edition (1945) of his book, Saltykov [18] derived the relations

$$E[B_A] = (\pi/4)S_V,$$

$$E[N_A] = (1/2)L_V.$$

Note that both estimators and estimands are ratios (or densities). This novel but not entirely satisfactory subscript notation for ratios is by now standard among stereologists.

SAMPLING

Since most spatial characteristics are in effect integrations over X, with all volume elements contributing equally regardless of location or orientation, sample planes, lines, quadrats, etc., should sample X in some uniform way. This implies that probes conform to the relevant invariant measures (probability distributions) of integral geometry (geometrical probability). Unbiased estimators generally result. Samples must often be multistage, e.g., a rectangular quadrat of a random plane section of a random cube obtained by dicing. Uniformity and unbiasedness are typically preserved in such sampling. As always, replication (to estimate MSEs) is desirable, but may be hampered by the destructive nature of sampling. Standard ad hoc sampling techniques such as stratified sampling*, importance sampling*, and the use of control variates and antithetic variates are all available (Miles and Davy [13]). A range of techniques has been devised to remove biases stemming from edge effects (Miles and Davy [13]; Miles [11]).

Conditions for the validity of the three above fundamental formulae are:

(i) area-weighted isotropic uniform random plane section of bounded X (Davy and Miles [2]); or

(ii) arbitrary plane section of unbounded X in the form of a homogeneous and isotropic two phase stochastic process in $\mathbb{R}^3$ (Miles [10]).

Estimator variances in (i) and (ii) are very different. Isotropy of the section in (i) or of the process in (ii) is not required for the validity of the Delesse relation $E[A_A] = V_V$.

CURVATURE

Embedded and exposed surfaces are of fundamental importance in geometric structure; hence the following resumé of their key features. For a smooth planar curve $\mathscr{C}$ given by $\phi = \phi(s)$ (s is the length along curve, ϕ is the tangent direction), the *curvature* $\kappa(s) = d\phi(s)/ds$ measures the local rate of turning. The integral of curvature

$$C = \int_{\mathscr{C}} \kappa(s)\, ds = [\phi]_{\mathscr{C}}$$

measures the net tangent vector rotation along $\mathscr{C}$. Thus at all points of a circle of radius r, $\kappa = \pm r^{-1}$; while for all smooth closed curves bounding convex regions, $C = \pm 2\pi$. The sign $\pm$ depends upon the chosen topological orientation.

Planar curvature leads to corresponding local and global characterization of smooth spatial surfaces, as follows. Consider a point ξ of $\mathscr{S}$ and a planar section $\mathscr{S} \cap T$ containing ξ. Then one has the classical differential geometric relation: The curvature of $\mathscr{S} \cap T$ at ξ is

$$\kappa(\theta,\phi) = (\kappa_1 \cos^2\phi + \kappa_2 \sin^2\phi)/\sin\theta,$$

where (θ,ϕ) are spherical polar coordinates of the normal to T relative to a coordinate frame, one of whose axes is parallel to the normal to $\mathscr{S}$ at ξ (θ is the angle between normals). This important relationship reflects the locally quadratic nature of smooth surfaces, and is generally associated with the names of Euler and Meusnier. κ_1, κ_2 are the *principal curvatures* of $\mathscr{S}$ at ξ. For example, at all points of a sphere of radius r, κ_1 and

κ_2 both equal r^{-1}, while at all points of a circular cylinder of radius r, κ_1 and κ_2 are 0 and r^{-1} (in some order). Of particular importance are the *mean curvature* $k = \frac{1}{2}(\kappa_1 + \kappa_2)$ and the *Gaussian curvature* $g = \kappa_1\kappa_2$. k is constant over a surface in equilibrium under surface tension. Surfaces partition into *elliptic regions* where $g > 0$ (e.g., mountain summit) and *hyperbolic regions* where $g < 0$ (e.g., mountain saddle). Integrating over a smooth (or polyhedral, by limit) surface, the *integral of mean curvature* $K = \int_{\mathscr{S}} k\, dS$ and the *integral of Gaussian curvature* $G = \int_{\mathscr{S}} g\, dS$ are both global characteristics of fundamental importance.

K: When $\mathscr{S}$ bounds a convex region, K is 2π times its "mean caliper diameter" M (i.e., mean linear projection onto isotropic random line). K_V may be estimated by

$$E[C_A] = K_V,$$

due to Cahn and DeHoff, which, like the other fundamental formulae, stems from classical integral geometric relations.

G: For simple closed surfaces $\mathscr{S}$, G is $4\pi(1 - n)$, where n is the number of "handles" possessed by $\mathscr{S}$. (For the topologically minded, G is 2π times the topological invariant called the *Euler–Poincaré characteristic* of $\mathscr{S}$.) Thus if α comprises an aggregate of disjoint particles without holes or handles, G_V is 4π times their spatial density. A_A, B_A, and C_A estimate V_V, $(\pi/4)S_V$, and K_V, respectively, but the estimation of G_V from sections evidently requires further assumptions. DeHoff [4, pp. 291–325] has described how laborious serial sectioning (equispaced parallel plane sections), with construction of equivalent graphs based on adjacent sections, may be used to estimate the two contributions to G_V, viz. total numbers of disjoint particles and handles.

PARTICLE AGGREGATES

Typically one wishes to estimate spatial particle density N_V and aggregate mean values like $\overline{V}$, $\overline{S}$, $\overline{K}$, and $\overline{G}$ defined in a natural manner. A serious stumbling block is that a plane may section a given particle in disjoint planar regions, with no means of determining from the section which regions correspond to the same particle! This problem disappears in the important case of convex particles. For a planar section of a homogeneous and isotropic convex particle process,

$$N_A = \overline{M}N_V$$

(see DeHoff [4, pp. 129–131]). Moreover, with double overbar relating to the aggregate of convex planar intercepts,

$$\overline{\overline{A}} = \overline{V}/\overline{M} \quad \text{and} \quad \overline{\overline{B}} = (\pi/4)\overline{S}/\overline{M};$$

for a line transect (section),

$$\overline{\overline{L}} = 4\overline{V}/\overline{S} \quad \text{and} \quad \overline{\overline{L^4}} = (12/\pi)\overline{V^2}/\overline{S}$$

(Miles [8]).

Particles may be said to have size and shape*, so that the ratio of volumes of two particles of the same shape with sizes s_1, s_2 is $(s_1/s_2)^3$. Often in practice looser definitions are used, e.g., the distribution of L in a line transect is referred to as the "aggregate size distribution." Similarly, many ad hoc empirical shape factors or characteristics have been defined, based upon shape characteristics in planar sections.

More is possible when specific particle shapes are assumed, e.g., ellipsoid, spheroid (ellipsoid with two semiaxes equal) or sphere (Moran [14]). Then shape and size are specified by the values of a few parameters, ϕ say. Planes section the above as ellipses, ellipses, and circles, respectively, with similar parametrization ψ. Thus a given (unknown) spatial density $f(\phi)$ gives rise to an (observable) planar density $g(\psi)$, with typically an integral equation determining f in terms of g. Wicksell [23] considered the spheres and spheroids cases, and subsequently much effort and sophistication has been applied towards efficient statistical solution of such integral equations, especially for spheres (Watson [22]; Jakeman and Anderssen [5]). Incidentally, these shape assumptions permit direct estimation of N_V.

ANISOTROPY

Many materials appear to be isotropic—that is, in a stochastic sense, invariant under arbitrary rotations—while just as many others are patently anisotropic (not isotropic). The former may be probed by planes of a single orientation, unlike the latter. Anisotropy may be characterized by means of orientation distributions, e.g., for embedded surfaces $\mathscr{S}$ the distribution of normal direction at a uniform random point of $\mathscr{S}$. Sometimes the coordinate frame is irrelevant, so that orientation distributions differing by a rigid body rotation are considered identical. Philofsky and Hilliard [16] have described the use of surface and curve probes of given orientation distribution, to determine coefficients in harmonic expansions of orientation distributions of curves and surfaces, respectively, in space.

PROJECTIONS

Consider a uniform probing beam of a penetrable bounded solid specimen X, with all rays of the beam parallel to the z axis. If $\lambda(x,y,z)$ is the (isotropic) "resistivity" at (x,y,z) to the beam, then beam intensity I typically conforms to the differential relation

$$I(x,y,z+dz) = I(x,y,z) - \lambda(x,y,z)\,dz,$$

so that

$$I(x,y,z) = I_0 \exp\left\{ -\int_{-\infty}^{z} \lambda(x,y,\zeta)\,d\zeta \right\}$$

(Beer's law). An image plane z = constant, perpendicular to the beam, records the transmitted beam. Thus, in principle, for any direction d, the value of $\int_{\mathscr{L}} \lambda(l)\,dl$ for all line transects $\mathscr{L}$ parallel to d may be estimated. Estimation of these values for closely and equispaced directions d all perpendicular to a fixed direction d_0 yields sufficient information for λ to be estimated at all points of X, essentially by Fourier inversion (Radon [17]). Such is the theoretical basis of the expensive modern-day computerized medical scanners. Often much more limited information, such as projection in a single direction, is available. For example, electron microscope sections are laminar and exceedingly thin, with projection only possible perpendicular to their plane faces (or at a limited range of tilt angles).

Randomness* of the material or randomness of the section reintroduces a stochastic element. The most general result relates to the homogeneous and isotropic two-phase material comprising Poisson convex sets (alternatively known as a Boolean scheme). That is, random interpenetrating convex sets sampled from some population, given independent isotropic orientations, and translated so that their centroids conform to a homogeneous spatial Poisson point process of intensity ρ. The two phases are the union α of the convex sets and its complement β. α (β) is supposed opaque (transparent) to the beam, so that perpendicular projection through a laminar section onto an image plane is itself two-phase (α', β'). Exact expressions exist for V_V, S_V, K_V, G_V and the observable A'_A, B'_A, C'_A for all lamina thicknesses t, leading to a variety of estimators for V_V, S_V, K_V, G_V (Miles [9]). When ρ is sufficiently small, most Poisson sets are disjoint, and simplifying approximations may be made to correct for the so-called Holmes effect in particle counts when overlapping may occur. Stereological exploitation of the tangential projections of embedded surfaces is also possible (Miles [12]).

IMAGE ANALYSIS

One necessary auxiliary stereological technique is the mensuration of planar images. A simple and cheap method of estimating some image characteristics is by "point-counting." For example, to estimate A_A, a transparent overlay marked with a point lattice may be randomly superposed on the image, and the image assessed and counted at each lattice point. Then A_A, and hence V_V, is estimated by the corresponding point ratio P_P. If the overlay has marked on it a regular and preferably isotropic (see Merz [7]) curve system, counting of the intersection points of this curve system with $\mathscr{S} \cap T$ allows estimation of B_A and hence S_V.

At the other extreme, where there is sufficient gradation of grey level, mensuration may be swiftly and accurately achieved by means of expensive computerized automatic image analysers, e.g., Quantimet, Leitz Texture Analyser. With these, interesting "logical" transformations of the image may be effected, to facilitate extraction of desired image characteristics (Serra [20]; a more elementary treatment is to be found in Coster and Chermaut [1a]). Typically grey levels are sufficiently wide in metallurgy and petrography, unlike biology. In the latter, pattern recognition* by a skilled observer applying qualitative stereology is often necessary before measurements can be made. Hence the prevalance of semiautomatic methods.

Of course, great accuracy in a single stage of what is essentially multistage sampling should be taken with a pinch of salt. For instance, the stereologist might be better advised to direct his effort to section replication, especially with anisotropic materials.

HISTORICAL

Despite Delesse's early (1847) contribution it was not until 1961 that a number of scientists in widely differing fields met to discuss their common interest, coined the term "stereology," and founded the International Society for Stereology. The main society activity is the holding of quadrennial international congresses, at which papers are largely applied in nature. References to the proceedings of these conferences are listed separately in the Bibliography. Also listed separately in the Bibliography and annotated in refs. 4 and 18 are treatises in a similar vein (an introduction to mathematical stereology has been written by Coleman [1]). The *Journal of Microscopy* is a recognized outlet for work in stereology, and *Acta Stereologica* is an international journal originating from the Medical Faculty of the E. Kardelj University of Ljubljana, Yugoslavia (Vol. 1, 1982).

Since 1961, statisticians have gradually taken more interest in the area, and proper mathematical foundations have been laid. However, as yet, statistical methods have been mainly standard (Nicholson [15]), not intrinsically geometric. Hopefully, sampling theorists will broaden their horizons to encompass geometric sampling theory, for they surely have much to offer!

References

[1] Coleman, R. (1979). An Introduction to Mathematical Stereology. *Memoirs No. 3*, Dept. of Theoretical Statistics, University of Aarhus, Aarhus, Denmark.

[1a] Coster, M. and Chermaut, J.-L. (1985). *Précis d'Analyse d'Images*. Éditions CNRS, Paris, France.

[2] Davy, P. J. and Miles, R. E. (1977). Sampling theory for opaque spatial specimens. *J. R. Statist. Soc. Ser. B*, **39**, 56–65.

[3] Delesse, M. A. (1847). Procédé mécanique pour déterminer la composition des roches. *C. R. Acad. Sci. (Paris)*, **25**, 544–545.

[4] DeHoff, R. T. and Rhines, F. N., eds. (1968). *Quantitative Microscopy*. McGraw-Hill, New York. (Treatise of an applied nature.)

[5] Jakeman, A. J. and Anderssen, R. S. (1975). Abel type integral equations in stereology. *J. Microscopy*, **105**, 121–133, 135–153.

[6] Kendall, M. G. and Moran, P. A. P. (1963). *Geometrical Probability*. Hafner, New York.

[7] Merz, W. A. (1967). Die Streckenmessung an gerichteten Strukturen im Microskop und ihre Anwendung zur Bestimmung von Oberflächen-Volumen-Relationen im Knochengewebe. *Mikroskopie*, **22**, 132–142.

[8] Miles, R. E. (1972). Multidimensional perspectives on stereology. *J. Microscopy*, **95**, 181–196.

[9] Miles, R. E. (1976). Estimating aggregate and overall characteristics from thick sections by transmission microscopy. *J. Microscopy*, **107**, 227–233.

[10] Miles, R. E. (1978). In *Geometrical Probability and Biological Structures: Buffon's 200th Anniversary. Lect. Notes Biomath.*, **23**, Springer, Berlin, Germany, pp. 115–136.

[11] Miles, R. E. (1978). The sampling, by quadrats, of planar aggregates. *J. Microscopy*, **113**, 257–267.

[12] Miles, R. E. (1980). The random tangential projection of a surface. *Adv. Appl. Prob.*, **12**, 425–446.

[13] Miles, R. E. and Davy, P. J. (1977). On the choice of quadrats in stereology. *J. Microscopy*, **110**, 27–44.

[14] Moran, P. A. P. (1972). The probabilistic basis of stereology. *Adv. Appl. Prob.*, **4** (Suppl.), 69–91.

[15] Nicholson, W. L. (1978). Application of statistical methods in quantitative microscopy. *J. Microscopy*, **113**, 223–239.

[16] Philofsky, E. M. and Hilliard, J. E. (1969). The measurement of the orientation distribution of lineal and areal arrays. *Quart. Appl. Math.*, **27**, 79–86.

[17] Radon, J. (1917). Über die Bestimmung von Funktionen durch ihre integralwerte längs gewisser Mannigfaltigkeiten. *Ber. Saechs. Akad. Wiss. Leipzig, Math.-Phys. Kl.*, **69**, 262–277.

[18] Saltykov, S. A. (1958). *Stereometric Metallography*, 2nd ed. State Publishing House for Metals and Sciences, Moscow, USSR. (Treatise of an applied nature.)

[19] Santaló. L. A. (1976). *Encyclopedia of Mathematics and its Applications*, Vol. 1. Addison-Wesley, Reading, MA.

[20] Serra, J. (1982). *Image Analysis and Mathematical Morphology*. Academic, London, England.

[21] Solomon, H. (1978). Geometric Probability. *SIAM Regional Conf. Ser. Appl. Math.*, **28**. Philadelphia, PA.

[22] Watson, G. S. (1971). Estimating functionals of particle size distributions. *Biometrika*, **58**, 483–490.

[23] Wicksell, S. D. (1925, 1926). The corpuscle problem. *Biometrika*, **17**, 84–99 (Part I); **18**, 151–172 (Part II).

Bibliography

Treatises of an Applied Nature

Underwood, E. E. (1970). *Quantitative Stereology*. Addison-Wesley, Reading, MA.

Weibel, E. R. (1979, 1980). *Stereological Methods*, Vols. I (Practical Methods for Biological Morphometry) and II (Theoretical Foundations). Academic, London, England.

Proceedings of International Congresses for Stereology

1st (Vienna, 1963): Haug, H., ed. (1963). Congressprint, Wien, Austria.

2nd (Chicago, 1967): Elias, H., ed. (1967). Springer-Verlag, New York.

3rd (Berne, 1971): Weibel, E. R., Meek, G., Ralph, B., and Ross, R., eds. (1972). *Stereology 3*. Blackwell, Oxford, England. [Also published in *J. Microscopy*, **95** (1972).]

4th (Gaithersburg, MD, 1975): Underwood, E. E., de Wit, R., and Moore, G. A., eds. (1976). *Special Publication 431*, National Bureau of Standards (USA).

5th (Salzburg, 1979): Adam, H., Bernroider, G., and Haug, H., ed. (1980). *Mikroskopie*, **37** (Suppl.), 1–495.

(GEOMETRIC PROBABILITY THEORY
PARTICLE-SIZE STATISTICS
RANDOM TESSELATIONS
SERRA'S CALCULUS
SPATIAL DATA ANALYSIS
SPATIAL SAMPLING)

R. E. MILES

STIELTJES FRACTIONS *See* PADÉ AND STIELTJES TRANSFORMATIONS

STIRLING DISTRIBUTIONS

There are two families of discrete distributions that are closely related to Stirling numbers* and named accordingly: Stirling distribution of the first kind (SDFK) and Stirling distribution of the second kind (SDSK). Stirling numbers are those numbers appearing in the sums relating the factorials of a variable to its powers, and vice versa. Riordan [10, p. 33] defines these numbers as follows. If $N^{(0)} = N^0 = s(0,0) = S(0,0) = 1$ and

$$N^{(n)} = N(N-1)\cdots(N-n+1) = \sum_{k=0}^{n} N^k s(n,k), \tag{1}$$

$$N^n = \sum_{k=0}^{n} N^{(k)} S(n,k), \tag{2}$$

then $s(n,k)$ and $S(n,k)$ are the Stirling numbers of the first and second kind, respectively, Note that for given n or given k, the numbers of the first kind alternate in sign.

STIRLING DISTRIBUTION OF THE FIRST KIND

We define the random variable (RV) Y to have the SDFK with parameters n and θ if its probability function (PF) is given by

$$\Pr(Y = y) = h(n, y)\frac{\theta^y}{y!}\Big/ f_n(\theta), \qquad y = n, n+1, \ldots, \tag{3}$$

where $f_n(\theta) = (-\log(1-\theta))^n$, with $0 < \theta < 1$, and $h(n, y) = n!|s(y, n)|$, where $s(y, n)$ is the Stirling number of the first kind with arguments y and n, defined by (1).

The SDFK arises as follows. Let $X_1, \ldots, X_n$ be n independent, identically distributed (i.i.d.) RV's, each having the logarithmic series distribution (LSD)* with parameter θ, i.e., having a PF

$$\Pr(X = x) = a\theta^x/x; \qquad x = 1, 2, \ldots, \tag{4}$$

where a is a normalizing constant: $1/a = -\log(1-\theta)$, $0 < \theta < 1$. The sum Y is distributed according to the SDFK (see, e.g., Patil and Wani [8]).

The SDFK (3) is a member of the class of power series distributions* (PSD), i.e. its PF has the form

$$\Pr(X = x) = a_x\frac{\theta^x}{x!}\Big/ g(\theta); \qquad x = 0, 1, \ldots, \tag{5}$$

where $\theta > 0$ is the parameter, $a_x > 0$ does not involve θ, and $g(\theta) = \Sigma a_x\theta^x$. This class of discrete distributions was introduced by Noak [5] in 1950, and has since been extensively studied, especially the slightly more general concept of a generalized PSD (GPSD) (see Patil [7]).

The probability generating function* (PGF) of the SDFK is

$$G(z) = \left[\frac{\log(1-\theta z)}{\log(1-\theta)}\right]^n.$$

The first two moments are

$$\text{mean} = \mu = na\theta/(1-\theta), \tag{6}$$

$$\text{variance} = \sigma^2 = na\theta(1-a\theta)/(1-\theta)^2, \tag{7}$$

respectively, where a is the constant appearing in (4). The fact that the SDFK belongs to the class of PSD can be exploited to set up various recurrence relations for moments and cumulants (cf. Patil and Wani [8]).

We now turn to estimation. As already noted, the SDFK with parameters n and θ arises as the sampling distribution of the sum of n observations from the LSD with parameter θ. Indeed, the sample sum is a complete, sufficient statistic* for θ. Therefore, the primary parameter in the SDFK is θ, while n is secondary; discussion of estimation in the literature has focused on θ.

Estimability* conditions for certain functions of the parameter of the PSD were given by Patil [7], and we may benefit by these general results in the case of inference for the LSD and SDFK. To indicate some results briefly, if $x_1, \ldots, x_n$ is a sample of size n from the LSD (4), then $y = \Sigma x_i$ follows the SDFK as defined by (3). When deriving the minimum variance unbiased* (MVU) estimate of some parametric function $\varphi(\theta)$, the argument used is based on the identity of two power series expansions. Thus if $\varphi(\theta) = \theta^\alpha$ (α a positive integer), and if $\hat{\varphi}(y)$ is unbiased, then

$$\sum_{y=n}^{\infty} \hat{\varphi}(y)h(n, y)\frac{\theta^y}{y!} = \theta^\alpha f_n(\theta),$$

where $h(y, n)$ and $f_n(\theta)$ are defined in (3). This leads to a MVU estimate in the form of a ratio (cf. Roy and Mitra [11])

$$\hat{\varphi}(y) = \begin{cases} 0 & \text{if } y \leqslant \alpha, \\ h(n, y-\alpha)/h(n, y) & \text{if } y > \alpha. \end{cases} \tag{8}$$

Other parametric functions, such as the PF at any point x, have been considered from the point of view of unbiased estimation (see Shanmugam and Singh [12]).

To obtain a maximum likelihood* (ML) estimate of the parameter θ in the LSD or SDFK is, in principle, straightforward. However, both the ML and the MVU estimates tend to be cumbersome to compute. The main difficulty with the MVU estimate (8), even in treating particular cases like

$\alpha = 1$, lies in deriving explicit or computationally attractive expressions. The MVU estimate involves Stirling numbers, which become too large too soon. In view of these difficulties, one may resort to a graphical estimation method suggested by Ord [6]. The basic idea behind Ord's technique is to use sample relative frequencies to approximate the theoretical PF.

The LSD was introduced by Fisher et al. [4] in an ecological context (see references contained in the separate LSD entry). Patil and Wani [8] report work on the SDFK. Cacoullos [2] studies truncated versions of the LSD and derives recurrence relations for generalized Stirling numbers, which facilitate computational work; see also Shanmugam and Singh [12].

STIRLING DISTRIBUTION OF THE SECOND KIND

The RV Y is said to have the SDSK with parameters n and θ if

$$\Pr(Y = y) = n!S(y, n)\frac{\theta^y}{y!}\bigg/(e^\theta - 1)^n, \qquad y = n, n+1, \ldots, \quad (9)$$

$\theta > 0$, where $S(y, n)$ is the Stirling number of the second kind (2) with arguments y and n.

This family of discrete distributions, which was first considered by Tate and Goen [13], arises as follows. Let Y be the sum of n i.i.d. RV's, $Y = X_1 + \cdots + X_n$, each having the zero-truncated, or positive Poisson distribution* (PPD) with parameter θ, i.e.,

$$\Pr(X = x) = \frac{\theta^x}{x!}\bigg/(e^\theta - 1), \qquad x = 1, 2, \ldots. \quad (10)$$

Then the sum Y has the SDSK with parameters N and θ.

Just as the SDFK, the SDSK belongs to the class of PSD or GPSD. The generating function associated with the SDSK is given by

$$G(z) = \left[(e^{\theta z} - 1)/(e^\theta - 1)\right]^n. \quad (11)$$

The mean and the variance are readily obtained as

$$\mu = n\theta/(1 - e^{-\theta}), \quad (12)$$

$$\sigma^2 = \mu(1 + \theta - \mu/n), \qquad \mu \text{ as in (12)}, \quad (13)$$

respectively.

Let us now briefly consider estimation of the PPD and SDSK. By and large, the remarks made in the discussion of the LSD and SDFK apply here too. Thus the sample sum is a complete sufficient statistic for the parameter θ in the Poisson, as well as in the PPD and the SDSK families, considered here.

In the case of n observations, $x_1, \ldots x_n$, with sum $y = \Sigma x_i$ made on the PPD, Tate and Goen [13] obtained the ratio of Stirling numbers of the second kind;

$$\hat{\theta}_y = \begin{cases} 0 & \text{if } y = n, \\ yS(n, y-1)/S(n, y) & \text{otherwise,} \end{cases} \quad (14)$$

as the MVU estimate of θ. Equivalently, the ratio (14) may be regarded as the MVU estimate of the parameter in the SDSK (9) based on a single observation. Tables for calculating the estimate (14) have been published [Tate and Goen [13] for $n = 3(1)50$ and $y = 2(1)n - 1$].

The simple MVU result (14) has been extended to more general parametric functions and to cases where the Poisson distribution is truncated at a point $c > 1$. In treating these more general cases, generalized Stirling numbers arise (see, e.g., Tate and Goen [13] and Ahuja [1]).

Estimation problems relating to the PPD have been discussed extensively from the point of view of ML estimation (Cohen [3]). A very simple estimate proposed by Plackett [9] is

$$\theta^* = (y/n)(1 - n_1/y),$$

where n_1 is the number of observations equal to 1.

A situation in which the PPD would occur is when it is desired to fit a distribution to Poisson-like data consisting of numbers of individuals in certain groups, possessing a given attribute, but in which a group cannot be sampled unless at least one of its members has the attribute.

EXTENSIONS

Cacoullos [2] introduces multiparameter Stirling distributions of both kinds, thereby extending the concept as presented here. He derives the distribution of the sufficient statistic for a one-parameter PSD truncated on the left at several known or unknown points. Multiparameter Stirling distributions of both kinds arise as special cases in this study.

References

[1] Ahuja, J. C. (1971). *Aust. J. Statist.*, **13**, 133–136.

[2] Cacoullos, T. (1975). In *Statistical Distributions in Scientific Work*, Vol. 1, G. P. Patil et al., eds. Reidel, Dordrecht, The Netherlands, pp. 19–30.

[3] Cohen, A. C. (1954). *J. Amer. Statist. Ass.*, **49**, 138–168.

[4] Fisher, R. A., Corbet, A. S., and Williams, C. B. (1943). *J. Animal Ecol.*, **12**, 42–58.

[5] Noak, A. (1950). *Ann. Math. Statist.*, **21**, 127–132.

[6] Ord, J. K. (1972). *Families of Frequency Distributions*. Hafner, New York.

[7] Patil, G. P. (1962). *Ann. Inst. Statist. Math.*, **14**, 179–182.

[8] Patil, G. P. and Wani, J. K. (1957). *Sankhyā A*, **27**, 369–388.

[9] Plackett, R. L. (1953). *Biometrics*, **9**, 485–488.

[10] Riordan, J. (1958). *An Introduction to Combinatorial Analysis*. Wiley, New York.

[11] Roy, J. and Mitra, S. (1957). *Sankhyā A*, **18**, 371–378.

[12] Shanmugam, R. and Singh, J. (1981). In *Statistical Distributions in Scientific Work*, Vol. 4, C. Taillie et al., eds. Reidel, Dordrecht, The Netherlands, pp. 181–187.

[13] Tate, R. F. and Goen, R. L. (1958). *Ann. Math. Statist.*, **29**, 755–765.

(LOGARITHMIC SERIES DISTRIBUTION
POISSON DISTRIBUTIONS
POWER SERIES DISTRIBUTIONS
STIRLING NUMBERS)

SVEN BERG

STIRLING MATRIX

FIRST KIND

The (infinite) matrix

$$\begin{pmatrix} 1 & 0 & 0 & 0 & 0 & \cdots \\ 0 & 1 & 0 & 0 & 0 & \cdots \\ 0 & -1 & 1 & 0 & 0 & \cdots \\ 0 & 2 & -3 & 1 & 0 & \cdots \\ & & \vdots & & & \end{pmatrix}$$

is a Stirling matrix of the first kind. Denoting by $S_{i,j}$ the element in the ith row and jth column,

$$S_{i,j} = S_{i-1,j-1} - (i-1)S_{i-1,j}.$$

Values of $S_{i,j}$ up to $i, j = 10$ are given in ref. 1.

SECOND KIND

The (infinite) matrix

$$\begin{pmatrix} 1 & 0 & 0 & 0 & 0 & \cdots \\ 0 & 1 & 0 & 0 & 0 & \cdots \\ 0 & 1 & 1 & 0 & 0 & \cdots \\ 0 & 1 & 3 & 1 & 1 & \cdots \\ & & \vdots & & & \end{pmatrix}$$

is a Stirling matrix of the second kind. It is the inverse of the Stirling matrix of the first kind.

Reference

[1] Morrison, N. (1969). *Introduction to Sequential Smoothing and Prediction*. McGraw-Hill, New York.

(STIRLING NUMBERS)

STIRLING NUMBERS

Stirling numbers, which were introduced by Stirling in 1730 [10], occupy a central position in combinatorial mathematics. The numbers appear in the sums relating powers of a variable to its factorials and vice versa. The classical book by Riordan [7] devotes several pages to the Stirling numbers. The more recent book by Comtet [3] contains a chapter on these numbers and a rich bibliography. The report by Charalambides and Singh [2] reviews the Stirling numbers together with generalizations, applications, and an extensive list of references.

Adopting Riordan's notation, we write $s(n, k)$ for the Stirling numbers of the first kind and $S(n, k)$ for those of the second kind. The numbers are defined by

$$x^{(n)} = \sum_{k=0}^{n} s(n, k)x^k,$$

$$x^n = \sum_{k=0}^{n} S(n, k)x^{(k)},$$

respectively, where $x^{(n)}$ is the falling factorial:

$$x^{(n)} = x(x-1)\cdots(x-n+1).$$

Note that both numbers are nonzero only for $k = 0, 1, \ldots, n$, $n > 0$. Furthermore, for given n or given k, the numbers of the first kind alternate in sign, whereas those of the second kind are always positive. Indeed, since $(-x)^{(n)} = (-1)^n x(x+1)\cdots(x+n-1)$, it follows at once from the defining relation that $(-1)^{n+k}s(n, k)$ is always positive. As an illustration, Table 1 shows Stirling numbers with argument $n = 7$.

Stirling numbers have been tabulated in a number of places (e.g., David et al. [4], Abramowitz and Stegun [1]). With increasing arguments, Stirling numbers of both kinds grow very quickly, and even for moderate values of the arguments they become troublesome to handle numerically. Jordan [6] obtained the simple approximations for large n,

$$|s(n+1, k)| \cong n!(\log n + C)^k/k!,$$

where $C = 0.57721$ is Euler's constant, and

$$S(n, k) \cong k^n/k!,$$

respectively. A great number of authors have discussed other approximations.

If Δ is the forward difference* operator, then the Stirling number of the second kind can be written in operator notation

$$S(n, k) = \frac{1}{k!}\Delta^k 0^n,$$

$\Delta^k 0^n$ being the kth difference of $f(y) = y^n$, computed at zero. Similarly, for a Stirling number of the first kind we can use operator symbols

$$s(n, k) = \frac{1}{k!}D^k 0^{(n)},$$

where $D^k 0^{(n)}$ is an abbreviation for the kth derivative of $x^{(n)}$, evaluated at zero.

Using the symbolic expansion

$$\Delta^k = (E-1)^k = \sum_{0}^{k}\binom{k}{r}(-1)^{k-r}E^r,$$

where E is the (unit) shift operator, together with the defining formula, we derive the simple summation formula for Stirling numbers of the second kind,

$$S(n, k) = \frac{1}{k!}\sum_{0}^{k}\binom{k}{r}(-1)^{r-k}r^n,$$

$$k = 0, 1, \ldots, n;\ n = 0, 1, 2, \ldots.$$

There does not exist an analogous single summation formula for Stirling numbers of the first kind.

The basic recurrence relation for the Stirling numbers of the second kind is

$$S(n+1, k) = kS(n, k) + S(n, k-1),$$

$$k = 1, 2, \ldots, n+1;\ n = 0, 1, 2, \ldots,$$

Table 1 Stirling Numbers of the First and Second Kind with $n = 7$

k	1	2	3	4	5	6	7
$s(7, k)$	720	−1764	1624	−764	175	−21	1
$S(7, k)$	1	63	301	350	140	21	1

with initial conditions $S(0,0) = 1$ and $S(n,0) = S(0,k) = 0$ for $n > 0$ and $k > 0$. A corresponding recurrence relation for the Stirling numbers of the first kind is

$$s(n+1,k) = s(n,k-1) - ns(n,k),$$
$$k = 1,2,\ldots,n+1,\ n = 0,1,2,\ldots,$$

with initial conditions $s(0,0) = 1$ and $s(n,0) = s(0,k) = 0$ for $n > 0$ and $k > 0$.

The Stirling numbers are *orthogonal* in the sense that

$$\sum_k S(n,k)s(k,m) = \delta_{n,m},$$

with $\delta_{n,m}$ a Kronecker delta ($\delta_{n,n} = 1$, $\delta_{n,m} = 0$ for $n \neq m$), and where the sum is over all values of k for which $S(n,k)$ and $s(k,m)$ are nonzero. Stirling numbers satisfy inverse relations (see Riordan [8] for further details).

Stirling numbers of the second kind arise naturally in simple occupancy problems*. Thus the number of ways of placing n different objects into k different cells, with no cell empty, is

$$\begin{aligned} k!S(n,k) &= \Delta^k 0^n \\ &= k^n - \binom{k}{1}(k-1)^n + \cdots \\ &\quad + (-1)^{k-1}\binom{k}{k-1}. \end{aligned}$$

The formula follows from a straight application of the method of inclusion-exclusion*. In the same manner, the number of ways of putting n different objects into N different cells so that k cells are occupied and $N - k$ are empty is $N^{(k)}S(n,k)$. For further applications of Stirling numbers in combinatorial theory see, e.g., Riordan [7] or Comtet [3].

The following formulas connect the factorial moments* $\mu_{(n)} = E[X^{(n)}]$ and the (power) moments* $\mu_k = E(X^k)$ of a random variable X:

$$\mu_{(n)} = \sum_{k=0}^{n} s(n,k)\mu_k,$$

$$\mu_n = \sum_{k=0}^{n} S(n,k)\mu_{(k)}.$$

For a further statistical application (actually a paraphrase of the occupancy problem mentioned above), consider simple random sampling* with replacement (SRSWR) from a population consisting of N units. The probability of obtaining exactly k different units in a SRSWR of size n is given by

$$p(k) = N^{(k)}S(n,k)/N^n,$$
$$k = 1,2,\ldots,\min(n,N),\ n = 1,2,\ldots,$$

the classical occupancy distribution known in the literature as the Stevens–Craig or Arfwedson's distribution*; see Johnson and Kotz [5, p. 251]. The Stevens–Craig distribution has been used for inference in capture–recapture* experiments, in which specimens are caught one at a time, marked, and released. In this application the parameter of interest is N, the size of the animal population, and is unknown. The minimum variance unbiased estimator* of N involves a ratio of Stirling numbers of the second kind; Seber [9] provides further details.

Stirling numbers occur in two further statistical applications. Let $x_1, x_2, \ldots, x_n$ be a random sample of size n from the logarithmic distribution* with parameter θ. Then the sample sum, $x_1 + x_2 + \cdots + x_n$, has the Stirling distribution* of the first kind with parameters θ and n. There is a corresponding Stirling distribution of the second kind, which arises as follows. Let $x_1, x_2, \ldots, x_n$ now be a random sample of size n from the zero-truncated Poisson distribution*. The sample sum, $x_1 + x_2 + \cdots + x_n$, then has a Stirling distribution of the second kind. The reader is referred to the entry STIRLING DISTRIBUTIONS for further information.

References

[1] Abramowitz, M. and Stegun, I. A. (1965). *Handbook of Mathematical Functions*. Dover, New York.

[2] Charalambides, Ch. A. and Singh, J. (1986). *Technical Report No. 40*, School of Business Administration, Temple University, Philadelphia, PA.

[3] Comtet, L. (1974). *Advanced Combinatorics*. Reidel, Dordrecht, The Netherlands.

[4] David, F. N., Kendall, M. G., and Barton, D. E. (1966). *Symmetric Functions and Allied Tables*. Cambridge University Press, London, England.

[5] Johnson, N. L. and Kotz, S. (1969). *Discrete Distributions*. Houghton-Mifflin, Boston, MA.

[6] Jordan, C. (1933). *Tohuku Math. J.*, **37**, 254–278.

[7] Riordan, J. (1958). *An Introduction to Combinatorial Analysis*. Wiley, New York.

[8] Riordan, J. (1968). *Combinatorial Identities*. Wiley, New York.

[9] Seber, G. A. F. (1982). *The Estimation of Animal Abundance and Related Parameters*, 2nd ed. Griffin, London, England.

[10] Stirling, J. (1730). *Methodus Differentialis sine Tractatus de Summatione et Interpolatione Serierum Infinitarium*. [English translation: F. Holliday (1749). *The Differential Method*. London.]

(COMBINATORICS
OCCUPANCY PROBLEMS
SHANMUGAN NUMBERS
STIRLING DISTRIBUTIONS)

SVEN BERG

STIRLING'S FORMULA

This is employed, in the form

$$n! = (2\pi)^{1/2} n^{n+1/2} e^{-n}, \quad (1)$$

to provide approximations to binomial*, negative binomial*, multinomial*, hypergeometric*, and related probabilities.

More precise forms are

$$\Gamma(x+1) = (2\pi)^{1/2} x^{x+1/2} \times \exp\left(-x + \tfrac{1}{12}\theta x^{-1}\right), \quad x > 0, 0 < \theta < 1, \quad (2)$$

and

$$\begin{aligned}\log \Gamma(x) &= \left(x - \frac{1}{2}\right)\log x - x + \frac{1}{2}\log(2\pi) + \frac{1}{12x} - \frac{1}{360x^3} + \frac{1}{1260x^5} - \frac{1}{1680x^7} + \cdots \\ &= \left(x - \frac{1}{2}\right)\log x - x + \frac{1}{2}\log(2\pi) + \sum_{j=1}^{n} \frac{B_{2j}}{2j(2j-1)x^{2j-1}} + R_n(x), \quad (3)\end{aligned}$$

where B_r is the rth Bernoulli number*; the remainder term $R_n(x)$ is smaller in absolute value than, and has the same sign as, the next term in the summation [1].

Stirling's original result [10] corresponds more closely to the form

$$n! = (2\pi)^{1/2}\left(n + \tfrac{1}{2}\right)^{n+1/2} e^{-(n+1/2)}, \quad (4)$$

whereas (3) is the form derived by de Moivre* [2]; see Tweddle [11] for an interesting historical review and discussion of (1) and (4) (note that the latter employs different notation for the B_{2r}'s). Stirling determined the constant $\sqrt{2\pi}$ using Wallis' formula.

Derivations of (1) are given by Feller [4], Robbins [9], and Diaconis and Freedman [3], who give a short but complete version of an argument that is due to Laplace*. A similar derivation was given by Glivenko in 1939 [5], following the method of Goursat [7, Chap. VI, Sec. 141] in his classical treatise on mathematical analysis. For inequalities giving upper and lower bounds on $n!$ based on refinements to (1) see Feller [4] and Mitrinovic [8, p. 185].

Although the second form in (3) is especially useful when x is large and positive, Good [6] shows that it can give remarkably good approximations to $\log \Gamma(x)$ for quite small values of x, taking n fairly small and ignoring $R_n(x)$. For example, when $x = 0.5$, taking $n = 3$ gives an error of less than 2%.

References

[1] Abramowitz, M. and Stegun, I. A., eds. (1964). *Handbook of Mathematical Functions*. National Bureau of Standards, Washington, DC, p. 257.

[2] de Moivre, A. (1730). *Miscellanea Analytica de Seriebus et Quadraturis*. London, England.

[3] Diaconis, P. and Freedman, D. (1986). *Amer. Math. Monthly*, **93**, 123–125.

[4] Feller, W. (1968). *An Introduction to Probability Theory and its Applications*, 3rd ed. Wiley, New York, Chap. 2.

[5] Glivenko, V. I. (1939). *Kurs Teorii Veroyatnostei*. GOSOBNT Izdat, Moscow, USSR.

[6] Good, I. J. (1985). *J. Statist. Comput. Simulation*, **21**, 84–86.

[7] Goursat, E. (1904). *A Course in Mathematical Analysis*, Vol. I, E. R. Hedrick, transl. Athenaeum Press, Boston, MA. (From the 1902 French edition.)

[8] Mitrinovic, D. S. (1970). *Analytic Inequalities*. Springer-Verlag, New York, p. 185.

[9] Robbins, H. (1955). *Amer. Math. Monthly*, **62**, 26–29.

[10] Stirling, J. (1730). *Methodus Differentialis sine Tractatus de Summatione et Interpolatione Serierum Infinitarium*. [English translation: F. Holliday (1749). *The Differential Method*.]

[11] Tweddle, I. (1984). *Amer. J. Phys.*, **52**, 487–488.

(COMBINATORICS)

STOCHASTICALLY LARGER VARIABLE

The random variable X is *stochastically larger* than the random variable Y if

$$\Pr[X > a] \geqslant \Pr[Y > a]$$

for all real values a. Sometimes the notation $X \succ Y$ is used.

(STOCHASTIC ORDERING)

STOCHASTIC ANALYSIS AND APPLICATIONS

Stochastic Analysis and Applications was founded by V. Lakshmikantham and G. S. Ladde of the University of Texas at Arlington in 1983 and is published by Marcel Dekker, Inc., New York. It appears four times a year, four issues consisting a volume. The size of each issue of the journal is approximately 120 pages. The current editorial address is V. Lakshmikantham and G. S. Ladde, Department of Mathematics, University of Texas at Arlington, Arlington, TX 76019.

The editorial policy of the journal is to achieve a balance between theory, methods, and applications, and to provide for the rapid publication of papers covering any aspect of the development and applications of stochastic analysis techniques in all areas of scientific endeavor. The utility of the journal is enhanced by its direct reproduction format, which allows papers to be published within just 10 weeks of their acceptance, ensuring readers of the most current information available.

The problems of a modern dynamic society are complex, interdisciplinary, and stochastic in nature. The necessity of a theory of stochastic analysis for the description of natural phenomena in a conceptually, analytically, and computationally tractable form has been well understood and accepted. The steady trend in the development of the theory from deterministic to stochastic is regarded as a primary goal of current and future research. Since random phenomena arise in every field of scientific endeavor, the theory of stochastic analysis developed so far has found applications in biological, engineering, medical, physical, social, and technological sciences. These areas of applications, in turn, are contributing to further develop the theory. There is every reason to believe from the accomplishments thus far achieved in this journey into the random world, therefore, that many more surprises are in store, that several vistas will open up for mathematical scientists, and that major breakthroughs are there waiting and offering an exciting prospect. It is in this spirit that we see the importance of this journal.

Stochastic Analysis and Applications will publish important research and expository contributions directly concerned with solving problems connected with random phenomena. Papers that are principally concerned with the theory of stochastic analysis should contain significant mathematical results. Authors of such papers are encouraged to indicate possible relevance to applications. Those papers that are involved with the probabilistic methods applicable in solving stochastic problems should either contain new techniques or more suitable and illuminating methods than those already known. Papers emphasizing applications of probabilistic analysis should consist of valid stochastic models and important real world problems, preferably with some analysis of data. Papers that tend to integrate and interrelate theory and applications within the scope of the journal will be particularly welcomed.

The Editors also encourage the submission of expository papers whose aim is to inform specialists about the state of the art of a particular topic of current interest, and of preliminary communications of outstanding developments in the field of stochastic analysis and applications.

V. LAKSHMIKANTHAM
G. S. LADDE

STOCHASTIC AND UNIVERSAL DOMINATION

One of the main aspects of theoretical statistics is to develop criteria to compare procedures. Stochastic domination and universal domination are two such criteria in estimation theory.

For any vector a let $|a|_D$ denote the generalized Euclidean length with respect to a nonnegative definite matrix D, i.e., $|a|_D = (a'Da)^{1/2}$. When D is the identity matrix, the generalized Euclidean length is the Euclidean length and is denoted by $|a|$. Let $\delta_1(X)$ and $\delta_2(X)$ be two estimators for some unknown parameter θ, where X is the data whose distribution is characterized by θ. The estimator δ_1 is said to *stochastically dominate* δ_2 under the generalized Euclidean error with respect to D if $|\theta - \delta_1(X)|_D$ is stochastically smaller than or equal to $|\theta - \delta(x)|_D$ for every θ and $|\theta - \delta_1(X)|_D$ is stochastically smaller than $|\theta - \delta_2(X)|_D$ for some θ. (For any two one-dimensional random variables Y_1 and Y_2, Y_1 is *stochastically smaller than or equal to* Y_2 if

$$P(Y_1 \geqslant c) \leqslant P(Y_2 \geqslant c), \qquad \forall c. \quad (1)$$

If in addition to (1), the first probability is strictly smaller than the second probability for some c, Y_1 is said to be *stochastically smaller than* Y_2.)

Notions similar to stochastic domination have attracted wide attention among statisticians. Pitman [14] defined a similar criterion under a fiducial distribution*. He called an estimator "the best" (and argued that it is undeniably the best on p. 401) if it has stochastically smallest Euclidean error under a fiducial distribution. In estimating the mean of a normal sample, he then showed that the sample average is "the best" estimator. The results were also extended by Rukhin [15, 16] in the Bayesian context.

Savage [17, criterion 3, p. 224] introduced a one-dimensional criterion that is even stronger than stochastic domination. He proposed comparing $P(\delta(X) \in (a, b))$ for all $a < 0 < b$. In contrast, the one-dimensional stochastic domination compares the same quantity only for all the symmetric sets (a, b), where $a = -b$. For any one-parameter monotone likelihood ratio* family, Lehmann [13, p. 83] constructed (in Savage's sense) a best median unbiased* estimator.

Cohen and Sackrowitz [9, Theorem 6.1] considered two independently normally distributed random observations X_1 and X_2 with means θ_1 and θ_2 and known variances. If it is known that $\theta_1 > \theta_2$, X_1 can be stochastically dominated by an estimator of θ_1 based on both X_1 and X_2.

It may seem as if stochastic domination is unrelated to Wald's domination criterion [20], which is defined for a specific loss function. According to Wald, δ_1 is as good as δ_2 if the risk function (i.e., the expected loss) of δ_1 is not greater than that of δ_2. Further, if these two risk functions are not identical, δ_1 is said to be better than δ_2. Hwang [11], however, showed that under an arbitrary generalized Euclidean error, stochastic domination is in fact equivalent to "universal domination": We say that an estimator δ_1 *simultaneously dominates* δ_2 with respect to a class of loss functions $\mathscr{D}$ if (in the sense of Wald) for any loss $L(\theta, \delta) \in \mathscr{D}$, δ_1 is as good as δ_2, and for at least one of these loss functions δ_1 is better than δ_2. Simultaneous domination with respect to all loss functions $L(|\theta - \delta|_D)$, where $L(\cdot)$ is nondecreasing, is called *universal domination under the generalized Euclidean error* with respect to D. Therefore stochastic domination (and universal domination) implies that Wald's domination criterion holds for a very broad class of loss functions. Note that the class is "universal" with respect to D, since any reasonable loss function $L(|\theta - \delta|_D)$ must have a nondecreasing L (and hence is included in the class) so that a statistician will not pay more if his estimate is closer to θ. Simultaneous domination, stochastic domination, and universal domination are attractive, because they avoid the exact specification of a loss, which may be difficult in practice (see, for example, Berger [2, p. 69] for more detailed comments).

A classic example of simultaneous domination is the Rao–Blackwell theorem*. For an estimator not a function of a sufficient statistic*, an estimator based on the sufficient statistics simultaneously dominates the given estimator under the class of all convex loss functions. Several other results are discussed below.

Brown [5] and Shinozaki [18] dealt with the problem of improving upon the intuitive estimator for a p-variate ($p \geqslant 3$) normal mean simultaneously under a class of quadratic losses with variable weights. Their results, in particular, imply the famous Stein effect* [19]. Ghosh and Auer [10] also established similar results relating to several intuitive estimators for exponential families*. In these problems, the intuitive estimators can be simultaneously dominated if and only if the weights of the quadratic losses are suitably restricted. For a slightly different class of loss functions, estimators simultaneously dominating the maximum likelihood* estimator for several means of independent Poisson* populations were also constructed in Clevenson and Zidek [8, Theorem 3.1].

Brown et al. [6] considered a class of loss functions similar to the "universal class." They assume that the observations have a distribution belonging to a one-parameter monotone likelihood ratio* family (or that there is a nontrivial sufficient statistic). For any nonmonotone procedure (or any procedure not a function of the sufficient statistic), they provided by an explicit algorithm a monotone procedure (or a procedure depending on the observations only through the sufficient statistic) that simultaneously dominates the given procedure. More interestingly, the simultaneously improved procedure can be both monotone and a function of the sufficient statistic in a problem where both monotonicity and sufficiency assumptions are met. Their formulation is more general than the estimation framework and applies to hypothesis testing* problems.

The names "stochastic domination" and "universal domination" for estimators were first coined in Hwang [11]. (Rukhin [15, 16] used a similar term, "universal Bayes," to denote the estimator simultaneously Bayes with respect to the "universal class of losses.")

We next discuss some results regarding the general linear model* $\mathbf{X}_{n\times 1} = \mathbf{A}_{n\times p}\,\boldsymbol{\theta}_{p\times 1} + \boldsymbol{\epsilon}_{n\times 1}$. When $\boldsymbol{\epsilon}$ has a spherical distribution, the least-squares* estimator $(\mathbf{A}'\mathbf{A})^{-1}\mathbf{A}'\mathbf{X}$ stochastically (or universally) dominates any other linear unbiased estimator for Euclidean error or any generalized Euclidean error. (These results were independently proved by Ali and Ponnapalli [1] and Hwang [11]. See also Berk and Hwang [3] for extensions.) Further results concerning the U-admissibility of the least-squares estimator are described below.

When $\boldsymbol{\epsilon}$ has a finite fourth moment and when the dimension p of θ is 1 or 2, the least-squares* estimator is U-admissible for any generalized Euclidean error, i.e., it cannot be universally dominated. This follows easily from the fact that, under the sum of squared error loss, the least-squares estimator is admissible, a property stronger than U-admissibility (see Hwang [11]).

For $p \geqslant 3$, we first consider the simplified situation $\mathrm{X} = \theta + \epsilon$. Whether the least-squares estimator is U-admissible depends on the specific distribution assumed. Suppose that ϵ has a t-distribution with N

(known) degrees of freedom having the probability density function

$$f(\epsilon) = \text{constant}\left(1 + \frac{1}{N}|\epsilon|^2\right)^{-(N+p)/2}$$

Then (see Hwang [11]) for $p \geqslant 3$ there exist estimators of the form

$$\delta_a(X) = \left(1 - \frac{a}{|X|^2}\right)^+ X$$

that stochastically dominate the least-squares estimator X for the Euclidean error. The estimator $\delta_a(X)$ is the well known James–Stein* [12] positive part estimator.

For the normal case, the picture seems to be completely different. Below we assume that ϵ has a p-variate normal distribution with a zero mean and an identity covariance matrix. For this case, it was shown in Brown and Hwang [7] that X is U-admissible for the Euclidean error.

In the last two paragraphs, we have assumed $A = I$ in the linear model, which is, in fact, equivalent to the assumption $A'A = I$. Below we discuss some other results of Brown and Hwang [7] about more general $A'A$ under the Euclidean error. (The more general case using a generalized Euclidean error can be reduced to the Euclidean error by a linear transformation.) When $p \leqslant 4$, it is again shown that the least-squares estimator $(A'A)^{-1}A'X$ is U-admissible. However, the same result does not hold in the higher-dimensional case. Let $q_1 \geqslant q_2 \geqslant \cdots \geqslant q_p$ be the eigenvalues of $A'A$. When q_1 is strictly greater than q_2, there exist estimators that universally dominate the least-squares estimator when p is "large enough." How large is "large enough" is in general unknown. However, if $q_1 > q_2 = \cdots = q_p$, then the least-squares estimator can be universally dominated if and only if $p \geqslant 4$. The results are surprising.

References

[1] Ali, M. M. and Ponnapalli, R. (1983). An Optimum Property of Gauss–Markoff Estimate When Errors are Elliptically Distributed, Tech. Report, Univ. of Western Ontario, London, Canada. (A paper presented in Cincinnati Annual Statistics Meeting, Aug. 1982.)

[2] Berger, J. O. (1985). *Statistical Decision Theory and Bayesian Analysis*, 2nd ed. Springer-Verlag, New York.

[3] Berk, R. and Hwang, J. T. (1984). Optimality of the Least Squares Estimator. *Tech. Rep. No. 84-8*, Cornell Statistics Center, Cornell University, Ithaca, NY.

[4] Birnbaum, Z. W. (1948). *Ann. Statist.*, **19**, 76–81.

[5] Brown, L. D. (1975). *J. Amer. Statist. Ass.*, **70**, 417–427.

[6] Brown, L. D., Cohen, A., and Strawderman, W. E. (1976). *Ann. Statist.*, **4**, 712–722.

[7] Brown, L. D. and Hwang, J. T. (1986). Universal Domination and Stochastic Domination: *U*-Admissibility and *U*-Inadmissibility of the Least Squares Estimator. Tech. Report, Cornell Statistical Center, Cornell University, Ithaca, NY.

[8] Clevenson, M. L. and Zidek, J. V. (1975). *J. Amer. Statist. Ass.*, **70**, 698–705.

[9] Cohen, A. and Sackrowitz, H. B. (1970). *Ann. Statist.*, **41**, 2021–2034.

[10] Ghosh, M. and Auer, R. (1983). *Ann. Inst. Statist. Math.*, **35A**, 379–387.

[11] Hwang, J. T. (1985). *Ann. Statist.*, **13**, 295–314.

[12] James, W. and Stein, C. (1960). In *Fourth Berkeley Symp. Math. Statist. Prob.*, Vol. 1. University of California Press, Berkeley, CA, pp. 361–379.

[13] Lehmann, E. L. (1986). *Testing Statistical Hypotheses*, 2nd. ed. Wiley, New York.

[14] Pitman, E. J. G. (1938). *Biometrika*, **30**, 391–421.

[15] Rukhin, A. L. (1978). *Ann. Statist.*, **6**, 1345–1351.

[16] Rukhin, A. L. (1984). *J. Multivariate Anal.*, **14**, 135–154.

[17] Savage, L. J. (1954). *The Foundations of Statistics*. Wiley, New York.

[18] Shinozaki, N. (1980). *J. Amer. Statist. Ass.*, **75**, 973–976.

[19] Stein, C. (1956). In *Proc. Third Berkeley Symp. Math. Statist. Prob.*, Vol. 1. University of California Press, Berkeley, CA, pp. 197–206.

[20] Wald, A. (1950). *Statistical Decision Functions*. Wiley, New York.

Bibliography

Baranchik, A. J. (1970). *Ann. Math. Statist.*, **41**, 642–645.

Ki, F. and Tsui, K. W. (1986). *Commun. Statist. A*, **15**, 2159–2173. (Results of Hwang [11] were generalized to the positive part Baranchik (1970) estimators.)

(ADMISSIBILITY
ESTIMATION, POINT
JAMES–STEIN ESTIMATORS
STEIN EFFECT)

JIUNN TZON HWANG

STOCHASTIC APPROXIMATION

INTRODUCTION

Stochastic approximation refers to a class of iterative procedures that estimate the optima or roots of functions that are only observable up to some statistical noise. Many of these procedures can be viewed as stochastic counterparts of iterative numerical analysis* procedures for optimizing known evaluatable functions. Stochastic approximation procedures have been applied to problems in adaptive control, simulation*, sequential estimation*, and quantal response*.

ROBBINS–MONRO AND KIEFER–WOLFOWITZ* PROCESSES

The origins of stochastic approximation are in sequential design, but the pioneering work is by Robbins and Monro [31]. They considered determining θ such that $M(\theta) = m_0$, where $M(x)$ is a real-valued function. However, $M(x)$ is not observable and all that can be observed for each x are realizations of a random variable $Y(x)$, where it is assumed that $EY(x) = M(x)$ for all x. Denote the cumulative distribution function (CDF) of $Y(x)$ by $H(y|x)$. Define the recursive sequence of random variables $\{X_n\}$, $n \geqslant 1$, by

$$X_{n+1} = X_n - a_n(Y(X_n) - m_0),$$

where $\{na_n\}$ is positive and bounded, X_1 is arbitrary and has finite variance, and $Y(X_n)$ given $X_1, \dots, X_n$ has the CDF $H(y|X_n)$. Suppose $Y(x)$ is uniformly bounded with probability 1 (w.p.1) for all x. Under the conditions that $M(x)$ is nondecreasing, there exists θ such that $M(\theta) = m_0$ and $M'(\theta) > 0$; Robbins and Monro proved that $\lim_{n\to\infty} E(X_n - \theta)^2 = 0$ and, consequently, $X_n \xrightarrow{P} \theta$. The Robbins–Monro (RM) process can be viewed as a stochastic derivativeless variant of the Newton–Raphson* algorithm of numerical analysis, the latter being a method for finding the root of an evaluatable and observable function. (See Ortega and Rheinboldt [30, Sec. 7.1].)

Weaker conditions were subsequently found by Blum [3, 4], Kallianpur [18], and Schmetterer [37] for proving convergence in probability of the RM process, Kallianpur giving an expression for the order of magnitude of $E(X_n - \theta)^2$. An exact expression for $E(X_{n+1} - \theta)^2$ can be obtained in a *linear regression** setting. Suppose

$$Y(X_i) = m_0 + \beta(X_i - \theta) + \epsilon_i, \qquad i = 1, \dots,$$

where the ϵ_i are i.i.d. with $E(\epsilon_i) = 0$ and $\operatorname{var} \epsilon_i = \sigma^2$. If $a_n = a/n$, $a > 0$, then

$$E(X_{n+1} - \theta)^2 = a^2\sigma^2\left[\sum_{i=1}^{n-1}\left\{i^{-2}\prod_{k=i+1}^{n}(1 - \beta a/k)^2\right\} + n^{-2}\right],$$

where X_1 is a constant.

Kiefer and Wolfowitz [21] were the first to give a stochastic approximation procedure for estimating the maximum of a function $M(x) = EY(x)$, where only $Y(x)$ can be observed. Suppose $M(x)$ is strictly increasing for $x < \theta$ and strictly decreasing for $x > \theta$. Define the recursive sequence of random variables $\{X_n\}$, $n \geqslant 1$, by

$$X_{n+1} = X_n + a_n\{Y(X_n + \delta_n) - Y(X_n - \delta_n)\}/\delta_n,$$

where $\{a_n\}$, $\{\delta_n\}$ are suitable sequences of real numbers (e.g., $a_n = n^{-1}$, $\delta_n = n^{-1/4}$); var $Y(x) \leqslant C$, for all x; $Y(X_n + \delta_n)$ and $Y(X_n - \delta_n)$ given X_n are conditionally independent random variables. Assuming certain regularity conditions on $M(x)$, Kiefer and Wolfowitz showed $X_n \xrightarrow{P} \theta$.

Some weaker conditions for the convergence of the Kiefer–Wolfowitz (KW) process were found by Blum [3], Burkholder [6], Krasulina [23], Kushner and Clark [25], and Dupač [10], who also considered optimal choices of a_n and δ_n. Dvoretsky [12] defined a much more general stochastic approximation procedure that includes as special cases both the RM and the KW processes, and obtained convergence in probability and in L^2 for his process. Blum [5] introduced a generalized RM process, which Krasulina [22] later showed was also a special case of a slightly generalized version of Dvoretsky's process. For the KW process, Dupač [10] found the order of magnitude of $(E(X_n - \theta)^2)$ as $n \to \infty$; under suitable conditions and if $a_n = a/n$, $\delta_n = d/n^{1/4}$, $a > a_0$, and $d > 0$, then $E(X_n - \theta)^2 = O(n^{-1/2})$. A further examination of rates of convergence using weak convergence* theory is in Kushner [24].

ASYMPTOTIC NORMALITY*

With certain assumptions, both the RM and the KW processes have an asymptotic distribution that is normal with mean θ. Chung [7], employing a method of moments* argument, was the first to obtain the asymptotic normality of the RM process. This approach was refined and applied to the KW and more general stochastic approximation processes by Burkholder [6], Derman [8], and Dupač [10]. Sacks [34] gave fairly general conditions for the asymptotic normality of both the RM and the KW processes. His methods are based on the Lindeberg–Feller central limit theorem* and a characteristic function* argument. Let $\{X_n\}$ be the RM process with $a_n = a/n$. Assume

$$M(x) = \alpha + \beta(x - \theta) + \delta(x, \theta),$$

where $\delta(x, \theta) = o(|x - \theta|)$ as $x \to \theta$, and where $\beta > 0$. Further assume that as $x \to \theta$, var $Y(x) \to \sigma^2 > 0$. Under these and a number of other assumptions on $M(x)$ and on the distribution of $Y(x)$, Sacks showed that $n^{1/2}(X_n - \theta)$ is asymptotically normally distributed with mean 0 and variance $a^2\sigma^2(2a\beta - 1)^{-1}$, where $a > (2\beta)^{-1}$. A similar type of result was given by Sacks for the KW process. Kersting [19] provides an approximation of the RM process by a weighted sum of i.i.d. random variables, so that asymptotic properties are readily derivable. Under much more general conditions on the noise distribution, Ruppert [33] has extended Kersting's technique to include also the KW process. Basically Ruppert approximates the RM and the KW processes by weighted sums of dependent random variables, so that central limit theorems for dependent random variables can be applied.

MODIFICATIONS OF THE RM AND KW PROCESSES

There has been relatively little published research on optimal stopping rules* for stochastic approximation procedures. Farrell [15] considered the problem of finding fixed width* confidence intervals* for θ and provided under certain conditions such a stopping rule for the RM process. Sielken [38] provided a different stopping rule for this problem.

A number of variations of the RM and the KW processes have been studied. Kesten [20] introduced accelerated processes based on the idea that infrequent fluctuations in the sign of $(X_n - X_{n-1})$ indicate X_n is likely not to be near θ and that frequent fluctuations indicate that $|X_n - \theta|$ is likely to be small. Consider $\{c_n\}$ such that $c_n > 0$, $\Sigma c_n = \infty$, $\Sigma c_{n^2} < \infty$, and $c_{n+1} \leqslant c_n$. Define

$$k(n) = 2 + \sum_{i=3}^{n} S[(X_i - X_{i-1}) \times (X_{i-1} - X_{i-2})],$$

where $S(t) = 1$, if $t \leqslant 0$ and 0 otherwise. For the RM process, where a_n are the random variables $c_{k(n)}$, Kesten shows under certain conditions that $X_n \to \theta$ w.p.1. He also obtains similar accelerated analogues of the KW process and the Dvoretsky process.

Fabian [13] considers a modification of the KW process that takes an additional number of observations at each step and provides increased speed of convergence. Venter [39] observed from Sacks' asymptotic normality result for the RM process that the choice of $a = \beta^{-1}$ minimizes the asymptotic variance. He introduces a variant of the RM process that estimates β, the slope of $M(x)$ at θ, and chooses an estimator of the value a, accordingly. Venter obtains the L^2 convergence for this process, which can also be viewed as a true stochastic analogue of the Newton–Raphson algorithm applied to $M'(x)$.

Dupač [11] considers a modification of the RM process when the desired root of $M(x)$ changes or drifts during the approximation process. Let $\{\theta_n\}$ be a suitable sequence of numbers with $M(\theta_1) = m_0$. Set $M_1(x) = M(x)$ and $M_n(x) = M(x - \theta_n + \theta_1)$, for $n > 1$. Based upon observing $Y_n(x)$ with mean $M_n(x)$ Dupač modifies the RM process and shows $E(X_n - \theta_n)^2 \to 0$, as $n \to \infty$. Another stochastic approximation procedure for when the root $M(x)$ changes over time is described by Ruppert [32].

Lai and Robbins [26] consider adaptive design schemes that provide good estimates of θ at "reasonable cost," where cost at stage n is defined by them to be $\Sigma_{i=1}^{n}(X_i - \theta)^2$. Their adaptive schemes include as special cases a number of the previously noted stochastic approximation schemes.

A stochastic approximation procedure using the RM process and isotonic regression* has been discussed by Hanson and Mukerjee [16].

MULTIVARIATE PROCESSES AND THEIR ASYMPTOTIC NORMALITY

A multivariate version of the RM process has been described by Blum [4]. Let $\mathbf{Y}(\mathbf{x})$ be a p-dimensional random variable observable for $\mathbf{x} \in \mathbb{R}^p$ and define $\mathbf{M}(\mathbf{x}) = E\mathbf{Y}(\mathbf{x})$. The multivariate RM process estimates $\boldsymbol{\theta}$, the unique root to the p simultaneous equations $\mathbf{M}(\boldsymbol{\theta}) = \mathbf{m}_0$. Under certain conditions the multivariate RM process defined by

$$\mathbf{X}_{n+1} = \mathbf{X}_n - a_n(\mathbf{Y}(\mathbf{X}_n) - \mathbf{m}_0),$$

where $\{a_n\}$ is a suitable sequence of positive real numbers, converges to $\boldsymbol{\theta}$ w.p.1.

Blum [4] has also given a multivariate version of the KW process. Let $Y(\mathbf{x})$ be a random variable observable for $\mathbf{x} \in \mathbb{R}^p$, where $M(\mathbf{x}) = EY(\mathbf{x})$. The goal is to find the unique maximum $\boldsymbol{\theta}$ of $M(\mathbf{x})$. Let $\mathbf{e}_1, \dots, \mathbf{e}_p$ be an orthonormal basis for $\mathbb{R}^p$. For each $\mathbf{x} \in \mathbb{R}^p$ and each positive number δ, generate $k + 1$ independent random variables $Y(\mathbf{x})$, $Y(\mathbf{x} + \delta\mathbf{e}_1), \dots, Y(\mathbf{x} + \delta\mathbf{e}_p)$, and define

$$\mathbf{Y}(\mathbf{x}, \delta) = \left(Y(\mathbf{x} + \delta\mathbf{e}_1) - Y(\mathbf{x}), \dots, Y(\mathbf{x} + \delta\mathbf{e}_p) - Y(\mathbf{x})\right)'.$$

The multivariate KW process is defined by

$$\mathbf{X}_{n+1} = \mathbf{X}_n + (a_n/\delta_n)\mathbf{Y}(\mathbf{X}_n, \delta_n),$$

where $\{a_n\}$, $\{\delta_n\}$ are sequences of real numbers satisfying certain conditions. Blum proved that under the appropriate conditions, $\mathbf{X}_n \to \boldsymbol{\theta}$ w.p.1.

For the multivariate RM process as described by Blum, it has been shown by Sacks [34] that $\sqrt{n}(\mathbf{X}_n - \boldsymbol{\theta})$ has an asymptotic normal distribution with mean $\mathbf{0}$. Suppose for the multivariate RM process that

$$\mathbf{M}(\mathbf{x}) = \boldsymbol{\alpha} + \mathbf{B}(\mathbf{x} - \boldsymbol{\theta}) + o(|\mathbf{x} - \boldsymbol{\theta}|)$$

and $\operatorname{cov}(\mathbf{Y}(\mathbf{x})) \to \boldsymbol{\Sigma}$ as $\mathbf{x} \to \boldsymbol{\theta}$. Write $\mathbf{B} = \boldsymbol{\Gamma}\mathbf{D}\boldsymbol{\Gamma}'$, where $\mathbf{D}$ is a diagonal matrix of the ordered eigenvalues $d_1 \geqslant \cdots \geqslant d_p$ of $\mathbf{B}$ and $\boldsymbol{\Gamma}$ is an orthonormal matrix. Then under certain assumptions, the (i, j)th element of the covariance matrix of the asymptotic distribution of $\sqrt{n}(\mathbf{X}_n - \boldsymbol{\theta})$ is given by $ap_{ij}(a(d_i + d_j) - 1)^{-1}$, where p_{ij} is the (i, j)th element of $\boldsymbol{\Gamma}'\boldsymbol{\Sigma}\boldsymbol{\Gamma}$, and $a_n = a/n$. The representation of Ruppert [33] can be applied to the multivariate RM process and also to the KW process to obtain asymptotic normality under quite mild conditions on the noise sequence.

CONSTRAINED OPTIMIZATION PROCESSES

Stochastic approximation procedures have been developed for *constrained optimization** problems. Many of these can be viewed as stochastic versions of deterministic *nonlinear programming** algorithms. (See Avriel [2] or Mangasarian [28] for the subject of nonlinear programming.) Consider minimizing $M(\mathbf{x})$ where either $\mathbf{x} \in C_1 \equiv \{\mathbf{x} : \phi_i(\mathbf{x}) = 0, i = 1, \ldots, s\}$ or $x \in C_2 \equiv \{\mathbf{x} : \psi_i(\mathbf{x}) \leqslant 0, i = 1, \ldots, s\}$, respectively, called the equality constrained case or the inequality constrained case. Again $M(\mathbf{x})$ is observable only up to stochastic noise. Kushner and Clark [25] review stochastic approximation procedures for both the equality constrained case and the inequality constrained case. Their methods are primarily based on penalty-multiplier-type algorithms or Lagrangian algorithms. Conditions for the convergence of these processes involve fairly general conditions on $Y(\mathbf{x})$ and typical deterministic conditions on $M(\mathbf{x})$ and C_1 or C_2. Kushner and Clark's projection algorithm for the inequality constrained case generates feasible $\mathbf{X}_n$ for each n, i.e., $\mathbf{X}_n \in C_1$ or C_2 for each n; under relatively weak conditions $\mathbf{X}_n$ converges with probability 1 to a Kuhn–Tucker point. Also Kushner and Clark consider the more general problem when the constraint set itself has a stochastic element, in particular, the case where for a given $\mathbf{x}$, $\phi_i(\mathbf{x})$ or $\psi_i(\mathbf{x})$ is not evaluatable, but either function can be observed with additive stochastic noise.

The first discussions of continuous versions of stochastic approximation algorithms were given by Driml and Hans [9] and Sakrison [35], with further discussions and examinations of their relationships to differential equations presented by Kushner and Clark [25] and by Ljung [27]. Extensions to more general spaces have been considered, these results being related to random fixed point theorems.

APPLICATIONS

Stochastic approximation methods have been applied in a number of settings. Good discussions of diverse applications are given by Albert and Gardener [1], Wasan [40], and Wetherill [41]. Albert and Gardener and Wasan also provide good general discussions of stochastic approximations. Examples of other applications, to multivariate estimation, and to quantile estimation* are given, respectively, by Sampson [36] and Hanson and Russo [17]. Martin and Goodfellow [29] have used the RM process to construct robust estimates of location and have compared by simulation these estimates to M-estimates.

An interesting application has been to quantal response* data, where a laboratory animal is given a certain amount of a drug or chemical and responds or does not respond. Response can be the animal's death or achieving a prespecified physiological response level, e.g., lowering of blood pressure by at least 25%. It is assumed that $M(x)$, the probability of response to amount x of the given drug or chemical, is a suitable unknown monotone increasing function of x. The usual goal is to estimate the dose θ such that $M(\theta) = 0.5$; depending on the response variable, θ is often called the LD_{50} or ED_{50}. Within this context stochastic approximation procedures provide iterative estimators of θ.

The stochastic approximation procedures developed for optimizing noisy functions over constrained spaces are useful in simulation problems. In engineering and economics modeling, the system response to a control setting of $\mathbf{x}$ can often only be simulated. The simulation response for a setting $\mathbf{x}$, $Y(\mathbf{x})$, can then be viewed as an estimator of the true system response $M(\mathbf{x})$. The objective is to find θ such that $M(\boldsymbol{\theta})$ is an optimum.

References

[1] Albert, A. and Gardner, L. (1967). *Stochastic Approximation and Nonlinear Regression*. MIT Press, Cambridge, MA. (Good reference text.)

[2] Avriel, M. (1976). *Nonlinear Programming*. Prentice-Hall, Englewood Cliffs, NJ.

[3] Blum, J. R. (1954). *Ann. Math. Statist.*, **25**, 382–386.

[4] Blum, J. R. (1954). *Ann. Math. Statist.*, **25**, 737–744.

[5] Blum, J. R. (1958). *Proc. Amer. Math. Soc.*, **9**, 404–407.

[6] Burkholder, D. L. (1956). *Ann. Math. Statist.*, **27**, 1044–1059.

[7] Chung, K. L. (1954). *Ann. Math. Statist.*, **25**, 463–483.

[8] Derman, C. (1956). *Ann. Math. Statist.*, **27**, 532–536.

[9] Driml, M. and Hans, O. (1960). *Trans. Second Prague Conf. Inf. Theory, Statist. Decision Funct. Random Process*. Czechoslovak Academy of Sciences, Prague, Czechoslovakia, pp. 113–122.

[10] Dupač, V. (1957). *Casopis Pest. Mat.*, **82**, 47–75. (In Czechoslovak with English and Russian summaries.)

[11] Dupač, V. (1965). *Ann. Math. Statist.*, **36**, 1695–1702.

[12] Dvoretzky, A. (1956). *Proc. Third Berkeley Symp. Math. Statist. Prob.*, Vol. 1. University of California Press, Berkeley, CA, pp. 39–55.

[13] Fabian, V. (1967). *Ann. Math. Statist.*, **38**, 191–200.

[14] Fabian, V. (1971). *Optimizing Methods in Statistics*, J. Rustagi, ed. Academic, New York, pp. 439–470. (Good review paper.)

[15] Farrell, R. (1962). *Ann. Math. Statist.*, **33**, 237–247.

[16] Hanson, D. and Mukerjee, H. (1981). Unpublished manuscript.

[17] Hanson, D. and Russo, R. (1981). *Zeit. Wahrsch. verw. Geb.*, **56**, 145–163.

[18] Kallianpur, G. (1954). *Ann. Math. Statist.*, **25**, 386–388.

[19] Kersting, G. (1977). *Ann. Prob.*, **5**, 954–965.

[20] Kesten, H. (1958). *Ann. Math. Statist.*, **29**, 41–59.

[21] Kiefer, J. and Wolfowitz, J. (1952). *Ann. Math. Statist.*, **23**, 462–466.

[22] Krasulina, T. (1962). *Theory Prob. Appl.*, **7**, 108–113.

[23] Krasulina, T. (1969). *Theory Prob. Appl.*, **14**, 522–526.

[24] Kushner, H. (1978). *SIAM J. Optimization Control*, **16**, 150–168.

[25] Kushner, H. and Clark, D. (1978). *Stochastic Approximation Methods for Constrained and Unconstrained Systems*. Springer-Verlag, New York. (Good mathematical reference text.)

[26] Lai, T. and Robbins, H. (1979). *Ann. Statist.*, **7**, 1196–1221. (Good paper for those intending to apply stochastic approximation.)

[27] Ljung, L. (1977). *IEEE Trans. Automat. Control*, **22**, 551–575.

[28] Mangasarian, O. (1969). *Nonlinear Programming*. McGraw-Hill, New York.

[29] Martin, R. and Goodfellow, D. (1984). *Commun. Statist. Simul. Comp.*, **13**, 1–46.

[30] Ortega, J. and Rheinboldt, W. (1970). *Iterative Solution of Nonlinear Equations in Several Variables*. Academic, New York.

[31] Robbins, H. and Monro, S. (1951). *Ann. Math. Statist.*, **22**, 400–407.

[32] Ruppert, D. (1981). *Ann. Math. Statist.*, **9**, 555–566.

[33] Ruppert, D. (1982). *Ann. Prob.*, **10**, 178–187.

[34] Sacks, J. (1958). *Ann. Math. Statist.*, **29**, 373–405.

[35] Sakrison, D. (1964). *Ann. Math. Statist.*, **35**, 590–599.

[36] Sampson, A. (1976). *J. Multivariate Anal.*, **6**, 167–175.

[37] Schmetterer, L. (1961). *Proc. Fourth Berkeley Symp. Math. Statist. Prob.*, Vol. 1. University of California Press, Berkeley, CA, pp. 587–609. (Good early review paper.)

[38] Sielken, R. (1973). *Zeit. Wahrsch. verw. Geb.*, **26**, 67–75.

[39] Venter, J. (1967). *Ann. Math. Statist.*, **38**, 181–190.

[40] Wasan, M. (1969). *Stochastic Approximation*. Cambridge University Press, Cambridge, England. (Good reference text.)

[41] Wetherill, G. (1966). *Sequential Methods in Statistics*. Methuen, London, England.

Acknowledgment

The work of the author is sponsored by the Air Force Office of Scientific Research under Contract F49620-79-C-0161.

(KIEFER–WOLFOWITZ PROCEDURE
NUMERICAL ANALYSIS
SEQUENTIAL ANALYSIS)

ALLAN R. SAMPSON

STOCHASTIC CONTROL *See* OPTIMAL STOCHASTIC CONTROL

STOCHASTIC CONVERGENCE *See* CONVERGENCE OF SEQUENCES OF RANDOM VARIABLES

STOCHASTIC DEMOGRAPHY

Stochastic demography is the theoretical and empirical study of random variation in demographic processes.

The most fundamental model of demography*, the *life table**, due to Graunt and Halley in the seventeenth century, is essentially probabilistic: it gives 1 minus the *cumulative distribution function* of the duration of life. Yet the practical use of the life table long preceded its interpretation in terms of probability theory. The statistical analysis of the sampling variability of life expectancy, computed from a life table based on a finite number of deaths, is recent (Wilson [105]). Elandt-Johnson and Johnson [23] describe present statistical techniques of life table analysis.

Another early model of stochastic demography, now called the *branching process**, was analyzed by Bienaymé* in 1845. It was independently formulated by Galton and Watson in 1873 to study the extinction of familial lines of descent. The theory of branching processes has undergone enormous mathematical diversification and development. Among demographic users of branching processes, Wachter et al. [103, Chap. 7] measure social mobility in seventeenth and eighteenth century England by comparing extinctions of families holding baronetcies with the predictions of branching processes and of a pure death process (*see* BIRTH AND DEATH PROCESSES).

The modern field of stochastic demography applies equally to human and nonhuman populations, to historical and evolutionary time scales. It emerges largely from the work (1939–1949) of Bartlett [6, 7], Feller [26], and Kendall [50–52]. M'Kendrick [78], in a remarkable but isolated early paper, analyzes pure birth processes in one and two dimensions. He derives, among other results, a partial differential equation for mortality in age-structured populations later attributed to von Foerster, as well as a probabilistic interpretation of the renewal equation for population growth*.

Keyfitz [53], Feichtinger [24], Pollard [87], Ludwig [66], Keiding [48], Menken [75], and Kurtz [55] give reviews. Smith and Keyfitz [95] reprint extracts from classic papers. The mathematics that supports virtually all of stochastic demography appears in Karlin and Taylor [46, 47].

Currently, the interpretation of deterministic population models (*see* MATHEMATICAL THEORY OF POPULATIONS) in terms of modern probability theory continues the task started by M'Kendrick (Hoem [42, 44]). However, this review emphasizes stochastic models for *population projection** developed since 1965. Other stochastic models in demography will be cited or reviewed selectively. Many stochastic models in other social and biomedical sciences that are relevant to demography are omitted entirely. Also largely omitted here are the statistical problems of measuring and estimating the parameters in stochastic models.

POPULATION PROJECTION MODELS

Stochastic models are needed for population projection because deterministic models fail to account for the variability of historical demographic data and to provide probabilistically meaningful estimates of the uncertainty of demographic predictions. Stochastic population projection models may include migration* and mortality as well as fertility*.

Lee [65] reviews comprehensively new techniques for projecting fertility. He concludes that stochastic models of the internal structure of fertility behavior provide a better basis for forecasting than do methods based on behavioral theories or on data about fertility expectations. McDonald [71] supports this conclusion.

Among stochastic projection models, one can distinguish, not always sharply, between structural models* and time-series* models. Structural models attempt to represent some underlying mechanism of population growth. Time-series models apply to demographic data general techniques in which the form of the model need not be based on a theory of demographic processes.

Structural projection models usually describe either or both of two sources of random fluctuation: demographic variation and environmental variation. Additional sources of variation that have been less fully considered in stochastic projection models include the heterogeneity of probability intensities among individuals (but see Keyfitz and Littman [54], Vaupel et al. [101], and Menken et al. [76]) and dependence between individuals (as in marriage; see below) or between factors affecting individuals (Peterson [83], Manton and Stallard [67]).

Demographic variation arises in ensembles of populations from the stochastic operation of mechanisms with fixed vital rates. For example (Pollard [85]), in populations each of $N = 1{,}000{,}000$ people, if the probability of dying within one year is $q = 0.002$, uniformly and homogeneously for all people, then the (binomial*) variance among populations in the number who die after one year in $Nq(1-q) = 1996$. This demographic variability should not be confused with sampling variability (though it is unfortunately sometimes given the same name), which is the variation in the properties of a sample of individuals randomly selected from a population. Whereas sampling variability arises from the procedures for observing a population, demographic variability arises intrinsically in an ensemble of populations each governed by a stochastic process*.

Environmental variation arises when the demographic rates themselves are governed by a stochastic process. To continue the example, if q is a random variable with mean 0.002, as before, and with standard deviation 0.0001, then, ignoring demographic variation, the variance in the expected number of deaths among populations of a million people is $\text{var}(Nq) = N^2\,\text{var}(q) = 10{,}000$. In this example, the purely environmental variation is more than five times the demographic variation, although the standard deviation of q, the fraction who die, is only 5% of the mean.

Generally, in large populations, fluctuations in vital rates cause fluctuations in population size that appear to dominate the fluctuations arising from demographic variation.

POPULATION PROJECTION WITH DEMOGRAPHIC VARIATION

Deterministic projections of populations closed to migration commonly use the recurrence relation

$$y(t+1) = L(t+1)\,y(t), \qquad t = 0, 1, 2, \ldots, \quad (1)$$

where $y(t)$ is a vector in which the ith component is the number of females in age class i at time t, $i = 1, \ldots, k$, $y(0)$ is a given initial age census* of the female population, and $L(t)$ is a $k \times k$ nonnegative matrix, conventionally called a *Leslie matrix*. All elements of L are 0 except those in the first row and those just below the diagonal. $L_{1i}(t+1)$ specifies the effective fertility of females in age class i at time t (the average number of daughters born between t and $t+1$ who survive to time $t+1$, per female aged i at t) and $L_{i+1,\,i}(t+1)$ gives the proportion of females in age class i at time t who survive to age class $i+1$ at time $t+1$ (Keyfitz [53]).

Taking L to be independent of t, i.e., constant in time, Pollard [84] reinterprets (1) as a multitype branching process. $L_{1i}(t+1)$ becomes the probability that a female in age group i at time t will give birth during the time interval $(t, t+1)$ to a single daughter, and that this daughter will be alive in age group 1 at time $t+1$; and $L_{i+1,i}(t+1)$ becomes the probability that a female from age group i at time t will survive to be in age group $i+1$ at time $t+1$. The survival and fertility of each female are assumed independent of each other and of the survival and fertility of all other females. Then $y(t)$ in (1) can be interpreted as the expectation of the age census at time t. A linear recurrence relation that uses the direct or Kronecker product* of two matrices describes the variances and covariances of each census of females. Goodman [35] computes the probability in Pollard's model that the line of descendants of an individual of any given age will eventually become extinct.

The multitype branching process model does not require that $L(t)$ in (1) have the form of a Leslie matrix. See Goodman [36] for a linear treatment of two sexes; Breev and Staroverov [12] for labor force migration; Wu and Botkin [106] for elephants. Deistler and Feichtinger [21] show that the multitype branching process model may be viewed as a special case of a model of additive errors proposed for population dynamics by Sykes [98].

Mode [80] develops population projection models using renewal theory* rather than matrix methods. Underlying these models are discrete-time versions of the Crump–Mode–Jagers age-dependent branching process.

The continuous-time stochastic theory analogous to what has just been described is presented by Keiding and Hoem [49] and Braun [11], with extensions to parity-dependent birth rates and multiregional populations.

Branching processes have been criticized as models of human and nonhuman populations. The criticisms have been directed at both assumptions and predictions. The best studied branching processes share two assumptions: stationarity and independence. Stationarity means that the underlying rates (e.g., of survival or of giving birth) are constant in time, though they may change with, e.g., age or parity; in branching processes in random environments (e.g., Pollard [85]), survival and birth rates may fluctuate, but the fluctuations are controlled by a stochastic process that is stationary in time. Independence means that the life history of one individual is independent of the life history of every other individual, though fecundity and survival may interact within one individual's life history and both fecundity and survival may depend on a fluctuating environment.

Of the assumptions, stationarity is the easiest target for criticism. For example, it has been objected that even a few human generations span a period of historical time in which social, political, and economic systems change so markedly and migration is so influential that any model based on unchanging rates or a stationary pattern of environmental fluctuations must be irrelevant. This objection would be valid if the long-run properties of branching processes occurred only when rates or environmental fluctuations had been stationary for all time. However, stochastic weak ergodic theorems* (Cohen [17]) suggest by analogy that even if the stochastic process governing vital rates is not stationary (but is ergodic), the behavior of a multitype branching process (specifically, the inhomogeneous product of the mean value matrices) should (under suitable conditions) depend on the recent past much more than on the remote past. Thus, loosely speaking, if the rate of historical change is slow relative to the smoothing effects of demographic processes, it may be adequate to model the present and recent past as if demographic processes were stationary, as long as no one assumes that today's conditions will extend indefinitely into the future. Thus the assumption of stationarity is not a priori disabling if its limitations are respected and if it leads to confirmed predictions.

The assumption of independence between individuals seems ineluctable in branching processes (but see Staroverov [96] for an example of how it can be modified). Independence seems easiest to defend in large populations and more difficult to defend in small ones because interactions seem more apparent in small populations, yet the demographic variation described by branching processes is most relevant to small populations and nearly irrelevant to large. Independence precludes a description of monogamous mating in a two-sex population. In studies of fish and wildlife* populations, it is widely assumed that population size must be stationary, on the average, over a long period of time, in part because of interactions between individuals that adjust birth and death rates. The assumption of independence is probably a greater obstacle to the success of branching process models than the assumption of stationarity.

If their assumptions are not grounds for dismissing branching processes a priori as population models, the empirical task remains of evaluating the predictions. Since existing populations are not extinct by definition, a supercritical branching process is usually chosen as a model and the asymptotic theory is applied (for a counterexample, see Wachter et al. [103, Chap. 11]).

In Pollard's model, the branching process is supercritical if the dominant eigenvalue $r(L)$ of the expected value matrix L exceeds 1 (so that the population asymptotically increases exponentially in size). In this case, if, for some t, every element of L^t is positive, then, with probability 1, the random vector that gives the number of females in each age class, divided by $(r(L))^t$, asymptotically becomes proportional to the stable age structure of the deterministic model with projection matrix L. It is obvious that no real population can forever grow exponentially and that $t \to \infty$ is never observed. At finite times large enough for the predictions of the model to be relevant, it appears that many real age censuses for human and nonhuman populations (e.g., elephant: Wu and Botkin [106]) deviate markedly from the stable age structure implied by current vital rates. This finding suggests that the age census is influenced by sources of variation in addition to purely demographic ones.

Projections of the Norwegian population as a multitype branching process give estimates of uncertainty that Schweder [91] considers unrealistically low.

Independently of Pollard [84], Staroverov [96] considers exactly the same model. Because the model variances are implausibly small compared to the historical variation in Soviet birth rates, Staroverov replaces the assumption that each individual evolves independently with the assumption that groups of c individuals evolve as units, independently of other groups. As c increases, the variance of numbers in each age group increases while the means remain unaltered. A comparison of observed and projected births from 1960 to 1973 suggests that even $c =$ 100,000 is too small, and that it is necessary to allow for temporal variation in the fertility and mortality parameters.

POPULATION PROJECTION WITH ENVIRONMENTAL VARIATION

In a large population, the effects of demographic variation are normally negligible compared to those of apparent changes in vital rates. In a model of multiplicative errors, Sykes [98] supposes that, given $L(t+1)$, $y(t)$ determines $y(t+1)$ exactly, but that there is no correlation between $L(t)$ and $L(s)$, $s \neq t$. He assumes an arbitrary covariance structure for the elements within a matrix, subject to the constraint that $L(t)$ be a Leslie matrix. Sykes computes the means and covariances of the age censuses, allowing the means and covariances of the sequence $\{L(t)\}$ to be inhomogeneous in time. Seneta [92] pursues the computation of variances in the models of Sykes [98] and Pollard [84]. Lee [63] discusses the numerical example that Sykes gives.

LeBras [61] considers populations satisfying (1) in which the sequence $\{L(t)\}$ is governed by a stationary stochastic process* with independence between Leslie matrices sufficiently distant in time. Under additional conditions on a finite sample space of Leslie matrices, he argues that for every sample path, $\lim_{t \to \infty}[t^{-1}\log y_1(t)]$ is a constant independent of the sample path, i.e., that the number of births asymptotically changes exponentially in every sample path.

Assuming that $\{L(t)\}$ is determined by a finite-state irreducible Markov chain of arbitrary fixed finite order, LeBras [61] computes the moments of $y(t)$ of all orders, for both finite t and as $t \to \infty$. He argues that the distribution of the number $y_1(t)$ of births at a given large time t is approximately log-normal*.

Independently of LeBras, Cohen [16] proves that when the Leslie matrices $L(t)$ are chosen from a denumerable set according to a Markov chain that is not necessarily stationary or homogeneous in time, the moments of age structure eventually become independent of the initial age census $y(0)$ and initial vital rates $L(1)$.

Cohen [16] also points out the relevance to stochastic demography of products of *random matrices** (Furstenberg and Kesten [29]). Under exactly stated conditions more general than those of LeBras, Furstenberg and Kesten prove theorems that imply that $y(t)$ changes asymptotically exponentially and that the elements of $y(t)$ are, for large t, asymptotically log normal.

Cohen [17] generalizes (1) to allow $L(t)$ to be *contractive operators* on the space of age structures, chosen by a first-order irreducible aperiodic Markov chain from a general (i.e., possibly uncountably infinite) state space of operators. When the Markov chain is time-homogeneous, the solution of a linear renewal integral equation* gives the limiting probability distribution of age structure. Even when the Markov chain that chooses $L(t)$ is inhomogeneous but suitably ergodic, the probability distribution of age structure asymptotically becomes independent of initial conditions. This weak stochastic ergodic theorem is the probabilistic analog of the deterministic weak ergodic theorem of Coale and Lopez (Pollard [87, pp.51–55]).

The use of products of random matrices to model environmental variability in age-structured populations is generalized and refined by, among other, Cohen [18], Lange [58], and Tuljapurkar and Orzack [99].

An elementary but important observation emerging from these studies is a distinction between two measures of the long-run rate of growth of a population in a stochastic environment. One measure, studied by Furstenberg and Kesten [29], is the average of the long-run rates of growth along each sample path,

$$\log \lambda = \lim_{t \to \infty} t^{-1} E(\log y_1(t)).$$

Another measure is the long-run rate of growth of the average populations,

$$\log \mu = \lim_{t \to \infty} t^{-1} \log E(y_1(t)).$$

For deterministic models $\lambda = \mu$, but in general, in stochastic models, $\lambda \leq \mu$ with strict inequality in most examples.

DEMOGRAPHIC AND ENVIRONMENTAL VARIATION

Demographic variation and environmental variation both exist in reality. If the probabilities of giving birth and of surviving in a multitype branching process are themselves random variables (Pollard [85], Bartholomew [3]), the moments of the numbers of individuals in each age class can be computed from a modification of a recurrence relation derived by Pollard [84].

For a multitype branching process such that the offspring probability generating functions* at all times are independently and identically distributed, Weissner [104] gives some necessary and some sufficient conditions for almost sure extinction of the population (see also Namkoong [81]). In a multitype branching process with probability generating functions determined by a stationary metrically transitive process (subject to certain bounds), the Furstenberg–Kesten limit of the product of expectation matrices determines whether the probability of extinction of all types is 1 or less than 1 (Athreya and Karlin [2]). Weissner, Athreya, and Karlin do not discuss the application of these results to age-structured populations.

A thorough empirical test of the merits for prediction of (1) when $L(t)$ has sequentially dependent, e.g., Markovian, random variation, in comparison with deterministic methods (Siegel [94]) of *population projection**, has yet to be performed. An outstanding example of the evaluation of a projection procedure, though it predicts only total population size, is given by Henry and Gutierrez [41]. In evaluating stochastic projections of age-structured populations, it will be necessary to consider, in addition to demographic variation and environmental variation, the uncertainty in specifying the form of a model that governs the $\{L(t)\}$ sequence and the uncertainty in estimating the parameter values of the model (Schweder [91], Hoem [43], Bartholomew [3]).

Time-Series Models

The application of modern stochastic *time-series* methods to demographic data originates with Lee [62] and Pollard [86]. Lee (see ref. 64 for summary) uses long time series, for example, of births and marriages or of mortality and wages, to test alternative historical theories of demographic and economic dynamics. Pollard [86] develops a second-order autoregressive model of the growth rate of total population size for Australia.

Lee's [63] analysis of births from 1917 to 1972 in the United States demonstrates that the distinction between structural models and time-series models is not sharp. Equation (1) implies that each birth may be attributed to the fertility of the survivors of some preceding birth cohort. Hence the sequence of births $\{y_1(t)\}$ is described by a renewal equation. By a sequence of approximations to this renewal equation, Lee transforms the residuals of births from their long-run trend into an autoregressive process for which variations in the net reproduction rate are the error term. Among the several stochastic models Lee considers for the net reproduction rate, a white-noise process and a first-order autoregressive (Markovian) model lead to poorer descriptions of births than a second-order autoregressive process.

Independently of Lee, Saboia [89] develops autoregressive moving average* (ARMA) models using *Box–Jenkins** techniques for the total population of Sweden. Based on data from 1780 to 1960 at five-year intervals, his projections for 1965 compare favorably with some standard demographic projections.

Saboia [90] relates ARMA models to the renewal equation for forecasting births. In these models, the age-specific vital rates can vary over time; migration is recognized. Using the female birth time series for Norway, 1919–1975, he gives forecasts with confidence intervals up to 2000. However, Saboia's [90] models are not the simplest required to describe the data (McDonald [70]). McDonald [69] describes the relationships among the renewal equation model, with migration added, structural stochastic econometric* models, and ARMA models.

Using Australian data from 1921 to 1965, he finds that the number of females aged 15 to 39 years does not help explain the number of births given the time series of past births, but that some additional explanatory power is obtained from the number of females aged 20 to 34 years. The ARMA models do not predict a sharp decline that occurred in the number of Australian births after 1971. McDonald suggests that exogenous, perhaps economic, variables will have to be invoked to explain this turning point. Land [56] similarly suggests incorporating exogenous variables in structural stochastic projection models with environmental variation.

The long-term forecasts of the time-series models have very wide confidence intervals* (e.g., McDonald [69], McNeil [74]). In view of the uncertainty of the demographic future, policy that depends on population size and structure should be flexible enough to allow for different possible futures.

In addition to spectral methods and *Box–Jenkins* techniques, other recent approaches to population time-series modelling include a stochastic version of the logistic equation (McNeil [74]) as a model of United States Census total population counts; the *Karhunen–Loève prodecure* (Basilevsky and Hum [8]) for quarterly records of births on two Jamaican parishes, 1880 to 1938; and an age- and density-dependent structural model, estimated by use of the *Kalman–Bucy filter* (Brillinger et al. [13]) for age-aggregated counts of the sheep *blow-fly*. In the study of blow-flies, even after the best of seven models for death rates had been fitted, the weighted residuals of the time series of deaths revealed substantial autocorrelation. Applying an ARMA model to these residuals improved the description of the data (Brillinger et al. [13, p. 75]). Hybrid models like this one, which combine demographic theory with general-purpose statistical descriptions, deserve further study in conjunction with efforts to determine empirically what exogenous nondemographic variables need to be incorporated. Granger and Newbold [38, Chap. 8.2] analyze the combination of forecasts in an economic context, which is relevant here.

Evaluation of Stochastic Population Projections

Prediction is the ultimate test of scientific understanding. Good population projection procedures might be found faster if two procedures were more systematically exploited: historical pseudo-experiments and multiple criteria.

To illustrate the meaning of historical pseudo-experiments, suppose one has data, demographic and otherwise, from year a to year b, and one wishes to forecast one or more components of the data. Using all the data from a to b to forecast for years $b + 1, b + 2, \ldots$ permits publication long before the model can be rejected. Using part of the data, from a to $b_1 < b$, at least permits a test of predictions against what happened in years $b_1 + 1, b_1 + 2, \ldots, b$. Why not pick a subinterval of the data for the years a_1 to b_1, $a \leq a_1 \leq b_1 \leq b$, fit the model to the years a_1 to b_1 and project forward, and then systematically vary *both* a_1 and b_1? The different tests of the model will not be independent, but one will have squeezed more information about the model's behavior out of the data. By varying a_1 for each value of b_1, one will learn how much knowledge of the past is relevant to a good prediction of the future, and whether the amount of the past that is relevant to the future itself changes over time.

Multiple criteria would help in deciding which models are good for which purposes under what conditions. One model may predict births well for the next five years; another, retirement age classes 20 years in the future. Instead of comparing a family of models by one criterion only, e.g., mean-squared deviations of observed from predicted age structure, why not look at median-absolute and maximum as well as mean-squared deviations, for each age class individually, for a whole flock of different intervals projected into the future? Flexibility is needed here to discover what are good criteria for evaluation.

Many of the general issues that arise in evaluating economic forecasts (Granger and Newbold [38, Chap. 8.3] arise equally here.

STOCHASTIC MODELS OF SPECIFIC PROCESSES

Several components of population change have been studied through the use of stochastic models. We shall sketch some models of human reproduction and of marriage. In addition to these, stochastic models have been developed to describe, among other topics, social mobility and the work force (Bartholomew [4], Bartholomew and Forbes [5], McClean [68]); the succession of rulers in an atoll society (Frauenthal and Goldman [28]); changes of residence (Ginsberg [32–34]); the population composition of the descendants of a collection of identical clones (Blackwell and Kendall [9], Cohen [15]);

cause-specific but age-aggregated mortality indices (Land and McMillen [57]); property crime rates (Cohen et al. [19]); dose–mortality curves in radiation biology (Turner [100], Miller [77]); and the frequencies of various *kinship* relations (Goodman et al. [37]; Feichtinger and Hansluwka [25], Wachter [102]).

Human Reproduction

In human reproduction a woman of appropriate age may be supposed to move from a state (state 1) of susceptibility to conception into one of two states: a state of conception (state 2) that does not end in a live birth and a state of conception (state 3) that does end in a live birth. From state 2, the woman returns to state 1, and from state 3, the woman enters a nonsusceptible state of postpartum amenorrhea, from which she eventually returns to state 1. Durations of stay in each state need not be exponentially distributed and rates of transition among states need not be homogeneous across women. This schema may be modeled as a renewal process (Sheps and Menken [93], who also refer to the pioneering work of Gini, Pearl, Henry, and Vincent; Das Gupta [20], Ginsberg [30, 31], Mode [79], Lange and Johnson [59]). This class of models has been used to analyze interbirth intervals and to evaluate the demographic consequences of contraception and abortion (Potter [88]). For example, models indicate that an effective contraceptive used by a small proportion of a population reduces birth rates more than a less effective contraceptive used by a much higher proportion of the population.

Braun [11] analyzes three sets of historical data on birth interval sequences using a different approach based on linear regression*.

Marriage

The standardized distribution by age of the frequency of first marriage* in a female cohort is well approximated by the convolution of a normal distribution of age of entry into a marriageable state and three exponentially distributed delays: the delay until the woman starts to keep frequent company with the eventual husband; the delay until engagement; and the delay until marriage (Coale and McNeil [14]).

Given the age distributions of brides and grooms separately in any period, how can the correlation in age of spouses be explained? Henry [40] models the two-way contingency table* of marriage frequencies categorized by age of bride and age of groom as a sum of contingency tables each with independence between rows and columns. Each summand is supposed to describe age-independent marriage choices within "panmictic circles."

Other probabilistic marriage models are developed by Hajnal [39], McFarland [72, 73], and Asmussen [1].

As Keiding and Hoem [49] point out, the assumption of stochastic independence between individuals is crucial to their probabilistic formulation of stable population theory, as well as to many other stochastic population models. A tractable and realistic stochastic formulation of marriage, which must drop the assumption of independence between individuals, is an open challenge.

COMPUTER SIMULATIONS

In addition to models that can be written down in what is accepted as "closed form," complex stochastic models of demographic processes are embodied in computer simulations. Menken [75] reviews simulation and other models. Wachter et al. [103, Chap. 1–5 and 11] use population simulations in historical demography; Orcutt et al. [82] in economics; Dyke and MacCluer [22] in population genetics; Howell [45] in anthropology; Feichtinger and Hansluwka [25] and Suchindran et al. [97] in demography proper.

Such simulations reveal the behavior of models when "realistic" assumptions preclude mathematical analysis. In addition, simulation of relatively simple models whose asymptotic behavior is understood may shed useful light on their transient behavior. For example, in simulations of the critical multitype branching process treated analytically by Pollard [84], Wachter et al. [103, Chap. 11] find that the interquartile range* in the size of individual age groups is very nearly steady, after a simulated generation or so, for more than a century. This so-called pre-asymptotic stochastic plateau is not described by asymptotic theory.

Simulations can also help test the field-worthiness of demographic estimation procedures derived from deterministic models (Wachter [102]).

Novel techniques make it possible to improve the precision (i.e., reduce the variance) of estimates derived from population simulations by replacing random numbers drawn independently for each replication or run with random numbers having a carefully chosen dependence (Fishman [27]).

ASSESSMENT AND PROSPECTS

A major lesson that has been learned from the testing against data of stochastic models for population projection and for other demographic processes is that it is frequently unsatisfactory to assume constancy over time and homogeneity over individuals in the vital rates or in the other forces of transition that directly affect individuals. Yet the purpose of theory, as Einstein said, is to make nature stand still when our back is turned. So constancy must be assumed at some level, probably far deeper than the rates that affect individuals, perhaps only in the form of causal relations. The question is: where?

The sheer diversity of the stochastic models that have been cited indicates part of the answer. There is a great need to integrate demographic with biological, social, and economic models, and with each other. For example, the fertility rates that serve as parameters in demographic models are the objects to be explained by biological models of fecundity in conjunction with sociological models of marriage and family formation and with economic models, e.g., of health services or food supply. Perhaps when the component models are correctly chosen and integrated with others correctly chosen, the total amount of variation remaining that must be attributed to pure randomness will be reduced. Even the empirical study of which local models of particular demographic processes are correct might be more successful if each model were not approached in isolation from all others.

References

[1] Asmussen, S. (1980). On some two-sex population models. *Ann. Prob.*, **8**, 727–744.

[2] Athreya, K. B. and Karlin, S. (1971). On branching processes with random environments: I. Extinction probabilities. *Ann. Math. Statist.*, **42**, 1499–1520.

[3] Bartholomew, D. J. (1975). Errors of prediction for Markov chain models. *J. R. Statist. Soc. Ser. B*, **37**, 444–456.

[4] Bartholomew, D. J. (1982). *Stochastic Models for Social Processes*, 3rd ed. Wiley, Chichester, England.

[5] Bartholomew, D. J. and Forbes, A. F. (1979). *Statistical Techniques for Manpower Planning*. Wiley, Chichester, England.

[6] Bartlett, M. S. (1955, 1966). *An Introduction to Stochastic Processes with Special Reference to Methods and Applications*. 1st ed., 2nd ed. Cambridge University Press, London, England.

[7] Bartlett, M. S. (1960). *Stochastic Population Models in Ecology and Epidemiology*. Methuen, London, England.

[8] Basilevsky, A. and Hum, D. P. J. (1979). Karhunen–Loève analysis of historical time series with an application to plantation births in Jamaica. *J. Amer. Statist. Ass.*, **74**, 284–290.

[9] Blackwell, D. and Kendall, D. G. (1964). The Martin boundary for Pólya's urn scheme and an application to stochastic population growth. *J. Appl. Prob.*, **1**, 284–296.

[10] Braun, H. I. (1978). Stochastic stable population theory in continuous time. *Scand. Actuarial J.*, **61**, 185–203.

[11] Braun, H. I. (1980). Regression-like analysis of birth interval sequences. *Demography*, **17**, 207–223.

[12] Breev, B. D. and Staroverov, O. V. (1977). O metode ucheta faktorov pri prognoze dvizheniia

naseleniia i trudovykh resursov. *Ekon. Mate. Metody*, **13**, 489–499. [English translation: Determinants in forecasting the movement of population and labor resources. *Matekon*, **14**, 80–97 (Spring, 1978).]

[13] Brillinger, D. R., Guckenheimer, J. Guttorp, P., and Oster, G. (1980). Empirical modelling of population time series data: The case of age and density dependent vital rates. *Lect. Math. Life Sci.*, **13**. Amer. Math. Soc., Providence, RI, pp. 65–90.

[14] Coale, A. J. and McNeil, D. R. (1972). The distribution by age of the frequency of first marriage in a female cohort. *J. Amer. Statist. Ass.*, **67**, 743–749.

[15] Cohen, J. E. (1976). Irreproducible results and the breeding of pigs; or nondegenerate limit random variables in biology. *Biosci.*, **26**, 391–394.

[16] Cohen, J. E. (1976). Ergodicity of age structure in populations with Markovian vital rates, I: Countable states. *J. Amer. Statist. Ass.*, **71**, 335–339.

[17] Cohen, J. E. (1977). Ergodicity of age structure in populations with Markovian vital rates, II: General states; III: Finite-state moments and growth rates; illustration. *Adv. Appl. Prob.*, **9**, 18–37, 462–475.

[18] Cohen, J. E. (1980). Convexity properties of products of random nonnegative matrices. *Proc. Nat. Acad. Sci. USA*, **77**, 3749–3752.

[19] Cohen, L. E., Felson, M., and Land, K. C. (1980). Property crime rates in the United States: A macrodynamic analysis, 1947–1977; with ex ante forecasts for the mid 1980's. *Amer. J. Sociol.*, **86**, 90–118.

[20] Das Gupta, P. (1973). A Stochastic Model of Human Reproduction. Population Monograph 11, *Institute of International Studies*, University of California, Berkeley, CA. Reprinted (1976) Greenwood Press, Westport, CT.

[21] Deistler, M. and Feichtinger, G. (1974). The linear model formulation of a multitype branching process applied to population dynamics. *J. Amer. Statist. Ass.*, **69**, 662–664.

[22] Dyke, B. and MacCluer, J. W. (1974). *Computer Simulation in Human Population Studies*. Academic, New York.

[23] Elandt-Johnson, R. C. and Johnson, N. L. (1980). *Survival Models and Data Analysis*. Wiley, New York.

[24] Feichtinger, G. (1971). *Stochastische Modelle demographischer Prozesse* (Stochastic Models of Demographic Processes). Springer, Berlin, Germany.

[25] Feichtinger, G. and Hansluwka, H. (1977). The impact of mortality on the life cycle of the family in Austria. *Zeit. Bevölkerungswiss.*, **4**, 51–79.

[26] Feller, W. (1939). Die Grundlagen der Volterraschen Theorie des Kampfes ums Dasein in Wahrscheinlichkeitstheoretischer Behandlung (The foundations of Volterra's theory of the struggle for existence probabilistically treated). *Acta Biotheor.*, **5**, 11–40.

[27] Fishman, G. S. (1979). Variance reduction for population growth simulation models. *Operat. Res.*, **27**, 997–1010.

[28] Frauenthal, J. C. and Goldman, N. (1977). Demographic dating of the Nukuoro society. *Amer. Math. Monthly*, **84**, 613–618.

[29] Furstenberg, H. and Kesten, H. (1960). Products of random matrices. *Ann. Math. Statist.*, **31**, 457–469.

[30] Ginsberg, R. B. (1972). A class of doubly stochastic processes: With an application to the effects of lactation on the postpartum anovulatory period. In *Population Dynamics*, T. N. E. Greville, ed. Academic, New York, 297–331.

[31] Ginsberg, R. B. (1973). The effect of lactation on the length of the post partum anovulatory period: An application of a bivariate stochastic model. *Theor. Popul. Biol.*, **4**, 276–299.

[32] Ginsberg, R. B. (1978). The relationship between timing of moves and choice of destination in stochastic models of migration. *Environ. Plann. A*, **10**, 667–679.

[33] Ginsberg, R. B. (1979). Timing and duration effects in residence histories and other longitudinal data. I. Stochastic and statistical models. II. Studies of duration effects in Norway, 1965–1971. *Reg. Sci. Urban Econ.*, **9**, 311–331, 369–392.

[34] Ginsberg, R. B. (1979). Tests of stochastic models of timing in mobility histories: Comparison of information derived from different observation plans. *Environ. Plann. A*, **11**, 1387–1404.

[35] Goodman, L. A. (1967). The probabilities of extinction for birth-and-death processes that are age-dependent or phase-dependent. *Biometrika*, **54**, 579–596.

[36] Goodman, L. A. (1968). Stochastic models for the population growth of the sexes. *Biometrika*, **55**, 469–487.

[37] Goodman, L. A., Keyfitz, N., and Pullum, T. W. (1974, 1975). Family formation and the frequency of various kinship relationships. *Theor. Popul. Biol.*, **5**, 1–27; **8**, 376–381.

[38] Granger, C. W. J. and Newbold, P. (1977). *Forecasting Economic Time Series*. Academic, New York.

[39] Hajnal, J. (1963). Concepts of random mating and the frequency of consanguineous marriages. *Proc. Roy. Soc. Lond. Ser. B*, **159**, 125–177.

[40] Henry, L. (1972). Nuptiality. *Theor. Popul. Biol.*, **3**, 135–152.

[41] Henry, L. and Gutierrez, H. (1977). Qualité des prévisions démographiques à court terme. Étude de l'extrapolation de la population totale des départements et villes de France, 1821–1975 (Quality of short-term demographic forecasts. Study of the extrapolation of the total population of the departments and cities in France, 1821–1975). *Popul. (Paris)*, **32**, 625–647.

[42] Hoem, J. M. (1972). Inhomogeneous semi-Markov processes, select actuarial tables, and duration-dependence in demography. In *Population Dynamics*, T. N. E. Greville, ed. Academic, New York, pp. 251–296.

[43] Hoem, J. M. (1973). Levels of Error in Population Forecasts. Article 61, Statistisk Sentralbyrå, Oslo, Norway.

[44] Hoem, J. M. (1976). The statistical theory of demographic rates: A review of current developments. *Scand. J. Statist.*, **3**, 169–185.

[45] Howell, N. (1979). *Demography of the Dobe: Kung*. Academic, New York.

[46] Karlin, S. and Taylor, H. M. (1975). *A First Course in Stochastic Processes*, 2nd ed. Academic, New York.

[47] Karlin, S. and Taylor, H. M. (1981). *A Second Course in Stochastic Processes*. Academic, New York.

[48] Keiding, N. (1975). Extinction and exponential growth in random environments. *Theor. Popul. Biol.*, **8**, 49–63.

[49] Keiding, N. and Hoem, J. M. (1976). Stochastic stable population theory with continuous time. I. *Scand. Actuarial J.*, **59**, 150–175.

[50] Kendall, D. G. (1949). Stochastic processes and population growth. *J. R. Statist. Soc. Ser. B*, **11**, 230–264.

[51] Kendall, D. G. (1950). Random fluctuations in the age distribution of a population whose development is controlled by the simple "birth-and-death" process. *J. R. Statist. Soc. Ser. B*, **12**, 278–285.

[52] Kendall, D. G. (1952). Les processus stochastiques de croissance en biologie (Stochastic growth processes in biology). *Ann. Inst. Poincaré*, **13**, 43–108.

[53] Keyfitz, N. (1968). *Introduction to the Mathematics of Population*. Addison-Wesley, Reading, MA.

[54] Keyfitz, N. and Littman, G. (1979). Mortality in a heterogeneous population. *Popul. Stud.*, **33**, 333–342.

[55] Kurtz, T. G. (1981). *Approximation of Population Processes*. Society for Industrial and Applied Mathematics, Philadelphia, PA.

[56] Land, K. C. (1980). Modeling macro social change. In: *Sociological Methodology*, Karl F. Schuessler, ed. Jossey-Bass, San Francisco, CA, pp. 219–278.

[57] Land, K. C. and McMillen, M. M. (1980). A macrodynamic analysis of changes in mortality indexes in the United States, 1956–1975: Some preliminary results. *Social Indicators Res.*, **7**, 1–46.

[58] Lange, K. (1979). On Cohen's stochastic generalization of the strong ergodic theorem of demography, *J. Appl. Prob.*, **16**, 496–504.

[59] Lange, K. and Johnson, N. J. (1978). Renewal processes with random numbers of delays: Application to a conception and birth model. *J. Appl. Prob.*, **15**, 209–224.

[60] LeBras, H. (1971). Éléments pour une théorie des populations instables (Elements of a theory of unstable populations). *Popul. (Paris)*, **26**, 525–572.

[61] LeBras, H. (1974). Populations stables aléatoires (Stochastic stable populations). *Popul. (Paris)*, **29**, 435–464.

[62] Lee, R. D. (1970). Econometric Studies of Topics in Demographic History. Ph.D. dissertation. Harvard University, Cambridge, MA.

[63] Lee, R. D. (1974). Forecasting births in post-transition populations: Stochastic renewal with serially correlated fertility. *J. Amer. Statist. Ass.*, **69**, 607–617.

[64] Lee, R. D. (1978). Models of preindustrial population dynamics with application to England. *In Historical Studies of Changing Fertility*, Charles Tilly, ed. Princeton Univ. Press, Princeton, NJ, pp. 155–207.

[65] Lee, R. D. (1978). New methods for forecasting fertility: An overview. *Popul. Bull. U.N.*, **11**, 6–11.

[66] Ludwig, D. (1978). *Stochastic Population Theories*. Springer, New York.

[67] Manton, K. G. and Stallard, E. (1980). A stochastic compartment model representation of chronic disease dependence: Techniques for evaluating parameters of partially unobserved age inhomogeneous stochastic processes. *Theor. Popul. Biol.*, **18**, 57–75.

[68] McClean, S. I. (1978). Continuous-time stochastic models of a multigrade population. *J. Appl. Prob.*, **15**, 26–37.

[69] McDonald, J. (1979). A time series approach to forecasting Australian total live-births. *Demography*, **16**, 575–601.

[70] McDonald, J. (1980). Births time series models and structural interpretations. *J. Amer. Statist. Ass.*, **75**, 39–41.

[71] McDonald, J. (1981). Putting Fertility Explanations in Perspective. Project Forecast: Dynamic Modelling. *Research Report No.* 22, Flinders University, S. Australia.

[72] McFarland, D. D. (1970). Effects of group size on the availability of marriage partners. *Demography*, **7**, 411–415.

[73] McFarland, D. D. (1975). Models of marriage formation and fertility. *Social Forces*, **54**, 66–83.

[74] McNeil, D. R. (1974). Pearl–Reed type stochastic models for population growth. *Theor. Popul. Biol.*, **5**, 358–365.

[75] Menken, J. A. (1977). Current status of demographic models. *Popul. Bull. U.N.*, **9**, 22–34.

[76] Menken, J. A., Trussell, J., Stempel, D., and Babakol, O. (1981). Proportional hazards life table models: An illustrative analysis of socio-demographic influences on marriage dissolution in the United States. *Demography*, **18**, 181–200.

[77] Miller, D. R. (1970). Theoretical survival curves for radiation damage in bacteria. *J. Theor. Biol.*, **26**, 383–398.

[78] M'Kendrick, A. G. (1925). Applications of mathematics to medical problems. *Proc. Edinburgh Math. Soc.*, **44**, 98–130.

[79] Mode, C. J. (1975). Perspectives in stochastic models of human reproduction: A review and analysis. *Theor. Popul. Biol.*, **8**, 247–291.

[80] Mode, C. J. (1976). Age-dependent branching processes and sampling frameworks for mortality and marital variables in nonstable populations. *Math. Biosci.*, **30**, 47–67.

[81] Namkoong, G. (1972). Persistence of variances for stochastic, discrete-time population growth models. *Theor. Popul. Biol.*, **3**, 507–518.

[82] Orcutt, G. H., Caldwell, S., and Wertheimer, R. (1976). *Policy Exploration Through Microanalytic Simulation*. Urban Institute, Washington, DC.

[83] Peterson, A. V. (1976). Bounds for a joint distribution function with fixed sub-distribution functions: Application to competing risks. *Proc. Nat. Acad. Sci. USA*, **73**, 11–13.

[84] Pollard, J. H. (1966). On the use of the direct matrix product in analyzing certain stochastic population models. *Biometrika*, **53**, 397–415.

[85] Pollard, J. H. (1968). A note on multi-type Galton–Watson processes with random branching probabilities. *Biometrika*, **55**, 589–590.

[86] Pollard, J. H. (1970). On simple approximate calculations appropriate to populations with random growth rates. *Theor. Popul. Biol.*, **1**, 208–218.

[87] Pollard, J. H. (1973). *Mathematical Models for the Growth of Human Populations*. Cambridge University Press, London, England.

[88] Potter, R. G. (1977). Use of family-building models to assess the impact of population policies upon fertility in the developing countries. *Popul. Bull. U.N.*, **9**, 47–51.

[89] Saboia, J. (1974). Modeling and forecasting populations by time series: The Swedish case. *Demography*, **11**, 483–492.

[90] Saboia, J. (1977). Autoregressive integrated moving average (ARIMA) models for birth forecasting. *J. Amer. Statist. Ass.*, **72**, 264–270.

[91] Schweder, T. (1971). The precision of population projections studied by multiple prediction methods. *Demography*, **8**, 441–450.

[92] Seneta, E. (1972). Population projection variances and path analysis. *J. Amer. Statist. Ass.*, **67**, 617–619.

[93] Sheps, M. C. and Menken, J. A. (1973). *Mathematical Models of Conception and Birth*. University of Chicago Press, Chicago, IL.

[94] Siegel, J. S. (1972). Development and accuracy of projections of population and households in the United States. *Demography*, **9**, 51–68.

[95] Smith, D. and Keyfitz, N. (1977). *Mathematical Demography: Selected Papers*. Springer, New York.

[96] Staroverov, O. V. (1976). Interval'nyi prognoz struktury naseleniia. *Ekon. Mat. Metody*, **12**, 56–71. [English translation: Interval estimates of population structure. *Matekon*, **13**, 42–67 (Winter 1976–77).]

[97] Suchindran, C. M., Clague, A. S., and Ridley, J. C. (1979). Estimation of stochastic variation in vital rates: A simulation approach. *Popul. Stud.*, **33**, 549–564.

[98] Sykes, Z. M. (1969). Some stochastic versions of the matrix model for population dynamics. *J. Amer. Statist. Ass.*, **64**, 111–130.

[99] Tuljapurkar, S. D. and Orzack, S. H. (1980). Population dynamics in variable environments, I. Long-run growth rates and extinction. *Theor. Popul. Biol.*, **18**, 314–342.

[100] Turner, M. E. (1975). Some classes of hit-theory models. *Math. Biosci.*, **23**, 219–235.

[101] Vaupel, J. W., Manton, K. G., and Stallard, E. (1979). The impact of heterogeneity in individual frailty on the dynamics of mortality. *Demography*, **16**, 439–454.

[102] Wachter, K. W. (1980). The sisters' riddle and the importance of variance when guessing demographic rates from kin counts. *Demography*, **17**, 103–114.

[103] Wachter, K. W., Hammel, E. A., and Laslett, P. (1978). *Statistical Studies of Historical Social Structure*. Academic, New York.

[104] Weissner, E. W. (1971). Multitype branching processes in random environments. *J. Appl. Prob.*, **8**, 17–31.

[105] Wilson, E. B. (1938). The standard deviation of sampling for life expectancy. *J. Amer. Statist. Ass.*, **33**, 705–708.

[106] Wu, L. S.-Y. and Botkin, D. B. (1980). Of elephants and men: A discrete, stochastic model for long-lived species with complex life histories. *Amer. Nat.*, **116**, 831–849.

(ACTUARIAL STATISTICS—LIFE
BIRTH AND DEATH PROCESSES
BRANCHING PROCESSES
DEMOGRAPHY
ECONOMETRICS
EPIDEMIOLOGICAL STATISTICS
FERTILITY
GALTON–WATSON PROCESS
GENETICS, STATISTICS IN
HUMAN GENETICS, STATISTICS IN
LIFE TABLES
MANPOWER PLANNING
MARKOV PROCESSES
MARRIAGE
MATHEMATICAL THEORY OF POPULATION
MIGRATION
POPULATION GROWTH MODELS
POPULATION PROJECTION
RANDOM MATRICES
RENEWAL THEORY
SIMULATION
TIME SERIES)

JOEL E. COHEN

STOCHASTIC DIFFERENTIAL EQUATIONS

A stochastic differential equation (SDE) is a differential equation of the form

$$dx/dt = f(x, t, \omega), \tag{1}$$

where ω varies in a probability space. A solution of (1) is therefore a stochastic process*.

A SDE was introduced first by Langevin* [24] to describe the motion of small grains or particles of colloidal size immersed in a fluid. The chaotic perpetual motion of such a particle was described mathematically by Einstein in his famous papers on Brownian motion* [5, 6]. This motion is the result of the random collisions of the particle (which is referred to as a *Brownian particle*) with the molecules of the surrounding medium. The frequency of such collisions is so high that the small changes in the particle's path are too fine to be discerned by the observer. Thus, the exact path of the particle cannot be followed in any detail and has to be described statistically. Since the particle under consideration is much larger than the particles of the surrounding fluid, the collective effect of the interaction between the Brownian particle and the fluid may be considered to be of hydrodynamical character.

Langevin decomposed the force acting on the Brownian particle into two parts: the first one is the hydrodynamical drag force, which is proportional to the velocity v of the particle, and the second is a force $f(t)$ due to the individual collisions with the particles of the surrounding fluid. The hydrodynamical drag force is given by $-\beta v$, where β is the friction coefficient, which, following Einstein, was taken to be given by Stokes' law $\beta = 6\pi a \eta / m$ (a is the radius of the Brownian particle, m is its mass, and η is the coefficient of hydrodynamical viscosity of the surrounding fluid). The fluctuating force $f(t)$ is described mathematically as "white noise," i.e., $\int_0^t f(s)\,ds = \sqrt{2q}\,w(t)$, where $w(t)$ is the standard three-dimensional Brownian motion and $2q$ is its "intensity." Thus, Newton's second law of motion was written in Langevin's form as

$$dv/dt = -\beta v + f(t). \tag{2}$$

The solution of (2) is given by

$$v(t) = v_0 \exp(-\beta t) + \int_0^t \exp[-\beta(t-s)]\, f(s)\,ds \tag{3}$$

or

$$v(t) = v_0 \exp(-\beta t) + \sqrt{2q} \times \int_0^t \exp[-\beta(t-s)]\,dw(s), \tag{4}$$

where the integral in (4) is a stochastic integral* in the sense of Itô or Stratonovich [15, 35]. The process $v(t)$ is called the Ornstein–Uhlenbeck process* [36] (it was first introduced by Lord Rayleigh [30]). We have

$\operatorname{var} v(t) = q[1 - \exp(-\beta t)]/\beta$. The value of the constant q is found by matching the transition probability density $p(v, t, v_0)$ with the Maxwell–Boltzmann law of statistical mechanics. This states that for a fluid in thermodynamical equilibrium the probability density of velocities is zero mean Gaussian, with variance kT/m, where T is absolute temperature, k is Boltzmann's constant, and m is the mass of the fluid particle. It follows that

$$q = \beta kT/m. \tag{5}$$

Relation (5) is called the *fluctuation–dissipation principle* [5, 6]. Thus,

$$p(v, t, v_0) = \left\{\frac{m}{2kT(1 - \exp(-2\beta t))}\right\}^{3/2} \times \exp\left[\frac{-m|v - v_0|^2}{2kT(1 - \exp(-2\beta t))}\right]. \tag{6}$$

The displacement process of the free Brownian particle is given by

$$\begin{aligned} x(t) &= x_0 + \int_0^t v(s)\,ds \\ &= x_0 + v_0\frac{[1 - \exp(-\beta t)]}{\beta} \\ &\quad + \frac{2q}{\beta}\int_0^t [1 - \exp(-\beta t)]\,dw(s). \end{aligned} \tag{7}$$

It follows that the transition probability density of $x(t)$ is given by

$$\begin{aligned} p(x, t, x_0, v_0) &= \left\{\frac{m\beta^2}{2\pi kT[2\beta t - 3 + 4\exp(-\beta t) - \exp(-2\beta t)]}\right\}^{3/2} \\ &\quad \times \exp\left\{\frac{-m\beta^2|x - x_0 - v_0 \times [1 - \exp(-\beta t)]/(2\beta)|^2}{2kT[2\beta t - 3 + 4\exp(-\beta t) - \exp(-2\beta t)]}\right\}. \end{aligned} \tag{8}$$

Averaging with respect to the stationary density of v_0 we obtain for large t

$$p(x, t, x_0) = \frac{1}{(4\pi Dt)^{3/2}}\exp\left[\frac{-|x - x_0|^2}{4Dt}\right], \tag{9}$$

where

$$D = kT/(m\beta) = kT/(6\pi a\eta). \tag{10}$$

Formula (10) is the celebrated Einstein relation [5, 6] that relates the diffusion coefficient to the physical properties of the Brownian particle and of the surrounding fluid. It is easy to see that $p(v, t, v_0)$ satisfies the Fokker–Planck equation*

$$\begin{aligned} \frac{\partial p}{\partial t} &= -\operatorname{div}(vp) + \frac{\beta kT}{m}\operatorname{div}\operatorname{grad} p, \\ p(v, t, v_0) &\to \delta(v - v_0) \quad \text{as } t \to 0. \end{aligned} \tag{11}$$

Similarly, the joint density $p(x, v, t, x_0, v_0)$ satisfies the Fokker–Planck equation

$$\begin{aligned} \partial p/\partial t &= -v\operatorname{grad}_x p \\ &\quad + \beta\operatorname{div}_v[vp + (kT/m)\operatorname{grad} p], \\ p(x, v, t, x_0, v_0) &\to \delta(x - x_0)\delta(v - v_0) \\ &\qquad \text{as } t \to 0. \end{aligned} \tag{12}$$

For large t

$$p(x, v, t, x_0, v_0) \sim p_1(v, t, v_0)p_2(x, t, x_0), \tag{13}$$

where p_1 satisfies (12) and p_2, given by (3), satisfies the diffusion equation

$$\begin{aligned} \partial p_2/\partial t &= D\operatorname{div}\operatorname{grad} p_2, \\ p_2(x, t, x_0) &\to \delta(x - x_0) \quad \text{as } t \to 0. \end{aligned} \tag{14}$$

For large t

$$E|x(t) - x_0|^2 = kTt/(3\pi a\eta) \quad \text{(Einstein)}, \tag{15}$$

while for small t, (8) shows that

$$E|x(t) - x_0|^2 = E|v_0|^2t^2 = 3KTt^2/m \quad \text{(Smoluchowski)}. \tag{16}$$

Equation (11) was first written by Lord Rayleigh [30] in the context of the random flights problem. This equation, together with (12) and (14), were derived again by

Einstein [5, 6], Smoluchowski [33], and Fokker [9] for special cases of molecular motion. The general nonlinear case is discussed after equation (28).

The theory of linear SDEs was extended by Smoluchowski [33], Furth [11], Wang and Uhlenbeck [39], and Uhlenbeck and Ornstein [36].

A nonlinear Langevin equation arises in case of Brownian motion (BM) in a field of forces. If the potential of the forces is $U(x)$ the Langevin equation takes the form

$$x''(t) + \beta x'(t) + U'(x(t)) = \sqrt{2\beta kT}\, w'(t) \tag{17}$$

or

$$\begin{aligned} x'(t) &= y(t) \\ y'(t) &= -\beta y - U'(x(t)) \\ &\quad + \sqrt{2\beta kT}\, w'(t). \end{aligned} \tag{18}$$

The general form of a SDE, driven by white noise, is the Itô [15] or Stratonovich [35] system (*see* DIFFUSION PROCESSES; STOCHASTIC INTEGRALS, and ref. 32),

$$dx = a(x,t)\,dt + b(x,t)\,dw(t), \quad x(t_0) = x_0, \tag{19}$$

where x_0 is a random variable, $a(x,t)$ is a vector, $b(x,t)$ is a matrix, and $w(t)$ is a vector of independent Brownian motions. If (19) is given the Stratonovich sense, the equivalent Itô form of (19) is

$$\begin{aligned} dx_i = \Bigg[& a_i(x,t) \\ & + \tfrac{1}{2}\sum_{j,k} b_{kj}(x,t)\,\partial b_{ij}(x,t)/\partial x_k \Bigg] dt \\ & + \sum_j b_{ij}(x,t)\,dw_j, \end{aligned} \qquad (i = 1,\ldots,n). \tag{20}$$

The second term in brackets is the Wong–Zakai [40] correction. To clarify the difference between the Itô and Stratonovich equations consider an example where "colored noise" (Ornstein–Uhlenbeck* process) $v_k(t)$ is used instead of white noise in a linear SDE. The process $v_k(t)$ is governed by the SDE

$$dv_k = -kv_k\,dt + k\,dw.$$

The autocorrelation function of $v_k(t)$ is given by

$$r(t) = k^2\exp(-k^2|t|) \to \delta(t) \quad \text{as } k \to \infty;$$

thus, the correlation time of $v_k(t)$ is finite, though very short, if k is large. Since the collision process is finitely correlated (the motion is perfectly correlated between collisions), $v_k(t)$ is a more physically acceptable model of the collision process than its white noise limit $w'(t)$. Note that $v_k(t)$ has continuous paths and finite power output, while $w'(t)$ is not even measurable with respect to t and has infinite power output. Consequently, the equation

$$x'(t) = a(x) + b(x)v_k(t) \tag{21}$$

is an ordinary differential equation, whose solution is differentiable. Letting $k \to \infty$, we obtain the limiting equation

$$dx = a(x)\,dt + b(x)\,dw \tag{22}$$

in the sense of Stratonovich, not Itô. Thus, the Stratonovich equation should be used as an idealized model for physical phenomena in the limit of vanishing correlation time of the fluctuations. Note, that, if $b(x,t)$ is independent of x, the Wong–Zakai correction vanishes, so that (19) is the same in either case.

EXISTENCE, UNIQUENESS, AND LOCALIZATION THEOREMS

If $a(x,t)$ and $b(x,t)$ are Lipschitz continuous with respect to x in each sphere $|x| < N$, uniformly in $t > 0$, and $|a| + |b| < K(1 + |x|)$ for all x and t, x_0 is independent of $w(t)$, and $Ex_0^2 < \infty$, then there exists a unique solution to (19) in the sense that, if x_1 and x_2 are any two solutions of (19), then

$$P\left[\sup_{0<t<T} |x_1(t) - x_2(t)| = 0\right] = 1 \quad \text{for all } T > 0. \tag{23}$$

The following localization principle holds. Let

$$a_1(x,t) = a_2(x,t)$$

and

$$b_1(x,t) = b_2(x,t)$$

for $|x| < N$, $t > 0$, and let $\tau_i = \inf[t \| x_i(t)| > N]$, $i = 1, 2$. Then

$$P(\tau_1 = \tau_2) = 1$$

and

$$P\Big(\sup_{0<t<\tau_1} |x_1(t) - x_2(t)| = 0\Big) = 1$$

(see ref. 12).

SDE AND DIFFUSION PROCESSES: THE TRANSITION PROBABILITY DENSITY

The solution of the Itô equation (19) is a diffusion process* with drift vector $a(x,t)$ and diffusion matrix $\sigma = bb*(x,t)$. A partial converse is also true: If the diffusion matrix σ is strictly and uniformly positive definite, the drift a and the diffusion matrix σ are sufficiently smooth, then $x(t)$ is a solution of (19) with b such that $\sigma = bb*$.

If (19) is given the Itô sense, the transition probability density

$$p(x,t,x_0,s)\,dx = P[x(t) \in x + dx \,|x(s) = x_0] \quad (24)$$

exists under certain smoothness assumptions on the coefficients, and satisfies with respect to the "forward" variables (x, t) the Fokker–Planck or the forward Kolmogorov equation [20, 29]

$$\partial p/\partial t = -\sum_i \partial[a_i(x,t)p]/\partial x_i + \tfrac{1}{2}\sum_{i,j} \partial^2[\sigma_{ij}(x,t)p]/\partial x_i\,\partial x_j \quad (25)$$

for $t > s$ and

$$p(x,t,x_0,s) \to \delta(x - x_0) \quad \text{as } t \to s.$$

The density $p(x, t, x_0, s)$ satisfies with respect to the "backward" variables (x_0, s) the backward Kolmogorov equation

$$\partial p/\partial s = -\sum_i a_i(x_0,s)\,\partial p/\partial x_{0i} - \tfrac{1}{2}\sum_{i,j}\sigma_{ij}(x_0,s)\,\partial^2 p/\partial x_{0i}\,\partial x_{0j} \equiv -Lp \quad (26)$$

for $s < t$, and

$$p(x,t,x_0,s) \to \delta(x - x_0) \quad \text{as } s \uparrow t.$$

The Fokker–Planck equation for the Stratonovich equation (1) is given by

$$\partial p/\partial t = -\sum_i \partial[a_i(x,t)p]/\partial x_i + \tfrac{1}{2}\sum_{i,j,k} \partial\big[b_{ik}(x,t) \times \partial(b_{jk}(x,t)p)/\partial x_i\big]/\partial x_j \quad (27)$$

and the initial conditions are the same as in (25). If a and b are independent of t, the density p depends on $t - s$, so that (26) can be written in the form

$$\partial p/\partial t = L^* p, \quad (28)$$

where L^* is the formal adjoint of L [cf. (26)] with the initial conditions of (25). In this case the conditional expectation of any function $f(x(t))$,

$$F(x,t) = E[f(x(t))|x(0) = x], \quad (29)$$

satisfies (28) with the initial condition $F(x, 0) = f(x)$.

Equation (25) was derived by Planck [29] from an arbitrary master equation assuming only that the jumps are small. For the case of the nonlinear equation (17) Kramers has derived the forward equation [21]. A rigorous mathematical derivation of the forward and backward equations (25) and (26) was given by Kolmogorov in [20].

BOUNDARY BEHAVIOR

Let D be a domain in R with sufficiently smooth boundary S. We say that S is an *absorbing boundary* for (19) if $P(x(t) \in S) = 1$ for all $t > \tau$, where τ is the first time that $x(t) \in S$. If S is an absorbing boundary, then

$$p(x, t, x_0, s) = \delta(x - x_0) \quad \text{for } x_0 \text{ in } S \text{ and } t > s. \tag{30}$$

Thus, (25) and (27) are an initial-boundary value problem of parabolic type.

We say that S is a *reflecting boundary* if $x(t)$ is instantly reflected at S, i.e., if a and b are extended across S by reflection, and the solution $x(t)$ of the extended system defines $x(t)$ in D by reflection in S. The reflection process is defined near S only, since the evolution of $x(t)$ in D is governed by (19). If S is a reflecting boundary then

$$\partial p(x, t, x_0, s)/\partial \nu = 0 \quad \text{for } x \text{ in } S \text{ or } x_0 \text{ in } S, \tag{31}$$

where ν is the normal to S. For other boundary behavior see refs. 8, 10, and 27.

The solution of a SDE can be used to represent the solutions of initial and boundary value problems for partial differential equations. Thus, the transition probability density is a representation of the solution of (25) with the specified initial and boundary conditions. Other representations for solutions of partial differential equations as functionals of $x(t)$ are given by Kolmogorov's representation formulae, the Feynman–Kac representation formula (*see* DIFFUSION PROCESSES), and formulae in the next section. They are the basis for a Monte Carlo* procedure for solution of partial differential equations.

FIRST PASSAGE TIMES

Let $\tau(x, s)$ be the first time the solution $x(t)$ of (19) ($x(s) = x$ in D) hits S, i.e.,

$$\tau(x, s) = \inf[t > s | x(t) \in S, \quad x(s) = x \text{ in } D]. \tag{32}$$

The random time $\tau(x, s)$ plays an important role in the theory and applications of SDE. The distribution and moments of $\tau(x, s)$ can be determined as solutions of partial differential equations [4, 17, 32]:

(i) Let $q(x, s, t)$ be the solution of the backward terminal value problem

$$\begin{aligned} \partial q/\partial t + Lq &= 0 \quad \text{for } x \text{ in } D,\ T > t > s, \\ q &= 1 \quad \text{for } x \text{ in } S,\ T > t > s, \\ q(x, s, T) &= 0 \quad \text{for } x \text{ in } D. \end{aligned} \tag{33}$$

with L defined in (26).
Then

$$p[\tau(x, s) < T] = q(x, s, s).$$

(ii) If $u(x, s, t)$ is the bounded solution of the backward boundary value problem

$$\begin{aligned} \partial u/\partial t + Lu &= -1 \quad \text{for } x \text{ in } D,\ t > s, \\ u(x, s, t) &= 0 \quad \text{for } x \text{ in } S, \end{aligned} \tag{34}$$

then

$$E\tau(x, s) = s + u(x, s, s) \quad (\text{see [29a]}).$$

(iii) Green's function $G(x, s, y, t)$ for the backward problem

$$\begin{aligned} \partial u/\partial t + Lu &= 0 \quad \text{for } x \text{ in } D,\ t > s, \\ u(x, t) &= f(x) \quad \text{for } x \text{ in } S,\ t > s, \end{aligned} \tag{35}$$

is the conditional density of points y where the solution $x(t)$ of (19) hits S for the first time, given that $x(s) = x$ in D, i.e.,

$$P[x(\tau(x, s)) \in A] = \int_A G(x, s, y, s)\, dS_y,$$

where $A \subset S$. If a and b in (19) are independent of t, (35) takes the form $Lu = 0$ and $G = G(x, y)$. Thus, the expected exit time $\bar{\tau}(x)$ of the BM $w(t)$ from an interval (a, b), given $w(0) = x$ in (a, b), is the solution of $\frac{1}{2}\bar{\tau}''(x) = -1$ for $a < x < b$, $\bar{\tau}(a) = \bar{\tau}(b) = 0$; hence $\bar{\tau}(x) = (x - a)(b - x)$. Note that if $a = -\infty$, then $\bar{\tau}(x) = \infty$, i.e., the mean first passage time of a BM through a point is infinite, although $P[\tau(x) < \infty] = 1$. Indeed, the probability $P[\tau(x) < t] = q(x, t)$, in case $a = -\infty$, is the solution of

$$\begin{aligned} &\partial q/\partial t = \tfrac{1}{2}\partial^2 q/\partial x^2, \qquad x < b,\ t > 0, \\ &q(b, t) = 1, \qquad q(x, 0) = 0. \end{aligned} \tag{36}$$

It is given by $q(x, t) = \operatorname{erf}[(b - x)/\sqrt{2t}]$;

hence

$$P[\tau(x) < \infty] = \lim_{t \to \infty} q(x, t) = 1.$$

The probability

$$u(x) = P[w(\tau) = a | w(0) = x]$$

(of hitting a before b, starting at x) is the solution of $u''(x) = 0$ for $a < x < b$, and $u(a) = 1$ and $u(b) = 0$; hence, $u(x) = (b - x)/(b - a)$ [7].

The ruin problem is a first passage problem*. Thus, if $x(t)$ in the one-dimensional equation (19) represents the assets process of an insurer, then ruin occurs when $x(t) = 0$ for the first time. The function q, defined by (33) is then the probability of ruin by time t. In the simplest model, let there be x claims on the average per unit time and let the average claim be of size m, with second moment α. Assume the insurer collects a surplus kmx (k is the loading constant) on the average above the expected amount of claims mx per unit time. The variance of the claims amount is $x\alpha$ per unit time. Thus, $a(x) = kmx$ and $b(x) = \alpha\sqrt{x}$ in (19). For $k > 0$, the probability of ruin is given by $\lim_{t \to \infty} q(x, t) = \exp[-kmx/\alpha]$ (see ref. 31).

STOCHASTIC STABILITY

The concept of stability of SDE is more complicated than that of deterministic differential equations. Thus, if $\mathbf{x} = \mathbf{0}$ is a globally asymptotically stable point of the system

$$\mathbf{x}' = \mathbf{a}(\mathbf{x}) \tag{37}$$

$[\mathbf{a}(\mathbf{0}) = \mathbf{0}]$, i.e., every solution $\mathbf{x}(t)$ of (37) satisfies $\mathbf{x}(t) \to \mathbf{0}$ as $t \to \infty$, even the smallest stochastic perturbation of (37),

$$d\mathbf{x}(t, \epsilon) = \mathbf{a}(\mathbf{x}(t, \epsilon))\, dt + \sqrt{2\epsilon}\, d\mathbf{w}(t) \tag{38}$$

($\epsilon \ll 1$), will cause the solution $\mathbf{x}(t, \epsilon)$ to leave every compact domain in finite time. This is the consequence of (34), which has a bounded solution for any $\epsilon > 0$.

There are several definitions of stochastic stability [13]. We say the origin is *RMS stable* for (19) if $a(0, t) = 0$, $b(0, t) = 0$, and for every $\epsilon > 0$ and t there exists $\delta > 0$ and $t_1 > t$ such that

$$E\left[|\mathbf{x}(t)|^2 | \mathbf{x}(t_0) = \mathbf{x}_0\right] < \epsilon \tag{39}$$

for all $|\mathbf{x}_0| < \delta$ and $t > t_1$. It is *asymptotically RMS stable* if there exists $\delta > 0$ such that

$$\lim_{t \to \infty} E\left[|\mathbf{x}(t)|^2 | \mathbf{x}(t_0) = \mathbf{x}_0\right] = 0 \tag{40}$$

for all $|\mathbf{x}_0| < \delta$. It is *globally asymptotically RMS stable* if (40) holds for all x_0. The origin is stochastically stable for (19) if for every $\epsilon_1 > 0$ and $\epsilon_2 > 0$ there exists t_1 such that

$$P\left[\sup_{t > t_1} |\mathbf{x}(t)| < \epsilon_1 \middle| |\mathbf{x}(t_0)| < \delta\right] > 1 - \epsilon_2 \tag{41}$$

(*weak stability*) or

$$\lim_{|\mathbf{x}| \to 0} P\left[\sup_{t > t_1} |\mathbf{x}(t)| > \epsilon \middle| \mathbf{x}(t_0) = \mathbf{x}\right] = 0 \tag{42}$$

(*strong stability*). It is *asymptotically stochastically stable* if

$$\lim_{|\mathbf{x}| \to 0} P\left[\lim_{t \to \infty} |\mathbf{x}(t)| = 0 \middle| \mathbf{x}(t_0) = \mathbf{x}\right] = 1 \tag{43}$$

and *globally asymptotically stochastically stable* if

$$P\left[\lim_{t \to \infty} \mathbf{x}(t) = \mathbf{0}\right] = 1 \tag{44}$$

for all solutions of (19).

If, for example, (19) is autonomous (i.e., $\mathbf{a}$ and $\mathbf{b}$ are independent of t), $\mathbf{a}(\mathbf{0}) = \mathbf{0}$, and $\mathbf{b}(\mathbf{0}) = \mathbf{0}$, then a sufficient condition for stochastic stability is the existence of a *Lyapunov function* at $\mathbf{x} = \mathbf{0}$ for (19): $V(\mathbf{x})$ is a Lyapunov function for (19) at $\mathbf{x} = \mathbf{0}$ if (i) $V(\mathbf{x})$ is defined and continuous in a neighborhood N of the origin, (ii) $V(\mathbf{0}) = 0$ and $V(\mathbf{x}) > 0$ for x in $N - \{\mathbf{0}\}$, and (iii) $LV < 0$ in N, where L is defined in (26). If, in addition, $V(\mathbf{x}) \to \infty$ as $|\mathbf{x}| \to \infty$, the asymptotic stability is global. Thus, $LV <$

$-kV$ for some $k > 0$ implies global asymptotic stability.

Consider, for example, an Itô system of homogeneous linear equations

$$d\mathbf{x} = \mathbf{A}(t)\mathbf{x}\,dt + \sum_{i=1}^{n} \mathbf{B}_i(t)\mathbf{x}\,d\mathbf{w}_i(t), \quad (45)$$

where $\mathbf{w}_i(t)$ are independent BMs. Taking expectations in (45), we obtain for $\mathbf{y}(t) = E[\mathbf{x}(t)|\mathbf{x}]$ the linear equation

$$\mathbf{y}'(t) = \mathbf{A}(t)\mathbf{y}(t);$$

hence, $\mathbf{y}(t) = \mathbf{0}$ must be a stable solution if $\mathbf{x}(t) = \mathbf{0}$ is to be stable for (45). Also, the covariance matrix of $\mathbf{x}(t)$ should not increase as $t \to \infty$ if the origin is to be stable in the sense of (43) or (44). This happens if the origin is stable for the system

$$\mathbf{Z}'(t) = \mathbf{A}(t)\mathbf{Z}(t) + \mathbf{Z}(t)\mathbf{A}^T(t) + \sum_{i=1}^{n} \mathbf{B}_i(t)\mathbf{Z}(t)\mathbf{B}_i^T(t), \quad (46)$$

where

$$\{\mathbf{Z}(t)\}_{ij} = E\left[x_i(t)x_j(t)|\mathbf{x}\right].$$

If $\mathbf{A}$ and $\mathbf{B}$ are independent of time, then there is stability in (46) if the system of algebraic equations

$$\mathbf{AZ} + \sum_i \mathbf{B}_i\mathbf{Z}\mathbf{B}_i^T = -\mathbf{C}$$

has a symmetric positive definite solution $\mathbf{Z}$ for some symmetric positive definite matrix $\mathbf{C}$.

A stochastic perturbation of a deterministically unstable system $x_i' = b_i x_i$ $(i = 1, 2)$ with $b_1 > 0 > b_2$ becomes stable if stochastically perturbed, as follows:

$$dx_i = b_i x_i\,dt + \sigma \sum_{j=1}^{2} x_j\,dw_j, \qquad i = 1, 2$$

(in the Stratonovich sense), with

$$1 + 2b_1/\sigma^2 < k\left[1 - I_1(k/2)/I_0(k/2)\right],$$

where $k = (b_1 - b_2)/2$ and I_0 and I_1 are Bessel functions* of a pure imaginary argument [13]. This fact is one of the reasons for introducing jitter into dynamical systems (e.g., stabilizing a broomstick balanced on its tip by juggling it at high frequency).

DIFFUSION APPROXIMATION TO MARKOV CHAINS

The construction of the BM as a limit of a rescaled random walk* can be generalized to a class of Markov chains (*see* MARKOV PROCESSES). Assume that for each N a Markov chain $X[N, t(n, n)]$ is given on the real line. More specifically, let $x(N, 1)$, $x(N, 2), \ldots$ be the possible states of the Nth Markov chain, and

$$P\{X[N, t(N, n)] = x(N, j)\,|\, X[N, t(N, n-1)] = x(N, i)\} = P(i, j, N, n),$$

where $t(N, 1) < t(N, 2) < \cdots$. Assume that

$$\max_n Dt(N, n) \equiv \max_n \left[t(N, n+1) - t(N, n)\right] \to 0 \quad \text{as } N \to \infty.$$

We define a piecewise constant interpolation of $X[N, t(N, n)]$ by setting

$$x(N, t) = X\left[N, t(N, n)\right] \quad \text{for } t(N, n) < t < t(N, n+1).$$

Assume that

$$E\left[x(N, t + Dt) - x|x(N, t) = x\right]/(Dt) \to a(x, t)$$

as $N \to \infty$, $Dt = Dt(N) \to 0$,

$$\operatorname{var}\left[x(N, t + Dt) - x|x(n, t) = x\right]/(Dt) \to b(x, t),$$

$$E\left[x(N, t + Dt) - x|x(n, t) = x\right]^{2+\delta}/(Dt) \to 0,$$

and that convergence is sufficiently rapid. Furthermore, assume that $a(x, t)$ and $\sqrt{b(x, t)}$ are bounded and continuous functions in $R \times [0, T]$. If the SDE

$$dx(t) = a(x(t), t)\,dt + \sqrt{b(x(t), t)}\,dw(t)$$

has a unique solution, then $x(N, t) \to x(t)$

as $N \to \infty$ in the weak sense; that is, for each continuous function $f(x)$,

$$E[f(x(N,t))|x(N,0) = x] \to E[f(x(t))|x(t) = x] \quad \text{as } N \to \infty$$

(see ref. 22). Diffusion approximations to Markov chains or Markov processes are useful in mathematical genetics* [3, 7, 14 and 17], in statistical mechanics [32, 37], in filtering theory [16], and so on.

SDES IN MATHEMATICAL PHYSICS

The *Langevin equation* [24] for a Brownian particle in a force field is very often used to model physical phenomena. It is given by

$$\mathbf{x}' = \mathbf{y},$$

$$\mathbf{y}' = -\beta\mathbf{y} - \mathbf{f}(\mathbf{x}) + \sqrt{2\beta kT/m}\,\mathbf{w}'. \quad (47)$$

The vector field $\mathbf{f}(\mathbf{x})$ represents the external forces acting on the particle. These may include electrostatic forces, chemical bonds or lattice bonds in a crystal, gravitational forces, and so on. Note that the fluctuation–dissipation principle (9) is used in (47). In case the force field is conservative it can be derived from a potential $\mathbf{f}(\mathbf{x}) = -\operatorname{grad} U(\mathbf{x})$.

For large β the Smoluchowski–Kramers approximate equation holds [21, 32, 33]:

$$\beta\mathbf{x}' = -\mathbf{f}(\mathbf{x}) + \sqrt{2\beta kT/m}\,\mathbf{w}' \quad (48)$$

[i.e., the inertial term $\mathbf{y}'$ is much smaller than the other terms in (47)]. Thus, for example, the phenomenon of sedimentation (variation of concentration with altitude of particles suspended in the atmosphere) is modeled by (48) with $\mathbf{f}(\mathbf{x}) = (0, 0, c)$, $c = (1 - \rho_0/\rho)g$, where ρ_0 is the density of the atmosphere, ρ is the density of the suspended Brownian particles and g is the gravitational constant. The Fokker–Planck equation in this case takes the form

$$\partial p/\partial t = -\operatorname{div}\mathbf{j}, \quad (49)$$

where $\mathbf{j} = -\mathbf{f}p + D\operatorname{grad} p$ is the diffusion current, and D is given by (16). The boundary condition for (49) is $j = 0$ at the bottom $z = 0$. The steady-state solution of (49) is

$$p = (c/D)\exp(-cz/D).$$

This is the law of *isothermal atmospheres*.

In the other extreme case, if β is very small, the following limits are obtained. If (47) with $T = 0$ has a stable equilibrium at $x = 0$, say, then averaging the Fokker–Planck equation over constant energy contours $[|y|^2/2 + U(\mathbf{x}) = E]$ yields the approximate Stratonovich equation [35, p. 115]

$$dE = -\beta E\,dt + \sqrt{2\beta kTE}\,dw. \quad (50)$$

A nonlinear self-excited oscillator driven by noise can be described by the second-order SDE

$$\mathbf{x}'' + \beta\mathbf{x}'h(\mathbf{x}, \mathbf{x}') + \mathbf{f}(\mathbf{x}) = \sqrt{2\beta kT}\,\mathbf{w}', \quad (51)$$

which is assumed to have a stable limit cycle S for $T = 0$. For small β the method of averaging yields the approximate Itô equation

$$dE = -\beta[b(E)/a'(E) - kT/2]\,dt + [2\beta kTa(E)/a'(E)]^{1/2}\,dw, \quad (52)$$

where

$$a(E) = \oint_E \mathbf{x}'\,d\mathbf{x} \oint_E \frac{d\mathbf{x}}{\mathbf{x}'}$$

and

$$b(E) = \oint_E \mathbf{x}'h(\mathbf{x}, \mathbf{x}')\,d\mathbf{x} \oint_E \frac{d\mathbf{x}}{x'}.$$

A case of particular interest is that of the damped physical pendulum driven by constant torque I, i.e., $f(x) = I - \sin x$ in (47). This is a model for the dynamics of the Josephson junction, charge density waves, ionic conductivity in crystals, etc. [1]. If $0 < \beta < \pi I/4 < \pi/4$, (47) has stable equilibrium states at $x = \arcsin I + 2n\pi$, $y = 0$, and a stable limit line S,

$$y = I/\beta + (\beta/I)\cos(x + \beta^2/I) + O\left[(\beta/I)^3\right].$$

The fluctuations about stable equilibrium states are governed by (50), the stationary density of which is the Boltzmann density. The stationary distribution of fluctuations about S is given by

$$p = \left[1/\sqrt{2\pi kT}\right]\exp\left[-(A - A_0)^2/(2kT)\right], \tag{53}$$

where A at a point (x, y) in phase space is the "mean action"

$$A = A(B) = (2\pi)^{-1}\int_0^{2\pi} y_B\,dx = I/B.$$

Here $y_B = y_B(x)$ is the limit line for (47), which passes through (x, y) obtained by replacing β with B, $A_0 = A(\beta) \cong I/\beta$. Equation (53) is valid in the domain $A > A_0 \cong 4/\pi$. Further limit theorems can be found in refs. 19 and 28.

THE EXIT PROBLEM

In many applications the noise term in (19) is small so that (19) can be written in the general form

$$d\mathbf{x} = \mathbf{a}(\mathbf{x}, t)\,dt + \sqrt{2\epsilon}\,\mathbf{b}(\mathbf{x}, t)\,d\mathbf{w}, \tag{54}$$

where ϵ is a small positive parameter. In particular, if the deterministic system (54) with $\epsilon = 0$ has a stable solution $\mathbf{x}_0(t)$ in a domain D (e.g., $\mathbf{x}_0(t)$ is a stable constant solution corresponding to a stable equilibrium or a stable periodic solution corresponding to a stable limit cycle), then (54) describes diffusion against the flow. In this case (34) and (35) become singular perturbation problems for which the solutions of the reduced problems ($\epsilon = 0$) are $+\infty$ and a constant C_0, respectively, The constant C_0 determines the limit of the probability density

$$p(\mathbf{x}, \mathbf{y}, \epsilon) = P\left[\mathbf{x}(\epsilon, \tau) = \mathbf{y} | \mathbf{x}(\epsilon, 0) = \mathbf{x}\right]$$

of exit points of the solution $\mathbf{x}(\epsilon, t)$ of (54) on the boundary of D as $\epsilon \to 0$. Indeed, by (35), $p(\mathbf{x}, \mathbf{y}, \epsilon)$ is the Green function, so that the solution of (35) [for autonomous (54)] is

$$C_0 = \lim_{\epsilon \to 0} u(\mathbf{x}, \epsilon) = \lim_{\epsilon \to 0} \oint_{\partial D} p(\mathbf{x}, \mathbf{y}, \epsilon) f(\mathbf{y})\,dS_{\mathbf{y}}. \tag{55}$$

In the special case that $\mathbf{b}$ is the identity matrix and $\mathbf{a}(\mathbf{x}) = -\operatorname{grad} U(\mathbf{x})$, the value of C_0 is given by

$$C_0 = \lim_{\epsilon \to 0}\left[\frac{\oint_{\partial D}\exp(-U/\epsilon) f(\mathbf{y})\mathbf{a}\cdot\mathbf{y}\,dS_{\mathbf{y}}}{\oint_{\partial D}\exp(-U/\epsilon)\mathbf{a}\cdot\mathbf{y}\,dS_{\mathbf{y}}}\right], \tag{56}$$

assuming $\mathbf{a}\cdot\boldsymbol{\nu} > 0$ on D ($\boldsymbol{\nu}$ is the outer unit normal). Hence, using the Laplace expansion of the integrals in (56) we obtain

$$p(\mathbf{y}) = \lim_{\epsilon \to 0} p(\mathbf{x}, \mathbf{y}, \epsilon) = \frac{\Sigma_k \Pi_i \omega_{ik} \mathbf{a}\cdot\boldsymbol{\nu}\delta(\mathbf{y} - \mathbf{x}_k)}{\Sigma_k \Pi_i \omega_{ik} \mathbf{a}\cdot\boldsymbol{\nu}(\mathbf{x}_k)}, \tag{57}$$

where ω_{ik} are the frequencies of vibration in the principal stable directions at the lowest saddle points $\mathbf{x}_k$ of the potential $U(\mathbf{x})$ on D. The mean exit time $\bar{\tau}(\epsilon) = E[\bar{\tau}(\epsilon)|\mathbf{x}]$, starting at a point $\mathbf{x}$ in D, is given by

$$\bar{\tau}(\epsilon) = \Omega^{-1}\exp\left(\min_{\partial D} U/\epsilon\right), \tag{58}$$

where the attempt frequency Ω is determined by the frequencies of vibration at the stable equilibrium in D (i.e., at the bottom of the potential well) and by the frequencies ω_{ik}. The quantity $\bar{\tau}(\epsilon)$ is asymptotically equal to the principal eigenvalue of the Fokker–Planck operator if ϵ is small. The distribution of $\tau(\epsilon)$ is asymptotically exponential. For further results see ref. 32.

The quantities $\bar{\tau}(\epsilon)$ and $p(\mathbf{y})$ determine many important physical parameters. Thus, in Kramers' diffusion model of a chemical reaction [21], $U(\mathbf{x})$ represents the potential of the chemical bonds and $\epsilon = \beta kT/m$ as in (47). The chemical reaction rate constant κ is the proportion of particles breaking the chemical bond per unit time; thus $\kappa = 1/\{2\bar{\tau}(\epsilon)\}$. The Arrhenius law can be derived from (47) in the form

$$\kappa = \tfrac{1}{2}\Omega\exp(-E/(kT)), \tag{59}$$

where E is the height of the potential barrier (the so-called activation energy).

Atomic migration in crystals can be described as the random walk of an impurity atom, whose motion consists of jumps between the potential wells of the crystalline lattice. These jumps are due to the random (thermal) vibrations of the crystal. Using (47) with a periodic potential $U(x)$, we obtain $\bar{\tau}$ as the mean time between jumps. The diffusion approximation to such a random walk in one dimension is given by

$$\partial p/\partial t = (\lambda/(2T))\, \partial^2 p/\partial x^2, \quad (60)$$

where λ is the width of the lattice cell. Thus, the diffusion coefficient for atomic migration in crystals is given by

$$D = \lambda\kappa \quad (61)$$

with κ given by (59). In higher dimensions the jumps have different probabilities in different directions. Using the exit distribution $p(y)$, defined in (57), in a diffusion approximation to a multidimensional random walk, a diffusion tensor* is obtained,

$$D(i, j) = \left[\sum_k P(k) z(k, i) z(k, j)\right]\kappa, \quad (62)$$

where $P(k)$ is the exit probability through the kth saddle point and $\mathbf{z}(k)$ is the jump vector.

Ionic conductivity in crystals is due to thermally induced jumps of the ions over the potential barrier in a crystal subjected to a uniform electrostatic field. Using (47) with $U(x) = -Eqx + qK \sin \omega x$, where q is the ionic charge, the definition of conductivity as

$$\sigma = \partial I/\partial E, \quad (63)$$

where I is the current and the the definition of current as

$$I = Cq[1/\bar{\tau}(R) - 1/\bar{\tau}(L)], \quad (64)$$

where C is the concentration of ions, $\bar{\tau}(R)$ and $\bar{\tau}(L)$ are mean jump times to the right and to the left, respectively, one obtains from (58), (63), and (64), for $E\omega K \ll 1$, that

$$\sigma = \{Cq^2K/(2\pi kTm)\}\exp(-2qK/kT). \quad (65)$$

Using (61) the Nernst–Einstein formula is obtained:

$$\sigma = Cq^2D/(kT). \quad (66)$$

In higher dimensions a conductivity tensor is obtained if (62) is used in (66) [32]. For further applications see refs. 1, 3, 28, and 32.

The exit problem was considered by Kramers [21] in the context of chemical reactions. He solved it in a one-dimensional model. A higher-dimensional model was discussed by Landauer and Swanson [25]. The problem in the form (54) was partially solved in a particular case by Ventzel and Freidlin [38], and in the more general case presented here by Ludwig [26] and by Matkowsky and Schuss (see ref. 32).

NONLINEAR FILTERING THEORY

In communications theory a random signal $x(t)$ (e.g., speech) is often modeled as a solution of an SDE. The signal is subjected to a linear or nonlinear transformation (modulation), and then it is transmitted through a communication channel (e.g., radio). The measured signal is usually corrupted by noise, so that the measurement process can be modeled by another SDE,

$$dy = g(x, t)\, dt + \sqrt{N}\, dw, \quad (67)$$

where $g(x, t)$ is the modulated signal, N is the noise level, and w is a BM independent of the noise driving the SDE of $x(t)$. The filtering problem is to estimate the process $x(t)$ given the measurements $[y(s), 0 < s < t]$. The optimal estimator $X(t)$ of $x(t)$, which minimizes the conditional estimation error variance

$$\langle e^2 \rangle = E\left[|X(t) - x(t)|^2 \middle| y(s), 0 < s < t\right], \quad (68)$$

is given by

$$X(t) = E[x(t) | y(s), 0 < s < t]. \quad (69)$$

In general, there is no sequential realization of (69). The conditional probability density $p(x, t) = p[x(t) = x | y(s),\ 0 < s < t]$ satisfies Kushner's equation [23, 32, 34], which in one dimension takes the form

$$dp = Lp\, dt + [g(x, t) - \langle g(x, t)\rangle] \times [dy - \langle g(x, t)\rangle\, dt]\, p/N, \quad (70)$$

where L is the Fokker–Planck operator for

$x(t)$, and $\langle \cdot \rangle$ represents the expectation conditioned on the measurements. In case $x(t)$ and $y(t)$ are governed by linear SDEs, (70) can be solved and (69) can be realized sequentially (*see* KALMAN FILTERING). For example, AM transmission of a linear signal $x(t)$ corresponds to

$$g(x, t) = x \sin \omega t$$

and can be estimated optimally. In contrast, FM transmission corresponds to

$$g(x(t), t) = C \sin\left[\omega t + D\int_0^t x(s)\, ds\right],$$

i.e., the measurements process is nonlinear. Equation (70), which is a nonlinear integro-partial-differential stochastic equation, can be transformed into a linear partial differential stochastic equation. This is achieved by changing the measure in function space defined by $x(t)$ and $y(t)$, to a measure defined by $x(t)$ and a pure noise measurements process by using Girsanov's formula [12] for the Radon–Nikodym derivative* q of the respective measures. It was shown by Zakai [41] that q satisfies the Itô SDE

$$dq = Lq\, dt + (gq/N)\, dy \qquad (71)$$

and $p(x, t) = q(x, t)/\int q(x, t)\, dx$. Equations (70) and (71) were used to obtain approximate solutions to the nonlinear filtering problem [2, 16, 18, 34].

References

[1] Ben-Jacob, E., Bergman, D. J., Matkowsky, B. J., and Schuss, Z. (1982). *Phys. Rev. A*, **26**, 2805–2816.

[2] Bobrovsky, B. Z. and Zakai, M. (1981). *Stochastic Systems, the Mathematics of Filtering and Identification and Applications. NATO-ASI Series*, Hazewinkel and Willems, eds. Reidel, Dordrecht, The Netherlands, p. 573.

[3] Crow, J. F. and Kimura, M. (1970). *An Introduction to Population Genetics*. Harper and Row, New York. (A textbook; contains many applications of SDEs in genetics.)

[4] Dynkin, E. B. (1965). *Markov Processes*, Vols. I and II. Springer, New York. (A classical advanced text.)

[5] Einstein, A. (1905). *Ann. Phys.*, **17**, 549 (in German).

[6] Einstein, A. (1906). *Ann. Phys.*, **19**, 371. (This and ref. 5 are two articles of historical importance in mathematics and physics.)

[7] Feller, W. (1966). *An Introduction to Probability Theory and Its Applications*, Vols. I and II. Wiley, New York. (A classical text.)

[8] Feller, W. (1957). *Ill. J. Math.*, **1**, 459. (A complete discussion of boundary behaviour of solutions of SDEs in one dimension.)

[9] Fokker, A. D. (1913). Thesis, Leyden, The Netherlands. [Also *Ann. Phys. (4)*, **43**, 810 (1914)].

[10] Friedman, A. (1979). *Stochastic Differential Equations*. Academic, New York. (An abstract theoretical treatise.)

[11] Fürth, R. (1917). *Ann. Phys.*, **53**, 177. [Also *Phys. Zeit.*, **19**, 421 (1918) and **20**, **21** (1919).]

[12] Gihman, I. I. and Skorohod, A. V. (1972). *Stochastic Differential Equations*. Springer, Berlin, Germany. (An abstract theoretical treatise.)

[13] Hasminsky, R. Z. (1980). *Stochastic Stability of Differential Equations*. Sijthoff and Noordhoff, Alphen aan den Rijn, The Netherlands. (A theoretical text, stressing the Lyapunov approach.)

[14] Iglehart, D. L. (1969). *J. Appl. Prob.*, **6**, 285.

[15] Itô, K. and McKean, H. P., Jr. (1965). *Diffusion Processes and Their Sample Paths*. Springer, New York.

[16] Jazwinsky, A. H. (1970). *Stochastic Processes and Filtering Theory*. Academic, New York.

[17] Karlin, S. and Taylor, H. M. (1981). *A Second Course in Stochastic Processes*. Academic, New York.

[18] Katzur, R., Bobrovsky, B. Z., and Schuss, Z. (1984). *SIAM J. Appl. Math.*, **44**, 591; 1176.

[19] Keller, J. B. (1977). *Statistical Mechanics and Statistical Methods in Theory and Applications*. V. Landman, ed. Plenum, New York. (An important collection of papers about the method of averaging.)

[20] Kolmogorov, A. (1931/1933) *Math. Ann.*, **104**, 415; **108**, 149.

[21] Kramers, H. A. (1940). *Physica*, **7**, 284. (A classical paper on mean first passage times.)

[22] Kushner, H. (1974). *Ann. Prob.*, **2**, 40.

[23] Kushner, H. (1964). *J. Math. Anal. Appl.*, **8**, 332.

[24] Langevin, P. (1908). *C. R. Acad. Sci. (Paris)*, **146**, 530.

[25] Landauer, B. and Swanson, J. A. (1961). *Phys. Rev.*, **121**, 1668.

[26] Ludwig, D. (1975). *SIAM Rev.*, **17**, 605.

[27] Mandl, P. (1968). *Analytical Treatment of One Dimensional Markov Processes*. Springer, New York. (An enhancement of Feller's work.)

[28] Papanicolaou, G. (1977). Modern Modeling of Continuous Phenomena, R. C. DiPrima, ed. *Lect. Appl. Math.*, **15**, 103.

[29] Planck, M. (1917). *Sitzungsber. Preuss. Akad. Wissens.*, 324.

[29a] Pontryagin, L. S., Andronov, A. A., and Vitt, A. A. (1933). *J. Exp. Theor. Phys. (Moscow)*, **3**(3), 165.

[30] Lord Rayleigh (1891). *Philos. Mag., Ser. 5*, **32**, 424; *Scient. Papers*, **III**, 473.

[31] Ruohonen, M. (1980). *Scand. Actu. J.*, **63**, 113.

[32] Schuss, Z. (1980). *Theory and Applications of Stochastic Differential Equations*. Wiley, New York.

[33] Smoluchowski, M. (1906/1915). *Ann. Phys.*, **21**, 756; **48**, 1105.

[34] Snyder, D. L. (1969). *The State Variable Approach to Continuous Estimation with Applications to Analog Communication Theory*. MIT Press, Cambridge, MA. (An SDE approach to filtering theory.)

[35] Stratonovich, R. L. (1967). *Topics in the Theory of Random Noise*, Vols. I and II. Gordon and Breach, New York. (An outstanding collection of problems and applications of SDEs.)

[36] Uhlenbeck, G. E. and Ornstein, L. S. (1930). *Phys. Rev.*, **36**, 823.

[37] van Kampen, G. (1981). *Stochastic Processes in Physics and Chemistry*. North-Holland, Amsterdam, The Netherlands. (A comprehensive review of SDEs and other processes in the physical sciences.)

[38] Ventzel, A. D. and Freidlin, M. I. (1984). *Random Partitions of Dynamical Systems*. Springer, Berlin.

[39] Wang, M. S. and Uhlenbeck, G. E. (1945). *Rev. Mod. Phys.*, **17**, 323.

[40] Wong, E. and Zakai, M. (1965). *Ann. Math. Statist.*, **36**, 1560.

[41] Zakai, M. (1969). *Zeit. Wahrsch. verw. Geb.*, **22**, 230.

Bibliography

Arnold, L. (1974). *Stochastic Differential Equations: Theory and Applications*. Wiley, New York. (A theoretical textbook with some applications.)

Chandrasekhar, S. (1943). *Rev. Mod. Phys.*, **15**, 1. (An excellent review.)

Doob, J. L. (1953). *Stochastic Processes*. Wiley, New York. (A classical advanced textbook.)

Einstein, A. (1956). *Investigations on the Theory of the Brownian Movement*. Dover, New York. (A historical collection of Papers.)

McKean, H. P., Jr. (1969). *Stochastic Integrals*. Academic, New York. (An elegant exposition of the Itô integral and some basic notions in SDEs.)

Wax, N. (1954). *Selected Papers on Noise and Stochastic Processes*. Dover, New York.

(BROWNIAN MOTION
DIFFUSION PROCESSES
FOKKER–PLANCK EQUATION
KALMAN FILTERING
MARKOV PROCESSES
ORNSTEIN–UHLENBECK PROCESS
RANDOM WALK
STOCHASTIC INTEGRALS
STOCHASTIC MECHANICS
STOCHASTIC PROCESSES)

Z. Schuss

STOCHASTIC DIFFERENTIAL EQUATIONS: APPLICATIONS IN ECONOMICS AND MANAGEMENT SCIENCE

There are two major strands of application of stochastic differential equations* in economics and management science*. The first strand involves modelling phenomena as stochastic differential equations and analyzing the solutions. The second involves the description of the state of a system as a stochastic differential equation and optimizing a criterion function subject to it.

A stochastic differential equation (SDE) may be regarded as a generalization of a deterministic ordinary differential equation

$$dx = m(t, x(t))\,dt \qquad (1)$$

through the addition of a stochastic term $v(t, x(t))\,dz$, where dz is the increment of a stochastic process $z(t)$ that obeys Brownian motion* or white noise* or a Wiener process. For a Wiener process $z(t)$ and for any partition $t_0, t_1, t_2, \ldots,$ of the time interval, the random variables $z(t_1) - z(t_0)$, $z(t_2) - z(t_1)$, $z(t_3) - z(t_2), \ldots,$ are independently and normally distributed with mean zero and variances $t_1 - t_0, t_2 - t_1, t_3 - t_2, \ldots,$ respectively. The SDE generalization of (1) is

$$dx = m(t, x(t))\,dt - v(t, x(t))\,dz. \quad (2)$$

Thus, over a short interval of time dt, the change dx in $x(t)$ is normally distributed with mean $m(t, x(t))\,dt$ and variance $v^2(t, x(t))\,dt$. The term $m(t, x(t))$ is often referred to as the *drift* of the stochastic process* describing the movement of dx through time and the term $v(t, x(t))$ is referred to as the diffusion.

As in the case of the deterministic differential equation (1), the solution to the SDE (2) is a function

$$x = F(t, z) \quad (3)$$

whose differential is (2). This solution to the SDE (2) is called a *diffusion process**. In order for the differential of (3) to coincide with (2), however, the differential elements dt and dz must obey the following multiplication table:

	dz	dt
dz	dt	0
dt	0	0

Thus, the differential of (3), where $z(t)$ is a Wiener process, will include a second partial derivative. In particular, taking the Taylor series expansion of (3) yields.

$$dx = F_t\,dt + F_z\,dz + F_{tt}(dt)^2/2 + F_{tz}\,dt\,dz + F_{zz}(dz)^2/2 + \text{high-order terms}, \quad (4)$$

that upon use of the above multiplication table yields as the differential of (3):

$$dx = (F_t + F_{zz}/2)\,dt + F_z\,dz. \quad (5)$$

Likewise, if

$$y = G(t, x), \quad (6)$$

where x obeys (2), then using Taylor's theorem and the above multiplication table yields the stochastic differential of (6):

$$dy = G_t\,dt + G_x\,dx + G_{xx}(dx)^2/2. \quad (7)$$

The formation of the stochastic differential of (6) as (7) is known as *Itô's lemma**. Expression (4) is of course the special case of (7) when $x \equiv z$. Substitution from (2) into (7) and employment of the multiplication table yields

$$dy = (G_t + G_x m + G_{xx}v^2/2)\,dt + G_x v\,dz. \quad (8)$$

A rigorous derivation of Itô's lemma is presented in Arnold [1].

It is Itô's lemma that is most widely employed in the application of SDEs in economics and management science. Equation (2) is often referred to as *Itô's stochastic differential equation*. The behavior of the stochastic variable $z(t)$ is assumed to follow either arithmetic or geometric Brownian motion. In arithmetic Brownian motion with zero mean, the probability of a unit increment or decrement of the stochastic variable is the same and independent of the value of the variable. In geometric Brownian motion the probability of a 1% increment or decrement of the stochastic variable is the same, and independent of the value of the variable.

OPTION PRICING

An option is a contract for the purchase or sale of an asset at a specified price on or before a specified date. The contract need not be exercised. The contracted price at which the asset may be purchased or sold is the *exercise* or *strike* price and the date at which the contract expires is the *exercise* or *maturity* date. A *call option* is a contract to buy an asset and a *put option* is a contract to sell an asset. A *European call option* is a contract that can only be exercised at the maturity date, while an *American call option* is a contract that can be exercised on or before the maturity date. Options can themselves be bought and sold and in fact there are a number of markets or exchanges for this purpose.

The question that has puzzled economists for some time is what the price of an option should be. A variety of answers were proposed, an excellent history of which is presented by Smith [13], but it was not until the work of Black and Scholes [2] that a satisfactory answer was provided.

The price of a call option should be a function of the price of the asset on which the option can be exercised and time to maturity of the option. Obviously, the price of the option will be zero at its maturity date if the price of the asset is below the exercise price, and will be equal to the difference between the price of the asset minus the exercise price otherwise. At the maturity date of the option the price of the asset is known. Prior to the maturity date, however, the price of the asset is uncertain. If the asset price follows some stochastic process it will induce a stochastic process on the call option price. Thus, determining the price of a call option amounts to finding the diffusion process that it obeys, given the stochastic process that governs the price of the asset.

Black and Scholes [2] envision a simple situation in which the asset is a stock that pays no dividends; the return from possession of the stock is based solely on the change in its price. There also exists a European call option that can be bought or sold short. Thus, an investor can hold a portfolio of shares of the stock and of units of the call option. The investor can also hold units of a riskless asset, say, a short term government security, that yields a return of r dollars per unit time.

Let P_s denote the price of a unit of the stock and S the number of units of the stock. Likewise, let P_c denote the price of one call option and C the number of call options. The value of an investor's portfolio consisting of units of the stock and the call option is denoted by V and is equal to

$$V = SP_s + CP_c. \tag{9}$$

The stock price is assumed to change according to a pattern described by geometric Brownian motion,

$$dP_s = mP_s\,dt + vP_s\,dz, \tag{10}$$

where m, a constant, is the instantaneous average return on the stock, and v^2, also a constant, is the instantaneous variance of the return on the stock.

Now the change in the value of the portfolio resulting from changes in the price of the stock and the price of the call option alone is

$$dV = S\,dP_s + C\,dP_c. \tag{11}$$

The sought after function is

$$P_c = P_c(t, P_s). \tag{12}$$

Thus, by Itô's lemma,

$$dP_c = \frac{\partial P_c}{\partial t}\,dt + \frac{\partial P_c}{\partial P_s}\,dP_s + \frac{1}{2}\frac{\partial^2 P_c}{\partial P_s^2}(dP_s)^2. \tag{13}$$

Now from (10) and the multiplication table it follows that

$$(dP_s)^2 = P_s^2 v^2\,dt. \tag{14}$$

Thus, substitution from (10), (13), and (14) into (11) yields the SDE

$$dV = \left[\frac{\partial P_c}{\partial t} + \frac{1}{2}\frac{\partial^2 P_c}{\partial P_s^2}P_s^2 v^2\right]C\,dt + \left[S + C\frac{\partial P_c}{\partial P_s}\right]dP_s, \tag{15}$$

where the coefficient of dt represents the average instantaneous return on the portfolio and the coefficient of dP_s, its instantaneous variance. Now the major insight of Black and Scholes is that instantaneous variance of the return on the portfolio can be made equal to zero by holding the number of shares of stock and the number of call options in the ratio

$$S/C = -\partial P_c/\partial P_s. \tag{16}$$

But then the portfolio becomes riskless as long as the investor maintains the ratio (16) and should earn the same rate of return per unit time as can be realized on a riskless asset. Thus, if (16) is satisfied

$$dV/V = r\,dt \tag{17}$$

or

$$dV = rV\,dt. \tag{18}$$

Without loss of generality, set $S = 1$ in (16) and substitute from (15) and (16) into (18) to get the partial differential equation for P_c,

$$\frac{\partial P_c}{\partial t} = r\left(P_c - P_s\frac{\partial P_s}{\partial P_c}\right) - \frac{1}{2}P_s^2 v^2\frac{\partial^2 P_c}{\partial P_s^2}, \tag{19}$$

with terminal condition

$$P_c = \max[0, P_s - P_\epsilon] \quad \text{at } t^*, \tag{20}$$

where t^* is the maturity date of the option and P_ϵ is the exercise price. The solution to (19) is

$$P_c = P_s \cdot N(y) - P_\epsilon e^{-rT} \cdot N(y), \tag{21}$$

where $N(y)$ denotes the log-normal distribution*,

$$y = \left[\ln(P_s/P_\epsilon) + (r + v^2/2)T\right] \big/ (v\sqrt{T}) \tag{22}$$

and T is the time to expiration of the option.

It can be shown that: $\partial P_c/\partial P_s > 0$, the price of the call option increases with the price of the stock; $\partial P_c/\partial P_\epsilon < 0$, the price of the option decreases as the exercise price increases; $\partial P_c/\partial T > 0$, the price of the option decreases as the expiration date of the option approaches; $\partial P_c/\partial r > 0$, the price of the option increases as the riskless rate of interest increases; and $\partial P_c/\partial v^2 > 0$, the price of the option increases as the variance of the price of the stock increases. This last result is perhaps the most interesting because it appears to be counterintuitive in the sense that it suggests that a holder of an option would prefer more variance in the price of a stock to less. Some reflection discloses, however, that this is precisely so because with a higher variance the chance that the stock price will exceed the exercise price is greater.

STOCHASTIC OPTIMAL CONTROL

The typical stochastic optimal control problem has the form

$$\max_{u(t)} E\left[\int_{t_0}^{T} f(t, x(t)u(t))\,dt + s(x(T), T)\right]$$

$$\text{subject to (s.t.) } dx = m(t, x(t), u(t))\,dt + v(t, x(t), u(t))\,dz,$$

$$x(t_0) = x_0, \tag{23}$$

where $u(t)$ is the *control variable*, $x(t)$ the *state variable*, and $z(t)$ is a Wiener process. The state of the system is described by the SDE above. The necessary conditions obeyed by the optimal control $u(t)$ are derived through Bellman's principle of optimality. According to this principle an optimal path has the property that whatever the initial conditions and control values over some initial period, the control over the remaining period must be optimal for the remaining problem, with the state resulting from the early decisions considered as the initial condition. Thus, define the optimal value function $J(t_0, x_0)$ as the maximum expected value obtainable for a problem of the form (23), starting at time t_0 in state $x(t_0) = x_0$:

$$J(t_0, x_0) = \max_u E\left[\int_{t_0}^{T} f\,dt + s(x(T), T)\right]$$

$$\text{s.t. } dx = m\,dt + v\,dz. \tag{24}$$

Now, according to the principle of optimality, (24) can be rewritten as

$$J(t_0, x_0) = \max_u E\left[\int_{t_0}^{t_0+\Delta t} f\,dt + \max_u \int_{t_0+\Delta t}^{T} f\,dt + s\right]$$

$$\text{s.t. } dx = m\,dt + v\,dz,$$

$$x(t_0 + \Delta t) = x_0 + \Delta x, \tag{25}$$

where Δt is taken as small and positive. Since Δt is supposed small, the first integral can be approximated by $f(t_0, x_0, u)\,\Delta t$ and the control can be assumed to be constant over this interval. Then (25) can be rewritten as

$$J(t_0, x_0) = \max_u E\left[f(t_0, x_0, u)\,\Delta t + J(t_0 + \Delta t, x_0 + \Delta x)\right]. \tag{26}$$

Assuming that $J(t, x)$ is twice differentiable, the function on the right is expanded about (t_0, x_0),

$$\begin{aligned} J(t_0 + \Delta t,\ x_0 + \Delta x) &= J(t_0, x_0) \\ &\quad + J_t(t_0, x_0)\,\Delta t + J_x(t_0, x_0)\,\Delta x \\ &\quad + \tfrac{1}{2} J_{xx}(t_0, x_0)(\Delta x)^2 \\ &\quad + \text{higher-order terms.} \end{aligned} \tag{27}$$

From the differential constraint in (23),

$$\Delta x = m\,\Delta t + v\,\Delta z, \tag{28}$$

and, therefore, by use of the multiplication table above,

$$(\Delta x)^2 = v^2\,\Delta t + \text{higher-order terms}. \tag{29}$$

Substituting from (29) and (27) into (26) yields

$$J(t_0, x_0) = \max_u E\big[f\Delta t + J + J_t\,\Delta t + J_x m\,\Delta t + J_x v\,\Delta z + \tfrac{1}{2}J_{xx}v^2\,\Delta t + \text{higher-order terms}\big]. \tag{30}$$

Taking the expectation in (30), observing that the only stochastic term is Δz and that its expectation is zero by assumption, subtracting $J(t_0, x_0)$ from both sides, dividing by Δt, and letting $\Delta t \to 0$, yields the basic necessary condition for the stochastic optimal control problem,

$$-J_t(t, x) = \max\big(f(t, x, u) + J_x m(t, x, u) + \tfrac{1}{2}v^2 J_{xx}(t, x)\big), \tag{31}$$

with boundary condition

$$J(t, x(T)) = s(x(T), T). \tag{32}$$

The expression holds for every (t, x) combination along the optimal path and not just at (t_0, x_0).

The basic necessary condition simplifies for problems of the form

$$V(x_0) = \max_u E\int_{t_0}^{\infty} e^{-r(t-t_0)}f(x, u)\,dt$$

$$\text{s.t. } dx = m(x, u)\,dt - v(x, u)\,dt, \quad x(t_0) = x_0 \tag{33}$$

(these are called *autonomous problems* by economists). Here

$$J(t, x) = e^{-rt}V(x) \tag{34}$$

and, therefore, upon substitution into (31) yields

$$rV(x) = \max_u\big(f + V''g + \tfrac{1}{2}V^2V''\big). \tag{35}$$

See Kamien and Schwartz [5] for a fuller demonstration of these results.

One of the earliest applications of stochastic optimal control was by Merton [7]. The question addressed was how a consumer would allocate personal wealth among current consumption, investment in a riskless asset, and investment in a risky asset so as to maximize the expected present value of utility over an infinite time horizon. Letting W represent total wealth, w the fraction of wealth in the risky asset, s the return on a sure asset, a the expected return on risky asset, with $a > s$, v^2 the variance per unit time on the risky asset, and c the consumption, and assuming that the utility function has the form $U(c) = c^b/b$ with $b < 1$, the problem is

$$\max_{c,w} E\int_0^{\infty}\big(e^{-rt}c^b/b\big)\,dt$$

$$\text{s.t. } dW = \big[s(1-w)W + awW - c\big]\,dt + wWv\,dz, \qquad w(0) = W_0, \tag{36}$$

where r is the rate of interest and dz follows a Wiener process. The state of the system is wealth $W(t)$ at each point in time that changes in a stochastic fashion. The deterministic portion of the change (the coefficient of dt) is composed of the return on the funds in the sure asset, plus the expected return on the funds in the risky asset, less consumption.

Application of the necessary condition (35) to this problem yields

$$c(t) = W(Ab)^{1/(b-1)},$$

$$w = (a - s)/(1 - b)v^2, \tag{37}$$

where A is a constant of integration that can be determined for specific values of the other parameters. According to (37) the individual optimality consumes a constant fraction of wealth at each moment. This fraction varies directly with the discount rate r, and with the riskiness of the risky asset v^2. The optimal division of wealth between two kinds of assets is a constant, independent of the total wealth. The fraction devoted to the risky asset varies directly with its expected return and inversely with its variance. This model was extended by Merton [8] to include several risky assets, and then was employed by him to formulate a dynamic version of the capital asset pricing model (Merton [9]). Breeden [3] simplified and extended this work of Merton.

Merton [10] also provided a stochastic version of the Ramsey problem. In its deterministic form the Ramsey problem requires a social planner to find a consumption plan to maximize society's total welfare, where total welfare is the integral of the utility of per capita consumption and all individuals are identical. This consumption plan is limited by the productive capacity of the economy, and that depends on the levels of capital and labor available. Only one good is produced and it can be consumed or reinvested to build up the stock of capital, and that will in turn allow for greater production in the future.

Formally the Ramsey problem can be posed as

$$\min_c \int_0^\infty (B - U(c(t))] \, dt$$

$$\text{s.t } \frac{dk}{dt} = f(k) - c - bk - kn,$$

$$k(0) = k_0, \, k \geqslant 0, \tag{38}$$

where $k(t)$ is the capital–labor ratio, $c(t)$ is per capita consumption, $f(k)$ is per capita output, b is the rate at which capital depreciates, B is the bliss level of utility, and n is the rate of population growth*. Merton transforms this problem into a stochastic optimal control problem by supposing that the rate of population growth is stochastic. Specifically, he assumes that the population changes according to the SDE

$$dL = nL\,dt + vL\,dz, \tag{39}$$

where dz is a Wiener process and nL and v^2L^2 are the instantaneous mean and variance per unit time, respectively. Applying Itô's lemma and (39) one transforms the deterministic state equation in (38) into a SDE. Formally, the new problem is

$$\max_{c(t)} E \int_0^T U(c(t)) \, dt$$

$$\text{s.t. } dk = \left[f(k) - c - (n + b - v^2)k \right] dt - vk\,dz,$$

$$k(0) = k_0, \, k(t) \geqslant 0, \tag{40}$$

where $U(c(t))$ is a strictly concave von Neumann–Morgenstern utility* function. Merton applies (35) and then lets $T \to \infty$ to derive a steady-state distribution for $k(t)$ (this distribution degenerates to a point in the deterministic case) and then derives a differential equation for the optimal savings rate

$$\left(\tfrac{1}{2}v^2k^2fU''\right) ds/dk = \left(fU' - \tfrac{1}{2}v^2k^2U''f'\right)s + \tfrac{1}{2}v^2k^2U''f' - U'(n + b - v^2) + U - B. \tag{41}$$

When $v^2 = 0$, expression (41) reduces to the condition for the optimal savings rate obtained in the solution to the deterministic Ramsey problem.

Tapiero [14] provides a stochastic version of a deterministic model of advertising and good will introduced by Nerlove and Arrow [11]. In the deterministic version, a firm faces the problem of determining an advertising plan $I(t)$ through time so as to maximize the present value of its profits, where profits $R(x)$ are a function of its current level of "good will" $x(t)$. Good will is accumulated through advertising. Formally, the problem is

$$\max_{J(t)} \int_0^\infty e^{-rt} [R(x(t)) - I(t)] \, dt$$

$$\text{s.t. } x' = I(t) - bx(t), \qquad x(0) = x_0, \tag{42}$$

where b is the rate at which good will decays. Tapiero's stochastic generalization is

$$\max_{I(t)} E \int_0^\infty e^{-rt} [R(x) - I(t)] \, dt$$

$$\text{s.t. } dx = (I - bx) \, dt + (I + bx)^{1/2} \, dz, \tag{43}$$

where dz follows a Wiener process. Tapiero's analysis of the necessary conditions for this problem discloses that, if the firm is risk neutral, then the optimal advertising policy coincides with the one followed in the deterministic case. When the firm is risk averse, then the optimal advertising policy followed under uncertainty depends importantly on the rate of decay of good will b. If b is large, the optimal policy under uncertainty approaches the optimal policy under certainty.

Gonedes and Lieber [4] present a stochastic version of a production planning problem often studied. In the deterministic production planning problem, a firm seeks to minimize the total cost of production that includes the manufacturing cost and the cost of holding inventory. The costs associated with inventory are storage costs when inventory is positive and stock-out costs in the form of lost sales or other penalties associated with not having the product on hand. In the deterministic version of this problem the demand for the product at each point in time is known. In the Gonedes and Lieber stochastic version the demand for the product is unknown. Formally, the problem is

$$\min E \int_0^T [F(p(t)) + n(I(t))]\, dt$$

$$\text{s.t. } dI = p(t)\, dt - dz, \qquad (44)$$

where $F(p(t))$ is the cost of production per unit time when the rate of production is $p(t)$, $n(I(t))$ is the inventory holding cost per unit time when the inventory level is $I(t)$, $z(t)$ is cumulative demand at time t, and dz is a Wiener process. They are able to convert this problem into a deterministic equivalent and analyze it by methods of optimal control.

FINAL REMARKS

The application of stochastic differential equations to problems in economics and management science is a recent event. The realm of application is growing beyond those mentioned here; see Mallioris and Brock [6], who also provide an excellent introduction to the subject.

References

[1] Arnold, J. L. (1974). *Stochastic Differential Equations: Theory and Application*. Wiley, New York.

[2] Black, F. and Scholes, M. (1975). The pricing of options and corporate liabilities. *J. Pol. Econ.*, **81**, 637–654.

[3] Breeden, D. T. (1979). An intertemporal asset pricing model with stochastic consumption and investment opportunities. *J. Financial Econ.*, **7**, 265–296.

[4] Gonedes, N. J. and Lieber, Z. (1974). Production planning for a stochastic demand process. *Operat. Res.*, **22**, 771–787.

[5] Kamien, M. and Schwartz, N. L. (1981). *Dynamic Optimization: The Calculus of Variations and Optimal Control in Economics and Management*. North-Holland, New York.

[6] Malliaris, A. G. and Brock, W. A. (1982). *Stochastic Methods in Economics and Finance*. North-Holland, New York.

[7] Merton, R. C. (1969). Lifetime portfolio selection under uncertainty: the continuous time case. *Rev. Econ. Statist.*, **51**, 247–257.

[8] Merton, R. C. (1971). Optimal consumption and portfolio rules in a continuous time model. *J. Econ. Theory*, **3**, 373–413.

[9] Merton, R. C. (1973). An intertemporal capital asset pricing model. *Econometrica*, **41**, 867–887.

[10] Merton, R. C. (1975). An asymptotic theory of growth under uncertainty. *Rev. Econ. Studies*, **42**, 375–394.

[11] Nerlove, M. and Arrow, K. J. (1962). Optimal advertising policy under dynamic conditions. *Econometrica*, **29**, 129–142.

[12] Ramsey, F. (1928). A mathematical theory of savings. *Econ. J.* [Reprinted in AEA, *Readings in Welfare Economics*, K. J. Arrow and T. Scitovsky, eds. Irwin, Homewood, IL (1969).]

[13] Smith, C. W., Jr. (1976). Option pricing. *J. Financial Econ.*, **3**, 3–51.

[14] Tapiero, C. (1978). Optimum advertising and goodwill under uncertainty. *Operat. Res.*, **26**, 450–463.

(FINANCE, STATISTICS IN MANAGEMENT SCIENCE, STATISTICS IN OPTIMAL STOCHASTIC CONTROL)

MORTON I. KAMIEN

STOCHASTIC DOMINATION *See* STOCHASTIC AND UNIVERSAL DOMINATION

STOCHASTIC FAILURE

Concepts defined and studied by Singh and Deshpande [1] extend those of increasing failure rate (IFR), decreasing failure rate (DFR), new better than used (NBU), and new worse than used (NWU) defined in RELIABILITY, PROBABILISTIC.

Let a random variable (RV) Y have life distribution F with finite mean μ. Let $X_0 = 0$ and $X_1, X_2, \ldots$ be a sequence of independent exponentially distributed RVs with common mean μ, and independent of Y. Then

1. Y or F has the *stochastically increasing failure rate* (SIFR) property if

$$\Pr\left(\sum_{i=0}^{k} X_i \leqslant Y < \sum_{i=0}^{k+1} X_i | Y \geqslant \sum_{i=0}^{k} X_i\right) \geqslant \Pr\left(\sum_{i=0}^{k-1} X_i \leqslant Y < \sum_{i=0}^{k} X_i | Y \geqslant \sum_{i=0}^{k-1} X_i\right), \quad k = 1, 2, \ldots. \quad (1)$$

2. Y or F has the *stochastically decreasing failure rate* (SDFR) property if (1) holds with the inequality reversed.
3. Y or F is *stochastically new better than used* (SNBU) if

$$\Pr\left(Y \geqslant \sum_{i=0}^{k+1} X_i | Y \geqslant \sum_{i=0}^{k} X_i\right) \leqslant \Pr\left(Y \geqslant \sum_{i=0}^{k+1} X_i\right), \quad k = 1, 2, \ldots. \quad (2)$$

4. Y or F is *stochastically new worse than used* (SNWU) if (2) holds with the inequality reversed.

The class of SNBU (SNWU) distributions contains and is strictly larger than both the NBU (NWU) class and SIFR (SDFR) class. The chain of implications appears in Fig. 1 of ref. 1.

Reference

[1] Singh, H. and Deshpande, J. V. (1985). *Scand. J. Statist.*, **12**, 213–220.

(HAZARD RATE AND OTHER CLASSIFICATIONS OF DISTRIBUTIONS
RELIABILITY, PROBABILISTIC)

STOCHASTIC GAMES

A stochastic game is a multistage game in which play proceeds from position to position according to probabilities controlled jointly by the players. An example in the two-person zero-sum case shows some of the features (*see* GAME THEORY for background).

Example 1. Consider a matrix game of the form

$$G_1 = \begin{pmatrix} 0 & 1 \\ 3 & G_2 \end{pmatrix},$$

where G_2 is some other game. In such a game, player I chooses a row and simultaneously player II chooses a column. If the entry in G_1 corresponding to the chosen row and column is a number, then II pays I that amount and the game ends. If the chosen entry is G_2, then the players play G_2. If the value of G_2 is known, say $v_2 = \text{val}(G_2)$, then the value $v_1 = \text{val}(G_1)$ may be found by replacing G_2 in the matrix by v_2 so that

$$v_1 = \text{val}\begin{pmatrix} 0 & 1 \\ 3 & v_2 \end{pmatrix}.$$

Of course, G_2 may itself have other games as components, in which case the above method must be iterated to find the solution. Moreover, one of these games may have G_1 as a component, in which case G_1 is defined in terms of itself and we may call G_1 a *recursive game*. Suppose that G_2 is the game G_1 with the roles of the players reversed,

$$G_2 = -G_1^T = \begin{pmatrix} 0 & -3 \\ -1 & G_1 \end{pmatrix}.$$

If the value v_1 of G_1 exists, then clearly $v_2 = -v_1$ and v_1 satisfies the equation

$$v_1 = \text{val}\begin{pmatrix} 0 & 1 \\ 3 & -v_1 \end{pmatrix}.$$

This has the unique solution $v_1 = \sqrt{7} - 2$. An optimal (mixed) strategy for I in G_1 is $(x_1, x_2) = (5 - \sqrt{7}, 1)/(6 - \sqrt{7})$, where x_i is the probability that I chooses row i. Similarly, $(y_1, y_2) = (3 - \sqrt{7}, 3)/(6 - \sqrt{7})$ is optimal for II in G_1, where y_j is the probability that II chooses column j. Use of these strategies in G_1 (interchanged when playing G_2) guarantees each player v_1, so that v_1 is the value of G_1.

In recursive games, play might proceed forever, so that to complete the description of the game, the payoff must be defined in the case of infinite play. In Example 1, it does not matter with optimal play of the players, but in other cases it might. Everett [9] has developed a general theory of recursive games of this type. Other types of multistage games include Milnor and Shapley's *games of survival* [20], a game-theoretic generalization of gambler's ruin, where two players repeatedly play the same matrix game until, if ever, one of them goes broke, and Blackwell's *multicomponent attrition games* [3], where resources can only decrease and the first player to run out of any resource is the loser.

A model of appropriate generality containing these games was proposed in the seminal paper of Shapley [25] under the title *Stochastic Games*. A detailed account may be found in the recent monograph by Vrieze [29]. A stochastic game **G** is a collection of matrix games $\{G^1, \ldots, G^K\}$, together with rules describing the stochastic movement from one game to the next. The set $S = \{1, \ldots, K\}$ is called the *state space*. If the game **G** is in state $k \in S$, the game G^k is played once with player I choosing a row i and player II choosing a column j, and with I receiving from II an immediate reward g_{ij}^k, the (i, j) entry of G^k. Then with probability s_{ij}^k the game stops, while with probability p_{ij}^{kl} the game continues by making a transition to state l, where $s_{ij}^k + \sum_{l=1}^K p_{ij}^{kl} = 1$ for all i, j, k. Symbolically, we may write

$$G^k = \left(g_{ij}^k + \sum_{l=1}^{K} p_{ij}^{kl} G^l \right).$$

A strategy for I is a sequence $\sigma = (\mathbf{x}_1, \mathbf{x}_2, \ldots)$ of vectors $\mathbf{x}_t = (\mathbf{x}_t^1, \ldots, \mathbf{x}_t^K)$, where $\mathbf{x}_t^k$ is a mixed strategy in game G^k for $k = 1, \ldots, K$ and $t = 1, 2, \ldots$. The interpretation of σ is that, if at stage t game G^k is to be played, then I will choose his row according to the probability distribution $\mathbf{x}_t^k$. The choice of $\mathbf{x}_t$ for I may depend upon the past history of the game up to stage t. If the choice of $\mathbf{x}_t$ is independent of the past and independent of t, say $\mathbf{x}_t = \mathbf{x}$, we say the strategy $\sigma = (\mathbf{x}, \mathbf{x}, \ldots)$ is *stationary*; similarly for strategies $\tau = (\mathbf{y}_1, \mathbf{y}_2, \ldots)$ for player II. The main theoretical problems for stochastic games center on the existence of the value and of stationary optimal strategies for various payoff criteria, and on methods of computing optimal stationary or ϵ-optimal stationary strategies when they exist.

When all stop probabilities are positive, $\min_{i,j,k} s_{ij}^k > 0$, the game will stop with probability 1 no matter what strategies the players use. Shapley [25] treats this case, using total reward as the payoff criterion. The main tool is the operator $T\mathbf{v} = \mathbf{w}$, which maps one set of possible values $\mathbf{v} = (v^1, \ldots, v^K)$ into another $\mathbf{w} = (w^1, \ldots, w^K)$ by the formula $w^k = \text{val}(G^k(\mathbf{v}))$, where $G^k(\mathbf{v})$ is the matrix game whose (i, j) component is $g_{ij}^k + \sum_{l=1}^K p_{ij}^{kl} v^l$. Shapley shows that T is a contraction operator and that the value $\hat{\mathbf{v}}$ of the game exists and is the unique solution to $T\hat{\mathbf{v}} = \hat{\mathbf{v}}$. Moreover, both players have optimal stationary strategies, which in state k are optimal for $G^k(\hat{\mathbf{v}})$. For any choice of $\mathbf{v}_0$, the iteration $\mathbf{v}_n = T\mathbf{v}_{n-1}$ converges to $\hat{\mathbf{v}}$ at an exponential rate.

SPECIAL CASES

Case 1. If $K = 1$, the game is equivalent to an infinitely repeated game with the criterion of minimaxing the return per unit time $\mathbf{xGy}/\mathbf{xSy}$, where $\mathbf{S}$ is the matrix of stop probabilities and s_{ij} now represents the time required to receive reward g_{ij}. An extension of this problem to stochastic games with $K > 1$ has been treated by Aggarwal et al. [1].

Case 2. If each game G^k is a game of perfect information, then **G** is a game of perfect information and both players have pure stationary optimal strategies.

Case 3. If each G^k is $m_k \times 1$ (player II has no choices), the problem reduces to a Markov decision* problem or dynamic program*.

Case 4. If all stop probabilities are the same, say $s_{ij}^k = 1 - \beta > 0$ for all i, j, k, then the problem is equivalent to a stochastic game with zero stop probabilities and transition probabilities $\hat{p}_{ij}^{kl} = p_{ij}^{kl}/\beta$, in which future rewards are discounted by β.

Subsequent research has focused on stochastic games with zero stop probabilities using either the total discounted reward or the limiting average reward per stage as the payoff criteria. In a finite discounted Markov decision process, it has been shown by Blackwell [4] that there is a stationary strategy independent of the discount β that is simultaneously optimal for all β sufficiently close to one and for the limiting average reward criterion. That this does not hold for stochastic games is seen in the following example.

Example 2. Let G^1 and G^2 be the games

$$G^1 = \begin{pmatrix} 1 + G^1 & 0 + G^1 \\ 0 + G^2 & 1 + G^2 \end{pmatrix} G^2 = (1 + G^1).$$

If, when G^1 is played, player I uses row 1 (row 2), then G^1 (resp. G^2) is played next. After G^2 is played, G^1 is played next. For the problem discounted by $\beta < 1$, the values $v_i = \text{val}(G^i)$ satisfy

$$v_1 = \text{val}\begin{pmatrix} 1 + \beta v_1 & \beta v_1 \\ \beta v_2 & 1 + \beta v_2 \end{pmatrix},$$

$$v_2 = \text{val}(1 + \beta v_1).$$

The solution gives

$$(1 - \beta) v_1 = (1 + \beta)/(2 + \beta)$$

and

$$(1 - \beta) v_2 = 2/(2 + \beta).$$

The optimal stationary strategies are $\mathbf{x} = (\frac{1}{2}, \frac{1}{2})$ for I and $(1 + \beta, 1)/(2 + \beta)$ for II. The values for the limiting average return criterion are found as the limit as $\beta \to 1$ of $(1 - \beta)v_1$ and $(1 - \beta)v_2$, namely, $v_1 = v_2 = \frac{2}{3}$.

Since the existence of values and stationary optimal strategies for discounted stochastic games follows from Shaply [25], the remaining theoretical problems concern the *limiting average return criterion.* Gillette [13] has identified two main cases in which the game has a value.

1. **The Cyclic Case.** If the transition probabilities are such that for every stationary strategy pair the resulting Markov chain on the state space is irreducible, then the game has a value and both players have stationary optimal strategies. The condition can be weakened; see Sobel [26] and Federgruen [10].
2. **The Perfect Information Case.** If each G^k is a game of perfect information, then the game has a value and the players have pure stationary optimal strategies. An error in Gillette's proof is corrected in Liggett and Lippman [17].

That not every stochastic game with *limiting average* payoff can have stationary optimal strategies can be seen in the following example, due to Gillette.

Example 3. *The big match* is the game

$$G = \begin{pmatrix} 1 & 0 \\ 0^* & 1^* \end{pmatrix}$$

in which starred entries represent absorbing states. Players I and II simultaneously choose a row and column, respectively, and there is an immediate reward for I of the corresponding numerical entry of G. If the entry is not starred, the game G is repeated. If the entry is starred, then all *future* rewards are fixed at that entry. Suppose the players are restricted to stationary strategies. If II always uses the mixed strategy $(y, 1 - y)$ in G, then I can receive an expected limiting average payoff of $\max(y, 1 - y)$ so that $y = \frac{1}{2}$ is the minimax* choice for II, with (upper) value $\frac{1}{2}$. If I uses a stationary strategy $(x, 1 - x)$ in G, then II can keep the limiting average payoff to zero almost surely by playing column 1 always if $x > 0$, and column 2 always if $x = 0$. Thus, the big match does not have a value in stationary strategies.

Yet the game does have a value of $\frac{1}{2}$, as shown by Blackwell and Ferguson [5] by exhibiting for every $\epsilon > 0$ an ϵ-optimal strategy for I, necessarily nonstationary. This result was generalized by Kohlberg [16] who showed that every repeated matrix game with absorbing states has a value.

The problem of whether every stochastic game with limiting average payoff has a value was open for many years. Bewley and Kohlberg [2] proved that the value $\mathbf{v}(\beta)$ of the β-discounted game has a convergent expansion in fractional powers of $(1 - \beta)$ for β close to 1, and that the limit $\mathbf{v}_\infty = \lim_{\beta \to 1}(1 - \beta)\mathbf{v}(\beta)$, exists and is equal to the limit of the average values for the n-stage game. Monash [21] has shown that stochastic games with limit average *expected* payoff have value $\mathbf{v}_\infty$. That every stochastic game with limiting average payoff has a value equal to $\mathbf{v}_\infty$ was finally proved in 1981 by Mertens and Neyman [19]. Their result shows in addition that *for every* $\epsilon > 0$, *there is a strategy for each player that is simultaneously* ϵ*-optimal in all sufficiently long finite games, and in the infinite game, and in all discounted games with discount sufficiently close to 1.*

Partial extensions of the general theory to metric state and action spaces, to nonzero-sum games and to games with more than two players have been made by Himmelberg et al. [14], Federgruen [10], and Couwenbergh [7].

An extension to games of *information lag*, when there is a time lag in informing one or both players of past transitions, has been made by Scarf and Shapley [24]. The prime example is the bomber–battleship game solved by Dubins [8] and others. See Ferguson [11], Ling [18], and Burrow [6] for further references.

COMPUTATION

The iteration $\mathbf{v}^{(n)} = T\mathbf{v}^{(n-1)}$ of Shapley may be used to approximate the value and optimal strategies of a stochastic game. Each iteration consists in solving a set of K matrix games. An improved method with extension to the limiting average case was described by Hoffman and Karp [15]. The method involves strategies rather than values. An arbitrary strategy $\mathbf{u}^0$ is chosen for I and the optimal reply by II, and the values $\mathbf{v}^0$ are computed by solving a Markov decision problem. Then an improved strategy $\mathbf{u}^1$ for I is found as the optimal strategy for the games $G^k(\mathbf{v}^0)$, and the process is iterated. Thus more work is required per stage but convergence to the solution is generally more rapid.

Both methods are convergent exponentially. An interesting method related to Newton–Raphson*, which gives second-order convergence when it converges, has been described by Pollatschek and Avi-Itzhak [23]. An arbitrary $\mathbf{v}^0$ is chosen, and the optimal strategies are found for the games $G^k(\mathbf{v}^0)$. Then the value $\mathbf{v}^1$ of the game when the players use these strategies repeatedly is found by solving the appropriate linear equations and the process is repeated. Thus a single iteration requires solving K games and solving a set of K linear equations. The method requires less work than Hoffman and Karp's and is of second-order convergence. However, van der Wal [27] has shown that the method does not always converge, so the status of the method as a general procedure is unsettled. Other computational methods have been suggested by van der Wal [28].

Unlike Markov decision problems, no finite computational method of solving a general stochastic game exists. This can be seem from Example 1, which has an irrational value even though all the coefficients are rational. A problem whose solution is to be found in the same ordered field as the coefficients is said to have the *order-field property*. Parthasarathy and Raghavan [22] have noted that stochastic games in which only one of the players controls transition probabilities (such as Example 2) has the order-field property and in fact can be solved by solving a linear program. This result is generalized by Filar [12] to stochastic games in which the player who controls transitions changes from state to state; computational methods for this case may be found in Vrieze [29].

References

[1] Aggarwal, V., Chandrasekaran, R., and Nair, K. P. K. (1980). *SIAM J. Alg. Disc. Meth.*, **1**, 201–210.

[2] Bewley, T. and Kohlberg, E. (1976). *Math. Operat. Res.*, **1**, 197–208, 321–336.

[3] Blackwell, D. (1954). *Naval Res. Logist. Quart.*, **1**, 327–332.

[4] Blackwell, D. (1962). *Ann. Math. Statist.*, **33**, 719–726.

[5] Blackwell, D. and Ferguson, T. (1968). *Ann. Math. Statist.*, **39**, 159–163.

[6] Burrow, J. L. (1978). *J. Optim. Theory Appl.*, **24**, 337–360.

[7] Couwenbergh, H. A. M. (1980). *Int. J. Game Theory*, **9**, 25–36.

[8] Dubins, L. (1957). *Ann. Math. Stud.*, **39**, 231–255.

[9] Everett, H. (1957). *Ann. Math. Stud.*, **39**, 47–78.

[10] Federgruen, A. (1978). *Adv. Appl. Prob.*, **10**, 452–471.

[11] Ferguson, T. S. (1966). *Proc. Fifth Berkeley Symp. Math. Statist. Prob.*, Vol. 1. University of California Press, Berkeley, CA, pp. 453–462.

[12] Filar, J. A. (1981). *J. Optim. Theory Appl.*, **34**, 503–515.

[13] Gillette, D. (1957). *Ann. Math. Stud.*, **39**, 179–187.

[14] Himmelberg, C. J., Parthasarathy, T., Raghavan, T. E. S., and van Vleck, F. S. (1976). *Proc. Amer. Math. Soc.*, **60**, 245–251.

[15] Hoffman, A. J. and Karp, R. M. (1966). *Manag. Sci.*, **12**, 359–370.

[16] Kohlberg, E. (1974). *Ann. Statist.*, **2**, 724–738.

[17] Liggett, T. and Lippman, S. A. (1969). *SIAM Rev.*, **11**, 604–607.

[18] Ling, K. (1974). *Ann. Statist.*, **2**, 988–999.

[19] Mertens, J.-F. and Neyman, A. (1981). *Int. J. Game Theory*, **10**, 53–66.

[20] Milnor, J. and Shapley, L. S. (1957). *Ann. Math. Stud.*, **39**, 15–45.

[21] Monash, C. A. (1979). Ph.D. thesis, Harvard University, Cambridge, MA.

[22] Parthasarathy, T. and Raghavan, T. E. S. (1981), *J. Optim. Theory Appl.*, **33**, 375–392.

[23] Pollatschek, M. A. and Avi-Itzhak, B. (1969). *Manag. Sci.*, **15**, 399–415.

[24] Scarf, H. E. and Shapley, L. S. (1957). *Ann. Math. Stud.*, **39**, 213–229.

[25] Shapley, L. S. (1953). *Proc. Nat. Acad. Sci.*, **39**, 1095–1100.

[26] Sobel, M. J. (1971). *Ann. Math. Statist.*, **42**, 1930–1935.

[27] van der Wal, J. (1978). *J. Optim. Theory Appl.*, **25**, 125–138.

[28] van der Wal, J. (1980). *Int. J. Game Theory*, **9**, 13–24.

[29] Vrieze, C. J. (1983). *Stochastic Games With Finite State and Action Spaces*. Mathematische Centrum, Amsterdam, The Netherlands.

(GAME THEORY
MARKOV DECISION PROCESSES
MINIMAX DECISION RULES)

T. S. FERGUSON

STOCHASTIC INDEPENDENCE *See* STATISTICAL INDEPENDENCE

STOCHASTIC IN DER SCHULE *See* *TEACHING STATISTICS*

STOCHASTIC INTEGRALS

Stochastic integral is the name given to an integral of the type $\int_0^1 \phi(t)\, dX(t)$, where $\phi(t)$ and $X(t)$ are stochastic processes*. This was introduced by Wiener for a Brownian motion* X and a deterministic function ϕ. Doob extended this to include processes X with orthogonal increments and used this to obtain spectral representations of stationary processes (wide sense) in discrete and continuous time [2]. Itô defined $\int_0^1 \phi(t)\, dX(t)$ for a Brownian motion X and a large class of stochastic processes ϕ (to be described later) and used it to construct diffusion processes* from the heuristic discription of their behaviour in an infinitesimal time interval [2, 6]. In many statistical inference problems for models of continuous time phenomena, stochastic integrals appear naturally, as can be seem from the following simple example.

Consider the model

$$Y(t) = \int_0^t h(\theta, u)\, du + W(t),$$
$$0 \leqslant t \leqslant 1,\ \theta \in \mathbb{R}, \quad (1)$$

where W is a Brownian motion, h is a known function, and θ is the unknown parameter to be estimated on the basis of the observations $\{Y(t) : 0 \leqslant t \leqslant 1\}$. To use the maximum likelihood* method of estimation, we first need to evaluate the likelihood function. To make a guess at it, let us look at the following discrete analogue of (1):

$$Z_n(k) = \sum_{i=0}^{k-1} h\left(\theta, \frac{i}{n}\right) \cdot \frac{1}{n} + W\left(\frac{k}{n}\right),$$
$$0 \leqslant k \leqslant n, \quad (2)$$

for a fixed integer n. The likelihood $L_n(\theta)$ of θ for (2) based on the observations $\{Z_n(k) : 0 \leqslant k \leqslant n\}$ w.r.t. the distribution of $\{W(k/n) : 0 \leqslant k \leqslant n\}$ is given by

$$L_n(\theta, z_n) = \exp\left[\sum_{i=0}^{n-1} h\left(\theta, \frac{i}{n}\right) \times \left(Z_n\left(\frac{i+1}{n}\right) - Z_n\left(\frac{i}{n}\right)\right) - \frac{1}{2}\sum_{i=0}^{n-1} h^2\left(\theta, \frac{i}{n}\right)\frac{1}{n}\right]. \quad (3)$$

From this, it is natural to expect that the likelihood $L(\theta, Y)$ of θ for (1) based on the observations $\{Y(s) : 0 \leqslant s \leqslant 1\}$ w.r.t. the distribution of $\{W(s) : 0 \leqslant s \leqslant 1\}$ is given by

$$L(\theta, Y) \approx \exp\left[\sum_{s=0}^{1} h(\theta, s) \times \{Y(s + ds) - Y(s)\} - \tfrac{1}{2}\sum_{s=0}^{1} h^2(\theta, s)\, ds\right] \quad (4)$$

i.e.,

$$L(\theta, Y) = \exp\left[\int_0^1 h(\theta, s)\, dY(s) - \tfrac{1}{2}\int_0^1 h^2(\theta, s)\, ds\right], \quad (5)$$

where the stochastic integral $\int_0^1 h(\theta, s)\, dY(s)$ is to be suitably interpreted. Indeed, if $\int_0^1 h(\theta, s)\, ds < \infty$, (5) is the Cameron–Martin formula and the stochastic integral is to be taken as the Wiener integral [4].

Even when $h(\theta, s)$ is random, (3) is valid if $h(\theta, i/n)$ depends only on $\{W(u) : 0 \leqslant u \leqslant i/n\}$ and hence we expect that (5) is also true if $h(\theta, s)$ depends only on (to be understood as measurable w.r.t. the σ field generated by) $\{W(u) : 0 \leqslant u \leqslant s\}$. In this case, h is called a *nonanticipative functional* of W. If h satisfies some integrability conditions, then (5) is correct and is known as the *Girsanov formula* [4].

Let the processes $\phi(t)$, $X(t)$, $0 \leqslant t \leqslant 1$, be defined on a probability space $(\Omega, \underline{\underline{A}}, P)$, so that ϕ, X are real valued functions on $\overline{\Omega} = [0, 1] \times \Omega$. For an $\omega \in \Omega$, the mapping $t \to X(t, \omega)$ is called the ω *path* of the process X. In most cases of interest (as in the above example) the ω paths of the process X are of unbounded variation and hence it is not possible to define $[\int_0^1 \phi(t)\, dX(t)](\omega)$ (for a fixed $\omega \in \Omega$) as the Riemann–Stieltjes integral of $\phi(t, \omega)$ w.r.t. $X(t, \omega)$.

As a first step, if g is a simple function of the form

$$g(t, \omega) = \sum_{i=0}^{n-1} 1_{(t_i, t_{i+1})}(t) g_i(\omega), \quad (6)$$

where $0 = t_0 < t_1 < \cdots < t_n = 1$, then it is natural to define

$$\left[\int_0^1 g(t)\, dX(t)\right](\omega) = \sum_{i=0}^{n-1} g_i(\omega)[X(t_{i+1}, \omega) - X(t_i, \omega)]. \quad (7)$$

The next step is to extend the mapping $g \to \int_0^1 g\, dX$ to a larger class of function-preserving continuity in some sense.

In the remaining part, we will outline this extension for a Brownian motion X. For details, see refs. 4, 7, and 9. For a discussion and details of the extension as well as properties of the integral in the general case of a semimartingale, see refs. 3, 5, 8, and 13 and references therein.

Let $(\Omega, \underline{\underline{F}}, P)$ be a complete probability space and let $\underline{\underline{F}}_t$, $0 \leqslant t \leqslant 1$, be an increasing family of sub σ-fields of $\underline{\underline{F}}$ such that

$$\underline{\underline{F}}_t = \bigcap_{s>t} \underline{\underline{F}}_s, \qquad 0 \leqslant t < 1, \tag{8}$$

and

$$\underline{\underline{F}}_0 \text{ contains all } P\text{-null sets in } \underline{\underline{F}}. \tag{9}$$

Let $W(t)$ be a Brownian motion on $(\Omega, \underline{\underline{F}}, P)$ such that

$$W(t) \text{ is } \underline{\underline{F}}_t \text{ measurable} \tag{10}$$

and

$$\{W(t+s) - W(t) : 0 \leqslant s \leqslant 1-t\} \text{ is independent of } \underline{\underline{F}}_t. \tag{11}$$

Given the Brownian motion $W(t)$, we can get $\underline{\underline{F}}_t$ to satisfy (8), (9), (10), and (11), as follows. Let

$$\underline{\underline{G}}_t' = \sigma(W(s) : s \leqslant t), \qquad \underline{\underline{G}}_t = \sigma(\underline{\underline{G}}_t \cup \underline{\underline{N}}),$$

where $\underline{\underline{N}} = \{A \in \underline{\underline{F}} : P(A) = 0\}$ and let $\underline{\underline{F}}_t = \bigcap_{s>t} \underline{\underline{F}}_s$, $0 \leqslant t < 1$, $\underline{\underline{F}}_1 = \underline{\underline{G}}_1$. Let $L^2(\overline{\Omega})$ be the class of functions $f : \overline{\Omega} = [0,1] \times \Omega \to \mathbb{R}$ satisfying:

$$f \text{ is } \underline{\underline{B}} \otimes \underline{\underline{F}} \text{ measurable, where } \underline{\underline{B}} \text{ is the Borel } \sigma\text{-field on } [0,1]. \tag{12}$$

$$\text{for all } t \geqslant 0,\ f(t,\cdot) \text{ is } \underline{\underline{F}}_t \text{ measurable,} \tag{13}$$

and

$$\|f\|_*^2 \equiv E\int_0^1 f^2(t,\omega)\,dt < \infty. \tag{14}$$

If $g \in L^2(\overline{\Omega})$ is a simple function, then $\int_0^1 g(t)\,dW(t)$ defined by (6) and (7) (with $W = X$) satisfies

$$E\left[\int_0^1 g(t)\,dW(t)\right]^2 = \|g\|_*^2. \tag{15}$$

It can be shown [7, 9] that for $f \in L^2(\overline{\Omega})$, there exists a sequence $\{f_n\} \subseteq L^2(\overline{\Omega})$ of simple functions such that

$$\|f_n - f\|_*^2 \to 0 \quad \text{as } n \to \infty. \tag{16}$$

If $\{f_n\}$ satisfies (16), then in view of (15), the sequence of random variables $Z_n = \int_0^1 f_n(t)\,dX(t)$ is Cauchy in $L^2(\Omega, \underline{\underline{F}}, P)$. Also, the L^2 limit of Z_n is independent of the choice of the sequence $\{f_n\}$ satisfying (16), so that we can define

$$\int_0^1 f(t)\,dX(t) = L^2 - \lim_{n\to\infty} Z_n. \tag{17}$$

Clearly, the mapping $f \to \int_0^1 f(t)\,dX(t)$ is linear. The relation (15) continues to hold for all $g \in L^2(\overline{\Omega})$. It should be observed that $\int_0^1 f(t)\,dX(t)$ is not defined for each $\omega \in \Omega$ separately, but is defined as an L^2 limit.

For $f \in L^2(\overline{\Omega})$ and $0 \leqslant s \leqslant 1$, let

$$Z_1(s) = \int_0^s f(t)\,dX(t) \equiv \int_0^s 1_{[0,t]}(s) f(t)\,dX(t).$$

The process $Z_1(s)$ admits a continuous modification, i.e., there exists a process $Z(t)$ such that for all $\omega \in \Omega$, the ω path $t \to Z(t,\omega)$ is continuous and for all $t \geqslant 0$, $Z_1(t) = Z(t)$ a.s., so that

$$Z(s) = \int_0^s f(t)\,dX(t).$$

One of the important properties of the stochastic integral is that for $f \in L^2(\overline{\Omega})$,

$$\int_0^s f(t)\,dX(t) \tag{18}$$

and

$$\left[\int_0^s f(t)\,dX(t)\right]^2 - \int_0^s f^2(t)\,dt \tag{19}$$

are martingales*. This makes it possible to use powerful results from martingale theory in studying stochastic integral properties.

In some cases, it is possible to obtain an explicit expression for a stochastic integral. Let n be an integer and λ be a real number.

Then

$$\int_0^s X(t)\,dX(t) = \tfrac{1}{2}X^2(s) - s, \tag{20}$$

$$\int_0^s [X(t)]^n\,dX(t) = \frac{1}{n+1}[X(s)]^{n+1} - \frac{n}{2}\int_0^s [W(t)]^{n-1}\,dt, \tag{21}$$

and

$$\int_0^s \exp(\lambda X(t))\,dX(t) = \frac{1}{\lambda}[\exp(\lambda X(s)) - 1] - \frac{\lambda}{2}\int_0^s \exp(\lambda X(t))\,dt. \tag{22}$$

Observe that the last term in these formulae would have been absent if $X(t)$ were a function of bounded variation. These formulae are special cases of the celebrated Itô formula and the second term in each of them is called *Itô's correction* [7].

If X is a continuous semimartingale, the stochastic integral $\int g\,dX$ can be defined along these lines (see ref. 11).

As remarked earlier, in general it is not possible to give a pathwise formula for $(\int_0^s g(t)\,dX(t))$, i.e., a formula expressing $[\int_0^s g(t)\,dX(t)](\omega)$ in terms of $\{g(u, \omega), X(u, \omega) : u \leqslant s\}$. However, such a pathwise formula can be given if we assume in addition that for all ω, $g(t, \omega)$ is a right continuous function with left limits (see refs. 1 and 10). The pathwise formula is important from the point of view of statistical applications. In the example discussed at the beginning, it shows how to compute the likelihood function $L(\theta, Y)$ for a given observation path Y.

For applications of stochastic integration to statistical problems, see refs. 9 and 12.

References

[1] Bicheler, K. (1981). *Ann. Prob.*, **9**, 49–89.

[2] Doob, J. L. (1953). *Stochastic Processes.* Wiley, New York.

[3] Dellacherie, C. and Meyer, P. A. (1982). *Probabilities and Potential B.* North-Holland, Amsterdam, The Netherlands.

[4] Friedman, A. (1975). *Stochastic Differential Equations and Applications*, Vol. I. Academic, New York.

[5] Ikeda, N. and Watanabe, S. (1981). *Stochastic Differential Equations and Diffusion Processes.* North-Holland, Amsterdam, The Netherlands.

[6] Itô, K. (1951). On Stochastic Differential Equations. *Mem. Amer Math. Soc.*, **4**.

[7] Itô, K. (1961). *Lectures on Stochastic Processes.* Tata Institute of Fundamental Research, Bombay, India.

[8] Jacod, J. (1979). Calcul Stochastique et Problèmes de Martingales. *Lect. Notes Math.*, **714**. Springer, New York.

[9] Kallianpur, G. (1980). *Stochastic Filtering Theory.* Springer, New York.

[10] Karandikar, R. L. (1981). *Sankhyā*, **43**, 121–132.

[11] Karandikar, R. L. (1983). *Stoch. Proc. Appl.*, **15**, 203–209.

[12] Lipster, R. Sh. and Shiryayev, A. N. (1977). *Statistics of Random Processes*, Vol. I. Springer, New York.

[13] Meyer, P. A. (1976). Un Cours sur les Integrales Stochastique. *Lect. Notes Math.*, **511**. Springer, New York.

(BROWNIAN MOTION
DIFFUSION PROCESSES
STOCHASTIC DIFFERENTIAL EQUATIONS
STOCHASTIC PROCESSES)

RAJEEVA L. KARANDIKAR

STOCHASTIC MATRIX

1. A matrix with elements (p_{ij}) such that for each row (i), $p_{i1} + p_{i2} + \cdots = 1$, and $0 \leqslant p_{ij}$.
2. A matrix with elements, some of which, at least, are random variables.

The first meaning is the more common one.

STOCHASTIC MECHANICS

Stochastic mechanics is an interpretation of solutions of the equations of quantum mechanics* in terms of stochastic processes*. Thus it establishes a correspondence between two very different subjects, a link that may be useful in giving probabilists an insight into modern physics.

Under consideration is the nonrelativistic quantum mechanics of particles, the only part of quantum mechanics that is presently understood from the point of view of rigorous mathematics. This subject cannot be understood in terms of classical probability theory, where random variables are defined on a sample space, corresponding to all possible outcomes of an experiment. In quantum mechanics the various random variables (usually called *observables* in this context) are typically defined on different sample spaces. The idea is that the choice of experiment to be performed defines the sample space*, and the choice of one experiment precludes the choice of another. Thus two or more random variables need not have a joint distribution. For example, in the usual interpretation of quantum mechanics, the random variables $X(t_1), \ldots, X(t_n)$ corresponding to the position of a particle at times $t_1, \ldots, t_n$ do not have a joint distribution. This makes it impossible to conceive of trajectories of the particle.

In stochastic mechanics the individual position random variables $X(t_1), \ldots, X(t_n)$ have the same distributions as in quantum mechanics. However, their relation to each other is quite different; in particular they are all defined on the same sample space and have a joint distribution. The trajectory X of a particle is a random continuous path, governed by a Markov diffusion process* with transition probabilities that depend on time in a complicated way.

BACKGROUND

Stochastic mechanics has its origin in work of Fényes, Nelson, and others. The book by Nelson [6] is an excellent introduction. His lectures [7] contain additional references. Dankel [2] extended the theory to include spin. Shucker [8] has analyzed position measurement. Many other authors have contributed; references to some of this work may be found in recent conference proceedings [1, 4].

Alternatives to quantum mechanics have been a source of speculation ever since the invention of quantum mechanics. Stochastic mechanics is an example of a class of alternative theories called *nonlocal hidden variable theories* [5]. Many physicists have objected to such theories on both experimental and philosophical grounds.

In the case of stochastic mechanics experimental disproof is difficult. The reason is that stochastic mechanics and quantum mechanics agree in their predictions of position measurements at fixed time. The most common type of measurement is a scattering experiment, and such a measurement may be interpreted as a measurement of whether the position of the particle is in a cone specified by a solid angle from the scatterer [3]. It is thus difficult to distinguish the two theories. In view of the spectacular success of quantum mechanics, an experimental refutation of either theory would be a surprise.

The philosophical objection is that it is difficult or impossible to observe hidden variables, and so for economy of thought they must be presumed to be meaningless or not to exist. This is a satisfying position for those who are content with the rather mysterious world view suggested by quantum mechanics.

One can avoid these questions by taking the attitude that stochastic mechanics is an appealing mathematical structure, quite apart from questions of physical reality. There is no doubt that Markov processes* have given insights into the very classical subject of potential theory. They may turn out to be equally relevant to a deeper understanding of quantum mechanics.

EQUATIONS OF MOTION

The process that describes the motion of a particle in stochastic mechanics is determined by a complex function ψ of space and time. This function is a solution of the Schrödinger equation of quantum mechanics. The parameters in this equation are a mass $m > 0$ and Planck's constant $\hbar > 0$ (a measure of the size of fluctuations). One must also specify a potential energy function ϕ, a real function of space (and perhaps also of time). The Schrödinger equation is

$$i(\partial\psi/\partial t) = -\tfrac{1}{2}(\hbar/m)\nabla^2\psi + (1/\hbar)\phi\psi. \tag{1}$$

The interpretation of the solution ψ is somewhat mysterious, but

$$\rho = |\psi|^2 \tag{2}$$

has a clear meaning in both quantum mechanics and stochastic mechanics; it is the probability density of the particle position at given time.

Write

$$u + iv = (\hbar/m)\nabla\psi/\psi. \tag{3}$$

This defines two time-dependent real vector fields u and v, called *osmotic* and *current velocities*. The osmotic velocity u satisfies

$$u = \tfrac{1}{2}(\hbar/m)\nabla\rho/\rho, \tag{4}$$

where ρ is the position probability density (2). It is easy to see that ρ satisfies an equation of continuity

$$\partial\rho/\partial t = -\nabla\cdot(v\rho) \tag{5}$$

involving the current velocity v.

The last two equations may be combined to give a diffusion equation

$$\partial\rho/\partial t = -\nabla\cdot((u+v)\rho) + \tfrac{1}{2}(\hbar/m)\nabla^2\rho. \tag{6}$$

This suggests defining the stochastic process as the Markov diffusion process* with constant diffusion parameter $\hbar$/m and variable drift $u + v$.

The process itself is a random function X of time. For every t, $\mathbf{X}(t)$ is a vector random variable that represents the position of the diffusing particle at time t. The process satisfies the Itô stochastic differential equation*

$$dX = (u(X) + v(X))\,dt + dW, \tag{7}$$

where dW is the increment of the Wiener process* with mean zero and diagonal covariance $(\hbar/m)\,dt$. Thus in each time step $dt > 0$ the particle takes a deterministic step given by the sum of the osmotic and current velocities and a random step given by the increment of the Wiener process. The random step is independent of the past, so the process is Markovian.

The Itô calculus for stochastic differentials is based on a Taylor expansion* to second order and the replacement of quadratic terms in dW by terms linear in dt, so

$$dW\cdot dW = (\hbar/m)\,dt. \tag{8}$$

For every smooth function f of position the Itô calculus and (7) give

$$\begin{aligned} d(f(X)) &= (\nabla f)(X)\cdot dX \\ &\quad + \tfrac{1}{2}\nabla\nabla f(X)\cdot dX\cdot dX \\ &= \nabla f(X)\cdot(u(X)+v(X))\,dt \\ &\quad + \tfrac{1}{2}(\hbar/m)\nabla^2 f(X)\,dt \\ &\quad + \nabla f(X)\cdot dW \end{aligned} \tag{9}$$

Taking expected values,

$$\begin{aligned} dE[f(X)]/dt &= E[\{(u+v)\cdot\nabla \\ &\quad + \tfrac{1}{2}(\hbar/m)\nabla^2\}f(X)]. \end{aligned} \tag{10}$$

Let ρ be the density of X at fixed time. Since for every function f,

$$E[f(X)] = \int f\rho\,d^n x, \tag{11}$$

(10) is equivalent to the diffusion equation (6). This shows that if the process X is started with the quantum mechanical density ρ at some fixed time, then it will continue to have the quantum mechanical density ρ in the future.

Since the current and osmotic velocities v and u play the fundamental role in stochastic mechanics, it is of interest to write the Schrödinger equation with these dependent variables. The result is

$$\begin{aligned} \{\partial/\partial t + v\cdot\nabla\}v - \{u\cdot\nabla + \tfrac{1}{2}(\hbar/m)\nabla^2\} \\ = -\nabla\phi/m. \end{aligned} \tag{12}$$

Equations (4), (5), and (12) form a complicated nonlinear system. In order to interpret these equations, it is useful to look at the two extreme cases where only the current velocity v is nonzero (the classical limit) and where only the osmotic velocity u is nonzero (stochastic equilibrium).

CLASSICAL MECHANICS

Classical mechanics is the limiting case of quantum mechanics (or of stochastic mechanics) when the parameter $\hbar$ approaches zero. It is evident from (4) that in this limit either u approaches zero or ρ becomes singular. For the moment we suppose that the limit of u is zero. (In the next section there is an example where u is nonzero except at a point, but ρ becomes concentrated at that point.)

With the assumptions $\hbar = 0$ and $u = 0$, the equation (7) for the path of the particle reduces to

$$dX/dt = v(X). \tag{13}$$

The current velocity v is determined by the limiting version of the Schrödinger equation (12), which is

$$\partial v/\partial t + (v \cdot \nabla)v = -\nabla\phi/m. \tag{14}$$

The last two equations may be combined to give

$$dv(X)/dt = -\nabla\phi(X)/m. \tag{15}$$

This is Newton's law of motion for a particle being carried along in a fluid with velocity field v. The only force acting on the fluid is the force $-\nabla\phi$ derived from the potential energy function ϕ.

STOCHASTIC EQUILIBRIUM

In stochastic mechanics the equilibrium state is complicated enough to be interesting in its own right. Assume that the potential energy ϕ is independent of time. Assume also that the solution ψ has the trivial time dependence

$$\psi = \psi_0 \exp(-i\,Et/\hbar), \tag{16}$$

where ψ_0 is independent of time and E is a real constant (the energy). The Schrödinger equation becomes

$$-\tfrac{1}{2}(\hbar/m)\nabla^2\psi_0 + (1/\hbar)(\phi - E)\psi_0 = 0. \tag{17}$$

The drift is independent of time and so the process has time-independent transition probabilities. The density

$$\rho = |\psi_0|^2 \tag{18}$$

is also independent of time, so the process is stationary.

Specialize further by assuming that the function ψ_0 is real. Then the current velocity v is zero and the drift is the osmotic velocity

$$u = (\hbar/m)\nabla\psi_0/\psi_0. \tag{19}$$

In this case the time-reversed process is the same process. This case typically arises in quantum mechanics at the lowest-energy stationary state. In stochastic mechanics it is an equilibrium maintained by fluctuations and time-reversible drift.

In this situation the Schrödinger equation may be written in terms of the osmotic velocity as

$$u \cdot \nabla u + \tfrac{1}{2}(\hbar/m)\nabla^2 u = \nabla\phi/m. \tag{20}$$

It follows from the Itô calculus and the Schrödinger equation (20) that

$$\begin{aligned}
du(X) &= dX \cdot \nabla u(X) \\
&\quad + \tfrac{1}{2} dX \cdot dX \cdot \nabla\nabla u(X) \\
&= u(X) \cdot \nabla u(X)\,dt \\
&\quad + \tfrac{1}{2}(\hbar/m)\nabla^2 u(X)\,dt \\
&\quad + dW \cdot \nabla u(X) \\
&= \nabla\phi(X)/m\,dt + (dW \cdot \nabla)u(X).
\end{aligned} \tag{21}$$

This says that the drift (in this case the osmotic velocity) tends to develop a component opposed to the force. (Recall that the force derived from a potential energy ϕ is $-\nabla\phi$, with a minus sign.) This seems reasonable for a particle coming into purely stochastic equilibrium. As the particle drifts toward equilibrium, it is no longer necessary to have a substantial drift component along the force.

In the equilibrium situation there is an especially intimate relation between the Schrödinger equation (17) and the diffusion process. In fact (17) and (19) imply that for every smooth function f,

$$\left\{\tfrac{1}{2}(\hbar/m)\nabla^2 - (1/\hbar)(\phi - E)\right\}(\psi_0 f) = \psi_0\left\{\tfrac{1}{2}(\hbar/m)\nabla^2 + u\cdot\nabla\right\}f. \quad (22)$$

In words, multiplication by ψ_0 is an isomorphism between the generator of the stochastic mechanics process (involving the drift u) and the Schrödinger operator (involving the potential ϕ). Thus many technical questions about Schrödinger operators have their exact counterpart in diffusion theory.

Example. The simplest concrete example is the lowest-energy state of the quantum mechanical harmonic oscillator. The potential energy is quadratic;

$$\phi = \tfrac{1}{2}m\omega^2x^2, \quad (23)$$

where $\omega > 0$ is a constant. The force $-\nabla\phi = -m\omega^2x$ is linear. The lowest-energy solution of the Schrödinger equation is given by (16) with

$$\psi_0 = C\exp\left(-\tfrac{1}{2}(m/\hbar)\omega x^2\right), \quad (24)$$

where C is a normalization constant and $E = \hbar\omega/2$. The stationary density ρ is Gaussian with mean zero and variance $\tfrac{1}{2}(\hbar/m)(1/\omega)$; explicitly,

$$\rho = C^2\exp\left(-m\omega x^2/\hbar\right). \quad (25)$$

The drift is the osmotic velocity

$$u = -\omega x, \quad (26)$$

which is also linear in the displacement.

The process satisfies the stochastic differential equation

$$dX = -\omega X\,dt + dW. \quad (27)$$

This is the Gaussian Markov process often referred to as the Ornstein–Uhlenbeck* velocity process. The simplest example of a stationary state in quantum mechanics corresponds exactly to the simplest example of a stationary Markov diffusion process.

References

[1] Albeverio, S., Combe, Ph., and Sirugue-Collin, M., eds. (1982). *Stochastic Processes in Quantum Theory and Statistical Physics*. Springer, Berlin, Germany. (Conference proceedings containing several papers on stochastic mechanics.)

[2] Dankel, T. G. (1970). *Arch. Rational Mech. Anal.*, **37**, 192–221. (Important technical paper developing stochastic mechanics on manifolds, thereby including spin in the theory.)

[3] Faris, W. (1982). *Foundations Phys.*, **12**, 1–26. (Discussion of spin measurement in stochastic mechanics, in connection with the Einstein–Rosen–Podolsky experiment.)

[4] Guerra, F. (1981). *Phys. Rep.*, **77**, 263–312. (General discussion of stochastic mechanics, with many references.)

[5] Jammer, M. (1974). *The Philosophy of Quantum Mechanics*. Wiley, New York. (Treatise on many aspects of quantum mechanics, including alternative hidden variable theories.)

[6] Nelson, E. (1967). *Dynamical Theories of Brownian Motion*. Princeton University, Princeton, NJ. (A delightful book. It begins with Brownian motion as a physical theory, develops the mathematics of stochastic motion, and concludes with stochastic mechanics as an alternative to quantum mechanics.)

[7] Nelson, E. (1982). In *Collected Works of Norbert Wiener*, Vol. 3, P. Masani, ed. MIT Press, Cambridge, MA. (Survey article containing more recent references on stochastic mechanics and stochastic electrodynamics.)

[8] Shucker, D. S. (1981). *J. Math. Phys.*, **22**, 491–494. (Analysis of position measurement in stochastic mechanics.)

(DIFFUSION PROCESSES
MARKOV PROCESSES
QUANTUM MECHANICS (various entries)
STOCHASTIC DIFFERENTIAL EQUATIONS
STOCHASTIC INTEGRALS
STOCHASTIC PROCESSES)

WILLIAM G. FARIS

STOCHASTIC MODELS

The journal *Stochastic Models* started publication in 1985 under the editorship of Marcel F. Neuts, and is published by Marcel Dekker, Inc. of New York. The editorial address is Marcel F. Neuts, Editor, *Stochastic Models*, Department of Systems and Industrial Engineering, University of Arizona, Tucson, AZ 85721 (USA).

In its first years of publication, *Stochastic Models* will appear in three issues, each of approximately 150 pages. It is anticipated that the number of issues will increase in later years. The journal is devoted to the dissemination of innovative material on all aspects of probabilistic models, including new theoretical contributions, algorithmic methods, numerical-experimental findings and applications of probability theory to new and to traditional areas of science and technology. Its Editorial Board reflects the international character of the research community in probabilistic modeling; its current members are S. Asmussen (Denmark), P. Brockwell (USA), F. P. Kelly (United Kingdom), D. Konig (German Democratic Republic), G. Letac (France), M. F. Neuts (USA), V. Ramasawami (USA), S. I. Resnick (USA), R. Rubinstein (Israel), Y. Takahashi (Japan), and R. Tweedie (Australia).

Upon receipt, all manuscripts submitted for *Stochastic Models* are sent to an editor for refereeing by one or two referees. When the manuscript is accepted for publication (possibly after revisions), the author is asked to prepare a final manuscript in a form suitable for photoreproduction. Specific recommendations on the preparation of final manuscripts may be found in the first issue (Instructions to Authors) or will be mailed by the Editor upon request. The Editors strongly encourage clear and informative communication between authors and readers. Authors are therefore requested to choose the level of abstraction appropriate to a rigorous but widely readable account of the subject, to include clarifying remarks where these may be helpful, to supply numerical and graphical material where appropriate, and to report on their experiences with algorithmic and experimental procedures in ways that may be helpful to readers faced with similar problems. In all cases, correct mathematical reasoning and procedure are essential.

The brief history of *Stochastic Models* does not yet permit a statistical analysis of its principal areas of content. A number of papers in the theory of queues, which continues to be one of the most productive areas of applied probability, have appeared. In many of these, the emphasis on implementable algorithmic methods and their immediate relationship to communications engineering and computer performance modeling have added a fresh element of style and a new perspective. The journal has also attracted contributions of a general theoretical nature as well as manuscripts in reliability theory, biomedical models, and the mathematics of simulation. As a general rule, the journal will not publish extensive computer codes, but a special corner devoted to the announcement of computer programs and discussions of their performance is planned for inclusion in future issues. The Editorial Board also welcomes the use of *Stochastic Models* as a vehicle for the publication of papers presented at symposia or specialized conferences in probability theory and its applications. This is with the understanding that the articles have not appeared elsewhere and that all material will be refereed prior to publication in the journal. Organizers of conferences who wish to avail themselves of this possibility are requested to write to the Editor, preferably prior to the conference, so that arrangements for speedy refereeing and publication can be worked out.

The contents of the first two issues of *Stochastic Models* are as follows:

"The relaxation time of two queueing systems in series" by J. P. C. Blanc.

"Sequencing two servers on a sphere" by A. R. Calderbank, E. G. Coffman, Jr., and L. Flatto.

"Efficient algorithms for solving the nonlinear matrix equations arising in phase type queues" by D. M. Lucantoni and V. Ramaswami.

"A state–age dependent policy for a shock process" by R. M. Feldman and N. Y. Joo.

"A survey of functional laws of the iterated logarithm for self-similar processes" by M. S. Taqqu and C. Czado.

"Simple bounds on system reliability function" by I. Gertsbakh.

"Stationary waiting time distribution in queues with phase type service and in quasi-birth-and death processes" by V. Ramaswami and D. M. Lucantoni.

"An exponential semi-Markov process, with applications to queueing theory" by G. Latouche.

"A stochasticproduction/inventory system with all-or-nothing demand and service measures" by A. G. DeKok and H. C. Tijms.

"Integer-valued self-similar processes" by K. van Harn and F. W. Steutel.

"A probabilistic analysis of multiprocessor list scheduling: the Erlang case" by O. J. Boxma.

"The queue GI/GI/k: finite moments of the cycle variables and uniform rates of convergence" by H. Thorisson.

"Certain Volterra integral equations arising in queueing" by D. Jagerman.

"Waiting times in the non-preemptive priority M/M/c queue" by O. Kella and U. Yechiali.

MARCEL F. NEUTS

STOCHASTIC ORDERING

With a general definition, a stochastic ordering is any way to compare probability distributions, i.e., any binary relation in a set of probability distributions. The most common interpretation is that one distribution in some sense attaches more probability to larger values than the other does. A second interpretation is that one distribution is more spread out or dispersed than the other; see related entries on ordering distributions in dispersion*, convexity*, majorization*, and star-shaped ordering*. Other interpretations combine these notions. Each field provides its own special flavor; see Barlow et al. [2], Barlow and Proschan [3], Bawa [4], Marshall and Olkin [14], Stoyan [24], Ziemba and Vickson [3], and ORDERING OF DISTRIBUTIONS, PARTIAL.

We discuss a standard, more restricted definition, requiring that the probability distributions be defined on a common underlying sample space S that is endowed with a partial order relation $\leqslant$, e.g., the real line or Euclidean space R^k. We say that one probability distribution P_1 is *stochastically less than or equal* to another P_2 and write $P_1 \leqslant_{\mathrm{st}} P_2$ if

$$\int_S h\,dP_1 \leqslant \int_S h\,dP_2 \tag{1}$$

for all nondecreasing real-valued functions h on S for which the expectations (integrals) are well defined. When the sample space S is the real line, then (1) holds if and only if

$$P_1((t,\infty)) \leqslant P_2((t,\infty)) \quad \text{for all } t. \tag{2}$$

For random variables or random elements on S, $X_1 \leqslant_{\mathrm{st}} X_2$ if their associated probability distributions satisfy (1), i.e., if $Eh(X_1) \leqslant Eh(X_2)$ for all nondecreasing real-valued functions h on S for which the expectations are well defined. Obviously, if $X_1 \leqslant_{\mathrm{st}} X_2$, then $h(X_1) \leqslant_{\mathrm{st}} h(X_2)$ for nondecreasing h.

Stochastic ordering (1) on the real line obviously implies that the means are ordered, but not conversely. The main idea is to have a stronger comparison than can be provided by a single-number description of location* such as the mean, median, or mode*. As a consequence, every pair of distributions need not be ordered.

The history of stochastic ordering is not well documented. The notion based on (2) was used in 1947 by Mann and Whitney [13] to characterize the alternatives when testing the equality of two distributions. Serious theoretical investigation seems to have been initiated in 1955 by Lehmann [11]; this is the first frequently cited reference. However, the closely related concept of utility* goes back to Bernoulli* in 1713; see ref. 4. For the history of other related concepts, see Marshall and Olkin [14].

STOCHASTIC ORDERINGS ON GENERAL SPACES

When the sample space is the real line, (1) and (2) are equivalent, but (1) is the preferred definition because it extends conveniently and appropriately to more general sample spaces. On Euclidean space R^k, (1) implies both

$$P_1((-\infty, t_1] \times \cdots \times (-\infty, t_k]) \geqslant P_2((-\infty, t_1] \times \cdots \times (-\infty, t_k]) \quad (3)$$

for all $(t_1, \ldots, t_k) \in R^k$ and

$$P_1((t_1, \infty) \times \cdots \times (t_k, \infty)) \leqslant P_2((t_1, \infty) \times \cdots \times (t_k, \infty)) \quad (4)$$

for all $(t_1, \ldots, t_k) \in R^k$, but for $k > 1$ neither (3) nor (4) implies the other and both together do not imply (1).

A subset A of S is said to be *increasing* if $y \in A$ whenever $x \in A$ and $x \leqslant y$ in S. Under general conditions on $(S, \leqslant)$, (1) is equivalent to

$$P_1(A) \leqslant P_2(A) \quad (5)$$

for all increasing open subsets A; see Kamae et al. [8].

Stochastic ordering of probability distributions on more general partially ordered spaces is of interest to compare stochastic processes*. For stochastic processes, S is a space of possible sample paths, such as the function spaces R^∞, $C[0, \infty)$, and $D[0, \infty)$; see ref. 8. To illustrate, consider $D[0, \infty)$, the space of right-continuous real-valued functions on $[0, \infty)$ with left limits, with elements $x_i \equiv \{x_i(t), t \geqslant 0\}$, where $x_1 \leqslant x_2$ if $x_1(t) \leqslant x_2(t)$ for all t. As stated before, $X_1 \leqslant_{st} X_2$ for stochastic processes $X_i \equiv \{X_i(t), T \geqslant 0\}$ in $D[0, \infty)$ if $Eh(X_1) \leqslant Eh(X_2)$ for all nondecreasing real-valued functions h on $D[0, \infty)$ for which the expectations are well defined. This means that the associated probability distributions on $D[0, \infty)$ satisfy (1). It is well known that the probability distribution P_i of X_i is determined by its finite-dimensional distributions. It turns out that stochastic order for such stochastic processes is also determined by the finite-dimensional distributions, i.e., $X_1 \leqslant_{st} X_2$ if and only if

$$[X_1(t_1), \ldots, X_1(t_k)] \leqslant_{st} [X_2(t_1), \ldots, X_2(t_k)] \quad (6)$$

for all $(t_1, \ldots, t_k) \in R^k$ and all k; see ref. 8. Usually the functions in $D[0, \infty)$ are real valued, but for stochastic orderings they can also take values in a general partially ordered space [8].

Stochastic orderings have also been defined and investigated for probability distributions on sample spaces endowed with an intransitive ordering; see Fishburn [6] and Whitt [27].

COUPLING: CONSTRUCTING ORDERED RANDOM ELEMENTS

A convenient device for obtaining stochastic orderings is a coupling, which is a special construction, given $P_1 \leqslant_{st} P_2$, of two random elements X_1 and X_2 mapping a common underlying probability space Ω into the space S such that $X_1(\omega) \leqslant X_2(\omega)$ for all $\omega \in \Omega$ with each X_i having distribution P_i. Strassen [25] showed that such a coupling always exists if $P_1 \leqslant_{st} P_2$ in the sense of (1) when S is a complete separable metric space endowed with a closed partial order; see ref. 8. However, the proof that employs the Hahn–Banach theorem is not constructive, so that in general such a coupling is hard to produce.

When S is the real line, the coupling is easy to construct using inverse CDFs (cumulative distribution functions)*. If F_i is the CDF of P_i and $P_1 \leqslant_{st} P_2$, then the coupling is obtained by setting $X_i = F_i^{-1}(U)$ for each i, where U is a random variable uniformly distributed on $[0, 1]$ and $F_i^{-1}(x) = \inf\{t: F_i(t) > x\}$; X_i so constructed has CDF F_i and $F_1^{-1}(x) \leqslant F_2^{-1}(x)$ for all x by virtue of (1). This result is contained in the early paper by Lehmann [11].

Stochastic orderings for stochastic processes via couplings were first obtained by O'Brien [18], primarily for Markov chains. For some continuous-time stochastic processes, couplings and associated stochastic orderings can be constructed by appropriately thinning a Poisson process*; see Sonderman [23] and Whitt [29]. Sometimes the coupling can be achieved by an ingenious modification of the model. For example, Wolff [32] compares different rules for assigning customers to servers in a multiserver queue*, and shows that the first-come–first-served rule minimizes congestion, by assigning successive service times in the order customers begin service instead of the order of arrival. This approach is further exploited by Smith and Whitt [22] to study the efficiency of combining separate service systems.

There are several other couplings for different purposes. One is the Skorohod representation theorem*, which replaces convergence in distribution by convergence with probability 1; see Wichura [31] and PROCESSES, EMPIRICAL. Another helps study the convergence to steady state by having a nonstationary process eventually coupled to a stationary version; see Lindvall [12].

ESTABLISHING STOCHASTIC ORDER

It is not difficult to establish new stochastic orderings from given ones using various preservation theorems [8]. For example, suppose that $X_k \leqslant_{st} Y_k$ for each k. First, $(X_1, \ldots, X_n) \leqslant_{st} (Y_1, \ldots, Y_n)$ for each n and thus $X_1 + \cdots + X_n \leqslant_{st} Y_1 + \cdots + Y_n$ if the random elements $X_1, \ldots, X_n$ are independent and the random elements $Y_1, \ldots, Y_n$ are independent. Second, $X \leqslant_{st} Y$ if X_k converges in distribution to X and Y_k converges in distribution to Y as $k \to \infty$. The related issue of continuity for the subset of undominated elements in a stochastic ordering is investigated by Goroff and Whitt [7].

For Markov chains and other stochastic processes, it is also important to make comparisons when the marginal distributions are dependent. As shown by Veinott [26], this can be done using conditional distributions. If $X_1 \leqslant_{st} Y_1$ and

$$(X_k | X_1 = x_1, \ldots, X_{k-1} = x_{k-1}) \leqslant_{st} (Y_k | Y_1 = y_1, \ldots, Y_{k-1} = y_{k-1}) \quad (7)$$

with probability 1 whenever $x_k \leqslant y_k$ for all k, $2 \leqslant k \leqslant n$, then $(X_1, \ldots, X_n) \leqslant_{st} (Y_1, \ldots, Y_n)$. O'Brien [18] also established this result via a coupling. Kamae et al. [8] showed that each random element can take values in a general space.

STOCHASTIC ORDERING AND CONDITIONING

Stochastic ordering is useful in conjunction with conditioning. A *uniform conditional stochastic ordering* holds between two distributions P_1 and P_2 if the associated conditional distributions, conditioning on some subset, are stochastically ordered for all subsets in some class; see Keilson and Sumita [10], Simons [21], and Whitt [28]. The conditioning events might correspond to new information; interesting applications in economics are contained in Milgrom [16] and Milgrom and Weber [17]. The conditioning might also correspond to the histories of a stochastic process, which of course also corresponds to new information systematically evolving in time; see Arjas [1].

When the sample space is the real line, uniform conditional stochastic order, given all subsets or just all subintervals, corresponds to the *monotone likelihood ratio* ordering**, which has many applications; see ref. [10]. A multivariate extension is introduced and investigated by Karlin and Rinott [9]. Connections to uniform conditional stochastic order are discussed in Whitt [30].

APPLICATIONS

Stochastic orderings have numerous applications, as the references testify; e.g., Stoyan [24] gives a comprehensive account of applications in queueing theory*. A typical application there is to make the idea that congestion increases when customers arrive more quickly or are served more slowly precise. Other important application areas are reliability* theory (see El-Neweihi [5]) and stochastic scheduling (see Ross [19]). Stochastic orderings also have an important place in the foundations of statistics, economics, and operations research*; see Savage [20] and Massé [15, p. 206]. In decision-making contexts the functions h in (1) typically represent utility functions, which are often assumed to be increasing and concave. This stochastic ordering is often called *stochastic dominance* [4]. When alternatives are stochastically ordered in this sense, the preferences are determined for all risk averse decision makers, without having to identify their specific utility functions. Alternatives that are stochastically dominated in this way are inadmissible and need not be considered further.

References

[1] Arjas, E. (1981). *Math. Operat. Res.*, **6**, 263–276. (Discusses uniform conditional stochastic orderings and applications in reliability theory.)

[2] Barlow, R. E., Bartholomew, D. J., Bremner, J. M., and Brunk, H. D. (1972). *Statistical Inference Under Order Restrictions*. Wiley, New York. (The standard stochastic ordering plays a relatively minor role here, but this is an important reference for order in statistics.)

[3] Barlow, R. E. and Proschan, F. (1975). *Statistical Theory of Reliability and Life Testing*. Holt, Rinehart and Winston, New York. (The various notions of aging can be viewed as stochastic orderings.)

[4] Bawa, V. S. (1982). *Manag. Sci.*, **28**, 698–712. (An extensive bibliography focusing especially on economics, utility theory and finance.)

[5] El-Neweihi, E. (1981). *Commun. Statist. A*, **10**, 1655–1672. (Uses stochastic orderings to characterize aging properties of multivariate life distributions in reliability theory.)

[6] Fishburn, P. C. (1978). *Manag. Sci.*, **24**, 1268–1277. (Introduces a stochastic ordering on a sample space with an intransitive order relation.)

[7] Goroff, D. and Whitt, W. (1980). *J. Econ. Theory*, **23**, 218–235. (Studies continuity of stochastic orderings: Determines conditions so that if two sets of probability distributions are close, the subsets of undominated distributions are also close.)

[8] Kamae, T., Krengel, U., and O'Brien, G. L. (1977). *Ann. Prob.*, **5**, 899–912. (Reviews and extends the theory of stochastic order for probability distributions on a complete separable metric space endowed with a closed partial order. A basic reference for stochastic ordering of stochastic processes.)

[9] Karlin, S. and Rinott, Y. (1980). *J. Multivariate Anal.*, **10**, 467–498. (Introduces and investigates multivariate monotone likelihood ratio orderings.)

[10] Keilson, J. and Sumita, U. (1983). *Canad. J. Statist.*, **10**, 181–198. (Discusses uniform conditional stochastic orderings.)

[11] Lehmann, E. (1955). *Ann. Math. Statist.*, **26**, 399–404. (A seminal paper.)

[12] Lindvall, T. (1979). *J. Appl. Prob.*, **16**, 502–505. (Describes a coupling, also exploited by D. Griffeath and J. W. Pitman, to study convergence to the steady state.)

[13] Mann, H. B. and Whitney, D. R. (1947). *Ann. Math. Statist.*, **18**, 50–60. (Uses stochastic ordering to characterize the alternatives in hypothesis testing.)

[14] Marshall, A. W. and Olkin, I. (1979). *Inequalities: Theory of Majorization and Its Applications*. Academic, New York. (The primary focus is on ordering distributions in dispersion, but there also is much related to standard stochastic orderings.)

[15] Massé, P. (1962). *Optimal Investment Decisions*. Prentice-Hall, Englewood Cliffs, NJ. (A broad clear view of economics and operations research.)

[16] Milgrom, P. R. (1981). *Bell J. Econ.*, **12**, 380–391. (Uses uniform conditional stochastic orderings to describe responses to new information in economics models.)

[17] Milgrom, P. R. and Weber, R. J. (1982). *Econometrica*, **50**, 1089–1122. (Applies uniform conditional stochastic orderings and multivariate monotone likelihood ratio to study auctions and competitive bidding.)

[18] O'Brien, G. L. (1975). *Ann. Prob.*, **3**, 80–88. (Establishes conditions for stochastic orderings of Markov chains and other stochastic processes in terms of orderings of initial distributions and transition kernels. Constructs a coupling for stochastic processes directly and simply.)

[19] Ross, S. (1983). *Introduction to Stochastic Dynamic Programming*. Academic, New York. (Chapter 6 contains a nice introduction to stochastic scheduling, where stochastic orderings are applied.)

[20] Savage, L. J. (1972). *The Foundations of Statistics*, 2nd ed. Dover, New York. (Stochastic orderings are not discussed, but their relevance is evident in this framework.)

[21] Simons, G. (1980). *Ann. Statist.*, **8**, 833–839. (Exposes the relationships between monotone likelihood ratio and uniform conditional stochastic order and gives applications to likelihood ratio tests.)

[22] Smith, D. R. and Whitt, W. (1981). *Bell Syst. Tech. J.*, **60**, 39–55. (Exploits couplings and monotone likelihood ratios to discuss the efficiency of combining separate service facilities.)

[23] Sonderman, D. (1980). *Math. Operat. Res.*, **5**, 110–119. (Obtains couplings and comparisons for semi-Markov processes by applying ref. 18 and appropriately thinning Poisson processes.)

[24] Stoyan, D. (1983). *Comparison Methods for Queues and Other Stochastic Models*, D. J. Daley, ed. Wiley, New York. (An English translation of the 1977 German edition. A basic reference for queueing theory.)

[25] Strassen, V. (1965). *Ann. Math. Statist.*, **36**, 423–439. (Establishes in Theorem 11 the existence of an ordered coupling on general partially ordered spaces.)

[26] Veinott, A. F., Jr. (1965). *Operat. Res.*, **13**, 761–778. (Establishes stochastic orderings for vectors in terms of conditional distributions and gives applications in inventory theory.)

[27] Whitt, W. (1979). *Manag. Sci.*, **25**, 505–511. (Studies stochastic orderings on a sample space with an intransitive order relation.)

[28] Whitt, W. (1980). *J. Appl. Prob.*, **17**, 112–123. (Introduces and investigates uniform conditional stochastic orderings.)

[29] Whitt, W. (1981). *Adv. Appl. Prob.*, **13**, 207–220. (Discusses stochastic orderings for stochastic point processes and applications to queues.)

[30] Whitt, W. (1982). *J. Appl. Prob.*, **19**, 695–701. (Discusses the connection between multivariate monotone likelihood ratio and uniform conditional stochastic order.)

[31] Wichura, M. J. (1970). *Ann. Math. Statist.*, **41**, 284–291. (Extends the Skorohod representation theorem, which is a different kind of coupling.)

[32] Wolff, R. W. (1977). *J. Appl. Prob.*, **14**, 884–888. (Uses an ingenious coupling to compare different rules for assigning customers to servers in a multiserver queue.)

[33] Ziemba, W. T. and Vickson, R. G. (1975). *Stochastic Optimization Models in Finance*. Academic, New York. (An anthology in finance related to stochastic ordering, which is referred to as stochastic dominance.)

(GEOMETRY IN STATISTICS: CONVEXITY
INEQUALITIES ON DISTRIBUTIONS: BIVARIATE AND MULTIVARIATE
MAJORIZATION AND SCHUR CONVEXITY
ORDERING DISTRIBUTIONS BY DISPERSION
ORDERING OF DISTRIBUTIONS, PARTIAL
ORDERING, STAR-SHAPED
UTILITY THEORY)

WARD WHITT

STOCHASTIC POPULATION MODEL

See BIRTH AND DEATH PROCESSES; STOCHASTIC DEMOGRAPHY

STOCHASTIC PROCESSES

INTRODUCTION

[In this entry the following abbreviations are used: a.e. = almost everywhere; i.i.d. = independent and identically distributed; w.p.1 = with probability 1.]

According to the most general definition, a stochastic process is any family of random variables. The word *stochastic* is of Greek origin (στοχαστικοξ) and means "to guess." The expressions *chance process* and *random process* are sometimes used as synonymous with the expression *stochastic process*.

More precisely, a stochastic process is a family $X = (X_t,\ t \in \mathscr{I})$, where $\mathscr{I}$ is an index set, of random variables defined on a probability space $(\Omega, \mathscr{F}, \mathbb{P})$ and taking values in the real line $\mathbb{R}$, or a countable set; the stochastic process is then said to have a continuous or a discrete *state space*, respectively. The term "stochastic process" is usually applied to an infinite family. When $\mathscr{I}$ is denumerable, the process is a *discrete parameter process*, for example, a random sequence $(X_n, n = 0, 1, \dots)$. When $\mathscr{I}$ is noncountable, the process is a *continuous parameter process*, for example, a random function $(X_t, 0 \leqslant t < \infty)$. In the case of *random fields**, t is a vector.

For a fixed point $\omega \in \Omega$, a function of t defined by $X_\cdot(\omega) \equiv X(\cdot, \omega)$ is a *sample function* (a sample sequence if $\mathscr{I}$ is denumerable) or a *path*, a *trajectory*, or a *realization* of the process X.

In physical applications, t usually represents time, and the empirical approach describes the stochastic process as any process in nature whose evolution (in time) depends on chance*. Then, random variables represent observations ("states" prevailing) at time t, with values $X_t(\omega) \equiv X(t, \omega)$. Therefore, the principal objects of investigations are properties of random variables X_t and of trajectories of a process. Probabilistic conditions are imposed on random variables X_t to characterize special classes of processes and to investigate behavior of sample functions.

Let $X = (X_t, 0 \leqslant t < \infty)$ be a real-valued stochastic process on a probability space $(\Omega, \mathscr{F}, \mathbb{P})$. The joint distribution functions for any finite collection of random variables of the family X are defined by

$$F_{t_1, \dots, t_n}(x_1, \dots, x_n) = \mathbb{P}(X_{t_1} \leqslant x_1, \dots, X_{t_n} \leqslant x_n)$$

for any finite collection of indices $t_1, \dots, t_n$ and for any real numbers $x_1, \dots, x_n$. The following comments should be made here. Suppose, conversely, that a family of such finite-dimensional distributions is given. Does there exist a stochastic process whose joint distributions coincide with those of the family? A positive answer, under certain conditions, is provided by the Daniell–Kolmogorov extension theorem*. However, these finite-dimensional distributions are not sufficient to determine probabilities of sets defined by a noncountable number of random variables; for example, a set of continuous sample functions or a set where $\lim_{s \to t} X_s$ exists. Additional assumptions, like separability of a process, must be introduced in order to determine probabilities of such events. For further discussion of these delicate points, see Mathematical Interest below.

ILLUSTRATIONS

The following well-known examples illustrate basic concepts already mentioned and those to be mentioned in later sections. These classical processes have long interesting histories and represent results of the work of many researchers. The reader will find the complete discussion in almost every text on stochastic processes (see the Bibliography).

Example 1. Random Walk*. Let Z_n, $n = 1, 2, \dots$, be a family of i.i.d. random variables (RVs) assuming values 1 and -1 with probability p and q, respectively. Define partial sums

$$X_n = Z_1 + \cdots + Z_n \quad \text{for } n \geqslant 1, \text{ and } X_0 = 0.$$

The stochastic process

$$X = (X_n, n = 0, 1, \dots)$$

has a discrete state space (set of all integers), a discrete parameter (it is customary to use the letter n here instead of t) and its RVs X_n represent a position at the nth step of the walk (starting from 0), with Z_n representing a single step to the right or to the left. A typical sample sequence may look like this: $0, 1, 2, 1, 0, -1, 0, -1, -2, \dots$.

It is known that X is a discrete parameter Markov chain, with one-step transition probabilities

$$\mathbb{P}(X_{n+1} = i + 1 | X_n = i) = p,$$
$$\mathbb{P}(X_{n+1} = i - 1 | X_n = i) = q$$

for $i = 0, \pm 1, \pm 2, \dots$. Hence

$$\mathbb{E}(X_{n+1} | X_n) = X_n + (p - q),$$

and X is a martingale* when $p = q$.

The random walk X is transient when $p \neq q$ (and then will drift to $\pm\infty$), but is recurrent null when $p = q$ (and will then oscillate); *see* MARKOV PROCESSES. In fact

$$\lim_{n\to\infty} X_n = \begin{cases} \infty & \text{w.p.1 when } p > q, \\ -\infty & \text{w.p.1 when } p < q, \end{cases}$$

and for every integer i,

$$\mathbb{P}(X_n = i \text{ infinitely often}) = 1 \qquad \text{when } p = q.$$

See ref. 8 (Vol. 1, Chap. XIV and Vol. 2, Chap. XII).

Example 2. Poisson Process*. This is an integer-valued stochastic process $X = (X_t, 0 \leqslant t < \infty)$, which is a time-homogeneous Markov chain (with a discrete state space and a continuous parameter) specified by transition probabilities

$$p_{ij}(t) = \mathbb{P}(X_{d+s} = j | X_s = i), \qquad t \geqslant 0,\ i, j = 0, 1, 2, \ldots,$$

of the form

$$p_{ij}(t) = \begin{cases} \dfrac{(\lambda t)^{j-i}}{(j-i)!} e^{-\lambda t}, & j \geqslant i, \\ 0, & j < i \end{cases}$$

(where λ is a constant), and by the initial distribution $\mathbb{P}(X_0 = i) = \pi(i)$. The finite-dimensional distributions take the form

$$\mathbb{P}(X_{t_1} = j_1, \ldots, X_{t_n} = j_n) = \mathbb{P}(X_{t_1} = j_1) \prod_{k=2}^{n} p_{j_{k-1} j_k}(t_k - t_{k-1}),$$

where

$$\mathbb{P}(X_t = j) = \sum_i \pi(i) p_{ij}(t).$$

Evidently, for $t \geqslant 0$,

$$\mathbb{E}(X_{t+s} | X_s) = \lambda t + X_s,$$

and it follows that the process $(X_t - \lambda t, 0 \leqslant t < \infty)$ is a martingale. Moreover, the increments $X_{t+s} - X_s$ (for nonoverlapping time intervals) are independent RVs with the Poisson distribution with mean λt. Hence, taking $X_0 = 0$ with probability 1, X is a process with independent increments (an alternative definition of a Poisson process takes this property as the defining requirement).

The following is the basic result: For any state i and for $h \geqslant 0$,

$$\mathbb{P}(X_s = i, t \leqslant s \leqslant t + h | X_t = i) = e^{-\lambda h}.$$

The preceding event is determined by a noncountable number of coordinates (time points s); therefore, the joint distributions at a countable number of points are not sufficient to determine its probability; hence the notion of separability must be invoked.

Sample functions $X(\cdot, \omega)$ of the (separable) Poisson process are almost all nondecreasing step functions with jumps of magnitude 1 (and may be taken to be right continuous with left limits), and constant between jumps. Furthermore, for each t, $X_s \to X_t$ with probability 1 as $s \to t$, so almost all sample functions are continuous at each t (with an exceptional set depending on t). Yet, as noted above, the Poisson process is not sample continuous for all t.

In applications, it is convenient to refer to jumps of sample functions as *events*. Then X_t represents the number of events (arrivals, renewals, births, etc.) occurring up to time t. Let $T_1, T_2, T_3, \ldots$ be the enumeration of the consecutive jumps. Then, $T_1, T_2 - T_1, T_3 - T_2, \ldots$ are independent identically distributed RVs having an exponential distribution with parameter λ. In this sense, the Poisson process is a point process (*see* STOCHASTIC PROCESSES, POINT). Alternatively, the renewal process with exponential lifetimes yields the Poisson process for the number X_t of renewals up to time t. See refs. 1 (Sec. 23) and 7, (Chap. VIII, Sec. 4).

Example 3. Brownian Motion*. This is a real-valued stochastic process $X = (X_t, 0 \leqslant t < \infty)$ with independent Gaussian increments $\Delta_t X(s) = X(t+s) - X(s)$ with

$$\mathbb{E}\Delta_t X(s) = 0, \qquad \mathbb{E}|\Delta_t X(s)|^2 = \sigma^2 |t|,$$

where σ is a positive constant. It is further assumed that the process starts initially from zero, with probability 1. Hence *Brownian motion* is a Markov process and also a martingale, and the transition density of $X(t+s)$, given $X(s)$, coincides with the Gaussian density of the increment $\Delta_t X(s)$ for $t > 0$. The transition functions (for real x and y) are

$$\mathbb{P}(X(t+s) \leqslant x | X(s) = y) = \int_{-\infty}^{x} f_t(z|y)\, dz,$$

where

$$f_t(z|y) = (2\pi t)^{-1/2}\sigma^{-1} \times \exp\{-(z-y)^2/(2\sigma^2 t)\}, \quad t > 0.$$

The mathematical theory of Brownian motion (also called the *Wiener process*) was developed by Wiener, who showed that almost all sample functions of a (separable) Brownian motion are continuous and nowhere differentiable. Actually, Wiener used the finite-dimensional joint distributions to construct a probability measure (the Wiener measure) on the space of all continuous paths of the process. For full discussion and applications see refs. 1 (Sec. 37), 4 (Chap. 4), and 7 (Chap. VIII); also refs. 14 and 9.

CLASSIFICATIONS

Classifications of stochastic processes may be arranged according to diverse criteria, which basically involve

1. probabilistic structure,
2. special processes,
3. methods and problems,
4. applications.

The notorious division into pure and applied fields is predominant in stochastic processes, although this distinction is somewhat blurred if only mathematical tools are considered, but is clearly visible in the object of interest. Here only the most typical stochastic processes will be mentioned briefly; for details see the appropriate entries in this encyclopedia.

For more precise definitions, events associated with a process should be specified. This is done as follows.

Take for $\mathscr{I}$ the positive real line $\mathbb{R}_+ = [0, \infty)$, and let $(\mathscr{F}_t,\ 0 \leqslant t < \infty)$ be an increasing family of σ fields of events in $(\Omega, \mathscr{F})$, i.e.,

$$\mathscr{F}_s \subset \mathscr{F}_t \subset \mathscr{F} \quad \text{for } s \leqslant t.$$

The family is right continuous when $\mathscr{F}_t = \mathscr{F}_{t+}$ for all T. The process $(X_t,\ 0 \leqslant t < \infty)$ is *adapted* to the family $(\mathscr{F}_t,\ 0 \leqslant t < \infty)$ if X_t is $\mathscr{F}_t$ measurable for every $t \in \mathbb{R}_+$. $\mathscr{F}_t$ is the *σ field of events prior to t*, and the family $(\mathscr{F}_t)$ is often referred to as a *history*. In particular, every process (X_t) is adapted to the family of *natural σ fields* $\mathscr{F}_t^0 = \sigma(X_s,\ s \leqslant t)$ generated by random variables of the process. In general, a process is taken with respect to a specified history, so in symbols $X = (X_t, \mathscr{F}_t, 0 \leqslant t < \infty)$; in the case of natural σ fields reference to $\mathscr{F}_t^0$ may be omitted from the notation. The conditional expectations $\mathbb{E}(X_{t+h}|\mathscr{F}_t)$ and conditional probabilities $\mathbb{P}(X_{t+h} \leqslant x|\mathscr{F}_t)$ are defined in the usual way (*see* CONDITIONAL PROBABILITY AND EXPECTATION), and in the case of natural σ fields the simpler notation, $\mathbb{E}(X_{t+h}|X_s,\ s \leqslant t)$ and $\mathbb{P}(X_{t+h} \leqslant x|X_s,\ s \leqslant t)$, respectively, is used.

The main types of stochastic processes are Markov processes, martingales*, stationary processes, and processes with independent increments. This classification is not exclusive; note that those mentioned in Examples 1–3 are all Markovian and have independent increments. The basic definitions may now be stated as follows.

A process $X = (X_t, \mathscr{F}_t, 0 \leqslant t < \infty)$ is said to be:

(i) a *Markov process* when

$$\mathbb{P}(X_t \leqslant x|\mathscr{F}_s) = \mathbb{P}(X_t \leqslant x|X_s) \quad \text{a.e.}$$

for all $s \leqslant t$ and x;

(ii) a *super-martingale* when all X_t are integrable and

$$\mathbb{E}(X_t|\mathscr{F}_s) \leqslant X_s \quad \text{a.e.}$$

for $s \leqslant t$ (X is a martingale if equality holds above);

(iii) a *stationary process* when

$$\mathbb{P}(X_{t_i} \leqslant x_i, i = 1, \ldots, n) = \mathbb{P}(X_{t_i + h} \leqslant x_i, i = 1, \ldots, n)$$

for any finite collection $t_1, \ldots, t_n$ and any h;

(iv) a *process with independent increments* when, for any $t_1 < \cdots < t_n$, the differences

$$X_{t_2} - X_{t_1}, X_{t_3} - X_{t_2}, \ldots, X_{t_n} - X_{t_{n-1}}$$

are mutually independent random variables.

It may be of interest to give a brief characterization of each of these types.

Markov Processes*

The definition of a Markov process stated above may be put into several equivalent forms suitable for specific purposes. Thus a process is said to be a Markov process when σ fields $\mathscr{F}_s$ and $\sigma(X_u, u \geqslant s)$ are conditionally independent given X_s, i.e.,

$$\mathbb{P}(M \cap W|X_s) = \mathbb{P}(M|X_s)\mathbb{P}(W|X_s) \quad \text{a.e.}$$

for every event $M \in \mathscr{F}_s$ and $W \in \sigma(X_u, u \geqslant s)$. When the Markov process is taken relative to its natural σ fields $\sigma(X_u, u \leqslant s)$, then the main definition simplifies to

$$\mathbb{P}(X_t \leqslant x|X_u, u \leqslant s) = \mathbb{P}(X_t \leqslant x|X_s) \quad \text{a.e.}$$

for any $t \geqslant s$ and x. Moreover, σ fields $\sigma(X_u, u \leqslant s)$ and $\sigma(X_u, u \geqslant s)$ appear symmetrically in the equivalent definition (involving sets M and W), expressing thus an intuitive meaning of the Markov property: The future $\sigma(X_u, u \geqslant s)$ and the past $\sigma(X_u, u \leqslant s)$ are independent, given the present X_s. Still another equivalent definition of a Markov process requires that

$$\mathbb{E}(f(X_t)|\mathscr{F}_s) = \mathbb{E}(f(X_t)|X_s) \quad \text{a.e.}, t \geqslant s,$$

for any bounded (measurable) function f.

Suppose that there exists a family of regular conditional probabilities $P_{s,t}(x, A)$ [i.e., for each $s \leqslant t$, $P_{s,t}(x, A)$ is a probability measure on Borel sets A of the real line (for fixed x), and is a measurable function of x (for fixed A)] such that

$$P_{s,t}(X_s, A) = \mathbb{P}(X_t \in A|\mathscr{F}_s) = \mathbb{P}(X_t \in A|X_s)$$

(in the sense that the term on the extreme left is a selected version of conditional probabilities indicated). The Markov property implies that the following Chapman–Kolmogorov equation* holds:

$$P_{s,t}(x, A) = \int P_{s,u}(x, dy) P_{u,t}(y, A) \qquad s \leqslant u \leqslant t.$$

The Markov process is then said to have $P_{s,t}(x, A)$ as its *transition function*, and

$$P_{s,t}(x, A) = \mathbb{P}(X_t \in A|X_s = x)$$

is the conditional probability that X_t takes a value in set A, given that $X_s = x$ when $s \leqslant t$. The Chapman–Kolmogorov equation expresses the intuitive fact that a transition from $X_s = x$ to $X_t \in A$ is decomposed into disjoint transitions from $X_s = x$, through intermediary position $X_u \in dy$, to $X_t \in A$, and that the probability of transition from y to A does not depend on x and s (by Markov property). It should be remarked that a transition function is a much simpler object than a Markov process, and that there exist Markov processes that do not possess transition functions. Moreover, it should be stressed that the Chapman–Kolmogorov equation is a consequence of a Markov property; indeed, there exist non-Markovian processes that satisfy the Chapman–Kolmogorov equation.

In analytic discussion of Markov processes, and especially in applications, one usually deals with transition functions, and the Chapman–Kolmogorov equation serves as a starting point for derivation (under additional assumptions) of equations for transition functions. Note that finite-dimensional distributions are now of the form

$$\mathbb{P}(X_{t_1} \leqslant x_1, \dots, X_{t_n} \leqslant x_n) = \int_{-\infty}^{\infty} \pi(dy_0) \int_{-\infty}^{x_1} P_{0, t_1}(y_0, dy_1) \cdots \int_{-\infty}^{x_n} P_{t_{n-1}, t_n}(y_{n-1}, dy_n),$$

where π is an initial distribution. Conversely, given a family of transition functions $P_{s,t}(x, A)$ and a distribution π, the Daniell–Kolmogorov theorem assures the existence of a Markov process whose transition functions coincide with those given.

When transition functions depend only on differences $t - s$, $P_{s,t}(x, A) = P_{t-s}(x, A)$, the Markov process is *time homogeneous*. The Chapman–Kolmogorov equation then takes the simpler form

$$P_{t+h}(x, A) = \int P_t(x, dy) P_h(y, A).$$

It is convenient to regard $P_t(X, A)$ as a kernel of a transformation P_t acting on bounded functions, according to the expression

$$P_t f(x) = \int P_t(x, dy) f(y),$$

where (probabilistically) $P_t f(x) = \mathbb{E}^x f(X_t)$. Analogously, one defines a dual transformation of measures μP_t. The family $(P_t, t \geqslant 0)$ forms a semigroup, and the Chapman–Kolmogorov equation expresses the semigroup property $P_{t+h} = P_t P_h$. Consequently, one approach to studying Markov processes uses techniques available from the theory of semigroups.

Classification of Markov processes may be arranged according to various criteria based on a state space, transition functions, and trajectories. A Markov process is a *Feller process* if the operator P_t maps continuous functions into continuous functions, and if $P_t f \to f$ (in supremum norm) when $t \to 0$. When a transition function possesses a density, so that

$$P_t(x, A) = \int_A p_t(x, y)\, dy,$$

it is convenient to work with equations for $p_t(x, y)$; typical examples are diffusion processes* (see Example 3). When a state space is discrete, it is customary to speak about a Markov *chain*, but this term is also used for a discrete parameter Markov process. For Markov chains with a continuous parameter, the operator $\mathbf{P}_t$ is a matrix

$$\mathbf{P}_t = (p_{ij}(t))$$

whose elements are transition probabilities

$$p_{ij}(t) = \mathbb{P}(X_{t+h} = j | X_h = i),$$

so that

$$P_t(i, A) = \sum_{j \in A} p_{ij}(t).$$

Typical examples are provided by the birth-and-death processes* (in particular, the Poisson process).

Sample functions of a Markov process exhibit a wide range of behavior, depending on additional assumptions (especially those on transition functions). In particular, even a Markov chain may have discontinuities that are not jumps. A (separable) Feller process may be selected in such a way that almost all its sample functions are right continuous with left limits. For a Hunt process, this property is taken as a part of the definition.

Properties of Markov processes mentioned here are discussed at length in the books [2, 4] devoted to the subject; see also refs. 7 (Chaps. V and VI), 11 (Chaps. 7 and 8), 12 (Chap. XII), and 14 (Chap. III). For an analytic treatment and examples, see refs. 8 (Chap. X), and 10 (Chap. 4).

Martingales*

Recall that a super-martingale $X = (X_t, \mathscr{F}_t, 0 \leqslant t < \infty)$ is always defined relative to some (increasing) family of σ fields $\mathscr{F}_t$. In the case of natural σ fields, the definition simplifies to

$$\mathbb{E}(X_t|X_u, u \leqslant s) \leqslant X_s \quad \text{a.e.} \ (s \leqslant t).$$

A super-martingale is thus a stochastic process that "shrinks in conditional mean." A martingale should not be confused with a Markov process; these two types of processes may however overlap (see for example the Brownian motion in Example 3). Defining inequalities may be put into the following equivalent form:

$$\int_B X_t \, d\mathbb{P} \leqslant \int_B X_s \, d\mathbb{P}$$

for every pair $s \leqslant t$ and every $B \in \mathscr{F}_s$ (with = for a martingale). A process X is a *submartingale* when $-X$ is a super-martingale*; hence, for simplicity, attention is here restricted to super-martingales.

The fundamental theorem of Doob asserts that under the "usual" conditions or under separability a super-martingale $X = (X_t, \mathscr{F}_t)$ has a modification (a "version") whose sample functions are all right continuous (and have left limits) if and only if the mapping $t \to \mathbb{E}X_t$ is right continuous; cf. refs. 4 (Theorem 3, Sec. 1.4), 7 (Theorems 11.3 and 11.5); and 6 (Theorem VI.4). Hence for all practical purposes one can restrict attention to such super-martingales.

The (super-) *martingale convergence theorem* shows that when $\sup_t \mathbb{E}|X_t| < \infty$, then random variables X_t converge a.e. to an integrable random variable X_∞ as $t \to \infty$. The above condition is satisfied when random variables $(X_t, \ 0 \leqslant t < \infty)$ are uniformly integrable. Then the super-martingale X is closed on the right, i.e., $(X_t, \mathscr{F}_t, 0 \leqslant t \leqslant \infty)$ is a super-martingale with $\mathbb{E}(X_\infty|X_s) \leqslant X_s$ a.e. (and $s \leqslant \infty$).

A particularly important example of a martingale is obtained by defining

$$X_t = \mathbb{E}(Z|\mathscr{F}_t), \qquad 0 \leqslant t < \infty,$$

where Z is an integrable random variable. Then X_∞ exists, $(X_t, \ 0 \leqslant t < \infty)$ are uniformly integrable, and $(X_t, \mathscr{F}_t, 0 \leqslant t \leqslant \infty)$ is also a martingale with $\mathbb{E}(Z|\mathscr{F}_\infty) = X_\infty$ a.e.

Of special interest (in connection with probabilistic potential theory) is the following form of the Riesz decomposition. Let $X = (X_t, \mathscr{F}_t)$ be a right-continuous positive super-martingale, hence necessarily convergent ($X_t \to X_\infty$ a.e.). Then, X has the unique (up to modification) decomposition

$$X_t = M_t + U_t,$$

where $M = (M_t, \mathscr{F}_t)$ is a uniformly integrable martingale with $M_t = \mathbb{E}(X_\infty|\mathscr{F}_t)$, and $U = (U_t, \mathscr{F}_t)$ is a positive super-martingale such that $U_t \to 0$ a.e. If (X_t) are uniformly integrable, so are (U_t) and then also $\mathbb{E}U_t \to 0$; see ref. 6 (Theorem VI.8). For further discussion of super-martingales see refs. 4 (Sec. 1.4), 6 (Chap. VI), 7 (Chap. VII), 12 (Sec. 36), 13 (Sec. IV.5), and 14 (Chap. II.2); see also ref. 10 for many interesting examples.

Stationary Processes

The definition of a stationary process* given above indicates that finite-dimensional distributions of a process depend only on time differences and not on the absolute time. This concept is taken in the sense of a *strictly stationary* process (in contrast to that of a *wide-sense* or *weakly stationary* process defined by covariance). In particular, this implies that for a stationary process all one-dimensional distribution functions are identical, i.e., $\mathbb{P}(X_t \leqslant x) = F(x)$ cannot depend on t. And, if they exist, then $\mathbb{E}X_t$ are constant, and the second joint moment $R(s) = \mathbb{E}X_{t+s}X_t$, known as a *covariance function*, depends only on s.

A (strictly) stationary process X may be regarded as generated by a translation semigroup $(T_t, \ 0 \leqslant t < \infty)$ of probability preserving transformations in the sense that for each t

$$X_t = T_t X_0.$$

A set $M \in \mathscr{F}$ is *invariant* if for each t the sets M and $T_t^{-1}M$ are $\mathbb{P}$ equivalent; the invariant sets form a σ field $\mathscr{L} \subset \mathscr{F}$. A random variable Y is *invariant* if for each t the Y and $T_t Y$ are equivalent. A (strictly) stationary process $(X_t,\ 0 \leqslant t < \infty)$ or its semigroup $(T_t,\ 0 \leqslant t < \infty)$ is said to be *ergodic* if the only invariant sets are those of probability 0 or 1; that is, if the only invariant random variables are equivalent to constants. The strong law of large numbers for a (strictly) stationary process (a special case of the Birkhoff ergodic theorem*) asserts that if a process is (strictly) stationary and X_0 is integrable, then the following limit exists with probability 1:

$$\lim_{t \to \infty} (1/t) \int_0^t X_s\, ds = \mathbb{E}(X_0 | \mathscr{L}).$$

In particular, if the process is also ergodic, then the value of the limit reduces to a constant $\mathbb{E} X_0$. This is a form of *ergodic hypothesis* according to which space averages and time averages coincide. For ergodic theorems see refs. 5 (p. 151), 7 (p. 515), and 12 (p. 412).

As an example, consider a time-homogeneous Markov process $X = (X_t)$ with a transition function $P_t(x, A)$, and suppose that μ is an invariant distribution in the sense that for any Borel set A,

$$\mu(A) = \int \mu(dx) P_t(x, A).$$

Taking this μ as an initial distribution for the process, one finds that the Markov process is stationary (hence, time homogeneity is not sufficient for stationarity). The process is ergodic if there is only one invariant distribution μ; see ref. 7 (p. 459).

Studies of stationary processes concerning properties of the covariance function $R(s)$ and its spectral representation essentially involve methods of harmonic analysis. Stochastic integrals* (with respect to processes with orthogonal increments) lead to spectral representations of random variables X_t. For a full discussion, see books devoted to the subject [5, 15] and appropriate chapters in refs. 7, 9, 11, and 12.

Process With Independent Increments

This process has been defined by a requirement that its increments $X_{t+h} - X_h$ for disjoint time intervals be independent. The formal expression

$$X_t - X_0 = \int_0^t dX_s$$

(in analogy with sums of independent random variables) suggests other names: an *additive*, *differential*, or *decomposable process*. Examples of such a process are specified by distributions of $X_{t+h} - X_h$, and not necessarily by those of X_t. To assign distributions to individual X_t, it is convenient to assume that $X_0 = 0$ (so that all sample functions vanish at the origin). Then $X = (X_t,\ 0 \leqslant t < \infty)$ is a Markov process and if $\mathbb{E}(X_{t+h} - X_h) = 0$, it is also a martingale.

A fundamental theorem of Lévy asserts that a (separable) process $X = (X_t)$ with independent increments is the sum of three components (not necessarily all present):

$$X_t = c(t) + X_t^d + X_t^c,$$

where $c(t)$ is a nonrandom "centering function," (X_t^d) is a discontinuous component with almost all sample functions continuous except at a countable number of fixed discontinuities, and (X_t^c) is an a.e. continuous component, with almost all sample functions continuous, except for a countable number of jumps. See refs. 7 (p. 417) and 12 (p. 540) and also the section Mathematical Interest below (on discontinuities).

If the distribution of $X_{t+h} - X_h$ depends only on t, a process X is said to have *stationary independent increments*. Such a process cannot have fixed discontinuities. Assuming that the centering function vanishes, then the distribution function F_t of the increments $X_{t+h} - X_h$ (for an a.e. continuous process) is shown to be infinitely divisible*. Its characteristic function* has the form

$$\int_{-\infty}^{\infty} e^{i\tau z}\, dF_t(z) = e^{t\psi(\tau)}$$

with the exponent (in Lévy form)

$$\psi(\tau) = im\tau - \tfrac{1}{2}\sigma^2\tau^2 + \int_{-\infty}^{\infty} \left(e^{i\tau x} - 1 - \frac{i\tau x}{1 + x^2} \right) dL(x),$$

where m is real and $\sigma \geqslant 0$; the function L is defined for all x, except $x = 0$, is nondecreasing for $x < 0$ and for $x > 0$ with $L(-\infty) = L(+\infty) = 0$, and satisfies

$$\int_{-1}^{1} x^2 \, dL(x) < \infty.$$

(The bar that crosses the above integrals excludes $x = 0$ from the domain of integration.) For the derivation, see refs. 7 (p. 419), 12 (p. 537), and 8 (p. 295). The Lévy function L has the following probabilistic interpretation. Let $v_t(x)$ be a number of jumps of sample functions in the time interval from 0 to t, of magnitude less than $x < 0$ or at least equal to $x > 0$. Then

$$\mathbb{E} v_t(x) = t|L(x)|;$$

see refs. 7 (p. 423), 12 (p. 546), and 8 (p. 295). Furthermore, the mean and variance of the interval distribution are now proportional to the length of the interval

$$\mathbb{E}(X_{t+h} - X_h) = (m + a)t,$$

$$\operatorname{var}(X_{t+h} - X_h) = (\sigma^2 + b^2)t,$$

where a and b^2 are constants depending on L.

For example, as noted earlier, Brownian motion and the Poisson process have stationary independent increments; $m = 0$ and $L(x) = 0$ for $x \neq 0$, so that $\psi(\tau) = -\tfrac{1}{2}\sigma^2\tau^2$ for Brownian motion; $m = \lambda$ and $L(x) = 0$ for $x < 0$ and for $x \geqslant 1$, but $L(x) = -\lambda$ for $0 < x < 1$, so that $\psi(\tau) = \lambda(e^{i\tau} - 1)$ for the Poisson process.

A process X with stationary independent increments, regarded as a Markov process, is both time and space homogeneous, with transition functions invariant under translations

$$P_{h,t+h}(x, A) = P_t(0, A - x),$$

where $A - x$ means $y - x$ for all $y \in A$. Such a process is also called a *Lévy process**. Space homogeneity implies that transition probabilities coincide with the distribution function of increments. Taking $A = (-\infty, z]$, one has

$$\begin{aligned} P_t(0, A) &= P_t(x, A + x) \\ &= \mathbb{P}(X_{t+h} \leqslant x + z | X_h = x) \\ &= \mathbb{P}(X_{t+h} - X_h \leqslant z) = F_t(z). \end{aligned}$$

The Chapman–Kolmogorov equation (Markov semigroup property) then becomes the convolution relation for the distribution functions F_t:

$$F_{t+h}(z) = \int_{-\infty}^{\infty} F_t(z - y) \, dF_h(y), \qquad t \geqslant 0, \ h \geqslant 0.$$

Convolution semigroups received considerable attention in the literature because of their interesting properties and many applications. For example, their generators may be expressed in terms of the Lévy function L; cf. ref. 8 (p. 293).

For further discussion of processes with independent increments see refs. 7 (Chap. VIII), 12 (Sec. 37), and 8 (Chap. 9).

Other Processes

Besides the four main types already discussed, there are other well known stochastic processes characterized by special properties. Some of these were developed to a great extent, each process forming a class of its own. These include renewal processes*, point processes, random walks*, branching processes*, diffusions, birth-and-death processes*, regenerative processes*, semi-Markov processes*, various extensions of martingales (amarts, quasi-martingales, semi-martingales, local martingales, martingales-in-the-limit, games fairer with time), etc. Moreover, depending on the form of distribution, one distinguishes special processes like the Poisson process, compound Poisson processes, Gaussian processes*, Cauchy and Bessel processes, stable processes, extremal processes, etc.

The reader should consult the Bibliography for definitions and properties. Here only a brief remark will be made on point processes, for the purpose of illustration.

A *point process* is obtained when one considers a sequence of events occurring in continuous time, individual events being distinguished only by their position in time. Let T_n, $n = 0, 1, 2, \ldots$, be a sequence of positive random variables such that

$$T_0 = 0, \qquad T_n < T_{n+1}, \qquad T_n \to T_\infty \leqslant \infty,$$

T_n representing the instant of occurrence of the nth event. The total number of events in the interval from 0 to t is a random variable N_t that may be written in the form

$$N_t = \sum_{n=0}^{\infty} I_{(T_n \leqslant t)},$$

where I_A is the indicator of a set A. The family $N = (N_t,\ 0 \leqslant t < \infty)$ is a *counting process*, and

$$N_t = n \quad \text{if} \quad T_n \leqslant t < T_{n+1}.$$

Note that N is an increasing process and its sample functions are right-continuous step functions with upward jumps of magnitude 1. Each of the random sequence (T_n) and the counting process (N_t) is known as a *point process*. Typical examples include the Poisson process, the birth process, the renewal process, and a process with positive independent increments (a *subordinator*). The definition of a point process stated here is a special case of a more general definition involving random measures (*see* STOCHASTIC PROCESSES, POINT).

The *intensity* of a point process N is a family of random variables $\Lambda(t)$ such that, relative to a history $(\mathscr{F}_s)$,

$$\mathbb{E}(N_t - N_s | \mathscr{F}_s) = \mathbb{E}\left(\int_s^t \Lambda(u)\, du | \mathscr{F}_s \right), \qquad s \leqslant t,$$

which implies that the random variables

$$M_t = N_t - \int_0^t \Lambda(u)\, du$$

form a martingale relative to $(\mathscr{F}_t)$.

For further discussion see ref. 3 (Chap. II).

Other Forms of Classification

Another form of classification of stochastic processes proceeds according to problems of interest and methods used for study. The latter range from analytic techniques to a measure theoretic approach.

Methods and problems include the general theory of processes (some aspects of which are given below in the section Mathematical Interest), analysis of sample functions behavior, convergence problems, random times (optional and predictable processes), passage problems, transformations of processes (stopping, random time change, etc.), distributional properties, etc. Of great interest are problems that exhibit connections with other branches of mathematics, like potential theory (harmonic functions, the Dirichlet problem), boundary theory (Martin boundary, Feller boundaries), stochastic equations (stochastic integrals*, Itô calculus). Other techniques of interest are harmonic analysis (spectral theory), ergodic theory, functional representation (the Choquet theorem), recurrent events, ladder variables*, the Wiener–Hopf technique, and the study of abstract-valued (e.g., Banach space) random variables. Of special character are control and optimization theories and the rather new field of stochastic approximation*. For discussion of these topics, see the Bibliography.

APPLICATIONS

Stochastic processes are found in applications to various fields where effects of randomness must be considered. These include many branches of engineering, physics and chemistry, economics, social and biological sciences, operations research, and others. The general form of such investigations is the analysis of stochastic models representing real-life situations; numerical methods and simulation complement analytical considerations. Intensive developments in applications grew into separate disciplines, rich in their own problems and methods. Queueing*, reliability* and survival analysis*, control, prediction and filtering*, and learning models are the best known examples.

Only a few illustrative examples will be given. For excellent surveys, see refs. 8 and 10 and the Bibliography.

Queueing

In particular, queueing theory*, is a source of many interesting examples. The input to a queuing system is described by a renewal process, or more generally by a point process. The number of customers present in the system at time t, represented by X_t, leads to a process $(X_t, \ 0 \leqslant t < \infty)$ that is non-Markovian in general. The ergodic distributions of such a process is of interest in typical applications. In the waiting system $GI/G/1$, the waiting process $(W_t, \ 0 \leqslant t < \infty)$, where W_t is the waiting time of a customer arriving at instant t, is non-Markovian. However, the imbedded process $(W_n, \ n = 0, 1, 2, \dots)$ with $W_n = W_{t_n - 0}$, evaluated at regeneration points t_n, is Markovian, and its transition probabilities are time and space homogeneous.

Other Processes

Point processes are applied to situations involving counting, like insurance claims, accidents, arrivals, departures, etc. Similarly, renewal processes are used in reliability theory and survival analysis. Stochastic processes, under the name of time series*, have been used in many applications, especially in economics. In communication theory*, signal and noise analysis provide many important examples of stochastic processes. Martingales also found their way into applications; see for example ref. 3.

Statistics

The distinction between stochastic processes and statistics is rather a matter of form and not of substance, and reflects different aspects of the same problem. Stochastic processes are basically concerned with structure and properties of models, and not with inferences* from real data. In recent years, however, statistical analysis of stochastic processes became a newly developing subject (see Bibliography D). This includes on one hand the use of classical methods of estimation and testing (e.g., distributions in Markov processes, spectral analysis* in stationary processes), and on the other hand the use of theoretical developments in stochastic processes (e.g., the advantages offered by martingale methods; large deviations*, empirical processes*). For this purpose, theoretical considerations of stochastic processes are indispensable.

Historical Aspects

Stochastic processes still wait for a historian. At present, historic notes (often biased or inaccurate) are scattered throughout the literature, in particular in works devoted to special processes and applications. Only a few undisputed names (mentioned in this article) are generally accepted. The following remarks represent folklore of the subject. It is pointless to pinpoint exact references; the reader should consult the annotations to works listed in the Bibliograpny.

Although the origin of the theory of stochastic processes may be traced back to developments in statistical mechanics (Gibbs, Boltzmann, Poincaré, Smoluchowski, Langevin), and later to pioneering work on Brownian motion (Einstein, Wiener, Lévy) and on teletraffic (Erlang, Molina), the theoretical foundations were laid down by Doob, Kolmogorov, and others, only after the Kolmogorov axioms* were formulated (the word "stochastic" went into use around that time). The introduction to birth-and-death processes by Feller, applications to physics (and potential theory) by Kac and Kakutani, developments in stationary processes (Cramer, Lévy) and in queueing (Pollaczek, Chinchin, Palm), and the pioneering work of Doob on Markov processes and martingales are further early milestones. These developments culminated in the wide spectrum of topics ranging from the recently formulated general theory of processes (see Bibliography A) to numerous applications in various fields (see Bibliography C).

MATHEMATICAL INTEREST

Stochastic processes, notwithstanding their many applications, form an object of mathematical interest. As concepts from the "general theory of processes" find their way every day to many applications, brief comments on basic ideas may be of interest to mathematically minded readers.

In such considerations, the state space in which random variables of a process take their values is a measurable space $(E, \mathscr{E})$, where E is a topological space with its Borel field $\mathscr{E}$ (generated by open sets). In most applications, $\mathscr{E}$ is a (locally compact separable) metric space, in particular the Euclidean space $\mathbb{R}^d$ of dimension d. For reasons of simplicity, in the following discussion E will be the real line $\mathbb{R} = (-\infty, \infty)$ and $\mathscr{E}$ will denote the class of its Borel sets. A process $X = (X_t, 0 \leqslant t < \infty)$ is said to be continuous in probability or almost everywhere or in the pth mean at a point t according as X_s converges to X_t in probability or almost everywhere or in the pth mean, respectively, as $s \to t$. A process X is *continuous* in the above sense if it is continuous at every t. A process X is *sample continuous* if its trajectories are continuous functions; it is *a.e. sample continuous* if the set of discontinuous trajectories is $\mathbb{P}$ negligible. Similar definitions refer to right-continuous and left-continuous processes, and to the property that a process has left-hand and right-hand limits (a right-continuous process with left-hand limits is said to be *cad-lag*). In the same fashion one considers differentiability and integrability properties.

In general, analytic properties of X relative to sample functions are stronger than the cooresponding properties relative to a space of random variables. For example, a.e. sample continuity of X implies a.e. continuity of X, but the converse need not be true. Indeed, the process X is continuous a.e. at a point t when the set N_t of sample functions that are discontinuous at t, has probability 0. Since the set of discontinuous sample functions $N = \bigcup_t N_t$ may have positive probability, so the continuity a.e. need not imply a.e. sample continuity of X. An example of such a situation is provided by the Poisson process (Example 2). When X is not continuous a.e. at a point t, that is, when the probability of N_t is strictly positive, then t is a *fixed discontinuity point* of X. A discontinuity point $t = t(\omega)$ of a sample function $X(\cdot, \omega)$, which is not a fixed discontinuity point of X, is a *moving* (with ω) *discontinuity point* of X. Thus a.e. sample continuity excludes fixed and moving discontinuity points outside a null set, whereas a.e. continuity excludes only fixed discontinuity points. Two stochastic processes $X = (X_t)$ and $Y = (Y_t)$ with the same index set $\mathscr{I}$ and state space $(E, \mathscr{E})$, defined on the same probability space $(\Omega, \mathscr{F}, \mathbb{P})$, are *indistinguishable* if for almost all $\omega \in \Omega$, their sample functions coincide for all t; the processes X and Y are a *modification* of each other if

$$X_t = Y_t \quad \text{a.e. for each } t.$$

The processes X and Y defined on different probability spaces with probabilities $\mathbb{P}$ and $\mathbb{P}'$ are *equivalent* if

$$\mathbb{P}(X_{t_1} \in A_1, \ldots, X_{t_n} \in A_n)$$
$$= \mathbb{P}'(Y_{t_1} \in A_1, \ldots, Y_{t_n} \in A_n)$$

for every finite collection of instants $t_1, \ldots, t_n$ and sets $A_1, \ldots, A_n$ of $\mathscr{E}$.

Consider now the function space $E^{\mathscr{I}}$ of paths, and introduce the product Borel field $\mathscr{E}^{\mathscr{I}}$ on $E^{\mathscr{I}}$. Let $\mathbb{P}_X$ be the image of $\mathbb{P}$ induced on $\mathscr{E}^{\mathscr{I}}$ by a process X, and let Y_t be a coordinate mapping of index t on $E^{\mathscr{I}}$. In other words, regard the process X as a mapping that with each point ω associates a sample function $X(\cdot, \omega) = \omega'$, considered as a point in the product space $E^{\mathscr{I}}$. Then, a coordinate mapping Y_t is simply the projection on the tth coordinate

$$Y_t(\omega') = X(t, \omega)$$

and the induced probability $\mathbb{P}_X$ is defined in the usual way as $\mathbb{P}_X(S) = \mathbb{P}(X^{-1}S)$ for sets S in the product σ field $\mathscr{E}^{\mathscr{I}}$. The stochastic process $(Y_t, t \in \mathscr{I})$ on the probability space $(E^{\mathscr{I}}, \mathscr{E}^{\mathscr{I}}, \mathbb{P}_X)$ is the *canonical process associated with* the process $(X_t, t \in \mathscr{I})$ on $(\Omega, \mathscr{F}, \mathbb{P})$. Processes (X_t) and $Y_t)$ are equivalent.

When $\mathscr{I}$ is uncountable, the canonical process is not sufficient because the product σ field $\mathscr{E}^{\mathscr{I}}$ is determined by at most a countable number of coordinates, so many important sets (e.g., the set of all continuous paths) would not be measurable, and thus would have no probability defined. Regarding $E^{\mathscr{I}}$ as a topological space, the probability measure on its Borel field (which is much larger than $\mathscr{E}^{\mathscr{I}}$) could be induced by $\mathbb{P}$ in the usual way, but even this construction is not really satisfactory (especially for sample functions analysis).

As noted in the Introduction, a process $X = (X_t)$ on a probability space $(\Omega, \mathscr{F}, \mathbb{P})$ defines a family of n-dimensional distributions F. Conversely, by the Daniell–Kolmogorov extension theorem, any collection of such functions F, satisfying natural consistency conditions, uniquely determines a probability measure on a function space $(E^{\mathscr{I}}, \mathscr{E}^{\mathscr{I}})$ such that its restrictions to finite-dimensional subspaces coincide with the given finite-dimensional distribution functions (projective system of probability measures). This is equivalent to construction of a canonical process (Y_t). For the complete discussion of the Daniell–Kolmogorov extension theorem, and its inadequacy for continuous parameter processes, see refs. 1 (Sec. 36), 12 (Sec. 4.3), 13 (Sec. III.3), and 14 (p. 42). I.-Tulcea extended this theorem to the arbitrary spaces E under a condition of existence of a family of regular conditional distributions (for the case of independence, validity of the theorem for arbitrary spaces E was shown earlier by von Neumann); see refs. 12 (Secs. 8.3 and 4.2) and 13 (Sec. V.I).

The basic concept of separability of a process was introduced by Doob in order to overcome difficulties in assigning probability measure to a wider class of sets than the product σ field $\mathscr{E}^{\mathscr{I}}$ (in modern treatment, separability has been overshadowed by stressing other properties of a process that imply separability).

A real-valued stochastic process $X = (X_t, 0 \leqslant t < \infty)$ is *separable* if there exists a countable set $S \subset \mathscr{I}$ (called a *separant*) and a $\mathbb{P}$-null set N that for every $\omega \in N^c$,

$$X(t, \omega) \in \bigcap_{t \in I} \overline{X(IS, \omega)}$$

for every $t \in \mathscr{I}$, where I runs over the class of open intervals in $\mathscr{I}$; here $\overline{X(K, \omega)}$ is the closure in $\overline{\mathbb{R}}$ of the image of $K \subset \mathscr{I}$ under the mapping $t \to X(t, \omega)$. In effect, the set determined by an uncountable number of coordinates is assigned the same probability as an appropriate set in $\mathscr{E}^{\mathscr{I}}$ containing it. Separability leads to important consequences listed in the following theorems due to Doob:

(i) Any stochastic process X has a separable modification $\tilde{X}$.

(ii) If the separable process is continuous in probability, then any countable dense set is a separant.

(iii) If the process X is separable and a.e. continuous, then it is measurable.

(iv) If the process X is continuous in probability, then there exists a separable and measurable modification $\tilde{X}$.

Note that separability is no restriction on finite-dimensional distributions; and according to **(i)** if the original process is not separable, it can always be replaced by a separable process. Moreover, as a consequence of separability, sequential limits $X_{s_n} \to X_t$ (along $s_n \in S$) can be replaced by limits $X_s \to X_t$ as $s \to t$. This last property is sometimes taken as an equivalent definition of separability. For discussions of separability see refs. (Sec. 38), 6 (Sec. IV.24), 7 (Sec. II.2), 12 (Sec. 35.2), and 13 (Sec. III.4).

In conclusion, it may be remarked that a stochastic process may be looked at in three different ways. In the customary interpretation, a stochastic process X is regarded as a mapping

$$t \to X_t$$

from the index set $\mathscr{I}$ into a vector space $\mathscr{K}$ of E-valued random variables. For example, $\mathscr{K}$ may be a space of random variables possessing moments of specified order (e.g., square integrable random variables), or a space of bounded random variables, etc. In effect, one then deals with classes of equivalent random variables and indistinguishable processes. In another interpretation already mentioned, a stochastic process X is regarded as a mapping.

$$\omega \to X(\cdot, \omega)$$

from Ω into a function space of paths (the product space) $E^{\mathscr{I}}$.

In theoretical considerations, a third interpretation is preferable, in which a stochastic process X is regarded as a mapping

$$(t, \omega) \to X(t, \omega)$$

from a product space $\mathscr{I} \times \Omega$ into a state space E. The process X is *measurable* if this mapping is measurable relative to a product σ field $\mathscr{U} \otimes \mathscr{F}$, where $\mathscr{U}$ is a σ field on $\mathscr{I}$,

$$X^{-1}(\mathscr{E}) \subset \mathscr{U} \otimes \mathscr{F}.$$

A *random set* is a subset M of $\mathscr{I} \times \Omega$ whose indicator I_M, as a function of (t, ω), is a stochastic process; thus, M is a measurable random set if the process $(I_M(t),\ t \in \mathscr{I})$ is measurable.

A process $X = (X_t, \mathscr{F}_t, 0 \leqslant t < \infty)$ with values in $(E, \mathscr{E})$ is said to be *progressively measurable* (or progressive with respect to a history) if for every $t \in \mathbb{R}_+$, the mapping,

$$(s, \omega) \to X(s, \omega)$$

of $[0, t] \times \Omega$ into $(E, \mathscr{E})$ is measurable with respect to the σ field $\mathscr{U}_t \otimes \mathscr{F}_t$, where $\mathscr{U}_t$ is a Borel field on $[0, t]$. Clearly, a progressive process is adapted and measurable. A process (X_t) adapted to a family $(\mathscr{F}_t)$ and assuming values in a metrizable space $(E, \mathscr{E})$ and whose paths are right continuous (or left continuous) is always progressively measurable.

A set M in the product space $\mathbb{R}_+ \times \Omega$ is *progressively measurable* if the real-valued process $(I_M(t), 0 \leqslant t < \infty)$ is progressively measurable. The progressive sets form a σ field $\mathscr{M}$ (the progressive field) included in $\mathscr{U} \otimes \mathscr{F}$, and a process X is progressive if it is measurable with respect to $\mathscr{M}$.

In the above definitions, the form of a probability measure $\mathbb{P}$ did not intervene. Probability structure must be imposed, however, for further analysis. It is usual to assume that $(\Omega, \mathscr{F}, \mathbb{P})$ is a complete probability space, and that each $\mathscr{F}_t$ contains all null sets of $\mathscr{F}$; the σ field $\mathscr{U} \otimes \mathscr{F}$ is also completed with respect to the product measure $\lambda \times \mathbb{P}$ (where λ is a measure on $\mathscr{U}$, in particular the Lebesgue measure).

The above remarks simply give a glimpse into the growing area of general theory of stochastic processes. For a complete discussion, see ref. 6 (Vol. 1).

References

The literature on stochastic processes is voluminous: Here a selection has been made to a few

(a) well-known works devoted to stochastic processes, in general;

(b) textbooks and introductory works;

(c) works devoted to special processes, but containing general material;

(d) works of historical interest.

Entries in the References and Bibliography are labeled according to this system. The † denotes an advanced mathematical text.

For specialized works devoted to special processes (Markov, martingales, queues, etc.) see the appropriate articles. The reader should also consult Proceedings of Berkeley Symposia, Proceedings of Conferences on Stochastic Processes and their Applications,

Proceedings of Seminars on Stochastic Processes, and Springer-Verlag Lecture Notes.

[1] Billingsley, P. (1986). *Probability and Measure* 2nd ed. Wiley, New York (b, c).

[2] Blumenthal, R. M. and Getoor, R. K. (1968). *Markov Processes and Potential Theory*. Academic, New York (c†).

[3] Bremaud, P. (1981). *Point Processes and Queues*. Springer-Verlag, New York (c†).

[4] Chung, K. L. (1980). *Lectures from Markov Processes to Brownian Motion*. Springer-Verlag, New York (c†).

[5] Cramér, H. and Leadbetter, M. R. (1967). *Stationary and Related Stochastic Processes*. Wiley, New York (c).

[6] Dellacherie, C. and Meyer, P. A. (1978/1982). *Probabilities and Potentials*, Vols. 1 and 2. North-Holland, Amsterdam, The Netherlands (a†).

[7] Doob, J. L. (1953). *Stochastic Processes*. Wiley, New York (a†).

[8] Feller, W. (1966). *An Introduction to Probability Theory and its Applications*, Vols. 1 and 2. Wiley, New York (a, b).

[9] Kac, M. (1959). *Probability and Related Topics in Physical Sciences*. Wiley-Interscience, New York (c).

[10] Karlin, S. and Taylor, H. M. (1975/1981). *First Course in Stochastic Processes*, 2nd ed.; *Second Course in Stochastic Processes*. Academic, New York (a, b).

[11] Lamperti, J. (1974). *Stochastic Processes*. Springer-Verlag, New York (c). (A survey of mathematical theory.)

[12] Loève, M. (1977). *Probability Theory*. Springer, New York (b, c).

[13] Neveu, J. (1965). *Mathematical Foundations of the Calculus of Probability*. Holden-Day, San Francisco, CA (b, c).

[14] Williams, D. (1979). *Diffusions, Markov Processes and Martingales*, Vol. 1. Wiley, New York (c).

[15] Yaglom, A. M. (1962). *Theory of Stationary Random Functions*. Prentice-Hall, Englewood Cliffs, NJ (c).

Bibliography

A. General Theory

In addition to refs. 1–14, see:

Doob, J. L. (1942). What is a stochastic process? *Amer. Math. Monthly*, **49**, 648–653 (d).

Doob, J. L. (1984). *Classical Potential Theory and its Probabilistic Counterpart*. Springer-Verlag, New York (c†).

Doob, J. L. and Ambrose, W. (1940). On two formulations of the theory of stochastic processes depending upon a continuous parameter. *Ann. Math.*, **41**, 737–745.

Kussmaul, A. U. (1977). *Stochastic Integration and Generalized Martingales*. Pitman, London, England (c†).

Meyer, P. A. (1966). *Probability and Potentials*. Blaisdell, Waltham, MA (c†, d). (This is the old edition of ref. 6.)

B. Analytic Methods

In addition to refs. 5, 8, 9, and 10, see:

Bartlett, M. S. (1962). *Stochastic Processes*. Cambridge University Press, Cambridge, England (b).

Çinlar, E. (1975). *Introduction to Stochastic Processes*. Prentice-Hall, Englewood Cliffs, NJ (b).

Cox, D. R. and Miller, H. D. (1968). *The Theory of Stochastic Processes*. Methuen, London, England (b).

Ephremides, A., ed. (1975). *Random Processes*, Vols. 1 and 2. Dowden, Hutchinson and Ross, Stroudsburg, PA (c). (Collection of reprints.)

Lévy, P. (1965). *Processus Stochastiques et Mouvement Brownien*. Gauthier Villars, Paris, France (c, d).

Parzen, E. (1962). *Stochastic Processes*. Holden-Day, San Francisco, CA (b).

Prabhu, N. U. (1965). *Stochastic Processes*. MacMillan, New York (b).

Takács, L. (1966). *Stochastic Processes (Problems and Solutions)*. Methuen, London, England (b).

Takács, L. (1967). *Combinatorial Methods in the Theory of Stochastic Processes*. Wiley, New York (c).

C. Applications

Bailey, N. T. J. (1964). *The Elements of Stochastic Processes with Applications to the Natural Sciences*. Wiley, New York (b).

Bartholomew, D. J. (1973). *Stochastic Models for Social Processes*, 2nd ed. Wiley, New York (b).

Moyal, J. L. (1949). Stochastic processes and statistical physics. *J. R. Statist. Soc. Ser., B*, **11**, 150–210 (c, d).

Ross, S. M. (1980). *Introduction to Probability Models*, 2nd ed. Academic, New York (b).

Syski, R. (1979). *Random Processes (A First Look)*. Dekker, New York (b).

Syski, R. (1980). *Congestion Theory*. North-Holland, Amsterdam, The Netherlands (c).

Wax, N., ed. (1954). *Noise and Stochastic Processes*. Dover, New York (c). (Collection of reprints.)

Wiener, N. et al. (1966). *Differential Space, Quantum Systems and Prediction*. MIT Press, Boston, MA (c).

D. Statistics

Billingsley, P. (1961). *Statistical Inference for Markov Processes*. University of Chicago Press, Chicago, IL (c).

Freidlin, M. I. and Wentzell, A. D. (1980). *Random Perturbations of Dynamical Systems*. Springer-Verlag, New York (c).

Jacobsen, M. (1982). *Statistical Analysis of Counting Processes*, Springer, New York.

Liptser, R. S. and Shiryayev, A. N. (1978). *Statistics of Random Processes*. Springer, New York (c).

Slud, E. V. (1987). *Martingale Methods in Statistics*. Wiley, New York.

(BIRTH-AND-DEATH PROCESSES
BRANCHING PROCESSES
BROWNIAN MOTION
DANIELL–KOLMOGOROV THEOREM
DIFFUSION PROCESSES
EMBEDDED PROCESSES
ERGODIC THEOREMS
GAUSSIAN PROCESSES
IMMIGRATION–EMIGRATION PROCESSES
KALMAN FILTERING
MARKOV PROCESSES
MARTINGALES
ORNSTEIN–UHLENBECK PROCESS
POINT PROCESS, STATIONARY
POISSON PROCESSES
PREDICTION AND FILTERING, LINEAR
PROCESSES, DISCRETE
PROCESSES, EMPIRICAL
QUEUEING THEORY
RANDOM WALK
RENEWAL THEORY
SEMI-MARKOV PROCESSES
STOCHASTIC DIFFERENTIAL EQUATIONS
STOCHASTIC PROCESSES, POINT)

R. SYSKI

STOCHASTIC PROCESSES AND THEIR APPLICATIONS

The history of the journal *Stochastic Processes and Their Applications* is best described by explaining the circumstances that led to starting the series of conferences with the same title. We quote extensively from a recent article on this topic by Prabhu [1].

> In the late 1960's a group of applied probabilists in the U.S.A. became concerned about the directions in which the field of applied probability was developing, and about the status of applied probabilists in the general scientific community. Applied probabilists seem to be caught between pure mathematicians concerned with abstraction and operations researchers preoccupied with problem-solving. In an effort to improve this unhappy situation Professor Julian Keilson of the University of Rochester, New York, invited a group of 20 probabilists to an informal meeting on stochastic processes and their applications at Rochester in 1971.

It became immediately clear that the concerns of this original informal group were widely shared by the international community of probabilists. The Rochester conference was followed by annual conferences in Australia, Canada, Europe, Israel, and the U.S.A. They were open to all participants, and invited as well as contributed papers were presented. To direct the planning of the conferences a committee was set up in 1973 with representation from all parts of the world. In 1975 this committee was affiliated with the Bernoulli Society* for Mathematical Statistics and Probability, which is a section of the International Statistical Institute*.

> At about the same time as the idea of these conferences was developing, a strong need was also felt for a journal that would meet more adequately the publication needs of a large group of probabilists whose areas of interest cover both the theory and applications of stochastic processes. By a fortunate coincidence, the North-Holland Publishing Company was also interested in the publication of such a journal. Negotiations resulted in an agreement to establish the journal, *Stochastic Processes and Their Applications*, in 1973 as a publication of North-Holland, with a Board of Editors consisting of Professors J. W. Cohen, J. Keilson, N. U. Prabhu, and R. Syski, the principal editors being Keilson and Prabhu. In 1980 the Journal became an official publication of the Bernoulli society, with Prabhu as the sole editor.

THE SCOPE OF THE JOURNAL

The journal publishes papers on the theory and applications of stochastic processes*. It is concerned with concepts and techniques, and covers a broad spectrum of mathematical, scientific, and engineering interests. The journal is interested in papers dealing with characterization, structural properties, inference, and control of stochastic processes. Every effort is made to promote innovation, vitality, and communication between disciplines. Papers dealing with new areas of application are especially welcome and are judged on the basis of their scientific interest rather than their mathematical novelty.

The contents of a recent issue [Vol. 23, No. 2 (1986)] were:

L. STETTNER, Discrete time adaptive impulsive control theory
K. YAMADA, A stability theorem for stochastic differential equations with application to storage processes, random walks and optimal stochastic control problems
E. ORSINGHER, Brownian fluctuations in space-time with applications to vibrations of rods
G. ALSMEYER and A. IRLE, Asymptotic expansions for the variance of stopping times in nonlinear renewal theory
J. G. SHANTHIKUMAR and D. D. YAO, The preservation of likelihood ratio ordering under convolution
T. NAKAGAWA, Convergence of critical multitype Galton–Watson branching processes
S. M. BERMAN, The supremum of a process with stationary independent and symmetric increments
D. T. PHAM, The mixing property of bilinear and generalised random coefficient autoregressive models
R. M. BURTON, A. R. DABROWSKI and H. DEHLING, An invariance principle for weakly associated random vectors
T. C. BROWN, B. G. IVANOFF and N. C. WEBER, Poisson convergence in two dimensions with application to row and column exchangeable arrays
T. HAYASHI, Laws of large numbers in self-correcting point processes
D. DEHAY and R. MOCHÉ, Strongly harmonizable approximations of bounded continuous random fields
R. J. WILSON, Weak convergence of probability measures in spaces of smooth functions (Short Communication)
R. GRÜBEL, A note on the distance of ladder height distributions (Short Communication)
J. B. MITRO, A note on discontinuous time changes of Markov processes

The Editor is assisted by an international board of associate editors. Papers submitted to the journal should be written in English or French. All papers are refereed, normally by two referees. The size of the journal has grown from a volume of 400 pages published in 1973 to three volumes of 1080 pages in all, published in 1987.

In January 1984 Professor C. C. Heyde became the Editor of the journal. His address is Department of Statistics, Institute of Advanced Studies, Australian National University, Canberra, ACT 2601. Under Professor Heyde the editorial policy remains substantially the same as described above.

Reference

[1] Prabhu, N. U. (1982). Conferences on stochastic processes and their applications. *Stochastic Process. Appl.*, **12**, 115–116.

N. U. PRABHU

STOCHASTIC PROCESSES, POINT

A point process is a random distribution of indistinguishable points in an underlying space such that only finitely many fall in any bounded (relatively compact) set. In the earliest examples the space was the positive half-line and points represented times of events, for example, arrivals of calls at a telephone exchange or customers at a queueing system. Initial applications to spatial processes were analyses of flying bomb hits on London and cluster structure of stars and galaxies. Since then, point processes on sets as complicated as function spaces and spaces of measures have emerged as fundamental to description of the structure of stochastic processes* such as Markov processes* and infinitely divisible random measures, while there has been a plethora of applications to phenomena as diverse as survival times, optical communication, precipitation, and flows in networks of queues*.

APPROACHES TO POINT PROCESSES

There exist two approaches to analysis of point processes: a *random counting measure* approach applicable on general spaces and a *counting process* approach valid only for point processes on $\mathbb{R}_+$; in the latter case the two are complementary rather than substitutes.

From the random counting measure viewpoint a point process on a (locally compact, separable) space E is a random integer-valued measure $N = (N(A): A \subset E)$ that is finite on compact sets; the random variable $N(A)$ represents the number of points in the set A. A point process N admits a representation as a sum of randomly located point masses, i.e., $N = \sum \epsilon_{X_i}$, where the X_i are random elements of E and ϵ_x denotes the point mass (Dirac measure) at x [$\epsilon_x(A) = 1$ or 0 according as $x \in A$ or $x \notin A$]; the number of summands is also random. Thus for each set A, $N(A) = \sum 1(X_i \in A)$, where 1 denotes an indicator function, and for f a function on E, $N(f) = \int f\,dN = \sum f(X_i)$ is the integral of f with respect to N. In a *simple* point process the X_i are distinct (i.e., there are no coincident points). A point process N on $\mathbb{R}_+ = [0, \infty)$ can be described by the (arrival) *counting process* (N_t), where N_t is the number of points (often termed arrivals) in $[0, t]$, by the *arrival times* $T_k = \inf\{t: N_t = k\}$, the time of the kth arrival, or by the *interarrival times* $U_i = T_i - T_{i-1}$ ($T_0 = 0$ is not an arrival time). Structure may be imposed via (N_t), as for nonhomogeneous Poisson processes*, via (T_k), which is rare, or via (U_i), as for renewal* processes.

The random counting measure context is especially suited to distributional theory; important descriptors are the *probability distribution* $\mathscr{L}_N(\cdot) = P\{N \in (\cdot)\}$; the *Laplace functional*

$$L_N(f) = E\left[\exp\left(-\int f\,dN\right)\right] = E\left[\exp\left(-\sum f(X_i)\right)\right];$$

and the *zero-probability functional*

$$z_N(A) = \Pr[N(A) = 0] = \lim_{t\to\infty} L_N(t1_A).$$

Moment measures (if they exist)

$$E\left[\int N(dx_1)\cdots\int N(dx_k)h(x_1,\ldots,x_k)\right] = E\left[\sum\cdots\sum h(X_{i_1},\ldots,X_{i_k})\right],$$

where h is a function on E^k, summarize key properties but do not determine $\mathscr{L}_N$. The two most important moment measures are the *mean measure* μ_N satisfying

$$\int h(x)\mu_N(dx) = E[N(h)] = E\left[\sum h(X_i)\right]$$

and the *covariance measure* ρ_N given by

$$\int h(x)g(y)\rho_N(dx, dy) = \operatorname{cov}(N(h), N(g)).$$

The Laplace functional is the fundamental distributional descriptor since it always determines the probability law and since moment measures are calculated from it by differentiation. The distribution is determined by the zero-probability functional in case N is simple (Kallenberg [5]), but calculation of moment measures from the zero-probability functional is often difficult or impossible.

In the random counting measure approach a key tool is *convergence in distribution*. Given point processes $N, N_1, N_2, \ldots$ on E, (N_n) converges in distribution to N if the probability distributions converge weakly. In particular, $N_n \to N$ in distribution if and only if the Laplace functionals converge for every positive, continuous function f on E with compact support. Many important classes of point processes, especially Cox processes and infinitely divisible point processes, admit characterizations as the class of all point processes arising as limits in distribution under some rarefaction mechanism.

Structure of a point process N under conditioning or partial observation is elucidated through *Palm distributions*. Given $x_1, \ldots, x_k$ in E, there exists (Kallenberg [5]) the reduced Palm distribution $Q_N^k(x, \cdot)$, a probability measure with the heuristic interpretation (it can be made precise using limits) as the distribution of $N - \Sigma_1^k \epsilon_{x_i}$ conditional on the (typically null) event that points of N are located at $x_1, \ldots, x_k$. In analysis of stationary point processes* the (unreduced) Palm distribution

$$P_0\{\cdot\} = Pr[N \in (\cdot)|N(\{0\}) = 1]$$

plays a central role. Palm distributions are crucial to state estimation for Cox processes and to comparison of different forms of stationary behavior for queueing systems. In this latter case the Palm distribution corresponds to initiating observation at the time of an arrival.

The key modern development in the theory of point processes on $\mathbb{R}$ is the martingale approach. For a large class of point processes N on $\mathbb{R}_+$ there exists a *stochastic intensity* process (λ_t) that is predictable (for practical purposes, left continuous; see Jacod [4]) and such that the process

$$M_t = N_t - \int_0^t \lambda_s\, ds$$

is a martingale* (with respect to the σ-algebras $\mathscr{F}_t^N = \sigma\{N_u : 0 \leqslant u \leqslant t\}$). Heuristically, with $\Delta N_t = N_t - N_{t-}$ the size (necessarily zero or one) of the jump of N at t,

$$\lambda_t\, dt = E[\Delta N_t | N_u : u < t]$$

$$= \Pr[\Delta N_t = 1 | N_u : u < t];$$

note well the strict inequality in the conditioning event. M is the *innovation martingale*, so termed because the increment $dM_t = dN_t - \lambda_t\, dt$ is the new information not predictable from knowledge of the strict past of N, that results from observation of N "at" t. Within the martingale context the most important tools are the *martingale representation theorem* and *martingale central limit theorems*. Concerning the former, let N be a point process on $\mathbb{R}_+$ with stochastic intensity (λ_t). Then the martingale M carries all the innovation in that *every* (suitably regular) $\mathscr{F}^N$-martingale $\tilde{M}$ is a stochastic integral* (Jacod [4]):

$$\tilde{M}_t = \tilde{M}_0 + \int_0^t Y_s\, dM_s$$

$$= \tilde{M}_0 + \int_0^t Y_s (dN_s - \lambda_s\, ds),$$

where Y is a predictable process, the *filter gain*; the name arises from applications where signals are filtered to remove noise. This theorem is vital to recursive state estimation and to construction of martingale estimators. Martingale central limit theorems (Rebolledo [10]) are utilized to establish asymptotic normality* of sequences of martingales, especially martingale estimators, in effect by reduction to a weak law of large numbers.

TRANSFORMATIONS OF POINT PROCESSES

Point processes are combined or transformed in several ways to produce additional point processes, among which the most important are the following. The *superposition* of point processes N_1, N_2 is the sum $N_1 + N_2$. Given a point process N with points X_i on E, a point process $\bar{N}$ with points (X_i, Z_i) on a product space $E \times E'$ is a *marked point process* based on N; the mark Z_i represents additional characteristics of the point located at X_i. (The arrival time T_i of a customer at a queueing system may be marked, e.g., by its service time Z_i, or a charged particle in space by the magnitude of its charge.) Typically construction of $\overline{N}$ from N entails introduction of additional randomness; for *position-dependent marking*, e.g., the Z_i are conditionally independent given N with the distribution of Z_i depending only on X_i. By first constructing a marked point process $\overline{N}$, one can construct *randomly translated* point processes with points $X_i + Z_i$, *compound point processes*, which are not point processes in general but rather purely atomic random measures, and in particular *thinned point processes* $N' = \sum Z_i \epsilon_{X_i}$, where the Z_i take only the values 0 and 1, so that N' is a point process whose points are a subset of those of N. If the Z_i are independent and identically distributed and independent of N with $\Pr[Z_i = i] = p$, then N results from independent random deletions of points of the original process N and is called a *p-thinning* of N.

IMPORTANT CLASSES OF POINT PROCESSES

Given a measure ν on E, a point process N is a Poisson process* with mean measure ν if N has *independent increments*; for disjoint sets $A_1, \ldots, A_k$ the random variables $N(A_1), \ldots, N(A_k)$ are independent, and if for each A, $N(A)$ has a Poisson distribution with mean $\nu(A)$. Homogeneous Poisson processes on $\mathbb{R}_+$ are the most important examples; here ν is a multiple of Lebesgue measure. The superposition of independent Poisson processes is Poisson. Ubiquity of Poisson processes in applications stems from Poisson limit theorems; under mild conditions the superposition of a large number of independent and uniformly sparse point processes is approximately a Poisson process (see Kallenberg [5]). Complicated stochastic processes such as (many) Markov processes* and Poisson cluster processes can be represented using Poisson processes on spaces of functions or measures (Çinlar and Jacod [1]; Matthes et al. [8]).

An important application of point processes and Poisson limit theorems is to *level crossings* of stochastic processes: if the paths of the process (X_t) are not too irregular, then for each a the random set of times at which X crosses the level a from below is a point process. For example, X_t may be the rate of flow in a river and a the flood level, so that the point process represents times of floods. As $a \to \infty$ this process is increasingly sparse; under simultaneous time and space rescaling it often converges to a Poisson process.

In a renewal process* N on $\mathbb{R}_+$ the interarrival times U_i are independent and identically distributed; principal features are asymptotic properties not only of the counting process (N_t) but also of solutions of renewal equations, permitting calculation of limit distributions of numerous functionals of renewal processes. Renewal processes have been applied extensively in queueing* and reliability theory*. Theoretical importance of renewal processes ensues in part from their being embedded* within other stochastic processes termed regenerative processes*. More generally, Markov renewal processes can be formulated as marked point processes, and inference for them performed using martingale methods (Gill [3]).

A *Cox process* is a Poisson process with the mean measure made random; important applications include precipitation and optical communication. Given a point process N and a random measure M, N is a Cox process *directed* by M if conditional on M; N is a Poisson process with mean measure M. A p-thinned Cox process is a Cox process; more generally (Mecke [9]) a point process N is a Cox process if and only if for each $p \in (0, 1]$ there is a point process N_p such that N has the same distribution as the p-thinning of N_p. In addition (Kallenberg [5]), Cox processes are all possible limits in distribution of sequences of p-thinned point processes in which the retention probabilities converge to zero. The class of mixed Poisson processes*, with directing measures $M = Y\nu$, where Y is a positive random variable and ν a fixed measure on E, coincides (Kallenberg [5]) with that of point processes *symmetrically distributed* with respect to ν in the sense that whenever $A_1, \ldots, A_k$ are disjoint sets with $\nu(A_1) = \cdots = \nu(A_k)$, the random variables $N(A_1), \ldots, N(A_k)$ are exchangeable*. Of special interest for Cox processes is the *state estimation* problem of minimum mean squared error reconstruction of an unobservable directing measure M from the Cox process N. Optimal state estimators are expressible via the Palm distributions of N (Karr [6]). For Cox processes on $\mathbb{R}_+$ methods are available for recursive *computation* of state estimators, which as integrals of a filter gain process with respect to the innovation martingale can be updated as additional observations are obtained, rather than recalculated from scratch.

Given a Poisson process $\tilde{N}$ with points X_i and a marked point process $\overline{N}$ with marks N_i that are themselves point processes, obtained by position-dependent marking, the point process $N = \Sigma N_i$ is a *Poisson cluster process*. Interpretation: To each cluster center X_i are associated cluster members comprising the point process N_i, and N consists of all the cluster members; for example X_i may be the "center" of a galaxy composed of stars represented by N_i. A widely applied (e.g., to models of precipitation) example is the Neyman–Scott cluster process, which is a Cox process with directing intensity

$$X_t = ab\int_0^t e^{-b(t-u)}\, d\tilde{N}_u.$$

Under mild integrability restrictions (Matthes et al. [8]) a point process N is a Poisson cluster process if and only if it is infinitely divisible in that for each n there exist independent, identically distributed point processes $N_{n1}, \ldots, N_{nn}$ such that $N = N_{n1} + \cdots + N_{nn}$ in distribution. The class of Poisson cluster processes coincides with that of limits in distribution of row sums of null arrays (N_{nk}) of point processes [i.e., the N_{nk} are independent in k for each n and are uniformly sparse: $\max_k \Pr[N_{nk}(B) > 0] \to 0$ for every bounded set B].

For E a finite-dimensional Euclidean space a point process N with points X_i is *stationary* (in law) if for each x, N has the same distribution as the translated point process with points $X_i - x$. The first moment measure is translation invariant, hence a scalar multiple of Lebesgue measure; the multiplier ν, the *intensity* of N, under regularity conditions can be recovered from local behavior as well (Leadbetter [7]). For stationary point processes* the *Palm measure* $\tilde{P}(\cdot) = \nu \Pr[N \in (\cdot) | N(\{0\}) = 1]$ plays a fundamental role. Analysis of stationary point processes focusses on ergodic theorems*, first and second moment measures, the covariance measure, the spectral measure and spectral density function, and linear prediction problems.

A *self-exciting point process* is described by an explicit specification of its stochastic intensity (λ_t), typically as

$$\lambda_t = a + \int_0^t h(t-s)\, dN_s,$$

where $a > 0$ and h is a positive function satisfying integrability restrictions. Examples include the Neyman–Scott cluster process.

INFERENCE FOR POINT PROCESSES

Statistical inference for point processes has four main facets: empirical inference, martingale inference, exploitation of special structure, and state estimation.

Empirical inference applies to observation of independent, identically distributed copies N_i of a point process N, and stresses and exploits relationships to empirical processes* on general spaces. Objects such as empirical probability laws, empirical Laplace functionals, and histogram estimators of Palm distributions have been examined (Karr [6]). Not only is use made of the theory of empirical processes, but also certain results for empirical processes can be derived from point process analogs.

Martingale inference pertains to statistical models of point processes on $\mathbb{R}_+$ in which the stochastic intensity is stipulated to exist and to depend on unknown, perhaps infinite-dimensional, parameters. The best compromise between generality and tractability is the *multiplicative intensity model*: for α a positive, integrable function on $[0,1]$, under the probability measure P_α the point process N has stochastic intensity $\alpha_t\lambda_t$, where (λ_t) is a baseline stochastic intensity that is observable together with N. For example, let $X_1, \ldots, X_n$ be independent, identically distributed survival times whose distribution function F admits hazard function $h = F'/(1-F)$ and let $N_t = \Sigma 1(X_i \leqslant t)$ be the number of deaths in $[0, t]$. Then the model is a multiplicative intensity model with $\lambda_t = n - N_{t-}$ and $\alpha_t = h(t)$. Typically (Jacod [4]) the log-likelihood function

$$L(\alpha) = \int_0^1 (1-\alpha_t)\lambda_t\, dt + \int_0^1 \log \alpha_t\, dN_t$$

$$= \int_0^1 (1-\alpha_t)\lambda_t\, dt + \sum \log \alpha_{T_i}$$

is unbounded above, but maximum likelihood estimation* can be effected indirectly (Karr [6]). The *martingale method* of inference regards a martingale as noise and estimates a process

$$B_t(\alpha) = \int_0^t Y_s \alpha_s 1(\lambda_s > 0)\, ds,$$

which, in the most important special case that $Y = 1$, is simply an approximation to $\int_0^t \alpha_s\, ds$, using the *martingale estimator*

$$\hat{B}_t = \int_0^t Y_s \lambda_s^{-1} 1(\lambda_s > 0)\, dN_s$$

$$= \sum_{T_i \leqslant t} Y_{T_i} \lambda_{T_1}^{-1} 1(\lambda_{T_i} > 0).$$

The error process $(\hat{B}_t - B_t(\alpha))$ is a P_α-martingale for every α; the martingale representation theorem implies that there are no other allowable error processes. Application has been made to product limit estimators for uncensored and censored data*, systems of Markov processes, the Cox regression model*, and medical statistics.

Exploitation of special structure has been carried out for parametric and nonparametric models of Poisson processes, renewal processes, Cox processes (especially mixed Poisson processes), and stationary point processes.

State estimation, the minimum mean squared error reconstruction of unobservable portions of a point process or of an associated but unobservable random mechanism, is best understood for Cox processes and for point processes on $\mathbb{R}$ with stochastic intensities, although some results are known for other special cases, e.g., linear state estimation for stationary point processes.

Point processes on the plane $\mathbb{R}^2$ are sometimes analyzed using *distance methods*, focussing on either the distance from a point of $\mathbb{R}^2$ to the nearest point of the process N or the distance from a point of N to its nearest neighbor* in N.

APPLICATIONS OF POINT PROCESSES

Serious applications of point processes, in addition to those already mentioned, include forestry, photography, earthquake occurrence and magnitude, reliability of complex systems, road traffic, hydrology*, precipitation scavenging of aerosol particles, and counters and spike trains of signals in neurons. The bibliography of Daley and Milne [2] provides access to these and others.

References

[1] Çinlar, E. and Jacod, J. (1981). In *Seminar on Stochastic Processes, 1981*, E. Çinlar, K. L. Chung, and R. K. Getoor, eds. Birkhäuser, Boston, MA.

[2] Daley, D. J. and Milne, R. K. (1973). *Int. Statist. Rev.*, **41**, 183–201.

[3] Gill, R. D. (1980). *Zeit. Wahrsch. verw. Geb.*, **53**, 97–116.

[4] Jacod, J. (1975). *Zeit. Wahrsch. verw. Geb.*, **31**, 235–253.

[5] Kallenberg, O. (1983). *Random Measures*, 3rd ed. Akademie-Verlag, Berlin, Germany.

[6] Karr, A. F. (1986). *Point Processes and Their Statistical Inference*. Dekker, New York.

[7] Leadbetter, M. R. (1972). *Proc. Sixth Berkeley Symp. Math. Statist. Prob.*, Vol. 3. University of California Press, Berkeley, CA, pp. 449–462.

[8] Matthes, K., Kerstan, J., and Mecke, J. (1978). *Infinitely Divisible Point Processes*. Wiley, New York.

[9] Mecke, J. (1968). *Zeit. Wahrsch. verw. Geb.*, **11**, 74–81.

[10] Rebolledo, R. (1980). *Zeit. Wahrsch. verw. Geb.*, **51**, 269–286.

Bibliography

Andersen, P. K. and Gill, R. D. (1982). *Ann. Statist.*, **10**, 1100–1120. (Martingale inference for the Cox regression model.)

Bartfai, P. and Tomko, J., eds. (1981). *Point Processes and Queueing Problems*. North-Holland, Amsterdam, The Netherlands. (Conference proceedings; papers on theory and applications.)

Bartlett, M. S. (1963). *J. R. Statist. Soc. Ser. B*, **25**, 264–296. (Spectral analysis for stationary point processes.)

Brémaud, P. (1981). *Point Processes and Queues: Martingale Dynamics*. Springer-Verlag, New York. (Stochastic intensities; filtering problems; application to queues and networks of queues.)

Brémaud, P. and Jacod, J. (1977). *Adv. Appl. Prob.*, **9**, 362–416. (Excellent survey; extensive bibliography.)

Brillinger, D. (1978). In *Developments in Statistics*, Vol. 1, P. R. Krishnaiah, ed. Academic, New York. (Analogies between inference for point processes and that for time series.)

Çinlar, E. (1972). In *Stochastic Point Processes*, P. A. W. Lewis, ed. Wiley, New York. (Poisson approximation; inverse problems.)

Cox, D. R. (1962). *Renewal Theory*. Methuen, London, England. (General source on renewal processes.)

Cox, D. R. and Isham, V. (1980). *Point Processes*. Chapman and Hall, London, England. (Modern introduction at an elementary level.)

Cox, D. R. and Lewis, P. A. W. (1966). *The Statistical Analysis of Series of Events*. Chapman and Hall, London, England. (Key reference on pre-martingale inference; many specific models.)

Daley, D. J. and Milne, R. K. (1973). *Int. Statist. Rev.*, **41**, 183–201. (Annotated bibliography of references prior to 1973.)

Feller, W. (1966). *An Introduction to Probability Theory and Its Applications*, Vol. II. Wiley, New York. (Poisson processes; concise, elegant discussion of renewal processes.)

Franken, P., König, D., Arndt, U., and Schmidt, V. (1980). *Point Processes and Queues*. Akademie-Verlag, Berlin, Germany. (Stationary point processes; Palm measures; application to queueing systems.)

Gaenssler, P. (1984). *Empirical Processes*. Institute of Mathematical Statistics, Hayward, CA. (Empirical processes on general measurable spaces.)

Gill, R. D. (1980). *Censoring and Stochastic Integrals*. Mathematical Centre, Amsterdam, The Netherlands. (Martingale analysis of product limit estimators for censored data.)

Grandell, J. (1976). *Lect. Notes Math.*, **529**, 1–234. (Linear and nonlinear state estimation for Cox processes.)

Grigelionis, B. (1980). *Scand. J. Statist.*, **7**, 190–196. (Short but very informative description of martingale inference.)

Jacobsen, M. (1982). *Lect. Notes Math.*, **12**, 1–226. (The martingale method from an algebraic viewpoint that minimizes measure-theoretic technicalities.)

Kallenberg, O. (1983). *Random Measures*, 3rd ed. Akademie-Verlag, Berlin, Germany. (*The* reference on the random counting measure approach.)

Karr, A. F. (1986). *Point Processes and Their Statistical Inference*. Dekker, New York. (Theory, nonparametric inference, and state estimation on general spaces.)

Khintchine, A. Y. (1960). *Mathematical Methods in the Theory of Queueing*. Griffin, London, England. (First rigorous analysis of many questions.)

Krickeberg, K. (1982). *Lect. Notes Math.*, **929**, 205–313. (Expository treatment of inference for Poisson, Cox, and stationary point processes.)

Leadbetter, M. R., Lindgren, G., and Rootzén, H. (1982). *Extremes and Related Properties of Random Sequences and Processes*. Springer-Verlag, New York. (Point processes arising from level crossings.)

Lewis, P. A. W., ed. (1972). *Stochastic Point Processes: Statistical Analysis, Theory and Applications*. Wiley, New York. (Conference proceedings; excellent coverage of the state of the field in 1972.)

Liptser, R. S. and Shiryayev, A. N. (1978). *Statistics of Random Processes*, Vols. I and II. Springer-Verlag, New York. (State estimation for point processes; martingale representation theorems.)

Matthes, K., Kerstan, J., and Mecke, J. (1978). *Infinitely Divisible Point Processes*. Wiley, New York. (Difficult but very complete.)

Neveu, J. (1977). *Lect. Notes Math.*, **598**, 249–447. (Nice presentation of stationary point processes.)

Neyman, J. and Scott, E. L. (1958). *J. R. Statist. Soc. Ser. B*, **20**, 1–43. (Analysis of distribution of stars and galaxies using clustered point processes.)

Rebolledo, R. (1977). *Lect. Notes Math.*, **636**, 27–70. (Very general martingale models in statistics; associated asymptotics.)

Ripley, B. D. (1980). *Spatial Statistics*. Wiley, New York. (Distance methods, among others; oriented to data analysis rather than theory.)

Shiryayev, A. N. (1981). *Int. Statist. Rev.*, **49**, 199–233. (Survey of recent developments in martingale theory; applications to inference for stochastic processes.)

Snyder, D. L. (1975). *Random Point Processes*. Wiley-Interscience, New York. (Inference and state estimation for Poisson and Cox processes on the real line; engineering examples.)

(MARKOV PROCESSES
MIXED POISSON PROCESS
POINT PROCESS, STATIONARY
POISSON PROCESSES
PROCESSES, DISCRETE
STOCHASTIC PROCESSES)

ALAN F. KARR

STOCHASTIC PROGRAMMING MODEL *See* LINEAR PROGRAMMING; MATHEMATICAL PROGRAMMING

STOCHASTIC RISK THEORY *See* RISK THEORY

STOCHASTICS

Stochastics is an international journal covering the theory and applications of stochastic processes*. It is published by Gordon and Breach Science Publishers, New York, and is currently appearing at the rate of two volumes per year, each volume consisting of four issues of about 80 pages per issue. The subscription rate in 1983 was $250 per volume. It was founded in 1973 by R. S. Bucy; the present editorial board (1983) was constituted in 1979, the editor-in-chief being M. H. A. Davis (editorial address: Department of Electrical Engineering, Imperial College, London SW7 2BT, England).

The journal's statement of scope reads partially as follows:

> Articles are published dealing with all aspects of stochastic systems—analysis, characterization problems, stochastic modelling and identification, optimization, filtering and control—and with related questions in the theory of stochastic processes. Also solicited are articles dealing with significant applications of stochastic process theory to problems in engineering systems, the physical and life sciences, economics and other areas.

In practice the coverage of the journal centres around stochastic differential equations*, stochastic control and optimization, and other questions related to stochastic systems. There is a bias towards "modern" probability theory, a large proportion of the papers involving Itô calculus, martingale* theory, and the like.

Stochastics publishes only substantial research or review articles. All contributions are reviewed by one or more referees. Most articles are in English, a significant minority in French; these are the journal's official languages. The average time for submission to publication is about a year or slightly less, with occasional outliers.

Sample table of contents (Vol. 10, No. 2):

"Stochastic equations and Krylov's estimates for semimartingales" by A. V. Mel'nikov.

"A finite fuel stochastic control problem" by S. D. Jacka.

"On a stochastic process determined by the conditional expectation and the conditional variance" by A. Plucińska.

"Theorème central limite de renormalisation pour des procéssus d'Ornstein–Uhlenbeck generalisés" by A. Benassi.

"On the non-uniqueness of the limit points of diffusion with a small parameter" by T. S. Chiang and C. R. Hwang.

M. H. A. DAVIS

STONE'S REJECTION CRITERION

One of the earliest test procedures for rejection of outliers*, suggested by Stone [3, 4]. It is very similar to Chauvenet's criterion*. See Barnett and Lewis [1] and Rider [2] for more details.

References

[1] Barnett, V. and Lewis, T. (1984). *Outliers in Statistical Data*, 2nd ed. Wiley, New York.

[2] Rider, P. R. (1933). *Washington Univ. Stud.—New Ser., Sci. Tech.*, **8**, 3–23.

[3] Stone, E. J. (1868). *Monthly Notices R. Astron. Soc.*, **28**, 165–168.

[4] Stone, E. J. (1873). *Monthly Notices R. Astron. Soc.*, **34**, 9–15.

(CHAUVENET'S CRITERION
OUTLIERS
PEIRCE'S CRITERION
WRIGHT'S REJECTION PROCEDURE)

STOPPED DISTRIBUTION

See CONTAGIOUS DISTRIBUTIONS; RANDOM SUM DISTRIBUTIONS

STOPPING DISTRIBUTION

See CONTAGIOUS DISTRIBUTIONS; RANDOM SUM DISTRIBUTIONS

STOPPING NUMBERS AND STOPPING TIMES

In a classical statistical inference* problem, based on a given set of (n) observations, one wants to draw statistical conclusions in an optimal (or efficient) manner. However, such a procedure may not turn out to be optimal in a more general setup where the sample number n may not be fixed in advance. Often, the experimental setup or the statistical protocol may give rise to a random sample size. Indeed, in sequential analysis*, the sample number is not predetermined but is treated as a random variable (N), adapted to the accumulating data in the sense that N takes on nonnegative integer values in such a way that the event $[N = n]$ depends only on the data observed up to the nth stage, for every $n \geqslant 0$. In the context of hypothesis testing* problems, Wald [10] showed that a sequential procedure requires on an average a smaller number of observations than in a fixed sample size procedure having the same probabilities of the first and second kind of errors. This remarkable feature of achieving superefficiency through random sample sizes has subsequently been observed in sequential point estimation theory (Robbins [5]), and in a more general context in optimal control theory (viz. Shiryayev [8], where other references are cited). Basically, in each case, one can define a *history process* $\mathscr{H} = \{\mathscr{H}_n; n \geqslant 0\}$ (where $\mathscr{H}_n$ depends only on the outcome up to the nth stage, $n \geqslant 0$), such that the sample number N is a nonnegative random variable adapted to $\mathscr{H}_n$ in the sense that the event $[N = n]$ depends only on $\mathscr{H}_n$, for every $n \geqslant 0$. Then N is termed a *stopping number*. In sequential analysis (and elsewhere), a stopping number N is dictated by a *stopping rule* that tells the statistician how to observe values from the population and when to stop sampling; for more details *see* OPTIMAL STOPPING RULES.

For illustration, we consider the following example. Suppose that we want to estimate the mean (μ) of a normal distribution with unknown variance σ^2. Based on n observations, the sample mean $\overline{X}_n$ is an optimal estimator of μ, and we may consider the risk function

$$\begin{aligned} R(n;a,c) &= aE\left(\overline{X}_n - \mu\right)^2 + cn \\ &= an^{-1}\sigma^2 + cn, \\ &\qquad a > 0,\ c > 0, \end{aligned}$$

where c represents the cost per unit sampling. If σ were known, this risk would have been a minimum for

$$n = N_0 = n_0(\sigma) = \left[(a/c)^{1/2} \cdot \sigma\right] + 1.$$

However, for unknown σ, no fixed sample size estimator may achieve this minimum risk for every $\sigma > 0$. Intuitively, one may consider the sample variances s_n^2, $n \geqslant 2$, and set

$$N = \min\left\{n \geqslant 2 : n \geqslant (a/c)^{1/2} \cdot s_n\right\}.$$

Thus N is a positive integer valued random variable and the event $[N = n]$ depends only on the partial sequence $\{s_k^2;\ k \leqslant n\}$, so that the history process is generated by the s_k^2, $k \geqslant 2$. A similar situation arises when one wants to provide a confidence interval* for μ with a coverage probability $1 - \gamma$ $(0 < \gamma < 1)$ and a prescribed bound $2d$ $(d > 0)$ on the width of the interval (*see* FIXED-WIDTH AND BOUNDED-LENGTH CONFIDENCE INTERVALS). Here also, for unknown σ, no fixed sample size procedure exists (Stein [9]). One may define

$$N = \min\left\{ n \geqslant 2 : s_n^2 \leqslant nd^2/\tau_{\gamma/2}^2 \right\},$$

where τ_ϵ is the upper $100\epsilon\%$ point of the standard normal distribution. Then N is a positive integer-valued random variable adapted to the same history process as in the point estimation case; see Chow and Robbins [2] for further details. Further, as in Wald [10], we may consider a sequential probability ratio test (SPRT) for a simple null vs. a simple alternative hypothesis, and the stopping number N may be defined in terms of the history process involving the (log-)likelihood ratios for the successive sample sizes. Stopping numbers of more general nature arising in the context of sequential estimation* as well as testing problems have been discussed in Sen [6, Chaps. 9 and 10; 7].

In the formulation of stopping numbers we should not exclude the possibility of the event $[N = \infty]$ occurring with a positive probability. If, however, $P\{N < \infty\} = 1$ (i.e., the sampling terminates with probability 1), then N is a *proper stopping number*. Also EN, whenever it exists, is termed the *average sample number* (ASN*). In the SPRT and other related sequential tests, both in the parametric and nonparametric setups, the optimality of stopping numbers is judged by the smallness of their ASNs, although in an asymptotic setup the distributions of stopping numbers themselves convey the same picture (see Sen [6, Chap. 9]). In the sequential point estimation problem, however, the optimality of the stopping number is judged by the smallness of the risk, and the solution may not necessarily lead to the minimum ASN. In general, a stopping number may be defined in a more general context by reference to some objective function, and an optimal stopping number maximizes this function. For details of such general formulations see Chow et al. [3], Shiryayev [8], and also OPTIMAL STOPPING RULES.

In the literature, the terms "stopping numbers" and "stopping times" have often been used synonymously, although stopping times need not be integer valued random variables. For example, if the history process $\mathscr{H} = \{\mathscr{H}(t),\ t \in R\}$ is a continuous time-parameter process, the stopping time adapted to $\mathscr{H}$ is nonnegative, but not necessarily integer valued. Thus, we may define a stopping time τ as a nonnegative random variable, adapted to a history process $\mathscr{H}$, in such a way that for every $t \geqslant 0$, the event $\{\tau \leqslant t\}$ depends only on $\{\mathscr{H}(s),\ s \leqslant t\}$.

For illustration, consider n items under life testing*; let $(X_{n:0} = 0 <) X_{n:1} < \cdots < X_{n:n} (< X_{n:n+1} = \infty)$ be the failure times, and for every $t > 0$, let $k_n(t)$ denote the number of failures within the time interval $[0, t]$, so that $k_n(t)$ is nonnegative integer valued and nondecreasing in t. The history process $\mathscr{H} = \{\mathscr{H}(t),\ t \geqslant 0\}$ is generated by $\{X_{n:r},\ r \leqslant k_n(t);\ k_n(t)\},\ t \geqslant 0$. In a type I censoring scheme, for some prefixed T $(0 < T < \infty)$, the experiment is stopped at time T, so that $k_n(T)$ is an integer valued stopping number, while in a type II censoring scheme, for some prefixed k $(k \leqslant n)$, the stopping time τ is defined by $\min\{t : k_n(t) = k\}$, so that $\tau = X_{n:k}$ is not necessarily integer valued. A more complicated situation arises in the context of progressive censoring schemes*, where we encounter proper stopping times that need not be stopping numbers in the sense described before. This latter illustration also pertains to time-sequential analysis*, where stopping numbers/stopping times need not be integer valued, and the concept of optimal stopping times requires a more general formulation than in the case of the ASN. In a general setup, stopping numbers or stopping times are associated with appropriate stopping rules (related to optimization of some objective functions), and need not be in the context of sequential or time-sequential sampling schemes. Stopping numbers and stopping times have also been extensively used in martingale* theory for both the discrete and continuous index cases, and in particular, in various distributional (and moment) inequalities; in the

Skorokhod–Strassen embedding of Wiener processes for martingales*, such stopping numbers and stopping times play a fundamental role; see Doob [4] for basic ideas in this area.

References

[1] Chernoff, H. (1972). *Sequential Analysis and Optimal Design*. SIAM, Philadelphia, PA.

[2] Chow, Y. S. and Robbins, H. (1965). *Ann. Math. Statist.*, **36**, 457–462.

[3] Chow, Y. S., Robbins, H., and Siegmund, D. (1970). *Great Expectations: The Theory of Optimal Stopping*. Houghton Mifflin, Boston, MA.

[4] Doob, J. L. (1967). *Stochastic Processes*, 2nd ed. Wiley, New York.

[5] Robbins, H. (1959). In *Probability and Statistics*. H. Cramér Volume. Almqvist and Wiksells, Uppsala, Sweden, pp. 235–245.

[6] Sen, P. K. (1981). *Sequential Nonparametrics: Invariance Principles and Statistical Inference*. Wiley, New York.

[7] Sen, P. K. (1985). *Theory and Applications of Sequential Nonparametrics*. SIAM, Philadelphia, PA.

[8] Shiryayev, A. N. (1977). *Optimal Stopping Rules*. Springer-Verlag, New York.

[9] Stein, C. (1945). *Ann. Math. Statist.*, **16**, 243–258.

[10] Wald, A. (1947). *Sequential Analysis*. Wiley, New York.

[11] Woodroofe, M. (1982). *Nonlinear Renewal Theory in Sequential Analysis*. SIAM, Philadelphia, PA.

(AVERAGE SAMPLE NUMBER (ASN)
FIXED-WIDTH AND BOUNDED-LENGTH CONFIDENCE INTERVALS
MARTINGALES
OPTIMAL STOPPING RULES
PROGRESSIVE CENSORING SCHEMES
RANDOM SUM DISTRIBUTIONS
SEQUENTIAL ANALYSIS
SEQUENTIAL ESTIMATION
STOCHASTIC APPROXIMATION
WALD'S EQUATION)

P. K. SEN

STOPPING RULES *See* OPTIMAL STOPPING RULES; RANDOM SUM DISTRIBUTIONS; STOPPING NUMBERS AND STOPPING TIMES

STORAGE THEORY *See* DAM THEORY; INVENTORY THEORY

STP$_2$ (STRICTLY TOTALLY POSITIVE FUNCTIONS OF ORDER 2)

Totally positive functions are Pólya frequency functions (*see* PÓLYA TYPE 2), where the argument is the difference between two variables. They were first studied by Gantmacher and Krein.

A function $p(x, y)$ is *strictly totally positive* of order $k(\mathrm{STP}_k)$ if and only if $p(x, y) > 0$ for all $x \in X$, $y \in Y$, and the determinant

$$p\begin{pmatrix} x_1, x_2, \ldots, x_m \\ y_1, y_2, \ldots, y_m \end{pmatrix} = \begin{vmatrix} p(x_1, y_1) & \cdots & p(x_1, y_m) \\ \vdots & & \vdots \\ p(x_m, y_1) & \cdots & p(x_m, y_m) \end{vmatrix} > 0, \tag{1}$$

where $x_1 < x_2 < \cdots < x_m$, $y_1 < y_2 < \cdots < y_m$, $x_i \in X$, and $y_j \in Y$, and $1 < m < k$. It is assumed in (1) that the rows and columns of the determinant are ordered so that x_i and y_j are arranged by increasing values (see Karlin [14], Karlin and Rubin [22], and Barlow and Proschan [3, 4]). Typically, X and Y are intervals of the real line or countable sets of discrete values on the real line, such as the set of all integers.

If $p(x, y)$ is STP$_k$, where X and Y represent open intervals on the real line, and all the indicated derivatives exist, then [17, p. 284]

$$\begin{vmatrix} p(x,y) & \frac{\partial}{\partial y}p(x,y) & \cdots & \frac{\partial^{m-1}}{\partial y^{m-1}}p(x,y) \\ \frac{\partial}{\partial x}p(x,y) & \frac{\partial^2}{\partial x\,\partial y}p(x,y) & \cdots & \frac{\partial^m}{\partial x\,\partial y^{m-1}}p(x,y) \\ \vdots & \vdots & & \vdots \\ \frac{\partial^{m-1}}{\partial x^{m-1}}p(x,y) & \frac{\partial^m p(x,y)}{\partial x^{m-1}\,\partial y} & \cdots & \frac{\partial^{2m-2}p(x,y)}{\partial x^{m-1}\,\partial y^{m-1}} \end{vmatrix} > 0 \tag{2}$$

for all $m \leqslant k$, $x \in X$, $y \in Y$. For $m = 2$, this implies that the determinant

$$\begin{vmatrix} p(x, y) & \dfrac{\partial}{\partial y} p(x, y) \\ \dfrac{\partial}{\partial x} p(x, y) & \dfrac{\partial^2 p(x, y)}{\partial x\, \partial y} \end{vmatrix} > 0,$$

which means that for $x_1 < x_2$ the function $p(x_2, y)/p(x_1, y)$ is increasing in y, i.e., $p(x, y)$ has a strictly monotone likelihood ratio*. Note that, even if $p(x, y)$ is STP_k, the determinant (2) for special choices may be zero.

The determinant $p(x, y)$ in (1) is written as $p^*(x, y)$ when $x_1 \leqslant x_2 \leqslant \cdots \leqslant x_m$ and $y_1 \leqslant y_2 \leqslant \cdots \leqslant y_m$, i.e., when some of the x's and y's may be equal. Karlin [18] proposes the definition of total positivity in terms of derivatives of $p(x, y)$, which can be regarded as the counterpart of the concept of strong monotonicity. The function $p(x, y)$ is said to be *extended totally positive* of order r in the x variable [abbreviated as $\text{ETP}_r(x)$] if

$$p^*\begin{pmatrix} x, x, \ldots, x \\ y_1, y_2, \ldots, y_m \end{pmatrix} > 0$$

for all $x \in X$, $y_1 < y_2 < \cdots < y_m$; $y_j \in Y$.

Similarly one may define the concept of $\text{ETP}_r(y)$. The emphasis is on strict inequality, since the case of inequality with equalities always holds for TP_r functions. Finally, the function $p(x, y)$ is extended totally positive of order r in both variables, ETP_r, if

$$p^*\begin{pmatrix} x, x, \ldots, x \\ y, y, \ldots, y \end{pmatrix} > 0 \qquad \text{for all } x \in X \text{ and } y \in Y,$$

where determinants are of size m, $1 < m < r$. For these definitions to be meaningful, it is assumed that $p(x, y)$ is sufficiently continuously differentiable.

Karlin [18] proved that if $p(x, y)$ is $\text{ETP}_r(x)$ or $\text{ETP}_r(y)$, then $p(x, y)$ is STP_r. Thus, there are four definitions of total positivity, namely, TP_r, STP_r, $\text{ETP}_r(x)$ and $\text{ETP}_r(y)$, and ETP_r, each of which implies the preceding property. In general, these four levels are not equivalent. In some special cases, STP_r and ETP_r are equivalent, e.g., if $p(x, y) = f(x - y)$, where $f(x)$ is r-fold continuously differentiable [14].

Schoenberg [13] has made an extensive study of the variation diminishing property (VDP) of TP_k kernels, where the variables x and y occur in translation form. Karlin [16–18] extended their applications in characterizing best statistical procedures for a variety of decision problems.

If $p(x, y)$ is STP_r and if differentiations under the integral sign are permissible, then the VDP can be strengthened as follows: Let $V(h)$ denote the number of changes of sign of $h(x)$. Let $Z(g)$ denote the number of zeros, counting multiplicities of $g(x)$. Then if $p(x, y)$ is STP_r we have $Z(g) \leqslant V(h)$. The concept of STP being intermediate between TP and ETP(x) possesses a VDP that is likewise intermediate between these results [14].

It is much easier to work with STP_r functions rather than with functions that are merely TP_r. Karlin [16] shows that TP_r functions, under some conditions, can be approximated by STP_r functions.

Brown et al. [5] adopt a more direct approach. They take the variation reducing (VR) property as basic and give criteria for a family of distributions to have the VR. Their method emphasizes the importance of the VR property for statistics. Karlin's [14, Sec. 5.3.1] theorem asserts the equivalence of STP_2 and SVR_2, which is equivalent to strict monotone likelihood ratio* (MLR). In this case, Karlin's result states that the expectation $\gamma(\theta) = E_\theta g(x)$ of a monotone function g will itself be monotone in θ.

More generally, Brown et al. [5] prove the equivalence of STP_r and SVR_r. Let $g = (g_1, g_2, \ldots, g_m) \in R^n$. Let $S^-(g)$ denote the number of sign changes of the sequence $g_1, g_2, \ldots, g_n$, ignoring zeros, with $S^-(0) = -1$. Let $S^+(g)$ denote the maximum number of sign changes of the sequence that can be obtained by counting zeros as either + or −1. Clearly $S^+(g) = \overline{\lim}_{h\to g} S^-(h)$. If $\{f_\theta(x)\}$ is a family of probability distributions, let $\gamma(\theta) = E_\theta\{g(x)\}$. We say that the kernel $\{f_\theta(x)\}$ reduces the variation of g if $S^-(\gamma) \leqslant S^-(g)$. Suppose

$$f_\theta(x): \Theta \times X \to [0,\infty], \qquad \Theta \subset R,\ X \subset R.$$

We say that f is $\mathrm{SVR}_{n+1}(x, \Theta)$ if $S^-(g) \leqslant n$ implies $S^+(\gamma) \leqslant S^-(g)$. If also $S^+(\gamma) = S^-(g)$, then $IS^+(\gamma) = IS^-(g)$, where $IS^+(\gamma)$ is the initial sign of g.

The basic theorem of Brown et al. [5] is

$$f \in \mathrm{SVR}_{n+1}(x, \Theta)$$
$$\Leftrightarrow f \in \mathrm{SVR}_{n+1}(x_{n+1}, \Theta_{n+1}).$$

This is used to prove that the exponential families* are SVR and to establish the symmetry of SVR and its equivalence with STP. They also show how basic statistical results can be derived using the SVR property.

The SVR_r property has been applied in many statistical contexts since Karlin's work. Examples are the study of the combination of independent one-sided test statistics (van Zwet and Oosterhoff [34]), scale families of symmetric and one-sided stable densities (Kanter [13]), the structure of optional screening and classification* procedures (Marshall and Olkin [25]), qualitative properties of the power function of the F test* (Farrell [9]), inequalities on the multinormal integral useful in optimal design (Rinott and Santner [30]), comparison of large deviation rates for differing distribution functions (Lynch [24]), and theory of hypothesis tests (Cohen [6]; Meeden [26]).

Karlin and Rinott [21] have reviewed several multivariate totally positivity properties of absolute value multinormal variables with applications to multivariate analysis and simultaneous statistical inference. These authors have extended the results of Kemperman [23], Barlow and Proschan [3], Abdel-Hameed and Sampson [1], Perlman and Olkin [27], Jogdeo [11], Sarkar [31], Esary et al. [8], Holley [10], Preston [29], Das Gupta et al. [7], and Pitt [28]. Many of these results involve positive dependence and associated inequalities for the multinormal, multivariate t* and Wishart distributions*, and are related to STP functions.

References

[1] Abdel-Hameed, M. and Sampson, A. R. (1978). *Ann. Statist.*, **6**, 1360–1368.

[2] Barlow, R. E. Marshall, A. W., and Proschan, F. (1963). *Ann. Math. Statist.*, **34**, 375–389.

[3] Barlow, R. E. and Proschan, R. (1965). *Mathematical Theory of Reliability*. Wiley, New York.

[4] Barlow, R. E. and Proschan, F. (1975). *Statistical Analysis of Reliability and Life Testing*. Holt, Rinehart, and Winston, New York.

[5] Brown, L. D., Johnstone, I. M., and MacGibbon, K. B. (1981). *J. Amer. Statist. Ass.*, **76**, 824–832.

[6] Cohen, A. (1965). *Ann. Math. Statist.*, **36**, 1185–1206.

[7] Das Gupta, S., Olkin, I., Perlman, M., Savage, L. J., and Sobel, M. (1972). *Proc. Sixth Berkeley Symp. Math. Statist. Prob.*, Vol. 2. University of California Press, Berkeley, CA, pp. 241–265.

[8] Esary, J. D., Proschan, F., and Walkup, D. W. (1967). *Ann. Math. Statist.*, **38**, 1466–1474.

[9] Farrell, R. H. (1968). *Ann. Math. Statist.*, **39**, 1978–1994.

[10] Holley, R. (1974). *Commun. Math. Phys.*, **36**, 227–231.

[11] Jogdeo, K. (1977). *Ann. Statist.*, **5**, 495–504.

[12] Johnson, N. L. and Kotz, S. (1970). *Continuous Univariate Distributions*, Vol. 1. Wiley, New York.

[13] Kanter, M. (1975). *Ann. Prob.*, **3**, 697–707.

[14] Karlin, S. (1968). *Total Positivity*, Vol. I. Stanford University Press, Stanford, CA.

[15] Karlin, S., Proschan, F., and Barlow, R. E. (1961). *Pacific J. Math.*, **11**, 1023–1033.

[16] Karlin, S. (1956). *Proc. Third Berkeley Symp. Math. Statist. Prob.*, Vol. 1. University of California Press, Berkeley, CA, pp. 115–128.

[17] Karlin, S. (1957). *Ann. Math. Statist.*, **28**, 281–308.

[18] Karlin, S. (1964). *Trans. Amer. Math. Soc.*, **111**, 33–107.

[19] Karlin, S. and McGregor, J. (1957). *Trans. Amer. Math. Soc.*, **86**, 366–400.

[20] Karlin, S. and Proschan, F. (1960). *Ann. Math. Statist.*, **31**, 721–736.

[21] Karlin, S. and Rinott, Y. (1981). *Ann. Statist.*, **9**, 1035–1049.

[22] Karlin, S. and Rubin, H. (1956). *Ann. Math. Statist.*, **27**, 272–299.

[23] Kemperman, J. H. B. (1977). *Indag. Math.*, **39**, 313–331.

[24] Lynch, J. (1979). *Ann. Prob.*, **7**, 96–108.

[25] Marshall, A. W. and Olkin, I. (1968). *J. R. Statist. Soc., Ser. B*, **30**, 407–435.

[26] Meeden, G. (1971). *Ann. Math. Statist.*, **42**, 1452–1454.

[27] Perlman, M. D. and Olkin, I. (1978). *Ann. Statist.*, **8**, 1326.

[28] Pitt, L. D. (1977). *Ann. Prob.*, **5**, 470.

[29] Preston, C. J. (1974). *Commun. Math. Phys.*, **36**, 233.

[30] Rinott, Y. and Santner, T. T. (1977). *Ann. Statist.*, **5**, 1228–1234.

[31] Sarkar, T. K. (1969). *Tech. Report No. 124*, Department of Operations Research and Statistics, Stanford University, Stanford, CA.

[32] Shaked, M. (1975). Ph.D. dissertation, University of Rochester, Rochester, NY.

[33] Schoenberg, I. J. (1951). *J. d'Analyse Math. Jerusalem*, **1**, 33.

[34] van Zwet, W. R. and Oosterhoff, J. (1967). *Ann. Math. Statist.*, **38**, 659–680.

(PÓLYA TYPE 2 FREQUENCY
DISTRIBUTIONS
SSR_2
TOTAL POSITIVITY)

B. RAJA RAO

ST. PETERSBURG PARADOX, THE

The St. Petersburg paradox is the most famous and fruitful of all the puzzles and paradoxes of probability. In the eighteenth century, Gabriel Cramer (1704–1752) and Daniel Bernoulli (1700–1782) invented the idea of numerical utility* as a way to resolve the paradox, and succeeding centuries have produced a fascinating variety of alternative resolutions. Even today the paradox inspires debate among statisticians, economists, and philosophers. It is called the St. Petersburg paradox because it came to the attention of the learned public through Daniel Bernoulli's famous paper [2] on utility*, *Specimen Theoriae Novae De Mensura Sortis*, published in 1738 in the journal of the St. Petersburg Academy. It runs as follows. Peter tosses a fair coin repeatedly until it lands heads. He agrees to pay Paul \$1.00 if it lands heads on the first toss, \$2.00 if it first lands heads on the second toss, \$4.00 if it first lands heads on the third toss, and so on; each time the coin lands tails the payment is doubled. What is Paul's expectation? What, in other words, is a fair price for Paul to pay for the game?

Peter pays Paul 2^{n-1} dollars if the coin first lands heads on the nth toss, and the chance of this is $(\frac{1}{2})^n$; so Paul's expectation is

$$\sum_{n=1}^{\infty} 2^{n-1}\left(\tfrac{1}{2}\right)^n = \sum_{n=1}^{\infty} \tfrac{1}{2} = \infty.$$

Paul should pay an infinite price. But this is paradoxical. Paul is certain to receive only a finite payment in return, so it hardly seems fair to ask him to pay an infinite price. In fact, he is almost certain to receive a very small payment—with probability 7/8, for example, he will receive at most \$4.00—and so it does not even seem fair to ask him to pay a substantial finite price.

NICHOLAS AND DANIEL BERNOULLI

The paradox was first studied by Daniel Bernoulli's older cousin Nicholas Bernoulli* (1687–1759), who discussed it (in a slightly different form, involving a die rather than a coin) in a series of letters [3] to Montmort* from 1713 to 1716. Nicholas' way of resolving the paradox was that large values of n should be assigned zero probability instead of merely small probability; it is morally certain that these values will not occur. But he was dissatisfied with this reasoning and wanted other opinions.

In order to understand why the paradox was so important to Nicholas, it is necessary to recognize that the early theory of games of chance* was based on problems and concepts of equity. The correspondence of Pascal* (1623–1662) and Fermat*

(1601–1665) was largely inspired by the problem of fairly dividing the stakes in an uncompleted game of chance (*see* PROBLEM OF POINTS), and Huygens*, in his *De Ratiociniis In Ludo Aleae* (1657), defined expectation in terms of equity, not in terms of probability. For Nicholas, it was fundamental that the expectation of a game was its fair price. So he was very disturbed by an example where this price did not seem fair.

In 1728 Gabriel Cramer, a student of Nicholas' uncle, John Bernoulli, sent Nicholas a proposed resolution of the paradox. Cramer argued that money should be valued in proportion, not to its quantity, but to the use one can make of it, or the pleasure it gives. He offered two hypotheses under which this viewpoint would render Paul's expectation finite after all: (1) Perhaps there is a point after which yet larger amounts of money can be of no further use or afford no further pleasure to a person. If, for example, a windfall greater than 2^{24} dollars is reckoned of no more value than a windfall of only 2^{24} dollars, then the expectation becomes

$$\sum_{n=1}^{24} 2^{n-1}\left(\tfrac{1}{2}\right)^n + \sum_{n=25}^{\infty} 2^{24}\left(\tfrac{1}{2}\right)^n = 13.$$

So Paul should pay \$13. (2) Perhaps the pleasure afforded by a large amount of money can be indefinitely large but is only proportional to the square root of the amount. Then the expected pleasure should be reckoned as

$$\sum_{n=1}^{\infty} (\sqrt{2})^{n-1}\left(\frac{1}{2}\right)^n = \frac{1}{2}\sum_{n=1}^{\infty}\left(\frac{1}{\sqrt{2}}\right)^{n-1}$$

$$= \frac{1}{2-\sqrt{2}},$$

which is the same as the pleasure afforded by the amount $(1/(2-\sqrt{2}))^2 \approx 2.9$. So Paul should pay \$2.90.

Still dissatisfied, Nicholas wrote to Daniel in St. Petersburg asking him to think about the problem. Daniel responded, in 1731, with a draft of his subtle and elegant *De Mensura Sortis*. In this paper Daniel begins, as Cramer had, with the idea that the value of a sum of money might be less than proportional to the amount. But whereas Cramer had suggested setting the value equal to the square root of the amount, Daniel gave an argument for setting it proportional to the logarithm. It is probable, Daniel argued, that the value or utility to a person of a small amount of money is inversely proportional to the amount he already has. This means an increase dx in a person's fortune will cause an increase

$$du = k(1/x)\,dx$$

in its utility. Integration gives

$$u = k \log x + c;$$

the utility of the fortune x is proportional to $\log x$.

Daniel then reasoned that Paul's payment should depend on his initial fortune. If Paul begins with a fortune a and pays Peter z for the game, he will have $a - z + 2^{n-1}$ in the end. So his expected final utility will equal his initial utility only if z satisfies

$$\sum_{n=1}^{\infty}\left[k\log(a-z+2^{n-1})+c\right]\left(\tfrac{1}{2}\right)^n$$

$$= k \log a + c$$

or

$$\sum_{n=1}^{\infty}\left[\log(a-z+2^{n-1})\right]\left(\tfrac{1}{2}\right)^n = \log a.$$

This equation does not involve the constants k and c, but it does involve the initial fortune a. If $a = \$10$, then $z \approx \$3$. If $a = \$1000$, then $z \approx \$6$. In general, z increases with a; the richer Paul is, the more he should be willing to pay for a small chance of an immense gain.

Nicholas was still not satisfied. For him, the problem was one of equity and had nothing to do with how much either player valued the money. Suppose Peter agreed to the game for some nonmonetary reason, and then the two are prevented from playing and must settle up. How much does Peter owe

Paul? Daniel had not, Nicholas felt, answered this question.

Nicholas' perplexity can be resolved only by facing up to the fact that there is no price for the St. Petersburg game that is fair in every respect. There is a sense in which it is fair to have to pay a substantial amount for a tiny chance of winning a large fortune, but also a sense in which it is unfair to have to pay a substantial amount when it is nearly certain that there will be very little return.

MANY OPINIONS

In the two centuries following the publication of Daniel's paper in 1738, almost every author who wrote on probability in a philosophical vein expressed an opinion on the St. Petersburg paradox.

One of the most cogent and interesting of the eighteenth-century writers is the famous naturalist Buffon (1707–1788). In his *Essai d'Arithmétique Morale*, published in 1777, Buffon gave a number of arguments to the effect that Paul's expectation should be valued at about $5. One of these arguments was Nicholas' contention that Paul should disregard small probabilities. Whereas Nicholas thought that a probability of 1/32 or less should be disregarded and so obtained an expectation of $2, Buffon obtained $5 by placing the cutoff at 1/10,000, the probability of a 56-year old man dying in a given day. He also supported his value of $5 by having a child repeat the game 2,048 times; the total return to Paul was $10,057. Buffon admitted that a longer run of games would give a greater average return, but he argued that 2,000 games is about as many as Peter and Paul would have time to play. Buffon also agreed with Daniel that the diminishing value of large amounts of money limits the value of Paul's expectation.

Few authors were as catholic as Buffon. Most settled on a single resolution of the paradox. A few, like Laplace* (1749–1827), wholeheartedly adopted Daniel's solution. A number of others, from Fontaine (1704–1777) and Poisson* (1781–1840) to twentieth-century authors such as Fry [7], insisted that the correct resolution is to take account of Peter's limited ability to pay.

One resolution that became more popular with time is Buffon's argument that a large or infinite price is unfair to Paul because he does not have time to repeat the game often enough for his average winnings to approach the price. This resolution was adopted by Condorcet (1743–1794), and it became increasingly popular with the rise of the frequentist view of probability in the mid-nineteenth century. It is the resolution given, for example, by Jakob Friedrich Fries (1773–1843) in *Versuch einer Kritik der Prinzipien der Wahrscheinlichkeitsrechnung* (1842) and by John Venn (1834–1923) in *The Logic of Chance* (1866).

Some authors gave much more complicated resolutions. William Allen Whitworth (1840–1905), for example, in the second edition of *Choice and Chance* (1870), argued that the price Paul should pay for the gamble depends on the funds he has available for speculation. The fraction of those funds he should pay should be such that if he repeats the gamble an infinite number of times, with the price he pays and the amounts Peter may pay back at each step scaled to his funds at that step, and if the frequencies of the outcomes come out proportional to their probabilities, then he will break even. This gives the same results as Daniel's logarithmic utility, with Paul's funds available for speculation substituted for his total fortune.

DISADVANTAGEOUS FAIR GAMES

Even within the frequentist view, the resolution advanced by Fries and Venn fails to do full justice to the paradox. It is not clear that Paul's inability to play the game sufficiently many times is the only reason the price seems unfair.

In his famous textbook on probability (1st ed., 1950), William Feller (1906–1970) clarified the sense in which expectations can, from the frequentist point of view, be regarded as fair prices. As he showed, even a finite expectation may fail to be an entirely fair price. If a game has a finite expectation μ, then the law of large numbers tells us only that if a person plays the game sufficiently many times then his average winnings will approach μ with probability 1. In symbols,

$$S_n/n \approx \mu, \tag{1}$$

where S_n denotes the total winnings after the nth play. Fries and Venn's point was that the number n of games required to give (1) large probability may be unreasonably large. But (1) also tells us only that the accumulated gain or loss $S_n - n\mu$ is likely to be small relative to n. If the variance for the winnings from each game is finite, then the central limit theorem* tells us more: The accumulated net gain is likely, for large n, to be of the order of magnitude $\sqrt{n}$ and is about equally likely to be positive or negative. But if the variance is not finite, we cannot say this. Feller constructed games where the probability tends to one that the player will have a net loss exceeding $n/\log n$. Such games may be technically "fair," but they are clearly disadvantageous.

Feller also suggested a new way of pricing the St. Petersburg game. Suppose Paul plays the game repeatedly and pays a greater price each time; he pays $\$\frac{1}{2}(\log_2 k)$ for the kth game. Then, as Feller proved, the ratio of Paul's total winnings to his total payment approaches 1 with probability 1 as the number of games played increases. This result is completely analogous to (1), which says that if Paul repeatedly plays a game with finite expectation μ each time, then $S_n/(n\mu)$, the ratio of his total winnings to his total payment, approaches 1. So in the frequentist view this scheme of increasing payments is fair to Paul in exactly the same limited sense that repeatedly paying a finite expectation is always fair.

EXPECTED UTILITY

Despite the interest shown in the St. Petersburg paradox, Daniel's idea of expected utility did not play a large role in statistics or economics in the two centuries following the publication of his paper. The English utilitarians had already made their broad conception of utility influential in the early nineteenth century, and in the 1870s William Stanley Jevons (1835–1882), Carl Menger (1840–1921), and Léon Walras (1834–1910) gave marginal utility a central role in economic theory. But probability and hence expected utility did not play an important role in these thinkers' work.

According to Oskar Morgenstern, a 1934 article [12] on the St. Petersburg paradox by Karl Menger (son of Carl Menger) inspired the axiomatization of utility under risk given in von Neumann and Morgenstern's *Theory of Games and Economic Behavior* (1944). This axiomatization deals with preferences between gambles, a gamble being any arrangement that pays various amounts of money with various known probabilities (*see* UTILITY THEORY). Von Neumann and Morgenstern's results had a profound influence. They helped inspire the work by L. J. Savage* and others that made subjective probabilities and utilities important to statisticians, and they also revived the idea of numerical utility within economics.

Within the modern expected utility theories the St. Petersburg paradox has retained its vitality because the question of whether infinite expected utilities are allowed to arise has implications for the choice of the axioms. In order to understand the role of the paradox here, we must notice a point made by Menger in 1934: gambles with infinite expected utility can be designed whenever the utility function is unbounded; and if there is one such gamble, there will be others. A second gamble that always pays more will, for example, also have infinite expected utility. But this poses a problem. Paul should presumably prefer a gamble that always pays more, yet the expected utility theorem says that Paul must be indifferent between two gambles with the same expected utility.

This problem can be handled in different ways. In theories such as Savage's, where arbitrary gambles are allowed and the axioms require Paul to prefer a gamble that always pays more, it is simply a theorem that the utility function is bounded. Other theories, such as Richard Jeffrey's [8], sharply restrict the gambles considered and thus allow unbounded utilities. (Jeffrey's theory requires that utilities be unbounded if they are to be determined up to changes in origin and scale.) Which approach one prefers depends, in part, on one's attitude towards the St. Petersburg paradox. Are infinite expected utilities to be ruled out because no one really has the means to offer us unbounded rewards or because our utility for rewards is bounded? Thus we find in current economic literature arguments about the paradox that mirror those of eighteenth- and nineteenth-century writers.

THE PSYCHOLOGICAL PERSPECTIVE

One might think that the argument between Nicholas and Daniel has finally been decided in favor of Daniel: We should reduce Paul's expectation by discounting the payoffs of the St. Petersburg game, not by discounting the probabilities. Recent work in psychology has made it clear, however, that expected utility theory does not adequately describe people's attitudes towards gambles. In addition to replacing money values by utilities, people also tend to neglect small differences between probabilities, and may tend, as Nicholas suggested, to neglect very small probabilities of gains.

It is clear, even from the historical record, that many considerations enter into people's reactions to the St. Petersburg paradox. Keynes [10] put it well:

> We are unwilling to be Paul, partly because we do not believe Peter will pay us if we have good fortune in the tossing, partly because we do not know what we should do with so much money or sand or hydrogen if we won it, partly because we do not believe we should ever win it, and partly because we do not think it would be a rational act to risk an infinite sum or even a very large finite sum for an infinitely larger one, whose attainment is infinitely unlikely.

LITERATURE

Nicholas Bernoulli's correspondence [3] on the St. Petersburg paradox has been published with commentary; the correspondence is in French, the commentary, by O. Speiss, is in German. Daniel Bernoulli's *Specimen Theoriae Novae de Mensura Sortis* [2] appeared in 1738; an English translation was published in 1954, and was reprinted, together with an earlier German translation, in 1967.

Todhunter's *History of the Theory of Probability* [16] is still the best source of information on late eighteenth-century discussion of the paradox. Todhunter reviews in detail the contributions of D'Alembert, Beguelin, Buffon, Condorcet, and Laplace. Stigler's masterly essay [15] on the history of utility puts the paradox in the context of the history of economic thinking. The more recent review by Samuelson [14] includes references to the recent economic literature.

Feller's original work [5] on fair games appeared in 1945. Further study of disadvantageous fair games is reported by Klass and Teicher [11]; see also Robbins [13].

Menger's article [12] on the paradox was originally published in German; an English translation appeared in 1967. The boundedness of the utility function is discussed by Blackwell and Girshick [4], Arrow [1], and Fishburn [6]. For a review and synthesis of empirical work on people's attitudes towards gambles, see Kahneman and Tversky [9].

References

[1] Arrow, K. J. (1971). *Essays in the Theory of Risk Bearing*. Markham, Chicago, IL, pp. 63–69.

[2] Bernoulli, D. (1738). Specimen theoriae novae de mensura sortis. *Commentarii Academiae Scientiarum Imperialis Petropolitanae*, **5**, 175–192. [English transl. (1954). *Econometrica*, **22**, 23–36. Reprinted (1967), Gregg Press, Franborough, Hampshire, England.]

[3] Bernoulli, N. (1975). *Die Werke von Jakob Bernoulli*, Vol. 3, B. L. van der Waerden, ed. Birkhaüser, Basel, Switzerland, pp. 555–567.

[4] Blackwell, D. and Girshick, M. A. (1954). *Theory of Games and Statistical Decisions*. Wiley, New York, pp. 104–110.

[5] Feller, W. (1945). *Ann. Math. Statist.*, **16**, 301–304. [See also *An Introduction to Probability Theory and Its Applications*, 1st ed., Vol. 1. Wiley, New York, Chap. 10, Secs. 3 and 4 (1950).]

[6] Fishburn, P. C. (1970). *Utility Theory for Decision Making*. Wiley, New York, pp. 206–207.

[7] Fry, T. C. (1928). *Probability and Its Engineering Uses*. Van Nostrand, New York, pp. 194–199.

[8] Jeffrey, R. (1983). *The Logic of Decision*, 2nd ed. University of Chicago Press, Chicago IL.

[9] Kahneman, D. and Tversky, A. (1979). *Econometrica*, **47**, 263–291.

[10] Keynes, J. M. (1921). *A Treatise on Probability*. Macmillan, New York, pp. 318-319.

[11] Klass, M. and Teicher, H. (1977). *Ann. Prob.*, **5**, 861–874.

[12] Menger, K. (1934). *Zeit. Nationaloekonomie*, **5**, 459–485 (in German). [English transl. (1967). In *Essays in Mathematical Economics in Honor of Oskar Morgenstern*, M. Shubik, ed. Princeton University Press, Princeton, NJ, Chap. 16.]

[13] Robbins, H. (1961). *Ann. Math. Statist.*, **32**, 187–194.

[14] Samuelson, P. (1977). *J. Econ. Lit.*, **15**, 24–55.

[15] Stigler, G. J. (1950). *J. Polit. Econ.*, **58**, 307–327, 373–396. [Reprinted (1965). In *Essays in the History of Economics*. University of Chicago Press, Chicago, IL, Chap. 5.]

[16] Todhunter, I. (1865). *History of the Theory of Probability*. [Reprinted (1965), Chelsea, New York.]

(BERNOULLIS, THE
GAMBLING, STATISTICS IN
GAME THEORY
UTILITY THEORY)

GLENN SHAFER